# SOURCES AND CONTROL
# OF AIR POLLUTION

# SOURCES AND CONTROL OF AIR POLLUTION

Robert Jennings Heinsohn, PhD, PE, DEE
*Emeritus Professor of Mechanical Engineering*
*The Pennsylvania State University*

Robert Lynn Kabel, PhD, PE
*Emeritus Professor of Chemical Engineering*
*The Pennsylvania State University*

PRENTICE HALL,
Upper Saddle River, New Jersey    07458

**Library of Congress Cataloging-in-Publication Data**

Heinsohn, Robert Jennings.
    Sources and control of air pollution / Robert Jennings Heinsohn,
  Robert Lynn Kabel
        p.    cm.
    Includes bibliographical references and index.
    ISBN 0-13-624834-9
    1. Air Pollution.   2. Air Purification.   I. Kabel, Robert Lynn
  II. Title.
    TD883. H4b4   1999
    628.5'3--dc21                                97-31971
                                                      CIP

Acquisitions Editor: Bill Stenquist
Editor-in-Chief: Marcia Horton
Director of production and manufacturing: David W. Riccardi
Production Manager: Bayani Mendoza de Leon
Production Editor: Ann Marie Longobardo
Manufacturing Buyer: Pat Brown
Cover Designer: Bruce Kenselaar
Editorial Assistant: Meg Weist

Cover Photo: Robert L. Kabel

 © 1999 by Prentice-Hall, Inc.
Division of Pearson Education
Upper Saddle River, NJ 07458

Printed in the United States

10  9  8  7  6  5  4  3

ISBN 0-13-624834-9

Prentice-Hall International (UK) Limited, *London*
Prentice-Hall of Australia Pty. Limited, *Sydney*
Prentice-Hall Canada Inc., *Toronto*
Prentice-Hall Hispanoamericana, S.A., *Mexico*
Prentice-Hall of India Private Limited, *New Delhi*
Prentice-Hall of Japan, Inc., *Tokyo*
Prentice-Hall Asia Pte. Ltd., *Singapore*
Editora Prentice-Hall do Brasil, Ltda., *Rio de Janeiro*

Mathcad is a registered trademark of Mathsoft Inc.

*To our friends and colleagues and
to Penn State, our academic home
for thirty-five years, all of whom
have made possible our rewarding
careers and rich friendship*

# Contents

**Preface**                                                                                    **xii**

# PART I
## Setting the Stage

## 1. INTRODUCTION                                                                              3

1.1 Prevention and Control of Air
     Pollution, 3
1.2 Voluntary and Involuntary Risks, 6
1.3 Risk Management and Assessment, 9
1.4 Liability, 17
1.5 Pollution Control Systems, 18
1.6 Fundamental Calculations, 27
1.7 Rate Concepts, 33

1.8 Chemical Equilibrium, 35
1.9 Chemical Kinetics, 43
1.10 Chemical Equilibrium and Steady State, 46
1.11 Professional Literature, 48
1.12 Closure, 49
Nomenclature, 50
References, 51
Problems, 53

## 2. ECOLOGY AND EXPONENTIAL GROWTH                                                            59

2.1 Biological Diversity, 59
2.2 Exponential Growth, 60
2.3 Population Growth and Energy
     Consumption, 67
2.4 Climate, 72
2.5 Lumped-Parameter Radiation
     Model, 73

2.6 Greenhouse Effect, 75
2.7 Interaction between Oceans
     and Atmosphere, 79
2.8 Closure, 81
Nomenclature, 82
References, 83
Problems, 84

# PART II
## Basics and Constraints

## 3. AIR POLLUTION LEGISLATION AND REGULATIONS                                                 89

3.1 Legislative Record, 89
3.2 Clean Air Act Amendments of 1990:
     Alternatives RACT and MACT, 100
3.3 Rule-Making, 110
3.4 Electronic Bulletin Boards, 111
3.5 State Regulations, 112

3.6 Hazardous Chemicals, 118
3.7 ISO 14000, 121
3.8 Closure, 122
Nomenclature, 122
References, 123
Problems, 124

## 4. EFFECTS OF POLLUTION ON THE RESPIRATORY SYSTEM                                            127

4.1 Physiology, 127

4.2 Respiratory Fluid Mechanics, 137

4.3  Analytical Models of Heat and Mass Transfer, 145
4.4  Toxicology, 159
4.5  Dose–response Characteristics, 168
4.6  Risk Analysis, 176
4.7  Closure, 179
Nomenclature, 180
References, 182
Problems, 185

## 5.  AESTHETICS                                                     187

5.1  Acid Rain, 187
5.2  Vegetation, 188
5.3  Soil, 192
5.4  Bodies of Water, 192
5.5  Surface Deterioration, 195
5.6  Visibility, 198
5.7  Plume Opacity, 201
5.8  Odors, 206
5.9  Noise, 207
5.10 Closure, 211
Nomenclature, 212
References, 213
Problems, 215

## 6.  ATMOSPHERIC CHEMICALS: SOURCES, REACTIONS, TRANSPORT, AND SINKS                                        217

6.1  Nature and Composition of the Atmosphere, 218
6.2  Biogenic and Anthropogenic Sources of Pollutants, 221
6.3  Natural Removal of Pollutants, 223
6.4  Carbon Cycle, 225
6.5  Sulfur Cycle, 228
6.6  Nitrogen Cycle, 228
6.7  Other Chemical Species, 230
6.8  Atmospheric Chemistry of Inorganic Species: Acid Rain, 240
6.9  Atmospheric Chemistry of Organic Species, 242
6.10 Atmospheric and Interfacial Transport, 254
6.11 Interfacial Equilibrium, 258
6.12 Closure, 268
Nomenclature, 269
References, 270
Problems, 274

# PART III
# Engineering

## 7.  FORMATION AND CONTROL OF POLLUTANTS IN COMBUSTION SYSTEMS                                                  281

7.1  Combustion Devices and Power Cycles, 281
7.2  Free-Radical Chemistry in Flames, 289
7.3  Formation of Carbon Oxides, 296
7.4  Formation of Nitrogen Oxides, 300
7.5  Formation of Sulfur Oxides, 308
7.6  Unburned Hydrocarbons, 312
7.7  Combustion Controls for Reciprocating Engines, 313
7.8  Combustion Controls for Stationary Combustors, 319
7.9  Fluidized-Bed Combustion, 320
7.10 Closure, 324
Nomenclature, 324
References, 325
Problems, 326

## 8. UNCONTROLLED POLLUTANT EMISSION RATES    329

8.1  Introduction, 329
8.2  Measurements of Process Gas Streams, 330
8.3  Pollutant Material Balance, 332
8.4  AP-42 Emission Factors, 336
8.5  Empirical Equations, 337
8.6  Evaporation and Diffusion, 338
8.7  Diffusion through Stagnant Air, 342
8.8  Evaporation of Single-Component Liquids, 346
8.9  Single-Film Theory for Multicomponent Liquids, 350
8.10 Two-Film Evaporation of Multicomponent Liquids, 354
8.11 Evaporation in Confined Spaces, 361
8.12 Drop Evaporation, 362
8.13 Leaks, 367
8.14 Closure, 376
Nomenclature, 377
References, 379
Problems, 381

## 9. ATMOSPHERIC DISPERSION    386

9.1  Box Model, 386
9.2  Global Atmospheric Circulation, 389
9.3  Lapse Rate, 397
9.4  Atmospheric Stability, 401
9.5  Wind-Speed Profiles, 405
9.6  Mixing Height, 407
9.7  Actual Atmospheric Temperature Variations, 408
9.8  Appearance of Plumes, 409
9.9  Gaussian Plume Model, 413
9.10 Plume Rise, 423
9.11 Building Exhaust Stacks, 426
9.12 Instantaneous Point Source: Puff Diffusion, 428
9.13 Continuous Elevated Line Sources, 431
9.14 Numerical Dispersion Models, 431
9.15 Closure, 431
Nomenclature, 432
References, 434
Problems, 435

## 10. CAPTURING GASES AND VAPORS    438

10.1 Condensation, 439
10.2 Adsorption, 440
10.3 Absorption, 453
10.4 Absorption and Chemical Reaction, 471
10.5 Thermal Oxidation Processes, 474
10.6 Summary of Operational Costs of Three Methods to Remove Diethylamine, 482
10.7 Thermal Reduction of $NO_x$, 482
10.8  Flue Gas Desulfurization, 485
10.9  Bioscrubbers, Biofilters, and Trickle-Bed Reactors, 488
10.10 Ultraviolet-Ozone Oxidation, 499
10.11 Supercritical Water Oxidation, 505
10.12 Closure, 507
Nomenclature, 508
References, 510
Problems, 513

## 11. MOTION OF PARTICLES    515

11.1 Drag, 515
11.2 Physical Properties of Aerosols, 518
11.3 Overall Collection Efficiency, 532
11.4 Equations of Particle Motion, 536
11.5 Freely Falling Particles in Quiescent Media, 539

11.6  Horizontal Motion in Quiescent
Air, 543
11.7  Gravimetric Settling in Chambers, 544
11.8  Gravimetric Settling in Ducts, 545
11.9  Clouds, 549
11.10  Stokes Number, 552

11.11  Inertial Deposition in Curved
Ducts, 553
11.12  Closure, 558
Nomenclature, 558
References, 560
Problems, 561

## 12. CAPTURING PARTICLES                                                           567

12.1  Cyclone Collectors, 567
12.2  Impaction between Moving
Particles, 575
12.3  Filtration, 591
12.4  Electrostatic Precipitators, 608
12.5  Engineering Design: Selecting and
Sizing Particle Collectors, 617

12.6  Closure, 619
Nomenclature, 619
References, 622
Problems, 623

## 13. COST OF AIR POLLUTION CONTROL SYSTEMS                                         627

13.1  Total Initial Cost and Total Annual
Cost, 628
13.2  Utility Costs, 632
13.3  Fabric Filters, 634
13.4  Electrostatic Precipitator, 637
13.5  Comparison of Costs to Remove
Particles from the Exhaust of a
Lime Kiln, 640
13.6  Carbon Adsorber, 640

13.7  Scrubbers, 643
13.8  Thermal Oxidizers, 646
13.9  Comparison of Costs to Remove
Diethylamine, 648
13.10  Closure, 649
Nomenclature, 649
References, 651
Problems, 651

## APPENDIX                                                                          652

Table A-1.1  AP-42 Emission Factors for
Uncontrolled Bituminous and Subbituminous
Coal Furnaces, 652
Table A-1.2  AP-42 Emission Factors for
Uncontrolled Anthracite Furnaces, 653
Table A-1.3  AP-42 Emission Factors for
Uncontrolled Emissions from Fuel Oil
Furnaces, 653
Table A-1.4  AP-42 Emission Factors for
Uncontrolled Emissions from Natural Gas
Furnaces, 653
Table A-1.5  AP-42 Emission Factors for
Uncontrolled Refuse Incinerators, 654
Table A-2.1  AP-42  Emissions From
Aircraft, 654

Table A-2.2  AP-42 Emission Factors for
Gasoline and Diesel Industrial Engines (20–250
HP), 655
Table A-2.3  Emission Factors for Automobiles
and Trucks, 655
Table A-3.1  AP-42 Hydrocarbon Emissions for
Loading and Transportation of Liquid Petroleum
Products, 656
Table A-3.2  AP-42 Emission Factors for
Fugitive Nonmethane Hydrocarbons From Piping
Systems, 656
Table A-4.1  AP-42  Emission Factors
for Uncontrolled Metallurgical
Processes, 657

Table A-5.1   AP-42 Emission Factors for Uncontrolled Emissions From Mineral Processes, 658

Table A-6.1   AP-42 Emission Factors for Fugitive Emissions From Vehicles Traveling On Roadways, 661

Table A-7   Emission Factors for Indoor Processes, Activities, and Furnishings, 662

Table A-8   Henry's Law Constants (H') and Diffusion Coefficients of Pollutants in Air and Water, 664

Table A-9   Critical Temperature, Pressure, and PEL values for Common Pollutants, 667

Table A-10   Thermophysical Properties of Air and Water, 668

Table A-11   Odor Threshold and OSHA PEL Values for Common Industrial Materials, 669

Table A-12   Vapor Pressure and OSHA PEL Values for Industrial Volatile Liquids, 673

Table A-13   Fresh Air Requirements for Ventilation, 675

Table A-14   Modified Euler Method, 676

Table A-15   S. I. Prefixes, Factors, and Symbols, 677

Table A-16   Common Physical Constants and Conversions, 677

# INDEX

**680**

# Preface

*What is it that is not a poison? All things are poison and nothing is without poison. It is the dose only that makes a thing a poison.*

*Paracelsus, 1493–1541*

The atmosphere may be our most precious resource. Accordingly, the balance between the use and protection of it is a high priority for our civilization. Potential "poisons," whether natural or manufactured must be understood, quantified, and controlled if we are to strike the proper balance. This is the essence of *Sources and Control of Air Pollution*. We believe that this book is the most comprehensive yet written in addressing not only rigorous engineering principles of air pollution control but also the human, ecological, and environmental context for the application of these principles. In fact, the first six chapters are devoted to this context. They provide a solid basis for bringing to a common understanding individuals and institutions of quite disparate views on atmospheric quality.

When writing this book we kept in mind undergraduate seniors and first-term graduate students in engineering and meteorology who are interested in understanding the sources of air pollution and applying engineering principles to prevent or control it. The book uses knowledge contained in typical courses in accredited programs of mechanical, chemical, civil, and environmental engineering. Material in the book is developed with mathematical rigor commensurate with that required for the accomplishment of engineering tasks in professional practice.

The primary objectives of the book are to enable engineers to understand natural and anthropogenic sources of air pollution and to learn ways to prevent or minimize pollution by the application of various control practices. Mass and energy balances are used to relate pollutant removal efficiency to process input parameters. In this way, engineers can alter the design and operation of industrial processes to minimize pollutant emission rather than merely attempting to correct processes after they are in operation.

We believe that a text should contain a substantial body of knowledge, much of which readers will learn simply by reading and self-study. To accomplish this, the presentations of important concepts are followed by examples that apply these principles. Consistent nomenclature is employed throughout the book and each chapter is followed by a complete list of the symbols, descriptions, and dimensions used in that chapter. Reference to all citations, including many general references works, facilitate the pursuit of additional breadth and depth by the user. Numerous homework problems offer the opportunity for application of material presented in the chapter. Finally, a detailed and comprehensive index covers the entire book.

Mathematical equations requiring trial-and-error solutions, or sets of coupled ordinary differential equations requiring Runge–Kutta (or other numerical) techniques, are often encountered. Accordingly, it is important that students and practitioners master commercially available mathematical software. Many examples were solved using *Mathcad* ®. Any number of other commercially available mathematical software programs can also be used. All the Mathcad programs used in the Examples are available on a Web page for the book (see the Prentice Hall Web site). Of special interest to the classroom instructor is the availability of "class notes" (handouts that free students from the need to transcribe everything said in a lecture). The class notes are also available on the Web page for the book (see the Prentice Hall Web site). We expect that readers will discover errors in the text that have escaped us. We encourage you to inform us of them so that we can post a complete list of errata on the web page.

Throughout the text we have cited Internet Web sites where readers can find thermodynamic data, government regulations, and so on relevant to the topic being discussed. We realize that Web sites are added from the Internet at a rapid rate and that some of the those cited will have limited longevity. But even if specific sites disappear, it is highly likely that they will be replaced by more comprehensive ones. The issue is not to provide a comprehensive and up-to-date list of Web sites, but rather, to initiate students to the browsing of Web sites, acquisition of confidence in doing so, and the desire to remain current via journals and conversations with colleagues.

Books in environmental science that present air pollution in qualitative terms may motivate and inform readers, but they are inadequate as engineering texts. Empathy concerning environmental pollution is no substitute for the competency to alter the processes that produce pollution. One should never hesitate to formulate issues in quantitative terms. Indeed, quantification improves the quality of thought. Not only should an engineer's work be quantitative, but the mathematical models employed must reflect the underlying physical and chemical processes that govern the production of the pollutants.

The book has thirteen chapters organized into three sections. Sections I and II (six chapters in all) will be informative to people from all disciplines of science if they have an interest in those things influencing and influenced by air pollution. An appreciation of these sections is also critical to the engineer proceeding through the final seven chapters (Section III) on engineering.

## Section I: Setting the Stage

Section I defines the boundaries on the subject and assists readers in developing perspectives with which to anticipate risk and formulate ways to prevent and, failing that, reduce pollution. *Chapter 1* discusses risk in quantitative terms and introduces the concept of voluntary and involuntary risk so that readers can act rationally about the host of risks to which the public is exposed. The chapter asks that readers be pragmatic concerning perceived risks and actual risks and ultimately to ask whether the benefits accrued from expenditures are commensurate with the risks that are ameliorated. The chapter concludes with a review of fundamental concepts from fluid mechanics, thermodynamics, kinetics, and chemistry that are used in subsequent chapters. *Chapter 2* shows that the exponential growth in the world's population, and its concomitant consumption of energy, is the principal factor underlying the generation of pollutants.

## Section II: Basics and Constraints

Section II presents the basic physical principles describing how pollution affects the earth's atmosphere and human well-being and describes the regulations which engineers will be expected to understand and comply with. *Chapter 3* summarizes the history of governmental action to reduce air pollution and ends with current regulations and how engineers can keep abreast of current and pending air pollution regulations. *Chapter 4* describes the effects of pollutants on health. The subjects of physiology, toxicology, and epidemiology are presented in the ways that they are taught to students of environmental health engineering in schools of public health. *Chapter 5* describes the aesthetic impact of pollution, which defines the quality of life. *Chapter 6* introduces the concept of formation and fate cycles to describe the interaction between pollutants and components of the environment. It is important that engineers comprehend the magnitude of natural and anthropogenic sources of pollution so that they can understand the impact of the industrial processes for which they are responsible. Chapter 6 also describes the physical processes and chemical reactions that govern the production of pollutants in the atmosphere. The presence of certain natural and anthropogenic compounds and the unique role of solar energy that drives photolytic reactions involving these chemical species is emphasized.

## Section III: Engineering

Section III develops the fundamental engineering relationships needed to prevent and/or control air pollution. The section contains equations describing how pollutants are formed in combustion systems and transported in the atmosphere, how control devices operate to capture particles, gases, and vapors, and how to estimate the costs associated with these systems. *Chapter 7* describes the principles governing pollutant production from stationary and mobile combustion systems. The chapter describes the chemical kinetic mechanisms associated with the oxides of nitrogen, sulfur, and carbon, the understanding of which may be used to alter the combustion process and prevent, or at least minimize, the generation of pollutants. *Chapter 8* describes how to model processes that give rise to pollutants and how to estimate the rate at which pollutants are produced for a variety of industrial operations. The chapter contains empirical relationships, emission factors, and methods to predict evaporation from conventional mass and energy balances common in chemical engineering. *Chapter 9* describes how pollutants are transported in the atmosphere. Mixing is described in terms of the well-mixed model, Gaussian plume models, and EPA dispersion models. The objective of the chapter is to enable engineers to predict pollutant concentrations at arbitrary points $(x, y, z)$ downwind of continuous point and line sources and of instantaneous point sources. These models offer a basis to understand more sophisticated methods found in the research literature. *Chapter 10* describes methods of separating gases and vapors from a process gas stream. Generic classes of gas removal systems

using each of these methods of separation are described in terms of the underlying physical principles and used as the basis for models of the performance of the removal system. *Chapter 11* describes the motion of solid and liquid particles in a moving gas stream. *Chapter 12* presents the physical process underlying the removal of particles from a process gas stream. Chapters 10 and 12 describe each generic class of vapor and particle collection systems, respectively, so that the removal efficiency can be predicted as a function of the available input parameters and design variables. *Chapter 13* addresses the engineering economics of generic modes of air pollution control. The chapter contains the rationale, equations and numerical parameters used by the EPA in estimating costs for RACT permitting procedures.

Each chapter is self-contained and requires minimal material presented in previous chapters (with the exception of the fundamental concepts in Chapter 1 and a coherent sequence of examples in Chapters 10, 12 and 13). There is no need for instructors to cover every chapter or to present chapters in a serial order. The Appendix contains tabular data needed for all topics and homework problems covered in the text.

We believe that a text on air pollution should contain the fundamental engineering principles needed to design and operate processes. In addition, it should educate engineers about the relationships between pollution and health, economics, governmental regulation, and philosophy, all of which raise and address questions about risks to public health and safety. To treat air pollution as just another set of physical principles to be applied misses the mark and ignores one of the reasons why pollution is an issue of public concern today. To pursue sound public policy, engineers must learn the social consequences of industrial activity.

A text should be more than a set of class notes strung together; instead, it is our hope that readers may hear a voice (witty or opinionated, but factual). Phrases, ideas, and opinions rattle around readers' heads and are recalled with fondness years into the future. Our good teachers evoked such responses in us, and we hope to do so with you.

Was Paracelsus right 500 year ago? Read on and decide for yourself.

## Epilogue

For over 40 years each, the authors have practiced engineering as teachers, researchers, and consultants to industry. We have seen interest in the environment emerge as legislation and regulation and have seen it absorbed in peoples lives as an element in a personal code of conduct. We wish to share our views for the future with the hope that readers can carry on this work without losing perspective or impeding the nations ability to participate successfully in the world's economy.

We note also that forms of pollution are often chemical and mechanical, even if their origins may be quite different. While the resolution of environmental dilemmas are often chemical and mechanical as well, we acknowledge with pleasure the contributions of, and our interactions with, other engineers, physical scientists, and thoughtful people of all persuasions. As professional engineers in the mechanical and chemical disciplines, we take pleasure in collaborating on this book.

# I

# Setting the Stage

- Chapter 1    Introduction
- Chapter 2    Ecology and Exponential Growth

To "set the stage" for subsequent detailed consideration of air pollution, we introduce the matter of risk and put it into an ecological context. We also identify methods of pollution control and put forth fundamental engineering principles and conventions that pervade subsequent chapters.

The concept of voluntary and involuntary risk is presented so that readers can think quantitatively and act rationally about the host of risks to which they are exposed daily. A pragmatic attitude concerning perceived and actual risks is suggested, which turns on the question of whether the benefits accrued from expenditures on pollution control are commensurate with the amelioration of risks. Fundamental concepts from fluid mechanics, stoichiometry, thermodynamics, kinetics, and chemistry that are used throughout are reviewed. The generic classes of pollution control systems are introduced. We end the section by showing that the exponential growth in the world's population, and its concomitant consumption of energy, are the principal factors underlying the generation of pollution. This chapter completes the big picture.

# 1

# Introduction

---

In this chapter you will learn:

- How to anticipate and rank risks quantitatively
- The legal definition of air pollution
- Generic classes of air pollution control systems
- How to compute pollutant concentrations
- How to compute properties of ideal mixtures of gases and liquids
- How to compute the composition of chemically reacting gases

## 1.1 Prevention and control of air pollution

The environment affects our health and we, and our activities, affect the environment. Since the number of humans on Earth increases exponentially, our technological activities have an increasing impact on the environment no matter how efficient our technology may be. The relationship between an accidental release of toxic material and injury is immediate

and recognition of the hazard is obvious. The relationship between environmental pollutants and illness is often delayed, obscured by natural causes of disease, and far more difficult to document. Numerous studies on air pollution epidemiology (Lipfert and Wyzga, 1995) have shown associations between premature mortality and air pollutant concentrations well below the national ambient air quality standards. It is difficult, however, to interpret the data and formulate appropriate public policies. A factor complicating the analysis is that the air quality inside residences, vehicles, and public buildings may be more important than the air quality outdoors. The goal of this book is to prepare engineers to:

- Understand quantitatively the relationship between technological activity and the environment.
- Design devices, processes, and systems to maintain a safe and healthy environment.

As a first step, engineers must determine (predict or measure) whether processes and devices are in compliance with air pollution regulations, and if they are not in compliance, how to alter the process or to

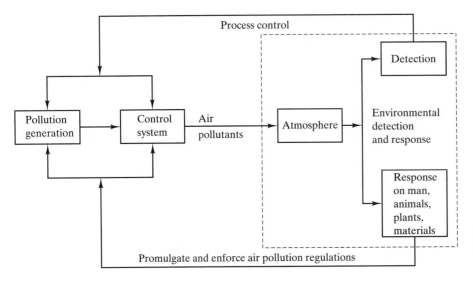

**Figure 1-1** Elements of air pollution generation, prevention and control

design control devices to achieve compliance. Figure 1-1 shows the elements of air pollution, prevention, and control. From left to right, the *source of pollution* is a process, building, machine, or activity that is generating pollutants. A *control system* governs the process and/or removes molecular species from the exit gas stream to bring emissions into compliance with regulations. *Air pollutants*, materials whose concentrations must be in compliance with regulations, pass on to the *atmosphere*, the medium that receives pollutants and in which solar radiation may initiate reactions that generate new pollutants. *Atmospheric detectors* monitor pollutant concentrations and initiate actions to reduce emissions by feeding back to the pollutant generator and downstream control system. *Receptors* are humans, animals, plants, or materials that are adversely affected by pollutants. Receptor response results in legislative actions and enforcement to reduce emissions.

Air pollution is defined by federal and state laws. In Pennsylvania, air pollution is defined as

> the presence in the outdoor atmosphere of any form of contaminant including but not limited to the discharging from stacks, chimneys, openings, buildings, structures, open fires, vehicles, processes, or any other source of any smoke, soot, fly ash, dust, cinders, dirt, noxious or obnoxious acids, fumes, oxides, gases, vapors, odors, toxic or radioactive substances, waste, or any other matter in such place, manner, or concentration inimical or which may be inimical to the public health, safety, or welfare or which is, or

may be injurious to human, plant or animal life, or to property, or which unreasonably interferes with the comfortable enjoyment of life or property.

Other states have comparable definitions. Aside from the amazing length of a single sentence, readers should note the breadth of human activities falling within the jurisdiction *air pollution*. Polluting materials may be gas, vapor, liquid, or solid and may be emitted by stationary and mobile sources. Air pollution comprises materials in the air that are inimical or possibly injurious to the public health and safety, injurious to plant or animal life or that unreasonably interfere with the enjoyment of life, or property.

It should be noted that the government's jurisdiction covers both health (plant, animal, and human) and aesthetics. From the definition of air pollution given above, it should be obvious that mere emission to the atmosphere of materials that some people find objectionable does not constitute pollution. For emissions to constitute pollution they must embody concentrations, duration, or mass flow rates exceeding prescribed amounts.

It is important to note that emissions to the indoor environment (homes, industrial environment, or enclosed public areas) are not included in the above. Indoor emissions that have the potential to escape to the outdoors are included, however. Within the indoor industrial environment, occupational exposures are covered by laws and regulations administered by the Occupational Safety and Health

Administration. There is a tendency among some people to believe that the environment (however it is defined) must be *preserved*. By definition, preservation is a static phenomenon, whereas the environment is certainly not static. *Conserving* the environment seems a more appropriate term.

If there is a cardinal rule that engineers should follow to abate pollution, it is to prevent pollutants from being produced in the first place. The logic is obvious, but as so often happens, the obvious is often overlooked. Whether it is retrofitting an existing process or designing a new process, engineers should seek ways to eliminate (or at least minimize) the use of materials which, emitted to the atmosphere, will be called pollutants. Such a position may be difficult to implement because management may wish to preserve practices that were successful in the past, practices whose costs are known to be low, or practices with which workers are familiar. If steps recommended to prevent pollution increase production or overhead costs without increasing the value of the product, engineers can expect their ideas to encounter resistance. Typically, engineers are asked is to design an "add-on" device to capture pollutants. Such a Band-Aid fix is more appropriately the engineers' last resort. Seeking other corrective measures takes courage and persuasion because old ways die hard.

Reducing air pollution calls for activity at three levels, each ranked in the order in which action should be taken and in order of increasing cost. The following *pollution prevention* strategies comprise a hierarchy of actions used by engineers in occupational health to abate pollution in the indoor air in an industrial environment.

- *Category 1: Administrative controls*
  a. Establish *work rules* and practices that specify how process are to be operated to prevent (or at least minimize) the generation of pollutants.
  b. Establish *maintenance*, *housekeeping*, and *waste disposal* practices that prevent (or at least minimize) the generation of pollutants.
  c. Establish *administrative procedures* and a *managerial structure* to implement actions (a) and (b).
- *Category 2: Engineering controls*
  a. *Eliminate* the use of polluting materials.
  b. *Substitute* less polluting or nonpolluting materials.
  c. *Enclose* the process so that polluting materials will not be emitted to the atmosphere.

  d. *Alter the process* to require minimal amounts of polluting materials.
  e. *Change the product* so that polluting materials are not needed.
- *Category 3: Pollution control systems*. Air pollution control systems are "tailpipe" devices to remove pollutants from a gas stream prior to its discharge to the atmosphere. Tailpipe solutions are employed if Categories 1 and 2 are insufficient. Control systems are engineering activities to remove pollutant gases and vapors from a discharge gas stream to satisfy government or more stringent company emission standards.

### EXAMPLE 1.1   SURFACE COATING AND CLEANING PROCESSES

Consider the airborne solvents [volatile organic compounds (VOCs)] generated by coating, treating, and cleaning surfaces in the following applications.

- *Furniture:* gluing, staining, finishing
- *Automotive vehicles, houses, aircraft, boats:* painting
- *Parts:* cleaning, degreasing, stripping
- *Textiles:* dying and treating, dry cleaning

Propose engineering controls (see Category 2 left) to prevent organic compounds (VOCs) from being emitted to the atmosphere.

*Solution*

a. *Eliminate the use of polluting materials.* Use water-based coatings instead of solvent-based coatings; eliminate solvent degreasers by cleaning with supercritical carbon dioxide, or acetic, citric, or tannic acid washes.
b. *Substitute less-polluting materials.* Use coatings with lower solvent content and higher solids content; use solvents with lower vapor pressures (lower emission rates).
c. *Enclose the process.* Use surface cleaning devices (dip tanks, vapor degreasers, etc.) that totally enclose the cleaning process and require parts to enter and leave through airlocks.
d. *Alter the process.* Use dry powder coating and infrared ovens that generate zero or minimal hydrocarbons, use water-based coatings.
e. *Change the product.* Use plastics or stainless steel that do not require coating or degreasing; coat the

product overseas and assemble the finished product in this country (readers may find this practice repugnant, but it is commonly used by industries around the world).

If the controls noted above are insufficient, engineers will need to design *pollution control systems* to limit atmospheric emissions. If it is necessary to retain the conventional solvent-based coatings, they should be applied inside ventilated paint booths with airborne solvents removed by activated carbon adsorbers or thermal oxidizers.

# 1.2 Voluntary and involuntary risks

Inflated prose is inimical to rational debate, and if we are to set economically sound public policies we must choose our words carefully. Discussions of air pollution often include the words *hazard*, *risk*, and *noxious* as if they were synonyms. They are not synonymous and the differences are important. A *hazard* has the potential to produce conditions that endanger safety and health, but it does not express the likelihood that it will occur. In the United States, 93 and 6300 people per year die from being struck by lightning and automobiles, respectively. By this measure, the likelihood, probability, or *risk* of death from an automobile strike is 68 times greater than from a lightning strike. The first dictionary definition for *noxious* is "physically harmful or destructive to living beings." Butyl mercaptan has a disagreeable, skunklike odor that is detectable at concentrations 1000 times smaller than concentrations that pose a risk to health. That may be obnoxious. Odorless carbon monoxide (CO) is a highly toxic gas whose high risk can be predicted accurately. Similarly, hydrogen cyanide (HCN) has a pleasant odor of burnt almonds at concentrations comparable to those that pose a serious risk to health. CO and HCN are certainly noxious.

To appreciate risks associated with human endeavors, Starr (1969) suggested that it was useful to categorize risks as voluntary and involuntary. *Voluntary risks* are taken by people of their own free will. Examples include risks associated with recreational sports and flying private aircraft. *Involuntary*

*risks* are imposed because of circumstances beyond our control. Examples include risks associated with using elevators in tall buildings and traveling in commercial aircraft. Definitions become cloudy concerning private automobiles because people voluntarily drive their own automobiles, yet are at risk from accidents caused by others.

Starr (1985) offers an elementary illustration of voluntary and involuntary risks in slicing bread. If you hold both the knife and bread, the distance between knife and bread will be small, yet if you hold the bread for someone else, you can bet that the distance will be greater. The example illustrates that confidence is at the root of public acceptance of risk not directly under one person's control.

Figure 1-2 shows how voluntary and involuntary risks vary with the perceived benefit of the risk. For a first approximation the risk is represented by the number of fatalities per hour of personal exposure associated with the activity. The perceived benefit of each activity is converted into a dollar equivalent as a measure of the activity's value to the individual. For voluntary activities, the amount of money spent on the activity by the average person was taken to be proportional to its benefit. In the case of involuntary activities, the contribution of the activity to the person's annual income was taken to be proportional to its benefit. While the approximations are crude, the large difference between the bands representing the voluntary and involuntary risks suggests that people *accept* higher risks in voluntary activities but *expect* lower risk in involuntary activities. For example, the risk due to disease for the entire U.S. population is $10^{-6}$ with no identifiable benefit. Note in Figure 1-2 that this risk is the same as for commercial aviation. Risks associated with the environment are considered involuntary and the public expects risks to be no larger than those associated with naturally occurring events.

The use of wood-burning stoves and political action to enact smoking ordinances shows the complexity of voluntary and involuntary risks. The concentration of carbon monoxide, *total suspended particles* (TSPs), and several polycyclic aromatic hydrocarbons (including benzo[$\alpha$]pyrene) in homes using wood-burning stoves (Traynor et al., 1987) is in general comparable to the concentration of these materials in public places where smoking is allowed

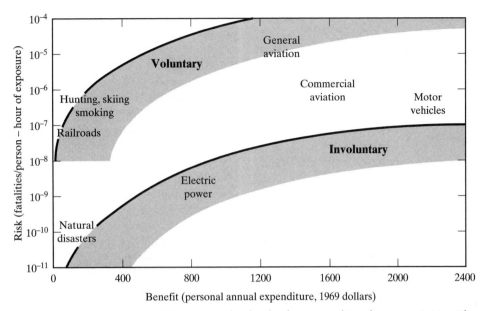

**Figure 1-2**   Risk versus benefit for various kinds of voluntary and involuntary activities. The risk of disease for the entire U.S. population is approximately $10^{-6}$ (abstracted from Starr, 1969)

(Repace and Lowery, 1980; Sterling et al., 1982). Indoor pollution from wood-burning stoves is a voluntary risk to the homeowner, whereas tobacco smoke in public buildings is an involuntary risk to the nonsmoker. While the risks to health may be comparable, many people object to the involuntary risk, yet accept with equanimity the voluntary risk. Some observers would consider this hypocrisy, others would consider it freedom of choice. The issue is even more complex, because in some regions of the United States there are times when up to one-half the outdoor total suspended particulates (TSP) comes from wood-burning stoves. The irony is that the voluntary risk to homeowners using wood-burning stoves becomes an outdoor air pollution involuntary risk to their neighbors.

Readers should not confuse ambient air quality standards with those in the workplace. The distinction between the two is dictated by the populations affected and the body's natural recuperative powers. Ambient *air quality standards* (AQSs) are set by a deliberative procedure in which experts in various fields identify concentrations above which the human population will suffer deleterious effects. Their recommendations include maximum values for both 24-hour (or less) and annual averages. Workplace air

quality standards, called *threshold limit values* (TLVs) and *permissible exposure limits* (PELs), are set by a similar process which specifies limits that healthy workers can be exposed to during working hours without deleterious effects. Workplace standards are based on the obvious fact that infants, aged, and infirm persons are not in the workplace and the logic that workers leave the workplace after eight to ten hours and the body's recuperative powers operate. For the population generally there is no escape from outdoor ambient air. Thus ambient air quality standards are more stringent than workplace standards.

When considering the health effects of air pollution, it is useful to have benchmarks for comparing environmental disease and injury statistics to those of the U.S. population (260 million) as a whole. Within the United States each year people die (National Safety Council, 1994) from the following:

- 720,862 heart attacks
- 514,657 cancer (150,000 lung cancer, 85% related to smoking)
- 500,000 (estimated) strokes
- 390,000 estimated diseases related to smoking (approximately one in every six deaths)
- 100,000 (estimated) alcohol

- 43,536 motor vehicle accidents
  - 17,900 vehicle–vehicle
  - 11,900 vehicle–fixed object
  - 6200 vehicle–pedestrian
  - 2394 motorcycles
  - 800 vehicle–bicycle
  - 600 vehicle–train
- 30,810 suicides (approximately 18,526 from firearms)
- 29,555 AIDS
- 9100 work, nonmotor vehicle, and 22,500 accidents in the home
  - 12,662 falls (7000 at home)
  - 4120 fires (3200 at home)
  - 2900 suffocation–ingestion (1900 at home)
  - 4621 drowning (800 at home, 312 in the bathtub)
- 26,513 homicides, 5000 unsolved murders
- 1833 carbon monoxide (suspected suicide, 50%)
- 93 lightning
- 90 venomous animals and plants

To put information about environmental health hazards into perspective, consider government *estimates* that *up* to 3500 people *may* die from radon or secondary tobacco smoke. It is helpful to realize that this is one-half of the *actual deaths* due to falls in the home. Motorcycles are fewer in number and are driven far fewer miles than automobiles, but they lead to a fatality rate per mile—16 times greater than that of other motor vehicles.

The Bureau of Labor Statistics (BLS) of the U.S. Department of Labor compiles the most reliable data on occupation-related deaths, although it only counts about three-fourths of the nation's workforce. Adjusted to include the entire nation's workforce, the Congressional Office of Technology Assessment (OTA) found that approximately 6000 deaths occurred annually from occupational injuries, or about 25 deaths for each working day. In 1912 the workforce was only half as large as now, yet an estimated 18,000 to 21,000 workers lost their lives. Table 1-1 shows the number of deaths and disabling injuries from several industries. On the basis of the on-the-job death rates (deaths/100,000 workers), agriculture and mining, followed by construction, were the riskiest occupational fields. Additional BLS data show that about half the fatal occupational injuries involve motor vehicles, off-road industrial vehicles, and falls. Surprisingly, murder associated with robbery makes taxicabs drivers (26.9/100,000 workers) and convenience store clerks (1600 total homicides) risky occupations.

**Table 1-1**  Work Accidents in 1993

| Industry division | Workers,[a] 1993 (millions) | Deaths, 1993 (percent change from 1986) | Death rates,[b] 1993 (percent change from 1986) | Disabling injuries (thousands) |
|---|---|---|---|---|
| All industry | 118.7 | 9100[c] (−15) | 8 (−20) | 3200[c] |
| Agriculture[d] | 3.1 | 1100 (−35) | 35[c] (−32.7) | 130 |
| Mining[d] | 0.6 | 200 (−5) | 33 (−34) | 20 |
| Construction | 5.9 | 1300 (−30) | 22 (−33) | 280 |
| Manufacturing | 17.8 | 700 (−36.4) | 4 (−33) | 560 |
| Trade[d] | 27.5 | 1300 (18.2) | 5 (−25) | 710 |
| Services[d] | 39.4 | 1300 (−13.3) | 3 (−40) | 780 |
| Government | 18.4 | 2000 (53.8) | 11 (37.5) | 490 |
| Transportation and public utilities | 6 | 1200 (−20) | 20 (−25.9) | 230 |

*Source:* National Safety Council (1994) estimates (rounded) based on data from the National Center for Health Statistics (p. 34), state vital statistics departments and state industrial commissions. Numbers of workers are based on Bureau of Labor Statistics data and include persons aged 14 and over.

[a] All persons gainfully employed, including owners, managers, paid employees, self-employed, and unpaid family workers, but excluding private household workers.

[b] Events per 100,000 workers in each group.

[c] About 3500 deaths and 200,000 of the injuries involved motor vehicles.

[d] Agriculture includes forestry and fishing. Mining includes quarrying and oil or gas extraction. Trade includes wholesale and retail trade. Services includes finance, insurance, and real estate.

Environmental illness is difficult to define for three reasons: (1) many environmental diseases are indistinguishable from diseases generally, (2) the relationship between a disease and the environment is not always recognized, and (3) many diseases have long latency periods and occur after exposure has ceased or the person has moved to another part of the country. The last reason is the most troubling factor inhibiting accurate assessments of risk. Serious diseases such as respiratory and neurological disorders and cancers are not generally captured in BLS records of work-related illness, but after considerable debate it is generally agreed that approximately five percent of the national cancer deaths are related to occupational exposures.

Environmental catastrophes such as occurred in Bhopal, India in 1984 (Elsom, 1987) are rare and should be defined as industrial accidents, not air pollution. Air pollution pertains to pollutant concentrations resulting from normal industrial operations that may be amplified from time to time by unusual meteorological conditions. Deaths related directly to air pollution occur rarely and comprise far fewer deaths than occur in the home due to accidents. Two notorious air pollution episodes resulting in the loss of life were in Donora, Pennsylvania in 1948, when 20 people died when $SO_2$ and particle emissions accumulated in the Monongahela river valley during a five day period, and the London fog of 1952, when several thousand people are believed to have died from large concentrations of $SO_2$ that accumulated over a period of approximately a week.

For the most part, air pollution is one of many *causal factors* involved in the death of older people suffering from other debilitating diseases (chronic bronchitis, heart disease, emphysema, asthma, bronchopneumonia, etc.) and not the immediate cause of death as in the case of the industrial accidents noted above. Air pollution assumes importance because (1) risks associated with the air we breath comprise the ultimate involuntary risk, (2) risks of unwanted climate change are involuntary risks for which remedies take decades to implement, and (3) risks to Earth's forests and waters require decades to correct. The exponential growth of population and its related exponential increase in pollutant production represents a clear and present danger that must be addressed while we have time to do so. Pollution is the unwanted consequence of deliberate decisions about industrial activity and the allocation of financial resources. Political institutions have it within their power to make rational decisions and reduce the emission of pollutants to the atmosphere.

# 1.3 Risk management and assessment

Health and safety are moral sentiments, and like other sentiments such as peace, freedom, and happiness, they are not absolutes or moral imperatives, however desirable they may be to us. They are, in fact, measured intangibly by the absence of their undesirable consequences. Nevertheless, they are of such evident importance that they must be assessed and managed. *Risk management* is the goal of personal and government policy, while *risk assessment* is an activity that estimates the spectrum and frequency of accidents and other negative events. Consider the automobile. If an environmental impact statement were prepared today it would reveal highly probable risks accounting for approximately 50,000 deaths per year. Do such risks banish automobiles? Of course not. Human intervention (either personal or governmental) manages the risks and dictates public policy. "How safe is safe enough?" is a rhetorical question that depends more on the management of risk than it does on quantitative statements derived from risk assessment. In the final analysis, public acceptance of risk depends more on public confidence in risk management than on informed use of the quantitative estimates of the consequences, probabilities, and magnitudes on undesirable consequences.

Air pollution involves familiar pollutants and new materials (gases, vapors, or particles). Familiar pollutants are ones whose deleterious effects are well known. Whether new materials produce benign or deleterious effects is generally not known, although experts may anticipate these effects. New materials may not even be subject to regulations. Nevertheless, professional responsibility requires engineers to assess risk in one way or another.

Entrepreneurs strive to compress the time between invention and commercial development to a few years rather than decades as in years past. Governmental regulatory procedures exist for premarket testing of new materials and devices, yet the number of new products is too large for an orderly testing of all of them. Our political system recognizes this fact and tempers its regulatory responsibility with the realization that the nation's future economy depends on sustaining a preeminent role in the development of new products and maintaining or reassuming a position of leadership in process technology (Thurow, 1987). The dilemma has been phrased aptly (Nichols and Zeckhauser, 1985) as, "Where should we spend whose money to undertake what programs to save which lives with what probability?"

A decision to develop new products is often cast in terms of the *cost–benefit ratio*. If the ratio is small, conventional wisdom argues that the venture should proceed. All things being equal, it is financially prudent to pursue the venture that has the lower ratio. A low cost–benefit ratio is not a basic moral norm that is intrinsically "good," however, because it sidesteps the fundamental question: "Who benefits … who pays?" Alternatively, it ignores the question of which political constituency derives the benefit and which political constituency bears the costs. The majority of large societal issues such as pollution and poverty are issues where the two political constituencies are not the same (Wilson and Crouch, 1987; Ames et al., 1987; Slovic, 1987; Russell and Gruber, 1987; Lave, 1987; Okrent, 1987; Dewees, 1987).

If a person plans to buy a personal computer with his or her own funds, the person buying it also profits from its use. Thus if two machines have the same capabilities, buying the one with the lowest cost–benefit ratio has obvious virtue. On the other hand, consider the citizens of a small community troubled by a proposed municipal waste incinerator. Defining costs and benefits is a dilemma. The costs and benefits for the community whose waste will be treated are easier to calculate than the costs and benefits borne by the community in which the facility will be located. Citizens generating the waste benefit, but those in the community where the facility will be located bear the burden of additional truck traffic and anxiety about stack emissions. Securing operating permits is designed to assure the smaller community that they will not be subjected to unhealthy conditions, but that often does not remove their anxiety particularly when they believe their political influence is insignificant compared to that of the community whose wastes they will treat. Possible benefits to the smaller community may be employment and tax revenue.

To illustrate the difficulty in assessing risk, consider the creation of new materials. The age of metals gave

way to the age of petrochemicals, which today is giving way to the age of ceramics and composite materials. The emerging materials industry is quite different from past industries in several significant ways: (1) unusual chemical compositions and manufacturing processes are created before the health and environmental implications can be understood fully; (2) the pace of these developments is faster than the pace at which health tests can be conducted; (3) new products are used by a diverse group of small manufacturers that are difficult to identify or monitor; and (4) only small amounts of these new materials are actually used in final products. The building blocks of advanced products are metals, polymers, ceramics, semiconductors and composites that are assembled sometimes molecule by molecule by unique processes many of which are not automated and may indeed be labor intensive. Some steps in the process involve exotic hazardous materials such as carcinogenic organics, highly toxic gases, and submicron particles to whose surface highly active organics have been added.

The majority of new materials are not invented by Fortune 400 companies but by small entrepreneurial firms that are labor intensive. The number of the exposed people is small but the exposure is likely to be a large percentage of their workforce. These firms are the least likely to have personnel whose sole occupational duty is health and safety. Thus there are no specialists to call upon as can be called upon in Fortune 400 companies. The volatility of these small firms and the mobility of their employees prevents defining the exposed population accurately, which in turn inhibits the accuracy of epidemiological studies.

The greatest barrier to maintaining a safe and healthy environment is the inadequacy of data to estimate the health risk of new materials. For the most part, new materials are not such obvious hazards as pesticides, herbicides, or explosives. There are approximately 60,000 chemicals and 2,000,000 mixtures in commercial use. Each year more than 1000 new chemical compounds are synthesized. The majority of these materials do not reach the public as end products but are used as intermediary materials in the production of finished products. These intermediate chemicals are often called *chemicals of commerce*.

Actions presently followed by regulatory agencies are based on the following assumptions:

1. Every perceived risk cannot be evaluated at the same time.
2. Plans must be developed to determine which materials pose the largest potential risk.
3. The actual risk of materials having the largest potential risk will be determined first.

Evidence of risk involves several components. *Acute* (short-term) hazards are the easiest to diagnose because symptoms are immediate and sometimes dramatic. Toxins such as carcinogens, teratogens, neuropathogens, and mutagens produce the following *chronic* (long-term) effects and are more difficult to assess:

- Disorders of the pulmonary, cardiovascular, and immune systems
- Disorders of the skeletal system, blood, and bone marrow
- Disorders of the skin and mucous membranes
- Hypersensitivity

Assessing risk requires knowledge of the actual amounts of the material used and the percent discharged to the environment. Merely listing materials used can create a false sense of risk. To begin with, engineers should conduct an *environmental audit* to identify the amounts of material used over a period of time. In addition, one must estimate the amounts remaining in the product, the amounts removed as waste and by difference the amounts that escape to the environment. Such audits are huge mass balances to define processes and identify materials posing a risk.

Responding to the *Toxic Substances Control Act* (TSCA), the EPA cataloged more than 56,000 manufactured or imported substances used in manufacture. The list, called the TSCA *Inventory*, excludes classes of materials regulated under other federal statues; for example, note 8627 food additives, 1815 prescription and nonprescription drugs, 3410 cosmetic ingredients, and 3350 pesticides in Figure 1-3.

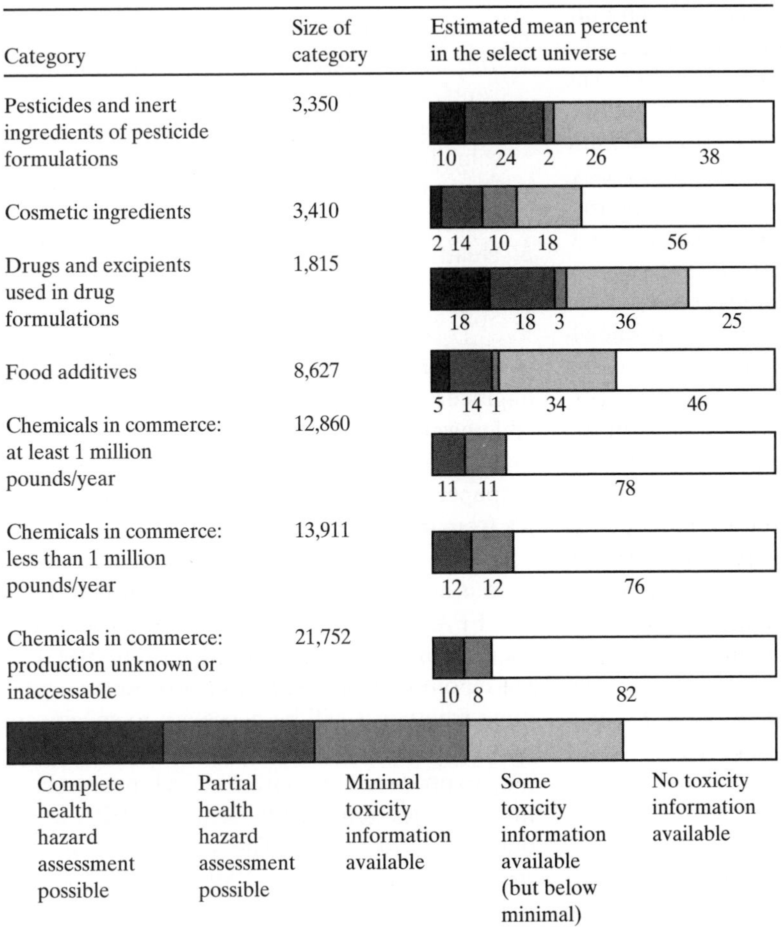

| Category | Size of category | Estimated mean percent in the select universe |
|---|---|---|
| Pesticides and inert ingredients of pesticide formulations | 3,350 | 10  24  2  26  38 |
| Cosmetic ingredients | 3,410 | 2  14  10  18  56 |
| Drugs and excipients used in drug formulations | 1,815 | 18  18  3  36  25 |
| Food additives | 8,627 | 5  14  1  34  46 |
| Chemicals in commerce: at least 1 million pounds/year | 12,860 | 11  11  78 |
| Chemicals in commerce: less than 1 million pounds/year | 13,911 | 12  12  76 |
| Chemicals in commerce: production unknown or inaccessable | 21,752 | 10  8  82 |

Legend (left to right):
- Complete health hazard assessment possible
- Partial health hazard assessment possible
- Minimal toxicity information available
- Some toxicity information available (but below minimal)
- No toxicity information available

**Figure 1-3**   Percent of materials for which health hazards are known (abstracted from Steering Committee on Identification of Toxic and Potentially Toxic Chemicals for Consideration by the National Toxicity Program, Board on Toxicology and Environmental Hazards, Commission on Life Sciences, National Research Council, "Toxicity Testing, Strategies to Determine Needs and Priorities", National Academy Press, Washington DC, 1984)

Identifying properties of chemicals that require special handling can be a complicated matter. Lists of chemicals are published by numerous governmental, trade, and professional agencies. The following are five familiar lists and the agency that created each one:

1. *189 Hazardous Air Pollutants* (HAPs) (Title III, Clean Air Act Amendments 1990)
2. *Registry of Toxic Effects of Chemical Substances* (RTECS) [National Institute for Occupational Safety and Health (NIOSH)]
3. *List of Toxic and Hazardous Substances* [Occupational Safety and Health Administration (OSHA)]
4. *Pocket Guide to Chemical Hazards* [National Institute for Occupational Safety and Health (NIOSH)]
5. *Threshold Limit Values for Chemical Substances and Physical Agents in the Workroom Environment with Intended Changes* (published and updated yearly) [American Conference of Governmental and Industrial Hygienists (ACGIH)]

The *Registry* (item 2 above), is published and updated by NIOSH every few years in compliance with the 1970 Occupational Safety and Health Act. The *Registry* is the most comprehensive source to consult. It contains toxicity data extracted from the scientific and professional literature for a fraction of the approximately 65,000 chemical compounds listed in the registry. The toxicity data should not be considered a definition of values for describing a safe dose for human exposure. Figure 1-3, taken from a study of a representative group of chemicals, shows the limited knowledge about the toxic properties of chemicals listed in RTECS. Most alarming is the fact that there is no toxicological information about nearly 75% of all of the chemicals in the three "chemicals in commerce" production categories which represent the majority (both in number and volume) of chemicals used in the United States. TSCA also requires NIH and EPA to create a computer database and search programs for chemicals listed in RTECS. The database is called the *Chemical Information Service* (CIS) and is an iterative, on-line service that enables users to retrieve toxicological data for chemicals identified by their RTECS numbers. In addition, CIS enables users to search for chemicals with specific toxicological properties, structures, dose, and so on. The American Chemical Society created a registry that uniquely identifies specific compounds by a *Chemical Abstract Service* (CAS) number. The CAS inventory contains many substances posing no hazard to health and is considerably larger than RTECS. Many professional journals require authors to list the CAS numbers for all chemicals included in their article. Chemicals listed in the RTECS and CIS also cite the CAS numbers.

Other compilations of this information can be found in engineering handbooks such as the *Handbook of Environmental Data* (Verschueren, 1983), *Kirk–Othmer Encyclopedia of Chemical Technology* (Prugh, 1982), and *Dangerous Properties of Industrial Materials* (Sax, 1979). Also, some states, including Massachusetts and California, publish a "Right-to-Know" list of substances. It is not the shortage, but the plethora of lists that causes confusion.

Two procedures required by the federal government help to identify hazardous substances and to handle them properly. Whenever chemicals are transferred between buyer and seller, OSHA requires that the transfer be accompanied by a *material safety data sheet (MSDS)* that specifies the following properties of the material:

1. Manufacturer's name and chemical synonym
2. Hazardous ingredients (pigments, catalysts, solvents, additives, vehicle)
3. Physical data (boiling temperature, vapor pressure, solubility evaporation rate, percent volatile material)
4. Fire and explosion data (flash point, flammability limits, ignition temperature, firefighting procedures)
5. Health hazard (threshold limit value, effects of overexposure, first-aid procedures)
6. Spill and leak procedures, waste disposal methods

Although the MSDS may not have all the data an engineer wants, they serve to alert engineers to possible hazards. Companies are obliged to file MSDSs and to make them available to workers, thus creating a "paper trail" that can be followed to trace materials from creation to destruction.

The second procedure guides the public about hazardous new materials. The TSCA legislation requires the EPA to evaluate new substances and methods used to manufacture them, determine potential release points, estimate potential exposures, and determine whether it will be necessary to specify specific procedures to minimize exposure. Companies planning to produce or import chemicals not on the TSCA inventory are required to notify the EPA at least 90 days prior to action. The notification is called a *Premanufacture Notice* (PMN). The document requires the formula, chemical structure, use, and details about the production process so that points of release can be anticipated and estimates prepared of the emissions and human exposure. Other data required are the physical and chemical properties of the substance: vapor pressure, solubility in water or solvents, normal melting and boiling temperatures, particle size if it is a powder, Henry's law constant, pH, flammability, volatilization from water, and any toxicological data that may be available. New substances are screened by the EPA and assigned a risk category that obliges the company to control release (general ventilation, protective clothing, respirators, glove boxes, etc.) or test the compound for its toxic or environmental effects. Toxicity is assessed by examining the toxicity of analogous substances (i.e., comparable molecular structures and physical properties). Both MSDS and PMN procedures were created to minimize environmental exposures, prevent chronic exposures of people living near manufacturing plants, and reduce acute exposure of people affected by transportation accidents.

Key elements in risk assessment are the fields of epidemiology and toxicology. *Epidemiology* is the branch of medicine that investigates the cause of disease; *toxicology* is the study of the adverse effects of chemical agents on biologic systems (Klaasen et al., 1986). Epidemiology begins with the symptoms that human subjects display and relates them to certain causative factors that can be shown to be statistically significant. In a sense, epidemiology looks backward to what has already happened. Thus it is inherently limited in its ability to assess the effects of new materials and new manufacturing processes. There are several limitations to epidemiology in environmental health:

1. Information is often lacking about the duration and concentration of the exposure.
2. There is frequently a long time between exposure and symptoms of the disease. Induction periods of five to 50 years are common in lung disease and cancer.
3. Workers change occupations and/or employer and it is difficult to define exposure.
4. The number of people in the sample is often small, which complicates determining biologically significant elevations in risk.
5. Multiple exposures to several chemicals in complex industrial settings makes it difficult to determine causative agents.
6. Agents affecting health (smoking, alcohol, drugs, etc.) to which people are exposed are not identified fully, or at all.

Toxicological studies require considerable time and involve large subject populations. Because the subjects are often animals with different physiology, extrapolating the results to humans who might be subjected to low doses is fraught with debate and controversy. Occupational exposures that produce chronic effects require tests over an animal's lifetime. Occupational exposures related to pregnancy, offspring and fertility of offspring require tests over several generations of animals. There are very few tests for neurological damage.

In addition to an assessment of risk in quantitative terms, it is also necessary to place risks in *perspective:* to view risks in relation to one another (Wilson and Crouch, 1987; Ames et al., 1987; Slovic, 1987; Russell and Gruber, 1987; Lave, 1987; Okrent, 1987) and to personal activities in which citizens engage voluntarily and involuntarily. If perspective is lost, risk data acquire a false concreteness in which priorities are apt to be set foolishly.

We live in an era when *perceptions*, including those about risk, become so important they are apt to be taken as "reality" (Cox and Strickland, 1988). If perceptions were based on fact and presented dispassionately, there might be little need to worry. Unfortunately, this is not the case. We live in an era when propaganda is ubiquitous (e.g., advertising and political discussion). *Propaganda* (Orwell, 1968; Ellul, 1965) is the dissemination of information designed to provoke a certain response by arousing our emotions rather than engaging our minds. Perceptions manipulated by skillful propaganda impede formulating sound public policies. For example, terrorist bombing of commercial airliners is abominable and threatens political stability. News of it captivates the nation's attention from time to time, yet since 1976, an average of only 61 people have died per year throughout the world. Americans lose sight of the fact that they are far more likely to drown in their bathtubs (an average of 353 per year).

People may overreact to risks of low probability but large visibility (Viscusi, 1992). As noted above, terrorist bombers (aircraft, buildings, car bombs, etc.) captivate the public's attention far in excess of the actual risk they pose because of the political challenge these acts pose, as well as their unpredictability. Risks arising from acts of commission draw the public's attention, whereas acts of omission are hardly reported. When the bearers of risk do not share in the cost of reducing the risk, extravagant remedies are apt to be adopted. Overreaction to small risks often impedes the technical progress that historically improves health. The fixation of a small portion of the public against water chlorination ignores the widely accepted benefit of chlorination. Manufactured carcinogens produce a strong reaction, whereas natural carcinogens are tolerated.

To acquire perspectives about risk, one must first understand the units with which data are presented (e.g., to ask "where are the zeros?" or if the values are small, "where is the decimal point?"). For example, if an air contaminant is described as having a concentration of 0.000001 mole fraction, or is described as being present at 1 ppm or 1000 ppb, laypersons are apt to respond quite differently unless they know that these concentrations are identical. The location of zeros enables one either to excite or to lull without logical persuasion (i.e., 0.000001 sounds small, whereas 1000 ppb sounds large). Presenting data in this way perverts logical debate, but unfortunately, it is often employed. Even the units themselves tax comprehension; for example, trichloroethylene in drinking water at a concentration of 1 ppb may sound ominous, but such a ratio is comparable to the ratio of one Chinese citizen to the entire population of China, or

one-sixth of an aspirin tablet (325 mg) dissolved in a railroad tank car filled with water (16,000 gal).

Shown in Table 1-2 are the *risks* (annual fatality rate) for persons at risk in different activities. One must be careful in using Table 1-2 since the types of people affected are quite different. For the most part, police killed in the line of duty are healthy adults, mountaineers are healthy young people, while those killed in falls at home are elderly. Nonetheless, important differences of several orders of magnitude exist between incurring cancer from smoking cigarettes, eating peanut butter, or drinking water containing chloroform or trichloroethylene.

An example in which perspective has been debated is the *removal of asbestos* from schoolrooms. The probability of children contracting lung cancer (mesothelioma) is estimated (Lave, 1987) to be five per million lifetimes, less than 1/5000 the chance of death faced by other events in children's lives. The risk to a building occupant for a 10-year exposure is less than one-fiftieth (1/50) the risk of a highway fatality resulting from commuting by car to and from the building (Dewees, 1987). In addition, improper removal of asbestos poses major risks to workers, their children, and to the population as a whole (Mossman

**Table 1-2**  Annual Fatality Rates for Persons at Risk

| Action | Annual fatality rate[a] | Uncertainty |
|---|---|---|
| Motorcycling | $2.0 \times 10^{-2}$ | |
| Aerial acrobatics (airplanes) | $5 \times 10^{-3}$ | |
| All cancers | $2.8 \times 10^{-3}$ | 10% |
| Cancer related to cigarettes (1 pack/day) | $3.6 \times 10^{-3}$ | Factor of 3 |
| Electrocution | $1.1 \times 10^{-4}$ | 5% |
| Police killed in line of duty (total) | $2.2 \times 10^{-4}$ | 20% |
| Motor vehicle accident (total) | $2.4 \times 10^{-4}$ | 10% |
| Mountaineering (mountaineers) | $6 \times 10^{-4}$ | 50% |
| Hang gliding | $8 \times 10^{-4}$ | |
| One diet drink per day (saccharin) | $10^{-5}$ | |
| Alcohol, light drinker | $2 \times 10^{-5}$ | Factor of 10 |
| Motor vehicle accident (pedestrian) | $4.2 \times 10^{-5}$ | 10% |
| Death due to home falls | $3.5 \times 10^{-6}$ | |
| 4 tablespoons peanut butter/day | $8 \times 10^{-6}$ | Factor of 3 |
| Lightning | $5 \times 10^{-7}$ | |
| Drinking water with EPA limit of chloroform | $6 \times 10^{-7}$ | Factor of 10 |
| Drinking water with EPA limit of trichloroethylene | $2 \times 10^{-9}$ | Factor of 10 |

*Source:* Abstracted from Wilson and Crouch (1987) and Lehr (1992).

[a] Fatality rate is the number of fatalities per year associated with an activity per number of people engaged in the activity.

et al., 1990). Consequently, it has been decided to leave the asbestos in many buildings until the building undergoes major renovation or demolition since removal may not reduce already low concentrations and a much larger improvement in public health could be bought by spending the money on other programs.

With regard to the carcinogenicity of certain chemical agents, a technique to improve our perspective has been proposed by Ames et al. (1987) and Gold et al. (1992). While animal cancer tests cannot be used to predict human risk with absolute certainty, the tests can be used to produce an index for setting priorities that reflect the carcinogenic hazard potential of certain chemical agents. A comparison of hazards from carcinogens ingested by humans should reflect the vastly different potency that various carcinogens produce in humans. The usual measure of potency $(TD_{50})$ is the daily dose rate (in milligrams of intake per kilogram of body weight) at which 50% of any

species will be tumor-free at the end of a standard lifetime. Data from which $TD_{50}$ values can be calculated for humans are rarely available. To arrive at an index, Ames defined a term that expressed the human daily lifetime dose (in milligrams per kilogram of body weight) as a percentage of the rodent $TD_{50}$ dose (in milligrams per kilograms) as a percentage for each carcinogen. Ames called the index the *human exposure dose/rodent potency dose* (HERP):

$$HERP\,(\%)$$

$$= 100 \times \frac{\text{daily human dose (mg/kg) over a lifetime}}{\text{rodent } TD_{50}\,(\text{mg/kg})}$$

$$(1.1)$$

The rodent $TD_{50}$ values can be taken from a database for 975 chemicals and typical human exposures can be estimated from data reported in the professional literature. Shown in Table 1-3 are selected HERP

**Table 1-3**  HERP Index

| Typical examples of daily human exposure | Carcinogen [dose (μg) for a 70-kg person] | HERP (%) |
|---|---|---|
| Chlorinated tap water | Chloroform (83) | 0.001 |
| Contaminated well water (1 l) | Trichloroethylene (267) | 0.0004 |
| (Woburn, Mass.) | Chloroform (12) | 0.0002 |
|  | Tetrachloroethylene (21) | 0.0003 |
| Swimming pool (1 hr/child) | Chloroform (250) | 0.008 |
| Conventional home air (14 h/day) | Formaldehyde (598) | 0.4 |
|  | Benzene (155) | 0.04 |
| Mobile home air (h/day) | Formaldehyde (2200) | 1.4 |
| Pesticide residue on food | PCB[a] (0.2) | 0.0002 |
|  | DDE[b] (2.2) | 0.0003 |
|  | Ethylene dibromide (0.42) | 0.0004 |
| Cooked bacon (100 g) | Dimethylnitrosamine (0.3) | 0.003 |
| Peanut butter (32 g) | Aflatoxin (0.064) | 0.03 |
| Brown mustard (5 g) | Allyl isothiocyanate (4600) | 0.07 |
| Basil (1 g, dried leaf) | Estragoll (3800) | 0.1 |
| Raw mushroom (15 g) | Mix of hydrazines | 0.1 |
| Diet cola (12 oz) | Saccharin (95,000) | 0.6 |
| Beer (12 oz) | Ethyl alcohol (18 ml) | 2.8 |
| Wine (250 ml) | Ethyl alcohol (30 ml) | 4.7 |
| Worker average daily intake | Formaldehyde (6100) | 5.8 |
| Phenobarbital (1 sleeping pill) | Phenobarbital (60,000) | 16 |
| High-exposure farmworker | Ethylene dibromide (150,000) | 140.0 |

*Source:* Abstracted from Ames et al. (1987).

[a] PCB, polychlorinated biphenyls.

[b] DDE, principal metabolite of DDT.

values. Low values of $TD_{50}$ imply high potency, and high exposures increase the hazard. Thus higher HERPs suggest greater possible hazard. For example, the $TD_{50}$ for exposure of mice to chloroform is 90 mg/kg. Thus a daily dose of 83 µg of chloroform over the lifetime of a 70-kg human leads to an HERP of 0.00132%.

Contaminated wells pose a considerably smaller risk than diet cola, wine, beer, or several natural foods. Chlorine added to water kills bacteria but also interacts with organic matter to produce chloroform; nevertheless, the amount of chloroform in tap water results in a lower HERP index than in common soft drinks and natural foods. Pesticide residues on food that once caused a great deal of anxiety are seen to result in a HERP of no particular concern. With respect to ethylene dibromide, the exposure of agricultural workers is huge compared to exposures associated to residues on foods. Plants produce their own toxins to combat a variety of insects and fungi; unfortunately, some of these possess significant HERP. The aflatoxin in peanut butter is 2 ppb, which corresponds to a HERP of 0.03% for a single daily peanut butter sandwich. Of course, you wouldn't eat a peanut butter sandwich every day of your life, so the hazard from peanut butter is much less than implied by the HERP. Still you have to eat something. The formaldehyde exposure for average U.S. workers is considerably higher than the dietary intake.

It would be a mistake to use HERP data as an absolute estimate of human hazard because of the uncertainty of applying rodent cancer tests to humans. At low dose rates, human susceptibility may differ systematically from rodent susceptibility and the shapes of the dose–response curves are not known. In the final analysis the HERP index is not a scale of human risks but only a tool for estimating relative risks and to help set priorities. No society is risk-free, and to believe that life today is more risky than in the past is doubtful and certainly unproved.

## EXAMPLE 1.2   INVOLUNTARY AND VOLUNTARY RISKS

It was recently announced that the drinking water in your town contained 100 ppb (molar basis) of chloroform $CHCl_3$, $M = 119$. (The EPA maximum

allowable concentration is 100 ppb.) At a public meeting several people are alarmed at the (involuntary) risk of tumors associated with drinking tap water. You assert that there is more (voluntary) risk from the saccharin in one 12-oz can of diet cola per day than all the tap water one could drink in a day! The audience is incredulous and hoots you down. Use data from Table 1-3 to estimate the number of 12-oz glasses of tap water per day to produce a risk equivalent to one 12-oz can of diet cola per day.

***Solution***   The concentration of $CHCl_3$ in water, $c$ (µg/kg water) is,

$$c = \frac{10^{-7}\,\text{mol CHCl}_3}{\text{mol H}_2\text{O}}\left(\frac{119}{18}\right)$$

$$= 661.1 \times 10^{-9}\,\text{kg CHCl}_3/\text{kg H}_2\text{O}$$

$$= 661.1\,\mu\text{g CHCl}_3/\text{kg H}_2\text{O}$$

One 12-oz glass contains 354 g of water. Thus the mass of $CHCl_3$ per glass of water $(m)$ is

$$m = \frac{(354)(661.1)}{1000} = 234.02\,\mu\text{g of CHCl}_3/\text{glass}$$

In Table 1-3, 83 µg of $CHCl_3$ is used as the basis of comparison; thus 83 µg of $CHCl_3$ will be ingested after the consumption of 0.354 glass of tap water (i.e., $83/234.02 = 0.354$). In terms of HERP, the number of glasses of water equivalent to one 12-oz can of diet cola can be found from the following:

- HERP(diet cola)/HERP(tap water) = $0.6\%/0.001\% = 600$
- 1 can of diet cola/0.354 glass of water $\Rightarrow 600$
- 1 can diet cola $\Rightarrow (0.354)(600) = 212.4$ 12-oz glasses of water

Thus consuming one 12-oz can of diet cola per day poses a risk equivalent to drinking 212 12-oz glasses of public water per day. The average consumption of water for an adult is 1 liter (approximately three 12-oz glasses) per day. Turning the argument upside down, the daily consumption of tap water poses a threat to cancer equivalent to consuming 1.4% of a can of diet cola! If the tap water contained only 10% of the EPA maximum allowable amount of $CHCl_3$ (10 ppb), one can of diet cola produces a risk equivalent to 2120 glasses of water! This is not to say that diet soda is hazardous, only that it has a higher hazard potential than that of chlorinated drinking water. Are you curious about diet drinks sweetened with Nutrasweet (aspartame, identified as containing phenylketonurics and phenylalanine)?

These data illustrate the relatively low involuntary risk associated with public drinking water compared to the larger voluntarily risk associated with diet cola. Do you think the audience would have become as excited if it had been reported that the CHCl₃ concentration in their water was 0.1 ppm?

---

To achieve perspective, citizens need to hear and see alternative points of view regarding pollution prevention/control. This is often difficult, owing to competing and entwined interests. *Competing interests* are apt to exaggerate, advance half-truths, and misrepresent an opposing interest to further its own gain. Examples of *conflicting interests* include (a) industry, which seeks to minimize expenditures on pollution control in order to remain competitive; (b) individuals, who wish freedom to act without governmental interference; and (c) government, which implements air pollution legislation passed by elected officials. Each of the above may operate at cross purposes to the other. *Entwined interests* occur when the goals of various independent elements of society reinforce one another and create a new and powerful constituency. To many Americans, an example of entwined interests in air pollution has the following elements:

- The news media contributes to the public's anxiety regarding possible adverse health effects of pollution.
- Elected governmental officials try to satisfy the public's concerns and enhance their chances of reelection.
- Governmental agencies do the bidding of their elected officials.
- The research community provides governmental sponsors with information on which further requests for support can be based.
- Environmental activists sustain the public's interest in combating pollution.

Regardless of conflicting and entwined interests, the objective of pollution prevention/control is first and foremost to improve public health and safety.

Policies must be measured in terms of the cost to the public and benefit derived by the public (Lehr, 1992) (Arrow et al., 1996). In the final analysis, one has to answer with candor, *Are the benefits derived commensurate with the costs incurred?*

## 1.4 Liability

In Section 1.3, risk was discussed in terms of injury to individuals. There is another kind of risk of which engineers must be aware, the risk incurred in designing products whose performance (or lack thereof) may cause injury to others. In the event of such injury, the engineer may be liable (i.e., legally bound to make good on losses or damages incurred by the other party). Liability is a legal concept described by the theory of torts. A *tort* is a wrongful or injurious act for which civil action is brought. A full discussion of the theory of torts is beyond the scope of this book, but engineers should be aware of two classes of product liability: *negligence* and *strict liability*.

To prevail in a product liability suit, the plaintiff (person initiating the suit) must prove the following:

- The product contained a defect.
- The product left the manufacturer containing the defect.
- The defect was a significant factor causing the injury (*proximate cause*).
- Compensable damage arose because of the injury.

Defects arise from two categories of errors: production errors and design errors. *Production errors* pertain to the manufacturer alone and occur when the manufacturer delivers a product that does not meet the manufacturer's standards. To prevail in a damage suit based on production errors, the plaintiff must show (1) that the manufacturer failed to satisfy its own standards, and (2) that such a failure led to the injury. *Design errors* (or defects) involve events external to the manufacturer and suits based on design errors must pass one of the following tests:

- *Negligence:* tests the defendant's conduct
- *Strict liability:* weighs utility against risk
- *Express warrantee and misrepresentation:* tests the performance against explicit claims made for the product

*Negligence* is the failure to use a reasonable amount of care that results in injury or damage to another. The negligence standard concentrates on whether the engineer was careful, prudently trained, and properly supervised. Common law negligence exists if the plaintiff can prove "the violation of a statute which is intended to protect the class of persons to which the plaintiff belongs against the risk is the type of harm which has in fact occurred" (Rothstein, 1983). Liability due to negligence requires that the plaintiff show that there is a causal relationship between the violation and the injury.

Of the three forms of liability, *strict liability* is the most elusive. Strict liability weighs utility against risk. All products have utility, but they also have attendant risks. The issue is determining the balance between the two. If utility outweighs risks, it is not "defective"; that is, it is not unreasonably dangerous. If there are reasonable things the manufacturer did not do to make the product safer, the product could be ruled unreasonably dangerous. Strict liability turns on the issue of whether the product is *unreasonably dangerous* (or reasonably safe). Strict liability does not depend on whether a product can cause injury, but only that the risks inherent in its use are those that a reasonable user ought to understand and avoid based on the following:

- Usefulness and desirability of the product
- Availability of other and safer products that meet the same needs
- Likelihood of injury and its probable seriousness
- Obviousness of the danger
- Common knowledge and normal public expectation of danger
- Potential for injury despite careful use that is in accord with written instructions and warnings
- Ability to eliminate danger without seriously impairing the product's usefulness or making it unduly expensive.

The courts have ruled that manufacturers must be as knowledgeable about their products as experts in the field. The manufacturer must know all the requirements, standards, and codes that have been imposed by statute, issued by government agencies, published by technical and industry associations, and even those things known as *good engineering practice* that change with the state of the art. The standards to provide the necessary information are supposed to provide guidance, but unfortunately, they sometimes are contradictory, omit specific details, and may appear so innocuous as to be useless. Responsible engineers do not design products that are intentionally defective, but they are nonetheless liable for technical inadequacies they are not aware of but are known by competing companies or by experts in the field, and if such inadequacies cause injury to persons or property and could have been avoided.

Sections 1.1 through 1.4 addressed the milieu within which we practice air pollution control. Chapters 2 through 6 continue this discussion in depth. At this point we wish to introduce, for perspective, the types of air pollution control systems covered in Chapters 7 through 12. We conclude this chapter with a presentation of fundamental calculations that form a basis for the quantitative materials in all subsequent chapters.

# 1.5  Pollution control systems

Air pollution control systems can be grouped generically into the following classes, which reflect the physical processes used to separate pollutants from the carrier gas.

**1.** Particle control systems
- Settling chambers (gravity)
- Cyclones (inertial separation)
- Filtration (inertial separation and diffusion)
- Electrostatic precipitators (electrostatic forces)
- Wet scrubbers (inertial separation and diffusion)

**2.** Gas and vapor control systems
- Wet scrubbers (absorption)
- Activated charcoal (adsorption)
- Thermal destruction (chemical oxidation): direct flame or catalytic
- Biological oxidation (biofiltration and bioscrubbers)
- Advance oxidation (chemical reactions initiated by ultraviolet light (UV) augmented by ozone and hydrogen peroxide)

*Gravimetric settling devices*, also called *dropout boxes* (Figure 1-4), are elementary devices that remove large particles. *Dropout boxes* are enlarged portions of a duct system in which the gas velocity is

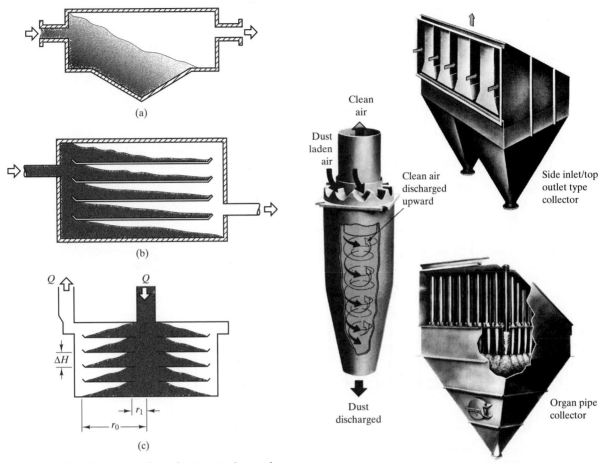

**Figure 1-4** Gravimetric settling devices (redrawn from Crawford, 1976)
(a) Dropout box
(b) Settling chamber with five trays
(c) Radial flow settling chamber with five trays

**Figure 1-5** Multiclone dust collectors containing several reverse-flow cyclones with helical inlets arranged in parallel. (Figure courtesy of Airotech, Inc.)

low and large particles settle out. *Settling chambers* incorporate several horizontal trays that enable smaller particles to fall through small distances to be collected. Settling devices should be equipped with easily accessible cleanout ports to enable workers to empty them on a routine basis. The pressure drop through settling chambers is small and they can be designed to accommodate high-temperature gases. Settling chambers are often placed upstream of other collectors to remove large particles that foul and reduce the performance of high-efficiency collectors.

*Reverse-flow cyclones* (Figure 1-5) and *straight-through cyclones* (Figure 1-6) are rugged *inertial separators* that remove particles under a variety of temperatures and pressures. Curvilinear flow produces centrifugal force that causes particles to travel radially outward, where they can be drawn off with bleed air or collected in a hopper. The removal efficiency is proportional the square of the tangential gas velocity and inversely proportional to the cyclone radius. Consequently, most inertial collectors consist of many individual cyclones carrying only a fraction of the flow arranged in parallel.

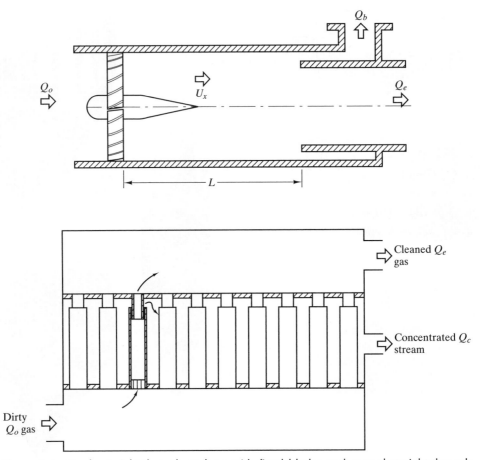

**Figure 1-6**    Single straight-through cyclone with fixed blades and several straight-through cyclones arranged in parallel as a particle concentrator (redrawn from Crawford, 1976).

Cyclones are often used upstream of high-efficiency devices to remove large particles, allowing high-efficiency devices to remove the smaller particles. The pressure drop is modest and cyclones can accommodate high-temperature gases. Because of their low collection efficiency, neither inertial nor gravimetric separators are considered stand-alone air pollution control devices.

*A baghouse* (Figure 1-7) is the name given to a high-efficiency particle collection device consisting of individual fabric filters arranged in parallel. Baghouses are divided into three categories that reflect the manner in which the filters are cleaned. *Shaker* and *reverse-flow baghouses* consist of several modular units arranged in parallel. Each module contains numerous fabric filters (bags) arranged in parallel. The filters bags are closed at the top and mounted on a cell plate (tube sheet) at the bottom. The tube sheet causes the dust-laden air to enter the bags and inflate them. The air passes radially outward and creates a dust cake on the inside of the bags. The modular units are arranged in parallel because each module is taken off-line for a short period of time when the dust cake is removed. Once the dust cake has been removed, the module is put back in line to clean the air. Bag cleaning is staggered so that for the majority of time, all modules are cleaning air but each module is at a different point of its operating cycle (i.e., one module has just come back on-line while another is about to go off-

**Mechanical shaking**

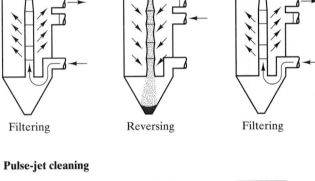

Filtering          Shaking          Filtering

**Reverse-flow cleaning**

Filtering          Reversing          Filtering

**Pulse-jet cleaning**

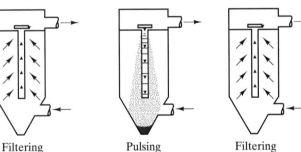

Filtering          Pulsing          Filtering

**Figure 1-7** Particle collectors using filters (baghouses) are classified by the method to remove the collected dust. (Figure provided by Wheelabrator Air Pollution Control, Inc.)

line to remove its dust cake). *Shaker filters* are cleaned by shaking the upper portion of the bag by a mechanical eccentric. *Reverse-flow filters* are cleaned by isolating the module and blowing cleaned air into the region outside the bags which flattens them and causes large fragments of the dust cake to fall below into the hopper. *Pulse jet filters* collect particles on the outside of fabric filters supported on wire frames to keep their cylindrical shape. Dust collected on the outside of the bag is removed by a pulse of high-pressure air discharged through a nozzle mounted at the upper end of the bag. As the pulse travels downward, it flexes the bag and dislodges *dust cake* on the outside of the bags, which falls to the hopper below. The bags are made of felt and cleaned every few minutes. Filtration systems have a high efficiency that is virtually independent of the volumetric flow rate, but they are limited by high temperatures and gas streams near the dew point.

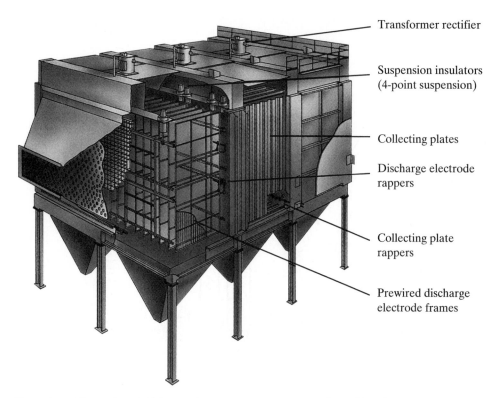

Transformer rectifier

Suspension insulators
(4-point suspension)

Collecting plates

Discharge electrode
rappers

Collecting plate
rappers

Prewired discharge
electrode frames

**Figure 1-8**    Plate-wire, rigid frame electrostatic precipitator with tumbling hammer rapping.
(Figure provided by Wheelabrator Air Pollution Control, Inc.)

*Electrostatic precipitators* (ESPs; Figure 1-8) are high-efficiency particle collectors that remove small particles from an airstream because of an electric charge placed on the particles. The plate-wire ESP shown in Figure 1-8 consists of hundreds of vertically mounted small-diameter wires placed midway between dozens of large vertically mounted plates that are electrically grounded. High-voltage applied to the wire produces a corona that generates electrons that are transferred to airborne particles. The electrically charged wire and plates also establish an electric field that causes the charged particles to migrate laterally and attach themselves to the plates. The plates are rapped mechanically, which causes large fragments of the deposited dust to slack off and fall to hoppers below. The performance of ESPs is very sensitive to variations in the gas volumetric flow rate and the electrical resistance of the collected dust.

They are commonly used to clean large steady volumetric flow rates of gas, such as in coal-fired electric utility boilers and lime and cement kilns.

While normally thought of as a gas and vapor removal system, *packed bed, bubble cap* and *transverseflow wet scrubbers* of a variety of designs can be used to remove particles from process gas streams. *Venturi scrubbers* (Figure 1-9) dispense with packing and cause high-speed airborne dust particles to impact and adhere to liquid droplets in the throat section. Packed bed scrubbers (Figure 1-10) are used to remove gases and vapors from a process gas stream. Airborne pollutants are transferred to the scrubbing liquid as the two streams encounter one another in the packing. The packing creates a large surface area per unit volume. Pollutants transferred to the scrubbing liquid must be removed by independent means before the liquid is recycled to the top of

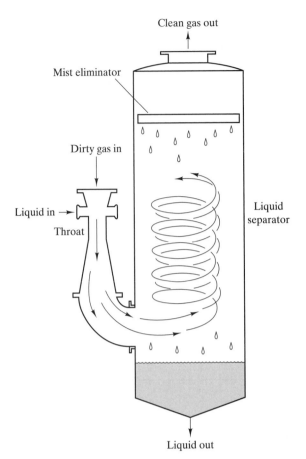

**Figure 1-9** Schematic diagram of a venturi scrubber. The cyclone separator and mist eliminator collect drops of scrubbing liquid

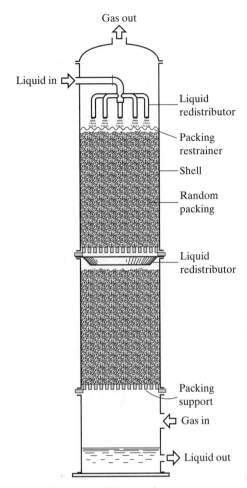

**Figure 1-10** Schematic diagram of a counter-current scrubber containing random packing (taken from Treybal, 1955)

the tower. *Transverse-flow scrubbers* (Figure 1-11) achieve the necessary gas–liquid interaction in a long length of packing in the horizontal direction if insufficient headroom is available. *Bubblecap scrubbers* (Figure 1-12) use numerous trays through which the gas is bubbled to achieve the gas–liquid interaction needed to remove particle and gaseous pollutants. At the exit of all wet scrubbers is a *demister* to capture small drops of the scrubbing liquid suspended in the cleaned gas.

If a pollutant is worth recycling or if the pollutant is unusually toxic, it can be captured in *adsorption* systems of a variety of designs. Adsorption is a batch process, so provision must be made to regenerate (or

replace) the adsorbent and recover or destroy the captured pollutant. The system shown in Figure 1-13 uses a rotary wheel to adsorb pollutant for a portion of its revolution and desorb for the remainder of its travel. The desorbed pollutant is burned in a thermal oxidizer in which the energy released by combustion is used to preheat the air used in desorption. If the pollutant has financial value, it can be recovered after desorption by a variety of methods.

*Thermal oxidation* (Figure 1-14) processes incorporating energy recovery techniques can be used to destroy pollutants chemically. *Regenerative thermal oxidizers* transfer combustion energy to one of several packed beds which heat incoming air at a later

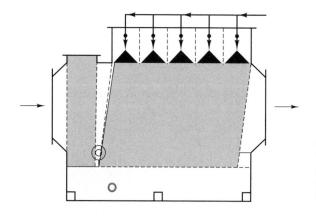

**Figure 1-11** Transverse (cross-flow) wet scrubber with random packing for gas or vapor removal (provided by Mystaire, Heat Systems Inc.)

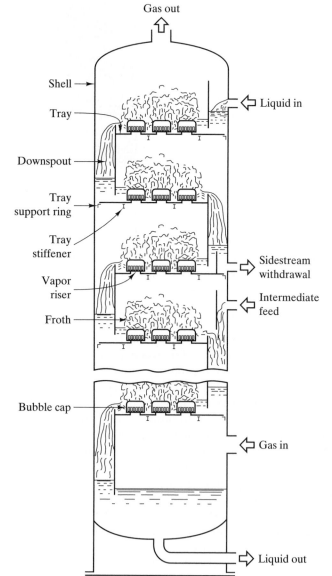

**Figure 1-12** Schematic diagram of a bubble-cap, plate scrubber for gas and vapor removal (taken from Treybal, 1955)

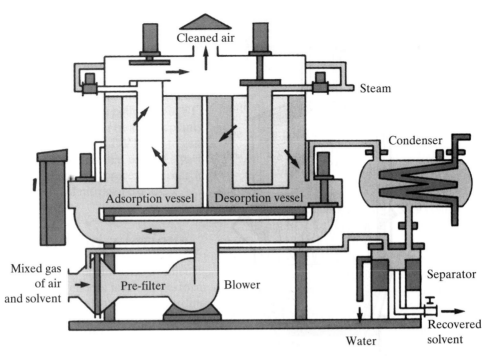

**Figure 1-13**  Solvent recovery system using activated carbon fiber adsorbent with regeneration with steam. (Figure courtesy of Met Pro Systems)

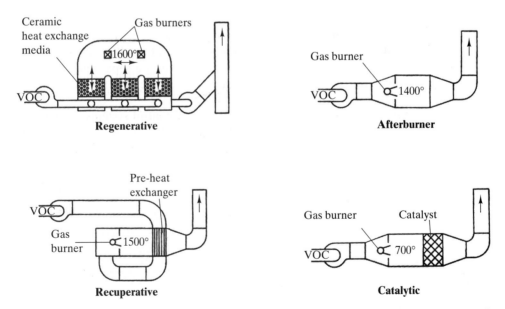

**Figure 1-14**  Four types of thermal oxidizers (a) regenerative (alternating beds of random packing), (b) recuperative (fin-tube heat exchanger), (c) afterburner (no energy recovery), (d) catalytic oxidizer (may incorporate energy recovery). (Figure provided by Ross Inc.)

*Nitrogen:* (1.3)(3.76) (8.6753 kmol/s)

   (28 kg $N_2$/kmol $N_2$)   = 1187.3363 kg $N_2$/s

   total mass leaving boiler = 1663.0227 kg/s

The difference between input and output mass is 6.96 kg/s (0.4%). The small discrepancy occurs because the average molecular weight (28.97) used to compute the input air flow includes gases not accounted for in the output.

## 1.6.2 Mass and volumetric flow rates

As a general proposition, fluid flow should always be described in terms of mass flow rate since there is no confusion about what kg/s means. Owing to ease of measurement, volumetric flow rates have been used in engineering for quite some time. For liquids, the volumetric flow rate is not confusing since the density of a liquid is often nearly constant. Gas flows are a different matter, however, and two terms are widely used to describe volumetric flow rates. In engineering units the two terms are *standard cubic feet per minute* (scfm) and *actual cubic feet per minute* (acfm). The unit of time, minutes, is sometimes replaced by seconds or hours and poses no confusion. On the other hand, the relationship between the two volumetric flow rate designations needs to be established.

The mass flow rate of gas through a duct is given by

$$\dot{m} = \int_A \rho U \, dA \qquad (1.2)$$

where $U$ is the velocity perpendicular to the differential area $dA$ and $\rho$ is the actual density of the gas at the temperature and pressure $(T, P)$. The ideal (perfect) gas law gives

$$\rho = \frac{PM}{ZR_u T} = \frac{PM}{R_u T} \qquad (1.3)$$

as process gas streams encountered in air pollution are for the most part at temperatures and pressures where the compressibility factor $(Z)$ is unity. The symbol $M$ is the molecular weight of the gas and $R_u$ is the *universal gas constant*. If the density $(\rho)$ does not vary with respect to the area $A$, Equation. 1.2 becomes

$$\dot{m} = \rho \int_A U \, dA = \rho Q \qquad (1.4)$$

where $Q$ is the *actual volumetric flow rate* at the actual temperature and pressure $(T, P)$ of the gas. Typical units for $Q$ are acfm (actual cubic feet per minute) or actual cubic meters per second (m³/s). It is common practice in air pollution to speak of a standard volumetric flow rate in such units as scfm (standard cubic feet per minute), which is a hypothetical volumetric flow rate equal to what the actual volumetric flow rate would be at standard temperature and pressure:

Standard pressure = 1 atm = 101.325 kPa

= 14.7 psia

Standard temperature = 25°C = 298.5 K       (1.5)

= 77°F = 537°R

Thus the *standard volumetric flow rate* is a hypothetical or corrected or reference volumetric flow rate; it is not the actual volumetric flow rate. Since the mass flow rate is independent of temperature and pressure,

$$Q_{STP} = Q_{actual} \frac{\rho_{actual}}{\rho_{STP}}$$

$$= Q_{actual} \frac{P_{actual}}{P_{STP}} \frac{T_{STP}}{T_{actual}} \qquad (1.6)$$

In the commonly used engineering units of cubic feet per minute, Equation 1.6 becomes

$$Q(\text{scfm}) = Q(\text{acfm}) \frac{P_{actual}}{P_{STP}} \frac{T_{STP}}{T_{actual}} \qquad (1.7)$$

There are times when engineers are casual and speak of the volumetric flow rate in the ambiguous terms cfm, which leaves listeners in a quandary since they do not know what temperature and pressure is implied. If the actual temperature and pressure are close to the standard values, the ambiguity is inconsequential. Engineers should always explicitly indicate the actual or standard volumetric flow rate. When using SI units, confusion seldom arises because the modifying terms *standard* and *actual* are generally not used. In SI units a volumetric flow rate is understood to be the actual value even though the modifying word *actual* is not used. Whenever possible, one should use the mass flow rate in appropriate units and avoid any confusion.

1. The term *acfm* is incomplete without specifying the actual (absolute) pressure and temperature

where the measurement is made. Thus one should always say that the volumetric flow rate is $Q(\text{acfm})$ at the pressure $P(\text{psia})$ and temperature $T(°R)$, for example.

2. The term *scfm* has a unique meaning and there is no need to specify the actual pressure and temperature. If the user wants to know the actual fluid velocity, it will be necessary to specify the actual pressure and temperature of interest and use Equation 1.7.

---

**EXAMPLE 1.4   DESCRIBING THE FLOW OF AIR INSIDE A DUCT**

Air passes through a 10 inch-ID duct at 250°F and 100 psig. The velocity is found to be uniform and equal to 25 ft/s. Compute the mass flow rate ($\text{lb}_m/\text{min}$) and the volumetric flow rate in acfm and scfm.

*Solution*   The air density is given by Equation 1.3:

$$\rho = \frac{PM}{RT} = \frac{(100 + 14.7)(144)(28.97)}{(1545)(250 + 460)}$$

$$= 0.436 \text{ lb}_m/\text{ft}^3$$

The mass flow rate is given by Equation 1.2:

$$\dot{m} = \rho A U = 0.436 \left[\frac{\pi(10)^2}{(4)(144)}\right](25)$$

$$= 5.94 \text{ lb}_m/\text{s} = 356.5 \text{ lb}_m/\text{min}$$

The volumetric flow rate in acfm is

$$Q(\text{acfm}) = \frac{\dot{m}}{\rho} = \frac{356.5}{0.436} = 817.6 \text{ acfm}$$

The volumetric flow rate in the units scfm is given by Equation 1.7:

$$Q(\text{scfm}) = 817.6\left(\frac{100 + 14.7}{14.7}\right)\left(\frac{77 + 460}{250 + 460}\right)$$

$$= 4825.1 \text{ scfm}$$

Clearly, it is foolhardy to specify a volumetric flow rate in the confusing units cfm. One must be precise and use either acfm or scfm. Better yet, avoid the confusion and use the mass flow rate.

## 1.6.3  Contaminant concentration

Air contaminants may be particles or gases. *Particles* may be solid and/or liquid. *Gaseous contaminants* may be truly gases or they may be potentially condensable vapors. Both contaminant gases and vapors that obey the ideal gas law and Dalton's law of partial pressures will be called simply *gas* throughout this book and their individual compressibility factors will be taken to be unity. *Fume* is the generic name for contaminants generated by exothermic processes, often metallurgical processes. Fumes are a mixture of gases, vapors, and particles in varying amounts. The particles are tiny (characteristic dimensions are generally considerably less than 1 μm), and for the most part are condensates of vapors produced by the exothermic process. Fumes may also contain small solid particles that never changed phase.

The mass of particles per unit volume of carrier gas is called the *mass concentration* ($c$). Often just the term *concentration* is used and readers must be careful to understand what units are implied since some users will be thinking of the number of particles per unit volume of carrier gas, while others will be thinking of the mass of particles per volume of carrier gas. Careful attention to the units will signify which definition is being used. There is no consensus on the phrase concentration because theoretical and experimental advantages accompany different definitions. To clarify the issue, users should demand that the units of concentration be specified. A variety of units are used for the mass concentration, but milligrams per cubic meter is common and is used in this book unless the powers of 10 suggest more convenient units.

The mass concentration of gaseous contaminants has the same meaning as for particles. If the molecular weight of the contaminant is known, the concentration can be given as a mole fraction. A relationship between mole fraction and mass concentration can be developed from the *ideal gas law* and *Dalton's law of partial pressure*. A single species of a gas satisfies

$$PV = Z\frac{m}{M}R_uT = Zn_tR_uT \tag{1.8}$$

where

$$m = \text{mass of the gas}$$

$$n_t = \text{total number of moles}$$

$$R_u = \text{universal gas constant}$$

$$Z = \text{compressibility factor}$$

The density of dry air $(\rho)$ is

$$\rho = \frac{m}{V} = \frac{PM}{ZR_uT} \qquad (1.9)$$

If the total pressure is well below the critical pressure $(P \ll P_c)$, the compressibility factor $(Z)$ is unity and Equation 1.9 reduces to Equation 1.3. If the air contains water vapor, the *density of moist air* can be computed from

$$\rho_{\text{moist}} = \frac{PM}{ZR_uT} \left( 1 - \frac{M_v}{M_a} \Phi f \frac{P_v}{P} \right) \qquad (1.10)$$

where $M_a$ and $M_v$ are the molecular weights of air and water, $P_v$ is the saturation pressure of water at temperature $T$, and $\Phi$ is the relative humidity (used as a fraction). Jennings (1988) claims that the constant $f$ is equal to 1.004 for air over a wide range of temperatures and pressures.

For a *mixture of perfect gases*, the ith molecular species satisfies the law of *partial pressures:*

$$P_iV = \frac{m_i}{M_i} R_uT = n_iR_uT \qquad (1.11)$$

The *mole fraction* of gaseous species $i$ can be written as

$$y_i \equiv \frac{n_i}{n_t} = \frac{P_i}{P_t} = \frac{V_i}{V_t} \qquad (1.12)$$

where $P_i$ and $V_i$ are the partial pressure and partial volume. By definition of the mole fraction $(y_i)$, a value of $y_i$ of $10^{-6}$ would be one part $i$ of species $i$ per million total parts. We call this ratio ppm:

$$1 \text{ ppm} \equiv \frac{1}{10^6} \frac{\text{part}}{\text{parts}}; \qquad \text{ppm} = 10^6 y_i \qquad (1.13)$$

For the very low concentrations often associated with the term *parts per million* the ratio of the number of molecules of species $i$ per million molecules of carrier gas is essentially equal to the ppm. That is, $1/(1 + 10^6) \simeq 1/10^6$. The value 1 ppm can also be used to mean one gram of a pollutant per million grams of surrounding medium. To eliminate confusion, the abbreviation ppm should always be defined with modifying words:

- *ppm based on volume:* ratio of moles of pollutant to $10^6$ total moles
- *ppm based on mass:* ratio of grams of pollutant to $10^6$ g of total material

Because the concentration of air pollutants is often expressed as a mole fraction $(y_i)$, in this book we define the abbreviation ppm consistently on the basis of volume alone. The words *on the basis of volume* are omitted for the sake of brevity. On those occasions when a ratio based on mass is appropriate, the abbreviation ppm will *never* be used. Rather, the ratio of the grams of pollutant per million grams of medium (or micrograms of pollutant per gram of medium) will be stated explicitly.

The mass concentration of gaseous species $i$ per total volume of carrier gas $(c_i)$ and the mole fraction of species $i$ are related as follows:

$$c_i = \frac{m_i}{V} = \frac{P_i}{T} \frac{M_i}{R_u} = y_i \frac{P}{T} \frac{M_i}{R_u} \qquad (1.14)$$

At *standard temperature and pressure* (STP) the mass concentration in the units mg/m$^3$ is

$$c_i(\text{mg/m}^3)_{\text{STP}} = \frac{y_iM_i}{24.5 \times 10^{-6}} \simeq \frac{(\text{ppm})M_i}{24.5} \qquad (1.15)$$

where $y_i$ is the mole fraction, ppm $\times 10^6$ is its virtual equivalent, and 24.5 is a conversion factor. If the actual pressure $P$ and temperature $T$ are not STP, the mass concentration is

$$c_i(\text{mg/m}^3)_{\text{actual}} = \frac{y_iM_i}{24.5 \times 10^{-6}} \frac{P_{\text{actual}}}{P_{\text{STP}}} \frac{T_{\text{STP}}}{T_{\text{actual}}}$$

$$\simeq \frac{(\text{ppm})M_i}{24.5} \frac{P_{\text{actual}}}{P_{\text{STP}}} \frac{T_{\text{STP}}}{T_{\text{actual}}} \qquad (1.16)$$

The absolute and actual pressure and temperature must be expressed in compatible units. Since the mole fraction is defined as a ratio of numbers of molecules in Equation 1.12, the mole fraction is independent of temperature and pressure. Similarly independent is composition expressed in parts per million:

$$\text{ppm}_{\text{STP}} = \text{ppm}_{\text{actual}} \qquad (1.17)$$

There are times when the concentration is so small that it is inconvenient to use the molar concentration in the units kmol/m$^3$ or even gmol/m. Examples of such small concentrations are the atmospheric concentrations of hydroxyl or oxygen free radicals. In these cases, the units molecules/m³ are often used and by making use of Avogadro's number, the mole fraction can be expressed as

$$y_i \equiv \frac{n_i}{n_t} = \frac{(n_i)6.02 \times 10^{23}}{(n_t)6.02 \times 10^{23}} \qquad (1.18)$$

At STP (101 kPa, 298 K) the number of molecules/m³ is $2.46 \times 10^{25}$. The mol fraction expressed as a percent (%) is also called the *percent by volume*. Recalling that the number of molecules in a gram mole is equal to *Avogadro's number* ($6.02 \times 10^{23}$), Equation 1.16 becomes

$$c_i(\text{molecules/m}^3)_{\text{actual}} = 2.457 \times 10^{25} \, y_i \frac{P_{\text{actual}}}{P_{\text{STP}}} \frac{T_{\text{STP}}}{T_{\text{actual}}}$$

$$\simeq 2.457 \times 10^{19} (\text{ppm}) \frac{P_{\text{actual}}}{P_{\text{STP}}} \frac{T_{\text{STP}}}{T_{\text{actual}}} \qquad (1.19)$$

---

**EXAMPLE 1.5   POLLUTANT CONCENTRATION AND EMISSION RATE**

An exhaust vent from a plastics manufacturing plant discharges 100 acfm of air at 200°C, 90 kPa. A continuous emissions monitor shows that the discharge contains 5 ppm of phenol vapor. Table A-11 in the Appendix indicates that 5 ppm is close to the odor threshold and the community can be expected to detect the medicinal, sweet odor of phenol. The molecular weight of phenol is 94.1. What is the actual phenol concentration in the exhaust stream, and what is the emission rate of phenol to the atmosphere?

*Solution*   From Equation 1.16, the actual concentration of phenol is

$$c(\text{mg phenol/m}^3 \text{ air})_{\text{actual}} = \frac{(5)(94.1)(90)(298)}{(24.5)(101)(473)}$$

$$= 10.8 \text{ mg/m}^3$$

The emission rate of phenol is

$$\dot{m}(\text{mg/h})_{\text{phenol}}$$
$$= c(\text{mg phenol/m}^3 \text{ air})_{\text{actual}} Q(\text{m}^3 \text{ air/h})_{\text{actual}}$$
$$= \frac{(10.8 \text{ mg/m}^3)(100 \text{ ft}^3/\text{min})(60 \text{ min/h})}{(3.28 \text{ ft/m})^3}$$
$$= 1836.3 \text{ mg/h}$$

## 1.6.4 Mixtures of ideal liquids

An *ideal solution* is one in which no energy is generated (or consumed) when a number of individual species are mixed. Thus the total volume ($V_t$) of a mixture containing $N$ distinct molecular species is equal to the sum of the volumes of the individual components:

$$V_t = \sum_i^N V_i \qquad (1.20)$$

For brevity, the limits on the summation are omitted in the material to follow.

The total mass ($m_t$) of the mixture and the total number of moles ($n_t$) of the mixture can be expressed as

$$m_t = \sum m_i \qquad (1.21)$$

$$n_t = \sum n_i = \sum \frac{m_i}{M_i} \qquad (1.22)$$

The average molecular weight, $M_{\text{avg}}$, can be defined by

$$M_{\text{avg}} = \frac{m_t}{n_t} = \frac{\sum_i n_i M_i}{n_t} = \sum_i \frac{n_i}{n_t} M_i \qquad (1.23)$$

The mole fraction of species $i$ in the liquid phase is

$$x_i = \frac{n_i}{n_t} \qquad (1.24)$$

Hence

$$M_{\text{avg}} = \sum_i x_i M_i \qquad (1.25)$$

Define *average density* $\rho_{\text{avg}}$ or *total density* $\rho_t$ by dividing the total mass ($m_t$) by the total volume ($V_t$):

$$\rho_{\text{avg}} = \rho_t = \frac{m_t}{V_t} = \sum \frac{m_i}{V_t} = \sum \bar{\rho}_i \qquad (1.26)$$

where $\bar{\rho}_i$ is defined as the *partial density*, which can be thought of as the density species $i$ would have if it alone occupied the entire volume. The volume occupied by each species ($V_i$) is

$$V_i = \frac{m_i}{\rho_i} \qquad (1.27)$$

The total volume ($V_t$) of an ideal mixture is

$$V_t = \sum V_i = \sum \frac{m_i}{\rho_i} \qquad (1.28)$$

The reciprocal of the average density $(1/\rho_{avg})$ can be expressed as

$$\frac{1}{\rho_{avg}} = \frac{V_t}{m_t} = \sum \frac{m_i}{m_t}\frac{1}{\rho_i} = \sum \frac{f_i}{\rho_i} \qquad (1.29)$$

where $f_i$ is the mass fraction of species $i$. The *average molar concentration* is equal to the total number of moles $(n_t)$ divided by the total volume $(V_t)$. The *average molar concentration* $(c_{avg})$ is also the *total molar concentration* $(c_{avg})$:

$$c_{avg} = c_t = \frac{n_t}{V_t} = \sum \frac{n_i}{V_t} = \sum \frac{m_i}{M_i V_t}$$

$$= \sum \frac{m_i}{V_t}\frac{1}{M_i} \qquad (1.30)$$

$$= c_t = \sum \frac{\bar{\rho}_i}{M_i} \sum \bar{c}_i \qquad (1.31)$$

The term $\bar{c}_i$ is defined as the *partial molar concentration*, which can be interpreted as the concentration species $i$ would have if it alone occupied the entire volume. Equations 1.26, 1.29, 1.30, and 1.31 suggest methods for calculating the total density and total concentration from pure component data. For ideal solutions these averaging methods are rigorous (i.e., $\rho_{avg} = \rho_t$ and $c_{avg} = c_t$). One may choose to calculate $\rho_{avg}$ and $c_{avg}$ this way for nonideal solutions, but one should not expect the averaged values to equal the true total values.

For a single species, the molar concentration is related to the density by

$$c = \frac{n}{V} = \frac{m}{M}\frac{1}{V} = \frac{m}{V}\frac{1}{M} = \frac{\rho}{M} \qquad (1.32)$$

Thus

$$M = \frac{\rho}{c} \qquad (1.33)$$

Written for a mixture, Equation 1.33 provides an alternative definition to Equation 1.23 for the *average molecular weight* $(M_{avg})$. Hence

$$M_{avg} = \frac{\rho_t}{c_{avg}} = \frac{\rho_t}{c_t} = \frac{m_t/V_t}{\sum(m_i/M_i)(1/V_t)} \qquad (1.34)$$

$$\frac{1}{M_{avg}} = \sum \frac{m_i}{m_t}\frac{1}{M_i} = \sum \frac{f_i}{M_i} \qquad (1.35)$$

where $f_i$ is the *mass fraction*, defined as

$$f_i = \frac{m_i}{m_t} = \frac{m_i}{\sum m_i} = \frac{m_i/V_t}{\sum(m_i/V_t)} = \frac{\bar{\rho}_i}{\rho_t} \qquad (1.36)$$

The *mole fraction in the liquid phase* $(x_i)$ is defined as

$$x_i = \frac{n_i}{n_t} = \frac{n_i}{\sum n_i} = \frac{n_i/V_t}{\sum(n_i/V_t)} = \frac{\bar{c}_i}{c_t} \qquad (1.37)$$

The mole fraction and mass fraction are related by

$$f_i = \frac{m_i}{m_t} = \frac{m_i/M_i}{(m_t/M_{avg})(M_{avg}/M_i)}$$

$$= \frac{n_i}{n_t}\frac{M_i}{M_{avg}} = x_i \frac{M_i}{M_{avg}} \qquad (1.38)$$

## EXAMPLE 1.6    COMPOSITION OF A SOLVENT MIXTURE

The label on a gallon $(3.788 \times 10^{-3}\ m^3)$ of a commercial product states the mole fractions of its contents; isopropyl alcohol (P), methyl alcohol (M), isobutyl alcohol (B), and water (W). The mole fractions $(x_i)$ and densities of the pure species $(\rho_i)$ are shown in the table below.

**a.** Compute the average molecular weight $(M_{avg})$ and average density $(\rho_{avg})$ of the mixture.
**b.** Compute the mass $(m_i)$, mass fraction $(f_i)$, and number of moles $(n_i)$ of each species.

| *Species* | $x_i$ | $M_i$ | $\rho_i$ (kg/m³) | $f_i$ |
|-----------|-------|-------|------------------|-------|
| P | 0.1 | 60 | 785 | 0.144 |
| M | 0.2 | 32 | 790 | 0.153 |
| B | 0.3 | 74 | 800 | 0.531 |
| W | 0.4 | 18 | 1000 | 0.172 |
|   | 1.0 |    |    | 1.000 |

**Solution**

**a.** From Equation 1.25,

$$M_{avg} = (0.1)(60) + (0.2)(32)$$
$$+ (0.3)(74) + (0.4)(18) = 41.8$$

The mass fraction $f_i$ can be found from Equation 1.38. The average density can be found from Equation 1.29:

$$\frac{1}{\rho_{avg}} = \sum \frac{f_i}{\rho_i}$$

$$= \frac{0.144}{785} + \frac{0.153}{790} + \frac{0.531}{800} + \frac{0.172}{1000}$$

$$= 0.00121286 \text{ m}^3/\text{kg}$$

$$\rho_{avg} = 824.49 \text{ kg/m}^3$$

Thus the mass of a gallon is

$$m_t = (824.49 \text{ kg/m}^3)(3.788 \times 10^{-3} \text{ m}^3)$$

$$= 3.1231 \text{ kg}$$

The total number of moles $(n_t)$ is

$$n_t = \frac{m_t}{M_{avg}} = \frac{3.123}{41.8} = 0.0741 \text{ kmol}$$

**b.** The mass of each species $(m_i)$ can be found from $m_i = f_i m_t$, and the number of moles of each species can be found from $n_i = m_i/M_i$.

| Species | $m_i$ (kg) | $n_i$ (kmol) |
|---------|-----------|--------------|
| P | 0.4497 | 0.007495 |
| M | 0.4778 | 0.01493 |
| B | 1.6584 | 0.02241 |
| W | 0.5371 | 0.02983 |
|   | 3.123 | 0.074665 |

The total molar concentration is equal to $\rho_{avg}/M_{avg}$ ($824.49/41.8 = 19.725$), or it can be found from the following:

$$V_i = \frac{m_i}{\rho_i} = \frac{m_i}{M_i}\frac{M_i}{\rho_i} = \frac{n_i M_i}{\rho_i}$$

$$V = \sum V_i = \sum \frac{n_i M_i}{\rho_i}$$

$$\frac{1}{c_t} = \frac{V}{n_t} = \sum \frac{n_i}{n_t}\frac{M_i}{\rho_i} = \sum \frac{x_i M_i}{\rho_i}$$

Upon substitution, we obtain

$$c_t = \left( \sum \frac{x_i M_i}{\rho_i} \right)^{-1}$$

$$= \left[ \frac{(0.1)(60)}{785} + \frac{(0.2)(32)}{790} \right.$$

$$\left. + \frac{(0.3)(74)}{800} + \frac{(0.4)(18)}{1000} \right]^{-1}$$

$$= 19.726 \text{ kmol/m}^3$$

# 1.7 Rate concepts

**Rate concept and mass balances.** Effective analysis of chemical processes requires recognition that chemical reactions are rate processes. The term *rate* is often used loosely or misunderstood. Therefore, a careful definition of the rate concept is essential at the outset.

For reasons of formalism in chemical reaction engineering (Fogler, 1992) it is preferable to think of the process rate in a chemically reacting system as the rate of *formation* of a species of interest. Then the more familiar term *rate of reaction* implies the rate of *disappearance* of that species of interest. Accordingly, for the formation of sulfuric acid from sulfur trioxide and water in the atmosphere,

$$SO_3 + H_2O \rightleftharpoons H_2SO_4 \tag{1.39}$$

the term $r_{H_2SO_4}$ represents the rate of formation of $H_2SO_4$ and $-r_{H_2SO_4}$ represents the rate of disappearance (or reaction) of $H_2SO_4$. From the stoichiometry of the simple reaction, it can be seen that

$$r_{H_2SO_4} = -r_{SO_3} = -r_{H_2O} \tag{1.40}$$

That is, the rate of formation of sulfuric acid is equal to the rates of reaction of sulfur trioxide and water.

**Rate of change and process rate.** The dictionary defines *rate* as "a quantity, amount, or degree of something measured per unit of something else." The "something" in this definition may be anything: for example, temperature, position, or concentration. The "something else" is often time, but it can also be position or some other independent variable. This definition corresponds to the *rate of change*, familiar from calculus. The rate of change may or may not be important in chemical processes. The *process rate*, also known as the *rate of reaction*, is always important. Rate of change and process rate are contrasted in Table 1-4.

To illustrate the distinction between a rate of change and a process rate, consider the *continuous-flow stirred tank reactor* (CSTR) (also called a *well-mixed reactor*) system shown in Figure 1-17. An unsteady-state mass balance on component A is expressed by

**Table 1-4**    Comparison of Rate of Change and Process Rate

| Rate of change | Process rate |
|---|---|
| Follows dictionary definition | Just a concept |
| Measured (examples: change of position with time, change of temperature with altitude or latitude) | Detemined from measurement and theory (examples: rate of reaction, rate of heat transfer) |
| Has a derivative character $$\lim \Delta x/\Delta t \rightarrow dx/dt$$ as $\Delta t \rightarrow 0$ | Just a concept usually denoted by $r$ for chemical reactions, $q$ for heat transfer, etc. |

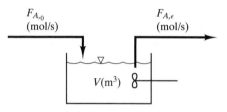

**Figure 1-17**    Molar flux of species A entering and leaving a continuous stirred tank reactor (CSTR)

$$F_{A,0} - F_{A,e} + r_A V = \frac{dn_A}{dt} \qquad (1.41)$$

where $n_A$ is the total number of moles of A in the reactor, and $F_{A,0}$ and $F_{A,e}$ are the molar flow (feed) rates of species A into and exiting the reactor. The rate of formation of A by reaction, $r_A$, is a process rate. The derivative, $dn_A/dt$, is the rate of change of moles of A with time.

**Deduction of the rate of reaction.** Equation 1.41 can be solved for the process rate, $r_A$, as a function of the rate of change, $dn_A/dt$, and other variables in the equation. A more familiar expression is obtained by restricting attention to the steady-state case, in which $dn_A/dt$ is equal to zero. Then Equation 1.41 can be written as

$$-r_A = \frac{F_{A,0} - F_{A,e}}{V} \qquad (1.42)$$

or alternatively,

$$r_A = \frac{F_{A,e} - F_{A,0}}{V} \qquad (1.43)$$

These equations show how the rate of reaction may be determined experimentally in a CSTR operated at steady state. Note, however, that a mass balance was

required to arrive at this conclusion. One does not measure process rates directly but only arrives at their values by some combination of measurement and theory of what occurs in the reactor.

A mass balance on a differential element of a plug flow reactor (Figure 1-18) is (Kabel, 1985)

$$-r_A \, dV = F_{A,0} \, dX \qquad (1.44)$$

where $F_{A,0}$ is the molar feed rate of species A and $dX$ is the differential fractional conversion in the differential volume, $dV$. The *fractional conversion, X,* is defined as moles of A reacted at any point in space or time per mole of A fed to the reactor. Equation 1.44 can be written as

$$-r_A = \frac{dX}{d(V/F_{A,0})} \qquad (1.45)$$

to express the rate of reaction $(-r_A)$. It can be seen that the process rate in this case is equal to the rate of change of conversion $(X)$ with $V/F_{A,0}$. This relation between a process rate and a rate of change is frequently used to determine rates of reaction in

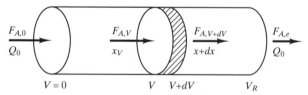

**Figure 1-18**    Molar flux of species A entering and leaving a plug flow reactor containing a differential control volume
$F_{A,0}, F_{A,e}$ - molar feed rate of molecular species A (kmol/s)
$F_{A,V}$ - molar feed rate entering the differential control volume
$Q_0$ - volumetric flow rate at inlet $(m^3/s)$
$X_V$ - fractional conversion of mol of A reacted at V per mol of A fed
$V$ - reactor volume downstream of the inlet $(m^3)$

tubular reactors. The term $V/F_{A,0}$ is often referred to as the *reciprocal space velocity*. When $F_{A,0}$ is expressed as a function of the volumetric flow (feed) rate $(Q_0)$, the term $V_R/Q_0$ is referred to as the *space time*, where $V_R$ is the total volume of the reactor.

As a final example, consider the case of a batch reactor. The mass balance given by Equation 1.41 applies, but because there is no flow of material into or out of the reactor, $F_{A,0}$ and $F_{A,e}$ are zero. Thus the mass balance on a batch reactor indicates that the rate of reaction $(-r_A)$ may be determined from the negative of the rate of change of $n_A$ with time, divided by the reactor volume, as shown by the first equality in Equation 1.46:

$$-r_A = -\frac{1}{V}\frac{dn_A}{dt} = -\frac{1}{V}\frac{d(Vn_A/V)}{dt}$$
$$= -\frac{1}{V}\frac{d(Vc_A)}{dt} \qquad (1.46)$$

Here again the theory concerning a mass balance enables one to deduce a process rate from a measured rate of change. For the special case in which a batch reaction is conducted at constant volume, the rate of reaction can be seen from Equation 1.46 to equal the negative of the rate of change of concentration, $c_A$, with time. This familiar result is by no means a definition of the reaction rate, but rather, a direct result of a mass balance for a very special case. Excellent references on these ideas are Dixon (1970), Churchill (1974), and Kabel (1981).

**Measurement of rates.** It has been shown how the rate of reaction may be calculated from measured variables. The simplest case is a CSTR where an algebraic collection of easily measured parameters suffices (see Equation 1.42). The simple result is one reason why CSTRs are often used for experimental measurement of reaction rates. A much more critical problem arises when one wishes to determine the rate of reaction from a rate of change. See Churchill (1974) and Bisio and Kabel (1985) for details.

**Objectives.** Having established the concept of reaction rate, we now explore what happens when the rate of reaction is zero (a definition of chemical equilibrium) and nonzero (a definition of chemical kinetics). Consider a mixture of gases A, B, C, and D contained in the well-mixed reactor shown in Figure

1-17. Assume that a reaction occurs when the temperature and pressure are changed to $T$ and $P$:

$$\nu_A A + \nu_B B \rightleftharpoons \nu_C C + \nu_D D \qquad (1.47)$$

Equation 1.47 is a *stoichiometric equation* and the term $\nu_i$ is the *stoichiometric coefficient* associated with the $i$th molecular species. The species on the left side of Equation 1.47 are called *reactants,* and the species on the right are called *products.* For this more general stoichiometry the process rates are related as follows:

$$-\frac{r_A}{\nu_A} = -\frac{r_B}{\nu_B} = \frac{r_C}{\nu_C} = \frac{r_D}{\nu_D} \qquad (1.48)$$

Two questions arise that are important to engineers:

1. What are the mole fractions $(y_i)$ of the components at equilibrium?
2. What are the mole fractions $(y_i)$ of the components at any instant of time?

# *1.8 Chemical equilibrium*

Chemical reaction thermodynamics give a limit on the maximum extent of chemical reaction. They also give important clues toward the effects of reaction conditions on reaction progress. A rather general development will be provided that highlights the importance of attending to units and reaction stoichiometry. Equation 1.47 expresses a general equilibrium reaction. For any reaction, Equation 1.49 states the relationship between the thermodynamic equilibrium constant and the standard change of Gibbs free energy for the reaction

$$\Delta G_T^0 = -R_u T \ln K_a \qquad (1.49)$$

The origin of this equation is given in thermodynamics texts. The equilibrium constant for the reaction is defined in terms of thermodynamic activities $(a_i)$:

$$K_a = \frac{a_C^{\nu_c} a_D^{\nu_d}}{a_A^{\nu_a} a_B^{\nu_b}} = \frac{(\bar{f}_C/f_C^0)^{\nu_c}(\bar{f}_D/f_D^0)^{\nu_d}}{(\bar{f}_A/f_A^0)^{\nu_a}(\bar{f}_B/f_B^0)^{\nu_b}} \qquad (1.50)$$

Values for $\Delta G_T^0$ are obtained from thermodynamic data in a manner similar to the enthalpy change of reaction and apply to the conversion of reactants in their standard states to products in their standard states. The standard states for reactants and products are defined in terms of the species as pure components at the temperature of interest and 1 atm pressure. An important ramification of this specification of the standard state is that the standard free-energy change $\left(\Delta G_T^0\right)$ may be a function of temperature, but it is not a function of the total pressure $\left(P_t\right)$. Therefore, $K_a$ in Equation 1.49 will also be independent of total pressure.

The activity $a_i$ of component $i$ in solution in the gas phase is defined as the ratio of the *fugacity* of that component in solution to the *fugacity* of that component in its standard state. This definition is used to express the right-hand equality of Equation 1.50. Standard-state fugacities are conventionally set at 1 atm:

$$f_i^0 \equiv 1 \text{ atm} \qquad (1.51)$$

so that they can be dropped from Equation 1.50. Obviously, the remaining component fugacities also have units of atmospheres. Thus an equilibrium constant $K_f$ based on solution fugacities $\overline{f}_i$ is equal to

$$K_f = \frac{\overline{f}_C^{v_c}\overline{f}_D^{v_d}}{\overline{f}_A^{v_a}\overline{f}_B^{v_b}} \, [=] \, (\text{atm})^{(v_C+v_D)-(v_A+v_B)} \quad (1.52)$$

The symbol $[=]$ means *has the units of*. Clearly $K_f$ will have the same numerical value as $K_a$. However, $K_f$ will have units of pressure (in atm) as indicated by Equation 1.52, whereas $K_a$ will always be dimensionless.

Fugacity is a difficult parameter to determine, indeed to comprehend physically. For an ideal gas, the fugacity of molecular species $i$ is equal to its partial pressure $\left(P_i\right)$. If the gas is not ideal, its fugacity is related to its partial pressure by its *fugacity coefficient* $\left(\phi_i\right)$:

$$\overline{f}_i = \phi_i P_i \qquad (1.53)$$

Fugacity coefficients $\left(\phi_i\right)$ can be obtained from generalized fugacity coefficient charts in thermodynamics texts. Substituting Equation 1.53 into Equation 1.52 yields

$$K_f = \left[\frac{\phi_C^{v_c}\phi_D^{v_d}}{\phi_A^{v_a}\phi_B^{v_b}}\right]\left[\frac{P_C^{v_c}P_D^{v_d}}{P_A^{v_a}P_B^{v_b}}\right]$$

$$= K_\phi K_P \, [=] \, (\text{atm})^{(v_C+v_D)-(v_A+v_B)} \quad (1.54)$$

where $K_\phi$ and $K_P$ are defined analogously to $K_a$ and $K_f$ by the bracketed terms in Equation 1.54. The partial pressure $\left(P_i\right)$ of molecular species $i$ is related to its mole fraction $\left(y_i\right)$:

$$P_i = y_i P_t \qquad (1.55)$$

where $P_t$ is the total pressure. Substituting Equation 1.55 into the definition of $\left(K_P\right)$ from Equation 1.54 results in the following:

$$K_P = \frac{y_C^{v_c}y_D^{v_d}}{y_A^{v_a}y_B^{v_b}} P_t^{(v_C+v_D)-(v_A+v_B)}$$

$$= K_y P_t^{(v_C+v_D)-(v_A+v_B)} \, [=] \, (\text{atm})^{(v_C+v_D)-(v_A+v_B)}$$

$$(1.56)$$

Equation 1.56 yields a relationship for an equilibrium constant, involving the measurable quantities total pressure $\left(P_t\right)$ and mole fraction $\left(y_i\right)$. Combining Equations 1.54 and 1.56 yields an expression (Equation 1.57) that relates all the variously defined equilibrium constants and the total pressure. It should be emphasized that $K_f$ has the units of pressure but is numerically equal to $K_a$, which is computed from thermodynamic data. Equation 1.57, like Equation 1.54, shows the relationship between $K_f$ and $K_\phi$ and $K_P$:

$$K_f = K_\phi K_P = K_\phi K_y P_t^{(v_c+v_d)-(v_a+v_b)} \quad (1.57)$$

The above also provides an alternative expression relating $K_\phi$, $K_y$, and the total pressure $\left(P_t\right)$. Clearly, $K_y$ is the term from which the equilibrium composition can be calculated, and $K_\phi$ is the correction for nonideal gases.

Several general conclusions can be drawn from the foregoing development.

**1.** Tiny values of $K_y(T)$ imply that the equilibrium mole fractions of the products are very small relative to the reactants, and conversely if the values of $K_y(T)$ are huge the product mole fractions are relatively large. If values of $K_y(T)$ are in the range, for example, $0.01 < K_y(T) < 100$, one can expect comparable amounts of both reactants and products in the final equilibrium mixture.

**2.** The equilibrium composition for an ideal gas (all $\phi_i = 1$) is independent of pressure if the sums of the stoichiometric coefficients on the right and left of Equation 1.47 are the same. If they are not the same, increasing the pressure at constant temperature increases the composition of the species on the side of the equation where the sum of stoichiometric coefficients is the smallest.

**3.** If one finds values of $K_P(T)$ for a reaction written in the reverse order of Equation 1.47, that is,

$$v_C C + v_D D = v_A A + v_B B \quad (1.58)$$

then

$$K_P(T)_{Eq.\ 1.58} = \frac{1}{K_P(T)_{Eq.\ 1.47}} \quad (1.59)$$

Since an equilibrium equation can be written as either Equation 1.47 or 1.58, readers must be careful with the words *products* and *reactants* since they only have meaning in terms of the way the equilibrium equation is written.

**4.** Sometimes readers can only find values of $K_P(T)$ for an equation where all stoichiometric coefficients are $n$ times the values in Equation 1.47, that is,

$$n v_A A + n v_B B = n v_C C + n v_D D \quad (1.60)$$

In this case,

$$K_P(T)_{Eq.\ 1.60} = [K_P(T)_{Eq.\ 1.47}]^n \quad (1.61)$$

When the gas mixture is assumed to be composed of ideal gases, the mole fraction $(y_i)$ can be expressed in terms of the number of moles $(n_i)$ or in terms of the partial pressure $(P_i)$:

$$y_i = \frac{n_i}{n_t} = \frac{P_i}{P_t} \quad (1.62)$$

where $n_t$ is the total number of moles and $P_t$ is the total pressure. The mole fractions in Equation 1.56 can be replaced and $K_P(T)$ rewritten as

$$K_P(T) = \frac{n_C^{v_C} n_D^{v_D}}{n_A^{v_A} n_B^{v_B}} \left(\frac{P_t}{n_t}\right)^{v_C + v_D - v_A - v_B} \quad (1.63)$$

For an ideal gas with partial pressures expressed in terms of atmospheres, $K_P = K_f$. Equilibrium data with pressure units other than atmospheres (e.g., mm Hg) are often reported as $K_P$ values, in which case the units come from the defining equation

$$K_P(T) = \frac{P_C^{v_C} P_D^{v_D}}{P_A^{v_A} P_B^{v_B}} \quad (1.64)$$

A simple representation of equilibrium constants for several reactions related to combustion and air pollution is shown in Figure 1-19.

**EXAMPLE 1.7   WATER-GAS SHIFT EQUILIBRIUM REACTION**

Consider the well-known water-gas shift reaction $H_2O + CO \rightleftharpoons H_2 + CO_2$. The equilibrium con-

stant $(K_P)$ as a function of temperature is shown below.

| $T(K)$ | $K_P(T)$ |
|---|---|
| 1000 | 1.44 |
| 1250 | 0.65 |
| 1500 | 0.39 |
| 1750 | 0.29 |
| 2000 | 0.22 |
| 2250 | 0.18 |
| 2500 | 0.16 |

(As a matter of interest, compare the values of $K_P$ above to the representation in Figure 1-19.) Assume initially that a mixture of the gases above exists in a closed vessel in the presence of 1 kmol of $N_2$. The table below shows the initial amounts (kmol) of each gas. The temperature is raised to the temperatures noted above. Compute and plot the $H_2$ concentration (ppm) at equilibrium over the range of temperatures 1000 to 2500 K.

| Time | $H_2O$ | CO | $H_2$ | $CO_2$ | $N_2$ | Total number of moles, $n_t$ |
|---|---|---|---|---|---|---|
| 0 | 0.3 | 0.5 | 0.1 | 0.2 | 1.0 | 2.1 |
| $\infty$ | $0.3 - \alpha$ | $0.5 - \alpha$ | $0.1 + \alpha$ | $0.2 + \alpha$ | 1.0 | 2.1 |

***Solution***   If the pressure is low enough that the gas is ideal (a likely circumstance for this equimolar reaction), Equation 1.56 can be used to relate $K_P(T)$ to gas composition in mole fractions.

$$K_P(T) = \left(\frac{0.1 + \alpha}{n_t}\right)\left(\frac{0.2 + \alpha}{n_t}\right)$$

$$\left(\frac{0.3 - \alpha}{n_t}\right)^{-1}\left(\frac{0.5 - \alpha}{n_t}\right)^{-1}$$

$$= \frac{(0.1 + \alpha)(0.2 + \alpha)}{(0.3 - \alpha)(0.5 - \alpha)}$$

The above can be reduced to a second-order algebraic equation and $\alpha$ found by solving the quadratic equation. Figure E1-7 shows the hydrogen concentration versus temperature. The MathCAD program producing Figure E1-7 can be found in the textbook's two Web sites identified at the end of this chapter.

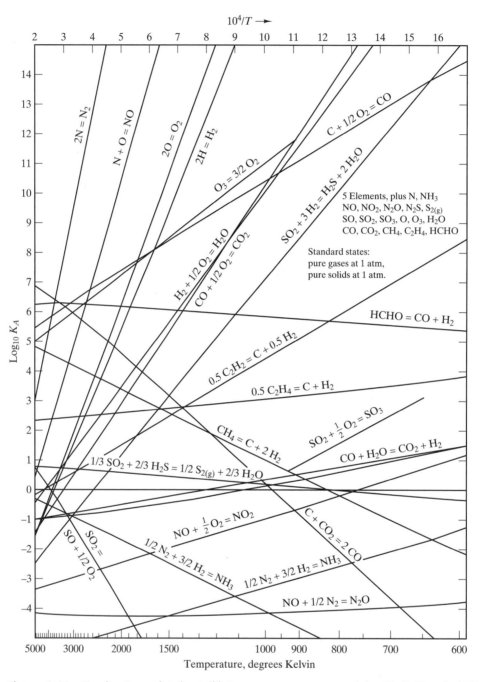

**Figure 1-19**   Combustion related equilibrium constants, prepared by H C Hottel, MIT (Balzhiser, Samuels and Eliassen, Prentice Hall, 1972)

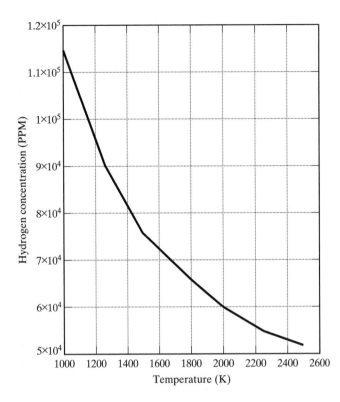

**Figure E1-7** Hydrogen concentration (ppm) versus temperature (K) for the water-gas shift reaction

In the absence of tabular or graphical values of the equilibrium constant at any temperature, $K_a$ can be computed from Equation 1.49 by computing the standard change of Gibbs free energy, $\Delta G_T^0$:

$$\Delta G_T^0 = \sum_{\text{products}} \nu_j \bar{g}_{i,T}^0 - \sum_{\text{reactants}} \nu_i \bar{g}_{i,T}^0 \quad (1.65)$$

where $\bar{g}_{i,T}^0$ is the molar Gibbs free energy of formation of species $i$ at temperature $T$. By convention, the standard Gibbs free energies of formation of elements in their natural states are assigned the value of zero at the reference temperature, 298 K. A tabulation of $\bar{g}_{i,T}^0$ for over 1000 chemical species at various temperatures can be found in the JANAF tables (Stull and Prophet, 1971). In the absence of JANAF tables, values of the Gibbs free energies of formation of species $i$ can be found from the definition of the Gibbs free energy,

$$\bar{g}_{i,T}^0 = \bar{h}_{i,T}^0 - T\bar{s}_{i,T}^0 \quad (1.66)$$

where $\bar{h}_{i,T}^0$ and $\bar{s}_{i,T}^0$ are the enthalpy and entropy of formation of species $i$ at temperature $T$ obtained from such sources as Perry et al. (1984) and Smith and van Ness (1975). Thus the equilibrium coefficient can be calculated from Equation 1.49 using

$$\Delta G_T^0 = \Delta H_T^0 - T\,\Delta S_T^0 \quad (1.67)$$

where the values of $\Delta H_T^0$ and $\Delta S_T^0$ are evaluated using a summation formulation similar to Equation 1.65.

On some occasions, values of entropy and enthalpy of formation are not available as functions of temperature, but their values at the standard temperature of 298 K and equations of specific heat $\bar{c}_p(T)$ for a species $i$ are available:

$$\frac{\bar{c}_p(T)}{R_u} = a_1 + a_2T + a_3T^2 + a_4T^3 + a_5T^4 \quad (1.68)$$

where $a_1, a_2, a_3, a_4,$ and $a_5$ are empirical constants (see, e.g., Kee et al., 1991, or this book's Web page). The entropy and enthalpy of formation at any temperature can be computed from the following:

$$\bar{h}_{i,T}^0 = \bar{h}_{i,298}^0 + \int_{298}^{T} \bar{c}_p(T)\,dT$$

$$\bar{s}_{i,T}^0 = \bar{s}_{i,298}^0 + \int_{298}^{T} \frac{\bar{c}_p(T)\,dT}{T} \quad (1.69)$$

The properties $\bar{c}_p(T)$, $\bar{h}^0_{i,298}$, and $\bar{s}^0_{i,298}$ are commonly referred to as *thermal data*. Combining Equations 1.49 and 1.67 enables one to compute the equilibrium constant solely on the basis of thermal data:

$$K_a(T) = \exp\left(\frac{-\Delta H^0_T}{R_u T}\right) \exp\left(\frac{\Delta S^0_T}{R_u}\right) \quad (1.70)$$

At this point use is made of Equations 1.68 and 1.69 to complete the calculation. If no data are available for the specific heats and enthalpies or entropies of formation, predictive methods for these properties are available in Reid et al. (1987).

A frequently used general chemical equilibrium computer code capable of handling over 400 different chemical species is the NASA Chemical Equilibrium Code (Gordon and McBride, 1976) called CE86. The code is updated from time to time, so the "86" will be replaced by the date of updating.

### EXAMPLE 1.8   EQUILIBRIUM COEFFICIENT FROM THERMAL DATA

You are away from your place of work, without your books, computer, and so on. All you have is a textbook on combustion (Turns, 1996). Compute the equilibrium coefficient of the water-gas shift reaction, $H_2O + CO \rightleftharpoons H_2 + CO_2$, over the range of temperatures 600 to 2000 K. Compare the results with the tabulated values, $K_p(1000) = 1.44$, $K_p(1500) = 0.39$, and $K_p(2000) = 0.22$.

*Solution*   From the textbook on combustion, the following thermodynamic data can be obtained.

| | $\bar{h}^0_{f,298}$ | $\bar{s}^0_{f,298}$ | $\bar{c}_p(T)/R_u$ (300 to 1000 K) |
|---|---|---|---|
| $H_2O$ | −241,845 | 188.715 | $3.386 + 0.00348\,T - 0.06354 \times 10^{-4}\,T^2 + 0.06968 \times 10^{-7}\,T^3 - 0.02506 \times 10^{-10}\,T^4$ |
| $CO$ | −110,541 | 197.548 | $3.262 + 0.00151\,T - 0.03881 \times 10^{-4}\,T^2 + 0.05582 \times 10^{-7}\,T^3 - 0.02475 \times 10^{-10}\,T^4$ |
| $H_2$ | 0 | 130.595 | $3.298 + 0.00825\,T - 0.08143 \times 10^{-5}\,T^2 + 0.09475 \times 10^{-9}\,T^3 + 0.04135 \times 10^{-11}\,T^4$ |
| $CO_2$ | −393,546 | 213.736 | $2.275 + 0.00992\,T - 0.10409 \times 10^{-4}\,T^2 + 0.06866 \times 10^{-7}\,T^3 + 0.02118 \times 10^{-10}\,T^4$ |

The units of $\bar{h}^0_{f,298}$, $\bar{s}^0_{f,298}$, and $\bar{c}_p$ are kJ/kmol, kJ kmol$^{-1}$ K$^{-1}$, and kJ kmol$^{-1}$ K$^{-1}$, respectively.

Calculate $\Delta H^0_T$ and $\Delta S^0_T$:

$$\Delta H^0_T = \bar{h}^0_{H_2,298} + \bar{h}^0_{CO_2,298} - \bar{h}^0_{H_2O,298} - \bar{h}^0_{CO,298}$$

$$+ \int_{298}^{T} \bar{c}_{p,H_2}(T)\,dT + \int_{298}^{T} \bar{c}_{p,CO_2}(T)\,dT$$

$$- \int_{298}^{T} \bar{c}_{p,H_2O}(T)\,dT - \int_{298}^{T} \bar{c}_{p,CO}(T)\,dT$$

$$\Delta S^0_T = \bar{s}^0_{H_2,298} + \bar{s}^0_{CO_2,298} - \bar{s}^0_{H_2O,298} - \bar{s}^0_{CO,298}$$

$$+ \int_{298}^{T} \frac{\bar{c}_{p,H_2}(T)\,dT}{T} + \int_{298}^{T} \frac{\bar{c}_{p,CO_2}(T)\,dT}{T}$$

$$- \int_{298}^{T} \frac{\bar{c}_{p,H_2O}(T)\,dT}{T} - \int_{298}^{T} \frac{\bar{c}_{p,CO}(T)\,dT}{T}$$

Select a temperature $(T)$, integrate the above, obtain values of $\Delta H^0_T$ and $\Delta S^0_T$, and substitute into

$$K_a(T) = \exp\left(\frac{-\Delta H^0_T}{R_u T}\right) \exp\left(\frac{\Delta S^0_T}{R_u}\right)$$

Repeat the calculation over the range of temperatures.

| $T$(K) | Calculated $K_a(T)$ | Tabulated $K_a(T)$ | Difference (%) |
|---|---|---|---|
| 600 | 28.547 | | |
| 700 | 9.509 | | |
| 800 | 4.27 | | |
| 900 | 2.335 | | |
| 1000 | 1.463 | 1.44 | 1.6 |
| 1200 | 1.011 | | |
| 1300 | 0.751 | | |
| 1400 | 0.589 | | |
| 1500 | 0.412 | 0.39 | 5.6 |
| 1600 | 0.363 | | |
| 1700 | 0.330 | | |
| 1800 | 0.308 | | |
| 1900 | 0.297 | | |
| 2000 | 0.296 | 0.22 | 34.5 |

At 1000 K the calculated and tabulated values are in good agreement. The disagreement becomes increasingly worse at temperatures above 1000 K. Since the empirical equation for the specific heats is accurate only for 300 to 1000 K, this fact is not surprising. The disparity above 1000 K underscores the need to use an empirical equation for the specific heat that is valid over the range of temperatures of interest. It is possible, however, that the discrepancy is due to an error in the value of enthalpy or entropy of formation.

The water-gas shift reaction of Example 1.7 illustrates an equilibrium computation in which the sum of the stoichiometric coefficients on the right- and left-hand sides of the equation are the same (i.e., *equimolar* reaction). Consider now an equilibrium computation in which the sum of the stoichiometric coefficients on the right- and left-hand sides of the equation are not the same.

## EXAMPLE 1.9   $SO_2/SO_3/O_2$ EQUILIBRIUM

To illustrate Equation 1.56, consider known amounts of $SO_2$, $O_2$, $SO_3$, and helium placed in a vessel. The temperature and pressure are changed to $T$ and $P$ and maintained until equilibrium occurs. The introduction of helium (or any other diluent) illustrates the effect of materials that do not participate in the reaction but merely dilute the species. Compute the equilibrium mole fractions if the reaction is expressed by the following equation.

$$SO_2 + \tfrac{1}{2}O_2 \rightleftharpoons SO_3$$

**Solution**   From thermodynamic literature, the equilibrium constants are

| $T$(K) | $K_p(T)$ (atm)$^{-1/2}$ |
|--------|------------------------|
| 298    | $2.6 \times 10^{12}$   |
| 1000   | 1.8                    |
| 2000   | $5.6 \times 10^{-3}$   |

(Check these values of $K_p$ with data from Figure 1-19. Do you notice anything curious?) You should anticipate the following effects of temperature, pressure, and dilution on this equilibrium.

**1.** *Temperature.* At 298 K, $SO_3$ will be the principal constituent, whereas at 2000 K, $SO_2$ and $O_2$ will be the principal constituents. At 1000 K, the mole fractions of all three species will have comparable values.

**2.** *Pressure.* From Equation 1.54, if the gaseous mixture is ideal, $K_\phi = 1$ and $K_p$ is independent of the total pressure. Since the sum of the stoichiometric coefficients of the reactants exceeds the sum for the products $\left( \nu_{SO_2} + \nu_{O_2} - \nu_{SO_3} = 1 + \tfrac{1}{2} - 1 = \tfrac{1}{2} \right)$, the ratio of mole fractions is proportional to the square root of the pressure:

$$\frac{y_{SO_3}}{y_{SO_2}\sqrt{y_{O_2}}} \propto \sqrt{P}$$

Thus at any temperature, the $SO_3$ mole fraction can be increased by increasing the pressure.

**3.** *Diluent.* Since the sum of the stoichiometric coefficients of the reactants exceeds the sum for the products, Equation 1.63 shows that the ratio of moles is related to the total number of moles as follows:

$$\frac{n_{SO_3}}{n_{SO_2}\sqrt{n_{O_2}}} \propto \sqrt{n_t}$$

Thus at any temperature and pressure, the reaction can be driven toward $SO_3$ somewhat by increasing the amount of diluent. In short, the effects of increasing pressure or adding diluent are in the same direction but not of the same magnitude.

Construct a table listing the initial number of moles of each species and the number of moles after equilibrium is achieved. Also tabulate the total number of moles of the mixture, $n_t$.

| Time | $SO_2$ | $O_2$ | $SO_3$ | He | $n_t$ |
|------|--------|-------|--------|-----|-------|
| 0 | $a$ | $b$ | $c$ | $d$ | $a + b + c + d$ |
| $\infty$ | $a - \alpha$ | $b - \alpha/2$ | $c + \alpha$ | $d$ | $a + b + c + d - \alpha/2$ |

*Note:* The moles of $SO_3$ at equilibrium are written as $(c + \alpha)$, where $\alpha$ is an unknown that will be determined by computation. Once the molar value of $SO_3$ is represented as $(c + \alpha)$, the remaining molar values in the table are dictated by a mole balance. The upper and lower bounds of $\alpha$,

$$-c \leqq \alpha \leqq a$$

are dictated by the fact that the number of moles of any of the species can never be negative. (Note that $O_2$ could further limit the bounds if $b < a/2$.) The mole fraction of each species is as follows:

$$y_{SO_2} = \frac{a - \alpha}{n_t} \qquad y_{O_2} = \frac{b - \alpha/2}{n_t}$$

$$y_{SO_3} = \frac{c + \alpha}{n_t}$$

Substitute these values in Equation 1.63 and simplify:

$$K_P(T) = \frac{c + \alpha}{a - \alpha} \left( \frac{a + b + c + d - a/2}{b - \alpha/2} \right)^{1/2} P^{-1/2}$$

At this point, readers can solve for $\alpha$ using any one of many commercially available mathematical software programs (e.g., MathCad) for known values of $T, P, a, b, c$ and $d$. Even if such programs are used, readers will find it useful to be able to estimate $\alpha$ to detect if errors have been made in programming the equation. In some cases these approximations can even yield an accurate solution. The following cases illustrate useful approximations that readers should master.

*Initial conditions:*

- *Case I: $P = 2$ atm, excess amount of one reactant but no product or diluent, $a = 1, b = 100, c = 0, d = 0$
- *Case II: $P = 2$ atm, equal amounts of reactants but no product or diluent, $a = b = 1, c = d = 0$
- *Case III: $P = 2$ atm, equal amount of product and a diluent but no reactants, $a = b = 0, c = d = 1$

*Case I:* Initial conditions; excess amount of one of the reactants but no product or diluent at $P = 2$ atm, $n_{SO_2,0} = 1$, $n_{O_2,0} = 100$, $n_{SO_3,0} = n_{He,0} = 0$ ($a = 1, b = 100, c = d = 0$). Substitute the values and obtain

$$K_P(T) = \frac{\alpha}{1 - \alpha} \left( \frac{101 - \alpha/2}{100 - \alpha/2} \right)^{1/2} 2^{-1/2}$$

Since $0 \leq \alpha \leq 1$, the middle square-root term involving $\alpha$ is approximately unity:

$$2^{1/2} K_P(T) = \frac{\alpha}{1 - \alpha}$$

Hence,

$$\alpha = \left[ 1 + \frac{1}{K_P(T) \sqrt{2}} \right]^{-1}$$

*Results case I: $a = 1, b = 100, c = 0, d = 0$, $n_t = 101 - \alpha/2$*

| T(K) | $\alpha$ | $y_{SO_3}$ | $y_{SO_2}$ | $y_{O_2}$ |
|------|----------|-----------|-----------|-----------|
| 298 | $\approx 1$ | $\approx 0.01$ | $\approx 0$ | 0.99005 |
| 1000 | 0.717 | 0.007 | 0.003 | 0.99006 |
| 2000 | 0.0008 | $\approx 0$ | $\approx 0.01$ | 0.99010 |

While the trends discussed above concerning the prominence of $SO_3$ and $SO_2$ at 298 K and 2000 K, respectively, remain true, the large amount of $O_2$ dilutes the other components.

*Case II:* Initial conditions—equal amounts of reactants but no product or diluent at $P = 2$ atm, $n_{SO_2,0} = n_{O_2,0} = 1$, $n_{SO_3,0} = n_{He,0} = 0$ ($a = b = 1$, $c = d = 0$). Substitute the values and obtain

$$K_P(T) \sqrt{2} = \frac{\alpha}{1 - \alpha} \left( \frac{4 - \alpha}{2 - \alpha} \right)^{1/2}$$

Determine that the limits of $\alpha$:

$$0 \leq \alpha \leq 1$$

Since $K_P(298 \text{ K}) = 2.6 \times 10^{12}$, $\alpha \approx 1$. At $T = 1000$ K trial-and-error techniques or MathCAD will be needed and one finds that $\alpha = 0.62$. At 2000 K, $K_P(2000 \text{ K})$ is small and one can expect that $0 < \alpha \ll 1$, with the result that

$$\alpha \approx K_P(2000) = 5.63 \times 10^{-3}$$

*Results case II: $a = 1, b = 1, c = 0, d = 0$, $n_t = 2 - \alpha/2$*

| T(K) | $\alpha$ | $y_{SO_3}$ | $y_{SO_2}$ | $y_{O_2}$ |
|------|----------|-----------|-----------|-----------|
| 298 | $\approx 1$ | 0.67 | $\approx 0$ | 0.33333 |
| 1000 | 0.62 | 0.37 | 0.22 | 0.40828 |
| 2000 | $5.63 \times 10^{-3}$ | $\approx 0$ | $\approx 0.50$ | 0.49929 |

*Case III:* Initial conditions—equal amounts of products and diluent, no reactants, $n_{SO_2,0} = n_{O_2,0} = 0$, $n_{SO_3,0} = n_{He,0} = 1$ ($a = b = 0, c = d = 1$). Substituting in the above, one finds that

$$2^{1/2} K_P(T) = \frac{1 + \alpha}{-\alpha} \left( \frac{4 - \alpha}{-\alpha} \right)^{1/2}$$

Determine the limits on $\alpha$:

$$-1 \leqq \alpha \leqq 0$$

Surprisingly, the limits on $\alpha$ are negative. But once committed to the table of initial and final number of moles, one has to adhere to the limits it stipulates. What this means physically is that the reaction goes in reverse. At 298 K, the huge value of $K_P(298\,\text{K})$ implies that $\alpha \approx 0$. At 2000 K, the small value of $K_P(2000\,\text{K})$ implies that $\alpha \approx -1$. At 1000 K, a graphical trial-and-error solution of Case III indicates that $\alpha = -0.55$.

*Results Case III:* $a = 0$, $b = 0$, $c = 1$, $d = 1$, $n_t = 2 - \alpha/2$

| $T(\text{K})$ | $\alpha$ | $y_{SO_3}$ | $y_{SO_2}$ | $y_{O_2}$ |
|---|---|---|---|---|
| 298 | $\approx 0$ | $\approx 0.50$ | $\approx 0$ | $\approx 0$ |
| 1000 | $-0.55$ | $0.197$ | $0.241$ | $0.12088$ |
| 2000 | $\approx -1$ | $\approx 0$ | $0.40$ | $\approx 0.20$ |

There are situations when several equations define a state of equilibrium. When a set of simultaneous equations define equilibrium, the analysis is similar to what has been presented before but separate equations similar to Equation 1.56 must be written for each chemical reaction. For each molecular species appearing in several equilibrium equations, a new unknown variable needs to be defined. To illustrate the analysis, consider the equilibrium of involving $N_2$, $O_2$, $NO$, and $NO_2$, given by the following simultaneous equations in which $K_{P,NO}(T)$ and $K_{P,NO_2}(T)$ are the designated equilibrium constants:

$$\tfrac{1}{2}N_2 + \tfrac{1}{2}O_2 \rightleftharpoons NO$$

equilibrium constant $K_{P,NO}(T)$     (a)

$$NO + \tfrac{1}{2}O_2 \rightleftharpoons NO_2$$

           (1.71)

equilibrium constant $K_{P,NO_2}(T)$     (b)

A table is constructed similar to the second table in Example 1.9 in which the initial and final number of moles of each species are specified. Define two variables, $\alpha$ and $\beta$, indicating the unknown number of moles converted by each of the two equilibrium equations.

| Time | $N_2$ | $O_2$ | $NO$ | $NO_2$ | $n_t$ |
|---|---|---|---|---|---|
| 0 | $a$ | $b$ | $c$ | $d$ | $a + b + c + d$ |
| $\infty$ | $a - \alpha/2$ | $b - \alpha/2 - \beta/2$ | $c + \alpha - \beta$ | $d + \beta$ | $a + b + c + d - \beta/2$ |

Write an equation involving $K_{P,NO}(T)$ and $K_{P,NO_2}(T)$ in the form of Equation 1.56. The mole fraction of each species is defined by the number of moles in the table divided by $n_t$:

$$K_{P,NO}(T)$$
$$= \frac{c + \alpha - \beta}{(a - \alpha/2)^{1/2}(b - \alpha/2 - \beta/2)^{1/2}} \quad (1.72)$$

$$K_{P,NO_2}(T)$$
$$= \frac{d + \beta}{c + \alpha - \beta} \left( \frac{a + b + c + d - \beta/2}{b - \alpha/2 - \beta/2} \right)^{1/2} P^{-1/2} \quad (1.73)$$

The initial mole numbers, $a$, $b$, $c$, and $d$, are known and are all positive or zero. The equations above should be solved by numerical means for $\alpha$ and $\beta$. An initial guess for the numerical solution should be performed following the practice of determining the limits of $\alpha$ and $\beta$ and on the magnitudes of $K_{P,NO}$ and $K_{P,NO_2}$. Readers should be careful to ensure that the numerical methods do not obtain unreal values of $\alpha$ and $\beta$. See Chapter 7 for details and how to solve Equations 1.72 and 1.73.

## 1.9 Chemical kinetics

Chemical equilibrium establishes the maximum extent that a reaction may progress to form products. Although this information is useful, what engineers really need to know is how far the reaction actually progresses. Chemical kinetics characterizes the rate at which chemical species appear or disappear. The first step in performing kinetic analysis is to postulate the *kinetic (molecular) mechanism* describing a reaction. A kinetic mechanism is the group of individual molecular reactions that describe a chemical reaction. Often, the stoichiometric equation used to perform an equilibrium calculation represents the overall reaction process but does not reflect the reaction mechanism of the chemical species that actually react to produce a product and the path the reaction takes. For example, the equation describing the $SO_2$, $O_2$, $SO_3$ reaction in Example 1.9 does not describe the actual species that react. Steps a, b, and c in the complete chemical kinetic mechanism that involve the species $SO_2$ and $SO_3$ are postulated to be

(a)    $SO_2 + O\cdot + M \rightleftharpoons SO_3 + M$

   rate constants, $k_a$ and $k_{-a}$        (1.74)

(b)    $SO_3 + O\cdot \rightarrow SO_2 + O_2$

   rate constant $k_b$        (1.75)

(c)    $SO_3 + H\cdot \rightarrow SO_2 + OH\cdot$

   rate constant $k_c$        (1.76)

Clearly, mechanistic steps accounting for species like $O_2$, $O\cdot$, $H\cdot$, and $OH\cdot$ are required to complete the mechanism. Readers should be careful and realize that rate constants ($k_i$) are really not "constants" but rate coefficients that are strong functions of temperature. Nevertheless, engineers use the phrase *rate constant* understanding that this is the case, and the phrase is used in this book. The symbol $\rightleftharpoons$ used herein denotes two equations involving the forward (left-to-right) reaction which has a kinetic rate constant $k_a$ and a reverse (right-to-left) reaction which has a kinetic rate constant $k_{-a}$. The appearance of a single arrow implies an irreversible reaction. The rates of formation of $SO_2$ and $SO_3$ can be expressed by the *law of mass action*:

$$r_{SO_3} = k_a[SO_2][O\cdot][M] - k_{-a}[SO_3][M]$$
$$-k_b[SO_3][O\cdot] - k_c[SO_3][H\cdot]    \quad (1.77)$$

$$r_{SO_2} = -k_a[SO_2][O\cdot][M] + k_{-a}[SO_3][M]$$
$$+k_b[SO_3][O\cdot] + k_c[SO_3][H\cdot]    \quad (1.78)$$

It should be noted that the net rate of formation of $SO_2$ (i.e., $r_{SO_2}$) is equal to the sum of the ways that $SO_2$ is formed minus the sum of the ways in which $SO_2$ is destroyed. In doing this, the reader must consider all the equations in the kinetic mechanism involving $SO_2$. The net rate of formation of $SO_3$ is quantified in a similar way. The symbol $[SO_3]$ denotes the molar concentration of $SO_3$ in such units as $kmol/m^3$ or $molecules/cm^3$. The kinetic rate constants $k_a$, $k_{-a}$, $k_b$, and $k_c$ are functions of temperature and given in the form

$$k(T) = AT^n \exp\left(\frac{-E_a}{R_u T}\right)    \quad (1.79)$$

called an *Arrhenius equation*, where $A$ is a *preexponential factor* (sometimes called the Arrhenius constant), $n$ is a known constant (often taken to be zero), and $E_a$ is the *activation energy*, which represents the amount of energy a molecule requires before it can react. For simplicity, the term $E_a/R_u$ is often combined into a single constant to eliminate confusion about the units of energy. Rate constants vary considerably in magnitude because some reactions are very fast while others are very slow. Therefore, readers should not be surprised to find $k$'s differing by many orders of magnitude. The units of the rate constants depend on whether the reaction is first, second, or third order. An example of a first-order reaction is a spontaneous decomposition,

$$-r_A = k[A]    \quad (1.80)$$

Equation (1.75) is a second-order reaction, and Reaction (1.74) is a third-order (or three-body reaction) in the forward direction. The units of $k$ follow from the units by which the rate of reaction ($mol/m^3 \cdot s$, ppm/min, etc.) and the concentrations ($mol/m^3 \cdot s$, ppm, etc.) are specified. In this book it is presumed that chemical kineticists have defined the operable kinetic mechanism and specified the kinetic rate constants.

The chemical mechanism applies to a small region of fluid wherein a statistical ensemble of molecules, free radicals, and so on, come together. Even at constant temperature and pressure, species concentrations may vary depending on the configuration and state of mixing in the reacting medium. As explained earlier, the process rate, $r_A$, is quantified by a mass balance. For a well-mixed system with continuous flow at steady state,

$$-r_A = \frac{F_{A,0} - F_{A,e}}{V}    \quad (1.81)$$

If there is no flow, the volume is constant, and the species are well mixed, Equation 1.46 leads to

$$r_A = \frac{(1/V)dn_A}{dt} = \frac{d(n_A/V)}{dt} = \frac{d[A]}{dt} \quad (1.82)$$

where $[A]$ and $c_A$ are common alternative symbols for the concentration of A. Three cases of different kinetics (first-order, second-order, and simultaneous reactions) are considered below for the case of a constant-volume well-mixed parcel of air and pollutants. Such models are often invoked without sufficient attention to atmospheric movement and changes in conditions. These issues are addressed more fully in Chapter 9.

**First-order reactions.** A common example of a first-order reaction is the dissociation of a species such as $H_2O_2$:

$$H_2O_2 \rightarrow OH\bullet + OH\bullet \quad (1.83)$$

For the generic compound AB, the rate of disappearance of AB is related to the rate of appearance of A and B:

$$AB \rightarrow A + B \quad (1.84)$$

$$r_A = \frac{d[A]}{dt} = r_B = \frac{d[B]}{dt}$$

$$= -r_{AB} = \frac{-d[AB]}{dt} \quad (1.85)$$

For a first-order reaction,

$$-r_{AB} = \frac{-d[AB]}{dt} = k[AB] \quad (1.86)$$

When integrated, the instantaneous concentration of AB is

$$[AB] = [AB]_0 \exp(-kt) \quad (1.87)$$

If $[A]_0 = 0$, the instantaneous concentration of A is

$$[A](t) = [AB]_0[1 - \exp(-kt)] \quad (1.88)$$

**Second-order reactions.** Most pollution reactions are bimolecular and proceed as the result of binary collisions such as

$$O_3 + CO \rightarrow CO_2 + [O\bullet] + [O\bullet] \quad (1.89)$$

For generic second-order bimolecular reactions written as

$$A + B \rightarrow C + D \quad (1.90)$$

the appearance of C and D accompanies the disappearance of A and B:

$$r_C = \frac{d[C]}{dt} = r_D = \frac{d[D]}{dt}$$

$$= -r_A = \frac{-d[A]}{dt} = -r_B = \frac{-d[B]}{dt} \quad (1.91)$$

For second-order kinetics,

$$-r_A = \frac{-d[A]}{dt} = k[A][B] \quad (1.92)$$

When Equation 1.92 is integrated, the instantaneous concentrations of A and B are related to each other and to time by

$$\frac{\ln[B]}{[A]} = \frac{k([B]_0 - [A]_0)t + \ln[B]_0}{[A]_0} \quad (1.93)$$

**Simultaneous reactions.** For a system of simultaneous reactions, it is important to be able to write and solve a set of simultaneous equations and to note some useful approximations that can be made depending on the magnitude of kinetic rate constants and the reactivity of certain species. Consider a kinetic reaction mechanism consisting of

$$A \rightarrow B \quad \text{rate constant, } k_a \quad (1.94)$$

$$B \rightarrow D \quad \text{rate constant, } k_b \quad (1.95)$$

Initially, assume that only species A is present at a concentration $[A]_0$ ($[B]_0 = 0$ and $[D]_0 = 0$). Write the rate equations,

$$-r_A = k_a[A] \quad (1.96)$$

$$r_B = k_a[A] - k_b[B] \quad (1.97)$$

$$r_D = k_b[B] \quad (1.98)$$

Combining the results of the rate equations for all species in the constant-volume batch reactor, the mass balance produces the following expressions:

$$\frac{d[A]}{dt} = -k_a[A] \quad (1.99)$$

$$\frac{d[B]}{dt} = k_a[A] - k_b[B] \quad (1.100)$$

$$\frac{d[D]}{dt} = k_b[B] \quad (1.101)$$

To compute the concentrations at any time, one integrates the differential equations based on known initial concentrations. The first equation can be integrated directly:

$$\int_{[A_0]}^{[A]} \frac{d[A]}{[A]} = -\int_0^t k_a \, dt$$

$$[A] = [A]_0 \exp(-k_a t) \qquad (1.102)$$

The second differential equation is a nonhomogeneous first-order differential equation and can be solved in closed form or by using Laplace transforms:

$$\frac{d[B]}{dt} + k_b[B] = k_a[A]_0 \exp(-k_a t)$$

$$[B] = \frac{[A]_0 k_a}{k_b - k_a [\exp(-k_a t) - \exp(-k_b t)]} \qquad (1.103)$$

The last differential equation can be solved directly:

$$\int_0^{[D]} \frac{d[D]}{[A]_0 k_a / (k_b - k_a)}$$

$$= \int_0^t k_b [\exp(-k_a t) - \exp(-k_b t)] \, dt \qquad (1.104)$$

$$[D] = \frac{[A]_0 - [A]_0 \exp(-k_a t) - [A]_0 k_a}{k_b - k_a [\exp(-k_a t) - \exp(-k_b t)]} \qquad (1.105)$$

Consider the consequence of what happens if the intermediate species B is highly reactive, so that it disappears as quickly as it forms. Such a case exists if B corresponds to the free radical $[O\cdot]$ or $[H\cdot]$ in Equations. 1.74 and 1.75. If the species disappears as quickly as it forms, one can assume that any instant

$$\frac{d[B]}{dt} \approx 0 \qquad (1.106)$$

This state of affairs is called a *pseudo steady state* for species B. Thence species B is said to possess its *steady-state concentration*, $[B]_{ss}$.

$$0 = -k_b[B] + k_a[A]$$

$$[B]_{ss} = \frac{k_a}{k_b}[A] = \frac{k_a}{k_b}[A]_0 \exp(-k_a t) \qquad (1.107)$$

If Equation 1.103 is examined, the above is true any time that $k_b$ is much greater than $k_a$ (i.e. $k_b \gg k_a$).

The larger consequence of the above is that if a kinetic mechanism contains a highly reactive species [B] such that its removal kinetic rate constant is large compared to the rate constant at which it is formed, one may set the derivative $d[B]/dt$ equal to zero and solve for the resulting value of $[B]_{ss}$ and then use this value in any other differential equation in which B appears. The advantage of this steady-state approximation to solving the set of differential equations is enormous.

## 1.10 Chemical equilibrium and steady state

All engineers are accustomed to identifying steady state as that circumstance when the partial time derivative is equal to zero (i.e., $\partial(\cdot)/\partial t = 0$). However, this circumstance also exists at chemical equilibrium, leading to potential confusion. While it is true that a system at chemical equilibrium is also at steady state, the reverse is not true. For example, in a flow system, steady state can be achieved, but if the reaction continues, chemical equilibrium does not exist. To pull the leg of our thermodynamicist friends, chemical equilibrium is the uninteresting state in which nothing happens. On the other hand, pragmatic engineers love steady state, in which desired changes in variables are achieved along with excellent control.

The concept of steady state requires one to focus attention on the time variations of a fluid property $[\cdot]$ at a particular *point in space* $(x, y, z)$. This is called the *Eulerian point of view*, in which a change with time is represented as $(\partial[\cdot]/\partial t)_{x,y,z}$. Thus a steady-state condition requires that at a point in space $(x, y, z)$ nothing changes with time [i.e., $(\partial[\cdot]/\partial t)_{x,y,z} = 0$]. However, at another point in space, $(x + \delta x, y, z)$, the value of the property $[\cdot]$ may be different than at $(x, y, z)$ even though again it does not change with time.

The concept of thermodynamic equilibrium requires one to focus attention on time variations of a property $[\cdot]$ of a *particular mass* of fluid. This is called the *Lagrangian point of view*, in which a change with time is represented by the total derivative, as one moves along with the fluid velocity. This special time derivative is called the *substantial derivative*, $D[\cdot]/Dt$, as opposed to the more general total time derivative $d[\cdot]/dt$,

$$\frac{D[\cdot]}{Dt} = \frac{\partial[\cdot]}{\partial t} + \frac{U\partial[\cdot]}{\partial x} + \frac{V\partial[\cdot]}{\partial y} + \frac{W\partial[\cdot]}{\partial z} \quad (1.108)$$

and $U, V$, and $W$ are the three components of the fluid velocity. Thus thermodynamic equilibrium requires that the property $[\cdot]$ of the fluid parcel does not change with time, irrespective of where the fluid parcel may be located. A particularly important case of the thermodynamic equilibrium is chemical equilibrium.

To illustrate these distinctions, consider changes occurring in the gas mixture shown in Figure 1-20 (note that this is the same gas mixture as used in Example 1.7 on the water-gas shift reaction). We wish to examine how the hydrogen concentration changes in closed and open systems. Initially, the temperature is sufficiently low that no reaction occurs.

- *Closed system.* A well-mixed constant-volume vessel initially contains the gases shown in Figure 1-20a.
- *Open system.* A duct with excellent radial mixing is fed steadily, with the same mixture supplying the inlet composition shown in Figure 1-20b.

In both cases the gas mixture suddenly experiences a step-function rise in temperature. Consider the hydrogen concentration as a function of time in the closed system and as a function of location in the open system.

**Closed system (well-mixed container).** Since there are no gradients in a well-mixed system, only $\partial[\cdot]/\partial t$ may exist. When chemical equilibrium is reached, $\partial[\cdot]/\partial t = 0$, and therefore $D[\cdot]/Dt = 0$. Thence the property $[\cdot]$ will no longer vary with time (see Figure

Closed system (well mixed, constant volume)

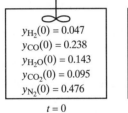
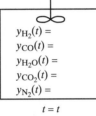
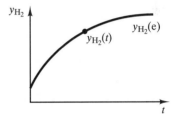

Open system (one dimensional, steady flow, no transverse gradients)

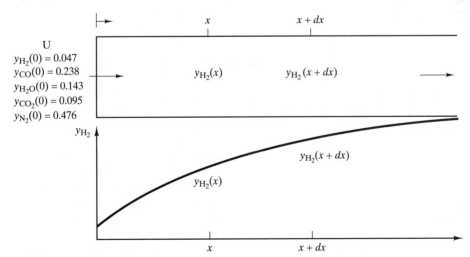

**Figure 1-20** Comparing steady state and chemical equilibirum

1-20a). Thus chemical equilibrium is mathematically equivalent to steady state for a closed system.

**Open system (steady flow through a duct).** Initially, fluid of the inlet composition fills the duct. Reaction begins after the temperature is raised. A transient period ensues, $(\partial[\cdot]/\partial t)_{x,y,z} \neq 0$, and eventually a steady state is achieved, $(\partial[\cdot]/\partial t)_{x,y,z} = 0$. Steady state requires that $(\partial[\ ]/\partial t)_{x,y,z} = 0$ but not that $D[\cdot]/Dt = 0$. Thus

$$\frac{D[\cdot]}{Dt} = \frac{U\partial[\cdot]}{\partial x} + \frac{V\partial[\cdot]}{\partial y} + \frac{W\partial[\cdot]}{\partial z} \neq 0 \quad (1.109)$$

Since there are no transverse gradients, $\partial[\cdot]/\partial y$ and $\partial[\cdot]/\partial z$ are zero. Since $U$ and $\partial[\cdot]/\partial x$ are not zero, $D[\cdot]/Dt$ is not zero and there is a variation of the hydrogen concentration along the duct as shown in Figure 1-20b. No matter how long you wait, only steady state, and not equilibrium, will be achieved. If the duct were infinitely long, or the reaction were infinitely fast, equilibrium would be achieved at the duct outlet. This situation is impractical in most circumstances and steady state is distinct from equilibrium. Thus for an open system flow, steady state does not ensure chemical equilibrium. Only when both the time and spatial derivatives are zero does steady state become equivalent to chemical equilibrium.

**Conclusion.** In a closed system, steady state and chemical equilibrium are conceptually distinct but mathematically equivalent. In an open system, chemical equilibrium ensures steady state, but steady state does not ensure chemical equilibrium.

## 1.11  Professional literature

The following is a brief list of textbooks and handbooks on environmental pollution. Listed also are materials from the related fields of industrial hygiene, industrial ventilation, and toxicology since the principles by which contaminants are controlled and their toxic effects are the same.

### Ecology

Elsom (1992)
Williamson (1973)

### Health effects and industrial hygiene

Cralley and Cralley (1979)

### Atmospheric chemistry and physics

Finlayson-Pitts and Pitts (1986)
Flagan and Seinfeld (1988)
Seinfeld and Pandis (1997)

### Aerosols

Friedlander (1977)
Hidy and Brock (1970)
Hinds (1982)
Willeke and Baron (1993)

### Air pollution control

Bunicore and Davis (1992)
Calvert and Englund (1984)
Cooper and Alley (1986)
Crawford (1976)
deNevers (1995)
Licht (1980)
Mycok et al. (1995)
Theodore and Bunicore (1994)
Wark and Warner (1981)

### Control of contaminants in the indoor environment

American Conference of Governmental and Industrial Hygienists (1988)
Burgess et al. (1989)
Heinsohn (1991)
Wadden and Scheff (1987)

### Key chemical engineering references

Allen and Rosselot (1997)
Bisio and Kabel (1985)
Fogler (1992)
McCabe et al. (1993)
Smith and Van Ness (1975)

**Professional societies.**  The following publications of professional societies contain material relevant to environmental pollution control.

*Aerosol Science and Technology*
*American Industrial Hygiene Association Journal*
*Annals of Biomedical Engineering*
*Annals of Occupational Hygiene*
*Applied Industrial Hygiene*
*Atmospheric Environment*
*Environmental Science and Technology*
*Journal of Aerosol Science*

*Journal of the Air and Waste Management Association*

*Transactions of the American Society of Heating, Refrigerating, and Air-Conditioning Engineers*

*Transactions of the American Society of Mechanical Engineers*

**Governmental agencies.** The following agencies of the U.S. government have specific missions devoted to the control of contaminants and publish technical reports expressly related to environmental pollution control. Their Web sites should be consulted to keep abreast of changes in regulations and standards.

Department of Energy (DOE)
Environmental Protection Agency (EPA)
Mine Safety and Health Administration (MSHA)
National Institute of Occupational Health (NIOSH)
Occupational Safety and Health Administration (OSHA)

**Internet.** A vast amount of information relating to topics in the book is available on the Internet and engineers can expect that the amount will continue growing at a rapid rate. Browsing (surfing) the Internet for information is a skill that engineers need to master. Web sites are added and updated so frequently that it is impractical to provide a comprehensive list. We have identified certain Web sites when it is appropriate to do so, but readers can expect that with the passage of time they will be superseded by new sites containing greater amounts of data. Our objective is to initiate readers to using the Internet to obtain information pertinent to air pollution, to acquire confidence doing so, and to impart the desire to remain current via professional journals and conversations with colleagues. The address for the Web site for this book is,

**http://www.engr.psu.edu/cde/me470**

The book's Web site can also be found by selecting a link on the Prentice Hall home page.

**http://www.prenhall.com**

Readers are urged to examine the Web page for the book to obtain solutions to Examples within chapters that use MathCAD, Internet Web sites containing government regulations on air pollution, addresses of Web pages that provide pollutant thermophysical property data, and Class Notes for each of the chapters.

**Symposia, Conferences, and Professional Meetings.** Various national and international symposia, conferences, workshops, and so on, are held every year. These meetings usually prove to be more valuable than one anticipates. Chance meetings with people in the hall, dinner, and so on, inevitably prove to be stimulating. The proceedings of many conferences are published but are unlikely to appear in university or corporate libraries. International meetings are of particular importance because the Japanese and Europeans follow practices that in many respects are superior to those used in the United States.

Engineers should subscribe to trade journals and magazines in environmental pollution control, which can be received by engineers working in the field at no cost. These "freebie" publications contain primarily advertisements about new equipment, products and services and written articles about general topics in the field. Subscribing to these publications is an excellent way that educators can keep abreast of new equipment and products that can be conveyed to their students. For engineers in industry the publications are essential.

## *1.12 Closure*

The overriding concern of engineers is to design and operate processes in ways to prevent discharging to the atmosphere materials that are injurious to health, the vitality of our natural resources, or our enjoyment of these natural resources. This might seem a banal statement, but the old chestnut "that an ounce of prevention is better than a pound of cure" applies to air pollution as well as it does for anything else. Consequently, preventing pollutants from being produced in the first place is cheaper than removing them from a discharge process gas stream or attempting to remove them from the atmosphere. Only as a last resort should engineers resort to removing pollutants in a discharge process gas stream. Nevertheless, engineers can be assured that they will be called on to perform this task. To spend public and private funds wisely, benefits must be commensurate with the reduction of perceived risk. To do otherwise is imprudent. Assessing risks is an activity involving many components of society. Engineers are one of those components. Not only must risks be assessed rationally, but we must never lose sight of risks that are entered into voluntarily but that affect others involuntarily.

# Nomenclature

| Symbol | Description (Dimensions*) |
|---|---|
| $\rightleftharpoons$ | simultaneous forward and reverse reactions |
| $\rightarrow$ | direction of a chemical reaction |
| $=$ | empirical stoichiometric equation |
| $[=]$ | has the units of |
| $[j]$ | concentration of species $j$ $(M/L^3, N/L^3)$ |
| $a, b, c, d$ | constants defined by equation |
| $A$ | area; preexponential factor in an Arrhenius equation |
| $a_i$ | activity of molecular species $i$ |
| $c, c_i$ | mass or molar concentration, total or molecular species $i$ $(M/L^3, N/L^3)$ |
| $\overline{c}_{p,i}$ | molar specific heat at constant pressure of molecular species $i$ (Q/NT) |
| $\overline{c}_i$ | partial concentration of molecular species $i$ $(M/L^3, N/L^3)$ |
| $f$ | constant |
| $f_i$ | mass fraction of molecular species $i$ |
| $F_i$ | molar flow rate of molecular species $i$ (N/t) |
| $\overline{f}_i$ | fugacity of molecular species $i$ |
| $f_i^0$ | standard state fugacity of molecular species $i$ |
| $E_a$ | activation energy (Q/N) |
| $\Delta G_T^0$ | standard change of Gibbs free energy (Q/N) |
| $\overline{g}_{i,T}^0$ | Gibbs free energy of formation of species $i$ at temperature $T$ (Q/N) |
| $\overline{h}_{i,T}^0$ | enthalpy of formation of species $i$ at temperature $T$ (Q/N) |
| $\Delta H_T^0$ | standard change of enthalpy (Q/N) |
| $k_1, k_{-1}$ | kinetic rate constant of forward and reverse equation 1, units depend on reaction |
| $k_f, k_r$ | kinetic rate constant of forward and reverse reaction, units depend on reaction |
| $K_a(T), K_f(T), K_P(T), K_\phi(T), K_y(T)$ different | ways to express equilibrium constants |

*Q, energy; F, force; L, length; M, mass; N, moles; t, time; T, temperature.

| Symbol | Description (Dimensions*) |
|---|---|
| $\dot{m}$ | mass flow rate (M/t) |
| $m_i$ | mass of molecular species $i$ (M) |
| $M_i$ | molecular weight of molecular species $i$ (M/N) |
| $n, n_i$ | number of moles of molecular species $i$ (N) |
| $n_i$ | mols of molecular species $i$ (N) |
| $n_t$ | total number of moles (N) |
| $P$ | total or overall pressure $(F/L^2)$ |
| $P_i$ | partial pressure of molecular species $i$ $(F/L^2)$ |
| $Q$ | volumetric flow rate $(L^3/t)$ |
| $r_A$ | rate of formation of species A $(N/t\ L^3)$ |
| $R_u$ | universal gas constant (Q/NT) |
| $\overline{s}_{i,T}^0$ | entropy of formation of species $i$ at temperature $T$ (Q/NT) |
| $\Delta S_T^0$ | standard change of entropy (Q/NT) |
| $t$ | time (t) |
| $T$ | temperature (absolute) (T) |
| $U$ | velocity (L/t) |
| $V$ | total or overall volume $(L^3)$ |
| $V_i$ | volume of molecular species $i$ $(L^3)$ |
| $x_i$ | mole fraction of molecular species $i$ in liquid phase |
| $X$ | fractional conversion of reactant A, moles of A reacted per mole of A fed |
| $y_i$ | mole fraction of molecular species $i$ in gas phase |
| $Z$ | compressibility factor |

*Greek*

|  |  |
|---|---|
| $\alpha$ | constant defined by equation |
| $\beta$ | constant defined by equation |
| $\nu_i$ | stoichiometric coefficient of molecular species $i$ |
| $\rho$ | density $(M/L^3)$ |
| $\overline{\rho}_i$ | partial density of molecular species $i$ $(M/L^3)$ |

| Φ | relative humidity |
| $\phi_i$ | fugacity coefficient of molecular species $i$ |

*Subscripts*

| $(\cdot)_{actual}$ | actual thermodynamic properties |
| $(\cdot)_{avg}$ | average |
| $(\cdot)_c$ | critical conditions |
| $(\cdot)_e$ | exit conditions |
| $(\cdot)_i$ | molecular species $i$ |
| $(\cdot)_{ss}$ | steady-state conditions |
| $(\cdot)_{STP}$ | standard temperature and pressure (298 K, 101 kPa) |
| $(\cdot)_t$ | total |
| $(\cdot)_0$ | initial conditions or inlet conditions |

*Abbreviations*

| acfm | actual cubic feet per minute |
| ACGIH | American Conference of Governmental Industrial Hygienists |
| AIHA | American Industrial Hygiene Association |
| AMCA | Air Movement and Control Association |
| ANSI | American National Standards Institute |
| APCA | Air and Waste Management Association (formerly Air Pollution Control Association) |
| AQS | air quality standards |
| ASHRAE | American Society of Heating, Refrigerating, and Air-Conditioning Engineers |
| ASME | American Society of Mechanical Engineers |
| BLS | Bureau of Labor Statistics |
| CAS | Chemical Abstract Service |
| CIS | Chemical Information Service |
| CSTR | continuous-flow stirred tank reactor |
| DOE | Department of Energy |
| fpm | feet per minute |

| EPA | Environmental Protection Agency |
| ESP | electrostatic precipitator |
| HAP | hazardous air pollutant |
| HERP | human exposure dose/rodent potency dose |
| ID | inside diameter |
| MSDS | Material Safety Data Sheet |
| MSHA | Mine Safety and Health Administration |
| NIH | National Institutes of Health |
| NIOSH | National Institute for Occupational Safety and Health |
| NSC | National Safety Council |
| OSHA | Occupational Safety and Health Administration |
| OTA | Office of Technology Assessment |
| PEL | permissible exposure limit |
| PMN | premanufacture notice |
| ppb | parts per billion |
| ppm | parts per million |
| RTECS | Registry of Toxic Effects of Chemical Substances |
| scfm | standard cubic feet per minute |
| STP | standard temperature (298 K) and pressure (101 kPa) |
| $TD_{50}$ | daily dose rate to halve the percent of tumor-free animals |
| TLV | threshold limit value |
| TSCA | Toxic Substances Control Act |
| TSP | total suspended particles |
| UV | ultraviolet radiation |
| VOC | volatile organic compounds |

# References

Allen, D. T., Rosselot, K. S., 1997. *Pollution Prevention for Chemical Processes*, Wiley-Interscience, New York.

*American Conference of Governmental and Industrial Hygienists, Committee on Industrial Ventilation*, ACGIH, 1988. *Industrial Ventilation: A Manual of Recommended Practice*, 20th ed., Lansing, MI.

Ames, B. N., Magaw, R., and Gold, L. S., 1987. Ranking possible carcinogenic hazards, *Science*, Vol. 236, April 17, pp. 271–280.

Arrow, K. J., Cropper, M. L., Eads, G. C., Hahn, R. W., Lave. L. B., Noll, R. G., Portney, P. R., Russell, M., Schmalensee, R., Smith, V. K., and Stavins, R. N., 1996. Is there a role for benefit–cost analysis in environmental, health, and safety regulation? *Science*, Vol. 272, pp. 221–224.

Balzhiser, R. E., Samuels, M. R., and Eliassen, J. D., 1972. *Chemical Engineering Thermodynamics*, Prentice Hall, Upper Saddle River, NJ.

Bisio, A., and Kabel, R. L., 1985. *Scaleup of Chemical Processes*, Wiley, New York, 699 pp.

Bunicore, A. J., and Davis, W. T. (Eds.), 1992, *Air Pollution Engineering Manual*, 3rd ed., Air and Waste Management Association, Pittsburgh, PA.

Burgess, W. A., Ellenbecker, M. J. and Treitman, R. D., 1989, *Ventilation for Control of the Work Environment*, Wiley-Interscience, New York.

Calvert, S., and Englund, H. M., 1984, *Handbook of Air Pollution Technology*, Wiley-Interscience, New York.

Churchill, S. W., 1974. *The Interpretation and Use of Rate Data: The Rate Concept*, McGraw-Hill, New York, 510 pp.

Cooper, C. D., and Alley, F. C., 1986. *Air Pollution Control: A Design Approach*, PWS Engineering, Boston.

Cox, G. V., and Strickland, G. D., 1988. Risk is normal to life itself, *American Industrial Hygiene Association Journal*, Vol. 49, pp. 223–227.

Cralley, L. J., and Cralley, L. V., (Eds.), 1979. *Patty's Industrial Hygiene and Toxicology*, Vols I, II, and III, Wiley, New York.

Crawford, M., 1976. *Air Pollution Control Theory*, McGraw-Hill, New York.

deNevers, N., 1995. *Air Pollution Control*, McGraw-Hill, New York.

Dewees, D. N., 1987. Does the danger from asbestos in buildings warrant the cost of taking it out? *American Scientist*, Vol. 75, May–June 1987, pp. 285–288.

Dixon, D. C., 1970. The definition of reaction rate, *Chemical Engineering Science*, Vol. 25, pp. 337–338.

Ellul, J., 1965. *Propaganda*, Vintage Books, New York.

Elsom, D. M., 1992. *Atmospheric Pollution: A Global Problem*, 2nd ed., Blackwell, Oxford.

Finlayson-Pitts, B. J., and Pitts, J. N., 1986. *Atmospheric Chemistry: Fundamentals and Experimental Techniques*, Wiley-Interscience, New York.

Flagan, R. C., and Seinfeld, J. H., 1988. *Fundamentals of Air Pollution Engineering*, Prentice Hall, Upper Saddle River, NJ.

Fogler, H. S., 1992. *Elements of Chemical Reaction Engineering*, 2nd ed., Prentice Hall, Upper Saddle River, NJ, 838 pp.

Friedlander, S. K., 1977. *Smoke, Dust and Haze*, Wiley-Interscience, New York.

Gold, L. S., Stone, T. H., Stern, B. R., Manley, N. B., and Ames, B., 1992. Rodent carcinogens: setting priorities, *Science*, Vol. 258, pp. 261–265.

Gordon, S., and McBride, B. J., 1976. *Computer Program for Calculation of Complex Chemical Equilibrium Compositions, Rocket Performance, Incident and Reflected Shocks, and Chapman–Jouguet Detonations*, NASA SP-273.

Heinsohn, R. J., 1991. *Industrial Ventilation: Engineering Principles*, Wiley-Interscience, New York.

Hidy, G. M., and Brock, J. R., 1970. *The Dynamics of Aerocolloidal Systems*, Pergamon Press, Tarrytown, NY.

Hinds, W. C., 1982. *Aerosol Technology*, Wiley-Interscience, New York.

Jennings, S. G., 1988. The mean free path, *Journal of Aerosol Science*, Vol. 19, No. 2, pp. 159–166.

Kabel, R. L., 1981. Rates, *Chemical Engineering Communications*, Vol. 9, pp. 15–17.

Kabel, R. L., 1985. Homogeneous reactions systems, in *Scaleup of Chemical Processes*, Bisio, A. L., and Kabel, R. L., (Eds.), Wiley, New York, pp. 124–128.

Kee, R. J., Rupley, F. M., and Miller, J. A., 1991. *The Chemkin Thermodynamic Data Base*, Sandia Report SAND87-8215B, reprinted March.

Klaassen, C. D., Amdur, M. O. and Doull, J. (Eds.), 1986. *Casarett and Doull's Toxicology*, 3rd ed., Macmillan, New York.

Lave, L. B., 1987. Health and safety risk analyses: information for better decisions, *Science*, Vol. 236, April 17, pp. 291–295.

Lehr, J. H., 1992. *Rational Readings on Environmental Concerns*, Van Nostrand Reinhold, New York.

Licht, W., 1980. *Air Pollution Control Engineering*, Marcel Dekker, New York.

Lipfert, F. W., and Wyzga, R. E., 1995. Air pollution and mortality: issues and uncertainties, *Journal of the Air and Waste Management Association*, Vol. 45, pp. 949–966.

McCabe, W. L., Smith, J. C., and Harriott, P., 1993. *Unit Operations of Chemical Engineering*, 5th ed., McGraw-Hill, New York, 1130 pp.

Mossman, B. T., Bignon, J., Corn, M., Seaton, A., and Gee. J. B. L., 1990. Asbestos: scientific developments and implications for public policy, *Science*, Vol. 247, pp. 294–301.

Mullen, J. F., 1988. Consider fluid-bed incineration for hazardous waste destruction, *Chemical Engineering*, Vol. 95, No. 10, pp. 22–26.

Mycok, J. C., McKenna, J. D., and Theodore, L., 1995. *Handbook of Air Pollution Control Engineering and Technology*, Lewis Publishers, Boca Raton, FL, 416 pp.

National Research Council, Steering Committee on Identification of Toxic and Potentially Toxic Chemicals for Consideration by the National Toxicology Program, Board on Toxicology and Environmental Health Hazards, Commission on Life Sciences, 1984. *Toxicity Testing: Strategies to Determine Needs and Priorities*, National Academy Press, Washington DC.

National Safety Council, 1994. *Accident Facts, 1994 Edition*, NSC, Ithaca, IL.

Nichols, A., and Zeckhauser. R., 1985. *The Dangers of Caution: Conservatism in Assessment and Management of Risk*, paper E-85-11, Harvard University, Cambridge, MA, November.

Okrent, D., 1987. The safety goals of the nuclear regulatory commission, *Science*, Vol. 236, April 17, pp. 296–300.

Orwell, G., 1968. Politics and the English language, *The Collected Essays*, in Vol. 4, *Journalism and Letters of George Orwell*, S. Angus and I. Angus (Eds.), Harcourt Brace Jovanovich, New York, pp. 127–139.

Perry, R. H., Green, D. W., and Maloney J. G., 1984. *Perry's Chemical Engineers Handbook*, 6th ed., McGraw-Hill, New York.

Prugh, R. W., (Ed.), 1982. *Kirk–Othmer Encyclopedia of Chemical Technology*, 3rd ed., Plant safety, Vol. 18, No. 60.

Reid, R. C., Prausnitz, J. M., and Poling, B.E., 1989. *The Properties of Gases and Liquids*, 4th ed., McGraw-Hill, New York.

Repace, J. L., and Lowery, A. H., 1980. Indoor air pollution, tobacco smoke, and public health, *Science*, Vol. 208, May, pp. 464–472.

Rothstein, M. A., (Ed.), 1983. *West's Handbook Series: Occupational Safety and Health Law*, 2nd ed., West Publishing, St. Paul, MN.

Russell, M., and Gruber, M., 1987. Risk assessment in environmental policy-making, *Science*, Vol. 236, April 17, pp. 286–290.

Sax, N. I., 1979. *Dangerous Properties of Industrial Materials*, 5th ed., Van Nostrand Reinhold, New York.

Seinfeld, J. H., 1986. *Atmospheric Chemistry and Physics of Air Pollution*, Wiley, New York.

Slovic, P., 1987. Perception of risk, *Science*, Vol. 236, April 17, pp. 280–285.

Smith, J. M., and Van Ness, H. C., 1975. *Introduction to Chemical Engineering Thermodynamics*, 3rd ed., McGraw-Hill, New York, 632 pp.

Starr, C., 1969. Social benefit versus technological risk, *Science*, Vol. 165, pp. 1232–1238.

Starr, C., 1985. Risk management, assessment and acceptability, *Risk Analysis*, Vol. 5, No. 2, pp. 97–102.

Sterling, T. P., Dimich, H., and Kobayashi, D., 1982. Indoor byproduct levels of tobacco smoke: a critical review of the literature, *Journal of the Air Pollution Control Association*, Vol. 32, No. 3, pp. 250–259.

Stull, D. R., and Prophet, H., 1971. *JANAF Thermochemical Tables*, 2nd ed., ASRDS-NBS 37, National Bureau of Standards, Washington, DC, June.

Theodore, L., and Bunicore, A. J., 1994. *Air Pollution Control Equipment*, Springer-Verlag, New York.

Thurow, L. C., 1987. A weakness in process technology, *Science*, Vol. 238, December 18, pp. 1659–1663.

Traynor, G. W., Apte, M. G., Carruthers, A. R., Dillworth, J. F., Grimsrud, D. T. and Gundel, L. A., 1987. Indoor air pollution due to emissions from wood-burning stoves, *Environmental Science and Technology*, Vol. 21, No. 7, pp. 691–697.

Turns, S., 1996. *An Introduction to Combustion*, McGraw-Hill, New York.

Verschueren, K., 1983. *Handbook of Environmental Data on Organic Chemicals*, 2nd ed., Van Nostrand Reinhold, New York.

Viscusi, W. K., 1992. *Fatal Tradeoffs: Public and Private Responsibilities for Risk*, Oxford University Press, New York.

Wadden, R. A. and Scheff, P. A., 1987. *Engineering Design for the Control of Workplace Hazards*, McGraw-Hill, New York.

Wark, K., and Warner, C. F., 1981. *Air Pollution*, Harper & Row, New York.

Willeke, K., and Baron, P. A. (Eds.), 1993. *Aerosol Measurement*, Van Nostrand Reinhold, New York.

Williamson, S. J., 1973. *Fundamentals of Air Pollution*, Addison-Wesley, Reading, MA.

Wilson, R., and Crouch, E. A. C., 1987. Risk assessment and comparisons: an introduction, *Science*, Vol. 236, April 17, pp. 267–270.

# Problems

*For Problems 1.1 to 1.3, students may find it useful to consult the web page http://webbook.NIST.Gov or the following reference books.*

*Lewis, R. J., Sr., Sax's Dangerous Properties of Industrial Materials, Van Nostrand Reinhold, New York, 1993*

*Lewis, R. J. Sr., Rapid Guide to Hazardous Chemicals in the Workplace, Van Nostrand Reinhold, New York, 1994*

*Norback, C. T., Hazardous Chemicals on File, Van Nostrand Reinhold, New York, 1988*

*Pohanish, R. P., and Greene, S. A., (Eds.), Hazardous Substances Resource Guide, Gale Research, Detroit, MI, 1993*

*Stecher, P. G. (Ed.), The Merck Index, 9th ed., Merck Corporation, Rahway, NJ, 1976*

*Weiss, G. (Ed.), Hazardous Chemicals Data Book, Noyes Data Corporation, Park Ridge, NJ, 1986*

**1.1.** An article in a professional journal lists the following chemicals and their CAS numbers: pyridine, CAS 110-86-1; methyl chloride, CAS 74-87-3; methylene chloride,

CAS 75-09-2; and acrolein, CAS 107-02-8. For each of these compounds, find the following.

(a) RTECS number
(b) Permissible exposure limit (PEL)
(c) Chemical formula
(d) Vapor pressure at 25°C
(e) Phase at 25°C, 1 atm
(f) Lower explosion limit (LEL)
(g) Health hazard symptom
(h) Recommended personal protection equipment and sanitation

**1.2.** You plan to conduct an experiment that requires the use of the following chemicals: dimethylamine, trimethylamine, and pentaborane. Consult one the chemical registries and find the following for each compound.

(a) CAS and RTECS numbers
(b) Permissible exposure limit (PEL)
(c) Chemical formula
(d) Vapor pressure at 25°C
(e) Phase at 25°C, 1 atm
(f) Lower explosion limit (LEL)
(g) Health hazard symptom
(h) Recommended personal protection equipment and sanitation

**1.3.** Prior to demolishing of an old building, workers find containers of liquid chemicals whose labels are nearly obliterated. The RTECS number on each can are legible: UH8225000, ZE2100000, and KX4550000. Consult one of the chemical registries and find the following for each can.

(a) CAS number
(b) Permissible exposure limit (PEL)
(c) Chemical formula
(d) Vapor pressure at 25°C
(e) Phase at 25°C, 1 atm
(f) Lower explosion limit (LEL)
(g) Health hazard symptoms
(h) Recommended personal protection equipment and sanitation

**1.4.** Visit the supply room in your institution and ask the attendant to show you the Material Safety Data Sheets (MSDS) for materials your institution uses that you know to be hazardous. (See Tables 3-1 and 3-2 for materials that OSHA considers hazardous.)

**1.5.** Adding chlorine to drinking water kills microscopic disease-causing organisms but it also reacts with organic matter and produces small amounts of chloroform. The issue is more serious with treatment of surface water supplies because groundwater contains negligible amounts of plant or animal matter. The U.S. average chloroform concentration in tap water is 83 μg/L. Chlorinating swimming pools also produces chloroform, and while as a rule children don't drink the water, they ingest some of it and breath air above the water surface containing chloroform and receive on the average a total dose of 250 μg per hour of exposure. Assess the cancer risk of playing in a pool compared with drinking tap water. Specifically, how many 8-oz glasses of tap water would a child have to drink to receive a chloroform dose equivalent to one hour of play in a pool? The average consumption of water for an adult is 1 L/day. (See Larson, et al., *Science*, April 8, 1994 for additional discussion.)

**1.6.** How many hours of exposure to formaldehyde in a conventional home produces cancer risks equivalent to drinking one 12-oz can of beer?

**1.7.** How many sandwiches containing 32 g of peanut butter produce cancer risks equivalent to formaldehyde exposure inside a conventional home for 14 h?

**1.8.** People who smoke outdoors are sometimes criticized for polluting the atmosphere. Compare the amount of CO generated by smoking to that generated by automobiles. Assume that all autos satisfy the 1990 Clean Air Act Amendments and produce no more that 9 g of CO per mile. The CO introduced to the atmosphere by cigarettes is on the average (inhaled plus sidestream smoke) 50 mg/cigarette. If smokers consume on the average 10 cigarettes ($\frac{1}{2}$ pack) a day and that on the average autos accumulate 20,000 miles per year, estimate how many smokers generate the same amount of CO as one automobile in a year.

**1.9.** A smoker typically inhales one cigarette over an elapsed (integrated) period of time equal to 20 s at an average volumetric flow rate of 5 L/min. The concentration of smoke particles in the inhaled air is $10^{15}$ particles/m$^3$. Consider a nonsmoker seated in a room with smokers. The average smoke concentration in the room is 0.3 mg/m$^3$ of air. Assume that the nonsmoker inhales air at a volumetric flow rate of 4 L/min. Assume that smoke particles are spherical with a uniform diameter $(D_p)$ of 0.1 μm and a density $(\rho_p)$ of 800 kg/m$^3$.

(a) What is the mass concentration of smoke particles inhaled by the smoker $(mg/m^3)$?

(b) How long would the nonsmoker have to remain in the room to inhale a mass of smoke equivalent to smoking one cigarette?

**1.10.** An operating permit application states that the $SO_2$ concentration is 1000 ppm in an air stream at 600°C, 100 kPa. The volumetric flow rate of air is 2000 scfm. What is the mass flow rate of $SO_2$ (kg/h)?

**1.11.** The concentration of carbon dioxide $(CO_2)$ in a flowing gas stream is 1000 ppm. The temperature and pressure of the gas stream are 375 K and 200 kPa. The duct containing the gas stream has a diameter equal to 1.5 m and the average air velocity is 10 m/s. Compute the following values.

    **(a)** (Actual) $CO_2$ concentration in $mg/m^3$ at 375 K, 200 kPa

    **(b)** (Actual) $CO_2$ concentration in $molecules/m^3$ at 375 K, 200 kPa

    **(c)** $CO_2$ concentration in $mg/m^3$ at STP

    **(d)** Mass flow rate of $CO_2$ in kg/s

    **(e)** Actual total gas volumetric flow rate in the units of acfm

    **(f)** Total gas volumetric flow rate in the units of scfm

**1.12.** One hundred (100) g of isopropyl alcohol ($M = 60.1$, SG = 0.79) and 100 g of glycol ($M = 74.1$, SG = 1.12) are added to 1 $m^3$ of water. If the mixture is ideal at STP, compute the mass fraction and mole fraction of each species.

**1.13.** Ten milliliters (10 mL) of the following liquids are added to 1 L of water,

    **(a)** MEK ($M = 72.1$, SG = 0.81)

    **(b)** Acetone ($M = 58.1$, SG = 0.79)

    **(c)** Diethylamine ($M = 73.1$, SG = 1.71)

    **(d)** Formaldehyde ($M = 30$, SG = 1.08)

Assuming that the mixture is ideal, compute the mass and mole fraction of each species.

**1.14.** Fly ash particles of constant diameter $(D_p)$ 10 $\mu$m, $\rho_p = 850$ $kg/m^3$ are present in flue gas at a temperature of 350°C. If there $10^6$ particles per cubic meter of gas at this temperature, find the mass concentration of particles corrected to STP (i.e., find $c_{particles}$ ($kg/m^3$ corrected to STP)].

**1.15.** Nickel carbonyl, $Ni(CO)_4$, is a highly toxic gas that is believed to decompose by the following mechanism:

$$Ni(CO)_4 \underset{k_{1r}}{\overset{k_{1f}}{\rightleftharpoons}} Ni(CO)_3 + CO \quad (1)$$

$$Ni(CO)_3 + O_2 \underset{k_{2r}}{\overset{k_{2f}}{\rightleftharpoons}} Ni(CO)O + CO + CO_2 \quad (2)$$

The reverse reaction for (2) is a three-body reaction and has a very small kinetic rate constant, $k_{2f} \gg k_{2r}$. One of the products, $Ni(CO)_3$, is a highly reactive species such that $k_{2f} \gg k_{1r}$ and the pseudo steady state approximation can be made [i.e., $d[Ni(CO)_3]/dt \approx 0$. Initially, a vessel of air at STP contains 100 ppm of $Ni(CO)_4$ and 10,000 ppm of CO and $O_2$. Show that at any time later,

$$\frac{[Ni(CO)_4]}{[Ni(CO)_4]_0} = \exp(-tk_{1f})$$

(See Stedman, D. H., Hikade, D. A., Pearson, R., Jr., and Yalvac, E. D., Nickel carbonyl: decomposition in air and related kinetics, *Science*, Vol. 208, pp. 1029–1030, 1980.)

**1.16.** Compute the annual fatality rates for persons at risk for the following accidents and place the values on

Figure 1-2 and Table 1-2. Are these fatalities voluntary or involuntary?

    **(a)** 20,500 accidental deaths in the home

    **(b)** 3900 fires in the home

    **(c)** 3000 accidental CO asphyxiations

    **(d)** 2400 suffocation–ingestion deaths

    **(e)** 48,000 motor vehicle accidental deaths

**1.17.** Molecular species W and Z are formed when species R disintegrates:

$$R \overset{k_0}{\longrightarrow} W + Z$$

where $k_0$ is the overall rate constant. The kinetic mechanism for the appearance of W and Z depends on two reactions, in which R reacts with a neutral species M which is in large abundance, to form a highly reactive state R*, which in turn disintegrates to form W and Z:

$$R + M \underset{k_{1r}}{\overset{k_{1f}}{\rightleftharpoons}} R^* + M \quad (1)$$

$$R^* \overset{k_2}{\longrightarrow} W + Z \quad (2)$$

where $k_{1f}$ and $k_{1r}$ are the reaction rate constants for the forward and reverse reactions in (1) and $k_2$ is the rate constant for reaction (2). Assuming that $k_{1f}$, $k_{1r}$, and $k_2$ are known, prove that (a) the rates of formation of W and Z are zeroth order in R if $k_{1r}[M] > k_2$, and (b) the rates of formation of W and Z are first order in R if $k_{1r}[M] \ll k_2$:

$$\frac{d[R]}{dt} = -k_0[R] \qquad \text{where } k_0 = k_{1f}[M]$$

$$\text{if } k_{1r}[M] \ll k_2$$

$$\frac{d[R]}{dt} \simeq -\gamma \qquad \text{where } \gamma \text{ is a very small constant,}$$

$$0 < \gamma \ll 1 \qquad \text{if } k_{1r}[M] > k_2$$

**1.18.** In fuel-lean flames, the vast majority of sulfur is oxidized to $SO_2$. A fraction is further oxidized to $SO_3$. At steady state, approximately 10% of the sulfur is in the form of $SO_3$. Sulfur trioxide is formed by the kinetic mechanism

$$SO_2 + O\cdot + M \overset{k_1}{\longrightarrow} SO_3 + M$$

Sulfur trioxide also undergoes two- and three-body removal reactions:

$$SO_3 + O\cdot + M \overset{k_2}{\longrightarrow} SO_3 + O_2 + M$$

$$SO_3 + H\cdot \overset{k_3}{\longrightarrow} SO_2 + OH\cdot$$

In fuel-lean combustion, the total moles of sulfur oxide remains constant (i.e., $S_t = SO_2 + SO_3 = $ constant).

    **(a)** Write a differential equation describing how the $SO_3$ concentration varies with time [i.e., $[SO_3] = f(t)$].

**(b)** Assuming that O• and H• remain constant, show that

$$[SO_3] = \frac{B}{A}\left[1 - \exp\left(\frac{-t}{\tau}\right)\right] + [SO_3]_0 \exp\left(\frac{-t}{\tau}\right)$$

where

$[SO_3]_0$ = known initial concentration (may be zero)

$B = k_1[O•][M][S_t]$

M = third body

$A = k_1[O•][M] + k_2[O•][M] + k_3[H•]$

$\tau = \dfrac{1}{A}$

**(c)** What is the steady-state $SO_3$ concentration, $[SO_3]_{ss}$?

**1.19.** $H_2O_2$ vapor in air dissociates spontaneously to form two $[OH•]$ radicals with a first order rate coefficient $k$:

$$H_2O_2 \xrightarrow{k} OH• + OH•$$

If $[OH•]_0$ and $[H_2O_2]_0$ are known, write an expression for the $[OH•](t)$ at any instant of time $> 0$.

**1.20.** Consider the reaction $NO + O_3 \rightarrow NO_2 + O_2$, in which the reaction rate constant is stated as $k(cm^3\ molecule^{-1}\ s^{-1}) = 2.2 \times 10^{-12}\exp(-1430/T)$. What is the reaction rate constant in the units $ppm^{-1}\ min^{-1}$ and $m^3\ gmol^{-1}\ s^{-1}$?

**1.21.** Find the ratio of the mass of air used per day by autos in a large city to the mass of air inhaled per day by the entire U.S. population.

- *U.S. population*: 220 million
- *Human air consumption*: 20 kg/day
- *Autos in the city*: 4.9 million
- *Gasoline use per day*: 2 gal/day
- *Gasoline density*: 6 $lb_m$/gal
- *Auto air–fuel (mass)*: 15

**1.22.** [*Design Problem*] A young health-conscious and environmentally responsible recent college graduate works for the EPA in Washington, DC. Since driving and parking in Washington is aggravating, he uses a motorcycle that consumes very little gasoline per mile. A routine day of work followed by a trans-continental flight is.

| Activity | Schedule (EST) |
|---|---|
| Awake | 6 A.M. |
| Exercise, bicycle laps on city streets | 6–7 A.M. |
| Breakfast and dress for work | 7–8 A.M. |
| Commute to work, motorcycle | 8–9 A.M. |
| Government work (including lunch) | 9 A.M.–5 P.M. |
| Commute to home, motorcycle | 5–6 P.M. |
| Dinner (1 h) | 6–7 P.M. |
| Drive to airport, auto | 7–8 P.M. |
| Fly to Los Angeles (3500 miles) on commercial airline | 8 P.M.–midnight |
| Taxi (auto) to motel | Midnight–1 A.M. |

The employee proposes that other health-conscious environmentally-responsible people in your company follow the same day-time routine. You are employed in the Environmental, Health and Safety Office in the company firm and have been asked by your superiors to prepare a company response to this person's proposal.

A review of National Safety Council data reveals the following data about the fatalities associated with the foregoing activities. The U.S. population is 262 million.

- *Bicycle–auto fatalities*: 800 deaths/yr; bicycle use $15 \times 10^9$ h/yr
- *Suffocation by food ingestion*: 1900 deaths/yr
- *Motorcycle–auto fatalities*: 2171 deaths/yr, motorcycle use $9.9 \times 10^9$ miles/yr, average speed 30 mph
- *Government work*: Table 1-1
- *Total auto fatalities*: 1.83 deaths/$10^8$ miles, average speed 30 mph
- *Commercial aircraft*: 0.07 death/$10^8$ miles

**1.23.** [*Design Problem*] The combustion of natural gas produces fewer pollutants than gasoline. Politicians and journalists jump to the conclusion that natural gas (assume $CH_4$) should be used as an automotive fuel, particularly for automobile fleets in cities (i.e., U.S. Postal Service, taxies, buses, trucks, and other vehicles with regular routes that return to a central location each day for maintenance and refueling). No one performs elementary calculations regarding the size and weight of the vessel needed to store high-pressure natural gas. An 1800-lb, four-door sedan typically has a 20-gallon molded plastic fuel tank weighing approximately 5 $lb_m$ located in the rear of the car and a passenger compartment that is 5 $m^3$.

**(a)** Estimate the volume and weight of:

- *Type A*: steel tank, storing high-pressure natural gas at an initial pressure of 200 atm, weighing 10 times the mass of the natural gas it contains. (*Note*: Be sure to consider the compressibility factor of natural gas.) The gas is stored at 300 K.

- *Type B*: fiberglass tank, storing low-pressure natural gas at an initial pressure of 5 atm, weighing 0.3 lbm for each cubic foot of gas it contains.

Assume the following fuel properties:

- *Higher heating value of gasoline: 48,256 kJ/kg*
- *Higher heating value of natural gas: 55,496 kJ/kg*
- *Density of gasoline: 900 kg/m$^3$*
- *Critical pressure and temperature of natural gas (methane): 4.64 MPa, 191.1 K*

**(b)** Compare the size and weight of types A and B fuel tanks with the volume and weight of the automobile. List procedures that you think need to be taken to cope with the following events:

- Filling of the tanks with high-pressure natural gas
- Natural gas leaks for the vehicles stored in a garage overnight
- Gas leaks, fire, and explosion attendant to using natural gas as a fuel in a vehicle traveling in cities and on open roads

(For more details, see Running on methane, *Mechanical Engineering*, Vol. 112, No. 5, pp. 66–71, May 1990.)

*Problems 24 to 30 are included to illustrate the use of MathCAD, or other equivalent software. The solution to many problems in this book using MathCAD are available on web page.*

**1.24.** The settling velocity ($v_t$) of small fly ash particles of diameter $D_p$ and density $\rho_p$ (1800 kg/m$^3$) in quiescent air is given by $v_t = D_p^2 \rho_p g / 18\mu$. The viscosity ($\mu$) is a function of temperature given by Equation 10.3 and $g$ is the acceleration of gravity (9.8 m/s$^2$). Compute and plot the settling velocity ($v_t$ in cm/s) versus particle diameter (in μm) for, 0.1 μm $< D_p <$ 100 μm, and for 300 K, 400 K, and 500 K.

**1.25.** Using the roots function in Mathcad, compute and plot the equilibrium concentration of CO (in ppm) versus the temperatures shown below. Initially 1 gmol of each of CO, $CO_2$, $O_2$, and argon are contained in a constant pressure vessel at 25°C and 2 atm. The temperature is raised to a constant value and held until equilibrium occurs. The equilibrium constants for the reaction

$$CO_2 \rightleftharpoons \tfrac{1}{2}O_2 + CO$$

are as follows:

| $T(K)$ | $K_p$ |
|--------|-------|
| 1600 | $1.98 \times 10^{-5}$ |
| 1800 | $2.04 \times 10^{-4}$ |
| 2000 | $1.313 \times 10^{-3}$ |
| 2200 | $5.976 \times 10^{-3}$ |
| 2400 | $2.106 \times 10^{-2}$ |
| 2600 | $6.075 \times 10^{-2}$ |

**1.26.** The objective of this problem is to become proficient with repetitive computations and the roots function to obtain trial-and-error solutions. Exhaust gas is discharged from a smoke stack of effective stack height ($H$) equal to 163 m. The mass emission rate ($\dot{m}$) of $SO_2$ is 160 g/s. At this height, the wind speed ($U$) is 6 m/s. The ground-level $SO_2$ concentration downwind of the stack is given by

$$c_{GL}(x) = \frac{\dot{m}}{2U\pi\sigma_y\sigma_z} \exp\left(\frac{-H^2}{2\sigma_z^2}\right)$$

where at any downwind distance ($x$) in km,

$$\sigma_y(m) = \left(\frac{1000}{2.15}\right) x \tan(T)$$

$$T = 24.167 - 2.5334 \ln(x)$$

$$\sigma_z(m) = 453.85 x^{2.1166}$$

**(a)** Compute and plot $c(\text{mg/m}^3)$ at various values of $x$ downwind of the stack, showing that as $x$ increases, $c(x)$ rises to a maximum and then falls.

**(b)** Using the roots function, find the distance ($x$) where the ground-level concentration is 0.1 mg/m$^3$.

**1.27.** The purpose of this problem is to illustrate Mathcad's solve block function. See the tutorial for details. You wish to predict the fugitive emissions from a long pile of coal that has a semicircular cross-sectional area of radius ($a$) 10 m. The axis of the pile is in the $z$-direction. Air approaches the pile in the negative $x$-direction with a velocity $U_x = -5$ m/s perpendicular to the axis of the pile. The equation of an air streamline of magnitude $A$ passing over the pile is given by

$$A = 5 \sin \theta \left(r - \frac{a^2}{r}\right) \quad \text{for } r > a$$

and the velocity components at any point ($r, \theta$) near the pile are

$$U_r = -5 \cos \theta \left[\left(\frac{a}{r}\right)^2 - 1\right]$$

$$U_\theta = 5 \sin \theta \left[ 1 + \left( \frac{a}{r} \right)^2 \right]$$

where $r$ is the radius from the center of the pile and the angle $\theta$ is measured between the $x$-axis and the radius.

(a) Find the coordinates $(r, \theta)$ of the point (or points) where

$$U_r = 1.32 \text{ m/s} \quad \text{and} \quad U_\theta = 4.25 \text{ m/s}$$

(b) Find the coordinates of all points along the streamline for which $A = 41.67 \text{ m}^2/\text{s}$ and plot the streamline using the Mathcad plotting function.

**1.28.** The emission rate $(\dot{m})$ of $SO_2$ from a stack varies with time as follows:

| $t(h)$ | $\dot{m}(t)\,(g/h)$ |
|--------|---------------------|
| 0.25   | 0.1   |
| 1.0    | 0.4   |
| 2.0    | 9.5   |
| 3.0    | 20.0  |
| 4.0    | 20.0  |
| 5.0    | 15.0  |
| 6.0    | 11.0  |
| 7.0    | 8.0   |
| 8.0    | 5.5   |
| 10.0   | 5.5   |
| 14.0   | 4.0   |
| 20.0   | 0.8   |

(a) Using a MathCAD curve-fitting function, derive and plot an empirical equation that describes the data.

(b) Using MathCAD's integration function, compute the total mass (g) of $SO_2$ emitted during the elapsed time period.

**1.29.** The relative humidity on summer day is 68% and the atmospheric (dry bulb) temperature is 90°F. The wet bulb temperature is 80°F. Using a psychrometric chart (see Perry and Chilton, 1973) the absolute humidity is found to be 0.02 lb$_m$ $H_2O$ vapor per lb$_m$ of dry air. The atmospheric air is drawn into a 24-in.-ID pipe with an average velocity of 1000 fpm.

(a) Compute the $H_2O$ concentration in mg/m$^3$ (based on 90°F) and in ppm.

(b) Compute the volumetric flow rate of the moist air in terms of acfm and scfm.

(c) The air is heated to various high temperatures and the water vapor dissociates:

$$H_2O \rightleftharpoons H_2 + \tfrac{1}{2}O_2$$

The equilibrium constants for dissociation are:

| $T(K)$ | $K_P = P_{H_2}P_{O_2}^{1/2}/P_{H_2O}$ |
|--------|----------------------------------------|
| 1400   | 0.00175 |
| 1600   | 0.00562 |
| 1800   | 0.01398 |
| 2000   | 0.02901 |
| 2200   | 0.05276 |
| 2400   | 0.08689 |

Using MathCAD, compute and plot the $H_2$ concentration (ppm) as a function of temperature.

**1.30.** A vessel contains 10 ppm of $N_2O_5$ and 5 ppm of $O_3$ in dry atmospheric air. The components react according to the following reaction mechanism in the absence of light:

(1) $N_2O_5 \rightarrow NO_2 + NO_3$
   $k_1(\text{ppm}^{-2}\text{min}^{-1}) = 3.4 \times 10^{16}\exp(-10{,}660/T)$

(2) $NO_2 + NO_3 \rightarrow N_2O_5$
   $k_2(\text{cm}^3 \text{ molecule}^{-1}\,\text{s}^{-1}) = 4.7 \times 10^{-13}\exp(259/T)$
   $k_2(\text{ppm}^{-2}\text{min}^{-1}) = 5.6 \times 10^3$

(3) $NO_2 + O_3 \rightarrow NO_3 + O_2$
   $k_3(\text{ppm}^{-2}\text{min}^{-1}) = 1.8 \times 10^2\exp(-2450/T)$

(4) $O_3 + NO \rightarrow NO_2 + O_2$
   $k_4(\text{ppm}^{-2}\text{min}^{-1}) = 3.1 \times 10^3\exp(-1450/T)$

(5) $NO + NO_3 \rightarrow 2NO_2$
   $k_5(\text{ppm}^{-2}\text{min}^{-1}) = 1.3 \times 10^4$

(6) $NO_2 + NO_3 \rightarrow NO + NO_2 + O_2$
   $k_6(\text{ppm}^{-1}\text{min}^{-1}) = 1.2 \times 10^1$

(a) Write equations for the rates of formation of $N_2O_5$, $NO_3$, $NO_2$, $O_3$, and $O_2$.

(b) Assuming that $NO_2$ and $NO_3$ achieve a state of pseudo steady state, write equations for their concentrations.

(c) Using MathCAD, compute and plot the concentration of $N_2O_5$ as a function of time.

# 2

# Ecology
# and Exponential Growth

---

In this chapter you will learn:

- About the human existence in the biosphere
- The nature of exponential growth
- The impact of energy consumption
- About radiant energy exchange between sun, earth, and space
- The influence of oceans on climate

There are countless geologic and biogenic pathways in which materials called pollutants enter and leave the environment. In this chapter we introduce the concept of *formation and fate cycles* to describe the interaction between pollutants and elements of the environment. It is important that engineers understand the magnitude of natural and anthropogenic sources of pollution so that they can understand the impact of the industrial processes they design and operate. We believe that readers should acquire a broad understanding of pollution early in the book so that they can better appreciate detailed discussions of individual topics given in subsequent chapters.

## 2.1 Biological diversity

The environment is a dynamic system (Schwartz, 1989) consisting of renewable and nonrenewable components. The objective of this chapter is to learn the processes and major pollutants affecting the atmospheric environment and the principal processes affecting their generation and removal. In short, who are the major players, and what are their roles? These processes and compounds are discussed in sequential fashion. Each is described in elementary terms to identify the principal roles they play. It is important that the reader become familiar with these players because specific aspects about pollution are discussed in detail in subsequent chapters, and unless readers retain an understanding about the basic relationships, they'll be overwhelmed by detail.

The relationship between organisms and their environment is *ecology*. The combination of organisms and the environment in which they live is an *ecosystem*. In terms of engineering thermodynamics, an ecosystem is an *open system (*control *volume)* in which energy and material are stored and across

These two aspects are illustrated by examples in three areas:

**1.** *Chemicals.* Polychlorinated biphenyls (PCBs) are halogenated hydrocarbons invented for their unusual stability and physical properties needed in electrical transformers, refrigerants, foaming agents, solvents, and so on. They were put into use without examining fully their carcinogenic properties and it was not until they were discarded and had migrated through land and water that the danger became evident. The immediate benefits of DDT became the basis to set goals for new pesticides before we fully understood the deleterious consequences of DDT on wildlife. Concern about saccharin, however, generated unwarranted anxiety (Cohen, 1978).

**2.** *Medical advances.* Pharmaceuticals such as thalidomide and intrauterine devices were created for laudable reasons, but their long-term effects were not fully understood and unfortunate consequences followed. Policies to remove asbestos in buildings (Dewees, 1987) were undertaken before we fully understood the economic consequences and limited benefits of such action.

**3.** *Communications.* Not all developments lead necessarily to technologies with undesirable consequences. Consider the history of communications (i.e., telegraph, telephone, radio, and television). Each advance has exceeded its creators' expectations. The need for governmental facilities to communicate with each other in the event that conventional systems were destroyed led to the creation of the *Internet* in the early 1960s. Decades later the network was made available to the public. Along with personal computers (which did not exist in the early 1960s) this communications network has evolved into what we call *cyberspace.* The limits of cyberspace are beyond the comprehension of most people.

The question implied in the examples above is not "right" or "wrong" but the more difficult question of "ought to" or "ought not to." The greater the urgency surrounding a technological undertaking, the more apt the full array of potential undesirable (or desirable) consequences will go undetected. Where was the prudence that should have delayed the *Challenger* spacecraft mission when there was an identified high probability that O-rings would fail under freezing conditions. On the other hand, there is not unlimited time to explore all the potential consequences of every undertaking. Owing to the accelerating pace of social change, decreasing amounts of time and money are available before decisions have to be made. But how much time can a society afford to spend contemplating unintentional consequences? There are no simple technological answers; the questions are political and we must look to the political process to resolve them.

The major event in the eighteenth century that altered people's lives was the industrial revolution, based on the invention of a portable supply of power. Prior to 1750, changes progressed *linearly* as an *arithmetic progression.* If $A$ is an activity that changes linearly with respect to time,

$$\frac{dA}{dt} = k_1 \tag{2.1}$$

$$A(t) = A_0 + k_1 t \tag{2.2}$$

$$\frac{A(t)}{A_0} = 1 + \frac{t}{t_d} \tag{2.3}$$

where $A_0$ is the initial value of $A$, $k_1$ is the coefficient of change, and $t_d$ is a linear *doubling time* related to $k_1$ by

$$t_d = \frac{A_0}{k_1} \tag{2.4}$$

Note, however, that this linear time depends on the starting point, $A_0$. Humans encounter countless linear relationships that we comprehend intellectually and cope with psychologically. Consider the following mundane linear relationships between cause and effect or stimulus and response.

- *Springs.* The deflection of a solid object is linearly proportional to the force applied to it (i.e., rubber bands, etc.).
- *Thermal comfort.* The comfort (or discomfort) we experience is linearly proportional to the temperature or velocity of air (or water) in which we are immersed.
- *Automobiles.* Under normal traffic conditions, the speed of an automobile is proportional to the deflection of the accelerator, or the angle of a turn is proportional to the angular deflection of the steering wheel.

Figures 2-2 and 2-3 show that changes in the world's population ($P$) and the per capita energy consumption ($c_p$) are no longer linear. After the

$t_d = 1 \text{ day}$

$\alpha = 0.693/$

The objective of the
ine the relation betwe
tion of energy, and the
There are equally imp
land and water, but the
The relationship betw
the atmosphere must
terms because it is on
that government can
success of its policies. F
tionships when one h
sponsible. Avoiding q
to confusion and the c
flourish.

An example of the
exponential growth is
Thomas Malthus was
fessor of history and p
the arguments of Rou
tion espousing the pe
published anonymous
show the infinite hun
was wishful thinking.
postulates: (1) that foc
tence, and (2) that the
necessary and will ren
Malthus was an eco
poverty as inescapable
population would incr
while the means of su
arithmetic progression
expand to the limit of
thereafter by famine, v

**EXAMPLE 2.2   MAL**

To illustrate Malthus's
on an island that has l
food and other require
is 1000 and 3000 acre
total tillable land on t
land yields 500 lb of f

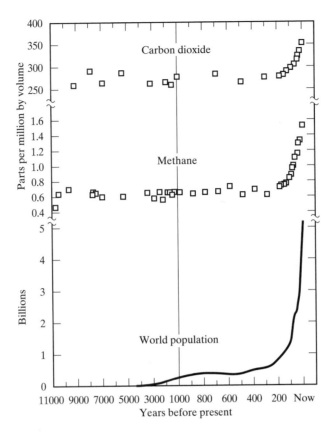

**Figure 2-2**  Ambient
concentration (ppm) of $CO_2$
and $CH_4$ and world
population versus time.
$CO_2$ and $CH_4$
concentrations were
deduced from
measurements of air
trapped in polar ice and
modern air samples. (Firor
and Jacobsen, 1993, Air
and Waste Management
Association, Vol 43, 1993)

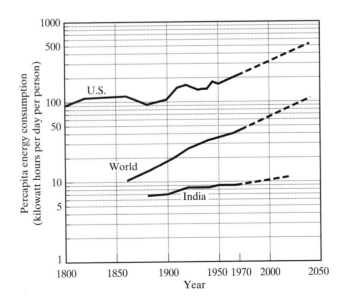

**Figure 2-3**  Trends in per
capita energy consumption
in the United States, India
and the world. (Barus,
1972)

eighteenth centu
increases with tir

$$\frac{dP}{dt} \neq$$

Change of this so
raises an interesti
to support the bel
an accelerating n
basically linear cr
consequences of
time, how can ind
sequences of tech
nological accomp
accelerating mani

*Exponential ch
sion*, occurs wher
portional to the ir

where $k_2$ is the cc
ing with time yiel

$$A(t)$$

where $A_0$ is the va
is an exponential
physical interpret
tion 2.6, where the
is proportional to

To a first appro
grew linearly, but

$$\frac{dP}{dt} =$$

where $\alpha$ is the ex
tion. The *doublin
is the elapsed ti
doubles [i.e., $P(t$

$$t_c$$

where $\alpha$ is expres
case of exponenti
not depend on the

Readers may n
type of change fr

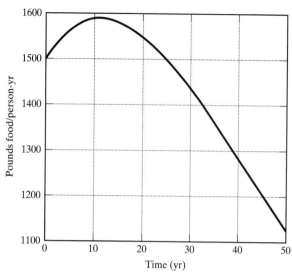

**Figure E2-2**   Malthusian dilemma, food production ($f_{h,p}(t)$, $lb_m$/person-yr) versus time (yr)

grows more rapidly than land is brought under cultivation, and the per capita food production begins to
decline. Food remains abundant, but after 25 years
the per capita food production falls below the minimum 1500 $lb_m$ per year. The society will fail because
it does not bring land under cultivation fast enough,
not because it runs out of land. At 25 years only 6000
acres, $3000 + (120)(25)$, out of 50,000, are under
cultivation. Clearly, two *temporary* technological
solutions are available for the Mathusian dilemma
and both have been used successfully in our society.
These are (1) to bring land under cultivation more
rapidly, and (2) to make each acre more productive.
Ponder irrigation, fertilization, herbicides, pesticides,
and hybrid seeds, but don't neglect biological diversity or forget the "dust bowl."

Malthus's theory was crude and presumed erro-
neously that the food produced per acre would
remain constant, failing to realize that agricultural
technology would make it possible to increase the
yield per acre and to bring more land into production
rapidly. While Malthus's theory lacked sophistication, his theory had empirical validity and established
a theoretical basis for economic theories developed
by many others after him. Malthus's postulation was
pessimistic, but the pessimism remains today, albeit

of a different form. In an evocative essay entitled
"The Tragedy of the Commons," Hardin (1968) (see
also Crowe, 1969) set forth the proposition that the
fundamental problem facing contemporary society is
the exponential growth in population. In colonial
America, the commons was the community's pasture
on which people grazed their livestock. The commons
can be sustained only if the population density is low.
As the human population increases, the shared commons is overgrazed and loses its value. Endorsing a
concept from Hegel that "freedom is the recognition
of necessity," Hardin held that we must recognize the
necessity to abandon the notion that breeding is a
"commons." Freedom to breed will bring ruin to all,
and no technical solution can rescue us from the
eventual misery of overpopulation.

In the book *The Limits to Growth*, Meadows et al.
(1972) argued that finite supplies of nonrenewable
resources posed an upper limit to a population that
expanded exponentially. Once again the analysis
lacked sophistication. By definition, nonrenewable
resources cannot replenish themselves, but while
finite limits exist, no one knows exactly how large
they are. Thus while nonrenewable resources are
finite, accessible ones simply become more expensive
to obtain. Consequently, the limit to nonrenewable
resources is an economic limit long before it becomes
a physical limit.

The writings of Malthus, Meadows, et al. belong to
a genre called *antiutopian* that is an integral part of
Western culture. The literature serves a valuable
function by examining positivist philosophy in quantitative terms, for example by emphasizing the finite-
ness of an ecosystem and the balance that must be
achieved for an ecosystem to sustain itself. The writing of mechanistic models, in which natural constraints (such as shortages of food, energy and water,
epidemics, pestilence, volcanic eruptions, genetic
variability of viruses, climate change, etc.) dictate the
maximum supportable population of the earth, might
describe nonhuman populations accurately but are
poorly suited for human populations. Human beings
are not just mouths to feed but have hands that work
and minds that shape the natural surroundings.
While the world's natural resources are finite, the
human imagination is limitless. Cohen (1995), introduces the concept of the earth's *human carrying
capacity*, $K(t)$. The carrying capacity is the maxi-

mum supportable population at any instant of time. The carrying capacity is a function of the natural constraints above and the human capacity to make individual and collective choices concerning the distribution of material goods and services, acceptable levels of well-being, technological innovation, political institutions, economic arrangements, family structure, migration, and other demographic parameters. Cohen suggests that the population growth rate, $dP(t)/dt$, can be described by

$$\frac{dP(t)}{dt} \propto P(t)[K(t) - P(t)] \qquad (2.12)$$

where the change in carrying capacity, $K(t)$, depends on the population growth rate,

$$\frac{dK(t)}{dt} = c\frac{dP(t)}{dt} \qquad (2.13)$$

where $c$ is a positive constant of proportionality. When $c$ exceeds unity, each person increases the carrying capacity by an amount greater than what they consume and the population grows. When $c$ is unity, the population grows in an exponential fashion. In times of scarcity, $c$ is less than unity and people contribute less than what they consume and the carrying capacity increases more slowly than the population increases.

The physical shortage of materials is less harsh than one might expect. Stimulated by economic incentives, alternative or new materials replace expensive natural materials in short supply. This is not to say that every material in short supply can be replaced, or that replacements will always be available when they are needed, but economic incentives stimulate inventive minds. Technology can ameliorate antiutopian predictions through more efficient use of energy, materials, land, and water (Ausubel, 1996). Within market economies, innovation creates a stream of alternatives that reduce the harsh notion that societies must always be pitted against each other, squabbling over a finite number of marbles.

**1.** *Energy*. Over time, we convert one form of energy to another with higher efficiency, thus requiring proportionally less fuel for a given output. Second, the ratio of hydrogen to carbon, H/C, in fuels increases, thus producing proportionally less carbon dioxide per Btu consumed. The decarbonization of fuel is already evident in the replacement of coal and oil with natural gas. Consider the ultimate in decarbonization, the "hydrogen economy."

**2.** *Land*. Up to 1940, the agricultural yield per acre of most crops advanced little. During the last half century, however, owing to improvements in fertilizer, drainage, equipment, seed stocks, and irrigation, yields of the world's major grains (maize, rice, soybeans, and wheat) improved rapidly. For example, the yields of wheat per acre are larger today in Ireland, Egypt, and India than in the United States. Thus not only are current agricultural yields per acre capable of sustaining the world's population, but there is opportunity to divert croplands to other purposes. World famine is currently a problem in distribution, not production.

**3.** *Materials*. Over time, economic incentives reduced both the amount and types of materials used to perform a variety of consumer and industrial functions. These changes also reduced the energy needed to manufacture materials and the mass of discarded waste. For example, compact disks selling for less than $100 can contain 90 million home phone numbers, which is equivalent to telephone books costing $60,000 and weighing 5 tons.

**4.** *Water*. The per capita withdrawal of water in the United States (for all purposes) quadrupled between 1900 and 1970. Because of economic incentives and governmental regulations, the per capita consumption now falls at a rate of 1.3% per year. Owing to improved industrial efficiencies, water purification, wastewater treatment, and desalinization, the per capita withdrawal of groundwater will continue to decrease.

On one hand, the exponential population growth is cause for concern. Yet on the other hand, human ingenuity is expansive; while energy and mass are conserved, there is no conservation of human ingenuity. Technology is one aspect of ingenuity and one reason to offset dour antiutopian predictions.

## 2.3 Population growth and energy consumption

The purpose of this section is to analyze how population and energy consumption affect the environment. Growth will be simplified by linear and exponential growth expressions in which the growth parameters are constant. Once understood, the analysis can be repeated using more accurate growth parameters that vary with time.

Figure 2-2 shows that the world's population has not increased in the gradual fashion that one might assume based on history prior to the industrial revolution. The industrial revolution marked the beginning of a period in which population can be described as

$$P(t) = P_0 \exp[\alpha(t - t_0)] \qquad (2.14)$$

where $P_0$ is the population at a reference date $t_0$ and $\alpha$ will be called the exponential population growth rate $(yr^{-1})$. Shown in Table 2-1 are the population growth rates of the world and the United States. Population growth rates are commonly expressed as a percent, for example 1.09% per year, and users must be careful to express this exponent as $0.0109 \ yr^{-1}$.

The distinguishing feature of the industrial revolution was the invention of *portable mechanical power*. In its elemental form, mechanical power is the transmission of energy through rotating shafts and gears that ultimately raise weights. The mechanical means to transmit and use power were firmly established prior to 1750. All society lacked was abundant, inexpensive power that could be installed wherever it was needed. Prior to 1750 portable power was provided by men and beasts. The ratio of the mass of the power supplier to net output power was large and there were working environments that men and beasts could not tolerate. Thus portable power possessed severe constraints. Power derived from moving water and wind was abundant only in selected geographic areas.

The invention of the steam engine by Newcomen and Watt, and its commercial exploitation by Boulton and others, marked the beginning of portable, inexpensive power that was limited only by the availability of fuel. The ratio of mass of the power supply to net output power was smaller than from beasts and man and the power could be provided under conditions inhospitable to men and beasts (i.e., at the bottom of coal mines, in foul air, on wheeled vehicles, or on ships). The availability of cheap portable power enabled industrial ventures to be undertaken on a scale and with profitability never seen before. With this power grew the political strength and prosperity of Europe and the United States. With prosperity, the population grew at an unprecedented rate.

The net useful mechanical energy is related to the total input energy by the overall *thermal cyclic efficiency* ($\eta$):

$$\eta = \frac{\text{net useful mechanical energy}}{\text{total input energy}}$$

$$= 1 - \frac{Q_{out}}{Q_{in}} \leqq 1 - \frac{T_{cold}}{T_{hot}} \qquad (2.15)$$

where $T_{cold}$ is the absolute temperature of the reservoir into which $Q_{out}$ is discharged and $T_{hot}$ is the absolute temperature of the reservoir from which $Q_{in}$ was obtained.

From 1750 to the present the overall thermal cyclic efficiency improved from approximately 10% to over 30%. The upper limit is established by the laws of thermodynamics (Equation 2.15), not by human inadequacy. Despite the improvement in thermal cyclic efficiency, total energy consumption has increased in an exponential fashion. Input energy is provided by the combustion of a fuel. Since the total mechanical energy used by society has grown exponentially, one expects that the consumption of fuel has also grown exponentially. The total consumption of energy in the world and in the United States can be expressed in an exponential fashion,

$$C(t) = C_0 \exp[\gamma(t - t_0)] \qquad (2.16)$$

---

**Table 2-1**   Population (millions) and Annual Population Growth Rate ($\alpha$)

|                | 1970 | 1980 | 1990 | 2000[a] |
|----------------|------|------|------|---------|
| World          | 3678; 1.90% | 4415; 1.81% | 5275; 1.76% | 6199; 1.56% |
| United States  | 204; 1.09% | 222; 0.83% | 243.6; 0.91% | 262; 0.63% |

*Source:* Extracted from Mauldin (1980), and Horiuchi (1992).

[a] Estimates for the year 2000 are understated since the Census Bureau reports that the U.S. population was 262.6 million in 1995.

where $\gamma$ is the *exponential energy consumption growth rate*. Shown in Table 2-2 are the total energy consumption and yearly growth rates of the world and the United States assuming that conservation measures are adopted. Once again the rate is expressed as a percent per year, and users must be careful to follow procedures described previously for the population growth rate.

It is instructive to examine the energy consumed per person (per capita energy consumption, $c_p$) since 1750. The *per capita energy consumption rate* is

$$c_p = \frac{C(t)}{P(t)} \qquad (2.17)$$

Figure 2-3 shows that the per capita energy consumption rate is not a constant but increases exponentially:

$$c_p = c_p(0) \exp[\beta(t - t_0)] \qquad (2.18)$$

Using Tables 2-1 and 2-2, representative exponential growth factors $\alpha$ and $\gamma$ for the United States and the world are shown in Table 2-3.

Combining Equations 2.16 and 2.17 gives us

$$
\begin{aligned}
C(t) &= P(t)c_p(t) \\
&= P_0 \exp[\alpha(t - t_0)]c_p(0) \exp[\beta(t - t_0)] \\
&= C_0 \exp[(\alpha + \beta)(t - t_0)] \\
&= C_0 \exp[\gamma(t - t_0)] \qquad (2.19)
\end{aligned}
$$

Hence

$$\gamma = \alpha + \beta \qquad (2.20)$$

and corresponding values for $\beta$ are shown in Table 2-3.

It is instructive to view the world's energy consumption to be driven by the *double exponential* ($\gamma$), composed of the exponential population growth rate ($\alpha$) plus per capita energy consumption exponential growth rate ($\beta$). Presented in this way, it reveals the compound nature of the problem facing the world. Even if we as a nation, or collectively as a global community, decide to conserve energy and maintain a constant per capita energy consumption rate, the total energy consumed in the world will continue to grow simply because the number of people grows exponentially.

The double exponential also reveals the difficult problem facing attempts to reduce the world's consumption of energy. Even if it was possible to reduce the per capita consumption of energy (improving the overall energy conversion efficiency, new technologies, etc.), it is quite another task to reduce population growth, or indeed, even to maintain it constant. Reductions in the per capita energy consumption are amenable to technology, while reducing the population growth is a sensitive issue within the realm of religion and politics.

Efforts to use energy frugally and to eschew wasteful activities have taken place for decades since OPEC attempted to control world oil prices. Figure 2-4 shows the per capita energy consumption versus the per capita gross national product (GNP). While there is some variability between nations of comparable affluence, there is a difference of several orders of magnitude in per capita energy consumption between the developed and underdeveloped nations.

**Table 2-3**    Average Exponential Growth Rates in Population and Energy Consumption Rate, 1970–2000

|  | Population, $\alpha$ (%/yr) | Per capita energy consumption, $\beta$ (%/yr) | Total energy consumption, $\gamma$ (%/yr) |
|---|---|---|---|
| World | 1.76 | 1.98 | 3.74 |
| United States | 0.86 | 0.51 | 1.37 |

**Table 2-2**    Total Energy Consumption Rate (Quads/yr)[a] and Annual Energy Consumption Growth Rate ($\gamma$)

|  | 1970 | 1980 | 1990 | 2000 |
|---|---|---|---|---|
| World | 140, 5.79% | 250, 3.36% | 350, 2.51% | 450, 3.28% |
| United States | 68, 1.11% | 76, 1.11% | 85, 1.62% | 100, 1.65% |

*Source:* Starr et al. (1992), Science, Vol. 256, pp. 98, 287, 1992

[a] Quad = $10^{15}$ Btu.

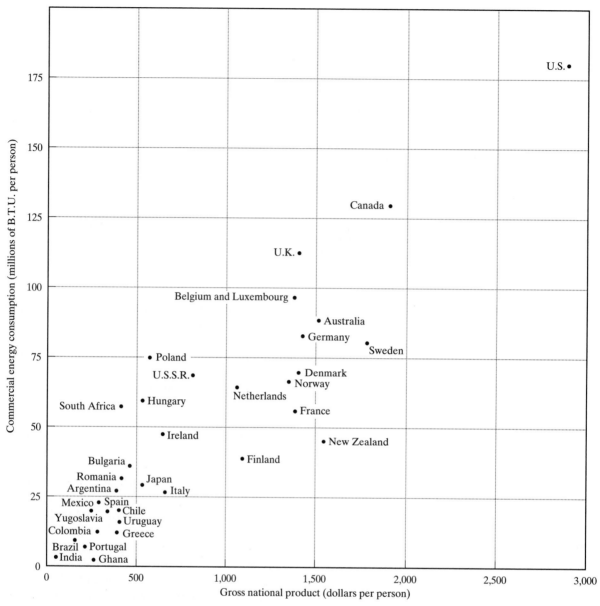

**Figure 2-4**    Per capita energy consumption versus per capita gross national product of different nations (Cook, 1971)

On the basis of Figure 2-4, it is unrealistic to believe that the United States or other advanced nations can reduce their per capita energy consumption without experiencing a substantial reduction in a standard of living.

In 1970 the world's energy consumption was 140 quads/yr (a quad is $10^{15}$ Btu). Figure 2-5 predicts future yearly energy consumption rates for three different exponential energy consumption growth rates:

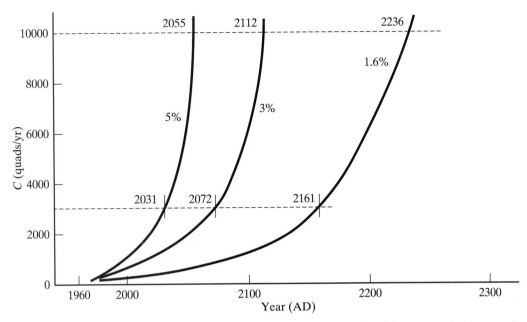

**Figure 2-5** World energy consumption rate (quads/yr.) versus time for different population growth rates and per capita energy consumption growth rates.

1. If the world consumes energy at a growth rate of 5%/yr
2. If the world adopts a rate of 1.6%/yr
3. If the world adopts a middle value of 3%/yr

| **Table 2-4** | Date When Climate Changes Can Be Expected | | |
|---|---|---|---|
| | $\gamma = 5\%/yr$ | $\gamma = 3\%/yr$ | $\gamma = 1.6\%/yr$ |
| 3,000 quads | 2031 | 2072 | 2161 |
| 10,000 quads | 2055 | 2112 | 2236 |

The values of 3000 quads/yr and 10,000 quads/yr have been identified (Table 2-4) because there is a general belief that when the energy discharged to the atmosphere achieves these values, there will be perceptible effects on the world's climate. A heat transfer rate of 10,000 quads/yr is equivalent to the 0.2% of the sun's energy impinging on the earth. It is important to note that 3000 and 10,000 quads can be achieved within the reader's life span, or at least within the life span of the reader's grandchildren or great grandchildren.

Disparities in per capita GNP (Figure 2-4) are thought by some to be the basis of international political unrest. Following this logic, it can be argued that if the developed and underdeveloped nations maintained their per capita energy growth rates (0.51%/yr. for the United States, and 1.98%/yr for the underdeveloped nations), the per capita energy consumption rates and per capita GNP would become equal at some time in the future. If the per capita GNPs were the same, one source of political unrest would no longer exist. Equation 2.18 can be used to predict this future date and shows that the world and U.S. per capita energy consumption rates will be equal in the year 2079. This date is of little solace since Figure 2-5 shows that in 2079 the world's total energy consumption would be large enough to be likely to alter the climate.

# 2.4 Climate

The exponential growth in the consumption of energy is the dominant anthropogenic factor affecting climate. The world's energy consumption produces two major effects on climate. One effect is the emission into the atmosphere of gases, vapors, and particles that affect the radiant heat transfer to and from earth. This phenomenon, often referred to as the *greenhouse effect*, has been widely (and sometimes wildly) discussed in the popular literature. The second effect has not received the attention it deserves and involves the direct heating of the earth's air and water by the energy discharged by human activities.

### EXAMPLE 2.3  TOO MANY PEOPLE IN A SWIMMING POOL

Consider a swimming pool 8 m × 16 m × 1.5 m (192 m³) and examine whether the water temperature will rise $(dT/dt)$ as a function of the number of people $(N_p)$ in the pool. The metabolic rate $(\dot{M})$ of an active person generates approximately 500 kcal/h·person, energy that is transferred to the water.

***Solution***  Consider the mass of water $(m_w)$ as a closed system, ignore work per unit time done on the water $(\dot{W})$, and assume that the only significant heat transfer $(\dot{Q})$ is equal to heat transferred from people's bodies $(\dot{Q} = N_p\dot{M})$. Applying the first law of thermodynamics, we obtain

$$\frac{dU}{dt} = \dot{Q} - \dot{W}$$

$$\frac{m_w c\, dT}{dt} = \dot{Q} = N_p\dot{M}$$

$$(192\text{ m}^3)(1000\text{ kg/m}^3)(1\text{ kcal/K·kg})\frac{dT}{dt}$$

$$= N_p \times 500 \text{ kcal/h·person}$$

$$\frac{dT}{dt} = N_p(0.002604)(\text{K/h})$$

When a single person is in the pool $(N_p = 1)$, the temperature rise is insignificant (0.0026 K/h). If, however, the pool is crowded and contains 1 person/m² $(N_p = 128)$, then

$$\frac{dT}{dt} = 0.33 \text{ K/h}$$

The model that follows is an elementary lumped-parameter model that illustrates the essential energy balance affecting climate. More refined models abandon lumped parameters and account for spatial variations in temperature and composition of the earth's atmosphere and ocean. Nonetheless, a lumped-parameter analysis is instructive since it illustrates clearly the manner in which the key parameters interact with one another.

In the biosphere, society's current energy addition is small with respect to the energy inventory of the biosphere (mass times heat capacity), and no significant temperature rise has occurred so far. On the other hand, since society's energy discharge increases in an exponential fashion, the situation will begin to resemble people in the swimming pool, and a biosphere temperature rise can be expected. Whether it is undesirable is another question, but its detection can be expected.

For purpose of analysis it will be assumed that all the energy consumed by humans is, after some delay, transferred to the atmosphere, hydrosphere, or space. This may appear like a rash assumption, but if one analyzes different scenarios that begin with the consumption of a fuel and then consider the energy discharged later by the products and services the original energy was used for, the assumption will be seen to be reasonable. Before considering several scenarios, recall Equation 2.15 for the overall thermal cyclic efficiency of devices producing power and the fact that only approximately 30% of the total input energy is transformed into useful output work. Thus 70% of the consumed fuel's energy is immediately transferred directly to the earth's air and water. Of the 30% that is used for productive purposes, consider the chain of energy-consuming products and services that lead to warming of the aerial and aqueous environments.

- Generated electricity is used to heat or cool homes, smelt metal, and rotate shafts of devices that ultimately convert mechanical energy to heat.

- Raw materials are converted into feed stocks that are ultimately manufactured into goods. At each step, heat is generated.
- Energy used for transportation is used to overcome frictional forces that ultimately result in heat being dissipated into the surroundings.
- Water is used directly for cooling exothermic processes.

Analyses of this sort lead one to conclude that it is reasonable to believe that within a short period of time, all the energy contained in fuels and in harnessed natural resources such as moving water and wind will be transformed into products and services whose end product will be heat transferred to earth's water and air. In terms of the second law of thermodynamics, fuel with an organized molecular structure obtained from well-defined locations on earth becomes energy transferred to earth's water and air, where it resides in a widely dispersed state. No energy is lost in the process; only its location has changed, and with each change in location, it becomes increasingly harder to extract any of this energy for future useful purposes. The process can be characterized as an irreversible process wherein energy has been conserved but the entropy of the earth and of its air and water has increased.

## 2.5  *Lumped-parameter radiation model*

Just as a swimming pool is not an infinite sink when there are a large number of swimmers, so the earth's atmosphere and waters are not infinite when the discharge of energy is sufficiently large. Consider the following radiation model patterned after one used by Hansen et al. (1981). To understand the relationship between heat transfer and the earth's temperature, model the earth as a solid mass possessing a thin spherical shell, hereafter called the *biosphere*, that contains the land, air, and water in which humans exist. Heat transfer between the sun, the earth's core, and space is shown in Figure 2-6.

Energy released by the combustion of fuels, $\dot{Q}_{human}$, will be treated as heat transfer to the biosphere. The sun radiates energy to the biosphere, the biosphere absorbing a portion of this radiant energy and reflecting the rest. Energy is also transferred to

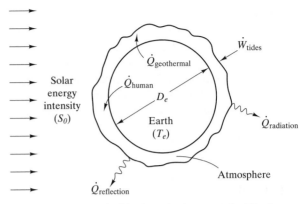

**Figure 2-6**  Blackbody radiation model of Earth

the biosphere by geothermal mechanisms within the earth's core. Work is done on the biosphere by the moon as it drives the tides, and the biosphere radiates energy to space. The biosphere will be treated as a closed (fixed-mass) system. Neglecting changes in kinetic and potential energy, the first law of thermodynamics is

$$\dot{Q} - \dot{W} = \frac{dU}{dt}$$

$$\dot{Q}_{sun} - \dot{Q}_{reflected} - \dot{Q}_{radiation} + \dot{Q}_{geothermal}$$

$$+ \dot{Q}_{human} - \dot{W} = \frac{dU}{dt} \quad (2.21)$$

where

$\dot{Q}_{sun}$ = radiant heat transfer from the sun to the biosphere

$\dot{Q}_{reflected}$ = radiant energy from the sun that is reflected by the biosphere

$\dot{Q}_{radiation}$ = radiant heat transfer from the biosphere to space

$\dot{Q}_{geothermal}$ = geothermal energy transfer from the earth's core to the biosphere, $32 \times 10^{12}$ W (956.32 quads/yr)

$\dot{Q}_{human}$ = energy consumed (and discharged) by humans

$\dot{W}$ = work done by the earth

$dU/dt$ = rate of change of the internal energy of the biosphere

The biosphere does no work on its surroundings, but the moon performs work $3 \times 10^{12}$ W (89.65 quads/yr) on the biosphere by driving the tides. Thus $\dot{W}_{\text{tides}} = -3 \times 10^{12}$ W ($-89.65$ quads/yr). The energy consumed by humans $(\dot{Q}_{\text{human}})$ increases in the exponential fashion discussed in Section 2.4. It will be assumed that the biosphere is in quasistatic equilibrium. Thus

$$\frac{dU}{dt} = 0 \qquad (2.22)$$

and

$$\dot{Q}_{\text{sun}} - \dot{Q}_{\text{reflected}} - \dot{W}_{\text{tides}} + \dot{Q}_{\text{geothermal}}$$
$$+ \dot{Q}_{\text{human}} - \dot{Q}_{\text{earth}} = 0 \qquad (2.23)$$

The radiation from the sun will be modeled as a collimated beam whose average solar intensity is $S_0$,

$$S_0 = 1387.5 \text{ W/m}^2 (440 \text{ Btu h}^{-1} \text{ ft}^{-2}) \quad (2.24)$$

and the biosphere as a radiating body whose *effective radiating temperature* is $T_e$. Following these assumptions, Equation 2.21 becomes

$$S_0 \pi R_e^2 - \rho S_0 \pi R_e^2 - \dot{W}_{\text{tides}} + \dot{Q}_{\text{geothermal}} + \dot{Q}_{\text{human}}$$
$$= \sigma 4\pi R_e^2 T_e^4$$

$$T_e = \left[ \frac{S_0(1 - \rho)}{4\sigma} + \frac{\dot{Q}_{\text{geothermal}} + \dot{Q}_{\text{human}} - \dot{W}_{\text{tides}}}{4\pi\sigma R_e^2} \right]^{0.25} \quad (2.25)$$

where

$T_e$ = effective radiating temperature of the biosphere (K)

$\sigma$ = Stefan–Boltzmann constant = $5.66 \times 10^{-8}$ $\text{W m}^{-2}\text{K}^{-4}$ ($0.1714 \times 10^{-8} \text{Btu h}^{-1} \text{ft}^{-2} \,^\circ\text{R}^{-4}$)

$R_e$ = earth radius = 6378 km

$\rho$ = earth's reflectivity, also called the *albedo*

It should be noted that the sun's energy strikes only the biosphere's projected area $(\pi R_e^2)$, whereas radiation to space occurs over the entire surface $(4\pi R_e^2)$. It is seen that the effective radiation temperature is governed by two parameters, the albedo $(\rho)$ and the world's energy consumption rate $(\dot{Q}_{\text{human}})$, which is expected to increase with time in an exponential manner. The *albedo* accounts for the portion of the sun's radiation that is reflected from earth. The albedo depends on the earth's cloud cover, particles in the atmosphere, and features of the earth's surface such as ice caps, ocean temperature, deserts, and forests. Table 2-5 summarizes predictions from Equation 2.25 for a variety of values of $\dot{Q}_{\text{human}}$ and albedo $(\rho)$. $\delta T_e$ is the temperature rise associated with heat transfer produced by industrial activity $(\dot{Q}_{\text{human}})$:

$$\delta T_e = \left[ T_e(\dot{Q}_{\text{human}} = 0) - T_e(\dot{Q}_{\text{human}}) \right]_{\rho=\text{const}} \quad (2.26)$$

A review of Table 2-5 shows the effect of the world's energy consumption rate and albedo. In simple terms, industrial activity causes the earth's effective radiation temperature to increase, but an increase in the albedo lowers the value of $T_e$. It is reasonably certain that the world's energy consumption will rise. The effect of industrial activity on the albedo is difficult to predict because there are compensating effects and the change may not be uniform over the earth's surface. In any event it can be concluded that when the energy consumption rate lays between 3000 and 10,000 quads, changes in climate can be expected.

The earth's *mean surface temperature* $(T_s)$ is approximately 288 K, and the earth's effective radiating temperature $(T_e)$ is approximately 255 K. Hansen et al. (1981) suggest that the difference $(T_s - T_e)$ can be thought of as *greenhouse heating* caused by the radiant energy exchange between the

**Table 2-5**   Lumped-Parameter Radiation Temperature

| $\dot{Q}_{\text{human}}$ (quads/yr) | Albedo $(\rho)$ | $T_e$(K) | $\delta T_e$(K) |
|---|---|---|---|
| 0 | 0.30 | 255.9 | 0 |
| | 0.32 | 254.1 | 0 |
| | 0.34 | 252.2 | 0 |
| | 0.36 | 250.3 | 0 |
| 3,000 | 0.30 | 256.0 | 0.1 |
| | 0.32 | 254.1 | 0 |
| | 0.34 | 252.3 | 0.1 |
| | 0.36 | 250.3 | 0 |
| 10,000 | 0.30 | 256.1 | 0.2 |
| | 0.32 | 254.3 | 0.2 |
| | 0.34 | 252.4 | 0.2 |
| | 0.36 | 250.5 | 0.2 |

earth's surface and clouds and gases in the atmosphere. Water, carbon dioxide, and other atmospheric gases capture infrared radiation (that would otherwise be lost to space) and return it to the earth's surface, thus inhibiting cooling. It has been observed that on the average, the temperature of the troposphere decreases at a rate of approximately 6.6 K/km. For a tropospheric thickness of 5 km, the difference, $T_s - T_e = 288 - 255 = 33$ K, produces the observed negative temperature gradient $-6.6°C/km$.

Cast in terms of quads, it is seen that energy discharges affecting climate are independent of the process producing the energy discharge. For example, using nuclear energy to produce electricity will eliminate the discharge of $SO_2$, and the greenhouse gases $CO_2$ and hydrocarbons, particles, and water vapor that affect radiant heat transfer between the earth and its atmospheric blanket, but nuclear energy sources do not change the fact that the energy generated is ultimately discharged to the environment. An interesting side issue is to consider the amount of energy consumed to produce uranium fuel from its ore.

There is little chance that the limited natural usable energy sources, such as wind, tides, and dams, can produce sufficient electricity to satisfy the world's power needs. Second, natural sources only eliminate the discharge of energy associated with the *generation* of electricity. Subsequent *uses* of the electricity discharge the same amount of energy to the earth's air and water as if natural sources had not been used. For many decades to come, gas, oil, and coal will be able to satisfy our power needs, but in time these materials will become too expensive and alternative supplies of energy will be needed. Energy from natural resources already occupy unique market niches in many local regions of the United States. Examples include solar, wind, geothermal, tides, small-scale hydroelectricity, and biomass conversion. Technological advances in these fields will expand their importance in the decades ahead, but energy generated from combustion and nuclear processes is not likely to be eliminated.

**1.** *Solar energy.* This is particularly attractive because no greenhouse gases are produced and there is no net addition of energy to the biosphere. Unfortunately, only arid regions of the United States (Nevada, New Mexico, Arizona, southern California, Utah, and Colorado) have reliable enongh incident annual power densities $(2300 \text{ Wh/m}^2 \cdot \text{yr})$ to make them attractive for large-scale solar energy power production (Service, 1996). The use of solar energy for low-power photovoltaic applications (i.e., power for environmental monitoring instruments, domestic electricity far from power lines, power for agricultural activities, etc.) will continue to grow.

**2.** *Wind.* The midwestern states from North Dakota and Montana south to New Mexico and west to California offer the most attractive potential for wind-power generation (Gustavson, 1979). However, the power density is such that wind power is attractive only as an intermittent resource. Although no greenhouse gases are produced, the generated electricity ultimately becomes a human energy addition to the biosphere.

**3.** *Geothermal energy.* Most of the land west of the Missouri River is underlain by hydrothermal or hot, dry rock resources. Geothermal power development is attractive only for rock deposits with thermal gradients of at least $70°C/km$. Evidence exists that the extraction of steam in certain California geothermal power sites exceeds the rate at which steam is formed naturally underground (there are too many "straws in the glass") and that steps must be taken to increase the return of condensate in order to sustain the desired extraction rates. Geothermal steam often contains gaseous sulfur compounds that pose unique pollution control problems (Axtmann, 1975).

**4.** *Biomass conversion.* The midwest is ideal for growing short-duration woody crops and other combustible vegetation suitable for fuel for power generation. Such fuels are carbonaceous and produce the full array of pollution problems associated with coal gas or oil. As the reserves of coal, gas and oil decrease, biomass fuels become economically attractive. Furthermore, such biomass products as paper and lumber are suitable for recycling.

## 2.6 *Greenhouse effect*

There is considerable concern that the products of combustion ($CO_2$, particles, hydrocarbons, etc.) will affect climate. Although there are plausible arguments to support concern and we will illustrate some of them, it is important to recall that significant fluctuations in climate have occurred prior to the establishment of human communities. The spreading and recession of continental ice sheets during the Pleistocene and the climatic changes in the Middle East (Issar, 1995) are dramatic examples that one can cite. Global temperatures computed from data obtained from air bubbles in ice cores (Raynaud et al., 1993; Mayewski et al., 1996), deep-sea marine sediment (Keigwin, 1996; Winograd et al., 1992), tree rings (Jacoby et al., 1996), and so on, indicate that the planet has experienced fluctuations in temperature before, during, and after the establishment of civilizations. It has been proposed that these events could have been produced by variations in solar activity (Haigh, 1996; Kerr, 1996), variability in oceanic circulation (Hurrell, 1995) and ocean surface temperatures (Lubin, 1994), periodic variations in the earth's orbit, tilt and rotation that changed solar insolation, sulfate aerosols (Kiehl and Briegleb, 1993), clouds generated from volcanoes (Minnis et al., 1993; Zielinski et al., 1994), and the impaction of meteors with earth. North Africa and Asia Minor, now arid, were once savannas supporting wildlife and early human civilizations. Over the centuries prior to Columbus, the temperature of the southwestern United States increased and the availability of rain for crops decreased, causing the migration of pueblo people who had maintained permanent communities there for thousands of years. What caused these significant climate changes is not known fully, but it should be clear that periodic changes in climate have occurred that had little or nothing to do with human activity. Having happened before, there is no reason to believe that such changes may not occur again.

The *greenhouse effect* (Schneider, 1989) is the anticipated warming of the earth produced by discharging increasing amounts of energy, pollutants, and combustion products to the atmosphere. The greenhouse effect occurs because gases discharged to the atmosphere block the radiation of infrared energy from earth to space but allow the transmission of incoming ultraviolet radiation from the sun.

Energy is radiated to the earth from the sun over a band of short (visible) wavelengths and radiated from earth to space over another band of longer (IR) wavelengths. A blackbody of temperature $T$ radiates energy over a range of wavelengths in a continuous fashion. The amount of energy at any wavelength is called the *spectral emissive power*, $E_{b,\lambda}$. The blackbody spectral emissive power is given by the Planck distribution law,

$$E_{b,\lambda} = \frac{2\pi hc_0^2}{\lambda^5}\left[\exp\left(\frac{hc_0}{\lambda kT}\right) - 1\right]^{-1} \quad (2.27)$$

where (in typical units)

$E_{b,\lambda}$ = spectral emissive power $\left(\text{Wm}^{-2}\,\mu\text{m}^{-1}\right)$
$h$ = Planck's constant $\left(6.6256 \times 10^{-34}\right)(\text{J}\cdot\text{s})$
$c_0$ = speed of light $\left(2.998 \times 10^8\right)(\text{m/s})$
$\lambda$ = wavelength of the radiant energy $(\mu\text{m})$
$k$ = Boltzmann constant $\left(1.3805 \times 10^{-23}\right)(\text{J/K})$
$T$ = absolute temperature of the radiating body (K)

A *blackbody* is a hypothetical body that radiates energy in the manner given by Equation 2.27. It can be shown that a blackbody is also one that absorbs all radiant energy incident upon it. In the physical world most materials are considered to be *graybodies* since they emit only a fraction of what would be emitted by a blackbody at the same temperature. Graybodies also reflect a portion of the energy incident upon them. Knowing the sun's radiant energy flux $\left(S_0, 1387.5 \text{ W/m}^2\right)$ incident on the earth, it is convenient to model the sun as a blackbody at 5800 K. Similarly, the earth may be treated as a blackbody radiating energy at the temperatures shown in Table 2-5. Figure 2-7 portrays Planck's distribution law (Equation 2.27) for a body radiating at 5800 K and at 245 K. The essential feature of the curves to note is that the dominant wavelengths of the sun's radiation occur in the visible (0.3 to 0.7 $\mu$m) and ultraviolet wavelengths, while the earth's radiant energy occurs over a range of wavelengths of 3 to 100 $\mu$m in the infrared range.

Radiant energy passing through the atmosphere is subject to preferential absorption because gases comprising the atmosphere are transparent to radiation at some wavelengths but opaque to radiation at other wavelengths. Each constituent gas has its own spectral absorptivity, some of which are shown in Figure 2-7. The net energy absorbed is the summation of the spectral absorptivities weighted by the concentrations of gases in the atmosphere.

A *greenhouse gas* is a gas that is transparent in the ultraviolet (UV) but opaque in the infrared (IR).

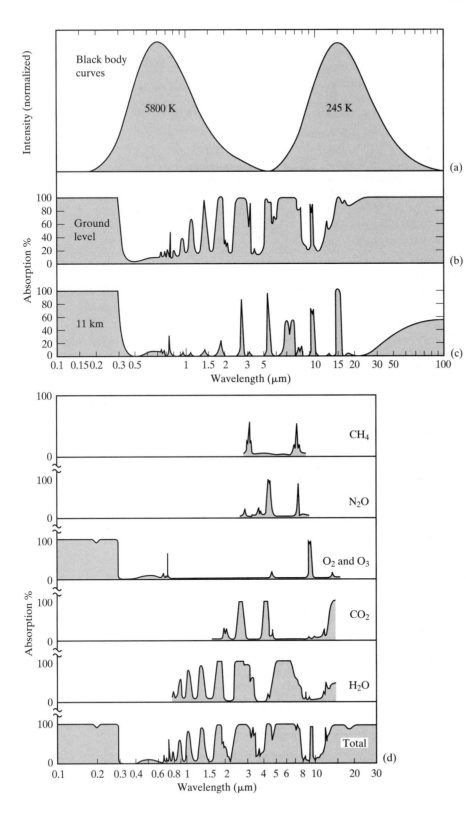

**Figure 2-7** (a) Blackbody emission spectra for earth and sun, (b) spectral absorptivity for solar energy reaching the earth's surface (c) atmospheric spectral absorptivity for solar energy at an altitude of 11 km, (a-c prepared by Wiliamson) (d) spectral absorptivity of several atmospheric gases (original source, R M Goody, "Atmospheric Radiation", Oxford Univ.. Press, 1964; R G Fleagle and J S Businger, "Introduction to Atmospheric Physics", Academic Press, NY 1963) (material taken from Williamson, p 86)

Thus greenhouse gases allow the sun's ultraviolet to pass through the atmosphere but selectively absorb the earth's infrared radiation toward space, making the atmosphere a selective filter. In inhibiting infrared radiation, the atmosphere acts as an insulator. The primary greenhouse gases are $CO_2$, $H_2O$, $O_3$, $N_2O$, $CH_4$, and halogenated hydrocarbons. Ozone is present in the atmosphere, particularly in the stratosphere, and is also an absorber of ultraviolet radiation, thus protecting against skin cancers in humans.

The distribution of energy associated with the radiant exchange between sun, earth, and space is shown in Figure 2-8. The consequence of increasing the concentration of greenhouse gases can be inferred from Figure 2-8. For purposes of simplicity the values of energy are shown as percentages (%) of the incoming solar radiation intensity 1387.5 W/m² (440 Btu/h·ft²).

Table 2-6 describes part of the energy balance. Outgoing infrared radiation to space from the earth and its atmosphere equals $25 + 29 + 12 + 4 = 70$. From the atmosphere (see the right-hand cloud in Figure 2-8) we see that only 12 of the 100% of the infrared radiation captured by the atmosphere is transmitted on to space. The other 88% is recycled back to earth (the greenhouse effect).

**Table 2-6**    Energy Exchange between Sun, Atmosphere, and Space

**Incoming energy from sun** (% of solar constant $S_0$)

| Absorbed radiation | | Reflected radiation | |
|---|---|---|---|
| By earth surface | 45 | By earth surface | −5 |
| By atmosphere | 25 | By atmosphere | −25 |
| Total absorbed | 70 | Total reflected | −30 |

**Outgoing energy from earth** (% of solar constant $S_0$)

| | |
|---|---|
| Radiation from earth surface to atmosphere | 100 |
| Radiation from earth surface to space | 4 |
| Evaporation of earth's surface water | 24 |
| Convection to atmosphere (thermals) | 5 |
| Output by earth and atmosphere | 133 |

Two effects of the world's energy consumption are to increase the amount of $CO_2$ and aerosols (solid or liquid particles) in the atmosphere. Carbon dioxide produced by oxidation processes has a long half-life (15 yr) and one can expect its concentration to increase with time as the world's energy consumption increases, thus contributing to global warming. The decrease in reception of solar radiation due to reflection by aerosols (part of the rationale for "nuclear

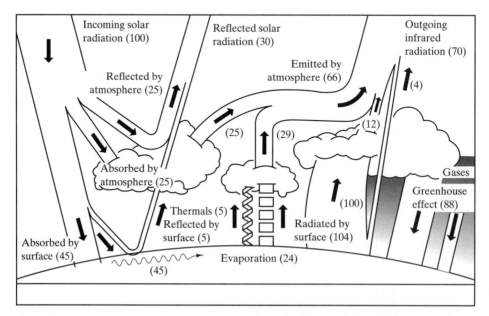

**Figure 2-8**    Energy exchange between sun, earth, atmosphere and space. The numbers represent the percentage of the average incident solar energy flux (adapted from Schneider, 1989)

winter") is estimated (Schwartz and Andreae, 1996) to be comparable, but opposite in sign, to the increased retention of terrestrial infrared radiation enhanced by anthropogenic $CO_2$.

The lumped-parameter radiation heat transfer model is too crude to explain fully the effect of pollutants on the atmosphere. Energy is consumed in selected places on earth, and large concentrations of pollutants exist at selected locations. Computer-based climate models of increasing sophistication have been under continuous development since 1969 (Firor and Jacobsen, 1993). When our understanding of the temperatures and movement of ocean currents improves, these models will more accurately describe the air–water exchange of mass and energy and thus enhance climate modeling. Desperately needed also in the public debate about global warming is inclusion of the economics of governmental policies (Lave and Dowlatabadi, 1993).

*Feedback* is the name given to the phenomenon where an event produces a change that either offsets (*negative feedback*) the stimuli causing it or enhances (*positive feedback*) the stimuli. Systems possessing negative feedback are defined as stable, and systems possessing positive feedback are unstable.

**1.** $H_2O$ *(positive feedback)*. If the global temperature rises, more water will evaporate. This is inherently a cooling process and by itself would illustrate negative feedback. However, as the moisture condenses (into rain and clouds), the latent heat of evaporation is released and the troposphere is heated. A warmer troposphere holds more water vapor, which traps more infrared radiation and amplifies the greenhouse effect.

**2.** *Ice-snow (positive feedback)*. If the global temperature rises, the melting of sea ice and snow increase the earth's absorbtivity (decreasing the albedo) and enhance the absorption of solar energy, which amplifies the greenhouse effect. Warming will have a more noticeable effect near sea-ice margins in the polar oceans.

**3.** *Cloud feedback.* It is unclear how clouds influence the greenhouse effect. The $H_2O$ feedback effect suggests an increase in cloud cover, but it is not known if it will occur generally or locally. An increase in cloud cover increases the albedo and reflection of solar energy, but it also increases trapping infrared radiation, which in turn increases the greenhouse

effect. The two consequences oppose each other and it is unclear if they will truly offset each other or whether one will dominate the other. Volcanic eruptions (Toon and Pollack, 1980; Minnis et al., 1993) introduced small particles into the upper troposphere and lower stratosphere. These particles increased the albedo and produced a cooling effect. The cooling effect was strongest over areas where the cloud cover was small. Because of their small size, these particles are more effective at reflecting the short-wavelength (UV) light from the sun than attenuating the longer-wavelength (IR) radiation from the earth. Thus they reflect more of the sun's energy back to space than they radiate infrared energy back to the earth.

**4.** *Ocean–atmosphere interaction.* Owing to the size of the oceans, their temperature, velocity, and salinity distributions are complex. It is unclear if the ocean surface will actually become warmer due to an increase in atmospheric $CO_2$. If the oceans become warmer, it is also not clear whether it will increase or decrease $CO_2$ in the atmosphere. It is difficult to model $CO_2$ uptake by the oceans (Siegenthaler and Oeschger, 1978). Modeling aside, measurements show that despite periodic variations there is a clear trend that the atmospheric temperature (Figure 2-9), and concentration of carbon dioxide and methane (Figure 2-2) have increased with time and the human presence.

## 2.7 Interaction between oceans and atmosphere

The meteorology of the tropics is characterized by an enormous thermal circulation. Moisture-laden tropical air convected upward from the warmest regions of the earth condenses in rising, towering cumulonimbus clouds that extend to heights of 15 km. In a sense, the tropical atmosphere is a gigantic heat engine driven by the sun operating between the relatively high temperature of the ocean surface and the low temperature of the lower stratosphere, where energy is radiated to space. The tropical convective zone moves seasonally north and south in response to changes in surface temperature.

Any relationship between pollution, energy discharge, and climate must include the oceans. Modeling

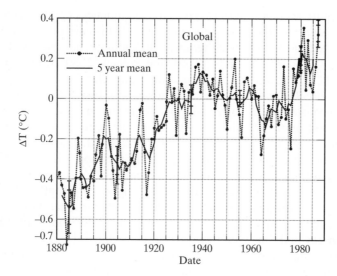

**Figure 2-9**   Global temperature vs. time (adapted from Schneider 1989)

global climate to include the effects of pollution and energy discharge will be no better than our ability to model ocean circulation (Semtner, 1995; Jin, 1996). Figure 2-10 shows that the oceans constitute a moving spherical shell of fluid containing unique global circulation currents of varying temperature and salinity (Post et al., 1990; Richardson, 1993). A deep current of cold, salty water from the north Atlantic travels south through the Atlantic, around the southern tip of Africa, and upwells in the equatorial Pacific and mid-Indian oceans. The resulting warm, less salty shallow currents travel across the Pacific, pass again below the southern tip of Africa and northwestward across the Atlantic, and downwell in the North Atlantic. Horizontal transport occurs in both the deep, cold ocean layers and in the warm surface waters.

A large body of research suggests (Robock, 1996) that surface temperature and precipitation anomalies can be predicted for months to a year in advance by accounting for the state of the sea surface temperatures in the tropical Pacific Ocean. The large surface temperature variations over North America in 1982–1983 were induced by the 1982 El Chichon volcanic eruption. Along the coast of Peru an upwelling region of cold water is moderated by a southward current of warm water in the early months of the year. The warm current is called *El Niño* because it appears around Christmas. Every few years it is exceptionally warm, penetrates unusually far south, and is accompanied by very heavy rains. El Niño is characterized by the following:

- High sea-surface temperatures
- Small differences in surface air pressure across the tropical Pacific
- Weak trade winds

The warm current extends thousands of kilometers offshore and is symptomatic of changes in circulation within the entire ocean basin in response to changes in the surface winds that drive the ocean. The historical record indicates that El Niño is associated with ecological and economic disasters, including devastating droughts over the western Pacific, torrential floods over the eastern Pacific, and unusual weather patterns in various parts of the world. The eastward expansion of the area of warm surface water during El Niño results in an eastward drift of the atmospheric tropical convective zone of the western tropical Pacific. This causes the eastern tropical Pacific to experience the El Niño effects above, while to the west of the dateline, the rainfall decreases and surface air pressure increases. El Niño is one part of a cycle of events called the *southern oscillation* (Philander, 1989), because after years of benign weather, a period of opposite conditions begins, called *La Niña*, characterized by:

- Unusually low sea-surface temperatures in the central and eastern Pacific
- Strong trade winds

La Niña conditions correlate with drought conditions over North America. During La Niña episodes,

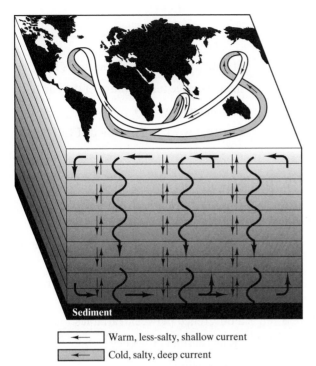

Warm, less-salty, shallow current

Cold, salty, deep current

**Figure 2-10**   The transport of oceanic carbon can be depicted as movements within and between oceanic compartments. Global oceanic circulation consists of an upwelling of deep waters in the northern Indian ocean and equatorial Pacific, and a downwelling in the North Atlantic. Horizontal transport of water occurs in both the deep, cold ocean layers and the warm surface waters. The wavy arrows represent biological pumping which transfers carbon to the deep waters. The thin vertical arrows represent the diffusion of carbon between the layers. It is estimated (Post, 1990) that the oceans absorbed between 26% and 34% of the fossil-fuel carbon discharged to the atmosphere between 1958 and 1980. (Post et al., 1990. Figure used with the permission of the American Scientist.)

ature from the South American coast to near the international dateline. From this point, the sea-surface temperature is high and it remains nearly constant to the Philippine Islands. Furthermore, the surface salinity of these waters differs. The salinity is higher in the central equatorial Pacific and lower in the warm western equatorial Pacific. The distribution of the sea-surface temperature is also reflected in the location of the thermocline. The *thermocline* is the boundary in the ocean between the wind-mixed and approximately constant-temperature surface waters and the colder and less active deep water. The thermocline is within 10 to 20 m of the surface of the sea at the Galápagos Islands and decreases to a depth of 100 to 150 m near the dateline, whereupon it remains at a nearly constant depth farther to the west.

In the western tropical Pacific, the waters above the thermocline have temperatures exceeding 28°C. This body of water is called the *Pacific warm pool.* Picaut et al. (1996) and Cane et al. (1997) contend that the Pacific warm pool generates the predominant ocean–atmosphere dynamic coupling that drives the world's intense atmospheric convection. The warm pool in the western equatorial Pacific is subject to strong east–west periodic migrations. The eastward movement of warm and less saline water from the central-western Pacific produces a well-defined salinity front at the eastern edge of the warm pool. The location of this zone is associated with El Niño–La Niña wind-driven surface current variations.

warm surface water contracts toward the western tropical Pacific so that heavy rainfall and low surface air pressures are confined to the western tropical Pacific and the eastern Pacific experiences reduced rainfall. The trade winds are intense during La Niña episodes.

The equatorial Pacific surface water is characterized by a steady increase in the sea-surface temper-

## 2.8   *Closure*

The single most important cause of air pollution is the expanding number of people in the world. An expanding population makes it essential that we reduce the per capita emission of pollutants. The consumption of energy is the underlying factor leading to the production of pollutants. Moreover, the same energy added to the biosphere affects directly and subtly the climate and our environment . The social and scientific issues are profound and complex. They require our best efforts at understanding and adaptation.

# Nomenclature

| Symbol | Description (Dimensions*) |
|--------|---------------------------|
| $A$ | area, parameter defined by equation $(L^2)$ |
| $c$ | specific heat of water (Q/MT) |
| $c_p$ | per capita energy consumption rate (Q/t) |
| $c_0$ | speed of light $(2.998 \times 10^8\,\text{m/s})$ (L/t) |
| $C(t)$ | total energy consumption at time $t$ (Q) |
| $D_p$ | diameter of a particle (L) |
| $E_{b,\lambda}$ | spectral emissive power $(Q/tL^3)$ |
| $f_{h,p}$ | per capita food harvested per year (M/t) |
| $F_h$ | mass of food harvested per year (M/t) |
| $F_p$ | mass of food required by population per year (M/t) |
| $h$ | Planck's constant $(6.6256 \times 10^{-34}\,\text{J·s})$ (Qt) |
| $k$ | Boltzmann constant $(1.3805 \times 10^{-23}\,\text{J/K})$ (Q/T) |
| $k_1, k_2$ | rate constants defined by equation |
| $M$ | molecular weight, neutral third body for a reaction |
| $\dot{M}$ | metabolic rate (Q/t) |
| $m_w$ | mass of water (M) |
| $N$ | number of items |
| $N_p$ | number of people |
| $P(t)$ | population, number of people at time $t$ |
| $\dot{Q}_j$ | heat transfer rate associated with source $j$ (Q/t) |
| $R_e$ | radius of the earth (6378 km) (L) |
| $S_0$ | solar intensity $(1387.5\,\text{W/m}^2, 440\,\text{Btu/h·ft}^2)$ $(Q/tL^2)$ |
| $t$ | time (t) |
| $T$ | temperature (T) |
| $t_d$ | doubling time, time for a variable to double (t) |

\*F, force; L, length; M, mass; N, mols; Q, energy; t, time; T, temperature.

|  |  |
|--------|-------------|
| $T_e$ | effective radiation temperature (T) |
| $T_s$ | mean surface temperature of the earth (T) |
| $U$ | internal energy (Q) |
| $\dot{W}$ | rate that work is performed (FL/t, Q/t) |

## Greek

|  |  |
|--------|-------------|
| $\alpha$ | population exponential rate of growth $(t^{-1})$ |
| $\beta$ | per capita energy exponential rate of growth $(t^{-1})$ |
| $\gamma$ | energy consumption exponential rate of growth, $\gamma = \alpha + \beta$ $(t^{-1})$ |
| $\delta, \Delta$ | a difference between variables |
| $\eta$ | cyclic thermal efficiency |
| $\lambda$ | wavelength (L) |
| $\rho$ | albedo, reflectivity |
| $\sigma$ | Stefan–Boltzmann constant $(5.66 \times 10^{-8}\,\text{Wm}^{-2}\,\text{K}^{-4})$ $(Q/tL^2T^4)$ |

## Abbreviations

|  |  |
|--------------|---------|
| ACGIH | American Conference of Governmental and Industrial Hygienists |
| Btu | British thermal unit |
| CFC | chlorofluorocarbon |
| EPA | Environmental Protection Administration |
| GNP | gross national product |
| OPEC | Organization of Petroleum Exporting Countries |
| PCB | polychlorinated biphenols |
| ppm | parts per million (by volume) |
| quad | $10^{15}$ Btu |
| STP | standard temperature and pressure (25°C, 101 kPa) |
| VOC | volatile organic compound |

# References

Radiant energy
energy flux $S_0 =$
  **2.4.** Assume
atmosphere inc
annual growth r
consumption of
$10^{15}$ Btu.
  **(a)** Estimate
perature betwee
stant at 0.35.
  **(b)** If energy
fashion,

and the albedo i
mate the chang
between 1970 ar
  Assume that
is used to produ
ing analysis:

_____

    **Ultimate a**
    **(mass fra**

    $C = 74.$
    $H_2 = 5.$
    $N_2 = 1.$
    $O_2 = 6.$
    $S = 0.7$

The heating val
mal efficiency w
  **(c)** What is t
  **(d)** How mu
year 2070?
  **(e)** What is
mulate during th
  **(f)** What is t
duced by combu
carbon is conver
  **2.5.** While t
earth's nonrene
products of hum
quoted as havin
sents hands (an
to feed." Thus
describe popula
senting advance

American Society of Mechanical Engineers, 1991. A closer look at the national energy strategy, ASME News, p. 2.

Ausubel, J. H., 1996. Can technology spare the earth? *American Scientist*, Vol. 84, pp. 166–178.

Axtmann, R. C., 1975. Environmental impact of a geothermal power plant, *Science*, Vol. 187, pp. 795–803.

Barus, C., 1972. *On the Limits to Energy Release and Implications for Present Policy*, Paper 72-WA/TS-1, annual meeting of ASME, November 26–30.

Cane, M. A., Clement, A. C., Kaplan, A., Kushnir, Y., Pozdnyakov, D., Seager, R., Zebiak, S. E., and Murtugudde, R., 1997. Twentieth-century sea surface temperature trends, *Science*, Vol. 275, pp. 957–960.

Cohen, B. L., 1978. Relative risks of saccharin and calorie ingestion, *Science*, Vol. 199, 3 March, p. 983.

Cohen, J. E., 1995. Population growth and earth's human carrying capacity, *Science*, Vol. 269, pp. 341–346.

Cook, E, "The Flow of Energy in Industrial Society" Sept. 1971 in "Readings from Scientific American," Rochlin, G.I. (ed), pp. 273–282, 1974.

Crowe, B. L., 1969. The tragedy of the commons revisited, *Science*, Vol. 166, pp. 1103–1107.

Dewees, D. N., 1987. Does the danger from asbestos in buildings warrant the cost of taking it out? *American Scientist*, Vol. 75, May–June, pp. 285–288.

Ehrlich, P. R., and Wilson, E. O., 1991. Biodiversity studies: science and policy, *Science*, Vol. 253, August 16, pp. 758–761.

Firor, J., and Jacobsen, J. E., 1993. Global climate change and sustainable development, *Journal of the Air & Waste Management Association*, Vol. 43, pp. 707–722.

Gustavson, M. R., 1979. Limits to wind power utilization, *Science*, Vol. 204, pp. 13–17.

Haigh, J. D., 1996. The impact of solar variability on climate, *Science*, Vol. 272, May 17, pp. 981–984.

Hansen, J., Johnason, D., Lacis, A., Lebedeff, S., Lee, P., Rind, D., and Russell, G., 1981. Climate impact of increasing atmospheric carbon dioxide, *Science*, Vol. 213, pp. 957–966.

Hardin, G., 1968. The tragedy of the commons, *Science*, Vol. 162, December 13, pp. 1243–1248.

Hart, H. L., 1972. Acceleration in social change, in *Technology and Society*, deNevers, N. (Ed.), Addison-Wesley, Reading, MA, pp. 51–79.

Horiuchi, S., 1992. Stagnation in the decline of the world population growth rate during the 1980s, *Science*, Vol. 257, pp. 761–765.

Hurrell, J. W., 1995. Decadal trends in the north atlantic oscillation: regional temperatures and precipitation, *Science*, Vol. 269, August 4, pp. 676–679.

Issar, A. S., 1995. Climatic change and the history of the middle east, *American Scientist*, Vol. 83, pp. 350–355.

Jacoby, G. C., D'Arrigo, R. D., and Davaajamts, T., 1996. Mongolian tree rings and 20th century warming, *Science*, Vol. 273, pp. 771–773.

Jin, F. F., 1996. Tropical ocean–atmosphere interaction, the pacific cold tongue, and the El Niño–southern oscillation, *Science*, Vol. 274, pp. 76–78.

Keigwin, L. D., 1996. The little ice age and medieval warm period in the Sargasso Sea, *Science*, Vol. 274, pp. 1504–1508.

Kerr, R. A., 1996. A new dawn for sun–climate links, *Science*, Vol. 271, March 8, pp. 1360–1361.

Kiehl, J. T., and Briegleb, B. P., 1993. The relative roles of sulfate aerosols and greenhouse gases in climate forcing, *Science*, Vol. 260, April 16, pp. 311–314.

Lave, L. B., and Dowlatabadi, H., 1993. Climate change: the effects of personal beliefs and scientific uncertainty, *Environmental Science and Technology*, Vol. 27, No. 10, pp. 1962–1972.

Lubin, D., 1994. The role of the tropical super greenhouse effect in heating the ocean surface, *Science*, Vol. 265, July 8, pp. 224–227

Malthus, T. R., 1972. An essay on the principle of population, as it affects the future improvement of society, in *Technology and Society*, deNevers, N. (Ed.), Addison-Wesley, Reading, MA, pp. 140–148.

Mauldin, W. P., 1980. Population trends and prospects, *Science*, Vol. 209, pp. 148–157.

Mayewski, P. A., Twickler, M. S., Whitlow, S. I., Meeker, L. D., Yang, Q., Thomas, J., Kreutz, K., Grootes, P. M., Morse, D. L., Steig, E. J., Waddington, E. D., Saltzman, E. S., Whung, P. Y., and Taylor, K. C., 1996. Climate change during last deglaciation in Antarctica, *Science*, Vol. 272, June 14, pp. 1636–1638.

Meadows, D. H., Meadows, D. L., Randers, J., and Behrens, W. W., III, 1972. *The Limits to Growth*, New American Library, New York.

Minnis, P., Harrison, E. F., Stowe, L. L., Gibson, G. G., Denn, F. M., Doelling, D. R., and Smith, W. L., Jr., 1993. Radiative climate forcing by the Mount Pinatubo eruption, *Science*, Vol. 259, pp. 1411–1415.

Philander, G., 1989. El Niño and La Niña, *American Scientist*, Vol. 77, September–October, pp. 451–459.

Picaut, J., Iou;
  McPhaden,
  placements
  ENSO, *Scie*
Post, W. M., Pe:
  V. H., and L
  cle, *America*
Raynaud, D.,
  Delmas, R.
  of greenh
  926–934.
Richardson, P.
  *Scientist*, M;
Robock, A., 19
  Vol. 272, M
Schneider, S. F
  public polic;
Schwartz, S.
  regional ph
  pp. 753–762
Schwartz, S. E
  mate chang
  24, pp. 1121
Semtner, A. J.
  Vol. 269, Se

## Problem

**2.1.** In an ;
1992, the EPA
as much smog-
ven for 50 mile
tant sources
Estimate how
car would hav
ating gas as th

- 83 × 10⁶ la;
  every year.
- A mower is
- 80 × 10⁶ car
- Average car

**2.2. (a)** Pr
the earth at t
energy is 12,0
transfer to the
by the moon t
the earth's alt

vehicles in a year $(L_a,$ miles/yr$)$ increases in an exponential fashion at a rate of 0.7% per year. If the total number of vehicles in the United States $(N_a,$ vehicles$)$ increases exponentially at a rate of 0.8% per year, how many years will elapse before the yearly NO emission rate (g/yr) will double?

**2.8.** The population of the world was $5.275 \times 10^9$ in 1990. Assume that the population increases in an exponential fashion at a rate of 1.76% per year. The radius of the earth is 6378 km and the average density of the earth is 5517 kg/m³.

**(a)** If the mass of the average person in the world is 55 kg, at what date in the future will the mass of the population equal the mass of the earth? (If you have friends who understand seismology, ask them what happens if all the people currently in the world, jump into the air to a height of 0.5 m?)

**(b)** Assuming that land covers only 25% of the earth's surface, what is the population density (people/km²) in 1990? What will be the overall population density in 1000 years? What will be the overall population density when part (a) is achieved?

**(c)** The population density of the developed portion of Hong Kong in 1973 was approximately 30,000 people/km². At the world growth rate above, how many years will elapse before the population density of the entire landmass of the world will equal 30,000 people/km²?

**(d)** The U.S. population is currently $252 \times 10^6$. The landmass of the United States covers $9.17 \times 10^6$ km². What is the current population density (people/km²)? If the growth rate of the U.S. population is 0.63%, in how many years will the population density (people/km²) equal that of Hong Kong?

**(e)** What steps do you think the United States or other nations could take to prevent the events noted above from occurring?

**2.9.** [*Design Problem*] Following graduation, you have been hired by a public utility company. Your supervisor has been asked to appear before a group of environmentalists who propose using solar power to satisfy the nation's energy needs and eliminate global warming produced by fossil-fuel combustion. You have been asked to provide some data for the presentation.

In 1981 the total consumption of energy in the United States was 74.4 quads $(74.4 \times 10^{15}$ Btu/yr$)$. The overall conversion efficiency between fuel consumed and useful energy (i.e., 74.4 quads) was 35% and will remain constant in the future. For this analysis assume that the future conversion efficiency between input solar energy to the solar panels and useful output energy is 15%. However, owing to reduced solar intensity due to clouds, solar panels out-of-service for repair and maintenance, etc., an overall conversion efficiency of only 8% can be relied upon. The average solar radiation energy flux in space is 440 Btu/ft²·h, of which only 48% actually reaches the earth and is available for absorption by solar panels. The per capita energy consumption rate in the United States increases in an exponential fashion at a rate of 0.7% per year and the U.S. population increases exponentially at a rate of 0.9% per year. The area of the 48 states is 3,037,951 miles², including:

- *Arizona*: 113,510 miles²
- *New Mexico*: 121,336 miles²
- *Nevada*: 109,895 miles²

Assume that the solar energy is available for only 12 h/day and that conventional 12-V lead–sulfuric acid batteries will be used at night. The solar panels must be large enough to provide energy during the day and at the same time to charge the batteries for nighttime use. Assume 70% conversion between energy absorbed by the solar panels and the energy stored in batteries. The batteries store an amount of energy equivalent to 150 A at 12 V for 30 s.

**(a)** What percentage of the 48 states would be under solar panels to provide U.S. energy requirements in the year 2000?

**(b)** Assume that a battery contains 10 lb$_m$ of lead and 5 lb$_m$ of acid and that a battery is replaced every two years. If only 50% of the lead and acid can be recycled, how much lead and acid must be produced per year to provide the power needed in 2000? (Note the world's lead production is approximately $2.5 \times 10^6$ tons/yr.)

**(c)** Suggest some questions for your supervisor to raise regarding the cost of buying land (assume $500/acre) needed for the solar panels, productive activities that can be performed beneath the solar panels, the magnitude of the disposal problem of discarded batteries, acid, and lead, or any mind-boggling problems you think should be included in the presentation.

# II

# Basics and Constraints

- Chapter 3   Air Pollution Legislation and Regulations
- Chapter 4   Effects of Pollution on the Respiratory System
- Chapter 5   Aesthetics
- Chapter 6   Atmospheric Chemicals: Sources, Reactions, Transport, and Sinks

Air pollution control is subject to certain basics and constraints of the physical and human environments in which engineering is practiced. Our physical surroundings determine the consequences of our actions. As human beings, we affect and are affected by our physical nature and environment.

Accordingly, we attempt to govern our behavior intelligently.

Current regulations and the history of governmental action to reduce air pollution are presented. The effects of pollutants on health, the quality of life, and aesthetic features of civilization are discussed. Formation and fate cycles are proposed to describe the interaction between pollutants and elements of the biosphere. It is important that everyone understand the magnitude of natural and anthropogenic sources of pollution in order to appreciate the impact of industrial processes for which they are responsible. Thus the milieu within which effective engineering must be practiced is defined.

# 3

# Air Pollution Legislation and Regulations

---

In this chapter you will learn:

- The chronology of U.S. air pollution legislation
- Titles and acronyms of important legislation, documents, criteria, and standards
- The character of state regulations
- How to estimate emission rates for compliance with state regulations
- How to access and use EPA electronic bulletin boards

## 3.1 Legislative record

### 3.1.1 Legislation before 1970

For the first hundred years of the country's existence (Stern, 1982), disputes about unwanted materials in the air were resolved by common law nuisance (public or private) or trespass litigation. After 1900, corrective action under the nuisance doctrine was replaced gradually by local governmental ordinances. The first municipal legislation was passed in Chicago and Cincinnati in 1881. By 1920, 175 municipalities had air pollution ordinances; by 1940 this number had grown to 200, 52 of which had smoke abatement enforcement agencies.

Problems with what constituted black, gray, or dense smoke led to the adoption of the *Ringelmann scale*, in which viewers compared a plume with the "blackness" of a Ringelmann 1 card, on which 20% of the area contained black dots. There were similar cards on which black dots comprised 40%, 60% and 80% of the area, and these were called Ringelmann 2, 3, and 4. Most municipalities prohibited smoke darker than Ringelmann 3. By the 1950s many municipal ordinances specified the maximum allowable particle mass concentration in a plume.

In 1910, Massachusetts became the first state to pass smoke control laws (for Boston). The first state to deal with air pollution other than black smoke from industrial stacks was California, which in 1947 authorized counties to regulate air pollution. Air that irritated the eyes and nose in Los Angeles became known as *smog* (smoke and fog). Political constituencies developed in other states after publicized air pollution episodes caused illness and even death. In *Donora, Pennsylvania*, a community on the Monongahela River south of Pittsburgh, an air pollution episode occurred in 1948 when industrial emissions and a persistent temperature inversion produced large $SO_2$, particle, and metal fume concentrations that resulted in the deaths of 20 people and respiratory illness to nearly half the population. In 1951, Oregon authorized a statewide agency to control air pollution. By 1960 there were 17 states with statewide air pollution activities, and eight had state air pollution control agencies whose authority transcended municipal boundaries.

What began as municipal ordinances grew to statewide authorities and then to a nationwide authority. Expansion of governmental authority was slow and there were many interruptions, owing to conflicting political constituencies that stood to gain or lose in the process. In 1950, President Truman instructed the Secretary of the Interior to organize the first U.S. *Technical Conference on Air Pollution*. Legislation urging national funds to support research was initiated by the California congressional delegation in 1950 but died in congressional committee. In 1952 the legislation passed the House but died in the Senate. In the 1955 state of the union address, President Eisenhower asked Congress to address the air pollution problems of the nation. In May the Senate passed the *Air Pollution Control Act* of 1955, which authorized the Public Health Service (PHS) of the Department of Health, Education and Welfare (HEW) to coordinate the annual dispersal of $3 \times 10^6$ for five years to state and local air pollution agencies for research, training, and technical assistance. In 1959, Congress passed the *Air Pollution Control Act Extension*, to extend the 1955 act for another four years and increase the annual appropriation to $5 \times 10^6$. Table 3-1 is a summary of principal federal air pollution legislation since 1955 (Stern, 1982).

Throughout the public debate was an interest to include emissions from both automobiles and stationary sources. In 1960, President Eisenhower signed the *Motor Vehicle Exhaust Study Act*, requiring the Secretary of HEW to report to the Congress on motor vehicles, air pollution and health. In 1962 President Kennedy signed the *Air Pollution Control Act Extension*, authorizing $5 \times 10^6$ annual expenditures for the Public Health Service and required PHS to include automotive emissions as a permanent part of their program.

## 3.1.2 Clean air act, 1963

Prior to 1963 air pollution had not become a truly national concern. Four conspicuous events in 1962 generated enthusiasm to expand national legislation: (1) publication of Rachel Carson's *Silent Spring*; (2) a series of smog episodes in London, England; (3) an air pollution episode in Birmingham, Alabama; and (4) a second National Conference on Air Pollution sponsored by the PHS. By 1963, a coalition of congressional leaders had formed that wished to increase the authority of the PHS with implementing legislation. The *Clean Air Act* of 1963 made permanent the Public Health Service's authority, directed HEW to prepare *criteria documents* summarizing scientific knowledge about the effects of air pollution on health, required semiannual reports to the Congress on motor vehicle emissions, established a technical committee to evaluate progress in the abatement of air pollution, and authorized a total appropriation of $65 \times 10^6$. The act was extended by the *Clean Air Act Amendments* (CAAA) of 1966. The 1963 and 1966 acts became landmark legislation and influenced developments in the United States for years to come. Prior to the 1963 act, individual(s) seeking relief from air pollution could only file suit for damages. Those filing suit had a difficult three-fold task: (1) prove that they had suffered financial losses, (2) prove that the losses were due to pollution, and (3) prove that the pollution was produced by the defendant. People of modest means found it very difficult to win cases against industries with considerable resources at their command. After 1963, the federal government took upon itself the task of abating pollution.

In 1965, Senator Muskie (Democrat, Maine) introduced legislation that was signed by President Johnson in October 1965 and known as the *Motor Vehicle Air Pollution Control Act*. The law allowed the Secretary of HEW to set emission standards for motor vehicles for the 1965 model year. The law also contained provisions to control air pollution between the United States and Canada and Mexico and called for research on $SO_2$ and motor vehicle exhausts.

**Table 3-1** Principal Federal Air Pollution Legislation

| Date | Legislation | Authorization |
|------|-------------|---------------|
| 1955 | Air Pollution Control Act | Provided Funds to local and state agencies for research and training |
| 1959 | Air Pollution Control Act, Extension | Extended the 1955 act |
| 1960 | Motor Vehicle Exhaust Study | Authorized PHS to study automotive emissions and health |
| 1962 | Air Pollution Control Act, Extension | Extended 1955 act and required PHS to include auto emissions in their program |
| 1963 | Clean Air Act | Provided:<br>Research at the federal level<br>Aid to states for training<br>Authority to abate interstate pollution |
| 1965 | Motor Vehicle Air Pollution Control Act | Provided:<br>Standards for auto emissions<br>Coordinated pollution control between United States, Canada, and Mexico<br>Research into $SO_2$ and auto emissions |
| 1966 | Clean Air Act, Amendments | Extended 1963 act<br>Provided grants-in-aid to state and local air pollution control programs |
| 1967 | Air Quality Act | Air Quality Control Regions (AQCRs)<br>Air Quality Criteria<br>Control Technology Documents<br>State Implementation Plans (SIPs)<br>Motor vehicle inspection program<br>Separate automotive emission standards for California |

President Nixon (1970) created the EPA as an agency within the Department of Health, Education and Welfare (later Department of Health and Human Services)

| Date | Legislation | Authorization |
|------|-------------|---------------|
| 1970 | Clean Air Act, Amendments | National Ambient Air Quality Standards (NAAQSs)<br>State Implementation Plans (SIPs) to achieve NAAQSs by 1973<br>New Stationary Source Emission Standards (NSPSs)<br>National Emission Standards for Hazardous Air Pollutants (NESHAPs)<br>Aircraft engine emission standards<br>Auto emission standards for 1975 models<br>States allowed to adopt air quality standards more stringent than federal standards<br>Citizen allowed to sue for air pollution violations |
| 1977 | Clean Air Act, Amendments | Geographic regions (class I, II, III) to preserve air quality<br>EPA-sanctioned emission offsets and emission banking within AQCRs<br>State permits that require prevention of significant deterioration (PSD) studies<br>Emission offsets and best available technology (BAT) in nonattainment regions<br>Lowest-achievable emission rate (LAER) in nonattainment regions<br>More stringent auto emissions standards<br>NSPS for fossil-fuel electric utility steam generating units |

*(continued)*

**Table 3-3**   *(continued)*

| Source | Maximum allowable emission rate |
|---|---|
| Ammonium sulfate manufacture | PM: 0.30 lb$_m$/ton product |
| Pressure-sensitive tape and label surface coating | VOC: 0.20 kg VOC/kg of coating |
| Asphalt roofing, saturator | PM: 0.04 kg/Mg of shingles |
| Beverage can surface coating<br>   Exterior coat<br>   Clear base coat<br>   Inside spraying | <br>VOC: 0.29 kg/L of coating solid<br>VOC: 0.46 kg/L coating solids<br>VOC: 0.89 kg/L coating solids |
| New residential wood heaters,<br>   manufactured after July 1990<br>   Catalytic combustor<br>   Noncatalyatic combustor | <br><br>PM: 4.1 g/h<br>PM: 15 g/h, if firing rate $<$ 1.5 kg/h<br>   18 g/h, if firing rate $>$ 1.5 kg/h |

*Source:* 40 CFR 60, 1989.

**PSD.** To preserve the foregoing classes, the 1977 act created a formal technical procedure called *prevention of significant deterioration* (PSD), in which an industry seeking an operating permit needed to show that changes in the air quality were consistent with the classifications above. PSD studies prevented industries from merely adopting minimal pollution control techniques that would worsen air quality but still be within NAAQSs.

**Economic incentives, offsets.** From its birth the EPA adopted a command-and-control philosophy to achieve its mission. Industries had no incentive to reduce pollution other than suffer financial penalities if they failed to do so. Tax relief offered some incentive, but the benefit was minimal. Guided by the writings of the Nobel laureate, F. Hayek (collected works, Bartley, 1989) astute observers of U.S. behavior suggested that a more effective way to reduce pollution was to generate *market financial incentives* in which industries could derive financial benefit by reducing pollution. Under this strategy, expenditures for pollution control that reduced emissions beyond what was required could sell the excess, called an *offset*, to other industries and generate profit. Thus rather than using conventional technology and merely complying with a standard, an industry could gain financially by improving the technology of pollution control. The dynamism of the U.S. economy rests on the creation of "property" that can be bought and sold in the market. Defining excess emission reductions as property that can be sold to others provides incentive far more attractive than tax relief. A command-and-control strategy produces begrudging compliance, whereas creating incentives serves both the interests of the public and industry. Incentives create a "win–win" situation that the United States finds compatible with its values.

By 1977 it was recognized that many regions in the United States (called nonattainment regions) had not attained NAAQSs despite conscientious efforts to do so and that further industrial development would only worsen the situation. Rather than quarantine such regions by prohibiting new construction and stunt economic development, the 1977 act created the notion of emission offsets. Under this policy new industrial development would be allowed provided that:

1. The new facility adopted the *best available technology* (BAT) to control emissions.
2. The company reduced emissions from other facilities that it operated in the region by an amount greater than what was allowed to be generated from its new operation.

Offsets allowed development, but as each new facility went on-line, another existing facility reduced its emission by a larger amount so that the total emissions in the area decreased. The offset policy was later liberalized to permit company X to buy (or "bank") reductions by another company, Y.

**LAER.** The 1977 act required states to revise their state implementation plans (SIPs) by 1979 and demonstrate that the NAAQSs would be attained by 1982. For areas of severe oxidant or carbon monoxide problems, states were required to submit a second revised SIP in 1982 that incorporated measures to satisfy NAAQSs by 1987. An alternative to the emission offset policy was later adopted if a state had an adequate program to reduce annual emissions in a manner that assured timely attainment of NAAQSs. The policy also required that new or modified major sources control emissions using a technology that achieved the *lowest-achievable emission rate* (LAER) based on current technology used by any state in the United States.

**Automotive emissions.** The 1977 act replaced the automobile emissions goals prescribed by the 1970 act by the following auto exhaust emission standards (g/mile):

|  | **Model year** | | |
| --- | --- | --- | --- |
|  | *1978–1979* | *1980* | *1981 and following* |
| Hydrocarbons | 1.5 | 0.41 | 0.41 |
| CO | 15 | 7 | 3.4 |
| $NO_x$ | 2.0 | 2.0 | 1.0 |

Automobiles using diesel engines and other innovative technology could receive a four-year waiver of the $NO_x$ standard to 1.5 g/mile under certain circumstances. The EPA was also empowered to raise the $NO_x$ standard to 7 g/mile for certain model lines in 1982–1983. States not in compliance with NAAQS auto emissions were granted permission to limit the sale of new cars to achieve the California emission standards. Warranties required on automobile thermal converters and catalytic reactors were relaxed from five years and 50,000 miles to 24 months and 24,000 miles. The requirement to develop a nationwide "high-altitude engine" was postponed until the 1984 model year.

All areas of the United States were classified as *attainment* or *nonattainment areas* with respect to each of the NAAQS Criteria Pollutants (Table 3-2). In 1975, 160 of the 247 U.S. air quality regions were nonattainment areas (i.e., were not in compliance with one or more of the NAAQSs). It was a common occurrence to find an area achieving compliance with respect to one or more pollutants (e.g., particles and $SO_2$) but not in compliance with another pollutant (e.g., ozone). To secure a permit for a new or modified source, engineers had to be cognizant of whether the source was located in an attainment or nonattainment area. If the new or modified source was located in an attainment area for a particular pollutant, a prevention of significant deterioration (PSD) study had to be conducted. If the new or modified source was located in a nonattainment area for another pollutant, an offset or LAER would have to be adopted for that pollutant.

## 3.1.5 Attainment areas: BACT

To secure an operating permit for a new or modified source located in an attainment area a PSD study is required if:

1. The new or modified source generates materials identified in the NSPS (Table 3-3), which has the *potential* to:
   a. Emit more than 100 (total) tons/yr
   b. Emit (or increase emissions by) 100 tons/yr of any pollutant regulated under NSPS or under standards for hazardous pollutants or regulated under mobile source control.
2. Any other new source emits 250 tons/yr of any pollutant.

Among other things, a PSD applicant must:

1. Monitor the ambient air for a period of one year to establish background pollutant levels prior to construction.
2. Negotiate with EPA on what pollution control method constitutes the *best available control technology* [BACT; 40 CFR 52.21(j)] for the process.
3. Using an EPA-approved air dispersion model, show that the estimated impact on air quality after the installation of the best available control technology (BACT) will be within the prescribed PSD increments for class I, II, or III regions.

The BACT analysis is performed for each new or modified source since there are no overall control strategies for an industry or generic group of sources. BACT analyses consist of evaluating several contemporary control technologies and estimating the reduction in emissions, broad environmental impact, energy consumed, and cost per mass of pollutant that is removed. Following the analysis, the technologies are ranked in order of overall efficacy. The reviewing authority then specifies an emission limitation for the source that reflects the maximum degree of reduction that can be achieved for each pollutant regulated under the act. In no case can a technology be recommended that does not meet NSPS standards. If the reviewing authority determines that there is no economically reasonable or technologically feasible way to measure emissions accurately, it may require the PSD applicant to design alternative equipment, or to establish new work practices or operational standards to reduce emissions to the maximum extent.

In summary, BACT is a *top-down process* that ranks in descending order the effectiveness of different control strategies. The PSD applicant first examines the most stringent, or "top" alternative. That alternative will define BACT for the source under consideration unless the applicant demonstrates that technical considerations, energy requirements, environmental impact, or economic impact justify a conclusion that the most stringent technology does not achieve the necessary emission reduction. If deemed "not achievable," the next most stringent alternative is considered, and so on.

## 3.1.6 Nonattainment areas: LAER and RACT

To secure an operating permit for a new or modified source in a nonattainment area, state implementation plans (SIPs) contain provisions for both an annual reduction of emissions from both existing sources and permitting procedures for new sources. SIPs were to achieve NAAQS by 1982 except for CO and oxidants, but under certain circumstances, compliance could be extended to 1987. In nonattainment regions, pollution control was required to achieve the lowest-achievable emission rate (LAER) for new or modified sources and *reasonable available control technology* (RACT) for existing sources. Selecting both LAER and RACT technology for any process is a matter of negotiation between the source operator and EPA, following procedures similar to those described in the BACT analysis described above.

In 1987, the EPA revised the particulate portion of the NAAQS to address the fact that small particles are more likely to enter the gas-exchange portion of the lung (pulmonary region) and are a more serious health risk than larger particles. *Particles whose diameter is less than 10 μm* were designated by the symbol ($PM_{10}$).

### EXAMPLE 3.1    CONTROL OF $SO_2$ AND $CO_2$ FROM AN ELECTRIC POWER STATION

You represent a public utility at a public hearing for the construction of a new electric power-generating facility. The power station will produce 1000 MW of electrical power with an overall thermal efficiency of 35%. The region is in compliance with the NAAQS for $SO_2$ and in accordance with BACT, the state air pollution agency requests that $SO_2$ be removed with a wet scrubber using lime (CaO). The power plant must satisfy the NSPSs and have an $SO_2$ removal efficiency of at least 90%, whichever is more stringent. To satisfy concerns about global warming, a group of citizens ask that 95% of the $CO_2$ should be removed with a second wet scrubber using caustic (NaOH). The overall scrubbing reactions are:

$$SO_2 + CaO + 2H_2O + \tfrac{1}{2}O_2$$
$$= CaSO_4 \cdot 2H_2O \text{ (solid)}$$
$$CO_2 + 2NaOH = H_2O + Na_2CO_3 \text{ (solid)}$$

You claim that capturing $CO_2$ will produce nearly 10 times as much solid waste as the mass of coal that is burned. The citizens scoff and hoot you down. To prove your point, compute the following ratios:

- $lb_m$ of CaO per $lb_m$ of coal
- $lb_m$ of $CaSO_4 \cdot 2H_2O$ per $lb_m$ of coal
- $lb_m$ of NaOH per $lb_m$ of coal
- $lb_m$ of $Na_2CO_3$ per $lb_m$ of coal
- Total $lb_m$ of solid waste per $lb_m$ of coal

The coal has the following composition:

sulfur = 3%      ash = 8%      carbon = 85%

HHV = 12,500 Btu/$lb_m$

**Solution**

1. Compute the coal consumption and firing rate.

$$\eta = \frac{W \text{ (net output)}}{Q \text{ (total input)}}$$

$$0.35 =$$

$$\frac{(1000 \text{ MW})(1000 \text{ kW/MW})(2545 \text{ Btu}/0.746 \text{ kWh})}{(\dot{m}_{coal} \text{ lb}_m/h)(12{,}500 \text{ Btu/lb}_m)}$$

$$\dot{m}_{coal} = 779{,}778 \text{ lb}_m \text{ coal/h}$$

$$Q_{firing} = (779{,}778 \text{ lb}_m/h)(12{,}500 \text{ Btu/lb}_m)$$

$$= 9.747 \times 10^9 \text{ Btu/h}$$

2. Compute the maximum allowable $SO_2$ emission rate. From the NSPSs (Table 3-3), the maximum allowable $SO_2$ emission rate is

$$\dot{m}_{SO_2} \text{ (NSPS max. allow.)}$$

$$= (0.8 \text{ lb}_m \text{ } SO_2/10^6 \text{ Btu})(9.747 \times 10^9 \text{ Btu/h})$$

$$= 7797.8 \text{ lb}_m \text{ } SO_2/h$$

The potential (uncontrolled) $SO_2$ emissions based on $S + O_2 = SO_2$ is

$$\dot{m}_{SO_2} \text{ (potential)} = (1 \text{ mol } SO_2/\text{mol S})$$

$$(64 \text{ lb}_m/\text{mol } SO_2)(\text{mol S}/32 \text{ lb}_m)$$

$$(0.03 \text{ lb}_m \text{ S/lb}_m \text{ coal}) \dot{m}_{coal}$$

$$= \left(\frac{64}{32}\right)(0.03)(779{,}778)$$

$$= 46{,}778 \text{ lb}_m/h$$

The capture efficiency of a device satisfying NSPSs:

$$\eta_{NSPS} = \frac{\dot{m}_{SO_2} \text{ (captured)}}{\dot{m}_{SO_2} \text{ (potential)}}$$

$$= 1 - \frac{\dot{m}_{SO_2} \text{ (NSPS max. allow.)}}{\dot{m}_{SO_2} \text{ (potential)}}$$

$$= 1 - \frac{7797.8}{46{,}788} = 0.833 = 83.3\%$$

The power plant is in an attainment area with respect to $SO_2$ and the EPA has accepted a BACT analysis indicating that a removable efficiency of 90% is possible. Thus the rate at which $SO_2$ is captured must be based on BACT.

$$\dot{m}_{SO_2, \text{ captured}} \text{ (BACT)} = (0.9)(46{,}788 \text{ lb}_m/h)$$

$$= 42{,}109 \text{ lb}_m \text{ } SO_2/h$$

The maximum $SO_2$ emission rate is

$$\dot{m}_{SO_2, \text{ max emission}} = 46{,}788 - 42{,}109$$

$$= 4679 \text{ lb}_m/h$$

Thus the mass of $SO_2$ that needs to be captured is 42,109 $lb_m/h$.

3. Compute the $lb_m$ of CaO per $lb_m$ of coal needed to achieve 90%.

$$\dot{m}_{CaO} = (\text{mol CaO/mol } SO_2)(56 \text{ lb}_m/\text{mol CaO})$$

$$(\text{mol } SO_2/64 \text{ lb}_m)(\dot{m}_{SO_2} \text{ lb}_m/h)$$

$$= \left(\frac{56}{64}\right)(42{,}109.2) = 36{,}845 \text{ lb}_m \text{ CaO/h}$$

$$\frac{\dot{m}_{CaO}}{\dot{m}_{coal}} = \frac{36{,}854.6}{779{,}778} = 0.0472$$

4. Compute the $lb_m$ of gypsum ($CaSO_4 \cdot 2H_2O$) produced per $lb_m$ of coal.

$$\dot{m}_{gyp} = (\text{mol gyp./mol } SO_2)(140 \text{ lb}_m/\text{mol gyp.})$$

$$(\text{mol } SO_2/64 \text{ lb}_m)(\dot{m}_{SO_2} \text{ lb}_m/h)$$

$$= \left(\frac{140}{64}\right)(42{,}109.2) = 92{,}113.9 \text{ lb}_m \text{ gyp./h}$$

$$\frac{\dot{m}_{gyp}}{\dot{m}_{coal}} = \frac{92{,}113.9}{779{,}778} = 0.118$$

5. Compute the $lb_m$ of NaOH per $lb_m$ of coal needed to remove $CO_2$. First compute the mass of $CO_2$ generated based on $C + O_2 = CO_2$.

$$\dot{m}_{CO_2} = (\text{mol } CO_2/\text{mol C})(\text{mol C}/12 \text{ lb}_m)$$

$$(44 \text{ lb}_m/\text{mol } CO_2)(0.85 \text{ lb}_m \text{ C/lb}_m \text{ coal})(\dot{m}_{coal})$$

$$= \left(\frac{44}{12}\right)(0.85)(779{,}778) = 2{,}430{,}308 \text{ lb}_m/h$$

For collection of 95% of the $CO_2$, the mass of NaOH is

$$\dot{m}_{NaOH} = (2\,NaOH/mol\,CO_2)(40\,lb_m/mol\,NaOH)$$
$$(mol\,CO_2/44\,lb_m)(0.95)(\dot{m}_{CO_2}\,lb_m/h)$$

$$= (2)\left(\frac{40}{44}\right)(0.95)(2,430,308)$$

$$= 4,197,805\,lb_m/h$$

$$\frac{\dot{m}_{NaOH}}{\dot{m}_{coal}} = \frac{4,197,923}{779,778} = 5.38$$

6. Compute the $lb_m$ of $Na_2CO_3$ produced per $lb_m$ of coal.

$$\dot{m}_{Na_2CO_3} = (1\,mol\,Na_2CO_3/2\,mol\,NaOH)$$
$$(mol\,NaOH/40\,lb_m)$$
$$(106\,lb_m/mol\,Na_2CO_3)(\dot{m}_{NaOH})$$

$$= \left(\frac{106}{40}\right)(4,197,805/2) = 5,562,092\,lb_m/h$$

$$\frac{\dot{m}_{Na_2CO_3}}{\dot{m}_{coal}} = \frac{5,562,248}{779,778} = 7.133$$

7. Compute the $lb_m$ of ash produced per $lb_m$ of coal. Assume that 99% of the ash is removed either as bottom ash or fly ash.

$$\dot{m}_{ash} = (0.99)(0.08\,lb_m\,ash/lb_m\,coal)(\dot{m}_{coal}\,lb_m/h)$$
$$(0.99)(0.08)(779,778) = 61,758\,lb_m/h$$

$$\frac{\dot{m}_{ash}}{\dot{m}_{coal}} = \frac{61,758}{779,778} = 0.0792$$

8. Compute the $lb_m$ of solid waste per $lb_m$ of coal.

$$\frac{\dot{m}_{total\,waste}}{\dot{m}_{coal}} = \frac{\dot{m}_{ash} + \dot{m}_{gyp} + \dot{m}_{Na_2CO_3}}{\dot{m}_{coal}}$$

$$= \frac{61,758 + 92,114 + 5,562,092}{779,778}$$

$$= \frac{5,715,990}{779,778} = 7.33$$

Thus your estimate of "nearly 10 times" was somewhat large, the actual mass of solid waste associated with removing 95% of the $CO_2$ is 7.33 times larger than the mass of coal consumed. Next time, be careful not to overstate your case.

In their concern about global warming, the citizens overlooked the fact that scrubbing $CO_2$ from the exhaust produces an enormous amount of solid waste. Such "pollution solutions" often substitute one kind of pollution for another. The important lesson to be learned from the above is that when pollutants ($SO_2$ or $CO_2$) are removed by combining with other material, the total mass of the waste material is considerably larger than the mass of pollutants.

# 3.2  Clean Air Act Amendments of 1990: alternatives RACT and MACT

By 1990 the air quality in several urbanized regions of the United States had improved only marginally and still did not satisfy NAAQS. Merely setting emissions standards failed to achieve the desired air quality. State agencies needed the opportunity to tailor emission controls to the air quality of certain areas and the unique features of the industrial source. The 1990 act revised the 1970 act substantially (Schultze, 1993) by adding or amending seven basic categories:

1. Nonattainment areas
2. Mobile sources
3. Air toxics
4. Acid rain
5. Permits
6. Stratospheric ozone and climate protection
7. Federal enforcement

**Title I: Nonattainment areas.** Title I classifies major urban areas by the amount that ozone, particulates, and CO exceed ambient air quality standards, In each nonattainment area, the EPA set deadlines for attaining reductions of these pollutants.

| Nonattainment level | Years to achieve attainment |
|---|---|
| Marginal | 3 |
| Moderate | 6 |
| Serious | 9 |
| Severe | 15 |
| Extreme (only Los Angeles) | 20 |

Any stationary source or group of stationary sources located within a contiguous area and under common control that emits or has the potential to emit, in the aggregate, 10 tons per year or more of any hazardous air pollutant or 25 tons per year or more of any combination of hazardous pollutants is defined as a *major source*. Any stationary source that is not a major source of hazardous air pollutants (HAPs) is an *area source*. All sources, new and old, will be required to secure *operating permits* that need to be renewed every five years. Permit fees will increase to provide revenue to support state environmental agencies. Until states achieve primary air quality standards, the states must revise their state implementation plans (SIPs). If states do not revise their SIPs, the government will impose a *federal implementation plan* (FIP). Depending on the severity of nonattainment, thousands of sources formerly too small and not subject to regulation will need to secure operating permits. For example, all gasoline dispensing systems will be required to capture vapors generated when tanks are filled.

**Title II: Mobile sources.** Today's cars and trucks account for nearly one-half of the emission of ozone precursors (VOCs and oxides of nitrogen) and 90% of the CO emissions in urban areas. Title II divides new mobile sources into passenger cars and trucks. Nitrogen oxides must be reduced by 35% (cars) and 69% (trucks) by 1997 through a phased-in plan. After 1993, automobiles and trucks must reduce CO emissions to 10 g/mile. Beginning in 1995, reformulated gasoline containing 15% less volatile organic compounds must be used in nine U.S. cities with the most severe ozone problems. The act establishes a *clean-fuel car* pilot program in California for 150,000 vehicles in model year 1996 and 300,000 by the model year 1999.

**Title III: Air toxics.** Title III establishes EPA's regulation over 189 specific chemicals (Table 3-4) called *hazardous air pollutants* (HAPs). The HAPs are a diverse collection (Kelly et al., 1994) of industrial chemicals and intermediates, pesticides, chlorinated and hydrocarbon solvents, metals, combustion by-products, polychlorinated biphenyls, and mixtures of chemicals such as coke oven gas. Some HAPs, principally VOCs, are ubiquitous ambient air contaminants. Some chemicals were defined as HAPs because they are well-known workplace toxins. About one-third of the HAPs are semivolatile, so that they may exist as both as vapors and solids. Emissions of HAPs will be limited to 10 tons/yr for a specific material or 25 tons/yr for combinations of these materials. The EPA will define *maximum achievable control technology* (MACT) techniques to be used in the operating permits that will have to be secured by all source operators. A list of 100 "extremely hazardous chemicals" was to be issued, with manufacturers handling these substances required to inform the public about possible hazards. The act establishes a *Chemical Safety and Hazard Investigation Board* to investigate accidents involving these chemicals.

**Title IV: Acid rain.** Title IV requires $NO_x$ emissions to be reduced by at least $2 \times 10^6$ tons/yr from 1980 levels. Emission standards for $NO_x$ were to be issued beginning in 1993. The act also called for $SO_2$ emissions to be reduced to $8.9 \times 10^6$ tons/yr by 2000. (Currently, approximately 25 million tons are emitted each year.) The first phase requires 110 utilities whose output exceeds 25 MW to reduce $SO_2$ emissions to "1.2 $lb_m$ $SO_2$/million Btu times their average 1985–1987 fuel use." In the second phase beginning 2000, a significantly greater number of utilities will need to reduce their $SO_2$ emissions to this level. An innovative *emissions trading plan* for $SO_2$ has been devised in which emissions below this value achieved over a period of time and in specified zones, called *emission credit or allowance*, can be traded or auctioned (bought and sold) as a commodity. The plan allows the forces of a market economy to change compliance from *command and control to market-based incentives*. It creates financial incentives to reduce emissions and to develop ingenious new technologies with performance better than what's required, with the hope that the superior performance can be sold to others rather than applying commonplace technology that merely satisfies emission standards. The Chicago Board of Trade has devised ways in which these credits can be bought and sold like other commodities (e.g., grain and stock contracts, futures, and options). The emission credit or allowance is not a "permit to pollute" because the EPA will determine what performance warrants an allowance and in what geographic region it can be

**Table 3-4** Hazardous Air Pollutants, Clean Air Act Amendments of 1990

| CAS number | Chemical name | CAS number | Chemical name |
|---|---|---|---|
| 75070 | Acetaldehyde | 334883 | Diazomethane |
| 60355 | Acetamide | 132649 | Dibenzofurans |
| 75058 | Acetonitrile | 96128 | 1,2-Dibromo-3-chloropropane |
| 98862 | Acetophenone | 84742 | Dibutylphthalate |
| 53963 | 2-Acetylaminofluorene | 106467 | 1,4-Dichlorobenzene |
| 1007028 | Acrolein | 91941 | 3,3-Dichlorobenzidene |
| 79061 | Acrylamide | 111444 | Dichloroethyl ether [bis(2-chloroethyl) ether] |
| 79107 | Acrylic acid | 542756 | 1,3-Dichloropropene |
| 107131 | Acrylonitrile | 62737 | Dichlorvos |
| 107051 | Allyl chloride | 111422 | Diethanolamine |
| 92671 | 4-Aminodiphenyl | 121697 | *N,N*-Diethylaniline |
| 62533 | Aniline | 64675 | Diethyl sulfate |
| 90040 | *o*-Anisidine | 119904 | 3,3-Dimethoxybenzidine |
| 1332214 | Asbestos | 60117 | Dimethyl aminoazobenzene |
| 71432 | Benzene | 119937 | 3,3'-Dimethyl benzidine |
| 92875 | Benzidine | 79447 | Dimethyl carbamoyl chloride |
| 98077 | Benzotrichloride | 68122 | Dimethyl formamide |
| 100447 | Benzyl chloride | 57147 | 1,1-Dimethyl hydrazine |
| 92524 | Biphenyl | 131113 | Dimethyl phthalate |
| 542881 | Bis(chloromethyl) ether | 77781 | Dimethyl sulfate |
| 117817 | Bis(2-ethylhexyl)phthalate | 534521 | 4,6-Dinitro-*o*-cresol and salts |
| 75252 | Bromoform | 51285 | 2,4-Dinitrophenol |
| 106990 | 1,3-Butadiene | 121142 | 2,4-Dinitrotoluene |
| 156627 | Calcium cyanamide | 123911 | 1,4-Dioxane (1,4-diethyleneoxide) |
| 105602 | Caprolactam | 122667 | 1,2-Diphenylhydrazine |
| 133062 | Captan | 106898 | Epichlorohydrin |
| 63252 | Carbaryl | | (1-chloro-2,3- epoxypropane) |
| 75150 | Carbon disulfide | 106887 | 1,2-Epoxybutane |
| 56235 | Carbon tetrachloride | 140885 | Ethyl acrylate |
| 463581 | Carbon sulfide | 100414 | Ethyl benzene |
| 120809 | Catechol | 51796 | Ethyl carbamate (urethane) |
| 133904 | Chloramben | 75003 | Ethyl chloride (chlorothane) |
| 57749 | Chlordane | 106934 | Ethylene dibromide (dibromoethane) |
| 7782505 | Chlorine | 107062 | Ethylene dichloride (1,2-dichloroethane) |
| 79118 | Chloroacetic acid | 107211 | Ethylene glycol |
| 532274 | 2-Chloroacetophenone | 151564 | Ethylene imine (aziridine) |
| 108907 | Chlorobenzene | 75218 | Ethylene oxide |
| 510156 | Chlorobenzilate | 96457 | Ethylene thiourea |
| 67663 | Chloroform | 75343 | Ethylidene chloride (1,1-dichloroethane) |
| 107302 | Chloromethyl methyl ether | 50000 | Formaldehyde |
| 126998 | Chloroprene | 76448 | Heptachlor |
| 1319773 | Cresols/cresylic acid (isomers and mixture) | 118741 | Hexachlorobenzene |
| 95487 | *o*-Cresol | 87683 | Hexachlorobutadiene |
| 108394 | *m*-Cresol | 77474 | Hexachlorocyclopentadiene |
| 106445 | *p*-Cresol | 67721 | Hexachloroethane |
| 98828 | Cumene | 822060 | Hexamethylene-1,6-diisocyanate |
| 94757 | 2,4-D, salts and esters | 680319 | Hexamethylphosphoramide |
| 3547044 | DDE | | |

*(continued)*

**Table 3-4** *(continued)*

| CAS number | Chemical name | CAS number | Chemical name |
|---|---|---|---|
| 100543 | Hexane | 75569 | Propylene oxide |
| 302012 | Hydrazine | 75558 | 1,2-Propylenimine (2-methyl aziridine) |
| 7647010 | Hydrochloric acid | 91225 | Quinoline |
| 7664393 | Hydrogen fluoride (hydrofluoric acid) | 106514 | Quinone |
| | | 100425 | Styrene |
| 123319 | Hydroquinone | 96093 | Styrene oxide |
| 78591 | Isophorone | 1746016 | 2,3,7,8-Tetrachlorodibenzo-p-dioxin |
| 58899 | Lindane (all isomers) | 79345 | 1,1,2,2-Tetrachloroethane |
| 108316 | Maleic anhydride | 127184 | Titanium tetrachloride |
| 67561 | Methanol | 108883 | Toluene |
| 72435 | Methoxychlor | 95807 | 2,4-Toluene diamine |
| 74893 | Methyl bromide | 584849 | 2,4-Toluene diisocyanate |
| 74873 | Methyl chloride (chloromethane) | 95534 | o-Toluidine |
| 71556 | Methyl chloroform (1,1,1-trichloroethane) | 8001352 | Toxaphene (chlorinated camphene) |
| 78933 | Methyl ethyl ketone (2-butanone) | 120821 | 1,2,4-Trichlorobenzene |
| 60344 | Methyl hydrazine | 79005 | 1,1,2-Trichloroethane |
| 74884 | Methyl iodide (iodomethane) | 79016 | Trichloroethylene |
| 108101 | Methyl isobutyl ketone (hexone) | 95954 | 2,4,5-Trichlorophenol |
| 624839 | Methyl isocyanate | 88062 | 2,4,6-Trichlorophenol |
| 80626 | Methyl methacrylate | 121448 | Triethylamine |
| 1634044 | Methyl tert-butyl ether | 1582098 | Trifluralin |
| 101144 | 4,4-Methylene bis(2-chloroaniline) | 540841 | 2,2,4-Trimethylpentane |
| 75092 | Methylene chloride (dichloromethane) | 108054 | Vinyl acetate |
| 101688 | Methylene diphenyl diisocyanate | 593602 | Vinyl bromide |
| 107779 | 4,4-Methylenedianiline | 75014 | Vinyl chloride |
| 91203 | Napthalene | 75354 | Vinylidene chloride (1,1-dichloroethylene) |
| 98953 | Nitrobenzene | 1330207 | Xylenes (isomers and mixture) |
| 92933 | 4-Nitrobiphenyl | 95476 | o-Xylene |
| 100027 | 4-Nitrophenol | 108383 | m-Xylene |
| 79469 | 2-Nitropropane | 106423 | p-Xylene |
| 684935 | N-Nitroso-N-methylurea | | Antimony compounds |
| 62759 | N-Nitrosodimethylamine | | Arsenic compounds (inorganic, including arsine) |
| 59892 | N-Nitrosomorpholine | | |
| 56382 | Parathion | | Beryllium compounds |
| 82688 | Pentachloronitrobenzene | | Cadmium compounds |
| 87865 | Pentachlorophenol | | Chromium compounds |
| 108952 | Phenol | | Cobalt compounds |
| 106503 | p-Phenylenediamine | | Coke oven emissions |
| 75445 | Phosgene | | Cyanide compounds |
| 7803515 | Phosphine | | Glycol ethers |
| 7723140 | Phosphorus | | Lead compounds |
| 85449 | Phthalic anhydride | | Manganese compounds |
| 1336363 | Polychlorinated biphenyls (aroclors) | | Mercury compounds |
| 1120714 | 1,3-Propane sulfone | | Fine mineral fibers $(D_p < \mu m)$ |
| 57578 | β-Propiolactone | | Nickel compounds |
| 123386 | Propionaldehyde | | Polycyclic organic matter |
| 114261 | Propoxur (baygon) | | Radionuclides (including radon) |
| 78875 | Propylene dichloride (1,2-dichloropropane) | | Selenium compounds |

traded. With respect to acid rain, the exact location of the $SO_2$ emitter within a certain geographical region makes no difference. Thus if one company can remove a ton of $SO_2$ cheaper than another company, the ton can be removed at the cheaper place.

**Title V: Permits.** Title V requires all current emitters of acid rain precursors and/or hazardous air pollutants and all new emitters subject to new sources regulations to secure operating permits. The permits issued to a facility will be for a fixed term of up to five years. The permits must contain provision for enforcing emission limitations, inspection, entry, monitoring, certification, scheduling, and reporting of compliance. Permit fees can be increased so as to provide revenue for the operation of state air pollution control agencies. Facility managers responsible for complying with Title V must realize that its monitoring, record-keeping, and reporting requirements are enormous. Compliance certification will be annual. Monitored data must be collected continuously and reported semiannually to the compliance agency, which will make the information public. If the data indicate that a facility is out of compliance, it could be subject to citizen lawsuits as well as EPA enforcement action. On the positive side, compliance with Title V places the facility within a *permit shield* which provides protection against third-party enforcement actions and lawsuits.

The EPA is far behind schedule on issuing MACT standards. If a facility is a major HAP's source, it is subject to MACT between the year of issuance and the year 2000. Even when a facility has been issued a permit, the permit can be reopened if it is due to expire in three or more years and the facility is not in compliance with current MACT standards. The EPA has proposed forming "MACT partnerships" with industrial groups in an attempt to meet its deadlines and set achievable standards. Once standards are set, the regulated community will have to live with them.

Under Section 112(g) of Title V, any construction, reconstruction, or modification of a major HAP's source must meet MACT standards *before these activities begin.* There are two ways existing major HAPs' sources may be exempt of Section 112(g), (1) employ offsets or (2) describe the source carefully so that certain activities are not classified as "modifications." For example, if flexible language is used to describe a process as "designed to accommodate" alternative fuels, raw materials, and operating scenarios, then changing fuels, raw materials, or operating scenarios do not constitute "modifications," whereas such changes might be considered "modifications" if flexible language had not been used. Facility managers should be careful and not inadvertently trap themselves. Securing offsets can become a convoluted regulatory process because the EPA has not specified fully how offsets are to be obtained. Another way to cope with Title V is to transform a major HAPs source into a *synthetic minor* source and place itself outside the jurisdiction of Title V. The process is still, however, subject to public notice and opportunity for public participation and EPA approval. A process can be transformed by replacing HAPs with nonpolluting materials. Managers must be careful; relief may be short-lived since future expansion of the process may place it within the jurisdiction of Title V.

Under Section 112(f), the EPA is required to set "residual risk standards" within eight years of issuing MACT standards. A *residual risk* is a potential impact on health and environment due to emissions that might persist after MACT standards are established. If it is determined that residual emissions do not afford an ample margin of safety, the EPA will issue additional controls. It is advantageous for industry groups to work with regulatory agencies to establish *de minimus* levels of emission.

**Title VI: Stratospheric ozone and climate protection.** Title VI brings about a steep increase in the price of ozone-destroying hydrocarbons by reducing their production. It also requires that chlorofluorocarbons (CFCs), Halons, and carbon tetrachloride may not be used after 2000. Methyl chloroform ($CH_3CCl_3$, 1,1,1-trichloroethane) will be banned in 2002. The production of hydrofluorocarbons (HCFCs) will be prohibited after 2030.

**Title VII: Federal enforcement.** Title VII increases the penalties for noncompliance. The act contains provision for citizens to seek penalties against violators.

## 3.2.1 Post-1990 CAAA developments

Measurements of pollution in southern California indicate the persistent inability to satisfy $O_3$ and CO ambient air quality standards even though auto per-

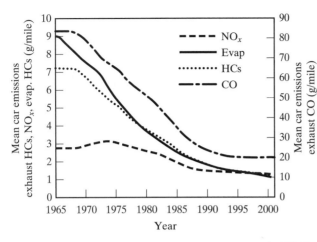

**Figure 3-1** NO$_x$, CO, hydrocarbon (HC), and evaporative losses from a light-duty vehicle fleet predicted by the EPA emissions inventory model (Calvert, et al., 1993)

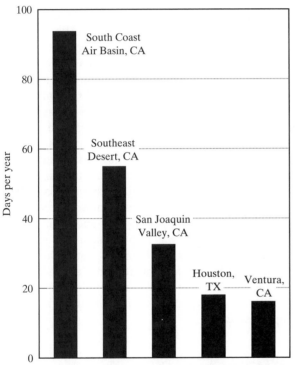

**Figure 3-2** The top five areas in the US not in compliance with the NAAQS ozone standard during 1993-1995. The rank reflects the average number of days/year each area was not in compliance with the 1-hr ozone standard (1 hr, 0.12 ppm). (Cooney C, 1996)

formance standards have become more stringent (Figures 3-1 and 3-2). Several explanations have been put forward which are broadly applicable to other urban areas and which might influence the promulgation of future regulations.

**1.** While auto emissions per mile have been reduced, the number of miles driven has risen and the population has increased (i.e., in 1970 the population of Los Angeles was 9 million, while in 1996 it was 13 million). In the last 25 years, the number of miles driven in the United States has doubled. The increasing popularity of heavy sport-utility vehicles also increases the emissions per mile.

**2.** The weakest link in predicting pollutant concentrations is the inaccurate emission inventory. Measurements suggest that the emission of CO and hydrocarbons has been understated by a factor of two.

**3.** Field measurements of vehicles on the highway indicate the presence *high emitters* (i.e., 50% of all the emitted CO and HC is generated by 10% of the vehicles). High emitters are vehicles displaying some or all of the following:
- Span all model years
- Are driven annually a great distance
- Undergo frequent accelerations and decelerations
- Have tampered emission controls (15 to 30% of all vehicles)

**4.** Evaporative losses on hot days have been underestimated.

**5.** Emissions from heavy-duty off-road vehicles used in construction activities are large and have not been subjected to the same scrutiny as passenger vehicles.

**6.** Vehicles with highly variable on-road tailpipe emissions are likely to be modern vehicles that have malfunctioning closed-loop emissions control systems that have not been tampered with (Bishop et al., 1996).

**7.** The mandated use of zero-emission vehicles (ZEVs), anticipated to be electric vehicles, may only shift pollutant generation from vehicles to electric utility boilers that will supply the energy and to industries that manufacture new batteries and dispose of used ones. The mix of pollutants may be different, but it is not obvious that the ultimate burden to the environment will be reduced.

In accordance with the CAAA 1990, the 1971 NAAQS standards (Table 3-2) must be reviewed every five years. In 1996, the ozone standard, 0.12 ppm for 1-h, was reviewed (Wolff, 1996a). Animal studies, controlled human chamber studies, field studies of ambient exposures, and hospital admission studies show that there is no scientific evidence to suggest the existence of a threshold concentration level. A small but statistically significant relationship exists between decreased lung function ($FEV_1$; see Chapter 4) at ozone concentrations below the NAAQS standard. Selecting a standard at the lowest-observable effects level and providing an "adequate margin of safety" is not possible. It was agreed that the new standard should be based on risk assessment for individuals whose preexisting respiratory diseases, such as asthma, places them at the highest risk. Furthermore, the new standard must be "robust" (i.e., account for periods of extreme meteorological conditions during which the concentration flips in and out of compliance). The EPA recommends that the old standard be replaced by an 8-h average within the range of 0.07 to 0.09 ppm, with one to five allowable *exceedances* per year averaged over a 3-year period. A contrary view recommends that the current air pollution warning infrastructure be expanded so that sensitive individuals can take appropriate *exposure avoidance* measures during periods of high ozone concentration.

An additional factor affecting the adoption of a new standard concerns the beneficial affect of tropospheric ozone to absorb harmful ultraviolet radiation (UV-B), which causes skin cancer and cataracts. Solar UV radiation of wavelengths less than 240 nm are absorbed by oxygen and ozone, but only ozone is effective for wavelengths between 240 and 320 nm. Wavelengths less than 320 nm produce deleterious biological effects. Reduced amounts of atmospheric ozone permit disproportionately large amounts of UV radiation to penetrate the atmosphere. For example, with an overhead sun and typical amounts of ozone, a 10% reduction in ozone results in a 20% increase of UV penetration at 305 nm, a 250% increase of UV penetration at 290 nm, and a 500% increase at 287 nm (Cicerone, 1987). It is estimated (Lutter and Wolz, 1997) that lowering tropospheric ozone increases the risk of UV-B, comparable to decreasing the risk on the respiratory system. Thus adoption of new standards must address both the adverse effects of ozone on respiratory system and the beneficial effects of ozone to absorb UV-B radiation.

In 1996 the ($PM_{10}$) particulate standard, 50 $\mu g/m^3$ annual average, 150 $\mu g/m^3$ 24-h average was reviewed (Wolff, 1996a). The EPA recommends a standards based on $PM_{2.5}$ $(D_p < 2.5 \ \mu m)$. Over the years a body of knowledge has accumulated (Schwartz et al., 1996) indicating that fine particles, $D_p < 2.5 \ \mu m$, associated with combustion sources induce pulmonary inflammation associated with a small but statistically significant increase in mortality of people over 65 suffering from chronic lung and heart disease. A review of the NAAQS particulate standard ended in 1996. There was consensus to retain the 24-h, $PM_{10}$ standard and to devise an additional $PM_{2.5}$ standard for fine particles (Wolff, 1996b). As with a review of the ozone standard, there was consensus that the particle standards should be robust. There was not a consensus (Vedal, 1997) regarding the maximum concentration and whether the standard should be a 24-hour and/or an annual standard. The EPA was to propose a new standard by June 1997. States will then have one year to submit modified state implementation plans (SIPs) and the EPA will review these plans within a year. After 2003 one can expect new control strategies to be implemented.

## EXAMPLE 3.2 AIR QUALITY REVIEW AND PERMIT FOR A NEW STEAM BOILER

A paper mill has an antiquated furnace to produce steam for the manufacture of high-grade paper. The company wishes to replace the boiler with a circulating fluidized boiler (CFB; see Figure E3-2) that produces high-pressure steam to generate electricity that will be sold to the local utility company and to use the discharged low-pressure steam to manufacture paper. The CFB boiler will consume low-quality

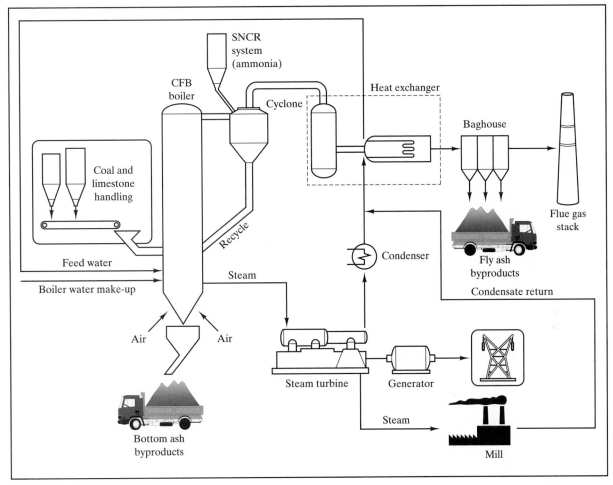

**Figure E3-2**  Components of a circulating fluidized bed (CFB) steam boiler

$(10{,}500$ Btu/lb$_m$) coal at the rate of 912,500 tons/yr $(1.916 \times 10^{13}$ Btu/yr, $9.58 \times 10^9$ Btu/h). This dual use of steam is called *cogeneration.* You have been asked to review permit applications prepared for the new boiler by a consulting firm. The plant is located in a mid-Atlantic state that is part of the Northeast Ozone Transport Region, parts of which are a nonattainment region with respect to ozone. The plant is located in a class II region that is an attainment region for all criteria pollutants except $O_3$. The permit application addresses the following issues:

- Review of BACT pollution control technologies
- Determination of PSD increment consumption
- Impact on a class II region
- Impact on NAAQSs
- Impact on nonattainment area
- Additional impacts (odor, visibility, soil, vegetation, commercial, residential, and industrial growth)

Pollutants will be removed by a variety of conventional means. Particles will be captured by a high-efficiency (99.9%) fabric filter (i.e., baghouse). Sulfur dioxide ($SO_2$) will be captured by injecting limestone into the fluidized bed of coal. Carbon monoxide (CO) and VOC's will be reduced to acceptable levels by the unique design of the combustion chamber. The oxides of nitrogen ($NO_x$) will be removed by selective noncatalytic reduction (SNCR) in which aqueous ammonia or urea is injected into the exhaust gas stream. Based on the performance of similar CFBs and their air pollution controls that have been permitted in the region, the following atmospheric emissions can be expected.

| Pollutant | Emission rate | NSPS (firing rate > $250 \times 10^6$ Btu/h) |
|---|---|---|
| $SO_2$ | 0.25 $lb_m$ $SO_2$/$10^6$ Btu | 0.8  $lb_m$ $SO_2$/$10^6$ Btu |
| Particles | 0.011 $lb_m$ particles/$10^6$ Btu | 0.1  $lb_m$ particles/$10^6$ Btu |
| $NO_x$ | 0.125 $lb_m$ $NO_x$/$10^6$ Btu | 0.70 $lb_m$ $NO_x$/$10^6$ Btu |
| CO | 0.15 $lb_m$ CO/$10^6$ Btu | |
| VOC | 0.004 $lb_m$ VOC/$10^6$ Btu | |

Thus the predicted emission from the CFB boiler and its controls are well within NSPS standards.

*$NO_x$ nonattainment region.* The company's present small boiler emits 900 tons $NO_x$/yr. A review of the EPA RACT/BACT/LAER Clearing House (Section 3.4) indicates that for cogeneration, BACT technology can be used but the much larger unit will emit 1437 tons $NO_x$/yr. If any source emits more than 100 tons $NO_x$/yr, PSD studies have to be conducted.

Second, the state was in a nonattainment region, and its air pollution agency committed itself to a state implementation plan (SIP) in which it agreed to reduce the total $NO_x$ emissions within the state by 15% each year. Thus whatever boiler the company adopted, it could no longer emit even 900 tons of $NO_x$/yr, and the company had to enable the state to reduce its $NO_x$ emission by an amount [called an *emission reduction credit* (ERC)] which is no less than 15% of the amount of $NO_x$ actually emitted by the CFB boiler.

The company chose to install the a CFB boiler that emitted 1437 tons $NO_x$/yr and to find other ways to enable the state to reduce its yearly $NO_x$ emission by (0.15)(1437) = 215 tons of $NO_x$/yr. To achieve these purposes, the company purchased a *pollution allowance* of 800 tons $NO_x$/yr from a nearby gas transmission company, GasTran Inc., which decided to power its compressors with electricity rather than steam generated by its own boiler using natural gas. Thus the NO*x* emissions from current equipment that will no longer be emitted was:

| | |
|---|---|
| Present boiler | 900  tons/yr |
| GasTran Inc. | 800  tons/yr |
| | 1700  tons/yr |

The actual reduction in $NO_x$ after new equipment was installed = 1700 − 1437 = 263 tons $NO_x$/yr. The

minimum reduction of $NO_x$ emissions the state needed to satisfy its SIP was 1652 − 1437 = 215 tons $NO_x$/yr. Thus by purchasing an allowance for $NO_x$ not emitted by the compressor station and installing the new CFB boiler, the state was able to achieve its goal of emitting less $NO_x$ (with 22% to spare) and the company generated steam for profit and continued to manufacture paper. Shown below are the current and proposed emission rates. Current emissions were actually measured. The proposed emissions are the product of the emission rates in the preceding table times the firing rate.

| Emission rate (tons/yr) | | | |
|---|---|---|---|
| Pollutant | Current | Proposed | Change (proposed-current) |
| $SO_2$ | 5310 | 2391 | −2919 |
| $PM_{10}$ | 209 | 105 | −104 |
| $NO_x$ | | | |
| Present boiler | 900 | | |
| GasTran Inc. | 800 | | |
| CFB boiler | | 1435 | −265 |
| VOC | 3.4 | 38 | 34.6 |
| CO | 32.1 | 1435 | 1402.9 |

Since the VOC emission rate is less than 50 tons/yr, no ERCs are needed. With the exception of CO, the CFB boiler and its controls has lower pollutant emission rates than the current company boiler and current GasTran boiler.

*PSD studies: air quality modeling.* The impact of the above on the region's air quality was evaluated by EPA-approved dispersion models (Chapter 9) following a protocol also approved by the EPA for the terrain of the proposed installation. A dispersion model is a computer model consisting of mathematical equations describing the transport of pollutants from the stack to points downwind. The models predict ground-level concentrations, $c_{GL}(x, y, 0)$, where $x$ is in the direction of the wind, $y$ is a distance transverse to the wind direction, and $z$ is the distance above the ground. The models predicted the ground-level concentrations that people at various points downwind of the stack $(x, y, 0)$ could experience over the period of a year. The models incorporated meteorological conditions that characterize the community over periods of several years, including conditions that produce both average and worst-case situations. Input data to the models were the stack height, exhaust temperature and velocity, pollutant and total gas mass flow rates, exit concentrations, and atmospheric stability parameters that described how the plume mixed in the atmosphere for different meteorological conditions.

The output (shown below) of air quality dispersion models serves several purposes. At the very least, the predicted concentrations must not exceed NAAQS values. In addition air quality models are used for two other determinations:

1. *Significant level:* predicted ground-level concentrations one can expect could be measured, compared to background values and confidently ascribed to the proposed source.

2. *PSD increment consumption:* ground-level concentrations below which new sources will be allowed to impact ambient air quality and which do not degrade existing background levels and/or produce adverse consequences. Thus the incremental increase in pollutant concentration cannot exceed the PSD incremental consumption.

The *significant level* is the predicted increase in ambient concentration due to the source that is expected to have statistical significance. The *PSD incremental consumption* is something quite different. Presuming that the region is in compliance with NAAQS, the difference $(c_{\text{NAAQS}} - c_{\text{predicted}})$ can be thought of as a valued environmental "asset" that will be allowed to be reduced by the proposed new source. Thus the issue is by what amount this asset can be "consumed" to benefit the community economically. In one sense any consumption is undesirable, yet the new source is an economic benefit for the community and one has to balance economic benefit against environmental cost.

The air quality modeling results for the class II region (concentrations, $\mu$g/m$^3$) are as follows:

| Pollutant | Predicted | Significance Level | PSD Incremental Level | NAAQS | Measured Background |
|---|---|---|---|---|---|
| SO$_2$ | | | | | |
| 3-h average | 113.5 | 25 | 512 | | 236 |
| 24-h average | 24.3 | 5 | 91 | 365 | 113 |
| Annual average | 4.9 | 1 | 20 | 80 | 26 |
| NO$_2$ (annual average) | 2.4 | 1 | 25 | 100 | 41 |
| CO | | | | | |
| 1-h average | 96.8 | 2,000 | | 40,000 | 13,000 |
| 8-h average | 50 | 500 | | 10,000 | 6,000 |
| Particles (PM$_{10}$) | | | | | |
| 24-average | 1.1 | 5 | 37 | 150 | 91 |
| Annual average | 0.2 | 1 | 19 | 50 | 32 |

All numerical values have the units $\mu$g/m$^3$.

The predicted particle (PM$_{10}$) and CO concentrations are so far below the significance level that further controls for these two pollutants are not necessary.

The incremental increases in SO$_2$ and NO$_x$ are within the PSD increment consumption, and the total concentrations of SO$_2$ and NO$_x$ after the installation of the CFB are predicted to be within the NAAQS. Thus the proposed permit to construct the CFB is in compliance with state and federal regulations.

## 3.3 *Rule-making*

Care taken in promulgating rules (Lippmann, 1987) reduces the number of changes required later following litigation. It is important for engineers to understand the federal process called *rule-making*. Legislation creating EPA defined its purpose and scope and authorized the creation of pollution control agencies in each state to enact and implement rules. Regulations are promulgated in a manner shown in Figure 3-3. The process takes considerable time since all interested parties must have the opportunity to make public comments (Melton, 1996).

Newly adopted regulations and their amendments, presidential proclamations, executive orders, federal agency documents having applicability and legal effect, and other federal agency documents of public interest are published in the *Federal Register*. The *Federal Register* is published daily by the National Archives Administration, Washington, DC 20408 to provide a uniform method for disseminating regulations and legal notices to the public. To cite any information in the *Federal Register*, one simply uses the volume and page number. For example, 51FR 1234 means *Federal Register*, volume 51, page 1234.

To keep track of all regulations relevant to a subject, such as particle emissions from municipal incinerators, an engineer would need to monitor every issue of the *Federal Register* and extract the relevant information. The task is formidable, but it is also unnecessary since it is accomplished by the *Code of*

Guiding principle: Protection of human health and the environment in conjunction with encouragement to reuse potential energy resources

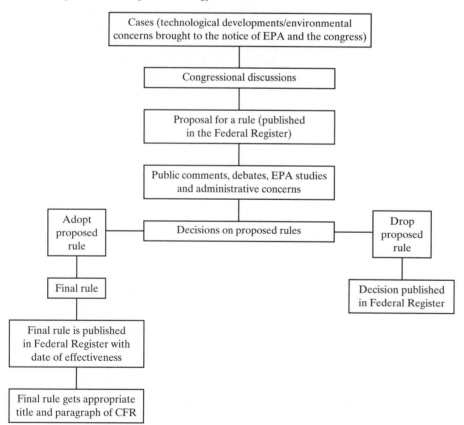

**Figure 3-3** Rulemaking, the process of promulgating environmental regulations

*Federal Regulations* (CFR), a compilation of all general and permanent rules and their amendments published by the *Federal Register* pertaining to certain issues or agencies. A complete searchable CFR database can be found in the textbook's Web site on the Prentice Hall Web page.

The code is divided into 50 titles representing broad areas subjected to federal regulation. Each title is divided into chapters which usually bear the name of the issuing agency, and each chapter is subdivided into parts covering specific regulatory areas. For example:

- *Title 29: Labor.* This title is composed of eight volumes; each volume contains parts grouped in ways to facilitate their use (e.g., Chapter XVII, parts 1900 to 1910, revised as of July 1, 1988, represents all regulations pertaining to the Occupational Safety and Health Administration, codified under this title as of July 1, 1988).
- *Title 40: Protection of environment.* This title is composed of 14 volumes; each volume contains parts grouped in ways to facilitate their use (i.e., Parts 1 to 51, Part 52, and Parts 53 to 60, revised as of July 1, 1989, represents all the regulations relating to the Environmental Protection Agency codified under this title as of July 1, 1989). Title 40 can be found from the textbook's Web site on the Prentice Hall Web page.

The CFR is kept to up date by the individual issues of the *Federal Register*. These two publications must be used together to determine the latest version of any rule. The CFR is revised at least once each calendar year. Each *Federal Register* lists the title number and parts of the CFR to which it makes reference. For example, a typical reference in the *Federal Register* could be 40CFR 270.70(a), which is interpreted as follows:

- Title 40 of the Code of Federal Regulations
- Part 270
- Paragraph 70(a)

Sometimes the *Federal Register* cites a reference simply as 266.34(b), which requires the reader to seek part 266, paragraph 34(b). A reference in a CFR may be listed as "reserved," which means that the reference will be or is planned to be incorporated in a future CFR.

# 3.4 Electronic bulletin boards

The Office of Air Quality Planning Standards (OAQPS) of the EPA provides information and technical support on air pollution control. These services include clearinghouses, conferences, reports, manuals, newsletters, support centers, workshops, classroom training, self-instructional courses, and an electronic technology transfer network (TTN). The TTN is a network of electronic bulletin boards that provides information and technology exchange in different areas of air pollution control, varying from emission test methods to regulatory air pollution models. The service is free except for the cost of using the telephone. Anyone in the world can exchange information, including personnel from the EPA and state agencies and people in the private sector and foreign countries. Individuals access the network from their computers through the use of a modem or via the Telnet network. The *Telnet access* address can be found on the textbook's Web site on the Prentice Hall Web page.

Individuals can download computer code for regulatory models, read a title summary of the 1990 Clean Air Act, find a course offered by the Air Pollution Training Institute, or request technical support in implementing an air pollution control program. Files with ".TXT" can be printed to the screen. Files with ".ZIP" or ".WPF" require the user to download a protocol "pkunzip <filename>" from the systems utilities to expand them. Individuals can transfer files, communicate with other users, leave questions for others to answer, or upload a file for others to use. The TTN presently contains the following 18 bulletin boards; additions and updating are made on a continual basis. In some cases files are compressed and special codes are needed before these codes can be used.

- *EMTIC (Emission Measurement Technical Information Center):* provides access to emission test methods and testing information for the development and enforcement of national, state, and local emission prevention and control programs.
- *AMTIC (Ambient Monitoring Technology Information Center):* provides information and all federal regulations pertaining to ambient air monitoring. Information on monitoring methodology, and field and laboratory studies is included.

- *AIRS (Aerometric Information Retrieval System):* provides information and documentation on the use and acquisition of air quality and emissions data from the AIRS mainframe computer.
- *BLIS (RACT/BACT/LAER Information Systems):* compilation of air permits from local, state, and regional air pollution control agencies. This bulletin board should be consulted to find out specific control technologies the EPA has accepted as BACT, RACT, and LAER.
- *NATICH (National Air Toxics Information Clearinghouse):* contains information submitted by the EPA, and state and local agencies regarding their air toxics programs to facilitate the exchange of information among governmental agencies.
- *COMPLI (Stationary Source Compliance):* provides stationary source and asbestos compliance policy and guidance information.
- *NSR (New Source Review):* offers guidance and technical information within the New Source Review permitting community.
- *SCRAM (Support Center for Regulatory Air Models):* provides regulatory air quality model computer codes, meteorological data, documentation, and modeling guidance. This bulletin board should be consulted for details about air quality models used in PSD studies.
- *CHIEF (Clearinghouse for Inventories/Emission Factors):* contains the latest information on air emission inventories and emission factors. The bulletin board provides access to numerical computational programs to estimate emissions of air pollutants for a variety of sources and performing air emission inventories for both criteria and toxic pollutants.
- *CAAA (Clean Air Act Amendments):* contains information on the Clean Air Act Amendments of 1990, regulatory requirements, implementation programs, criteria pollutants, and technical analyses.
- *APTI (Air Pollution Training Institute):* describes the current course offerings on air pollution, including curriculum, schedules, locations, costs, and up-to-date changes.
- *CTC (Control Technology Center):* offers free engineering assistance, a hot line, and technical guidance to state and local air pollution control agencies in implementing air pollution control programs.
- *USC (User Support Center):* offers information on modems, downloading, communication software, and other communications issues. It also provides a public message area for users to share information related to the use of the TTN.
- *ORIA (Office of Radiation and Indoor Air):* disseminates information to state and local governments, industry, professional groups, and citizens on how to reduce exposure to harmful levels of radiation and indoor air pollutants.
- *USCAN (USD/Canada Air Quality Agreement):* provides information for the exchange of permitting information between states on the U.S.–Canadian border.
- *OMS (Office of Mobile Sources):* provides information on mobile source emissions, including regulations, test results, models, and guidance.
- *AIRISC (Air RISC):* provides technical assistance and information primarily to state and local air pollution control agencies on health, risk, and exposure assessment on toxic and criteria pollutants.
- *SBAP (Small Business Assistance Program):* disseminates information to small business about new federal rules related to small business issues.

The EPA intends to make all TTNBBS bulletin boards accessible through the Web. At the time of writing, a CD-ROM containing the AP-42 Emission Factors was available from the Government Printing Office for a modest price. With the passage of time, other information useful to engineers will become available from the Printing Office. In this fast-changing area of electronic information transfer, it is incumbent upon engineers to be fluent in the latest information sources.

## 3.5  State regulations

State implementation plans (SIPs) approved by the EPA require states (or regional agencies) to devise regulations to reduce air pollution. State legislatures pass air pollution control acts defining the enforcement duties and responsibilities of their environmental departments. Legislation typically defines two other bodies to assist in this task, an *environmental quality board* and an *environmental hearing board*. Regulations to reduce air pollution are adopted by the state environmental quality board following rule-making procedures similar to those described for the federal government. The environmental hearing board has the power to rule on appeals and adjudicates appeals in conformity with administrative agency law. States use different titles for their environmental enforcement departments (e.g., Department of Environmental Resources, Department of Natural Resources, etc.). Contained in this section is an abridged summary of the regulations for Pennsylvania, which typify regulations from other states. In what follows the generic phrase *state environmental department* will be used for the name of the state or regional agency that enforces air pollution regulations.

States are divided into geographical regions. Engineers should contact the regional office at the inception of plans to modify an existing source or construct new sources to learn of new or impending changes in regulations. When plans to construct or modify a source have progressed to a certain level of refinement, an application should be submitted to the state environmental department for approval. The purpose of the application is to determine that the process contains the appropriate technology to ensure that the process is in compliance. Such approval will also detect bogus claims made by equipment vendors or source operators. The application requires the operator to:

- Describe components of the process, its operating cycle, and the maximum production rates.
- Estimate pollutant emission rates (per hour, day, and year) from the uncontrolled process under maximum operating conditions. These emissions are called *potential emissions.*
- Describe air pollution control devices and estimate their performance.
- Estimate pollution emission rates to the atmosphere (per hour, day, or year) after controls are installed.

Once approval is secured, source operators can enter into contractual agreements with confidence. Consummating contracts prior to state approval is risky. Table 3-5 summarizes the steps to be taken to secure operating permits for new or modified sources. Tables 3-6 to 3-8 are abridged versions of typical state standards for sources of air pollution. It must be remembered that securing permits is done on a case-by-case basis following a top-down process

---

**Table 3-5**   Process to Secure Permits for New or Modified Sources in Pennsylvania

1. *Best available technology.* Each new or modified source must show that the emissions will be the minimum attainable through the use of the *best available technology* (BAT). A control strategy is not acceptable merely because it reduces emissions to some low value.

2. *Case-by-case, top-down process.* Each new or modified source is evaluated by the state air pollution agency on a *case-by-case basis* following a *top-down strategy* (beginning with the most stringent technology) that reduces *potential emissions* by the largest amount. *Potential emissions* are the yearly emissions (tons/yr) that are apt to occur if there is no attempt to control emissions from a source at maximum operating capacity, 24 h/day and 365 days/yr.

3. *Emission of hazardous air pollutants* (HAPs). An acceptable control strategy must be one that brings about *maximum achievable control technology* (MACT). MACT applies if the potential emissions rate ($\dot{m}$)
   $\dot{m}$(single HAP) $\geqq$ 10 tons/yr $\qquad\qquad\qquad$ $\dot{m}$(combination of HAPs) $\geqq$ 25 tons/yr
   MACT is selected by process 2. Technologies are deemed to satisfy MACT by the EPA.

4. *Emissions from new or modified sources.* A *prevention of significant deterioration* (PSD) study is conducted if potential emissions exceed 100 tons/yr. If the source is located in an *attainment area* with respect to the pollutant, the agency will ask that *best available control technology* (BACT) be adopted. If the source is located in a *nonattainment area* with respect to the pollutant, the agency will ask for a technology providing the *lowest achievable emission rate* (LAER). The selections of BACT and LAER are by process 2. Technologies to satisfy BACT and LAER must have been approved by the EPA. If the technology does not achieve the required reduction (*offset*), the offset must be secured from another facility in the region.

5. *Emissions from an existing source.* If the potential emissions exceed 50 tons/yr for VOC or 100 tons/yr for $NO_x$, the agency requires a *reasonable available control technology* (RACT). The selection of RACT is by process 2 provided that the actual emissions are less than the amounts in 3. above and cost no more than $1500 per ton of reduction following procedures in the EPA Control Cost Manual (see Chapter 13).

---

**Table 3-6**   Abridged Summary of Pennsylvania $SO_2$ Emission Standards from Combustion Units[a]

---

*Nonbasin areas* (low pollution level, good atmospheric ventilation)

$$\text{Oil: } A < 4.0 \% \; S_{max} = \begin{cases} 0.5 \text{ for No. 2 fuel and lighter} \\ 2.8 \text{ for No. 4, 5, and 6 fuel and heavier} \end{cases}$$

Coal: If $E < 250$, then $A < 4.0$

$$\text{If } E > 250, \; A = \begin{cases} 3.7 \text{ thirty-day running average} \\ 4.0 \text{ daily average not to be exceeded more than 2 days in 30} \\ 4.8 \text{ daily average maximum never to be exceeded} \end{cases}$$

*Moderate air basin* (periodic poor atmospheric ventilation, i.e., Allentown, Easton, Johnstown, etc.)

$$\text{Oil: } A < 3.0 \% \; S_{max} = \begin{cases} 0.3 \text{ for No. 2 fuel and lighter} \\ 2.0 \text{ for No. 4, 5, and 6 fuel and heavier} \end{cases}$$

Coal: *If* $E < 250$, then $A < 3.0$

$$\text{If } E > 250, \; A = \begin{cases} 2.8 \text{ thirty-day running average} \\ 3.0 \text{ daily average not to be exceeded more than 2 days in 30} \\ 3.6 \text{ daily average maximum never to be exceeded} \end{cases}$$

*Critical air basin* (history of air pollution episodes, i.e., Monongahela River valley, Pittsburgh, etc.)

$$\text{Oil and coal: } \quad A = \begin{cases} 1 & \text{if } 2.5 < E < 50 \\ 1.7E^{-0.14} & \text{if } 50 < E < 2000 \\ 0.6 & \text{if } E > 2000 \end{cases}$$

---

[a] $A$, maximum emissions ($lb_m$ $SO_2/10^6$ Btu; $E$, firing rate ($10^6$ Btu/h); % $S_{max}$, maximum allowable percent sulfur in fuel oil used in the area.

---

beginning with the most stringent technology that reduces potential emissions by the largest amount. Depending on the pollutant and geographical region, the selection of the approved control strategy may include economic considerations. Thus some of the standards a company may be asked to satisfy may be more stringent than those seen in Tables 3-6 to 3-8.

An abridged summary of $SO_2$ emission standards for coal and oil combustion units (industrial boilers, electric power generation boilers, etc.) is shown in Table 3-6. It should be noted that more stringent standards apply in geographical region (basins) where history has shown that meteorological conditions have produced air pollution episodes in the past. Table 3-7 is an abridged summary of particulate emission standards for a variety of industrial processes. States define many standards for industries unique to their region containing detailed stipulations unique to the region. Depending on the potential emission rate and whether the region is an attainment or nonattainment region, BACT,

RACT, or LAER standards may be applied. State regulations also seek to abate fugitive particulate emissions, unpleasant odors, and the opacity of process gas streams as they enter the atmosphere (Table 3-8).

Solvents, or volatile organic compounds (VOCs) as they are more properly called, are the subject of considerable concern to EPA because of atmospheric reactions with ozone. A *volatile organic compound* (VOC) is defined as a hydrocarbon with a normal boiling point below 100°C (i.e., saturation pressure = 101 kPa, saturation temperature < 100°C). Because of its low reactivity, methane is not considered a VOC. The phrase *nonmethane hydrocarbon* (NMH) is sometimes used instead of VOC. VOC controls apply to numerous processes: surface coating and cleaning, dry cleaning, degreasing, petroleum refineries, fuel storage vessels, asphalt paving, printing, graphic arts, pharmaceutical manufacture, and so on. Each of these processes has unique standards and control methods. VOCs from surface coating processes are ubiquitous in U.S. industry. If a process

---

**Table 3-7**   Abridged Summary of Pennsylvania Particulate Emission Standards from Combustion Units and General Processes

---

*Combustion units* [$A$ = maximum emissions ($lb_m$ particles/$10^6$ Btu), $E$ = firing rate ($10^6$ Btu/h)]:

$$A = \begin{cases} 0.4 & \text{if } 2.5 < E < 50 \\ 3.6E^{-0.56} & 50 < E < 600 \\ 0.1 & \text{if } E > 600 \end{cases}$$

*Incinerators:* maximum particle concentration in exit gas = 0.1 grain/ft$^3$ corrected to STP and 12% $CO_2$

*General processes*
$$\dot{m}_p(lb_m/h) = (0.76)I^{0.42} \quad I \text{ is given by} \quad I(lb_m/h) = (F)(\dot{w})$$

where $F$ is the process factor (see chart below) and $\dot{w}$ is the process charging rate (production unit/h, where the production units are defined in the chart for $F$).

| Process | F-process factor (units) |
|---|---|
| Asphaltic concrete production | 6 (aggregate feed) |
| Coal dry cleaning | 2 (product) |
| Grain | |
|     Drying | 200 (product) |
|     Elevators, load and unloading | 90 (grain) |
|     Screening, cleaning | 300 (grain) |
| Iron foundry | |
|     Melting < 5 tons/h | 150 (iron) |
|     Melting > 5 tons/h | 50 (iron) |
|     Sand handling | 20 (sand) |
|     Shake-out | 20 (sand) |
| Lime calcining | 200 (product) |
| Portland cement manufacture | |
|     Clinker production | 150 (dry solids feed) |
|     Clinker cooling | 50 (product) |

---

**Table 3-8**   Pennsylvania Fugitive Emissions, Visible Emissions, and Odor Standards

---

*Fugitive emissions.* No person shall cause, suffer, or permit fugitive particulate emissions to be emitted into the outdoor atmosphere from any source or sources if such emissions are visible, at any time, at the point such emissions pass outside the person's property.

*Visible emissions.* No person shall cause, suffer, or permit the emission into the outdoor atmosphere of visible air contaminants in such a manner that the opacity (percent light attenuated) of the emissions is:

    (1) equal to or greater than 20% for a period or periods aggregating more than 3 min in any one hour; or
    (2) equal to or greater than 60% at any time.

Exempted are emissions from agricultural activity and emissions of condensed $H_2O$, commonly called "steam."

*Odors.* No person shall cause, suffer, or permit the emission into the outdoor atmosphere of any malodorous air contaminants from any source in such a manner that the malodorous contaminants are detectable outside the property of the person on whose land the source is being operated. Exempted are odors arising from the production of agricultural commodities in their unmanufactured state on the premises of the farm operation. If control of malodorous air contaminants is required, they shall be incinerated at a minimum of 1200°F (649°C) for at least 0.3 s prior to discharge into the outdoor atmosphere. Other control techniques may be used if such techniques are equivalent or better than incineration.

has the potential to generate less than 10 tons VOC/yr, the process is exempt from regulation. If the process has the potential to generate more than 50 tons VOC/yr and/or 500 $lb_m$ VOC/day, compliance with state regulations can be achieved in several ways.

1. $X < Y$. The manufacturer can use water-based coatings or use a coating ($X$) with a lower VOC/gallon value than the *compliant coatings* ($Y$) recommended (Table 3-9), or
2. $X > Y$. If the manufacturer selects a coating ($X$) with a higher VOC/gallon value than the *compliant coating* ($Y$) recommended, an air pollution control device must be installed to remove VOCs before they are emitted to the outdoor atmosphere. The minimum removal efficiency is given by

$$\% \text{ removal} = 100\left[1 - \frac{Y(Z-X)\text{TE}_{\text{proposed}}}{X(Z-Y)\text{TE}_{\text{BAT}}}\right] \quad (3.1)$$

where

$$\text{TE}_{\text{proposed}}, \text{TE}_{\text{BAT}} = \text{proposed and BAT spray transfer efficiency (Table 3-9)}$$
$$X = lb_m \text{ VOC/gallon of present or proposed coating (minus water)}$$
$$Y = \text{allowable VOC content (use values from Table 3-9)}$$
$$Z = \text{density of VOC (solvent) } (lb_m/\text{gal})$$

The spraying transfer efficiency refers to the percent of the spray that is actually deposited on the coated surface. Manufacturers may choose whatever spraying method they like, but Table 3-9 lists the EPA recommended particular methods for particular coating operations.

### EXAMPLE 3.3   PAINTING FLUORESCENT LIGHT FIXTURES

One of the steps in the manufacture of fluorescent light fixtures is to apply a surface coat of white enamel. For quite some time the plant has used a rotating electrostatic sprayer. The metal part is grounded and the paint particles are given a positive electric charge. The moving paint particles travel along the electric field lines and 95% of the paint lands on the intended metal part. Such a spraying

method has a transfer efficiency ($\text{TE}_{\text{proposed}}$) that exceeds the best available transfer technology for painting metal parts (40%). After being coated, the parts pass through an infrared oven and the remaining VOC is removed by air drying. The firm presently uses 100 gal/day of enamel containing 6.2 $lb_m$ VOC/gal. To obtain the quality of coating that is desired, workers thin the paint by 10% using the solvent isopropyl alcohol (6.6 $lb_m$ VOC/gal). Your employer asks you to prepare the RACT application required by Title V of the 1990 CAAA.

***Solution***   Based on the amount of VOCs used, the potential VOC emission rate is as follows.

$$\dot{m}_{\text{potential}} = (\dot{m}_{\text{enamel}} \text{ gal/day})(6.2 \text{ } lb_m \text{ VOC/gal})$$
$$+ (0.1)(\dot{m}_{\text{enamel}} \text{ gal/day})(6.6 \text{ } lb_m \text{ VOC/gal})$$
$$= (100 \text{ gal/day})(6.2 \text{ } lb_m \text{ VOC/gal})$$
$$+ (0.1)(100 \text{ gal/day})(6.6 \text{ } lb_m \text{ VOC/gal})$$
$$= 620 + 66 = 686 \text{ } lb_m \text{ VOC/day}$$

Since the potential emissions are greater than 500 $lb_m$/day, the company must adopt either strategy 1 or 2 mentioned to the left. You suggest four alternatives.

1. Use less paint or reduce the production rate to comply with the maximum VOC emission rate of 500 $lb_m$/day, 50 tons/yr.
2. Replace the present enamel with a compliant coating containing no more than 3.5 $lb_m$ VOC/gal.
3. Use a water-based enamel containing no VOCs.
4. Continue using the present materials and coating practices but purchase a VOC emission control system whose VOC removal efficiency satisfies Equation 3.1.

Management selects alternative 4 for several reasons. Management does not want to reduce the amount of enamel used on each fixture or reduce production rate. Exploration shows that there are no new solvent-based coatings or water-based coatings available that qualify as a compliant coating and produce the quality management expects. You point out that the cost of using water-based coatings is considerably less than purchasing and operating a VOC removal system, but management is adamant and insists that they wish to continue using their present coating practices and materials. Table 3-9 indicates that the best available spray transfer efficiency

**Table 3-9**  Abridged Summary of Pennsylvania Compliant Coatings for Selected Process and Recommended Coating Transfer Efficiencies

| Process | Compliant coating VOC content, $Y$ (lb$_m$/gal) | Transfer Efficiency, TE |
|---|---|---|
| Fabric coating | 2.92 | 0.95 |
| Paper coating | 2.92 | 0.95 |
| Automobile coating | | |
|   Prime coat | 1.92 | 0.4 |
|   Top coat | 2.83 | 0.4 |
|   Repair | 4.84 | 0.4 |
| Wood cabinet and furniture finishing | | |
|   Clear top coat | 5.92 | 0.65 |
|   Wash coat | 6.50 | 0.65 |
|   Final repair coat | 6.00 | 0.30 |
|   Semitransparent spray stains and toners | 6.84 | 0.65 |
|   Clear sealers | 6.17 | 0.65 |
|   All other coatings | 7.00 | 0.65 |
| Miscellaneous metal parts | | |
|   Heavy-duty trucks | 4.33 | 0.4 |
|   Extreme-performance coating | 3.50 | 0.4 |
|   Pail and drum interiors | 4.33 | 0.4 |
|   Air-dried coatings | 3.5 | 0.4 |
|   All other coatings | 3.00 | 0.4 |
| *Application method* | *Best available transfer technology,* TE$_{BAT}$ | |
| Air atomized | 0.25 | |
| Airless atomized | 0.40 | |
| Manual electrostatic spraying | 0.65 | |
| Nonrotational electrostatic spraying | 0.70 | |
| Rotational Head Electrostatic spraying | 0.80 | |
| Dip and flow | 0.9 | |
| Electrodeposition | 0.95 | |
| Roller Coating | 1.00 | |

(TE$_{BAT}$) is 0.4 and that the compliant coating ($Y$) for air-dried coatings has a VOC content of 3.5 lb$_m$ VOC/gal. The minimum removal efficiency of a VOC removal system is predicted by Equation 3.1.

$$\eta = 100\left[1 - \frac{(3.5)(6.6 - 6.2)(0.95)}{(6.2)(6.6 - 3.5)(0.4)}\right]$$

$$= 100\left(1 - \frac{1.33}{7.69}\right) = 82.7\%$$

Another part of the state implementation plans concerns *air pollution episodes*. An episode is the name given to the buildup of air pollutants due to particular meteorological conditions which has the potential to produce concentrations that could jeopardize health. All states are required to monitor the concentration of certain air pollutants on a continuous basis at selective sites throughout the state. The data are processed centrally and coupled with weather forecasts. If there is a possibility that pollutant concentrations in a region might lead to a substantial threat to the health, the state is required to inform the public and take action to minimize harm. There are five levels of air pollution episodes defined by the concentrations shown in Table 3-10.

**Table 3-10**    Air Pollution Episodes in Pennsylvania

| Level | Criteria [a] |
|---|---|
| Forecast | Meteorological conditions may cause pollutant concentration to increase above normal levels. |
| Alert | When one of following levels is reached and meteorological conditions are such that the concentrations can be expected to remain at the said level for 12 or more hours, or increase.<br>$SO_2$: 0.3 ppm, 6-h average<br>Particles: COH = 4.0, 6-h average<br>ppm ($SO_2$) $\times$ COH = 0.3, 24-h average<br>$NO_2$: 0.2 ppm, 24-h average |
| Warning | When one of the following levels is reached and meteorological conditions are such that the concentrations can be expected to remain at the said level for 12 or more hours, or increase.<br>$SO_2$: 0.5 ppm, 6-h average<br>Particles: COH = 6.0, 6-h average<br>ppm ($SO_2$) $\times$ (COH) = 0.9, 24-h average<br>CO: 30 ppm 8-h average<br>Oxidants: 0.25 ppm 4-h average<br>$NO_2$: 0.3 ppm, 24-h average |
| Emergency | When any one of the following levels is reached and meteorological conditions are such that the concentrations can be expected to remain at the said level for 12 or more hours.<br>$SO_2$: 0.6 ppm, 24-h average<br>particles: COH = 7.0, 24-h average<br>ppm ($SO_2$) $\times$ (COH) = 1.4, 24-h average<br>CO: 40 ppm, 8-h average<br>Oxidants: 0.35 ppm, 4-h average<br>$NO_2$: 0.4 ppm, 24-h average |
| Termination | Once declared, any level reached in accordance with these criteria should remain in effect until the criteria for the alert level are no longer met and meteorological conditions are such that pollutant concentrations can be expected to decrease. |

[a] The definition of COH is given by Equation 5.25.

The action the state is required to take is shown in Table 3-11.

The legal authority of state or regional environmental departments is established by state law yet there is provision to appeal actions taken by the department. Environmental departments may grant temporary *variances* from any standard or requirement upon petition, a public hearing, and review by the state environmental department. Actions taken concerning variances can be appealed to the environmental hearing board. For the sake of expediency, state environmental departments and source operators often enter into *consent agreements* to mitigate air pollution. Such consent agreements are not subject to appeal.

Fees, penalties, and fines collected for violations of environmental regulations are paid to a general fund disbursed by the state environmental department for purposes of eliminating air pollution. Penalties are typically called *summary offenses* and *misdemeanors*. Each pertains to the severity of the unlawful conduct and may contain fines and possible imprisonment.

## 3.6 *Hazardous chemicals*

Assessing risk requires knowledge of the actual amounts of the material used and the percent discharged to the environment. Merely listing materials used can create a false sense of risk. To begin with, engineers should conduct an *audit* which indicates the amounts of material used over a period of time. In addition, one must estimate the amounts remain-

**Table 3-11**   Actions in Response to Air Pollution Episodes in Pennsylvania

| Level | Action to be Taken |
|---|---|
| **Alert** | |
| General | Prohibit open burning |
| | Limit incineration and soot blowing |
| | Reduce space heating and cooling |
| | Defer facility alterations |
| | Eliminate unnecessary motor vehicle operations |
| Specific | Electric generating facilities |
| |    Switch to low-sulfur fuel |
| |    Shift load to new units |
| |    Shift load to facilities outside area |
| |    Switch to low-sulfur fuels for industrial boilers |
| | Reduce blast furnace operation |
| | Increase coking time of coke ovens |
| | Steelmaking furnaces |
| |    Reduce oxygen and steel production |
| |    Switch to low-sulfur fuels |
| **Warning** | |
| General | Cease incineration of all solid and liquid waste |
| | Cease, postpone, and delay facility alterations |
| | Minimize space heating and cooling |
| | Minimize use of electricity |
| | Encourage car pools |
| Specific | Electric generating facilities reduce power to outside users |
| | Maximum production reduction of blast furnaces |
| | Maximum reduction in coke oven operations |
| | Maximum reduction of steelmaking furnaces |
| | Maximum reduction of quarry operations |
| **Emergency** | |
| General | Close places of employment, with the exception of essential goods and services |
| Specific | Maximum reduction of power generation for outside area users |

ing in the product, the amounts removed as waste and lastly the amounts that may escape to the environment. Such audits are really huge mass balances and identify clearly processes and materials posing a health risk.

Responding to the *Toxic Substances Control Act* (TSCA), the EPA cataloged more than 56,000 manufactured or imported substances used in manufacture. The list, called the *TSCA inventory*, excludes classes of materials regulated under other federal statues: food additives (8627), prescription and nonprescription drugs (1815), cosmetic ingredients (3410), and pesticides (3350).

Identifying properties of chemicals that require special handling can be a complicated matter. Lists of chemicals are published by numerous governmental, trade, and professional agencies. The following are five familiar lists and the agency that created each one:

1. *189 Hazardous Air Pollutants* (HAPs) (Title III, Clean Air Act Amendments 1990)
2. *Registry of Toxic Effects of Chemical Substances* (RTECS) [National Institute for Occupational Safety and Health (NIOSH)]
3. *List of Toxic and Hazardous Substances* [Occupational Safety and Health Administration (OSHA)]
4. *Pocket Guide to Chemical Hazards* [National Institute for Occupational Safety and Health (NIOSH)]

**5.** *Threshold Limit Values for Chemical Substances and Physical Agents in the Workroom Environment with Intended Changes* (published and updated yearly) [American Conference of Governmental and Industrial Hygienists (ACGIH)]

The *Registry* is published and updated by NIOSH every few years in compliance with the 1970 Occupational Safety and Health Act. The *Registry* is the most comprehensive source to consult. It contains toxicity data extracted from the scientific and professional literature for a fraction of the approximately 65,000 chemical compounds listed in the registry. The toxicity data should not be considered a definition of values for describing a safe dose for human exposure.

TSCA also requires NIH and EPA to create a computer database and search programs for chemicals listed in RTECS. The database is called the *Chemical Information Service* (CIS) and is an iterative, on-line service that enables users to retrieve toxicological data for chemicals identified by their RTECS numbers. In addition, CIS enables users to search for chemicals with specific toxicological properties, structures, dose, and so on. The American Chemical Society created a registry that uniquely identifies specific compounds by a *Chemical Abstract Service* (CAS) number. The CAS inventory contains many substances posing no hazard to health and is considerably larger than RTECS. The properties of hazardous chemicals and their CAS, CIS, and RTECS numbers can be found in Norback (1988), Lewis (1986, 1994), and Pohanish and Greene (1993). This information can also be obtained from entries at the Web site, http://chemfinder.camsoft.com/. Many professional journals require authors to list the CAS numbers for all chemicals included in their article. Chemicals listed in the RTECS and CIS also cite the CAS numbers.

Two procedures required by the federal government help to identify hazardous substances and to describe how to handle them safely. First whenever chemicals are transferred between buyer and seller, OSHA requires that the transfer be accompanied by a *material safety data sheet* (MSDS) that specifies the following properties of the material:

1. Manufacturer's name and chemical synonym
2. Hazardous ingredients (pigments, catalysts, solvents, additives, vehicle)
3. Physical data (boiling temperature, vapor pressure, solubility evaporation rate, percent volatile material)
4. Fire and explosion data (flash point, flammability limits, ignition temperature, firefighting procedures)
5. Health hazard (threshold limit value, effects of overexposure, first-aid procedures)
6. Spill and leak procedures and waste disposal methods

Although the MSDSs often don't have all the data engineers want, they at least serve to alert engineers of possible hazards. Companies are obliged to file MSDSs and to make them available to workers, thus creating a "paper trail" that can be followed to trace materials from their creation to destruction.

The second procedure guides the public about hazardous new materials. The TSCA legislation requires the EPA to evaluate new substances and methods used to manufacture them, determine potential release points, estimate potential exposures, and determine whether it will be necessary to specify procedures to minimize exposure. Companies planning to produce or import chemicals not on the TSCA inventory are required to notify the EPA at least 90 days prior to action. The notification is called a *premanufacture notice* (PMN). The document requires the formula, chemical structure, use, and details about the production process so that points of release can be anticipated and estimates prepared of the emissions and human exposure. Other data required are the physical and chemical properties of the substance: vapor pressure, solubility in water or solvents, normal melting and boiling temperatures, particle size if it is a powder, Henry's law constant, pH, flammability, volatilization from water, and any toxicological data that may be available. New substances are screened by EPA and assigned a risk category that obliges the company to control release (general ventilation, protective clothing, respirators, glove boxes, etc.) or test the compound for its toxic or environmental effects. Toxicity is assessed by examining the toxicity of analogous substances (i.e., comparable molecular structures and physical properties). Both

MSDS and PMN procedures were created to minimize workplace exposures, prevent chronic exposures of people living near manufacturing plants, and reduce acute exposure of people affected by transportation accidents.

Outdoor and indoor air quality standards are often confused with one another. The National Ambient Air Quality Standards (NAAQS; Table 3-2) specify the maximum allowable concentration of pollutants in the outdoor environment of the United States. *Permissible exposure limits* (PELs) and *eight-hour time-weighted threshold limit values* (TWA-TLVs) specify the maximum allowable concentration of pollutants in the indoor workplace environment for periods of eight hour/day, forty hour a week. PELs are created by NIOSH and used by OSHA for enforcement purposes. TWA-TLVs are created by the American Congress of Governmental and Industrial Hygienists (ACGIH) to define unhealthy pollutant concentrations in the workplace. Particular pollutants may possess *short-time exposure limits* (STELs), which are 15-minute average allowable concentrations. A small selected set of pollutants are assigned *ceiling* (C) values that should never be exceeded. While PELs and TWA-TLVs are created by two different organizations, both values are designed to achieve the same objective. Thus for the majority of pollutants, TWA-TLVs and PELs are identical and those that are not, do not differ by much. For practical purposes either the PEL or the TWA-TLV can be used to define a safe indoor industrial environment. A list of the current PELs for many industrial chemicals is shown in Table A-11 in the Appendix.

The maximum allowable outdoor concentrations specified in the NAAQSs are generally more stringent than PELs and TWA-TLVs. The latter pertain to healthy men and women working in an indoor environment. At the end of the day they return to the outdoor environment, rest, and allow the body's recuperative power to restore itself. Thus workers can withstand higher maximum allowable concentrations for short periods of time. Outdoor standards pertain to a continuous period of time and must protect the entire U.S. population (i.e., infants, aged, and those suffering from life-threatening disease).

A clear illustration of the difference between NAAQS and PELs are the standards for $O_3$, $NO_2$, and NO (Seinfeld, 1984).

|  | $O_3$ (ppb) | $NO_2$ (ppb) | NO (ppb) |
|---|---|---|---|
| 8-h PEL | 100 | 1000 | 25,000 |
| NAAQS | 119 (1-h) | 53 (annual) |  |
| Polluted urban air | 100–500 | 50–250 | 50–750 |

In the atmosphere NO, $NO_2$, and $O_3$ are the principal stable molecular species which participate in the photochemical reactions that pollute urban air. Thus the NAAQS standards for ambient $NO_2$ are considerably more stringent than the PEL standards for air in the workplace. Since there is little direct sunlight in the workplace to initiate these photochemical reactions, only $O_3$ has a comparable standard because it is an active oxidizing agent in and of itself. Nitric oxide (NO) and nitrogen dioxide ($NO_2$) are only mildly toxic in and of themselves, and hence their PELs are similar to a large number of hazardous industrial materials.

## 3.7 ISO 14000

Standards established by the *International Standards Organization* (ISO) have an enormous impact on U.S. industry. ISO standards are not just quantitative performance criteria to be satisfied but an organizational structure and set of procedures, voluntarily created by companies which ensure that companies are engaged in continuous improvement. ISO 9000 was a set of managerial procedures for continuous improvement in the quality of industrial products. ISO 14000 is a set of managerial procedures for the continuous minimization of pollutant emission. U.S. manufacturing firms wishing to participate in the global economy will need ISO 14000 registration. ISO 14000 is designed to assist companies achieve environmental performance and to ensure that environmental issues will not become barriers to international trade. The voluntary standard is in essence the passport to international trade. Bankers and insurers must be able to assess a company's environmental performance record to make sound decisions on lending and liability. Governments may insist on ISO 14000 standards to approve of international trade agreements. ISO 14000 requires companies to engage in four major activities:

1. Create an environmental management system (EMS) that formalizes corporate policies and procedures.
2. Audit operations (*ecoaudits*) to ensure conformity; evaluate and document performance.
3. Label "environmental" products.
4. Perform *life-cycle assessments*.

ISO standards also include procedures to train and certify auditors who will establish compliance with ISO 14000 standards. The EPA welcomes ISO 14000 because the standards are compatible with EPA standards and the management systems that assist companies improve compliance with EPA regula-

tions. The EPA looks at ISO 14000 registration as evidence of "due diligence" for compliance with EPA regulations.

## 3.8 Closure

Over the last three decades the population of the Western world has accepted the belief that the discharge of pollutants possesses a risk to the health and well-being of the population. These beliefs have sustained the passage of increasingly stringent pieces of federal legislation designed to improve the quality of the atmosphere. U.S. legislation empowers states to impose standards limiting the emission of pollutants to the atmosphere. In the United States concern for the environment is deep-seated and supported by the public. There is interest in promulgating regulations in which the economic burden of improving air quality is commensurate with environmental risks to be ameliorated.

## Nomenclature

*Abbreviations*

| | |
|---|---|
| ACGIH | American Conference of Governmental and Industrial Hygienists |
| AQCR | air quality control regions |
| BACT | best available control technology |
| BAT | best available technology |
| CAAA | clean air act amendments |
| CAS | Chemical Abstract Service |
| CFB | circulating fluidized boiler |
| CFC | clorofluorocarbon |
| CFR | Code of Federal Regulations |
| CIS | Chemical Information Service |
| COH | coefficient of haze |
| dscf | dry (no water vapor) standard (25°C, 101 kPa) cubic foot |
| dscf | dry standard cubic foot |
| EMS | environmental management systems |
| EPA | Environmental Protection Agency |
| ERC | emission reduction credit |
| $FEV_1$ | forced expiratory volume in 1 second volume |
| FIP | federal implementation plan |
| grains | unit of mass, 7000 grains $= 1\,lb_m$ |

| | |
|---|---|
| HAP | hazardous air pollutant |
| HC | hydrocarbon |
| HCFC | hydrofluorocarbon |
| HEW | Health, Education and Welfare Administration |
| HHV | higher heating value |
| ISO | International Standards Organization |
| LAER | lowest-achievable emissions rate |
| MACT | maximum achievable control technology |
| MBtu | million Btu |
| MSDS | material safety data sheet |
| Mw | megawatt |
| NAAQS | national ambient air quality standards |
| NESHAP | national emission standards for hazardous air pollutants |
| NIH | National Institutes for Health |
| NIOSH | National Institute of Occupational Safety and Health |
| NMH | nonmethane hydrocarbon |
| NSPS | new source performance standards |
| OAQPS | Office of Air Quality Planning Standards |
| OSHA | Occupational Safety and Health Administration |
| PEL | permissible exposure limits |

| | | | | |
|---|---|---|---|---|
| PHS | Public Health Service | | STP | standard temperature (298 K) and pressure (101 kPa) |
| PM | particulate matter | | TE | transfer efficiency (surface coating spraying) |
| $PM_{2.5}$ | particulate matter less than 2.5 $\mu$m | | | |
| $PM_{10}$ | particulate matter less than 10 $\mu$m | | TLV | threshold limit value |
| PMN | premanufacture notice | | TSCA | Toxic Substances Control Act |
| PSD | prevention of significant deterioration | | TTN | technical transfer network |
| RACT | reasonable available control technology | | TTNBBS | technology transfer network bulletin boards |
| RTECS | Registry of the Toxic Effects of Chemical Substances | | TWA | time-weighted average |
| SIP | state implementation plans | | VOCs | volatile organic compounds |
| SNCR | selected noncatalytic reduction | | $X, Y, Z$ | variables defined by Equation 3.1 |
| STEL | short-time exposure limits | | ZEV | zero-emission vehicle |

## References

Bartley, W. W. ed., "Collected works F. A. Hayek", Univ. Chicago Press Chicago, 1989.

Bishop, G. A., Stedman D. H., and Ashbaugh, L., 1996. Motor vehicle emissions variability, *Journal of the Air and Waste Management Association*, Vol. 46, pp. 667–675.

Calvert, J. G., Heywood, J. B., Sawyer, R. F., and Seinfeld J. H., 1993. Achieving acceptable air quality: some reflections on controlling vehicle emissions, *Science*, Vol. 261, July 2, pp. 37–45.

Cicerone, R. J., 1987. Changes in stratospheric ozone, *Science*, Vol. 237, pp. 35–42.

Cooney, C., 1996. Untitled, *Environmental Science and Technology*, Vol. 30, p. 434.

Kelly, T. J., Mukund, R., Spicer, C. W., and Pollack, A. J., 1994. Concentrations and transformations of hazardous air pollutants, *Environmental Science and Technology*, Vol. 28, No. 8, pp. 378A–387A.

Lewis, R. J., Sr., 1986. *In Rapid Guide to Hazardous Chemicals in the Workplace*, N. Irving (Ed.), Van Nostrand Reinhold, New York.

Lewis, R. J. Sr., 1994. *Sax's Dangerous Properties of Industrial Materials*, 8th ed., Van Nostrand Reinhold, New York.

Lippmann, M., 1987. Role of science advisory groups in establishing standards for ambient air pollutants, *Aerosol Science and Technology*, Vol. 6, pp. 93–114.

Lutter, R., and Wolz, C., 1997. UV-B screening by tropospheric ozone: implications for the national ambient air quality standard, *Environmental Science and Technology*, Vol. 31, No. 3, pp. 142A–146A.

Melton, L., 1996. Industrial combustion coordinated rule-making, *Journal of the Air and Waste Management Association*, Vol. 46, pp. 769–777.

Norback, C. T., 1988. *Hazardous Chemicals on File*, J.C. Norback, New York.

Pohanish, R. P., and Greene, S. A., 1993. *Hazardous Substances Resources Guide*, Detroit Gale Research, Detroit, MI.

Schulze, R. H., 1993. The 20-year history of the evolution of air pollution control legislation in the U.S.A., *Atmospheric Environment*, Vol. 27B, No. 1, pp. 15–22.

Schwartz, J., Dockery, D. W., and Neas, L. M., 1996. Is daily mortality associated specifically with fine particles? *Journal of the Air and Waste Management Association*, Vol. 46, pp. 927–939.

Stern, A. C., 1982. History of air pollution legislation in the united states, *Journal of the Air Pollution Control Association*, Vol. 32, No. 1, pp. 44–61.

Vedal, S., 1997. Ambient particles and health: lines that divide, *Journal of the Air and Waste Management Association*, Vol. 47, pp. 551–581.

Wolff, G. T., 1996a. The scientific basis for a new ozone standard, *EM, Air and Waste Management Association Magazine for Environmental Managers*, September, pp. 27–32.

Wolff, G. T., 1996b. The scientific basis for a particulate matter standard", *EM, Air and Waste Management Association Magazine for Environmental Managers*, October, pp. 26–31.

# Problems

**3.1.** The emission rate of particles from a lime kiln is 2.8 $lb_m/h$. The volumetric flow rate of air is 100,000 scfm, the coal (12,500 Btu/$lb_m$) consumption rate is 10 tons/h, and limestone enters the kiln at a rate of 5 tons/h. Show that this particle emission rate is (or is not) in compliance with new source performance standards.

**3.2.** Answer briefly the following questions concerning legislation and regulations.

**(a)** What does it mean to say that a plume is a Ringlemann number 2?

**(b)** What does smog mean?

**(c)** What was the purpose of the Air Pollution Control Act of 1955?

**(d)** What legislation first explicitly addressed air pollution from motor vehicles?

**(e)** What are criteria documents? What legislation caused which agency to create them?

**(f)** What legislation was the first to establish emission standards for automobiles?

**(g)** What are state implementation plans, and which federal legislation asked for their creation?

**(h)** What is an air quality control region?

**(i)** When was the EPA created?

**(j)** What are national ambient air quality standards, and which legislation established them?

**(k)** What are NESHAPs?

**(l)** What is a class II region, and what are the annual $SO_2$ and particulate ambient air quality standards in a class II region?

**(m)** What is the objective and content of a PSD study?

**(n)** What is an emission offset, and what two conditions have to be met?

**(o)** Define and contrast BACT and LAER pollution control systems.

**(p)** What is the difference between national ambient air quality standards and new source performance standards?

**(q)** What are the NAAQS criteria pollutants?

**(r)** What does a standard for $PM_{10}$ mean?

**(s)** Your firm wishes to build a lime kiln in a region that is in compliance with respect to particles but not with respect to oxidants. What types of studies will the firm be expected to conduct, and what type of pollution control equipment will it be expected to employ?

**(t)** Under the CAAA of 1990, what is a "serious" no-attainment level, and when will a region be expected to achieve attainment?

**(u)** What is the difference between LAER and RACT pollution control strategies?

**(v)** What is the difference between a SIP and a FIP?

**(w)** Describe the BACT analysis.

**(x)** What are the differences between, LAER, BACT, and RACT?

**(y)** Define MACT and the sources for which it applies.

**(z)** What is emission trading? Describe how it operates.

**(aa)** What are HAPs as defined by the CAAA of 1990?

**(bb)** What is the difference between *Federal Register* and Code of Federal Regulations (CFR)?

**(cc)** Contrast the duties of an environmental quality board and an environmental hearing board.

**(dd)** What are air pollution episodes, and what actions must a company take if an emergency episode level exists?

**(ee)** What is a variance, and who grants it?

**3.3.** A new electric generating plant is planned that will burn bituminous coal (1% sulfur, 8% ash) and natural gas (0% sulfur and ash) in a cyclone furnace. Thirty percent (30%) of the input energy will be provided by gas and 70% by coal. The plant's electrical output is 800 MW and the overall thermal efficiency is 30%. The energy content of the coal and gas are 12,500 Btu/$lb_m$ and 975 Btu/$ft^3$, respectively. The cost of coal and a gas are $1.50/million Btu and $4.00/million Btu, respectively.

**(a)** Estimate the yearly fuel cost assuming that the boiler is operative 98% of the time throughout the year.

**(b)** Assuming that all the sulfur in coal is converted to $SO_2$, what percent reduction in $SO_2$ will be necessary to satisfy the new source performance standards for fossil-fuel electric utility steam generating units?

**3.4.** Answer each question briefly from information obtained from TTNBBS, and where possible, indicate which section or pages of the bulletin board you used to answer the question.

**(a)** *AMTIC bulletin board*

**(1)** You wish to monitor [CO] in ambient air. Select an instrument and indicate the range of concentrations that can be measured.

**(2)** What is the subject of CFR Part 50, section 50.8?

**(b)** *BLIS bulletin board*. Access BLIS and query the last 5 years of the database. Perform a standard search.

Select a process and company in your state. Discuss the details of emission controls for your process.

**(c)** *CAAA bulletin board.* Your antique 1973 Chevy Vega burns leaded gasoline. When will lead be banned from gasoline in the United States?

**(d)** *CHIEF bulletin board*

    **(a)** You wish to estimate the VOC emissions from a wastewater treatment plant. Is there a program in CHIEF you can use? If so, what is the name of the program?

    **(b)** You wish to estimate the fugitive emissions from an unpaved dusty road. Is there a program in CHIEF you can use? If so, what is the name of the program?

**(e)** *COMPLI bulletin board.* According to the COMPLI woodstove database, what is the efficiency , heat output, and emissions rate (g/h) of a woodstove that interests you.

**(f)** *EMTIC bulletin board.* You need to measure $NO_x$ from a stationary gas turbine. Which EPA emission testing method is designed specifically for this purpose?

**(g)** *ORIA bulletin board.* According to ORIA, how can humidifiers adversely affect indoor air? How can you reduce this risk?

**(h)** *SCRAM bulletin board*

    **(a)** You need to know mixing height data for a regulatory model. From which location(s) in your state can you get this information?

    **(b)** According to the SCRAM news section, what NTIS (National Technical Information Service) model would you use to address the accidental release of a substance over several minutes, such as puffs of dust from cement bags?

**3.5.** Answer the following questions regarding the Clean Air Act Amendments of 1990.

**(a)** A serious nonattainment area has how many years to reduce pollution and achieve ambient air quality standards?

**(b)** Do all sources, old and new, have to secure operating permits?

**(c)** After 1993, what is the maximum CO emissions (g/mile) for all new trucks and automobiles?

**(d)** MACT techniques must be applied if the emission of HAPs exceeds how many tons/yr?

**(e)** $SO_2$ emissions must be reduced to what value (tons/yr) by 2000?

**(f)** Will methyl chloroform be banned, and if so, when?

**(g)** In what title of CFR can air pollution regulations be found?

**3.6.** Answer the following questions with regard to your state's air pollution regulations.

**(a)** A lime calcining process produces 5 tons of lime/h, 8760 h/yr. What is the maximum allowable particle emission rate ($lb_m$/h).

**(b)** Is it true that neighbors can ask your state air pollution agency to prevent a farmer from spreading foul-smelling pig manure on his fields?

**(c)** A prime coating is applied to automobiles and trucks. The primer contains 5 $lb_m$/gal of VOC and the solvent used in the process contains 7 $lb_m$/VOC gal. What is the minimum removal efficiency of a VOC removal system?

**3.7.** Consider federal New Source Performance Standards.

**(a)** Hot asphalt is produced at a rate 5 tons/h. What is the maximum allowable particle concentration (mg/dry m$^3$)?

**(b)** An electric arc furnace produces steel at a rate of 3 tons/h. What is the maximum allowable particle concentration (mg/dry m$^3$)?

**(c)** What is the maximum allowable particle emission rate ($lb_m$/h) from a process that produces 1.6 tons/h of ammonium sulfate?

**(d)** What is the maximum allowable emission rate ($lb_m$/h) of $SO_2$ from a fossil-fuel electric utility furnace that burns 300 tons/h of coal that contains 2.5% sulfur and has a heating value of 12500 Btu/$lb_m$?

**3.8.** A coal-fired electric utility steam generating unit burns bituminous coal (12,500 Btu/$lb_m$) at a rate of 400 tons/h. At the stack exit, the mass flow rate of exhaust gas (assume air) is 2000 kg/s, and the temperature and pressure are 380 K and 95 kPa. The generating unit generates $NO_x$ (assume $NO_2$) at a rate given by emission factors for utility boilers, 17 kg $NO_x$ per Mg of coal (Mg $\doteq 10^6$ g).

**(a)** To satisfy the 1980 New Source Performance Standards for $NO_2$, the EPA requires the installation of a suitable $NO_2$ control device. What is the minimum efficiency of this device?

**(b)** If the actual $NO_2$ emission rate is 590 g/s, what is the concentration of $NO_2$ (in ppm) in the exhaust gas?

**3.9.** [*Design Problem*] Call the office of your state's regional air pollution agency and find out their Web site address. Examine the regulations for your state's air pollution control agency. Answer briefly each question below and *indicate the section(s) or page(s)* of the regulations on which your answer is based.

**(a)** Define air pollution.

**(b)** What is a contamination source?

**(c)** Does the agency

    **(1)** Enter a premise only if it has a search warrant?

**(2)** Have the right to inspect and sample a contamination source?

**(3)** Require that records be kept and have the right to inspect these records?

**(4)** Set standards for automobile emissions?

**(d)** What is a permit, and when is it required?

**(e)** Can municipalities enact air pollution ordinances that are more stringent than the state agency?

**(f)** What is a variance? What body grants it? Are all variances temporary, or are some permanent?

**(g)** What is a fugitive air contaminant? What is particulate matter?

**(h)** Define a volatile organic compound (VOC)?

**(i)** Are odors from spreading manure on farm fields not belonging to the person who raised the animals subject to agency's regulations?

**(j)** Beyond what potential emission rate is it necessary to install a vapor collection system on an existing surface coating process that uses VOCs?

**(k)** How does your state define air pollution episodes, and what actions must industries take during these episodes?

**(l)** How does your state decide when RACT or LAER standards have to be used?

# 4

# Effects of Pollution on the Respiratory System

In this chapter you will learn:

- Names and functions of components of the respiratory system and how pollutants affect these functions
- How to model the motion of air in the respiratory system
- How to model gas exchange in the respiratory system
- Primary occupational lung diseases
- Dose and risk relationships

The large surface area of the lung, its airways, and the thin membrane separating air space and capillaries make the lung the primary organ, not only for gas exchange but also for the absorption of toxins. Table 4-1 lists common industrial toxicants and the lung diseases they produce. Engineers interested in air pollution and industrial hygiene should learn the fundamental concepts of toxicology and physiology as they pertain to the respiratory system. It is recommended that engineers become familiar with Cralley

and Cralley (1979), Guyton (1986), and Klaasen et al. (1986). Several chapters (Ultman, 1985, 1988; Slutsky et al., 1985; Paiva, 1985; Engle, 1985) in research monographs and three physiology texts used by first-year medical students (West, 1974; Slonim and Hamilton, 1987; Levitzky, 1986) are particularly suited for users of this book since they describe the motion of gas in the lung in terms congenial to engineers. The purpose of this chapter is to describe the respiratory system and how air contaminants affect its function.

## 4.1 Physiology

The lungs expand and contract by the downward and upward movement of the diaphragm and by the expansion and contraction of the rib cage. Normal quiet breathing is accomplished with the diaphragm. The expansion and contraction of the lungs causes the gas pressure within the alveoli to become negative or positive with respect to atmospheric pressure. The pressure difference for normal breathing is approximately

**Table 4-1** Selected Industrial Toxicants Producing Lung Disease through Inhalation

| Toxicant | Chemical composition | Occupational source | Common name of disease | Site of action | Acute effect | Chronic effect |
|---|---|---|---|---|---|---|
| Aluminium | $Al_2O_3$ | Manufacture of abrasives, smelting | Aluminosis | Upper airways, alveolar interstitium | Cough, short-ness of breath | Interstitial fibrosis |
| Ammonia | $NH_3$ | Ammonia production, manufacture of fertilizers, chemical production, explosives | | Upper airway | Immediate upper and lower respiratory tract irritation, edema | Chronic bronchitis |
| Arsenic, | $As_2O_3$, $AsH_3$ (arsine), $Pb_3(AsO_4)_2$ | Manufacture of pesticides, pigments, glass, alloys | | Upper airways | Bronchitis | Lung cancer, bronchitis, laryngitis |
| Asbestos | Fibrous silicates (Mg, Ca, and others) | Mining construction, shipbuilding, manufac-ture of asbestos containing materials | Asbestosis | Parenchyma | | Pulmonary fibrosis, pleural calcification lung cancer, pleural mesothelioma |
| Berylium | Be, $Be_2Al_2(SiO_3)_4$ | Ore extraction, manufacture of alloys, ceramics | Berylliosis | Alveoli | Severe pulmonary edema, pneumonia | Pulmonary fibrosis, progressive dyspnea, interstitial granulomatosis, corpulmonale |
| Chlorine | $Cl_2$ | Manufacture of pulp and paper, plastics, chlorinated chemicals | | Upper airways | Cough, hemoptysis, dyspnea, tracheo-bronchitis, broncho-pneumonia | |
| Chromium (IV) | $Na_2CrO_4$ and other chromate salts | Production of Cr compounds, paint pigments, reduction of chromite ore | | Nasopharynx, upper airways | Nasal irritation bronchitis | Lung tumors and cancers |
| Coal dust | Coal plus $SiO_2$ and other minerals | Coal mining | Pneumo-coniosis | Lung parenchyme, lymph nodes, hilus | | Pulmonary fibrosis |

(continued)

**Table 4-1**  (continued)

| Toxicant | Chemical composition | Occupational source | Common name of disease | Site of action | Acute effect | Chronic effect |
|---|---|---|---|---|---|---|
| Coke oven emissions | Polycyclic hydrocarbons, $SO_x$, $NO_x$, and particulate mixtures of heavy metals | Coke production | | Upper airways | | Tracheobronchial cancers |
| Hydrogen flouride | HF | Manufacture of chemicals, photographic film, solvents, plastics | | Upper airways | Respiratory irritation, hemorrhagic pulmonary edema | |
| Iron oxides | $Fe_2O_3$ | Welding, foundry work, steel manufacture, hematite mining, jewelry making | Siderotic lung disease: silver finisher's lung, hematite miner's lung, arc welder's lung | Silver finisher's: pulmonary vessels and alveolar walls; hematite miner's: upper lobes, bronchi and alveoli; art welder's; bronchi | Cough | Silver finisher's: subpleural and perivascular aggregations of macrophages; hematite miner's diffuse fibrosis-like pneumoconiosis; arc welder's; bronchitis |
| Kaolin | $Al_2O_3 \cdot 2SiO_2 \cdot 2H_2O$ plus crystalline $SiO_2$ | Pottery making | Kaolinosis | Lung parenchyme, lymph nodes, hilus | | Pulmonary fibrosis |
| Nickel | NiCO (nickel carbonyl), Ni, $Ni_2S_3$ (nickel subsulfide), NiO | Nickel ore extraction, nickel smelting, electronic electroplating, fossil fuel | | Parenchyma (NiCO), nasal mucosa ($Ni_2S_3$), bronchi (NiO) | Pulmonary edema delayed by two days (NiCO) | Squamous cell carcinoma of nasal cavity and lung |
| Oxides of nitrogen | NO, $NO_2$, $HNO_3$ | Welding, silo filling, explosive manufacture | | Terminal respiratory bronchi and alveoli | Pulmonary congestion and edema | Emphysema |

(continued)

**Table 4-1** *(continued)*

| Toxicant | Chemical composition | Occupational source | Common name of disease | Site of action | Acute effect | Chronic effect |
|---|---|---|---|---|---|---|
| Ozone | $O_3$ | Welding, bleaching flour, deodorizing | | Terminal respiratory bronchi and alveoli | Pulmonary edema | Emphysema |
| Perchloroethylene | $C_2Cl_4$ | Dry cleaning, metal degreasing, grain fumigating | | | Pulmonary edema | |
| Phosgene | $COCl_3$ | Production of plastics, pesticides, chemicals | | Alveoli | Edema, pulmonary edema | Bronchitis |
| Silica | $SiO_2$ | Mining, stone cutting construction, farming, quarrying | Silicosis, pneumoconiosis | Lung parenchyma, lymph nodes, hilus | | Pulmonary fibrosis |
| Sulfur dioxide | $SO_2$ | Manufacture of chemicals refrigeration, bleaching, fumigation | | Upper airways | Bronchoconstriction cough, tightness in chest | |
| Talc | $Mg_3Si_4O_{10}(OH)_2$ | Rubber industry, cosmetics | Talcosis | Lung parenchyma, lymph nodes | | Pulmonary fibrosis |
| Tin | $SnO_2$ | Mining, processing of tin | Stanosis | Bronchioles and pleura | | Widespread mottling of x-ray without clinical signs |
| Toluene 2.4-diisocyanate (TDI) | $CH_3C_6H_3(NCO)_2$ | Manufacture of plastics (polyurethane) | | Upper airways | Acute bronchitis, bronchospasm, pulmonary edema | |
| Xylene | $C_6H_4(CH_3)_2$ | Manufacture of resins, paints, varnishes, other chemicals, general solvent for adhesives | | Lower airways | Pulmonary edema | |

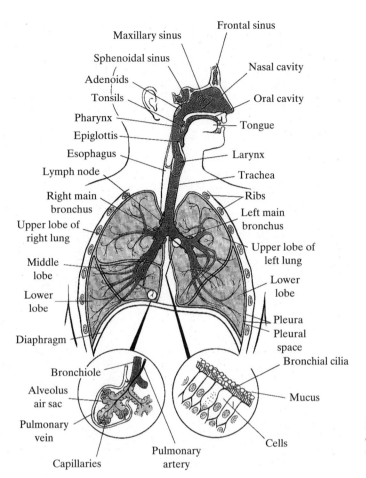

**Figure 4-1** Components of the respiratory system (courtesy of the American Lung Association)

±1 mm of mercury. Figure 4-1 shows the three regions of the respiratory system: (1) nasopharyngeal, (2) tracheobronchial, and (3) pulmonary.

## 4.1.1 Nasopharyngeal region

The nasopharyngeal region lies between the nostrils and larynx; it is also called the *upper or extrathoracic airways*. There are large convoluted surfaces in the nose called *nasal turbinates* containing blood vessels. The nasal turbinates are covered by a mucous membrane that transfers energy and water to fresh air during inhalation and from exiting gas during exhalation. Nasal turbinates are the primary sites for the absorption of water-soluble contaminants in air. Under quiet breathing most of the air passes through these nasal airways. Under vigorous exercise the mouth becomes the primary airway for inhalation and exhalation. The *switching point* is on average at

a respiratory rate (also called minute volume) of 34.5 L/min. The upper airway volume is approximately 50 mL and the flow path between lips and glottis is approximately 17 cm long and 22 cm between the nose to glottis. The nasal passages are lined with cellular tissue and mucous glands (vascular mucous epithelium). Downstream of the *pharynx* is the *larynx*, which constricts the flow of air, producing a *vena contracta*, enabling humans to utter sounds. The cross-sectional area of the adult trachea is approximately 2.54 cm², while the cross-sectional area within the larynx varies with the volumetric flow rate;

$$15 \text{ L/min} = 0.88 \text{ cm}^2$$

$$30 \text{ L/min} = 1.40 \text{ cm}^2$$

$$60 \text{ L/min} = 2.42 \text{ cm}^2$$

Instabilities and vortices are produced in the airstream passing through the apertures of the larynx. Particle deposition in the larynx is large and a cause of malignant tumors (Martonen et al., 1993). During inspiration, air passing through the folds and vocal cords forms a high-velocity airstream (*laryngeal jet*) that affects the velocity of air and particle size distribution entering the tracheobronchial tree.

## 4.1.2 Tracheobronchial region

The tracheobronchial region (Figure 4-2) consists of the *trachea, primary and secondary bronchi*, and *terminal bronchiole*. The diameter and length of the trachea are 1.8 cm and 12.0 cm, respectively. The diameter and length of the right and left main bronchi are 1.22 cm and 4.76 cm, respectively. The angle separating the right and left bronchi is 70°. The

tracheobronchial region is also called the *conducting airway*. All the passageways, from the nose to the terminal bronchiole, are lined with ciliated epithelium and coated with a thin layer of *mucus* from mucus-secreting cells. This surface is called the *mucosal layer* or *mucous membrane*. Mucus production is about 10 mL per 24 hours In humans, the thickness of the mucous layer is from 5 to 10 μm. *Cilia* are hair-like organs protruding 3 to 4 μm above the surface of the cell (Figures 4-3 and 4-4). As many as 200 cilia per cell protrude from special epithelial cells lining air passages. *Mucociliary escalation* is the whiplike movement of cilia that occurs in the nasal cavity and respiratory tract. Once in the pharynx, the mucus passes through the digestive system. Cilia move forward with a sudden rapid stroke 10 to 20 times a second, bending sharply so that they remain within the mucous layer. They move backward slowly. The

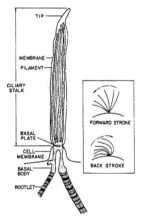

**Figure 4-3** Structure and function of the cilium (adapted from Guyton 1986)

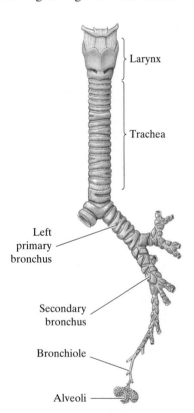

**Figure 4-2** Schematic diagram of branching airways. The trachea divides into two primary bronchi, one to each lung. Each bronchus divides 22 more times, and finally terminates in a cluster of alveoli. (Diagram taken from Silverthorn, 1998)

**Figure 4-4** Scanning electron micrograph of a portion of a bronchial passage showing cilia interspersed with mucus-secreting cells whose surfaces are covered with microcilli. (Photograph taken from Silverthorn, 1998)

whiplike movement propels mucus toward the pharynx at a rate of approximately 1 cm/min, but the slow return motion does not displace the mucus. The mechanism of cilia and mucus that transfers material from the lung to the digestive tract is called *lung clearance.*

The trachea is called the *first generation respiratory passage.* The trachea divides into the *right and left bronchi* at a point called the *carnia.* The right and left main bronchi are called the *second-generation respiratory passages.* Each division of bronchi thereafter is an additional generation. There are between 20 and 25 generations before air finally reaches alveoli. The final few generations are less than 1 to 1.5 mm in diameter and are called *bronchiole.* All the intermediate passageways between the trachea and bronchiole are called *bronchi.* To keep the trachea from collapsing, multiple rings of cartilage surround five-sixths of the circumference. In the bronchi, smaller amounts of cartilage provide partial rigidity. The bronchiole have no rigidity and expand and contract along with alveoli. The walls of the trachea and bronchi are composed of cartilage and smooth muscle. Bronchiole walls are entirely smooth muscle. The upper and conducting airways are relatively rigid and have a volume of 150 to 200 mL. The upper conducting airways are called *anatomic dead space* because very little gas exchange occurs.

Under *quiet breathing* (12 to 15 breaths/min) the pressure drop between alveoli and trachea is approximately 1 mm of mercury. The bronchi and trachea are very sensitive to touch, light, particles, and certain gases or vapors. The larynx and carnia are particularly sensitive. Nerve cells in these passageways initiate a series of involuntary actions that produce coughing. *Coughing* begins with inspiring about 2.5 L of air followed by the spontaneous closing of the *epiglottis,* tightening the vocal cords and traping air in the lungs. The abdominal muscles then contract, forcing the diaphragm to contract the lung, collapse the bronchi, constrict the trachea, and increase the air pressure in the lungs to values as high as 100 mm Hg. The epiglottis and vocal cords suddenly open and air under pressure is expired. The rapidly moving air carries foreign matter upward from the bronchi and trachea. *Sneezing* is another involuntary reflex similar to the cough, except that the *uvula* (portion of the soft palate above the back of the tongue) is depressed and large amounts of air pass rapidly through the nose and mouth.

### 4.1.3 Pulmonary region

The pulmonary region is also called the *respiratory airspace.* The pulmonary region consists of tiny sacs called *alveoli* clustered in groups and interconnected

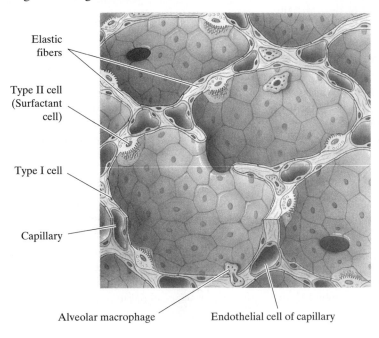

Elastic fibers

Type II cell (Surfactant cell)

Type I cell

Capillary

Alveolar macrophage

Endothelial cell of capillary

**Figure 4-5** Alveoli in the pulmonary system. The alveoli are composed of type I cells for gas exchange and type II cells that synthesize surfactant. Macrophage lying on the alveolar membrane ingest foreign material that reach the alveoli. (Diagram taken from Silverthorn, 1998)

by openings called *alveolar ducts*. Alveoli (Figure 4-5) are thin-walled polyhedral pouches of characteristic width of 250 to 350 μm and which have at least one side open to either a *respiratory bronchiole* or an *alveolar duct*. Each terminal bronchiole supplies air for a segment of the lung called an *acinus*. The function of the alveoli is to provide a surface for the exchange of oxygen, carbon dioxide, and volatile metabolites between air and blood in the capillaries. The air side of alveoli consists of squamous epithelial cells and rounded septal cells. Mobile phagocytic cells and macrophage lie on the inner surface of the alveolus. *Macrophage* are white blood cells 7 to 10 μm in diameter that metabolize inhaled particulate material, spores, bacteria, and so on. The process is called *phagocytosis*. The metabolizing cells are also called *phagocytes*. Infectious particles, except for some chronic bacterial and fungal infections such as tuberculosis and some viral diseases, are usually killed by macrophage. The alveoli are served by a labyrinth of venous and arterial capillaries approximately 8 μm in diameter. Approximately 90 to 95% of the alveolar surface is served by capillaries. There are 300 to 500 million alveoli in the adult human. The mean thickness of the tissue between alveoli is about 9 μm.

The branching airway is an ordered structure in which each airway divides into two airways of similar, if not identical geometry. The *Weibel symmetric model* (Weibel, 1963) is widely used to describe the structure (Figure 4-6). The tracheobronchial region consists of generations 0 to 16. Generations 0 to 3 contain cartilage and are called *bronchi* (BR). Generations 4 to 16 contain no cartilage and are called *bronchiole* (BL). Generation 16 is called the *terminal bronchiole* (TBL). The *pulmonary region* consists of generations 17 to 23, which are subdivided into three generations of *partially alveolated respiratory bronchiole* (RBR), three generations of *fully alveolated alveolar ducts* (AD), and end in the dead-ended or terminal *alveolar sacs* (AS).

Figure 4-7 shows airway cross-sectional area ($A$), surface area ($S$), Reynolds number (Re) versus distance ($y$) from the nasal tip, and the Weibel airway generations. Once inside the lung, the airflow decelerates and the flow path decreases rapidly. While the Reynolds numbers are low, one should not assume that the flow is fully developed since bronchial bifurcations produce regions of recirculation and unsteadi-

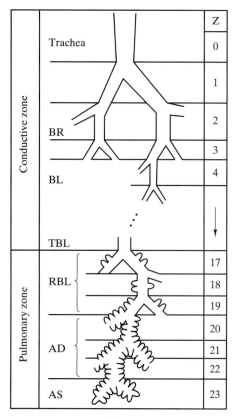

**Figure 4-6**  Symmetric Weibel model. Z is the airway generation, (BR) bronchi, (BL) bronchioles, (TBL) terminal bronchiole, (RBL) partially alveoliated respiratory bronchioles, (AD) fully alveoliated ducts, (AS) terminal alveolar sacs (redrawn from Weibel 1963, Ultman 1985)

ness. Fully developed flow does not occur until $L/L_e > 1$, which only occurs in the bronchioles (i.e., airway generations beyond 6).

The volume of the lung changes as air is inhaled and exhaled and the honeycomb–alveoli structure expands and contracts. The total *alveolar area* in the adult human is about 35 m$^2$ during expiration and 100 m$^2$ during deep inspiration. By comparison, the area of a singles tennis court is 195.7 m$^2$. Great debate exists about the existence and role of a thin (0.5 to 1 μm) liquid layer on the air side of alveoli. The aqueous layer is called a *surfactant* and is a complex mixture of proteins predominately dipalmitoyl lecithin. The surfactant forms a liquid monolayer on the alveolus–air interface capable of lowering the normal surface tension (72 nN/m) to virtually zero (Longo et al.,

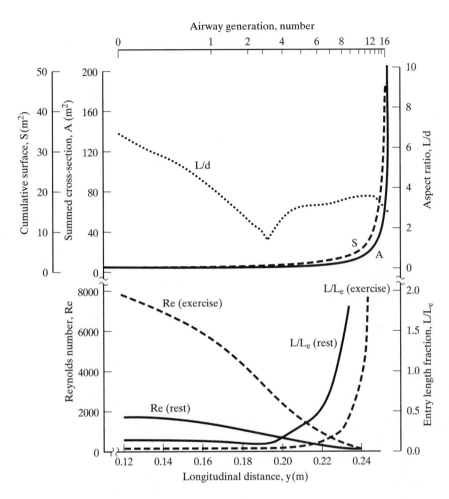

**Figure 4-7**  Geometric and aerodynamic characteristics of a symmetric Weibel model for two ventilation flow rates, rest (0.4 L/s) and exercise (1.6 L/s). $L_e$ is the length needed for the flow to become fully developed (redrawn from Ultman 1988)

1993). To accomplish this reduction, surfactant must adhere to the interface and be capable of maintaining a coherent, tightly packed monolayer that will not collapse even under high compression accompanying expiration. The collapse of the monolayer results in its expulsion from alveoli and an increase in surface tension. Deficiency and inactivation of the surfactant are contributing factors to a number of pulmonary diseases.

*Arteries* carry blood containing nutrients to tissue and *veins* carry blood containing waste products away from tissue. Each nutrient artery entering an organ divides six to eight times, after which the remaining vessels are called *arterioles.* Arterioles are approximately 40 μm in diameter. Arterioles subdivide two to five times and become *capillaries* with diameters of 8 or 9 μm. The body contains nearly

10 billion capillaries with a total surface area of 500 to 700 m². Nearly all cells are within 20 to 30 μm of a capillary. Blood flows through capillaries in an intermittent fashion. On the return leg, capillaries recombine to form *venules*, which in turn recombine to form *veins. Perfusion* is the phrase used to describe the flow of blood through the blood vessels in the lung. There is a large pressure drop associated with blood flow in arteries and arterioles. Thus the walls of these vessels are considerably stronger and thicker than veins and venules. Capillary walls are composed of unicellular endothelial cells surrounded by basement membrane. The thickness of the wall is about 0.5 μm. Material on the outside of capillaries is called *interstitial fluid.*

Materials are transferred through the capillary membrane by a variety of processes broadly called *diffusion*. Cells in the capillary wall are composed of materials in which different nutrients are soluble. If the nutrient, such as oxygen and carbon dioxide, is soluble in *lipids* (materials within cells that are soluble in fat solvents but not water), it will diffuse through the cells in the capillary membrane containing lipids. Water-soluble but lipid-insoluble materials such as sodium and chloride ions, glucose, and so on, pass through the membrane through slit pores (6 to 7 nm wide). The phrase *permeability* is used to describe the diffusing capacity of different materials to pass through the capillary membrane. The concentration difference is the driving potential.

Approximately one-sixth of body tissue is the space between cells and is called *interstitium*. Fluid in this space is generally called *interstitial fluid*. Figure 4-8 shows the elements of the interstitium. Solid material is composed of two major elements, collagen fiber bundles and proteoglycan filaments. *Collagen* fiber bundles have large tensile strength and provide tensile strength to the tissue. *Proteoglycan* filaments are small coiled protein molecules that form a "mat." Interstitial fluid is a plasma derived from capillaries. The combination of proteoglycan filaments and interstitial fluid is called *tissue gel*. After passing through the capillary wall, materials diffuse through the tissue gel, molecule by molecule, to cells receiving the material. Although virtually all fluid is bound in some way, small rivulets of "free fluid" are present. *Edema* is the name of a physical disorder that occurs when the amount of free fluid accumulates in pockets and the rivulets expand enormously.

Not all material can be transferred to the blood by diffusion (e.g., insoluble material, indigestible bacteria, and dust particles). The *lymphatic system* is an additional route by which these materials are transferred from the interstitial space to the blood. Many large inhaled particles find their way into the lymphatic system, where they pose a serious hazard to health. Small lymphatic vessels called capillaries carry proteins and large particles of foreign matter from interstitial spaces. About one-tenth of the fluids from the arteries passes through the lymphatic system rather than return through venous capillaries. Substances with large molecular weights that cannot diffuse may enter lymphatic capillaries. Figure 4-9 illustrates how the unique structure of lymphatic capillaries enables them to accept very large molecules and foreign particles. Anchoring filaments of the lymphatic capillary are attached to the connective tissue in the interstitial space between cells. Adjacent ends of endothelial cells (of the lymphatic capillary) overlap and produce a "flap." Large particles in the interstitial fluid are able to push these flaps open and flow directly into the lymphatic capillary.

Lymphatic fluid is derived from interstitial fluid but is richer in proteins. Lymphatic capillaries branch and merge, as do veins and arteries. Lymphatic channels merge at *lymph nodes*, where addi-

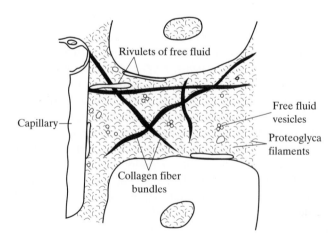

**Figure 4-8**   Structure of the interstitium (redrawn from Guyton 1986)

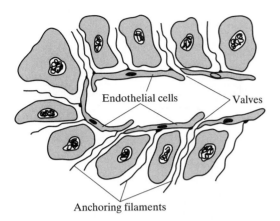

**Figure 4-9** Lymphatic capillaries showing structures that enable material of large molecular weight to enter circulation (redrawn from Guyton 1986)

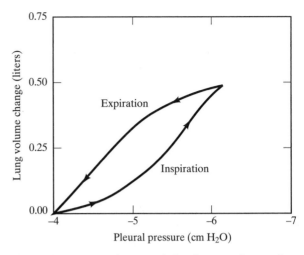

**Figure 4-10** Compliance of the lung (redrawn from Guyton 1986)

tional chemical processes occur. Approximately 100 mL of lymph flows through the thoracic duct of a resting person. An estimated additional 20 mL of lymph flows through other channels. Lymph is believed to be pumped by movement of muscles and by the numerous flaps that pass material to larger collecting lymphatic vessels.

## 4.2 *Respiratory fluid mechanics*

The flexible properties of the lung are due to elastic fibers within the lung tissue and the surface tension of the surfactant liquid lining the alveolar surfaces. The change in the volume of the lung with respect to pressure ($dV/dP$) is called *compliance*. Compliance is comparable to the elastic coefficient or Young's modulus in solid mechanics or isothermal expansion coefficient in thermodynamics. Figure 4-10 is a graph of the volume of air input and removed from the lung (within the thorax) as a function of the pressure in the space between the lung and rib cage (a space called the *pleura*). Two things are evident: (1) the slope is not constant and (2) the lung displays *hysteresis* in that expiration and inspiration do not produce coincident curves. Both effects are believed to be due to the viscoelastic properties of human tissue, the honeycomb structure of alveoli, and the surface tension of the sur-

factant. A linear relationship connecting the endpoints in Figure 4-10 is approximately 0.13 L/cm water pressure drop. If the lungs were removed from the thorax, the slope would be 0.22 L/cm of water. The difference is due to the elastic properties of the thoracic cage.

### 4.2.1 Spirometry

Air is drawn into and out of the lungs by positive or negative pressure in the lung cavity. *Spirometry* is the analysis of the volume and volumetric flow rate of air during respiration. Figure 4-11 shows an elementary way to measure the volume of respired air corrected to STP. The *tidal volume* $(V_t)$ is the volume of air (at STP) inhaled (and exhaled) during *normal breathing*. The *vital capacity* $(V_{vc})$ is the maximum volume of air one can expel from the lungs after first filling the lungs to their maximum extent and then expiring to the maximum extent. A typical value of the vital capacity is 4800 mL. The *residual volume* $(V_r)$ is the volume of air remaining in the lungs following the maximum expiration that one can produce. Typical residual volume is 1200 mL. The *anatomic dead space* $(V_d)$ is the volume of the conducting airways from the mouth to the respiratory bronchiole within which there is little gas exchange. The volume of the anatomic dead space increases with lung expansion, varying from 144 mL in a collapsed lung to 260 mL

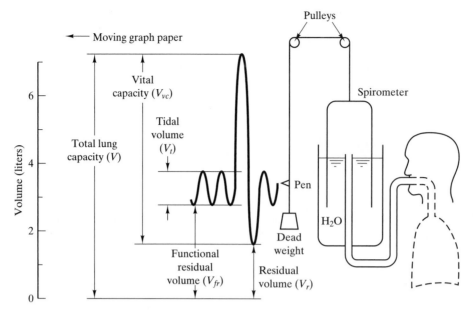

**Figure 4-11** Lung volumes and elements of spirometry. A pen records changes in the air volume on graph paper that moves to the left. The residual volume and functional residual volume cannot be measured with the spirometer (redrawn from Klaassen et al., 1986, and adapted from Silverthorn, 1998)

after maximum inspiration. Approximately 50% of the anatomic dead space volume is located in the upper airways. The *functional residual capacity* $(V_{fr})$ is the difference between the minimum of the tidal volume (about 2300 mL) and zero. The *total lung capacity (V)* is the maximum volume to which the lungs can be expanded, with the greatest possible inspiratory effort (about 5800 mL). All pulmonary volumes and capacities are about 20 to 25% less in women than in men. They tend also to be greater for athletic persons.

The *minute respiratory rate* (or more simply, the *ventilation rate*) $(Q_t)$, also called the *minute volume*, is equal to the tidal volume times the respiratory rate. The *normal tidal volume* is 500 mL and the normal respiratory rate is 12 to 15 breaths/min. Using 12 breaths/min, the *normal minute respiratory rate* is approximately 6 L/min. A person can live for short periods of time at two to four breaths per minute (1 to 2 L/min). Under unusual conditions the tidal volume can be as large as the vital capacity, or the respiratory rate can rise to as high as 40 to 50 breaths/min. At these rapid respiratory rates, a per-

son usually cannot sustain a tidal volume greater than one-half the vital capacity. The *alveolar ventilation rate* $(Q_a)$ (Figure 4-12) is the volumetric flow rate of fresh air that reaches the alveoli that is available for

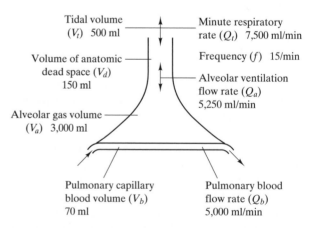

**Figure 4-12** Typical lung volumes and volumetric flow rates. There may be considerable individual differences in these values (redrawn from Klaassen et al., 1986)

gas exchange. It is equal to the breathing rate times the difference between the tidal volume and the (anatomic) dead volume; that is,

$$\text{alveolar ventilation} = (500\text{ mL} - 150\text{ mL})$$
$$(12\text{ breaths/min})$$
$$= 4200\text{ mL/min}$$

The rate at which a person inhales and exhales depends on the level of physical activity. A great deal of information about lung disease can be gained by measuring the rates of inhalation and exhalation. Of particular importance is the volume of gas exhaled in the first second called *forced expiratory volume at 1 second* $(\text{FEV}_1)$. This parameter is particularly reproducible and a sensitive indication of obstructions in lung airways (Figure 4-13).

Like any other fluid system, the volumetric flow rate depends on the pressure difference $(\delta P)$ between lung and atmosphere and the frictional forces within the air passageways. The frictional forces can be lumped together and called the *airway resistance (R)* and the volumetric flow rate $(Q)$ expressed as

$$Q = \frac{\delta P}{R} \qquad (4.1)$$

The airway resistance may vary throughout the respiratory cycle and the value for inspiration may be different than that for expiration (Figure 4-10).

If the instantaneous volume of air versus time is measured by the equipment shown in Figure 4-11, graphs similar to Figure 4-13 can be obtained. The normal lung has a vital capacity of approximately 4.6 L and a residual volume of 1.2 L. Note that a maximum

volumetric flow rate of about 400 L/min is achieved rather quickly after one begins to exhale. A test to classify respiratory abnormalities is to expel rapidly as much air as one can from maximally filled lungs. The volume so measured is called the *forced vital capacity* (FVC). Figure 4-13 shows the trace of normal lungs, obstructed lungs, and constricted lungs. People with obstructed lungs will not be able to inspire quite as much as unobstructed lungs, but most important, it will take a longer time to expel the air they have inhaled.

*Constricted lungs* (sometimes called *restricted*) cannot be fully expanded because they contain lesions or because abnormalities in the chest cage prevent people from inflating their lungs fully. The constricted lung is similar to the normal lung except that the vital capacity is smaller. Constricted lungs occur in people suffering from fibrosis of the lungs (black lung disease, silicosis, asbestosis), thoracic deformities, large tumors, or other problems that restrict the amount of air that can be put into the lungs.

People with *obstructed lungs* have a larger airway resistance and have to exert a larger pressure to expel air because the airways are narrower and cannot be emptied as easily as normal lungs. Thus the residual volume is also larger than the normal lung. Obstructed lungs occur in people suffering from chronic asthma, asthmatic bronchitis, emphysema, mucosal edema, and inflammation of the conducting air passages.

A common way to quantify respiratory disorders is to examine three parameters: the vital capacity (VC), *forced expiratory volume at 1 second* $(\text{FEV}_1)$, and the residual volume (VR):

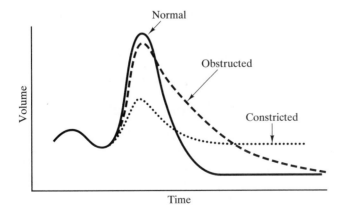

**Figure 4-13**  Spirometric measurements of normal, obstructed and constricted lung

*Normal lung*: $0.7 < \left(\dfrac{FEV_1}{VC}\right)_{normal} < 0.75$

*Constricted lung*: $FEV_{1,\,constricted} < FEV_{1,\,normal}$

*Obstructed lung*: $FEV_{1,\,obstructed} < FEV_{1,\,normal}$

$$\left(\frac{FEV_1}{VC}\right)_{constricted} > 0.75$$

$$\left(\frac{FEV_1}{VC}\right)_{obstructed} < 0.70$$

$$VC_{constricted} < VC_{normal}$$

$$VC_{obstructed} \approx VC_{normal}$$

$$VR_{constricted} > VR_{normal}$$

$$VR_{obstructed} \approx VR_{normal}$$

An additional technique to quantify respiratory disorders involves the concept of *moments* (Menkes et al., 1981; Permutt and Menkes, 1979) since it provides insight into the pathophysiology of disease and allows one to have quantitative means to discriminate between curves in Figure 4-13. The *arithmetic mean transient time* $(t_m)$ is the arithmetic average time it takes to expire an amount of air equal to the vital capacity. Assuming that $\rho$ is constant, the conservation of mass for air during the process of expiration is

$$Q = -\frac{dV}{dt} \qquad (4.2)$$

where $Q$ is the expiration volumetric flow rate, and $V$ is the volume. The negative sign is needed because the lung volume decreases during expiration. Multiply both sides by time $t$ and integrate over the time, $t_{vc}$, the time it takes to expire a volume of air equal to the vital capacity $(V_{vc})$:

$$\int_0^{t_{vc}} tQ\,dt = -\int_{V_{vc}}^0 t\,dV \qquad (4.3)$$

Define the *arithmetic mean transient time*:

$$t_m = \int_0^{V_{vc}} \frac{t\,dV}{V_{vc}} = \frac{1}{V_{vc}}\int_0^{V_{vc}} t\,dV \qquad (4.4)$$

Rearrange the above:

$$t_m = \int_0^{t_{vc}} \frac{Q}{V_{vc}} t\,dt = \frac{1}{V_{vc}}\int_0^{t_{vc}} Qt\,dt \qquad (4.5)$$

Define the $n$th moment as

$$\alpha_n = \int_0^{t_{vc}} \frac{Q}{V_{vc}} t^n\,dt \qquad (4.6)$$

The first moment $(n = 1$ or $\alpha_1)$ is the arithmetic mean transient time $(t_m)$ described in Equation 4.5:

$$t_m = \alpha_1 = \int_0^{t_{vc}} \frac{Q}{V_{vc}} t\,dt = \frac{1}{V_{vc}}\int_0^{t_{vc}} tQ\,dt \quad (4.7)$$

The second moment $(n = 2$ or $\alpha_2)$ is defined as

$$\alpha_2 = \int_0^{t_{vc}} \frac{Q}{V_{vc}} t^2\,dt = \frac{1}{V_{vc}}\int_0^{t_{vc}} t^2 Q\,dt \quad (4.8)$$

The square root of $\alpha_2$ is the root mean square of the transient time. Because of the nature of the tails in Figure 4-13, it can be seen that higher moments reflect the importance of the tails and smaller expired volumes of air. The arithmetic mean transient time $(t_m)$ provides an index of the average time it takes for air to leave a normal lung. It can be shown that

$$t_m = \alpha_1 = \sqrt{\frac{\alpha_2}{2}} \qquad (4.9)$$

For obstructed and constricted lungs, the relationship between $\alpha_1$ and the square root of $\alpha_2$ changes. Where there is an obstruction in the upper airway, Figure 4-13 shows that the lung volume does not decrease rapidly and the square root of $\alpha_2$ does not fall relative to $\alpha_1$. In cases of airway obstruction because of emphysema and age, $\sqrt{\alpha_2}$ rises relative to $\alpha_1$. Permutt and Menkes (1979) report that analysis of 57 nonsmoking males shows that $\alpha_1$, and $\alpha_2$, and age (in years) are related by

$$\sqrt{\alpha_2} = 0.229 + 1.165\alpha_1 + 0.00347(\text{age}) \qquad (4.10)$$

with a standard deviation of 0.078. The companion equation for a group of smokers is

$$\sqrt{\alpha_2} = 0.371 + 0.676\alpha_1 + 0.00981(\text{age}) \qquad (4.11)$$

with a standard deviation of 0.206.

## 4.2.2 Bohr model

The Bohr model (Ultman, 1985), shown in Figure 4-14, is a simple but useful analytical model that has been used for many years to illustrate the distribution of inspired air between the rigid conducting airways and the expandable alveolar region. The model con-

Ventilation volumetric flow rate
$$Q_t = fV_t$$

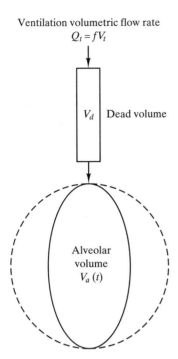

$V_d$ | Dead volume

Alveolar
volume
$V_a(t)$

**Figure 4-14** Bohr model

sists of a rigid volume called the *anatomic dead space* $(V_d)$ and an expandable volume called the *alveolar region* $(V_a)$. Gas inside the alveolar region is assumed to be well mixed, while the gas flowing through the dead space is assumed not to mix. During inspiration, a portion of the volume of inspired air [i.e., tidal volume $(V_t)$] will occupy the dead space $(V_d)$, and the remainder $(V_t - V_d)$ will mix with gases in the alveolar region. The product of the breathing frequency $(f)$ and the tidal volume is a volumetric flow rate called the *minute volume* or more simply the *ventilation rate* $(Q_t)$:

$$Q_t = fV_t \qquad (4.12)$$

The variation of the alveolar volume with time $(dV_a/dt)$ is called the *alveolar ventilation* $(Q_a)$. Thus the volume of air entering the alveolar region during inspiration is

$$V_a = (V_t - V_d) = V_t\left(1 - \frac{V_d}{V_t}\right) \qquad (4.13)$$

Differentiate with time and assume that the ratio $V_d/V_t$ is constant:

$$\frac{dV_a}{dt} = \frac{dV_t}{dt}\left(1 - \frac{V_d}{V_t}\right) \qquad (4.14)$$

Since the alveolar ventilation $(Q_a)$ is given by

$$Q_a = \frac{dV_a}{dt} \qquad (4.15)$$

and $dV_t/dt$ is the ventilation rate $(fV_t)$.

$$Q_a = fV_t\left(1 - \frac{V_d}{V_t}\right) = Q_t\left(1 - \frac{V_d}{V_t}\right) \qquad (4.16)$$

The Bohr model is useful in understanding the role of pulmonary and cardiac function under varying degrees of exercise. Table 4-2 shows values of ventilation rate $(Q_t)$ and blood flow rate $(Q_b)$ during exercise. Alveolar ventilation can be computed from Equation 4.16. The ratio of the alveolar ventilation rate $(Q_a)$ to the blood volumetric flow rate $(Q_b)$ is called the *ventilation perfusion ratio* $(R_{vp} = Q_a/Q_b)$. By measuring this ratio and monitoring the concentration of specific gases in the inspired air and in the blood, one can obtain quantitative estimates of the effectiveness of the lung to transfer the gas to the blood. This transfer is commonly called *gas uptake*. In the analysis above it has been assumed that the blood absorbs all the material diffusing through the alveolar barrier. If only a fraction of the material is really absorbed, the analysis can be modified to include the solubility constant of the material in blood (see the extended Bohr model in Section 4.3).

The elementary Bohr model shows that exercise increases the ventilation volumetric flow rate by a factor of 6.9, while the blood flow rate increases by only a factor of slightly over 3. The fraction of the inspired air actually reaching the alveolar region increases only slightly (0.66 to 0.84) and the ventilation–perfusion ratio $(Q_a/Q_b)$ increases by a factor of 2.63. Thus the uptake of gases by the blood is limited more by the supply of blood than it is of air reaching the alveolar region.

The greater the ratio of dead space to tidal volume $(V_d/V_t)$, the smaller the fraction of inspired air reaching the alveoli. Thus effective alveolar ventilation occurs only if the tidal volume $(V_t)$ exceeds the anatomical dead volume $(V_d)$. If the respiratory frequency is sufficiently high, there is evidence (Slutsky et al., 1985) that effective gas exchange occurs even when the tidal volume is smaller than the dead space.

**Table 4-2**   Ventilation, Blood Flow, and the Ventilation Perfusion Ratio $(R_{vp})$ during Exercise

|  | Rest | Light | Moderate | Heavy |
|---|---|---|---|---|
| Ventilation rate, $Q_t$ (L/min) | 11.6 | 32.2 | 50.0 | 80.4 |
| Frequency $(\text{min}^{-1})$ | 13.6 | 23.3 | 27.7 | 41.1 |
| Tidal volume, $V_t$ (L) | 0.85 | 1.38 | 1.81 | 1.96 |
| $V_d/V_t$ | 0.34 | 0.20 | 0.16 | 0.16 |
| Blood flow, $Q_b$ (L/min) | 6.5 | 13.8 | 18.4 | 21.7 |
| $R_{vp} = Q_a/Q_b$ | 1.18 | 1.87 | 2.28 | 3.11 |

*Source:* Abstracted from Ultman (1989, 1988).

The phenomenon is called *high-frequency ventilation.* An example is panting in dogs.

For years physiologists have contemplated the mechanisms by which gas exchange in the lung is maintained, given that the tidal volume (500 mL) is only a fraction of the vital capacity (4800 mL). The transport of gases between the mouth (or nose) and the alveoli can be divided into five modes (Haselton and Scherer, 1980; Ultman, 1985; Slutsky et al., 1985; Paiva, 1985; Engle, 1985): (1) direct alveolar ventilation by bulk convection, (2) convection by high-frequency "pendelluft," (3) convective dispersion due to asymmetric inspiration and expiratory velocity profiles, (4) Taylor-type dispersion, and (5) molecular diffusion. Each mode pertains to a different Weibel airway generation.

### 4.2.3 Bulk convection

*Bulk convection* is the conventional flow of air in a passageway. The *volumetric flow rate* is the cross-sectional area times the average velocity. The flow of air through the trachea is bulk convection. The development of boundary layers and the resulting modification of the velocity profile outside the boundary layer can be analyzed by conventional boundary layer theory. One must be cautious. The flow is pulsatile, the airways contain constrictions (such as the vocal cords) and bifurcations, and there is a moving layer of mucus on the walls. Thus fully established, steady flow does not exist.

### 4.2.4 Pendelluft

There is considerable *asymmetry* in the human bronchial tree. Because of the location of the heart and other organs, the right and left lungs are not

identical. The right lung is cleft by horizontal and oblique fissures into three lobes; the left lung is cleft into two lobes. When a tidal volume of air is inhaled, the fresh air may reach upper alveoli but not lower alveoli. The path of the inspired air and the number of alveoli immediately reached is a function on the level of physical activity, number of breaths per minute, and the size of the tidal volume. There is asynchronous filling and emptying of the lungs that may lead to an exchange of air between parallel lung units (Figure 4-15).

Inhalation and exhalation can be modeled crudely as two resistance–capacitance circuits in parallel (Otis, 1956) driven by a common square-wave power supply (Figure 4-16). The square-wave power supply

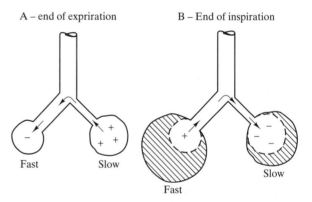

A – end of expiration     B – End of inspiration

Fast     Slow
Fast     Slow

**Figure 4-15**   When the breathing frequency is large, the lung resistance dominates the rate of filling and emptying for parallel lung units having different time constants $(\tau = RC)$. Case A - expired air from the slow unit is transferred to the fast unit. Case B - at the end of inspiration, the fast unit will transfer air to the slow unit. The plus (+) and the minus (−) signs indicate the pressure relative to atmospheric pressure (adapted from Chang 1984)

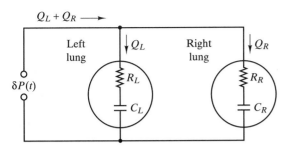

**Figure 4-16**  Schematic diagram to illustrate asynchronous filling and emptying of lungs, i.e. pendelluft.

corresponds to the positive and negative pressure difference produced by the diaphragm that fills and empties the lungs. The pressure drop ($\delta P$) to force air through a series of branching bronchiole can be thought to be a product of a resistance ($R$) and the volumetric flow rate ($Q$). Since the lung volume ($V$) changes, the pressure drop ($\delta P$) needed to fill the lung can be taken to be proportional to the lung volume ($V$) and inversely proportional to a lung capacitance ($C$). Thus for a single lung,

$$\delta P = RQ + \frac{V}{C} \qquad (4.17)$$

It is recognized that the lumped-parameter circuit in Figure 4-16 is simplistic since the lung resistance ($R$) and capacitance ($C$) are not truly constant. Nevertheless, assuming $R$ and $C$ constant affords an opportunity to analyze gross behavioral characteristics of pendelluft. Applying the law of conservation of mass for air entering a expandable lung unit, one finds that if the air density is constant, the rate of change of the lung volume is equal to the volumetric flow rate of air into the lung:

$$\frac{dV}{dt} = Q \qquad (4.18)$$

Combining these equations gives

$$\delta P = RQ + \frac{V}{C} = R\frac{dV}{dt} + \frac{V}{C} \qquad (4.19)$$

Then integrating gives

$$\int_{V_0}^{V(t)} \frac{dV}{\delta P/R - V/RC} = \int_0^t dt$$

Following algebraic manipulation, we obtain

$$\frac{V_f - V(t)}{V_f - V_0} = \exp\left(-\frac{t}{RC}\right) \qquad (4.20)$$

and

$$V(t) = \Delta V\left[\frac{V_f}{\Delta V} - \exp\left(-\frac{t}{RC}\right)\right] \qquad (4.21)$$

where $V_0$ is the initial lung volume, $V_f$ is the final volume of the inflated lung $(V_f = C\,\delta P)$ and $\Delta V = V_f - V_0$. The product $RC$ is a time constant ($\tau$) equal to the time required for the lung to acquire 64% of its final volume $[$i.e., $V(\tau)/V_f = 0.64]$. The volumetric flow rate of air at any instant entering the lung unit can be obtained from Equations 4.19 to 4.21:

$$Q = \frac{\delta P}{R} - \frac{V_f}{RC} + \frac{\Delta V}{RC}\exp\left(-\frac{t}{\tau}\right) \qquad (4.22)$$

If the right and left lung units having different final volumes $(V_f)$ and time constants ($\tau$) are linked in parallel as shown in Figure 4-16 and subjected to a common pressure difference ($\delta P$), one can compute the volume and volumetric flow rates for each lung. It is clear that the response characteristics of each lung will be different. At the end of a rapid expiration of air, the unit with the smaller time constant (the fast unit) on the left (e.g., Figure 4-15) is ready to fill, while the slower unit on the right is still emptying. Thus there is a flow of gas from the slower to the faster unit, where the breathing frequency is large. At the end of a rapid inspiration, air will flow from the fast unit to the slow unit, which is still filling. This "sloshing" between lung units is known as *pendelluft*.

### 4.2.5 Asymmetric velocity profiles

Experiments have shown that the velocity profiles in expiratory flow are flatter than those in inspiratory flow, which tend to resemble fully developed profiles. Figures 4-17 a and b provide a simple explanation (Haselton and Scherer, 1980). Upon inhalation, flow passes into the bronchial tree system as plug flow moving to the right. After a certain rightward displacement, the profile becomes parabolic if the flow is *laminar* (i.e., at low Reynolds numbers). Thus it moves rightward as *fully established*. Upon exhala-

Inspiration: Parabolic velocity profile, $u > 0$

$$U(r)/U_0 = \{1 - (r/R)^2\}$$
$$U(0) = U_0$$
$$U_{ave} = Q/A = U_0/2$$
$$r_{ave} = 0.707\,R$$

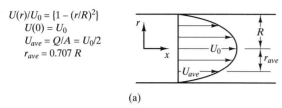

(a)

Expiration: Plug flow, $u < 0$

$$U_{ave} = Q/A$$

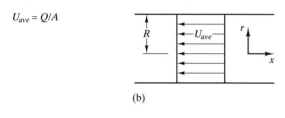

(b)

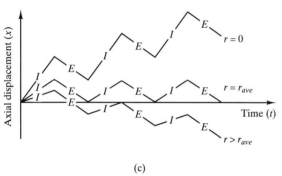

(c)

**Figure 4-17** (a) Inspiration: parabolic velocity profile, (b) Expiration: plug flow, (c) Displacement of a fluid element at three radial locations during inspiration (I) and expiration (E)

tion, flow emerges from the bronchial tree system as individual fully established profiles from each bronchi. Figure 4-7 shows that beyond the eighth airway generation, $L/L_e > 1$ and the flow is fully established. Upon merging, the many velocity profiles flatten into one such that the net profile is *uniform plug flow* moving to the left.

Consider a parcel of air at three radial locations in an airway:

**1.** Along the centerline, $r = 0$ and $U > U_{ave}$.
**2.** At the value $r_{ave}$, where $U(r_{ave}) = U_{ave}$ and $U_{ave} = Q/A$.
**3.** In the annular region, $r > r_{ave}$ and $U < U_{ave}$.

Trace the forward and backward displacement of the air parcel during several cycles of inhalation and exhalation. Neglect any transverse motion of the air parcel.

The displacement of the air parcel is equal to the air velocity, $U(r)$, integrated over the elapsed time. Figure 4-17c shows the result of this integration. Air in the center of the airway ($r = 0$) moves to the right (i.e., toward the alveoli), while air along the airway walls $(r > r_{ave})$, moves to the left (i.e., toward the pharynx). At a radius $r = r_{ave}$, the air has no net displacement. While the net volumetric flow rate through the airway as a whole during inspiration and expiration is zero, there is net flow to the right into the alveoli through the central portion of the airway and a net flow to the left out of the alveoli along the outer annular portion of the airway. Recall that cilia and their mucociliary escalation mechanism (Section 4.1.2) cause mucus to move along the bronchi walls in the same direction. Consequently, the flow in the conducting airways is both *pulsatile* and *countercurrent*.

## 4.2.6 Taylor dispersion

The transport of gases during breathing is a function of complex oscillating flows and a highly complex compliant (expanding/contracting) system of bronchioles. Taylor-type dispersion is the name given to this composite flow. To begin to understand Taylor-type dispersion, consider the original work of Taylor (1953) on a fully established laminar flow in a duct of constant radius $a$. The average velocity is one-half the maximum centerline velocity. At time zero, imagine that a diffusible material is injected (continuously) into the flow at a velocity equal to the local liquid velocity. The material will be convected downstream (longitudinally) but will also diffuse radially. Now imagine a reference system moving in the direction of flow at the average velocity $U$. Taylor showed that the radial transport of mass can be described as *virtual longitudinal diffusion* governed by Fick's law, in which the mass transport of material, $\dot{m}$, relative to a moving frame of reference is described by

$$\dot{m} = -D'A\frac{dc}{dx'} \tag{4.23}$$

where $x' = x - Ut$, and the molecular diffusion coefficient $D$ is related to the effective diffusion coefficient $D'$:

$$D' = \begin{cases} (\text{laminar}) = \dfrac{(aU)^2}{D} \\ (\text{turbulent}) = \text{proportional } (Ua) \end{cases} \quad (4.24)$$

Taylor (1954) also considered such dispersion in turbulent flow through a tube. This result is also shown in Equation 4.24. In reality, air in bronchioles moves in an oscillatory fashion, and one must also cope with inertial and viscous effects. The parameter used to characterize oscillatory flow is the *Womersley number* (Wo) (Slutsky et al., 1980):

$$\text{Wo} = a \sqrt{\frac{2\pi f}{v}} \quad (4.25)$$

where $f$ is the frequency of oscillation and $v$ is the kinematic viscosity of air. If Wo is less than unity, the flow can be analyzed as quasi-steady viscous flow; if Wo is considerably larger than unity, the flow must be analyzed as unsteady viscous flow. For laminar oscillatory flow (Wo $\gg$ 1) the dispersion coefficient, called $K$ by meteoroligists for dispersion in the atmosphere, is

$$K \propto \frac{\text{Re}^2}{\text{Wo}^n} \quad (4.26)$$

where $n$ is a constant and the Reynolds number is

$$\text{Re} = \frac{2aU_{\text{rms}}}{v} \quad (4.27)$$

where $U_{\text{rms}}$ is the root-mean-square axial velocity, $a$ is the radius of the bronchiole, and $n$ is approximately 3. Dispersion in the bronchiole is very complex. The flow is often unsteady, and it is clear that mass transport based solely on the mechanism of Taylor-type dispersion is inadequate.

### 4.2.7 Molecular diffusion

Molecular diffusion is the mechanism by which oxygen and carbon dioxide are transferred through the alveolar membrane to and from capillaries (Figure 4-18). The overall efficiency of gas exchange is a function of the five modes of gas transport. They are not mutually exclusive and certainly interact; however, for a given set of physical conditions, one mode may be dominant in a certain airway generation. Fig-

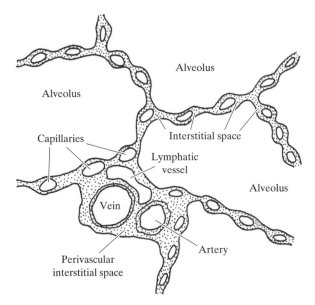

**Figure 4-18** Capillaries in the alveolar walls.

ure 4-19 is a summary of the dominant modes of transport in the lung. In the trachea and main-stem bronchi, turbulent Taylor-type dispersion should be important. If the tidal volume is large, convective flow can clear this portion of the dead space and ventilate some alveoli directly. In the medium-sized airways, large phase lags and oscillatory convective flow can occur. Mixing in the conducting zone of the lung occurs either by convective dispersion resulting from asymmetry of the lungs or out-of-phase bulk flow. In the small peripheral airways in the respiratory zone, out-of-phase oscillatory motion may be responsible for the ventilation of some lung units. Finally, in the alveoli and near the gas exchange surface, molecular diffusion is the dominant mode of gas exchange.

## 4.3 Analytical models of heat and mass transfer

Lung *permeability* is the rate at which materials penetrate the epithelial lining of the bronchial tree and lung respiratory surfaces. Air contaminants absorbed by the blood migrate to different body organs and affect them differently. Sulfur dioxide is soluble in water and is removed primarily by mucus in the upper respiratory system. Ozone has low solubility in water and penetrates the tracheobronchial region

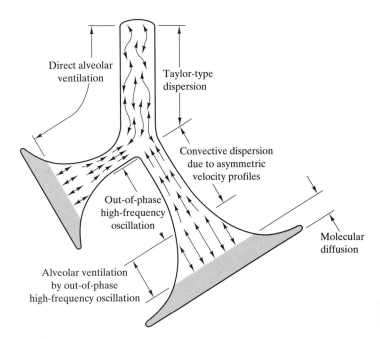

**Figure 4-19**  Modes of gas transport in the lung.

where much of it is absorbed and reacts chemically with mucus; however, a portion penetrates to attack underlying tissue. Carbon monoxide and volatile hydrocarbons have low solubility in water and do not react with mucus; thus they penetrate the pulmonary region, where they diffuse through the alveolar membrane and are absorbed in the blood. These examples illustrate widely different questions that the modeler must address in designing an analytical model.

- Is the contaminant soluble in water?
- Does the contaminant react with mucus or tissue?
- Can the contaminant be absorbed by blood?
- Do the effects of the contaminant vary with airway generation?
- Can well-mixed conditions be assumed?

The functional residual capacity $(V_{fr})$ of the lungs is approximately 2300 mL, but only 350 mL of inspired air is added to the alveolar volume and 350 mL of alveolar air is expired. Table 4-3 shows the typical composition of inspired air, gas in the alveoli, and the expired gas. Trace amounts of carbon tetrachloride and benzene produced in metabolism are also transferred to the expired gas. The large volume of alveolar air with respect to the volume of inspired gas ensures slowly changing conditions within the alveoli and prevents sudden changes in the gas exchange rates to the blood. An

elementary model of alveolar ventilation will be useful. Such a model, shown in Figure 4-20, assumes that the alveolar volume is a well-mixed region. At the quasi-steady state (i.e., averaged over many in- and exhalation cycles), inspired air enters at 4200 mL/min and mixes with 2300 mL of alveolar air, resulting in a 4200 mL/min departing steam of alveolar air. Oxygen is removed from the volume at a rate (rest conditions, 250 mL/min) dictated by the body's metabolic rate and carbon dioxide enters the volume at a rate (200 mL/min, rest conditions) also dictated by the metabolic rate. Note that 250 and 200 mL/min do not agree with the difference in oxygen and carbon dioxide entering and leaving the volume. The disparity arises because materials also

**Table 4-3**  Mole Fraction of Respiratory Gases That Enter and Leave the Lungs at Sea Level

| Species | Humid air[a] | Alveolar air | Expired air |
|---------|----------|--------------|-------------|
| $N_2$   | 0.7413   | 0.749        | 0.745       |
| $O_2$   | 0.1967   | 0.136        | 0.157       |
| $CO_2$  | 0.0004   | 0.053        | 0.036       |
| $H_2O$  | 0.0618   | 0.062        | 0.062       |

*Source:* Abstracted from Guyton (1986).
[a] Humid air: 25°C, 101 kPa, 50% relative humidity.

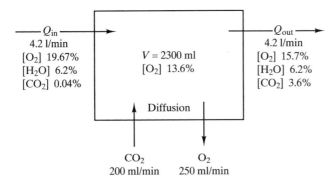

Figure 4-20 Control volume for alveolar gas exchange (adapted from Guyton 1986)

enter and leave the body as solids and liquids. Furthermore, steady-state concentrations of oxygen and carbon dioxide vary with the ventilation rate, as shown in Figures 4-21 and 4-22. These figures show that the normal alveolar partial pressures of $O_2$ and $CO_2$ occur at an alveolar ventilation rate of 4200 mL/min. If people exercise and produce a metabolic process requiring 1000 mL of oxygen and producing 800 mL of carbon dioxide per minute, the lung ventilation rates given by the dashed lines will be needed.

Figure 4-23 shows the components of the alveolar membrane through which gases are exchanged. The labyrinth of capillaries is so dense that the alveolar surface can be thought to be a sheet of blood contained within a thin membrane. The volume of blood

in the lung is between 60 and 140 mL and is contained within capillaries about 8 μm in diameter. The movement of blood through the myriad of blood vessels in the alveolar membrane is called *perfusion*. The rate of gas exchange is expressed in terms of a diffusing capacity times the difference in partial pressure of the diffusing gas in the blood and the gas within the alveoli. Under strenuous exercise (elevated metabolic rate), the pulmonary capillaries dilate, the blood flow rate increases, and the gas exchange rate increases. Thus the diffusing capacity may vary by a factor of 3. The diffusing capacity of carbon dioxide is considerably higher than for oxygen, 65 mL/min per mm Hg compared to 21 mL/min per mm Hg, respectively, under restful conditions. Thus if alveoli are damaged, individuals suffer from

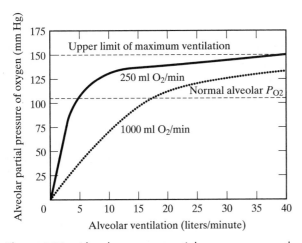

Figure 4-21 Alveolar oxygen partial pressure versus the alveolar volumetric flow rate for two oxygen absorption rates in the blood (adapted from Guyton 1986)

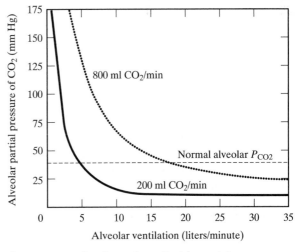

Figure 4-22 Alveolar carbon dioxide partial pressure versus alveolar volumetric flow rate for two carbon dioxide desorption rates in the blood (adapted from Guyton 1986)

for the layer, and $f$ is the fraction of the contaminant that remains after depletion because the contaminant may react chemically with materials in the layer. The solubility $k_s$ is a thermodynamic property of the contaminant and materials in the layers and is independent of the rate of transfer. The mass transfer coefficient $k_m$ is a transport property and depends on the rate of transfer, velocity field of the blood and/or air, geometry of the layer, and diffusivity of the species being transferred. The fraction of the contaminant that diffuses without depletion depends on the chemical kinetics of the contaminant and materials in the layer. If the contaminant is transferred without any reaction, the fraction is unity. For the *four-layer barrier* shown in Figure 4-25, the overall resistance to transfer is

$$\frac{1}{K} = \sum_i \left(\frac{f}{k_s k_m}\right)_i \quad (4.30)$$

where the subscript $i$ pertains to each of the four layers. For contaminants that do not react with materials in the four layers, the fraction $f$ is unity and the overall resistance is governed by the layer with the smallest product $(k_s k_m)$. The alveolar membrane consists of four layers, but mucus should be replaced by the pulmonary surfactant.

Consider an element of the capillary and the mass transferred into it. Transfer across the alveolar membrane can be written as

$$dm = \frac{KM(P_a - P_y)P' \, dy}{R_u T} \quad (4.31)$$

where $P'$ is the perimeter of the capillary into which mass is transferred, $P_y$ is the contaminant partial pressure at a variable point somewhere in the capillary, and $K$ is the overall mass transfer coefficient for the alveolar membrane.

Consider the blood flowing in the capillaries and write a mass balance for a differential element of the blood absorbing the contaminant.

$$\frac{dm}{M} + \frac{Q_b k_b P_y}{R_u T} = \frac{Q_b k_b (P_y + dP_y)}{R_u T}$$

$$\frac{dm}{M} = \frac{Q_b k_b \, dP_y}{R_u T} \quad (4.32)$$

where $k_b$ is the solubility of the transferred material in blood (moles absorbed per volume of blood) since it is assumed that the alveolar membrane does not absorb the contaminant. Eliminate $dm$ from Equations 4.31 and 4.32 and solve the resulting differential equation to find $P_y$:

$$\int_0^y \frac{KP' \, dy}{k_b Q_b} = \int_{P_v}^{P_y} \frac{dP_y}{P_a - P_y} \quad (4.33)$$

After simplification we have

$$P_a - P_y = (P_a - P_v)\exp\left(-\frac{yKP'}{k_b Q_b}\right) \quad (4.34)$$

Return to Equation 4.31, replace $P_y$ by Equation 4.34 and integrate:

$$\frac{R_u T}{MKP'(P_a - P_v)} \int_0^{\dot{m}} dm$$

$$= \int_0^L \exp\left(-\frac{yKP'}{k_b Q_b}\right) dy \quad (4.35)$$

$$\dot{m} = \frac{M(P_a - P_v)Q_b k_b}{R_u T}\left[1 - \exp\left(-\frac{LKP'}{k_b Q_b}\right)\right]$$

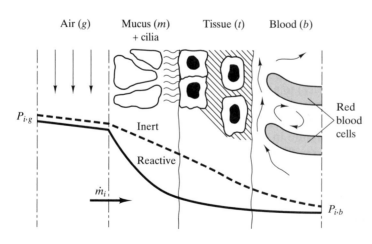

Air (g)   Mucus (m) + cilia   Tissue (t)   Blood (b)

Red blood cells

$P_{i,g}$

Inert

Reactive

$\dot{m}_i$

$P_{i,b}$

**Figure 4-25**   Schematic diagram of four-layer diffusion through the bronchial wall. In the alveoli, the mucus + cilia layer is replaced by a surfactant layer (adapted from Ultman 1988).

Now consider the alveolar region as the control volume and write a mass balance for the contaminant in the entire alveolar volume:

$$V_a \frac{dc_a}{dt} = Q_a c_0 - Q_a c_a - \frac{\dot{m}}{M} \quad (4.36)$$

where $V_a$ is the volume of the alveolar region and $c_a$ is the molar contaminant concentration in the alveolar region. Assuming quasi-steady-state conditions $(dc_a/dt = 0)$, we obtain

$$\frac{c_a}{c_0} = 1 - \frac{\dot{m}}{M Q_a c_0} \quad (4.37)$$

Assume that $P_a \gg P_v$, divide both sides of Equation 4.35 by $Q_a c_0$, and simplify. One obtains the ratio of the contaminant absorbed by the blood to the rate at which the contaminant is inhaled [i.e., uptake *absorption efficiency* $(\dot{m}/M Q_a c_0)$]:

$$\frac{\dot{m}}{M Q_a c_0} = \frac{1}{1 + R_{vp}/[1 - \exp(-N_D)]} \quad (4.38)$$

where

$$N_D \,(\text{diffusion parameter}) = \frac{KS}{Q_b k_b} \quad (4.39)$$

$S$ (total area over which mass is

$$\text{transferred}) = P'L \quad (4.40)$$

$R_{vp}$(ventilation–perfusion ratio

$$\text{modified by } k_b) = \frac{Q_a}{Q_b k_b} \quad (4.41)$$

Figure 4-26 is a graph of the uptake absorption efficiency versus the ventilation–perfusion ratio for several values of the diffusion parameter. The following general conclusions can be drawn regarding the influence of exercise on the absorption of contaminants.

**1.** Large values of $N_D$ imply good diffusion of gas through the alveolar–capillary barrier and the uptake absorption efficiency is uniformly high. The location of $N_D$ in the exponential term of Equation 4.38 ensures that at large values of $N_D$, uptake absorption efficiency is only a function of the $R_{vp}$. When $R_{vp}$ becomes large, the efficiencies decrease and approach approximately uniform values irrespective of the diffusion parameter $(N_D)$ since absorption is now limited by blood flow rate $(Q_b)$.

**2.** Designate a rest state corresponding to

*Rest:* $N_D = 1.0$   at which $R_{vp}$ falls slightly above 0.1

and locate the dot in Figure 4-26. Assume that the overall mass transfer coefficient $K$ and surface area $S$ remain constant, and use the data in Table 4-2 to quantify light, moderate, and heavy exercise. One finds that $N_D$ decreases with exercise since it is inversely proportional to the blood flow rate $Q_b$. Also, $R_{vp}$ increases due to increases in the ventilation–perfusion ratio. Thus the dots move down and to the right with increasing exercise. We see that as the rate of exercise increases, the uptake absorption efficiency decreases. From a health perspective, exercise does not increase the uptake absorption efficiency, although the uptake itself [i.e., the dose (kmol/s)] increases since the alveolar ventilation flow rate and blood flow rate increase (Nadel, 1985). The original presumption that the bodily dose is proportional to the ventilation rate is overly pessimistic regarding

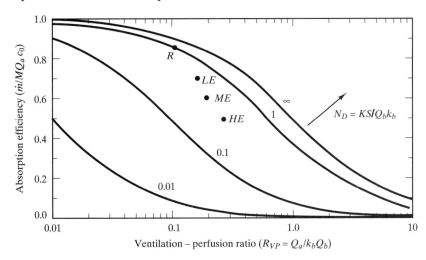

**Figure 4-26**  Absorption efficiency versus ventilation-perfusion ratio for different values of the diffusion parameter corresponding to rest (R), light exercise (LE), moderate exercise (ME) and heavy exercise (HE) (adapted from Ultman 1988)

injury by pollution. While the body dose increases, it is not strictly proportional to the ventilation rate.

**3.** Consider the transfer of oxygen through the alveolar membrane. People suffering from *emphysema* experience a reduction in the surface area for gas exchange ($S$) and hence a reduction in the diffusion parameter $N_D$. Thus they experience a reduction in absorption efficiency nearly proportional to the reduction in surface area irrespective of the ventilation–perfusion ratio.

### EXAMPLE 4.1    EFFECT OF EMPHYSEMA ON GAS UPTAKE

The reduction in alveolar surface area ($S$) accompanying emphysema affects the oxygen uptake (absorption) efficiency. Compute and plot the oxygen uptake efficiency versus age (after 18) of a person who suffers a reduction in alveolar surface area as follows:

$$\frac{S(t)}{S(18)} = 1 - \frac{t - 18}{120} \qquad \text{for } t > 18$$

Assume the following:

- $S(18) = 100 \text{ m}^2$.
- Solubility of oxygen in blood ($k_b$) is 10.7 L oxygen/L blood.
- Diffusion parameter at rest ($N_D$) at age 18 is 1.0.

***Solution*** From Table 4-2, the ventilation–perfusion ratio ($R_{vp}$), and the volumetric flow rates of air ($Q_a$) and blood ($Q_b$) for four levels of activity are:

| Activity | $Q_a$ (L/min) | $Q_b$ (L/min) | $R_{vp}=$ $Q_a/Q_b k_b$ |
|---|---|---|---|
| Rest (R) | 7.67 | 6.5 | 0.1103 |
| Light exercise (LE) | 25.81 | 13.8 | 0.1748 |
| Moderate exercise (ME) | 41.95 | 18.4 | 0.2131 |
| Heavy exercise (HE) | 67.49 | 21.7 | 0.2907 |

The overall mass transfer coefficient ($K$) can be found from Equation 4.39:

$$K = \frac{N_D(18)Q_b k_b}{S(18)}$$
$$= \frac{(1.0)(6.5 \text{ L/min})(10.7)(\text{m}^3/1000 \text{ L})(\text{min}/60 \text{ s})}{100 \text{ m}^2}$$
$$= 1.159 \times 10^{-5} \text{ m/s}$$

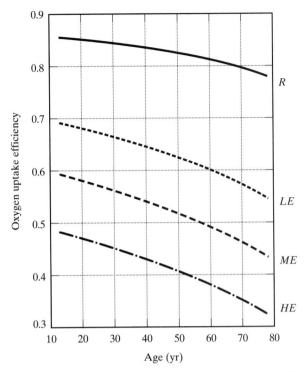

**Figure E4-1** Effect of emphysema on oxygen uptake efficiency versus time at four level of physical activity; R - rest, LE - light exercise, ME - moderate exercise, HE - heavy exercise

Figure E4-1 shows the absorption efficiency (Equation 4.38) as a function of age and physical activity. The MathCad program producing Figure E4-1 can be found in the textbook's Web site on the Prentice Hall Web page. In all cases the reduction in alveolar surface area from age 18 to 75 reduces the uptake of oxygen. At rest the reduction is only 8%, but under heavy exercise, the reduction is 30%. Such a reduction under heavy exercise subjects the body to a serious oxygen deficiency and prevents people from undertaking heavy exercise.

### 4.3.2 Distributed parameter model

There is considerable experimental evidence that ozone causes short-term biochemical functional changes in the lung. Specifically, there is a reduction in the 1-second forced expiratory volume ($FEV_1$) (Folinsbee et al., 1988; Lipmann, 1989). The distributed parameter model enables one to model these effects.

If one is interested in studying the effects of contaminants on portions of the bronchial tree as air pollutants flow in the longitudinal direction, the extended Bohr model is of little help and one must turn to a distributed parameter model. A single differential equation can be written for the transport of contaminants in the direction of flow. The bifurcations of the bronchial tree are included by modeling the airway as a single conduit resembling a *trumpet* (Figure 4-27), whose cross-sectional area and perimeter varies with longitudinal distance in accord with the Weibel model. Geometric data for the airways are provided by Figure 4-7. Mass transfer between passing air and blood vessels is modeled as mass transfer through several layers in series, as shown in Figures 4-23 and 4-25. The parameters describing the transfer across each layer may vary widely depending on the contaminant gas. Gases such as carbon monoxide diffuse to the blood and are not absorbed or reacted chemically with materials in the layers, gases such as ozone react chemically with the mucosal layer, and gases such as sulfur dioxide are mildly soluble in the mucosal layer.

The differential equation describing the longitudinal ($y$-direction) transport of contaminant species $i$ through the airway is

$$\frac{A}{R_u T}\frac{\partial P_i}{\partial t} = -\frac{Q}{R_u T}\frac{\partial P_i}{\partial y} + \frac{1}{R_u T}\frac{\partial\left[DA(\partial P_i/\partial y) - (P'\dot{m}_i/S)\right]}{\partial y} \quad (4.42)$$

where

$A, P'$ = cross-sectional area and perimeter of the airway; both vary with $y$ (Figure 4-7)

$Q$ = volumetric flow rate

$D$ = diffusion coefficient

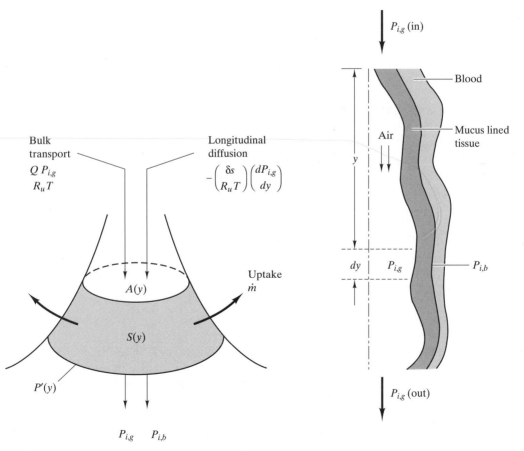

**Figure 4-27**  Schematic diagram of a distributed parameter model (redrawn from Ultman 1988).

$P_i$ = partial pressure of contaminant $i$

$S$ = total surface area over which mass is transferred to the blood

$\dot{m}$ = rate at which contaminant is transferred to the blood (i.e., uptake)

The left-hand term in Equation 4.42 represents the rate of accumulation and can be set to zero for quasi-steady-state solutions. The terms on the right-hand side correspond respectively to bulk transport, longitudinal diffusion, and uptake through the walls of the air passage. A differential equation for uptake is Equation 4.31, where the overall mass transport coefficient $K$ depends on properties of each of the four layers in Figure 4-25.

Equation 4.42 can be solved numerically. Miller et al. (1985) and Georgopoulos et al. (1997) have used such a model to describe the effects of ozone inhalation and predicted the total dose received in branches of the bronchial tree and the dosage penetrating to the tissue. Figure 4-28 shows the relative dose (dose/tracheal ozone concentration) versus location at the inside surface of the airway (mucus–air interface for airway generations up to 16 and surfactant–air interface in the pulmonary region) for several different first-order rate constants $(k_r)$ for the ozone–mucus reaction. Figure 4-29 shows the relative dose that penetrates the tissue for four levels of physical activity (rest to heavy exercise) similar but not identical to those in Table 4-2. The difference between the relative dosages in these two figures is the ozone that reacts in the mucus layer or surfactant layers. While the rate constants for the chemical reactions are not known with precision, Figure 4-28 shows that the dose at the mucus–air interface is roughly independent of airway generation, but Figure 4-29 shows that the net dose to tissue increases in the tracheobronchial region (TB). In both cases the dose decreases in the pulmonary region (P). Thus the tracheobronchial tissue is protected by *ozone absorption and chemical reaction with mucus* and the protection this provides tissue is superior to that provided by pulmonary surfactant. The tracheobronchial removal efficiency increases with the rate constant. The highest relative dose experienced by tissue occurs at the seventeenth airway generation for R, LE, and ME physical activity and at the twentieth for HE physical activity, conclusions borne out by exper-

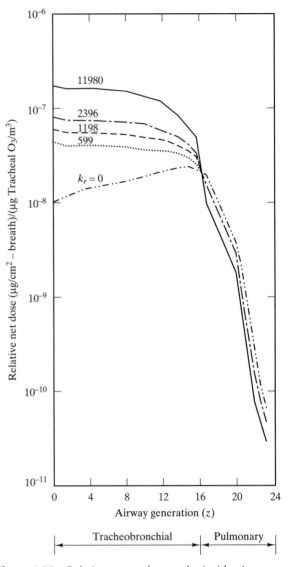

**Figure 4-28**   Relative ozone dose to the inside airway surface versus airway generation for several first-order rate constants $(k_r)$, for the ozone-mucus reaction (redrawn from, Miller et al, 1985)

iment. Tissue dose in the tracheobronchial region is affected only slightly by exercise, but the point of maximum tissue dosage penetrates into the pulmonary region.

The distributed parameter model has also been used by Hanna and Scherer (1986a–c) to model the transfer of energy and water vapor during inhalation and exhalation (Figures 4-30 to 4-32). The upper res-

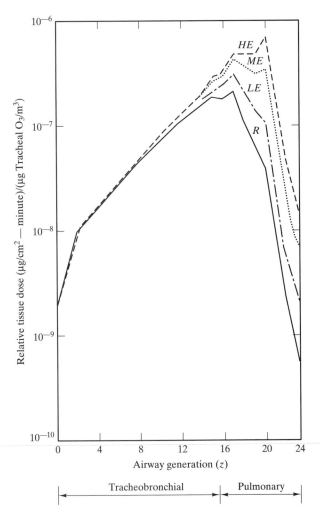

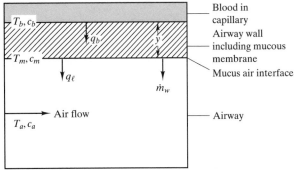

**Figure 4-30**  Transverse heat and mass transfer across layers of an airway passage during inspiration. Quantities are functions of the longitudinal distance along the airway (redrawn from Hanna and Sherer 1986 a,b,c)

**Figure 4-29**  Relative ozone dose to tissue versus airway generation for rest (R), light exercise (LE), moderate exercise (ME) and heavy exercise (HE) (redrawn from Miller et al, 1985)

piratory tract region between the nasal tip to the midtrachea is very susceptible to inflammation and infection. The upper respiratory tract is also the region within which the majority of energy is transferred to heat the incoming air to body temperature and to add water vapor to the incoming air (i.e., *conditioning inhaled air*). Some energy and water is recovered upon expiration. The amount of water and energy transferred to the inspired air depends on the temperature and humidity of the incoming air; the amount transferred during expiration is affected by

the blood temperature and blood flow rate in the nasal and oral cavity. The bronchi may constrict because of the loss of energy and water from the mucosal surface; thus it is important to be able to describe quantitatively the transfer of energy and water from the tracheobronchial passages.

To preserve the functioning of the alveoli, inspired air must be heated to the body temperature and contain the maximum amount of water vapor. This conditioning is performed upstream in the trachea and bronchi. The mucous membrane conditions incoming air and collects pathogenic organisms and particles. The particles are removed by ciliary motion. The mucus secretion rate is a function of air temperature and humidity. The rheological properties of the mucus are also a function of net loss of energy and water.

The majority of conditioning occurs in the nasal cavity where during normal breathing the temperature is raised to 70% of body core temperature (37°C) and the air becomes saturated with water vapor. The temperature of the mucosal surface is 32 to 33°C. After the first third of the bronchial tree (Figure 4-6) the temperatures of the air and mucosal surface are equal to the body core temperature. The energy associated with water evaporation in the respiratory system is approximately 85% of the entire body's heat loss. The difference in the humidity of the expired and inspired air is the largest source of water loss by the body. During expiration, 20 to 25%

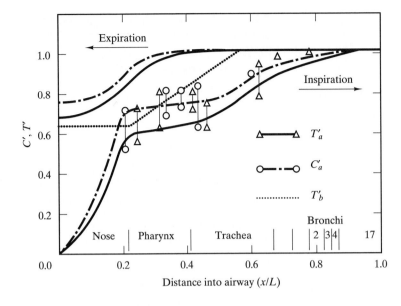

**Figure 4-31**
Dimensionless air temperature $(T_a')$, blood temperature $(T_b')$ and water vapor concentration $(c_a')$ as a function of the longitudinal distance in the bronchial system during inspiration and expiration in the rest condition (redrawn from Hanna and Sherer 1986 a,b,c)

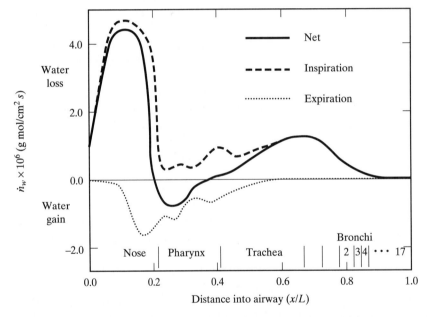

**Figure 4-32**   Molar transfer rate $(\dot{n}_w)$ of water to air as a function of the longitudinal distance in the bronchial system during inspiration and expiration in the rest condition. Positive values of $(\dot{n}_w)$ denote the body's loss of water (redrawn from Hanna and Sherer 1986 a,b,c)

of the energy lost by evaporation during inhalation is regained by condensation. For quiet breathing the net heat transfer to the inspired air per day is about 350 kcal, or 15% of the basal metabolic rate, and the net transfer of water is 250 to 400 mL/day.

Respiratory airflow is oscillatory, but for purposes of analysis the flow can be assumed to be quasi-static since the Womersley numbers are less than unity. Experimental measurements of air temperature in the nose and trachea show that steady-state temperatures are reached after 20% of the total time for inspiration and expiration. Measurements of relative humidity in air expired through the mouth show steady-state values after 20% of the total time. Measurements of the temperature at the interface between the mucus layer and wall of the air passage show that this temperature is constant throughout the respiratory cycle. Beyond the second or third division of the bronchial tree the temperature and humidity of the air do not vary over the respiratory cycle. This point in the bronchial tree is called the *isothermal saturation boundary* (ISB). The precise location of ISB varies with the ambient temperature tidal volume and breathing rate.

Within the mucus layer and tissue, steady-state conditions are achieved within characteristic times equal to the layer thickness divided by the thermal conductivity or mass diffusivity. Assuming that the thickness of the layer is 50 μm, these characteristic times are on the order of milliseconds, whereas the duration of inspiration and expiration is 2.5 seconds for a breathing rate of 12 breaths/min. In summary, assuming quasi-static conditions for inspiration and another set for expiration is a reasonable assumption for flow in the trachea and first two generations of the bronchial system.

Figures 4-31 and 4-32 show the results of an analysis based on an elementary quasi-static model of the transport of mass and heat in the respiratory tract. Air is assumed to pass uniformly through a duct whose walls are a composite of the mucous membrane, bronchial wall, and a layer composed of the tissue containing capillaries. The energy transferred from the blood through the mucus–air interface is assumed to be equal to the latent heat of evaporation of the evaporated water plus energy transferred to the air by convection. The blood is assumed to have constant temperature:

$$\frac{k_t}{y}(T_b - T_a) = h_c(T_m - T_a) + N_w M_w h_{fg} \quad (4.43)$$

where

$$
\begin{aligned}
x &= \text{direction of flow} \\
T_b &= \text{blood temperature (constant)} \\
T_m(x) &= \text{mucus–air interface temperature} \\
T_a(x) &= \text{air temperature} \\
M_w, h_{fg} &= \text{molecular weight and enthalpy of} \\
&\quad \text{vaporization of water} \\
N_w(x) &= \text{number of moles of water that} \\
&\quad \text{evaporates per unit time and area} \\
y &= \text{thickness of airway wall including} \\
&\quad \text{mucus membrane (constant)} \\
k_t(x), h_c(x) &= \text{conductivity of airway wall, convec-} \\
&\quad \text{tion heat transfer coefficient at} \\
&\quad \text{mucus–air interface}
\end{aligned}
$$

The number of moles of water that evaporate per unit time and area is equal to the mass transfer of water vapor to air:

$$N_w = k_c(c_m - c_a) \quad (4.44)$$

The parameter $k_c(x)$ is the convective mass transfer coefficient. The concentration of water vapor in air $c_a(x)$ is a variable, but the concentration of water vapor at the mucus–air interface is equal to its saturation value at temperature $T_m(x)$ and can be given by the empirical equation (Hanna and Scherer, 1986a)

$$c_m(x) = 22.4 \exp\left(-\frac{4900}{T_m}\right) \quad (4.45)$$

where $c_m(x)$ is in mol/cm$^3$. Alternatively, the Clausius–Clapeyron equation can be used. The diffusion of water vapor in the axial direction is small compared to bulk transport and can be neglected (Hanna and Scherer, 1986a). Thus the conservation of mass for water vapor is given by

$$U\frac{dc_a}{dx} = \frac{P k_c(c_m - c_a)}{A} \quad (4.46)$$

The conservation of energy is given by

$$U\frac{dT_a}{dx} = \frac{P'}{A\rho c_{p,a}}\left[h_c(T_m - T_a)\right.$$
$$\left. + c_{p,w} M_w N_w(T_m - T_a)\right] \quad (4.47)$$

where

$$
\begin{aligned}
U &= \text{velocity (assumed to be plug flow)} \\
A(x), P'(x) &= \text{airway cross-sectional area and} \\
&\quad \text{perimeter} \\
\rho &= \text{air density} \\
c_{p,w}, c_{p,a} &= \text{specific heat of water vapor and} \\
&\quad \text{air}
\end{aligned}
$$

The thermal conductivity of the airway wall and mucous membrane, $k_t$, can be taken as a constant and the transport parameters $h_c(x)$ and $k_c(x)$ are variables that can be evaluated (Hanna and Scherer, 1986a–c) by dimensionless relationships from the literature of heat and mass transfer.

Equations 4.43 to 4.47 constitute a set of coupled ordinary differential equations that can be solved numerically. The values of the blood temperature $T_b$, airway cross-sectional area $A(x)$, and perimeter $P'(x)$ can be taken as constants at the appropriate location $(x)$ in the bronchial tree. The air velocity $U$ is equal to the tidal volumetric flow rate divided by $A(x)$.

The computations can be performed for both inspiration and expiration associated with restful room air breathing. Figure 4-31 shows the predicted air temperatures and water vapor concentration, which are normalized with respect to the temperature of the inspired air $[T_a(\text{insp})]$,

$$T_a(x)' = \frac{T_a(x) - T_a(\text{insp})}{T_c - T_a(\text{insp})} \qquad (4.48)$$

where $T_c$ is the body-core temperature. The water vapor concentration is normalized in a similar fashion. Variations in the values of the transport coefficients (by 50%) are shown to have only moderate effect on the results for inspiration and little effect for expiration. Of all the parameters in the analysis, the blood temperature and volume of the nasal cavity were found to have the most importance. The evaporation rate of water is shown in Figure 4-32 for both inspiration and expiration.

Figures 4-31 and 4-32 show vividly that the nasal cavity is the primary organ that *conditions inspired air*. Within the nasal cavity the water vapor concentration in air nearly reaches equilibrium during both inspiration and expiration. During expiration, water vapor condenses on to the cooler nasal mucosa. The mucosa of the trachea and larynx lose water upon inspiration but regain some of it upon expiration. For the cycle of inspiration and expiration, the nasal cavity, lower trachea, and bronchial tree experience a net loss of water, but the pharynx has a net gain of water. Air is predicted to be expired at nearly the nasal blood temperature and nearly saturated. These results are confirmed by experiment. Downstream of the nasal cavity, the airstream reaches 60 to 70% of body-core temperature and is fully saturated during inspiration. Relatively little conditioning of the inspired air occurs within the pharynx and upper trachea. The upper portions of the bronchial tree fully condition inspired air in a relatively short distance. Conditioning in the tracheobronchial tree is the result of a large surface area of the numerous bifurcations. As the temperature of the inspired air decreases or the tidal volumetric flow rate increases,

less conditioning is accomplished in the nasal cavity and more is accomplished in the tracheobronchial tree.

The health implications of these studies are important. Water must be transported to the pharynx since the pharynx lacks cilia and mucus-secreting glands and cells. The pharynx is thus particularly vulnerable to drying, disease, organisms, bacterial infiltration, irritation, and assault by air pollutants. The conclusion is also reinforced by exercise experiments involving sulfur dioxide (Kleinman, 1984), in which the dose to the pharynx resulting from mouth breathing is larger than that from nasal breathing, owing to the efficient scrubbing that occurs in the nasal cavity.

In the outdoor environment, sulfur dioxide forms sulfuric acid, which may be partially neutralized by atmospheric ammonia $NH_3$ to form ammonium bisulfate $NH_4HSO_4$ and ammonium sulfate $(NH_4)_2SO_4$. It is reported (Hattis et al., 1987) that sulfuric acid particles are 10 times more potent than $(NH_4)_2SO_4$ and 33 times more potent than $NH_4HSO_4$. When inhaled, small particles contact various surfaces in the tracheobronchial region. Because of the buffering capacity and volume of mucus, it is believed (Hattis et al., 1987) that the pH of the tracheobronchial mucus does not change appreciably. However, the acid concentration in very small individual particles may be sufficient to produce localized *irritant signals* in the lower airways to increase mucus secretion and contribute to processes involved in chronic bronchitis. Depending on the exact pH depression to produce a signal, the minimum-size acid particle required to produce a signal is believed to be between 0.4 and 0.7 μm. Neutralization by $NH_3$ in the upper respiratory tract may have an affect as well.

To predict the deposition of aerosols or the absorption of gases and vapors in the various parts of the respiratory system, it will be necessary to develop accurate fluid mechanics mathematical models. These models must account for the asymmetry of the respiratory system, the expansion and contraction of the air passages, and the mucus ciliary motion. Attempts to secure models that explain observed phenomena have gone on for several decades and have improved steadily. The work of Nixon and Egan (1987), Miller et al. (1985), Muller et al. (1990), Gradon and Yu (1989), Gradon and Orlicki (1990), Eisner and Martonen (1989), Yu and Xu (1987), Yu and Neretnieks (1990), Xu and Yu (1987), Ultman (1985, 1988, and 1989), and Johannon (1991) should be consulted.

## EXAMPLE 4.2 LOSS OF BODY WATER THROUGH RESPIRATION

The body loses water through perspiration, elimination, and respiration. Estimate the net loss of water through respiration associated with different physical activities conducted in a dry environment where the ambient temperature is 25°C and the relative humidity is 5%. Assume that expired air has a relative humidity of 100% based on a temperature of 30°C.

**Solution** The saturation (vapor) pressure of water at 30 and 25°C are $P_v(30°C) = 4.246$ kPa and $P_v(25°C) = 3.169$ kPa. The relative humidity, $RH(T) = P_v$(actual at $T$)/$P_v$(saturation at $T$). Thus

$$P_v(\text{expired air, } 30°C) = 4.246 \text{ kPa}$$

$$P_v(\text{inspired air, } 25°C) = (0.05)(3.169)$$

$$= 0.158 \text{ kPa}$$

The water vapor concentration in inspired and expired air are

$$c_v(\text{inspired air, } 25°C)$$

$$= \frac{P_v(\text{inspired air, } 25°C)}{(R_u/M)T} = \frac{0.158 \text{ kPa}}{(8.314/18)(298)}$$

$$= 0.001147 \text{ kg/m}^3 = 1.147 \text{ g/m}^3$$

$$c_v(\text{expired air, } 30°C) = \frac{P_v(\text{expired air, } 30°C)}{(R_u/M)T},$$

$$= \frac{4.246}{(8.314/18)(303)} = 0.030339 \text{ kg/m}^3$$

$$= 30.339 \text{ g/m}^3$$

The rate at which the body loses water,

$$\dot{m}_{w,\text{loss}} = Q_t[c_v(\text{expired air, } 30°C)$$

$$- c_v(\text{inspired air, } 25°C)]$$

where $Q_t$ is the volumetric flow rate of air given in Table 4-2 for four levels of activity. For the rest condition,

$$\dot{m}_{w,\text{loss}}(\text{rest}) = (11.6 \text{ L/min})(\text{m}^3/1000 \text{ L})$$

$$(30.339 - 1.147)(\text{g/m}^3)(60 \text{ min/h}) = 20.32 \text{ g/h}$$

| Activity | $Q_t$ (L/min) | $\dot{m}_{\text{water}}$ (g/h) |
|---|---|---|
| Rest | 11.6 | 20.32 |
| Light exercise | 32.2 | 56.57 |
| Moderate exercise | 50.0 | 87.59 |
| Heavy exercise | 80.4 | 140.84 |

1. Normal day (8 h of rest and 16 h of light exercise)

$$\dot{m}_{w,\text{loss}}(\text{normal day}) = (8)(20.32)$$

$$+ (16)(56.57)(\text{g/h})[\text{L}/1000 \text{ g})] = 1.068 \text{ L/day}$$

2. 4-h round of golf (light exercise)

$$\dot{m}_{w,\text{loss}}(\text{4-h golf}) = \frac{(4)(56.57)}{1000} = 0.226 \text{ L/4-h golf}$$

3. 6-h hike (moderate exercise)

$$\dot{m}_{w,\text{loss}}(\text{6-h hike}) = \frac{(6)(87.59)}{1000}$$

$$= 0.525 \text{ L/6-h hike}$$

4. 3-h rock climbing (heavy exercise)

$$\dot{m}_{w,\text{loss}}(\text{3-h rock climbing}) = \frac{(3)(140.84)}{1000}$$

$$= 0.422 \text{ L/3-h rock climbing}$$

A rule of thumb is that people should consume approximately 1.5 L of water per day. On the basis of the calculation above, physically active people should consume more than this if they live in a dry region. (Of course, one does not need a baccalaureate degree in engineering to know this!)

## 4.4 Toxicology

Respiratory diseases and disorders produce symptoms that are grouped in several categories: hypoxia, cyanosis, dyspnea, and hypercapnia. *Hypoxia* is an inadequate supply of oxygen to support bodily functions and can be subdivided.

1. *Hypoxic hypoxia.* Sufficient oxygen does not reach the alveoli. Hypoxic hypoxia can be caused by environmental factors.

| **Table 4-4**  Effects of Low Oxygen Concentration | |
|---|---|
| **[O$_2$] (%)** | **Manifestations** |
| 20.9 | Normal oxygen concentration in air (78% nitrogen, 1% argon, and trace gases) |
| 17 | Hypoxia occurs with deteriorating night vision, increased heartbeat, and accelerated breathing |
| 14–16 | Very poor muscular coordination, rapid fatigue, and intermittent respiration |
| 6–10 | Nausea, vomiting, inability to perform, and unconsciousness |
| Below 6 | Spasmodic breathing and convulsive movements; death within minutes |

2. *Anemic hypoxia.* Inadequate hemoglobin prevents sufficient oxygen from reaching the cells.
3. *Circulatory hypoxia.* The blood flow rate carrying oxygen to the cells is insufficient.
4. *Histotoxic hypoxia.* Tissues cannot use oxygen properly.

*Cyanosis* refers to the blue hue acquired by skin because of excessive amounts of deoxygenated hemoglobin and may be a symptom of respiratory insufficiency. *Hypercapnia* is a condition of excess carbon dioxide in the blood. If the ventilation rate is abnormally high, both the oxygen and carbon dioxide concentrations become excessive. Excess carbon dioxide results in dyspnea. *Dyspnea* is also a state of mind (i.e., anxiety) related to the inability to provide the body with sufficient air. It is clear that the symptom may have several causes, such as hypoxia, hypercapnia, or purely emotional factors.

An *oxygen-deficient atmosphere* is defined by regulatory agencies as one in which the oxygen concentration is less than 19.5%. Safe practices dictated by OSHA are designed to prevent people from accidentally being exposed to an oxygen-deficient atmosphere. The effects of low oxygen concentration (Table 4-4) are very serious and cause many deaths each year.

Unlike the exchange of oxygen and carbon dioxide, the transport of toxic gases occurs throughout all parts of the respiratory tract (see Table 4-1). Throughout this section the term *gas* includes both gases and vapors. The rates at which gases are taken up and distributed to body organs vary considerably. On the one hand, anesthetics produce their effect rapidly. Toxic gases such as hydrogen cyanide and hydrogen sulfide are lethal within minutes. Hallucinogens and carbon monoxide take longer to produce physiological effects.

## EXAMPLE 4.3   AIR CONTAINING LARGE CONCENTRATIONS OF CO$_2$

You are an officer in the U.S. Navy and have volunteered for submarine duty. You are undergoing training and have been instructed that submarines have an air purification system that automatically generates oxygen to maintain an O$_2$/N$_2$ molar ratio of 0.25 and a water vapor mole fraction of 0.039. You are also told that meters record both CO and CO$_2$ on a continuous basis. The CO$_2$ meter indicates than an alarm will sound if the CO$_2$ concentration exceeds 250,000 ppm. A fellow officer knows that CO is dangerous and has a PEL of 35 ppm (see Table A-11 in the Appendix) but is puzzled why it is necessary to monitor CO$_2$ since there is no PEL for CO$_2$ and that even normal atmospheric air contains approximately 350 ppm of CO$_2$. Explain why a CO$_2$ concentration of 250,000 ppm inside the submarine is dangerous.

*Solution*  Assuming that air inside the submarine contains only O$_2$, N$_2$, H$_2$O, and CO$_2$, the mole fractions are related:

$$1 = y_{O_2} + y_{N_2} + y_{H_2O} + y_{CO_2}$$

The air purification system maintains

$$y_{H_2O} = 0.039 \qquad y_{N_2} = 4y_{O_2}$$

$$\frac{y_{N_2}}{y_{O_2}} = \frac{0.8}{0.2} = 4,$$

Substituting into the above results in

$$1 = y_{O_2} + 4y_{O_2} + 0.039 + y_{CO_2}$$

If the CO$_2$ monitor trips an alarm at $y_{CO_2} = 0.25$ (250,000 ppm), the oxygen mole fraction at this time will be

$$y_{O_2} = \frac{(1 - 0.039 - 0.25)}{5} = 0.1422 = 14.22\%$$

Thus if $y_{O_2}/y_{N_2}$ remains at 0.25, the large amounts of $CO_2$ displace oxygen and nitrogen and the crew will inhale air containing only 14.2% oxygen. At this concentration, their performance will be impeded, and under combat conditions this represents a dangerous situation. Thus the immediate risk is not from $CO_2$ "poisoning" but from insufficient $O_2$. As an engineer you should be looking into the function of the $CO_2$ removal system even before the alarm sounds.

Toxic gases either react directly on portions of the respiratory tract or are transported to other organs before their effect is registered. Examples of the first type are ozone and sulfur dioxide. Examples of the second type are carbon monoxide and hydrogen cyanide. Exposure to ozone for two hours reduces FVC and $FEV_1$ in healthy adults by small but statistically significant amounts (4%) and its effects persist for approximately 18 hours (Lipmann, 1987, 1989, 1991). Highly reactive agents soluble in water (e.g., anhydrous acids and strong oxidants) are apt to damage tissue, while less reactive gases such as nickel carbonyl diffuse through tissue to react with endothelial cells. Other gases may damage capillaries.

In Section 4.3 it was shown that the nasal cavity is the principal organ that transfers water and energy to and from respired air. Because the surface of the nasal cavity contains a great deal of water, any toxic gas soluble in water will be removed. Anhydrous acids and sulfur dioxide are more apt to be removed than ozone, owing to the latter's lower solubility in water. The nasal cavity is the body's most efficient wet scrubber. In the tracheobronchial region toxic gases encounter mucus lining the airways. Gases penetrating the mucous lining contact goblet or ciliated cells. Ciliated cells are generally more sensitive to toxins than goblet cells, and reducing the number of cilia per unit area of passageway impairs the clearance mechanisms.

The body's response to particles contained in inspired air is entirely different than its response to gases. *Clearance* is the process by which particles are removed from the lung. *Mucociliary clearance* is the process by which the conducting airways of the lung remove depositing particles and carry them to the lar-

ynx on surface mucus propelled by cilia. *Alveolar clearance* is the process by which particles are removed by nonciliated surfaces in the gas-exchange region of the lung. Clearance mechanisms include ingestion by macrophage followed by the migration from the lung and the gradual dissolution of the particle.

Figures 4-33 and 4-34 show that particles deposit themselves throughout all regions of the lung through a variety of processes (Hatch and Gross, 1964; Perra and Ahmed, 1979; Rothenberg and Swift, 1984; Martonen, 1992). In the *nasopharyngeal* region *inertial impaction* is the dominant mechanism and relatively large particles are removed. Particles removed in the *trachea* and *bronchial tree* are removed by a combination of *inertial* and *gravitational settling* (sedimentation) processes. The larynx affects particle motion because of turbulence created by air passing over the vocal cords. In addition, particles entrained in the air leaving the larynx contact the trachea at localized "hot spots" (Balashazy et al., 1990; Martonen et al., 1992).

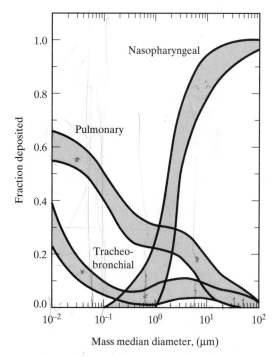

**Figure 4-33** Predicted regional deposition of particles in the respiratory system for a tidal volumetric flow rate of 21 L/min. Shaded area indicates the variation resulting from two geometric standard deviations, 1.2 and 4.5 (redrawn from Perra and Ahmed 1979)

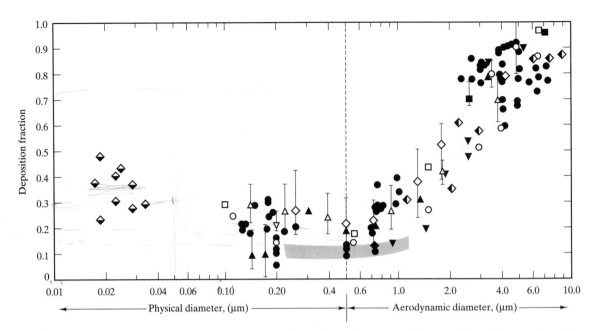

**Figure 4-34**   Deposition of monodisperse aerosols in the total respiratory tract for mouth breathing as a function of particle diameter. Below 0.5 μm refers to the actual physical diameter while above 0.5 μm refers to the aerodynamic diameter (from Seinfeld 1986)

Elsewhere in the lung, particles are deposited more uniformly over the airway surface. Submicron particles penetrate the *alveolar* region and are metabolized by macrophage. Particles that are not metabolized remain in the alveoli or may diffuse to other parts of the body, such as the lymphatic system. Particles removed in the nasopharyngeal region enter the digestive tract and pass through the body in a short time. Unless these materials enter the bloodstream through the digestive system, they are of little importance. Particles entering the trachea are removed by cilia and mucus and transported to the digestive tract. Particles penetrating the bronchial tree are flushed more slowly. Exactly which particles penetrate which region of the respiratory tract depends on the breathing rate and tidal volume. Thus a population of energetic workers in a dusty industrial environment may have more particles of a given size deposited to the bronchial tree and alveoli than do people in a nonindustrial environment under restful breathing. For slender particles of asbestos, cotton, and similar materials, deposition processes must take into account their unusual shape (Kasper et al., 1985; Gallily, 1986).

The terms *inhalable* and *respirable* are used to categorize particles that pose a hazard to health. The

EPA defines *inhalable* as particles having an aerodynamic diameter less than 10 μm, while ACGIH defines *respirable* as particles having an aerodynamic diameter less than 2.5 μm. Particles with $D_p < 2.5$ μm are called *fine particles* and those with $D_p > 2.5$ μm are called *coarse* particles. Fine particles produced directly by combustion processes are called *primary particles*, and particles produced by reactions in the atmosphere of gaseous products of combustion are called *secondary particles*. The *aerodynamic diameter* is the diameter $(D_p)$ of a spherical water drop that has the same settling velocity as the actual particle. To a first approximation, the actual and aerodynamic diameters are related by

$$D_p(\text{aerodynamic}) = D_p(\text{actual}) \sqrt{\frac{\rho_p}{\rho_w}} \quad (4.49)$$

The definitions of *respirable* and *inhalable* may change or new terms may be coined as understanding of the health effects of inhaled particles improves. Particles deposited in the alveoli are acted upon by one or a combination of four processes.

**1.** Particles may be phagocytized and passed up the tracheobronchial tree by the mucociliary escalator.

2. Particles may be phagocytized and transferred to the lymphatic drainage system.
3. Particle surface material may be dissolved and transferred to blood vessels or lymphatics.
4. Particles and some dissolved material may be retained in the alveoli permanently.

There is basic disagreement on the interpretation of the epidemiological data on health effects associated with increases in particle air pollution. Although a strong case has been presented that this association reflects a causal relationship, plausible alternative explanations have also been suggested (Vedal, 1997). Inhaled fine particles entering the pulmonary region induce *pulmonary inflammation* and increase the risk of hospitalization or death of sensitive individuals such as those with chronic lung and heart disease. Inflammatory lung disease is associated with an increase in mortality of people 65 and older. The portion of the increase due to chronic obstructive pulmonary disease is 3.3%, the portion due to ischemic heart disease is 2.1%, and the portion due to pneumonia is 4% (Schwartz and Andreae, 1996). Hypotheses describing how inflammation is induced depend on the chemical properties of the particles, such as acidity, the presence of transition metals, and the presence of ultrafine particles, $D_p < 0.02$ μm (20 nm). Transition metal ions such as ferric ions catalyze the production of hydroxyl radicals (OH·) via the Fenton reaction. Ultrafine particles, $D_p < 20$ nm, are taken up poorly by lung macrophage and are capable of penetrating the pulmonary epithelium and may pass into the interstitium.

Studies of particle trajectories in the bronchial tree (Balashazy and Hofmann, 1993) reveal that during inspiration, particles deposit themselves preferentially on the tissue dividing the bronchial airways into two passages (i.e., *carnial ridges*). The fraction of particles that are deposited is higher for large particles, owing to their larger inertia. Mucociliary clearance is smallest at these airway bifurcations; consequently, the epithelial cells on these carnial ridges suffer a larger dose than other cells on the airway passages. The phenomenon helps explain the high occurrence of bronchial carcinomas attributed to cigarette smoke and radon progeny.

The process of *phagocytosis* is one of the body's essential protective mechanisms. Organs of the body, including the alveoli, contain special cells called *macrophage* that attack foreign matter by a variety of processes. Phagocytosis begins when the macrophage membrane attaches itself to the surface of the particle. Receptors on the surface of macrophage bond with the particle. The edges of the membrane spread outward rapidly and attempt to engulf the particle (mechanism of *pinocytosis*). The macrophage membrane contracts and pulls the particle into its interior, where cell lysosome attach themselves to the particle. *Lysosome* are special digestive organelle within cells that contain enzymes called *hydrolase* that digest foreign matter. The engulfed particle is called a *vesicle*. Following digestion, the vesicle containing indigestible material is excreted through the macrophage membrane. Macrophage are generally thought to originate in bone marrow. How they migrate to different parts of the body, including alveoli, is not fully understood except that they are very flexible and capable of amoebalike expansion and contraction that enable them to pass through minute openings in tissue.

Many toxic materials enter the body through the respiratory system, but only some affect the lung directly; the others enter the blood system or lymphatic system and harm other body organs. Lung disease will be divided into five categories:

1. *Irritation*. Air passageways become irritated, leading to constriction and perhaps edema and secondary infection.
2. *Cell damage*. Cells lining the air passageways are damaged, resulting in necrosis, increased permeability, and edema within the airway.
3. *Allergies*. Materials in pollen, cat hair, and similar substances excite nerve cells that cause muscles surrounding bronchiole to contract, which constricts airways and taxes the cardiovascular system.
4. *Fibrosis*. Lesions consisting of stiff protein structures appear on lung tissue, inhibiting lung function. Fibrosis of the pleura may also occur, which restricts movement of the lung and produces pain.
5. *Oncogenesis*. Tumors are formed in parts of the respiratory system.

Table 4-1 lists the principal occupational diseases of the lung and describes the effects of exposure to several occupational toxic substances.

### 4.4.1  Airway irritation

The cross-sectional area of elements in the bronchial tree are affected by many industrial chemicals. Most chemicals reduce the cross-sectional area of the airway, but a selected few enlarge the airway. *Dyspnea* is the anxious feeling that one cannot breath deeply and rapidly enough to satisfy respiratory demand and may be caused by a narrowing of the airways. Gases such as ammonia, chlorine, and formaldehyde vapor (Alenandesson and Hedenstierna, 1989) are soluble in water and produce dyspnea. Exposure does not produce chronic respiratory damage, but high concentrations can result in death. *Asthma* is a chronic and debilitating disease, causing swollen and inflamed airways that are prone to sudden and violent constrictions and can increase lung resistance 20-fold. Asthmatic attacks are characterized by shortness of breath and wheezing and can be life threatening. Asthmatic airway inflammation is initiated by airborne proteins (*allergens*). *Hay fever* is a similar disorder induced by hypersensitivity that affects the upper respiratory tract.

### 4.4.2  Ozone exposure

Ozone is an air pollutant generated by complex atmospheric reactions initiated by the photolysis of nitrogen dioxide $(NO_2)$ in the troposphere. Ozone can be generated indoors by corona from electronic air cleaners and incorrectly operating photocopy machines. Ozone is a lung irritant producing demonstrable short-term (acute) effects and possible long-term (chronic) effects (Tilton, 1989; Lipmann, 1989, 1991). Depending on dose, the acute effects are reduced lung function ($FEV_1$) for a period up to 42 hours, irritated throat, chest discomfort, cough, and headache. Bronchoalveolar lavage (BAL) measurements show that ozone produces lung inflammation. There is also a positive association between ambient temperature, ozone concentration, and daily hospital admissions for pneumonia and influenza. For asthmatics, additional effects include increased use of medication and restricted physical activity. Data to support chronic effects are limited and are open to multiple interpretations. Chronic ozone exposure during the summer seems to reduce lung function, which persists for a few months but dissipates by the spring. There is general agreement, however, that ozone contributes to premature aging of lungs. Studies with rats and monkeys support this conclusion.

### 4.4.3  Cellular damage and edema

A variety of substances can damage cells of the bronchial tree and alveoli, which in turn release fluid into these passageways. The site of the damage depends on the solubility of the material in water; the more soluble materials damage the nasal cavity, and less soluble materials damage upper elements of the respiratory tree. The production of fluids (edema) may take considerable time to evidence itself. Phosgene irritates the nasal cavity and upper respiratory passages because it reacts with the abundant supply of water to form carbon dioxide and hydrochloric acid. Ozone and nitrogen dioxide, on the other hand, are less soluble and penetrate the bronchiole and alveoli. Cadmium oxide fume is a submicron particle that travels to the alveoli and produces edema. Sustained exposure results in an irreversible destruction of alveoli and reduction of oxygen uptake by the blood. Nickel oxide, nickel sulfide fume, and vapors of nickel carbonyl damage the cells of whatever surface they reside on. Hydrocarbon vapors such as xylene and perchloroethylene have low solubility in water and travel to alveoli, where they diffuse to capillaries and are transported to the liver and other organs. Oxygenated intermediaries produce pulmonary edema.

*Emphysema* is a debilitating disease that reduces vital capacity of the lung and taxes the cardiovascular system. The disease is the result of three events:

1. Chronic infection of alveoli or bronchial tissue, produced by tobacco smoke or other irritants, increases mucus excretion and paralyzes cilia.
2. Infected lung tissue, entrapped air, and obstructing fluids in alveoli and alveolar ducts destroy alveolar walls and their web of capillaries (Figure 4-35).
3. Airway resistance increases, diffusing capacity decreases, and ventilation–perfusion ratios decrease.

Chronic emphysema progresses slowly. Insufficient oxygen produces hypoxia and taxes the cardiovascular system.

The symptoms of *pulmonary tuberculosis* are persistent cough, fever, chills, night sweats, tiredness, appetite loss, weight loss, and spitting of blood. Tuberculosis can also cause infections of bone, kidney, brain, and the

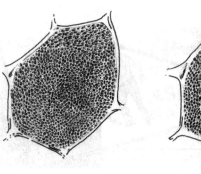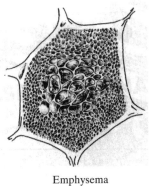

Normal               Emphysema

**Figure 4-35** Emphysema destroys normal alveolar walls and leaves large holes in the tissue.(Adapted from Occupational Health and Safety, Vol 58, No 3, March 1989)

lymphatic system. Tuberculosis is spread by droplets $(1 < D_p < 5 \; \mu m)$ containing the bacilli. Tuberculosis is highly contagious. The infection rate among close contacts or family members to an infected person is approximately 29%. A single laryngeal tuberculosis lesion may produce thousands to millions of organisms whenever an infected person sneezes, coughs, speaks, sings, or laughs. In a typical sneeze, approximately 1 million tuberculosis bacilli are contained in droplets that evaporate before the droplets reach the floor.

*Pneumonia* is a disease in which the alveoli are filled with fluid. A common pneumonia is caused by bacterial infection of the alveoli. Lung fluid, largely water but also containing red and white blood cells, enter alveoli and may, in time, fill the entire lung. Reduction of the total alveolar membrane and decrease of the ventilation–perfusion ratio is called *hypoxemia. Bronchitis* (Figure 4-36) is an inflammation of the mucous lining of the bronchial tree and results in an increased production of mucus.

### 4.4.4 Pulmonary fibrosis

*Pneumoconiosis* is the general class of disease of which pulmonary fibrosis is the central feature. Silica $(SiO_2)$ exists in several forms, of which *cristobalite* and *tridymite* induce fibrosis and quartz does not. *Silicosis*

is formation of silicotic nodules of concentric fibers of collagen 1 to 10 mm in diameter that appear in lymphatics around blood vessels beneath the pleura in the lungs and sometimes in mediastinal lymph nodes. The complete explanation of how pulmonary lesions are formed is lacking, even though silicosis has been recognized for centuries. Macrophage attach themselves to silica particles. It is believed that the lysosomal membrane of the macrophage ruptures and releases lysosomal enzymes that "digest" the macrophage. New macrophage repeat the process and the cycle is repeated. The damaged macrophage release some unknown material that stimulates the formation of *collagen* (a protein and major constituent in the intercellular connective tissue of meats that is not readily digested by most enzymes). The nodules may fuse and block blood vessels and reduce the flow of blood. Alveolar walls may be destroyed. The size of the alveolar sacs and ducts may enlarge, gas exchange can be reduced, and symptoms of emphysema occur.

*Asbestos* is the generic name for a group of hydrated silicates existing in the form of clustered fibers. The length, diameter, and number of fibrils (fibers of smaller diameter attached to the larger fiber) on each fiber vary. Exposure to asbestos can cause four types of disorders (Mossman et al., 1990):

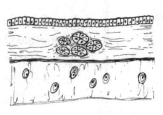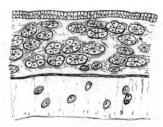

Normal               Chronic bronchitis

**Figure 4-36** Mucous gland layer in the normal human airway (left) contrasted with the airway from a patient with chronic bronchitis (right). Note the increase in size and number of mucous glands in chronic bronchitis. (Adapted from Occupational Health and Safety, Vol 58, No 3, March 1989)

1. Asbestosis
2. Lung cancer
3. Mesotheliomas of the pleura, pericardium, and peritoneum
4. Benign changes in the pleura

There are six forms of *asbestos*, but only three forms have been used to any extent in industry: *chrysotile*, *crocidolite*, and *amosite*, each is capable of causing disease. Chrysotile amounts to 95% of all the asbestos used in the United States. Concentrations of chrysotile dust particles in buildings rarely exceed 0.001 fiber/cm$^3$. There is evidence that chrysotile asbestos does not cause mesothelioma, even after heavy exposure. Other minerals—cummingtonite, grunerite , tremolite, anthophyllite, and actinolite—are defined as "asbestos" for regulatory purposes, but they have not been shown to cause disease in miners.

*Asbestosis* is pulmonary interstitial fibrosis associated with an excessive deposition of collagen that stiffens the lung and impairs gas exchange. *Lung cancers* are tumors in the tracheobronchial epithelial or alveolar cells. In general, lung cancers have been found in asbestos workers who are smokers and only rarely in nonsmokers. Diffuse malignant *mesothelioma* is a fatal tumor associated with mesothelial cells or underlying mesenchymal cells in the pleura, pericardium, and peritoneum. Mesothelioma is caused principally by crocidolite. Diagnosing mesotheliomas is difficult, as the tumor may resemble metastates of other tumor types and assume a wide variety of microscopic appearances that may be attributed to cancers of the gastrointestinal tract or other organs. A number of benign pleural changes may occur in asbestos workers that rarely cause functional impairment.

As with silicosis, a complete explanation describing how fibrosis occurs does not exist, although a great deal is known about some aspects of the process. The length and diameter of the fiber and the character of the fibrils comprising the total fiber are believed to influence fibrosis. Similar to silicosis, the incidence of disease is strongly enhanced by smoking. Bronchogenic carcinoma can be found in all portions of the bronchial tree. Interstitial fibrosis is most commonly found in the lower lobes of the lung. Asbestos itself is chemically inactive, but chemical carcinogens that reside on the fiber surface may also initiate cancer within the lung.

## 4.4.5 Respiratory allergies

An *allergy* is a hypersensitivity to a specific substance, called an *allergen*, which may be harmless to others. Many allergens are finely divided organic dust; others are proteins. An *antigen* is an enzyme, toxin, or other substance in the allergen to which the body reacts. More than 150 types of airborne allergens are currently recognized (Hamilton et al., 1992). Tree, grass and weed pollen, and molds are generated in the outdoor environment and may also enter the indoor environment through infiltration or ventilation. Allergens from house pets, insects, microorganisms, and aerosols from tobacco smoke, cooking, and other indoor activities are generated in the indoor environment. Allergens produce rhinitis, asthma, hay fever, and various reactive airway and alveolar diseases. The typical size of allergen particles is shown in Table 4-5.

**1.** *House pets.* Cat allergens are produced by sublingual salivary glands and hair root sebaceous glands of the domestic cat. Cat allergens bind strongly to clothing, carpet, upholstered furniture, or inert airborne dust and can be resuspended. Cat antigen is highly persistent and not affected by high temperatures or steam cleaning. Regular vacuuming does lit-

**Table 4-5**    Particle Size of Several Common Indoor Allergens

| Contaminant | $D_p(\mu m)$ |
|---|---|
| Pollen | 1–200 |
| Ragweed (oblate or prolate spheroids) | 20–30 |
| Dust mites (microscopic anthropods) | >10 |
| Guinea pig pelt | 0.8–4.9 |
| Rat urine | 0.8 |
| Fungal spores | 1–200 |
| Tobacco smoke | 0.01–1 |
| Viruses | 0.004–0.05 |
| Bacteria | 0.3–50 |
| Mycobacteria tuberculosis | 0.3–0.6 (length, 1–4 $\mu$m) |
| Legionella pneumophila | 0.4–0.7 (length, 30–50 $\mu$m) |

tle to reduce indoor antigen levels. Bird feathers and droppings shed albumins, proteins, and fungi that are allergens. The domestic dog produces complex protein allergens associated with hair, dander, saliva, feces, and urine. Dog allergens are highly persistent and remain stable in homes for extended periods.

2. *Household pests.* The most important source of allergens is the feces of dust mites that thrive in warm, moist conditions and are ubiquitous in human bedding. Dust mite antigens are excreted in fecal pellets; antigens are also associated with mite body debris. Dust mites have a life span of 3 to 3.5 months and thrive in a warm, humid environment. Dust mites consume human epithelial cells and animal dander in mattresses, pillows, carpets, and furniture. Skin flakes are shed by the abrasion of clothing and body parts rubbing against each other. An average skin flake has an equivalent diameter of 14 $\mu$m. It is estimated that the entire outer layer of skin is shed every day or two (Rothman, 1954) at a rate of 7 million skin flakes per minute (Clark and Cox, 1973). Tests of indoor environmental dust in homes and offices have shown it to be composed (70 to 90%) primarily of skin flakes (Clark and Cox, 1973; Clark, 1974). Assuming a density of 1 g/cm$^3$, 7 million skin flakes per minute corresponds to a mass emission rate of about 20 mg/min. Rat and mice allergens are generated from rodent hair, skin, feces, and urine. Allergens from cockroaches are produced from the exoskeleton and feces. Air pollution may also aggravate existing asthma, but air pollution is not believed (Cookson and Moffatt, 1997) to be responsible for the doubling of the disease in the last 20 years.

3. *Plant allergens.* During the peak pollen season, approximately 1 $\mu$g of pollen is inhaled each day. Many pollens release antigens within 30 seconds of hydration on mucous membranes or the respiratory tract. Allergic pollens can be produced by trees, scrubs, grasses, weeds, mosses, ferns, herbs, and many popular indoor plants. Ragweed allergens consist of pollen and fragments of leaves and stems. Ragweed pollen can transfer antigens to airborne inert dust, which can be resuspended. A variety of fungi and molds exist in the outdoor and indoor environment throughout the year.

4. *Bacteria.* Bacteria are generated as aerosols from humans, plants, animals, or water sources such as showers, urinals, and HVAC systems. Bacteria can accumulate quickly in the humidifier reservoirs. Soil bacteria are tracked into homes on the soles of shoes or attach themselves to airborne inert dust. Humans shed 7 million skin flakes per minute, each of which contains on average four viable bacteria. Sneezing, coughing, and singing generate airborne bacteria. Thermophilic actinomycetes are organisms that grow at elevated temperatures (104 to 158°F) and high humidity. Farm, sugarcane, tobacco, and mushroom workers are occupational groups at risk. Chronic hypersensitivity may result in inflammation of interstitial tissue leading to lung scarring and fibrosis. In severe cases the alveolar interstitium becomes fibrotic, bronchiole walls thicken, the alveoli widen, and edema and emphysema may occur.

Over 35 million Americans suffer from some kind of allergy. Fifteen million suffer from hay fever and 9 million have asthma. Asthma is associated with the death of over 4000 people annually. Certain types of white blood cells in allergic persons produce antibodies, which attach themselves to mast cells. *Mast cells* are usually found in the respiratory system, gastrointestinal tract, and the skin. When these antibodies detect an allergen, a large number of chemicals are produced, the best known of which is *histamine,* which produces watery eyes, runny nose, itching, and sneezing. The type of allergic response (e.g., asthma, hay fever or hives, etc.) depends on which part of the body interacts with the activated mast cells.

*Asthma* attacks may also be induced by viral infections or allergic reactions. Mast cells that induce the reaction are located in the bronchial tubes and lungs. Symptoms include swelling of the bronchial tubes and spastic contractions of the muscles surrounding the bronchial tubes. Both actions reduce the internal cross-sectional area of bronchial passages. Simultaneously, sensory nerve endings in the lung release chemicals that stimulate the secretion of excessive amounts of mucus. Asthma attacks are associated with wheezing and shortness of breath. Sulfur dioxide is a known respiratory irritant for asthmatics.

*Farmers lung* results from the inhalation of certain spores that inflame alveoli and produce fever and dyspnea. *Bagassosis* is a disease having similar symptoms inflicting workers handling dried and partially fermented sugarcane. *Byssinosis* is constriction of the bronchial tree, producing dyspnea in workers handling cotton, flax, and hemp. *Black lung* is a similar disease associated with coal mining. Numerous industrial chemicals, such as toluene diisocyanate (TDI) used in the manufacture of polyurethane plastic and methylisocyanate (MCI) used in the manufacture of insecticide, initiate allergic-like symptoms, and more serious symptoms if the exposure and concentration are sufficiently high.

## 4.5 Dose–response characteristics

A statement written by *Paracelsus* (Swiss physician, 1493–1541) is a basic tenant of modern toxicology: "What is it that is not a poison? All things are poison and nothing is without poison. It is the dose only that makes a thing a poison." This axiom has stood the test of time and should be recalled today to address the public's anxiety about substances that are in the air, water, and food we consume. The similarity between this quote and the line from Hamlet is striking: "There is nothing either good or bad but thinking makes it so."

Safe exposure limits are derived from experience and from experimental studies on humans and animals. People in the workforce comprise a population that is narrow compared to the general population. Such people are, on the whole, healthy adults in the prime of life who are exposed to contaminants for only several hours of the day and can return home to rest and breath cleaner air for the remainder of the day. Thus standards of occupational exposure should not be confused or equated with general environmental standards applicable for the general population. In setting standards for the outside environment, infants, the aged, and the ill and infirm have to be included in the population. There is no escape from the outside environment, so standards are more conservative.

Each person responds to the exposure of toxins in a unique way. Luckily, groups of people respond in ways that can be analyzed statistically and parameters can be defined that have statistical significance. Compilation of such data constitutes the basis for health standards and public policies prescribing the maximum exposure to which people may be subject in the workplace or the outside environment. Under-

lying these standards are the dose–response characteristics for specific toxins. The definitions of *dose* and *response* depend on the specific toxin but have the following general properties;

- *Dose:* total mass of toxin to which the body is subjected. The dose is a function of both the concentration of the toxin, duration of exposure, and the rate at which it is introduced to the body.
- *Response:* measurable physiological changes produced by the toxin.

Toxicity may be acute or chronic. An *acute toxicity* is sudden damage resulting from a single exposure to a large concentration of a toxin. *Chronic toxicity* is the accumulated damage from repeated exposure to small concentrations of a toxin over long periods of time. Reactions to these two different types of exposure have little resemblance to one another. Chronic toxicity is not predictable from knowledge of acute toxicity for the same chemical. For example, chemicals such as vitamin D, sodium fluoride, and chloride have a large acute toxicity but no chronic toxicity. On the other hand, a single ingestion of metallic mercury passes through the body without causing significant damage, whereas metallic mercury ingested in small amounts over a long period of time accumulates in the body and is very harmful.

Toxicity may also be local or systemic. *Systemic toxicity* is distributed throughout the body, while *local toxicity* pertains to particular organs. Entry to the body can be through the skin, by oral intake, or by inhalation. Figure 4-37 is a schematic diagram

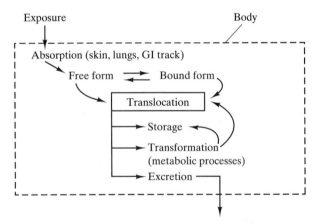

**Figure 4-37**    Pathways for the movement of toxic material

showing the various pathways that a toxin may take. Toxins can be categorized by the response they produce. Shown below are these categories and some example toxins.

- *Asphyxiant* (insufficient oxygen in blood): carbon monoxide
- *Irritant* (inflames tissue on contact): ammonia
- *Corrosive* (damages tissue on contact): chromic acid
- *Allergen* (material that affects the immune system): ragweed pollen
- *Carcinogen* (material that produces cancer): nickel carbonyl
- *Site-specific toxins*
  - *Pulmonary toxin* (lung): asbestos, beryllium, chromium
  - *Hepatoxin* (liver): carbon tetrachloride, nitriles
  - *Nephrotoxin* (kidney): kepone, lead, allyl chloride
  - *Brain toxin*: narcotics, ketones
  - *Skin toxin*: benzyl chloride
  - *Cardiotoxin* (heart): chloroethane
  - *Neurotoxin* (nervous system): mercury, malathion
  - *Ocular toxin* (eyes): methyl chloride, methanol, phenol
  - *Hematopoietic toxin* (blood stream): lead, benzene
  - *Bone toxin*: inorganic fluorides
- *Reproductive toxins* (materials that lead to low sperm production, genetic damage to egg or sperm cells, menstrual disorders): polychlorinated biphenyls
  - *Mutagen* (materials that alter the character of genetic material in cells): ethyleneimine
  - *Gametoxin* (material that damages sperm or ova): benzo[*a*]pyrene
  - *Teratogen* (material that interferes in the normal development of the fetus after conception and may result in miscarriage, visible birth defects, or defects not noticeable at birth): thalidomide
- *Transplacental carcinogen* (cancer-producing substance crossing the placenta that reaches the fetus): ethyl nitrosourea

Occupational exposure is traditionally assessed by comparing the concentration of the contaminant in inhaled air with concentration standards promulgated by professional health organizations. One could logically ask why one would measure the concentration in air presuming that it produces certain concentrations in the blood, urine, soft tissue, or bone, rather than measuring the concentration of the material or its by-products in these parts of the body. Certainly, measuring the concentration in blood and urine poses no technical difficulty. Such *biological monitoring* may replace air monitoring in the future, but until it does, Leung and Paustenbach (1988) recommend that a series of *biological exposure indices* (BEIs) should be developed. The biological exposure index (BEI) is the mass of contaminant per volume of the body component in which the contaminant is stored (i.e., blood, soft tissue, urine, etc.). Biological monitoring actually measures the concentration; the BEI infers the value based on measurement of external parameters.

*Pharmacokinetics* is the science that relates the rate processes of absorption, distribution, metabolism, and excretion of chemical substances in a biological system (Kreibel and Smith, 1990). To establish a BEI, one begins by establishing a relationship between the mass of contaminant in the body, *body burden*, and the concentration in inhaled air. For a first approximation, the relationship can be expressed as first-order kinetics. The mass rate of accumulation of a contaminant in the body is equal to its rate of absorption from inspired air minus its rate of removal (by all forms):

$$\frac{dm_{bb}}{dt} = FQc_a - k_r m_{bb} \qquad (4.50)$$

where

$m_{bb}$ = mass of contaminant in the body, referred to as the *body burden*

$Q$ = volumetric flow rate of inspired air

$c_a$ = contaminant mass concentration in air

$k_r$ = first-order rate constant describing the overall process that transforms the contaminant and/or removes the contaminant from the body

$F$ = *bioavailability* ratio of the mass of contaminant absorbed by the body to the mass of contaminant in inspired air

Integrating Equation 4.50, one obtains

$$m_{bb} = m_{bb,ss}[1 - \exp(-k_r t)] \qquad (4.51)$$

where the steady-state value of the body burden $(m_{bb,ss})$ is computed from

$$\frac{dm_{bb}}{dt} = FQc_a - k_r m_{bb} = 0 \qquad (4.52)$$

$$m_{bb,ss} = \frac{FQc_a}{k_r} \qquad (4.53)$$

By definition, the BEI concentration of the material in the blood is related to the body burden by

$$\text{BEI} = c_b = \frac{m_{bb}}{V_b} \qquad (4.54)$$

where $V_b$ is the volume of blood in which the material is stored. The blood may be the major component but it is not the only component; indeed, it may not even be the major component. Thus for any contaminant, $V_b$ represents an unknown that must be determined.

Independent experimental measurements can be conducted to reveal the values of $F$ and $k_r$. The bioavailability is determined by comparing the area under the concentration–time curves following equal doses of a chemical by intravenous injection and by inhalation. From Equation 4.51 it can be shown that the first-order rate constant is related to the half-life of the contaminant in the body:

$$k_r = \frac{0.693}{t_{1/2}} \qquad (4.55)$$

the time it takes the body burden to decrease by a factor of 2 after exposure to the contaminant in inhaled air has ceased.

---

**EXAMPLE 4.4    BEI OF ACETONE ASSOCIATED WITH ITS PEL**

Your superior asks you to estimate the BEI for a person exposed to acetone at a concentration equal to the PEL.

***Solution***    The 8-hour PEL for acetone in air is equal 750 ppm (1780 mg/m³). Not wanting to expose a person to the PEL, you conduct a series of experiments in which someone inhales air containing 545 ppm of

acetone (1293 mg/m³) at a ventilation volumetric flow rate $(Q_t)$ of 1.25 m³/h for 2 hours. The experiments reveal that after 2 hours, the concentration of acetone in the blood $(c_b)$ is 10 mg/L, the bioavailabilty $(F)$ is 45%, and the half-life $(t_{1/2})$ is 4 hours. The rate constant $k_r$ can be found from Equation 4.55:

$$k_r = \frac{0.693}{t_{1/2}} = \frac{0.693}{4} = 0.173 \text{ h}^{-1}$$

*Two-hour exposure.* The steady-state body burden for the 2-hours experiment $(m_{bb,ss\,2h})$ is found from Equation 4.53:

$$m_{bb,ss\,2h} = \frac{FQc_a}{k_r} = \frac{(0.45)(1.25 \text{ m}^3/\text{h})(1293 \text{ mg/m}^3)}{(0.173 \text{ h}^{-1})}$$

$$= 4204 \text{ mg}$$

The body burden after 2 h $(m_{bb\,2h})$ can be found from Equation 4.51:

$$m_{bb\,2h} = m_{bb,ss\,2h}[1 - \exp(-tk_r)]$$

$$= 4204\{1 - \exp[-(0.173)(2)]\} = 1230 \text{ mg}$$

From these experiments, the person's volume of blood can be estimated from Equation 4.54:

$$V_b = \left(\frac{m_{bb}}{c_b}\right)_{2h} = \frac{1230 \text{ mg}}{10 \text{ mg/L}} = 123 \text{ L}$$

*Eight-hour exposure.* The steady-state body burden corresponding to an 8-hours exposure to acetone at a concentration equal to its PEL (750 ppm, 1780 mg/m³) can be found from Equation 4.53:

$$m_{bb,ss\,8h} = \frac{FQc_a}{k_r} = \frac{(0.45)(1.25 \text{ m}^3/\text{h})(1780 \text{ mg/m}^3)}{0.173 \text{ h}^{-1}}$$

$$= 5788 \text{ mg}$$

The BEI equivalent to such an 8-hours exposure can be found from Equation 4.54:

$$\text{BEI}_{8h} = c_{b,8h} \frac{m_{bb,ss\,8h}}{V_b} = \frac{5788 \text{ mg}}{123 \text{ L}} = 47 \text{ mg/L}$$

Thus if a biological testing program reveals blood samples containing an acetone concentration of 47 mg/L or higher, it suggests that a person was exposed to acetone concentrations equal to or greater than the PEL and that corrective actions should be taken.

Similar correlations can be obtained for other air contaminants that are absorbed primarily in the blood. For contaminants that are absorbed in the urine, soft tissue, or bone, similar correlations can be obtained, but the task becomes progressively more difficult.

The body's response to a contaminant depends on how the contaminant affects various organs. In the simplest case, the contaminant is a mild irritant that is metabolized rapidly and the response is directly proportional to the ambient concentration $(c_a)$:

$$R_p = kc_a \qquad (4.56)$$

where $R_p$ is a measurable response and $k$ is a constant. At higher concentrations, the initial response may be superseded by a secondary acute or chronic response that may be cumulative. If the material accumulates in the body, the response can be expressed in a form called *Haber's law*:

$$R_p = kc_a t^n \qquad (4.57)$$

The coefficient $n$ is not necessarily unity, nor is it necessarily constant and may have higher values for progressive stages of the disease. The body burden $(m_{bb})$ is highest for fat-soluble organics, organic bases, and weak acids. Asbestos, chlorocarbons, and radioactive fluorides may be stored for life, while lead and DDT are retained for years. In contrast, the body burden for water-soluble compounds is low.

At the cellular level, the response is proportional to the rate of change with time of the effective local concentration $(c)$,

$$R_p = k \frac{dc}{dt} \qquad (4.58)$$

The *effective local concentration* is the concentration within tissue, in contrast to $c_a$, which is the concentration in air. The effective concentration at the site depends on the ambient concentration $(c_a)$, rate of metabolism, and rate of absorption. The response

$$R_p = k_{ab}c_a - k_r c \qquad (4.59)$$

where $k_{ab}$ is the rate of absorption and $k_r$ is the metabolic removal rate. For drugs, a moderate value of $k_r$ is sought so that all the therapeutic agent reaches its target and the dose can be kept low. The respiratory system is an efficient transfer system, and many con-

taminants can readily be absorbed and transmitted to certain organs. Carbon monoxide is a classic example.

*Biotransformation* can render a contaminant harmless or toxic. Formaldehyde is converted by the liver to formic acid, which is a normal metabolic by-product and harmless at modest concentrations. On the other hand, aromatic compounds such as benzene and benzopyrene are metabolized and form highly reactive products called *epoxides* that initiate tumors.

The response of a person to a given dose cannot be predicted accurately, owing to the unique characteristics of individuals. Groups of people, on the other hand, display dose–response characteristics that can be described statistically. Of great use are cumulative distribution functions, median values and standard deviations. Three commonly used ranges are easily determined by experiment:

1. A dose from zero to a threshold value $(c_t)$ at which a response is observed
2. A dose at which 100% of the subjects manifest a certain response
3. A selected midrange between ranges 1 and 2: for example, a median value of 50%.

It is common to express dose–response data as a cumulative distribution graph and to identify the median 50% response dosage and the standard deviation.

Arsenic and selenium are important trace metals needed by the body. If the daily dose is below some threshold, a deficiency occurs which may cause disease. As the dose increases above the necessary level, the body will eliminate it. If the dose exceeds the body's ability to metabolize or eliminate the material, it will accumulate and cause another type of disease. In pharmacology, the *Ehrlich index*, shown in Figure 4-38, is used to elucidate the safety of substances. The Ehrlich index is the difference in concentration between the 95% point of the curative dose–response curve and the 5% point of the toxic dose–response curve. The larger the Ehrlich index, the safer the material. In the case of mercury, which was once used to treat syphilis, the index is small. For anesthetics the index is larger but not large enough to be indifferent to their effects.

Experimental data are often acquired at large doses well in excess of normal exposures, followed by extrapolation of the data to estimate the response

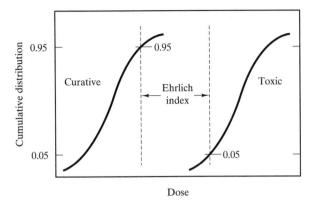

**Figure 4-38**   Ehrlich index

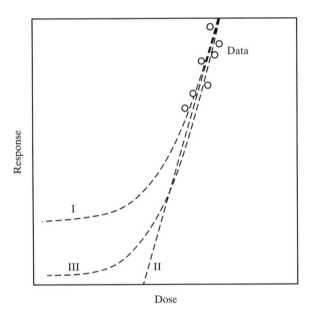

**Figure 4-39**   Methods to extrapolate dose-response data for three classes of toxins

at lower dosage. The way that extrapolations are made is critical and the cause for much controversy. Figure 4-39 shows three ways that extrapolations can be made. Certain highly toxic bacteria such as the tubercle bacillus and certain carcinogens are toxic at very small concentrations and low dosages. Curve I represents such cases. Other air contaminants are quite harmless at low concentrations and require a certain threshold dose before any response is observed. In this case a straight-line extrapolation

through the threshold (curve II) is reasonable. Naturally occurring materials such as CO, $CO_2$, $NH_3$, and HCHO represent intermediate cases (see curve III) because there is always a certain amount of such materials in the lungs and airways in excess of what is present in the atmosphere.

Since toxins produce different effects, it is difficult to generalize dose–response characteristics for all toxins. Sulfur dioxide constricts the tracheobronchial system. Thus the airway resistance is higher than normal (Colucci and Strieter, 1983). Second, sulfur dioxide irritates and inflames bronchial tissue. Two parameters are used to quantify exposure, dose rate and total dose:

$$\text{dose rate} = D_{\min} = Qc \qquad (4.60)$$

$$\text{total dose} = D_t \qquad (4.61)$$

The *total dose* is the cumulative (or time-averaged) amount of sulfur dioxide introduced to the body, but as an integrated quantity it obscures any possible distinction between slow breathing, length of exposure period, and concentration that may exist. Previous studies failed to establish a close correlation between response and total dose. The *dose rate* is the instantaneous assault of sulfur dioxide on the body. To study the effect of sulfur dioxide during exercise when inhaled volumetric flow rates vary considerably, the dose rate is the more attractive dose parameter. If the toxin was cumulative and the time for it to be transported to body organs was significant, the total dose might be more attractive. The use of dose rate is logical for toxins such as sulfur dioxide, ozone, PAN, and other irritants that produce an immediate response.

*Carbon monoxide* is a well-studied toxin (McCartney, 1990). It is toxic because *hemoglobin* absorbs carbon monoxide more readily than oxygen. The brain fails to receive sufficient oxygen and produces effects such as reduced visual acuity, psychomotor skill, and pulmonary function, and eventually, death. Carbon monoxide can aggravate angina, a pain in the chest and left arm caused by a sudden decrease in blood supply to the heart muscle. For sedentary individuals, breathing at approximately the same volumetric flow rate, the total dose is an attractive dose parameter. The response parameter could be altered by behavioral responses such as loss of psychomotor skills, but the direct measurement of *carboxyhemoglobin* in the blood is a more accurate measure of response. Figure 4-40 is a dose–response curve for carbon monoxide. It should be noted

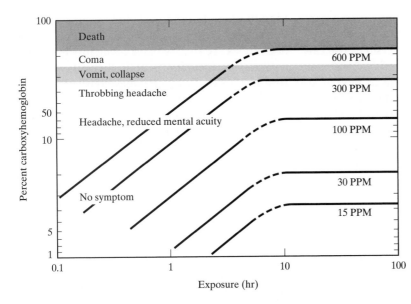

**Figure 4-40**  Response to carbon monoxide as a function of concentration (ppm) and time. The OSHA 8-hr PEL is 35 ppm and the EPA Primary Air Quality Standard is 9 ppm, (redrawn from Seinfeld 1986)

that CO toxicity is serious, resulting in approximately 800 accidental deaths per year in the United States.

Ozone is a highly reactive gas and strong oxidant. It reacts with body fluids and tissue to impair lung function in the short term (Figures 4-28 and 4-29). In simple terms, ozone limits the ability to take a deep breath. A demonstrative effect of ozone is a reduction in the forced expiratory volume at 1 second $(FEV_1)$. Figure 4-41 is a compilation (Tilton, 1989) of several studies and shows that the forced expiratory volume in 1 second is reduced for healthy males. Following exposure, lung function is partially restored in seven days, and complete restoration occurs in two weeks. In the long term, the effects on health are twofold:

there is a transient reduction in the resistance to infection, and it is believed that there may be chronic damage to the gas-exchange region in the lung.

The dose–response relationships for carcinogens are quite different from those for ozone, carbon monoxide, and sulfur dioxide. *Cancer* can be thought to progress through several stages: *initiation*, *promotion*, and *progression* (Ricci and Molton, 1985). Initiation is an irreversible lesion in the DNA that leads to cancer if further attack occurs. The attack can occur through exposure to chemicals or other agents, such as viruses. Promotion is a biochemical process that accelerates progression of the initiated cell to cancer. If a promoter attacks an uninitiated cell, the damage is thought to be reversible.

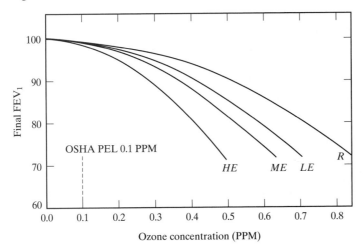

**Figure 4-41**  Final $FEV_1$, for healthy young adult males exposed to ozone for 2-hr consisting of alternating 15min periods of exercise and rest at four levels of exercise; light exercise (LE) 24 L/min $< Q_t < 43$ L/min; moderate exercise (ME) 44 L/min $< Q_t$ $< 63.1$ L/min; heavy exercise (HE) 64.1L/min $> Q_t$ (redrawn from Tilton 1989)

Cancer can be caused by a "single hit" of a toxicant on a DNA molecule in the target organ. The resulting *point mutation* may or may not be reversible. The carcinogen–DNA complex is called an *adduct* and may be removed from the DNA molecule as new unaffected DNA comes into being. If the DNA adduct is retained, it is likely to alter cellular control and initiate the cancer process. Adduct formation is an index of exposure but not necessarily an index of initiation, since continued attack on the DNA does not mean that cancer necessarily develops.

Some chemicals are transformed into carcinogens through human metabolism. For example, Benzo[*a*]pyrene must be metabolized to an epoxide which reacts with DNA in the cell to induce cancer. Other chemicals, such as bis(chloromethyl) ether, do not appear to need metabolic conversion to be reactive. The existence of a threshold for a cancer is a point of contention among professionals who assess risk. The argument *against a threshold* assumes that a single hit will lead to the uncontrolled growth of a somatic cell and eventually produce cancer. Arguments *for a threshold* are based on the existence of gene repair mechanisms and immune defenses within the body. The *one-hit model* assumes that a single hit causes irreversible damage to DNA and leads to cancer. In the *multistage model,* a cell line must pass through $N$ stages before a tumor is initiated irreversibly. The rate at which cell lines pass through one or more stages is a function of the dose rate. In the *multihit model,* $N'$ dose-related hits to sensitive tissue are required to initiate a cancer. The Weibel model assumes that these hits occur in a single cell line and that different cell lines compete independently to produce tumors. In all these models, it is assumed that the rate at which dose-related hits occur is a linear function of the dose rate. The most important difference between the multistage model and either the Weibel or multihit models is that in the Weibel and multihit models all hits result from the dose, whereas in the multistage model, passage through some of the stages can occur spontaneously.

### EXAMPLE 4.5 ONE-HIT MODEL, BENZENE AND RISK OF LEUKEMIA

Considerable evidence exists linking exposure to benzene vapor with leukemia. Benzene is metabo-

lized to phenols, quinones, and unstable oxygenated benzene derivatives. These oxygenated metabolic derivatives are reactive and covalently bind to tissue constituents, including genetic elements in bone marrow. In 1969 the government established a *maximum, average (8-hours) threshold limit value* (TWA-TLV) of 10 ppm. The standard produced considerable debate and legal challenge because more conservative interests wanted a TWA-TLV of 1 ppm. White et al. (1982) described how the one-hit model can be used to quantify risk for different benzene exposures and different dose rates, thus providing regulators a quantitative way to select exposure limits.

The one-hit model presumes no threshold value and is based on the assumption that a single encounter with a carcinogen triggers a series of biological events leading to cancer. The basis of White's analysis were 314 white male workers who were exposed to benzene in the production of a rubber film for a period exceeding 5 years during the years 1940–49. Between 1950 and 1975, 5 of the 314 (probability $1.592 \times 10^{-2}$) workers died of leukemia compared to the national average of 758 per 100,000 (probability $7.58 \times 10^{-3}$) of comparable workers in other industries during 1940–1949. Death rates from all types of leukemia were taken from the 1975 mortality tables for U.S. white males for 1937–1977, and a background probability $(P_0)$ was found to be 707 per 100,000 $(P_0 = 7.07 \times 10^{-3})$. The *standard mortality ratio* (SMR) is defined as

$$\text{SMR} = \frac{\text{observed deaths}}{\text{expected deaths}} = \frac{0.01592}{0.00758} = 2.10$$

Thus the rubber-film workers experienced twice as many leukemia cancers. Over the 1937–1975 period, the TWA-TLV was steadily lowered.

| Date | TWA-TLV (ppm) |
|---|---|
| 1937–1940 | 150 |
| 1941–1946 | 100 |
| 1947 | 35 |
| 1948–1956 | 25 |
| 1957–1962 | 25 |
| 1963–1968 | 25 |
| 1969–1975 | 10 |

Over the period of 1940–1949, the average value of the TWA-TLV was 83 ppm. Over the period 1937–1975, measurements of the actual benzene concentrations in the workplace was on the average 50 ppm. The mathematical expression for the one-hit model is

$$P_d = \frac{P_t - P_0}{1 - P_0} = \frac{P_t}{1 - P_0} - \frac{P_0}{1 - P_0}$$

$$= 1 - \exp(-BD)$$

where

$B =$ constant that needs to be determined from independent data

$D$ (dose) = (concentration) (exposure time) (ppm·yr)

$P_d =$ excess cancer probability due to benzene exposure

$P_t =$ total risk probability of developing leukemia $(1.59 \times 10^{-2})$

$P_0 =$ background risk probability in the absence of the exposure $(7.58 \times 10^{-3})$

Combine the two equations above:

$$\text{SMR} = \frac{P_t}{P_0}$$

$$= 1 + \frac{1 - P_0}{P_0}[1 - \exp(-BD)]$$

$$= 1 + P_d \frac{(1 - P_0)}{P_0}$$

$$= 1 + P_d \frac{1 - 7.58 \times 10^{-3}}{7.58 \times 10^{-3}}$$

$$= 1 + (130.9)(P_d)$$

The single-hit equation presumes that risk depends only on the benzene dose (ppm·yr) to which workers are exposed and that a certain level of risk is associated with the dose, irrespective of whether exposure occurs over a long period of time at low concentration or whether exposure occurs over a short period of time at high concentration. For purpose of calculation, estimate the constant $B$ based on two types of exposure: (1) actual 5-year exposure of 314 workers at 83 ppm and (2) 30-year exposure of workers at a TWA-TLV of 50 ppm.

The one-hit model equation above can be rearranged to compute the value of $B$,

$$B = \frac{-1}{D} \ln \frac{1 - \text{SMR}(P_0)}{1 - P_0}$$

Two values of $B$ can be found corresponding to $D_{5\text{yr}}$ and $D_{30\text{yr}}$:

**1.** *Five-year exposure, high concentration* $\left(83 \text{ PPM for 5 years}, D_{5\text{yr}} = 415 \text{ PPM}\cdot\text{yr}\right)$

$$B_{5\text{yr}} = \left(-\frac{1}{415} \text{ PPM}\cdot\text{yr}\right) \ln\left[1 - \frac{(2.1)(7.58 \times 10^{-3})}{1 - 7.58 \times 10^{-3}}\right]$$

$$= 2.034 \times 10^{-5} (\text{PPM}\cdot\text{yr})^{-1}$$

**2.** *Thirty-year exposure, low concentration* $\left(50 \text{ PPM for 30 years}, D_{30\text{yr}} = 1500 \text{ PPM}\cdot\text{yr}\right)$

$$B_{30\text{yr}} = \left(-\frac{1}{1500} \text{ PPM}\cdot\text{yr}\right) \ln$$

$$\left[1 - \frac{(2.1)(7.58 \times 10^{-3})}{1 - 7.58 \times 10^{-3}}\right]$$

$$= 5.626 \times 10^{-6} (\text{PPM}\cdot\text{yr})^{-1}$$

For $B$ values obtained in cases (1) and (2), one can estimate the excess leukemia risks $(P_d)$ associated with the two possible TWA-TLV values considered by NIOSH (i.e., 10 and 1 ppm) for people working over lifetimes of 45 and 25 years:

$$P_d = 1 - \exp(-BD)$$

*Case I: 45-year working lifetime.* For a TWA-TLV = 10 ppm and a working lifetime of 45 years, the standard mortality ratios (SMRs) are

$$P_{d,5\text{yr}}(45 \text{ yr}) = 1 - \exp(-B_{5\text{yr}}D)$$

$$= 1 - \exp\left[-(2.034 \times 10^{-5})(10 \text{ PPM})(45 \text{ yr})\right]$$

$$= 9.111 \times 10^{-3}$$

$$P_{d,30\text{yr}}(45 \text{ yr}) = 1 - \exp\left[-(5.626 \times 10^{-6})(10)(45)\right]$$

$$= 2.528 \times 10^{-3}$$

$$\text{SMR}_{5\text{yr}}(45 \text{ yr}) = 1 + (130.9)(9.111 \times 10^{-3}) = 2.19$$

$$\text{SMR}_{30\text{yr}}(45 \text{ yr}) = 1 + (130.9)(2.528 \times 10^{-3}) = 1.33$$

$$\text{Average SMR} = 1.76$$

For a TWA-TLV = 1 ppm and a working lifetime of 45 years, the SMRs are

$$P_{d,5yr}(45 \text{ yr}) = 1 - \exp(-B_{5yr}D)$$
$$= 1 - \exp[-(2.034 \times 10^{-5})(1)(45)]$$
$$= 9.149 \times 10^{-4}$$

$$P_{d,30yr}(45 \text{ yr}) = 1 - \exp[-(5.626 \times 10^{-6})(1)(45)]$$
$$= 2.531 \times 10^{-4}$$

$$\text{SMR}_{5yr}(45 \text{ yr}) = 1 + (130.9)(9.149 \times 10^{-4}) = 1.12$$

$$\text{SMR}_{30yr}(45 \text{ yr}) = 1 + (130.9)(2.53 \times 10^{-4}) = 1.03$$

$$\text{Average SMR} = 1.08$$

*Case II: 25-yr working lifetime.* For a TWA-TLV = 10 ppm and a working lifetime of 25 years, the SMRs are

$$P_{d,5yr}(25 \text{ yr}) = 5.07 \times 10^{-3}$$

$$P_{d,30yr}(25 \text{ yr}) = 1.45 \times 10^{-3}$$

$$\text{SMR}_{5yr}(25 \text{ yr}) = 1.66$$

$$\text{SMR}_{30yr}(25 \text{ yr}) = 1.19$$

$$\text{Average SMR} = 1.42$$

For a TWA-TLV = 1 ppm and a working lifetime of 25 years, the SMRs are

$$P_{d,5yr}(25 \text{ yr}) = 5.08 \times 10^{-4}$$

$$P_{d,30yr}(25 \text{ yr}) = 1.41 \times 10^{-4}$$

$$\text{SMR}_{5yr}(25 \text{ yr}) = 1.07$$

$$\text{SMR}_{30yr}(25 \text{ yr}) = 1.02$$

$$\text{Average SMR} = 1.04$$

*Conclusions.* For a 45-year working lifetime, reducing the TWA-TLV to 1 ppm results in risks to leukemia that are little different than for the population generally, whereas a 10 ppm standard produces a risk to leukemia comparable to what workers experienced in 1940–1949. Reducing the workplace benzene concentration by a factor of 10 reduces the risk to leukemia by 39% for a worker who was exposed for 45 years. If a more traditional working lifetime of 25 years is used as the basis of comparison, the 10-fold reduction of the TWA-TLV reduces the risk to leukemia by 14%.

This example demonstrates how risk-reduction strategies can be quantified. The question that remains is whether the benefits gained are commensurate with the cost to achieve them. While the engineers are com-fortable with quantitative discussions, the decision to set a new TWA-TLV is a political decision and engineers must learn how to cope with the emotional factors entering all political discussions. At times, people advocating a particular course of action deliberately mislead by using numerical data irresponsibly. For example, someone might suggest that reducing the TWA-TLV by a factor of 10 will reduce the risk to leukemia by a factor of 10. This is clearly not the case and it is hoped that engineers will rise to the occasion and correct such misleading claims.

## 4.6 Risk analysis

Products of incomplete combustion are a mixture of organic and inorganic compounds. Most studies of mutagenicity (DeMarini et al., 1996) involve extracting organic material from captured particles with solvents [i.e., *extractable (particulate) organic material* (EOM)]. Mutagenic activity is measured by the number of mutants produced in *Salmonella* per unit mass of EOM. Because *Salmonella* mutagenic assays use mutant cells that revert to phenotypically wild-type cells, the resulting mutants are called *revertants* (rev). Mutagenic potency (rev/μg of EOM) can also be expressed as a mutagenic emission factor in which mutagenic activity is expressed per mass of fuel or the amount of energy released in combustion. Table 4-6 shows *mutagenic emission factors* for various combustion processes. Expressing the data in these terms shows that open, uncontrolled burning has greater mutagenic activity than controlled burning in combustors designed for this purpose. Even with controlled combustors, mutagenic emission factors enable engineers to rate the carcinogenic risk from different fuels and combustors. Engineers need to know how these materials are diluted in the atmosphere and are transported downwind to reach a person's breathing zone. This is the subject of Chapter 9.

Detecting a contaminant's ability to cause acute effects is relatively straightforward since acute effects appear immediately. Assessing chronic effects are much more difficult to analyze since chronic effects manifest themselves over a long period of time, as much as decades after exposure. Figure 1-3 shows that no toxicity information is available for over 46,000 of the 65,000 chemicals used in industry. A

| | Table 4-6 | Mutagenic Emission Factors of Combustion Products | |
|---|---|---|---|
| **Source** | **Revertants/μg EOM** | **Revertants/kg fuel × 10⁵** | **Revertants/MJ** |
| Residential heating | | | |
| Wood | 1 | 50 | 250,000 |
| Oil | 2–5 | 1 | 2,500 |
| Coal-fired power plant | 0.5–9.4 | 0.06 | 230 |
| Vehicles | | | |
| Diesel | 1.4–15.1 | 40 | |
| Gasoline with catalyst | 8.6 | 1 | |
| Incineration | | | |
| Municipal waste | 1–27 | 1–12 | 300–38,000 |
| Medical/pathological waste | 1–2 | 0.7 | |
| Hazardous waste (pesticide) | 0.1–18 | 1–2 | |
| Rotary kiln | | | |
| Natural gas | 0.8 | 0.01 | 20 |
| Toluene | 20 | 8.97 | 13,730 |
| Polyethylene | 400 | 6.43 | 19,050 |
| Open burning | | | |
| Agricultural plastic | 0.2–1.8 | 100 | 250,000 |
| Scrap-rubber tires | 2–12 | 800 | |
| Cigarette smoke condensate | 0.5– 2.4 | | |
| Coke oven gas | 3.4–3.7 | | |
| Roofing coal tar | 0.5–78.1 | | |
| Urban air | 0.1–8.2 | | |

*Source:* DeMarini and Lewtas (1995).

complete assessment of health hazards (including chronic effects) has been made for 6400 (only 9.8%) of the total chemicals used in commerce. Furthermore, it has been estimated that the world's laboratories are capable of studying the toxicity of only 500 compounds a year, woefully less than the 700 to 1000 new chemicals that are introduced each year. Since a test for carcinogenicity for a single compound may take as long as three years and cost $250,000 or more, it is naive to believe that society will allocate money to establish the chronic risk of all the chemicals with indisputable certainty. Alternatively, tests to detect a chemical's genotoxicity can be conducted in a short period of time and at a considerably lower cost. There is a considerable body of knowledge to suggest that latent diseases such as cancer, birth defects, and genetic disease begin by altering DNA.

For potential carcinogens, the current regulatory attitude (Rich, 1990) is extremely conservative and assumes that even small concentrations can cause cancer. Such an attitude is contrary to the traditional regulatory approach for noncarcinogens, in which there is assumed to exist a threshold concentration below which there are no observable effects. For potential carcinogens a variety of mathematical models are used to extrapolate from high to low dose. The EPA currently favors a linearized multistage model, which provides a 95% upper bound estimate of cancer incidence at a given dose. The slope of the dose–response curve called the *carcinogen potency factor* or *unit risk factor* $(r_{u,j})$, is used to estimate the probability of cancer associated with a particular exposure. Table 4-7 lists the risk factors $(r_{u,j})$ for 17 common chemicals of commerce. The probability of developing cancer following exposure is called the *excess lifetime cancer risk* $(r_e)$ and is the estimate of increased risk of cancer above that occurring in the general population. The excess lifetime cancer risk is the product of the exposure and the unit risk for each constituent. Excess lifetime risk is a measure of the probability that a person might develop cancer resulting from exposure to a certain contaminant at a certain concentration over a 70-year exposure. It is calculated by multiplying

**Table 4-7** EPA Unit Risk Factors[a]

| Chemical (CAS number) | Unit risk factor $(\mu g/m^3)^{-1}$ |
|---|---|
| Acetaldehyde (75-07-0) | $3.96 \times 10^{-6}$ |
| Acrylonitrile (107-13-1) | $0.48 \times 10^{-4}$ |
| Arsenic (7740-38-2) | $27 \times 10^{-2}$ |
| Benzene (71-43-2) | $2.65 \times 10^{-5}$ |
| Benzo[a]pyrene | $1.75 \times 10^{-2}$ |
| Beryllium (7440-41-7) | $8.85 \times 10^{-4}$ |
| 1,3-Butadiene (106-99-0) | $0.19 \times 10^{-4}$ |
| Cadmium (7440-43-9) | $8.28 \times 10^{-3}$ |
| Carbon tetrachloride (56-23-5) | $9.44 \times 10^{-5}$ |
| Chloroform (67-66-3) | $1.12 \times 10^{-4}$ |
| Chromium VI (7440-47-3) | $2.55 \times 10^{-2}$ |
| 1,2-Dichloroethane (75-34-3) | $1.05 \times 10^{-4}$ |
| 1,1-Dichloroethylene (540-59-0) | $1.98 \times 10^{-4}$ |
| Epichlorohydrin (106-89-8) | $4.54 \times 10^{-6}$ |
| Ethylene dibromide (106-93-4) | $1.69 \times 10^{-3}$ |
| Ethylene oxide (75-21-8) | $1.80 \times 10^{-4}$ |
| Formaldehyde (50-00-0) | $1.60 \times 10^{-5}$ |
| Gasoline (marketing) | $1.69 \times 10^{-6}$ |
| Hexachlorobenzene | $5.71 \times 10^{-3}$ |
| Methylene chloride (75-09-2) | $1.63 \times 10^{-6}$ |
| Nickel-refinery dust (7440-02-0) | $2.36 \times 10^{-3}$ |
| Nickel subsulfide | $1.18 \times 10^{-3}$ |
| Perchloroethylene (127-18-4) | $3.93 \times 10^{-6}$ |
| Propylene oxide (75-56-9) | $8.79 \times 10^{-5}$ |
| Styrene (100-42-5) | $2.43 \times 10^{-6}$ |
| Trichloroethylene (79-01-6) | $9.28 \times 10^{-5}$ |
| Vinyl chloride (75-01-4) | $1.05 \times 10^{-5}$ |

[a] These factors are subject to change, and only up-to-date values should be used.

the unit risk factor by the long-term average concentration $c_j (\mu g/m^3)$:

$$r_e = r_{u,j} c_j \qquad (4.62)$$

where the subscript $j$ refers to the chemical in question. The *maximum lifetime individual risk* is calculated by multiplying the unit risk factor by the highest concentration to which a person is exposed:

$$r_e(\max) = r_{u,j} c_j(\max) \qquad (4.63)$$

For example, if the unit risk factor $(r_{u,j})$ is $4.0 \times 10^{-5}$ $(\mu g/m^3)^{-1}$ and the highest concentration is $3 \mu g/m^3$, the maximum individual lifetime risk would be

$1.2 \times 10^{-4}$. This means that there is little more than one chance in $10^4$ (risk $1 \times 10^{-4}$) that a person residing in the area will contract cancer as a result of exposure to this pollutant for a period of 70 years. Note that nationwide, the risk for all cancers is $2.8 \times 10^{-3}$ and cancer related to cigarettes is $3.6 \times 10^{-3}$ (Table 1-2).

For several constituents and exposures, the excess lifetime cancer risk is the sum of the products of the unit risk potency factor and exposure for each constituent:

$$r_e = \sum [(r_{u,j})(c_j)]_n \qquad (4.64)$$

where the subscript $(n)$ represents the total number of contaminants $(j)$ to which one is exposed.

An excess lifetime cancer risk $(r_e)$ of less than $10^{-7}$ corresponds to the likelihood that one person in 10 million is apt to contract cancer after a period of 70 years of continuous exposure. The involuntary risk of $10^{-6}$ is the risk of death by disease for the entire U.S. population. Figure 1-2 shows that a risk of $10^{-7}$ is comparable to the involuntary risk of fatality riding the railroad. An excess lifetime cancer risk of $10^{-7}$ is generally considered acceptable. Risks between $10^{-4}$ and $10^{-7}$ (Table 1-2) resulting from indentifiable causes are common targets for remediation.

The unit risk factor is a quantitative estimate of carcinogenic potency and expresses the chance of contracting cancer (though not necessarily life threatening) from continuous exposure to a contaminant of concentration of $1 \mu g/m^3$ in air for a period of 70 years. For contaminants in water or food, other units may be used. The unit risk is a plausible upper estimate of the risk referred to as the 95% confidence level. With such an estimate, the true risk is not likely to be higher but it could be lower.

Cancer risks may be reported in terms of risk estimates for a person (*excess lifetime individual risk*, $r_{u,j}$) or risk estimates for an exposed population (*aggregate risk*, $r_{a,j}$). Assessing a group's risk to cancer from exposure to a contaminant $(j)$ requires three pieces of data:

1. Carcinogenic potency of unit risk factor $(r_{u,j})$
2. Concentration $(c_j)$
3. Number of people exposed $(N)$

*Aggregate risk* $(A)$ pertains to all people within the exposed group. Aggregate risk is equal to the unit risk factor $(r_{u,j})$ times the sum of the products of the number of people multiplied by the estimated concentration to which they were exposed:

$$A_{a,j} = r_{u,j} \sum P_n c_n \qquad (4.65)$$

where $j$ is the contaminant in question and $P_n$ is the population that is exposed to contaminant $j$ whose concentration is $c_n$. If the aggregate risk is normalized for the entire population $(a_{a,j})$ it can be expressed as

$$a_{a,j} = \frac{A_{a,j}}{\sum P_n} = \frac{r_{u,j} \sum P_n c_n}{\sum P_n} \qquad (4.66)$$

### EXAMPLE 4.6   CANCER RISK TO A POPULATION LIVING IN THREE DIFFERENT COMMUNITIES

An industrial park contains several industries that use 1,2-dichloroethane, a colorless, flammable liquid with a pleasant sweet odor. It is used in the manufacture of plastics, as a solvent in resins, as a degreaser in textiles, and as an extracting agent for soybean oil and caffeine. For years the PEL was 50 ppm, 200 mg/m$^3$. The compound irritates the skin and eyes, and prolonged exposure is associated with liver and kidney damage. For decades, individual companies in the park discharged small amounts of the vapor to the atmosphere unaware that other industries were doing the same thing. Communities downwind of the industrial park experienced the following average annual concentrations.

| Community | Population | Annual average concentration ($\mu$g/m$^3$) |
|-----------|-----------|---------------------------------------------|
| A | 10,000 | 4.0 |
| B | 30,000 | 1.0 |
| C | 100,000 | 0.2 |

From Table 4-7, the unit risk factor $(r_u)$ for 1,2-dicholoroethane is found to be $1.05 \times 10^{-4} \, (\mu\text{g/m}^3)^{-1}$. The individual risk $(r_j)$ for each community is

$$r_{j,\text{A}} = (4 \, \mu\text{g/m}^3)(1.05 \times 10^{-4})(\mu\text{g/m}^3)^{-1}$$
$$= 4.2 \times 10^{-4}$$

$$r_{j,\text{B}} = (1 \, \mu\text{g/m}^3)(1.05 \times 10^{-4})(\mu\text{g/m}^3)^{-1}$$
$$= 1.02 \times 10^{-4}$$

$$r_{j,\text{C}} = (0.2 \, \mu\text{g/m}^3)(1.05 \times 10^{-4})(\mu\text{g/m}^3)^{-1}$$
$$= 0.21 \times 10^{-4}$$

The aggregate risk incurred by the three communities is

$$A_a = r_u \sum P_n C_n$$
$$= (1.05 \times 10^{-4})[(4)(10^4) + (1)(3 \times 10^4)$$
$$+ (0.2)(10^5)] = 9.45$$

Thus a total of 9.45 cases of cancer may occur if the 140,000 people were exposed to 1,2-dichloroethane for 70 years. The anticipated number of cases per year would be 9.45/70, or 0.135 case/yr (between one and two cases every 10 years). The anticipated risk $(a_a)$ for the population of these three communities is

$$a_a = \frac{0.135 \text{ case/yr}}{140,000 \text{ people}}$$
$$= 9.64 \times 10^{-7} \text{ case/person} \cdot \text{yr}$$

which is approximately three orders of magnitude smaller than the incidence of cancer for the entire U.S. population, where there are 454,000 cancer deaths a year:

$$a_a \, (\text{U.S. population}) = \frac{454,000}{250 \times 10^6}$$
$$= 1.82 \times 10^{-3}$$

Since many people move into and out of the communities, only a few people are exposed for 70 years, and the exposure of people moving into a community is often unknown.

## 4.7 Closure

Engineers are sometimes asked to explain how pollutants affect health. Replying that this is not your field may be true, but it also ducks the issue. Ducking the issue is not necessary since the rudiments of physiology are not difficult to learn; indeed, they are satisfying to learn. It is important that readers understand the components of the respiratory system, how they function, and how they may be affected by air pollutants. Material engineers should feel comfortable with the literature of physiology and have no hesitancy to consult it. The manner in which pollutants are transported in the respiratory system and absorbed by blood is inherently interesting to those who appreciate the thermal sciences and lends itself to applying the principles of mass transport.

## Nomenclature

| Symbol | Description Dimensions[*] |
|---|---|
| $[j]$ | concentration of species $j$ $(M/L^3, N/L^3)$ |
| $a$ | radius of airway (L) |
| $a_{a,j}$ | aggregate risk per person in a population $P_n$ |
| $A_{a,j}$ | aggregate risk from pollutant $j$ |
| $A(y)$ | total airway cross sectional area as a function of longitudinal distance $y$ $(L^2)$ |
| $B$ | constant defined by the one-hit equation $(t^{-1})$ |
| $c$ | concentration $(M/L^3)$ |
| $C$ | lung capacitance $(L^5/F)$ |
| $c_a$ | concentration in the alveolar region $(M/L^3)$ |
| $c_a, c_b, c_m$ | molar concentration in air, blood, and at mucus–air interface $(M/L^3)$ |
| $C_L, C_R$ | capacitance of left and right lung $(L^5/F)$ |
| $c_{p,w}, c_{p,a}$ | specific heat of water vapor and air $(Q/MT)$ |
| $c_y$ | concentration at an arbitrary location (y) in the capillary $(M/L^3)$ |
| $c_0$ | inlet concentration $(M/L^3)$ |
| $d$ | airway diameter (L) |
| $D$ | molecular diffusion coefficient $(L^2/t)$ |
| $D'$ | effective diffusion coefficient $(L^2/t)$ |
| $D_{min}$ | dose rate (M/t) |
| $D_{p,act}$ | actual diameter of a particle (L) |
| $D_{p,aero}$ | aerodynamic diameter of a particle (L) |
| $D_t$ | total dose (M) |
| $f$ | breathing frequency $(t^{-1})$; fraction of contaminant remaining after depletion by |

[*] F, force; L, length; M, mass, N, mols; Q, energy; t, time; T, temperature.

|  |  |
|---|---|
| | chemical reaction during mass transfer (Bohr model) |
| $F$ | bioavailability |
| $f_i$ | fraction of species $i$ depleted by chemical reaction |
| $h_c$ | convective heat transfer coefficient of mucus–air interface (Q/TtL) |
| $h_{fg}$ | enthalpy of vaporization (Q/M) |
| $k$ | constant of proportionality |
| $K$ | overall mass transfer coefficient (L/t) |
| $k_{ab}$ | absorption coefficient $(L^3/t)$ |
| $k_{b,i}$ | solubility of molecular species $i$ in the blood |
| $k_c$ | convective mass transfer coefficient (L/t) |
| $k_{m,i}$ | mass transfer coefficient of molecular species $i$ in the layer (L/t) |
| $k_r$ | first-order rate constant for the removal of any contaminant $(1/t)$ |
| $k_{s,i}$ | solubility constant of molecular species $i$ in a layer |
| $k_t$ | thermal conductivity of airway wall (Q/MtT) |
| $L$ | longitudinal length of airway (L) |
| $L/d$ | airway aspect ratio |
| $L_e$ | length of airway needed for fully developed flow (L) |
| $m$ | mass (M) |
| $m_{bb}$ | body burden (M) |
| $\dot{m}_i$ | mass transport of molecular species $i$ (M/t) |
| $M_i$ | molecular weight of molecular species $i$ (M/N) |
| $n$ | constant |
| $N$ | number of persons in a population |
| $N_D$ | diffusion parameter, defined by equation |

| | |
|---|---|
| $N_w$ | number mols of water per unit area and time $(N/L^2t)$ |
| $P$ | pressure $(F/L^2)$ |
| $P'$ | airway perimeter (L) |
| $P_a$ | partial pressure in alveolar region $(F/L^2)$ |
| $P_d$ | excess risk probability due to an atmospheric pollutant |
| $P_{i,g}, P_{i,b}$ | partial pressure of molecular species $i$ in gas or blood $(F/L^2)$ |
| $P_n$ | population |
| $P_r$ | partial pressure in arterial blood $(F/L^2)$ |
| $P_v$ | partial pressure in venous blood $(F/L^2)$ |
| $P_y$ | partial pressure at an arbitrary location ($y$) in the capillary $(F/L^2)$ |
| $P_0$ | inlet partial pressure, background probability $(F/L^2)$ |
| $\delta P$ | pressure drop $(F/L^2)$ |
| $Q$ | volumetric flow rate $(L^3/t)$ |
| $Q_a$ | alveolar ventilation, volumetric flow rate into (or out of) alveolar region (Bohr model) $(L^3/t)$ |
| $Q_b$ | blood volumetric flow rate $(L^3/t)$ |
| $q_l, q_b$ | total heat transfer from the liquid and blood (Q/t) |
| $Q_L, Q_R$ | volumetric flow rates in left and right lung $(L^3/t)$ |
| $Q_t$ | minute respiratory rate $(L^3/t)$ |
| $r$ | radius (L) |
| $R$ | airway resistance |
| $r_{ave}$ | radius at which the local velocity is equal to $U_{ave}$ (L) |
| $r_e$ | excess lifetime cancer risk |
| $R_L, R_R$ | resistance of left and right lungs $(Ft/L^5)$ |
| $R_p$ | response $(M/L^3)$; measurable response |
| $R_u$ | universal gas constant $(Q/NT)$ |
| $r_{u,j}$ | unit risk factor for species molecular $j$ $(L^3/M)$ |
| $R_{vp}$ | ventilation–perfusion ratio $(Q_a/Q_b k_b)$ |
| Re | Reynolds number |
| $S$ | capillary surface area, cumulative airway surface area $(L^2)$ |
| $t$ | time (t) |

| | |
|---|---|
| $t_{1/2}$ | half life (t) |
| $T_b, T_a, T_m$ | temperature of blood, air, and mucous (T) |
| $T_c$ | body-core temperature (T) |
| $t_n$ | arithmetic mean transient time (t) |
| $t_{vc}$ | time to expire volume of air equal to vital capacity (t) |
| $T_a(x)'$ | normalized air temperature, defined by equation; dimensionless temperature defined by equation |
| $U$ | velocity (L/t) |
| $U_{ave}$ | average velocity (L/t) |
| $U_{rms}$ | root-mean-square velocity (L/t) |
| $U_0$ | centerline velocity (L/t) |
| $V, V(t)$ | lung volume $(L^3)$ |
| $V_a(t)$ | alveolar volume $(L^3)$ |
| $V_b$ | pulmonary capillary blood volume $(L^3)$ |
| $V_d$ | anatomic dead space volume (volume of conducting airways) $(L^3)$ |
| $V_f$ | final volume of inflated lung $(L^3)$ |
| $V_{fr}$ | lung functional residual capacity $(L^3)$ |
| $V_r$ | lung residual volume $(L^3)$ |
| $V_t$ | tidal volume $(L^3)$ |
| $V_{vc}$ | vital capacity volume $(L^3)$ |
| $V_0$ | initial value of lung volume $(L^3)$ |
| $x, y, z$ | spatial coordinates (L) |
| $Z$ | airway generation number |

*Greek*

| | |
|---|---|
| $\alpha_1, \alpha_2, \alpha_n$ | constants defined by equation |
| $\Delta V$ | $V_f - V_0$ $(L^3)$ |
| $\mu$ | dynamic viscosity $(Ft/L^2)$ |
| $\nu$ | $(\mu/\rho)$ kinematic viscosity $(L^2/t)$ |
| $\pi$ | 3.14159 |
| $\rho$ | density $(M/L^3)$ |
| $\tau$ | lung time constant $(\tau = RC)$ (t) |

*Subscripts*

| | |
|---|---|
| $(\cdot)_a$ | air property |
| $(\cdot)_b$ | blood property |
| $(\cdot)_i$ | molecular species $i$ |

| | |
|---|---|
| $(\cdot)_m$ | mucus property |
| $(\cdot)_p$ | particle |
| $(\cdot)_{ss}$ | steady state |
| $(\cdot)_w$ | water property |
| $(\cdot)_0$ | initial value |

*Abbreviations*

| | |
|---|---|
| AD | fully alveolated alveolar ducts |
| ACGIH | American Conference of Governmental Industrial Hygienists |
| AS | alveolar sacs |
| BAL | bronchoalveolar lavage |
| BEI | biological exposure index |
| BL | bronchiole |
| BR | bronchi |
| DNA | deoxyribonucleic acid |
| EOM | particulate organic material |
| EPA | Environmental Protection Agency |
| $FEV_1$ | forced expiratory volume in 1 s |
| FVC | forced vital capacity |
| ISB | isothermal saturation boundary |
| MCI | methylisocyanate |
| NIOSH | National Institute of Occupational Safety and Health |
| OSHA | Occupational Safety and Health Administration |
| P | pulmonary region |
| PAN | peroxyacetylnitrate |
| PEL | permissible exposure limit |
| ppm | parts per million (mole fraction) |
| RBR | partially alveolated respiratory bronchiole |
| Re | Reynolds number |
| SMR | standard mortality ratio, defined by equation |
| $SR_{aw}$ | airway resistance |
| STP | standard temperature and pressure |
| TBL | tracheobronchial region |
| TDI | toluene diisocyanate |
| TL | terminal bronchiole |
| TWA-TLV | time-weighted average threshold limit value |
| VC | vital capacity |
| VOC | volatile organic compound |
| VR | residual volume |
| $W_o$ | Womersley number defined by equation |

# References

Alenandesson, R., and Hedenstierna, G., 1989. Pulmonary function in wood workers exposed to formaldehyde: a prospective study, *Archives of Environmental Health*, Vol. 44B, No. 1, pp. 5–11.

Balashazy, I., and Hofmann, W., 1993. Particle deposition onto indoor residential surfaces, *Environmental Science and Technology*, Vol. 24, No. 6, pp. 745–772.

Balashazy, I., Martonen, T. B., and Hofmann, W., 1990. Simultaneous sedimentation and impaction of aerosols in two-dimensional channel bends, *Aerosol Science and Technology*, Vol. 13, pp. 20–34.

Chang, H. K., 1984. Mechanisms of gas transport during ventilation by high-frequency oscillation, *Journal of Applied Physiology*, Vol. 56, No. 3, pp. 553–563.

Clark, R. P., 1974. Skin scales among airborne particles, *Journal of Hygiene (Cambridge)*, Vol. 72, pp. 47–51.

Clark, R. P., and Cox, R. N., 1973. The generation of aerosols from the human body, Article 95 in *Airborne Transmission and Airborne Infection*, Hers, J.F.P., and Winkler, K. C. (Eds.), Oosthoek Publishing, Utrecht, The Netherlands, pp. 413–426.

Colucci, A. V., and Strieter, R. P., 1983. Dose considerations in the sulfur dioxide exposed exercising asthmatic, *Environmental Health Perspectives*, Vol. 52, pp. 221–232.

Cookson, M.O.C.M., and Moffatt, M., 1997. Asthma: an epidemic in the absence of infection, *Science*, Vol. 275, pp. 41–42.

Cralley, L. J., and Cralley, L. V., (Eds.), 1979. *Patty's Industrial Hygiene and Toxicology*, Vols. I, II, and III, Wiley, New York.

DeMarini, D. M., and Lewtas, J., 1995. Mutagenicity and carcinogenicity of complex combustion emissions: emerging molecular data to improve risk assessment, *Toxicological and Environmental Chemistry*, Vol. 49, pp. 157–166.

DeMarini, D. M., Shelton, M. L., and Bell, D. A., 1996. Mutation spectra of chemical fractions of a complex mixture: role of nitroarenes in the mutagenic specificity of municipal waste incinerator emissions, in *Mutation Research Fundamental and Molecular Mechanisms of Mutagenesis*, Ashby, M., Gentile, J., Sankaranarayanan, K., and Glickman, B. (Eds.), Elsevier, Amsterdam. Vol. 349, pp. 1–20.

Eisner, A. D., and Martonen, R. T. B., 1989. Simulation of heat and mass transfer processes in a surrogate bronchial system developed for hygroscopic aerosol studies, *Aerosol Science and Technology*, Vol. 11, pp. 39–57.

Engle, L., 1985. *Intraregional gas mixing and distribution*, Chapter 7, in *Gas Mixing in the Lung*, Engle, L. A., and Paiva, M. (Eds), Marcel Dekker, New York, pp. 287–358.

Folinsbee, L. J., McDonnell, W. F., and Horstman, D. H., 1988. Pulmonary function and symptom responses after 6.6-hour exposure to 0.12 ppm ozone with moderate exercise, *Journal of the Air Pollution Control Association*, Vol. 38, No. 1, pp. 28–35.

Gallily, I., Schiby, D., Cohen, A. H., Hollander, W., Schless, D., and Stober, W., 1986. On the inertial separation of nonspherical aerosol particles from laminar flows. I. The cylindrical case, *Aerosol Science and Technology*, Vol. 5, pp. 267–286.

Georgopoulos, P. G., Walia, A., Roy, A., and Lioy, P. J., 1997. Integrated exposure and dose modeling and analysis system. 1. Formulation and testing of microenvironmental and pharmacokinetic components, *Environmental Science and Technology*, Vol. 31, No. 1 pp. 17–27.

Gradon, L., and Orlicki, D., 1990. Deposition of inhaled aerosol particles in a generation of the tracheobroncial tree, *Journal of Aerosol Science*, Vol. 21, No. 1, pp. 3–19.

Gradon, L., and Yu, C. P., Diffusional particle deposition in the human mouth, *Aerosol Science and Technology*, Vol. 11, pp. 213–220.

Guyton, A. C., 1986. Textbook of medical physiology, 7th ed., W.B. Saunders, Philadelphia.

Hamilton, R. G., Chapman, M. D., Platts-Mills, T. A. E., and Adkinson, N. F., 1992. House dust aeroallergen measurements in clinical practice: a guide to allergen-free home and work environments, *Immunology and Allergy Practice*, Vol. 14, No. 3, pp. 9–25.

Hanna, L. M., and Scherer, P. W., 1986a. A theoretical model of localized heat and water vapor transport in the human respiratory tract, *Transactions of the ASME, Journal of Biomechanics Engineering*, Vol. 108, pp. 19–27.

Hanna, L. M., and Scherer, P. W., 1986b. Regional control of local airway heat and water vapor losses, *Journal of Applied Physiology*, Vol. 61, No. 2, pp. 624–632.

Hanna, L. M., and Scherer, P. W., 1986c. Measurement of local mass transfer coefficients in a cast model of the human upper respiratory tract, *Transactions of the ASME, Journal of Biomechanics Engineering*, Vol. 108, pp. 12–18.

Haselton, F. R., and Scherer, P. W., 1980. Bronchial bifurcations and respiratory mass transport, *Science*, Vol. 208, pp. 69–71.

Hatch, T. F., and Gross, P., 1964. *Pulmonary Deposition and Retention of Inhaled Aerosols*, Academic Press, San Diego, CA.

Hattis, D., Wasson, J. M., Page, G. S., Stern, B., and Franklin, C. A., 1987. Acid particles and the tracheo-bronchial region of the respiratory system: an irritation-signaling model for possible health effects, *Journal of the Air Pollution Control Association*, Vol. 37, No. 9, pp. 1060–1066.

Johanson, G., 1991. Modeling of respiratory exchange of polar solvents, *Annals of Occupational Hygiene*, Vol. 35, No. 3, pp. 323–339.

Kasper, G., Nida, T., and Yang, M., 1985. Measurements of viscous drag on cylinders and chains of spheres with aspect ratios between 2 and 50, *Journal of Aerosol Science*, Vol. 16, No. 6, pp. 535–556.

Klaassen, C. D., Amdur, M. O., and Doull, J., (Eds.), 1986. *Casarett and Doull's Toxicology*, 3rd ed., Macmillan, New York.

Kleinman, M. T., 1984. Sulfur dioxide and exercise: relationships between response and absorption in upper airways, *Journal of Air Pollution Control Association*, Vol. 34, No. 1, pp. 32–37.

Kreibel, D., and Smith, T. J., 1990. A nonlinear pharmacologic model of the acute effects of ozone on the human lungs, *Environmental Research*, Vol. 51, pp. 120–146.

Leung, H. W., and Paustenbach, D. J., 1988. Application of pharmacokinetics to derive biological exposure indexes from threshold limit values, *American Industrial Hygiene Association Journal*, Vol. 49, No. 9, pp. 445–450

Levitzky, M. G., 1986. *Pulmonary Physiology*, 2nd ed., McGraw-Hill, New York.

Lippmann, M., 1987. Role of science advisory groups in establishing standards for ambient air pollutants, *Aerosol Science and Technology*, Vol. 6, pp. 93–114.

Lipmann, M., 1989. Health effects of ozone, a critical review, *Journal of the Air Pollution Control Association*, Vol. 39, No. 5, pp. 672–695.

Lipmann, M., 1991. Health effects of tropospheric ozone, *Environmental Science and Technology*, Vol. 25, No. 12, pp. 1954–1961

Longo, M. L., Bisagno, A. M., Zasadzinski, J. A. N., Bruni, R., and Waring, A. J., 1993. A function of lung surfactant protein SP-B, *Science*, Vol. 261, pp. 453–456.

Martonen, T. B., 1992. Deposition patterns of cigarette smoke in human airways, *American Industrial Hygiene Journal*, Vol. 53, pp. 6–18.

Martonen, T. B., Zhang, Z., and Yang, Y., 1992. Interspecies modeling of inhaled particle deposition patterns", *Journal of Aerosol Science*, Vol. 23, No. 4, pp. 389–406.

Martonen, T. B., Zhang, Z., and Lessmann, R. C., 1993. "Fluid dynamics of human larynx and upper tracheo-bronchial airways, *Aerosol Science and Technology*, Vol. 19, pp. 133–156.

McCartney, M. L., 1990. Sensitivity analysis applied to Coburn–Foster–Kane models of carboxyhemoglobin formation, *American Industrial Hygiene Association Journal*, Vol. 51, No. 3, pp. 169–177.

Menkes, H., Cohen, B., Permutt, S., Beatty, T., and Shelhamer, J., 1981. Characterization and interpretation of forced expiration, *Annals of Biomedical Engineering*, Vol. 9, pp. 501–511.

Miller, F. J., Overton, J. H., Jaskot, R. H., and Menzel, D. B., 1985. A model of the regional uptake of gaseous pollutants in the lung. 1. The sensitivity of the uptake of ozone in the human lung to lower respiratory track secretions and to exercise, *Journal of Toxicological Environmental Health*, Vol. 79, pp. 11–27.

Mossman, B. T., Bignon, J., Corn, M., Seaton, A., and Gee, J. B. L., 1990. Asbestos: scientific developments and implications for public policy, *Science* Vol. 247, pp. 294–301

Muller, W. J., Hess, G. D., and Scherer, P. W., 1990. A model of cigarette smoke deposition, *American Industrial Hygiene Association Journal*, Vol. 51, No. 5, pp. 245–256.

Nadel, E. R., 1985. Physiological adaptations to aerobic training, *American Scientist*, Vol. 73, pp. 334–342.

Nixon, W., and Egan, M. J., 1987. Modeling study of regional deposition of inhaled aersols with special reference to effects of ventilation asymmetry, *Journal of Aerosol Science*, Vol. 18, No. 5, pp. 563–579.

Otis, A. B., 1956. Mechanical factors in distribution of pulmonary ventilation, *Journal of Applied Physiology*, Vol. 8, pp. 427–443.

Paiva, M., 1985. Theoretical studies of gas mixing in the lung, Chapter 6, in *Gas Mixing in the Lung*, Engle, L. A., and Paiva, M., (Eds.), Marcel Dekker, New York, pp. 221–286.

Permutt, S., and Menkes, H. A., 1979. Spirometry, analysis of forced expiration within the time domain, in *Lung in the Transition Between Health and Disease*, Macklen, P. T., and Permutt, S., (Eds.), Marcel Dekker, New York, pp. 113–152.

Perra, F. P., and Ahmed, A. K., 1979. *Respiratory Particles*, Ballinger, Cambridge, MA.

Ricci, P. F., and Molton, L. S., 1985. Regulating cancer risks, *Environmental Science and Technology*, Vol. 19, No. 6, pp. 473–479.

Rich, G., 1990. A primer on risk calculations, *Pollution Engineering*, Vol. 22, No. 5, pp. 94–99.

Rothenberg, S. J., and Swift, D. L., 1984. Aerosol deposition in the human lung at variable tidal volumes: calculation of fractional deposition, *Journal of Aerosol Science*, Vol. 3, pp. 215–226.

Rothman, S., 1954. *Physiology and Biochemistry of the Skin*, University Chicago Press, Chicago.

Schwartz, S. E., and Andreae, M. O., 1996. Uncertainty in climate change caused by aerosols, *Science*, Vol. 272, May 24, pp. 1121–1122.

Seinfeld, J. H., 1986. *Atmospheric Chemistry and Physics of Air Pollution*, Wiley, New York.

Silverthorn, D.S., 1998. *Human Physiology, and Integrated Approach*, Prentice Hall Upper Saddle River, NJ.

Slonim, N. B., and Hamilton, L. H., 1987. *Respiratory Physiology*, 5th ed., C.V. Mosby, St Louis, MO.

Slutsky, A. S., Drazen, J. M., Ingram, R. H., Kamm, R. D., Shapiro, A. H., Fredberg, J. J., Loring, S. H., and Lehr, J., 1980. Effective pulmonary ventilation with small-volume oscillations at high frequency, *Science*, Vol. 209, pp. 609–610.

Slutsky, A. S., Kamm, R. D., and Drazen, J. M., 1985. Alveolar ventilation at high frequencies using tidal volumes smaller than the anatomical dead space, Chapter 4 in *Gas Mixing in the Lung*, Engle, L. A., and Paiva, M., (Eds.), Marcel Dekker, New York, pp. 137–176.

Taylor, G. I., 1953. Dispersion of soluble matter in solvent flowing slowly through a tube, *Proceedings of the Royal Society (London), Series A*, Vol. 219, p. 186.

Taylor, G.I., 1954. The dispersion of matter in turbulent flow through a pipe, *Proceedings of the Royal Society (London), Series A*, Vol. 220, pp. 440.

Tilton, B. E., 1989. Health effects of tropospheric ozone, *Environmental Science and Technology*, Vol. 23, No. 3, pp. 257–263.

Ultman, J. S., 1985. Gas transport in the conducting airways, Chapter 3 in *Gas Mixing in the Lung*, Engle, L. A., and Paiva, M., (Eds.), Marcel Dekker, New York, pp. 63–136.

Ultman, J. S., 1988. Transport and uptake of inhaled gases, in *Air Pollution, the Automobile, and Public Health*, Watson, A. Y., Bates, R. R., and Kennedy, D., (Eds.), Health Effects Institute, National Academy Press, Washington, DC, pp. 323–366.

Ultman, J. S., 1989. Exercise and regional dosimetry: factors governing gas transport in the lower airways, in *Susceptibility to Inhaled Pollutants*, ASTM STP 1024,

Utell, M. J., and Frank, R., (Eds.), American Society for Testing and Materials, Philadelphia, pp. 111–126.

Vedal, S., 1997. Ambient particles and health: lines that divide, *Journal of the Air and Waste Management Association*, Vol. 47, pp. 551–581.

Weibel, E., 1963. *Morphology of the Lung*, Academic Press, San Diego.

West, J. B., 1974. *Respiratory Physiology: The Essentials*, Williams & Wilkins, Baltimore, MD.

White, M. C., Infante, P. F., and Chu, K. C., 1982. A quantitative estimate of leukemia mortality associated with occupational exposure to benzene, *Journal for the Society of Risk Analysis*, Vol. 2, No. 3, pp. 195–204.

Xu, G. B., and Yu, C. P., 1987. Deposition of diesel exhaust particles in mammalian lungs, *Journal of Aerosol Science and Technology*, Vol. 7, pp. 117–123.

Yu, J.-W., and Neretnieks, I., 1990. Single-component and multicomponent adsorption equilibria on activated carbon of methylcyclohexane, toluene, and isobutyl methyl ketone, *Industrial Engineering Chemistry Research*, Vol. 29, No. 2, pp. 220–231.

Yu, C. P., and Xu, G. B., 1987. Predicted deposition of diesel particles in young humans, *Journal of Aerosol Science*, Vol. 18, No. 4, pp. 419–429.

# Problems

**4.1.** Unlike engineering, where knowledge is based on writing and using quantitative relationships, knowledge in physiology and toxicology is based on naming things and explaining their function. The following are *diseases, parts of the body, or physiological parameters* that you should be able to recall from memory. If the word is a *disease,* state its symptoms and explain the physiological impairment. If the word is *part of the body*, define it succinctly and explain its physiological function. If the word is a *physiological parameter*, describe what it represents.

(a) Aerodynamic particle diameter
(b) Allergic responses
(c) Alveoli
(d) Asthma, bronchitis and emphysema
(e) Bronchi, bronchiole and alveolar ducts
(f) Capillary permeability
(g) Cilia
(h) Collagen
(i) Convective flow, pendelluft, Taylor diffusion, molecular diffusion
(j) Dose and dose rate
(k) Dyspnea
(l) Edema
(m) Epithelium and endothelium cells
(n) Hypoxia
(o) Interstitium
(p) Lipids
(q) Lung compliance
(r) Lymphatics
(s) Lysosome and hydrolase
(t) Mucus and surfactant
(u) Phagocytosis, macrophage
(v) Pharynx, trachea, esophagus, and epiglottis
(w) Pneumonia and tuberculosis
(x) Respirable and inhalable particles
(y) Silicosis and asbestosis
(z) Spirometer, tidal volume, vital capacity, residual volume, anatomical dead space, functional residual capacity, forced expiratory volume in 1 s
(aa) Veins, arteries, arterioles, venules, capillaries
(bb) Ventilation perfusion ratio

**4.2.** Write and solve the differential equations that predict the volumetric flow rate in the right and left lungs $(Q_R \text{ and } Q_L)$ arranged in parallel as shown in Figures 4-15 and 4-16. Assume that during exhalation (when the pressure is positive) the pressure (cm water) varies as follows:

$$
P = \begin{cases}
\dfrac{6t}{0.1t_c} & 0 < t < 0.1t_c \\[2mm]
6 & 0.1t_c < t < 0.4t_c \\[2mm]
6\left(1 - \dfrac{t - 0.4t_c}{0.1t_c}\right) & 0.4t_c < t < 0.5t_c
\end{cases}
$$

where $t_c$ is the length of the cycle (inhalation followed by exhalation) in seconds. During inhalation, the cycle is repeated except that the pressure pulse is negative. The frequency of oscillation is to be varied over a range of 5 to 20 times the normal breathing frequency. The frequency for normal breathing is 13.6 cycles per minute $(t_c = 4.41 \text{ s})$. (For illustrative purposes, you may wish to allow the upper limit to be 300 cycles/min to simulate canine "panting.") For computation purposes assume that the capacitance $(C_L)$ and resistance $(R_L)$ of the left lung are 0.083 L/cm water and 0.8 cm of water per L/min, respectively. Assume that the right lung has a capacitance and resistance 30% larger that the left lung. Vary the breathing frequency from 13.6 to 50 cycles/min and show when pendelluft occurs by plotting the volumetric flow rates in the right and left lungs.

**4.3.** If the tail of the forced expiratory capacity curve (Figure 4-13) was an exponential function, show that

$$\sqrt{\alpha_2} = \sqrt{2\alpha_1}$$

**4.4.** Show that, in general, the square root of $\alpha_2$ is the root mean square of $\alpha_1$.

**4.5.** Describe the acute and chronic effects of the following toxicants on the pulmonary system. List the OSHA PEL for each of the substances.

(a) Ammonia
(b) Chlorine
(c) Hydrofluoric acid vapor
(d) Perchloroethylene vapor
(e) Silica
(f) Sulfur dioxide
(g) Toluene vapor

**4.6.** The following are characteristic sizes of various airborne particles.

- *Tobacco smoke:* 0.01 to 0.10 μm
- *Pollen:* 10 to 100 μm
- *Sea salt:* 0.1 to 1.0 μm
- *Bacteria:* 0.5 to 50 μm
- *Insecticide dusts:* 1 to 10 μm

On the basis of information in Figures 4-33 and 4-34, estimate the amount (in percentage of inlet particles) you would expect to be deposited in different parts of the respiratory system. If the ventilation rate (volumetric flow rate of inhaled air, m³/min) tripled, would you expect these values to change?

**4.7.** Beginning with Equation 4.31, derive fully the equations for the extended Bohr model in Section 4.3 to establish the validity of Equation 4.38.

**4.8.** Assume that people with impaired cardiovascular function suffer a 10% reduction in blood flow rate $(Q_b)$ (i.e., multiply $Q_b$ in the Table 4-2 by 0.9). Using the extended Bohr model, compute the percentage reduction in gas uptake efficiency $(E_u)$ between people with healthy and impaired hearts for four levels of activity assuming that all other parameters remain the same.

**4.9.** Many riding lawn tractors are equipped with exhaust pipes that discharge engine exhaust gases a few feet upwind of the operator's face. Operators inhale carbon monoxide at a concentration of 200 ppm $(228.6 \text{ mg/m}^3)$ over a period of time and may display the effects of carbon monoxide poisoning. Using the extended Bohr model, estimate the dose (kmol) of carbon monoxide an operator receives in 2 h:

$$\text{dose} = \int_0^2 E_u Q_a c_0 \, dt$$

where $E_u$ and $Q_a$ are the appropriate values for light exercise (LE).

**4.10.** Explain briefly the difference between the following terms, which are sometimes confused with one another.

(a) Chronic and acute response
(b) Mutagens and teratogens
(c) Asthma and bronchitis
(d) Tuberculosis and pneumoconiosis
(e) Pinocytosis and phagocytosis
(f) Obstructed and constricted lungs
(g) Respiratory and pulmonary system
(h) Residual volume and functional residual volume
(i) Vital capacity and tidal volume
(j) Epiglottis and esophagus
(k) Pharynx and trachea
(l) Fibrosis and pneumoconiosis
(m) Emphysema and edema and dyspnea
(n) Mucociliary clearance and alveolar clearance
(o) Biological exposure index and permissible exposure limit
(p) Dose and dose rate

**4.11.** The annual average ambient concentrations in PPB of several pollutants are measured in New York City $(\text{population} = 8 \times 10^6)$, Newark, NJ (400,000), and Jersey City, NJ (300,000).

| Pollutant (M) | New York City | Newark | Jersey City | PEL (ppm) |
|---|---|---|---|---|
| Benzene (78.1) | 18 | 10 | 5 | 10 |
| Carbon tetrachloride (153.8) | 3 | 12 | 10 | 2 |
| Chloroform (119.4) | 10 | 15 | 8 | 2 |
| Formaldehyde (30) | 20 | 12 | 15 | 3 |
| Perchloroethylene (166) | 5 | 8 | 10 | 25 |
| Styrene (104.2) | 5 | 20 | 18 | 50 |

(a) Which city has the highest excess lifetime cancer risk $(r_e)$ due to chloroform?

(b) Which city has the highest excess lifetime cancer risk $(r_e)$ for all the pollutants?

(c) Which pollutant poses the highest aggregate risk $(r_{a,j})$ to the total population of the three cities?

**4.12.** Repeat Problem 4.11 for St. Louis, MO (950,000) and East St. Louis, IL. (65,000) having the ambient concentrations in μg/m³ as given in the following table.

| Pollutant (M) | St. Louis | East St. Louis |
|---|---|---|
| Benzene (78.1) | 4.6 | 10.6 |
| Styrene (104.2) | 2.9 | 3.3 |
| Chloroform (119.4) | 0.3 | 0.5 |
| Trichloroethylene (131.4) | 3.5 | 6.7 |
| Carbon tetrachloride (153.8) | 0.8 | 1.4 |
| Formaldehyde (30.0) | 5.2 | 6.8 |

# 5

# Aesthetics

---

In this chapter you will learn:

- How air pollutants affect:
  - Vegetation and soil composition
  - Bodies of water and aquatic life
  - Atmospheric visibility
  - Surfaces of buildings, works of art, and personal property
- The relationship between pollution, noise, and odors

Air pollution regulation includes preventing injury to plants, animals, and property and "the comfortable enjoyment of life or property." While less obvious than its impact on health, the impact of air pollutants on objects that people find beautiful and that bring harmony to their lives cannot to be ignored. There is ample evidence that air pollutants affect plants, trees, and soil; bodies of water and aquatic life; atmospheric visibility; surfaces of buildings, automobiles, monuments, and works of art.

## 5.1 Acid rain

The acidification of rain and aquatic ecosystems is one of the major consequences of pollution. Acidification results from the accumulation of hydrogen ions $(H^+)$ in water. The degree of acidity or alkalinity is expressed numerically as pH with $[H^+]$ having the units of gmol/L.

$$pH = \log_{10} \frac{1}{H^+} \tag{5.1}$$

The pH scale varies from 0 to 14. The range of pH of several common materials is shown in Figure 5-1. The pH of unpolluted precipitation is less than 7.0 (neutral) owing to the presence of dissolved $CO_2$ (that forms carbonic acid in water) or sea salts.

$$CO_2 + H_2O \rightleftharpoons H_2CO_3 \rightleftharpoons H^+ \\ + HCO_3^- \rightleftharpoons 2H^+ + CO_3^{2-} \tag{5.2}$$

In parts of the world dissolved ammonia, volcanic ash or alkaline particulates from arid regions may cause the pH to exceed the 5.0-5.6 range, which is typical in "unpolluted precipitation" (see Figure 5-1). The

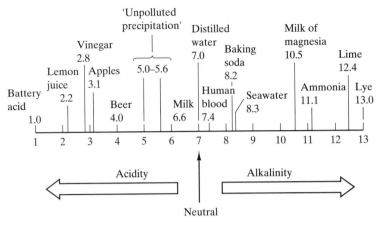

**Figure 5-1**  pH scale and common materials (used by Elsom, 1992)

principal precursors of acid rain are anthropogenic and natural emissions of sulfur and nitrogen oxides and to a lesser extent chlorine compounds. Through catalytic and photochemical reactions, these form $H_2SO_4$, $HNO_3$, and $HCl$ which comprise 62%, 32%, and 6% respectively of the acidity in acid rain. Sulfuric acid $\left(H_2SO_4\right)$ is formed from $SO_2$ by reactions catalyzed by metals contained within rain drops:

$$2SO_2 + O_2 + \text{metal catalyst} \rightarrow 2SO_3$$

$$SO_3 + H_2O \rightarrow H_2SO_4 \qquad (5.3)$$

Nitric acid $\left(HNO_3\right)$ is formed by OH• attack on $NO_2$ during the daytime,

$$NO_2 + OH\bullet + M \rightarrow HNO_3 + M \qquad (5.4)$$

and $NO_2$ and $O_3$ at night:

$$NO_2 + O_3 \rightarrow NO_3 + O_2 \qquad (a)$$

$$NO_2 + NO_3 \rightarrow N_2O_5 \qquad (b) \quad (5.5)$$

$$N_2O_5 + H_2O \rightarrow 2HNO_3 \qquad (c)$$

Hydrochloric acid (HCl) is transported down from the stratosphere, where it is formed through reactions of CFCs and O•:

$$CFC + h\nu(UV) \rightarrow Cl\bullet + \text{products} \qquad (a)$$

$$CFC + O\bullet \rightarrow ClO + \text{products} \qquad (b)$$

$$O\bullet + ClO \rightarrow Cl\bullet + O_2 \qquad (c) \qquad (5.6)$$

$$Cl\bullet + CH_4 \rightarrow HCl + CH_3\bullet \qquad (d)$$

Shown in Figure 5-2 are pH isopleths in North America. Sulfur dioxide emissions are high in Ohio, West Virginia, and Pennsylvania because of the large number of coal-fired power generating facilities located in the Ohio River valley. Not only is there liquid acidic precipitation in the form of rain and fog, but acidification also occurs in the form of snowpack. Aquatic ecosystems suffer a steep increase in acidification each spring because approximately 80% of the pollutants acquired by snowpacks are released in the first 30% of the meltwater.

# 5.2  *Vegetation*

The secondary national ambient air quality standards (NAAQSs; Table 3-2) are essentially values selected to protect vegetation. Damage to plants affects their aesthetic value adversely and reduces their economic value as food (Westenbarger and Frisvold, 1994) and fiber. Pollutants enter plants through their foliage as vapor, gases, and particulate matter, or by uptake through roots. *Hydrophilic* compounds are capable of combining with water and are more readily absorbed in solution through roots. *Hydrophobic* compounds are not capable of combining with water but may be adsorbed by or reacted in the soil; or if they are volatile, they may be absorbed by foliar lipids. The uptake of hydrophobic compounds can be modeled in ways similar to absorption of gases by solids (Paterson et al., 1991, 1994; O'Dell et al., 1977).

It is important that engineers recognize the types of injury plants suffer and the general causes for the damage. Visible injury to foliage can be grouped as follows:

- Collapse and patterns of dead (*necrotic*) leaf tissue
- Loss of chlorophyll (*chlorosis*) in specific portions of the leaf
- Growth abnormalities

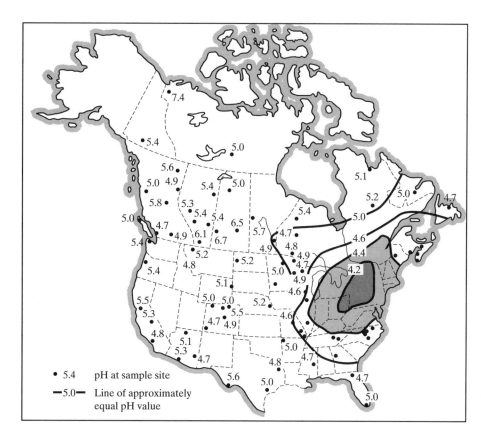

**Figure 5-2**  pH isopleths in the United States (used by Elsom, 1992)

In some cases the injury is characteristic of specific pollutants and diagnosis is easy. In other cases, injury produces symptoms similar to symptoms caused by disease, insects, soil nutrition, climate, light, temperature, humidity, soil water, and genetic history. Table 5-1 is a summary of pollutants, and specific damage produced in certain plants. These plants can be used as *indicator plants* (Simonich and Hites, 1995). Stern (1968) suggests a linear relationship between time (*t*), concentration (*c*), and plant injury (*I*):

$$I = t(c - c_t) \qquad (5.7)$$

where *I* refers to an acute quantifiable injury and $c_t$ is a threshold concentration below which there is no visible injury.

Sulfuric acid $(H_2SO_4)$ interacts with foliage in two ways. The acid forms a low-pH liquid film on the leaf or needle that erodes the cuticular wax which leads to desiccation and injury to the leaf or needle. Second, a portion of the acidity is transferred from the leaves to the soil. Sulfuric acid $(H_2SO_4)$ entering the stomata is neutralized by calcium stored in the leaf. Calcium in the leaf must be replaced by calcium carried in solution from the soil via fine roots as calcium bicarbonate. Calcium bicarbonate is formed in the soil by exchange with hydrogen from carbonic acid formed from water and $CO_2$ generated by root respiration.

At concentrations below 25 µg/m³, $H_2SO_4$ may not affect foliage directly, but wet and dry deposition of sulfuric acid (as well as other acidic sulfates) affects the soil. Airborne acidic sulfates deposited on foliage as wet media (rain, snow, and fog) and dry media (aerosols and gases) are washed to the forest floor by rain and snow. When sulfuric (or nitric) acid seeps into the soil, it may deplete soil minerals such as calcium, magnesium, potassium, and other trace nutrients essential to vegetation nutrition.

The rate at which nutrient cations are leached from the rooting zone depends on the mobility of the

---

**Table 5-1**    Effect of Pollutants on Plants

---

**Sulfur dioxide $(SO_2)$**
*Indicator plants*: alfalfa, barley, cotton, wheat, apple
*Sensitivity*: acute, $c > 0.3$ ppm; chronic, $0.1$ ppm $< c < 0.3$ ppm
*Sulfur dioxide markings*
    *Broad leaf*: marginal and interveinal white to straw blotched areas distributed over leaf
    *Grasses*: irregular necrotic streaks either side of midvein
    *Conifers*: brown necrotic tips of needles
*Resistant plants*: potato, onion, corn, maple
*Similar markings produced by other causes*
    *Broad leaf*: leafhopper injury, fungal diseases producing blotched leaves, high-temperature scorch on maples
    *Grasses*: bacterial blight
    *Conifers*: winter and drought injury

**Ozone $(O_3)$**
*Indicator plants*: tobacco, tomato, bean, spinach, potato, lilac, begonia
*Sensitivity*: 0.02 ppm for 4–8 h, 0.05 ppm 1–2 h.
*Ozone Markings*
    *Broad leaf*: white flecks uniformly distributed over leaf surface
    *Grasses*: scattered markings distributed over surface
    *Conifers*: brown necrotic needle tips (tip burn) similar to $SO_2$ damage but no separation between living and
        damaged tissue
*Resistant plants*: mint, geranium, gladiolus, pepper, maple
*Similar markings*
    *Broad leaf*: red spider, some rusts, coalesced regions resemble $SO_2$ damage
    *Grasses*: red spider and mite damage
    *Conifers*: any disease causing tip burn

**Ethylene $(C_2H_4)$**
*Indicator plants*: orchid blossom, tomato, cowpea, cotton
*Sensitivity*: orchids, 0.005 ppm; tomato, 0.1 ppm
*Ethylene markings*
    *Broad leaf*: epinasty (curled growth since upper surface grows more rapidly than lower surface)
    *Grasses*: retarded growth
    *Conifers*: abscission (thin layer of pithy cells at base of stem)
*Resistant plants*: lettuce
*Similar markings*
    *Broad leaf*: water stress, bacterial wilt, nematode and aphid injury
    *Grasses*: suppressed growth due to other causes
    *Conifers*: water stress

**Peroxyacetylnitrate (PAN)**
*Indicator plants*: petunia, romaine lettuce, pinto bean, bluegrass
*Sensitivity*: 0.01–0.05 ppm
*PAN markings*
    *Broad leaf*: collapse of underside of leaf, glazed or silvered or bronzed leaves
    *Grasses*: collapsed bleached leaf
    *Conifers*: nonspecific blight
*Resistant plants*: cabbage, corn, wheat, pansy
*Similar markings*
    *Broad leaf*: sun scald, various viral and fungal diseases

*(continued)*

**Table 5-1**   *(continued)*

*Grasses*: various viral and fungal diseases
*Conifers*: red spider and mite injury, drought, excessive salts

**Fluorides**
*Indicator plants*: gladiolus, pine, Chinese apricot, Italian prune
*Sensitivity*: 0.1 ppb for 5 weeks (gladiolus)
*Fluoride markings*
   *Broad leaf*: accentuated dark red-brown band between green leaf and necrotic border around entire leaf
   *Grasses*: tip burn, accentuated line between live and necrotic tissue
   *Conifers*: similar to grasses
*Resistant plants*: alfalfa, rose, tobacco, cotton, tomato
*Similar markings*
   *Broad leaf*: fungal diseases, wind, high temperature, drought
   *Grasses*: many fungal and bacterial scalds, blotches, streaks
   *Conifers*: winter and drought injury, $SO_2$ injury

**Nitrogen dioxide $(NO_2)$**
*Indicator plants*: tomato, tobacco and bean respond to $NO_2$
*Sensitivity*: acute injury at 2–5 ppm is similar to $SO_2$ injury

---

anions $NO_3^-$ and $SO_4^{2-}$ in the soil. In passing through the soil, each anion carries a cation with it. The major cations in soil are $Ca^{2+}$, $Mg^{2+}$, $K^+$, $Na^+$, $Al^{3+}$, $Mn^{2+}$, $NH_4^+$, and $H^+$. The number of cations depends on mineralization of humus and organic matter and from the weathering of minerals in the soil minus those taken up by feeder roots. As the cations are depleted, the soil becomes more acidic. Hydrogen ions react with aluminum or other aluminum-containing minerals in the soil and bring $Al^{3+}$ into aqueous solution:

$$Al(OH)_3 + 3H^+ \rightarrow Al^{3+} + 3H_2O \qquad (5.8)$$

Similar reactions may occur with iron and manganese and bring iron and manganese ions into aqueous solution. Nutrient assimilation in the root system depends on the availability of $Ca^{2+}$ ion and nutrition suffers when either the $Ca^{2+}/Al^{3+}$ ratio decreases or when the $Ca^{2+}$ concentration itself decreases. Tomlinson (1983) reports that the $Al^{3+}$ ion becomes toxic to the feeder root system when the ratio $Ca^{2+}/Al^{3+}$ falls below certain values for a period of time. As soil acidity increases, pH decreases and the concentration of $H^+$ increases, which leads to the production of the following metal ions:

$pH > 6.5$:
$$CaCO_3 \rightarrow Ca^{2+} \text{ and } CO_3{}^{2-} \qquad (a)$$

$5 < pH < 6.5$:
$$CaSiO_3 \rightarrow Ca^{2+} + SiO_3{}^{2-} \qquad (b)$$

$4.2 < pH < 5$:
$$\text{clay } (H_4Al_2Si_2O_9) \rightarrow Al^{3+} \qquad (5.9)$$
$$+ \text{ various anions} \quad (c)$$

$3 < pH < 4.2$:
$$Al(OH)_3 \rightarrow Al^{3+} + 3(OH)^- \qquad (d)$$

$pH < 3$:
$$Fe(OH)_3 \rightarrow Fe^{3+} + 3(OH)^- \qquad (e)$$

Thus if the pH $< 5$, there are few carbonates and silicates in the soil to produce $Ca^{2+}$, and incoming hydrogen ions produce $Al^{3+}$, which displaces $Ca^{2+}$ near the root system.

The major impact of $SO_2$ on the human environment is local. Approximately 17% of all anthropogenic sulfur emitted in the United States in 1975 was released in the northeastern United States over just 0.6% of the nation's land area. Electric power generation in the Ohio River valley constitutes 16% of all the electrical power generated in the United States but 47% of all the $SO_2$ released by U.S. power plants. Even though sulfate $SO_4^{2-}$ deposition in

Pennsylvania and West Virginia is the highest in the United States, (Figure 5-2), the limestone soils in these states buffer these acidic reactions and minimize damage. The granitic soils of northeastern North America lack this buffering capacity and suffer higher sulfate damage.

## 5.3 Soil

When protein in humus and other organic matter in the soil decomposes, nitric acid is formed, the soil becomes acidic, and the pH decreases:

$$R—NH_2 + 2O_2 \rightleftharpoons R—OH + H^+ + NO_3^-$$
(5.10)

where R represents one of many hydrocarbon chains. Humus decomposition also depends on soil temperature, moisture, oxygen content, and soil organisms. When the nitrate ion is absorbed by the feeder root system, it is converted to protein in the plant, driving Equation 5.10 in the reverse direction. If large quantities of $H^+$ are present in the soil, this reverse reaction is augmented. Thus acid-forming and acid-consuming reactions may occur in soils at different times and locations in a forest.

The earth's minerals are subject to continual attack by atmospheric acids that produce clays, which in turn undergo reactions to produce aluminum ions, $Al^{3+}$. The principal atmospheric acid is carbonic acid formed by the absorption and dissociation of $CO_2$ in rainwater (Equation 5.2). Pollutants such as $SO_2$ and $NO_x$, which eventually form $H_2SO_4$ and $HNO_3$, enhance weathering. Minerals needed by plants (phosphorus, potassium, and calcium) are produced in soil by weathering. Soil minerals (e.g., $K^+$) can be formed by *weathering* reactions such as the following between carbonic acid and the most abundant mineral, feldspar.

$$2KAlSi_3O_8 + 3H_2O + 2CO_2 \rightarrow H_4Al_2Si_2O_9$$
$$+ SiO_2 + 2KHCO_3$$
(5.11)

Potassium bicarbonate $(KHCO_3)$ dissociates into $K^+$ and $HCO_3^-$. Similar reactions involving calcium and magnesium compounds release $Ca^{2+}$ and $Mg^{2+}$ to the soil. Forest soils are modified by the short- and long-term transport of air pollutants, particularly sulfates. A summary of these effects is shown in Figure 5-3.

The effect of sulfates on the feeder root system of trees is closely related to the effect of sulfate on the production of protein humus in the soil.

Cadmium is believed to be the most dangerous form of trace element contamination in soil (Brown et al., 1996). The primary risk pathway associated with cadmium in soils is the soil–plant–human pathway resulting from the consumption of staple grains or garden crops grown on these soils. Cadmium is transferred to soil as a contaminant in phosphorus fertilizer, mining and smelting of zinc and lead ores, and application of biosolids as fertilizer. Factors affecting cadmium uptake in vegetables and fruits include plant species, cadmium concentration, pH, and soil factors such as organic matter, chloride concentration, zinc status, hydrous iron, manganese, and aluminum oxides.

## 5.4 Bodies of water

Many lakes in North America and Europe have been acidified by the deposition of atmospheric pollutants. Havens et al. (1993), found that the abundance of six common crustacean zooplankton taxa that are important links between the producers and consumers in the food chain in Lake Ontario decrease in number as the acidity increases. In the early spring when the snow melts rapidly, nitrates and sulfates that have accumulated in the snow pass directly into lakes as acids without coming into prolonged contact with the soil or vegetation. Newly borne fish are particularly vulnerable. When a lake is subjected to acid deposition, its pH declines slowly until the buffering capacity of the soil and water is depleted, whereupon the pH decreases rapidly. Acidification of bodies of water increases the concentration of the aluminum ion, $Al^{3+}$, by the same mechanisms as in soil. Unfortunately, aluminum ions irritate fishes' gills and cause the gills to produce a protective mucus that impedes the absorption of salts, which in turn impedes the ability of fish to breath. Ultimately, the fish suffocate. Amphibians react in a similar fashion, which affects other animals in the food chain. Similar results have been found in lakes throughout the Adirondack Mountains of New York.

Mercury is a toxic material whose ability to migrate through the environment occurs because of biological activity that occurs in bodies of water. An illustration of this is the age-old use of mercury to extract amalga-

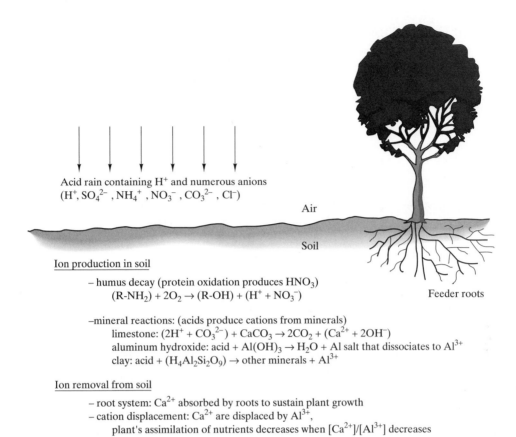

Acid rain containing $H^+$ and numerous anions
$(H^+, SO_4^{2-}, NH_4^+, NO_3^-, CO_3^{2-}, Cl^-)$

Air

Soil

Feeder roots

Ion production in soil

– humus decay (protein oxidation produces $HNO_3$)
$(R\text{-}NH_2) + 2O_2 \rightarrow (R\text{-}OH) + (H^+ + NO_3^-)$

–mineral reactions: (acids produce cations from minerals)
limestone: $(2H^+ + CO_3^{2-}) + CaCO_3 \rightarrow 2CO_2 + (Ca^{2+} + 2OH^-)$
aluminum hydroxide: acid $+ Al(OH)_3 \rightarrow H_2O + Al$ salt that dissociates to $Al^{3+}$
clay: acid $+ (H_4Al_2Si_2O_9) \rightarrow$ other minerals $+ Al^{3+}$

Ion removal from soil

– root system: $Ca^{2+}$ absorbed by roots to sustain plant growth
– cation displacement: $Ca^{2+}$ are displaced by $Al^{3+}$,
plant's assimilation of nutrients decreases when $[Ca^{2+}]/[Al^{3+}]$ decreases

**Figure 5-3**   Effect of $[H^+]$ on soil and feeder roots

mate gold from its ore, as practiced by small-time, aggressive miners in Indonesia and Brazil (Hoffmann, 1994) operating outside the law. Mercury forms an amalgam with gold particles dispersed in ore. The miners heat the amalgam with acetylene torches, which melts the gold and vaporizes ("fumes off") the mercury. The concentration of mercury vapor is dangerously large and jeopardizes the health of everyone in the vicinity. A portion of the escaping mercury enters the atmosphere as a vapor and the remainder condenses to small particles that settle on the soil, in dwellings, on clothing, and so on. The miners also leave considerable mercury in tailings that ultimately enters the soil and bodies of water. This use of mercury is an example of carelessness whose unfortunate consequences will become evident in the years to come.

Under ambient conditions, metallic mercury is quite stable but in the soil and water (Mason et al., 1996), biochemical reactions create organic mercury compounds, such as salts of monomethylmercury

$(CH_3Hg^+)$ that can be transported across membranes, stored in fat tissue, and ultimately damage the central nervous system. Figure 5-4 illustrates the formation and fate of mercury compounds in an aqueous, environment. Not shown in the figure are possible geologic sources of mercury, which Rasmussen (1994) indicates are significant and not to be ignored. If industrial effluents containing mercury enter aqueous systems, natural biologic processes convert mercury into the potent neurotoxin dimethylmercury, $(CH_3)_2Hg$. Because dimethylmercury is produced faster than natural organisms can degrade it, it accumulates in the flesh of fish and shellfish (Driscoll et al., 1994) in the ionic form, monomethylmercury $(CH_3Hg^+)$, and is then consumed by humans. Consuming contaminated fish and shellfish was the cause of a major well-publicized mercury poisoning in Minamata, Japan, in 1953 which began when industrial mercury wastes were discharged into Minimata bay (Kinjo et al., 1993).

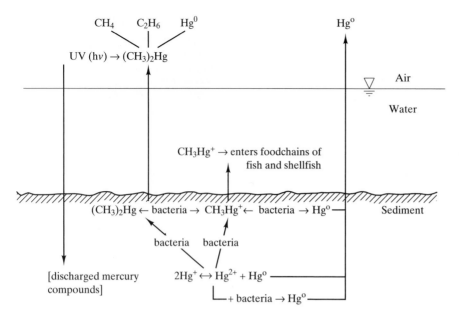

**Figure 5-4**  Biological cycle for mercury

Dimethylmercury poisoning results in a progressive disease beginning with impaired vision and speech and leading ultimately to uncontrolled muscular movement. Fetuses and young children are particularly vulnerable.

Transformation of inorganic mercury compounds is catalyzed by microorganisms. Aerobes oxidize such materials as HgS to form sulfites and later sulfates that ultimately form water-soluble $Hg^{2+}$. Once in solution, enzymes present in a number of bacteria reduce $Hg^{2+}$ to $Hg^0$. Within sediment, the following equilibrium occurs:

$$Hg_2^{2+} \rightarrow Hg^{2+} + Hg^0 \qquad (5.12)$$

Elemental mercury, $Hg^0$, has a sufficiently high vapor pressure to be transferred from sediment and the aqueous environment into the atmosphere. A second mechanism converts $Hg^{2+}$ into mono- and dimethylmercury. There are methylating agents available to make methyl radicals $(CH_3^+)\cdot$ available in biological systems. The coenzymes methylcorrinoids are known to participate in this reaction. Dimethylmercury is synthesized by a reaction involving $CH_3Hg^+$. Dimethylmercury is volatile and once in the atmosphere is photolyzed by UV to yield $Hg^0$ plus the methyl free radicals $(CH_3)\cdot$, which eventually become methane and ethane.

$$UV + (CH_3)_2Hg \rightarrow Hg^0 + 2CH_3\cdot \qquad (a)$$

$$CH_3\cdot + H\cdot \rightarrow CH_4 \qquad (b) \qquad (5.13)$$

$$CH_3\cdot + CH_3\cdot \rightarrow C_2H_6 \qquad (c)$$

Other microorganisms convert methylmercury to $Hg^0$ plus methane.

Fossil-fuel combustion, waste incineration, metal-ore roasting, and refining are significant anthropogenic sources of mercury emission to the atmosphere. Mercury released from the combustion of coal can be discharged as a vapor or retained in fly ash (Billings and Matson, 1972). Unfortunately, the concentration of mercury in bottom ash and boiler slag deposits is low, suggesting that most of the mercury leaves the boiler in the flue gas. Mercury can also be released from geothermal steam plants (Robertson et al., 1977). Galbreath and Zygarlicke (1996) suggest that greater than 50% of the $Hg^0(g)$ reacts with the oxidants in combustion flue gases. $Hg^0(g)$ is difficult to control and likely to enter the atmosphere because it is volatile (boiling point 357°C) and not soluble in water.

As the flue gas cools, a significant portion of mercury vapor, $Hg^0(g)$, condenses or is adsorbed on soot and fly ash particles. Small particles have a large surface-to-volume ratio and offer a greater opportunity to remove mercury (Natusch and Wallace, 1974; Natusch et al., 1974). The speciation of mercury in flue gas involves gaseous ($g$) and solid ($s$) phases of mercury. The three oxidation states of mercury are the elemental ($Hg^0$), mercurous cation ($Hg_2^{2+}$), and mercuric cation ($Hg^{2+}$). The two mercury cations appear as numerous inorganic and organic compounds; however, $Hg_2^{2+}$ compounds are generally unstable in the flue gas and the atmosphere. Inorganic $Hg^{2+}$ compounds are generally reactive and soluble in water and will be removed in a wet scrubber.

$Hg^0(g)$ is the thermodynamically stable form of mercury in the high temperature region of the combustor. With decreasing temperature, $Hg^0(g)$ reacts to form $Hg^{2+}$ compounds. In flue gas, thermochemical equilibrium calculations predict that $Hg^0(g)$, $HgCl_2(g)$, $HgO(s, g)$, and $HgSO_4$ are the predominant forms. At $100 < T < 270°C$, $HgSO_4(s)$ and $HgO(s)$ are the stable forms of mercury. At higher temperatures, mercury is predicted to react with flue gases and produce three compounds: $HgCl_2(g)$, $HgO(g)$, and $Hg^0(g)$. The $HgCl_2$ phase is dominant at temperatures less than about 450°C, but with increasing temperature $HgCl_2(g)$ reacts with $H_2O$ to produce $HgO(g)$:

$$HgCl_2(g) + H_2O \rightleftharpoons HgO(g) + 2HCl(g) \tag{5.14}$$

The half-life of airborne elemental mercury, $Hg^0(g)$, is on the order of hours for reactions with $Cl_2$ in the nocturnal atmosphere and for reactions with $H_2O_2$ in ambient air (Seigneur et al., 1994). The total mass of mercury in the atmosphere is currently thought to be approximately 6000 metric tons. (Douglas, 1994). Annual worldwide mercury emissions are thought to be approximately 6000 metric tons per year. Of these, some 2000 metric tons per year are generated by the oceans and another 600 to 2000 metric tons per year are generated by natural terrestrial sources. The remaining 2000 to 3400 metric tons per year are generated by human activities. The largest component of the anthropogenic emissions, about 1200 metric tons per year, is from nonutility industrial activities (i.e., disposal of batteries and fluorescent lamps) and approximately 600 metric tons per year from solid waste incineration. Approximately 300 metric tons per year are emitted by the world's electric utility boilers. U.S. utilities emit less than 60 metric tons per year, which is only 1% of the total global mercury emissions:

- Natural terrestrial sources    2000 metric tons/yr
- Natural marine sources    2000 metric tons/yr
- Anthropogenic sources    2000 metric tons/yr
  - Nonutility industrial    1200 metric tons/yr
  - Waste incineration    600 metric tons/yr
  - Global coal combustion    300 metric tons/yr

## 5.5 *Surface deterioration*

Pollutants discolor and erode buildings (Zappia et al., 1992, 1993; Sabbioni et al., 1992), cathedrals (Fobe et al., 1995; Nord et al., 1994), archaeological religious artifacts (Christoforou et al., 1996), historical monuments (Schiavon and Zhou, 1996), works of art, private property, and so on. The cost of cleaning soiled personal objects (cloths, homes, windows, cars, etc.), the cost of cleaning corporate buildings, and the cost of cleaning historic monuments and public buildings are costs borne ultimately by everyone. Soiling is very serious, and in the case of works of art and archaeological sites (Salmon et al., 1994), may be irreversible. Soiling occurs in both the indoor and outdoor environment and is produced from particles, pollutant gases, and vapors. All surface materials are adversely affected by ultraviolet radiation.

Metals, stone, painted surfaces, and textiles deteriorate when subjected to atmospheric pollutants in combination with sunlight, moisture, temperature, and air movement. Solid and liquid particles deposited on surfaces reduce their reflectance. In combination with water, some particles react chemically with the surface material. Some pollutants are inherently reactive and induce chemical reactions on surfaces. In combination with water, particles containing metals may initiate corrosive electrochemical reactions.

Pollutants affecting material surfaces are sulfur dioxide and sulfates, nitrogen oxides and nitrates, chlorides, carbon dioxide, and ozone (Butlin, 1991). Damage occurs from dry or wet deposition. Damage from particles, especially from diesel engine emissions, is of increasing importance. In buildings containing books, historical artifacts, works of art, and so on, pollutants entering the building from the outside environment or generated within the building cause significant damage (Hisham and Grosjean, 1991). Building materials that are the most sensitive to pollutants are calcareous building stone and ferrous metals. Soiling includes the loss of mass, changes in porosity, discoloration, and embrittlement. On indoor walls containing frescoes and mural paintings, Camuffo (1983) found that diurnal heating causes hygromteric cycles in which moisture is released or absorbed by the wall. Such breathing cycles cause art work to deteriorate.

## 5.5.1  Painted surfaces

Exposure to oxidants, either singularly or in combination, or exposure to particles results in substantial fading (Grosjean et al., 1993). Over 50% of the particles that soil painted surfaces (Figure 5-5) are coarse (i.e., over 10 μm). Such soiling can be quantified by measuring the surface's reflectance. For most people, a 30% reduction in reflectance justifies repainting the surface. In the absence of rain, Hamilton and Mans-feld (1993) report the change in reflectance, $R(\%)$, can be estimated by:

$$R(\%) = 1.413(tc_{\text{TSP}})^{0.5} \qquad (5.15)$$

where $t$ is the exposure in years and $c_{\text{TSP}}$ is the total suspended particulate in μg/m$^3$. When rainfall is taken into account, the authors report that:

$$R(\%) = 100[1 - \exp(-kt)] \qquad (5.16)$$

where $k$ depends on the concentration and type of particles. For total suspended particulates (TSP) and particles of elemental carbon (PEC):

$$k_{\text{TSP}}(\text{yr}^{-1}) = 0.0085c_{\text{TSP}}(\mu\text{g/m}^3) \qquad \text{(a)}$$
$$k_{\text{PEC}}(\text{yr}^{-1}) = 0.095c_{\text{PEC}}(\mu\text{g/m}^3) \qquad \text{(b)}$$
$$(5.17)$$

## 5.5.2  Stone surfaces

The patina on monuments composed of limestone (Fobe et al., 1995), sandstone, and marble (Guidobaldi and Mecchi, 1993) may contain material from deposited particles that leads to their deterioration (Figure 5-6). The surfaces may retain their contour but

**Figure 5-5**  Effect of air pollution on a painted surface (American Scientist, Jan/Feb 1976)

**Figure 5-6**  Polluted air reacts chemically with stone (marble, sandstone, limestone, etc.) monuments. The white surfaces are exposed to intense rain and wash away the softened stone detail. A black crust containing atmospheric dust and biological organisms accumulate on surfaces sheltered from rain. Detail of the Ospedale Civile in Venice. (Reprinted from Zappia G, Sabbioni C and Gobbi G, "Non-Carbonate Carbon Content on Black and White Areas of Damaged Stone Monuments", Atmospheric Environment, Vol 27A, No 7, pp 1117-1121, 1993. Reprinted with the kind permission from Elsevier Science Ltd, The Boulevard, Langford Lane, Kidlington OX5 1GB, UK)

become blackened, or they may remain white but lose their contour by the erosion of stone. Surface contours on carved stone are removed by rain on exposed surfaces and buildup known as *black gypsum crust* in sheltered areas. Monument stone, *calcite*, is a hexagonal crystalline form of $CaCO_3$. In the presence of carbonaceous aerosols, $SO_2$, and acidic compounds, calcite reacts to form gypsum (Nord et al., 1994):

$$CaCO_3 + H_2SO_4 \rightarrow$$
$$CaSO_4 + H_2O + CO_2 \quad (a)$$
$$CaCO_3 + SO_2 + 2H_2O\,(vapor) \rightarrow$$
$$CaSO_4 \cdot 2H_2O + CO_2 \quad (b)$$

(5.18)

Carbonaceous particles, especially those containing trace amounts of metal, are also believed to act as catalysts for the heterogeneous oxidation of $SO_2$ (Hutchinson et al., 1992a). When both HCl vapor and $SO_2$ are present, Hutchinson et al. (1992b) found that HCl is approximately 20 times more reactive than $SO_2$. In moist films on calcareous building stone, Haneef et al. (1992) found that oxidation by $SO_2$ is more significant than that by NO or $NO_2$. Carbonic acid formed from atmospheric carbon dioxide (Equation 5.2) is the main precursor for damage to concrete (highways, bridges, structures, etc.). There is evidence also that damage occurs due to sulfur dioxide and nitrogen oxides. The total carbon $[C]_t$ in the patina can be expressed as

$$[C]_t \rightarrow [C]_c + [C]_{nc} \quad (5.19)$$

where $[C]_c$ is the natural carbon in the original stone and $[C]_{nc}$ is the noncarbonate carbon contributed by pollution. Measurements of the carbon in damaged patinas (Zappia et al., 1993) show that the noncarbonated $[C]_{nc}$ far exceeds the natural carbon and is due entirely to atmospheric pollutants. Analysis shows that within the blackened patina, called *black gypsum crust*, 56 to 86% is composed of gypsum, which is far in excess of the gypsum in the natural rock. The white eroded portions of monuments correspond to surfaces that are unsheltered and subject to flushing by rain. The white eroded surfaces show less gypsum than the blackened areas but more than the underlying rock. Further analysis of the blackened patina shows the presence of minerals found in windblown soil particles.

### 5.5.3 Metal surfaces

Acid deposition produces corrosion via electrochemical reactions on metal surfaces (Knotkova and Barton, 1992). In the presence of water vapor and $SO_2$, solid particles containing trace metals create an environment in which electrochemical corrosion of ferrous metals occurs. The corrosion of iron and steel is caused primarily by oxygen and moisture that is accelerated by contaminants such as sulfates and chlorides. At low humidity, aluminum surfaces show little evidence of corrosion by $SO_2$. At higher humidity, a white powdery deposit of aluminum oxide appears:

$$2Al + SO_2 + H_2O \rightleftharpoons Al_2O_3 + H_2S \quad (5.20)$$

Copper alloys develop a thin stable surface film that inhibits further deterioration. Initially, a black sulfide or oxide film appears that is later replaced by the familiar green sulfate patina.

Not all the surface effects of copper are related to pollution. Variations in the color of the patina on the Statue of Liberty are not due to attack by acid rain (Livingston, 1991). The color patterns are related to the direction of the prevailing wind and rainwater striking the statue. The rainwater stabilizes the sulfate copper minerals over the chloride minerals produced by exposure to sea salt. Ironically, these patterns are more prominent now than in the past because ambient sulfur dioxide concentrations have decreased over the last few decades.

### 5.5.4 Wood

Aside from biological degradation and insects, the main factors degrading wood are UV light, moisture, and ozone. Acidity may lead to hydrolysis of cellulosic components or to minor increases in weathering rates. Sulfur dioxide and nitrogen oxides can lead to oxidation of the cellulose structure of wood and reduce its mechanical properties.

### 5.5.5 Textiles and elastomers

Air pollutants damage the underlying textile material and/or discolor the dye in fabrics. Cellulose fibers such as cotton, linen, hemp, and rayon are susceptible to losing their tensile strength due to acid damage. Animal fibers such as wool, fur, and hair are more tolerant of air pollutants since they contain nitrogen and sulfur compounds more resistant to a loss of tensile strength. Dyes, however, are subject to ultraviolet radiation damage from the sun and from damage due to nitrogen oxides, ozone, and sulfur oxides. Elastomers of the unsaturated type are attacked by ozone at the carbon–carbon double bond, which results in a loss of tensile strength.

## 5.6  Visibility

The appearance of industrial emissions and the degradation of scenic vistas (Zhang et al., 1994) are two characteristics of air pollution that humans object to. Reduction in visibility suggests worsening pollution levels (Molenar et al., 1994). Visibility is characterized by either its *visual range,* (i.e., the greatest distance an object can be seen) or by *opacity* (i.e., the attenuation of light passing through the polluted air). The visual range and opacity are inversely proportional to one another. Figure 5-7 illustrates the reduction in visibility due to pollution. The visual range is useful to describe the aesthetic quality of the landscape. As a parameter it is used in transportation safety (aircraft, ships, automobiles, etc.). Opacity is a technical parameter used for regulatory purposes to define unacceptable industrial stack emissions. Reductions in visual range of 10 km when the visual range is 300 km are unnoticeably small to most people and represents a useful baseline for comparison. An opacity of 7.5% is barely noticeable to trained observers and represents a baseline for regulatory compliance. Gazzi et al. (1994) suggest that visibility can be quantified by the luminance of a target. Pitchford and Malm (1994) propose a visual index that is linear with a parameter called *just noticeable changes in visibility* (JNCs) defined in terms of either attenuation of light or changes in visual range. By averaging over both space and time (one year), Malm et al. (1994) estimate that visibility in the United States could be improved by 21% if $SO_2$ emissions were reduced by 10 million tons. The largest improvement would be seen in the Ohio River valley.

With the exception of $NO_2$, pollutant gases and vapors do not absorb light (i.e., they are transparent). Nitrogen dioxide is unique because it absorbs nearly all green-blue light and produces the orange-brown haze associated with smog. Changes in the visual range and opacity are produced by particles in the atmosphere. *Hygroscopic* particles such as $H_2SO_4$, $CaHSO_4$, and $NH_4HSO_4$ absorb water vapor, increase their size, and attenuate light. The particle mass concentration may be very small and the particles pose no hazard to health, but they obscure vision. Consider the following natural visual phenomena that are based on aerosols:

- Fog
- Mountain haze (i.e., Blue Ridge Mountains of Virginia)
- Red sky at sunrise and sunset, blue sky during the daytime

The passage of light through air containing small particles is physically different from light passing through a continuous medium. Rather than light being described by such phrases as diffraction, reflection, and absorption, the effect of aerosols on light is defined as *scattering,* which integrates these phenomena. The attenuation of light intensity ($I$) by an aerosol is given by the Beer–Lambert law,

$$dI = -\sigma I \, dx \qquad \text{(a)}$$
$$\frac{I(L)}{I_0} = \exp(-\sigma L) \qquad \text{(b)}$$

$$(5.21)$$

where $L$ is the distance traveled between source and observer (i.e., the *optical path length*) and $\sigma$ is called the *scattering coefficient.* The scattering coefficient is a function of particle composition, size, and concentration.

The most dramatic attenuation of light occurs when the aerosol diameter ($D_p$) is comparable to the wavelength of light (i.e., $0.3 \, \mu m < D_p < 0.7 \, \mu m$). Such scattering is called *Mie scattering.* If the particles are smaller than the wavelength of light, the scattering is called *Rayleigh scattering. Mie scattering* is the principal phenomenon reducing visibility. For Mie scattering, the scattering coefficient ($\sigma$) is a function of the particle number concentration, $n_t$ (particles/m$^3$), and the particle surface area

$$\sigma = \frac{K n_t \pi D_p^2}{4} \qquad (5.22)$$

The parameter $K$ depends on the light wavelength and the particle's diameter and refractive index. For

**Figure 5-7** Reduction of visibility. View of Los Angles under clear and smoggy conditions.

an aerosol consisting of particles of different diameters, one sums over the range of particle diameters:

$$\sigma = \frac{\sum n_i K_i \pi D_{p,i}^2}{4} \quad (5.23)$$

where $n_i$ is the number of particles of diameter $D_{p,i}$ per unit volume of gas.

The opacity $(O)$ of an aerosol of optical path length $L$ is related to the scattering coefficient $(\sigma)$ as follows:

$$O = 1 - \frac{I(L)}{I_0} = 1 - \exp(-\sigma L) \quad (5.24)$$

## EXAMPLE 5.1  STACK OPACITY AND PARTICLE CONCENTRATION

The exhaust from a pulverized-coal electric-utility steam boiler is in compliance with the state's stringent emission standards and well within the federal New Source Performance Standards. At the time of the compliance tests the opacity was found to be 10% and the particle concentration $(c_p)$ in the exhaust was 0.46 g/dry standard m³, just within compliance standards. The state requires that the opacity must be monitored by a calibrated in-stack opacity monitor which records the opacity on a continuous basis. Over a period of several hours the opacity monitor records a 15% opacity, which is still in compliance with the state's visible emissions standard. On the basis of opacity, what particle concentration $(g/m^3)$ would you expect to exist during this period?

*Solution* Between disparate sources, there is no dependable relationship between opacity and particle concentration because the particle size distribution and optical properties may be vastly different from one source to another. Even the plumes from one steam boiler to another will vary. However, for a single source, such as the boiler in question, the particle size distribution is apt to be the same from day to day provided that the properties of the pulverized coal do not change. Thus while the engineer may not know the size distribution needed in Equation 5.24, there is no reason to believe that it changes appreciably. For the same reason, there is no reason to believe that the optical properties of the aerosol have changed. The optical path length and the wavelength of the light source used in the in-stack opacity moni-

tor certainly do not change. The particle number concentration $(n_t)$ is related to the particle mass concentration $(c_p)$:

$$c_p = n_t m_p$$

where $m_p$ is the mass of an average-sized particle. From Equation 5.23, the scattering coefficient for the two measurements can be expressed as

$$\sigma_1 = \frac{\sum K n_{i,1} \pi D_{p,i}^2}{4} = \frac{K n_{t,1} \sum \pi (n_{i,1}/n_{t,1}) D_{p,i}^2}{4}$$

$$\sigma_2 = \frac{\sum K n_{i,2} \pi D_{p,i}^2}{4} = \frac{K n_{t,2} \sum \pi (n_{i,2}/n_{t,2}) D_{p,i}^2}{4}$$

where $n_{t,1}$ and $n_{t,2}$ are the total particle concentrations for the two opacity measurements. The particle concentrations are related by

$$\frac{c_{p,2}}{c_{p,1}} = \frac{n_{t,2}}{n_{t,1}} \frac{m_{p,2}}{m_{p,1}} = \frac{\sigma_2}{\sigma_1}$$

The scattering coefficients are related to the opacity by

$$\sigma = -\frac{\ln(1 - O)}{L}$$

Thus

$$\frac{c_{p,2}}{c_{p,1}} = \frac{\ln(1 - O_2)}{\ln(1 - O_1)}$$

$$c_{p,2} = \frac{(0.46) \ln(1 - 0.15)}{\ln(1 - 0.10)} = 0.708 \text{ g/m}^2$$

Consequently, the boiler is just within compliance with respect to particle mass concentration when the opacity is 10% and is likely to be out of compliance at an opacity of 15% since the mass concentration has more than doubled.

---

When light impinges on an aerosol, light is scattered in all directions. The intensity of the light observed by a person depends on the angle between the observer and incident light. The angle between the source and observer is called the *scattering angle* $(\theta_s)$ (Figure 5-8):

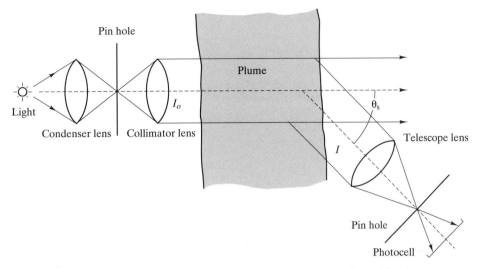

**Figure 5-8**   The intensity of light passing through an aerosol is affected by the scattering angle. Steam plumes appear dark and dense when sunlight passes through them (i.e. the sun is in the viewers face) while they appear white and less dense if the sun is behind the viewer.

- *Forward scattering*, $\theta_s = 0°$. Viewers looks directly into the incident light.
- *Back scattering*, $\theta_s = 180°$. Viewers look at the plume with light directly behind them.
- *90-degree scattering*, $\theta_s = 90°$. Viewers look at the plume perpendicular to the incident light.

Figure 5-9 shows that the intensity of light observed by the viewer is strongly dependent on the scattering angle, which is why EPA Method 9, describing how to estimate the opacity of plumes, requires viewers to observe plumes with the sun within a cone of an included angle of 140° to their rear.

Scattering also depends on the light's wavelength. Blue light has a smaller wavelength than red light and is scattered more easily than red light. At sunrise and sunset, the scattering angle is small (i.e., forward scattering) and the sky has a red hue because blue light is scattered away from the viewer's line of sight. During daylight hours, the myriad atmospheric particles scatter blue light sideways (i.e., 90° scattering) to produce the blue sky, unless of course one looks directly into the sun. *Skylight* is light scattered sideways by the particles in the atmosphere.

A parameter used in air pollution regulations to quantify the opacity of particles in air is called the *coefficient of haze* (COHs; see Table 3-10). The coefficient of haze is determined by ASTM Method D1704-16. An aerosol is drawn through filter paper at a prescribed volumetric flow rate and for a prescribed period of time. The filter paper is illuminated by light. The percent of light that is attenuated by particles on the filter paper is used to define the COHs:

$$\text{COHs} = 100 \log_{10} \frac{I_0}{I} \tag{5.25}$$

where $I$ and $I_0$ are the outlet and incident light intensities.

## 5.7  *Plume opacity*

A visible plume discharged from a smokestack (indeed, even the phrase *smokestack*) is a vivid statement that material from an industrial process is being added to atmosphere. The public objects to plumes for aesthetic reasons and because the presence of a plume suggests the possibility of a hazardous or noxious material being added to the atmosphere. As a result of the passage of air pollution legislation, not only is the mass emission rate (kg/s) of a source subject to regulation, but so is the opacity of the emission. The *opacity* is the percent of light that is attenuated by the plume (Connor and Hodkinson, 1967). Thus an opacity of 20% means that 80% of incident light passes through the plume.

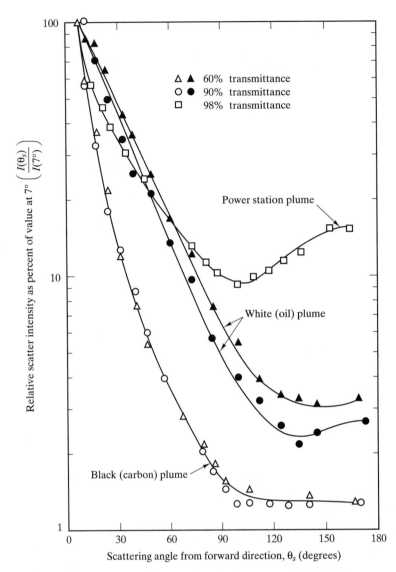

Relative scatter intensity as percent of value at 7° $\left(\frac{I(\theta_s)}{I(7°)}\right)$

Scattering angle from forward direction, $\theta_s$ (degrees)

△ ▲ 60% transmittance
○ ● 90% transmittance
□ 98% transmittance

Power station plume

White (oil) plume

Black (carbon) plume

**Figure 5-9** The intensity of light passing through the aerosol decreases with increasing scattering angle up to approximately 90 degrees. (figure taken from, Connor W D and Hodkinson J R, 1967)

In Pennsylvania no plume is permitted to have an opacity exceeding 15% except for three minutes within any hour, when it may not exceed 60%. Other states have similar standards. In virtually all industrial installations in the United States, optical instruments called *transmissometers* are mounted in the exhaust stack to monitor opacity on a continuous basis. To measure the opacity, people (*smoke readers*) are trained and certified to assess opacity visually following procedures defined in EPA Reference Method 9 (40 CFR Part 60).

Opacity produced by various forms of water are (appropriately) exempt from regulation. Thus water vapor that condenses into a highly visible and opaque steam plume is not subject to regulation. The public is generally unaware of this and may lump together plumes of steam and plumes containing particles. At night, steam plumes appear more ominous than in the day. It is important for people involved in air pollution control to (1) distinguish steam plumes from plumes containing particles, and (2) know how certified smoke readers assess opacity.

Technically speaking, *steam* is water vapor and thus is transparent. What we see and call a *steam plume* is the condensed phase consisting of small water droplets, similar to those comprising clouds.

Nevertheless, this visible discharge has been called steam by many people for a long time and is used in this book. If a discharge is produced from a cooling tower, dryer, or a wet scrubber, one expects to see a steam plume. In the absence of information on the source of the plume, there are two features of steam plumes that distinguish them from plumes composed of nonaqueous particles.

**1.** Steam plumes possess clearly visible boundaries. Steam plumes have convoluted surfaces but at any instant well-defined boundaries are clearly evident to the naked eye. Figure 5-10 is an example of a steam plume from a power plant that clearly illustrates these properties. Steam that is invisible at the stack exit (uncondensed vapor) but becomes opaque (condenses) a short distance downwind of the stack exit is called a *detached* steam plume, whereas a steam plume that appears immediately at the stack exit is called *attached*.

**2.** Steam plumes end abruptly when the water droplets evaporate. To be sure, the end of the steam plume contains transient isolated visible regions (steam) that exist for short periods of time before they evaporate, but nonetheless steam plumes end abruptly. A plume containing particles, on the other hand, does not end abruptly but persists and becomes diluted as the plume drifts downwind and mixes with ambient air. Figure 5-11a shows that the steam plume from a cooling tower ends abruptly, while the faint particle plume from a tall stack behind it persists as it drifts downwind. Steam plumes are normally white, but if they also contain particles, they appear tan (see darkening in the black and white Figure 5-11b). The opacity produced by the particles is subject to regulation and the certified observer must make observations at a location where the water droplets have evaporated completely, for example on the left one-third of Figure 5-11b.

EPA Reference Method 9 requires that opacity is assessed visually by certified observers who face the plume with the sun at their backs. Furthermore, the sun's rays must pass within a cone of included angle 140°, with the axis of rotation horizontal to the ground. If these procedures are followed, a plume with an opacity of 15% is barely visible to the untrained observer. Figures 5-12a and 5-12b illustrate how exaggerated an opaque steam plume appears when the viewer faces the sun. If viewed with the sun at one's back, even steam plumes are less dramatic. Photographs of a plume cannot be used to assess its opacity, although photographs obviously indicate the existence of the plume. Statements that photographs similar to Figures 5-12a and 5-12b demonstrate or even imply that plumes are not in

**Figure 5-10** Attached steam plume from a power plant. Note its well-defined boundaries and abrupt downstream disappearance often accompanied by transitory puffs of steam. (*photo by R.J. Heinsohn*).

**Figure 5-11a** Photograph of steam and particle plumes taken while facing the sun (Meaghen J F, et al. 1982)

**Figure 5-11b** Attached steam and particulate plume from a coal-washing facility. Note the persistence of the particulate plume after the steam plume evaporates. (*photo by R. J. Heinsohn*)

compliance with the visible emission standard are bogus. To begin with, steam plumes are exempt from regulation irrespective of their opacity, and second, in Figure 12b the photographer was facing in the wrong direction for accurate assessment. An egregious journalistic practice is to illustrate articles on air pollution with riveting pictures (Figure 5-12b) of voluminous, but benign, steam plumes taken by photographers facing the sun. The implication is that what is shown comprises pollution. Professionals concerned about air pollution must avoid this temptation and strive for accurate communication with policymakers and the public.

**Figure 5-12**  Plumes of condensed water vapor ("steam plumes") have vastly different appearances depending on the location of the sun. (a) In the top picture the sun is behind the viewer and the plume has the distinctive white (benign) appearance. (b) In the bottom picture the sun shines through the plume (back-lighting) and the innocuous steam plume acquires an ominous gray/black appearance. (photos by R.L. Kabel)

# 5.8 Odors

An *odor* is the physiological response to a particular airborne molecular species by the olfactory nerve cells, *olfactory neurons*, located at the top of the nose, just above the bridge of the nose. The concentration of a species that causes a person to detect an odor varies considerably between individuals. A person's response is affected by temperature, humidity, and exposure to simultaneous odors. While an average person might perceive an odor, only a few can identify the odor or even compare it to some other odor. People tend to become accustomed to odors, even those they find initially unpleasant. By far the majority of complaints about air pollution and inadequate ventilation concern odors (Shusterman, 1992). An odor may be harmless, indeed even pleasant (such as food), but if the odor persists, is intense, or inundates a person, the odor constitutes an involuntary intrusion and is a legitimate basis for compliant.

A *detection threshold* is the minimum concentration that produces a noticeable change in the odor of a person's environment. A more stringent concept of threshold is the *recognition threshold*, which is the minimum concentration at which the odor can be described. Even when detected it may take a 30 to 60% increase in the concentration before a panel identifies consistently that a higher concentration exists.

Although there is no satisfactory comprehensive theory of odors, chemicals that elicit an olfactory response tend to have low vapor pressures. Alcohols and acetates have odor thresholds that decrease with increasing carbon chain length and pungency thresholds that decrease exponentially with chain length (Cometto-Muniz and Cain, 1993). Thus odor threshold values reported in the literature (Ruth, 1986; Leonardos et al., 1969; Hellman and Small, 1974; Dravnieks et al., 1986; Verschueren, 1983; Stahl, 1978; Nagy, 1991) are apt to vary considerably. Table A-11 is a tabulation of odor threshold values that correspond to 100% recognition. Ruth (1986) tabulated lower concentrations that have been reported in the literature. The reader may find such values reported elsewhere and should be prepared for values that are several orders of magnitude lower than those in Table A-11.

If an odor is associated with toxic materials, the mere existence of the odor should be taken seriously.

Nearly 55% of the materials in Table A-11 have odor threshold values similar to their PEL or TLV values, 28% have odor threshold values above their PEL, and 17% have PEL greater than odor threshold values. Schaper (1993) found a strong correlation between TLV and materials whose odors produce a 50% reduction in the respiratory rate in mice. Shown below are several examples where the ratio of the odor threshold to the PEL change from very large to very small.

| Material | Odor threshold (ppm) | PEL (ppm) |
|---|---|---|
| Nickel carbonyl | 3 | 0.001 |
| Methyl bromide | 1,031 | 5 |
| Methyl alcohol | 20,485 | 200 |
| Methyl formate | 2,802 | 100 |
| Hydrogen cyanide | 4.5 | 4.7 |
| Methyl methacrylate | 0.3 | 100 |
| Xylidene | 0.005 | 2 |
| Butyl mercaptan | $9 \times 10^{-4}$ | 0.5 |
| Carbon monoxide | No odor | 35 |

Certain generalities can be seen from reviewing Table A-11.

- *Mercaptans:* unpleasant, repulsive materials that are detectable at concentrations well below PEL
- *Sulfides:* burnt, decayed matter, putrid
- *Acetates:* pleasant, fragrant, sweet, fruity
- *Aldehydes:* varying from sweet and fruity to pungent
- *Amines:* fishy, sharp, pungent

*Olfaction* is a sensitive physiological system that responds to contaminants at low concentrations, sometimes less than parts per billion. The olfactory system is capable of discriminating one odor from a background of different odors. Unfortunately, the olfaction system exhibits *fatigue* and individuals may become insensitive to an odor until the concentration changes. Odors elicit an involuntary response that either pleases or offends people. While no two individuals are alike, groups of people exhibit responses that can be quantified with statistical significance. There is a strong correlation between the senses of smell and taste since the neurological response involves the same organs. Food flavors are generally associated with aromas which are airborne mixtures of molecular species that stimulate the sensory organs. Practices followed in the foods and flavors industry in which the sensory response is described are also used to describe environmental odors. To *quantify* odors, at least three independent properties are needed:

1. *Quality* (or *character*): a description of the odor using familiar functional groupings
2. *Acceptability* (or *hedonic tone*): sensual pleasure, annoyance, or offense the odor evokes
3. *Intensity*: a quantitative response proportional to the concentration

The intensity or strength of a response can be described in terms of the *Weber–Fechner law*:

$$S_i = K \log(c_i) \qquad (5.26)$$

where $S_i$ is the magnitude (intensity) of the response, $c_i$ is the concentration of the material, and $K$ is a constant. Experiments to establish threshold values involve trained observers who ascribe a numerical value to diluted samples of the odor sometimes spanning six to eight orders of magnitude. Minimal detectable levels are established by diluting samples until the odor is no longer detected. Statistical methods are used to establish minimal detectable levels at which 50% and 100% of trained observers detect the odor. The most reliable parameter is the *recognition threshold*, which is defined as the lowest concentration 100% of a group of trained observers can positively identify an odor which is consistent with the response at all higher concentrations. The *detection threshold* is defined as the concentration that produces the first evidence of the odor above the background. Unfortunately, there are different detection thresholds depending on the background. An additional way to categorize odors is the *odor index*, which is the quotient of the actual vapor pressure to its partial pressure at the odor threshold value:

$$\text{odor index} = \frac{\text{vapor pressure}}{\text{threshold level (partial pressure)}}$$
$$(5.27)$$

Another way to express threshold values is called the *dilution factor* (Dravnieks et al., 1986), which is defined as the total number of volumes to which one volume of air saturated with the vapor or odorant must be diluted to reach the odor threshold. The odor index and dilution factor are the same and,

because the values are large, are often reported in terms of their logarithms. Both terms are useful because they the take into account the escaping tendency of the material (its vapor pressure) and the ability for people to recognize its odor.

## 5.9 Noise

Longitudinal pressure waves from 20 to 20,000 Hz are called *sound waves*. A vast number of superimposed sound waves in which there is no predetermined relationship between the frequencies and amplitudes of the waves is *noise*. Noise is unwanted sound and is divided into the following categories.

1. *Broadband noise.* Acoustical energy distributed over a broad range of wavelengths (typically, eight or more octaves) having no distinguishing tone is called broadband noise. Examples include the noise produced by a large crowd of people or by railroad trains.
2. *Narrowband noise.* Acoustical energy distributed over a narrow range of wavelengths and possessing a dominant tone is called narrowband noise. Examples include jet engines and circular saws.
3. *Impulse, or impact, noise.* Acoustical energy distributed over a short period of time (less than 1 s in duration) and generated at a frequency less than 200 periods/min is called impulse noise. Examples include a pneumatic jack hammer and a punch press.
4. *White noise.* Broadband noise in which the power density (acoustical power at a particular frequency) varies with time is called white noise. Examples include highway traffic noise.
5. *Pink noise.* Broadband noise in which the power densities are equal is called pink noise.

Hearing can be impaired permanently if people are exposed to sound or noise above certain amplitudes and for a certain duration. It is important for engineers to be able to predict the noise associated with their designs, take steps to reduce the generation of noise, add material to absorb it, or know when to recommend personal protective devices.

The square of the *speed of sound* (*a*) is quantified thermodynamically as the change of pressure with respect to density at constant entropys (s) across the sound wave:

$$a^2 = \left(\frac{\partial P}{\partial \rho}\right)_s \qquad (5.28)$$

For sound traveling in a gaseous medium that can be approximated as a perfect gas, the speed of sound can be shown to be

$$a = \sqrt{\frac{k R_u T}{M}} \qquad (5.29)$$

where

$k$ = ratio of specific heats $\left(c_p/c_v\right)$

$R_u$ = universal gas constant

$M$ = molecular weight of the gaseous medium

$T$ = absolute temperature

The speed of sound in air at STP is 344.5 m/s. Sound travels much faster in solids and liquids. The speed of sound in solids and liquids can be derived from Equation 5.28 and is equal to

$$a = \frac{1}{\sqrt{\beta\rho}} \qquad (5.30)$$

where $\beta$ is the isothermal expansion coefficient:

$$\beta = \frac{1}{\rho}\left(\frac{\partial \rho}{\partial P}\right)_T \qquad (5.31)$$

Consider a *point source* of sound located in free space. At a distance $(r)$ from the source, sound can be characterized by three parameters: *pressure ($P$ in force/area)*, *intensity ($I$ in power/area)*, and *power ($W$ in energy/time)*. Assuming that pressure waves travel through air without loss (which is consistent with the assumption of constant entropy), the power distributed over a spherical surface is equal to the acoustical power of the source. The power associated with the sound wave per unit area is called the *intensity ($I$)* and is related to the pressure by

$$I = \frac{P^2}{\rho a} \qquad (5.32)$$

Thus *sound power ($W$)* is related to sound intensity by

$$W = I\left(4\pi r^2\right) = \frac{P^2\left(4\pi r^2\right)}{\rho a} \qquad (5.33)$$

If the source of sound is not located in free space, and ideal reflecting surfaces direct the sound in pref-

erential directions, it is convenient to modify the above as follows:

$$W = \frac{I\left(4\pi r^2\right)}{Q} \qquad (5.34)$$

where $(Q)$ is called the *directivity factor*. Sound directivity is defined as the ratio of the sound power of a small omnidirectional hypothetical source to the sound power of an actual source that produces the same sound-pressure level at a point of measurement. If the sound source is bounded by acoustically reflecting surfaces,

- $Q = 1$: sound source in free space
- $Q = 2$: sound source located in an infinite reflecting plane
- $Q = 8$: sound source located at the intersection of three mutually perpendicular intersecting reflecting planes

It is convenient to define a *sound-pressure level ($L_P$)* that is related to the sound pressure ($P$):

$$L_P = 20 \log_{10}\left(\frac{P}{P_0}\right) \qquad (5.35)$$

The term $P_0$ is a reference value,

$$P_0 = 2 \times 10^{-5}\,\text{N/m}^2 \qquad (5.36)$$

which at one time was thought to be a threshold value for a typical young person at 1000 Hz. In a similar fashion it is useful to define a sound intensity level $(L_I)$:

$$L_I = 10 \log_{10} \frac{I}{I_0} \qquad (5.37)$$

where the reference value $I_0$ corresponds roughly to the reference pressure level $\left(P_0\right)$:

$$I_0 = 10^{-12}\,\text{W/m}^2 \qquad (5.38)$$

If Equation 5.32 is substituted into Equation 5.37, it can be seen that $L_P = L_I$. Thus sound pressure level is numerically the same as sound intensity level. Sound power ($W$) can also be expressed as a sound power level $(L_W)$,

$$L_W = 10 \log_{10} \frac{W}{W_0} \qquad (5.39)$$

where

$$W_0 = 10^{-12}\,\text{W} \qquad (5.40)$$

The units used to express the sound-pressure level, sound-intensity level, and sound-power level are called *decibels* (dB).

It is of advantage to manufacturers to reduce the acoustic power of equipment (not HiFi of course) they sell. Consequently, the acoustic power of equipment is information that is often available to the purchaser. The acoustic power can be used to predict the sound-pressure level $(L_P)$ at arbitrary locations from the equipment. Assume that the sound power $(W)$ remains constant, Equations 5.33 and 5.34 indicate that the sound pressure and sound intensity decrease as the distance $(r)$ between the source and listener increases. Combining Equations 5.32 to 5.39 and evaluating the constant terms, it can be shown that

$$L_I = L_P = L_W - 20 \log_{10} r(m) + 10 \log_{10} Q - 11$$

$$(5.41)$$

Using the above, it is simple to show that the sound-pressure level changes by 6 dB each time the distance between a listener and a source doubles or is halved.

---

### EXAMPLE 5.2   SOUND-PRESSURE LEVEL OF A JACKHAMMER

A pneumatic jackhammer is rated as having a free–field sound level of 85 dB at 10 m. What is the acoustic power of the equipment, and what is the sound-pressure level at workstations 0.5 m and 2.0 m from the jackhammer, used at the junction of the pavement and two perpendicular walls?

***Solution***  The acoustic-power level of the source $(L_W)$ is found by substituting $L_P = 85$ dB, $r = 10$, and $Q = 1$ in Equation 5.41. One finds that the acoustic-power level of the source $(L_W)$ is 116 dB. The acoustic power is

$$L_W = 116 = 10 \log_{10} \frac{W}{10^{-12} \text{ W}}$$

$$11.6 - 12 = -0.4 = \log_{10}(W) = \frac{\ln(W)}{2.303}$$

$$W = 0.398 \text{ W}$$

The sound-pressure level at $r = 0.5$ m and $r = 2.0$ m are found from the same equation using $L_W = 116$ dB and $Q = 8$.

$$L_p(r = 0.5 \text{ m}) = 120 \text{ dB}$$
$$L_p(r = 2.0 \text{ m}) = 108 \text{ dB}$$

---

Since noise consists of several simultaneous sound waves of different frequencies and intensities, the ear is subjected to a composite wave in which the total power is the sum of the powers of each sound wave. Thus the sound intensity at a point in space subject to *several sources* of noise is

$$\frac{I}{I_0} = \sum \frac{I_j}{I_0} \qquad (5.42)$$

where $j$ refers to the $j$th source of noise. The right-hand side of above can be evaluated by taking antilogs of Equation 5.37:

$$\frac{L_{I,j}}{10} = \log_{10} \frac{I_j}{I_0} \qquad (5.43)$$

The total sound-intensity level of several sources of noise can be expressed as

$$L_I (\text{dB}) = 10 \log_{10} \left( \sum 10^{n_j} \right)$$
$$n_j = \frac{L_{i,j}}{10} \qquad (5.44)$$

### EXAMPLE 5.3   SOUND-INTENSITY LEVEL FROM SEVERAL FANS

A company wishes to place six forced draft fans in a room. What is the total sound-intensity level produced by all six fans at a certain point $(A)$ in the room if at point $(A)$ each fan separately produces the following sound-intensity levels.

| Fan | Sound-intensity level (dBA) at point $A$ |
|-----|------------------------------------------|
| 1   | 85 |
| 2   | 92 |
| 3   | 90 |
| 4   | 84 |
| 5   | 93 |
| 6   | 87 |

***Solution***  Substitution in Equation 5.44 shows that the total sound level is 97 dBA:

$$L_I = 10 \log_{10}\left(10^{8.5} + 10^{9.2} + 10^{9.0} \right.$$
$$\left. + 10^{8.4} + 10^{9.3} + 10^{8.7}\right)$$
$$= 97 \text{ dB}$$

The human function of hearing depends on both frequency and sound level as shown in Figure 5-13. To simulate the ear it is convenient to define three sound levels (A, B, and C scales) in which there are different weighting factors for selected bands of frequencies shown in Figure 5-14. The OSHA Permissible Noise Exposures shown in Table 5-2 are defined

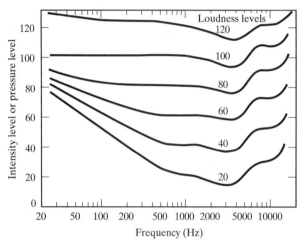

**Figure 5-13** Free-field equal loudness contours of pure tones, (redrawn from Zimmerman, 1984)

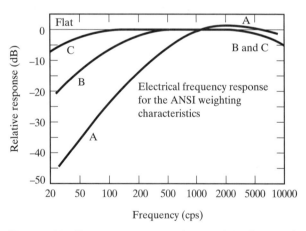

**Figure 5-14** Frequency-response characteristics for sound-level meters, (redrawn from NIOSH, "Occupational Diseases, A Guide to Their Recognition, 1977)

**Table 5-2** OSHA Threshold Limit Values for Noise

| Duration per day (hs) | Sound level (dBA) |
|:---:|:---:|
| 16 | 80 |
| 8 | 85 |
| 4 | 90 |
| 2 | 95 |
| 1 | 100 |
| $\frac{1}{2}$ | 105 |
| $\frac{1}{4}$ | 110 |
| $\frac{1}{8}$ | 115 |

*Source:* Abstracted from CFR 29, Parts 1900–1910.

in terms of the A scale. While the 8-h standard is 85 dBA, higher sound levels are allowed for shorter periods of time. Noise produced by impact should not exceed 140 dB. Figure 5-15 shows the correspondence between sound-pressure level and commonplace sources of noise.

When the daily exposure is composed of different periods of exposure to $n$ different sound levels, the hazard is evaluated by the following:

$$E_n = \sum_n \left(\frac{c}{t}\right)_i \qquad (5.45)$$

where the subscript $i$ refers to the sound level, $c_i$ is the duration of period of exposure at one sound level, $t_i$ is the total duration of exposure permitted at that level (Table 5-2), and the summation is over the number of periods of exposure. If the values of $E_n$ exceeds unity, the mixed exposure is considered to have exceeded the OSHA limit for safe conditions.

The jackhammer in Example 5.2 is so noisy that it should not be used in its present condition because, while workers using it may wear personal protective devices, other people in its proximity will be affected. The sound level (97dB) of the six fans in Example 5.3 clearly exceeds the 8-h OSHA standard and a person located at the point in question should not remain there for more than 1.5 h unless a personal hearing protective device is worn. *Hearing impairment* develops long after exposure and is irreversible. Incidence of hearing impairment is illustrated in Figure 5-16

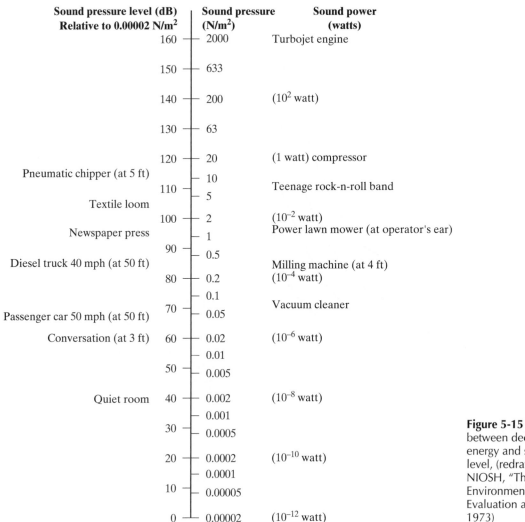

**Figure 5-15** Relationship between decibels, sound energy and sound pressure level, (redrawn from NIOSH, "The Industrial Environment - Its Evaluation and Control", 1973)

and shows that even at 90 dB, a significant portion of a working population will suffer impairment. We note that "incidence" is an accumulated result and the young may be just as susceptible to impairment as the old. Designers of pollution control systems must be sensitive to the noise produced by fans, air jets, or indeed any vortices, recirculation, or flow separation that is apt to occur. Not only must OSHA standards be met, but one must minimize noise in general.

## 5.10 Closure

Pollution that interferes with our enjoyment of the natural environment is within the jurisdiction of the Clean Air Act. The impact on the environment may not be as dramatic or immediate as the impact on health, but nonetheless, engineers may be called upon to devise ways to reduce these effects. Engineers need to understand how pollutants affect the surroundings and how to express these effects in quantitative terms.

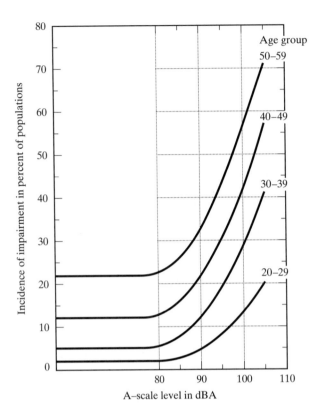

**Figure 5-16**  Prevalence of impaired hearing and sound-levels for different age groups, (redrawn from NIOSH, Occupational Diseases, A Guide to Their Recognition, 1977)

# Nomenclature

| Symbol | Description (Dimensions*) |
|---|---|
| $\rightleftharpoons$ | simultaneous forward and reverse reaction |
| $\rightarrow$ | direction of a chemical reaction |
| $=$ | stoichiometric equation |
| $[j]$ | concentration of species $j$ $(M/L^3, N/L^3)$ |
| $[=]$ | has units of |
| $a$ | speed of sound (L/t) |
| $c$ | concentration $(M/L^3)$ |
| $c_p$ | particle concentration $(M/L^3)$ |
| $D_p$ | particle diameter (L) |
| $E_n$ | sound exposure index, defined by equation |

*F, force; L, length; M, mass; N, mols; Q, energy; t, time; T, temperature.

|  |  |
|---|---|
| $I$ | acute quantifiable injury, light intensity, sound intensity |
| $k$ | rate constant defined by equation |
| $k$ | ratio of specific heats, $c_p/c_v$ |
| $K$ | parameter defined by equation |
| $L$ | optical path length (L) |
| $L_P, L_I, L_W$ | sound pressure, intensity, power level |
| $M$ | molecular weight (M/N) |
| $m_p$ | mass of a particle (M) |
| $n_j$ | number of particles of diameter $D_{p,j}$ per unit volume of gas $(L^{-3})$ |
| $n_t$ | total number of particles per unit volume of gas $(L^{-3})$ |
| $O$ | opacity |
| $P$ | pressure $(F/L^2)$ |
| $Q$ | sound directivity factor |
| $r$ | radius to noise source (L) |

| $R_u$ | universal gas constant (Q/NT) |
|---|---|
| $S_i$ | intensity or strength of an odor |
| $t$ | time (t) |
| $T$ | temperature (T) |
| $W$ | sound power $(Q/L^2)$ |

*Abbreviation*

| ASTM | American Society of Testing and Materials |
|---|---|
| cps | cycles per second |
| dB | decibel |

| COHs | coefficient of haze |
|---|---|
| JNC | just noticeable change |
| pH | defined by equation |
| UV | ultraviolet |

*Greek*

| $\beta$ | isothermal expansion coefficient $(L^3/F)$ |
|---|---|
| $\Theta_s$ | scattering angle |
| $\sigma$ | scattering coefficient |
| $\rho$ | density $(M/L^3)$ |
| $\Phi$ | relative humidity |

# References

Billings, C.E., and Matson, W.R., 1972. Mercury emissions from coal combustion, *Science*, Vol. 176, pp. 1232–1233.

Brown, S. L., Chaney, R. L., Lloyd, C. A., Angle, J. S., and Ryan, J. A., 1996. Relative uptake of cadmium by garden vegetables and fruits grown long-term biosolid-amended soils, *Environmental Science and Technology*, Vol. 30, No. 12, pp. 3508–3511.

Butlin, R. N., 1991. Effects of air pollutants on buildings and materials, *Proceedings of the Royal Society of Edinburgh*, Vol. 97B, pp. 255–272.

Camuffo, D., 1983. Indoor dynamic climatology: investigations on the interactions between walls and indoor environment, *Atmospheric Environment*, Vol. 17, No. 9, pp. 1803–1809.

Christoforou, C. S., Salmon, L. G., and Cass, G. R., 1996. Fate of atmospheric particles within the Buddhist cave temples at Yungang, China, *Environmental Science and Technology*, Vol. 30, No. 12, pp. 3425–3434.

Cometto-Muniz, J. E., and Cain, W. S., 1993. Efficacy of volatile organic compounds in evoking nasal pungency and odor, *Archives of Environmental Health*, Vol. 48, No. 5, pp. 309–314.

Connor, W. D., and Hodkinson, J. R., 1967. *Optical Properties and Visual Effects of Smoke Stack Plumes*, EPA report NTIS PB 174-705, National Technical Information Service, Springfield, VA.

Douglas, J., 1994. Mercury and the global environment, *EPRI Journal*, pp. 14–21. Jan/Feb.

Dravnieks, A., Schmidsdorff, W., and Meilgaard, M., 1986. Odor thresholds by forced-choice dynamic triangle olfactometry: reproducibility and methods of calculation, *Journal of the Air Pollution Control Association*, Vol. 36, No. 8, pp. 900–905.

Driscoll, C. T., Yan, C., Schofield, C. L., Munson, R., and Holsapple, J., 1994. The mercury cycle and fish in the Adirondack lakes, *Environmental Science and Technology*, Vol. 28, No. 3, pp. 136A–143A.

Elsom, D. M., 1992. *Atmospheric Pollution*, 2nd ed., Blackwell, Oxford.

Fobe, B. O., Vleugels, G. J., Roekens, E. J., Van Grieken, R. E., Hermosin, B., Ortega-Calvo, J. J., Del Junco, A. S., and Saiz-Jimenez, C., 1995. Organic and inorganic compounds in limestone weathering crusts from cathedrals in southern and western Europe, *Environmental Science and Technology*, Vol. 29, No. 6, pp. 1691–1701.

Galbreath, K. C., and Zygarlicke, C. J., 1996. Mercury speciation in coal combustion and gasification flue gases, *Environmental Science and Technology*, Vol. 30, No. 8, pp. 2421–2426.

Gazzi, M., Vicentini, V., and Bonafe, U., 1994. A field experiment on contrast reduction law, *Atmospheric Environment*, Vol. 28, No. 5, pp. 901–907.

Grosjean, D., Grosjean, E., and Williams, E. L., II, 1993. Fading of artists' colorants by a mixture of photochemical oxidants, *Atmospheric Environment*, Vol. 27A, No. 5, pp. 765–772.

Guidobaldi, F., and Mecchi, A. M., 1993. Corrosion of ancient marble monuments by rain: evaluation of pre-industrial recession rates by laboratory simulations, *Atmospheric Environment*, Vol. 27B, No. 3, pp. 339–351.

Hamilton, R. S., and Mansfield, T. A., 1993. The soiling of materials in the ambient atmosphere, *Atmospheric Environment*, Vol. 27A, No. 8, pp. 1369–1374.

Haneef, S. J., Johnson, J. B., Dickinson, C., Thompson, G. E., and Wood, G. C., 1992. Effect of dry deposition of NO$_x$ and SO$_2$ gaseous pollutants on the degradation of calcareous building stones, *Atmospheric Environment*, Vol. 26A, No. 16, pp. 2963–2974.

Havens, K. E., Yan, N. D., and Keller, W., 1993. Lake acidification: effects on crustacean zooplankton populations, *Environmental Science and Technology*, Vol. 27, No. 8, pp. 1621–1624.

Hellman, T. M., and Small, F. H., 1974. Characterization of the odor properties of 101 petrochemicals using sensory methods, *Journal of the Air Pollution Control Association*, Vol. 24, No. 10, pp. 979–982.

Hisham, M. W. M., and Grosjean, D., 1991. Air pollution in southern California museums: indoor and outdoor levels of nitrogen dioxide, peroxyacetyl nitrate, nitric acid and chlorinated hydrocarbons, *Environmental Science and Technology*, Vol. 25, No. 5, pp. 857–862.

Hoffmann, R., 1994. Winning gold, *American Scientist*, Vol. 82, pp. 15–17.

Hutchinson, A. J., Johnson, J. B., Thompson, G. E., and Wood, G. C., 1992a. The role of fly-ash particulate material and oxide catalysts in stone degradation, *Atmospheric Environment*, Vol. 26A, No. 15, pp. 2795–2803.

Hutchinson, A. J., Johnson, J. B., Thompson, G. E., and Wood, G. C., 1992b. Stone degradation due to dry deposition of HCl and SO$_2$ in a laboratory-based exposure chamber, *Atmospheric Environment*, Vol. 26A, No. 15, pp. 2785–2793.

Kinjo, Y., Higashi, H., Nakano, A., Sakamoto, M., and Sakai, R., 1993. Profile of subjective complaints and activities of daily living among current patients with Minamata disease after 3 decades, *Environmental Research*, Vol. 63, pp. 241–251.

Knotkova, D., and Barton, K., 1992. Effects of acid deposition on corrosion of metals, *Atmospheric Environment*, Vol. 26A, No. 17, pp. 3169–3177.

Leonardos, G., Kendall, D., and Barnard, N., 1969. Odor threshold determinations of 53 odorant chemicals, *Journal of the Air Pollution Control Association*, Vol. 19, No. 2, pp. 91–95.

Livingston, R. A., 1991. Influence of the environment on the patina of the Statue of Liberty, *Environmental Science and Technology*, Vol. 25, No. 8, pp. 1400–1408.

Malm, W. C., Trijonis, J., Sisler, J., Pitchford, M., and Dennis, R. L., 1994. Assessing the effect of SO$_2$ emission changes on visibility, *Atmospheric Environment*, Vol. 28, No. 5, pp. 1023–1034.

Mason, R. P., Reinfelder, J. R., and Morel, F. M. M., 1996. Uptake, toxicity, and tropic transfer of mercury in a coastal diatom, *Environmental Science and Technology*, Vol. 30, pp. 1835–1845.

Meaghen, J. F., Bailey, E. M., and Luria, M., 1982. The impact of mixing cooling tower and power plant plumes on sulfate aerosol formation, *Journal of the Air Pollution Control Association*, Vol. 32, No. 4, pp. 389–391.

Molenar, J. V., Malm, W. C., and Johnson, C. E., 1994. Visual air quality simulation techniques, *Atmospheric Environment*, Vol. 28, No. 5, pp. 1055–1063.

Nagy, G. Z., 1991. The odor impact model, *Journal of the Air & Waste Management Association*, Vol. 41, No. 10, pp. 1360–1362.

Natusch, D. F. S., and Wallace, J. R., 1974. Urban aerosol toxicity: the influence of particle size, *Science*, Vol. 186, November, 22, pp. 695–699.

Natusch, D. F. S., Wallace, J. R., and Evans, C. A., Jr., 1974. Toxic trace elements: preferential concentration in respirable particles, *Science*, Vol. 183, pp. 202–204.

Nord, A. G., Svardh, A., and Tronner, K., 1994. Air pollution levels reflected in deposits on building stone, *Atmospheric Environment*, Vol. 28, No. 16, pp. 2615–2622.

O'Dell, R. A., Taheri, M., and Kabel, R. L., 1977. A model for uptake of pollutants by vegetation, *Journal of the Air Pollution Control Association*, Vol. 27, No. 11, pp. 1104–1116.

Paterson, S., Mackay, D., Bacci, E., and Davide, C., 1991. Correlation of the equilibrium and kinetics of leaf–air exchange of hydrophobic organic chemicals, *Environmental Science and Technology*, Vol. 25, No. 5, pp. 866–871.

Paterson, S., Mackay, D., and McFarlane, C., 1994. A model of organic chemical uptake by plants from soil and the atmosphere, *Environmental Science and Technology*, Vol. 28, No. 13, pp. 2259–2266.

Pitchford, M. L., and Malm, W. C., 1994. Development and applications of a standard visual index, *Atmospheric Environment*, Vol. 28, No. 5, pp. 1049–1054.

Rasmussen, P. E., 1994. Current methods of estimating atmospheric mercury fluxes in remote areas, *Environmental Science and Technology*, Vol. 28, No. 13, pp. 2233–2241.

Robertson, D. E., Crecelius, W. A., Fruchter, J. S., and Ludwick, J. D., 1977. Mercury emissions from geothermal power plants, *Science*, Vol. 196, pp. 1094–1097.

Ruth, J. H., 1986. Odor thresholds and irritation levels of several chemical substances: a review, *Journal of the American Industrial Hygiene Association*, Vol. 47, March, pp. A-142 to A 151.

Sabbioni, C., Zappia, G., and Gobbi, G., 1992. Carbonaceous particles on carbonate building stones in a simulated system, *Journal of Aerosol Science*, Vol. 23, Suppl. 1, pp. S921–S924.

Salmon, L. G., Christoforou, C. S., and Cass, G. R., 1994. Airborne pollutants in the Buddhist cave temples at the

Yungang Grottoes, China, *Environmental Science and Technology*, Vol. 28, pp. 805–811.

Schaper, M., 1993. Development of a database for sensory irritants and its use in establishing occupational exposure limits, *American Industrial Hygiene Association Journal*, Vol. 54, No. 9, pp. 488–544.

Schiavon, N., and Zhou, L., 1996. Magnetic, chemical, and microscopical characterization of urban soiling on historical monuments, *Environmental Science and Technology*, Vol. 30, No. 12, pp. 3624–3629.

Seigneur, C., Wrobel, J., and Constantinou, E., 1994. A chemical kinetic mechanism for atmospheric inorganic mercury, *Environmental Science and Technology*, Vol. 28, No. 9, pp. 1589–1597.

Shusterman, D., 1992. Critical review: the health significance of environmental odor pollution, *Archives of Environmental Health*, Vol. 47, No. 1, pp. 76–91.

Simonich, S. L., and Hites, R. A., 1995. Organic pollutant accumulation in vegetation, *Environmental Science and Technology*, Vol. 29, pp. 2905–2914.

Stahl, W. H., (Ed.), 1978. *Compilation of Odor and Taste Threshold Data*, ASTM DS 48A, American Society for Testing and Materials, Philadelphia.

Stern, A. C. (Ed.), 1968. *Air Pollution*, Vols. I, II, and III, 2nd ed., Academic Press, San Diego, CA.

Tomlinson, G. H., II, 1983. Air pollutants and forest decline, *Environmental Science and Technology*, Vol. 17, No. 6, pp. 246A–256A.

Verschueren, K., 1983. *Handbook of Environmental Data on Organic Chemicals*, 2nd ed., Van Nostrand Reinhold, New York.

Westenbarger, D. A., and Frisvold, G. B., 1994. Agricultural exposure to ozone and acid precipitation, *Atmospheric Environment*, Vol. 28, No. 18, pp. 2895–2907.

Zappia, G., Sabbioni, C., and Pauri, M. G., 1992. Damage induced by atmospheric aerosol on ancient and modern building materials, *Journal of Aerosol Science*, Vol. 23, Suppl. 1, pp. S917–S920.

Zappia, G., Sabbioni, C., and Gobbi, G., 1993. Non-carbonate carbon content on black and white areas of damaged stone monuments, *Atmospheric Environment*, Vol. 27A, No. 7, pp. 1117–1121.

Zhang, X., Turpin, B. J., McMurry, P. H., Hering, S. V., and Stolzenburg, M. R., 1994. Mie theory evaluation of species contributions to 1990 wintertime visibility reduction in the Grand Canyon, *Journal of the Air & Waste Management Association*, Vol. 44, February, pp. 153–162.

Zimmerman, N.J., "Principles of Occupational Safety and Health Engineering," Instructor's Guide published by NIOSH, PO NO 81-3030, March 1984.

# Problems

**5.1.** Mercury in coal vaporizes during combustion. Ash from coal combustion is removed from the furnace as "bottom ash" and suspended in the exhaust gas as "fly ash". As the exhaust gases cool, 10% by weight of the vapor condenses on the surface of fly ash particles and 90% remains as vapor in the exhaust gas (Billings and Matson, 1972). These emissions might become subject to regulation in the future (see the *New York Times*, August 26, 1991 for details).

**(a)** Estimate the emission of mercury vapor (tons/yr) to the atmosphere from a 1000-MW electric generating station that has an overall thermal efficiency of 33%, burns coal with a heating value of 12,500 Btu/lb$_m$, and contains 1 mg of mercury per kilogram of coal. Assume that 99.5% by weight of the fly ash particles are captured by a fabric filter or electrostatic precipitator and that only 0.5% of the particles are emitted to the atmosphere. See Example 3.1 for guidance.

**(b)** Estimate the mass of mercury leaving the plant in the form of collected fly ash.

**(c)** Estimate the total emission of mercury (tons/yr) in the United States at the present time if 25% of our total energy consumption is provided by coal with the mercury content noted above. In 1970, the total U.S. consumption of energy was 70 quads/yr $\left(\text{quad} = 10^{15} \text{ Btu}\right)$ and the yearly consumption of energy increased exponentially at a rate of 1.6% per year.

**5.2.** You work in a plant where the chemicals listed below are used. You believe that detecting the chemical's odor during working hours (40 h/week) indicates hazardous conditions.

- Acetic acid
- Amyl acetate
- Butyl acetate
- Butyl mercaptan
- Carbon disulfide
- Carbon tetrachloride
- Cyclohexane
- Ethyl ether
- Ethylene oxide
- Formaldehyde
- Methyl alcohol
- Methyl bromide

- Methylene chloride
- Nickel carbonyl
- Pentaborane
- Phenol
- Styrene
- Trimethylamine
- Vinyl chloride

**(a)** Which of the chemicals are HAPs?

**(b)** What are the odor threshold concentrations (ppm)? How do people characterize each odor?

**(c)** What are the OSHA PELs?

**(d)** Identify the chemicals whose detectable odor for 40 h/week constitutes hazardous conditions.

**(e)** Identify the chemicals that could be present at hazardous concentrations for 40 h/week and whose odor would not be detectable below hazardous concentrations.

Tabular data about some of these materials can be found in the Appendix; other information can be obtained from the following:

Norback, C. J., *Hazardous Chemicals on File*, Van Nostrand Reinhold, New York, 1988

Lewis, R. J., Sr., *Sax's Dangerous Properties of Industrial Materials*, Von Nostrand Reinhold, New York, 1993

Lewis, R. J., Sr., *Rapid Guide to Hazardous Chemicals in the Workplace*, Von Nostrand Reinhold, New York, 1994 (web site) http://chemfinder.camsoft.com/

**5.3.** An automobile shredding machine is rated as having a free-field sound pressure level of 85 dB at 10 m. What is the acoustic power output (watts)? What is the sound pressure level $(L_p)$ for workers standing 2 m from the machine when the shredder is placed next to a high sound reflecting wall used to shield the machine from public view?

**5.4.** Six fans will be located on the ground outside a plant. The sound pressure level at point P for each fan is found to be 85, 92, 90, 84, 93, and 87 dBA. What is the sound pressure level (dBA) at point P when all the machines are operating at the same time?

**5.5.** A plant manager buys five fans, each of which is rated as having an acoustic power output of 0.01 W. The five fans will be located on the ground 5 m apart and mid-way along the long wall of a building. The building dimensions are 30 m × 200 m × 20 m high. Estimate the sound pressure level experienced by a person standing 0.2 m away from the center fan.

**5.6.** Three identical fans are located on the floor in the center of a large building. The layout of the fans is in the form of an equilateral triangle 2 m on a side. The free-field sound pressure level of each machine is 85 dBA at a distance of 0.1 m from the machine. When all three fans are in use, what sound pressure level (dBA) can be anticipated for a person standing in the center of the equilateral triangle?

**5.7.** Estimate the sound pressure level (dBA) experienced by an aircraft ground chief who guides a four-engine jet aircraft into its berth on an airport runway. Each engine produces 10,000 W of acoustical power. At the closest point to the aircraft, the chief stands directly in front of the aircraft's nose 10 m from the inboard engines and 20 m from the outboard engines. Neglect sound reflected from any buildings.

**5.8.** Derive Equation 5.41.

**5.9.** Derive Equation 5.44.

**5.10.** Show that the sound pressure level decreases by 6 dBA when a listener doubles the distance to the source.

**5.11.** [*Design Problem*] Aircraft flying tourists over the Grand Canyon have become so numerous that the noise they generate (collectively) has become a source of irritation to visitors to the park. Such noise "unreasonably interferes with the comfortable enjoyment" of the park and thus constitutes air pollution following the definition in Section 1.1. The Park Service wishes to curtail their activity to reduce the noise to acceptable levels. On the average, the sound power level $(L_w)$ of aircraft is 140 dBA. FAA flight rules require aircraft to maintain a distance no less than $\frac{1}{2}$ mile from each other.

**(a)** *Noise from a single aircraft.* Consider a single aircraft that initially is directly overhead but flies away at 60 mph at an altitude of 1000 m. Compute and plot the sound pressure level, $L_p(t)$, as a function of time for a period of 60 s.

**(b)** *Noise from many aircraft.* Recommend a minimum altitude that all aircraft must maintain so that if a large number of aircraft are flying the minimum distance apart, the sound pressure level $(L_p)$ to people on the ground never exceeds 50 dBA.

# 6

# Atmospheric Chemicals: Sources, Reactions, Transport, and Sinks

In this chapter you will learn:

- About the formation and fate cycles of carbon, sulfur, and nitrogen compounds
- About interfacial mass transfer between air pollutants and bodies of water and solid surfaces
- About atmospheric photochemical reactions

In the chapters on exponential growth (2), the human respiratory system (4), and aesthetics (5), we introduced the interconnection of air purity and pollution with all aspects of the global environment. In this chapter we are more explicit about biogenic and anthropogenic processes that inject chemicals into the atmosphere. Natural processes that occur are called *biogenic processes*, and processes associated with human activities are called *anthropogenic processes*. We look in detail at the physical and chemical processes that occur in the atmosphere and change its chemical constitution. This defines the condition of the atmosphere and enables consideration of its impact on the other media at its boundaries. Finally, we indicate how fluxes of chemical species between the atmosphere and the bounding media (water bodies, vegetation, and soil) may be quantified. Having characterized the atmosphere and its impact on the surrounding media and with consideration of the long- and short-term risks to the biosphere (Chapter 1) and the legal consequences (Chapter 3), we can address the need for, and the desirable degree of, control of air pollution in subsequent chapters. In Sections 6.1 to 6.7 we describe how pollutants are formed by biogenic and anthropogenic processes, how they travel through the biosphere, and their ultimate fate (i.e., *formation and fate cycles*). Sections 6.8 to 6.11 present in quantitative terms the chemical and physical processes underlying the formation and fate cycles, thus allowing engineers to aid industry and government in satisfying emission and ambient air quality standards.

# 6.1 Nature and composition of the atmosphere

Before we examine anthropogenic pollution of the atmosphere, we need to understand the natural conditions and constituents of the atmosphere and how these influence the *formation and fate cycles* of chemical species in the entire biosphere. Readers will not be able to comprehend human influence on the environment until they understand these natural processes first. We begin with a physical description of the atmosphere and identify regions of differing characteristics. As one goes higher in the atmosphere, the related physical properties of pressure, mass density, and molecular density (or concentration) all decrease, as shown in Figure 6-1. We think of the atmospheric temperature as decreasing with altitude, and it usually does in the lower atmosphere, but Figure 6-1 shows

that it is not that simple. The layers of the atmosphere, based primarily on temperature, that will be used throughout this book are as follows:

**1.** *Troposphere* is the layer between the ground and up to approximately 15 km at the equator and 10 km at the poles. In the troposphere the temperature typically decreases at the rate of 6.5°C/km, the air is relatively well mixed, and thermally driven convection is relatively strong. The troposphere is the layer that contains most of our weather.

**2.** *Stratosphere* is the layer between the troposphere and approximately 50 km. There is little bulk mixing in the stratosphere. The temperature is relatively constant except in the upper stratosphere, where the absorption of UV radiation by ozone causes the temperature to rise to approximately 0°C (270 K).

**3.** *Mesosphere* is the layer between 50 and 85 km within which the temperature decreases almost linearly to approximately 175 K.

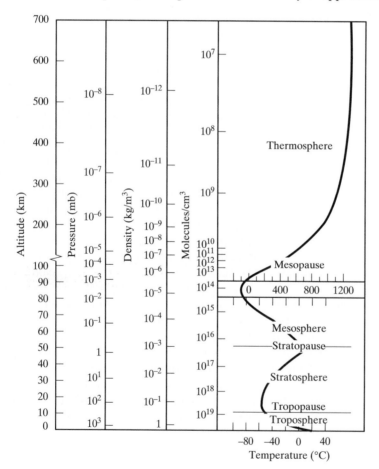

**Figure 6-1** Temperature profile in atmospheric layers (Williamson, 1973)

4. *Thermosphere* is the uppermost layer of the atmosphere. The temperature increases to approximately 1500 K, but it could not be measured with a thermometer because the molecular concentrations are very small. At its greatest density, the thermosphere has a molecular concentration on the order of $10^{13}$ molecules/cm$^3$. By comparison the molecular concentration at the earth's surface is $2.5 \times 10^{19}$.

The transition regions between the layers are known as the *tropopause, stratopause,* and *mesopause.*

---

**EXAMPLE 6.1   ATMOSPHERIC PRESSURE AND DENSITY**

If the molecular concentration at the earth's surface is $2.5 \times 10^{19}$ molecules/cm$^3$, use the ideal gas law to calculate the pressure (kPa) and density $\left(\text{kg/m}^3\right)$ if the temperature is 293 K.

**Solution**   The molecular weight of air is 28.97. Avogadro's number is $6.023 \times 10^{23}$ molecules/gmol.

$$\rho = \left(2.5 \times 10^{19} \text{ molecules/cm}^3\right)$$
$$\left(100 \text{ cm/m}\right)^3 \left(28.97 \text{ g/gmol}\right)$$
$$\left(\text{gmol}/6.023 \times 10^{23} \text{ molecules}\right)\left(\text{kg/1000 g}\right)$$
$$= 1.202 \text{ kg/m}^3$$

Assuming that air is an ideal gas yields

$$P = \rho R_u T = \left(1.202 \text{ kg/m}^3\right)\left(8.314 \text{ kJ/kmol} \cdot \text{K}\right)$$
$$\left(298 \text{ K}\right)\left(\text{kmol}/28.97 \text{ kg}\right)\left(\text{kPa} \cdot \text{m}^3/\text{kJ}\right)$$
$$= 102.79 \text{ kPa}$$

These calculated values are essentially consistent with Figure 6-1.

---

Although the alchemists of antiquity thought of "air" as one of the four basic elements of nature (the others were earth, water, and fire), today we realize that "clean air," for example measured on a remote Pacific island or the Antarctic content, contains a wide variety of chemical species. A list of these is given in Table 6-1 along with their average concentrations and approximate residence times. There is a great deal to learn from the information in this table.

Molecules of nitrogen and oxygen comprise 990,300 of 1 million molecules of clean air. The ratio of nitrogen to oxygen is almost four to one. Argon brings the total to 999,640, and adding carbon dioxide accounts for 999,990 molecules. If we keep adding the other minor constituents, we find that the total exceeds 1 million by about 0.002%, which implies that there is some uncertainty in the numbers. We would be fortunate indeed if our uncertainty in other matters of atmospheric conditions were so minimal. Note that water is not included in these calculations because of its great variability from place to place and time to time. Spatial and temporal variability in other species is possible as well, especially as we consider a polluted atmosphere.

The third column of Table 6-1 gives approximate values of the residence times for many of the species in the table. Without becoming mathematically exact, the residence time is a measure of how long it takes for the atmospheric inventory of a given species to be turned over or replaced. That is, the residence time can be thought of as the length of time that an average molecule resides in the atmosphere before it is somehow removed at a boundary or loses its identity by reaction. For example, the residence times of nitrogen and oxygen are $10^6$ and 10 years, respectively. The reasons for this are the relative chemical inertness of nitrogen and the fact that combustion (oxidation), metabolism, and photosynthesis results in a large exchange of oxygen, often involving carbon dioxide. Carbon dioxide is much more stable than oxygen and has an accordingly longer residence time. By contrast to $CO_2$, carbon monoxide is quite reactive and has a residence time of the order of weeks rather than years. The extreme inertness of $N_2O$ and CFC11 leads to very long residence times. We will address these species in more detail later.

---

**EXAMPLE 6.2   WATER VAPOR CONCENTRATION IN THE ATMOSPHERE**

On a given summer day a newspaper reports the weather in Washington, DC and Albuquerque, New Mexico, to be as follows:

|  | Washington, D.C. | Albuquerque N. Mex. |
|---|---|---|
| $T$ (ambient) | 95°F (35°C) | 95°F (35°C) |
| $P_0$ (ambient) | 101 kPa | 95 kPa |
| Relative humidity (RH) | 70% | 4% |

**Table 6-1**   Composition of Atmospheric Air

| Species | | Average concentration (ppm) | Approximate residence time |
|---|---|---|---|
| Major species | $N_2$ | 780,840 | $10^6$ yr |
| | $O_2$ | 209,460 | 10 yr |
| | $H_2O$ | Variable | |
| Inert gases | Ar | 9,340 | |
| | Ne | 18 | |
| | He | 5.2 | |
| | Kr | 1.1 | |
| | Xe | 0.09 | |
| Trace species | $CO_2$ | 350 | 15 yr |
| | $CH_4$ | 1.72 | 10 yr |
| | $H_2$ | 0.58 | 10 yr |
| | $N_2O$ | 0.33 | 150 yr |
| | CO | 0.05–0.2 | 65 days |
| | $NH_3$ | 0.01 | 20 days |
| | $NO/NO_2$ | | |
| |    Remote regions | < 0.00004 | 1 day |
| |    Rural continental United States | 0.0002–0.0005 | 1 day |
| |    Urban | 0.001 | 1 day |
| | $O_3$ (troposphere) | 0.02–0.05 | < 1yr |
| | $H_2O_2$ | 0.001 | |
| | $HNO_3$ | 0.001–0.0001 | 1 day |
| | HCs (Table 6-7) | 0.001–0.050 | |
| | HCHO | < 0.0005–0.00075 | |
| | HCOOH | > 0.02 | |
| | $CH_3OH$ | 0.04–0.06 | |
| | CFC 11 | 0.003 | 65 yr |
| Sulfur compounds | $SO_2$ | 0.0002 | 40 day |
| | COS | 0.0005 | $> 0.3 \times 10^5$ h |
| | $CS_2$ | 0.00001–0.0002 | $> 1.8 \times 10^5$ h |
| | $CH_3SH$ | | 3–13 h |
| | $(CH_3)_2S$ | | 31 h |
| | $H_2S$ | | 53 h |
| Free radicals | OH· | $1$–$10 \times 10^6$ molecules/cm$^3$ | |
| | $HO_2$· | $1 \times 10^9$ molecules/cm$^3$ | |

*Source:* Seinfeld (1986).

Compute the water vapor concentration (ppm) in Washington and Albuquerque.

***Solution***   The saturation pressure of water vapor, $P_v$ (saturation at 35°C), is 5.628 kPa. The relative humidity is the ratio of the actual partial pressure of the water vapor, $P_v$ (actual), divided by the partial pressure of the water vapor if it existed as a saturated vapor, $RH = 100[P_v(\text{actual})/P_v(\text{saturation})]$.

*Washington, D.C.* The actual partial pressure of water vapor, $P_v$ (actual) = 0.7(5.628) = 3.9396 kPa. Assuming the actual air and water vapor to be ideal gases, the water vapor mole fraction in Washington is

$$y_{\text{Washington}} = \frac{P_v(\text{actual})}{P_0} = \frac{3.9363}{101} = 0.039006$$

$$= 39,006 \text{ ppm}$$

*Albuquerque, N. Mex.* The actual partial pressure of water vapor, $P_v$ (actual) $= 0.04(5.628) = 0.22512$ kPa. The water vapor mol fraction in Albuquerque is

$$y_{\text{Alburquerque}} = \frac{P_v(\text{actual})}{P_0} = \frac{0.22512}{95} = 0.002369$$

$$= 2369 \text{ ppm}$$

From Table 6-1 it is evident that water vapor is the third most prevalent species in the air of Washington, D.C., approximately 18% of the oxygen concentration. In Albuquerque, water vapor is the fourth most prevalent species in air and less common than argon! The overall density of air in Albuquerque is less than it is in Washington, because the altitude is higher (hence the disparity in ambient pressure, $P_0$). For people with tuberculosis (TB), the light dry air in Albuquerque was more beneficial to recuperation than the heavy humid air in Atlantic coast cities. This fact (and low population densities) led the U.S. government to locate many TB sanatoria in the Southwest in the early part of the twentieth century prior to the development of pharmaceutical cures for TB.

# 6.2 Biogenic and anthropogenic sources of pollutants

Now that we have identified numerous constituents of clean air, we move on to identify biogenic and anthropogenic sources of such chemical species. Table 6-2 gives background concentrations, major sources and sinks, and estimates of emissions of several familiar gaseous pollutants. It can be seen that the combustion of fossil fuels produces far more $SO_2$ than volcanoes. On the other hand, natural sources (especially biological decay) of the reduced form of sulfur, $H_2S$, far outweigh anthropogenic sources. Of course, the $H_2S$ is readily oxidized to $SO_2$ in the atmosphere, giving an approximate balance between natural and anthropogenic sulfur. Enormous amounts of CO are generated by incomplete combustion in motor vehicle engines. Nevertheless, the CO concentration in the atmosphere has not risen over the years, whereas the concentration of $CO_2$ has increased (see Figures 2-2 and 6-2). No doubt oxidation of CO to $CO_2$ is one reason that CO does not accumulate.

More important, perhaps, is the dwarfing effects of many very large natural sources of CO.

The four nitrogen-bearing compounds in Table 6-2 are the reduced form, ammonia ($NH_3$), and three increasingly oxidized forms: $N_2O$, NO, and $NO_2$. Nitrous oxide ($N_2O$) is inert and has been used as an anesthetic, *laughing gas*. The inertness plus a high rate of natural generation explain its high background concentration. Nitric oxide (NO) and nitrogen dioxide ($NO_2$) are both produced in large amounts by the oxidation of the $N_2$ in air during high-temperature combustion and analogously by lightning. Nitrogen compounds in fossil fuels are also oxidized during combustion. Nevertheless, natural sources of these two nitrogen oxides (critical players in the formation of photochemical smog) are a factor of 10 higher than anthropogenic production. Ammonia ($NH_3$) itself is of little consequence, except perhaps locally, as an atmospheric pollutant, but it plays an important role in atmospheric chemistry for its part in the formation of aerosols.

Ozone ($O_3$), is easily recognized by its familiar odor following lightning storms. It is also formed in the stratosphere and transported downward into the troposphere. Ozone is formed by complex atmospheric photochemical reactions. At ordinary concentrations, ozone is not hazardous but at concentrations 10 times background levels, the effects on humans begin with irritation of mucous membranes and escalate to bronchial irritation, breathing difficulties, and chest pains. Reactive hydrocarbons, largely olefins and aromatics, have both natural and anthropogenic origins. Like ozone and the nitrogen oxides, reactive hydrocarbons are major participants in atmospheric photochemical reactions that produce smog.

Nonreactive hydrocarbons, largely methane ($CH_4$) but also other saturated hydrocarbons, are rather inert (do not participate in atmospheric photochemical reactions) and are not considered major pollutants or contributors to pollution. Very large quantities of methane are generated as *swamp gas* produced by biological decay, by ruminant animals (Wahlen et al., 1989), and in rice paddies (Khalil et al., 1991; Venkataramani and Subbaraya, 1993). One might wonder whether such agricultural activities should be considered natural or human-made since they are related to human activities. On the other hand, if it weren't for human beings, the great herds of bison might be roaming the plains instead.

**Table 6-2** Sources, Background Concentrations, and Sinks of Familiar Gaseous Pollutants

| Pollutant | Major source | | Estimated emission (kg/yr) | | Background concentration ($\mu g/m^3$) | Major identified sinks |
|---|---|---|---|---|---|---|
| | Anthropogenic | Natural | Anthropogenic | Natural | | |
| $SO_2$ | Combustion of coal and oil | Volcanoes | $130 \times 10^9$ | $2 \times 10^9$ | 1–4 | Scavenging, chemical reactions, soil and surface water absorption, dry deposition |
| $H_2S$ | Chemical processes, sewage treatment | Volcanoes, biological decay | $3 \times 10^9$ | $100 \times 10^9$ | 0.3 | Oxidation to $SO_2$ |
| $N_2O$ | None | Biological decay | None | $590 \times 10^9$ | 460–490 | Photodissociation in stratosphere, surface water and soil absorption |
| $NO$ | Combustion | Bacterial, action in soil, photo-dissociation of $N_2O$ and $NO_2$ | $53 \times 10^9$ combined with $NO_2$ | $768 \times 10^9$ | 0.25–2.5 | Oxidation to $NO_2$ |
| $NO_2$ | Combustion | Bacterial action in soil, oxidation of NO | | | 1.9–2.6 | Photochemical reactions, oxidation to nitrate, scavenging |
| $NH_3$ | Coal burning, fertilizer, waste treatment | Biological decay | $4 \times 10^9$ | $170 \times 10^9$ | 4 | Reaction with $SO_2$, oxidation to nitrate, scavenging |
| $CO$ | Auto exaust and other combustion processes | Oxidation of methane, photodissociation of $CO_2$, forest fires, oceans | $360 \times 10^9$ | $3000 \times 10^9$ (?) | 100 | Soil absorption, chemical oxidation |
| $O_3$ | None | Tropospheric reactions and transport from stratosphere | None | (?) | 20–60 | Photochemical reactions; absorption by and surfaces (soil and vegetation) and surface water |
| Nonreactive hydrocarbons | Auto exhaust, combustion of oil | Biological processes in swamps, ruminant emissions | $70 \times 10^9$ | $300 \times 10^9$ | $CH_4 = 1000$ Non-$CH_4$ < 1 | Biological action |
| Reactive hydrocarbons | Auto exhaust, combustion of oil | Biological processes in forests | $27 \times 10^9$ | $175 \times 10^9$ | < 1 | Photochemical oxidation |

*Source:* Adapted from Rasmussen et al. (1974b).

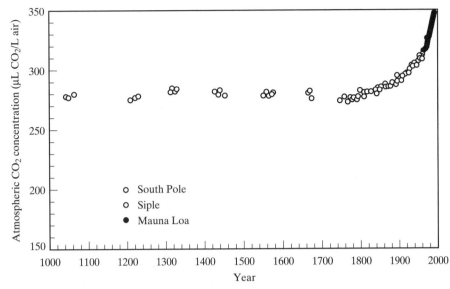

**Figure 6-2** Atmospheric $CO_2$ began increasing in the 18th century and direct measurements made at Mauna Loa Observatory in Hawaii since 1958 indicate that the increase has accelerated. In 1988 the atmospheric carbon reservoir was estimated at 748 gigatons, equivalent to a $CO_2$ concentration of 351 microliters $CO_2$ per liter air which is larger than at any time during the past 160,000 years. Post, et al. 1990)

It should be noted that the largest background concentrations occur for constituents that are chemically inert and least soluble in water. The importance of this will become more evident as the chapter continues. If the reader compares Table 6-2 to other tables that appear in the popular and professional literature, differences may be found. Where differences occur, they are due more to the gross uncertainty in the estimates than to atmospheric changes over time. Nevertheless, we offer such data because it is important that readers be able to rank the sources and appreciate their orders of magnitude.

# 6.3 *Natural removal of pollutants*

Table 6-2 also indicates major sinks for gaseous pollutants. In this section the natural processes by which pollutants are eliminated or converted are considered. There are two main categories of removal processes, as indicted in Table 6-3. *Bulk processes*

occur within the atmosphere and have a volumetric character. By contrast, *interfacial processes* occur at the interfaces of the atmosphere with specific surface features and have the character of an area.

We noted earlier that ozone was formed in the stratosphere and transported downward to the troposphere. Conversely, the degradation of the ozone layer hinges on the transport of certain species upward from the earth's surface to the stratosphere. Within our current level of understanding, one might think of the stratosphere as an upper boundary. But who can foresee what the future holds? Experience shows that it is worth trying to anticipate the future impact of today's action.

## 6.3.1 Bulk processes

*Chemical reactions in the atmosphere* are probably the best-known removal processes, owing to the much deserved attention given to *photochemical smog*. Smog is by no means the only important consequence of atmospheric reactions. The oxidation of $H_2S$ to $SO_2$ and CO to $CO_2$ mentioned earlier are simple and meaningful examples because the relative insolubility of $H_2S$ and CO renders other removal processes insignificant. Later we devote an entire section to the vast topic of atmospheric chemistry.

Two methods of *precipitation scavenging* exist. In the first the pollutant is incorporated into droplets and/or particles during the nucleation and growth phases of cloud formation. The pollutant may even

**Table 6-3**   Removal Processes

| Bulk | Interfacial |
|---|---|
| Atmospheric reactions | Vegetation |
| Precipitation scavenging | Soil |
| Dry deposition | Water |
| Aerosol scavenging | Stone |

contribute to this process. Second, when solid or liquid precipitation falls, gases and particles are collected during the vertical passage through the polluted region. The relative importance of these two scavenging mechanisms is not quantitatively defined, but the phenomenon of acid rain is well known. The pollution of surface regions, especially water bodies, via precipitation also occurs. Thus for certain contaminants this effective cleansing mechanism of air becomes, at the same time, a polluter of water.

The term *dry deposition* has been used to imply all removal processes at the earth's surface except precipitation scavenging and is much too broad to be useful. Therefore, we will identify several interfacial processes separately in the next section. Consider the following mechanistic illustration of one important bulk process that might accurately be called dry deposition. A common air pollutant, $SO_2$, is oxidized and absorbed in a water droplet to produce $H_2SO_4$. The oxidation occurs either before or after the absorption. The resulting droplet is a natural site for the absorption of ammonia, another common pollutant that is inherently basic. Since $NH_3$ is very soluble, its absorption can also precede that of the sulfur dioxide. In either case the result is that the uptake of one component is enhanced by the presence of the other. If the water in the droplet now evaporates $\left(\text{as }(NH_3)_2SO_4\right.$ and $NH_4HSO_4$ are not particularly hygroscopic$\left.\right)$, a dry particle of ammonium sulfate, $(NH_3)_2SO_4$, or ammonium bisulfate, $(NH_4HSO_4)$, remains and can eventually work its way to the ground.

The previous illustration shows how ammonia can be scavenged by sulfuric acid aerosol. Of course, *aerosol scavenging* need not be restricted to aqueous solutions of absorbed gases, as the potential variety of aerosols is staggering. Considerable attention is given to aerosol characteristics and transport in Chapter 10, albeit in the context of their removal from gases produced by industrial processes.

## 6.3.2 Interfacial processes

The ability of *vegetative surfaces* to exchange large quantities of gaseous materials with their surroundings is dramatically evidenced by the photosynthetic process involving $CO_2$ and $O_2$ and by the evapotranspiration of $H_2O$. Table 6-4 shows the uptake rates of several ordinary pollutants in alfalfa (Hill, 1971).

**Table 6-4**    Solubility in Water and Uptake Rate of Pollutants

| Pollutant | Uptake rate in alfalfa[a] $(gmol/m^2 \cdot s)$ | Solubility at 20°C $(g/100\ g)$ |
|---|---|---|
| CO | 0 | 0.00234 |
| NO | $2.1 \times 10^{-9}$ | 0.00625 |
| $O_3$ | $34.7 \times 10^{-9}$ | 0.052 |
| $NO_2$ | $39.6 \times 10^{-9}$ | Decomposes |
| $SO_2$ | $59.0 \times 10^{-9}$ | 10.8 |

[a] Concentration of the gas in the chamber was $2 \times 10^{-6}$ gmol/m$^3$.

Also demonstrated (Kabel, 1979) is the striking trend of increasing pollutant uptake, with its increasing solubility in water. Because of its substantial solubility in water, direct absorption of $SO_2$ from the atmosphere can supply much needed nutrients for plant metabolic processes, especially in areas where the soil is deficient in sulfur. This is a reminder that like most aspects in the balance of life, acidic gases are not entirely bad.

*Soil* can be a very effective sink for atmospheric pollutants, especially if the soil is moist and the pollutants are soluble in water. However, CO is virtually insoluble in water, yet is taken up in huge quantities by microbiological agents in the soil. Were it not for this enormous sink, the residence time of CO in the atmosphere could be 30 times greater than it is. Not surprisingly, the uptake rate of an acidic gas like $SO_2$ is greater when the pH of the soil is higher (i.e., is more basic).

*Water* is a critical factor in the majority of pollutant removal processes. In addition to mechanisms discussed earlier, it plays a major role via the direct absorption of water-soluble pollutants in over the three-fourths of the earth's surface covered by water. To illustrate, consider a local-scale example involving $NH_3$, which is six times more soluble in water than $SO_2$. Cattle feedlots generate large quantities of $NH_3$, as anyone passing by has noticed. Hutchinson and Viets (1969) showed that the amount of $NH_3$ removed from the atmosphere by precipitation scavenging was "insignificant compared to the amount absorbed directly from the air by aqueous surfaces in the vicinity of the cattle feedlots."

*Stone* does not appear to be a significant sink for any pollutants, but as discussed in Chapter 5, some pollutants have a significant impact on stone. Sulfur dioxide in the atmosphere causes inestimable damage to frescos, monuments, and other edifices throughout the world. The damage is the result of enhanced weathering rates caused by the attack of $H_2SO_4$ on the carbonate matrix of limestone and sandstone to form gypsum,

$$CaCO_3 + H_2SO_4 + H_2O \rightleftharpoons CaSO_4 \cdot 2H_2O + CO_2 \tag{6.1}$$

Because the resulting salt is more soluble in water than is the carbonate, gypsum is leached more readily from the stone surface and accelerates the deterioration.

## 6.4  Carbon cycle

A mass balance on any chemical species in the atmosphere (input − output = accumulation) enables an accounting of fluxes and an inventory of that species. In this section we focus on carbon, particularly $CO_2$, but also on the related species CO and $CH_4$. A review of Table 6-1 shows that other hydrocarbons in the atmosphere, while possibly important in regards to smog, are present at considerably lower concentrations than CO and $CH_4$ and do not have to be included in the carbon inventory. Sources and sinks of CO and $CH_4$ have been discussed previously. At the end of this section we consider the global carbon cycle, encompassing land and oceans as well as the atmosphere.

Carbon dioxide is the principal constituent of the atmosphere, following $N_2$, $O_2$, $H_2O$ vapor, and Ar. Figure 6-2 shows that while the atmospheric concentration of $CO_2$ had been relatively constant for centuries, it has risen noticeably since 1750 and continues to do so today. (See Figure 6-8 for a record of $CO_2$ concentrations for the last 150,000 years.) Reservoirs (Post et al., 1990) for carbon (expressed in teragrams of carbon, 1 Tg = $10^{12}$g) are:

- *Oceans:* 34.5 × $10^6$ Tg (84.1%)
- *Rocks and minerals:* 3.632 × $10^6$ Tg (8.9%)
- *Vegetation and soil:* 1.816 × $10^6$ Tg (4.5%)
- *Atmosphere:* 0.6792 × $10^6$ Tg (1.7%)

Reporting the mass of carbon in this fashion masks its molecular form and availability for chemical reactions. The mass of carbon in any molecular form is equal to the mass of the carbon-bearing molecule times the ratio of the molecular weight of carbon (12) to the molecular weight ($M$) of the molecular species. The ocean stores carbon in three forms:

1. *Inorganic carbon in solution* (dissolved gas; bicarbonate ions, $HCO_3^-$; and carbonate ions, $CO_3^{2-}$): 33.6 × $10^6$ Tg
2. *Dissolved organic compounds such as carbohydrates and proteins:* 0.908 × $10^6$ Tg
3. *Particles of dead plants and animals:* 0.02724 × $10^6$ Tg

While the public is often concerned with $CO_2$ in the atmosphere, the list above shows that the atmosphere is the smallest reservoir for carbon, while carbon in the oceans is by far the largest. Owing to the small amount of carbon in the atmosphere, however, any global event that alters the exchange of $CO_2$ between the atmosphere and water can significantly affect (increase or decrease) the concentration of $CO_2$ in the atmosphere. Atmospheric $CO_2$ concentrations are expected (Schneider, 1989) to continue their exponential rise, that is, in proportion to the consumption of fossil fuels.

Figure 6-3 shows that compounds containing carbon pass through the environment in several ways. Oceanic carbon is the largest reservoir and carbon exchange at the air–sea interface is the dominant mechanism affecting atmospheric $CO_2$. Between 30 and 50% of the $CO_2$ generated by fossil fuel is removed from the atmosphere by the oceans and transported to depth as dissolved inorganic carbon and biogenic carbon produced by biological organisms in the ocean (Rivkin et al., 1996). Biological processes maintain high levels of carbon compounds in the oceans. Polar oceans are sites of abundant biological activity. Phytoplankton flourish in both the Arctic and Antarctic Oceans but are highly seasonal. Deep-ocean water, is richer in dissolved inorganic carbon than warmer surface water, in which the dissolved $CO_2$ is consumed by photosynthesis in plant life near the surface. Plants at the surface eventually die and sink, taking their carbon with them. Surface water absorbs or desorbs $CO_2$, depending on temperature, wind speed, and liquid- and air-phase concentrations. Carbonate chemistry in the oceans is also affected by ocean salinity. Figure 6-3 represents an approximate global carbon cycle in which all forms of carbon and all media are considered (Kokoszka and Kabel, 1986). The numerical values

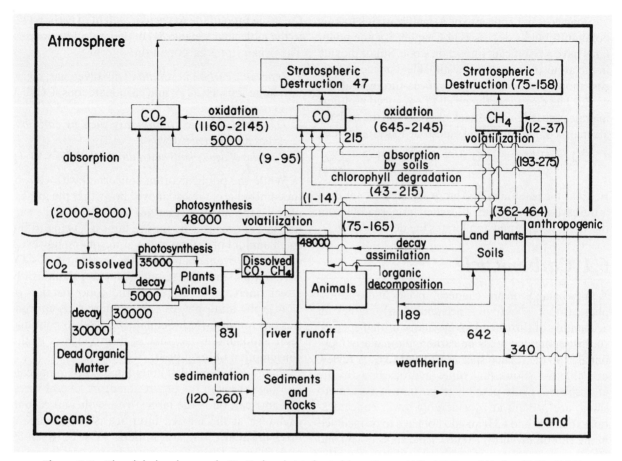

**Figure 6-3**    The global carbon cycle (Tg Carbon/yr) (adapted from Freyer H D, 1979; and Holland HD, 1978)

are highly uncertain, but the diagram provides insight into the movement of carbon through earth's solid, liquid, vapor, and living media. While actual numerical values may be disputed, the order of magnitude of the terms is important and identifies the important exchange mechanisms in the global environment.

The most dramatic oceanic response to global warming is an increase in ocean temperature and a reduction in oceanic circulation. Sarmiento and LeQuere (1996) suggest that global warming reduces the oceanic uptake of $CO_2$ due to the elevated temperatures and weakened circulation. Offsetting this, however, is an increase in the downward flux of biogenic carbon. In deep water, dissolved inorganic carbon formed by the mineralization of biogenic materi-

als accumulates since its return to the surface is impeded by decreased oceanic circulation. On balance, not enough is known about oceanic biology to conclude whether the oceanic uptake of $CO_2$ is helped or hindered by increases in atmospheric $CO_2$ associated with global warming.

## EXAMPLE 6.3    CARBON DIOXIDE IN THE AIR AND OCEAN

Examine the carbon cycle (Fig 6-3).

**a.** Draw a diagram that isolates atmospheric $CO_2$ and identify all the sources of this $CO_2$ and the amount each source contributes (Tg/yr). Identify all the sinks of atmospheric $CO_2$ and the amounts each sink removes (Tg/yr). Test to see if the total of the sources equals the total of the sinks.

b. Draw a diagram that isolates $CO_2$ dissolved in the oceans and identify all the sources and sinks and determine if the totals of each are equal.

**Solution** $CO_2$ in the atmosphere:

Sources (Tg/yr)

| | |
|---|---:|
| Biological decay of land plants | 48,000 |
| Anthropogenic activity | 5,000 |
| Oxidation of CO | 1,160–2,145 |
| Total of sources | 54,000–55,000 |

Sinks (Tg/yr)

| | |
|---|---:|
| Photosynthesis | 48,000 |
| Absorption in bodies of water | 2,000–8,000 |
| Weathering | 340 |
| Total of sinks | 50.000–56,000 |

$CO_2$ dissolved in bodies of water:

Sources (Tg/yr)

| | |
|---|---:|
| Biological decay of organic sediment | 30,000 |
| Absorption from atmosphere | 2,000–8,000 |
| Biological decay of plants and animals | 5,000 |
| River runoff | 831 |
| Total of sources | 38,000–44,000 |

Sinks (Tg/yr)

| | |
|---|---:|
| Photosynthesis water plants | 35,000 |
| Total of sinks | 35,000 |

The atmospheric mass balance is reasonable; however, the mass balance in bodies of water is not. There must be additional sinks in bodies of water that have not identified, or perhaps aquatic photosynthesis is larger than shown in Figure 6-3.

The concentrations of CO and $CO_2$ in the atmosphere and their residence times are shown in Table 6-1. Carbon dioxide is formed naturally by oxidation processes. Anthropogenic CO and $CO_2$ are produced by combustion processes related to the world's energy consumption rate. Thus they can be expected to increase at a similar rate (Khalil and Rasmussen,

1984; Woodwell et al., 1983). Carbon monoxide is converted to $CO_2$ in the atmosphere by reacting with OH· radicals:

$$CO + OH\cdot \rightarrow CO_2 + H\cdot$$

$$k(m^3/gmol \cdot s) = 4.4 \times T^{1.5} \exp\left(\frac{372}{T}\right) \quad (6.2)$$

## EXAMPLE 6.4 ATMOSPHERIC [OH·] BASED ON CO RESIDENCE TIME

Free radicals such as OH· are very reactive and it is very difficult to measure their concentration. Estimates of their concentration can be inferred by measuring the concentration of specific atmospheric molecular species with which they are known to react. Estimate the concentration of OH· in the atmosphere at $T = 280$ K if CO disappears by Equation 6.2 and the half-life (for this closed batch system) is taken as a surrogate for the average residence time of 65 days $= 5.616 \times 10^6$ s (in an open flow system) shown in Table 6-1. Compare the estimated value with the value shown in Table 6-1, $1–10 \times 10^6$ molecules/cm³.

**Solution** The rate of change of the CO concentration can be expressed as

$$\frac{d[CO]}{dt} = -k[CO][OH\cdot]$$

$$\frac{d[CO]}{[CO]} = -k[OH\cdot]\, dt$$

$$\int_{[CO]_0}^{1/2[CO]_0} \frac{d[CO]}{[CO]} = -k[OH\cdot] \int_0^{t_{1/2}} dt$$

$$= \ln(0.5) = -k[OH\cdot]t_{1/2}$$

The value of the rate constant is

$$k(280\ K) = (4.4)(280^{1.5})\exp\left(\frac{372}{280}\right)$$

$$= 77,836\ m^3/gmol \cdot s$$

Assuming that [OH·] is constant yields

$$\ln(0.5) = -0.693 = -k[OH\cdot]\, t_{1/2}$$

$$[OH\cdot](gmol/m^3) = \frac{0.693}{(77,836)(5.616 \times 10^6)}$$

$$= 1.585 \times 10^{-12}$$

$$[OH\cdot] = 1.585 \times 10^{-12}$$

$$(gmol/m^3)(m^3/10^6\ cm^3)$$

$$(6.023 \times 10^{23}\ molecules/gmol)$$

$$= 0.955 \times 10^6\ molecules/cm^3$$

$$\simeq 1 \times 10^6\ molecules/cm^3$$

The concentration shown in Table 6-1 is $1$–$10 \times 10^6$ molecules/cm$^3$. The predicted value is on the low side of the tabulated value, but owing to the uncertainty in the rate constant, the agreement is nonetheless good.

## 6.5  Sulfur cycle

To enhance the reader's perspective, we would like to quantify the movement of sulfur through the environment, arriving finally at an indication of the global sulfur cycle. Owing to the attention given to the generation of $SO_2$ during the combustion of coal, the public has heard a lot about the presence of this acidic gas in the atmosphere. It is less aware, however, of the following forms of sulfur generated naturally by biological processes in the oceans and soil:

- $CS_2$ (carbon disulfide)
- $H_2S$ (hydrogen sulfide)
- COS (carbonyl sulfide)
- $CH_3SCH_3$ (dimethyl sulfide)
- $CH_3SSCH_3$ (dimethyl disulfide)
- $CH_3SH$ (methyl mercaptan)

Typically, 90% of the sulfur in humid regions is organic. Aerobic soils contain less organic sulfur than do anaerobic soils. Several mechanisms remove sulfur from the soil.

1. Soluble sulfates are leached from the soil by rainwater and are carried to the seas via rivers.
2. Bacteria metabolize sulfur in anaerobic soil and produce gaseous $H_2S$, which migrates to the atmosphere.
3. In aerobic soils, compounds such as dimethyl sulfide $(CH_3SCH_3)$, dimethyl disulfide $(CH_3SSCH_3)$, and methyl mercaptan $(CH_3SH)$ are broken down to gaseous products by reaction with the hydroxyl radical $(OH\cdot)$.

There are several biogenic and anthropogenic terrestrial sources of sulfur of comparable magnitude, but there is little evidence that sulfur species are released to the atmosphere from them (Shinn and Lynn, 1979).

- *Sulfur fertilizers:* 0.908 Tg/yr
- *Natural erosion of denuded rock:* 0.636 Tg/yr
- *Acid mine drainage:* 1.816 Tg/yr
- *Mine-spoil erosion (strip mines):* 0.908 Tg/yr
- *Sulfate sludge from power plant scrubbers:* 0.908 Tg/yr

Bacterial volatilization of sulfur compounds in soil generates sulfur compounds that enter the atmosphere, while an equally large source of sulfur is released by sea spray and bacterial volatilization in the ocean.

Figure 6-4 shows the biogenic and anthropogenic mass fluxes of sulfur in the environment. While volcanoes and anthropogenic sources are just two of many sources, the sulfur they generate rises to high altitudes, enabling it to travel great distances and for long periods of time, with the result that the sulfur returns to earth as sulfates. Table 6-5 summarizes estimates of the principal fluxes in the sulfur cycle (Kokoszka and Kabel, 1986). These estimates are keyed to Figure 6-4, an approximate global sulfur cycle in which all forms of sulfur and all media are considered. The numerical values are highly uncertain but their orders of magnitude are important and identify the important exchange mechanisms in the global environment.

In the atmosphere, $H_2S$ and other reduced forms of sulfur react and ultimately become $SO_2$. The initial oxidation step is believed to be

$$H_2S + OH\cdot \rightarrow HS\cdot + H_2O \qquad (6.3)$$

Sulfur dioxide is transformed into the sulfate ion, which is removed as either sulfuric acid (rainout) or as particles when the acid reacts with ammonia:

$$SO_2 + \tfrac{1}{2}O_2 + H_2O$$
$$+ \ catalyst \rightleftharpoons H_2SO_4 \quad (a)$$
$$H_2SO_4 + 2NH_3 \rightleftharpoons (NH_4)_2SO_4 \quad (b) \quad (6.4)$$
$$H_2SO_4 + NH_3 \rightleftharpoons NH_4HSO_4 \quad (c)$$

## 6.6  Nitrogen cycle

As with carbon and sulfur, sources and sinks of nitrogen compounds in the atmosphere were discussed earlier. To round out the picture, we add information on the transport of and conversion among various forms

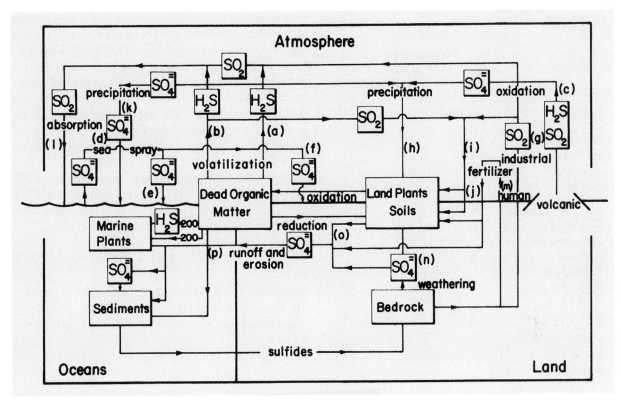

**Figure 6-4**    The global sulfur cycle (Tg Sulfur/yr), see Table 6.5 for values of specific fluxes, (adapted from Moss M R, 1978)

of nitrogen within the environment, arriving finally at an indication of the global cycle. Since both NO and $NO_2$ occur simultaneously and interactively in the atmosphere, the symbols $NO/NO_2$ or simply $NO_x$ are used to represent the compounds. Nitrogen exists in many forms in the land, water, and air. Figure 6-5 shows the terrestrial, aqueous, and atmospheric cycles.

Nitrous oxide $(N_2O)$ is the dominant oxide of nitrogen in the stratosphere, where it is an important component in the UV–$O_3$ reaction mechanism. Nitrous oxide is an absorber of infrared radiation and a major component in the greenhouse effect. The concentration of $N_2O$ in the stratosphere is approximately 310 ppb and is reported to be increasing at a rate of 0.25 to 0.31% per year (Czepiel et al., 1996). Total global $N_2O$ emissions are estimated to be 21 ± 2 Tg/yr. Biogenic sources are estimated to be 60% of the total emissions and include undisturbed soils, oceanic waters, and possibly atmospheric reactions. Anthropogenic sources

include cultivated soils, biomass burning, combustion, and treatments of high-organic municipal and livestock wastewaters (Czepiel et al., 1996).

Biogenic processes producing $N_2O$ are *nitrification* and *denitrification* processes. Nitrification is an aerobic two-step process in which $NH_4^+$ is oxidized to $NO_2^-$, which is followed by oxidation to $NO_3^-$. Nitrous oxide is a by-product of these reactions. Denitrification occurs under anaerobic conditions in which bacteria react with $NO_2^-$ and $NO_3^-$ to produce NO, $N_2O$, or $N_2$. The availability of $O_2$ is the determining factor in $N_2O$ production.

A major anthropogenic source of $N_2O$ is the application of fertilizer on cultivated soil. Fertilizer in the form of ammonia $(NH_3)$, ammonium nitrate $(NH_4NO_3)$, or urea $[CO(NH_2)_2]$ reacts rapidly with water and the enzyme urease in the soil to form $(NH_4)_2CO_3$. This form of nitrogen is not susceptible to denitrification until it is oxidized to nitrite or

**Table 6-5** Principal Fluxes in the Sulfur Cycle

| Flux | Corresponding letter of flux in Fig. 6-4 | Flux [Tg(S)/yr] |
|---|---|---|
| Biological decay (land) | (a) | 50–100 |
| Biological decay (ocean) | (b) | 30–100 |
| Volcanic activity | (c) | 1–3 |
| Sea spray (total) | (d) | 44 |
| To ocean | (e) | 40 |
| To land | (f) | 4 |
| Anthropogenic emission | (g) | 50–70 |
| Precipitation (land) | (h) | 50–90 |
| Particulate deposition (land) | (i) | 10–20 |
| Absorption (vegetation) | (j) | 15–30 |
| Precipitation and particulate deposition (ocean) | (k) | 70–75 |
| Absorption (ocean) | (l) | 25 |
| Fertilization | (m) | 10–25 |
| Rock weathering | (n) | 15–40 |
| Pedosphere runoff | (o) | 50–90 |
| Total runoff | (p) | 70–140 |

*Source:* Moss (1978).

nitrate by soil microorganisms. Nitrous oxide $(N_2O)$ is released by the soil during *nitrification* of ammonium-producing fertilizers under aerobic conditions and during *denitrification* of nitrate under anaerobic conditions. Field measurements indicate that approximately 1.3% of the fertilizer nitrogen applied is released in the form of $N_2O$ (Firestone et al., 1980). Nitrification processes in the soil and water produce $N_2O$, which is photolyzed in the stratosphere to produce $N_2$ and oxygen radicals. In the troposphere, $N_2O$ is virtually inert, as we have seen before.

Figure 6-5 shows that NO and $NO_2$ are emitted to the atmosphere directly or formed by the reaction of $NO_2$ and oxygen free radicals. Once in the atmosphere, $NO_x$ reacts with a variety of molecular species, depending on location. The eventual fate of $NO_x$ is the formation of nitrates as either particles or acid. In urban areas where hydrocarbons are generated, a vast array of photochemical reactions occurs to form oxygenated nitrogen-bearing compounds. Anthropogenic $NO_x$ produced by combustion processes is much smaller than the $NO_x$ produced by natural processes. Unfortunately, anthropogenic $NO_x$ is con-

centrated in urban areas, where high atmospheric concentrations react photochemically with oxygen and hydrocarbons to produce urban smog. Figure 6-5 represents an approximate global nitrogen cycle in which all forms of nitrogen and all media are considered (Kokoszka and Kabel, 1986). The numerical values of each flux are highly uncertain, but the orders of magnitude of the fluxes are important and identify the important exchange mechanisms in the global environment.

Nitrogen compounds are removed from the atmosphere in a variety of ways. The higher oxides of nitrogen are photolytic and ultimately become $NO_2$, which reacts with OH· to become nitric acid,

$$NO_2 + OH\cdot \rightarrow HNO_3 \qquad (6.5)$$

Ammonia $(NH_3)$ is soluble in water and is removed by absorption in surface water or water droplets in the atmosphere, where it may react with sulfuric acid (Equation 6.4). In any event, ammonia compounds are removed from the atmosphere as rainout.

## 6.7 Other chemical species

One can envision diagrams similar to Figures. 6-3 to 6-5 for any chemical element. In the following paragraphs we examine several more air pollutants of interest.

### 6.7.1 Photochemical oxidants and ozone

Photochemical oxidants are pollutants that irritate mucous membranes (eyes, nasal cavity, tracheobronchial system). They are generated in the atmosphere by sunlight acting on less irritating pollutants emitted by human activity. The principal photochemical oxidants are ozone, certain aldehydes, and peroxyacetyl nitrate, often referred to by its acronym PAN. Figure 6-6 summarizes reactions that produce these photochemical oxidants. The mechanism typifies the way that chemical kinetics are described in air pollution, and engineers need to become familiar with sketches of this sort. The first step in understanding Figure 6-6 is to realize that it consists of three smaller mechanisms:

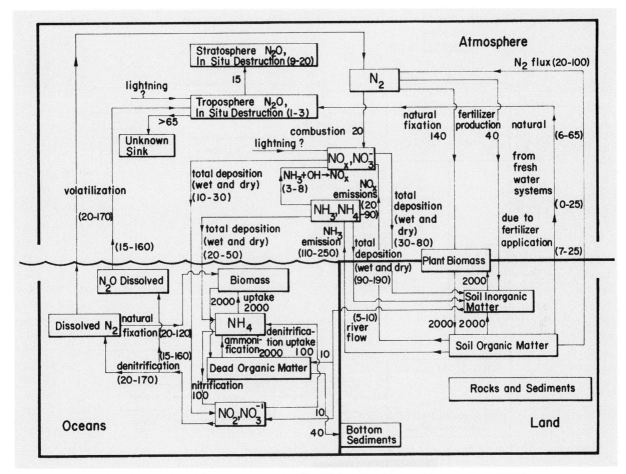

**Figure 6-5**   The global nitrogen cycle, (Tg Nitrogen/yr) (adapted from Bolin B, 1979)

1. NO/NO$_2$ and O$_3$ photolysis
2. Hydrocarbon oxidation reactions initiated by O$_3$
3. Gas-phase reactions that produce aerosols

Combustion reactions generate NO/NO$_2$ and a variety of volatile organic hydrocarbons (olefins and aldehydes in particular). The box above and to the left of the sun describes the photodissociation of NO$_2$, which in turn leads rapidly to the generation of ozone. The two right-hand boxes at the top of Figure 6-6 describe a series of reactions between ozone and aldehydes and olefin hydrocarbons that generate new and additional aldehydes, some of which irritate mucous membrane. The lower boxes in Figure 6-6 summarize a number of complex reactions that generate hydrocarbons, which condense and produce the very small particles present in smog.

## 6.7.2 Radon

The radiation dose from the daughters of radon is a significant part of natural background radiation, constituting about half of the total dose the general population receives from natural radiation (Berger,

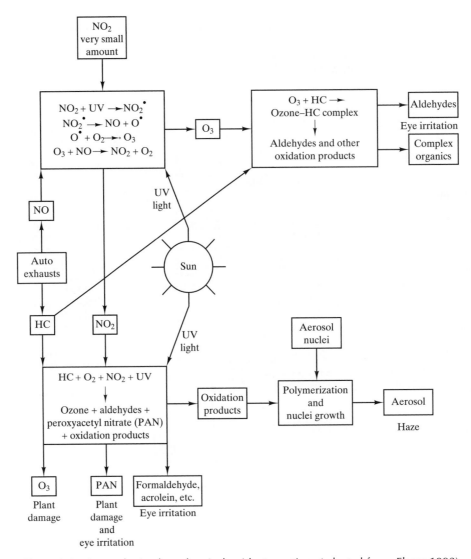

**Figure 6-6**   Atmospheric photochemical oxidant reactions (adapted from, Elsom 1992)

1990). The greatest fraction of natural radiation to which humans are exposed occurs from the inhalation of the short-lived decay products of radon ($^{222}$Rn and $^{220}$Rn) which occur in ambient air and, in possibly higher concentrations, in air inside buildings. Radon is a gas generated by the radioactive decay of uranium-238 in the earth's crust. Radon 220, also called thoron, is generated in a similar fashion from thorium 232. Radon is ubiquitous and its concentration varies throughout the United States. Its concentration is large where the amounts of uranium and

thorium are large. Because of the movement of atmospheric air and mixing through the troposphere, the out-of-doors radon concentration is not large. For most of the United States the estimated average radon concentration is 0.25 pCi/L (Nero et al., 1986). Inside buildings the concentration may be large if the source is strong and ventilation within the building is poor (Nazaroff and Teichman, 1990).

A diagram of the radioactive decay of uranium and thorium leading to the production of radon is shown in Figure 6-7. Once radon is produced, it

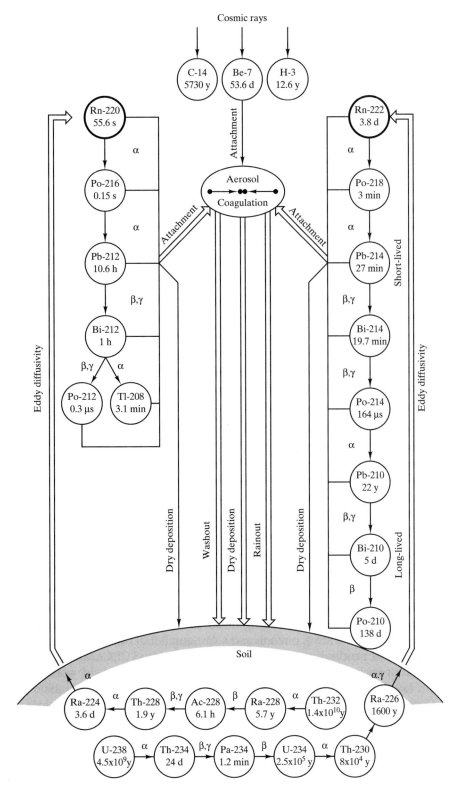

**Figure 6-7** Radioactive decay leading to the production of radon, y - years, d - days, h - hours, min - minutes, s - seconds (Portsendorfer, 1994)

**Table 6-8**   Estimated Particle Emissions

| Source | Tg/yr | Source | Tg/yr |
|---|---|---|---|
| **Global** | | Transportation | |
| | | Motor vehicles | |
| Anthropogenic | | Gasoline | 0.381 |
| Particles | 83.5 | Diesel | 0.236 |
| Gas-conversion particles | | Aircraft | 0.027 |
| $SO_2$ | 133.5 | Watercraft | 0.136 |
| $NO_x$ | 27.2 | Nonhighway | |
| Photochemical | 24.5 | Agriculture | 0.072 |
| Natural | | Commercial | 0.011 |
| Soil | 181.6 | Construction | 0.003 |
| Gas-conversion particles | | Other | 0.024 |
| $H_2S$ | 185.2 | | |
| $NO_x$ | 392.3 | | |
| $NH_3$ | 224.3 | | |
| Photochemical | 181.6 | Incineration | |
| Volcanoes | 3.6 | Municipal | 0.089 |
| Forest fires | 2.7 | On-site | 0.168 |
| Sea salt | 908 | Open dump | 0.557 |
| Global Total: 2348 | | | |
| **United States** | | | |
| Natural | 57.2 | Other | |
| Fires | | Tire particles (highways) | 0.272 |
| Forest fire | 33.6 | Cigarette smoke | 0.209 |
| Slash burning | 5.4 | Aerosols from spray cans | 0.354 |
| Accumulated litter | 10 | Sea spray | 0.309 |
| Agricultural burning | 2.2 | | |
| | | United States Total: 111.2 | |

*Source:* Fennelly (1976).

# 6.8 Atmospheric chemistry of inorganic species: acid rain

In Chapter 5 we introduced the concept of acid rain and indicated the impact that it has on our environment. The primary precursors that lead to acid rain are inorganic species $CO_2$, $SO_2$, $NO_x$, and chlorinated hydrocarbons which produce $H_2CO_3$, $H_2SO_4$, $HNO_3$, and $HCl$, respectively. The latter species are deposited with precipitation. Here we give more details on the atmospheric chemistry of each of these species, all of which lead to acid rain.

CO has a short lifetime, on the order of weeks, because it reacts with hydroxyl radicals ($OH\cdot$) to form $CO_2$:

$$CO + OH\cdot \rightarrow CO_2 + H\cdot \qquad (6.7)$$

Atmospheric $CO_2$, on the other hand, has a long lifetime, on the order of 15 years. Both CO and $CO_2$ are formed by biogenic and anthropogenic processes, although $CO_2$ is formed at a higher rate than CO. Carbon monoxide (CO) is formed naturally by the reaction between $CH_4$ and $OH\cdot$ in the soil and atmosphere:

$$CH_4 + OH\cdot \rightarrow CH_3\cdot + H_2O \qquad (6.8)$$

This second-order reaction is quite slow and determines the overall rate of production of CO, since subsequent oxidation steps (see Equations 6.19) leading to CO are much more rapid.

## EXAMPLE 6.5  CO GENERATION BY $CH_4$ OXIDATION

Carbon monoxide is generated mole for mole from methane by Equation 6.8 and subsequent rapid reactions. Calculate the global production of CO in Tg C/yr by this mechanism.

***Solution***  Rasmussen et al. (1974a) give $k = 5.5 \times 10^{-12} \exp(-1900/T)$, where $k$ $[=]$ cm$^3$ molecule$^{-1} \cdot$s$^{-1}$ and $T$ $[=]$ K for the rate-determining step. For the background concentrations of methane $= 1000$ $\mu$g/m$^3$ (see Table 6-2) and hydroxyl radicals $= 5 \times 10^6$ molecules/cm$^3$ (see Table 6-1), we can calculate the rate of production of CO at any temperature of interest, in this case $-40°$C (233 K) and $+20°$C (293 K).

$$-r_{CH_4} = r_{CO} = k[CH_4][OH\cdot]$$

$$k_{233K} = 5.5 \times 10^{-12} \exp(-1900/233)$$
$$= 1.6 \times 10^{-15} \text{ cm}^3 \text{ molecule}^{-1} \text{ s}^{-1}$$

$$k_{293K} = 5.5 \times 10^{-12} \exp(-1900/293)$$
$$= 8.4 \times 10^{-15} \text{ cm}^3 \text{ molecule}^{-1} \text{ s}^{-1}$$

$$[CH_4] = (1000 \ \mu g/m^3)(m^3/10^6 \text{ cm}^3)$$
$$(g/10^6 \ \mu g)(gmol/16 \text{ g})$$
$$= 6.25 \times 10^{-11} \text{ gmol/cm}^3$$

$$[OH\cdot] = (5 \times 10^6 \text{ molecules/cm}^3)$$
$$(gmol/6.02 \times 10^{23} \text{ molecules})$$
$$= 8.3 \times 10^{-18} \text{ gmol/cm}^3$$

$$r_{CO,233K} = (1.6 \times 10^{-15})(6.25 \times 10^{-11})$$
$$(8.3 \times 10^{-18})(6.02 \times 10^{23})$$
$$= 5.0 \times 10^{-19} \text{ gmol cm}^{-3} \text{ s}^{-1}$$

$$r_{CO,293K} = (8.4 \times 10^{-15})(6.25 \times 10^{-11})$$
$$(8.3 \times 10^{-18})(6.02 \times 10^{23})$$
$$= 2.6 \times 10^{-18} \text{ gmol cm}^{-3} \text{ s}^{-1}$$

The result appears imperceptibly small, and it is, unless we sum over the volume of the troposphere and put it on an annual basis.

$$V_{tr} = S_{earth} H_{tr} = (510 \times 10^6 \text{ km}^2)$$
$$(10 \text{ km})(10^{15} \text{ cm}^3/\text{km}^3)$$
$$= 5.1 \times 10^{24} \text{ cm}^3$$

$$r_{CO,233K} = (5.0 \times 10^{-19} \text{ gmol cm}^{-3} \text{ s}^{-1})$$
$$(5.1 \times 10^{24} \text{ cm}^3)(3.2 \times 10^7 \text{ s/yr})$$
$$= 8.16 \times 10^{13} \text{ gmol CO/yr}$$

$$r_{CO,293K} = (2.6 \times 10^{-18})(5.1 \times 10^{24})(3.2 \times 10^7)$$
$$= 4.24 \times 10^{14} \text{ gmol CO/yr}$$

To convert these values to Tg C/yr, multiply by (1 gmol C/gmol CO)(12 g G/gmol C)(Tg/$10^{12}$ g) to obtain

$$r_{CO,233K} = 979 \text{ Tg C/yr}$$

$$r_{CO,293K} = 5088 \text{ Tg C/yr}$$

These calculated values are in the same range (645 to 2145 Tg C/yr) as shown in Figure 6-3 for the oxidation of methane to carbon monoxide.

---

In the atmosphere, $H_2S$ is oxidized by $O(^1D)\cdot$ and $O_3$ to $SO_2$, which is then transformed to $H_2SO_4$ by several atmospheric reactions and by heterogeneous reactions within marine fog (Gundel et al., 1994).

$$2SO_2 + O_2 + \text{catalyst} \rightarrow 2SO_3 \quad \text{(a)}$$
$$SO_3 + H_2O \rightarrow H_2SO_4 \quad \text{(b)} \qquad (6.9)$$

There is also evidence (Seinfeld, 1986) that $SO_2$ reacts with $OH\cdot$ and $HO_2\cdot$ in the presence of NO and $NO_2$ to form $H_2SO_4$:

$$SO_2 + OH\cdot \rightarrow HOSO_2\cdot \quad \text{(a)}$$
$$HOSO_2\cdot + O_2 \rightarrow HO_2\cdot + SO_3 \quad \text{(b)}$$
$$HO_2\cdot + NO \rightarrow NO_2 + OH\cdot \quad \text{(c)} \qquad (6.10)$$
$$SO_3 + H_2O \rightarrow H_2SO_4 \quad \text{(d)}$$

and increasing evidence (Faust, 1994) of aqueous-phase photochemical reactions involving $H_2O_2$ mediated oxidation of $SO_2$ to $H_2SO_4$.

Nitric acid ($HNO_3$) in the atmosphere is generated by reactions of NO with the hydroxyl and hydroperoxyl radical:

$$NO + HO_2\cdot \rightarrow NO_2 + OH\cdot \quad \text{(a)}$$
$$NO_2 + OH\cdot + M \rightarrow HNO_3 + M \quad \text{(b)} \qquad (6.11)$$

Hydrochloric acid (HCl) can be formed by any number of reactions involving chlorinated hydrocarbons and free radicals. For example, in the stratosphere the following reaction mechanism produces

HCl, which is transported downward into the troposphere. Reactions with CFCs begin with the oxygen radical that forms chlorine (and bromine) oxides ClO (and BrO). Alternatively, CFCs may be photolyzed by ultraviolet radiation to form the Cl· (and Br·) radicals. The oxide ClO reacts with the oxygen radical to produce Cl·, which consumes ozone and generates ClO. The ultimate fate of chlorine is HCl (Cicerone, 1987):

$$
\begin{aligned}
\text{CFC} + \text{O·} &\rightarrow \text{ClO} + \text{products} &\text{(a)}\\
\text{CFC} + hv &\rightarrow \text{Cl·} + \text{products} &\text{(b)}\\
\text{O·} + \text{ClO} &\rightarrow \text{Cl·} + \text{O}_2 &\text{(c)}\\
\text{O}_3 + \text{Cl·} &\rightarrow \text{O}_2 + \text{ClO} &\text{(d)}\\
\text{Cl·} + \text{CH}_4 &\rightarrow \text{HCl} + \text{CH}_3\text{·} &\text{(e)}\\
\text{OH·} + \text{HCl} &\rightarrow \text{H}_2\text{O} + \text{Cl·} &\text{(f)}
\end{aligned}
\qquad (6.12)
$$

The result of the acidification implied by the mechanisms above is the distribution of pH in precipitation over the United States and Canada (see Figure 5-2). The pH of rain in Europe is around 4.0 to 4.5. A mass balance on nitrogen in the Netherlands shows far more nitrate $(NO_3^-)$ in the water and soil than can be accounted for by deposition (Freemantle, 1995). The counterintuitive explanation is that inherently, basic ammonia $(NH_3)$, an agricultural chemical, is oxidized in the air and by microorganisms in the soil to form nitric acid $(HNO_3)$:

$$
\begin{aligned}
\text{NH}_3 + 2\text{O}_2 &\rightarrow \text{NO}_3^- + \text{H}^+ + \text{H}_2\text{O} &\text{(a)}\\
\text{NH}_4^+ + 2\text{O}_2 &\rightarrow \text{NO}_3^- + 2\text{H}^+ + \text{H}_2\text{O} &\text{(b)}
\end{aligned}
\qquad (6.13)
$$

Nature often has surprises for us as we try to take command of the environment.

# 6.9 *Atmospheric chemistry of organic species*

Many reactions occur in the atmosphere on a molecular basis (as shown below) or an ionic basis as the products below dissociate in solution:

$$
\begin{aligned}
\text{SO}_3 + \text{H}_2\text{O} &\rightarrow \text{H}_2\text{SO}_4 &\text{(a)}\\
\text{NH}_3 + \text{H}_2\text{O} &\rightarrow \text{NH}_4\text{OH} &\text{(b)}
\end{aligned}
\qquad (6.14)
$$

The mechanisms of other reactions, such as the oxidation of $SO_2$ to $SO_3$, as discussed in Chapter 1, are more complex, occurring via free radicals.

## 6.9.1 Photochemical reactions

A distinctive feature of atmospheric pollution is the initiation of reactions by solar radiation. Such reactions are called *photochemical reactions* and the process is called *photolysis*. For particular species, the reaction rate depends on the frequency and intensity of radiation. The chief function of photochemical reactions is to generate radicals, which in turn initiate other reactions. The important photochemical reactions relevant to pollution involve oxygen and ozone, several oxides of nitrogen, and numerous aldehydes.

*Free radicals* (sometimes merely called radicals) are active molecular fragments that initiate atmospheric reactions. Free radicals are combinations of atoms (or fragments of compounds) in which there is an unbound electron that is free to bond with electrons of other molecules. Several common radicals of importance in air pollution are:

1. OH· (hydroxyl)
2. $HO_2$· (hydroperoxyl)
3. $CH_3$· (methyl)
4. $O(^1D)$· (single oxygen)
5. H· (hydrogen)

Radicals should not be confused with *ions* (such as $OH^-$), which are molecules or molecular fragments in which there is an excess (or deficit ) of electrons. Radicals exist for only a short period of time; nevertheless, at any given temperature and pressure it is possible to calculate their equilibrium concentrations. Precisely which combinations of atoms exist as radicals is a matter of chemistry. For engineers interested in air pollution, it is only necessary to become familiar with the radicals one expects to exist at any temperature and pressure and to be able to determine their concentrations and kinetic rate constants for various reactions.

It is important to understand the principal atmospheric radicals that exist at standard temperatures and pressure. Atmospheric reactions occur in parallel but at different rates. Thus one should not look for a single "trigger" that starts a chain reaction. Researchers may designate one species or reaction as the trigger, but in reality all reactions occur simultaneously, and at any instant many atmospheric radicals exist at various concentrations, some insignificantly small. Shown in Table 6-1 are the two principal atmospheric radicals and typical concentrations one can expect in clean tropospheric conditions (Seinfeld, 1986). It is convenient to group free radical kinetics into four types:

1. Reactions that initiate radicals
2. Propagation reactions that consume one radical but produce two radicals
3. Branching reactions that consume one radical but produce a different radical
4. Terminating reactions that annihilate radicals

In the atmosphere initiation reactions are generally photolytic reactions. Consider for the moment only one hydrocarbon, $CH_4$, and its products. The following illustrates reactions that may occur depending on altitude, temperature, and radiation wavelength.

1. Initiation (photolytic) reactions that create radicals:

$$O_3 + h\nu \rightarrow O\cdot + O_2 \qquad (a)$$

$$NO_2 + h\nu \rightarrow O\cdot + NO \qquad (b)$$

$$O_2 + h\nu \rightarrow O\cdot + O\cdot \qquad (c) \quad (6.15)$$

$$H_2O + h\nu \rightarrow O\cdot + H_2 \qquad (d)$$

$$HCOH + h\nu \rightarrow HCO\cdot + H\cdot \qquad (e)$$

2. Propagation (doubling) reactions that consume one radical but produce two radicals:

$$O\cdot + H_2O \rightarrow OH\cdot + OH\cdot \qquad (a)$$

$$O\cdot + O_3 \rightarrow O\cdot + O\cdot + O_2 \qquad (b)$$

$$O\cdot + H_2 \rightarrow OH\cdot + H\cdot \qquad (c) \quad (6.16)$$

$$H\cdot + O_2 \rightarrow OH\cdot + O\cdot \qquad (d)$$

$$O\cdot + CH_4 \rightarrow CH_3\cdot + OH\cdot \qquad (e)$$

where M is any molecular species that participates in the reaction but does not react chemically.

3. Branching reactions that consume one radical but produce another radical:

$$OH\cdot + H_2 \rightarrow H\cdot + H_2O \qquad (a)$$

$$H\cdot + H_2O \rightarrow OH\cdot + H_2 \qquad (b)$$

$$H\cdot + O_2 + M \rightarrow HO_2\cdot + M \qquad (c)$$

$$HO_2\cdot + O_3 \rightarrow OH\cdot + 2O_2 \qquad (d) \quad (6.17)$$

$$HO_2\cdot + NO \rightarrow OH\cdot + NO_2 \qquad (e)$$

$$OH\cdot + CO \rightarrow H\cdot + CO_2 \qquad (f)$$

$$OH\cdot + CH_4 \rightarrow CH_3\cdot + H_2O \qquad (g)$$

4. Terminating reactions that annihilate radicals:

$$H\cdot + H\cdot + M \rightarrow H_2 + M \qquad (a)$$

$$O\cdot + O\cdot + M \rightarrow O_2 + M \qquad (b)$$

$$H\cdot + OH\cdot + M \rightarrow H_2O + M \qquad (c)$$

$$O\cdot + M + O_2 \rightarrow O_3 \qquad (d)$$

$$O\cdot + N_2O \rightarrow NO + NO \qquad (e) \quad (6.18)$$

$$O\cdot + NO + M \rightarrow NO_2 + M \qquad (f)$$

$$OH\cdot + NO \rightarrow HONO \qquad (g)$$

$$OH\cdot + NO_2 + M \rightarrow HNO_3 + M \qquad (h)$$

The methyl radical $(CH_3\cdot)$ reacts with numerous species to generate formaldehyde (HCHO), formyl radical $(HCO\cdot)$, carbon monoxide (CO), and carbon dioxide $(CO_2)$:

$$CH_3\cdot + O_2 \rightarrow HCOH + OH\cdot \qquad (a)$$

$$HCOH + OH\cdot \rightarrow HCO\cdot + H_2O \qquad (b)$$

$$HCO\cdot + OH\cdot \rightarrow CO + H_2O \qquad (c) \quad (6.19)$$

$$HCO\cdot + O_2 \rightarrow HO_2\cdot + CO \qquad (d)$$

$$CO + OH\cdot \rightarrow CO_2 + H\cdot \qquad (e)$$

## 6.9.2 Oxygen and ozone

There are two oxygen radicals, $O(^1D)\cdot$ and $O(^3P)\cdot$. The radical $O(^1D)\cdot$ is an excited singlet form of the oxygen molecule capable of initiating a variety of chemical reactions. The second radical, $O(^3P)\cdot$, is less reactive. In the stratosphere, the UV portion of $(\lambda < 200\text{ nm})$ solar radiation and molecular oxygen produce ozone, oxygen radicals, and hydroxyl radicals. The oxygen radicals, $O(^1D)\cdot$, react with molecular oxygen and a third body (M) in the reaction to form ozone. Ozone is photolytic and over a broad range of UV ($240 < \lambda < 320$ nm) produces oxygen radicals. Ozone also reacts with oxygen radicals to annihilate one another.

**Stratosphere**

$$H_2O + h\nu \rightarrow H_2 + O(^1D)\cdot \tag{a}$$

$$O_2 + h\nu \rightarrow \begin{cases} O(^1D)\cdot + O(^3P)\cdot & \lambda < 175 \text{ nm} \\ O(^1D)\cdot + O(^1D)\cdot & \lambda < 137 \text{ nm} \end{cases} \tag{b}$$

$$\tag{6.20}$$

$$O_3 + h\nu \rightarrow \begin{cases} O(^1D)\cdot + O_2 \\ O(^3P)\cdot + O_2 \end{cases} \tag{c}$$

$$O(^3P)\cdot + O_2 + M \rightarrow O_3 + M \tag{d}$$

As a result, solar radiation and $O_2$ engage in a series of simultaneous reactions in the stratosphere that both form and destroy $O_3$, thus establishing a steady-state $O_3$ concentration. These reactions are exothermic and cause the temperature to rise (i.e., an inversion) in the stratosphere (Figure 6-1). The presence of the reactive $O(^1D)\cdot$ radical in the stratosphere gives rise to several important reactions that generate the reactive hydroxyl radical OH·, the reactive ClO· radical from chlorofluorocarbons (CFCs), and NO from the normally unreactive $N_2O$:

$$O(^1D)\cdot + H_2O \rightarrow OH\cdot + OH\cdot \tag{a}$$

$$O(^1D)\cdot + CH_4 \rightarrow OH\cdot + CH_3\cdot \tag{b}$$

$$O(^1D)\cdot + N_2O \rightarrow NO + NO \tag{c} \quad (6.21)$$

$$O(^1D)\cdot + CFCs \rightarrow \begin{array}{l} \text{reactive products} \\ \text{such as ClO} \end{array} \tag{d}$$

The stratospheric inversion acts as a lid or ceiling to impede the upward convection of tropospheric gases. For this reason, any weakening of the stratospheric inversion, such as would occur by the depletion of stratospheric ozone, affects upward convection, weather, and climate. In addition, depleting the ozone layer increases the intensity of UV radiation reaching the earth, which affects adversely the photosynthetic processes of terrestrial and aquatic plant life and increases the chances of skin cancer in humans.

---

**EXAMPLE 6.6   THE OZONE LAYER: A SUCCESS STORY IN ATMOSPHERIC CHEMISTRY**

In the early years of refrigeration, the hazardous gases $SO_2$ and $NH_3$ were the refrigerants of choice. So the arrival of inert chlorofluorocarbons (CFCs) looked like a godsend. In 1971, Crutzen (Zurer, 1995b) showed how relatively inert $N_2O$ diffuses slowly up into the stratosphere, where some if it is converted to $NO_x$, a known catalyst for the destruction of $O_3$. In 1973, Rowland and Molina (Zurer, 1995b) postulated that (human-made) CFCs follow the same transport path to the stratosphere as $N_2O$. Once there, the CFCs are attacked by UV radiation, generating chlorine atoms which attack ozone (see Figure 6-14 for details of these reactions and transport processes).

Ninety percent of all the ozone is in the stratosphere, with a maximum partial pressure at 14 to 23 km above the earth. Despite an increase in ozone near the ground in recent years, the total amount of ozone has decreased substantially. The health hazard associated with this decrease is that ozone absorbs harmful solar UV-B radiation (radiation of wavelengths less than 320 nm). The UV-B radiation that reaches the surface of the earth has the potential to cause "eye cataracts, an increase of non-melanoma skin cancers, damage to generic DNA and suppression of the efficiency of the immune system" (Bojkov, 1995). Seasonal thinning of the ozone layer (*ozone hole*) was first observed over the Antarctic and later in Arctic regions. When the upper stratosphere ozone decreases, the UV-B radiation at the earth's surface increases. In recent years the ozone hole has been penetrating farther into the midlatitude, where the bulk of the population lives.

For two decades, academic, industrial, and government scientists from around the world pursued full understanding of this global issue. In the late 1980s and early 1990s, international agreements limiting the production and use of CFCs were implemented. Correction of the problem is measured in decades, with nature adding to the confusion along the way. In 1996, Solomon (Zurer, 1996b) showed that aerosols from volcanic eruptions can accelerate the degradation of the ozone layer. Nevertheless, the agreed-upon action caused the atmospheric concentrations of some CFCs to begin to decrease in 1991, with more CFCs decreasing every year (Zurer, 1995a). Detectable ozone recovery is expected in 2005 to 2010. There were plenty of arguments about whether and when we should act while scientific, health, and economic issues were being debated. Are we, or will we be, better off for addressing this issue when we did? Yes, Were Crutzen, Sherwood, and Molina right? In 1995, the three were awarded the Nobel Prize in Chemistry (Zurer, 1995b).

Will we be as successful in assessing and addressing "global warming?" See also Chapter 2.

---

In the troposphere, ozone is formed primarily by photolytic reactions of $NO_2$. The oxygen radical reacts with $O_2$ via Equation 6.20d to produce $O_3$. The $O(^1D)\cdot$ radical reacts with water vapor to produce hydroxyl radicals or with methane to form methyl and hydroxyl radicals.

**Troposphere**

$$NO_2 + h\nu \rightarrow \begin{cases} NO + O(^1D)\cdot & \lambda < 244 \text{ nm} \\ NO + O(^3P)\cdot & \lambda < 398 \text{ nm} \end{cases} \quad (a)$$

$$O(^1D)\cdot + H_2O \rightarrow OH\cdot + OH\cdot \quad (b)$$

$$O(^3P)\cdot + O_2 + M \rightarrow O_3 + M \quad (c)$$

$$(6.22)$$

$$O(^1D)\cdot + CH_4 \rightarrow OH\cdot + CH_3\cdot \quad (d)$$

$$O_3 + h\nu \rightarrow O(^1D)\cdot + O_2 \quad \lambda < 320 \text{ nm} \quad (e)$$

$$O(^1D)\cdot + O_3 \rightarrow O_2 + O(^3P)\cdot + O(^3P)\cdot \quad (f)$$

The oxygen radical $O(^1D)\cdot$ generates $OH\cdot$ radicals (Equation 6.22), which in turn react with hydrocarbons in the troposphere, as exemplified by the following reaction, with methanol $(CH_3OH)$ ultimately producing $HNO_3$.

$$CH_3OH + OH\cdot \rightarrow \begin{cases} CH_2OH + H_2O & (a) \\ CH_3O\cdot + H_2O & (b) \end{cases}$$

$$CH_2OH + O_2 \rightarrow HCHO + HO_2\cdot \quad (c)$$

$$CH_3O\cdot + O_2 \rightarrow HCHO + HO_2\cdot \quad (d) \quad (6.23)$$

$$HO_2\cdot + NO \rightarrow NO_2 + OH\cdot \quad (e)$$

$$OH\cdot + NO_2 + M \rightarrow HNO_3 + M \quad (f)$$

The reaction with methanol above shows that it forms formaldehyde (HCHO), which reacts to generate new hydroxyl radicals, which in turn react with $NO_2$ to produce nitric acid. Oxidation mechanisms for olefins (which react with ozone as well as $OH\cdot$ radicals), aldehydes (which react with $OH\cdot$ and also photodissociate into free radicals), and aromatics are more complex, but again have similar chain initiation, branching, and termination steps (Finlayson-Pitts and Pitts, 1993).

In the planetary boundary layer, $O_3$ is formed via the photolytic reactions 6.22 because combustion sources generate $NO_x$. In the *marine boundary layer* (lower troposphere over the oceans) very little $NO_x$ is to be found. The small amounts of $O_3$ that are encountered disappear via photolysis and reactions involving CO:

$$CO + OH\cdot \rightarrow H\cdot + CO_2 \quad (a)$$

$$H\cdot + O_2 + M \rightarrow HO_2\cdot + M \quad (b) \quad (6.24)$$

$$HO_2\cdot + O_3 \rightarrow OH\cdot + 2O_2 \quad (c)$$

## 6.9.3 Oxides of nitrogen

The nitrogen oxides $(NO \text{ and } NO_2)$ are the principal species in tropospheric chemistry. In an urban troposphere, NO is formed by combustion processes and produces $NO_2$ via reactions with ozone, $OH\cdot$, $HO_2\cdot$, $O(^3P)\cdot$. Once formed, $NO_2$ is photolyzed to produce NO, which begins the following reaction scheme:

$$NO_2 + h\nu \rightarrow \begin{cases} NO + O(^1D)\cdot & \lambda < 244 \text{ nm} \quad (a) \\ NO + O(^3P)\cdot & \lambda < 398 \text{ nm} \quad (b) \end{cases}$$

$$HONO + h\nu \rightarrow NO + OH\cdot \quad \lambda < 591 \text{ nm} \quad (c)$$

$$NO + O_3 \rightarrow NO_2 + O_2 \quad (d)$$

$$(6.25)$$

$$NO + HO_2\cdot \rightarrow NO_2 + OH\cdot \quad (e)$$

$$NO + O(^3P)\cdot + M \rightarrow NO_2 + M \quad (f)$$

$$NO + OH\cdot \rightarrow HONO \quad (g)$$

$$NO_2 + OH\cdot + M \rightarrow HNO_3 + M \quad (h)$$

At night, the UV reactions cease and nitrogen trioxide $(NO_3)$ is formed from the reaction of nitrogen dioxide and ozone:

$$NO_2 + O_3 \rightleftharpoons NO_3 + O_2 \quad (6.26)$$

Nitrogen trioxide $(NO_3)$ is photolytic and decomposes rapidly during the daytime to form NO and $NO_2$:

$$NO_3 + h\nu \rightarrow \begin{cases} NO_2 + O(^3P)\bullet & \text{(a)} \\ NO + O_2 & \text{(b)} \end{cases} \quad (6.27)$$

The first reaction generates oxygen radicals, which react with molecular oxygen to generate ozone via Equation 6.22(c).

### 6.9.4 Aldehydes

Many hydrocarbons react with hydroxyl and oxygen radicals, but aldehydes and ketones are different in that they are also photolytic. Photochemical reactions with aldehydes yield methyl and acetyl radicals, which in turn react with molecular oxygen and NO to form oxygenated radicals. Two common aldehydes that illustrate this are formaldehyde (HCOH) and acetaldehyde $(CH_3CHO)$:

$$HCOH + h\nu \rightarrow HCO\bullet + H\bullet \quad \text{(a)}$$
$$CH_3CHO + h\nu \rightarrow CH_3\bullet + HCO\bullet \quad \text{(b)} \quad (6.28)$$

The acetyl radical $(HCO\bullet)$ and methyl radical $(CH_3)\bullet$ react with oxygen and $OH\bullet$ to form $CO_2$ via Equation 6.19. In all cases the reaction rate is proportional to the intensity of the radiation, which varies throughout the day and with the season of the year.

### 6.9.5 Urban air pollution models and the ozone ridge

An important technique in EPA's pollution abatement mission is the *empirical kinetic modeling approach* (EKMA), which seeks to relate changes in atmospheric hydrocarbon and $NO_x$ emissions to changes in the maximum ozone concentrations (Seinfeld, 1989). The EKMA consists of numerical computational models (also called *airshed models or air basin models*) that describe a polluted urban air basin containing $NO_x$, nonmethane organic compounds in which sunlight generates oxidants [typically, ozone and peroxyacetylnitrate (PAN)]. The goal of modeling is to provide government regulators with a rational way to set emission standards. Indeed, without knowledge gained from these models, implementing certain pollutant emission standards could inadvertently increase the production of oxidants. Shown in Table 6-9 are the range of pollutant concentrations

**Table 6-9**  Range of Pollutant Concentrations in Urban Regions

| Pollutant | Range (ppb) |
|---|---|
| CO | 1000–10,000 |
| $SO_2$ | 20–200 |
| $O_3$ | 100–500 |
| $HNO_3$ | 3–50 |
| $NO_2$ | 1–500 |
| NMHC | 500–1200 ppb C[a] |
| HCHO | 20–50 |
| TSP—24 h | 5–1500 $\mu g/m^3$ |
| Lead | 0.0001–10 $\mu g/m^3$ |

*Source:* Abstracted from Seinfeld (1986), Finlayson-Pitts and Pitts (1986), and Flagan and Seinfeld (1988).
[a] Nonmethane hydrocarbons, reported as carbon.

found in different cities of the world whose air is considered polluted. Because the mix of pollutants in one air basin is different from the mix in another air basin, an abatement policy that is appropriate for one air basin could be entirely inappropriate for another. Even in a single air basin, the mix of pollutants varies between seasons (indeed, even between night and day; Russell et al., 1986), suggesting that abatement strategies could be tailored to the season. Once the underlying kinetic relationships are understood, regulatory agencies can set emission and ambient air quality standards for specific pollutants to minimize the generation of oxidants in a cost-effective fashion. Modeling urban air pollution is a long-standing field of research involving complex chemical kinetics and sophisticated computational techniques. Even if engineers do not perform these tasks themselves, it is important that they understand how these models are designed, the numerical techniques used to solve the set of stiff nonlinear ordinary differential equations (ODEs), and the experimental smog chamber experiments used to corroborate results predicted by the analytical models (Hess et al., 1992).

Volatile organic compounds (VOCs), generated by biogenic and anthropogenic processes in the outdoor environment, participate in photolytic reactions that ultimately form ozone and other oxidants that irritate the mucous membrane. Different urban and rural airsheds contain a variety of VOCs (Russell et al., 1995). The reactivities of these compounds vary considerably and the hydrocarbon free radicals that they generate are similarly varied. To illustrate airshed modeling, we assume that this complex mixture can be characterized by a generic hydrocarbon RH, a generic aldehyde RCHO, and two hydrocarbon free radicals:

1. $RO_2\cdot$ (alkyl peroxy radical)
2. $RC(O)O_2\cdot$ (peroxyacetyl radical)

We postulate that the pollution follows the kinetic mechanism shown in Figure 6-10 and Table 6-10. The airshed is taken to be well mixed and contains known initial concentrations, $[NO]_0$, $[NO_2]_0$, $[RH]_0$, and $[RCHO]_0$. The mixture is exposed to sunlight, and the concentration of these species are computed throughout the period of exposure. There is no external addition of these species to the air mixture after it is irradiated.

Figure 6-10 is useful because it enables readers to recognize which species are reactants, which are products, and which are intermediates that are formed, consumed, and form again. The lower por-

tion of Figure 6-10 involving reactions 1, 2, and 3 is a self-sustaining cycle. The photolysis of $NO_2$ to NO and O· leads to the creation of $O_3$. The regeneration of $NO_2$ occurs via reaction 3. The upper portion of Figure 6-10, involving organic compounds, aldehydes, and hydrocarbon radicals present in an urban airshed, is another cycle that generates $HO_2\cdot$, which in turn reacts with NO to replenish OH· to sustain the cycle. The boxes represent species of interest immersed in a sea of oxygen, water vapor, and free radicals OH·, $HO_2\cdot$, and O·. The hydrocarbon (RH) reacts with hydroxyl radical (OH·) to form the peroxyalkyl radical $(RO_2\cdot)$, eventually leading to molecular species CO, $CO_2$, and PAN. The aldehyde (RCHO) reacts by photolysis and OH· attack and ultimately forms PAN through the intermediate peroxyacetyl radical $(RC(O)O_2\cdot)$. Aldehydes are generated in the process as well but through photolysis and OH· attack they ultimately disappear. The most dominant player in the kinetic mechanism is the hydroxyl radical OH· followed by sunlight. In the presence of reactive hydrocarbons (RH and RCHO) that produce $RO_2\cdot$ and $RC(O)O_2\cdot$, NO reacts to

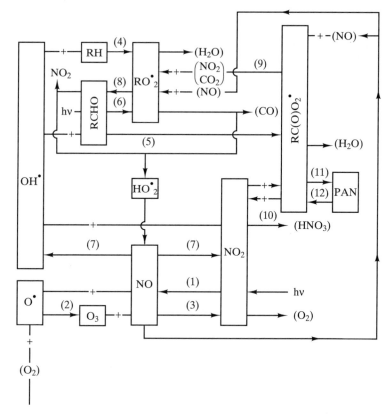

**Figure 6-10** Diagram of generalized kinetic mechanism for urban air pollution, Table 6.10

**Table 6-10**   Generalized Reaction Mechanism

| Reaction | Rate constant | k (T=298K) |
|---|---|---|
| (1) $NO_2 + h\nu \rightarrow NO + O\bullet$ | | (assumed) 0.533[a] |
| (2) $O\bullet + O_2 + M \rightarrow O_3 + M$ | $6.0 \times 10^{-34} (T/300)^{-2.3}$ cm$^6$ molecule$^{-2}$ s$^{-1}$ | $2.183 \times 10^{-5}$[b] |
| (3) $NO + O_3 \rightarrow NO_2 + O_2$ | $2.2 \times 10^{-12} \exp(-1430/T)$ cm$^3$ molecule$^{-1}$ s$^{-1}$ | 26.59[b] |
| (4) $RH + OH\bullet \rightarrow RO_2\bullet + H_2O$ | $1.68 \times 10^{-11} \exp(-559/T)$ cm$^3$ molecule$^{-1}$ s$^{-1}$ | $3.775 \times 10^3$[b] |
| (5) $RCHO + OH\bullet \rightarrow RC(O)O_2\bullet + H_2O$ | $6.9 \times 10^{-12} \exp(250/T)$ cm$^3$ molecule$^{-1}$ s$^{-1}$ | $2.341 \times 10^4$[b] |
| (6) $RCHO + h\nu \rightarrow RO_2\bullet + HO_2\bullet + CO$ | | (assumed) $1.91 \times 10^{-4}$[a] |
| (7) $HO_2\bullet + NO \rightarrow NO_2 + OH\bullet$ | $3.7 \times 10^{-12} \exp(240/T)$ cm$^3$ molecule$^{-1}$ s$^{-1}$ | $1.214 \times 10^4$[b] |
| (8) $RO_2\bullet + NO \rightarrow NO_2 + RCHO + HO_2\bullet$ | $4.2 \times 10^{-12} \exp(180/T)$ cm$^3$ molecule$^{-1}$ s$^{-1}$ | $1.127 \times 10^4$[b] |
| (9) $RC(O)O_2\bullet + NO \rightarrow NO_2 + RO_2\bullet + CO_2$ | $4.2 \times 10^{-12} \exp(180/T)$ cm$^3$ molecule$^{-1}$ s$^{-1}$ | $1.127 \times 10^4$[b] |
| (10) $OH\bullet + NO_2 \rightarrow HNO_3$ | $1.1 \times 10^{-11}$ cm$^3$ molecule$^{-1}$ s$^{-1}$ | $1.613 \times 10^4$[b] |
| (11) $RC(O)O_2\bullet + NO_2 \rightarrow RC(O)O_2NO_2$ | $4.7 \times 10^{-2}$ cm$^3$ molecule$^{-1}$ s$^{-1}$ | $6.893 \times 10^3$[b] |
| (12) $RC(O)O_2NO_2 \rightarrow RC(O)O_2\bullet + NO_2$ | $1.95 \times 10^{16} \exp(-13{,}543/T)$ s$^{-1}$ | $2.143 \times 10^{-2}$[a] |

*Source:* Seinfeld (1986).
(a)  k[=] min$^{-1}$
(b)  k[=] ppm$^{-1}$ min$^{-1}$

increase the production of PAN as well as to produce $HO_2\cdot$ radicals that react with NO (reaction 7) to produce $NO_2$ and generate $OH\cdot$.

To model the system, ODEs are written for all species of interest. The steady-state approximation is applied to the differential equations for the free radicals and $O_3$ to obtain expressions for their "local" steady-state values. These values are substituted in the remaining ODEs and where possible, an order-of-magnitude analysis is performed to eliminate negligible terms. The resulting ODEs are solved by numerical computational techniques for the species of interest (RH, RCHO, NO, $NO_2$, $O_3$, etc.), obtaining the concentrations as a function of time over a 10-h irradiation period. Figure 6.11 is an example of

the results that are obtained for one set of initial conditions. It should be noted that $[O_3]$ rises throughout the irradiation period. The calculation may be repeated using different values for $[NO_x]_0$ and $[VOC]_0$. A three-dimensional surface depicting the ozone concentration at the end of 10 h ($[O_3]_{10\,h}$ plotted along the $z$-axis) may be constructed for a range of $[NO_x]_0$ and $[VOC]_0$ lying along the $x$ and $y$ axes in the horizontal plane. The three-dimensional surface generated is an *ozone ridge* beginning at the origin and rising in height along a diagonal "ridge line" drawn from the origin. Figure 6-12 is a top view of the ozone ridge, consisting of a series of nested $[O_3]_{10\,h}$ isopleths. The ridge line in Figure 6-12 has the value $[VOC]_0/[NO_x]_0 = 5$. If the rate constants in the

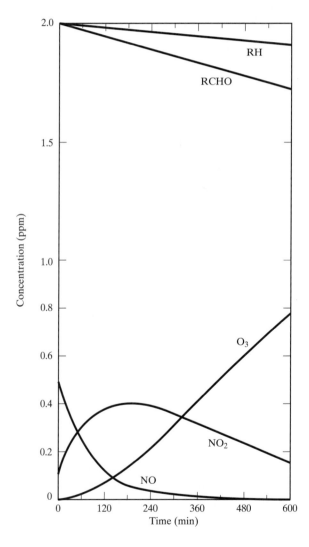

**Figure 6-11** Urban pollutant species vs time using generalized reaction mechanism, Table 6.10 (Seinfeld, 1986), 10-hr reaction time, $[RH]_0 = 2$ ppm, $[RCHO]_0 = 2$ ppm, $[NO]_0 = 0.5$ ppm, $[NO2]_0 = 0.1$ ppm

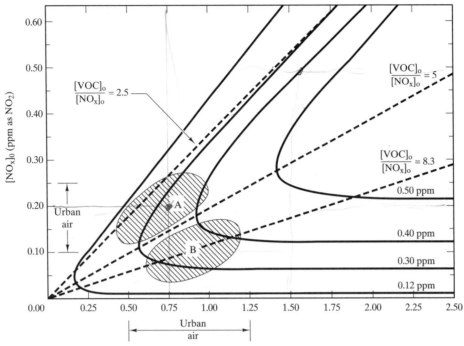

**Figure 6-12**   Ozone isopleths The maximum [$O_3$] achieved during a fixed period of irradiation of a mixture whose initial concentrations are [$NO_x]_0 = [NO]_0 + [NO_2]_0$ and [$VOC]_0$, where the latter is measured as ppm Carbon. The general features of the graph are similar in virtually all urban smog systems, although the actual location of the $O_3$ isopleths depends on the specific conditions such as solar intensity, organic species involved, etc. (adapted from Seinfeld, 1986)

kinetic mechanism are changed, a similar ozone ridge and ridge line will still be generated.

Ozone ridges (similar to Figure 6-12) appear frequently in the pollution literature and it is important that engineers be able to interpret them and draw conclusions to aid regulators in setting policies. Cost-effective policies to reduce oxidant concentration in any airshed depend on knowing whether the community lies above or below the ridge line. In the words of the authors (Finlayson-Pitts and Pitts, 1993) regulators should set emission standards resulting in [$NO_x]_0$ and [$VOC]_0$ that allow a region to "roll down the ozone ridge."

The first step is to plot the range of [$VOC]_0$ and [$NO_x]_0$, describing each air basin on Figure 6-12. For discussion purposes, two hypothetical air basins, urban region A and rural region B, are shown in Figure 6-12 with diagonals describing their [$VOC]_0/[NO_x]_0$ ratios. Region A has a [$VOC]_0/[NO_x]_0$ less than the ridge line and much less than urban region B, owing

to the greater amount of $NO_x$ generated by combustion sources in urban regions and the many more VOCs generated by biogenic sources in rural region B.

Examination of Figure 6.12 shows the upper diagonal corresponding to an urban [$VOC]_0/[NO_x]_0$ of 2.5 and the lower diagonal to a [$VOC]_0/[NO_x]_0 = 8.3$. Along the lower diagonal, ozone levels are limited by the availability of $NO_x$ to generate it. Along the upper diagonal, ozone levels are limited by the availability of the VOCs needed to generate the free radicals, which in turn generate ozone. Reducing both [$NO_x]_0$ and [$VOC]_0$ simultaneously while keeping their ratio constant also reduces ozone. If [$VOC]_0$ and [$NO_x]_0$ lie along the ridge line, abatement policies that reduce either [$NO_x]_0$ or [$VOC]_0$ will reduce ozone. If [$NO_x]_0$ and [$VOC]_0$ are reduced simultaneously, care should be taken to change the [$VOC]_0/[NO_x]_0$ ratio appropriately in order to roll down the ozone hill.

These consequences can also be explained in terms of reactions 1-3 in the kinetic mechanism in Table 6-10. Below and to the right of the ridge line, oxidant is limited by the amount of the precursor $NO_2$ that leads to ozone via photolysis. In this region, oxidants are more sensitive to decreases in an already low $NO_x$ than to the hydrocarbon free radicals produced by VOCs. Above and to the left of the ridge line, the formation of ozone is limited by the availability of organic free radicals (generated slowly by the hydrocarbons) that convert NO to $NO_2$. These results of the well-mixed model beg the question of the accurate location of the ozone ridge. Thus more rigorous analysis is required.

In an actual urban air basin, the matter is more complex. $NO_x$, alkenes, alkanes, aldehydes, alcohols, and ketones are continually added to the atmosphere, particularly during commuting time in the morning and early evening. Present also are a variety of biogenic VOCs of varying reactivity (Dimitriades, 1981; Grosjean et al., 1993) to make them players in the ozone kinetics. Air containing these pollutants and ozone is convected into and out of the urban air basin and the solar radiation varies throughout the day (and season).

More sophisticated analyses require that the generalized kinetic mechanism shown above must include the VOCs of high reactivity (Russell et al., 1995; Dimitrades, 1996) unique to the region's air basin. Many reactive species have been omitted from the mechanism of Table 6-10. For example, $H_2O_2$ should be added (Sakugawa et al.,1990; Sakugawa and Kaplan, 1993). A considerable body of knowledge concerning atmospheric photochemistry has been amassed (Whitten et al., 1980; Atkinson et al., 1992; Atkinson, 1990; Carter, 1990; Roberts, 1990; Dechaux et al., 1994). Improved EKMA models such as the Lagrangian model (McRae et al., 1982, 1983) have been devised to account for the transport of air into and out of the region. Sophisticated numerical techniques to solve the equations have been devised.

Reported values of $[O_3]$ and $[VOC]/[NO_x]$ in urban regions indicated that measured values exceeded values predicted by a variety of early airshed models by a considerable amount. It was concluded that the underestimation could arise from underestimating VOC emissions from mobile sources (i.e., inadequate estimates of automotive running and evaporative losses, autos not in compliance with emission standards, or autos whose emission controls had been tampered with). A second and more interesting source of VOCs could be biogenic VOCs that had been ignored previously. Chief among these VOCs were isoprenes from hardwood trees and $\alpha$-pinene from conifers. These two biogenic organics react rapidly with OH· and O· and $NO_3$ at night to generate ozone. On a larger scale, particularly in rural areas, highly reactive compounds such as isoprene and the terpenes must be included in airshed models.

Figure 6-13 shows the relative importance of various VOCs. These are the results of a study using a chemical kinetic mechanism containing 203 reactions and 91 reacting species, including 27 organics (Carter, 1990). The range of the potential to produce ozone is enormous; for example, 1 kg of formaldehyde (HCHO) will produce two orders of magnitude more ozone that 1 kg of ethane. Organic species such as alkanes and alcohols generate one order of magnitude less ozone than equal mass emissions of alkenes and aldehydes.

## EXAMPLE 6.7  LEIGHTON MECHANISM

The upper left box in Figure 6-6, called the *Leighton mechanism* (Leighton, 1961), describes the production of ozone in urban air containing the oxides of nitrogen. Baring the influence of hydrocarbons, the Leighton relationship predicts that the ozone concentration is directly proportional to the intensity of the solar radiation (which varies throughout the day) and the ratio of $NO_2$ to NO:

$$[O_3]_{ss} = k_1(I)\frac{[NO_2]}{k_3[NO]}$$

The Leighton mechanism consists of Equations 6.25b, 6.22c, 6.25d, renumbered for simplicity as

$$NO_2 + I(hv) \rightarrow NO + O\cdot$$

$$k_1 = f(I), \text{ typcally} \approx 0.533 \text{ min}^{-1}$$

$$O\cdot + O_2 + M \rightarrow O_3 + M$$

$$k_2(cm^6 \text{ molecule}^{-2} \text{ s}^{-1}) = 6 \times 10^{-34}\left(\frac{T}{300}\right)^{-2.3}$$

$$O_3 + NO \rightarrow NO_2 + O_2$$

$$k_3(cm^3 \text{ molecule}^{-1} \text{ s}^{-1}) = 2.2 \times 10^{-12}\exp\left(\frac{-1430}{T}\right)$$

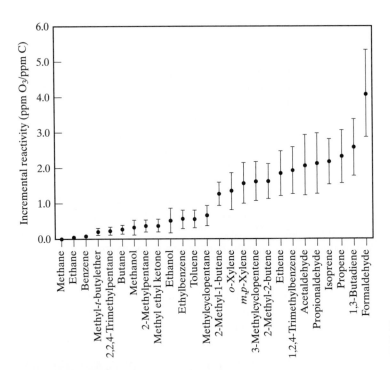

**Figure 6-13** Hydrocarbon reactivities (taken from Russell et al, 1995)

Assuming that $O_3$ and $O\cdot$ achieve their steady-state concentrations because they are so active, prove the Leighton relationship.

*Solution*  Write the differential mass balances for the $O_3$ and $O\cdot$ concentrations in a parcel of air and set the derivatives equal to zero because the species are very reactive

$$\frac{d[O_3]}{dt} = k_2[O\cdot][O_2][M] - k_3[O_3][NO] \simeq 0$$

$$[O_3]_{ss} = \frac{k_2[O\cdot][O_2][M]}{k_3[NO]}$$

$$\frac{d[O\cdot]}{dt} = k_1[NO_2] - k_2[O\cdot][O_2][M] \simeq 0$$

$$[O\cdot]_{ss} = \frac{k_1[NO_2]}{k_2[O_2][M]}$$

Insert the second equation into the first:

$$[O_3]_{ss} = \frac{k_2[O_2][M]}{k_3[NO]}\frac{k_1[NO_2]}{k_2[O_2][M]} = \frac{k_1}{k_3}\frac{[NO_2]}{[NO]}$$

$$= k_1(I)\frac{[NO_2]}{k_3[NO]}$$

Since $k_1$ depends on the solar radiation intensity $(I)$ while $k_3$ depends only on temperature, it can be seen that $[O_3]$ rises and falls in proportion to the solar intensity and the instantaneous ratio $[NO_2]/[NO]$. Measurements in urban air show that this is not the case; indeed, ozone reaches its peak sometime after the peak in $[NO_2]/[NO]$. Thus some other mechanism must exist that influences the production of urban ozone. This part is played by the reactions with hydrocarbons, as suggested by the other boxes in Figure 6-6.

## 6.9.6 Stratospheric ozone layer

We have already noted that $N_2O$ and CFC11 have very long residence times in the atmosphere, owing to their chemical inertness at tropospheric conditions. Accordingly, for $N_2O$ the ultimate destruction occurs by photodissociation in the stratosphere. Reactions in the stratosphere are summarized in Figure 6-14. Reactions with CFCs begin with the oxygen free radical that forms chlorine (and bromine) oxides, ClO (and BrO). Alternatively, CFCs are photolyzed by ultraviolet radiation to form the $Cl\cdot$ (and $Br\cdot$) radicals. The oxide ClO reacts with oxygen radicals to produce $Cl\cdot$, which consumes ozone and regenerates

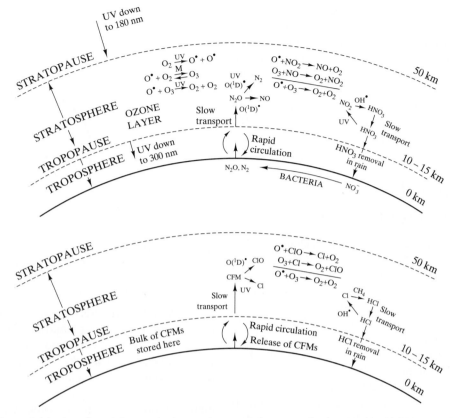

**Figure 6-14** (top figure) Stratospheric reactions contributing to the formation and destruction of ozone. (bottom figure) Simplified portrayal of stratospheric reactions involving chlorofluoromethanes (CFM's) affecting stratospheric ozone. (National Academy of Sciences, 1976)

ClO. The ultimate fate of chlorine is HCl (Cicerone, 1987). Equations 6.12 and 6.20 show the kinetic mechanisms that occur in the stratosphere.

---

### EXAMPLE 6.8 STEADY-STATE ClO IN THE STRATOSPHERE

Consider the CFC stratospheric reactions in Equations 6.12 and 6.20. Show that the mechanism is capable of sustaining a steady-state concentration of ClO at the expense of ozone and CFCs.

***Solution*** The concentration of ClO is governed by Equations 6.12a, c, and d.

$$\frac{d[\text{ClO}]}{dt} = k_{12a}[\text{CFC}][\text{O}\bullet] - k_{12c}[\text{O}\bullet][\text{ClO}]$$
$$+ k_{12d}[\text{O}_3][\text{Cl}\bullet]$$

If steady state exists, $[\text{ClO}] = [\text{ClO}]_{ss}$ and $d[\text{ClO}]/dt = 0$. Thus

$$[\text{ClO}_{ss}] = \frac{k_{12a}[\text{CFC}][\text{O}\bullet]}{k_{12c}[\text{O}\bullet]} + \frac{k_{12d}[\text{O}_3][\text{Cl}\bullet]}{k_{12c}[\text{O}\bullet]}$$

$$= \frac{k_{12a}}{k_{12c}}[\text{CFC}] + \frac{k_{12d}}{k_{12c}}\frac{[\text{O}_3][\text{Cl}\bullet]}{[\text{O}\bullet]}$$

UV radiation and $\text{O}_2$ and $\text{O}_3$ will produce a steady-state $[\text{O}\bullet]$ concentration. Thus by reaction 6.12a and the first term in the equation above, ClO will be produced as long as CFCs are added to the stratosphere even if no CFCs are photolyzed. If CFCs are photolyzed, Cl• will be finite, reaction 6.12d and the second term in the equation above will come into play, and $[\text{ClO}\bullet]_{ss}$ will increase.

The chain is capable of sustaining the concentration of ClO at the expense of ozone. It is convenient to think of CFCs as catalysts for the destruction of ozone. While ozone is generated in the stratosphere by the photolysis reaction Equation 6.20, the inclusion of an additional removal reaction lowers the steady state ozone concentration. Methane reacts with Cl• and OH• to generate a steady-state concentration of HCl which after many years is transported downward to the upper troposphere, where it is removed by rainfall.

# 6.10 *Atmospheric and interfacial transport*

So far in this chapter we have described characteristics of the atmosphere and indicated how chemical species enter, leave, and are changed by the atmosphere. We have shown how the troposphere interacts with the bounding media of water, land, living species, and the stratosphere. In all of these contexts we have assessed natural and human-made influences on the phenomena and species of interest. Now we wish to introduce the quantitative modeling of interfacial transport of chemical species (or pollutants) of interest.

The existence of atmospheric transport models that provide information on temperature, pressure, humidity, winds, pollutant concentrations, and so on, is presumed. Some of this information is available from the atmospheric dispersion models to be presented in Chapter 9, or from input information required by them. Many atmospheric transport models do not incorporate the injection and removal of materials at the earth's surface. It will be clear that the models described below (or similar models) could fulfill this role, as allowance for finite surface flux manifests itself simply as a boundary condition on any properly constructed atmospheric transport model. Sophisticated modeling is beyond the scope of this book.

## 6.10.1 Interfacial flux calculations

Consider the transfer of material from the gas phase across a gas–liquid interface and into the liquid phase (Figure 6-15). A macroscopic approach that has found widespread application in chemical engineering, meteorology, oceanography, and other fields is described below. The vertical flux of a material in a gas

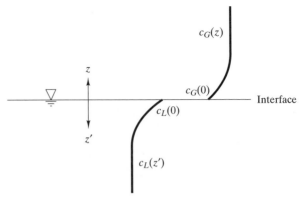

**Figure 6-15** Schematic diagram of material transfer across a gas-liquid interface

phase, $N_G(0)$, is proportional to a concentration driving force $[c_G(z) - c_G(0)]$ and will be expressed as

$$N_G(0) = k_G(z)[c_G(z) - c_G(0)] \qquad (6.29)$$

where $N_G(0)$ is the vertical molar flux at the interface (kmol/m$^2$·s), $c_G(z)$ and $c_G(0)$ are the molar concentrations at an elevation $z$ and at the interface ($z = 0$) (kmol/m$^3$), and $k_G(z)$ is the gas-phase mass transfer coefficient evaluated at elevation $z$ (m/s). If the gas-phase concentration is known by measurement or model prediction at some height and the gas-phase mass transfer coefficient is obtained for the same height from some predictive model, the surface flux can be calculated if the interfacial concentration is known. The $z$-dependence implies that the coefficient must be selected to be consistent with the height at which the pollutant concentration is measured. The choice is quite arbitrary, although 1 m and 10 m are commonly used in presenting meteorological data. Sometimes $c_G(0)$ is taken to be zero. However, this implies that the non-atmospheric side of the interface has an infinite capacity for the material and offers no resistance to the rate of mass transfer. In case of uptake by rock, soil, and/or vegetation, neither of these assumptions are likely to be valid. Indeed, it may be the atmospheric phase resistance to mass transfer, which is negligible by comparison. These solid phases may have to be characterized by a complex series of resistances according to their individual peculiarities.

By contrast, an aqueous (liquid) phase, such as a lake or ocean, can be dealt with similarly to the gas phase. The flux equation can be expressed as

$$N_L(0) = k_L(z')[c_L(0) - c_L(z')] \quad (6.30)$$

where $N_L(0)$ is the molar flux at the interface in the liquid phase ($kmol/m^2 \cdot s$), $c_L(0)$ and $c_L(z')$ are the concentrations in the liquid phase at the interface ($z' = 0$) and depth $z$, respectively ($kmol/m^3$), and $k_L(z')$ is the mass transfer coefficient in the liquid phase corresponding to some depth $z'$ at which the concentration is known (m/s).

Again, $c_L(z')$ is obtained by measurement or prediction and $k_L(z')$ is obtained from a predictive model. Correlations of the gas- and liquid-phase mass transfer coefficients with readily determined parameters are the subject of later parts of this section. In general, the interfacial concentration, $c_L(0)$, is not easily obtained.

At steady state, there is continuity in the transfer of mass across the interface. That is, the molar flux of pollutant through the gas must be equal to the molar flux of pollutant through the liquid:

$$N_G(0) = N_L(0) \quad (6.31)$$

Combining Equations 6.29 and 6.30 leaves only two unknowns, the gas and liquid interfacial concentrations. A relationship between them is supplied by the postulate that phase equilibrium exists at the interface. The validity of this postulate of interfacial equilibrium is well documented for many cases. For any particular case, solution thermodynamics can be used to obtain a quantitative expression for the interdependence of the gas and liquid interfacial concentrations generally:

$$c_G(0) = f[c_L(0)] \quad (6.32)$$

An example of a rigorous calculation for $SO_2$ in equilibrium with fresh water and with seawater is given by Rasmussen et al. (1974a). A commonly seen form of Equation 6.32 is $c_G(0) = Hc_L(0)$, known as Henry's law (see Chapter 8 for details about Henry's law). The calculation procedure is as follows:

1. $c_G(z)$ and $c_L(z')$ must be known or specified.
2. $k_G(z)$ and $k_L(z')$ must be predicted.
3. Equations 6.30 through 6.32 are solved simultaneously for $c_G(0)$, $c_L(0)$, and $N(0)$, the interfacial compositions and the desired absorption flux.

## 6.10.2  Gas-phase mass transfer coefficient

To calculate the deposition of a pollutant from the atmosphere to any medium at the earth's surface, it is necessary to quantify the mass transfer through air near the surface. As described above, this can be achieved if the gas-phase mass transfer coefficient $k_G(z)$ can be predicted. The starting point for the prediction is the Reynolds analogy, which expresses a similarity among mass, heat, and momentum transfer (more details are presented in Chapter 8). One result of the Reynolds analogy for mass transfer from air is the expression

$$k_G(z) = c_D U(z) = \frac{U_*^2}{U(z)} \quad (6.33)$$

where

$k_G(z)$ = mass transfer coefficient in air (m/s)

$U(z)$ = air velocity at an elevation $z$, (m/s)

$U_*$ = friction velocity (m/s)

$c_D$ = drag coefficient, $c_D = [U_*/U(z)]^2$ (6.34)

The left equality of Equation 6.33 says that the mass transfer coefficient, $k_G(z)$, is linearly proportional to the horizontal velocity, $U(z)$, which in turn is a function of the height, $z$, and that the coefficient of proportionality is the drag coefficient, $c_D$. Noting that $c_D = [U_*/U(z)]^2$, Equation 6.33 can be expressed as shown in terms of the *friction velocity*, $U_*$, which is related to the air shear stress, $\tau_0$, at the air–water interface,

$$U_* = \left(\frac{\tau_0}{\rho}\right)^{1/2} \quad (6.35)$$

The chemical engineering literature contains many manifestations of the Reynolds analogy for different flow configurations. One of the best known expressions is the Chilton and Colburn (1934) equation, which allows for the variation among chemical species by including the Schmidt number ($Sc = \mu/\rho D$):

$$k_G(z) = \frac{U_*^2}{U(z)Sc^{2/3}} \quad (6.36)$$

Owen and Thomson's (1963) research on heat transfer provided a somewhat different correlation to allow for the fact that different species may diffuse at different rates and especially to allow for the fact that surfaces may exhibit a bluff body character not accounted for by the Reynolds analogy. Their equation is

$$\frac{1}{k_G(z)} = \frac{1}{U_*}\left[\frac{U(z)}{U_*} + \frac{1}{B}\right] \qquad (6.37)$$

Owen and Thomson's equation has been accepted by many investigators who have tried to correlate $B$ with system parameters. One such equation was suggested by Dipprey and Sabersky (1963):

$$\frac{1}{B} = 10.25\text{Re}_0^{0.2}\text{Sc}^{0.44} - 8.5 \qquad (6.38)$$

where

$$\text{Re}_0 = \frac{U_* z_0}{\nu} \qquad (6.39)$$

The Reynolds number is based on a characteristic *roughness height* $z_0$ and friction velocity $U_*$. Sutton (1953) has tabulated values of $z_0$ for various types of underlying surfaces (see also Table 9-3). Charnock (1955) developed the following relationship for a water surface:

$$z_0 = \frac{bU_*^2}{g} \qquad (6.40)$$

where $g$ is the acceleration of gravity and $b$ is a constant for which various values have been given. The velocity profile $U(z)$ is required and can be given by the logarithmic velocity profile for neutral atmospheric conditions,

$$\frac{U(z)}{U_*} = \left(\frac{1}{0.4}\right)\ln\frac{z}{z_0} \qquad (6.41)$$

The constant 0.4 is called the *von Kármán constant.* For nonneutral atmospheric conditions (see Chapter 8 for details), corrections can be made to the above as required (Panofsky, 1963). To obtain a value of the gas-phase mass transfer coefficient, the calculation proceeds as follows.

**1.** For a value of the wind speed, $U(z)$, at a single height ($z$), Equations 6.40 or 6.42 and 6.41 can be solved for $z_0$ and $U_*$.

**2.** The velocity profile $U(z)$ can be obtained from the logarithmic velocity profile equation 6.41.

An alternative expression for directly predicting the friction velocity for airflow over a surface of water is given by Hicks (1973):

$$c_D = \left[\frac{U_*}{U(10)}\right]^2 = [0.65 + 0.07U(10)] \times 10^{-3} \qquad (6.42)$$

Knowing the appropriate physical properties for air and pollutant, the correction factor, $1/B$, can be obtained from Equation 6.38. From the known velocity profile (Equation 6.41) and the calculated values of the $U_*$ and $1/B$, the mass transfer coefficient, $k_G$, can be determined from Equation 6.37 for whatever height the pollutant concentration is known. Figure 6-16 shows the gas-phase mass transfer coefficient, at $z = 10$ m, as a function of roughness height, $z_0$, predicted from equations proposed by different researchers.

The development above is only illustrative. Many ideas have reached more sophisticated levels and almost classic form in a variety of special applications. For smooth surfaces (low $z_0$), model predictions are in rather good agreement. However, they diverge sharply as the roughness increases.

An empirical and theoretically more limited version of the gas-phase mass transfer coefficient is the *deposition velocity*, defined as $v_G = N_G(0)/c_G$. This quotient of the interfacial flux and the gas-phase concentration has the units of velocity, length/time, the conceptual charm of an analogy with falling particles, and the ready determination from measured fluxes and concentrations. By definition it assumes, reasonably for particles and not so for most gases, that the receiving phase has no influence on the interfacial transfer [i.e., $c_G(0) = 0$]. Thus the deposition velocity could not be successfully rationalized into a coherent predictive tool. Nevertheless, the literature is full of values for the deposition velocity. They are useful if the specific value that you find was determined for exactly the same conditions that you face. McMahon and Denison (1979) have surveyed empirical atmospheric deposition parameters. Selected values from the paper are collected in Table 6-11.

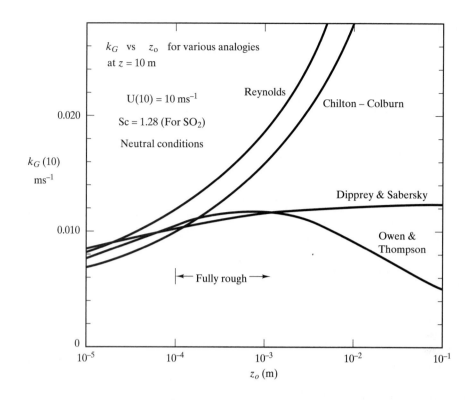

**Figure 6-16**   Mass transfer coefficients for different roughness heights (Kabel R L, 1975)

The figure shows: $k_G$ vs $z_0$ for various analogies at $z = 10$ m. $U(10) = 10$ ms$^{-1}$, Sc = 1.28 (For SO$_2$), Neutral conditions. Curves labeled Reynolds, Chilton – Colburn, Dipprey & Sabersky, Owen & Thompson, with a "Fully rough" region indicated. Vertical axis $k_G(10)$ ms$^{-1}$ with marks at 0.010, 0.020; horizontal axis $z_o$ (m) from $10^{-5}$ to $10^{-1}$.

**Table 6-11**   Selected Values of Deposition Velocities[a]

| Reference | $v_g$ (cm/s) | Gas | Surface | Comment |
|---|---|---|---|---|
| Aldaz (1969) | 2 | $O_3$ | Juniper bush | |
| | 0.6 | $O_3$ | Sand or dry grass | |
| | 0.16 | $O_3$ | Snow | |
| | 0.07 | $O_3$ | Fresh water | |
| | 0.04 | $O_3$ | Ocean | |
| | 0.02 | $O_3$ | Distilled water | |
| van Dop et al. (1977) | 0.13 | $O_3$ | Dry grass | |
| Garland et al. (1973) | 1.2 | $SO_2$ | Grass | Tracer method |
| | 0.8 | $SO_2$ | Grass | Gradient method |
| Owers and Powell (1974) | 2.6 | $SO_2$ | Grass | $U = 5.2$ m/s |
| | 0.7 | $SO_2$ | Grass | $U = 1.8$ m/s |
| Whelpdale and Shaw (1974) | 2.6 | $SO_2$ | Grass | |
| | 2.2 | $SO_2$ | Water | |
| | 0.5 | $SO_2$ | Snow | |
| Fowler (1978) | 0.3–1.5 | $SO_2$ | Wheat | Stomata open, $v_g = 0.8$ cm/s |
| | | | | Stomata closed, $v_g = 0.3$ cm/s |

*Source:* McMahon and Denison (1979).

[a] The reader is encouraged to examine the data in this table for trends in $v_g$ for (1) different receiving surface media for the same gas, (2) different gases for the same receiving surface medium, (3) one investigator and/or experimental method compared to another, and (4) differing meteorological and plant physiological conditions.

### 6.10.3 Liquid-phase mass transfer coefficient

Researchers in chemical engineering, oceanography, and other fields have measured liquid-phase mass transfer coefficients. Most often the data have been interpreted in terms of the thickness of a hypothetical stagnant layer or film and thus have resisted successful correlation. Danckwerts (1951) proposed a *surface renewal theory*, which is attractive conceptually but contained a surface renewal rate parameter that has proved hard to predict. A breakthrough occurred when Fortescue and Pearson (1967) and Lamont and Scott (1970) proposed *roll cell* mass transfer models based on "large eddy" and "eddy cell" characterizations of the turbulence in the liquid phase, respectively. Recent experiments by Asher and Pankow (1991) support these models.

Brtko and Kabel (1978) adapted the two roll cell models to the situation where the roll cells are the result of wind-induced turbulence in the liquid phase. A physical picture of a water body is helpful in appreciating these correlations. Figure 6-17 is a schematic diagram of a *stratified water body*. The lake or ocean is bounded vertically by an air–water interface and, of course, by the bottom of the lake. Between the interface and bottom it is assumed that there is a mixed region of approximately uniform temperature separated from a relatively stagnant region of slow-moving or motionless water by the *thermocline*. This stratification does not occur in all bodies of water, but in large lakes and oceans it does exist in many situations. Next, it has been observed that air flowing over a water surface induces *roll cell* behavior in an

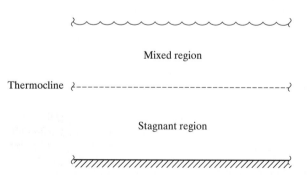

Mixed region

Thermocline

Stagnant region

**Figure 6-17** Schematic diagram of a stratified body of water (Kabel R L, 1975)

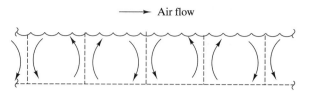

➝ Air flow

**Figure 6-18** Wind-induced roll cells in water at the air-water interface (Kabel R L, 1975)

upper layer of the water as illustrated in Figure 6-18. In this top layer of water, counterrotating cells are set up. Thus in some regions the water flows upward and in the adjacent regions the flow is downward.

The result for the *eddy cell model* under neutral conditions is

$$k_L(z') = 0.4\left(\frac{D}{\nu}\right)^{1/2}\left[\frac{\nu U_*^3}{kz'}\left(\frac{\rho_a}{\rho_w}\right)^{3/2}\right]^{1/4} \quad (6.43)$$

To use this model only the friction velocity in the air at the free liquid surface, the depth $z'$, in the liquid phase where the concentration is known, and the temperature of both phases are required. The friction velocity is obtained from equations described earlier. The temperatures enable estimation of the physical properties in the liquid, and $k$ is the von Kármán constant (0.4).

## 6.11 Interfacial equilibrium

Now that methods have been presented for the prediction of liquid and gas mass transfer coefficients, attention is turned to the remaining relationship required to complete the calculation of the gas absorption flux. That relationship depends on the equilibrium at the interface. The quantitative relationship depends on the specification of the particular chemical of interest. Nitrogen dioxide, a rather soluble material in water, will be chosen for illustration. Equilibrium at the interface is often essentially a solubility calculation. But in the case of a material like $NO_2$, which reacts with the aqueous phase forming a variety of ionic species and even producing nitric oxide (NO) which goes back into the atmosphere, this calculation becomes very complicated. Nitrogen dioxide $(NO_2)$ is important as an air pollutant, but it is also a water pollutant, since on absorption it forms a nitrate nutrient for a lake.

Actually, very few data are available for absorption of the nitrogen oxides in water. This is especially true at the low concentration levels that exist in the atmosphere. Thus most of the information presented in this section is taken from experience with the manufacture of nitric acid $(HNO_3)$, which is an altogether different situation since the concentrations are much higher. On the other hand, insofar as the matter is described by thermodynamics, the thermodynamic parameters that are correct for one case will also be correct for the other case. Reviewing the literature of $HNO_3$ manufacture yields the following seven steps, which are believed to occur in the absorption of $NO_2$ in water.

(1) $2NO_2(g) \rightleftharpoons N_2O_4(g)$

(2) $N_2O_4(g) + H_2O(l) \rightleftharpoons HNO_2(aq)$
$\quad\quad + HNO_3(aq)$

(3) $3HNO_2(aq) \rightleftharpoons HNO_3(aq)$
$\quad\quad + 2NO + H_2O$

(4) $HNO_3(aq) \rightleftharpoons H^+ + NO_3^-$ \hfill (6.44)

(5) $HNO_2(aq) \rightleftharpoons H^+ + NO_2^-$

(6) $H_2O \rightleftharpoons H^+ + OH^-$

(7) $NO(g) + \frac{1}{2}O_2(g) \rightleftharpoons NO_2(g)$

The steps above and the reversible reaction symbol indicate stoichiometry, not the steps of a chemical kinetic mechanism.

Reaction (1) is the dimerization of $NO_2$ to form nitrogen tetroxide $(N_2O_4)$. This step is important because it is believed that it is the dinitrogen form that actually is absorbed in water, as shown by reaction (2) to give nitrous acid $(HNO_2)$ and nitric acid $(HNO_3)$ in the aqueous phase. Reaction (3) shows a postulated conversion of the $HNO_2$ in the aqueous phase to $HNO_3$ and the liberation of water and the two molecules of NO. Nitric oxide (NO) is insoluble in water and would thus be liberated to the gas phase. Further, reactions (4) and (5) show the ionization of $HNO_3$ and $HNO_2$ acids to the nitrate $(NO_3^-)$, nitrite $(NO_2^-)$, and hydrogen $(H^+)$ ions. To be complete, one includes the ionization of water, reaction (6), although this term is usually negligible in most practical calculations. Finally, reaction (7) shows the oxidation in the gas phase of NO to $NO_2$. For purposes of illustration in what follows, the ionization of water,

reaction (6), will be ignored. Also, it will be assumed for simplification that the rate of reaction (7) is so slow that it does not equilibrate and in fact does not even occur in the gas phase. The equilibrium constants for reactions (6) and (7) are well known and could be incorporated if desired.

Shown below are the literature values for the equilibrium constants for reactions (1) through (5). The equilibrium constant $(K_1)$ for the dimerization of $NO_2$ is quite well known. On the other hand, the equilibrium constant $(K_2)$ for the absorption of $N_2O_4$

| Reaction | Interfacial equilibrium constant |
|---|---|
| (1) | $K_1 = 6.8 \;\; atm^{-1}$ |
| (2) | $K_2 = (3.6 \pm 0.5) \times 10^4 \;\; gmol/L^2 \; atm^{-1}$ |
| (3) | $K_3 = 30 \pm 1 \;\; atm^2 \; gmol/L^{-2}$ |
| (4) | $K_4 = 29 \pm 4 \;\; gmol/L$ |
| (5) | $K_5 = (4.5 \pm 0.1) \times 10^{-4} \;\; gmol/L$ |

in water is known with much less certainty. The equilibrium constant $(K_3)$ for the reaction of $HNO_2$ to form $HNO_3$ and NO is quite large and has about a 3% uncertainty. The equilibrium constants for the ionization of nitric and nitrous acids in the aqueous phases, $K_4$ and $K_5$, respectively, have considerable uncertainty. Nevertheless, these values can be used in the simultaneous solution of the defining equations for equilibrium of each of the reactions to obtain an equilibrium relationship between the gas- and liquid-phase interfacial concentrations of the respective species.

### EXAMPLE 6.9 INTERFACIAL FLUX CALCULATION FOR $NO_2$

The methods used to predict the rate of gas absorption at an air–water interface will be illustrated with an example of the absorption of $NO_2$ to show the kinds of data that are needed and the kinds of results that can be obtained.

*Solution* First in the calculation of the gas-phase mass transfer coefficient, $k_G(z)$, the only thing that is required to complete the calculation besides physical properties is $U(10)$, the airspeed at $z = 10$ m. In the case of this calculation, it is assumed that $U(10) = 5$ m/s.

The most direct approach for airflow over water is to use Equation 6.42 to estimate $c_D$, which is then substituted into Equation 6.33 to find $k_G(z)$. For $U(10) = 5$ m/s.

$$k_G(10) = c_D U(10)$$
$$= [0.65 + 0.07 U(10)] \times 10^{-3} U(10)$$
$$= [0.65 + 0.07(5)] 10^{-3}(5)$$
$$= 0.005 \text{ m/s} = 0.5 \text{ cm/s}$$

An alternative method of calculating $k_G(10)$, which could apply to surfaces other than water, follows for illustration and comparison. If we take values of $U_* = 0.21$ m/s and $z_0 = 0.02$ cm for a smooth sea from Table 9-3, $\nu = \mu/\rho = 1.5 \times 10^{-5}$ m$^2$/s, and Sc = 1.3; we calculate $Re_0 = 2.8$ and $1/B = 5.6$ from Equation 6.38 and $k_G(10) = 0.7$ cm/s from Equation 6.37. The two values of $k_G(10)$ are roughly equivalent and, like Fig 6-16, indicate the uncertainty in such calculations. These values for the air-phase mass transfer coefficient are reasonable in comparison with experimentally observed deposition velocities. The specified values of $\nu$ and Sc imply a diffusivity for NO$_2$ in air of

$$D = \frac{\nu}{\text{Sc}} = \frac{1.5 \times 10^{-5} \text{ m}^2/\text{s}}{1.3}$$
$$= 1.15 \times 10^{-5} \text{ m}^2/\text{s} = 0.115 \text{ cm/s}$$

This value is well below the values for similar compounds (CO$_2$, N$_2$O, and SO$_2$) in Table A-8 in the Appendix because NO$_2$ partially dimerizes to N$_2$O$_4$ in the gas phase.

The next step is to calculate the liquid mass transfer coefficient. Retaining the atmospheric phase values of $U(10) = 5$ m/s, $U_* = 0.16$ m/s (calculated from Equation 6.42), and $\rho_a = 1.2$ kg/m$^3$, we add aqueous phase properties $\rho_w = 10^3$ kg/m$^3$, $\nu = 10^{-2}$ cm$^2$/s, and $D = 10^{-5}$ cm$^{-2}$/s, giving Sc $= \nu/D = 10^3$. Estimating the diffusivity of NO$_2$ in water is fictitious, as Equation 6.44(2) shows the dimer of NO$_2$ to dissolve in water, forming the ionic species nitrous (HNO$_2$) and nitric (HNO$_3$) acids. So we merely choose $D = 10^{-5}$ cm$^2$/s as a value typical of binary diffusivities in the liquid phase. Taking 1 m as the depth at which aqueous phase concentrations are measured, the calculation yields

$$k_L = 1.1 \times 10^{-5} \text{ m/s} = 4.0 \text{ cm/h}$$

This value is also in the right range for typically observed values of the liquid-phase mass transfer coefficient.

The thing that remains to be done is to determine the reaction equilibria at the interface. The thermodynamic relationships relating the concentrations of reactants and products to the five equilibrium constants in the table above lead to five equations to be solved simultaneously. It was necessary to assume a gas-phase concentration of NO. The background concentration of NO in the atmosphere, 1.4 $\mu$g/m$^3$, was chosen (see Table 6-2). Algebraic manipulation of the remaining five equations with the unknowns involved there leads to two equations, which retain three unknowns. These equations are

$$c_{H^+}^2 = 2.26 \times 10^{10} P_{NO_2}^{3/2} \qquad \text{(a)}$$
$$+ 1.41 \times 10^{-7} P_{NO_2}^{1/2} + 10^{-14}$$

$$\sum c_{NO_2} = \left( 7.84 \times 10^8 + \frac{2.26 \times 10^{10}}{c_{H^+}} \right) P_{NO_2}^{3/2} \qquad \text{(b)}$$
$$+ \left( 3.16 \times 10^{-4} + \frac{1.41 \times 10^{-7}}{c_{H^+}} \right) P_{NO_2}^{1/2}$$

Equation (a) relates $c_{H^+}$, the hydrogen ion concentration (gmol/L) in the liquid phase to $P_{NO_2}$ (atm), the partial pressure of NO$_2$ in the gas phase. Equation (b) relates $\sum c_{NO_2}$, the total molar concentration (gmol/L) of the NO$_2$-bearing species (HNO$_2$, HNO$_3$, NO$_2^-$, and NO$_3^-$) in the liquid phase to the hydrogen ion concentration in the liquid phase and the partial pressure of NO$_2$ in the gas phase. Note that NO$_2$ does not actually exist in the liquid phase. The loading of nitrogen oxide species in the liquid phase, $\sum c_{NO_2}$, is what is meant by $C_L$ in what follows. Thus the combination of these two equations relates the gas and liquid-phase concentrations of NO$_2$, which is the objective of this relationship. It should be pointed out that this equilibrium relationship applies only at the interface and does not relate to concentrations at any other point in the liquid or gas phases.

Now the gas and liquid mass transfer rate equations and the interfacial equilibrium relationships can be combined into single calculation of the absorption flux of NO$_2$ into a freshwater lake. The specified gas and liquid concentrations are

$$c_G(10) = 20 \ \mu g \ NO_2/m^3$$

$$c_L(1) = 3 \times 10^{-2} \ gmol/L$$

The given concentration of gas-phase $NO_2$ was taken to be about 10 times higher than the background atmospheric concentration. The given liquid-phase $NO_2$ concentration for a depth of 1 m, which equals $1.4 \times 10^9 \ \mu g/m^3$, is taken as the liquid-phase concentration which would be in equilibrium with the background concentration of $NO_2$ in the atmosphere. In practice it would be the total measured $NO_2$ equivalents at a depth of 1 m. For these concentrations and the given wind speed, $U(10) = 5$ m/s, the calculations yield the interfacial flux and concentrations

$$c_G(0) = 2.31 \ \mu g/m^3$$

$$c_L(0) = 3 \times 10^{-2} \ gmol/L$$

$$N(0) = 8.7 \times 10^{-8} \ g/m^2 \cdot s$$

It should be noticed that the liquid-phase interfacial concentration is virtually identical to the liquid-phase concentration specified for 1 m. The implication of this is that:

1. The liquid-phase resistance is negligible and the gas-phase resistance controls the mass transfer process.
2. $NO_2$ is very soluble in water.

This is not surprising and an experienced engineer would have made the simplifying assumption of a perfect liquid-phase sink in the first place. Such an assumption would be quite poor for a less soluble gas. The gas-phase $NO_2$ concentration drops almost 90% from a height of 10 m to the interface. The perfect sink approximation of $c_G(0) = 0$ would imply a 10% error in the calculated flux. With the knowledge that the aqueous-phase resistance is nearly negligible for the absorption of $NO_2$, it is reasonable to compare $k_G(10) = 0.5$ cm/s to Whelpdale and Shaw's (1974) deposition velocity for $SO_2$ to water (Table 6-11), $v_g = 2.2$ cm/s. The magnitudes of $k_G(10)$ for $NO_2$ and $v_G$ for $SO_2$ are comparable and we would expect the transport of $NO_2 \left(N_2O_4\right)$ to be less than $SO_2$.

Based on these general methods and the sample calculations that have been performed, several conclusions can be drawn about our ability to predict mass transfer rates from the air to bodies of water.

1. Methods exist to predict interfacial fluxes for various species and physical conditions.
2. Mass transfer coefficients are not very species dependent.
3. Interfacial equilibrium and gas and liquid bulk concentrations are very species dependent.
4. In general, fluxes involve, in a complex but rational way, all the factors that have been discussed above.
5. Although any given predicted number contains uncertainty (and sometimes this uncertainty is large), the accuracy of the predictions will often be adequate to allow excellent approximation and problem simplification.

## 6.11.1 Mass transfer to vegetation

Long before the scientific concept of gaseous diffusion was formalized, vegetation was known to exchange large quantities of $CO_2$, $O_2$, and water vapor with the atmosphere. Leaves comprise the major plant sink for air pollutants. A cross-sectional view of a typical leaf is shown in Figure 6-19. The upper and lower surfaces of the leaf are composed of *epidermic* cells. Situated among the epidermic cells are stomatal openings, which are the major pathways for the plant to absorb gases. Guard cells line the stomatal openings. The stomata are elliptical or round vascular conduction tissues distributed within the upper, lower, or both sides of the leaf. The green *palisade* and spongy *mesophyll* cells contain the bulk of the leaf *chloroplasts* and serve as a sink for various pollutants. Because of the moist character of the mesophyll, the solubility properties of atmospheric pollutants affect the concentrations of solutes in the mesophyll and ultimately, their absorption rates.

When a polluted atmosphere comes into contact with a plant, the uptake mechanisms are both physical and chemical. Generally, the mechanism can be divided into the following steps:

1. Diffusion of pollutants from the bulk gas to the openings of the stomata
2. Further diffusion of the pollutants into the stomatal pores
3. Absorption and biochemical reaction of pollutants in the mesophyllic cells

The overall uptake rate of pollutants is the result of the steps above taking place in succession.

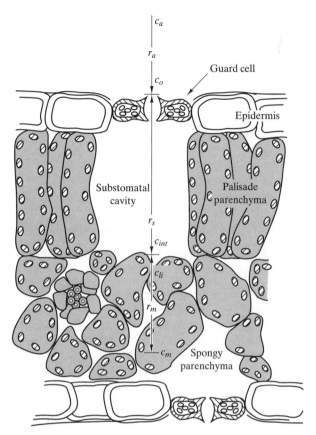

**Figure 6-19** Schematic cross-section of a dicotyledonous leaf showing gaseous pollutant concentrations and resistances (Adapted from Bennett, J H, et al., 1973)

---

**EXAMPLE 6.10   MEASUREMENT OF NO$_2$ UPTAKE BY POTATO PLANTS**

Many researchers measure pollutant uptake in 6 atch systems. However there may be advantages to using flow systems as illustrated below.

**Solution**   Sinn et al. (1984) measured the uptake rate of NO$_2$ by potato plants under laboratory conditions. Fifty-five three-week old potato plants were exposed to a continuous flow of NO$_2$ in air in a growth chamber of 5.66 m$^3$ total volume. It had been demonstrated earliar that the pot surfaces did not attract detectable amounts of NO$_2$. Further more, plastic bags were sealed over the pots at the base of the plant stems to prevent NO$_2$ absorption by the soil. Six experiments were performed. In each experiment a stable NO$_2$ concentration was established in the empty chamber and is referred to as the concentration of NO$_2$ in the

inlet line. Once the NO$_2$ concentration had equilibrated, 55 plants were transferred to the exposure chamber. Five to 10 minutes after plants were placed in the exposure chamber, a lower stable NO$_2$ concentration was recorded. At this time, the exposure to NO$_2$ for a measured period of 5-h began. At the conclusion of the 5-h exposure, all plants were removed from the chamber. With plants in the exposure chamber, the pollutant concentration was lower than in the empty chamber, indicating NO$_2$ uptake by the plants.

Nitrogen dioxide uptake rate was estimated for each experiment using the following statement of the conservation of mass in a well-mixed chamber.

$$Qc_{in} - Qc_{out} - Vr = 0$$

where

$Q$ = airflow into and out of the chamber (m$^3$/s)

$c_{in}$ = inlet concentration ($\mu$g/m$^3$ NO$_2$ in the empty chamber)

$c_{out}$ = outlet concentration ($\mu$g/m$^3$ NO$_2$ with plants in the chamber)

$V$ = chamber volume, 5.66 m$^3$

$r$ = uptake rate ($\mu$g NO$_2$/m$^3 \cdot$ s)

For the above, the uptake rate ($r$) is

$$r = \frac{Q}{V}(c_{in} - c_{out})$$

Nitrogen dioxide exposure concentrations and corresponding uptake rates are listed in the table below.

| Experiment | $c_{in}$ ($\mu$g/m$^3$) | $c_{out}$ ($\mu$g/m$^3$) | NO$_2$ uptake ($\mu$g/m$^3 \cdot$ s) |
|---|---|---|---|
| A | 380 | 228 | 0.343 |
| B | 380 | 247 | 0.299 |
| C | 475 | 304 | 0.386 |
| D | 665 | 475 | 0.428 |
| E | 950 | 703 | 0.557 |
| F | 1140 | 817 | 0.728 |

A linear increase in uptake rate accompanied the increase in NO$_2$ exposure concentration, as shown in Figure E6-19.

---

The following equation, familiar to engineers, is commonly used by plant physiologists to express the uptake rate:

$$\text{flux (kmol/s} \cdot \text{m}^2) = \frac{\text{driving force (kmol/m}^3)}{\text{resistance (s/m)}}$$

(6.45)

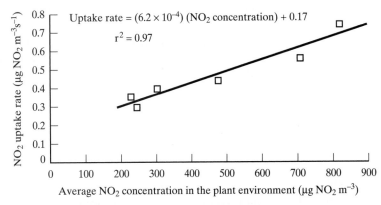

**Figure E6-19** Nitrogen dioxide uptake rate of three-week old potato plants as a function of nitrogen dioxide concentration in the plant environment (Sinn et al, 1984)

The driving force represents the concentration difference between points along the path from the air into the leaf. The resistance depends on the characteristics of each process, and one quantitative expression for the resistance of each of the above steps will have to be developed.

**Aerodynamic resistance, $r_a$.** The details and constraints of the derivation indicated below are found in O'Dell et al. (1977). The first resistance, pertaining to the transport of a gas to the surface of the leaf, is the aerodynamic resistance, $r_a$, the reciprocal to the gas-phase mass transfer coefficient, $k_G$. The aerodynamic resistance is analogous to the resistance to heat and mass transfer in a boundary layer near a flat plate. Following Pohlhausen (1921) for laminar flow across a flat plate, this equation will be taken to be

$$\frac{Lk_G}{D} = \frac{L}{Dr_a} = 0.664 Re^{1/2} Sc^{1/3} \qquad (6.46)$$

where $L$ is a characteristic length defined by the problem at hand and $D$ is the binary diffusivity of the pollutant in air. Two sets of data for water vapor transfer from a single leaf are compared to Equation 6.46 in Figure 6-20. The term $L$ is the leaf dimension in the direction of flow. The recommended correlation is seen to give reasonable representation of the data. Some experimenters have noted that at the wind speeds above about 1 m/s, the aerodynamic resistance becomes small compared to other resistances.

**Stomatal resistance, $r_s$.** The stomatal pore is represented by a narrow tube that opens into an absorptive chamber as seen in Figure 6.19. The diffusive resistance of such a pore can be estimated from mass transfer theory. The resistance of all pores must be summed in parallel over the entire leaf surface to yield a total stomatal resistance, $r_s$. The result is

$$r_s = \frac{4L_e}{D\pi abn_s} \qquad (6.47)$$

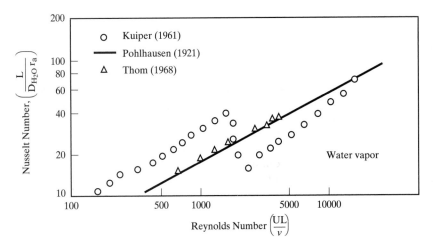

**Figure 6-20** Comparison of two workers' data on water vapor transfer from leaf models with Pohlhausen's correlation (adapted from O'Dell et al., 1977)

where

$a$ = stomatal slit length

$b$ = stomatal slit width

$n_s$ = stomatal density = total number of stomata/leaf area, both sides

$L_e$ = stomatal slit length plus the length of the major axis of the substomatal cavity

Values of $a$, $b$, and $n_s$ can be found by inspection of the leaf surface.

For a particular leaf, the stomatal population, $n_s$, effective stomatal length, $L_e$, and major axis of the stomatal slit, $a$, are constant. Hence $r_s$ is inversely proportional to the stomatal slit width $b$ and diffusivity $D$.

$$r_s \propto \frac{1}{bD} = \frac{p}{bD} \tag{6.48}$$

where $p$, the *stomatal proportionality constant*, is a function of characteristics of the leaf physiology:

$$p = \frac{4L_e}{\pi a n_s} \tag{6.49}$$

For a given leaf type, once the stomatal resistance has been determined for a single aperture size, $b$, and for a given gas with diffusivity, $D$, the stomatal constant, $p$, can be calculated. Then $r_s$ can be estimated for any aperture and any pollutant gas.

**Mesophyllic resistance, $r_m$.** After entering the substomatal cavity, the pollutant dissolves in the aqueous media of the spongy mesophyll cells and diffuses into the liquid. As the pollutant diffuses into the liquid, it also undergoes chemical reactions. A mass balance on a differential element in the mesophyll gives

$$\frac{D_m d^2 c_A}{dz^2} + k c_A = 0 \tag{6.50}$$

where $D_m$ is the effective diffusivity of pollutant $A$ in the mesophyll and $k$ is the first-order rate constant. Two boundary conditions are required to solve this differential equation. At the interface, $z = 0$, the mesophyllic concentration, $c_A$, is designated as the liquid-phase concentration $c_{li}$; thus the first boundary condition is

$$z = 0 \quad c_A = c_{li} \tag{6.51}$$

At the bottom depth of the mesophyll, $z = L_m$, the second boundary condition is $dc_A/dz = 0$. The solution of Equation 6.50, which predicts the flux of pollutant $A$ into the moist mesophyll can be found in Bird et al. (1960):

$$N_A = \frac{D_m c_{li}}{L_m} b_l \tanh b_l \tag{6.52}$$

where

$$b_l = \left( k L_m^2 D_m \right)^{1/2} \tag{6.53}$$

At the stomatal–mesophyllic interface the gas-phase and liquid-phase concentrations may be related by Henry's law, which often applies at equilibrium for very low concentrations.

$$c_i = H c_{li} \tag{6.54}$$

Working through the theory of diffusion with chemical reaction yields the following expression for the mesophyllic mass transfer resistance when the bulk average concentration in the mesophyll is significant:

$$r_m = \frac{1}{A_m n_s} \left\{ \left[ (k D_l)^{1/2} \tanh b_l \right]^{-1} - \frac{1}{k L_m} \right\} \tag{6.55}$$

If the reaction in the mesophyll is very fast, $k$ will be large and cause $1/k L_m$ to be negligible. Also, if $L_m$ is large, the bulk liquid concentration will approach zero and the term $1/k L_m$ will not appear.

**Overall mass transfer resistance, $R_{\text{leaf}}$.** The series of resistances discussed above controls the flux of gas into the leaf. To obtain an equation in terms of the ambient concentration and the bulk concentration in the mesophyll cell, we equate the fluxes through each of the resistances to obtain the following overall equation:

$$N_A = \frac{c_A - H c_m}{r_a + r_s + H r_m} \tag{6.56}$$

where $c_m$ is the bulk pollutant concentration in the mesophyllic liquid. Recall that the flux can be written in terms of an overall driving force and a total mass transfer resistance $R$ as

$$N_A = \frac{c_A - H c_m}{R_{\text{leaf}}} \tag{6.57}$$

Thus

$$R_{\text{leaf}} = r_a + r_s + H r_m \tag{6.58}$$

The equations above become even simpler for highly soluble gases (i.e., small value of $H$) such as HF, $SO_2$, $Cl_2$, and $NH_3$ because the term $Hr_m$ becomes negligibly small.

A comparison between predicted uptake and observed data for $SO_2$ over a relatively wide range of concentrations is seen in Figure 6-21. The validity of the model is demonstrated by the close agreement between the chamber data, field data, and the corresponding predictions. The sensitivity of the calculation to various values of the stomatal constant, $p$, is indicated by the parametric lines.

The foregoing pertains to mass transfer to a single leaf. This is quite a different matter from mass transfer to a collection of leaves in a *forest canopy*, a flower garden, or a grain field. In our treatment of uptake of pollutants by bodies of water, we were able to ignore the fact that wavy surfaces have more area than flat ones. For vegetative surfaces the variation of area is much larger and is accounted for by the leaf area index, $R$, which has units of square meter of leaf surface per square meter of horizontal cross section

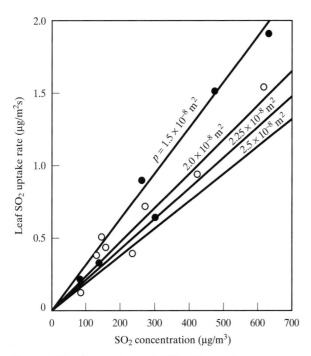

**Figure 6-21**   Comparison of Hill's $SO_2$ uptake rates measured in the field (o) and in the chamber (•) with predicted rates using four values of p (Hill A C, 1971)

of canopy. Accordingly, we estimate the mass transfer coefficient for a single leaf, $k_{leaf} = 1/R_{leaf}$, and then multiply it by the *leaf area index*, $R$, to allow for the increased surface area offered by many leaves. Values of $R$ are available in the plant physiology literature (Schulze, 1982; Pirson, 1993). Balancing this area factor somewhat is the fact that closely clustered leaves diminish the effectiveness of any one leaf to the exchange of gases with air. This phenomenon is accounted for by an effectiveness factor, β, Hence

$$k_v = \beta R k_{leaf} \qquad (6.59)$$

DeCaria (1981) analyzed a forest canopy by modeling it by analogy to a packed bed of a catalyst particles, using a method developed theoretically in chemical reaction engineering. The effectiveness factor, β, by which the flux into a single leaf should be multiplied to find the flux to an average leaf in a canopy can be obtained from

$$\beta = \frac{1}{\Phi_s} \tanh \Phi_s \qquad (6.60)$$

where

$$\Phi_s = \left( \frac{a_{v,\,avg} k_{leaf,\,avg} h_v^2}{D_{e,\,avg}} \right)^{1/2} \qquad (6.61)$$

in which

$a_{v,\,avg}$ = average leaf area density $\left(\text{m}^2 \text{ of leaf area per m}^3 \text{ of canopy}\right)$

$k_{leaf,\,avg}$ = average leaf mass transfer coefficient

$h_v$ = height of the plant canopy

$D_{e,\,avg}$ = average effective diffusivity

### EXAMPLE 6.11   MASS TRANSFER TO A LEAF AND A CANOPY OF LEAVES

We wish to make use of the foregoing principles of mass exchange with vegetation to calculate the uptake of $SO_2$ by a spruce forest, a matter of consequence in northern Europe.

***Solution***   The basis of the calculation is Equation 6.56:

$$N_A = \frac{c_A - Hc_m}{r_a + r_s + Hr_m}$$

We know that $SO_2$ is metabolized by many plants at a reasonable rate; hence $c_m$ and $r_m$ are expected to be

The flux of $CO_2$ absorption in the ocean, calculated above, may be put on a global basis for comparison with Figure 6-3 on the carbon cycle. The surface area of the oceans of the earth is $3.61 \times 10^8$ $km^2$. Thus

$$N_G(0) = (7.37 \times 10^{-6} \text{ g } CO_2/m^2 \cdot s)(10^6 \text{ m}^2/km^2)$$

$$(3.61 \times 10^8 \text{ km}^2)(3600 \text{ s/h}) \times (24 \text{ h/day})$$

$$(365 \text{ days/yr})(\text{Tg } CO_2/10^{12} \text{ g } CO_2)$$

$$(12 \text{ Tg C}/44 \text{ Tg } CO_2) = 23,000 \text{ Tg C/yr}$$

We see that this value falls reasonably between the 2000 and 8000 Tg/yr for mere absorption of $CO_2$ in the water and the massive 35,000 Tg C/yr oceanic photosynthetic loop. The fact that there is generation as well as consumption of $CO_2$ in the ocean suggests that the earlier limiting assumption of zero $CO_2$ concentration in the ocean at 10 m depth deserves reexamination. As $c_L(10)$ increases, the calculated $CO_2$ absorption flux will decrease.

The following is a summary of the results of several researchers and the calculation above. Converting the calculated flux of $7.37 \times 10^{-6}$ g $CO_2/m^2 \cdot s$ and Craig's $2.95 \times 10^{-5}$ g $CO_2/m^2 \cdot s$ and the value 2000 to 8000 Tg C/yr in Figure 6-3 into yearly fluxes of $CO_2$, one obtains

- *Brtko:* 84,000 Tg/yr
- *Craig:* 340,000 Tg/yr
- *Figure 6-3:* 7000 to 29,000 Tg/yr

Without performing a complete sensitivity analysis, it can be shown that Brtko's calculated flux could vary over the range of the values above, merely by changing $U(10)$ to other plausible values.

The environment, and especially the oceanic component, is like an amorphous, multidimensional puzzle. We try to make sense of it via a large number of small fragments of information. It will take all of the good science and engineering that we can muster to live in harmony with our surroundings.

## 6.12 Closure

There is enormous uncertainty in the material in this chapter. Knowledge of the atmosphere and processes occurring within it and at its boundaries increases both minutely and massively every day. One looks at a particular medium as a possible sink for some pollutant and discovers that it is, in fact, a source for this or even another pollutant. One spends years on a particular reaction mechanism only to discover a significant alternative pathway. The oceans represent a huge chemical inventory compared to the atmosphere. Their potential to influence air pollution is enormous and we know precious little about them. To be effective, engineers will need to be in continual close touch with the chemical, biological, physical, and applied sciences relevant to their particular interests.

Some of the information provided here, especially in connection with the global cycles of carbon, sulfur, and nitrogen and with residence times of atmospheric constituents, are necessarily averages over long periods of time. Such averaged information offers perspective on the larger picture. On the other hand, mechanistic information and rate equations apply to points in space and time. These offer the potential for the reconciliation of conflicting assumptions and objectives. In this way they comprise the basis for the solution of actual pollution problems.

It is crucial to acknowledge that the atmosphere is dynamic and to realize that pollution is inherently episodic. Atmospheric and biospheric conditions vary dramatically from summer to winter, between day and night, and with changing weather systems. Thus even with steady input of a chemical species, the impact of that species could be enormous (polluting) at one time and insignificant (benign) at another. Even if there were such a thing as a constant atmosphere, contaminant introduction is variable both spatially and temporally. Consider the fertilization of fields, production cycles at factories, rhythms of automotive use, application and later termination of use of an insecticide such as DDT, and the contrast between developed and developing societies. Thinking macroscopically in terms of averages and statistics is not enough. We must contemplate and act microscopically, quantitatively, in terms of cause and effect, and actions and impact.

# Nomenclature

| Symbol | Description (Dimensions*) |
|--------|--------------------------|
| $[j]$ | concentration of species $j$ $(M/L^3, N/L^3)$ |
| $[=]$ | has the units |
| $\rightarrow$ | direction of a chemical reaction |
| $\rightleftharpoons$ | simultaneous forward and reverse reactions |
| $=$ | empirical stoichiometric equation |
| $a$ | stomatal slit length (L) |
| $A_m$ | mesophyll surface area associated with a single stoma $(L^2)$ |
| $a_{v,\text{avg}}$ | average leaf area density |
| $b$ | stomatal slit width (L); constant defined by equation |
| $B$ | correction factor defined by equation |
| $b_1$ | constant defined by equation |
| $c_D$ | drag coefficient, defined by equation |
| $c_G(z), c_L(z')$ | pollutant molar concentrations in the gas and liquid phases at location $z$ and $z'$ $(M/L^3, N/L^3)$ |
| $c_i$ | pollutant concentration in the gas phase at the air–water interface $(M/L^3, N/L^3)$ |
| $c_{int}$ | pollutant concentration in the gas phase at the bottom of the substomatal cavity $(M/L^3, N/L^3)$ |
| $c_{li}$ | pollutant concentration in the liquid phase at the air–water interface $(M/L^3, N/L^3)$ |
| $D$ | diffusion coefficient $(L^2/t)$ |
| $D_{e,\text{avg}}$ | average effective diffusion coefficient $(L^2/t)$ |
| $D_m$ | effective diffusivity of the pollutant in the mesophyll cell $(L^2/t)$ |
| $D_p$ | particle diameter (L) |
| $g$ | acceleration of gravity $(L/t^2)$ |
| $H$ | Henry's law constant $(F/L^2)$ |
| $h_v$ | height of vegetation (L) |
| $h\nu$ | symbol denoting UV radiation |

*F, force; L, length; M, mass; N, mols; Q, energy; t, time; T, temperature.

| Symbol | Description (Dimensions*) |
|--------|--------------------------|
| $I$ | solar intensity $(Q/tL^2)$ |
| $k$ | reaction rate coefficient (units determined by the rate equation) |
| $K$ | reaction equilibrium constant |
| $k'$ | first-order rate constant for chemical reactions in mesophyll cells |
| $k_G(z), k_L(z')$ | mass transfer coefficients in the gas and liquid phases evaluated at $z$ and $z'$ $(L/t)$ |
| $k_{\text{leaf}}$ | mass transfer coefficient for a single leaf $(L/t)$ |
| $k_{\text{leaf, avg}}$ | average mass transfer coefficient for a canopy of leaves $(L/t)$ |
| $k_v$ | mass transfer coefficient for vegetation $(L/t)$ |
| $L$ | characteristic length defined by details of the application (L) |
| $L_e$ | effective pore length (L) |
| $L_m$ | depth of the mesophyll cavity (L) |
| $N$ | pollutant flux $(N/tL^2)$ |
| $N_G(0), N_L(0)$ | molar flux in the gas and liquid phases evaluated at air–liquid interface $(N/tL^2)$ |
| $n_s$ | stomatal density, number of stomata per area of leaf, front and back $(L^{-2})$ |
| $p$ | stomatal constant defined by equation |
| $P$ | total pressure $(F/L^2)$ |
| $P_v$ | partial pressure of water vapor $(F/L^2)$ |
| $Q$ | volumetric flow rate $(L^3/t)$ |
| $r$ | uptake rate by vegetation $(M/L^3t)$ |
| $(-r), r$ | rate of reaction, formation $(N/tL^3)$ |
| $R$ | leaf area index; overall mass transfer resistance, defined by equation $(L/t)$ |
| $r_a$ | aerodynamic mass transfer resistance $(L/t)$ |
| $R_{\text{leaf}}$ | overall mass transfer resistance for a single leaf $(L/t)$ |
| $r_m$ | mesophyllic mass transfer resistance $(L/t)$ |
| $r_p$ | mass transfer resistance through the pores of a leaf $(L/t)$ |

| | | | |
|---|---|---|---|
| $r_s$ | stomatal mass transfer resistance (L/t) | $\Delta_h$ | change in sea level (L) |
| $R_u$ | universal gas constant (Q/NT) | $\beta$ | constant defined by equation |
| $t$ | time $(t)$ | $\beta$ | effectiveness factor for plant canopy |
| $T$ | temperature (T) | $\lambda$ | radiation wavelength (L) |
| $U_*$ | friction velocity, defined by equation (L/t) | $\mu$ | viscosity $(Ft/L^2)$ |
| $U(z)$ | gas-phase velocity profile, air velocity at elevation $z$ (L/t) | $\nu$ | kinematic viscosity $(L^2/t)$ |
| $V$ | volume $(L^3)$ | $\rho$ | density $(M/L^2)$ |
| $y_i$ | mol fraction of molecular species $i$ | $\tau_0$ | air shear stress at air–water interface $(F/L^2)$ |
| $z, z'$ | vertical distance with respect to the air–water interface (L) | $\Phi_s$ | modulus for effectiveness factor, defined by equation |
| $z_0$ | roughness height, defined by equation or table (L) | | |

*Greek*

| | | |
|---|---|---|
| $\alpha$ | constant defined by equation | |

*Abbreviations*

| | |
|---|---|
| Re | Reynolds number, defined by equation |
| Sc | Schmidt number $(\mu/\rho D = \nu/D)$ |

# References

Aldaz, L., 1969. Flux measurements of atmospheric ozone over land and water, *Journal of Geophysical Research*, Vol. 74, pp. 6943–6946.

Anon., 1994. untitled, *EPRI Journal*.

Asher, W. E., and Pankow, J. F., 1991. Prediction of gas/water mass transport coefficients by surface renewal model, *Environmental Science and Technology*, Vol. 25, No. 7, pp. 1294–1300.

Atkinson, R., 1990. Gas-phase tropospheric chemistry of organic compounds: a review, *Atmospheric Environment*, Vol. 24A, No. 1, pp. 1–41.

Atkinson, R., Baulch, D. L., Cox, R. A., Hampson, R. F., Jr., Kerr, J. A., and Troe, J., 1992. Evaluated kinetic and photochemical data for atmospheric chemistry: supplement IV, *Atmospheric Environment*, Vol. 26A, No. 7, pp. 1187–1230.

Berger, R., 1990. The carcinogenicity of radon, *Environmental Science and Technology*, Vol. 24, No. 1, pp. 30–31.

Billings, C. E., and Matson, W. R., 1972. Mercury emissions from coal combustion, *Science*, Vol. 176, pp. 1232–1233.

Bird, R. B., Stewart, W. E., and Lightfoot, E. N., 1960. *Transport Phenomena*, Wiley, New York.

Bojkov, R. D., 1995. *The Changing Ozone Layer*, joint publication of the World Meteorological Organization and the United Nations Environment Programme, 26 pp.

Bolin, B., 1979. On the role of the atmosphere in biogeochemical cycles, *Quarterly Journal of the Royal Meteorological Society*, Vol. 105, pp. 25–42.

Bolin, B., Degens, E. T., Kempe, S., and Ketner. P. (Eds.), 1979. *The Global Carbon Cycle*, SCOPE Report 13, Wiley, New York, pp. 101–128.

Bennett, J.H., Hill, A. C., and Gates, D.M., 1973. A model for gaseous pollutant sorption by leaves, *Journal of the Air Pollution Control Assoc.*, Vol. 23, No. 11, pp. 957–962.

Brook, E. J., Sowers, T., and Orchardo, J., 1996. Rapid variations in atmospheric methane concentration during the past 110,000 years, *Science*, Vol. 273, pp. 1087–1090.

Brtko, W. J., 1976. Mass transfer at natural air–water interfaces, M.S. thesis, Pennsylvania State University, 135 pp.

Brtko, W. J., and Kabel, R. L., 1978. Transfer of gases at natural air–water interfaces, *Journal of Physical Oceanography*, Vol. 8, pp. 543–556.

Carter, W. P. L., 1990. A detailed mechanism for the gas-phase atmospheric reactions of organic compounds, *Atmospheric Environment*, Vol. 24A, pp. 481–518.

Charnock, H., 1955. Wind stress on a water surface, *Quarterly Journal of the Royal Meteorological Society*, Vol. 81, pp. 639–640.

Chilton, T. H., and Colburn, A. P., 1934. Mass transfer (absorption) coefficients, prediction from data in heat transfer and fluid friction, *Industrial and Engineering Chemistry*, Vol. 26, pp. 1183–1187.

Cicerone, R. J., 1987. Changes in stratospheric ozone, *Science*, Vol. 237, pp. 35–42.

Coles, D. G., Ragaini, R. C., Ondov, J. M., Fisher, G. L., Silberman, D., and Prentice, B. A., 1979. Chemical studies of stack fly ash from a coal-fired power plant, *Environmental Science and Technology*, Vol. 13, No. 4, pp. 455–459.

Craig, H., 1957. The natural distribution of radiocarbon and the exchange time of carbon dioxide between atmosphere and sea, *Tellus*, Vol. 9, No. 1.

Czepiel, P., Douglas, E., Harris, R., and Crill, P., 1996. Measurements of N$_2$O from composed organic wastes, *Environmental Science and Technology*, Vol. 30, No. 8, pp. 2519–2525.

Danckwerts, P. V., 1951. Significance of liquid-film coefficients in gas absorption, *Industrial Engineering Chemistry*, Vol. 43, pp. 1460–1467.

DeCaria, A. J., 1981. Aircraft measurements and mathematical modeling of the removal of gaseous air pollutants at natural air–land and air–water interfaces, Ph.D. thesis, Pennsylvania State University, pp. 444.

Dechaux, J. C., Zimmermann, V., and Nollet, V., 1994. Sensitivity analysis of the requirements of rate coefficients for the operational models of photochemical oxidant formation in the troposphere, *Atmospheric Environment*, Vol. 28, No. 2, pp. 195–211.

Dimitriades, B., 1981. The role of natural organics in photochemical air pollution, *Journal of the Air Pollution Control Association*, Vol. 31, No. 3, pp. 229–235.

Dimitriades, B., 1996. Scientific basis for the VOC reactivity issues raised by Section 183(e) of the Clean Air Act Amendments of 1990, *Journal of the Air & Waste Management Association*, Part C, Vol. 46, pp. 963–970.

Dipprey, D. F., and Sabersky, R. H., 1963. Heat and momentum transfer in smooth and rough tubes at various Prandtl numbers, *International Journal of Heat and Mass Transfer*, Vol. 6, pp. 329–353.

Elsom, D. M., 1992. *Atmospheric Pollution: A Global Problem*, 2nd ed., Blackwell, Oxford.

Faust, B. C., 1994. Photochemisry of clouds, fogs and aerosols, *Environmental Science and Technology*, Vol. 28, No. 5, pp. 217A–222A.

Fennelly, P. F., 1976. The origin and influence of airborne particulates, *American Scientist*, Vol. 64, January–February, 1976, pp. 46–56.

Finlayson-Pitts, B. J., and Pitts, J. N., 1986. *Atmospheric Chemistry: Fundamentals and Experimental Techniques*, Wiley-Interscience, New York.

Finlayson-Pitts, B. J., and Pitts, J. N., 1993. Atmospheric chemistry of tropospheric ozone formation: scientific and regulatory implications, *Journal of the Air & Waste Management Association*, Vol. 43, pp. 1091–1100.

Firestone, M. K., Firestone, R. B., and Tiedje, J. M., 1980. Nitrous oxide from soil denitrification: factors controlling its biological production, *Science*, Vol. 208, pp. 749–750.

Flagan, R. C., and Seinfeld, J. H., 1988. *Fundamentals of Air Pollution Engineering*, Prentice Hall, Upper Saddle River, NJ.

Fortescue, G. E., and Pearson, J. R. A., 1967. On gas absorption into a turbulent liquid, *Chemical Engineering Science*, Vol. 22, pp. 1163–1175.

Fowler, D., 1978. Dry deposition of SO$_2$ on agricultural crops, *Atmospheric Environment*, Vol. 12, pp. 369–373.

Freemantle, M., 1955. The acid test for Europe, *Chemical and Engineering News*, May 1.

Freyer, H. D., Atmospheric cycles of trace gases containing carbon, in *The Global Carbon Cycle*, SCORE Report 13, Bolin, B., Degens, E.T., Kempe, S., and Ketner, P. (Eds.), Wiley, New York.

Garland, J. A., Clough, W. S., and Fowler, D., 1973. Deposition of sulphur dioxide on grass, *Nature (London)*, Vol. 242, pp. 256–257.

Gribble, G. W., 1994. The natural production of chlorinated compounds, *Environmental Science and Technology*, Vol. 28, No. 7, pp. 310A–319A.

Grosjean, D., Williams, E. L., Grosjean, E., Andino, J. M., and Seinfeld, J. H., 1993. Atmospheric oxidation of biogenic hydrocarbons: reaction of ozone with β–pinene, D-limonene and *trans*-caryophyllene, *Environmental Science and Technology*, Vol. 27, No. 13, pp. 2754–2758.

Guenther, A., Zimmerman, P., and Wildermuth, M., 1994. Natural volatile organic compound emission rate estimates for U.S. woodland landscapes, *Atmospheric Environment*, Vol. 28, No. 6, pp. 1197–1210.

Gundel, L. A., Benner, W. H., and Hansen, A. D. A., 1994. chemical composition of fog water and interstitial aerosol in Berkeley, California, *Atmospheric Environment*, Vol. 28, No. 16, pp. 2715–2725.

Hess, G. D., Carnovale, F., Cope, M. E., and Johnson, G. M., 1992. The evaluation of some photochemical smog reaction mechanisms. I. Temperature and initial composition effects. II. Initial addition of alkanes and alkenes. III. Dilution emissions effects, *Atmospheric Environment*, Vol. 26A, No. 4, pp. 625–659.

Hicks, B. B., 1973. *The Dependence of Bulk Transfer Coefficients upon Prevailing Meteorological Conditions*, Radiological and Environmental Research Annual Report ANL-8060, Part IV, Argonne National Laboratory, Argonne, Ill.

Hill, A. C., 1971. Vegetation: a sink for atmospheric pollutants, *Journal of the Air Pollution Control Association*, Vol. 21, pp. 341–346.

Holland, H. D., 1978. *The Chemistry of the Atmosphere and Oceans*, Wiley, New York, pp. 259–283.

Husted, S., 1993. An open chamber technique for determination of methane emission from stored livestock manure, *Atmospheric Environment*, Vol. 27A, No. 11, pp. 1635–1642.

Hutchinson, G. L., and Viets, F. G. Jr., 1969. Nitrogen enrichment of surface water by absorption of ammonia volatilized from cattle feedlots, *Science*, Vol. 166, pp. 514–515.

Jeffree, C. E., Johnson, R. P. C., and Jarvis, P. G., 1971. Epicuticular wax in the stomatal antechamber of sitka spruce and its effect on the diffusion of water vapor and carbon dioxide, *Planta (Berlin)*, Vol. 98, pp. 1–10.

Kabel, R. L., 1975. Atmospheric impact on nutrient budgets, *Proceedings of the First Speciality Symposium on Atmospheric Contributions to the Chemistry of Lake Waters*, International Association of Great Lakes Reserach, September 28–October 1, pp. 114–126.

Kabel, R. L., 1979. Natural removal of gaseous pollutants, in *Advances in Environmental Science and Engineering*, Pfafflin, J. R., and Ziegler, E. N., Jr. (Eds.), Gordon and Breach, New York, pp. 86–102.

Khalil, M. A. K., and Rasmussen, R. A., 1984. Carbon monoxide in the earth's atmosphere: increasing trend, *Science*, Vol. 224, pp. 54–56.

Khalil, M. A. K., Rasmussen, R. A., Wang, M. X., and Ren, L., 1991. Methane emissions from rice fields in China, *Environmental Science and Technology*, Vol. 25, No. 5, pp. 979–981.

Kokoszka, L. C., and Kabel, R. L., 1986. A review of dry gaseous deposition and its relation to the biogeochemical cycles of carbon, nitrogen and sulfur, in Vol. 5, *Advances in Environmental Science and Engineering*, Pfafflin, J. R., and Ziegler, E. N. (Eds.), Gordon and Breach, Montreux, Switzerland, pp. 40–93.

Kuiper, P. J. C., 1961. The effects of environmental factors on the transpiration of leaves with special reference to stomatal light response, *Mededelingen Landbouwhogeschool Wageningen*, Vol. 61, No. 7, p. 1.

Lal, S., Venkataramani, V., and Subbaraya, B. H., 1993. Methane flux measurements from paddy fields in the tropical Indian region, *Atmospheric Environment*, Vol. 27A, No. 11, pp. 1691–1694.

Lamont, J. C., and Scott, D. S., 1970. An eddy cell model of mass transfer into the surface of a turbulent liquid, *AIChE Journal*, Vol. 16, pp. 513–519.

Landsberg, J. J., and Jarvis, P. G., 1973. A numerical investigation of the momentum balance on a spruce forest, *Journal of Applied Ecology*, Vol. 10, pp. 645–655.

Leighton, P. A., 1961. *Photochemistry of Air Pollution*, Academic Press, San Diego, CA.

Linton, R. W., Loh, A., Natusch, D. F. S., Evans, C. A., Jr., and Williams, P., 1976. Surface predominance of trace elements in airborne particles, *Science*, Vol. 191, pp. 852–854.

MacDonald, R. C., and Fall, R., 1993. Detection of substantial emissions of methanol from plants to the atmosphere, *Atmospheric Environment*, Vol. 27A, No. 11, pp. 1709–1713.

Marcinowski, F., and Napolitano, S., 1993. Reducing the risks from radon, *Journal of the Air and Waste Management Association*, Vol. 43, pp. 955–962.

McBride, J. P., Moore, R. E., Witherspoon, J. P., and Blanco, R. E., 1978. Radiological impact of airborne effluents of coal and nuclear plants, *Science*, Vol. 202, pp. 1045–1050.

McMahon, T. A., and Denison, P. J., 1979. Empirical atmospheric deposition parameters: a survey, *Atmospheric Environment*, Vol. 13, pp. 571–585.

McRae, G. J., Goodin, W. R., and Seinfeld, J. H., 1982. Development of a second-generation mathematical model for urban air pollution. I. Model formulation, *Atmospheric Environment*, Vol. 16, No. 4, pp. 679–696.

McRae, G. J., Goodin, W. R., and Seinfeld, J. H., 1983. Development of a second-generation mathematical model for urban air pollution. II. Evaluation of model performance, *Atmospheric Environment*, Vol. 17, No. 3, pp. 501–522.

Miller, A., and Thompson, J. C., 1970, *Elements of Meteorology*, Charles E. Merrill, Columbus, OH.

Moss, M. R., 1978. In *Sulfur in the Environment*, Nriagu, J. O. (Ed.), Vol. 1, Wiley, New York, pp. 23–50.

Mueller, J. A., and DiToro, D. M., 1993. Multicomponent adsorption of volatile organic chemicals from air stripper offgas, *Water Environment Research*, Vol. 65, No. 1, pp. 15–25.

Natusch, D. F. S., Wallace, J. R., and Evans, C. A. Jr., 1974. Toxic trace elements: preferential concentration in respirable particles, *Science*, Vol. 183, pp. 202–204.

Nazaroff, W. W., and Teichman, K., 1990. Indoor radon, *Environmental Science and Technology*, Vol. 24, No. 6, pp. 774–782.

Nero, A. V., Schwehr, M. B., Nazaroff, W. W., and Revzan, K. L., 1986. Distribution of airborne radon-22 concentrations in U.S. homes, *Science*, Vol. 234, pp. 992–997.

Novakov, T., Chang, S. G., and Harker, A. B., 1974. Sulfates as pollution particulates: catalytic formation on carbon (soot) particles, *Science*, Vol. 186, pp. 259–261.

O'Dell, R. A., Taheri, M., and Kabel, R. L., 1977. A model for uptake of pollutants by vegetation, *Journal of the Air Pollution Control Association*, Vol. 27, No. 11, pp. 1104–1116.

Owen, P. R., and Thomson, W. R., 1963. Heat transfer across rough surfaces, *Journal of Fluid Mechanics*, Vol. 15, pp. 321–334.

Owers, M. J., and Powell, A. W., 1974. Deposition velocity of sulfur dioxide on land and water surfaces using 35S tracer method, *Atmospheric Environment*, Vol. 8, pp. 63–67.

Panofsky, H. A., 1963. Determination of stress from wind and temperature measurements, *Quarterly Journal of the Royal Meteorological Society*, Vol. 89, pp. 85–94.

Perry, R. H., and Green, D. W., and Maloney, J. O., 1984. *Chemical Engineers' Handbook*, 6th ed., McGraw-Hill, New York.

Pirson, A. (Ed.), 1993. General index, *Encyclopedia of Plant Physiology*, Vol. 20, Springer-Verlag, Berlin.

Pohlhausen, E., 1921. Der Wärmeaustausch zwischen festen Körpern und Flüssigkeiten mit kleiner Reibung und kleiner Wärmeleitung, *Zeitschrift Angewandte Mathematik und Mechanik*, Vol. 1, p. 115.

Porstendorfer, J., 1994. Properties and behavior of radon and thoron and their decay products in the air, *Journal of Aerosol Science*, Vol. 25, No. 2, pp. 219–263.

Post, W. M., Peng, T. H., Emanuel, W. R., King, A. W., Dale, V. H., and DeAngelis, D. L., 1990. The global carbon cycle, *American Scientist*, Vol. 78, July-August, pp. 310–326.

Rasmussen, K. H., Taheri, M., and Kabel, R. L., 1974a. *Sources and Natural Removal Processes for Some Atmospheric Pollutants*, EPA Grant Report EPA–650/4-74-032.

Rasmussen, K. H., Taheri, M., and Kabel, R. L., 1974b. Global emissions and natural processes for removal of gaseous pollutants, *Water, Air, and Soil Pollution*, Vol. 4, pp. 33–64.

Raynaud, D., Jouzel, J., Barnola, J. M., Chappellaz, J., Delmas, R. J., and Lorius, C., 1993. The ice record of greenhouse gases, *Science*, Vol. 259, pp. 926–934.

Rivkin, R. B., Legendre, L., Deibel, D., Tremblay, J. E., Klein, B., Crocker, K., Roy, S., Silverberg, N., Lovejoy, C., Mesple, F., Romero, N., Anderson, M. R., Matthews, P., Savenkoff, C., Vezina, A., Therriault, P., Wesson, J., Berube, C., and Ingram, R. G., 1996. Vertical flux of biogenic carbon in the ocean: is there food web control? *Science*, Vol. 272, pp. 1163–1166.

Roberts, J. M., 1990. The atmospheric chemistry of organic nitrates, *Atmospheric Environment*, Vol. 24A, No. 2, pp. 243–287.

Robertson, D. E., Crecelius, W. A., Fruchter, J. S., and Ludwick, J. D., 1997. Mercury emissions from geothermal power plants, *Science*, Vol. 196, pp. 1094–1097.

Russell, A. G., Cass, G., and Seinfeld, J. H., 1986. On some aspects of nighttime atmospheric chemistry, *Environmental Science and Technology*, Vol. 20, No. 11, pp. 1167–1172.

Russell, A., Milford, J., Bergin, M. S., McBride, S., McNair, L., Yang, Y., Stockwell, W. R., and Croes, B., 1955. Urban ozone control and atmospheric reactivity of organic gases, *Science*, Vol. 269, pp. 491–496.

Sakugawa, H., and Kaplan, I. R., 1993. Comparison of $H_2O_2$ and $O_3$ content in atmospheric samples in the San Bernardino mountains, southern California, *Atmospheric Environment*, Vol. 27A, No. 9, pp. 1509–1515.

Sakugawa, H., Kaplan, I. R., Tsai, W., and Cohen, Y., 1990. Atmospheric hydrogen peroxide, *Environmental Science and Technology*, Vol. 24, No. 1, pp. 1452–1462.

Sarmiento, J. L., and LeQuere, C., 1996. Oceanic carbon dioxide uptake in a model of century-scale global warming, *Science*, Vol. 274, pp. 1346–1350.

Schneider, S. H., 1989. The greenhouse effect: science and public policy, *Science*, Vol. 243, February 10, pp. 771–781.

Schulze, E.-D., 1982. Plant life forms and their carbon, water, and nutrient relations, in *Encyclopedia of Plant Physiology, Physiological Plant Ecology II,* Vol. 12B, *Water Relations and Carbon Assimilation,* pp. 615–676, Springer-Verlag, Berlin.

Seinfeld, J. H., 1986. *Atmospheric Chemistry and Physics of Air Pollution*, Wiley, New York.

Seinfeld, J. H., 1989. Urban air pollution: state of the science, *Science*, Vol. 243, February 10, pp. 745–752.

Shinn, J. H., and Lynn, S., 1979. Do man-made sources affect the sulfur cycle of northeastern states? *Environmental Science and Technology*, Vol. 13, No. 9, pp. 1062–1067.

Sinn, J. P., Pell, E. J., and Kabel, R. L., 1984. Uptake rate of $NO_2$ by potato plants, *Journal of the Air Pollution Control Association*, Vol. 34, No. 6, pp. 668–689.

Sutton, O. G., 1953. *Micrometeorology*, McGraw-Hill, New York.

Takahashi, T., 1961. Carbon dioxide in the atmosphere and in atlantic ocean waters, *Journal of Geophysical Research*, Vol. 66, p. 477.

Tanner, R. L., and Zielinska, B., 1994. Determination of the biogenic emission rates of species contributing to VOC in the San Joaquin Valley of California, *Atmospheric Environment*, Vol. 28, No. 6, pp. 1113–1120.

Thom, A. S., 1968. The exchange of momentum, mass and heat between an artificial leaf and the airflow in a wind-tunnel, *Quarterly Journal of the Royal Meterological Society*, Vol. 94, p. 44.

van Dop, H., Guichiert, R., and Lanting, R. W., 1977. Some measurements of the vertical distribution of ozone in the atmospheric layer, *Atmospheric Environment*, Vol. 11, pp. 65–71.

Venkataramani, V., and Subbaraya, V., 1993. Methane flux measurements from paddy fields in a tropical region, *Atmospheric Environment*, Vol. 27A, No. 11, pp. 1691–1693.

Wahlen, M., Tanaka, N., Henry, R., Deck, B., Zeglen, J., Vogel, J. S., Southon, J., Shemesh, A., Fairbanks, R., and Broecker, W., 1989. Carbon-14 in methane sources and in atmospheric methane: the contribution from fossil carbon, *Science*, Vol. 245, pp. 286–245.

Wallington, T. J., Schneider, W. F., Worsnop, D. R., Nielsen, O. J., Sehested, Debruyn, W. J., and Shorter, J. A., 1994. The environmental impact of CFC replacements: HFC's and HCFC's, *Environmental Science and Technology*, Vol. 28, No. 7, pp. 320A–326A.

Whelpdale, D. M., and Shaw, R. W., 1974. Sulfur dioxide removal by turbulent transfer over grass, snow and water surfaces, *Tellus*, Vol. 26, pp. 195–204.

Whitten, G. Z., Hogo, H., and Killus, J. P., 1980. The carbon-bond mechanism: a condensed kinetic mechanism for photochemistry, *Environmental Science and Technology*, Vol. 14, No. 6, pp. 690–700.

Williamson, S. J., 1973. *Fundamentals of Air Pollution*, Addison-Wesley, Reading, MA.

Woodwell, G. M., Hobbie, J. E., Houghton, R. A., Melillo, J. M., Moore, B., Peterson, B. J., and Shaver, G. R., 1983. Global deforestation: contribution to atmospheric carbon dioxide, *Science*, Vol. 222, pp. 1081–1086.

Zurer, P., 1995a. Global monitoring shows ozone treaty is working, *Chemical Engineering News*, July 17, p. 7.

Zurer, P., 1995b. Chemistry Nobel prizes: three win for ozone depletion research, *Chemical Engineering News*, October 16, pp. 4–5.

Zurer, P., 1996. Volcanic aerosols accelerate loss, *Chemical Engineering News*, April 8, p. 7.

# Problems

**6.1.** The concentrations in Table 6-2 have units of $\mu g/m^3$. Use the molecular weights and ideal gas law to convert the concentrations to parts per million (ppm).

**6.2.** Use the ideal gas law and the appropriate pressure and temperature to calculate the air density at an altitude of 80 km. Compare your answer to Figure 6-1.

**6.3.** Identify the principal anthropogenic and natural sources of the following chemical species. If known, compute the ratio of total anthropogenic emission rate to total natural emission rate for these pollutants.

  **(a)** CO
  **(b)** $CO_2$
  **(c)** $NO_x$
  **(d)** $SO_2$
  **(e)** Radon
  **(f)** Photochemical oxidants

**6.4.** Write the major chemical reactions (and/or physical processes) that affect the following gases in the atmosphere.

  **(a)** CO
  **(b)** $H_2S$
  **(c)** $CH_4$
  **(d)** $SO_2$
  **(e)** $CO_2$
  **(f)** $NO_2/NO$
  **(g)** $NH_3$
  **(h)** VOCs.

Interpret your results in terms of observed atmospheric residence time.

**6.5.** Prepare short answers to the following questions.

  **(a)** Describe the formation and fate of CO.
  **(b)** What reactions involving CO occur in the atmosphere?
  **(c)** Describe the formation and fate of $CH_4$.
  **(d)** What reactions involving $CH_4$ occur in the atmosphere?
  **(e)** Describe the formation and fate of organic compounds (generally).
  **(f)** How will an increase in the temperature of land and water affect the transfer of $CO_2$?
  **(g)** What reactions in the atmosphere transform $SO_2$ to sulfate aerosols?
  **(h)** What are the principal anthropogenic and biogenic sulfur compounds?
  **(i)** What are the sources of atmospheric $NH_3$, $N_2O$, NO, $NO_2$, and higher-order oxides?
  **(j)** What nitrogen oxide exists and reacts in the stratosphere?
  **(k)** What nitrogen oxides exist and react in the troposphere?
  **(l)** Describe in kinetic terms the transformation of $NO_x$ to aerosol nitrates.

**6.6.** Rank the following types of particles in terms of increasing diameter

- Pollen
- Tobacco smoke
- Beach sand
- Insecticide dust
- $(H_2O)$ mist
- Viruses
- Human hair

**6.7.** Examine the carbon cycle (Figure 6-3).

**(a)** Draw a diagram that isolates atmospheric $CO_2$ and identify all the sources of this $CO_2$ and the amount that each contributes (Tg/yr). Identify all the sinks for atmospheric $CO_2$ and the amount that each removes (Tg/yr). Rank each of the sources and sinks by orders of magnitude (powers of 10). Do the total of the sources equal the total of the sinks, and if not, explain why.

**(b)** Draw a diagram that isolates $CO_2$ dissolved in the ocean and identify all the sources of this $CO_2$ and the amount that each contributes (Tg/yr). Identify all the sinks for this dissolved $CO_2$ and the amount that each removes (Tg/yr). Rank each of the sources and sinks by orders of magnitude (powers of 10). Do the total of the sources equal the total of the sinks? If not, explain why.

**6.8.** Examine the carbon cycle (Figure 6-3). Draw a diagram that isolates atmospheric CO and identify all the sources of this CO and the amount that each contributes (Tg/yr). Identify all the sinks for atmospheric CO and the amount that each removes (Tg/yr). Rank each of the sources and sinks by orders of magnitude (powers of 10). Do the total of the sources equal the total of the sinks, and if not, explain why. Do the oceans contain a large inventory of CO? Explain.

**6.9.** Examine the carbon cycle (Figure 6-3). Draw a diagram that isolates atmospheric $CH_4$ and identify all the sources of this $CH_4$ and the amount that each contributes (Tg/yr). Identify all the sinks for atmospheric $CH_4$ and the amounts that each remove (Tg/yr). Rank each of the sources and sinks by orders of magnitude (powers of 10). Do the total of the sources equal the total of the sinks? If not, explain why. Do the oceans contain a large inventory of $CH_4$? Explain.

**6.10.** Examine the sulfur cycle (Figure 6-4).

**(a)** Draw a diagram that isolates atmospheric $SO_2/SO_4^{2-}$ and identify all the sources of this $SO_2/SO_4^{2-}$ and the amount that each contributes (Tg/yr). Identify all the sinks for atmospheric $SO_2/SO_4^{2-}$ and the amount that each removes (Tg/yr). Rank each of the sources and sinks by orders of magnitude (powers of 10). Do the total of the sources equal the total of the sinks, and if not, explain why.

**(b)** Draw a diagram that isolates $SO_2/SO_4^{2-}$ dissolved in the ocean and identify all the sources of this dissolved $SO_2/SO_4^{2-}$ and the amount each contributes (Tg/yr). Identify all the sinks for the dissolved $SO_2/SO_4^{2-}$ and the amount that each contributes (Tg/yr). Rank each of the sources and sinks by orders of magnitude (powers of 10). Do the total of the sources equal the total of the sinks? If not, explain why.

**6.11.** Examine the sulfur cycle (Figure 6-4). Draw a diagram that isolates atmospheric $H_2S$ and identify all the sources of this $H_2S$ and the amount that each contributes (Tg/yr). Identify all the sinks for atmospheric $H_2S$ and the amount that each removes (Tg/yr). Rank each of the sources and sinks by orders of magnitude (powers of 10). Do the total of the sources equal the total of the sinks, and if not, explain why. Do the oceans contain a large amount of $H_2S$? Explain.

**6.12** Compare the $NO_2$ uptake calculated in Example 6.9 with that shown in Table 6-4. Should these values be equal? Why? Why not?

**6.13.** Apply the $NO_2$ uptake calculated in Example 6.9 to the 70% of the earth's surface covered by water to determine a global uptake rate in kg/yr. Compare this calculated value to $NO_2$ emission rates given in Table 6-2 and to the corresponding $NO_x$ deposition rate given in Figure 6-5. What do you conclude? Do such comparisons have any value? If so, what is it?

**6.14.** Examine the nitrogen cycle (Figure 6-5).

**(a)** Draw a diagram that isolates atmospheric $NO_x/NO_3^-$ and identify all the sources of this $NO_x/NO_3^-$ and the amount that each contributes (Tg/yr). Identify all the sinks for atmospheric $NO_x/NO_3^-$ and the amount each removes (Tg/yr). Rank each of the sources and sinks by orders of magnitude (powers of 10). Do the total of the sources equal the total of the sinks, and if not, explain why.

**(b)** Draw a diagram that isolates the oxides of nitrogen dissolved in the ocean and identify all the sources of dissolved nitrogen oxides and the amount that each contributes (Tg/yr). Identify all the sinks for dissolved nitrogen oxides and the amount that each removes (Tg/yr). Rank each of the sources and sinks by orders of magnitude (powers of 10). Do the total of the sources equal the total of the sinks? If not, explain.

**6.15.** A crude estimate of the OH· concentration in the atmosphere can be obtained from knowing the half life of $CH_4$ in the atmosphere. Assume that the OH· concentration is constant, that the half-life of $CH_4$ is equal to the residence time listed in Table 6-1, and that $CH_4$ reacts with OH· by Equation 6-4.

$$CH_4 + OH \cdot \rightarrow CH_3 \cdot + H_2O$$

$$\frac{d[CH_4]}{dt} = -k[CH_4][OH\cdot]$$

$$k(m^3 \, gmol^{-1} s^{-1}) = 3.5 \times 10^8 \exp\left(\frac{-4500}{T}\right)$$

Estimate the OH· concentration at 300 K and compare it with the value shown in Table 6-1.

**6.16.** Consider the thermal decomposition of peroxyacetylnitrate (PAN), $CH_3(O)O_2NO_2$, described by Equation (12) in Table 6-10. Estimate the half-life $(t_{1/2})$ of PAN if Equation (12) was the only reaction affecting its concentration. Considering the entire reaction mechanism in Table 6-10, is PAN a species that can be expected to persist for a reasonable period of time?

**6.17.** Consider the formation of $O_3$ given by Equation (2) in Table 6-10. How quickly is $O_3$ formed once $NO_2$ undergoes photolysis? Specifically estimate the half-life of $O\cdot$ if the pseudo rate constant $k'$ for the following reaction is $1.5 \times 10^{-14}$ cm$^3$ molecule$^{-1}$ s$^{-1}$

$$\frac{d[O\cdot]}{dt} = -k[M][O\cdot][O_2]$$

$$= -k'[O\cdot][O_2]$$

The oxygen concentration can be obtained from Table 6-1. Using a pseudo rate constant $k'$ that incorporates the concentration of a neutral species M is a practice often encountered in the chemical kinetics of air pollution.

**6.18.** Consider the CFC stratospheric reactions in Figure 6-14. Show that the mechanism is capable of sustaining a steady-state concentration of ClO at the expense of ozone and CFCs.

**6.19.** Consider the half-lives of two species found in an urban atmosphere.

**(a)** Estimate the half-life of PAN $[CH_3C(O)O_2NO_2]$ in the atmosphere if it disappears by a first-order thermal decay reaction in which the kinetic rate constant $(k)$ at STP is equal to $3.6 \times 10^{-4}$ s$^{-1}$

$$CH_3C(O)O_2NO_2 \rightarrow CH_3(O)O_2\cdot + NO_2$$

**(b)** Estimate the half-life of $CH_4$ in the atmosphere if the main removal mechanism is the following reaction with $OH\cdot$:

$$CH_4 + OH\cdot \rightarrow CH_3\cdot + H_2O$$

At STP, the second-order kinetic rate constant $(k)$ is equal to $8.4 \times 10^{-15}$ cm$^3$ molecule$^{-1}$ s$^{-1}$ and the $OH\cdot$ concentration is equal to $5 \times 10^5$ molecules/cm$^3$.

**6.20.** A major organic pollutant produced by the combustion of hydrocarbons is formaldehyde (HCHO). Formaldehyde is photolytic, reacts with $OH\cdot$, and generates photolytic reactions that occur in parallel with the Leighton mechanism (Problem 6.15). Seinfeld (1986) reports the following reactions and rate constants:

**(1)** $NO_2 + I(h\nu) \rightarrow NO + O\cdot$
$k_1 = 0.533$ min$^{-1}$

**(2)** $O\cdot + O_2 + M \rightarrow O_3 + M$
$k_2 = 6 \times 10^{-34} (T/300)^{-2.3}$ cm$^6$ molecule$^{-2}$ s$^{-1}$

**(3)** $O_3 + NO \rightarrow NO_2 + O_2$
$k_3 = 2.2 \times 10^{-12}$ exp $(-1430/T)$ cm$^3$ molecule$^{-1}$ s$^{-1}$

**(4a)** $HCHO + I(h\nu) \rightarrow 2HO_2\cdot + CO$
$k_{4a} = 1.6 \times 10^{-3}$ min$^{-1}$

**(4b)** $HCHO + I(h\nu) \rightarrow H_2 + CO$
$k_{4b} = 2.11 \times 10^{-3}$ min$^{-1}$

**(5)** $HCHO + OH\cdot \rightarrow HO_2\cdot + CO + H_2O$
$k_5 = 1.1 \times 10^{-11}$ cm$^3$ molecule$^{-1}$ s$^{-1}$

**(6)** $HO_2\cdot + NO \rightarrow NO_2 + OH\cdot$
$k_6 = 3.7 \times 10^{-12}$ exp $(240/T)$ cm$^3$ molecule$^{-1}$ s$^{-1}$

**(7)** $OH\cdot + NO_2 \rightarrow HNO_3$
$k_7 = 1.1 \times 10^{-11}$ cm$^3$ molecule$^{-1}$ s$^{-1}$

Species M is a neutral species, such as $N_2$. Assuming that species $O_3$, $OH\cdot$, and $OH_2\cdot$ are reactive species such that the pseudo-steady-state assumption applies (i.e., $d[O_3]/dt \approx 0$, $d[OH\cdot]/dt \approx 0$, and $d[HO_2\cdot]/dt \approx 0$), show that the NO, $NO_2$ and HCHO concentrations are governed by the following differential equations:

$$\frac{d[NO_2]}{dt} = \frac{2k_{4a}k_5[HCHO]^2}{k_7[NO_2]}$$

$$\frac{d[NO]}{dt} = -2k_{4a}\left(1 + \frac{k_5[HCHO]}{k_7[NO_2]}\right)[HCHO]$$

$$\frac{d[HCHO]}{dt} = -\left(k_{4a} + k_{4b} + 2k_{4a}\frac{k_5[HCHO]}{k_7[NO_2]}\right)[HCHO]$$

If the initial concentrations are, $[NO_2]_0 = 0.1$ ppm, $[NO]_0 = 0.01$ ppm, and $[HCHO]_0 = 0.1$ ppm, compute and plot the concentrations of HCHO, NO, $NO_2$, and $O_3$ as a function of time. You will need to solve the three, nonlinear differential equations above using a numerical method such as the fourth order Runge–Kutta method.

**6.21.** Table 6-1 gives the concentration of $CO_2$ in atmospheric air as 355 ppm. Assuming that falling rain equilibrates with this $CO_2$ and does not contact any other acid or basic gases, calculate the pH of such rain. Compare your calculated value to Figures. 5-1 and 5-2 and discuss the comparison. The concentration of $CO_2$ in a raindrop, $[CO_2]$, is related to the partial pressure of $CO_2$ in air, $P_{CO_2}$, by Henry's law constant, $H_{CO_2}$. At 35°C,

$$H_{CO_2} = \frac{P_{CO_2}}{[CO_2]} = 421 \text{ psia/gmol} \cdot \text{kg}$$

Once absorbed in water, the $CO_2$ dissociates and produces hydrogen ion, $H^+$, according to the following:

$$[CO_2] + [H_2O] \rightleftharpoons [H^+] + [HCO_3^-]$$

$$\frac{[H^+][HCO_3^-]}{[CO_2]} = 4.33 \times 10^{-7} \text{ gmol/kg}$$

$$[HCO_3^-] \rightleftharpoons [H^+] + [CO_3^{2-}]$$

$$\frac{[H^+][CO_3^{2-}]}{[HCO_3^-]} = 4.64 \times 10^{-11} \text{ gmol/kg}$$

$$[H_2O] \rightleftharpoons [H^+] + [OH^-]$$

$$[H^+][OH^-] = 10^{-14} \text{ gmol}^2/\text{kg}$$

In these equations liquid-phase concentrations have the units of gmol/kg of solution, the partial pressure of $CO_2$ has the units of psia, and the (nearly pure) water concentrations have been lumped with the equilibrium constants.

**6.22.** Consider the experimental determination of $NO_2$ uptake rate by Sinn et al. (1984) in Example 6.9. In their work they maintained a constant level of gas mixing in the growth chamber and they demonstrated by independent measurement with an "autoporometer" that the leaf diffusive resistance was essentially constant over all experiments and that it was not significantly affected by the presence or absence of $NO_2$ over the range of concentrations tested.

**(a)** Under these circumstances, do you consider it surprising that the uptake rate depends linearly with the exposure concentration? Explain.

**(b)** Suggest modifications of the experiment that you could run to evaluate and quantify independently the aerodynamic, stomatal, and mesophyllic resistances.

**6.23.** In Example 6.9 the wind velocity at 10 m was taken to be 10 m/s (22.5 mph). Conditions are not calm but this is hardly an exceptional wind for the North Atlantic Ocean. Repeat the calculation with winds over the range 0 to 80 mph. Based upon your results, comment on the steady state and episodic perspectives on gas uptake by large bodies of water. Equally ill-defined, and possibly more critical than the air velocity, is the concentration profile of $CO_2$ (or its equivalent) in the ocean. Visit your local geosciences library or post a query on the Internet to obtain such data. Use the new information to improve the accuracy of the calculation in Example 6.9. Compare the sensitivity of the calculated results to varying $U(10)$ and $c_L(10)$. Is it possible that $CO_2$ could move from the ocean to the atmosphere? If so, what might be the impact?

**6.24.** *[Design Problem]*

**(a)** A train is derailed among the wheat fields of South Dakota. A tank car of ammonia $(NH_3)$ develops a substantial leak and it is 3 days before the leak can be brought under control. As a member of the state's emergency environmental action team, you are asked for an immediate assessment of the impact of the $NH_3$ leak on the surrounding wheat as well as workers and inhabitants. You have 1 h to prepare for a meeting with officials in the vicinity of the derailment. What will you tell them?

**(b)** When the crisis is over you are expected to prepare a report detailing the impact on the crops of the exposure to ammonia. The farmers will want to know what to expect when harvest comes, and you know that your predictions will be scrutinized thoroughly at that time. What information (meteorological, physical, chemical, biological, theoretical, experimental, etc.) would you like to have? How will you get it? What will you do with it? Be as explicit as possible. Your reputation depends on it.

**6.25.** *[Design Problem]* A leak of bromine gas $(Br_2)$ occurs near a farmer's pond and a law suit erupts. You are asked to calculate the total uptake of bromine by the pond during the time of the leak. The molecular weight of $Br_2$ is 160 and its solubility at 20°C is 35.8 g $Br_2$ per liter of water. Begin your analysis by calculating the gas and liquid mass transfer coefficients, $k_G(10)$ and $k_L(1 \text{ m})$, for wind velocities of $U(10 \text{ m}) = 2$ and 10 m/s. There is more to the resolution of this problem than the calculation of these two coefficients, of course, but what can you say based on your results so far? What information do you need to proceed further?

**6.26.** *[Design Problem]* You are employed as an engineer in a regional air pollution control agency in your state. A consulting firm has modeled the air quality in your area and created the "ozone ridge" shown in Figure 6-12. Two regions of the state experience high ozone concentrations during the spring, summer, and fall and the governor is under pressure from voters and the EPA to recommend plans to reduce these high ozone concentrations.

- *Region 1:* $[NO_x]$ is $0.5 \pm 0.05$ ppm and $[VOC]$ is $1.5 \pm 0.005$ ppm.

- *Region 2:* $[NO_x]$ is $0.2 \pm 0.05$ ppm and $[VOC]$ is $0.75 \pm 0.1$ ppm.

A number of proposals have been recommended by interested parties.

**1.** There is nothing than can be done to reduce $O_3$.

**2.** Reduce $[VOC]$ and disregard $[NO_x]$.

**3.** Reduce both $[VOC]$ and $[NO_x]$ such that the ratio $[VOC]/[NO_x] \approx 2.5$.

**4.** Increase $[VOC]$ and disregard $[NO_x]$.

**5.** Reduce $[NO_x]$ but disregard $[VOC]$.

Recommend a tentative strategy for regions 1 and 2 (and its justification) and identify ancillary factors that need to be considered before a final recommendation can be made to the governor.

# III

# Engineering

- Chapter 7   Formation and Control of Pollutants in Combustion Systems
- Chapter 8   Uncontrolled Pollutant Emission Rates
- Chapter 9   Atmospheric Dispersion
- Chapter 10   Capturing Gases and Vapors
- Chapter 11   Motion of Particles
- Chapter 12   Capturing Particles
- Chapter 13   Cost of Air Pollution Control Systems

Having defined the circumstances in which we live and work, we move on the rigorous study of "engineering" in regards to air pollution control. We identify the major sources of human-made pollutants, how they move in the atmosphere, and how they might be prevented or removed. Not surprisingly, economics is a major player in decisions affecting the generation and amelioration of air pollution.

In this section we present chemical kinetic mechanisms associated with the generation of oxides of nitrogen, sulfur, carbon, and unburned hydrocarbons, with emphasis on stationary and mobile combustion systems. Mathematical models to quantify the rate of generation of pollutants and options for its prevention or minimization are introduced for a variety of industrial processes. The transport of pollutants in the atmosphere is described mathematically, and it is shown how to predict ground-level pollutant concentrations at points downwind of sources. Methods for separation and collection of gaseous and particulate pollutants from a process stram are modeled mathematically. This enables the prediction of the collection efficiency of control systems as a function of process input parameters and design variables. The section concludes the book with a discussion of engineering economics and presents ways to estimate the costs of purchasing and operating generic classes of air pollution control systems.

# 7

# Formation and Control of Pollutants in Combustion Systems

---

I n this chapter you will learn:

- To apply chemical equilibrium and kinetic relationships associated with CO and $CO_2$, NO and $NO_2$, and $SO_2$ and $SO_3$
- To understand the chemical kinetics associated with the combustion of hydrocarbons
- To identify the combustion parameters to minimize pollutant generation in stationary combustors, reciprocating engines, and gas turbines

## 7.1 Combustion devices and power cycles

Many pollutants are produced by high-temperature combustion processes. Therefore, it is worthwhile for readers to become familiar with power cycles used in mobile and stationary combustors. Throughout this chapter, devices that burn fuel to achieve various practical purposes will be referred to as *combustors*:

- *Stationary combustors*
  - Coal, oil, and gas combustion to generate steam (electricity, process steam, etc.)
  - Processes to remove $CO_2$ from ores (calcination) in kilns
  - Combustion of waste material in thermal incinerators
  - Dryers and ovens used in the manufacture bricks, foods, coated surfaces, etc.
- *Mobile combustors*
  - Combustion cycles to operate vehicles, aircraft, or portable power supplies

**Stationary combustors.** The objective of these devices is simply to produce a region of sufficiently high temperature to accomplish the task at hand. Electric utility boilers (Figure 7-1) burn pulverized coal and air to produce steam for steam turbines that generate electricity. Stoker boilers (Figure 7-2) burn coal on a vibrating or traveling grate to generate process steam. Kilns (Figure 7-3) are inclined cylindrical combustors

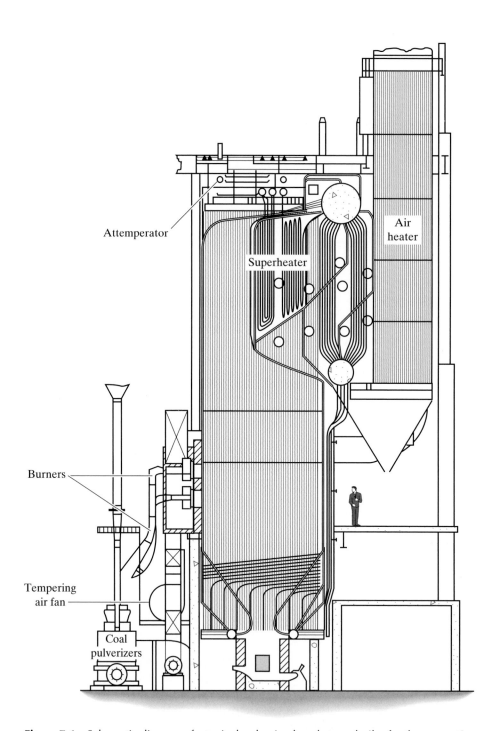

**Figure 7-1** Schematic diagram of a typical pulverized coal steam boiler for the generation of electrical power. (Babcock and Wilcox, 1978)

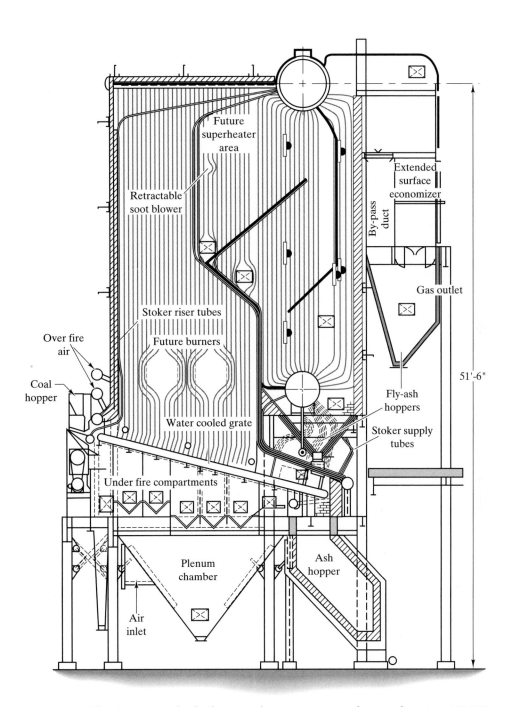

**Figure 7-2** Vibration grate stoker boiler to produce process steam for manufacturing, 120,000 lbm steam/hr, 450 psig saturated steam, bituminous coal. (With the permission of the Kverner Pulping, Inc.)

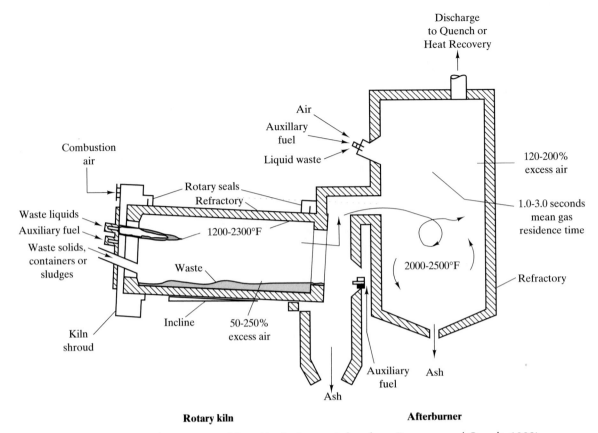

**Figure 7-3** Liquid waste rotary kiln with afterburner (taken from Dempsey and Oppelt, 1993)

that rotate slowly. A long luminous flame is produced along its axis, heating materials (limestone, wastes, etc.) that tumble down the cylinder. Incinerators (Figure 7-4) burn waste materials and emit as few intermediate products as possible. Packaged oil combustors (Figure 7-5) produce process steam for industrial processes. The purity of the fuel and ease of control make gas-fired combustors less polluting except perhaps for the production of $NO_x$.

In stationary combustors fuel (solid, liquid, or gas) and air enter the combustor as separate streams, mix, and burn at atmospheric pressure. Particulate matter and pollutants are removed and the remaining products of combustion are discharged to the atmosphere. In the case of high-temperature processes, steps must be taken to ensure that the discharge of the oxides of sulfur and nitrogen are within limits dictated by state and federal environmental agencies.

**Mobile combustors.** Mobile combustors are unique because they burn fuel carried by the vehicle. Commonly, the fuel and air burn as a diffusion flame, but the pressure is generally well above 1 atm. The products of combustion are discharged to the atmosphere. Mobile combustion processes are optimized to achieve the desired vehicle performance (i.e., acceleration, torque, fuel economy, starting, etc.) and to produce high *thermal cyclic efficiency* ($\eta$):

$$\eta = \frac{\text{net output mechanical work}}{\text{total input thermal energy}} \quad (7.1)$$

It is convenient to model mobile combustors as cyclic devices in which air is the operating fluid. The gas flow is simplified as a path in which the discharged air is reintroduced as inlet air. It is instructive to depict three basic cycles used for mobile power shown on *P–v* and *T–s* diagrams (Figure 7-6).

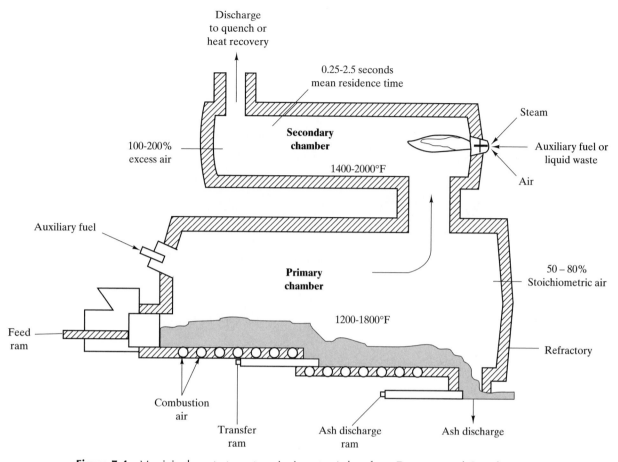

**Figure 7-4** Municipal waste two-stage incinerator (taken from Dempsey and Oppelt, 1993)

1. *Otto cycle*: spark ignition (SI) cycle used widely in automobiles, boats, and small-horsepower consumer products
2. *Diesel cycle*: compression ignition (CI) cycle used widely in utility vehicles requiring high-torque and low-speed performance such as trucks, farm tractors, and small ships
3. *Brayton cycle*: used in aircraft jet engines (Figure 7-7) and ground-based military vehicles such as tanks

Figure 7-8 is a diagram of a four-stroke (one power stroke/two revolutions) spark or compression ignition engine. A *reciprocating* engine is a spark or compression ignition engine that uses reciprocating pistons. It is seen that the maximum temperature in these cycles $(T_3)$ occurs after combustion. Assuming that the working fluid (air) is an ideal gas with constant specific heats (*air standard cycle*), it can be shown that the thermal cyclic efficiencies can be expressed as follows,

*Otto cycle*: $\quad \eta = 1 - \left(\dfrac{1}{R_v}\right)^{k-1}$

$$R_v = \text{compression ratio} = \frac{V_1}{V_2} \quad (7.2)$$

*Diesel cycle*: $\quad \eta = 1 - \left(\dfrac{1}{R_v}\right)^{k-1} \dfrac{R_c^k - 1}{k(R_c - 1)} \quad (7.3)$

$$R_v = \text{compression ratio} = \frac{V_1}{V_2}$$

$$\text{and } R_c = \text{cutoff ratio} = \frac{V_3}{V_2}$$

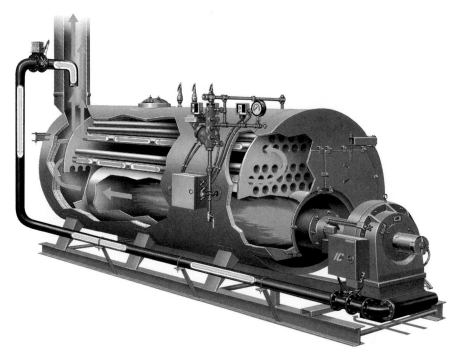

**Figure 7-5**    Oil or gas, two-pass low NOX steam boiler incorporating flue gas recirculation, with permission from Industrial Combustion)

*Brayton cycle:* $\eta = 1 - \left(\dfrac{1}{R_p}\right)^{(k-1)/k}$

$$R_p = \text{compression ratio} = \frac{P_2}{P_1} \quad (7.4)$$

It can be shown that the thermal efficiency of the Brayton and Otto cycles can be written as

$$\eta = 1 - \frac{T_4}{T_3} = 1 - \frac{T\,(\text{end of the expansion})}{T\,(\text{peak})}$$

$$(7.5)$$

---

**EXAMPLE 7.1    COMPRESSION RATIO, PEAK TEMPERATURE, AND THERMAL EFFICIENCY**

**a.** Prove Equations 7.2 and 7.5.

**b.** Compute and plot the peak cycle temperature $(T_3)$ and cycle thermal efficiency $(\eta)$ versus the compression ratio $(R_v)$ for an air standard spark ignition (Otto) cycle when the heat added by combustion $(Q_{in})$ is 1,000 kJ/kg. The air begins the cycle at $T_1 = 300$ K, $k = 1.4$, and $c_v = 0.7165$ kJ/kg·K.

**Solution**

**a.** Between any two points (a) and (b) in an isentropic process for an ideal gas with constant specific heats,

$$\frac{T_a}{T_b} = \left(\frac{P_a}{P_b}\right)^{(k-1)/k} = \left(\frac{\rho_a}{\rho_b}\right)^{(k-1)} \quad \text{but} \quad \frac{\rho_a}{\rho_b} = \frac{v_b}{v_a}$$

From the first law of thermodynamics for a cyclic process $\left(\Sigma Q_{in} = \Sigma Q_{out}\right)$ and Equation 7.1,

$$\eta = 1 - \frac{Q_{out}}{Q_{in}} = 1 - \frac{c_v(T_4 - T_1)}{c_v(T_3 - T_2)}$$

$$= 1 - \frac{T_1\left[(T_4/T_1) - 1\right]}{T_2\left[(T_3/T_2) - 1\right]}$$

From Figure 7-6, it is seen that $v_4 = v_1$ and $v_3 = v_2$. From the isentropic relationships,

$$\frac{v_4}{v_3} = \left(\frac{T_3}{T_4}\right)^{1/(k-1)} \quad \text{and} \quad \frac{v_1}{v_2} = \left(\frac{T_2}{T_1}\right)^{1/(k-1)}$$

Thus $T_2/T_1 = T_3/T_4$, which can be rearranged as $T_4/T_1 = T_3/T_2$. Substituting this into the equations

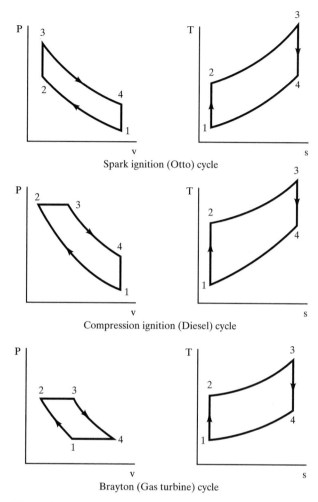

Spark ignition (Otto) cycle

Compression ignition (Diesel) cycle

Brayton (Gas turbine) cycle

**Figure 7-6** P-v and T-s diagrams of spark ignition, compression ignition and Brayton cycles

for $\eta$, one finds that the cyclic thermal efficiency of an air standard spark ignition (Otto) cycle can be expressed as

$$\eta = 1 - \frac{1}{R_v^{k-1}} \quad \text{and} \quad \eta = 1 - \frac{T_4}{T_3}$$

**b.** The heat added by combustion is $Q = c_v(T_3 - T_2)$. But $T_2 = T_1(v_1/v_2)^{k-1} = R_v^{k-1}$. Thus

$$T_3 = \frac{Q}{c_v} + T_1 R_v^{k-1} = \frac{1000}{0.7165} + 300(R_v)^{0.4}$$

Graphs of $\eta$ and $T_3$ for a variety of compression ratios $R_v$ are shown in Figures E7.1a and b. It is seen that both the peak cycle temperature $(T_3)$ and the thermal cyclic efficiency $(\eta)$ are strongly dependent on the compression ratio.

Designers of engines based on these cycles attempt to maximize the thermal cyclic efficiency $(\eta)$ by maximizing the peak temperature $(T_3)$, but they are faced with a series of difficult compromises. High efficiency related to high compression ratio, cutoff ratio, and pressure ratio produce low concentrations of CO and unburned hydrocarbons but simultaneously produce high concentrations of nitrogen oxides. The cycles shown in Figure 7-6 enable one to predict how the gas temperature varies with piston volume but do not show how fast temperatures change with time $(dT/dt)$. In analyses that follow it will be necessary to estimate how long it takes for the combustion products to expand during the power stroke (3–4) and the rate of change of temperature with time during the power stroke. For the Brayton cycle, a typical residence time $(t_r)$ during which combustion occurs can be expressed as

$$t_r \equiv \frac{V_{\text{burner}}}{Q_{\text{actual}}} = \frac{(A_{\text{burner}})(L_{\text{burner}})}{Q_{\text{actual}}} \quad (7.6)$$

Typical values of $t_r$ for jet engines are

$$10^{-4}s < t_r < 10^{-3}s$$

**Figure 7-7** Pratt and Whitney F100-PW-229 aircraft jet engine. (Photo provided by the Pratt & Whitney, Corp.)

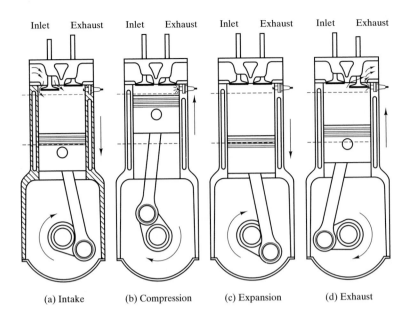

**Figure 7-8** Schematic diagram of a four-stroke (1 power stroke/2 revolutions) for a spark ignition engine (redrawn from Heywood, 1988)

(a) Intake    (b) Compression    (c) Expansion    (d) Exhaust

For the *four-stroke* spark ignition (Otto) cycle in Figure 7-8, there is one power stroke for every two revolutions. The residence time $(t_r)$ is a conceptually useful measure of the time it takes to complete the power stroke and is related to the engine rpm as follows:

$$t_r = \frac{\text{time (min)}}{\text{stroke}} = (\text{min/rev.})(\text{rev./2 strokes})$$

$$= (2\,\text{rpm})^{-1} \tag{7.7}$$

For an engine running at 3000 rpm, the residence time $(t_r)$ is approximately 0.01 s.

If the gas mixture is treated as an ideal gas with constant specific heats, the time rate of change of temperature during the expansion stroke (3–4) can be found from the isentropic expansion from the peak cyclic temperature $(T_3)$.

$$T(t) = T_3 \left[\frac{V_3}{V(t)}\right]^{k-1} \tag{7.8}$$

Differentiating with respect to time, one can show that

$$\frac{dT}{dt} = \left[(1-k)T_3 V_3^{k-1}\right]V^{-k}A_p v_p \tag{7.9}$$

where $A_p$ is the cross-sectional area of the piston and $v_p$ is the average piston speed. The average piston speed is related to engine rpm,

$$v_p = \frac{\omega L_p}{\pi}$$

$$v_p(\text{m/min}) = 2(\text{rpm})L_p(\text{m}) \tag{7.10}$$

where $\omega$ is the engine rotational speed (radians per unit time) and $L_p$ is the length of the stroke.

Shown in Figure 7-9 are the predicted concentrations of the equilibrium products of combustion in a spark-ignition engine using isooctane and air at different fuel–air ratios and temperatures at 30 atm pressure. The fuel–air ratio is expressed as an *equivalence ratio* $(\phi)$:

$$\phi = \frac{(F/A)_{\text{actual}}}{(F/A)_{\text{stoichiometric}}} \tag{7.11}$$

While this high pressure is only applicable to spark-ignition engines, the manner in which stable and free-radical species vary with temperature and fuel–air ratio is valid for combustors in general. Figure 7-9 can be used when readers perform an *order-of-magnitude* analysis to simplify a set of kinetic equations.

$T = 1750$ K:   fuel-lean flames, $\phi < 1$,

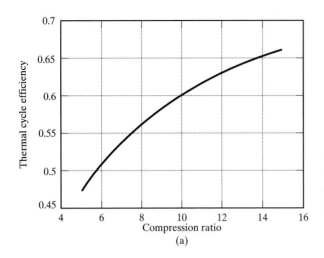

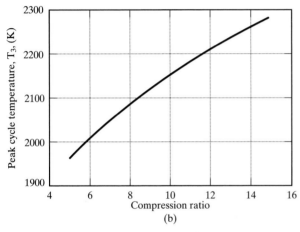

**Figure E7-1** (a) Thermal cyclic efficiency, versus Otto cycle compression ratio. (b) Peak cycle temperature, $T_3$ (K), versus Otto cycle compression ratio

$$y_{OH\cdot} \approx 1.5 \times 10^{-4},$$

negligible $y_{O\cdot}$ and $y_{H\cdot}$.

fuel-rich flames, $\phi > 1$,

negligible free radicals \hfill (7.12)

$T = 2250$ K: fuel-lean flames, $\phi < 1$,

$$y_{OH\cdot} \approx 10^{-3},$$

$$y_{O\cdot} < 3 \times 10^{-4}, \ y_{H\cdot} \approx 0$$

fuel-rich flames, $\phi > 1$,

$y_{OH\cdot}$ decreases with $\phi$ while

$y_{H\cdot}$ rises with $\phi$, $y_{O\cdot} \approx 0$ \hfill (7.13)

$T = 2750$ K: fuel-lean flames, $\phi < 1$,

$$y_{OH\cdot} \approx 10^{-2}, \ y_{O\cdot} < 3 \times 10^{-3},$$

$y_{H\cdot}$ rises to $10^{-3}$ at $\phi = 1.0$

fuel-rich flames, $\phi > 1$,

$y_{OH\cdot}$ and $y_{H\cdot} \approx 10^{-3},$

falling $y_{O\cdot}$. \hfill (7.14)

where $y$ expresses the concentration as a mole fraction. It is clear that free-radical concentrations increase with temperature and OH· radicals are favored in fuel-lean flames

## 7.2 Free-radical chemistry in flames

A prerequisite to understanding how pollutants are formed by combustion is knowledge of how free radicals govern individual chemical reactions that sustain combustion. The reader is advised to review material in Chapter 6 describing how free radicals govern the production of pollutants in the atmosphere. The role of free radicals is the same, but the high temperatures, pressures, and concentrations in the cyclic processes cause the reactions to proceed more rapidly. Chemical kinetic rate constants needed for this analysis are obtained from a variety of sources (Carter, 1990; Roberts, 1990; Atkinson, 1990; Atkinson et al., 1992; Seinfeld, 1986; Finlayson-Pitts and Pitts, 1986).

*Diffusion flames* are commonly used in practical combustors because it is easier and safer to introduce pure fuel into the combustor, where it mixes and burns with air. Figure 7-10 shows the essential features of a diffusion flame surrounding a burning solid or liquid drop. When combustible solid or liquid particles are introduced into a region of high temperature, the fuel volatizes and the hydrocarbon vapors travel radially outward and mix with air traveling radially inward. At some radius, the composition of air and fuel vapor reach *stoichiometric proportions*, (or are within narrow values either side of stoichiometric proportions) and a flame is produced. A portion of the heat generated in combustion travels radially inward to volatilize additional fuel. The products of combustion diffuse radially

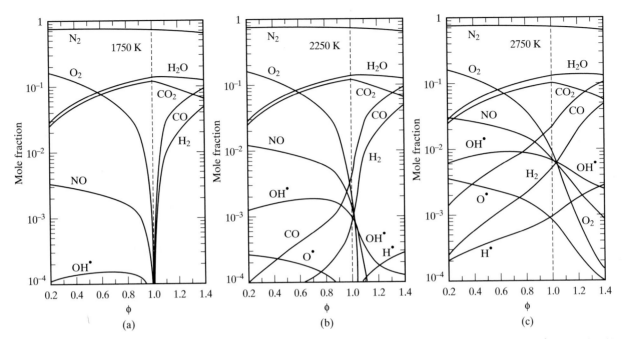

**Figure 7-9** Mol fractions of equilibrium products of isooctane-air mixtures as a function of fuel/air equivalence ratio ($\phi$) at 30 atmospheres and (a) 1750 K, (b) 2250 K and (c) 2750 K (figure taken from Heywood, 1988)

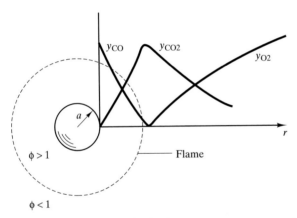

**Figure 7-10** Diffusion flame showing temperature and concentration profiles of liquid drop of radius a (redrawn from Flagan & Seinfeld, 1988)

outward. The combustion zone is associated with radiant energy in the visible spectrum and produces the orange and blue region we call the flame. Diffusion flames can also be produced by gaseous fuels except that it is not necessary for heat to be transferred to volatilize the fuel.

Diffusion flames are inherently safe because the fuel and oxygen travel toward one another and com-

bustion occurs only in the region where they are in combustible proportions. In *premixed flames* the fuel and air are mixed prior to the flame. Also, the premixed flames exist in close proximity to a solid surface, called a *flame holder*, to which they transfer energy. Shown in Figure 7-11 is a *flat flame* produced by a Meeker burner used in elementary chemistry laboratories. The flame is a thin flat disk several millimeters above a screen (flame holder) mounted at the exit of the burner. The screen acts as a heat sink to "hold" the flame. Were it not for the heat sink, the flame would propagate upstream into the premixed fuel and air, transform itself into a detonation wave, and eventually produce an explosion. Because sufficient heat is transferred to the flame holder, the temperature of the premixed gases flowing through the screen is too low to sustain a propagating flame. Thus the solid body called the *flame holder* literally holds or secures the flame at particular location. A screen surrounding a candle flame was used by Sir Humphry Davy in 1815 when he invented the mine safety lamp that prevented candle flames from igniting methane in coal mines.

To appreciate the role of free radicals in the production of pollutants associated with combustion, consider the flat flame in Figure 7-11. It is easier to

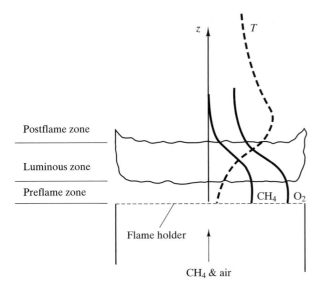

**Figure 7-11** Premixed flame showing zones, temperature and concentration profiles

illustrate combustion kinetics with a premixed methane $(CH_4)$–air flat flame than a diffusion flame. If the fuel–air ratio is too large, the flame will have a large luminous orange top and a flat blue disk at its base. As the fuel–air ratio approaches stoichiometric proportions, the orange region disappears, and remaining is a thin blue circular disk approximately 1 mm thick with a diameter equal to the diameter of the burner mouth. Imagine that an experiment is conducted in which the temperature and concentrations of various molecular species are measured along the vertical axis of the flame. Figure 7-11 shows the resulting profiles. It is convenient to divide the flame into three regions: *preflame zone, luminous zone*, and *postflame zone*. Within each zone the following reactions occur.

**1.** *Preflame zone.* The preflame zone is the transparent region a few millimeters thick between the luminous zone and the flame holder. Within the preflame zone, the temperature rises because heat is transferred from the luminous zone to the relatively cool flame holder. Within this zone $CH_4$ either *pyrolizes* (by reacting with a neutral species M) or is attacked by free radicals $(H\cdot, OH\cdot, O\cdot, HO_2\cdot)$. The free radicals are generated in the flame and propagate back into the incoming premixed fuel and air.

Thus within the premixed region, methane is converted to methyl free radicals $(CH_3\cdot)$.

$$CH_4 + M \rightarrow CH_3\cdot + H\cdot + M \tag{7.15}$$
$$k_+(m^3\ gmol^{-1}\ s^{-1}) = 1.41 \times 10^{11} \exp(-44{,}535/T)$$
$$CH_4 + H\cdot \rightarrow CH_3\cdot + H_2 \tag{7.16}$$
$$k_+(m^3\ gmol^{-1}\ s^{-1}) = 2 \times 10^8 \exp(-5950/T)$$
$$CH_4 + OH\cdot \rightarrow CH_3\cdot + H_2O \tag{7.17}$$
$$k_+(m^3\ gmol^{-1}\ s^{-1}) = 3.5 \times 10^8 \exp(-4500/T)$$
$$CH_4 + O\cdot \rightarrow CH_3\cdot + OH\cdot \tag{7.18}$$
$$k_+(m^3\ gmol^{-1}\ s^{-1}) = 2 \times 10^7 \exp(-3450/T)$$
$$CH_4 + HO_2\cdot \rightarrow CH_3\cdot + H_2O_2 \tag{7.19}$$

**2.** *Luminous zone.* The bottom of the luminous zone corresponds to the point where the temperature rises to a value commonly referred to as the *ignition temperature*, the temperature below which it is not possible to sustain combustion. Within the luminous zone, temperatures reach their maximum values and vigorous chemical activity occurs that results in a net production of free radicals. *Branching reactions* occur in which one free radical gives birth to two free radicals, thus doubling the number of free radicals.

$$H\cdot + O_2 \rightarrow OH\cdot + O\cdot \tag{7.20}$$
$$k_+(m^3\ gmol^{-1}\ s^{-1}) = 1.2 \times 10^{11} T^{-0.91} \exp(-8310/T)$$
$$O\cdot + H_2 \rightarrow OH\cdot + H\cdot \tag{7.21}$$
$$k_+(m^3\ gmol^{-1}\ s^{-1}) = 1.5 \times 10^1 T^2 \exp(-3800/T)$$
$$O\cdot + H_2O \rightarrow OH\cdot + OH\cdot \tag{7.22}$$
$$k_+(m^3\ gmol^{-1}\ s^{-1}) = 1.5 \times 10^4 T^{1.14} \exp(-8680/T)$$

Within the luminous zone there is also an exchange of free radicals.

$$OH\cdot + H_2 \rightarrow H\cdot + H_2O \tag{7.23}$$
$$k_+(m^3\ gmol^{-1}\ s^{-1}) = 1.0 \times 10^2 T^{1.6} \exp(-1660/T)$$
$$O\cdot + HO_2\cdot \rightarrow OH\cdot + O_2 \tag{7.24}$$
$$k_+(m^3\ gmol^{-1}\ s^{-1}) = 2 \times 10^7$$
$$H\cdot + HO_2\cdot \rightarrow OH\cdot + OH\cdot \tag{7.25}$$
$$k_+(m^3\ gmol^{-1}\ s^{-1}) = 1.5 \times 10^8 \exp(-500/T)$$
$$H\cdot + H_2O \rightarrow OH\cdot + H_2 \tag{7.26}$$

$$H\cdot + O_2 + M \rightarrow HO_2\cdot + M \tag{7.27}$$

$$k_+\left(m^6\ gmol^{-2}\ s^{-1}\right) = 1.5 \times 10^3\ exp\left(-500/T\right)$$

$$O\cdot + H\cdot + M \rightarrow OH\cdot + M$$

$$k_+\left(m^6\ gmol^{-2}\ s^{-1}\right) = 1 \times 10^4 \tag{7.28}$$

where M is any neutral molecular species that retains its identity. There are also reactions involving $H_2O_2$.

$$OH\cdot + OH\cdot + M \rightarrow H_2O_2 + M \tag{7.29}$$

$$k_+\left(m^6\ gmol^{-2}\ s^{-1}\right) = 9.1 \times 10^2\ exp\left(2550/T\right)$$

$$H_2O_2 + M \rightarrow OH\cdot + OH\cdot + M \tag{7.30}$$

$$k_+\left(m^3\ gmol^{-1}\ s^{-1}\right) = 1.3 \times 10^{10}\ T^{-2}$$

$$H_2O_2 + OH\cdot \rightarrow HO_2\cdot + H_2O \tag{7.31}$$

$$k_+\left(m^3\ gmol^{-1}\ s^{-1}\right) = 1.9 \times 10^{12}\ exp\left(-187/T\right)$$

Within the luminous zone reactions occur that lead to the production of nitrogen oxides and carbon oxides.

$$HCHO + OH\cdot \rightarrow HCO\cdot + H_2O \tag{7.32}$$

$$HCO\cdot + OH\cdot \rightarrow CO + H_2O$$

$$k_+\left(m^3\ gmol^{-1}\ s^{-1}\right) = 3 \times 10^7 \tag{7.33}$$

$$HCO\cdot + O_2 \rightarrow HO_2\cdot + CO \tag{7.34}$$

$$k_+\left(m^3\ gmol^{-1}\ s^{-1}\right) = 3.3 \times 10^{12}\ exp\left(\pm 150/T\right)$$

$$CO + OH\cdot \rightarrow CO_2 + H\cdot$$

$$k_+\left(m^3\ gmol^{-1}\ s^{-1}\right) = 4.4T^{1.5}\ exp\left(372/T\right) \tag{7.35}$$

$$H\cdot + CO_2 \rightarrow CO + H\cdot + O\cdot$$

$$k_+ = \text{very small} \tag{7.36}$$

$$CO + O_2 \rightarrow CO_2 + O\cdot \tag{7.37}$$

$$k_+\left(m^3\ gmol^{-1}\ s^{-1}\right) = 2.5 \times 10^6\ exp\left(-24,060/T\right)$$

$$CO + O\cdot + M \rightarrow CO_2 + M \tag{7.38}$$

$$k_+\left(m^6\ gmol^{-2}\ s^{-1}\right) = 5.3 \times 10^1\ exp\left(2285/T\right)$$

$$CO + HO_2\cdot \rightarrow CO_2 + OH\cdot \tag{7.39}$$

$$k_+\left(m^3\ gmol^{-1}\ s^{-1}\right) = 1.5 \times 10^8\ exp\left(-11,900/T\right)$$

$$N\cdot + O_2 \rightleftharpoons NO + O\cdot \tag{7.40}$$

$$k_+\left(m^3\ gmol^{-1}\ s^{-1}\right) = 1.8 \times 10^4 T\ exp\left(-4680/T\right)$$

$$k_-\left(m^3\ gmol^{-1}\ s^{-1}\right)$$

$$= 3.8 \times 10^3 T\ exp\left(-20,820/T\right) \tag{7.41}$$

$$O\cdot + N_2 \rightleftharpoons NO + N\cdot$$

$$k_+\left(m^3\ gmol^{-1}\ s^{-1}\right)$$

$$= 1.8 \times 10^8\ exp\left(-38,370/T\right) \tag{7.42}$$

$$k_-\left(m^3\ gmol^{-1}\ s^{-1}\right)$$

$$= 3.8 \times 10^{-7}\ exp\left(-425/T\right) \tag{7.43}$$

$$O_2 + M \rightleftharpoons O\cdot + O\cdot + M$$

$$k_+\left(m^3\ gmol^{-1}\ s^{-1}\right)$$

$$= 1.1 \times 10^{12} T^{-1/2}\ exp\left(-59,386/T\right) \tag{7.44}$$

$$k_-\left(m^6\ gmol^{-2}\ s^{-1}\right) = 1 \times 10^5 T^{-1} \tag{7.45}$$

$$N\cdot + OH\cdot \rightleftharpoons NO + H\cdot$$

$$k_+\left(m^3\ gmol^{-1}\ s^{-1}\right)$$

$$= 7.1 \times 10^7\ exp\left(-450/T\right) \tag{7.46}$$

$$k_-\left(m^3\ gmol^{-1}\ s^{-1}\right)$$

$$= 1.7 \times 10^8\ exp\left(-24,560/T\right) \tag{7.47}$$

Equations 7.33 to 7.39 show that CO is formed before $CO_2$. If flames lack sufficient OH·, the oxidation of CO to $CO_2$ will be impeded. Equations 7.40 and 7.47 describe how NO is formed from $O_2$ and $N_2$; such NO is called *thermal NO*.

**3.** *Postcombustion zone.* Downstream of the luminous zone the temperature decreases and terminating reactions occur in which free radicals annihilate themselves.

$$H\cdot + H\cdot + M \rightarrow H_2 + M$$

$$k_+\left(m^6\ gmol^{-2}\ s^{-1}\right) = 6.4 \times 10^5 T^{-1} \tag{7.48}$$

$$O\cdot + O\cdot + M \rightarrow O_2 + M$$

$$k_+\left(m^6\ gmol^{-2}\ s^{-1}\right) = 1.0 \times 10^5 T^{-1} \tag{7.49}$$

$$H\cdot + OH\cdot + M \rightarrow H_2O + M$$

$$k_+\left(m^6\ gmol^{-2}\ s^{-1}\right) = 1.4 \times 10^{11} T^{-2} \tag{7.50}$$

$$HO_2\cdot + HO_2\cdot \rightarrow H_2O_2 + O_2$$

$$k_+\left(m^3\ gmol^{-1}\ s^{-1}\right) = 2 \times 10^7 \tag{7.51}$$

$$OH\cdot + OH\cdot + M \rightarrow H_2O_2 + M$$

$$k_+\left(m^6\ gmol^{-2}\ s^{-1}\right) = 1.3 \times 10^{10} T^{-2} \tag{7.52}$$

$$H\cdot + HO_2\cdot \rightarrow H_2 + O_2 \tag{7.53}$$

$$k_+\left(m^3\ gmol^{-1}\ s^{-1}\right) = 2.5 \times 10^7\ exp\left(-350/T\right)$$

$$OH\cdot + HO_2\cdot \rightarrow H_2O + O_2$$

$$k_+\left(m^3\ gmol^{-1}\ s^{-1}\right) = 2 \times 10^7 \tag{7.54}$$

A common approximation is to assume equilibrium values of OH·, O·, and H· based on the following equilibrium reactions:

$$\tfrac{1}{2}O_2 \rightleftharpoons O\cdot$$

$$[O\cdot]_e = K_{P,O}[O_2]^{1/2} \tag{7.55}$$

$$K_{P,O}(\text{atm}^{1/2}) = 3030 \exp(-30{,}790/T) \tag{7.56}$$

$$\tfrac{1}{2}H_2O + \tfrac{1}{4}O_2 \rightleftharpoons OH\cdot$$

$$[OH\cdot]_e = K_{P,OH}[H_2O]^{1/2}[O_2]^{1/4} \tag{7.57}$$

$$K_{P,OH}(\text{atm}^{1/4}) = 166 \exp(-19{,}680/T) \tag{7.58}$$

$$\tfrac{1}{2}H_2O \rightleftharpoons H\cdot + \tfrac{1}{4}O_2$$

$$[H\cdot]_e = K_{P,H}[H_2O]^{1/2}[O_2]^{-1/4} \tag{7.59}$$

$$K_{P,H}(\text{atm}^{3/4}) = 44{,}100 \exp(-42{,}500/T) \tag{7.60}$$

$$\tfrac{1}{2}N_2 \rightleftharpoons N\cdot$$

$$[N]_e = K_{P,N}[N_2]^{1/2} \tag{7.61}$$

$$K_{P,N}(\text{atm}^{1/2}) = 3030 \exp(-57{,}830/T) \tag{7.62}$$

The kinetics describing the combustion of higher-order aliphatic hydrocarbons are complex. For example, the kinetics of ethane $(C_2H_6)$ combustion (Notzold and Algermissen, 1981) and other alkanes is similar to methane but include additional pyrolysis reactions that dissociate ethane and produce two $CH_3\cdot$ radicals which subsequently react with ethane to form ethyl radicals $(C_2H_5\cdot)$ and additional aliphatic hydrocarbons. Additional reactions occur between pyrolysis products and oxygen to produce aldehydes. Pyrolysis products as well as unstable intermediates interact with oxygen $(O_2)$ to form OH· and $HO_2\cdot$ radicals. Chain branching occurs as it does for methane, but there are additional reactions that abstract hydrogen from ethane.

A description of the combustion kinetics for aromatic hydrocarbons is a formidable task. A commonly followed practice is to represent the individual reaction by hypothetical reactions that replicate the rate at which the overall reaction progresses. These reactions are called *global or overall mechanisms* and their kinetics are characterized by *global rate constants*. Global reactions are useful for modeling the performance of combustors because engineers need to be able to predict the rate at which heat is generated by combustion, the rate at which pollutants are formed, and so on. One such formulation is

a *two-step overall reaction*, the first of which describes rapid hydrocarbon combustion expressed as

$$C_nH_m + (n/2 + m/4)O_2$$
$$\xrightarrow{k_{\text{fuel}}} nCO + (m/2)H_2O \tag{7.63}$$

The rate at which the hydrocarbon is consumed $(\text{gmol m}^{-3}\,\text{s}^{-1})$ is expressed as

$$r_{\text{fuel}} = -A \exp\left(-\frac{E_a}{R_uT}\right)[C_nH_m]^a[O_2]^b \tag{7.64}$$

Westbrook and Dryer (1981) determined values of $A$, $E_a$, and the exponents $a$ and $b$ for a variety of aliphatic and aromatic hydrocarbons. Table 7-1 summarizes the values of $A$, $E_a$, $a$, and $b$ for several reactions reported by Flagan and Seinfeld (1988).

The second reaction describes the slow conversion of CO to $CO_2$:

$$CO + \tfrac{1}{2}O \xrightarrow{k_{CO}} CO_2 \tag{7.65}$$

which proceeds at a rate $(\text{gmol m}^{-3}\,\text{s}^{-1})$

$$-r_{CO} = A \exp\left(-\frac{E_a}{R_uT}\right)[H_2O]^c[O_2]^b[CO] \tag{7.66}$$

The conversion of CO is expressed as a function of $H_2O$ because the dominant reaction affecting the conversion of CO involves the reaction with OH·, which to a first approximation can be estimated from the dissociation of $H_2O$. Dryer and Glassman

---

**Table 7-1** Rate Parameters for Two-Step Global Kinetics[a]

| Fuel | $A \times 10^{-6}$ | $(E_a/R_u) \times 10^{-3}$ | $a$ | $b$ |
|---|---|---|---|---|
| $CH_4$ | 2,800 | 24.4 | −0.3 | 1.3 |
| $C_2H_4$ | 75 | 15 | 0.1 | 1.65 |
| $C_3H_6$ | 15 | 15 | −0.1 | 1.85 |
| $C_3H_8$ | 31 | 15 | 0.1 | 1.65 |
| $C_6H_6$ | 7 | 15 | −0.1 | 1.85 |
| $C_8H_{18}$ | 18 | 15 | 0.25 | 1.5 |
| $CH_3OH$ | 117 | 15 | 0.25 | 1.5 |
| $C_2H_5OH$ | 56 | 15 | 0.15 | 1.6 |

[a] Convention: $A \times 10^{-6} = 2800 \Rightarrow A = 2800 \times 10^6$; $(E_a/R_u) \times 10^{-3} = 24.4 \Rightarrow E_a/R_u = 24.4 \times 10^3$.
For consistent units in Equation 7.64, use gmd,m³, s, and K.
*Source*: Abstracted from Flagan and Seinfeld (1988).

(1973) propose the following data for a global reaction for CO:

$$-r_{CO}(\text{gmol m}^{-3}\text{ s}^{-1}) = 1.3 \times 10^{10}$$

$$\exp\left(-\frac{20{,}140}{T}\right)[CO][H_2O]^{1/2}[O_2]^{1/4} \qquad (7.67)$$

The concentrations $[CO]$, $[H_2O]$, and $[O_2]$ should be expressed in gmol/m³. The rate of reaction predicted by global kinetics is useful in estimating the rate at which energy is released during combustion and the time it takes for combustion to occur.

---

**EXAMPLE 7.2   DURATION OF COMBUSTION IN A SPARK IGNITION ENGINE**

The idealized Otto (SI) cycle in Figure 7-6 presumes that combustion occurs instantly during the constant-volume process after the piston achieves its topmost position (i.e., top dead center). Examine this assumption by computing the time required to burn the fuel if the rate of combustion is described by the global reaction Equation 7.64. For this inquiry, assume that the Otto cycle has a compression ratio $(R_v)$ of 8. Premixed air and fuel enter the engine at a temperature $(T_1)$ and pressure $(P_1)$ of 300 K, 100 kPa. The fuel is isooctane $(C_8H_{18}, M = 114)$ and it is burned with 25% excess air. The heating value of isooctane is 44,788 kJ/kg of fuel.

**Solution**

*Stoichiometry.* The stoichiometric equation for complete combustion is

$$C_8H_{18} + (1.25)(12.5)[O_2 + 3.77N_2]$$

$$= 8CO_2 + 9H_2O + 3.13O_2 + 58.91N_2$$

The following molar ratios will be needed in subsequent calculations:

$$\frac{n_{\text{fuel}}}{n_{\text{air}}} = \frac{n_{\text{fuel}}}{n_{O_2}} \frac{n_{O_2}}{n_{\text{air}}}$$

$$= \frac{1}{(1.25)(12.5)}\left(\frac{1}{4.77}\right) = 0.0134$$

$$\frac{n_{\text{fuel}}}{n_{\text{total}}} = \frac{n_{\text{fuel}}}{n_{\text{fuel}} + n_{\text{air}}}$$

$$= \frac{n_{\text{fuel}}/n_{\text{air}}}{(n_{\text{fuel}}/n_{\text{air}}) + 1} = \frac{0.0134}{1 + 0.0134} = 0.0132$$

$$\frac{n_{\text{air}}}{n_{\text{total}}} = \frac{n_{\text{air}}}{n_{\text{fuel}} + n_{\text{air}}}$$

$$= \frac{1}{(n_{\text{fuel}}/n_{\text{air}}) + 1} = \frac{1}{0.0134 + 1} = 0.9868$$

$$\frac{n_{O_2}}{n_{\text{total}}} = \frac{n_{O_2}}{n_{\text{fuel}}} \frac{n_{\text{fuel}}}{n_{\text{total}}}$$

$$= (12.5)(1.25)(0.0132) = 0.2062$$

*Temperature and pressure at point 2.* The temperature and pressure at point 2 are

$$T_2 = T_1(R_v)^{k-1} = 300(8)^{0.4} = 689.2\text{ K}$$

$$P_2 = (R_v)^k = 100(8)^{1.4} = 1837.9\text{ kPa}$$

The total molar specific volume $(n_{\text{total}}/V)$ at point 2 is

$$\frac{n_{\text{total}}}{V} = \frac{P_2}{R_u T_2}$$

$$= \frac{1837.9\text{ kPa}}{(8.314\text{ kJ/kmol·K})(689.2\text{ K})}\left(\frac{\text{kJ}}{\text{kPa·m}^3}\right)$$

$$= 0.3207\text{ kmol/m}^3 = 320.7\text{ gmol/m}^3$$

The molar specific volumes of fuel and oxygen volumes at point 2 are

$$\frac{n_{\text{fuel}}}{V} = \frac{n_{\text{total}}}{V} \frac{n_{\text{fuel}}}{n_{\text{total}}}$$

$$= (320.7\text{ gmol/m}^3)(0.0132) = 4.233\text{ gmol/m}^3$$

$$\frac{n_{O_2}}{V} = \frac{n_{\text{total}}}{V} \frac{n_{O_2}}{n_{\text{total}}}$$

$$= (320.7\text{ gmol/m}^3)(0.2062) = 66.128\text{ gmol/m}^3$$

*Temperature at point 3.* The energy released by combustion per mole of air is related to the heating value:

$$q\,(\text{kJ/kmol air}) = (0.0134\text{ kmol fuel/kmol air})$$

$$(114\text{ kg fuel/kmol fuel})$$

$$(44{,}788\text{ kJ/kg fuel})$$

$$= 68{,}418\text{ kJ/kmol air}$$

Since the combustion from 2–3 occurs at constant volume, the temperature $T_3$ is related to the energy added by combustion by

$$q = 68,418 = \int_{T_2}^{T_3} c_v \, dT$$

If $c_v$ is assumed to be constant and equal to 0.7165 kJ/kg·K, the computed value of $T_3$ is 3981.93. For the high temperatures associated with combustion, it would be advisable to use a value of $c_v$ that is a function of temperature. Assume that the $c_p$ of air and nitrogen are the same. The constant-volume specific heat $(c_v)$ is equal to $c_p - R_u$, and the constant-pressure specific heat $(c_p)$ can be found in van Wylen and Sonntag (1985) and Turns (1996).

$$c_v(\text{kJ/kmol air·K}) = c_p - R = 39.06 - 512.79\theta^{-1.5}$$

$$+ \ 1072.7\theta^{-2} - 820.4\theta^{-3} - 8.314$$

where $\theta = T/100$ and $q = 100 \int c_v \, d\theta$:

$$68,418 = 100 \int_{6.892}^{\theta_3} (30.746 - 512.79\theta^{-1.5}$$

$$+ \ 1072.7\theta^{-2} - 820.4\theta^{-3}) \, d\theta$$

The above can be integrated and $\theta_3$ determined by trial and error. A value of $T_3 = 3226$ K is obtained.

*Rate of temperature change during combustion.* The rate at which the temperature rises from point 2 to 3 is related to the rate of reaction (Equation 7.64) and the first law of thermodynamics applied to a constant-volume process:

$$r(\text{gmol fuel m}^{-3}\ \text{s}^{-1})(44,788\ \text{kJ/kg fuel})$$

$$(114\ \text{kg fuel/kmol fuel})$$

$$(\text{kmol/1000 gmol})V(\text{m}^3) = n_{\text{total}} c_v \frac{dT}{dt}$$

$$\frac{r(44,788)(114)V}{1000} = 5105.83 r V\ (\text{kJ/s})$$

$$= n_{\text{total}} c_v \frac{dT}{dt}(\text{kJ/s})$$

where $n_{\text{total}}$ is the total moles of material within the cylinder, $V$ is the volume, and $c_v(T)$ is given by the

foregoing function of temperature. The volume is constant between points 2 and 3; thus

$$n_{\text{total}} = \frac{PV}{R_u T} \qquad P = P_2 \frac{T}{T_2}$$

Replace $n_{\text{total}}$ and simplify:

$$5105.83 r = \frac{P_2}{R_u T_2} c_v \frac{dT}{dt}$$

Replace $r$ by Equation 7.64 and substitute numerical data from Table 7-1 for $C_8H_{18}$:

$$A \exp\left(-\frac{E_a}{R_u T}\right)\left(\frac{n_{\text{fuel}}}{V}\right)^a\left(\frac{n_{O_2}}{V}\right)^b (5105.83)$$

$$= \frac{P_2}{T_2 R_u} c_v \frac{dT}{dt}$$

$$18 \times 10^6 \exp\left(-\frac{15,000}{T}\right)$$

$$(4.232)^{0.25}(66.128)^{1.5}(5105.83)$$

$$= \frac{1837.9}{(689.2)(8.314)} c_v \frac{dT}{dt}$$

$$7.0848 \times 10^{13} \exp\left(-\frac{15,000}{T}\right) = 0.3207 c_v \frac{dT}{dt}$$

In terms of $\theta$, the above becomes

$$2.2092 \times 10^{12} \exp\left(-\frac{150}{\theta}\right) = c_v \frac{d\theta}{dt}$$

$$\int_0^t dt = t = 100 \int_{6.892}^{32.26}$$

$$\frac{(30.756 - 512.79\theta^{-1.5} + 1072.7\theta^{-2} - 820.4\theta^{-3})\,d\theta}{2.2092 \times 10^{12} \exp(-150/\theta)}$$

$$= 0.01\ \text{s} = 10\ \text{ms}$$

At the end of the compression stroke (point 2) the piston is at top dead center, whereupon it immediately begins its downward motion during the power stroke. If it takes 40 ms to complete a stroke and if it takes 10 ms for combustion, it is necessary to initiate combustion (produce the spark) before the piston reaches top dead center. Thus the spark is set (advanced) to initiate combustion momentarily before the piston reaches top dead center and combustion continues as the piston moves downward during the power stroke. The assumption that combustion occurs instantaneously is a useful idealization for comparing different power cycles, but it is an idealization nevertheless. The Mathcad program producing the results above can be found on the textbook's Web site on the Prentice Hall Web page.

## 7.3 Formation of carbon oxides

To understand the formation of the carbon oxides in combustion processes, one needs to realize that chemical equilibrium sets an upper limit on what products can be formed from the reactants, whereas chemical kinetics dictates how fast products are formed. Given a mixture of CO, $CO_2$, and $O_2$, the value of [CO]/[CO2] at equilibrium is dictated by

$$CO_2 \rightleftharpoons CO + \tfrac{1}{2}O_2 \qquad (7.68)$$

The natural logarithm of the equilibrium constants based on concentration $\left(\ln K_a\right)$ are shown in Table 7-2. On the basis of the values in Table 7-2, one expects high temperatures to encourage the formation of CO. Once $CO_2$ exists, it is highly unlikely that CO will be formed as the gas cools. Kinetics relates the rate of the reverse reaction to the rate of cooling.

**Table 7-2**  Equilibrium Constant for $CO_2 \rightleftharpoons CO + \tfrac{1}{2}O_2$

| $T(K)$ | $\ln K_a$ |
|---|---|
| 1000 | −23.529 |
| 1200 | −17.871 |
| 1400 | −13.842 |
| 1600 | −10.830 |
| 1800 | −8.497 |
| 2000 | −6.635 |
| 2200 | −5.120 |
| 2400 | −3.860 |
| 2600 | −2.801 |

## EXAMPLE 7.3    EQUILIBRIUM CONCENTRATIONS OF CO AND $CO_2$

Assume that an exhaust gas mixture at STP is composed of CO, $CO_2$, and $O_2$ with the following initial molar composition: 18% $CO_2$, 3.3% $O_2$, and 78.7% diluent (predominately $N_2$ and $H_2O$). The temperature is raised to a high value and maintained at this value. Compute and plot the equilibrium concentration of CO (ppm) for pressures ($P$), 0.5, 1.0, and 2.0 atm over the range of temperatures 1600 to 2600 K if equilibrium is governed by Equation 7.68.

**Solution**    The material in Section 1.7 describes how to set up a table to perform the calculation. Assume 1 gmol of the gas mixture and let $z$ be the number of gmol of CO at equilibrium. If no $CO_2$ is converted to CO, $z = 0$; if all the $CO_2$ is converted to CO, $z = 0.18$. Thus the bounds on $z$ are $0 < z < 0.18$ because there can never be more CO produced than there was $CO_2$ to start with (i.e., each molecule of CO began as a molecule of $CO_2$).

| Time | $CO_2$ | CO | $O_2$ | Diluent | $n_t$ (total gmol) |
|---|---|---|---|---|---|
| 0 | 0.18 | 0 | 0.033 | 0.787 | 1.0 |
| ∞ | $0.18 - z$ | $z$ | $0.033 + z/2$ | 0.787 | $1.0 + z/2$ |

Assuming perfect gas behavior, Equations 1.50 to 1.57 can be used to relate the equilibrium composition to the equilibrium coefficients $K_a$ (Table 7-2) by the following:

$$K_a = K_f = K_y P^{1/2} = \frac{y_{CO} \, y_{O_2}^{1/2}}{y_{CO_2}} P^{1/2}$$

where the mole fractions are expressed as $y$. Substituting data from the stoichiometric table results in

$$K_a = \frac{(z/n_t)\left[(0.5z + 0.033)/n_t\right]^{1/2} P^{1/2}}{(0.18 - z)/n_t}$$

$$= \frac{z}{0.18 - z}\left(\frac{0.5z + 0.033}{1 + 0.5z}\right)^{1/2} P^{1/2}$$

The equation above can be solved by trial and error using one of several commercial mathematical computer programs to find the value $z$ for any temperature within the range 1600 to 2600 K. The CO concentration in ppm is related to $z$ as follows:

$$CO(ppm) = \frac{10^6 z}{1 + 0.5z} \approx (z)(10^6)$$

because the values of $z$ are small (i.e., $z \ll 1$). At high temperature the assumption is weak and the trial-and-error solution is more accurate. A summary of the calculations follows.

| $T$(K) | $K_a$ | $P = 0.5$ atm [CO] (ppm) | $P = 1$ atm [CO] (ppm) | $P = 2$ atm [CO] (ppm) |
|---|---|---|---|---|
| 1,600 | $1.98 \times 10^{-5}$ | 28 | 20 | 14 |
| 1,800 | $2.04 \times 10^{-4}$ | 285 | 212 | 143 |
| 2,000 | $1.313 \times 10^{-3}$ | 1,796 | 1,279 | 909 |
| 2,200 | $5.976 \times 10^{-3}$ | 7,580 | 5,506 | 3,972 |
| 2,400 | $2.106 \times 10^{-2}$ | 22,190 | 16,790 | 12,530 |
| 2,600 | $6.075 \times 10^{-2}$ | 46,920 | 37,420 | 29,290 |

Figure E7-3 shows the equilibrium CO mole fraction in ppm over the range 1600 to 2600 K. It is clear that high temperatures and low pressures favor the production of CO. The Mathcad program producing Figure E7-3 can be found on the textbook's Web site on the Prentice Hall Web page.

Example 7.3 shows that $CO/CO_2$ equilibrium is very sensitive to temperature. Increasing the temperature from 1600 K to 1800 K increases the CO concentration by an order of magnitude. While the amount of CO that is created increases with temper-

ature, only 3.9% of the $CO_2$ has a been converted to CO at 2600 K.

From Equations 7.32 to 7.39, it is clear that $CO_2$ is formed by the oxidation of CO in the high-temperature luminous portion of flames. Carbon monoxide is formed by reactions of hydrocarbon radicals such as HCO· and hydroxyl radical OH·. Carbon dioxide is formed later by the reaction between CO and OH·. In flames with insufficient $O_2$ there is insufficient OH· and one can expect larger amounts of CO. If equilibrium existed as the exhaust gases cool, one would find much lower concentrations of CO. Unfortunately, if

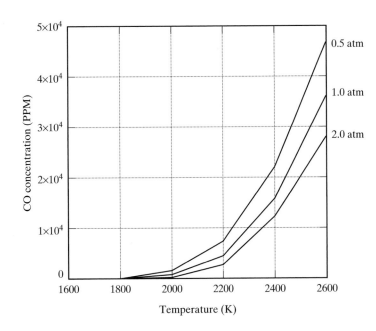

**Figure E7-3** Equilibrium concentration of CO (ppm), versus temperature (K) at three pressures 0.5, 1.0 and 2.0 atm. Initial conditions in percent by volume:
$[CO_2](0) = 18\%$,
$[O_2](0) = 3.3\%$, [diluent] = 78.7%

cooling occurs too rapidly, the low equilibrium values of CO at the cool temperatures may not be achieved.

The first step in predicting the rate at which CO is converted to $CO_2$ is to identify the kinetic mechanism (i.e., identify all the reactions that consume and produce CO and the respective rate constants for each reaction). Equations 7.35 to 7.39 will be selected since it will be assumed that CO has been formed by combustion and that only its conversion to $CO_2$ is of concern.

1. Write Equations 7.35 to 7.39.

2. Write the mass balance for the net rate of appearance of CO (i.e., the sum of the rates at which CO is formed minus the sum of the rates at which it is removed):

$$r_{CO} = k_{36}[H\bullet][CO_2] - [CO](k_{35}[OH\bullet]$$
$$+ k_{37}[O_2] + k_{38}[M][O\bullet] + k_{39}[HO_2]) \quad (7.69)$$

It may be necessary to write mass balances for other species if those species appear as unknown quantities on the right-hand side of the equation. In the case of the mechanism above this is not necessary since all the species on the right-hand side are constants or can be evaluated in terms of other known quantities. In some cases the species concentration need not be truly constant if the variation in concentration is small and can be neglected. For example, if $O_2$ and $H_2O$ are considerably larger than the maximum amount of CO that is consumed, $[O_2]$ and $H_2O$ can be assumed to be constant.

3. Compute the value of all the rate constants at the temperature(s) under study.

4. Estimate the concentration of free radicals and intermediate species on the right-hand side of the equation above. In lieu of other information, the expressions in Table 7-3 can be used. The table expresses the concentrations of OH$\bullet$ and O$\bullet$ from the dissociation of $H_2O$. The other species are related to $O_2$ and $H_2O$ through kinetics described by Flagan and Seinfeld (1988) involving OH$\bullet$.

5. Determine the order of magnitude of the product of the rate constant and species concentration for each term in Equation 7.69. Be careful; sometimes the product of a small rate constant but a large concentration may be larger than the product of another term involving a large rate constant but small species concentration.

6. Simplify the equation by retaining terms of large magnitude and ignore terms of smaller magnitude.

Following the steps above and using Equations 7.37 to 7.39, the reader will find that for the conversion of CO to $CO_2$, the only reaction of any significance is Equation 7.35 (i.e., the second term on the right-hand side):

$$-r_{CO, 35} = k_{35}[CO][OH\bullet] \quad (7.70)$$

Assuming that OH$\bullet$ is a highly reactive species, the pseudoequilibrium assumption can be employed and the value of OH$\bullet$ replaced by its equilibrium value given in Table 7-3.

$$-r_{CO, 35} = \left[ 4.4T^{1.5} \exp\left(\frac{372}{T}\right) \right] \quad (7.71)$$
$$\left\{ 166[H_2O]^{1/2}[O_2]^{1/4} \exp\left(-\frac{19,680}{T}\right) \right\}[CO]$$

An equivalent empirical equation, called a *global rate* (Dryer and Glassman, 1973) equation, for CO conversion is derived from experiment.

$$-r_{CO} = (k_+ K_{c, OH})[CO][H_2O]^{1/2}[O_2]^{1/4}$$
$$- (k_- K_{c, H})[H_2O]^{1/2}[O_2]^{-1/4}([C_t] - [CO]) \quad (7.72)$$

**Table 7-3** Approximate Equilibrium Concentrations of Intermediate Species Associated with Combustion

| Reaction | Equilibrium Constant | Reference |
|---|---|---|
| $\frac{1}{2}N_2 \rightleftharpoons N\bullet$ | $K_{P, N}(atm^{-1/2}) = 3030 \exp(-57,830/T)$ | Flagan and Seinfeld (1988) |
| $\frac{1}{2}O_2 \rightleftharpoons O\bullet$ | $K_{P, O}(atm^{-1/2}) = 3030 \exp(-37,790/T)$ | Flagan and Seinfeld (1988) |
| $\frac{1}{2}H_2O + \frac{1}{4}O_2 \rightleftharpoons OH\bullet$ | $K_{P, OH}(atm^{1/4}) = 166 \exp(-19,680/T)$ | Flagan and Seinfeld (1988) |
| $\frac{1}{2}H_2O \rightleftharpoons H\bullet + \frac{1}{4}O_2$ | $K_{P, H}(atm^{3/4}) = 44,100 \exp(-42,500/T)$ | Flagan and Seinfeld (1988) |
| $\frac{1}{2}H_2O + \frac{3}{4}O_2 \rightleftharpoons HO_2\bullet$ | $K_{P, HO_2}(atm^{-1/4}) = 0.073 \exp(-16,920/T)$ | Unknown |
| $H_2O + \frac{1}{2}O_2 \rightleftharpoons H_2O_2$ | $K_{P, H2O_2}(atm^{-1/2}) = 8.26 \times 10^{-4} \exp(-12,685/T)$ | Unknown |

where $k_+K_{c,OH}$ and $k_-K_{c,H}$ pertain to Equations 7.57 and 7.59. Neglecting the reverse reaction, Equation 7.72 has the same functional form as Equation 7.71:

$$-r_{CO} = 1.3 \times 10^{10} \exp\left(-\frac{20,140}{T}\right)$$

$$[CO][H_2O]^{1/2}[O_2]^{1/4} \quad (7.73)$$

where the concentration is expressed in gmol/m³. From this equation it can be seen that the oxidation of CO to $CO_2$ depends on the presence of $O_2$, which agrees with experimental measurements. From Equation 7.70 one can see that in fuel-lean flames when the temperature and the OH• concentration are apt to be large, the conversion is rapid. In fuel-rich flames the OH• concentration and temperatures are apt to be lower and the conversion is slower.

To determine [CO] as a function of time in a closed system, a mass balance is required. If the closed system has constant temperature and volume, the mass balance gives the following differential equation, which can be combined with the rate equation and integrated in a closed form:

$$r_{CO} = \frac{1}{V}\frac{dN_{CO}}{dt} = \frac{d[CO]}{dt} \quad (7.74)$$

### EXAMPLE 7.4    RATE OF OXIDATION OF CO TO CO₂ FOLLOWING A STEP-FUNCTION DECREASE IN TEMPERATURE

Assume an exhaust gas mixture maintained at constant pressure (1 atm) at 2000 K in a closed system. Initially it has the following molar composition, 18% $CO_2$, 3.3% $O_2$, 12% $H_2O$, and the rest is a diluent, predominately $N_2$. Example 7.3 indicates that at 2000 K, the initial CO concentration is 1280 ppm. The gas mixture now undergoes an instantaneous step-function decrease in temperature to a final constant temperature of either 1100, 1050, 1000, and 900 K. The objective is to predict [CO] as a function of time.

**Solution**    At 2000 K, the total molar concentration $(n_t)$ is

$$n_t = \frac{P}{R_u T} = \frac{1\ \text{atm}}{(0.082\ \text{L·atm/gmol K})(2000\ \text{K})}$$

$$= 0.00609\ \text{gmol/L} = 6.09\ \text{gmol/m}^3$$

Thus the initial molar concentrations of $H_2O$ and $O_2$, are

$$[H_2O] = (0.12)(6.09) = 0.732\ \text{gmol/m}^3$$

$$[O_2] = (0.033)(6.09) = 0.201\ \text{gmol/m}^3$$

Since the amounts of $H_2O$ and $O_2$ are large compared to CO, it can be assumed that the gmol values of $O_2$ and $H_2O$ do not change. Since the mixture can be assumed to be an ideal gas, the molar densities $(\text{gmol/m}^3)$ of $H_2O$ and $O_2$ are functions of the final temperature $(T)$:

$$[H_2O]_{final} = (0.732)\left(\frac{2000}{T}\right) = \frac{1464}{T}\ (\text{gmol/m}^3)$$

$$[O_2]_{final} = (0.201)\left(\frac{2000}{T}\right) = \frac{402}{T}\ (\text{gmol/m}^3)$$

From Equations 7.73 and Equation 1.82 the rate of change of CO concentration is equal to

$$-\frac{d[CO]}{dt} = -r_{CO} = [CO][H_2O]^{1/2}[O_2]^{1/4}$$

$$\left[1.3 \times 10^{10} \exp\left(\frac{-20.140}{T}\right)\right]$$

Substituting for the water and oxygen concentrations and separating variables yields

$$\frac{d[CO]}{[CO]} = -\left(\frac{1464}{T}\right)^{1/2}\left(\frac{402}{T}\right)^{1/4}$$

$$\left[1.3 \times 10^{10} \exp\left(\frac{-20,140}{T}\right)\right]dt$$

Integrating with respect to time and expressing the CO concentration in ppm results in the following expression:

$$CO(\text{ppm}) = 1280 \exp\left\{-t\left(\frac{1,464}{T}\right)^{1/2}\right.$$

$$\left.\left(\frac{402}{T}\right)^{1/4}\left[1.3 \times 10^{10} \exp\left(-\frac{20,140}{T}\right)\right]\right\}$$

The results, shown in Figure E7-4, indicate that the conversion of CO to $CO_2$ proceeds slowly for low final temperatures. At the high temperatures encountered in combustion processes, the conversion of CO to $CO_2$ is seen to occur more rapidly.

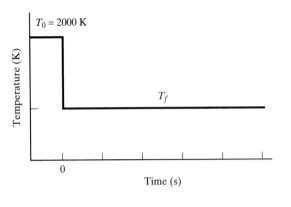

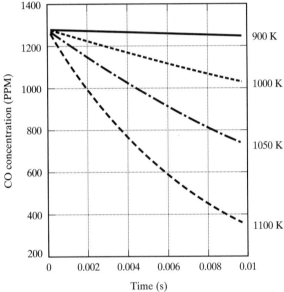

**Figure E7-4** CO concentration (ppm) versus time following a step-decrease in temperature for several final temperatures $T_f$ (900, 1,000, 1050 and 1,100 K)

Initial composition by volume:

$[CO]_0 = 1280$ ppm, $[CO_2]_0 = 18\%$, $[O_2]_0 = 3.3\%$, $[H_2O]_0 = 12\%$, $[\text{diluent}]_0 = 60.7\%$

While we recognize that practical combustion systems do not operate at constant pressure and temperature, the results above provide useful qualitative indications of performance. The elapsed time during the expansion stroke of a reciprocating engine or during the expansion through a turbine of a Brayton cycle are between 0.001 and 0.010 s; consequently, there is sufficient time at the higher temperatures for significant conversion. It must be emphasized that the reduction in temperature

experienced by exhaust gases in these cycles is not a step-function decrease, nor is the pressure constant; nevertheless, the computed results reveal the sensitivity of the kinetics of conversion to the final temperature. Since $CO/CO_2$ kinetics are fast, it is highly likely that some CO will be converted to $CO_2$ during the expansion portion (point 3 to 4) of the power cycles in Figure 7-6. The Mathcad program producing Figure E7-4 can be found in the textbook's Web site on the Prentice Hall Web page.

# 7.4 Formation of nitrogen oxides

The oxides of nitrogen associated with combustion are nitric oxide (NO) and nitrogen dioxide $(NO_2)$. Because they appear together, the oxides of nitrogen $(NO/NO_2)$ will hereafter be referred to as $NO_x$. Nitrogen oxides of higher order $(NO_3, N_2O_5,$ etc.) form in the atmosphere but are only of importance at night since they are photolytic and are reduced quickly to $NO_x$ in the presence of sunlight. Anthropogenic $NO_x$ is generated by stationary and mobile power supplies. Nitrogen oxides originate from two sources of nitrogen:

1. Oxidation of diatomic nitrogen $(N_2)$ in air, hereafter called atmospheric $N_2$
2. Conversion of molecular nitrogen compounds in fuel

Nitrogen dioxide is formed at low temperature by the exothermic reactions of NO. The major mechanistic reactions are slow reactions such as

$$NO + NO + O_2 \rightarrow NO_2 + NO_2 \qquad (7.75)$$

$$k_{+1}(m^6\,\text{gmol}^{-2}\,\text{s}^{-1}) = 1.2 \times 10^{-3} \exp(530/T)$$

$$NO + O\cdot + M \rightarrow NO_2 + M \qquad (7.76)$$

$$k_{+2}(m^6\,\text{gmol}^{-2}\,\text{s}^{-1}) = 1.5 \times 10^3 \exp(940/T)$$

$$NO_2 + M \rightarrow NO + O\cdot + M \qquad (7.77)$$

$$k_{-2}(m^3\,\text{gmol}^{-1}\,\text{s}^{-1}) = 1.1 \times 10^{10} \exp(-33,000/T)$$

$$NO + HO_2\cdot \rightarrow NO_2 + OH\cdot \qquad (7.78)$$

$$k_{+3}(m^3\,\text{gmol}^{-1}\,\text{s}^{-1}) = 2.1 \times 10^6 \exp(240/T)$$

**Table 7-4** Equilibrium Constants for NO and NO$_2$

| $T$(K) | $\frac{1}{2}N_2 + \frac{1}{2}O_2 \rightleftharpoons NO$<br>$K_{P,\,NO} = (P_{NO})/[(P_{N_2})^{1/2}(P_{O_2})^{1/2}]$ | $NO + \frac{1}{2}O_2 \rightleftharpoons NO_2$<br>$K_{P,\,NO_2}(\text{atm}^{1/2}) = (P_{NO_2})/[(P_{NO})(P_{O_2})^{1/2}]$ |
|---|---|---|
| 300 | 0 | $2.63 \times 10^6$ |
| 600 | $6.072 \times 10^{-8}$ | 25.6 |
| 900 | $2.59 \times 10^{-5}$ | 0.547 |
| 1200 | $5.348 \times 10^{-4}$ | 0.080 |
| 1500 | 0.003 | 0.025 |
| 1800 | 0.011 | 0.012 |
| 2100 | 0.026 | 0.007 |
| 2400 | 0.050 | 0.004 |
| 2700 | 0.83 | 0.003 |

NO$_2$ formation is favored by rapid cooling in the presence of substantial O$_2$ and occurs after the exhaust gas enters the atmosphere.

Combustion produces two types of nitric oxide. *Thermal NO* is nitric oxide formed by the high-temperature oxidation of N$_2$ in the combustion air, and *prompt NO* is nitric oxide formed by reactions of N$_2$ and hydrocarbon free radicals generated in flames. If the fuel contains nitrogen-bearing compounds, free-radical reactions and nitrogen atoms in the fuel produce NO$_x$ called *fuel-NO$_x$*. Approximately 85% of the NO$_x$ in combustion products is thermal NO$_x$ and the remainder is prompt or fuel-NO$_x$.

## 7.4.1 Thermal NO$_x$

We begin by considering reaction equilibria and then move on to kinetics and rates. The processes that produce thermal NO$_x$ are also called *thermal fixation of atmospheric nitrogen*. The overall chemical equilibria involving NO, NO$_2$, and atmospheric oxygen and nitrogen can be expressed as

$$\tfrac{1}{2}N_2 + \tfrac{1}{2}O_2 \rightleftharpoons NO$$

$$K_{P,\,NO} = 4.71 \exp(-10{,}900/T) \quad (7.79)$$

$$NO + \tfrac{1}{2}O_2 \rightleftharpoons NO_2$$

$$K_{P,\,NO_2}(\text{atm}^{-1/2}) = 2.5 \times 10^{-4} \exp(6923/T) \quad (7.80)$$

Tabulated in Table 7-4 are defining equations and values of the equilibrium constants for Equations 7.79 and 7.80.

### EXAMPLE 7.5 EQUILIBRIUM CONCENTRATIONS OF NO AND NO$_2$

Consider a gaseous mixture containing 3.3% O$_2$ and 77% N$_2$. The remainder of the mixture is inert diluents, predominately CO$_2$ and H$_2$O. This mixture simulates the exhaust products of an internal combustion engine. Compute and plot the equilibrium concentrations of NO and NO$_2$ at several elevated temperatures ($T$) and atmospheric pressure ($P = 1$ atm). Neglect any reactions with CO$_2$ and H$_2$O.

**Solution** Assume that initially there is 1 gmol of the gas mixture. The unknowns $x$ and $y$ represent the number of gmol of NO and NO$_2$ that are formed via Equations 7.79 and 7.80.

| Time | N$_2$ | O$_2$ | NO | NO$_2$ | Diluent | $n_t$ (total gmol) |
|---|---|---|---|---|---|---|
| 0 | 0.77 | 0.033 | 0 | 0 | 0.197 | 1 |
| $\infty$ | $0.77 - x/2$ | $0.033 - x/2 - y/2$ | $x - y$ | $y$ | 0.197 | $1 - y/2$ |

From Equation 1.56, the appropriate expressions for equilibrium, noting that $P_i = y_i P$, where $y_i$ is the mole fraction of component $i$, are

$$K_{P,\text{NO}} =$$

$$\frac{(x - y)/n_t}{\left[(0.77 - x/2)/n_t\right]^{1/2}\left[(0.033 - x/2 - y/2)/n_t\right]^{1/2}} P^{(1-1)}$$

$$K_{P,\text{NO}} = \frac{x - y}{(0.77 - x/2)^{1/2}(0.033 - x/2 - y/2)^{1/2}}$$

$$K_{P,\text{NO}_2} =$$

$$\frac{y/n_t}{\left[(x - y)/n_t\right]\left[(0.033 - x/2 - y/2)/n_t\right]^{1/2}} \left(\frac{1}{P}\right)^{1/2}$$

$$K_{P,\text{NO}_2} = \frac{y}{x - y}\left(\frac{1 - y/2}{0.033 - x/2 - y/2}\right)^{1/2}\left(\frac{1}{P}\right)^{1/2}$$

where $P = 1$ atm. The NO and $NO_2$ concentrations in ppm are

$$\text{NO(ppm)} = 2 \times 10^6 \frac{x - y}{2 - y}$$

$$\text{NO}_2\text{(ppm)} = 2 \times 10^6 \frac{y}{2 - y}$$

These equations can be solved numerically by trial and error. Figure E7-5 and the summary below show

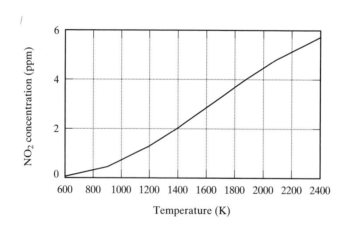

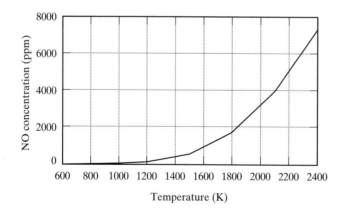

**Figure E7-5** Equilibrium concentrations (ppm) of $NO_2$ and NO versus temperature (K). Initial conditions % by volume: $[N_2]_0 = 77\%$, $[O_2]_0 = 3.3\%$, [diluent] $= 19.7\%$, Pressure $= 1$ atm

the equilibrium concentrations of NO and $NO_2$ versus temperature.

| T(K) | $(NO)_e$ (ppm) | $(NO_2)_e$ (ppm) |
|------|------|------|
| 600 | 0.01 | 0.04 |
| 900 | 4.1 | 0.4 |
| 1200 | 85.2 | 1.2 |
| 1500 | 522.2 | 2.4 |
| 1800 | 1736 | 3.7 |
| 2100 | 4045 | 4.8 |
| 2400 | 7512 | 5.8 |

From the equilibrium calculations above it is seen that $NO_2$ is slightly larger than NO at low temperatures, but at high temperatures there is considerably more NO than $NO_2$. Thus it would not be surprising to find NO produced by combustion at peak cycle temperatures. The Mathcad program producing Figure E7-5 can be found in the textbook's Web site on the Prentice Hall Web page.

Measurements of NO and $NO_2$ concentrations in the exhaust gases of a variety of combustors indicate that they are not governed by their equilibrium values based on the measured exhaust temperature. The NO concentration exceeds the equilibrium value predicted by the measured exhaust gas temperature by a considerable amount. The measured exhaust concentrations are closer to the equilibrium values one would expect based on the peak cycle temperature, $T_3$, in Figure 7-6. The conclusion to draw is that as the gas expands and cools during the power stroke, there is insufficient time for an adjustment to a new equilibrium to occur. In succinct terms, the formation of NO seems to be dictated by chemical kinetics rather than equilibrium and the reactions are too slow for equilibrium to be achieved. Engineers should be suspicious of conclusions drawn from analyses based solely on chemical equilibrium thermodynamics.

Thermal nitric oxide (NO) formation is governed by free-radical reactions referred to as the extended *Zeldovitch mechanism* (Bowman and Kesten, 1973; and Bowman 1973, 1975)

$$N_2 + O\cdot \rightleftharpoons NO + N\cdot \qquad (7.81)$$

$$k_{+1}(\text{m}^3 \text{ gmol}^{-1} \text{ s}^{-1}) = 1.8 \times 10^8 \exp(-38{,}370/T)$$

$$k_{-1}(\text{m}^3 \text{ gmol}^{-1} \text{ s}^{-1}) = 3.8 \times 10^7 \exp(-425/T) \quad (7.82)$$

$$N\cdot + O_2 \rightleftharpoons NO + O\cdot \qquad (7.83)$$

$$k_{+2}(\text{m}^3 \text{ gmol}^{-1} \text{ s}^{-1}) = 1.8 \times 10^4 T \exp(-4680/T)$$

$$k_{-2}(\text{m}^3 \text{ gmol}^{-1} \text{ s}^{-1}) = 3.8 \times 10^3 T \exp(-20{,}820/T) \qquad (7.84)$$

$$N\cdot + OH\cdot \rightleftharpoons NO + H\cdot \qquad (7.85)$$

$$k_{+3}(\text{m}^3 \text{ gmol}^{-1} \text{ s}^{-1}) = 7.1 \times 10^7 \exp(-450/T)$$

$$k_{-3}(\text{m}^3 \text{ gmol}^{-1} \text{s}^{-1}) = 1.7 \times 10^8 \exp(-24{,}560/T) \quad (7.86)$$

The rate of formation of NO and N· are

$$r_{NO} = k_{+1}[N_2][O\cdot] - k_{-1}[N\cdot][NO]$$
$$+ k_{+2}[N\cdot][O_2] - k_{-2}[NO][O\cdot]$$
$$+ k_{+3}[N\cdot][OH\cdot] - k_{-3}[NO][H\cdot] \quad (7.87)$$

$$r_{N\cdot} = k_{+1}[N_2][O\cdot] - k_{-1}[N\cdot][NO]$$
$$- k_{+2}[N\cdot][O_2] + k_{-2}[NO][O\cdot]$$
$$- k_{+3}[N\cdot][OH\cdot] + k_{-3}[NO][H\cdot] \quad (7.88)$$

Equations similar to the above can be written for the other species (i.e., H·, $O_2$, and $N_2$), but it is not necessary to do so. An equation for N· was written since the free radical appears in Equation 7.87. Since N· is an active species, it is logical to assume the pseudo-steady-state assumption (see Section 1.7.2): set $r_{N\cdot} = 0$, solve for the steady-state value of $[N\cdot]$, and designate the value by the symbol $[N\cdot]_{ss}$:

$$[N\cdot]_{ss} =$$
$$\frac{k_{+1}[N_2][O\cdot] + k_{-2}[NO][O\cdot] + k_{-3}[NO][H\cdot]}{k_{-1}[NO] + k_{+2}[O_2] + k_{+3}[OH\cdot]} \quad (7.89)$$

At this point users should perform an *order-of-magnitude* analysis for each term on the right-hand side of the combined equation and eliminate (cautiously) terms that are several orders of magnitude smaller than others. The following are guidelines for performing an order-of-magnitude analysis.

**1.** Select a typical temperature that one expects to encounter in the process and determine the order of magnitude of each kinetic rate constant.

**2.** Define the order of magnitude of the concentrations of each stable and free-radical species on the right-hand side of the combined equation. Use data from Figure 7-9 for guidance. Keep in mind what kind of flame you are studying. Fuel-lean flames will contain $O_2$ in their exhausts, perhaps with mole fractions of 0.03 to 0.05. Fuel-rich flames contain much smaller concentrations of $O_2$. Free radical concentrations are several orders of magnitude smaller still and depend on temperature.

**3.** Since [NO] is the unknown that is sought, be hesitant to assign a value to it.

**4.** Stable species such as $[N_2]$ and $[H_2O]$ will probably have very large concentrations that change little during the combustion process.

**5.** Each term on the right-hand side of the combined equation is the product of three elements, a rate constant (or product of rate constants) and the concentrations of two species, one or both of which may be a stable species or a free radical. Don't be hasty; while a rate constant may be small, the remaining terms may be large (or vice versa), such that the product is significant.

Most combustion systems are operated with excess air (fuel-lean combustion) and oxygen $(O_2)$ is typically found in the exhaust products. In the analysis that follows it will be assumed that there is excess oxygen. For this case Wark and Warner (1981) assert that Reactions 7-85 and 7-86 are unimportant. Heywood (1988) supports this position for fuel-lean combustion (for example at fuel-air equivalence ratios less than 0.6) but makes it clear that this assumption would lead to substantial errors in stoichiometric and fuel-rich mixtures. Thus the last term in the numerator and denominator of Equation 7.89 can be neglected. Substitute $[N\bullet]_{ss}$ into the equation for $r_{NO}$. After manipulation the combined equation can be written as

$$r_{NO} = \frac{2[O\bullet]\left\{k_{+1}[N_2] - \left(\dfrac{k_{-1}k_{-2}}{k_{+2}[O_2]}\right)[NO]^2\right\}}{\left\{1 + \dfrac{k_{-1}[NO]}{k_{+2}[O_2]}\right\}}$$

(7.90)

Consider the change in NO concentration resulting from a step decrease or increase in temperature.

To achieve a familiar dimensionless form for this equation, we define a non-dimensionalizing parame-

ter $[NO]_e$ as the NO concentration that would exist at equilibrium following a step change in temperature which is the context of this analysis. At equilibrium, Equations 7.81 to 7.84 can be rewritten as follows:

$$\frac{k_{+1}}{k_{-1}} = \frac{[NO]_e[N\bullet]_e}{[N_2]_e[O\bullet]_e}$$

(7.91)

$$\frac{k_{+2}}{k_{-2}} = \frac{[NO]_e[O\bullet]_e}{[N\bullet]_e[O_2]_e}$$

(7.92)

Solving Equation 7.92 for $[N\bullet]_e/[O\bullet]_e$ and substituting into Equation 7.91 gives

$$\frac{k_{+1}}{k_{-1}}\frac{k_{+2}}{k_{-2}} = \frac{[NO]_e^2}{[O_2]_e[N_2]_e}$$

(7.93)

The right-hand side of Equation 7.93 is seen to be the square of the equilibrium constant for Equation 7.79. Thus

$$K_{P,NO}^2 = \frac{[NO]_e^2}{[O_2]_e[N_2]_e} = \frac{k_{+1}}{k_{-1}}\frac{k_{+2}}{k_{-2}}$$

(7.94)

and

$$K_{P,NO} = 4.71 \exp\left(-\frac{10{,}900}{T}\right)$$

(7.95)

Because $K_{P,NO}$ and the corresponding K based on concentration are dimensionless, they may be used interchangeably as long as consistent units are used for all concentrations. Otherwise, the ideal gas law would have to be used to convert the equilibrium constant to appropriate units. Using the relationships above, Equation 7.90 can be rewritten as

$$r_{NO} = \frac{2k_{+1}[O\bullet][N_2]\left\{1 - \dfrac{[NO]^2}{K_{P,NO}[N_2][O_2]}\right\}}{\left\{1 + \dfrac{k_{-1}[NO]}{k_{+2}[O_2]}\right\}}$$

(7.96)

The concentrations of $N_2$ and $O_2$ in fuel-lean combustion products are known quantities and decrease only slightly when NO is produced. Since $[O\bullet]$ is a highly reactive species, it can be replaced by its local equilibrium value $[O\bullet]_e$. At the high temperatures of combustion $[\text{O·}]_e$ is dictated by the high-temperature dissociation of $O_2$ (Equation 7.55):

$$\tfrac{1}{2}O_2 \rightleftharpoons \text{O·}$$

$$K_{P,\text{O}}(\text{atm}^{1/2}) = 3030 \exp\left(-\frac{30{,}790}{T}\right) \quad (7.97)$$

Thus

$$[\text{O·}] = [\text{O·}]_e = \frac{[O_2]^{1/2} K_{P,\text{O}}}{(R_u T)^{1/2}} \quad (7.98)$$

The expression for $[\text{O·}]_e$ above can be substituted into Equation 7.96 to obtain $r_{\text{NO}}$ as a function of the concentrations of molecular species only. To generalize Equation 7.96, it is useful to define several substitute parameters.

$$Y = \frac{[\text{NO}]}{[\text{NO}]_e} \quad (7.99)$$

$$C = \frac{k_{-1}(K_{P,\text{NO}})[N_2]^{1/2}}{k_{+2}[O_2]^{1/2}} \quad (7.100)$$

$$M = \frac{4k_{+1}K_{P,\text{O}}[N_2]^{1/2}}{(R_u T)^{1/2} K_{P,\text{NO}}}$$

$$= 5.7 \times 10^{15} T^{-1} P^{1/2} \exp\left(-\frac{56{,}000}{T}\right) \quad (7.101)$$

with $M$ in seconds$^{-1}$, $P$ in atmospheres, and $T$ in Kelvin. Thus Equation 7.96 becomes

$$r_{\text{NO}}[\text{NO}]_e = \frac{M(1 - Y^2)}{2(1 + CY)} \quad (7.102)$$

The reciprocal of $M$ is a time constant ($\tau$) that characterizes the speed with which the reaction proceeds. For a particular pressure, the value of $M$ is strongly dependent only on temperature, rising by a factor of 10 every 200 K.

| $T$(K) | $M$(s$^{-1}$) | $\tau = 1/M$(s) |
|---|---|---|
| 1,800 | 0.098 | 10.251 |
| 1,900 | 0.475 | 2.104 |
| 2,000 | 1.971 | 0.507 |
| 2,100 | 7.120 | 0.140 |
| 2,200 | 22.839 | 0.044 |
| 2,300 | 66.072 | 0.015 |
| 2,400 | 174.633 | 0.006 |
| 2,500 | 426.323 | 0.002 |
| 2,600 | 970.212 | 0.001 |

The value of $C$ is weakly dependent on the final temperatures and since the concentrations of $[N_2]$ and $[O_2]$ are essentially equal to their values in the original gas, $C$ is essentially a constant and $0 < C < 1$.

To illustrate the generation of thermal $\text{NO}_x$, consider known initial amounts of $N_2$ and $O_2$ in a container that is suddenly subjected to a step increase in temperature and predict the $\text{NO}_x$ concentration as a function of time. Such a process is a crude simulation of the generation of $\text{NO}_x$ in the gases in a reciprocating engine during the heat addition path prior to the power stroke. Admittedly, the gas temperature and pressure decrease as the gas expands during the power stroke, but the results of the analysis for a step rise in temperature provides a conservative estimate of the maximum $\text{NO}_x$ that can be generated at the peak cycle temperature. For a constant-volume batch system, the mass balance for NO generation is

$$r_{\text{NO}} = \frac{1}{V}\frac{dN_{\text{NO}}}{dt} = \frac{d[\text{NO}]}{dt} \quad (7.103)$$

Combining the mass balance (Equation 7.103) with the rate equation (Equation 7.102) gives the ordinary differential equation

$$\frac{dY}{dt} = \frac{M(1 - Y^2)}{2(1 + CY)} \quad (7.104)$$

Integrating Equation 7.104,

$$\int_{Y_0}^{Y(t)} \frac{(1 + CY)\,dY}{1 - Y^2} = \int_{Y_0}^{Y(t)} \frac{dY}{1 - Y^2}$$

$$+ \int_{Y_0}^{Y(t)} \frac{CY\,dY}{1 - Y^2} = \int_0^t \frac{M\,dt}{2} \quad (7.105)$$

becomes

$$\left(\frac{1 - Y}{1 - Y_0}\right)^{C+1}\left(\frac{1 + Y}{1 + Y_0}\right)^{C-1} = \exp(-Mt) \quad (7.106)$$

where $Y_0$ is based on the initial NO concentration.

Example 7.5 illustrated how sensitive the equilibrium concentrations of NO and $NO_2$ are to temperature. With regard to combustion cycles in Figure 7-6, it is useful to inquire if there is sufficient time for equilibrium to occur during the compression portion of the cycle. A first step in such an analysis is to consider how fast the concentrations of NO and $NO_2$

respond to an instantaneous increase in temperature, or "step-up" change in temperature. For the step-up problem presented above, when the temperature rises instantaneously from 300 K($Y_0 \approx 0$) to an elevated temperature, Equation 7.106 reduces to the following:

$$(1 - Y)^{C+1}(1 + Y)^{C-1}$$

$$= \exp(-Mt) = \exp\left(-\frac{t}{\tau}\right) \quad (7.107)$$

### EXAMPLE 7.6 GENERATION OF NO IN AN EXHAUST GAS MIXTURE SUBJECTED TO A STEP INCREASE IN TEMPERATURE

To simulate the rate at which NO is generated at the peak combustion cycle temperature, consider an oxygen–nitrogen mixture in which the molar composition is 77% $N_2$, 3.3% $O_2$ (the remainder being diluents that do not participate in the chemical kinetics) contained in a constant-pressure vessel at STP. Such a composition is similar to the products of combustion in a fuel-lean flame. The temperature is suddenly raised and maintained at a constant value for a period of time. The pressure ($P$) is assumed to be constant (1 atm). Using the analytical model above, predict and plot the NO concentration in ppm versus time for selected high final temperatures. The simulation is limited in that once the gas reaches its peak cycle temperature in a reciprocating engine, the temperature and pressure decrease during the power stroke. Nonetheless, the example illustrates how important small changes in peak cycle temperatures are in the kinetics of NO generation.

***Solution*** To predict the instantaneous concentration of NO, one solves Equation 7.107 to find the time ($t$) for selected values of $Y$. However, since a graph of concentration versus time is sought, it is easier to compute the time ($t$) for selected values of $Y$ and plot $t$ versus $Y$. Figure E7-6 shows the NO concentration (ppm) as a function of time for different final temperatures. Figure E7-6 is a graph of the computed results. Several important consequences are apparent.

**1.** A review of the equilibrium $[NO]_e$ as function of temperature (Figure E7-4) shows that within 1500

to 2100 K, the increase in NO concentration is approximately 50 ppm per 10 K. Small changes in temperature bring about large changes in $[NO]_e$.

**2.** The next question to face is whether there is sufficient time for [NO] to achieve (or even approach) its equilibrium value. A review of the Figure E7-6 shows that if the final temperature is 1900 K, it takes nearly 10 s to reach $Y \simeq 1$, (i.e., $[NO]_e \simeq 3000$ ppm), while at 2300 K it takes approximately 0.1 seconds to reach $Y \simeq 1$ (i.e., $[NO]_e \simeq 6000$ ppm). Thus the kinetics at 1900 K are considerably slower than they are at 2300 K (i.e., a 400° increase in temperature decreases the time it takes to achieve equilibrium by a factor of 100).

In conclusion, NO kinetics are very sensitive to peak temperature. Reducing peak temperatures 100° results in large reductions in NO concentration and only minimal reductions in overall thermal efficiency of the Otto cycle (Equation 7.5). The Mathcad program predicting Figure E7-6 can be found in the textbook's Web site on the Prentice Hall Web page.

Figure E7-6 illustrates the amount of NO formed at the end of the compression stoke at which time combustion occurs. In addition, Equation 7.106 can be used to predict instantaneous values of NO concentration following a step-function *decrease* in temperature. In this case the initial value of NO concentration, $[NO]_0$, will be assumed to be $[NO]_e$, as predicted previously on the basis of the peak combustion temperature (see Figure E7-5). In the case of a step-down problem, when the temperature decreases instantaneously from an initial elevated temperature, $Y_0$ is significant since it represents the initial NO concentration at the elevated temperature. Once again, if the final temperature is below 2200 K, the kinetics of NO reactions are very slow and little NO can be removed within the 0.01 s of the expansion process. Consequently, the NO concentration in the exhaust gases is basically equal to $[NO]_e$ based on the peak cycle temperatures, $T_3$, in Figure 7-6.

In either spark (SI) or compression ignition (CI) reciprocating engines, combustion occurs within 0.001 to 0.01 s and the postcombustion processes occur within 0.01 to 0.1 s. If peak cycle temperatures exceed approximately 2500 K, chemical kinetics are

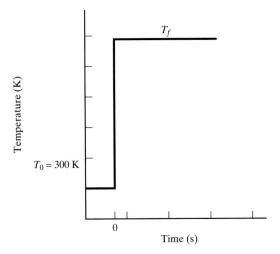

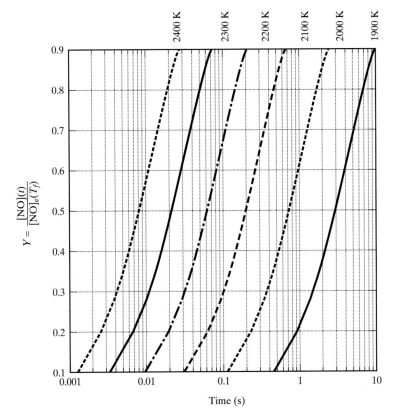

**Figure E7-6** NO concentration in an exhaust gas mixture subjected to a step-increase in temperature $(T_f)$. Initial temperature, $T_0 = 300$ K, initial concentrations by volume: $[N_2]_0 = 77\%$, $[O_2]_0 = 3.3\%$, [diluent] $= 19.7\%$

fast and large NO concentrations are nearly equal to their equilibrium values at the peak cycle temperature. If, however, peak combustion temperatures are below 2000 K, chemical kinetics are sufficiently slow to prevent NO from approaching its equilibrium value. Clearly, peak temperature is the most important variable affecting thermal $NO_x$.

Example 7.6 provides insight into the design of low-$NO_x$ combustors. High peak cycle temperatures produce large equilibrium values of NO, and cooling the products of combustion by expansion or dilution results in sluggish chemical kinetics that does not consume NO rapidly. Thus one can expect NO concentrations in combustion exhaust gases to be considerably larger than what one would expect based on the temperature of the exhaust. It is convenient to state that the concentration of NO is *frozen* at its equilibrium value, $[NO]_e$, based on the peak cycle temperature.

### 7.4.2  Prompt NO

Prompt NO is nitric oxide formed from reactions between diatomic nitrogen $(N_2)$ and hydrocarbon free radicals. Prompt NO occurs in fuel-rich regions where the hydrocarbon free radicals are more likely to be found. Since the activation energy of these reactions is low, the reactions occur "promptly" in the preflame zone rather than in the luminous or postflame zones. Considerably more NO is formed by thermal fixation (Equations 7.81 to 7.86) than by the generation of prompt NO. If the combustor geometry is modified to reduce peak temperatures and reduce the production of thermal NO, prompt NO may become more important.

### 7.4.3  Fuel $NO_x$

Fuel oils may contain nitrogen-bearing compounds having N — H bonds and C — N bonds in the form of pyridine and pyrrole structures (Figure 7-12). Typical nitrogen concentrations in fuel oil vary between 0.1 and 0.5%. In coal there may be 1.2 to 1.6% nitrogen. Early in the combustion process these ring structures break and eventually yield HCN, $NH_3$, or free radicals such as $NH_2\cdot$ and $CN\cdot$, all of which ultimately form $NO_x$.

# 7.5  *Formation of sulfur oxides*

Coal and some fuel oils contain sulfur-bearing compounds, including inorganic sulfides and organic structures called thiophenes and thiols (Figure 7-13). The majority of human-made emissions of sulfur oxides are associated with coal-fired stationary combustors (i.e., electric utility boilers, kilns, etc.). Sulfur oxide emissions from engines using distillate fuels (gas turbines, spark and compression ignition engines) are very small and arise only from trace amounts of sulfur in additives to lubricants. A review of the emission factors (Section 8.4) shows that mobile power supplies produce relatively small amounts of sulfur oxides.

### 7.5.1  $SO_2/SO_3$ equilibrium

Exhaust gases from stationary combustors typically contain excess oxygen. Sulfur is present in the form $SO_2$ along with trace amounts of $H_2S$ and COS. Exhaust gases also contain various amounts of $SO_3$, depending on the maximum combustion temperature and how rapidly the exhaust gases cool. Sulfur trioxide $(SO_3)$ in combustion systems is of interest because its presence leads to corrosion. To appreciate the relationship between $SO_2$, $SO_3$, and $O_2$ consider the chemical equilibrium

$$SO_2 + \tfrac{1}{2}O_2 \rightleftharpoons SO_3$$

$$K_{P,SO_3}(\text{atm}^{-1/2}) = \frac{P(SO_3)}{P(SO_2)P(O_2)^{1/2}}$$

$$= 1.53 \times 10^{-5} \exp\left(\frac{11{,}750}{T}\right) \quad (7.108)$$

**EXAMPLE 7.7   CONCENTRATIONS OF $SO_2$ AND $SO_3$ AT CHEMICAL EQUILIBRIUM**

Prior to studying the chemical kinetics associated with $SO_2$ and $SO_3$, consider the equilibrium between

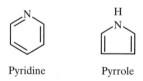

Pyridine          Pyrrole

**Figure 7-12**   Structure of two nitrogen-bearing compounds in fuels (from Flagan & Seinfeld, 1988)

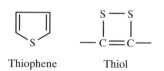

Thiophene          Thiol

**Figure 7-13**   Structure of two common sulfur compounds in coal (from Flagan & Seinfeld, 1988)

these species governed by Equation 7.108. Assume that 0.0002 gmol of $SO_3$, 0.001 gmol of $SO_2$, 0.2 gmol of $O_2$, and diluents are placed in a vessel and raised to an elevated constant temperature and at constant pressure until equilibrium occurs. This gas mixture was chosen because it simulates the exhaust gas that might be produced by a coal-fired boiler burning a sulfur-bearing coal. Compute and plot the equilibrium concentrations, $[SO_2]_e$ and $[SO_3]_e$, in ppm, and the ratio $C = [SO_2]_e/[SO_3]_e$ for a range of temperatures, 1000 K $< T <$ 2000 K, for three pressures $(P)$, 0.5, 1.0, and 2.0 atm.

**Solution** The computation will use the table described in Example 1.9. Assuming an initial gas mixture of 1 gmol total, it will be assumed that $a = 0.0002$, $b = 0.001$, $c = 0.2$, and the diluent $= 0.7988$. The total number of moles $(n_t)$

$$n_t = (0.001 - x) + \left(0.2 - \frac{x}{2}\right)$$
$$+ (0.0002 + x) + 0.7988 = 1 - \frac{x}{2}$$

From Equation 1.52,

$$K_{P, SO_3} = \frac{(0.0002 + x)/n_t}{[(0.001 - x)/n_t][(0.2 - x/2)/n_t]^{1/2}} P^{-1/2}$$
$$= \frac{0.0002 + x}{0.001 - x} \left(\frac{2 - x}{0.4 - x}\right)^{1/2} P^{-1/2}$$

where

$$K_{P, SO_3}(\text{atm}^{-1/2}) = 1.53 \times 10^{-5} \exp\left(\frac{11{,}750}{T}\right)$$

The bounds on $x$ are $-0.0002 < x < 0.001$. The concentration of $SO_2$ and $SO_3$ at equilibrium in the units of ppm are

$$[SO_2]_e(\text{ppm}) = \frac{2(0.001 - x)10^6}{2 - x}$$

$$[SO_3]_e = \frac{2(0.0002 + x)10^6}{2 - x}$$

Solve the equations involving $K_{P, SO_3}$ for $x$ by trial and error. Compute and plot the equilibrium values, $[SO_2]_e$ and $[SO_3]_e$, in ppm as a function of temperature over the range 1000 to 2000 K and pressures of 0.5 to 2.0 atm. Also plot the parameter $C$, where $C \equiv [SO_2]_e/[SO_3]_e$. The results are as follows:

| | P = 0.5 atm | | | P = 1.0 atm | | | P = 2.0 atm | | |
|---|---|---|---|---|---|---|---|---|---|
| T(K) | $[SO_2]_e$ | $[SO_3]_e$ | C | $[SO_2]_e$ | $[SO_3]_e$ | C | $[SO_2]_e$ | $[SO_3]_e$ | C |
| 1000 | 744 | 456 | 1.6 | 643 | 557 | 1.2 | 539 | 661 | 0.8 |
| 1100 | 991 | 209 | 4.7 | 924 | 275 | 3.4 | 844 | 356 | 2.4 |
| 1200 | 1104 | 96 | 11.6 | 1069 | 131 | 8.2 | 1023 | 177 | 5.8 |
| 1300 | 1153 | 47 | 24.5 | 1135 | 65 | 17.4 | 1110 | 90 | 12.3 |
| 1400 | 1175 | 25 | 46.8 | 1165 | 35 | 33.1 | 1151 | 49 | 23.4 |
| 1500 | 1185 | 14 | 81.9 | 1180 | 20 | 57.9 | 1171 | 29 | 40.9 |
| 1600 | 1191 | 9 | 133.6 | 1187 | 12 | 94.5 | 1181 | 18 | 66.8 |
| 1700 | 1194 | 6 | 205.8 | 1192 | 8 | 145.5 | 1188 | 12 | 102.9 |
| 1800 | 1196 | 4 | 301.9 | 1194 | 6 | 213.6 | 1192 | 8 | 151.1 |
| 1900 | 1197 | 3 | 425.9 | 1196 | 4 | 301.1 | 1194 | 6 | 212.4 |
| 2000 | 1198 | 2 | 580.3 | 1197 | 4 | 410.3 | 1196 | 4 | 290.1 |

where $[SO_2]_e$ and $[SO_3]_e$ are in ppm.

It is seen that increasing temperature favors the production of $SO_2$ and the disappearance of $SO_3$. In all cases the ratio of $SO_2$ to $SO_3$ $(C)$ increases with temperature. An increase in pressure by a factor of 4 results reducing C by a factor of two at all temperatures. The results for $P = 1$ atm are seen in Figure E7-7. The

Mathcad program producing Figure E7-7 can be found in the textbook's Web site on the Prentice Hall Web page.

---

The doubling of C with a quadrupling of P is a direct result of the $P^{1/2}$ factor in the equation $K_{P, SO_3}$. This is a manifestation of *Le Châtelier's principle*,

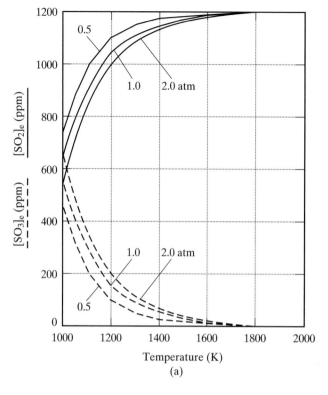

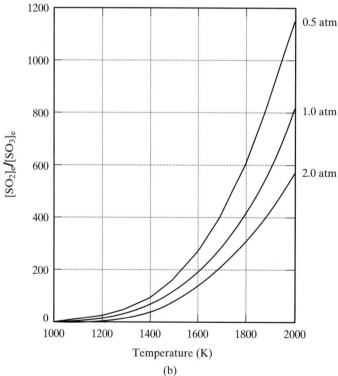

**Figure E7-7** (a) Equilibrium concentrations (ppm) of $SO_2$ and $SO_3$ versus temperature (K). (b) The ratio, $C = [SO_2]_e/[SO_3]_e$, at equilibrium for various temperatures (K)

relieve that stress. In this case the reaction shifts toward $SO_3$ (i.e. fewer moles) to relieve the stress of pressure. The table of $[SO_2]_e$ and $[SO_3]_e$ versus temperature shows that $[SO_2]_e$ is the predominant product at temperatures above 1500 K, while $[SO_3]_e$ becomes comparable in magnitude at 1000 K. Thus on the basis of equilibrium, one expects the ratio $C = [SO_2]_e/[SO_3]_e$ in flames and hot exhaust products from coal-fired boilers to be very high, and it is.

## 7.5.2 $SO_2/SO_3$ chemical kinetics

Following combustion, the hot exhaust gases pass through a heat exchanger and transfer energy to incoming combustion air. The exhaust products cool in the presence of excess oxygen. The equilibrium values tabulated in Example E7.7 would suggest that much of the $SO_2$ would be converted to $SO_3$. In fact, if we examine the value of $K_{P,500K} = 2.6 \times 10^5$, which is vastly higher than the value of 1.939 at 1000 K, we see that at low temperatures $SO_3$ becomes the predominant equilibrium product. This is *not* observed. To understand why, one must consider the rates of reactions as well. To predict the rate at which $SO_3$ is formed in the cooler environment, one needs to analyze the kinetic mechanism describing the conversion of $SO_2$ to $SO_3$.

$$SO_2 + O\cdot + M \rightarrow SO_3 + M$$

$$k_{+1}(m^6/gmol^2 \cdot s) = 8.0 \times 10^4 \exp(-1400/T) \tag{7.109}$$

$$SO_3 + O\cdot + M \rightarrow SO_2 + O_2 + M$$

$$k_{+2}(m^6/gmol^2 \cdot s) = 7.04 \times 10^4 \exp(785/T) \tag{7.110}$$

$$SO_3 + H\cdot \rightarrow SO_2 + OH\cdot$$

$$k_{+3}(m^3/gmol \cdot s) = 1.5 \times 10^7 \tag{7.111}$$

where [M] is the concentration of a neutral species. The net rate of $SO_3$ formation can be written as

$$r_{SO_3} = k_{+1}[SO_2][O\cdot][M] - k_{+2}[SO_3][O\cdot][M] + k_{+3}[SO_3][H\cdot] \tag{7.112}$$

If sulfur exists only as $SO_2$ and $SO_3$, the number of moles of sulfur $[S_t]$ will be constant:

$$[S_t] = [SO_2] + [SO_3] = \text{constant} \tag{7.113}$$

and Equation 7.112 can be written as

$$r_{SO_3} = k_{+1}[O\cdot][M]([S_t] - [SO_3])$$

$$- [SO_3](k_{+2}[O\cdot][M] - k_{+3}[H\cdot])$$

$$r_{SO_3} = k_{+1}[O\cdot][M][S_t] - [SO_3](k_{+1}[O\cdot][M]$$

$$+ k_{+2}[O\cdot][M] + k_{+3}[H\cdot]) \tag{7.114}$$

The value of $[SO_3]$ at equilibrium (see Section 1.9) can be found by setting the left-hand side $(r_{SO_3})$ of Equation 7.114 equal to zero:

$$[SO_3]_e = \frac{k_{+1}[O\cdot][S_t][M]}{\{[O\cdot][M](k_{+1} + k_{+2}) + k_{+3}[H\cdot]\}} \tag{7.115}$$

For simplicity, assume that cooling is accomplished by a step-function decrease to a constant low final temperature. For this invariant final temperature, the rate coefficient and the concentrations of $[O\cdot]$ and $[H\cdot]$ are constant. Assuming a constant-volume batch reactor, the mass balance gives for the instantaneous decrease in temperature,

$$r_{SO_3} = \frac{d[SO_3]}{dt} \tag{7.116}$$

Combining Equations 7.116 and 7.114 and defining $a$ and $\tau$ as

$$a = k_{+1}[O\cdot][M][S_t] \tag{7.117}$$

$$\tau = (k_{+1}[O\cdot][M] + k_{+2}[O\cdot][M] + k_{+3}[H\cdot])^{-1} \tag{7.118}$$

Equation 7.116 becomes

$$\frac{d[SO_3]}{dt} = a - \frac{1}{\tau}[SO_3] \tag{7.119}$$

The resulting equation can be integrated directly:

$$\int_{[SO_3]_0}^{[SO_3]} \frac{d[SO_3]}{a - [SO_3]/\tau} = \int_0^t dt \tag{7.120}$$

Integrating and simplifying, we obtain

$$[SO_3] = \frac{k_{+1}[S_t][O\cdot][M]}{k_{+1}[O\cdot][M] + k_{+2}[O\cdot][M] + k_{+3}[H\cdot]}$$

$$(1 - e^{-t/\tau}) + [SO_3]_0 e^{-t/\tau} \tag{7.121}$$

where $[SO_3]_0$ is the initial $SO_3$ concentration. To check the integration partially, we let $t$ go to infinity and observe that the remainder of Equation 7.121 is identical to Equation 7.115 for $(SO_3)_e$. Mathematically, $\tau$ is the elapsed time for the quantity

$1 - \exp(-t/\tau)$ to achieve the value 0.632. In a physical sense, the larger the value of $\tau$, the longer it takes to consume the same amount of $SO_3$, all else remaining the same.

One could calculate the conversion of $SO_2$ to $SO_3$ during cooling in ways similar to what was done concerning the oxides of carbon and nitrogen. Alternatively, a partial understanding can be gained by analyzing the time constant ($\tau$) and $[SO_3]_e$ given by Equations 7.118 and 7.115. For fuel-lean flames when $[H\cdot]$ is small, the equations above can be approximated as

$$[SO_3]_e = \frac{[S_t]}{1 + k_{+2}/k_{+1}}$$

$$= \frac{[S_t]}{1 + 0.88 \exp(2185/T)} \quad (7.122)$$

and

$$\frac{1}{\tau} = [O\cdot][M](k_{+1} + k_{+2})$$

$$= 10^4[M][O\cdot]\left[ 8.0 \exp\left( -\frac{1400}{T} \right) \right.$$

$$\left. + 7.04 \exp\left( \frac{785}{T} \right) \right] \quad (7.123)$$

where $[M]$ and $[O\cdot]$ are expressed in $gmol/m^3$. The concentration of the oxygen radical $[O\cdot]$ can be estimated by the dissociation of $O_2$ (Equation 7.98):

$$[O\cdot] = \frac{[O_2]^{1/2} K_{P,O}}{(R_u T)^{1/2}} \quad (7.98)$$

$$K_{P,O}(atm^{1/2}) = 3303 \exp\left( -\frac{30,790}{T} \right)$$

Consider final temperatures of 1200 K and 1000 K. Evaluation of Equation 7.123 reveals that

- $T_{final} = 1200$ K:   $[M] = 10.12$ gmol/m$^3$
  $[O\cdot] = 9.88 \times 10^{-8}$ gmol/m$^3$ and   $\tau = 6.2$ s
- $T_{final} = 1000$ K:   $[M] = 12.15$ gmol/m$^3$
  $[O\cdot] = 1.56 \times 10^{-9}$ gmol/m$^3$ and   $\tau = 302$ s

The significance of the above is that it requires from 6 to 302 s for 63.2% of the original $SO_2$ to be converted to $SO_3$. Consider now the elapsed time as the combustion gases leave the combustor and pass through elements of the exhaust system. Combustion

exhaust gases may spend as much as 6 s cooling, in which time some $SO_2$ may be converted to $SO_3$. Their final temperatures of 1000 K or lower are achieved in elapsed times far smaller than 302 s. Thus there is insufficient time for equilibrium to be achieved at the final temperatures, and $SO_3$ concentrations are far lower than would be expected from equilibrium theory alone.

## 7.6 Unburned hydrocarbons

An unburned hydrocarbon (HC) is the name given to an organic emission produced by incomplete combustion. It is important that readers are familiar with the classes of organic compounds that comprise the generic category *unburned hydrocarbons*. See Figure 7.14 for a summary of those compounds important to air pollution and a brief characterization of each class. Because stationary combustors operate in a steady-state fashion, engineers have the opportunity to optimize the design and operation of stationary combustors in ways that minimize the emission of unburned hydrocarbons. By contrast, the transient nature of combustion in reciprocating engines results in the generation of considerably more unburned hydrocarbons per unit of fuel burned.

Reciprocating engine exhaust gases contain a mixture of hydrocarbons, some more reactive than others. Typical spark ignition combustion (SI) products are paraffins, 33%; olefins, 27%; acetylene, 8%; and aromatics, 32% (Heywood, 1988). Regulatory agencies find it useful to categorize unburned hydrocarbons in terms of their *atmospheric reactivity* to produce $NO_2$ and urban smog. Table 7-5 summarizes the reactivity of several classes of hydrocarbons. See also Figure 6-13 *Oxygenated hydrocarbons* are organic compounds (hereafter called simply *oxygenates*) that contain oxygen. Oxygenates in an engine exhaust irritate the mucous membrane and also participate in atmospheric photochemical reactions. Particularly irritating oxygenates are carbonyls and phenols. Low-molecular-weight carbonyls such as aldehydes and aliphatic ketones are compounds of particular importance to pollution. Carbonyls represent approximately 10% of hydrocarbon emission from diesel automobile engines. Formaldehyde comprises approximately 20% of the carbonyls.

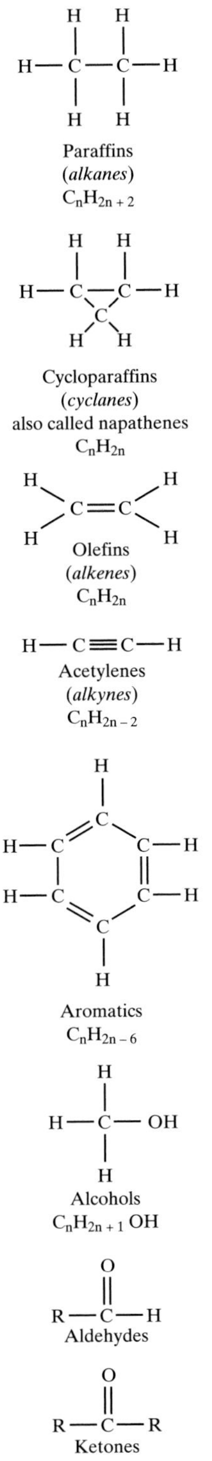

**Figure 7-14** Classes of organic compounds important to air pollution (abstracted from Heywood, 1988)

**Table 7-5** Reactivity of Hydrocarbon Emissions[a]

| Class of hydrocarbon | Relative reactivity |
|---|---|
| $C_1$–$C_4$ paraffins, acetylene, benzene | 0 |
| 2,3-Dimethyl-2-benzene | 1 |
| $C_4$ and higher-molecular-weight paraffins, monoalkyl benzenes, *ortho*- and *para*-dialkyl benzenes, cyclic paraffins | 2 |
| Ethylene, *meta*-dialkyl benzenes, aldehydes | 5 |
| 1-Olefins (except ethylene), diolefins, tri- and tetraalkyl benzenes | 10 |
| Internally bonded olefins | 30 |
| Internally bonded olefins with substitution at the double bond, cycloolefins | 100 |

*Source*: Abstracted from Heywood (1988).

[a] General Motors reactivity scale (0–100) based on the $NO_2$ formation rate for the hydrocarbon relative to the $NO_2$ formation rate for 2,3-dimethyl-2-benzene as the reference.

Diesel fuels are fuels used in compression ignition (CI) engines and are less volatile than fuels used in spark ignition engines. Consequently, they contain a larger percentage of high-molecular-weight species. Since the diesel fuels are sprayed into the combustion chamber, the particle size distribution, transport of particles inside the cylinder, and the eventual nonuniformity of the fuel (vapor)–air ratio affects the production of unburned hydrocarbons. The pyrolysis of diesel fuels is considerably more important than it is with spark ignition fuels. There is also a larger percentage of particulate unburned hydrocarbons (soot) in the exhaust of compression engines.

# 7.7 Combustion controls for reciprocating engines

A distinction will be made between *combustion control* and *air pollution controls*. Pollution controls are also called *tailpipe devices*. Combustion control is achieved through engineering design in which the temperature and velocity fields in the combustion region are established in ways that chemical reaction mechanisms will not produce pollutants, or if they are produced, that there is sufficient time for subsequent reactions to remove

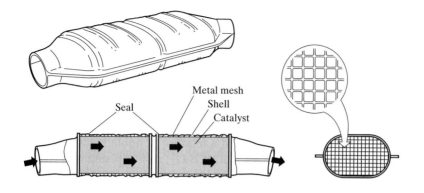

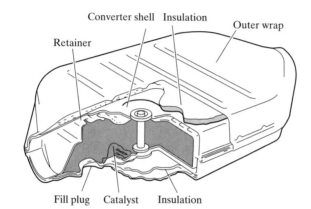

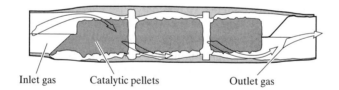

**Figure 7-18** Catalytic converters for spark-ignition engine emission control: (top) monolith design; (bottom) pelletized design (taken from Heywood, 1988)

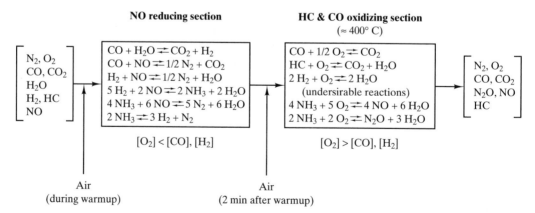

**Figure 7-19** Dual catalyst reactions (Heywood 1988)

**Table 7-6**  Typical Composition of SI Engine Exhaust (400–600°C, normal range)

| | | | |
|---|---|---|---|
| $CO_2$ | 12% (120,000 ppm) | HC | 1000–6000 ppm (methane equivalent, assumes all hydrocarbon is methane) |
| $H_2O$ | 12% (120,000 ppm) | | |
| $O_2$ | 1–5% (10,000–50,000 ppm) | NO | 100–2000 ppm |
| CO | 0.2–5% (2000–50,000 ppm) | $SO_2$ | $\approx 20$ ppm |
| | | Trace amounts of lead and phosphorus | |

*Source*: Heywood (1988).

reduce NO in the presence CO and $H_2$ mixtures. Under slightly rich conditions, the relative activity of noble metals is ruthenium (Ru) > rhodium (Rh) > palladium (Pd) > platinum (Pt). If the exhaust gas mixture is too fuel rich, there may be unfortunate reverse reactions that convert NO into $NH_3$ or HCN. Reducing catalysts are deactivated in the presence of sulfur.

# 7.8  Combustion controls for stationary combustors

Stationary combustors include pulverized coal electric utility boilers, process steam stoker boilers, gas turbines, kilns, glass furnaces, municipal waste incinerators, and so on. A detailed discussion of pollution associated with the incineration of hazardous wastes can be found in Dempsey and Oppelt (1993). The common feature of these combustors is that they operate steadily except for programmed changes in feed rate and output load. Steady-state operation allows engineers the opportunity to design combustors with unique geometry that minimizes production of $NO_x$, CO, and unburned hydrocarbons. There is little one can do to reduce the generation of sulfur oxides except to reduce the sulfur content in the fuel, since virtually all the sulfur in the fuel exits the combustor as sulfur oxides. Beyond using low-sulfur fuels, one must resort to tailpipe flue gas desulfurization (FGD) techniques, discussed in Chapter 10.

It is convenient to group combustion controls for stationary combustors into four categories: flue gas recirculation, low-$NO_x$ burners, staged combustion, and fuel blending. All but the last are techniques to lower the peak combustion temperatures to minimize NO production while providing sufficient time and oxygen to minimize the production of CO and unburned hydrocarbons.

*Flue gas recirculation* involves burners that recirculate a fraction of the flue gas (combustion exhaust) into the primary combustion air. Flue gas is relatively inert, reduces the concentration of the reacting species, and thus reduces the combustion temperature. This in turn reduces $NO_x$ emissions by 15 to 30%.

*Low-$NO_x$ burners* are combustors that incorporate techniques that allow fuel and air to mix and burn in a staged fashion so as to minimize pollutant production. Stoker boilers cannot accommodate low-$NO_x$ burners. Figure 7-20 typifies a pulverized coal burner that incorporates these features. A portion of the air is mixed with the coal so that the primary combustion zone produces a fuel-rich flame. The remainder of the air is added as an annular sheath that surrounds the fuel-rich primary zone. The partially combusted products in the primary zone complete their combustion in the secondary fuel-lean burnout zone. Figure 7-21 illustrates the concept of staged combustion in a *low-$NO_x$ gas turbine can burner*. By adding air through slots judiciously located in the burner periphery, peak temperatures can be kept within limits.

*Staged combustion* is the concept of initiating a primary fuel-rich flame and then injecting secondary air at particular locations downstream to complete the combustion in a fuel-lean fashion. In both the primary and secondary region the peak temperatures are less than what would have occurred if the fuel and air had been introduced at a single time. In reality, the low-$NO_x$ combustor (Figure 7-20) is an example of staged combustion. Reduction of $NO_x$ of 10 to 15% can be achieved in staged combustion. Staged combustion may also be accomplished by adding fuel in a staged fashion.

*Fuel blending* is a control strategy in which the fuel that has traditionally been used is augmented with other fuels. The clearest example of the strategy is blending low-sulfur coal with traditional coal so that the total amount of sulfur/Btu is within acceptable

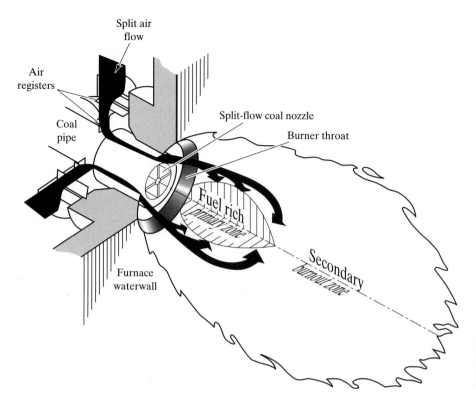

**Figure 7-20** Elements of a Low-NO$_x$ pulverized coal burner incorporating staged introduction of air.

limits. Alternatively, a secondary set of burners using natural gas (*overfiring*) could be installed and the coal feed rate reduced to maintain the same firing rate (Btu/h). Some boilers can be retrofitted with overfiring, while others can not. Accommodations may also have to be made because a coal flame has different radiant heat transfer properties than a natural gas flame.

## 7.9 *Fluidized-bed combustion*

Fluidized-bed combustion is an advanced combustion technology. The distinctive features of fluidized-bed combustion are illustrated by a graph of the gas and particle velocity versus bed expansion (Figure 7-22). At low velocity, gas passes upward but the particles in the bed do not move. As the gas velocity increases, particles in a bed become agitated and begin to move both horizontally and vertically and the volume of the bed increases. Such motion is called a *classic (bubbling) fluidized bed* because the air and particles resemble a *bubbling* medium. Figure 1-15 shows the basic components of a bubbling fluidized-bed combustor (see A in Figure 7-22).

At larger gas velocities, particles of burning coal become entrained in the rising combustion gases. The bed expands and there is an ill-defined freeboard above the bed. The bulk solids fraction, which in the case of a stationary bed is equal to the bulk density of the coal, decreases from the bottom to the top of the reactor. Burning coal particles become entrapped in rising column of combustion gas. The entrained particles are separated from the hot combustion gases by a cyclone. A fraction of these particles contain significant amounts of coal and are returned to the reactor. Solid material is returned to the reactor through a novel loop seal that prevents combustion gases from traveling in wrong direction. A reactor containing these features is called an (*atmospheric*) *circulating fluidized-bed* (CFB) *reactor*. Figure 7-23 illustrates the essential features of CFB (see also B in Figure 7-22). If combustion occurs at pressures in excess of atmospheric pressure, the reactor is called a *pressurized circulating fluidized-bed reactor*.

At still larger gas velocities, a point is reached at which elutriation occurs and the bed volume increases with a small change in gas velocity. Eventually, a bed ceases to exist and all particles are entrained in the

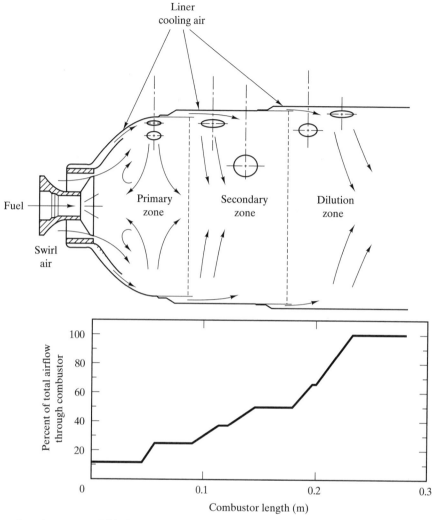

**Figure 7-21**  Staged combustion in an aircraft gas turbine (taken from Flagan and Seinfeld, 1988)

combustion gases. This is the transport reactor see C in Figure 7-22.

The numerical values and shape of the curves in Figure 7-22 depend on the mass flow rate at which solids are added to the reactor. A line drawn to the asymptotic value of the particle velocity when the bed material is transported is called the *mean gas velocity*. The difference between the mean gas velocity and particle velocity is called the *slip velocity*. This asymptotic velocity can be thought of as a settling velocity of the bed particles. Thus when the air velocity is equal to this value, the upward force on the bed is equal to the downward force of gravity on the particles in the bed and the bed levitates. At this point (C) the *transport reactor* behaves as a pulverized-coal combustor (Figure 7-1).

One of the virtues of fluidized-bed combustors is that they can burn low-quality (high-ash, low-heating-value) fuels such as lignites, anthracite culm, petroleum coke, oil shale, bark, peat, and industrial sludges that conventional traveling-grate stokers and pulverized coal combustors cannot use. At the present time there are literally mountains of coal-bearing materials discarded from anthracite coal mining activities of past generations. This material is called *culm*. The runoff from these culm piles is acidic and pollutes creeks and streams. Culm has very little economic value but can be used in CFBs to generate electricity economically.

Crushed limestone can be mixed with coal to remove $SO_2$ as soon as it is created in the bed (Badin and Frazier, 1985). Thus CFBs do not require the addi-

reduced in size, and mixed thoroughly in the bed to produce a high degree of fineness (100 to 300 μm) in the solid material leaving the combustion chamber. The long residence time for small particles enables sulfur to be removed with less limestone than in a bubbling bed furnace. In addition, the higher gas velocities of a CFB produce larger heat transfer rates than in the bubbling bed furnaces.

Fuel is fed directly into the CFB reactor, without a costly preparation and distribution system, through a combination of gravimetric feeders and chain conveyors that feed fuel into the loop seals. Limestone is fed to the bed by gravimetric feeders, through a rotary valve where transport blowers convey the limestone to the periphery of the combustion chamber. Combustion air is added in two stages: primary air enters the reactor through a nozzle grate at the bottom of the CFB reactor, and secondary air is added through the side of the reactor above where the fuel is added. Such staged introduction of air minimizes the production of nitrogen oxides. Crushed limestone needed to remove sulfur compounds is added near the bottom the combustor. Fuel in a CFB is burnt at approximately 850°C. Elutriated fuel, limestone, and lime exiting the furnace pass through a reverse-flow cyclone that removes the larger particles that are returned to the CFB for further reaction from very small particles that are removed later by a baghouse or ESP. A cleverly designed loop seal exists in the return leg between cyclone and CFB to maintain a seal to prevent bed material from entering the base of the cyclone.

Heat exchangers are located in two places. Conventional convection heat exchangers are located in the discharge from the cyclone to produce steam, to preheat air for combustion and to heat feed water. A novel heat exchanger to generate steam is embedded in a fluidized-bed chamber located below the cyclone. On leaving the recycling cyclone, a portion of the hot solids are diverted to the fluidized-bed heat exchanger, where they are cooled before being returned to the reactor. By embedding steam tubes in the bubbling bed, contact between the steam tubes and the agitated burning coal particles produces heat transfer rates $(kJ/s \cdot m^2)$ considerably larger than convective heat transfer in which only a hot combustion gases pass over the steam tubes.

## 7.10 Closure

Combustion systems are the primary source of anthropogenic $NO_x$ and $SO_x$. They are also a principal source of pollutants that enter the atmosphere and participate in photochemical reactions leading to urban air pollution. As shown in this chapter, the formation of the sulfur and nitrogen oxides is very sensitive to the peak temperatures in the combustion process. It is important that engineers understand how these sulfur and nitrogen oxides are formed so that they can design future combustion systems in ways that will not produce them. Such knowledge requires an understanding of chemical kinetics driven by free radicals.

## Nomenclature

| Symbol | Description (Dimensions*) |
|---|---|
| [j] | concentration of species $j$ $[M/L^3,\ ppm,\ \text{mole fraction}(N/L^3)]$ |
| = | empirical stoichiometric equation |
| [=] | has the units |
| → | direction of a chemical reaction |
| ⇌ | simultaneous forward and reverse reactions |

* F, force; L, length; M, mass; N, mols; Q, energy; t, time; T, temperature.

| | |
|---|---|
| $a, b$ | constants defined by equation |
| $A_s$ | total surface area of cylinder $(L^2)$ |
| $C$ | unitless parameter defined by equation |
| $[C_t]$ | total concentration of carbon $(N/L)$ |
| $D_p$ | piston bore (L) |
| $E_a$ | activation energy $(Q/NT)$ |
| $k$ | isentropic coefficient, $c_p/c_v$ |
| $k_{+1}$ | forward rate constant for equation number 1, units depend on the equation |
| $k_{-1}$ | reverse rate constant for equation number 1, units depend on the equation |

| | |
|---|---|
| $K_P$ | equilibrium constant based on partial pressure, units depend on the equation |
| $L_p$ | piston stroke (L) |
| $M$ | parameter defined by equation $(t^{-1})$ |
| $n_j$ | number of moles of molecular species $j$ |
| $n_t$ | total number of moles |
| $P$ | total pressure $(F/L^2)$ |
| $P_i$ | partial pressure of species $i$ $(F/L^2)$ |
| $Q$ | volumetric flow rate $(L^3/t)$ |
| r | reaction rate, defined by equation |
| $r_A, -r_A$ | rate of formation of species $A$, rate of reaction of $A$ $(N/L^3t)$ |
| $R_c$ | cutoff ratio in a diesel cycle, defined by equation |
| $R_p$ | compression ratio of the compressor in a Brayton cycle, $P_{max}/P_{min}$ |
| $R_u$ | universal gas constant $(Q/NT)$ |
| $R_v$ | compression ratio, $V_{max}/V_{min}$ |
| $[S_t]$ | total concentration of sulfur $(N/L)$ |
| $t$ | time (t) |
| $T$ | absolute temperature (T) |
| $t_r$ | residence time, defined by equation (t) |
| $V$ | volume $(L^3)$ |
| $v_p$ | piston speed $(L/t)$ |
| $x, y, z$ | unknowns |
| $Y$ | unitless concentration defined by equation |

| | |
|---|---|
| $y_i$ | mole fraction of species $i$ |

*Greek*

| | |
|---|---|
| η | cyclic thermal efficiency |
| τ | characteristic time of a chemical kinetic equation (t) |
| φ | equivalence ratio, actual fuel to air ratio/stoichiometric fuel–air ratio |
| ω | engine speed (rpm) $(t^{-1})$ |

*Abbreviations*

| | |
|---|---|
| CFB | circulating fluidized-bed |
| CI | compression ignition engine |
| EGR | exhaust gas recirculation |
| ETBE | ethyl tertiary butyl ether |
| FBC | fluidized-bed combustion |
| FGR | flue gas recirculation |
| HC | unburned hydrocarbons |
| MTBE | methyl tertiary butyl ether |
| RFG | reformulated gasoline |
| RPM | engine speed in revolutions per minute |
| RVP | Reid vapor pressure |
| SI | spark ignition engine |
| VOC | volatile organic compound |

# *References*

Atkinson, R., 1990. Gas-phase tropospheric chemistry of organic compounds: a review, *Atmospheric Environment*, Vol. 24A, No. 1, pp. 1–41.

Atkinson, R., Baulch, D. L., Cox, R. A., Hampson, R. F., Jr., Kerr, J. A., and Troe, J., 1992. Evaluated kinetic and photochemical data for atmospheric chemistry: supplement IV, *Atmospheric Environment*, Vol. 26A, No. 7, pp. 1187–1230.

Bobcock and Wilcox, 1978, "Steam/Its Generation and Use," 161 E. 42$^{nd}$ St., New York, NY.

Badin, E. J., and Frazier, G. C., 1985. Sorbents for fluidized-bed combustion, *Environmental Sciences and Technology*, Vol. 19, No. 10, pp. 895–901.

Bowman, C. T., 1975. Kinetics of pollutant formation and destruction in combustion, *Progress in Energy Combustion Science*, Vol. 1, pp. 33–45.

Bowman, C. T., and Kesten, A. S., 1973. Kinetic modeling of nitric oxide formation in combustion processes, Fall Meeting, Western Section of the Combustion Institute, October.

Calvert, J. G., Heywood, J. B., Sawyer, R. F., and Seinfeld, J. H., 1993. Achieving acceptable air quality: some reflections on controlling vehicle emissions, *Science*, Vol. 261, July 2, pp. 37–45.

Carter, W. P. L., 1990. A detailed mechanism for the gas–phase atmospheric reactions of organic compounds, *Atmospheric Environment*, Vol. 24A, pp. 481–518.

Dempsey, C. R., and Oppelt, E. T., 1993. Incineration of hazardous waste: a critical review update, *Journal of the Air & Waste Management Association*, Vol. 43, pp. 25–73.

Dryer, F. L., and Glassman, I., 1973. High temperature oxidation of CO and CH₄, *Proceedings of the 14th Symposium (International) on Combustion*, Combustion Institute, Pittsburgh, PA, pp. 987–1003.

Finlayson-Pitts, B. J., and Pitts, J. N., 1986. *Atmospheric Chemistry: Fundamentals and Experimental Techniques*, Wiley-Interscience, New York.

Flagan, R. C., and Seinfeld, J. H., 1988. *Fundamentals of Air Pollution Engineering*, Prentice Hall, Upper Saddle River, NJ.

Heywood, J. B., 1988. *Internal Combustion Engine Fundamentals*, McGraw-Hill, New York.

Notzold, D., and Algermissen, J., 1981. Chemical kinetics of the ethane-oxygen reaction. High temperature oxidation at ignition temperatures between 1400 K and 1800 K, *Combustion and Flame*, Vol. 40, pp. 293–313.

Odum, J. R., Jungkamp, T. P. W., Griffin, R. J., Flagan, R. C., and Seinfeld, J. H., 1997. The atmospheric aerosol–forming potential of whole gasoline vapor, *Science*, Vol. 276, pp. 96–99.

Roberts, J. M., 1990. Review article: the atmospheric chemistry of organic nitrates, *Atmospheric Environment*, Vol. 24A, No. 2, pp. 243–287.

Seinfeld, J. H., 1986. *Atmospheric Chemistry and Physics of Air Pollution*, Wiley, New York.

Turns, S., 1996. *An Introduction to Combustion*, McGraw-Hill, New York.

van Wylen, J., and Sonntag, R. E., 1985. *Fundamentals of Classical Thermodynamics*, 3rd ed., Wiley, New York.

Westbrook, C. K., and Dryer, F. L., 1981. Simplified reaction mechanisms for the oxidation of hydrocarbon fuels in flames, *Combustion Science and Technology*, Vol. 287, pp. 31–43.

# *Problems*

**7.1.** Consider an air-standard Otto cycle (air is assumed to be a perfect gas with constant specific heats). If $T_1 = 15°C$, $P_1 = 101$ kPa, the input of energy is constant and equal to 1800 kJ/kg air, and the compression ratio $(r_v)$ is variable, plot the thermal cyclic efficiency η versus maximum cycle temperature $(T_3)$ over the range 1500 K < $T_3$ < 2500 K. Compute $d\eta/dT_3$.

**7.2.** Consider an air-standard Diesel cycle (air is assumed to be a perfect gas with constant specific heats). If $T_1 = 15°C$, $P_1 = 101$ kPa, the input of energy is constant and equal to 1800 kJ/kg air, and the compression ratio $(r_v)$ and cutoff ratio $(r_c)$ are variables, plot the thermal cyclic efficiency η versus the maximum cyclic temperature $(T_3)$ over the range 1500 K < $T_3$ < 2500 K. Compute $d\eta/dT_3$.

**7.3.** Consider an air-standard Brayton cycle (air is assumed to be a perfect gas with constant specific heats). If $T_1 = 15°C$, $P_1 = 101$ kPa, and the input of energy is constant and equal to 720 kJ/kg air, plot the thermal cyclic efficiency η as a function of the maximum cyclic temperature $(T_3)$ over the range 1500 K < $T_3$ < 2500 K. Compute $d\eta/dT_3$.

**7.4.** In Section 7.7.3 the statement is made that for a given displacement volume, the minimum surface area occurs when the stroke is equal to the bore. Prove that the statement is true.

**7.5.** Refer to Example 7.3. Consider the combustion products from fuel-rich combustion in an Otto or Brayton cycle in which the initial CO and $CO_2$ concentrations are the following values (by volume): $[CO]_0 = 0.2\%$, $[CO_2]_0 = 15\%$, and $[O_2]_0 = 0.5\%$, $[H_2O] = 12\%$. The remainder is a diluent. Assuming that the only equilibrium to contend with is the $CO/CO_2$ equilibrium, compute and plot the equilibrium values of CO over the range 1600 K < $T$ < 2600 K.

**7.6.** Refer to Example 7.4. Begin with $CO/CO_2$ equilibrium concentrations at 2600 K using the data from Problem 7.5 as initial data. The temperature decreases in a stepwise fashion to a low value. Compute and plot the CO concentration as a function of time over the range 1200 K < $T$ < 1800 K.

**7.7.** Refer to Example 7.5. Consider combustion products from Otto or Brayton cycles equipped with exhaust gas recirculation (EGR) and fuel–air ratios near stoichiometric values in which the initial NO, $NO_2$, and $O_2$ concentrations have the following values (by volume): $[O_2]_0 = 0.5\%$, $[NO]_0 = 5$ ppm, $[NO_2]_0 = 10$ ppm. Assuming that the only equilibrium to contend with is $NO/NO_2$ equilibrium, compute and plot the equilibrium values of NO over the range 1500 K < $T$ < 2500 K.

**7.8.** Refer to Example 7.6. Consider a mixture of NO, $NO_2$, and other combustion gases at $T = 2100$ K in which $[NO]_0 = 4000$ ppm, $[O_2]_0 = 0.2\%$, $[N_2]_0 = 80\%$. The remainder of the gases are CO, $CO_2$, and $H_2O$, whose dissociation reactions will be ignored for the moment. Compute and plot the NO concentration as a function of time if the temperature decreases in a stepwise fashion to several final low temperatures in the range 1500 K $< T <$ 1800 K.

**7.9.** Refer to Example 7.7. Consider the combustion products from a coal-fired electric utility boiler incorporating fuel gas recirculation (FGR) such that the combustion products contain sulfur oxides with the initial concentrations $[SO_2]_0 = 100$ ppm, $[SO_3]_0 = 0.1$ ppm, $[O_2]_0 = 5\%$. The remainder of the gases will be treated as diluents for the moment. Compute and plot the equilibrium concentrations of $SO_2$ and $SO_3$ at the temperatures 1500 K $< T <$ 2500 K.

**7.10.** Assume that a cylinder in a novel high-compression spark ignition engine cycle has the following values:

- $T_3 = 2200$ K, $P_3 = 10,430$ kPa, $V_3 = 30$ cm$^3$
- Piston: stroke $(L_p) = 5$ cm, diameter $(D_p) = 10$ cm
- Displacement volume $V_d = V_4 - V_3 = L_p A_p = 392.5$ cm$^3$
- Engine speed $= 3000$ rpm
- $k = 1.4$

At top dead center the temperature and pressure are $T_3$ and $P_3$ and the $CO_2$ mole fraction is 0.18, the $O_2$ mole fraction is 0.033, the water vapor mole fraction is 0.12, and the CO mole fraction is 0.00128 (1280 ppm).

**(a)** Find the time $(t_4)$ it takes for the piston to travel from top dead center (crank angle $\theta = 0$ and $V = V_3$) to bottom dead center ($\theta = 180$ and $V = V_4$). Compute $T_4$.

**(b)** Using the global kinetic expression discussed in Example 7.4, compute and plot $[CO](t)/[CO]_0$ versus time from $t = 0$ to $t = t_4$. Note that $[CO]$, $[O_2]$, and $[H_2O]$ in Equation 7.64 are expressed in the units gmol/m$^3$ at the particular $T$ and $P$ in the gas. Since the differential equation is nonlinear, you will need to use the fourth-order Runge–Kutta method in a commercially available mathematical program or the modified Euler method in Table A-14. Compare the results of the calculation with Example 7.4.

**7.11.** One kilomole of each $SO_2$, $O_2$, and He are contained in a vessel. The contents are compressed to 3 atm and raised to a constant temperature of 1000 K. What is the concentration (ppm) of $SO_3$ at equilibrium if equilibrium is governed by Equation 7-108.

**7.12.** A constant-pressure vessel is filled with a homogeneous mixture of CO, $H_2O_2$ (vapor) and $N_2$. The mixture

is instantaneously raised to a temperature of 800 K and maintained at this temperature. The initial concentrations are

$$[CO]_0 = [H_2O_2]_0 = 7.62 \times 10^{-7} \text{gmol/cm}^3$$
$$[N_2]_0 = 1.37 \times 10^{-5} \text{gmol/cm}^3$$

The reaction is governed by the following kinetic mechanism:

$$H_2O_2 + M \xrightarrow{k_1} OH\bullet + OH\bullet + M$$

$$CO + OH\bullet \xrightarrow{k_2} CO_2 + H\bullet$$

where

$$M = \text{nitrogen}$$
$$k_1(\text{cm}^3 \text{ gmol}^{-1} \text{ s}^{-1}) = 1.17 \times 10^{17} \exp\left(-\frac{E_{a,1}}{RT}\right)$$
$$E_{a,1} = 45,500 \text{ cal/gmol}$$
$$k_2(\text{cm}^3 \text{ gmol}^{-1} \text{ s}^{-1}) = 5.5 \times 10^{11} \exp\left(-\frac{E_{a,2}}{RT}\right)$$
$$E_{a,2} = 1080 \text{ cal/gmol}$$
$$R = 1.987 \text{ cal/gmol}\cdot\text{K}$$

Compute the concentration $(\text{gmol/cm}^3)$ of the highly reactive species OH$\bullet$ after 10 ms.

**7.13.** Assume that a vessel contains 77% nitrogen $(N_2)$ and 3.3% oxygen $(O_2)$. The total pressure is 100 kPa. The mixture is raised to an initial temperature $(T_{\text{initial}})$ and maintained at this temperature until all species reach their equilibrium concentrations. The gas mixture is suddenly cooled to a final temperature $(T_{\text{final}})$ of 1800 K and maintained at this temperature until all species reach their equilibrium values. Compute and plot the NO concentration (ppm) as a function of elapsed time for several values of $T_{\text{initial}}$ (2400 K, 2200 K, and 2000 K). Assume that in Equations 7.100 and 7.101 the parameter $C$ is zero and that the value of $M$ (1800 K) is 0.0975 s$^{-1}$. Note the differential equations for a step-down process are the same as a step-up process except that the initial and final limits are different. If it takes approximately 0.01 s for combustion exhaust gases inside a reciprocating engine or gas turbine to be discharged to the atmosphere, what is the NO concentration as a function of time?

$$K_{P,\text{NO}}(2400 \text{ K}) = 2.386 \times 10^{-3}$$
$$K_{P,\text{NO}}(2000 \text{ K}) = 3.85 \times 10^{-4}$$
$$K_{P,\text{NO}}(2200 \text{ K}) = 1.043 \times 10^{-3}$$
$$K_{P,\text{NO}}(1800 \text{ K}) = 1.148 \times 10^{-4}$$

# 8.3  Pollutant material balance

There are occasions when pollutants are formed in a predictable fashion. In these situations the pollutant mass flow rate can be estimated from a simple mass balance.

**VOC from surface coating.**  When metal, wood, paper, textiles, and so on are coated, the coating contains solvents used to convey the pigment, solids, and so on, to the surface. The solvent may be water (water-based coatings) or it may be a volatile organic compound (VOC). Virtually all of the solvent leaves the coating between the time it is applied (spraying, dipping, etc.) and when it has dried. In terms of the instantaneous emission rate, it is important to know the rate at which solvent evaporates, but in terms of the total amount of solvent discharged, engineers only need to know the amount of solvent in the coating as it was applied and the rate (kg/s) at which coating was used. The key element that makes this elementary analysis valid is the knowledge that the coating retains a negligible amount of solvent; thus all the solvent in the coating ultimately enters the atmosphere.

**Halogenated hydrocarbons.**  As a first approximation, halogenated hydrocarbons do not react or become absorbed or adsorbed on particles in the troposphere. Thus there is no depletion or accumulation of them in the troposphere. The yearly emission rate can be estimated by the masses of each that are produced in a year minus the corresponding masses that industry recovers per year.

**Carbon dioxide from stationary and mobile sources.**  The carbon in hydrocarbon fuels consumed by stationary sources and mobile sources such as automobiles, buses, trucks, railroads, and aircraft are eventually converted to $CO_2$. Even unburned hydrocarbons and CO eventually react in the atmosphere and form $CO_2$. To a first approximation, the average yearly rate at which $CO_2$ is generated can be estimated from the mass of carbon in the fuels times the ratio (44/12) representing its conversion to $CO_2$:

$$C + O_2 = CO_2 \tag{8.3}$$

If other air pollutants enter the atmosphere directly without reacting, their emission can be estimated in ways similar to VOCs. Unfortunately, there are countless pollutants whose emission rates depend on complex chemical reaction mechanisms containing many intermediate chemical species that cannot be analyzed as easily. Thus the generation rate of particles, oxides of nitrogen, and many hazardous air pollutants (HAPs) have to be estimated by other means.

> ## EXAMPLE 8.2  POLLUTANTS GENERATED BY ELECTRIC POWER STATIONS: USING MASS BALANCES

An electric power generating plant burns bituminous coal in a wet-bottom, pulverized coal-fired furnace (also called a steam boiler) equipped with a wet scrubber to remove 98% of the $SO_2$ from the exhaust gases. Assume the following:

- Power plant output = 1000 MW
- Overall thermal cyclic efficiency ($\eta$) = 30%
- Coal assay: sulfur (by mass) = 2.1%

   higher heating value (HHV) = 12,500 Btu/lb$_m$

   average molecular weight = 12

   ash (by mass) = 8%

- Wet scrubber converts $SO_2$ to gypsum ($CaSO_4 \cdot 2H_2O$) by the following overall equation:

$$SO_2 + CaO + 2H_2O + \tfrac{1}{2}O_2 = CaSO_4 \cdot 2H_2O \downarrow$$

Estimate the following in tons/h:

**a.** Coal consumption rate $(\dot{m}_c)$
**b.** $SO_2$ mass flow rate entering the scrubber $(\dot{m}_{SO2})$
**c.** Total ash accumulation rate $(\dot{m}_{ash})$
**d.** CaO consumption rate $(\dot{m}_{CaO})$
**e.** Gypsum accumulation rate $(\dot{m}_g)$

**Solution**  To begin, recall the definition of overall thermal cyclic efficiency ($\eta$):

$$\eta = \frac{\text{net output power}}{\text{total input energy}} = \frac{\dot{W}_{net}}{q}$$

The total rate at which energy is released by burning coal is

$$q = \frac{\dot{W}_{net}}{\eta} = \left( \frac{1000 \text{ MW}}{0.30} \right) \left( \frac{2545 \text{ Btu/h}}{0.746 \text{ kW}} \right)$$

$$(1000 \text{ kW/MW}) = 1.1372 \times 10^{10} \text{Btu/h}$$

**a.** The mass flow rate of coal,

$$\dot{m}_c = (1.1372 \times 10^{10}\ \text{Btu/h})$$
$$(\text{lb}_\text{m}\ \text{coal}/12{,}500\ \text{Btu})(\text{ton}/2000\ \text{lb}_\text{m})$$
$$= 454.88\ \text{tons/h}$$

**b.** Irrespective of the chemical kinetics describing how sulfur is oxidized to $SO_2$, the overall stoichiometry can be expressed

$$S + O_2 = SO_2$$

Thus 1 mol of oxygen combines with 1 mol of sulfur to form 1 mol of $SO_2$:

$$\dot{m}_{SO_2} = (1.1372 \times 10^{10}\text{Btu/h})$$
$$(\text{lb}_\text{m}\ \text{coal}/12{,}500\ \text{Btu})$$
$$(\text{mol}\ SO_2/\text{mol}\ S)(64\ \text{lb}_\text{m}\ SO_2/\text{mol})$$
$$(\text{mol}/32\ \text{lb}_\text{m}\ S)(0.021\ \text{lb}_\text{m}\ S/\text{lb}_\text{m}\ \text{coal})$$
$$= 3.821 \times 10^4\ \text{lb}_\text{m}/\text{h}$$
$$= 19.10\ \text{tons/h}$$

and the mass flow rate of $SO_2$ entering the scrubber is 19.10 tons/h.

**c.** Ash appears in two places. The bulk accumulates as *bottom ash* in the base of the furnace and is removed as conventional ash. A smaller amount of the ash is released as *fly ash* that leaves the furnace with the flue gas and is removed by the wet scrubber (or other particle removal system). It is not possible to estimate the amount of fly ash from a material balance, and other techniques will be needed. Nevertheless, the total ash production is

$$\dot{m}_\text{ash} = \dot{m}_c(\text{mass of ash}/\text{mass of coal})$$
$$= (454.88\ \text{tons coal/h})$$
$$(0.08\ \text{lb}_\text{m}\ \text{ash}/\text{lb}_\text{m}\ \text{coal})$$
$$= 36.39\ \text{tons/h}$$

**d.** From the overall stoichiometric equation describing the removal of $SO_2$ by the wet scrubber, it is seen that 1 mol of CaO is needed to remove 1 mol of $SO_2$. The molecular weight of $SO_2$ is 64 and the molecular weight of CaO is 56. Thus

$$\dot{m}_\text{CaO} = \dot{m}_{SO_2}(\text{tons}\ SO_2/\text{h})$$
$$(\text{ton mol}\ SO_2\ \text{formed}/64\ \text{tons}\ SO_2)$$

$$(\text{mol CaO}/\text{mol}\ SO_2)$$
$$(56\ \text{tons CaO}/\text{ton mol CaO})$$
$$(0.98\ SO_2\ \text{removed}/SO_2\ \text{formed})$$
$$= (19.10)\left(\frac{56}{64}\right)(0.98) = 16.38\ \text{tons/h}$$

**e.** Similarly, 1 mol of gypsum is formed for each mole of $SO_2$ that is removed. The molecular weight of gypsum is 172. Thus the accumulation rate of gypsum is

$$\dot{m}_g = \dot{m}_{SO_2}(\text{ton mol}\ SO_2\ \text{formed}/64\ \text{tons})$$
$$(\text{mol gyp.}/\text{mol}\ SO_2)$$
$$(172\ \text{tons gyp.}/\text{ton mol gyp.})$$
$$(0.98\ SO_2\ \text{removed}/SO_2\ \text{formed})$$
$$= (19.10)\left(\frac{172}{64}\right)(0.98) = 50.30\ \text{tons/h}$$

Consider the amount of $CO_2$ formed by the combustion of coal. Assume that ultimately all the combustion products are converted to $CO_2$. Assume also that coal is all carbon except for the 8% that is ash and the 2.1% that is sulfur.

$$\text{mass coal} = \text{mass of carbon} + \text{mass of ash}$$
$$+ \text{mass of sulfur}$$

$$1 = \frac{\text{mass of carbon}}{\text{mass of coal}} + 0.08 + 0.021$$

$$\frac{\text{tons C}}{\text{tons coal}} = 1 - 0.08 - 0.021 = 0.899$$

The mass flow rate of $CO_2$ is

$$\dot{m}_{CO_2} = \dot{m}_c(\text{tons coal/h})$$
$$(\text{ton mol C}/12\ \text{tons carbon})$$
$$(1\ \text{mol}\ CO_2/\text{mol C})$$
$$(44\ \text{tons}\ CO_2/\text{ton mol}\ CO_2)$$
$$(0.899\ \text{ton C}/\text{ton coal})$$
$$= (454.88\ \text{tons coal/h})\left(\frac{44}{12}\right)(0.89)$$
$$= 1484.2\ \text{tons/h}$$

*Summary:*

| Output (tons/h) | | Input (tons/h) | |
|---|---|---|---|
| Ash | 36.39 | Coal | 454.88 |
| Gypsum | 50.30 | CaO | 18.72 |
| $CO_2$ | 1484.42 | | |
| | 1571.11 | | 473.60 |

For every ton of coal that is burned, 0.021 ton of sulfur enters the furnace. Thus in the example above, 9.55 tons/h enters the furnace. For every ton of sulfur that is removed, 1.96 tons of CaO is needed and 5.27 tons of gypsum is collected. Considering just the solids, 473.6 tons/h is consumed and 86.69 tons/h of dry material is discarded. Recalling that power plants usually store ash and scrubber sludge on the plant's property, one can appreciate the huge mass of material that power plants accumulate. Add to this the high cost of land near metropolitan areas and readers can appreciate why these plants might be interested in using natural gas for fuel or replacing fossil-fuel steam generators with nuclear steam generators. The analysis demonstrates the economic necessity to remove sulfur from coal by coal cleaning techniques so that it never enters the boiler in the first place. The size of the equipment and the mass of waste is far smaller when sulfur is removed before the coal enters the boiler rather than removing it afterward when it has combined with other materials. Removing sulfur as $CaSO_4 \cdot 2H_2O$ requires handling 50.3 tons/h, whereas removing 98% of the sulfur by coal cleaning requires handling only $(0.98)(0.021)(454.88) = 9.36$ tons/h.

The leakage of gases and vapors from valve packing, pump seals, flange gaskets, and so on, the escape of vapors or gases from contaminated soil, or the evaporation of vapor from the free surface of a liquid are difficult quantities to measure directly. A simple experimental method (Figure 8-2) of estimating these rates is to enclose the source in an airtight container of volume $V$, pass a clean stream of air through the container at a measured volumetric flow rate $Q_a$, mix the contents of the chamber thoroughly, and measure the concentration of the gas or vapor either leaving or inside the container. Enclosures used to make these measurements are called *flux chambers* (Eklund, 1992).

If the emission source strength ($S$) is constant (e.g., leaks in packing or flanges), the user should maintain a constant volumetric flow rate of clean air $(Q_a)$ and either monitor $c(t)$ or record its steady-

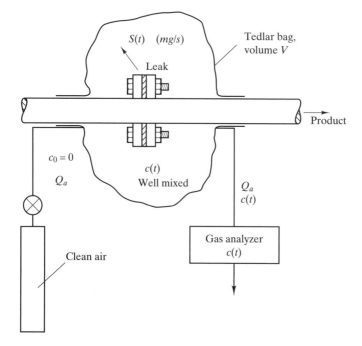

**Figure 8-2** Flux chamber used to measure leak rate in a flange and gasket

state value $(c_{ss})$ with an appropriate gas analyzer. If the source strength $(S)$ decreases with time (e.g., solvent evaporation from painted surfaces), the user may want to isolate the chamber $(Q_a = 0)$ and merely record $c(t)$ inside the chamber. In either event, the source strength $(S)$ can be determined by writing a mass balance for the escaping fluid,

$$V \frac{dc}{dt} = S - Q_a(c - c_0) \qquad (8.4)$$

**Measuring the steady-state concentration, $c_{ss}$.** If steady-state conditions are reached within a reasonable time, the source strength $(S)$ can be found by setting the left-hand side of Equation 8.4 to zero. If the volumetric flow rate $(Q_a)$ is controlled and the values of $c_{ss}$ and $c_0$ are measured, a constant source strength can be computed from

$$S = Q_a(c_{ss} - c_0) \qquad (8.5)$$

If the escaping gas or vapor is adsorbed by the walls of the container, Equation 8.4 must be modified to incorporate wall losses. The experiment should be repeated for several values of $Q_a$ to ensure that the right-hand side of Equation 8.4 is constant. No inaccuracy will occur if small values of $Q_a$ are used (as long as $Q_a$ can be measured accurately), but there is a danger that if $Q_a$ is too large, air velocities sweeping past the source may influence the rate of emission $S$ of the escaping gas or vapor. Large values of $Q_a$ produce small values of $c_{ss}$ which may be below the low-limit accuracy of the gas measuring equipment.

**Measuring the instantaneous concentration, $c(t)$.** If the source strength $S(t)$, varies with time and the chamber is isolated, $Q_a = 0$,

$$V_{iso} \frac{dc}{dt} = S(t) \qquad (8.6)$$

or in integral form,

$$\int_{c_0}^{c(t)} dc = c(t) - c_0 = \int_0^t \frac{S(t)}{V_{iso}} dt \qquad (8.7)$$

From the value of either the integral or the slope of the concentration versus time curve, the user can determine an algebraic expression for the source strength $(S)$.

## EXAMPLE 8.3 METHANE EMISSION RATE FROM RICE PADDIES

At a professional meeting an environmentalist suggests that methane is emitted to the atmosphere at a flux, $s$, of approximately 10 g acre$^{-1}$ h$^{-1}$ by rice grown in rice paddies. You wish to measure the emission rate to confirm or refute the assertion. To measure the methane emission rate, you construct a flux chamber (1 m high) over 3 m$^2$ in the center of the rice paddy. The chamber is covered with a transparent material that allows the rice in the chamber to receive the same amount of sunlight as the rest of the rice paddy. The soil temperature and air humidity inside the flux chamber are the same as outside the flux chamber. Ambient air containing 0.001 mg/m$^3$ of methane is passed into the chamber at a rate $(Q)$ of 0.01 m$^3$/s and the methane concentration is measured in the air exiting the chamber. A fan inside the chamber produces well-mixed conditions. The following are measurements of the methane concentration versus time.

| t (min) | $c_{CH4}$ (mg/m$^3$) |
|---------|---------------------|
| 0 | 0.0010 |
| 1.0 | 0.0272 |
| 2.0 | 0.0494 |
| 3.0 | 0.0677 |
| 4.0 | 0.0826 |
| 5.0 | 0.0948 |
| 6.0 | 0.1048 |
| 7.0 | 0.1130 |
| 8.0 | 0.1197 |
| 9.0 | 0.1252 |
| 10.0 | 0.1297 |
| 15.0 | 0.1425 |
| 20.0 | 0.1472 |
| 30.0 | 0.1496 |
| 60.0 | 0.1499 |

From the data above it can be seen that the steady-state concentration is approximately 0.15 mg/m$^3$. Thus from Equation 8.5, the methane emission rate is

$$S = s(\text{mg m}^{-2}\text{ s}^{-1})(A \text{ m}^2) = 3s = Q_a c_{ss} =$$
$$(0.01 \text{ m}^3/\text{s})(0.15 \text{ mg/m}^3) = 0.0015 \text{ mg/s}$$

$$s = S/A = 0.0005 \text{ mg m}^{-2} \text{ s}^{-1}$$
$$1.8 \text{ mg h}^{-1} \text{ m}^{-2}(43,560 \text{ ft}^2/\text{acre})(m/3.28 \text{ ft})^2$$
$$(g/1000 \text{ mg}) = 7.288 \text{ g acre}^{-1} \text{ h}^{-1}$$

Thus the assertion is reasonable.

# 8.4 AP-42 emission factors

In the absence of actually measuring the rate at which pollutants are generated, engineers often use emission factors. The original publication of the EPA emission factors was in a document noted as AP-42. The notation *AP-42* has become a well-used word in the vocabulary of the pollution control community; readers should not forget it. AP-42 emission factors express the amount of pollutant typically produced in an uncontrolled process per unit quantity of material being processed. If the process has no air pollution control system to remove pollutants, the rate at which pollutants are generated $(\dot{m}_g)$ is equal to the rate at which they are discharged to the atmosphere $(\dot{m}_d)$. Rarely, however, are processes uncontrolled, and when an air pollution control device with a removal (control or capture) efficiency $(\eta)$ cleans the exhaust process gas stream, the rate at which the pollutant is discharged to the atmosphere is

$$\dot{m}_d = (1 - \eta)\dot{m}_g \qquad (8.8)$$

To compute the pollutant generation $(\dot{m}_g)$, engineers need to know the rate at which raw materials are processed. The rate at which a product is manufactured (production rate) is a quantity the source operator is keenly aware of (i.e., fuel firing rate, production rates, mass of product being stored, etc.) since it is a key item to the company's profitability. It must be stressed that emission factors only typify the rate or amount of pollutant generated by a class of industrial operations and may not necessarily quantify any specific source in the category. Within a category, individual operations may generate quite different amounts. AP-42 emission factors have been compiled by the EPA for a large variety of industries. The EPA updates and expands emission factors on a regular basis. Abridged Lists of AP-42 emission factors are contained in the appendix tables as follows:

A-1.1 Furnaces using bituminous and subbituminous coal

A-1.2 Furnaces using anthracite coal

A-1.3 Furnaces using fuel oil

A-1.4 Furnaces using natural gas

A-1.5 Refuse incinerators

A-2.1 Aircraft

A-2.2 Industrial engines

A-3.1 Loading and transport of liquid petroleum products

A-4.1 Metallurgical processes

A-5.1 Mineral processes

A-6.1 Fugitive emissions from roadways

A full selection of AP-42 emission factors can be accessed through the Internet TTNBBS bulletin board called CHIEF. Also contained in the AP-42 are emission factors for industrial processes equipped with a variety of air pollution control devices. The data are of great value in preparing state permit applications and environmental impact statements. Professional journals (Miller et al., 1996) (Butcher and Ellenbecker, 1982) and handbooks (Bond and Staub, 1972) (ASHRAE, 1989; Orlemann et al., 1983, Shen, 1982, Struer & MacKay, 1984) publish data from time to time that, while not strictly AP-42 emission factors, fulfill the same function. For example, see Jenkins et al. (1996) for emission factors for polycyclic aromatic hydrocarbons from biomass burning. Table A-7 is a summary of emission data for a variety of indoor activities and sources. If the data are accurate, receive sufficient attention, and fulfill the same purpose, readers are encouraged to use them. The AP-42 emission factors are available in commercial microcomputer databases that are reviewed in professional journals (Bare, 1988). AP-42 emission factors are also available on a CD-ROM (AIR CHIEF, EFIG/EMAD/OAQPS/EPA) that can be purchased from the U.S. Government Printing Office at a modest cost.

## EXAMPLE 8.4 POLLUTANTS GENERATED BY AN ELECTRIC POWER STATION: USING AP-42 EMISSION FACTORS

Consider the electric power generating plant described in Example 8.2. Estimate the rate at which the following pollutants are generated using AP-42 emission factors.

**a.** Total particulate matter $(\dot{m}_p)$
**b.** Sulfur dioxide $(\dot{m}_{SO_2})$
**c.** Oxides of nitrogen $(\dot{m}_{NO_x})$
**d.** Carbon monoxide $(\dot{m}_{CO})$
**e.** Nonmethane VOC $(\dot{m}_{VOC})$
**f.** Methane $(\dot{m}_{CH_4})$

**Solution** A review of the AP-42 emission factors from Table A-1.1 shows the following in the units of kg of pollutant per Mg($Mg = 10^6$ g) of coal burned.

- Total particulate matter (fly ash):
  $(3.5)(\%A)$ kg/MG (where A = % ash in coal)
- $SO_2$: 19.5S (S = % sulfur in coal) kg/Mg
- $NO_x$: 17 kg/Mg
- CO: 0.3 kg/Mg
- Nommethane VOC: 0.04 kg/Mg
- Methane: 0.015 kg/Mg

From Example 8.2 the coal firing rate was 454.88 tons/h or 413 Mg/h. Thus

**a.** $\dot{m}_p$(fly ash) $= (413 \text{ Mg/h})(3.5)(8)$ kg/Mg
$= 11{,}564$ kg/h $= 12.74$ tons/h
**b.** $\dot{m}_{SO_2} = (19.5)(2.1)(413 \text{ Mg/h})$
$= 1.691 \times 10^4$ kg/h $= 18.63$ tons/h
**c.** $\dot{m}_{NO_x} = (17)(413) = 7021$ kg/h $= 7.73$ tons/h
**d.** $\dot{m}_{CO} = (0.3)(413) = 123.9$ kg/h
$= 0.136$ tons/h
**e.** $\dot{m}_{VOC} = (0.04)(413) = 16.52$ kg/h
$= 0.01819$ ton/h
**f.** $\dot{m}_{CH_4} = (0.015)(413) = 6.195$ kg/h
$= 0.0068$ ton/h

It must be emphasized that these values represent the pollutant generation rate, not the rate at which pollutants are emitted to the atmosphere. Note from the mass balance computation in Example 8.2 that the $SO_2$ emission rate was calculated to be 19.10 tons/h, which agrees well with the value above (18.63 tons/h). If a wet scrubber is used to remove $SO_2$, it will also remove some amount of the particulate matter but probably negligible amounts of the other pollutants unless the scrubbing liquid is designed especially to do so.

Except for $SO_2$, the AP-42 emission factors are the only way to estimate the generation rates of other pollutants since material balances are of little value,

owing to the complicated ways by which the pollutants are formed during combustion. Sulfur dioxide is the only pollutant that can be estimated reasonably by a material balance since it is known that virtually all the sulfur leaves the process in the flue gas. Readers should also note that the estimated generation rate for fly ash ($\dot{m}_p = 12.74$ tons/h) is smaller than the total generation rate of ash ($\dot{m}_a = 36.39$ tons/h) (i.e., fly ash = 35% of the total ash). Ash from the coal leaves the process as bottom ash and fly ash and AP-42 emission factors are the only way to estimate the generation rate of fly ash.

## 8.5 Empirical equations

Empirical equations lack the mathematical rigor students are accustomed to in engineering courses. Nonetheless, as engineering professionals, they will encounter empirical equations published in professional journals from time to time (e.g., Kumar et al., 1966) or developed for specialized purposes by their employers, vendors of equipment, or by federal and state agencies. When used for the purpose for which they were created and provided that the user carefully abides by the prescribed units, the results can be used with confidence.

The evaporation of a volatile liquid from an industrial spill is a situation that engineers may have to analyze. An *industrial spill* is a shallow pool in which the surface to volume ratio is large and heat transfer is sufficient to compensate for the cooling effect of evaporation. An expression (Kunkel, 1983) for cases where the wind speed is low and the Reynolds number based on length is less than 20,000 is

$$S_s = 0.3 \left(\frac{U_a}{T_a}\right)^{0.8}$$
$$A^{0.9} M_j P_v \left[\frac{(3.1 + (\rho_j/M_j)^{-0.33})^2}{T_a^{0.5}(1/29 + 1/M_j)^{0.5}}\right]^{-0.67} \quad (8.9)$$

where

$S_s$ = evaporation rate (kg/h)
$U_a$ = airspeed (m/s)
$A$ = area of spill $(\text{m}^2)$
$M_j$ = molecular weight of volatile liquid (species $j$)

$T_a$ = air temperature (K)

$\rho_j$ = average density of the volatile liquid $j$ $(g/cm^3)$

$P_v$ = vapor pressure of the volatile liquid at the liquid–air interface (mm Hg)

Kunkle (1983) discusses several analytical models that predict the evaporation rate for spills in which heat transfer is of varying importance.

---

**EXAMPLE 8.5   ESTIMATING THE EVAPORATION OF A VOLATILE MATERIAL USING AN EMPIRICAL EQUATION**

Ethyl mercaptan $(CH_3CH_2SH)$ is a liquid with an unpleasant skunklike odor. The material is a strong oxidizer, with a PEL value of 0.5 ppm. A 55-gal drum of the material is handled roughly and a small leak develops in a seam. A circular pool 10 m in diameter develops. You have been asked to model the atmospheric dispersion of the vapor, and as a first step you need to estimate the evaporation rate $(S_s,$ kg/h$)$ from the spill. The air temperature is 17.7°C and the wind speed is 3 m/s. The properties of ethyl mercaptan are

$$P_v = 400 \text{ mm Hg} \qquad M = 62.1 \text{ g/gmol}$$

$$\rho = 0.84 \text{ g/cm}^3$$

**Solution**   Substitute into Equation 8.9:

$$S_s = (0.3)\left(\frac{3}{290.7}\right)^{0.8}\left(\frac{100\pi}{4}\right)^{0.9}(62.1)(400)$$

$$\left[\frac{\left(3.1 + (0.84/62.1)^{-0.33}\right)^2}{K}\right]^{-0.67}$$

where

$$K = 290.7^{0.5}\left[\left(\frac{1}{29}\right) + \left(\frac{1}{62.1}\right)\right]^{0.5} = 3.8347$$

Thus

$$S_s = (0.3)(0.0258)(50.7)(24{,}840)$$

$$\left[\frac{(3.1 + 4.137)^2}{3.8347}\right]^{-0.67}$$

$$= 9747.7\left(\frac{52.24}{3.8347}\right)^{-0.67}$$

$$= (9747.7)(0.1718)$$

$$= 1674.4 \text{ kg/h}$$

---

This completes our treatment of the first four methods of estimating pollutant emission rates as they were enumerated at the beginning of the chapter. Often, unconventional applications and increasingly rigorous requirements necessitate mathematical modeling based on the principles of fluid flow, heat transfer, and diffusive and/or convective mass transport. Numerous practical examples of such modeling follow.

## 8.6 *Evaporation and diffusion*

A variety of volatile liquids are used in industrial operations. Unless precautions are taken, vapors from these liquids will enter the indoor or outdoor environment. To predict pollutant concentrations in air, it will be necessary to know the rate at which the vapors evaporate. In certain cases one can estimate these rates accurately from the principles of mass transfer and obtain more accurate rates than predicted by Equation 8.9 or AP-42 emission factors. Texts on the theory of mass transfer (Sherwood et al., 1975; McCabe et al., 1993; Treybal 1980; Bird et al., 1960; Hirschfelder et al., 1954) should be consulted to acquire a thorough understanding of the physical concepts and principles of mass transfer. Mackay and Paterson (1986) developed a comprehensive model describing the rates of mass transfer at the air–liquid interfaces for organic chemical volatilization, absorption in water, dissolution in rainfall, and wet and dry particle deposition.

To model the generation of a contaminant, one must know the thermodynamic and physical properties of air and the contaminant. The properties of air are readily available, but the properties of contaminants are less abundant. Table A-12 lists the vapor pressures of many organic liquids for a variety of temperatures. Primary references and handbooks (Vargaftik, 1975; Perry and Chilton, 1984; MacKay et al. 1982) should be consulted for data for specific contaminants or for more accurate data at different temperatures.

Values of vapor pressure $(P_{v2})$ of pure species at untabulated temperatures $(T_2)$ can be estimated from their values at $(P_{v1}, T_1)$ using the *Clausius–Clapeyron* equation

$$\ln \frac{P_{v2}}{P_{v1}} = \frac{h_{fg}}{R_u} \frac{T_2 - T_1}{T_2 T_1} \qquad (8.10)$$

where $R_u$ is the universal gas constant and $h_{fg}$ is the *enthalpy of vaporization per mole*, also called the *heat of evaporation* preferably between or at least near $T_1$ and $T_2$. The reference values $P_{v1}$ and $T_1$ are often 1 atm and the saturation temperature at 1 atm respectively. The saturation temperature at 1 atmosphere pressure is called the *normal boiling temperature*. The accuracy of this calculation is high if $T_1$ and $T_2$ are not vastly different.

### EXAMPLE 8.6   COMPUTATION OF THE VAPOR PRESSURE AT AN ARBITRARY TEMPERATURE

Compute the vapor pressure (in atmospheres) of toluene at 136.5°C and 31.8°C if the only information available is that the normal boiling temperature is 110.6°C and the heat of vaporization at the normal boiling point is 7997 cal/gmol.

*Solution*   The normal boiling temperature is the saturation temperature at 1 atm pressure. The heat of vaporization $(h_{fg})$ at the normal boiling point corresponds to a saturation pressure of 1 atm. Substituting a temperature 409.65 K (136.5°C) in the above, introducing constants to convert units, and neglecting the variation of $h_{fg}$ with temperature yields

$$\ln \frac{P_v}{1} = \frac{(7997)(4.186)(409.65 - 383.75)}{(8.314)(409.65)(383.75)}$$

$$P_v = 1.94 \text{ atm}$$

Substituting a temperature of 304.95 K (31.8°C) in the above, we obtain

$$\ln \frac{P_v}{1} = \frac{(7997)(4.186)(304.95 - 383.75)}{(8.314)(304.95)(383.75)}$$

$$P_v = 0.06646 \text{ atm (50.5 mm Hg pressure)}$$

Tabulated values of vapor pressures $(P_v)$ shows that:

1. $P_v(136.5°C) = 2.0$ atm; thus the computed value is 3.0% below the tabulated value.
2. $P_v(31.8°C) = 40$ mm Hg pressure; thus the computed value is 26.2% above the tabulated value.

The agreement at 136.5°C is quite good because the heat of vaporization was available at 110.6°C, which is quite close to the temperature of interest. In contrast, at 31.8°C the agreement is poor, owing to the disparity between the normal boiling point of 110.6°C and 31.8°C, the temperature of interest. A value of $h_{fg} = 9079$ cal/gmol is available for toluene at 25°C (Perry and Chilton, 1984). Using this value, which is closer to the temperature of interest (31.8°C), gives $P_v = 35.0$ mm Hg, which is 12.5% below the tabulated value. This show the sensitivity of predictions by the Clausius–Clapeyron equation to the selected value of the heat of vaporization.

Table A-8 contains a list of the *molecular diffusion coefficients* or *diffusivities* for a variety hydrocarbons in air and water at 0 to 25°C. Handbooks or source books should be consulted to find diffusion coefficients for other materials. *Gas-phase binary diffusion coefficients* $(D_{12}$ in cm²/s) can also be estimated by an equation proposed by Chen and Othmer and reported by Vargaftik (1975):

$$D_{12}(\text{cm}^2/\text{s}) = \qquad (8.11)$$

$$\frac{0.43 (T/100)^{1.81} \sqrt{1/M_1 + 1/M_2}}{P(T_{c1}T_{c2}/10{,}000)^{0.1405}[(v_{c1}/100)^{0.4} + (v_{c2}/100)^{0.4}]^2}$$

where $v_c$ and $T_c$ are *critical* volume (cm³/gmol) and temperature (K), $M_i$ is the molecular weight, and total pressure $(P)$ is in atmospheres. Other equations to predict the diffusion coefficient can be found in Sherwood et al. (1975), Perry and Chilton (1984), Treybal (1980), and Bird et al. (1960). Evaluations of the best methods are provided in Reid et al. (1987). Table A-9 lists values of critical temperature and pressure. Values of critical specific volume can be computed from

$$v_c = \frac{Z_c R_u T_c}{P_c} \qquad (8.12)$$

where $Z_c$ is the *compressibility factor* at the critical point, $(T_c, P_c)$. Values of $Z_c$ vary from about 0.23 to 0.33 for different substances, but a value of 0.27 is a good average for many substances encountered in air pollution.

The net molar transfer rate per unit area past the observer is $\mathbf{N}_a + \mathbf{N}_j$. The *molar average velocity* $(\mathbf{U}_m)$ is defined as

$$\mathbf{U}_m c_t = \mathbf{N}_a + \mathbf{N}_j = \mathbf{U}_a c_a + \mathbf{U}_j c_j$$

$$\mathbf{U}_m = \frac{\mathbf{N}_a + \mathbf{N}_c}{c_t} = \frac{\mathbf{U}_a c_a + \mathbf{U}_j c_j}{c_t} \quad (8.19)$$

where $c_t$ is the total molar concentration, $c_t = c_a + c_j$. The molar flux of species $j$ and air relative to the bulk gas can be written as

$$\mathbf{J}_j = c_j(\mathbf{U}_j - \mathbf{U}_m) \qquad \mathbf{J}_a = c_a(\mathbf{U}_a - \mathbf{U}_m) \quad (8.20)$$

Replace $\mathbf{J}_j$ and $\mathbf{J}_a$ by Equation 8.16 and $\mathbf{U}_m$ by Equation 8.19, rearrange, and obtain

$$\mathbf{N}_j = \frac{c_j}{c_t}(\mathbf{N}_j + \mathbf{N}_a) - D\,\mathbf{Grad}\,c_j$$

$$= \frac{c_j}{c_t}(\mathbf{N}_j + \mathbf{N}_a) - D\nabla c_j \quad (8.21)$$

$$\mathbf{N}_a = \frac{c_a}{c_t}(\mathbf{N}_a + \mathbf{N}_j) - D\,\mathbf{Grad}\,c_a$$

$$= \frac{c_a}{c_t}(\mathbf{N}_j + \mathbf{N}_a) - D\nabla c_a \quad (8.22)$$

Equations 8.21 and 8.22 relate concentration to molar flow rate. To obtain an equation involving the concentration only, consider a volume element in space through which a fluid composed of several molecular species flows. A mass balance results in the following expression, hereafter called the *species continuity equation*:

$$\frac{\partial c_j}{\partial t} + \mathbf{Div}(\mathbf{U}c_j) = \frac{\partial c_j}{\partial t} + \nabla\cdot(\mathbf{U}c_j) = D_{aj}\nabla^2 c_j + S_j$$
$$(8.23)$$

where **Div** is the vector operation of taking the divergence of the vector $(\mathbf{U}c_j)$ and $S_j$ represents the rate of production of molecular species $j$ by a chemical reaction (i.e., a chemical source of $j$). In a nonreacting gas mixture, $S_j$ is zero.

Consider the diffusion in the $z$-direction in which $\mathbf{N}_a$, $\mathbf{N}_j$, and $D_{aj}$ are all constant. Since there is only one spatial coordinate ($z$), vector quantities in the $z$-direction such as $N_{j,z}$ and $J_{j,z}$, and so on will be written simply as $N_c$ and $J_c$, and so on, for the sake of brevity and $\nabla c_j = dc_j/dz$. Equation 8.21 becomes

$$\int_{c_{j,1}}^{c_{j,2}} \frac{dc_j}{c_t N_j - c_j(N_a + N_j)} = -\int_{z_1}^{z_2} \frac{dz}{c_t D_{aj}} \quad (8.24)$$

which when integrated and rearranged becomes

$$N_j = \frac{N_j}{N_a + N_j}\frac{c_t D_{aj}}{z_2 - z_1}$$
$$\ln\frac{N_j/(N_j + N_a) - c_{j,2}/c_t}{N_j/(N_j + N_a) - c_{j,1}/c_t} \quad (8.25)$$

The general expression above can be used to predict the concentration of contaminants for several types of processes of interest: for example, evaporation of a pure liquid through stagnant air, or evaporation of a volatile species from a homogeneous liquid mixture through stagnant air.

## 8.7 Diffusion through stagnant air

Consider the diffusion of the vapor of a volatile liquid (denoted by subscript $j$) through a stagnant layer of air (denoted by subscript $a$) before entering the atmosphere. Such mass transfer is also called *pure diffusion* because no convective effects exist in the medium through which diffusion occurs. An example of such diffusion is the evaporation from a partially filled barrel of liquid that is open at the top (Figure 8-5). Assume that room air currents sweep the vapors away from the opening of the barrel (point 2) but do not induce any motion of the air and vapor inside the barrel. Assume also that air is not absorbed by the volatile liquid and that air currents carrying away vapors establish a vapor concentration at the mouth of the barrel that is negligibly small. The temperature and pressure will be assumed to be constant. Assume that radial gradients are zero and that there is only one spatial coordinate ($z$). Thus vector quantities will be understood to be in the $z$-direction. Since air is not absorbed by the liquid $N_a \ll N_j$, $N_j$ is essentially constant and $N_j/(N_j + N_a) \simeq 1$. Equation 8.25 becomes

$$N_j = \frac{c_t D_{aj}}{z_2 - z_1}\ln\frac{1 - c_{j,2}/c_t}{1 - c_{j,1}/c_t} \quad (8.26)$$

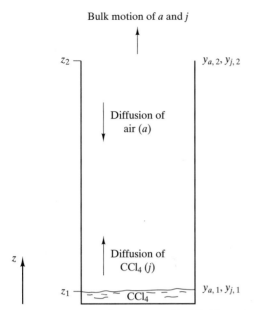

Bulk motion of $a$ and $j$

$z_2$

$y_{a,2}, y_{j,2}$

Diffusion of air ($a$)

$z$

Diffusion of $CCl_4$ ($j$)

$z_1$ — $CCl_4$ — $y_{a,1}, y_{j,1}$

**Figure 8-5** Evaporation from a partially filled barrel open at the top

Assuming that air and vapor can be described by the ideal gas equation, the mol fraction $(y_j)$ is related to concentrations and partial pressures:

$$P = c_t R_u T \qquad \text{(a)}$$

$$P_j = c_j R_u T \qquad \text{(b)}$$

$$y_j = \frac{c_j}{c_t} = \frac{P_j}{P} \qquad \text{(c)} \qquad (8.27)$$

$$c_t = c_a + c_j \qquad \text{(d)}$$

$$1 = \frac{c_a}{c_t} + \frac{c_j}{c_t} = y_a + y_j \qquad \text{(e)}$$

Equation 8.26 can be rewritten as

$$N_j = \frac{P D_{aj}}{R_u T (z_2 - z_1)} \ln \frac{y_{a,2}}{y_{a,1}} \qquad (8.28)$$

Define a *log mean mole fraction* $(y_{am})$ and *log mean partial pressure ratio* $(P_{am})$:

$$P_{am} = \frac{P_{a,2} - P_{a,1}}{\ln (P_{a,2}/P_{a,1})} \qquad y_{am} = \frac{y_{a,2} - y_{a,1}}{\ln (y_{a,2}/y_{a,1})} \qquad (8.29)$$

$y_{am}$ and $P_{am}$ are related as follows:

$$y_{am} = \frac{y_{a,2} - y_{a,1}}{\ln (y_{a,2}/y_{a,1})} = \frac{1}{P} \frac{P_{a,2} - P_{a,1}}{\ln (P_{a,2}/P_{a,1})} = \frac{P_{am}}{P} \qquad (8.30)$$

Since

$$y_{a,2} - y_{a,1} = (1 - y_{j,2}) - (1 - y_{j,1}) =$$

$$y_{j,1} - y_{j,2} \qquad (8.31)$$

$y_{am}$ can be written as

$$y_{am} = \frac{y_{j,1} - y_{j,2}}{\ln (y_{a,2}/y_{a,1})} \qquad (8.32)$$

The term, $\ln(y_{a,2}/y_{a,1})$ in Equation 8.28 can be replaced by the above, resulting in the molar flux $N_j$ expressed as

$$N_j = \frac{P D_{aj}}{R_u T (z_2 - z_1) y_{am}} (y_{j,1} - y_{j,2}) \qquad (8.33)$$

$$N_j = \frac{P^2 D_{aj}}{R_u T (z_2 - z_1) P_{am}} (y_{j,1} - y_{j,2}) \qquad (8.34)$$

Because of concepts that will be developed in subsequent sections, it is instructive to define the first terms on the right-hand side of Equations 8.33 and 8.34 as a *mass transfer coefficient* $k_G$. Thus for pure diffusion,

$$k_G = \frac{D_{aj}}{R_u T (z_2 - z_1) y_{am}} \qquad (8.35)$$

which because $P_{am} = y_{am} P$ can also be written as

$$k_G = \frac{P D_{aj}}{R_u T (z_2 - z_1) P_{am}} \qquad (8.36)$$

Convenient units of $k_G$ are kmol m$^{-2}$ s$^{-1}$ kPa$^{-1}$. Using the mass transfer coefficient $(k_G)$, the molar flux $N_j$ can be expressed as

$$N_j = k_G P (y_{j,1} - y_{j,2}) \qquad (8.37)$$

Alternatively, the driving potential $(y_{j,1} - y_{j,2})$ can be expressed in terms of a molar concentration difference $(c_{j,1} - c_{j,2})$ or as the difference in partial pressures $(P_{j,1} - P_{j,2})$. Using Equation 8.27, the molar flux $(N_j)$ can be written as

$$N_j = k_G R_u T (c_{j,1} - c_{j,2})$$

$$= k_G (P_{j,1} - P_{j,2}) \qquad (8.38)$$

Summary of mass transfer equations:

$$N_j = k_G (P_{j,1} - P_{j,2}) \qquad \text{(a)}$$

$$N_j = k_G P (y_{j,1} - y_{j,2}) \qquad \text{(b)} \qquad (8.39)$$

$$N_j = k_G R_u T (c_{j,1} - c_{j,2}) \qquad \text{(c)}$$

where for pure diffusion,

$$k_G = \frac{D_{aj}}{R_u T (z_2 - z_1) y_{am}} \quad \text{(d)}$$

Points 1 and 2 are chosen by the user to identify two different locations in the particular application. Often, points 1 and 2 correspond to the far-field location (subscript $\infty$) and the liquid–gas interface (subscript $i$).

When the interfacial contaminant vapor pressure is considerably less than the total pressure $(P_{c,i} \ll P)$ and the far-field contaminant vapor pressure is negligible $(P_{j,\infty} \simeq 0$ and $P_{a,\infty} \simeq P)$, $P_{am}$ may be difficult to evaluate. Help can be gained by expressing the logarithm of $P_{a,i}/P_{a,\infty}$ in a Taylor series:

$$\ln \frac{P_{a,i}}{P_{a,\infty}} = \ln \frac{P - P_{j,i}}{P}$$

$$= \ln \left(1 - \frac{P_{j,i}}{P}\right)$$

$$= -\frac{P_{j,i}}{P} \quad \text{(8.40)}$$

Thus

$$P_{am} = \frac{P_{a,i} - P_{a,\infty}}{\ln(P_{a,i}/P_{a,\infty})} \simeq \frac{(P - P_{j,i}) - P}{-P_{j,i}/P} \simeq P \quad \text{(8.41)}$$

Similarly, when $1 \gg y_{j,i} > y_{j,\infty} \simeq 0$, $y_{am} \approx 1$.

$$y_{am} = \frac{y_{a,i} - y_{a,\infty}}{\ln(y_{a,i}/y_{a,\infty})} \simeq \frac{y_{a,i} - 1}{\ln y_{a,i}} \simeq \frac{y_{a,i} - 1}{y_{a,i} - 1} = 1 \,(8.42)$$

In summary, when the contaminant vapor pressure at the liquid–air interface is small and the far-field contaminant partial pressure is negligible, it can be assumed that $P_{am} = P$.

The concentration profile that accompanies diffusion through a stagnant air layer can be found by substituting $N_a \simeq 0$ into Equation 8.21 and obtaining

$$N_j = -\frac{c_t D_{aj}}{1 - y_j} \frac{dy_j}{dz} \quad \text{(8.43)}$$

Since $N_j$ is constant,

$$\frac{dN_j}{dz} = \frac{d[-c_t D_{aj}(dy_j/dz)/(1 - y_j)]}{dz} = 0 \quad \text{(8.44)}$$

Integrating twice yields

$$-\ln(1 - y_j) = C_1(z) + C_2 \quad \text{(8.45)}$$

The two constants may be determined from knowledge of the concentration at $z_1$ and $z_2$:

$$y_j(z_1) = y_{j,1} \qquad y_j(z_2) = y_{j,2} \quad \text{(8.46)}$$

Upon rearranging, the concentration profile becomes

$$\frac{1 - y_j}{1 - y_{j,1}} = \left(\frac{1 - y_{j,2}}{1 - y_{j,1}}\right)^{(z-z_1)/(z_2-z_1)} \quad \text{(8.47)}$$

## EXAMPLE 8.8 EVAPORATION FROM AN EMPTY BARREL

Consider an open 55-gal drum containing a thin layer of carbon tetrachloride at the bottom. The drum is left uncovered. Estimate the rate at which carbon tetrachloride evaporates and enters the air.

- Drum cross-sectional area $(A) = 0.25 \text{ m}^2$.
- Drum height $(z_2 - z_1) = 0.813 \text{ m}$.
- Room temperature 80°F (26.7°C).
- Diffusion coefficient of carbon tetrachloride in air, $D = 0.62 \times 10^{-5} \text{ m}^2/\text{s}$.
- Carbon tetrachloride vapor pressure is 15.6 kPa.

*Solution* In the far field, $P_{a,2} = P$, since $P_{j,2} \approx 0$. At the liquid interface, $P_{j,1} = 15.6$ kPa, $P_{a,1} = P - P_{j,1}$. Using Equations 8.37 and 8.39 yields

$$N_j = k_G(P_{j,1} - P_{j,2})$$

$$= \frac{P D_{aj}}{R_u T (z_2 - z_1) P_{am}} (P_{j,1} - P_{j,2})$$

From Equation 8.29,

$$P_{am} = \frac{P_{a,2} - P_{a,1}}{\ln(P_{a,2}/P_{a,1})}$$

$$= \frac{P - (P - P_{j,1})}{\ln[P/(P - P_{j,1})]}$$

$$= \frac{P_{j,1}}{\ln(1/1 - P_{j,1}/P)}$$

Thus

$$N_j = \left(\frac{P D_{aj}}{R_u T (z_2 - z_1)}\right)$$
$$\left(\frac{(P_{j,1} - 0)\ln[(1/1) - (P_{j,1}/P)]}{P_{j,1}}\right)$$

$$= \frac{P D_{aj}}{R_u T (z_2 - z_1)} \ln \left[ \frac{1}{(1 - P_{j,1}/P)} \right]$$

$$= \frac{(101 \text{ kPa})(0.62 \times 10^{-5} \text{ m}^2/\text{s})}{(8.314 \text{ kJ kmol}^{-1} \text{ K}^{-1})(299.7 \text{K})(0.813 \text{ m})}$$

$$\ln \left( \frac{1}{1} - \frac{15.5}{101} \right)$$

$$= 5.19 \times 10^{-8} \text{ kmol m}^{-2} \text{ s}^{-1}$$

The rate of evaporation from the open drum is thus

$$\dot{m}_j = (5.19 \times 10^{-8} \text{ kmol m}^{-2} \text{ s}^{-1})(0.25 \text{ m}^2)$$

$$(154 \text{ kg/kmol}) = 1.99 \times 10^{-3} \text{ g/s} = 7.164 \text{ g/h}$$

If the barrel contained other volatile liquids, the evaporation rate could be found in a similar fashion. A compilation of evaporation rates, vapor pressures and diffusion coefficients for $CCl_4$ and other volatile liquids is shown below.

From the following table it may be noted that the evaporation rate increases with the vapor pressure but is relatively insensitive to the molecular diffusion coefficient.

| Hydrocarbon | Evaporation Rate (g/h) | Evaporation relative to $CCl_4$ | $P_v(T_{room})$ (mm Hg) | $D_{aj}(10^{-5} \text{ m}^2/\text{s})$ |
|---|---|---|---|---|
| Acetone | 7.95 | 1.11 | 180 | 0.83 |
| Carbonte trachloride | 7.16 | 1.00 | 91 | 0.62 |
| Benzene | 3.90 | 0.54 | 75 | 0.77 |
| Toluene | 1.20 | 0.17 | 20 | 0.71 |
| Acetic acid | 0.75 | 0.10 | 11 | 1.06 |
| Chlorobenzene | 0.23 | 0.03 | 12 | 0.62 |

### EXAMPLE 8.9   EMISSION OF HCL FROM AQUEOUS HCL SOLUTIONS

Consider the emission of HCl from drums of aqueous HCl solutions inadvertently left open. Assume the drums are half full and that the distance $(z_2 - z_1)$ is 50 cm. Assuming that room air sweeps the acid fumes away from the opening such that $y_{j,2}$ is very small with respect to unity, the emission of HCl can be estimated from Equation 8.39. In attempting to perform the calculations, two questions present themselves,

1. What is the diffusion coefficient of HCl in air?
2. Can the aqueous solution of HCl be considered to be homogeneous (i.e., is the acid concentration at the air–liquid interface equal to the overall concentration)?

The diffusion coefficient of polar compounds in air is difficult to predict, but for a first approximation,

Equation 8.11 will be used. The computation results in

$$D_{\text{HCl, air}} = 1.864 \times 10^{-5} \text{ m}^2/\text{s}$$

Uncertainty about the value of the acid concentration at the liquid–air interface is more difficult to resolve. The rigorous approach would be to do a simultaneous diffusional analysis on the transport in the liquid phase. In the case of strong electrolytes dissolved in water, however, the diffusion rates are those of the individual ions, which move rapidly (McCabe et al., 1993). Thus a good estimate of the maximum emission rate is to neglect diffusional resistance in the liquid film and assume that the liquid HCl concentration at the liquid–air interface is equal to the overall liquid concentration. The equilibrium partial pressure of HCl over an aqueous HCl solution can be obtained from Perry and Chilton (1984). The table below shows the result of using Equation 8.39

to estimate the HCl emission rate at 25°C for a variety of HCl concentrations:

| % HCl (by mass) | $P_{j,1}$ (mm Hg) | $N_j$ (kmol HCl m$^{-2}$ s$^{-1}$) |
|---|---|---|
| 20 | 0.32 | $6.30 \times 10^{-10}$ |
| 24 | 1.49 | $2.95 \times 10^{-9}$ |
| 28 | 7.05 | $1.40 \times 10^{-8}$ |
| 32 | 32.5 | $6.57 \times 10^{-8}$ |
| 36 | 142 | $3.11 \times 10^{-7}$ |
| 40 | 515 | $1.70 \times 10^{-6}$ |

To illustrate the sensitivity of HCl emission to temperature, the computations above can be repeated for a 30% acid concentration but varying temperature.

| $T$ (°C) | $P_{j,1}$ (mm Hg) | $N_j$ (kmol m$^{-2}$ s$^{-1}$) |
|---|---|---|
| 15 | 7.6 | $1.51 \times 10^{-8}$ |
| 20 | 10.6 | $2.11 \times 10^{-8}$ |
| 25 | 15.1 | $3.00 \times 10^{-8}$ |
| 30 | 21.0 | $4.21 \times 10^{-8}$ |
| 35 | 28.6 | $5.77 \times 10^{-8}$ |

From these calculations it should be apparent that the vapor pressure of the volatile component is the dominant factor controlling the emission rate of the contaminant. Thus whether it is the concentration or the temperature, anything that increases the vapor pressure increases the rate of evaporation.

# 8.8 Evaporation of single-component liquids

This section is devoted to the fundamental equations describing the evaporation of a volatile liquid consisting of a single molecular species (i.e., pure species or single-component liquid). Section 8.9 is devoted to two-component systems (i.e., the evaporation of a volatile liquid dissolved in another liquid, which may be water). Evaporation from two-component systems is also called *gas stripping* (Figure 8-6), *desorption*, or more generally, *distillation*. For the most part the liquid from which the volatile material evaporates is water, although the analysis is not necessarily

**Figure 8-6** Air stripping or gas scrubbing using non-clogging, fluidized non-spherical packing. Figure provided by Diversified Remediation Controls, Inc. Mpls, MN/Fluid Technologies (environmental) LTD

restricted to water. The opposite effect, pollutants absorbed by water, is discussed in Chapter 10.

Consider the evaporation of a single molecular species into air passing over it at a velocity $U_\infty$. When air passes over a volatile liquid the volatile material evaporates if its partial pressure in the gas phase is sufficiently low. Figure 8-7 is a diagram of the velocity and concentration profiles above the air–liquid interface. Evaporation is an endothermic process, but it will be assumed that there is sufficient heat transfer from the surroundings to maintain a constant liquid temperature. When evaporation occurs in conjunction with the motion of air over the liquid surface, the molar flux of molecular species $j$ normal to a unit area is expressed as

$$N_j = k_G(P_{j,i} - P_{j,\infty}) \qquad (8.48)$$

where (in typical units)

$N_j$ = molar flux $(\text{kmol m}^{-2}\,\text{s}^{-1})$

$k_G$ = *gas-phase mass transfer coefficient* $(\text{kmols m}^{-2}\,\text{s}^{-1}\,\text{kPa}^{-1})$

$P_{j,i}$ = partial pressure of species $j$ at the liquid–gas interface

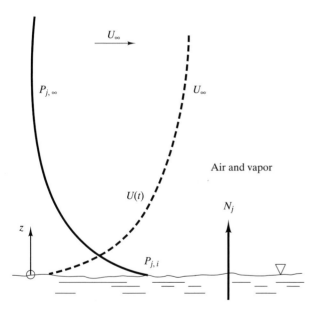

Single molecular species liquid

**Figure 8-7** Mass transfer through a single-film. Concentration and velocity profiles in air passing over a volatile (single molecular species) liquid.

$P_{j,\infty}$ = partial pressure of species $j$ in the far field

For a pure liquid $j$, $P_{j,i}$ will equal the vapor pressure $P_{v,j}$:

For mass transfer by bulk flow it is useful to express the gas-phase mass transfer coefficient in dimensionless form,

$$\text{Sh} = C_1(\text{Re})^{a1}(\text{Sc})^{b1} \qquad (8.49)$$

where $C_1$, $a1$, and $b1$ are dimensionless quantities that depend on the geometry of the evaporating surface and the range of Reynolds numbers under consideration.

$$\text{Sh (Sherwood number)} = \frac{k_G R_u T P_{am} L}{P D_{aj}} \qquad (8.50)$$

$$\text{Re (Reynolds number)} = \frac{LU\rho}{\mu} = \frac{LU}{\nu} \qquad (8.51)$$

$$\text{Sc (Schmidt number)} = \frac{\mu}{D_{aj}\rho} = \frac{\nu}{D_{aj}} \qquad (8.52)$$

The length $L$ is chosen by the user or defined in a particular empirical equation. It is the practice in heat transfer also to express the film coefficient in dimensionless form:

$$\text{Nu} = C_2(\text{Re})^{a2}(\text{Pr})^{b2} \qquad (8.53)$$

where $C_2$, $a2$, and $b2$ are dimensionless constants that depend on the geometry of the surface and the range of Reynolds numbers under consideration. The other parameters are dimensionless and are defined as

$$\text{Nu (Nusselt number)} = \frac{Lh}{k} \qquad (8.54)$$

$$\text{Pr (Prandtl number)} = \frac{\mu c_p}{k} \qquad (8.55)$$

In all of the equations above, the viscosity ($\mu$ and $\nu$), density ($\rho$), and thermal conductivity ($k$) pertain to air, not the liquid phase. Owing to the small contaminant concentration, the properties of pure air can be used.

From the theory of mass and heat transfer (see Treybal, 1980, and Incropera and DeWitt, 1990, for details), it can be shown that the Prandtl number (Pr) is proportional to the ratio of the local velocity boundary layer thickness $(\delta_U)$ to the local thermal boundary layer thickness $(\delta_T)$:

$$\mathrm{Pr}^n = \frac{\delta_U}{\delta_T} \qquad (8.56)$$

where $n$ is a positive number, typically 3. Similarly, the Schmidt number (Sc) is proportional to the ratio of the velocity boundary layer thickness $(\delta_U)$ to the concentration boundary layer thickness $(\delta_m)$:

$$\mathrm{Sc}^n = \frac{\delta_U}{\delta_m} \qquad (8.57)$$

The ratio of the Schmidt and Prandtl numbers is called the *Lewis number* (Le):

$$\mathrm{Le} = \frac{\mathrm{Sc}}{\mathrm{Pr}} = \left(\frac{\delta_T}{\delta_m}\right)^{1/n} \qquad (8.58)$$

Heat and mass transfer processes can be described by fundamental equations containing convection and diffusion terms of similar form. When these equations are made dimensionless (Incropera and Dewitt, 1990) one finds that each equation is related to the velocity through the Reynolds number (Re), while the Prandtl and Schmidt numbers depend on the temperature and composition of the phase of interest, albeit differently in detail. Because the dimensionless equations are similar, it is qualitatively useful to think of heat and mass transfer as analogous processes. Indeed, if the respective dimensionless parameters are equal, the solution of one equation is interchangeable with the solution of the other, provided that the boundary conditions are the same.

The *Reynolds analogy* is the name given to the condition when both the Prandtl and Schmidt numbers are unity and the solutions of both equations are identical. In actuality, the Schmidt and Prandtl numbers are not unity, nor even equal to each other. Nevertheless, the Reynolds analogy is useful as a starting point, in which case Equations 8.49 and 8.53 are comparable to each other:

$$C_1 = C_2 \qquad a1 = a2$$

$$b1 = b2 \qquad \mathrm{Sh} = \mathrm{Nu}\left(\frac{\mathrm{Sc}}{\mathrm{Pr}}\right)^{b1} \qquad (8.59)$$

and the gas-phase mass transfer coefficient $k_G$ can be written as

$$k_G = \mathrm{Nu}\,\frac{D_{ac}}{L}\left(\frac{\mathrm{Sc}}{\mathrm{Pr}}\right)^{b1}\frac{P}{P_{am}}\frac{1}{R_u T} \qquad (8.60)$$

In this way one can estimate mass transfer behavior from (more easily measured) heat transfer data. However, compare the Reynolds analogy to other correlations of the mass transfer coefficient in Figure 6-16 for a sobering dose of reality.

We have emphasized analogies between heat and mass transfer. Heat transfer correlations are often more readily available because research to generate them has been well supported and sustained. Secondly, experiments in heat transfer that generate correlations are easier to conduct. Table 8-1 represents a variety of heat transfer relationships and the text shows how these can be used to predict mass transfer behavior. Bird et al. (1960) analyze the equivalence of heat and mass transfer theory and point out that true equivalence requires:

1. Constant physical properties
2. A small rate of mass transfer
3. No chemical reactions in the fluid
4. No viscous dissipation
5. No emission or absorption of radiant energy
6. No pressure diffusion, thermal diffusion or forced diffusion

These restrictions will not be constraining in most air pollution calculations, but users of analogies should be aware of their limitations.

### EXAMPLE 8.10   ESTIMATING EVAPORATION RATE FROM FUNDAMENTAL PRINCIPLES

Repeat Example 8.5 using fundamental relationships. It will assumed that the partial pressure of the contaminant at the liquid–gas interface is equal to the vapor pressure of the contaminant $P_{v,j}$, whose value can be found in Table A-12. Also assume that the characteristic length in the problem is the diameter of the pool (i.e., $L = D$). The temperature is 17.7°C (290.7 K) and the total pressure is 1 atm (760 mm Hg).

| Ethyl mercaptan | Air |
|---|---|
| $P_{j,i} = P_{v,j} = 400\,\mathrm{mm\,Hg}$ (see Table A-12) | $v = 15.89 \times 10^{-6}\,\mathrm{m^2/s}$ |
| $D_{aj} = 0.9 \times 10^{-5}\,\mathrm{m^2/s}$ | $\mathrm{Pr} = 0.707$ |
| $M_j = 62.1$ | $U_\infty = 3\,\mathrm{m/s}$ |
| | $P_{a,i} = 760 - P_{j,i}$ $= 360\,\mathrm{mm\,Hg}$ |

The presence of significant amounts of methyl mercaptan in air just above the interface will influence the Prandtl (Pr) and Schmidt (Sc) numbers. For a first approximation this influence will be neglected because allowing for the presence of the mercaptan is time consuming, and as we have shown earlier, the property groups seldom dominate such calculations. Compute the following.

$$P_{am} = \frac{760 - 360}{\ln(760/360)} = \frac{400}{0.7472}$$

$$= 535.3 \text{ mm Hg}$$

$$\text{Re} = \frac{U_\infty D}{\nu}$$

$$= \frac{(10 \text{ m})(3 \text{ m/s})}{15.89 \times 10^{-6} \text{ m}^2/\text{s}} = 1.88 \times 10^6$$

The criteria in Table 8-1 show that the flow is turbulent. Table 8-1 indicates that the Nusselt number can be calculated from

$$\text{Nu} = [(0.037)\text{Re}^{0.8} - 871](\text{Pr})^{0.33}$$

$$= [(0.037)(1.88 \times 10^6)^{0.8} - 871]$$

$$(0.707)^{0.33} = 2.673$$

The Schmidt number is

$$\text{Sc} = \frac{\mu}{D_{aj}\rho}$$

$$= \frac{15.89 \times 10^{-6} \text{ m}^2/\text{s}}{0.9 \times 10^{-5} \text{ m}^2/\text{s}} = 1.765$$

**Table 8-1**   Convection Heat Transfer and Mass Transfer Relationships

1. *Air flowing over a flat surface*: Nu = $Lh/k$, Pr = $\mu c_p/k$, Re = $U_\infty Lr/\mu$
   *Laminar Flow*: Pr > 0.6; Re < $5 \times 10^5$
   Nu = $0.664 \text{ Re}^{0.5}\text{Pr}^{0.33}$
   *Turbulent Flow*: 0.6 < Pr < 60; $5 \times 10^5$ < Re < $10^8$
   Nu = $(0.037 \text{ Re}^{0.80} - 871)\text{Pr}^{0.33}$

2. *Air flowing through a cylindrical duct*: Sh = $k_G(R_u TLP_{am}/D_{aj}P)$
   Sh = $0.023 \text{ Re}^{0.83}\text{Sc}^{0.33}$, 4000 < Re < 60,000, 0.6 < Sc < 300

3. *Air flowing over a stationary sphere* ($L = D$)
   Nu = $2 + (0.4\text{Re}^{0.5} + 0.06\text{Re}^{0.67})(\mu_0/\mu_s)^{0.25}\text{Pr}^{0.4}$ if the following are satisfied:
   0.71 < Pr < 380
   3.5 < Re < $7.6 \times 10^4$
   1.0 < $\mu_\infty/\mu_s$ < 3.2
   The viscosity $\mu_\infty$ is evaluated at the far-field temperature and the viscosity $\mu_s$ is evaluated at the average surface temperature.

4. *External air flowing over plates and cylinders*: Nu = $C\text{Re}^m\text{Pr}^{1/3}$ (Pr > 0.7)

| | Re | C | m | |
|---|---|---|---|---|
| Horizontal cylinder | | | | |
| ($D = L$) | 0.4-4 | 0.989 | 0.330 | |
| | 4-40 | 0.911 | 0.385 | |
| | 40-4000 | 0.683 | 0.466 | |
| | 4000-40,000 | 0.193 | 0.618 | |
| | 40,000-400,000 | 0.027 | 0.805 | |
| Square ($L$ by $L$) | 5000-100,000 | 0.246 | 0.588 | flow 90° to flat surface |
| | 5000-100,000 | 0.102 | 0.675 | flow 45° to a flat surface |
| Semi-infinite vertical plate | | | | |
| (height $L$) flow 90° to flat surface | 4000-15,000 | 0.288 | 0.731 | |

5. *Air flowing through a fixed bed of pellets*: Sc = 0.6, $\varepsilon = 1 - (D_p a_s/6)$
   $a_s$ = total pellet surface area per volume of fixed bed
   Nu = $(2.06/\varepsilon)\text{Re}^{0.422}\text{PrSc}^{-0.67}$     90 < Re < 4000
   Nu = $(20.4/\varepsilon)\text{Re}^{0.185}\text{PrSc}^{-0.67}$     5000 < Re < 10,300

Abstracted from Bird et al. (1960) and Incropera and DeWitt (1990).

The mass transfer coefficient, $k_G$, can be computed from Equation 8.60, where $b_1 = 0.33$:

$$k_G = \text{Nu}\, \frac{D_{aj}}{D} \left(\frac{\text{Sc}}{\text{Pr}}\right)^{0.33} \frac{P}{P_{am}} \frac{1}{R_u T}$$

$$= (2673) \left(\frac{0.9 \times 10^{-5} \text{ m}^2/\text{s}}{10 \text{ m}}\right) \left(\frac{1.765}{0.707}\right)^{0.33}$$

$$\left(\frac{760}{535.3}\right) \left(\frac{1}{8.314 \text{ kJ/kmol·K}}\right) (290 \text{ K})$$

$$= 1.92 \times 10^{-6} \text{ kmol m}^{-2} \text{ s}^{-1} \text{ kPa}^{-1}$$

The molar flux, $N_j$, can be computed from Equation 8.39, assuming that the mercaptan partial pressure in the far field is zero $(P_{j,\infty} = 0)$:

$$N_j = k_G (P_{j,i} - P_{j,\infty})$$

$$= 1.92 \times 10^{-6} (\text{kmol m}^{-2} \text{ s}^{-1} \text{ kPa}^{-1})$$

$$\left[\left(\frac{400}{760}\right)(101)\text{kPa} - 0\right]$$

$$= 1.0208 \times 10^{-4} \text{ kmol s}^{-1} \text{ m}^{-2}$$

and the total evaporation rate from the pool is

$$\dot{m}_j = N_j A M_j$$

$$= 1.0208 \times 10^{-4} (\text{kmol s}^{-1} \text{ m}^{-2})$$

$$(62.1 \text{ kg/kmol}) \frac{100\pi}{4} (3600 \text{ s/h})$$

$$= 1790 \text{ kg/hr}$$

It is seen that this evaporation rate is larger than the value (1674.4 kg/h) computed from the empirical equation. An explanation for the discrepancy (5.4%) may lie in the fact that the equation used to compute the Nusselt number assumes that the velocity and concentration boundary layers begin at the leading edge of the pool. In the actual case air passing over the ground has a fully established boundary layer at the leading edge of the pool. Thus the equation for the Nusselt number does not describe accurately the actual case.

## 8.9 Single-film theory for multicomponent liquids

Consider air passing over a liquid that contains several molecular components of varying volatility. Let the characteristic far-field air velocity and temperature be denoted by $U_\infty$ and $T_\infty$. The most volatile components will evaporate fastest. For the single-film model it will be assumed that there is a mixing mechanism inside the liquid such that the liquid-phase concentration is uniform with regard to space but variable with respect to time.

$$c_j(x, y, z, t) = c_j(t) \qquad (8.61)$$

The mass transfer is similar to that shown in Figure 8-7 and those discussed in Section 8.8 except that now there are several molecular species ($j$) in the liquid phase.

The theory is called single film because the primary resistance to mass transfer is in the concentration boundary layer in the air phase. The task for the engineer is to estimate the mass transfer rate of any one of the molecular species and the total evaporation rate of all species as a function of time.

For the analysis that follows (Drivas, 1982; Stiver et al., 1989), it will be assumed that the liquid is an *ideal solution* (see Section 1.6.4) in the form of a pool of diameter $D_0$ and depth $h$. The depth decreases very slowly with time. *Raoult's law* is invoked to describe the equilibrium that exists at the interface between a multicomponent liquid and a multicomponent vapor. The partial pressures of the molecular species $j$ in the gas phase $(P_j)$ and at the interface $(P_{j,i})$ are

$$P_j = y_j P \qquad \text{(a)}$$

$$P_{j,i} = y_{j,i} P = x_{j,i} P_{v,j} \qquad \text{(b)} \qquad (8.62)$$

where $P_{v,j}$ is the vapor pressure (see Table A-12) or saturation pressure of pure species $j$ at the temperature of the liquid and $x_{j,i}$ is the mole fraction at the interface in the liquid phase. Note that $x_{j,i} = x_j$ because of the liquid-phase uniformity assumption made earlier. It will be assumed that the system is isothermal $(T_{\text{liquid}} = T_{\text{air}} = T_\infty = \text{constant})$, that the total number of moles per volume of liquid $(c_L)$ is constant, and that the far-field partial pressure of each volatile molecular species is zero $(P_{j,\infty} = 0)$.

The rate of evaporation of molecular species $j$ $(kmol\ s^{-1}\ m^{-2})$ can be found from Equation 8.48:

$$N_j = k_G(P_{j,i} - P_{j,\infty})$$
$$= k_G P_{j,i} = k_G x_j P_{v,j}$$
$$= k_G \frac{c_j}{c_L} P_{v,j} \qquad (8.63)$$

Be mindful that the subscript $i$ refers to conditions at the interface and particular molecular species are denoted by the subscript $j$ if multiple species are involved.

If the mole fraction of species $j$ in the liquid were unchanging, the evaporation rate of species $j$ would be constant. However, in a multicomponent liquid pool, the most volatile liquids will evaporate faster than the less volatile. If molecular species $j$ is the most volatile, its (spatially uniform) concentration will decrease with time and the liquid will become increasingly rich in the less volatile species. If the pool is of finite mass, the ratio of the mole fraction of the most volatile species to the mole fraction of the least volatile species will decrease with time. Consequently, while the total number of mols of liquid decreases slowly with time, the mole fraction of the least volatile increases slowly with time simultaneously with a slow decrease in the mole fraction of the most volatile species. The following analysis shows how to estimate the instantaneous evaporation rate of each molecular species $(N_j)$.

Define a control volume in a layer of liquid of cross-sectional area $A$ and thickness $L$. From the conservation of mass,

$$\frac{dc_j}{dt} = -\frac{k_G c_j P_{v,j}}{Lc_L} \qquad (8.64)$$

where (typical units)

$P_{v,j}$ = vapor pressure (atm) of pure species $j$ evaluated at temperature $T_\infty$

$c_j$ = number of moles of species $j$ per volume of liquid $(mol/m^3)$

$k_G$ = mass transfer coefficient $(mol\ m^{-2}\ atm^{-1}\ s^{-1})$; assumed to be the same for every species

$c_L$ = total number of moles per volume of liquid $(mols/m^3)$; assume constant

$L$ = thickness of liquid pool (m); assume to decrease very slowly with time

Integrating Equation 8.64 with time yields the concentration of species $j$ at any instant

$$c_j(t) = c_j(0) \exp\left(\frac{-k_G P_{v,j} t}{Lc_L}\right) \qquad (8.65)$$

where $c_j(0)$ is the initial molar concentration of species $j$ in the liquid.

For a mixture of liquid species $j$, it is useful to define the total mass of liquid per unit area $(m_t/A,\ kg/m^2)$:

$$\frac{m_t}{A} = \sum Lc_j M_j \qquad (8.66)$$

where the summation is taken over the total number of molecular species $(c_j)$ and $M_j$ is the molecular weight of contaminant species $j$. The total rate of evaporation per unit area $(\dot{m}_t/A)$ in which several species evaporate at individual rates is obtained by differentiating Equation 8.66 with time:

$$\frac{\dot{m}_t}{A} = -\frac{d(m_t/A)}{dt} = -\sum LM_j \frac{dc_j}{dt} \qquad (8.67)$$

Substituting Equation 8.64 gives

$$\frac{\dot{m}_t}{A} = \sum k_G P_{v,j} M_j \frac{c_j}{c_L} \qquad (8.68)$$

Replace the instantaneous concentration of species, $c_j(t)$, by Equation 8.65:

$$\frac{\dot{m}_t}{A} = \sum_j M_j k_G P_{v,i} \frac{c_j(0)}{c_L} \exp\frac{-k_G P_{v,i} t}{Lc_L} \qquad (8.69)$$

Multiply Equation 8.69 by the initial mass per unit area of liquid, $m_t(0)/A$, where

$$\frac{m_t(0)}{A} = Lc_L M_{avg}(0) \qquad (8.70)$$

$$M_{avg}(0) = \sum x_j(0) M_j \qquad (8.71)$$

where $x_j(0)$ is the initial mole fraction of species $j$. Rearrange, simplify, and obtain

$$\frac{\dot{m}_t}{A} = \frac{m_t(0)k_G}{c_L L A M_{avg}(0)}$$

$$\sum_j P_{v,i} x_j(0) M_j \exp\left(\frac{-k_G P_{v,i} t}{L c_L}\right) \quad (8.72)$$

Equation 8.72 reduces to

$$\frac{\dot{m}_t}{A} = k_G \sum_j P_{v,i} x_j(0) M_j \exp\left(\frac{-k_G P_{v,i} t}{L c_L}\right) \quad (8.73)$$

For *nonideal solutions*, Raoult's law can be modified by including the liquid *activity coefficients* $(\Gamma_j)$, so that Equation 8.64 can be rewritten as

$$\frac{dc_j}{dt} = -\frac{k_G c_j \Gamma_j P_{v,j}}{L c_L} \quad (8.74)$$

The remainder of the derivation proceeds in the same fashion, with the activity coefficient carried through as a known parameter.

The mass transfer coefficient $(k_G)$ can be obtained from Equation 8.60. For large spills of petroleum on bodies of water, spills on the order of hundreds of meters in diameter, Drivas (1982) recommends the expression

$$k_G = \frac{0.0292 U_\infty^{0.78}}{R_u T_\infty D_0^{0.11} Sc^{0.67}} \quad (8.75)$$

for all the evaporating species. The parameters in Equation 8.75 require the following units:

$$k_G = gmol\ m^{-2}\ atm^{-1}\ h^{-1}$$

$U_\infty$ = far-field air velocity (m/h)

$D_0$ = diameter of the spill (m)

$Sc$ = gas-phase Schmidt number (using a mass weighted average for the liquid mixture)

$R_u$ = universal gas constant $(8.206 \times 10^{-5}\ atm\ m^3\ gmol^{-1}\ K^{-1})$

$T_\infty$ = air temperature (K)

---

## EXAMPLE 8.11 EVAPORATION OF VOLATILE COMPOUNDS FROM A WELL-MIXED WASTEWATER LAGOON

A lagoon is 10 m in diameter $(D_0)$ and 3 m deep $(d)$. The land is flat surrounding the lagoon. Dry air (zero water vapor in the far field, $P_{water, \infty} = 0$) at 25°C passes over the surface of the lagoon at a velocity $U_\infty$ of 3 m/s. An agitator in the lagoon ensures that the composition in the liquid phase is uniform. Initially, the mole fractions of benzene $(C_6H_6)$, methyl alcohol $(CH_3OH)$, carbon tetrachloride $(CCl_4)$, and toluene $(C_6H_5CH_3)$ are each 0.05. The remainder of liquid is water. Estimate the evaporation rate of each species as a function of time. Let the subscript $j$ denote the particular species

*Solution* Some useful properties for air are: $Pr = 0.708$, $\nu = 1.5 \times 10^{-5}\ m^2/s$, and $\rho = 1.2\ kg/m^3$. The average molecular weight of the liquid is

| Species, $j$ | $M$ | $x_j(0)$ | $D_j$ in air $(m^2/s)$ | $Sc_j$ | $P_{v,i}(25°C)$ (kPa) |
|---|---|---|---|---|---|
| $C_6H_6$ | 78 | 0.05 | $0.77 \times 10^{-5}$ | 1.948 | 12.77 |
| $CH_3OH$ | 32 | 0.05 | $1.33 \times 10^{-5}$ | 1.128 | 15.72 |
| $CCl_4$ | 154 | 0.05 | $0.62 \times 10^{-5}$ | 2.419 | 14.40 |
| $C_6H_5CH_3$ | 92 | 0.05 | $0.71 \times 10^{-5}$ | 2.113 | 4.19 |
| $H_2O$ | 18 | 0.80 | $2.64 \times 10^{-5}$ | 0.568 | 3.17 |

$$M_{avg} = 0.05(78 + 32 + 154 + 92) + 0.8(18)$$
$$= 32.2\ kg/kmol$$

The initial mass fractions of each species, $f_j(0)$, in the liquid phase shown above are given by

$$f_j = \frac{x_j M_j}{M_{avg}}$$

To compute the total molar density $(c_L)$, it will be assumed that the mixture is an ideal liquid mixture. Assume 1000 kg of the liquid mixture. The mass fraction $f_j$, density of the pure species $\rho_j$, mass $m_j$, number of moles $n_j$, and volume $V_j$ of each species are as follows:

| Species, $j$ | $M_j$ | $f_j$ | $\rho_j \,(\text{kg/m}^3)$ | $m_j \,(\text{kg})$ | $n_j \,(m_j/M_j)$ | $V_j \,(\text{m}^3) = m_j/\rho_j$ |
|---|---|---|---|---|---|---|
| $C_6H_6$ | 78 | 0.1210 | 879 | 121 | 1.55 | 0.1376 |
| $CH_3OH$ | 32 | 0.0496 | 790 | 49.6 | 1.55 | 0.0628 |
| $CCl_4$ | 154 | 0.2391 | 1595 | 239.1 | 1.55 | 0.1499 |
| $C_6H_5CH_3$ | 92 | 0.1428 | 866 | 142.8 | 1.55 | 0.1649 |
| $H_2O$ | 18 | 0.4472 | 1000 | 447.2 | 24.84 | 0.4472 |
| | | | | | $n_t = \Sigma \, n_j = 31.04$ | $V_t = \Sigma \, V_j = 0.9624$ |

The total molar density $c_L$ is

$$c_L = \frac{n_t}{V_t} = \frac{31.04 \text{ kmol}}{0.9624 \text{ m}^3} = 32.25 \text{ kmol/m}^3$$

The mole-weighted Schmidt number for the pollutants in air (including water vapor),

$$\begin{aligned} Sc_{avg} = &\, [(0.05)(1.948) + (0.05)(1.128) \\ &+ (0.05)(2.419)(0.05)(2.113) \\ &+ (0.80)(0.568)] = 0.8349 \end{aligned}$$

A mole-weighted average diffusion coefficient of pollutants in air (including water vapor),

$$\begin{aligned} D_{avg} = &\, 10^{-5}[(0.05)(0.77) \\ &+ (0.05)(1.33) + (0.05)(0.62) \\ &+ (0.05)(0.71) + (0.80)(2.64)] \\ =&\, 2.284 \times 10^{-5} \text{ m}^2/\text{s} \end{aligned}$$

The Reynolds number (characteristic length $D_0 = 10$ m) of the air passing over the lagoon is

$$Re = \frac{U_\infty D_0}{\nu} = \frac{(3 \text{ m/s})(10 \text{ m})}{1.5 \times 10^{-5} \text{ m}^2/\text{s}} = 2 \times 10^6$$

From Table 8-1, such a Reynolds number indicates that the flow is turbulent and that the following Nusselt number correlation can be used for the air:

$$Nu = (0.037 \, Re^{0.8} - 871) Pr^{0.33}$$

$$= [(0.037)(2 \times 10^6)^{0.8} - 871](0.708)^{0.33} = 2847$$

The agitator in the lagoon ensures uniform concentration within the aqueous phase. Thus mass transfer from the liquid surface is governed by a single-film theory. The mass transfer coefficient can be found from Equation 8.60:

$$k_G = Nu \, \frac{D_{aj}}{D_0} \left( \frac{Sc}{Pr} \right)^{0.33} \frac{P}{P_{am}} \frac{1}{R_u T}$$

where $P_{am}$ is computed from Equation 8.29:

$$P_{am} = \frac{P_{a,\infty} - P_{a,i}}{\ln (P_{a,\infty}/P_{a,i})}$$

In the far field it has been assumed that there is no water vapor (RH = 0%); thus $P_{a,\infty} = 101$ kPa. At the air–liquid interface, $P_{a,i}$ is equal to the following:

$$\begin{aligned} P_{a,i} = &\, P_t - [(0.05)P_v(C_6H_6) \\ &+ (0.05)P_v(CH_3OH)(0.05)P_v(CCl_4) \\ &+ (0.05)P_v(C_6H_5CH_3) + (0.8)P_v(H_2O)] \\ =&\, 101 - 4.89 = 96.11 \text{ kPa} \end{aligned}$$

$$P_{am} = \frac{101 - 96.11}{\ln (101/96.11)} = \frac{4.89}{\ln (1.05)} = 98.53 \text{ kPa}$$

The mass transfer coefficient is

$$\begin{aligned} k_G = &\, 2847 \left( \frac{2.284 \times 10^{-5} \text{ m}^2/\text{s}}{10 \text{ m}} \right) \left( \frac{0.8349}{0.708} \right)^{0.33} \\ &\frac{101/98.53}{(8.314 \text{ kJ kmol}^{-1} \text{ K}^{-1})(298 \text{ K})} \\ =&\, 3.166 \times 10^{-6} \text{ kmol kPa m}^{-2} \text{ s}^{-1} \end{aligned}$$

The evaporation rate per unit area is given as Equation 8.73:

$$\frac{\dot{m}_t}{A} = k_G \sum x_j(0) M_j P_{v,j} \exp \left( -\frac{k_G P_{v,j} t}{L c_L} \right)$$

where $m_t(0)/A$ is the total mass per unit area of the lagoon and given by Eq. 8.70:

$$\begin{aligned} m_t(0) = &\, L c_L M_{avg} \\ =&\, (3 \text{ m})(32.25 \text{ kmol/m}^3)(32.2 \text{ kg/kmol}) \\ =&\, 3115.4 \text{ kg/m}^2 \end{aligned}$$

Substitute this value in the previous equation and obtain

$$\frac{\dot{m}_t}{A} \left( \text{kg m}^{-2} \text{ s}^{-1} \right) =$$

$$3.166 \times 10^{-6} \sum x_j(0) M_j P_{v,j} \exp \left( -\frac{k_G P_{v,j} t}{L c_L} \right)$$

The evaporation rate for each species $j$ is given by $k_G[c_j(0) - 0]M_j P_{v,j} x_j$. Replace $c_j(0)$ by Equation 8.65:

$$\frac{\dot{m}_j}{A} \left(\text{kg m}^{-2}\text{s}^{-1}\right) =$$

$$3.166 \times 10^{-6} x_j(0)\, M_j P_{v,j} \exp\left(-\frac{k_G P_{v,j} t}{L c_L}\right)$$

The evaporation rate of each species can be determined by repetitive computations using commercially available mathematical computer programs. The results are seen in Figure E8-11. The MathCAD program producing Figure E8-11 can be found in the textbook's Web site on the Prentice Hall Web page.

The equations above assume that $L$ and $c_L$ remain constant throughout the period of evaporation. Such an assumption is clearly untrue since $L$ has to decrease as materials evaporate from the pool and $c_L$ will also vary slightly, owing to the fact that the most volatile species evaporate fastest and their mole fractions and partial molar concentrations decrease to zero while the partial molar concentrations of the less volatile species are nonzero. To estimate the total evaporation rate for such a case, the user should use

Equation 8.73 to estimate the evaporation rate during a time step $t = \delta t$ and then update the values of $l$ and $c_L$ in Equation 8.73 to compute the evaporation rate for the next time step. By marching forward through a series of many time steps, the user can estimate the evaporation rate over a long period of time. The value of $L$ and $c_L$ at the beginning of each time step should be equal to the values at the end of the previous time step:

$$c_j(t) = c_j(t - \delta t)$$

$$L(t) = L(t - \delta t) - \frac{\delta t(\dot{m}_t)}{\rho(t - \delta t)}$$

$$c_L(t) = \sum c_j(t - \delta t)$$

## 8.10 Two-film evaporation of multicomponent liquids

Section 8.7 concerned the evaporation of a contaminant through a column of stagnant air. Such mass transfer is called *pure (molecular) diffusion*. If the air

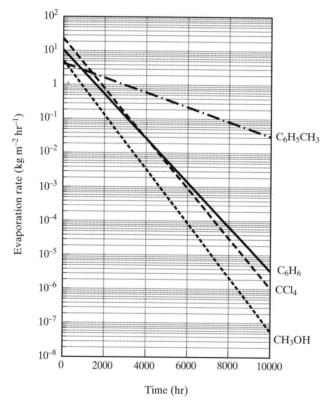

C6H5CH3

C6H6

CCl4

CH3OH

**Figure E8-11**   Rate of evaporation ($\text{kg m}^{-2}\text{ hr}^{-1}$) of volatile compounds from a well mixed, waste-water lagoon versus time (hr).

is moving, evaporation is more rapid because the moving air enhances mass transfer. Such mass transfer is called *convective diffusion*. Section 8.10 concerns convective diffusion.

*Desorption* is the name of the physical process in which a dissolved gas (or vapor) is transferred from the liquid phase to the gas phase. Carbon dioxide bubbles in carbonated soft drinks and beer are familiar examples of desorption. *Stripping* is the name of the process in which air, steam, or other stripping medium is used to enhance desorption. An example (Figure 8-6) is the removal of volatile organic compounds from contaminated groundwater.

Less familiar examples of desorption follow. Trace amounts of trichloroethylene (TCE) in tap water desorb from hot water passing through a bathroom shower fixture. Little (1992) found that the daily indoor inhalation exposure to materials in tap water for a 10-min shower in a home bathroom is equivalent to 1.5 times that incurred by ingesting 2 L (typical daily consumption) of the same water. Giardino et al. (1992) found that approximately 60% of TCE in water is volatilized from the spray drops and water deposited on the surface of the shower enclosure. Shepherd et al. (1996) found that chloroform was generated in washing machines that use bleach containing sodium hypochlorite. Once generated, chloroform entered the indoor environment. Chloroform in the discharge water could constitute a significant fraction of chloroform mass loading in a municipal wastewater plant. On the larger scale, Jones et al. (1996) found that volatile HAPs in municipal wastewaters escaping through curb drains and holes in manholes in municipal sewer systems are a potentially significant source of hazardous air pollutants (HAPs) in the urban environment.

In Section 8.9 it was assumed that the concentration in the liquid phase was spatially uniform. In general this is not the case; it certainly should not be assumed without evidence to support it. The evaporation of pollutants from a multicomponent liquid mixture is complicated by the fact that each species must be transported from the liquid interior through a concentration boundary layer in the liquid before it reaches the air–liquid interface to evaporate. Once leaving the surface as a vapor it must be transported through another boundary layer on the air side. Thus mass is transferred through *two films or resistances*,

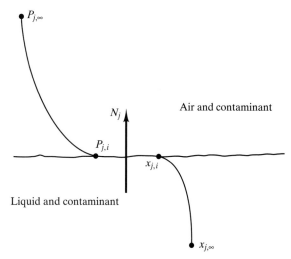

**Figure 8-8** Concentration profiles for two-film resistance

as shown in Figure 8-8. If one film offers much more resistance than the other, the analysis reduces to a single-film analysis, but until this is known for a fact, both films have to be dealt with as if both were equally important. As a general rule, the resistance of the liquid film controls mass transfer for very volatile contaminants (contaminants with large values of the Henry's law constant, $H$) while the resistance of the gas-phase film becomes increasingly important as the volatility decreases.

If the mass of liquid and air are large, such as evaporation of a volatile organic compound from an industrial spill or from a wastewater lagoon, a steady-state mass transfer rate can be achieved. If on the other hand, the mass of the liquid is small, the analysis is complicated by the depletion of material in the liquid phase and its accumulation in the gas phase, as shown in Figure 8-9. Only steady-state mass transfer through two films will be considered in the analysis that follows. As in single-film theory, isothermal conditions will be presumed. Physically, this means that the liquid phase is in contact with an infinite heat source that transfers sufficient energy to the liquid to replace the energy consumed in evaporation.

Before postulating equations to describe the mass transfer, it is essential to understand the equilibrium of a liquid air-pollutant system. Consider a cylinder equipped with a frictionless piston, shown in Figure 8-10. The cylinder contains water and air, each of

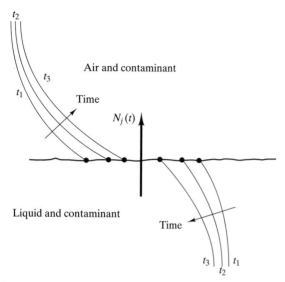

**Figure 8-9**   Instantaneous concentration profiles for two-film resistance showing depletion in the liquid and accumulation in the gaseous phase

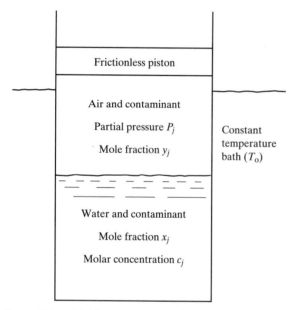

**Figure 8-10**   Multicomponent equilibrium for contamination in air and water

which contains a certain amount of a dissolved molecular species $j$. The vessel is immersed in a constant temperature bath $(T_0)$. Water containing species $j$ is placed in the vessel along with (pure) air. Some of the species evaporates from the water and enters the

air. Eventually, the system comes to equilibrium. At equilibrium, the mole fraction of species $j$ in water $(x_j)$ and the partial pressure of $j$ in air $(P_j)$ are measured. The experiment is repeated for a variety of mole fractions of $c$ in water at temperature $T_0$. A typical graph of the partial pressure versus the liquid mole fraction at equilibrium is shown in Figure 8-11. If the temperature is varied and the experiment is repeated, a series of equilibrium lines called *equilibrium isotherms* can be drawn. At any given partial pressure, the equilibrium liquid mole fraction goes down as the temperature goes up.

For purposes of understanding air pollution, in which solutes exist at very low concentrations, only the portion of the equilibrium curve in the vicinity of the origin will be considered. In this range the equilibrium isotherms satisfy *Henry's law*, which can be expressed in several forms, depending on the units chosen to represent the solute concentration:

$$P_j = Hc_j = Hx_jc_L = H'x_j \qquad (8.76)$$

$$Hc_L = H' \qquad (8.77)$$

The parameters $H$ and $H'$ are the slopes of the equilibrium curve and are called the *Henry's law constants*. The units of $H$ are atm m$^3$ kmol$^{-1}$ and the units of $H'$ are atm. Alternatively, the units of pressure could be kPa or N/m$^2$. The parameters $c_j$ and $c_L$ are the solute and total liquid molar concentrations. Henry's law constant is a physical property that reflects how a molecular species *partitions* itself between air and water. Chemicals with low values of

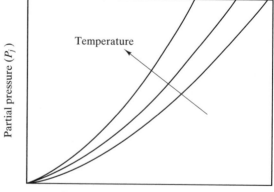

**Figure 8-11**   Absorption equilibrium isotherm for contaminant in air and water

Henry's law constant are those that partition more of the solute in the liquid phase than in the air phase, while higher values of the Henry's law constant imply the opposite. Henry's law constant can be predicted from thermodynamic data (Nirmalakhandan and Speece, 1988) or computed from experimental measurements of vapor pressure in air and solubility in water (Yaws et al. 1991; MacKay and Shiu 1981). Henry's law constants ($H$) for many organic compounds can be found in an entry by Schwarzenbach in Web site http://chemfinder.camsoft.com/. Table A-8 is a listing of $H'$ and diffusion coefficients ($D$) for many important pollutants.

## EXAMPLE 8.12 COMPUTATION OF HENRY'S LAW CONSTANT FROM SOLUBILITY DATA

Estimate the Henry's law constant for toluene in water.

**Solution**  From Perry et al. (1984), the solubility of toluene ($M = 92$) is reported to be 0.05 g of toluene per 100 g of water at 16°C and 1 atm. Thus the mole fraction in water is

$$x_j = \frac{0.05/92}{(100/18) + (0.05/92)} = 9.77 \times 10^{-5}$$

From Table A-12 the vapor pressure ($P_v$) of toluene is found to be 40 mm Hg at 31.8°C and 20 mm Hg at 18.4 °C. The vapor pressure at 16°C is not given. Since $H'$ depends on the vapor pressure ($P_{v,j}$), which in turn is keenly dependent on temperature, you should take the time to estimate the vapor pressure at 16°C carefully. Extrapolating linearly, $P_v(16°C)$ is

$$\frac{P_{v,j} - 20}{40 - 20} = \frac{16 - 18.4}{31.8 - 18.4}$$

$$P_{v,j}(16°C) = 20 + \frac{(-2.4)(20)}{(13.4)} = 16.4 \text{ mm Hg}$$

Better accuracy could be obtained using the Clausius–Clapeyron Equation 8.10 to estimate the vapor pressure at 16 °C.

Because the solubility condition implies equilibrium with the saturated vapor, the Henry's law constant at 16°C can now be estimated from Equation 8.77:

$$H' = \frac{P_j}{x_j} = \frac{P_{v,j}}{x_j}$$

$$\frac{(16.4 \text{ mm Hg})(\text{atm}/760 \text{ mm Hg})(1.01 \times 10^5 \text{N atm}^{-1} \text{m}^{-2})}{9.77 \times 10^{-5} \text{ mol/total mol}}$$

$$= 2.23 \times 10^7 \text{ N/m}^2$$

The value listed in Table A-8 is $3.72 \times 10^7 \text{ N/m}^2$.

The Henry's law constant has a substantial temperature dependence and the temperature for the datum from Table A-8 is unknown. The two values are in the same ballpark and the value calculated for 16 °C is probably the more reliable one.

---

Vapor–liquid equilibrium embodies a wide spectrum of chemical behavior. For components present in very small amounts, Henry's law is often a satisfactory approximation. For components in very large amounts (e.g., solvents), Raoult's law may suffice. In general an entire range of nonideal behavior would need to be quantified by methods available in the chemical engineering literature but beyond the scope of this book.

## EXAMPLE 8.13 COMPARISON AND WARNINGS ABOUT HENRY'S AND RAOULT'S LAWS

Given:

- Aqueous solution of toluene with mole fraction $x_{tol} = 9.77 \times 10^{-5}$
- Toluene vapor pressure 25°C = 29.8 mm Hg
- Henry's law constant = $3.72 \times 10^7 \text{ N/m}^2$

Calculate the partial pressure of toluene over the aqueous solution at 25°C using Henry's law and Raoult's law.

**Solution**  By Henry's law (Equation 8.76), partial pressure at the interface

$$P_j = H'x_j$$
$$= (3.72 \times 10^7 \text{ N/m}^2)(9.77 \times 10^{-5})$$
$$= 3634 \text{ N/m}^2$$

By Raoult's law (Equation 8.62), partial pressure at the interface

$$P_j = x_j P_{v,j}$$
$$= (9.77 \times 10^{-5}) \frac{29.8}{760} \text{atm}(10^5 \text{ N m}^{-2} \text{ atm}^{-1})$$
$$= 0.382 \text{ N/m}^2$$

This 10,000-fold discrepancy illustrates the dramatic errors that are possible when Raoult's law is applied to minor components in nonideal liquid solutions. At a mole fraction of $10^{-4}$, toluene certainly meets the criterion as a minor component. As an aromatic hydrocarbon ($C_7H_8$) in water, its solution will surely be nonideal. One characterization of nonideal solutions is activity coefficients ($\Gamma$) that differ from unity. Chemically, one might expect a toluene molecule, surrounded by water, to try to leave the solution (i.e., to exert a much greater partial pressure than that implied by its very low mole fraction). Equally gross errors may be expected when Henry's law is applied to components present in large proportion.

If a liquid containing a pollutant is brought into contact with air also containing the pollutant, the pollutant will desorb and transfer to air if the coordinates $(P_j, x_j)$ lie below the equilibrium line on Figure 8-12. If the coordinates lie above the equilibrium line, the pollutant will leave the air phase and be absorbed in the liquid. In terms of the variables shown in Figure 8-12, the mass transfer rate can be expressed as

$$N_j = k_L(x_{j,\infty} - x_{j,i})c_L = k_G(P_{j,i} - P_{j,\infty}) \quad (8.78)$$

Unfortunately, both the properties at the interface $(x_{j,i}$ and $P_{j,i})$ and the mass transfer coefficients $(k_L$ and $k_G)$ are unknown. An alternative expression for the mass transfer can be postulated that is more empirical but often more useful,

$$N_j = K_Lc_L(x_{j,\infty} - x_j^*) = K_G(P_j^* - P_{j,\infty}) \quad (8.79)$$

where $K_L$ and $K_G$ are called *overall mass transfer coefficients* in the liquid and gas phases, and the *star states* $(x_j^*$ and $P_j^*)$ are the hypothetical mole fraction and partial pressure values defined by the equilibrium diagram (Figure 8-12). $x_j^*$ is the hypothetical liquid-phase mole fraction, which would be in equilibrium with the actual far-field gas-phase partial pressure $P_{j,\infty}$ and $P_j^*$ is the hypothetical gas-phase partial pressure, which would be in equilibrium with the actual liquid-phase mole fraction $x_{j,\infty}$.

It is important to recognize that the star states are hypothetical states defined by the graphical relationship shown in Figure 8-12. The overall mass transfer coefficient in the liquid phase $(K_L)$ can be expressed using the following construction:

$$x_{j,\infty} - x_j^* = (x_{j,\infty} - x_{j,i}) + (x_{j,i} - x_j^*) \quad (8.80)$$

From Equation 8.77 it can be seen that Henry's law allows one to relate $x_{j,i}$ to $P_{j,i}$ and $x_j^*$ to $P_{j,\infty}$.

$$P_{j,i} = H'x_{j,i} \quad (8.81)$$

$$P_{j,\infty} = H'x_j^* \quad (8.82)$$

Thus

$$x_{j,\infty} - x_j^* = (x_{j,\infty} - x_{j,i}) + \frac{(P_{j,i} - P_{j,\infty})}{H'} \quad (8.83)$$

The differences contained in the three terms in parentheses above can be replaced by expressions

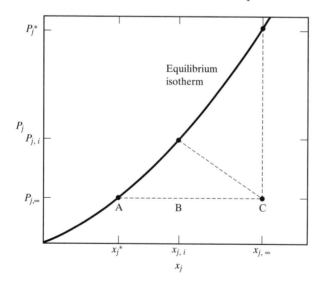

**Figure 8-12**   Relationship between star state and actual state

involving the molar flux $N_j$ using Equation 8.78. In so doing, one obtains

$$\frac{N_j}{c_L K_L} = \frac{N_j}{c_L k_L} + \frac{P_{j,i} - P_{j,\infty}}{H c_L} \quad (8.84)$$

After simplifying, we obtain

$$\frac{1}{K_L} = \frac{1}{k_L} + \frac{1}{H k_G} \quad (8.85)$$

It is useful to think of each one of the terms above as a resistance. Return to Equation 8.79, introduce Equations 8.81 and 8.77, and the mass transfer rate can be written as

$$\begin{aligned} N_j &= K_L c_L \left( x_{j,\infty} - x_j^* \right) \\ &= K_L \left( c_L x_{j,\infty} - c_L x_j^* \right) \\ &= K_L \left( c_{j,\infty} - \frac{c_L P_{j,\infty}}{H'} \right) \\ &= K_L \left( c_{j,\infty} - \frac{c_L P_{j,\infty}}{H c_L} \right) \\ &= K_L \left( c_{j,\infty} - \frac{P_{j,\infty}}{H} \right) \quad (8.86) \end{aligned}$$

It is useful to think of Equation 8.86 in the form of Ohm's law. The mass being transferred $(N_j)$ is equal to the driving potential $(c_{j,\infty} - P_{j,\infty}/H)$ divided by the resistance to mass transfer:

$$\begin{aligned} N_j &= \frac{c_{j,\infty} - P_{j,\infty}/H}{1/K_L} \\ &= \frac{c_{j,\infty} - P_{j,\infty}/H}{1/k_L + 1/H k_G} \\ &= \frac{c_{j,\infty} - P_{j,\infty}/H}{R_L + R_G} \quad (8.87) \end{aligned}$$

where $R_L$ is the liquid-film mass transfer resistance $(= 1/k_L)$ and $R_G$ is the gas-film mass transfer resistance $(= 1/H k_G)$.

Like any resistance circuit consisting of two resistances in series, if one resistance is much larger than the other, it dictates the current. Similarly, if mass transfer is governed by boundary layers (i.e., films) in the liquid and gas phases, a film possessing a substantially larger resistance than the other dictates the mass transfer rate. Quite often, the liquid film has the largest resistance since air passing over the liq-

uid surface convects vapors from the liquid in a vigorous manner. Counter-examples, however, are offered in Sections 6.10 and 6.11.

Equation 8.86 has been used to estimate the rate of evaporation of volatile organic hydrocarbons from ocean spills, industrial waste lagoons, and for stripping operations to remove dissolved pollutants from underground water. The mass transfer coefficient $(k_G)$, in units of gmol m$^{-2}$ s$^{-1}$ atm$^{-1}$, has been extracted from Mackay and Yeun's (1983) empirical expressions, which also included the mass transfer coefficient $(k_L)$ in the units of m/s:

$$k_G \left( \text{gmol m}^{-2} \text{ s}^{-1} \text{ atm}^{-1} \right)$$
$$= 4.1 \times 10^{-2} + 1.9 U^* \text{Sc}_G^{-0.67} \quad (8.88)$$

$$k_L(\text{m/s}) \begin{cases} = 1.0 \times 10^{-6} + 34.1 \times 10^{-4} U^* \text{Sc}_L^{-0.5} \\ \qquad \text{when } U^* > 0.3 \quad \text{(a)} \quad (8.89) \\ = 1.0 \times 10^{-6} + 144 \times 10^{-4} U^{*2.2} \text{ Sc}_L^{-0.5} \\ \qquad \text{when } U^* < 0.3 \quad \text{(b)} \end{cases}$$

The parameter $U^*$ (m/s) is called the *friction velocity* and is defined as

$$U^* = \left( \frac{\tau_a}{\rho_a} \right)^{0.5} \quad (8.90)$$

where $\tau_a$ is the shear stress of air passing over the liquid surface. In the outdoor environment where the liquid surface is large and evaporation is driven by the wind, Mackay and Yeun (1983) recommend that the friction velocity $(U^*)$ be expressed as

$$U^*(\text{m/s}) = U(10)[6.1 + 0.63\,U(10)]^{0.5} \quad (8.91)$$

where $U(10)$ is the wind speed (m/s) a distance 10 m above the liquid surface.

### EXAMPLE 8.14  EVAPORATION OF VOLATILE COMPOUNDS FROM A STAGNANT WASTE LAGOON

Estimate the evaporation rate (kg/h) of **(a)** benzene and **(b)** chloroform from a waste lagoon when the concentrations in water are 100 mg/L of benzene $(M = 78)$ and 100 mg/L of chloroform $(M = 119)$. The lagoon is 25 m wide × 40 m long × 3.5 m deep.

The air and liquid temperatures are 25°C and the wind speed is 1.7 m/s at $z = 10$ m.

**Solution** Schmidt numbers:

$$Sc(\text{benzene–water}) = 1000$$
$$Sc(\text{chloroform–water}) = 1100$$
$$Sc(\text{benzene–air}) = 1.76$$
$$Sc(\text{chloroform–air}) = 2.14$$

Henry's law constant (water), $H' = Hc_L$:

$$c_L = (1000 \text{ kg/m}^3)$$
$$(\text{kmol/18kg})(1000 \text{ gmol/kmol})$$
$$= 55.56 \times 10^3 \text{ gmol/m}^3$$

benzene: $H' = 3.05 \times 10^7 \text{ N/m}^2$

$$H = \frac{H'}{c_L} = 5.50 \times 10^{-3} \text{ atm m}^3 \text{ gmol}^{-1}$$

chloroform: $H' = 2.66 \times 10^7 \text{ N/m}^2$

$$H = \frac{H'}{c_L} = 4.78 \times 10^{-3} \text{ atm m}^3 \text{ gmol}^{-1}$$

The mass transfer rate is given by Equation 8.86:

$$N_j = K_{L,j}\left(c_{j,\infty} - \frac{P_{j,\infty}}{H}\right)$$

where $c_{j,\infty}$ is the concentration of the species in the liquid far-field and $P_{i,\infty}$ is the partial pressure of species $j$ in the gas far-field. The overall mass transfer coefficient $K_{L,j}$ is given by Equation 8-85:

$$\frac{1}{K_L} = \frac{1}{k_L} + \frac{1}{Hk_G}$$

and the mass transfer coefficients for the gas and liquid phases will be taken from Equations 8.88 and 8.89. The friction velocity $U^*$ can be found from Equation 8.91:

$$U^*(\text{m/s}) = U(10)[6.1 + 0.63U(10)]^{1/2}$$
$$= 1.7[6.1 + 0.63(1.7)]^{1/2} = 4.55 \text{ m/s}$$

From Equations 8.88 to 8.91,

$$k_L(\text{m/s}) = 10^{-6} + \frac{34.1 \times 10^{-4}(U*)}{Sc_L^{0.5}}$$

**a.** Benzene (subscript b):

$$k_{G,b} = 4.1 \times 10^{-2} + \frac{(1.9)(4.55)}{1.76^{0.67}}$$

$$= 5.9593 \text{ gmol m}^{-2} \text{ s}^{-1} \text{ atm}^{-1}$$

$$k_{L,b} = 10^{-6} + 34.1 \times 10^{-4}\frac{(4.55)}{1,000^{0.5}}$$

$$= 10^{-6} + 490.6 \times 10^{-6}$$

$$= 491.6 \times 10^{-6} \text{ m/s}$$

$$\frac{1}{K_{L,b}} = \frac{10^4}{4.916}(\text{s/m})$$

$$+ \frac{10^3}{(5.9593 \text{ gmol m}^{-2} \text{ s}^{-1} \text{ atm}^{-1})(5.5 \text{ atm m}^3 \text{ gmol}^{-1})}$$

$$= 2134.17 + 30.51 = 2064.7 \text{ s/m}$$

$$K_{L,b} = 4.841 \times 10^{-4} \text{ m/s}$$

**b.** Chloroform (subscript c):

$$k_{G,c} = 4.1 \times 10^{-2} + \frac{(1.9)(4.55)}{2.14^{0.67}}$$

$$= 5.2326 \text{ gmol m}^{-2}\text{s}^{-1} \text{ atm}^{-1}$$

$$k_{L,c} = 10^{-6} + \frac{34.1 \times 10^{-4}(4.55)}{1100^{0.5}}$$

$$= 10^{-6} + 467.8 \times 10^{-6}$$

$$= 4.688 \times 10^{-4} \text{ m/s}$$

$$\frac{1}{K_{L,c}} = \frac{10^4}{4.688}$$

$$+ \frac{10^3}{(5.2326 \text{ gmol m}^{-2} \text{ s}^{-1}\text{atm}^{-1})(4.78 \text{ atm m}^3 \text{ gmol}^{-1})}$$

$$= 2,133.1 + 39.98 = 2173.0 \text{ s/m}$$

$$K_{L,c} = 4.602 \times 10^{-4} \text{ m/s}$$

The mass transfer of species $j$ is given by Equation 8.86:

$$N_j = K_{L,j}\left(c_{j,\infty} - \frac{P_{j,\infty}}{H}\right)$$

In the liquid far-field (near the bottom of the lagoon),

$$c_{b,\infty} = c_{c,\infty} = 100 \text{ mg/L}$$

but in the gas far-field (distances far above the liquid–air interface there is only pure air),

$$P_{b,\infty} = P_{c,\infty} = 0$$

The mass transfer of benzene and chloroform from the lagoon (kg/h) of surface area 1000 m²:

$$m_b = N_b (1000 \text{ m}^2)$$
$$= (1000 \text{ m}^2)(K_{L,b} \text{ m/s})(C_{b,\infty} \text{ mg/l}) = (1000 \text{ m}^2)$$
$$(4.841 \times 10^{-4} \text{ m/s})(100 \text{ mg/L})$$
$$= 174.2 \text{ kg/h}$$

$$m_c = N_c (1000 \text{ m}^2)$$
$$= (1000 \text{ m}^2)(K_{L,c} \text{ m/s})(c_{c,\infty} \text{ mg/L})$$
$$= (1000 \text{ m}^2)(4.602 \times 10^{-4} \text{ m/s})(100 \text{ mg/L})$$
$$= 165.6 \text{ kg/h}$$

Review for a moment the resistance of the liquid and gas films to mass transfer. From the analysis above, it is clear that the liquid film resistance is nearly one hundred times larger than the gas film resistance.

# 8.11 Evaporation in confined spaces

An interesting use for the Equation 8.86 is to predict the rate of evaporation of the constituents from a multicomponent liquid when the liquid is contained in a confined space and the vapors accumulate in the air above the air-liquid interface. On rare occasions individuals enter confined spaces, i.e. utility workers enter underground vaults to repair power or telephone lines, farmers enter manure pits under animal confinement buildings to remove manure, chemical workers enter reactors to make repairs of mechanical components, etc. Entering confined spaces is a very hazardous activity. Workers often underestimate the danger since everything looks peaceful and they are unaware of the presence of toxic gas or inadequate oxygen. Inadequate oxygen can cause one to pass out and die in a matter of minutes. Workers entering the space to assist a comrade will face the same fate. Operating electrical equipment in the presence of combustible gases can cause an explosion even when there is no detectable odor. For these reasons, OSHA prescribes detailed procedures for entering a confined space. Individuals are risking their lives if these procedures are ignored. Equation 8.86 can be used to predict the equilibrium vapor concentration of volatile liquids in confined spaces and to predict the rate at which these equilibrium values are achieved.

## EXAMPLE 8.15 COMBUSTIBLE ORGANIC COMPOUNDS IN A CONFINED SPACE

A reactor is filled with a solution of water and ethyl alcohol (EA), mol fraction $(x_{EA,\infty})$ of 0.001. The temperature is 26°C. Initially 75% of the reactor is filled with the liquid and the remaining space is air. There is concern that eventually the lower explosion limit (LEL) may be reached in the air space and there is interest in how fast the alcohol desorbs. You have been asked to determine if the liquid evaporates entirely or if an equilibrium is established. If equilibrium is reached what is the partial pressure for the alcohol. Secondly you have been asked to derive an expression to predict the rate of desorption.

**Solution** Equation 8.86 specifies the desorption rate. Desorption occurs at a decreasing rate until equilibrium is achieved whereupon $N_{EA} = 0$. Thus,

$$c_{EA,\infty} = \left[ \frac{P_{EA,\infty}}{H_{EA}} \right]$$

Since 75% of the volume contains liquid, the mole fraction of alcohol in water $(x_{EA,\infty} = 0.001)$ remains essentially constant.

$$[P_{EA,\infty}] = H_{EA} c_{EA,\infty}$$
$$= H_{EA} x_{EA,\infty} c_L = H'_{EA} x_{EA,\infty}$$

where $c_L$ is the total molar concentration of the liquid phase, which is essentially water. From Table A-8,

$$H'_{EA} = 0.00456 \times 10^7 \text{ N/m}^2$$
$$[P_{c,\infty}] = (0.00456 \times 10^7 \text{ N/m}^2)(0.001)$$
$$= 0.00456 \times 10^4 \text{ N/m}^2$$
$$= (45.6 \text{ N/m}^2)(\text{kPa}/10^3 \text{ N m}^{-2})$$
$$= 0.0456 \text{ kPa}$$

Assuming an ambient pressure of 100 kPa, alcohol partial pressure corresponds to a gas-phase mole fraction of 0.000456 (456 ppm). Since the LEL of ethyl alcohol is 3.28% (mole fraction = 0.0328) and the OSHA PEL is 1000 ppm, there is no chance of fire or explosion nor asphyxiation in the air space above the liquid; nevertheless other confined entry procedures still have to be followed. Your conclusions might be different on a hot day.

At any instant the desorption rate of alcohol is given by Equation 8.86, where $P_{EA,\infty}$ refers to the partial pressure of alcohol in the far-field gas phase. It is obvious that while $P_{EA}$ is initially zero, it increases with time until the equilibrium value on the preceding page is achieved. To predict the partial pressure at any instant, define the air space above the liquid as a control volume $(V)$ and employ the conservation of mass to predict the change in the molar gas phase alcohol concentration $c_{EA,\infty}$, with time:

$$\frac{dm_{EA}}{dt} = \frac{V M_{EA} d c_{EA,\infty}}{dt}$$

$$= M_{EA} N_{EA} A$$

$$= A M_{EA} K_L \left[ c_{EA,\infty} - \frac{P_{EA,\infty}}{H} \right]$$

where A is the area of the liquid-air interface.

Replace $c_{EA,\infty}$ using the ideal gas law,

$$P_{EA,\infty} = c_{EA,\infty} R_u T$$

$$\frac{V M_{EA}}{R_u T} \frac{d P_{EA,\infty}}{dt}$$

$$= A M_{EA} K_L \left[ c_{EA,\infty} - \frac{P_{EA,\infty}}{H} \right]$$

Simplify, separate and integrate:

$$\int_0^{P_{EA}} \frac{d P_{EA}}{c_{EA,\infty} - P_{EA}/H} = \frac{K_L R_u T A}{V} \int_0^t dt$$

The instantaneous alcohol partial pressure in the air space and the desorption rate can now be expressed as

$$P_{EA,\infty}(t) = H c_{EA,\infty} \left[ 1 - \exp\left( -\frac{t A R_u T K_L}{H V} \right) \right]$$

$$= P_{EA,ss} \left[ 1 - \exp\left( -\frac{t A R_u T K_L}{H V} \right) \right]$$

$$N_{EA}(t) = K_L \left[ c_{EA,\infty} - \frac{P_{EA,\infty}(t)}{H} \right]$$

## 8.12 Drop evaporation

*Atomizers*, *spray nozzles*, and so on create drops of liquid and are used in countless industrial processes such as spray drying applications or coatings; transferring liquids; open-channel flow; scrubbing, washing, and cleaning; aeration; and, fuel atomization. Liquid drops are also generated naturally as sea spray and when gas bubbles traveling upward through seawater collapse and break through the liquid–air interface. Every drop formed by this process has a liquid–air interface through which the liquid or species dissolved in the liquid evaporates.

Consider the evaporation of drops of a liquid consisting of a single molecular species, moving or suspended in air at a constant temperature $(T_\infty)$. In the analysis that follows it is assumed that the number of drops per unit volume of air is low, such that one drop does not influence another. Thus the total evaporation rate from a cloud of drops is equal to the evaporation from each drop summed over the total number of drops. More complete analyses can be found in meteorology texts on the microphysics of clouds (Pruppacher and Klett, 1978). While the assumption of independent drops is simplistic, valuable intuition can be gained that will aid readers studying more complex issues such as clouds.

Before modeling the process, consider the physical processes that occur. Vapor escapes from the surface of the drop as shown in Figure 8-13 because the vapor pressure of the saturated liquid (based on the drop surface temperature) exceeds the partial pressure of the vapor in the far field. As the liquid evaporates, the drop diameter decreases, which in turn influences the rate of evaporation. Simultaneously, the evaporating liquid extracts energy from the drop (a process called *evaporative cooling*) that lowers the temperature of the drop, which in turn lowers the saturation vapor pressure at the drop–air interface. Because evaporation tends to lower the drop temperature, energy is transferred to the drop from the air by convection. Thus both mass and heat transfer are coupled and together control the rate at which the drop evaporates. The evaporation rate from a drop of diameter $D_p$ (where the subscript $p$ refers to properties of the particle) is

$$\dot{m}_p = N \pi D_p^2 M_p \qquad (8.92)$$

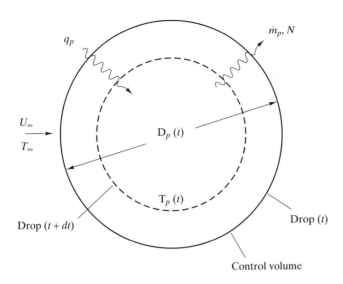

**Figure 8-13**   Control volume for evaporating drop

where the molar flux $N(\text{kmol m}^{-2}\,\text{s}^{-1})$ is

$$N = k_G R_u T_\infty (c_i - c_\infty) \qquad (8.93)$$

Since the drop contains only one species, of molecular weight $M_p$, the nomenclature can be simplified so that the concentration subscripts pertain only to the liquid–air interface (subscript $i$) and to the far-field concentration (subscript $\infty$). The molar concentrations can be converted to partial pressures using the perfect gas law. Thus

$$\dot{m}_p = \pi M_p D_p^2 k_G T_\infty \left( \frac{P_v}{T_{p,i}} - \frac{P_\infty}{T_\infty} \right) \qquad (8.94)$$

where $P_\infty$ is the pollutant partial pressure in the far-field where the temperature is $T_\infty$, and $P_v$ is the liquid saturation pressure based on the drop interface temperature $T_{p,i}$. The convection mass transfer coefficient $(k_G)$ can be found from heat transfer relationships using the Reynolds analogy described in Section 8.8. Combining Equations 8.93 and 8.60, in which the characteristic length $L$ is the drop diameter $(D_p)$, one obtains

$$\dot{m}_p = \pi M_p D_p \frac{D_{ac}}{R_u} \left( \frac{\text{Sc}}{\text{Pr}} \right)^{b1}$$

$$\frac{P}{P_{am}} \left( \frac{P_v}{T_{p,i}} - \frac{P_\infty}{T_\infty} \right) \text{Nu} \qquad (8.95)$$

The Nusselt number (Nu) for a sphere can be expressed as a function of the Reynolds and Prandtl numbers, and the Reynolds number is

$$\text{Re} = \frac{\rho D_p U_\infty}{\mu} \qquad (8.96)$$

and $U_\infty$ is the velocity of the drop relative to the air.

As the drop evaporates, its diameter decreases. To compute the instantaneous drop diameter $D_p(t)$, write the equation for the conservation of mass $(m_p)$ for the drop:

$$\dot{m}_p = -\frac{dm_p}{dt}$$

$$= \frac{d(\pi \rho_p D_p^3 / 6)}{dt} = -\frac{\rho_p \pi D_p^2}{2} \frac{dD_p}{dt} \qquad (8.97)$$

If the drop temperature at the liquid–air interface is known, the diameter can be found as a function of time by equating Equations 8.94 and 8.97. Unfortunately, one does not know the liquid drop interface temperature $(T_{p,i})$, nor can one assume that the drop surface temperature is necessarily equal to the air temperature. A separate calculation is needed to compute the drop temperature. From the field of heat transfer (Kreith, 1973; Incropera and DeWitt, 1990) the *Biot number* is defined as

$$\text{Bi} = \frac{hD_p}{2k_p} \qquad (8.98)$$

where $h$ is the convection heat transfer film coefficient and $k_p$ is the liquid thermal conductivity. If Bi is less than approximately 0.1, the drop temperature does not vary with radius. Thus if Bi $< 0.1$, the drop temperature is uniform throughout:

$$T_p(r, t) = T_{p,i}(t) = T_p(t)$$

Note that while the Biot and Nusselt numbers are composed of the same variables, the thermal conductivity $(k_p)$ in the Biot number refers to the liquid, while the value of $k$ in the Nusselt number refers to air. If the drop has a spatially uniform temperature, one can assume that the drop has a *lumped heat capacity*. Physically, this means that the resistance to conduction heat transfer within the drop is negligible, such that whenever heat is transferred to the drop it is distributed throughout instantly, so that the temperature of the drop does not vary with radius. For small drops that constitute an evaporating cloud, the assumption of a lumped heat capacity is generally valid and is adopted in this analysis.

The drop temperature, $T_p(t)$, can be found from writing the conservation of energy using the drop surface as the control surface. The integral form of the equation contains an unsteady term that is important because both the mass and internal energy within the control volume change with time. We neglect shear work and the kinetic energy of the evaporating drop. The energy equation expressing the heat transfer to the drop $(q_p)$ is

$$q_p = \frac{d(m_p u_f)}{dt} + \dot{m}_p h_g(T_p)$$

$$q_p = m_p c_v \frac{dT_p}{dt} + u_f(T_p)\frac{dm_p}{dt} + \dot{m}_p h_g(T_p) \quad (8.99)$$

where $h_g(T_p)$ is the enthalpy of the saturated vapor at the drop temperature. The change in internal energy of the saturated liquid $(u_f)$ is expressed as the liquid specific heat $(c_v)$ times the change in temperature, that is, $c_v dT_p$. The mass of the drop, at any instant $m_p(t)$, is

$$m_p(t) = \frac{\pi \rho_p D_p^3(t)}{6} \quad (8.100)$$

and its change with time is given by Equation 8.97. The rate of heat transfer $q_p(\text{watts})$ is given by the equation for convection heat transfer,

$$q_p = h\pi D_p^2(T_\infty - T_p) \quad (8.101)$$

Substituting, Equations 8.101 and 8.97 into 8.99 leads to

$$h\pi D_p^2(T_\infty - T_p) = \frac{\pi \rho_p D_p^3}{6} c_v \frac{dT_p}{dt} + \dot{m}_p[h_g(T_p) - u_f(T_p)] \quad (8.102)$$

The quantity $h_g(T_p) - u_f(T_p)$ is approximately equal to the enthalpy of the vaporization, $h_{fg}/(T_p)$, also called *heat of evaporation*. Replacing $dT_p/dt$ by $-d(T_\infty - T_p)/dt$, Equation 8.102 becomes

$$\frac{d(T_\infty - T_p)}{dt} = -6h(\rho_p D_p c_v)(T_\infty - T_p) + \frac{6\dot{m}_p h_{fg}}{\rho_p \pi D_p^3 c_v} \quad (8.103)$$

where $h_{fg}$ is taken to be constant and assumed to be equal to $h_{fg}(T_\infty)$. Equations 8.97 and 8.103 constitute a set of simultaneous differential equations whose solution yields the evaporation rate, drop temperature, and drop diameter as a function of time. Since the equations are coupled, numerical techniques should be used.

A significant simplification can be employed if the quantity $\rho_p D_p c_v/6h_h$ is small compared to characteristic time of the process. The quantity can be thought of as a time constant $(\tau)$ for a thermal $R$–$C$ circuit in which the reciprocal of the time constant is the product of a thermal resistance $(h\pi D_p^2/k_p)$ and a thermal capacitance $(6k_p/\rho_p c_v \pi D_p^3)$,

$$\tau = \frac{\rho_p D_p c_v}{6h} = \left(\frac{h\pi D_p^2}{k_p}\right)^{-1}\left(\frac{6k_p}{\rho_p \pi D_p^3 c_v}\right)^{-1} \quad (8.104)$$

If the time constant $(\tau)$ is small, the system responds quickly to changes in the thermal environment and the drop achieves a quasi-steady-state temperature. If such a steady state is achieved quickly, the left-hand side of Equation 8.103 can be assumed to be zero, and at any time a quasi-steady-state temperature can be assumed:

$$T_p(\text{steady state}) = T_\infty - \left(\frac{\dot{m}_p h_{fg} T_\infty}{\pi h D_p^2}\right) \quad (8.105)$$

The temperature difference $(T_p - T_\infty)$, is called the *temperature depression*. Finally, the evaporation rate of a drop can be found by solving Equation 8.95, in which the drop temperature is given by Equation 8.105.

## EXAMPLE 8.16 DROP EVAPORATION

Consider a cloud of water droplets in quiescent humid air at STP and analyze the evaporation of water droplets smaller than 500 μm. The settling velocity of a drop of water falling freely in quiescent air is the relative velocity $U_\infty$ and is given by the following,

$$v_t = \frac{\rho_p D_p^2 g}{18\,\mu} = U_\infty$$

Details about gravimetric settling and the settling velocity can be found in Section 11.5 and Figure 11-19. The relevant properties of air and water are:

- Air: $T_\infty = 25°C$
  $k = 0.0258$ W m$^{-1}$ K$^{-1}$
  Sc $= 0.5769$
  Relative humidity $= 80\%$
  $P_\infty = 101$ kPa,
  Pr $= 0.707$
  $\mu = 1.794 \times 10^{-5}$ kg m$^{-1}$ s$^{-1}$
  $\nu = 1.5 \times 10^{-5}$ m$^2$/s
  $D_{aw} = 0.26 \times 10^{-4}$ m$^2$/s

- Water: $k_p = 0.613$ W m$^{-1}$ K$^{-1}$
  $P_v(25°C) = 3.169$ kPa
  $h_{fg}(25°C) = 2442.3$ kJ/kg
  $\rho_p = 1000$ kg/m$^3$
  $P_{w,\infty} =$ partial pressure of water vapor in far field $= (0.8)(3.169) = 2.5352$ kPa
  $c_v = 4.18$ kJ kg$^{-1}$ K$^{-1}$

**a.** Compute and plot the evaporation rate $\dot{m}_p$(mg/s) of a drop of water. Initially, the drop is 500 μm in diameter. Assume that the temperature depression $(T_p - T_\infty)$ is zero. Thus $T_p = 25°C$.

**b.** Compute and plot the instantaneous drop diameter $D_p(t)$ (μm) versus time.

*Solution*  The Reynolds number is calculated on the basis of the settling velocity and is thus very small. Using the appropriate equation from heat transfer (Table 8-1), the Nusselt number can be computed for a sphere from which the convection heat transfer coefficient ($h$) can be computed. The Biot numbers are computed and are seen to be less than 0.1 for particles 100 μm and smaller. On the other hand, the computed values for $D_p$ equal to 500 μm are at the limit of an assumption in the theory. For all particles the lumped heat capacity assumption is reasonable. The time constant $(\tau = \rho_p D_p c_v / 6h)$ can be computed directly. The time constant is less than 0.1s for particles smaller than 100 μm, and one would expect that the drop temperature is equal to its equilibrium value predicted by Equation 8.105.

For purposes of reference, one can categorize the particles above as follows:

Fog:  $1(\mu m) < D_p < 10\,(\mu m)$
Mist:  $10\,(\mu m) < D_p < 100\,(\mu m)$
Drizzle: $100\,(\mu m) < D_p < 1000\,(\mu m)$
Rain:   $D_p > 1000\,(\mu m) = 1$ mm

The evaporation rate is given by Equation 8.95. The Nusselt number (Nu) is obtained from Table 8-1 for flow over a sphere. The instantaneous drop diameter, $D_p(t)$, can be obtained by integrating Equation 8.97. The calculations that follow are tedious and the reader must be very careful to check the units used in the equations. Throughout the calculations, the drop diameter $D_p$ will be expressed in the units of μm. The log mean pressure difference $(P_{am})$ is

$$P_{am} = \frac{P_{a,i} - P_{a,\infty}}{\ln\left(P_{a,i}/P_{a,\infty}\right)} =$$

| $D_p$ (μm) | $v_t$ (m/s) | Re | Nu | $h$ (W m$^{-2}$ K$^{-1}$) | Bi | $\tau$ (s) |
|---|---|---|---|---|---|---|
| 5 | 0.0008 | 0.001 | 2 | 10,320 | 0.042 | 0.0003 |
| 10 | 0.003 | 0.002 | 2 | 5,160 | 0.042 | 0.001 |
| 50 | 0.078 | 0.258 | 2.18 | 1,125 | 0.046 | 0.031 |
| 100 | 0.255 | 1.7 | 2.54 | 655 | 0.054 | 0.106 |
| 500 | 2.0 | 66.7 | 5.81 | 300 | 0.122 | 1.162 |

$$= \frac{(101 - 3.169) - (101 - 2.535)}{\ln(97.831/98.469)} = 98.157 \text{ kPa}$$

The Reynolds number

$$\text{Re} = \frac{\rho D_p v_t}{\mu} = \frac{D_p v_t}{\nu} = \frac{D_p}{\nu} \frac{g \rho_p D_p^2}{18\mu}$$

$$= (1000 \text{ kg/m}^3) \left(\frac{D_p}{10^6 \text{ m}}\right)^3$$

$$\frac{9.8 \text{ m/s}^2}{18(1.5 \times 10^{-5} \text{ m}^2/\text{s})(1.794 \times 10^{-5} \text{ kg m}^{-1} \text{s}^{-1})}$$

$$= 0.2023 \frac{D_p^3}{10^5}$$

**a.** From Equation 8.95, the evaporation rate $(\dot{m}_p)$ is equal to

$$\dot{m}_p = (\text{Nu})\pi M_p D_p \frac{D_{aw}}{R_u} \left(\frac{\text{Sc}}{\text{Pr}}\right)^{b1}$$

$$\frac{P}{P_{am}} \left(\frac{P_v}{T_p} - \frac{P_\infty}{T_\infty}\right)$$

$$\doteq \text{Nu}(D_p)\pi \frac{D_p}{10^6 \text{ m}}$$

$$\frac{(18 \text{ kg/kmol})(0.26 \times 10^{-4} \text{ m}^2/\text{s})}{8.314 \text{ kJ kmol}^{-1} \text{K}^{-1}(0.5769/0.707)^{-0.4}}$$

$$\left(\frac{101}{98.157}\right)\left(\frac{3.169}{298} - \frac{2.5352}{298}\right) \text{ kPa/K}$$

$$= \text{Nu}(D_p) \frac{18\pi D_p}{10^6} \left(\frac{0.03127}{10^4}\right)$$

$$(0.9219)(1.0289)(0.0021268)(10^6 \text{ mg/kg})$$

$$\dot{m}_p (\text{mg/s}) = \text{Nu}(D_p)(0.35654 \times 10^{-6}) D_p$$

The calculated results are shown in Figure E8-16a

**b.** The Nusselt number in the equation above is a function of the drop diameter, $\text{Nu}(D_p)$:

$$\text{Nu}(D_p) = 2 + [0.4(\text{Re})^{0.5} + 0.06(\text{Re})^{0.67}]\text{Pr}^{0.4}$$

$$= 2 + (0.707)^{0.4}$$

$$\left[0.4\left(\frac{0.2023 D_p^3}{10^5}\right)^{0.5} + 0.06\left(\frac{0.2023 D_p^3}{10^5}\right)^{0.67}\right]$$

The differential equation expressing the rate of change of the drop diameter is given by Equation 8.97:

$$\frac{d}{d(\pi\rho_p D_p^3/6)} = \pi\rho_p D_p^2/2 \frac{dD_p}{dt} = -\dot{m}_p$$

Rearranging yields

$$\frac{dD_p}{dt} = -\frac{2\dot{m}_p}{\pi\rho_p D_p^2}$$

$$= -(2\text{Nu})\frac{M_p}{\rho_p D_p} \frac{D_{aw}}{R_u} \left(\frac{\text{Sc}}{\text{Pr}}\right)^{0.4}$$

$$\frac{P}{P_{am}}\left(\frac{P_v}{T_p} - \frac{P_\infty}{T_\infty}\right)$$

$$= -\frac{2\text{Nu}(D_p)(18 \text{ kg/kmol})}{(1000 \text{ kg/m}^3)(D_p/10^6 \text{ m})}$$

$$\left(\frac{0.03127}{10^4}\right)(0.9219)(1.0289)$$

$$\times (0.0021268)(10^6 \text{ μm/m})$$

$$\frac{dD_p}{dt} (\text{μm/s}) = \frac{-0.227098 \times 10^3 \text{ Nu}(D_p)}{D_p}$$

Since the Nusselt number is a function of the drop diameter, the differential equation above cannot be integrated in closed form and will be solved numerically using a fourth-order Runge–Kuttta technique, available in several commercially available mathematical computer programs. Figure E8-16b shows the results of these computations, the instantaneous drop diameter as a function of time. The MathCAD program producing Figures E8-16a and b can be found in the textbook's Web site on the Prentice Hall Web page.

The analysis above can be applied to steam plumes produced by industry and the calculated lifetimes can be compared to those in steam plumes.

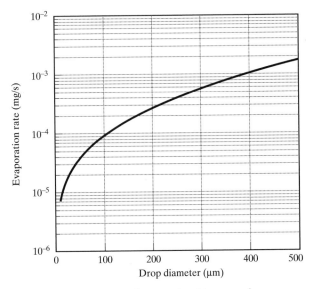

**Figure E8-16a** Evaporation rate (mg/s) versus drop diameter (μm)

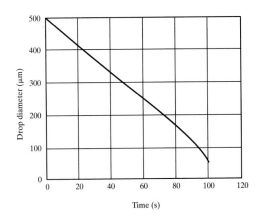

**Figure E8-16b** Drop diameter (μm) versus time (s).

## 8.13 Leaks

Hazardous materials may escape as leaks of gas, vapor or liquid from vessels, piping, pumps, flanges, valves, and so on. Figure 8-14 illustrates leaks that one may expect. Leaks may occur through holes in the vessels or piping or through cracks in valve packing, pump seals, or gaskets called *leakers*. Leakers are generally the result of broken seal joints and have a characteristic opening larger than 100 μm. Emissions may also occur through small pores 1 μm or less, by the process of capillary flow. For the same pressure drop across the leak site, the average emission rate is higher if the emitting fluid is a volatile liquid than if it is a permanent gas (Choi et al., 1992). The total emission rate ($Ib_m$/h) of an organic compound can be estimated by the EPA's *emission factors* shown in Tables A-3.2. Leak rates can also be measured experimentally using flux chambers (Section 8.1).

The term *aperture* is used in this section to denote leakers and openings (larger than 100 μm). Liquid leaks may produce conspicuous puddles, while escaping gases and vapors may produce a conspicuous noise. Often, however, no such symptoms are evident and the leak may be undetected for a considerable time. While leaks occur through openings of very small cross-sectional area, the length (in the direction of flow) of the hole or crack is often several times larger than the characteristic dimension perpendicular to the direction of flow. In a sense, leaks occur through a "conduit" and have both entrance and exit pressure losses as well as frictional losses through the conduit. The analysis in this section treats an aperture as a sharp-edged hole with loss coefficients associated with rapid contractions and expansions. A more sophisticated analysis should treat the aperture as a conduit. The upstream temperature and pressure are noted by $T_1$ and $P_1$ with the final atmospheric temperature and pressure noted by $T_a$ and $P_a$. The temperature and pressure of the leaking fluid at the exit plane of the aperture will be noted by $T_2$ and $P_2$. The pressure at the exit plane of the aperture will be equal to $P_a$ if the exit velocity is less than the speed of sound or possibly larger than $P_a$ if the exit velocity exceeds the speed of sound.

- *Subsonic flow:* $P_2 = P_a$
- *Sonic or supersonic flow:* $P_2 > P_a$

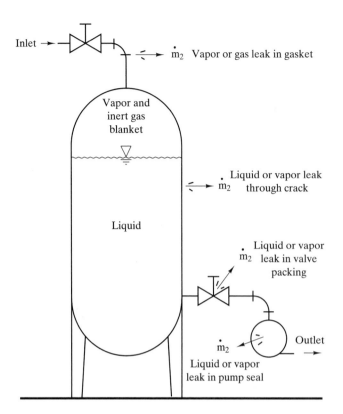

**Figure 8-14** Locations of potential leaks in a pressure vessel and its inlet and outlet piping (adapted from Crowl and Louvar, 1990)

Shown below is a convenient way to categorize leaks in terms of the hardware in which they occur and the mass flow rate of the leak.

|  | Isothermal slow leak | Adiabatic rapid discharge |
|---|---|---|
| Storage tank | $T_1$ = constant | $T_1$ and $P_1$ decrease |
| Transmission pipeline | $P_1$ and $T_1$ = constant, steady flow | Rupture |

These criteria define the thermodynamic and heat transfer analysis.

A *puncture* of a storage vessel or *rupture* of a transmission pipeline may produce a discharge with a large mass flow rate that is more properly called a *process dump* and will not be analyzed in this section. Leaks, consisting of small mass flow rates, will be addressed. Material will leak through an aperture when the upstream pressure $P_1$ exceeds atmospheric pressure $P_a$. Thermodynamics enters the analysis by relating the aperture temperature and pressure $(T_2, P_2)$ to the upstream temperature and pressure $(T_1, P_1)$ and to heat transferred into or out of the fluid.

High-pressure gases and vapors produce a jet as they escape and the escaping material can often be assumed to be an ideal gas as it enters the atmosphere. Liquids stored under pressure produce different types of leaks, depending on the relationship between the aperture temperature $(T_2)$ and the normal boiling temperature $(T_b)$.

A leak from the space above the liquid level can produce a stream of vapor or a two-phase stream composed of liquid and vapor. If the space above the liquid contains an inert blanket of gas, the inert gas may also be present in the leak. The rapid vaporization (boiling) of the liquid due to a sudden decrease in pressure is called *flash boiling*, or *flashing* for short. A leak below the liquid level produces a liquid stream that flashes (partially or totally) into vapor as it exits. If the exit temperature $(T_2)$ is less than $T_b$, flashing will not occur and the escaping liquid merely drips downward.

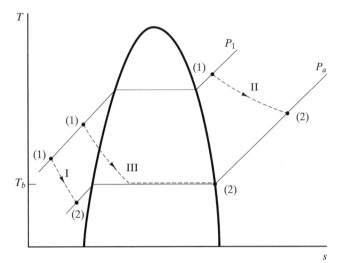

**Figure 8-15** Three types of adiabatic but irreversible leaks (adapted from Crowl and Louvar, 1990)
I - liquid leaks
II - gas or vapor leaks
III - adiabatic flashing leaks

Three types of leaks that engineers may encounter can be shown on a temperature–entropy diagram (Figure 8-15).

- I. *Liquid leak:* $P_1$, $T_1$ correspond to a compressed liquid and $T_2 < T_b$.
- II. *Gas or vapor leak:* $P_1$, $T_1$ correspond to a superheated vapor and $T_2 > T_b$.
- III. *Flash boiling:* $P_1$, $T_1$ correspond to a compressed liquid and $T_2$ is equal or greater than $T_b$.

A *compressed liquid* is a liquid at a temperature $(T_1)$ less than the saturation temperature $(T_{sat})$ corresponding to the pressure $(P_1)$. If there is heat transfer to the upstream fluid, point 1 will remain constant. If, however, the flow is adiabatic, point 1 in Figure 8-15 moves downward with time as the temperature $T_1$ decreases.

Leaks associated with the steady flow of a fluid in a pipeline are analyzed as either isothermal or adiabatic steady flow through an aperture. Leaks associated with storage vessels may be isentropic, isothermal, or nearly adiabatic (but not isentropic). On Figure 8-15 isothermal leaks map as horizontal lines and isentropic flow maps as vertical lines. Paths shown in Figure 8-15 correspond to leaks that are adiabatic but not isentropic. Owing to the small cross-sectional area of the aperture, a leak, by definition, has a small mass flow rate. If the leak is a liquid, the density and temperature of the fluid are essentially constant and no heat transfer issues arise. If the fluid is a gas or vapor or if flashing occurs, the temperature of the fluid passing through the aperture will decrease since there is inadequate time for heat to be transferred to the escaping fluid. If the mass flow rate is small, there may be adequate time for heat to be transferred from the vessel, pipe, or from outside agents to the upstream fluid. If this occurs, the upstream fluid may be assumed to be isothermal. If, however, a discharge occurs because of a dramatic event (e.g., worker drives a forklift vehicle into a high-pressure vessel, a pipe fractures due to thermal fatigue, etc.), the mass flow rate will be large and it is unwise to assume that there is adequate time for heat transfer. Such dramatic events are not discussed in this section, and the reader is encouraged to consult Crowl and Louvar (1990) for a more thorough discussion of the subject.

## 8.13.1 Liquid leaks

*Liquid leaks* refer to the discharge of a fluid that remains a liquid as it passes through an aperture. Liquid leaks pertain to incompressible flow (or more specifically constant density flow). Figure 8-16

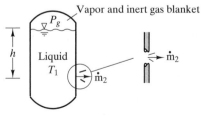

**Figure 8-16** Liquid leak, $T_1 = T_L < T_b$ (adapted from Crowl and Louvar, 1990)

shows the essential features of liquid leaks. The velocity of an incompressible liquid through an aperture is governed by

$$\frac{P_1}{\rho} + \frac{U_1^2}{2} + gz_1 = \frac{P_2}{\rho} + \frac{U_2^2}{2} + gz_2 + h_{LT} \quad (8.106)$$

where the subscript 1 designates upstream flow conditions and the subscript 2 designates conditions in the exit plane of the aperture. For liquid leaks, $P_2 = P_a = 1$ atm. The term $h_{LT}$ is the *total head loss*, which for the leak is equal to the *minor head loss* $h_{LM}$ (head loss comprising entry and exit losses) since frictional losses are assumed to be negligible:

$$h_{LM} = \frac{C_0 U_0^2}{2} \quad (8.107)$$

The term $C_0$ is a *loss coefficient* for the aperture and $U_0$ is a characteristic velocity defined in the table specifying $C_0$. In the absence of other data, one can assume that the aperture is a rapid contraction. The *ASHRAE Fundamental Handbook*, 1994, shows that for a rapid contraction ($\theta = 180°$) and a circular opening considerably smaller than the upstream area,

$$C_0 = 0.43 \quad (8.108)$$

and $U_0$ is the velocity through the opening ($U_2$). The mass flow rate of the leak ($\dot{m}_2$) is given by

$$U_2 = \left[ \frac{2(P_1 - P_2)}{\rho_2(1 + C_0)} \right]^{1/2} \quad (8-109)$$

$$\dot{m}_2 = \rho_2 A_2 U_2 \quad (8.110)$$

where $U_2$ is the velocity of the fluid leaking through the aperture. Crowl and Louvar (1990) suggest using a discharge coefficient to characterize flow through a sharp-edged orifice and another discharge coefficient to account for the vena contracta of the flow through the aperture. It is not obvious that this is more accurate, owing to the fact that the flow through the aperture is really flow through a narrow conduit of length considerably larger than the dimension perpendicular to the flow. In this section only the single coefficient $C_0$ is used.

---

### EXAMPLE 8.17 LEAK THROUGH VALVE PACKING

The packing around the stem of a shutoff valve mounted in a 6-inch pipe carrying 500 gal/min of trichloroethylene ($\rho_2 = 1460$ kg/m³) at 25°C and 2 atm pressure leaks at the rate of 1 kg/day. Estimate the area of the aperture.

*Solution* The elevation change ($z_1 - z_2$) is negligible. While the velocity inside the pipe is appreciable, the fluid located in the valve housing upstream of the leak is relatively stagnant, but its pressure is 2 atm. Thus

$$A_2 = \frac{\dot{m}_2}{\rho_2 U_2}$$

$$U_2 = \left[ \frac{2(P_1 - P_2)}{\rho_2(1 + C_0)} \right]^{0.5}$$

where $P_1$ is the pipe pressure (2 atm) and $P_2$ is the atmospheric pressure, $P_2 = P_a = 1$ atm.

$$U_2 = \left[ \frac{(2)(100 \text{ kPa})(1000 \text{ kg m}^{-1}\text{ s}^{-2}\text{ kPa}^{-1})}{(1460 \text{ kg/m}^3)1.43} \right]^{0.5}$$

$$= 9.78 \text{ m/s}$$

$$A_2 = \frac{1 \text{ kg/day}(\text{day/24 h})(\text{h/3600 s})}{(9.78 \text{ m/s})(1460 \text{ kg/m}^3)}$$

$$= 8.106 \times 10^{-10} \text{ m}^2 = 8.106 \times 10^{-4} \text{ mm}^2$$

If the aperture was a circular hole, its diameter would be 32 $\mu$m.

---

Suppose that a leak develops in a pressurized storage vessel containing an incompressible liquid protected by a blanket of inert gas at a pressure $P_g$ as shown in Figure 8-16. If the leaking liquid does not flash, Equation 8.109 can be used to compute its velocity. Because there is a hydrostatic head ($h$) between the surface of the liquid and the aperture in the vessel wall, the pressure difference in Equation 8.109 should be replaced by $(P_g - P_2) + g\rho_2 h$ [i.e., $P_1$(at the leak site) $= P_g + g\rho_2 h$] (see Equation 8.106). Thus the velocity of the leaking fluid is

$$U_2 = \left[ \frac{2\left(P_g - P_a\right) + g\rho_2 h}{\rho_2\left(1 + C_0\right)} \right]^{0.5} \quad (8.111)$$

Unique to the configuration in Figure 8-16 is the fact that as the leak progresses both the fluid level ($h$) and the pressure of the inert gas blanket $\left(P_g\right)$ decrease, which in turn reduces the leak mass flow rate. Leaks in storage vessels occur slowly and there is sufficient time for heat transfer to establish isothermal conditions; thus $T_g$ is constant. Applying the conservation of mass for the leaking liquid shows that at any instant, the liquid level $h(t)$ is

$$h(t) = h_0 - \frac{1}{\rho_2 A_2} \int_0^t \dot{m}_2 \, dt \quad (8.112)$$

At any instant, the pressure of the gas blanket,

$$P_g(t) = \frac{R_g T_g m_g}{V_t - Ah(t)} \quad (8.113)$$

where $T_g$ and $m_g$ are the inert-gas blanket temperature and mass and $V_t$ is the total volume of the vessel.

## 8.13.2 Vapor and gas leaks

For a fluid stored at a pressure $P_1$ several times larger than the atmospheric pressure $\left(P_1 > P_a\right)$, if the temperature $T_1$ is greater than $T_{\text{sat}}$ at $P_1$, the fluid will be a superheated vapor. If leaks occur in a pipe, pipe fittings, or a storage vessel containing gases or vapors under high pressure as shown in Figure 8-17, the density of the material will not remain constant as it escapes and the equations in Section 8.13.1 cannot be

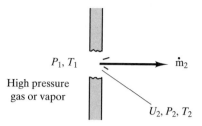

**Figure 8-17**   Vapor and gas leaks (adapted from Crowl and Louvar, 1990)

used. For analysis it will be assumed that the escaping material can be approximated as an ideal gas and that the flow through the aperture, up to the point it leaves the aperture exit plane, is isentropic. Such a leak can be depicted on the temperature–entropy diagram Figure 8-18.

If the leak is in the wall of a high-pressure vessel, the upstream velocity $U_1$ is zero. Even if the leak is in the wall of a pipe containing a flowing gas or vapor, the velocity $U_1$ is small compared to the sonic velocity in the aperture and can be neglected. For those familiar with compressible flow, it is to be noted that the upstream temperature and pressure $T_1$ and $P_1$ are the stagnation temperature and pressure $T_0$ and $P_0$. If

$$\frac{P_1}{P_a} > \left( \frac{\gamma + 1}{2} \right)^{\gamma/(\gamma-1)} \quad (8.114)$$

the flow will be choked and the aperture velocity $\left(U_2\right)$ will be equal to the speed of sound, $U_2 = a$.

$$a^2 = \left( \frac{\partial P}{\partial \rho} \right)_s \quad (8.115)$$

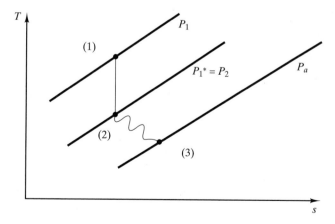

**Figure 8-18**   Isentropic flow of an ideal gas through an aperture with oblique shocks downstream of aperture (adapted from Crowl et al., 1990)

For an ideal gas it can be shown (Shapiro, 1953) that the speed of sound ($a$) is equal to

$$a = (\gamma R T_2)^{0.5} \qquad (8.116)$$

where $\gamma = c_p/c_v$ (isentropic coefficient). $T_2$ is the aperture temperature and is related to the upstream temperature $T_1$ by the following:

$$T_2 = T_1^* = \frac{2 T_1}{1 + \gamma} \qquad (8.117)$$

The pressure corresponding to choked conditions is noted as $P_1^*$ in Figure 8-18 and has the value

$$P_2 = P_1^* = \frac{P_1}{[(k+1)/2]^{\gamma/(\gamma-1)}} \qquad (8.118)$$

If the atmospheric pressure is less than $P_1^*$, there will be an irreversible pressure adjustment (oblique shocks) external to the aperture that has no bearing on the mass flow rate of the leak. If the atmospheric pressure is larger than $P_1^*$, the flow through the aperture will be compressible but subsonic and governed by another set of equations. Since gas and vapor leaks are generally associated with high-pressure sources ($P_1 > P_a$), choked flow is a common experience. The leak mass flow rate ($\dot{m}_2$) is given by Equation 8.110, where the density is the actual density of the leaking gas at the exit of the aperture ($\rho_2$). From the theory of isentropic flow (Shapiro, 1953),

$$\rho_2 = \rho_1^* = \frac{\rho_1}{[(\gamma+1)/2]^{1/(\gamma-1)}} \qquad (8.119)$$

---

### EXAMPLE 8.18   NATURAL GAS LEAK THROUGH CRACKED GASKET

A gasket in a flange in a high-pressure natural gas (assume methane) transmission pipeline develops a crack 10 μm in diameter ($7.85 \times 10^{-11}$ m²). The pressure ($P_1$) and temperature ($T_1$) in the pipeline are 20 atm and 25°C. Estimate the leak mass flow rate. From Perry and Chilton (1984), the molecular weight ($M$) and isentropic coefficient ($\gamma = c_p/c_v$) for $CH_4$ are 16.04 and 1.299, respectively.

***Solution***   Since, $P_1/P_a = 20 > [(\gamma+1)/2]^{\gamma/(\gamma-1)} = 1.8$, the gas velocity at the aperture exit will be equal

to the speed of sound. At the aperture exit plane the gas pressure is

$$P_1^* = \frac{(100 \text{ kPa})(20)}{[(1.299+1)/2]^{1.299/0.299}} = 1097 \text{ kPa}$$

Thus there is a considerable pressure adjustment to be achieved, and a complex shock pattern will occur downstream of the crack. A detectable noise can be expected. The gas density at the aperture exit plane can be found from Equation 8.119. The gas density at point 1,

$$\rho_1 = \frac{P_1}{RT_1} = \frac{2000}{(8.314/16.04)(298)}$$
$$= 12.948 \text{ kg/m}^3$$

The gas density and temperature at the exit plane of the aperture is

$$\rho_2 = \rho_1^* = \frac{12.948}{[(1.299+1)/2]^{1/0.299}} = 8.125 \text{ kg/m}^3$$

$$T_2 = T_1^* = \frac{(2)(298)}{2.299} = 259.2 \text{ K}$$

The exit gas velocity is equal to the speed of sound, thus

$$U_2 = a = \left[ \frac{(1.299)(8.314)}{(16.04)(1000)(259.2)} \right]^{0.5} = 417.5 \text{ m/s}$$

The leak rate $\dot{m}_2$ is

$$\dot{m}_2 = A_2 \rho_2 U_2$$
$$= (7.85 \times 10^{-11} \text{ m}^2)(8.125 \text{ kg/m}^3)(417.5 \text{ m/s})$$
$$= 2.662 \times 10^{-7} \text{ kg/s} = 0.958 \text{ g/h}$$

---

## 8.13.3 Flashing

Consider the high-pressure storage vessel in Figure 8-19 containing a compressed liquid at pressure $P_1$ and temperature $T_1$. If a hole enables a compressed liquid to escape, the liquid will undergo a rapid reduction in pressure as it passes through the aperture. If the aperture temperature ($T_2$) or the atmospheric temperature ($T_a$) is larger than the normal boiling temperature ($T_b$), the liquid will "flash" into the vapor phase. Flashing may not be instantaneous and one can expect some liquid droplets to exist for a short period of time before they evaporate. To model

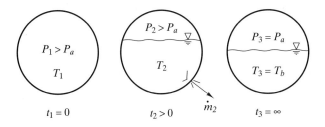

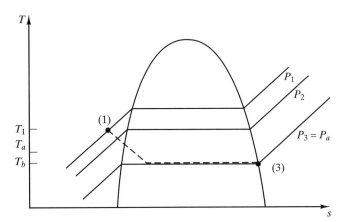

**Figure 8-19** Adiabatic flash boiling of a compressed liquid leaking from a storage vessel (adapted from Crowl at al., 1990)

the process it will be assumed that vaporization is complete and that the escaping vapor exists at the boiling temperature $(T_b)$. Eventually, the vapor will mix with atmospheric air and the temperature will rise to $T_a$. If one can assume that the velocity of the vapor leaving the aperture will be equal to the speed of sound (choked flow), $U_2 = a$.

To model flash boiling, assume that the process is adiabatic and that the energy to vaporize the leaking fluid is transferred from the compressed liquid upstream of the aperture. During the increment of time $(dt)$ when the mass flow rate of vapor escaping the vessel is $\dot{m}_2$, the mass of liquid inside the vessel changes by $dm$. Applying the conservation of mass of the compressed liquid in the vessel yields

$$dm = -\dot{m}_2 \, dt \qquad (8.120)$$

The energy required to vaporize this mass is equal to the enthalpy of vaporization $(h_{fg})$ based on the instantaneous temperature of the compressed liquid. The energy is obtained from the compressed liquid causing its temperature to change. Thus

$$dm \, h_{fg} = c_p m \, dT \qquad (8.121)$$

where $c_p$ is the specific heat of the compressed liquid. For liquids in general, $c_p \simeq c_v$ and is often written simply as $c$.

Integrating both equations from the time the leak begins $(T_1, P_1)$ to the time when the leak ceases $(P_1 = P_a \text{ and } T_3 = T_b)$ yields

$$m_1 - m_3 = -\int_0^{t_3} \dot{m}_2 \, dt \qquad m_2 = m_1 - m_3$$

$$\int_{m_1}^{m_3} \frac{dm}{m} = \int_{T_1}^{T_b} \frac{c_p}{h_{fg}} \, dT$$

$$\frac{m_3}{m_1} = \exp\left[-\frac{c_p(T_1 - T_b)}{h_{fg}}\right] \qquad (8.122)$$

This result assumes that $h_{fg}$ remains constant over the temperature range, $T_1$ to $T_b$

$$\frac{m_1 - m_3}{m_1} = f_{LK} = 1 - \exp\left[-\frac{c_p(T_1 - T_b)}{h_{fg}}\right]$$

where $f_{LK}$ is the fraction of the original liquid that ultimately leaks:

$$f_{LK} = \frac{m_2}{m_1} = \frac{m_1 - m_3}{m_1} \qquad (8.123)$$

The leak mass flow rate $(\dot{m}_2)$ can be found from the continuity equation. As long as the atmospheric pressure $(P_a)$ satisfies

$$\frac{P_1}{P_a} \gg \left(\frac{\gamma}{\gamma - 1}\right) \qquad (8.124)$$

one can expect that the flow through the aperture will be choked and $U_2$ will equal the speed of sound. Thus

$$\dot{m}_2 = A_2 a \rho_2 \qquad (8.125)$$

where $A_2$ is the aperture area and $\rho_2$ is the density of the escaping vapor, which will be assumed to be equal to the reciprocal of the specific volume of the saturated vapor based on the boiling temperature $T_b$. For a two-phase liquid–vapor mixture, Crowl and Louvar (1990) state that the speed of sound is

$$a = h_{fg}\left(\frac{g_c}{c_p T_2}\right)^{0.5} \qquad (8.126)$$

where $h_{fg}$ is the enthalpy of vaporization and $c_p$ is the specific heat of the saturated liquid.

The temperature, pressure, and mass of compressed liquid remaining in the vessel decrease with time as the leak continues. As soon as the tank pressure no longer satisfies Equation 8.124, the velocity through the aperture will be less than the speed of sound and another set of equations will have to be used. Thus ultimately the leaking mass flow rate $(\dot{m}_2)$ is not constant even though the flow may be choked initially.

### EXAMPLE 8.19 CHLORINE LEAK THROUGH A CRACKED VESSEL

Liquid chlorine is stored in a 1000-gal vessel at 80°F and 500 psia. A crack 1 mm in diameter $(A_2 = 7.85 \times 10^{-7} \text{ m}^2)$ develops in a gasket and chlorine escapes to the atmosphere. Determine the mass flow rate during the initial portion of the leak and estimate the fraction of the original chlorine that ultimately escapes to the atmosphere after a long period of time.

**Solution** From Perry et al. (1984), the following data on chlorine are obtained.

- $T_1 = 80°F$ (26.7°C), $P_1 = 500$ psia, chlorine is a compressed liquid. At $T_1$:
  - $P_{sat} = 116.46$ psia
  - Specific volume of saturated liquid, $v_f = 0.01154 \text{ ft}^3/\text{lb}_m$
  - $h_{fg} = 107.32 \text{ Btu/lb}_m$
  - Liquid chlorine specific heat: $c_p = 0.229$ cal g$^{-1}$ °C$^{-1}$ (0.7426 Btu lb$_m^{-1}$ °R$^{-1}$)

- $P_a = 14.7$ psia, chlorine is an ideal gas. At $P_a$:
  - $T_{sat} = -33.8°C$ (−28.8°F) (431°R)
  - Specific volume of saturated vapor, $v_g = 4.335 \text{ ft}^3/\text{lb}_m$
  - $h_{fg} = 123.67$ Btu lb$_m^{-1}$ °F$^{-1}$
  - Chlorine gas: $\gamma = c_p/c_v = 1.355$

The mass of chlorine originally in the vessel $(m_1)$ is

$$m_1 = \frac{V}{v_f} = \frac{1000 \text{ gal}(0.1336 \text{ ft}^3/\text{gal})}{0.01154 \text{ ft}^3/\text{lb}_m}$$

$$= 11,584 \text{ lb}_m$$

The fraction of chlorine that leaks is

$$f_{LK} = 1 - \exp\left[-\frac{c_p(T_1 - T_b)}{h_{fg}}\right]$$

$$= 1 - \exp\left[-\frac{0.7426[80 - (-28.8)]}{115.5 \text{ Btu lb}_m^{-1} °\text{F}^{-1}}\right]$$

$$= 0.503$$

The value 115.5 Btu/lb$_m$ is the average value of $h_{fg}$ between $T_1$ and $T_b$. During the initial period of escape when the liquid temperature is $T_1$, the acoustic velocity $(a)$ is

$$a = h_{fg}\left(\frac{g_c}{c_p T_b}\right)^{0.5}$$

$$= (115.5 \text{ Btu/lb}_m)$$

$$\left(\frac{(32.2 \text{ lb}_m \text{ ft lb}_f^{-1} \text{s}^{-2})(778 \text{ ft lb}_f \text{Btu}^{-1})}{(0.7426 \text{ Btu lb}_m^{-1} °\text{R}^{-1})(431°\text{R})}\right)^{0.5}$$

$$= 1022 \text{ ft/s } (311.5 \text{ m/s})$$

The mass flow rate will be a maximum during this period:

$$\dot{m}_2 = \frac{aA_2}{v_g}$$

$$= \frac{(311.9 \text{ m/s})(7.85 \times 10^{-7} \text{ m}^2)(3.28 \text{ ft/m})^3}{4.335 \text{ ft}^3/\text{lb}_m}$$

$$= 0.001993 \text{ lb}_m/\text{s}$$

If the flash boiling mass flow rate is small, the process may be isothermal and the leak mass flow rate and flash fraction $f_{LK}$ will need to be calculated by another set of equations.

## 8.13.4 Leaks in process and storage vessels

The objective of this section is to analyze leaks of vented (upper surface of liquid vented or open to the atmosphere and hence at atmospheric pressure throughout the leak) incompressible liquid from chemical storage and batch process vessels. Large vessels of varied geometry are used, such as spheres, upright vessels of constant circular cross-sectional area, and horizontal vessels of constant circular cross-sectional area.

To compute the leak rate, time to drain a vessel, and so on, you will need to compute the volume of the stored liquid as a function of the location of the leak in the vessel and the top surface of the liquid. The computation is complicated for several reasons:

- Cylindrical vessels of circular cross-sectional area have dome-shaped ends for strength.
- Vessels often contain heating or cooling coils, blenders, agitators, etc.
- A vessel may be totally, or less than totally, filled with liquid. If the liquid does not fill the vessel, there may be a blanket of inert gas above it.

Leaks occur because seams may rupture, the vessel may be accidentally punctured, or gaps may be present in valves, pipe fittings, and so on. If a leak occurs in an upright vessel of constant circular cross-sectional area containing no internal members, the time to drain the vessel from liquid level $z_1$ above the leak to $z_2$ can be computed to be

$$t = \left(\frac{\pi D^2}{A_2}\right)\left(\frac{2(1 + C_0)}{g}\right)^{1/2}\left[(z_1)^{0.5} - (z_2)^{0.5}\right] \quad (8.127)$$

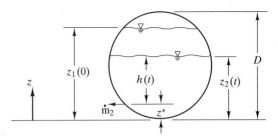

**Figure 8-20** Leak in a spherical vessel at a distance $z^*$ from the base of the sphere. (adapted from Crowl et al., 1990)

If, on the other hand, a leak occurs in a spherical or horizontal vessel of circular cross-sectional area, the task in more difficult (Crowl et al., 1990).

Consider the vented spherical vessel shown in Figure 8-20. Initially, the height of the liquid is $z_1(0)$; if a leak occurs, the liquid level is $z_2(t)$. A leak occurs through an opening of area $A_2$ located a vertical distance $z^*$ above the bottom of the vessel. At time $t$, let $h$ be the distance between the upper surface of the liquid and the leak opening,

$$h(t) = z_2(t) - z^* \quad (8.128)$$

As the leak progresses, $h$ decreases with time. Engineers need to know how to compute the leak mass flow rate at any instant and the amount of time it takes to either drain the vessel or for the liquid level to drop from $z_1$ to $z_2$. Assume that the tank is vented to the atmosphere. Consequently, the pressure above the liquid surface is always equal to the atmospheric pressure. From the conservation of mass and the expression for the velocity through the leak, one can write

$$\dot{m} = -\rho\frac{dV}{dt} = \rho A_2\left[\frac{2gh(t)}{1 + C_0}\right]^{1/2} \quad (8.129)$$

where $V$ is the volume of the liquid. The volume of a spherical segment defined by the upper free surface of liquid and a horizontal plane passing through the leak is

$$V = \frac{\pi h}{6}\left[6z^*(D - z^* - h) + 3hD - 2h^2\right] \quad (8.130)$$

When a leak occurs, the volume of liquid inside the tank decreases with time:

$$\frac{dV}{dt} = \left(z^*D - z^*z^* - 2hz^* + Dh - h^2\right)\pi\frac{dh}{dt} \quad (8.131)$$

To estimate the time ($t$) required for the liquid surface to fall from $z_1$ to $z_2$, combine Equations 8.129 and 8.131 and integrate the following:

$$\int_0^t dt =$$

$$-\int_{z_1}^{z_2} \frac{\pi \left[Dz^* - (z^*)^2 - 2hz^* + Dh - 2h^2\right] dh}{A_2 \left[2gh/(1 + C_0)\right]^{1/2}}$$

$$(8.132)$$

---

**EXAMPLE 8.20 LIQUID LEAK THROUGH A PUNCTURED VESSEL**

A vented spherical storage vessel 12 m in diameter is filled to capacity with a toxic hydrocarbon (density 800 kg/m³). A forklift operator inadvertently strikes the vessel but does not inform his supervisor. Sometime later, another worker notices hydrocarbon leaking through a hole 1.8 m above the bottom of the vessel and reports it to you. You need to submit a "spill" report to the state department of environmental resources. When reported to you, the upper level of liquid is 4.8 m below the top of the vessel and measurements reveal that the cross-sectional area of the puncture is 0.8 cm² and the current leak rate is 0.4 kg/s.

**a.** Find the volume of liquid that has been spilled.
**b.** Estimate the discharge coefficient $(C_0)$.
**c.** Estimate the leak discharge rate (kg/s) at the time of the puncture.

**Solution** In terms of the parameters shown in Figure 8-20 and described above, $D = 12$ m and $z^* = 1.8$ m:

$t = t_1 = 0$: $\quad z_1(0) = D = 12$ m

$$h_1 = z_1 - z^* = 12 - 1.8 = 10.2 \text{ m}$$

$t = t_2 \quad z_2(t) = (12 - 4.8)\text{m} = 7.2$ m

$$h_2 = z_2 - z^* = 7.2 - 1.8 = 5.4 \text{ m}$$

The initial volume of liquid is

$$V_1 = V(0) = \frac{\pi D^3}{6} = \frac{3.14(12)^3}{6} = 904.32 \text{ m}^3$$

The volume of liquid in the vessel when the spill was reported,

$$V_2 = \frac{\pi h_2}{6}\left[6z^*(D - z^* - h_2) + 3Dh_2 - 2h_2^2\right]$$

$$= \frac{\pi}{6}(5.4)\left[6(1.8)(12 - 1.8 - 5.4)\right.$$

$$\left. + 3(12)(5.4) - 2(5.4)^2\right] = 531.06 \text{ m}^3$$

**a.** The volume and mass of fluid spilled on the ground are

$$V_{\text{spill}} = V_1 - V_2 = 904.3 - 531.06 = 373.2 \text{ m}^3$$

$$m_{\text{spill}} = (373.2 \text{ m}^3)(800 \text{ kg/m}^3) = 29{,}776 \text{ kg}$$

**b.** The discharge coefficient $C_0$ can be estimated from Equation 8.129 using the measured mass flow rate at the time of the report and $h_2 = 5.4$ m:

$$\dot{m}_2 = \rho A_2 \left[2gh/(1 + C_0)\right]^{1/2}$$

$$0.4 \text{ kg/s} = (800 \text{ kg/m}^3)(0.8 \times 10^{-4} \text{ m}^2)$$

$$\left[\frac{(2)(9.8 \text{ m/s}^2)(5.4 \text{ m})}{1 + C_0}\right]^{1/2}$$

$$C_0 = 1.709$$

**c.** At the time of the puncture, the mass flow rate can be estimated in a similar way using $h_1 = 10.2$ m:

$$\dot{m}_1 = \rho A_2 C_0 (2gh_1)^{0.5}$$

$$(800 \text{ kg/m}^3)(0.8 \times 10^{-4} \text{ m}^2)(0.608)$$

$$\left[2(9.8 \text{ m/s}^2)(10.2)\right]^{1/2} = 0.905 \text{ kg/s}$$

---

## 8.14 Closure

It is essential that engineers be able to estimate the rates of emission of pollutants from a variety of sources. Securing operating permits from environmental regulatory agencies requires these estimates. In some cases it may be possible to measure the emission rates. In other cases, empirical equations can be used if the physical limitations on their use are satisfied. Emission factors developed by the EPA lack sophistication but are available for hundreds of different sources and processes and have been used by the EPA and industry for decades. Since state and federal environmental operating permits often require the use of emission factors, engineers must know where these values can be obtained and how to use them. Use of the EPA electronic bulletin board grows steadily and it is essential that engineers know how to use it. There are times when the previous methods are unavailable or when greater accuracy is needed. Accordingly, engineers must be able to model emission processes and estimate rates from the principles of mass transfer found in numerous reference works and textbooks. The professional journals regularly add to this literature and should be consulted.

# Nomenclature

| Symbol | Description (Dimensions*) |
|---|---|
| $\rightleftharpoons$ | simultaneous forward and reverse reaction |
| $\rightarrow$ | direction of a chemical reaction |
| [=] | has the units of |
| = | stoichiometric equation |
| $[j]$ | concentration of species $j$ $(M/L^3, N/L^3)$ |
| $A$ | area $(L^2)$ |
| $a_1, a_2$ | constants defined by equation |
| $a_s$ | total pellet surface area per volume of fixed bed |
| $b_1, b_2$ | constants defined by equation |
| $c_j$ | mass concentration of molecular species $j$ $(M/L^3)$ |
| $c_L$ | total molar concentration of the liquid phase $(N/L^3)$ |
| $c_p$ | specific heat at constant pressure $(Q/MT)$ |
| $c_v$ | specific heat at constant volume $(Q/MT)$ |
| $C_0$ | loss coefficient |
| $C_1, C_2$ | constants defined by equation |
| $D$ | diameter of tank (L) |
| $D_{ac}$ | diffusion coefficient of a contaminant in air $(L^2/t)$ |
| $D_p$ | diameter of a particle or drop (L) |
| $D_0$ | diameter of a spill (L) |
| $f_j$ | mass fraction of species $j$ |
| $f_{LK}$ | leak fraction, defined by equation |
| $g$ | acceleration of gravity $(L/t^2)$ |

*F, force; L, length; M, mass; N, mols; Q, energy; t, time; T, temperature.

| | |
|---|---|
| $g_c$ | gravitational constant $= 32.2\ \text{lb}_m\ \text{ft}\ \text{lb}_f^{-1}\ \text{s}^{-2}$ |
| $h$ | height of liquid in tank (L) |
| $h$ | heat transfer coefficients $(Q/tL^2)$ |
| $H, H'$ | Henry's law constant (variable) |
| $h_{fg}$ | enthalpy of vaporization (Q/M) |
| $h_g$ | enthalpy of a saturated vapor (Q/M) |
| $h_{LM}$ | minor head loss (L) |
| $h_{LT}$ | total head loss (L) |
| $J_j$ | molar flux of molecular species j per unit area $(N/L^2t)$ |
| $k$ | thermal conductivity (Q/tLT) |
| $K$ | constant defined by equation |
| $k_G, k_L$ | individual mass transfer coefficient for gas and liquid phase (variable) |
| $K_G, K_L$ | overall mass transfer coefficient for gas and liquid phase (variable) |
| $L$ | characteristic length or distance defined by equation (L) |
| $L$ | thickness of liquid pool (L) |
| $\dot{m}_d, \dot{m}_g,$ | mass flow rate of pollutant discharge, pollutant generation (M/t) |
| $m_g$ | mass of inert gas blanket (M) |
| $\dot{m}_i$ | mass flow rate of molecular species $i$ (M/t) |
| $M_i$ | molecular species $i$ |
| $m_t$ | total mass (M) |
| $n_j$ | total number of moles of molecular species $j$ (N) |
| $N_j$ | molar flux of molecular species $j$ $(N/L^2t)$ |
| $P$ | pressure $(F/L^2)$ |
| $P_{am}$ | log mean partial pressure difference, defined by equation $(F/L^2)$ |

| | | |
|---|---|---|
| $P_{c,\,sat}$ | saturation pressure of molecular species $c$ $(F/L^2)$ | |
| $P_{v,\,j}$ | vapor pressure of molecular species $j$ $(F/L^2)$ | |
| $P^*$ | pressure at choked flow $(F/L^2)$ | |
| $P_j^*$ | hypothetical (star state) partial pressure of molecular species $j$ $(F/L^2)$ | |
| $q$ | heat transfer rate $(Q/t)$ | |
| $Q$ | volumetric flow rate $(L^3/t)$ | |
| $q/A$ | heat transfer rate per unit area $(Q/tL^2)$ | |
| $R_u$ | universal gas constant $(M/N)$ | |
| $s$ | entropy $(Q/T)$ | |
| $S, s$ | source strength, source strength per unit area $(M/t, M/L^2t)$ | |
| $S_s$ | evaporation rate $(M/t)$ | |
| $t$ | time $(t)$ | |
| $t'$ | time to empty a leaking vessel $(t)$ | |
| $T, T_a$ | temperature$T$ $(T)$ | |
| $U, U_a$ | velocity $(L/t)$ | |
| $U^*$ | friction velocity, defined by equation $(L/t)$ | |
| $u_f$ | internal energy of a saturated liquid $(Q/M)$ | |
| $U_m$ | molar average gas velocity, defined by equation $(L/t)$ | |
| $U(10)$ | wind speed at an elevation of 10 m $(L/t)$ | |
| $V, V_j$ | volume, volume of molecular species $j$ $(L^3)$ | |
| $v$ | molar specific volume $(L^3/N)$ | |
| $v_{c,j}$ | specific volume of molecular species $j$ at its critical state $(L^3/M)$ | |
| $v_t$ | terminal settling velocity of a particle $(L/t)$ | |
| $x_j$ | mole fraction of molecular species $j$ in the liquid phase | |
| $\dot{W}_{net}$ | net output power $(Q/t)$ | |
| $x_j^*$ | hypothetical mole fraction of molecular species $j$ in the liquid phase | |
| $y_{am}$ | log mean mole fraction, defined by equation | |
| $y_i$ | mole fraction of molecular species $i$ in the gas phase | |
| $z$ | distance $(L)$ | |
| $Z$ | compressibility | |
| $z^*$ | distance between puncture and base of a spherical vessel $(L)$ | |

*Greek*

| | |
|---|---|
| $\gamma$ | isentropic constant |
| $\Gamma$ | activity coefficient |
| $\delta_U, \delta_T, \delta_m$ | thickness of the velocity, thermal, and concentration boundary layer $(L)$ |
| $\eta$ | efficiency |
| $\mu$ | dynamic viscosity $(M/Lt)$ |
| $\nu$ | kinematic viscosity $L^2/t$ |
| $\rho$ | density $M/L^3$ |
| $\tau_a$ | air shear stress $F/L^2$ |
| $\phi$ | association factor, defined by equation |

*Subscripts*

| | |
|---|---|
| $(\cdot)_a$ | air |
| $(\cdot)_{avg}$ | average value of a property |
| $(\cdot)_c$ | critical value |
| $(\cdot)_{c,i}$ | value of molecular species $c$ at the air–liquid interface |
| $(\cdot)_{c,0}$ | value of molecular species in the bulk gas |
| $(\cdot)_g$ | inert–gas blanket |
| $(\cdot)_{iso}$ | isolated |
| $(\cdot)_j$ | molecular species $j$ |
| $(\cdot)_L$ | leak or liquid |
| $(\cdot)_p$ | particle |
| $(\cdot)_s$ | evaluated at constant entropy |
| $(\cdot)_t$ | total property |
| $(\cdot)_0$ | initial value |
| $(\cdot)_\infty$ | evaluated in the far-field |

*Abbreviations*

| | |
|---|---|
| acfm | volumetric flow rate, actual cubic feet per minute |
| ACGIH | American Congress of Governmental Industrial Hygienists |
| AP-42 | compilation of EPA emission factors |
| ASHRAE | American Society of Heating, Refrigerating, and Air-Conditioning Engineers |
| ASTM | American Society of Testing and Materials |
| Bi | Biot number |
| CAAA (1990) | Clean Air Act amendments of 1990 |

| CFR | Code of Federal Regulations |
| EPA | Environmental Protection Agency |
| HAP | hazardous air pollutant |
| HHV | higher heating value |
| Le | Lewis number |
| LEL | lower explosion limit |
| Nu | Nusselt number |
| OSHA | Occupational Safety and Health Administration |
| PEL | permissible exposure limit |
| ppm | parts per million |
| Pr | Prandtl number |
| Re | Reynolds number |
| Sc | Schmidt number |
| Sh | Sherwood number |
| STP | standard temperature (25°C) and pressure (1 atm) |
| TCE | trichloroethylene |
| VOC | volatile organic compound |

# References

ACGIH, 1989. *Threshold Limit Values for Chemical Substances and Biological Exposure Indices from 1989–1990*, American Conference of Governmental and Industrial Hygienists, Cincinnati, OH.

ASHRAE, 1989. *Ventilation for Acceptable Indoor Air Quality*, ASHRAE Standard 62-1989, American Society of Heating, Refrigerating, and Air-Conditioning Engineers, Atlanta, GA.

ASHRAE, 1994. *ASHRAE Fundamental Handbook.* American Society of Heating, Refrigerating, and Air-Conditioning Engineers, Atlanta, GA.

Anon., 1991. Guidance on HAP early reduction compliance extensions, *The Air Pollution Consultant*, Vol. 1, No. 2, November–December, pp. 2.1–2.13.

Bare, J. C., 1988. Indoor air pollution source database, *Journal of the Air Pollution Control Association*, Vol. 38, No. 5, pp. 670–671.

Bird, R. B., Stewart, W. E., and Lightfoot, E. N., 1960. *Transport Phenomena*, Wiley, New York.

Bond, R., and Staub, C. P., (Eds.), Prober, R., (Coord. Ed.), 1972. *Handbook of Environmental Control*, Vol. I, *Air Pollution*, CRC Press, Cleveland, OH.

Butcher, S. S., and Ellenbecker, M. J., 1982. Particulate emissions for small wood stoves and coal stoves, *Journal of the Air Pollution Control Association*, Vol. 32, No. 4, pp. 380–384.

Choi, S. J., All, R. D., Overcash, M. R., and Lim, P. K., 1992. Capillary-flow mechanism for fugitive emissions of volatile organics from valves and flanges: model development, experimental evidence, and implications, *Environmental Science and Technology*, Vol. 26, No. 3, pp. 478–484.

Crowl, D. A., and Louvar, J. F., 1990. *Chemical Process Safety: Fundamentals with Applications,* Prentice Hall, Upper Saddle River, NJ.

Drivas, P. J., 1982. Calculation of evaporative emissions from multicomponent liquid spills, *Environmental Sciences and Technology*, Vol. 16, pp. 726–728.

Eklund, B., 1992. Practical guidance for flux chamber measurements of fugitive volatile organic emission rates, *Journal of the Air and Waste Management Association*, Vol. 42, pp. 1583–1591.

Giardino, N. J., Esmen, N. A., and Andelman, J. B., 1992. Modeling volatilization of trichloroethylene from a domestic shower spray: the role of drop-size distribution, *Environmental Science and Technology*, Vol. 26, No. 8, pp. 1602–1606.

Heinsohn, R. J., 1991. *Industrial Ventilation: Engineering Principles*, Wiley-Interscience, New York.

Hirschfelder, J. O., Curtiss, C. F., and Bird, R. B., 1954. *Molecular Theory of Gases and Liquids*, Wiley, New York.

Incropera, F. P., and DeWitt, D. P., 1990. *Fundamentals of Heat Transfer*, Wiley, New York.

Jenkins, B. M., Jones, A. D., Turn, S. Q., and Williams, R. B., 1996. Emission factors for polycyclic aromatic hydrocarbons from biomass burning, *Environmental Science and Technology*, Vol. 30, No. 8, pp. 2462–2469.

Jones, D. L., Burklin, C. E., Seaman, J. C., Jones, J. W., and Corsi, R. L., 1996. Models to estimate volatile organic hazardous air pollutant emissions from municipal sewer systems, *Journal of the Air and Waste Management Association*, Vol. 46, pp. 657–666.

Kreith, F., 1973. *Principles of Heat Transfer*, 3rd ed., Harper & Row, New York.

Kumar, A., Vatcha, N. S., and Schmelzle, J., 1996. Estimate emissions from atmospheric releases of hazardous substances, *Environmental Engineering World*, Vol. 2, No. 6, November–December, pp. 20–23.

Kunkel, B. A., 1983. *A Comparison of Evaporative Source Strength Models for Toxic Chemical Spills*, Report AFGL-TR-0307, Air Force Geophysics Laboratory, U.S. Air Force Systems Command, 50 pp.

Little, J. C., 1992. Applying the two-resistance theory to contaminant volatilization in showers, *Environmental Science and Technology*, Vol. 26, No. 7, pp. 1341–1349.

Mackay, D., and Paterson, S., 1986. Model describing the rates of transfer processes of organic chemicals between atmosphere and water, *Environmental Science and Technology*, Vol. 20, No. 8, pp. 810–816.

Mackay, D., and Shiu, W. Y., 1981. A critical review of Henry's law constants for chemicals of environmental interest, *Journal of Physical and Chemical Reference Data*, Vol. 10, No. 4, pp. 1175–1191.

Mackay, D., and Yeun, A. T. K., 1983. Mass coefficient correlations for volatilization of organic solutes from water, *Environmental Science and Technology*, Vol. 17, No. 4, pp. 211–217.

Mackay, D., Bobra, A., Chan, D. W., and Shiu, W. Y., 1982. Vapor pressure correlations for low-volatility environmental chemicals, *Environmental Science and Technology*, Vol. 16, No. 10, pp. 645–649.

McCabe, W. L., Smith, J. C., and Harriott, P., 1993. *Unit Operations of Chemical Engineering*, 5th ed., McGraw-Hill, New York.

McKone, T. E., 1987. Human exposure to volatile organic compounds from common household tap water: the indoor inhalation pathway, *Environmental Science and Technology*, Vol. 21, No. 12, pp. 1194–1201.

Miller, C. A., Ryan, J. V., and Lombardo, T., 1996. "Characterization of Air Toxic from an Oil-Tireture Boiler", *Journal of the Air Waste Managment Assoc.*, Vol. 46, pp. 742–748.

Nirmalakhandan, N. N., and Speece, R. E., 1988. QSAR model for predicting Henry's constant, *Environmental Science and Technology*, Vol. 22, No. 11, pp. 1349–1357.

Orlemann, J. A., Kalman, T. J., Cummings, J. A., Lin, E. Y., Jutze, G. A., Zoller, J. M., Wunderle, J. A., Zieniewski, J. L., Gibbs, L. L., Loudin, D. J., Pfetzing, E. A., and Ungers, L. J., 1983. *Fugitive Dust Control Technology*, Noyes Data Corporation, Park Ridge, NJ.

Perry, R. H., and Chilton, C. H., 1984. *Chemical Engineers' Handbook*, 6th ed., McGraw-Hill, New York.

Pruppacher, H. R., and Klett, J. D., 1978. *Microphysics of Clouds and Precipitation,* D. Reidel, Dordrecht, The Netherlands.

Reid, R. C., Prausnitz, J. M., and Poling, B. E., 1987. *The Properties of Gases and Liquids*, 4th ed., McGraw-Hill, New York.

Shapiro, A. H., 1953. *The Dynamics and Thermodynamics of Compressible Flow*, Vol. I, Ronald Press, New York.

Shen, T. T., 1982. Estimation of organic compound emissions from waste lagoons, *Journal of the Air Pollution Control Association*, Vol. 32, No. 1, pp. 80–82.

Shepherd, J. L., Corsi, R. L., and Kemp, J., 1996. Chloroform in indoor air and wastewater: the role of residential washing machines, *Journal of the Air and Waste Management Association*, Vol. 46, pp. 631–642.

Sherwood, T. K., Pigford, R. L., and Wilke, C. R., 1975. *Mass Transfer*, McGraw-Hill, New York.

Stiver, W., and Mackay, D., 1984. Evaporation rate of spills of hydrocarbons and petroleum mixtures, *Environmental Science and Technology*, Vol. 18, No. 11, pp. 834–840.

Stiver, W., Shiu, W. Y., and Mackay, D., 1989. Evaporation times and rates of specific hydrocarbons in oil spills, *Environmental Sciences and Technology*, Vol. 23, No. 1, pp. 101–105.

Treybal, R. E., 1980. *Mass-Transfer Operations*, 3rd ed., McGraw-Hill, New York.

U.S. Department of Health and Human Services, Public Health Service, Centers for Disease Control, National Institute for Occupational Safety and Health, 1985. *NIOSH Pocket Guide to Chemical Hazards*, U.S. Government Printing Office, Washington, DC.

U.S. Department of Health, Education, and Welfare, Public Health Service, Centers for Disease Control, National Institute for Occupational Safety and Health, 1977. *Occupational Diseases: A Guide to Their Recognition*, Publication 77-181, U.S. Government Printing Office, Washington, DC, revised.

U.S. Department of Labor, 1989. Air contaminants: permissible exposure limits, Title 29, *Code of Federal Regulations*, Part 1910.1000, U.S. Government Printing Office, Washington, DC.

U.S. EPA, AP-42: *Compilation of Pollutant Emission Factors*, Vol. I, *Stationary Point and Area Sources*, PB87-150959 Supplement A, October, 1986; PB82-101213 Third Edition Supplement 12,1981; PB81-244097 Third Edition Supplement 9; PB81-178014 Third Edition Supplement 11, October 1980; PB80-199045 Third Edition Supplement 10, February 1980; National Technical Information Service, Springfield, VA.

U.S. EPA, 1987. *User's Guide: Emission Control Technologies and Emission Factors for Unpaved Road Fugitive Emissions,* EPA/625/5-87/022, September. U. S. Government Printing Office, Washington, DC.

U.S. Office of the Federal Register, 1989. Protection of Environment, *Code of Federal Regulations (CFR) 40*, Part 60, Superintendent of Documents, U.S. Government Printing Office, Washington, DC, pp. 195–1013.

Vargaftik, N. B., 1975. *Tables on the Thermophysical Properties of Liquids and Gases*, 2nd ed., Hemisphere, Washington, DC.

Yaws, C., Yang, H. C., and Pan, X., 1991. Henry's law constants for 362 organic compounds in water, *Chemical Engineering*, Vol. 98, No. 11, pp. 179–185

# *Problems*

**8.1.** Using emission factors, estimate the contaminant generation rate in units of kg/s for the following cases.

**(a)** Your firm makes office furniture and paints the metal surfaces with an enamel paint (8.0 $lb_m$/gal). The consumption of paint is 5 gal/min. What is the hydrocarbon emission rate?

**(b)** You work for a slaughterhouse and in the fall of the year local hunters bring venison for smoking. Estimate the emission of particles (smoke) if meat is processed at the rate of 50 $lb_m$/h.

**(c)** You use an unvented kerosene space heater in your hunting cabin. Estimate the rate at which carbon monoxide is emitted while you sleep. Kerosene (density 7.5 $lb_m$/gal, heating value 40,000 kJ/kg) is burned at the rate of 0.1 gal/h.

**(d)** Six people smoke simultaneously during a meeting in a small conference room. Estimate the emission rate of carbon monoxide assuming only sidestream smoke.

**(e)** You are in charge of loading concrete mix into trucks. The space where this is done is small. Estimate the particle emission rate if trucks are loaded at a rate of one every 10 minute Each truck carries 10 tons of concrete mix.

**(f)** Estimate the emission rate of gasoline fumes from a filling station (having no fume controls) if on the average 10 gal of gasoline are added to each auto and if tanks are filled ("splash filling") at the rate of one every 5 min.

**8.2.** Using emission factors, estimate the contaminant generation rate in the units of kg/s for the following cases.

**(a)** Estimate the generation rate (kg/h) of particles, $SO_2$, $NO_x$, CO, VOCs, $CH_4$ from a hand-fired stoker (Figure 7-2) that burns anthracite coal at a rate of 5 tons/h. The coal assay is 12,550 HHV, 3% sulfur and 8% ash by mass. Compare the $SO_2$ generation rate with the value estimated by a mass balance.

**(b)** Estimate the hourly generation of particles, $SO_2$, CO, $NO_2$, and organics from a municipal multiple-chamber refuse incinerator (Figure 7-4) processing 10 tons refuse/h.

**(c)** Estimate the number of miles a new automobile must be driven to emit an amount of $NO_x$ equal to what is emitted by a Boeing 747 in one landing-and-takeoff cycle. Repeat for unburned hydrocarbons.

**(d)** Compare the fugitive emission rate ($lb_m$/h) from a flange in a 6-in high-pressure pipeline carrying $C_2H_6$ at 10 atm. Use data in Table A-3.1-2.

**(e)** Estimate the $NO_x$ generation rate from a rotary lime kiln and an asphalt mix plant each handling 10 tons/material per hour.

**(f)** What is the minimum particle collection efficiency needed to bring a plant that manufactures flat glass into compliance with the new source performance standards?

**(g)** Estimate the generation rate (kg $h^{-1}$ $mile^{-1}$) of fugitive dust from a 10-lane paved highway carrying 100 passenger vehicles/minute traveling at an average speed of 50 mph.

**8.3.** An electric generating station burns coal at a rate of 200 tons/h. The plant operates continuously 8760 h/yr. The coal has a higher heating value of 12,500 Btu/$lb_m$, 3% sulfur and 8% ash, and 0.01% mercury (by weight). An electrostatic precipitator (ESP) removes sufficient fly ash to satisfy the new source performance standards. If all the mercury condenses on minuscule flyash particles that are not captured by the ESP, estimate the mercury emissions (tons/yr).

**8.4.** Sour natural gas is methane containing odorous gases such as $H_2S$. If a sour natural gas (assume that $M = 18$) containing 5 mole % of $H_2S$ is burned at a rate of 10 kg/min, what is the $SO_2$ emission rate (kg/min) assuming all the sulfur is converted to $SO_2$?

**8.5.** In an automotive body shop, autos are refinished in a paint booth at a net rate that corresponds to 5 $m^2$/h. Refinishing consists of applying all the following materials:

- *Primer:* 100 g/$m^2$ per coat, one coat applied
- *Enamel:* 200 g/$m^2$ per coat, two coats applied
- *Lacquer:* 150 g/$m^2$, one coat applied

Exhaust from the paint booth is vented to the atmosphere. Using emission factors, estimate the rate (kg/h) at which hydrocarbon vapor enters the atmosphere.

**8.6.** Using the empirical equations in Section 8.5, estimate the rate at which trichlorethylene vapor will be generated (kg/h) when a conventional closed 55-gal drum is filled at a rate of 2 gal/min under conditions called *splash loading*.

**8.7.** Using the empirical equations in Section 8.5, estimate the rate at which gasoline (assume octane) vapor will be emitted to the atmosphere (tons/yr) when U.S. automobile gasoline tanks (assume 20 gal/car) are filled at a rate of 5 gal/min. Assume that $80 \times 10^6$ U.S. autos are filled with 5000 gal of gasoline each year.

**8.8.** Crushed limestone is transferred from one conveyor to another. To minimize the generation of fugitive dust, an enclosure is built to surround the operation. Using the emission factors for screening, conveying, and handling in stone quarries, estimate the mass of respirable dust generated per ton of stone. For calculation purposes, define respirable dust as particles 10 μm or less. Assume that the density of limestone is 105 $lb_m/ft^3$ and that the respirable mass fraction is 15%.

**8.9.** Trichlorethylene is a common nonflammable hydrocarbon used to clean metal surfaces. The chemical formula is $C_2HCl_3$ and the molecular weight is 131. The enthalpy of vaporization at 85.7°C is 57.24 cal/g. Shown below are saturation temperatures and pressures:

| P(mm Hg) | 1 | 5 | 10 | 20 | 40 | 60 | 100 | 200 | 400 | 760 |
|---|---|---|---|---|---|---|---|---|---|---|
| T(°C) | −43.8 | −22.8 | −12.4 | −1.0 | 11.9 | 20.0 | 31.4 | 48.0 | 67.0 | 86.7 |

The enthalpy of vaporization at 85.7°C is 57.24 cal/g. Using the Clausius–Clapeyron equation, estimate the vapor pressure at 25°C.

**8.10.** An open 55–gal drum contains a thin layer of trichlorethylene at the bottom. The drum is left uncovered in a storage room.

**(a)** Using first principles, estimate the rate (kg/h) at which vapor enters the environment.

**(b)** Using emission factors taken from the CHIEF (see Section 3.4), estimate the rate (kg/h) at which vapor enters the atmosphere.

**8.11.** Benzene (MW = 78) at 28°C (83°F) lays in a shallow pool in a flat, bottomed cylindrical reactor 3 m in diameter and 5 m high. The normal boiling temperature is 80°C, the heat of vaporization is 433.5 kJ/kg at 25°C and the diffusion coefficient of benzene in air is $0.77 \times 10^{-5} \ m^2/s$. The reactor has a circular opening (diameter 0.5 m) at the top through which materials are added. The lid on the opening has been removed and benzene vapor escapes to the workplace atmosphere (25°C). Estimate the rate (kg/h) at which benzene vapor escapes through the opening.

**8.12.** An artist is casual about handling chemicals in her small hot unvented studio. To save money the artist cleans her brushes with used allyl alcohol brought to her by a friend who obtains it from his place of work and is unaware of its toxicity. (Unfortunately neither the friend nor the artist are aware of MSDS literature!) Uncapped 1-gal cans (4 inch by 6.5 inch base, 9.5 inch high, 1-inch-diameter cylindrical opening, 0.5 inch high) with pools of alcohol at the bottom are stashed in corners of the room. Near her easel she keeps a 6-inch cup filled to the rim with alcohol for cleaning brushes. The artist uses a room fan to cool herself while working. Air passes over the open cup at a velocity of 3 m/s. Assume that the diffusion coefficient for allyl alcohol in air is $1 \times 10^{-5} \ m^2/s$. See Table A-12 for other properties of allyl alcohol.

**(a)** Estimate the vapor pressure (kPa) and enthalpy of vaporization ($kJ \ kmol^{-1} \ K^{-1}$) at 25°C.

**(b)** Using first principles, estimate the emission rate of fumes (kg/h) from an open 1-gal can assuming that:

**(1)** Evaporation is governed by diffusion through a stagnant air column 1in. in diameter, 9.5 in. high.

**(2)** The concentration of alcohol in the air inside the can is dictated by its vapor pressure. Diffusion through the opening can be approximated as diffusion through a stagnant air column $\frac{1}{2}$ inch high.

**(c)** Using first principles, estimate the emission rate of fumes (kg/h) from the cup.

**8.13.** A pie plate is filled with water. The upper diameter of the pie plate is 10 in. Water evaporates at a rate of 20 mg/s when the air and water are at 25°C and the air is absolutely dry (relative humidity is zero).

**(a)** What is the overall convection mass transfer coefficient?

**(b)** Estimate the evaporation rate if the relative humidity is 70%?

**8.14.** Cylindrical brass stock $\frac{1}{4}$ in. in diameter leaves a machine with a thin (0.1-mm) layer of water coating its surface. The speed with which the brass stock moves is 0.1 m/s. Room air at 25°C and 20% relative humidity passes over the brass perpendicular to the axis at a velocity of 10 m/s and causes the water to evaporate. The temperature of the water film remains constant (30°C) due to the large heat capacity and thermal conductivity of the brass. Estimate:

**(a)** The overall convection mass transfer coefficient $(h_m)$.

**(b)** The local evaporation rate $(kg\,s^{-1}\,m^{-2})$ 5 s after it has left the machine.

**(c)** The water film thickness in part (b).

**8.15.** Moth repellent (paradichlorobenzene) in the form of a long cylinder of circular cross-sectional area is hung in an airstream (at STP) moving with a uniform velocity $U$. The moth repellent has uniform chemical composition and a density $\rho$, and it sublimates.

**(a)** Show that the sublimation rate $(\dot{m}, kg/s)$ varies as $\dot{m} = $ constant $U^m$ where $m$ is a constant given in a table of heat transfer correlations.

**(b)** Show that the rate of change in diameter $(dD/dt)$ is $dD/dt = -$constant $D^m$.

**8.16.** Ethyl alcohol $(C_2H_5OH)$ is accidentally spilled on the floor of a store room in which the temperature and pressure are 23°C, 1 atm. The average air velocity is 2 m/s and the spill is 2 m in diameter. Estimate the rate of evaporation in $kg/m^2$ from first principles and compare the results with those obtained from empirical equations.

**8.17.** A large open vessel contains an aqueous mixture of volatile waste liquids. The vessel is 10 m in diameter, 3 m high, and filled to the rim. The vessel is stored outdoors in a remote part of an industrial plant at 25°C. Air passes over it at a velocity of 3 m/s parallel to the surface. Initially, the waste liquid consists of the following:

| Hydrocarbon | Mole fraction |
|---|---|
| Benzene | 0.05 |
| Methyl alcohol | 0.05 |
| Carbon tetrachloride | 0.05 |
| Toluene | 0.05 |

The remainder of the liquid is water. Estimate the evaporation rate (kg/h) of volatile materials after 10 h, 100 h, 1000 h, and so on, assuming that the wastes are agitated to make the concentration within the vessel uniform. Does the evaporation rate change with time? Does the composition of the remaining liquid remain constant?

**8.18.** Questions are sometimes raised (McKone, 1987) about the exposure that humans experience due to volatile organic compounds (VOCs) that evaporate from household tap water. If the concentration of trichloroethylene (TCE) in household tap water is 1 mg/L of water, estimate the rate of evaporation of TCE from a bathtub of water at 25°C using two-film theory. For a first approximation, neglect any TCE in the air.

**8.19.** Using two-film theory, estimate the evaporation rates (kg/h) of benzene and chloroform from a waste lagoon containing 100 mg/L of benzene $(M = 78)$ and 100 mg/L of chloroform $(M = 119)$. The lagoon is 25 m $\times$ 40 m and 3.5 m deep. The air and liquid temperatures are 25°C and the wind speed is 3.0 m/s. The Henry's law constants are:

- *Benzene:* $5.5 \times 10^{-3}$ atm m$^3$ gmol$^{-1}$
- *Chloroform:* $3.39 \times 10^{-3}$ atm m$^3$ gmol$^{-1}$

and the Schmidt numbers are:

- *Benzene and water:* 1000
- *Chloroform and water:* 1100
- *Benzene and air:* 1.76
- *Chloroform and air:* 2.14

**8.20.** Grain stored in silos, transported in ships' holds, and so on, is sprayed with a fumigant to prevent deterioration. A spray containing ethyl formate and other dissolved fumigant materials will be used. Your supervisor wishes to spray this material on grain manually as it is transferred to a vessel. You suspect that spraying will produce an unhealthy exposure to ethyl formate vapor. To prove your point you wish to estimate the concentration to which workers might be exposed. As a first step you need to estimate the evaporation occurring during the time a drop leaves a spray nozzle until it lands on the grain. Using Runge–Kutta techniques, solve the simultaneous differential equations and predict the diameter $(D_p)$, temperature $(T_p)$, and evaporation rate (g/s) of a drop as function of time. As a first step, consider only a single drop falling through motionless air at its terminal settling velocity (see Figure 11-15). Assume that the ethyl formate vapor concentration in air (far removed from the drop) is zero at all times. *Air:* is at 25°C, 101 kPa, 20% relative humidity. Initial drop diameter is 500 μm, initial drop temperature is 25°C; see the Appendix for ethyl formate vapor pressures. Estimate the enthalpy of vaporization $(h_{fg})$ with the Clausius–Clapeyron equation, and if other property data about ethyl formate cannot be found, assume that its properties are those of water.

**8.21.** Waste isopropyl alcohol $(M = 60)$ leaks into a pond 10 m in diameter. Because of its infinite solubility, the mole fraction of the alcohol in the aqueous phase is uniform and equal to 0.001. The air and pond are at a temperature of 23.8°C. Atmospheric air passes over the pond at a (far-field) velocity of 3 m/s. Estimate the alcohol evaporation rate $(mg\,m^{-2}\,s^{-1})$. The overall pressure is 101 kPa (760 mm Hg). For non air: $\nu = 1.589 \times 10^{-5}$ m$^2$/s, $\rho = 1.186$ kg/m$^3$, $\mu = 1.88 \times 10^{-5}$ kg m$^{-1}$ s$^{-1}$, and Pr = 0.707.

**8.22.** Imagine that the pure isopropyl alcohol is contained in a constant-pressure $(P_0)$, constant-temperature $(T_0)$ vessel. Initially, 75% of the vessel is filled with alcohol and the space above $(V_0)$ is filled with air. The surface area of the air–liquid interface is $A_i$. Initially, the evaporation rate will be equal to the value developed in the Example 8.15, but as the space above fills with alcohol vapor, the alcohol evaporation rate will change.

**(a)** What is the alcohol mole fraction in air at equilibrium?

**(b)** Write and integrate a differential equation describing how the alcohol evaporation rate varies with time. Assume that the mass transfer coefficient(s) are constant.

**8.23.** Xylene is to be transferred from one vessel to an overhead vessel. The volumetric flow rate is 100 gal/min, the temperature is 25°C, and the pressure inside the pump is 1.5 atm. The pump is inexpensive and the seal surrounding the pump shaft is poor and can be characterized as having a clearance (open) area of 0.001 mm². Will xylene leak through the pump seal? If so, estimate the leak rate ($lb_m/h$) and compare the value to the value estimated by emission factors. Does a leak constitute a potential fire hazard?

**8.24.** Ammonia $(NH_3)$ is used in a selective noncatalytic reduction process (SNCR) to remove $NO_x$ from the exhaust gas of an electric utility boiler. See Chapter 10 for details about the SNCR process. The ammonia gas is stored at 25°C in large high-pressure vessels at 2500 kPa. There is evidence that small holes $(10^{-6}\ mm^2)$ may occur due to corrosion, allowing the ammonia to escape to the atmosphere.

**(a)** What is the leak rate (g/h) at a time when the upstream temperature and pressure of the ammonia are 25°C, 2500 kPa?

**(b)** As the ammonia escapes to the atmosphere, the upstream pressure decreases. Since the leak is slow, there is adequate heat transfer to maintain the upstream temperature at 25°C. Write an expression to predict the leak rate as a function of time. Compute and plot the leak rate and upstream pressure as a function of time.

**8.25.** Hydrogen cyanide (HCN) is stored in high-pressure vessels at 25°C, 1000 kPa. Initially, the HCN in the vessel is roughly 80% liquid and 20% vapor (by mass) at this temperature and pressure. The vessels stand upright, each with a shutoff valve at the top. Unfortunately, a small leak develops in the valve seat of the shutoff valve (leak area = $10^{-6}\ mm^2$) and HCN escapes to the atmosphere. HCN is a lethal gas.

**(a)** By means of a *T–s* phase diagram, show the state of the HCN in the storage vessel.

**(b)** Estimate the leak rate (g/s) under these conditions.

**8.26.** [*Design Problem*] Automobile parts are forged from heated steel. After forging, the parts have a surface temperature of 150°C and are placed on an overhead conveyor, where they are transported to several machining operations. While on the conveyor, they cool by forced convection (air velocity 10 m/s) with room air at 25°C. After being cooled to 25°C and after several machining operations, the parts are dipped in trichloroethylene (TCE) to remove oils, cutting fluids, and so on. After being dipped in TCE, the parts have a 0.01-mm film of TCE over the entire surface. The parts are placed on an overhead conveyor. While being conveyed, TCE evaporates. You have been asked to estimate the initial heat transfer rate (kJ/s) after forging and the evaporation rate (kg/s) immediately after being removed from the TCE cleaning tank. The part can be assumed to resemble a sphere of diameter 0.5 m. Your supervisor claims that within 5 s 80% of the requisite cooling has occurred and 80% of the TCE film has evaporated. Is you supervisor correct?

**8.27.** [*Design Problem*] You work in a research facility that models rivers and flood control projects, tests the design of ship hulls, and so on. Topographically, scaled models are laid out on the floor of the building to simulate a river, bay, lake, impounded water behind a dam, and so on. Thus a large water surface is exposed to room air and evaporation occurs. At other times, a tank of water is used and scale models of ship hulls are towed through the water so that drag and wave patterns can be studied. In all these experiments certain values of the Reynolds, Prandtl, Froude, and so on, numbers are needed so that researchers can achieve proper similitude. To control the water's viscosity, special hydrocarbons are added to the water. Unfortunately, researchers do not read the MSDS literature and select volatile toxic materials. Toxic vapors escape into the room. Using two-film theory, estimate the evaporation rate $(kmol\ s^{-1}\ m^{-2})$ of toxic vapor from the water under the following conditions.

- The characteristic length of water surface is 100 m.
- Room air has an average velocity of 2 m/s.
- There is negligible water movement.
- Henry's law constant, diffusivities, and other physical properties of the additives are similar to perchloroethylene.
- The initial mole fraction $(x_{c,0})$ in water is 0.01.

What precautions should be taken to keep the vapors from entering the workplace?

**8.28.** [*Design problem*] Fiberglass is sprayed into molds to manufacture caps used to enclose the beds of pickup trucks. Periodically, workers flush their spray nozzles with xylene. To minimize vapor entering the workplace atmosphere, workers direct the spray downward into a container equipped with a lateral exhauster that prevents vapors from leaving the container. Despite this practice, you believe that too much xylene evaporates and enters the atmosphere. Your supervisor contends that only a small amount of vapor enters the atmosphere as long as the particles are captured by the exhauster within 0.5 s. Estimate:

**(a)** The initial evaporation rate per drop (g/s).

**(b)** The initial temperature depression in the drop.

**(c)** The percent of the original drop that enters the workplace before the drop enters the container (neglect the temperature depression).

**(d)** The time required for 99% of the drop to evaporate (neglect the temperature depression).

- Initial drop diameter, $D_p(0) = 100\ \mu m$.
- $M(\text{xylene}) = 106$.
- $h_{fg}(\text{xylene}) = 409\ \text{kJ/kg}$.
- $P_v$ (saturation) xylene (see Table A-8).
- Diffusion coefficient of xylene in air is $0.70 \times 10^{-5}\ \text{m}^2/\text{s}$.
- Drop velocity $= 0.261\ \text{m/s}$
- Use Table A-11 for the properties of air.

**(e)** Is this practice a hazard to the workers' health? If so, what precautions should be taken to eliminate, or at least minimize, the hazard?

**8.29.** [*Design Problem*] A citizen's group in your community knows that you're an engineering whiz and asks you to assist them in characterizing the emissions from a cogeneration plant that is proposed to be built in the community. The plant uses a spreader stoker boiler that burns bituminous coal to produce steam. The steam generates electricity and is then used for process steam in the manufacture of paper. To satisfy NSPS (Table 3-3), the exhaust from the boiler passes through a baghouse (refer to Figure 1-7) to remove fly ash particles. The flue gas then enters a lime scrubber (refer to Figure 10-29) to remove $SO_2$. The scrubber stoichiometry is expressed as

$$CaO + 2H_2O + \tfrac{1}{2}O_2 + SO_2 \rightarrow CaSO_4 \cdot 2H_2O(s) \downarrow$$

The pollution removal systems do not remove $CO_2$, $CO$, and $H_2O$. Water leaves the scrubber as a saturated vapor at 165°C since a portion of the scrubbing liquid (water) evaporates. The following characteristics of the proposed plant have become public knowledge:

- electrical output: 225 MW
- Overall thermal cyclic efficiency of the plant: 33%
- Air–fuel ratio: 15 $\text{lb}_m$ air/$\text{lb}_m$ coal
- Incoming combustion air: 25°C, 101 kPa, 30% relative humidity
- Operating time: 50 weeks/yr, 7 days/week, 24 h/day
- Stack exit conditions: $T_{\text{exit}} = 165°C$, $D_{\text{exit}} = 5$ ft
- Coal assay: HHV = 12,000 Btu/$\text{lb}_m$

$$S = 2.3\%\ (\text{by weight})$$
$$Ash = 8.3\%\ (\text{by weight})$$
$$H_2 = 3\%\ (\text{by weight})$$
$$Hg = 1\ \text{mg Hg/kg coal}$$

- Mass flow rate of water circulated in the scrubber $= 4000$ gal/min
- Mass flow rate of makeup fresh water $= 1.5$ times the amount of water lost by evaporation in scrubber

$SO_2$ scrubbing occurs at 80°C; thus water and mercury vapors leave the scrubber at their saturation pressures. The vapor pressure of mercury at 80°C is unknown, but Perry et al. (1984) report the following vapor pressure/temperature data:

- 1 mm Hg pressure at 126.2°C
- 5 mm Hg pressure at 164.8°C
- 10 mm Hg pressure at 184°C

You have asked the local state environmental agency for the latest uncontrolled emission factors for a stoker boiler and been given the following information.

*Particles:* 60 $\text{lb}_m$ particles/ton coal
*$SO_2$:* 39S (where S is the percent sulfur in coal) $\text{lb}_m$/ton coal
*$NO_x$:* 14 $\text{lb}_m$/ton coal
*CO:* 5 $\text{lb}_m$/ton coal
*Nonmethane hydrocarbons:* 0.07 $\text{lb}_m$/ton coal
*$CH_4$:* 0.03 $\text{lb}_m$/ton coal

You neighbors ask you to estimate:

**(a)** The following mass flow rates (tons/yr)

**(1)** $\dot{m}(CO_2)$

**(2)** $\dot{m}(H_2O$, including the water vapor entering with the combustion air)

**(3)** $\dot{m}(NO_x)$

**(4)** $\dot{m}(CO)$

**(5)** $\dot{m}$(nonmethane hydrocarbons)

**(6)** $\dot{m}(CH_4)$

**(7)** $\dot{m}$(Hg vapor)

**(8)** $\dot{m}$(generated particles entering the baghouse)

**(9)** $\dot{m}(SO_2$ generated and entering the scrubber)

**(10)** $\dot{m}$(NSPS maximum allowable particle rate discharged to the environment)

**(11)** $\dot{m}$(NSPS maximum allowable $SO_2$ rate discharged to the environment)

**(b)** Minimum baghouse collection efficiency

**(c)** Minimum $SO_2$ scrubber efficiency

**(d)** Total mass flow rate of gas and vapor leaving the stack exit (tons/yr)

**(e)** Average exit velocity of the stack gas (m/s)

**(f)** Total mass flow rate of ash (fly ash and bottom ash) and gypsum sludge (tons/yr)

# 9

# Atmospheric Dispersion

In this chapter you will learn;

- The concepts underlying global air movement
- To predict pollutant concentrations using box models
- To predict atmospheric velocity, temperature, and pressure gradients
- To predict pollutant concentrations downwind of ground-level and elevated sources
- About large commonly used numerical dispersion models

When pollutants enter the atmosphere, they mix with air and are transported downwind by prevailing winds. Atmospheric dispersion is the name of the body of knowledge that enables engineers to predict pollutant concentrations in the vicinity of the source and how the concentrations vary with time, type of weather, and season of the year. Such predictions are needed to assess the long-term consequences of a source of pollution and to ensure that the source is in compliance with state and federal regulations.

## 9.1 Box model

Pollutants mix in the atmosphere at finite rates such that the downwind concentrations vary with location $(x, y, z)$ and perhaps time if the source is unsteady. It is useful to predict concentrations as if the mixing were infinitely fast. Rarely is mixing infinitely fast, but calculations based on this assumption are easy to carry out and represent the upper limit to mixing that occurs at finite rates. Good engineering requires the engineer to bracket an answer between anticipated upper and lower limits. Box modeling represents the upper limit since it presumes infinitely fast mixing.

Suppose that engineers need to compute the ground-level pollutant concentration in a community containing sources of pollution. The first step is to define an area $(ws)$ and elevation $(h)$ that defines a "box" within which there is reason to believe that mixing occurs rapidly. The next step is to define the cross-sectional area $(hw)$ through which air enters and leaves. A sketch of the box model is shown in Figure 9-1. Physical examples where these assumptions are reasonable are automobile tunnels, streets

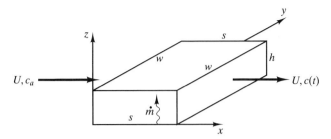

**Figure 9-1** Box model

bounded by tall buildings, a community located in a valley or basin, and indoor environments (homes, automobiles and aircraft passenger compartments, etc.).

In the outdoor environment the area of interest is usually defined by the space between impervious boundaries, ground, mountains, buildings, and so on. The height of the box is usually selected to be a mixing height, which is the altitude of an atmospheric layer through which pollutants do not penetrate. The layer may be an elevated inversion layer or the height of a region within which mixing is good. The equations for the box model shown in Figure 9-1 are the following:

*Conservation of mass of air:*

$$Q_{\text{in}} = Q_{\text{out}} = Q = Uhw \quad (9.1)$$

where $w$ and $h$ are the width and height of the box and $U$ is the average air speed perpendicular to the inlet area $(hw)$.

*Conservation of mass for the pollutant:*

$$whs\frac{dc}{dt} = \dot{m} + Uc_a hw - Uchw \quad (9.2)$$

where $\dot{m}$ is the pollutant *source strength* (mass/time) and $c_a$ is the pollutant concentration in the ambient air entering the box. It is useful to arrange the solution of Equation 9.2 in a form expressing a steady-state concentration $(c_{ss})$. The steady-state implies that $dc/dt = 0$. Thus

$$c_{ss} = c_a + \frac{\dot{m}}{Uhw} \quad (9.3)$$

Equation 9.2 can be integrated directly by the separation of variables,

$$\int_{c(0)}^{c(t)} \frac{dc}{(\dot{m} + Uhwc_a) - Uhwc} = \frac{1}{whs}\int_0^t dt \quad (9.4)$$

where the initial concentration in the box, $c(0)$, is assumed to be equal to the ambient concentration, $c_a$. If, $\dot{m}$, $U$, $h$, $w$, $s$, and $c_a$ are constant, the terms in the integrands of Equation 9.4 are constant and the concentration at any time $c(t)$ is

$$\frac{c(t) - c_a}{c_{ss} - c_a} = 1 - \exp\left(-\frac{Ut}{s}\right) \quad (9.5)$$

where $s$ is the length of the box in the direction of the wind. If the ambient concentration $(c_a)$ is zero, Equation 9.5 becomes

$$\frac{c(t)}{c_{ss}} = 1 - \exp\left(-\frac{Ut}{s}\right) \quad (9.6)$$

For readers familiar with the literature of industrial hygiene and environmental health engineering, Equation 9.5 is the same as that for the well-mixed model in industrial ventilation. If the source strength, $\dot{m}(t)$, and/or the wind speed, $U(t)$, varies with time, Equation 9.4 can be integrated using the modified Euler method (Table A-14) or Runge–Kutta techniques.

## EXAMPLE 9.1 POLLUTION IN CITY STREETS

A city street is 25 m wide $(w)$ and is bounded on either side by tall buildings 100 m high $(h)$ which trap pollutants below 100 m. During the evening an underground gas main leaks natural gas (methane) into the stormwater drainage system. Natural gas $(M = 16)$ escapes through manholes and sidewalk gutter drains in a stretch of street $(s)$ 1000 m long.

*Case I: Steady emission rate and steady wind speed.* Assume that the gas emission rate is 60 g s$^{-1}$ m$^{-1}$ of street (in the direction of traffic flow). If the wind is blowing steadily $(U = 0.5$ m/s), how much time elapses before explosive limits (methane, LEL = 5.4% by volume) are reached? It can be assumed that the ambient $(c_a)$ and initial $(c(0))$ concentrations of natural gas are zero. The total mass flow rate of escaping gas $[\dot{m}(g/s)]$ is

$$\dot{m} = (60 \text{ g s}^{-1} \text{ m}^{-1})(1000 \text{ m}) = 60,000 \text{ g/s}$$

Assuming STP, the lower explosive limit (LEL) of 0.054 (54,000 ppm) corresponds to a mass concentration (see Equation 1.15),

$$c(\text{mg/m}^3) = \frac{(\text{ppm})M}{24.5} = \frac{(54,000)(16)}{24.5}$$

$$= 35,260 \text{ mg/m}^3$$

Assuming that the ambient concentration of natural gas is negligible $(c_a = 0)$, the steady-state concentration is

$$c_{ss} = c_a + \frac{\dot{m}}{Uwh} = \frac{60,000 \text{ g/s}}{(0.5 \text{ m/s})(25 \text{ m})(100 \text{ m})}$$

$$= 48 \text{ g/m}^3 = 48,000 \text{ mg/m}^3$$

The time to reach LEL is

$$\frac{c(t)}{c_{ss}} = 1 - \exp\left(-\frac{Ut}{s}\right)$$

$$= \frac{35.26}{48} = 0.7346 = 1 - \exp\left(-\frac{0.5t}{1000}\right)$$

$$0.2654 = \exp(-0.0005t)$$

$$t = \frac{1.3264}{0.0005} = 2652.8 \text{ s} = 44.21 \text{ min}$$

The above can be thought of as a limiting case. If meteorological conditions caused the height $(h)$ of the box to change, or if the leak rate changed (indeed, if any of the parameters $\dot{m}$, $U$, $h$, $w$, $s$, and $c_a$ change), the steady-state solution above is invalid and techniques discussed in Case II need to be followed.

*Case II: Emission rate varies with time and the wind speed increases with time.* The volume $(V)$ of the box remains the same $2.5 \times 10^6$ $(\text{m}^3)$. The initial natural gas concentration is zero. Assume that the air speed $(U)$ increases steadily with time. Assume also that the emission rate increases with time, reaches a maximum, and then decreases with time. Predict the natural gas concentration as a function of time (minutes) if the emission rate and wind speed vary as follows:

$$\dot{m}(t)(\text{g/min}) = 3.6 \times 10^6 t^{2.533} \exp(-0.5t^{1.033})$$

$$U(t)(\text{m/min}) = 30 \exp(0.005t)$$

where $t$ is in minutes. The differential equation to predict the concentration as a function of time is

$$\frac{dc}{dt} = \frac{1}{whs}\{-75,000c \exp(0.005t)$$

$$+ 3.6 \times 10^6[t^{2.533} \exp(-0.5t^{1.033})]\}$$

Since the coefficients on the right-hand side vary with time $(t)$, the modified Euler method (Table A-14) or Runge–Kutta method should be used. The Runge–Kutta method is more accurate and can be accessed in several commercially available mathematical computational programs. Figure E9-1 shows how the air speed $(U)$, gas emission rate $(\dot{m})$, and gas concentration $(c)$ vary with time. The concentration increases with time, reaches a maximum, and then decreases. The increase with time is expected on the basis of the analysis in Case I. The concentration decreases because the increasing air velocity sweeps gas out of the box and because the emission rate eventually decreases. Potentially explosive conditions (LEL) exist between 4 and 12 min. Before 3 min, and after 15 min, the natural gas concentrations are below LEL. The Mathcad program producing Figure E9-1 can be found in the textbook's Web site on the Prentice Hall Web page.

---

If the source of pollution $(\dot{m})$ varies in magnitude over the land area $(ws)$ and engineers want to incorporate this into the model so as to obtain concentrations at selected points $c(x, y, z)$ within the region, it is possible to write equations similar to the above for each of a series of smaller boxes contained within the larger single box $(whs)$. Following this idea, engineers will write equations like Equation 9.2 for each box involving the dimensions of the smaller boxes and the wind speeds acting over the faces of the boxes which share common boundaries. Solving this set of equations requires knowledge of the velocity field along the faces defining the boxes. In the field of industrial ventilation this technique is called the *sequential box model*, and methods to solve the set of simultaneous linear differential equations can be found in Heinsohn (1991). Applying Runge-Kutta techniques, the sequential box model can even accommodate source strengths and wind speeds that vary with time.

The box model can also be used to model photochemical reactions in urban air and predict the con-

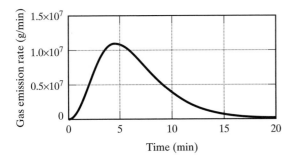

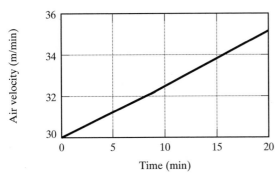

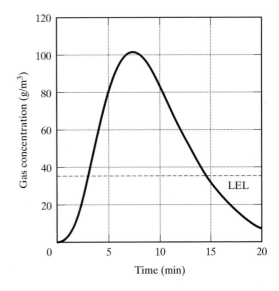

**Figure E9-1** Pollution in a city street. Natural gas concentration $(g/m^3)$ versus time (min), assuming a box model when the gas emission rate (g/min) and air speed (m/min) vary with time.

centrations of the reactive species with time (Jin and Demerjian, 1993). Separate source terms have to be written for each species. The source terms for photolytic species can be written to account for the time-varying nature of the solar flux.

## 9.2 Global atmospheric circulation

*General circulation* describes the macroscopic movement of air that manifests itself in such things as the trade winds, westerlies, and so on. Understanding this movement is needed to appreciate the long-distance transport of pollutants over the span of hundreds of kilometers. Readers who like fluid mechanics will find the subject fascinating and should consult Holton (1992) for a full exposition of the subject. Only selected portions of this body of knowledge will be included in this book and readers are encouraged to investigate this literature for themselves since many books have been written for readers of different levels of skill. Three scales are used to describe global atmospheric circulation:

1. *Microscopic scale:* distances less than 10 km
2. *Mesoscopic scale:* distances between 10 and 100 km
3. *Macroscopic scale:* distance greater than 100 km

To describe the circulation of the atmosphere, it is convenient to divide the earth's "air ocean" into two major layers, one of which can be subdivided (Figure 9-2):

1. *Geostrophic layer* $(z > 300$ to $500$ m). The geostrophic layer is an inviscid region, a region in which viscosity and shear stresses are relatively unimportant, and the air's movement is governed principally by horizontal pressure gradients and Coriolis forces.
2. *Planetary boundary Layer* $(0 < z < 300$ to $500$ m). The planetary boundary layer is a region immediately overhead that is affected by horizontal pressure gradients, viscosity, and Coriolis forces. The planetary boundary layer can be subdivided in two parts. The *surface layer* $(0 < z < 50$ to $100$ m) is the lower part of the

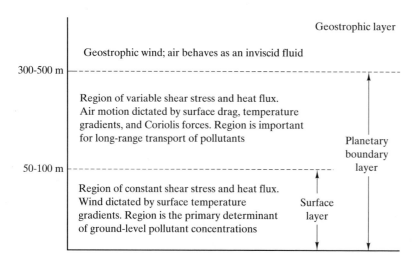

Figure 9-2   Surface layer (0 < z < 50-100 m) and planetary boundary layer (0 < z < 300-500 m) in the lower troposphere (adapted from Seinfeld, 1986)

planetary boundary layer in which the earth's topology and vertical temperature gradients strongly affect ground-level pollutant concentrations. For 100 < z < 500 m, Coriolis forces begin to affect the transport of air pollutants.

Figure 9-3 is a two-dimensional plot that shows the general circulation of the earth's atmosphere. The northeast and southeast trade winds travel from east to west while the westerlies travel in the opposite direction. Figure 9-4 is a three-dimensional plot which shows that above the equator Hadley cells cause the eastward trade winds and westerlies to

travel as "ribbons of air" that rotate in opposite directions as they travel across the globe.

The factors driving the macroscopic movement of the atmosphere are the sun's radiation, earth's rotation, and Coriolis forces. Intense solar heating in the tropics creates an updraft of air that divides and travels toward the poles, where it cools and descends (*subsides*). In the temperate regions, the rotation of the earth and air create smaller regions of circulating flow called *Hadley cells*.

An important force affecting the dynamics of air motion in the geostropic layer is the *Coriolis force*, which readers need to understand if they are to

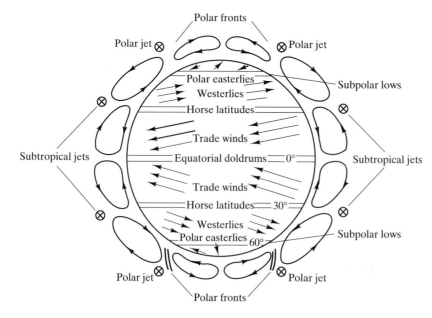

Figure 9-3   Global air circulation, circulation eddies below the subtropical and polar jets are called Hadley cells (Williamson, 1973)

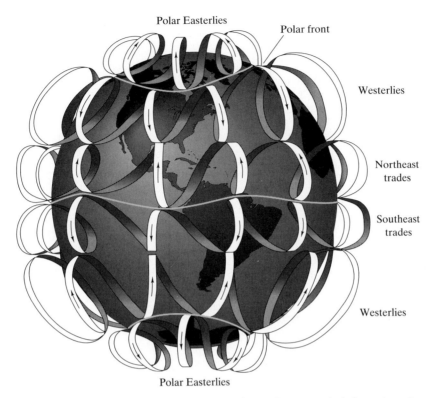

Polar Easterlies

Polar front

Westerlies

Northeast
trades

Southeast
trades

Westerlies

Polar Easterlies

**Figure 9-4** Global circulation of the atmosphere. Planetary winds form six major belts, three in each hemisphere, and distribute energy from an unequal distribution of solar radiation. Over the equator, humid air rises, moves toward both poles and sinks at about 30 degrees north and south latitude, creating a subtropical belt of high-pressure surface air. Some of this air moves toward the poles and forms the surface midlatitude westerly winds. Over the poles, cold air sinks and moves toward the equator and forms the polar front when it encounters the midlatitude westerlies. At this junction the polar jet stream is formed which is a necessary component of the nor'easterlies (Davis R E and Dolan R, 1993)

appreciate these smaller circulation patterns. The Coriolis acceleration $(a_c)$ is defined by the vector cross product,

$$\mathbf{a}_c = \frac{\mathbf{F}_c}{m} = -2(\mathbf{\Omega} \times \mathbf{U}) \qquad (9.7)$$

where $\mathbf{U}$ is the wind speed and $\mathbf{\Omega}$ is the earth's rotational speed. The acceleration can be understood as the Coriolis force per unit mass.

To understand the physical significance of the Coriolis force in the inviscid geostropic layer, consider two frames of reference. An *inertial reference system* is a reference system fixed with respect to the stars, and an *Eulerian reference system* is a reference system defined with respect to the earth. We begin

with Figure 9-5 and adopt the inertial reference system. Recall that viscous effects are negligible in the geostropic layer. At an instant of time $t$, consider a point $A$ (latitude $\alpha$, longitude $\beta$) in the air where $U_S$ is the southerly wind speed directed to point $B$, latitude $(\alpha - \delta\alpha)$, longitude $\beta$. After an elapsed time $\delta t$, the air has moved southward and westward (with respect to the earth's surface) to point $C$, latitude $(\alpha - \delta\alpha)$ and longitude $(\beta + \delta\beta)$. To explain this result, we return to the inertial frame of reference and note that in the absence of viscous effects and owing to inertia, the air and the earth rotate eastward with a velocity $U_E$ (in the inertial reference system):

$$U_E = R_e \Omega \cos \alpha \qquad (9.8)$$

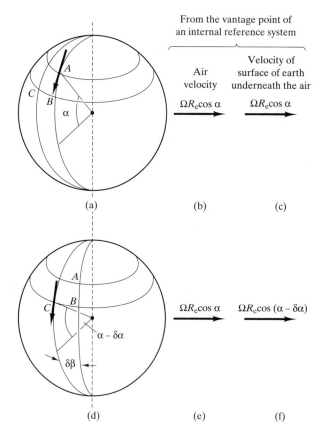

**Figure 9-5** Physical explanation of the Coriolis force (adapted from Williamson, 1973)

During the interval $\delta t$, the air will have moved southward to a lower latitude ($\alpha - \delta\alpha$), and the distance traveled toward the equator is

$$U_S \delta t = R_e \delta\alpha \qquad (9.9)$$

However, at the latitude $\alpha - \delta\alpha$, the earth moves faster (in the inertial reference system) than at latitude $\alpha$. At latitude $\alpha - \delta\alpha$ the eastward speed of the earth is

$$U'_E = R_e \Omega \cos(\alpha - \delta\alpha) \qquad (9.10)$$

For a southward wind, therefore, the eastward rotating air lags behind the earth increasingly as the wind moves toward lower latitudes.

The air that was originally moving southward (and still is in the inertial reference system) with the velocity $U_S$ no longer appears to be moving truly southward when observed from a point on earth. See, for example, Figure 9-6, in which the view is from a position directly over the north pole. As the earth rotates through the angle ($\Omega \delta t$), the velocity component $U_S$ contin-

ues, but to an observer on earth it appears that the air has acquired a westward velocity component, $U_S \Omega \delta t \sin \alpha$. The total westward velocity of the air relative to an observer on earth $U_W$ at the time $t + \delta t$ is

$$U_W = (U'_E - U_E) + U_S \Omega \delta t \sin \alpha \qquad (9.11)$$

In Equation 9.11 the term $U'_E - U_E$ represents the component of the apparent westward velocity, owing to the change in latitude and the last term represents the component of the westward component, owing to the apparent change in direction of the southward wind.

The westward acceleration $a_w$ of the air relative to an observer on earth is

$$a_w = \frac{U_w}{\delta t} \qquad (9.12)$$

Using the following algebraic manipulation

$$a_w = \frac{U'_E - U_E}{\delta t} + U_S \Omega \sin \alpha$$

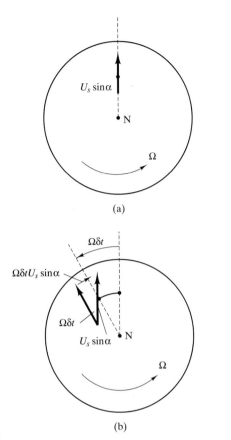

$U_s \sin\alpha$

N

$\Omega$

(a)

$\Omega\delta t$

$\Omega\delta t U_s \sin\alpha$

$\Omega\delta t$

$U_s \sin\alpha$ N

$\Omega$

(b)

**Figure 9-6** (a) Winds viewed from a point directly over the North Pole. The apparent magnitude of the wind speed is $U_s \sin\alpha$, (b) The wind moves in a straight path as the earth rotates to the east through an angle ($\Omega\delta t$). To an observer on earth, the wind appears to have a westward component ($U_s\Omega\delta t \sin\alpha$) observed from earth, (adapted from Williamson, 1973)

$$= \frac{[\cos(\alpha - \delta\alpha)]R_e\Omega}{\delta t} + U_S\Omega\sin\alpha$$

$$= \frac{[(\cos\alpha)(\cos\delta\alpha) + (\sin\alpha)(\sin\delta\alpha) - \cos\alpha]U_S\Omega}{\delta\alpha}$$

$$+ U_S\Omega\sin\alpha \tag{9.13}$$

and taking the limit as $\delta t$ and $\delta\alpha$ approach zero, one obtains

$$a_w = 2U_S\Omega\sin\alpha \tag{9.14}$$

This acceleration is called the *Coriolis acceleration*, and it represents a westward acceleration of air that is moving south toward the equator. When multiplied by the mass of the air parcel, the phenomenon appears to an earthbound observer to be a force, called the *Coriolis force*.

If the wind was initially moving northward, the Coriolis force would cause the air to advance ahead of the rotating earth and the air would acquire an eastward velocity component relative to an observer on earth. The Coriolis acceleration is always directed perpendicular to the wind velocity in the sense that observers on earth with their backs to the wind will observe it veer toward the right (in the northern hemisphere) and to the left in the southern hemisphere (see Figure 9-7).

On both sides of the equator warm air rises. This plume of warm equatorial air enters the stratosphere at the tropical tropopause, rises at tropical latitudes, divides, and travels toward both poles. The Coriolis force acting on these airmasses produces winds from the east, the *trade winds*. (Winds from the east moving to the west are called *east winds*. The name *east* or *west* refers to where the wind comes from, not where it is going.) At approximately 30° north and south latitude, stratospheric air cools (subsides), descends and returns to the tropopause (Figure 9-8). Warm air from temperate zones (latitudes between 30 and 60°) travels toward the poles, where it cools and subsides. Coriolis forces acting on these air-masses produce the *polar easterlies*. Between 40 and 55° north and south latitude is a region of turbulent air in which tropical and polar air interact and the Coriolis force produces westerly winds. In a belt between 30 and 35° north and south latitude (*horse latitudes*) lies a calm region characterized by light winds and high barometric pressure. The equatorial region is a band of gentle, and for considerable periods of time nonexistent, winds called the *doldrums*. The mean age of air in the stratosphere is approximately 5 years at altitudes above 25 km and 2 years at altitudes below 2 km.

The Coriolis force is also manifested in ocean currents. There is a clockwise rotation of the Gulf Stream and Japanese Current in the northern hemisphere. A counterclockwise current of water along the coast of Peru and northern Chile is another example (see Chapter 2). Surface water is also affected by surface winds. Thus the upwelling current of cold water west of Peru is driven westward by the southeast tradewinds, producing the well-known El Niño and La Niña effects.

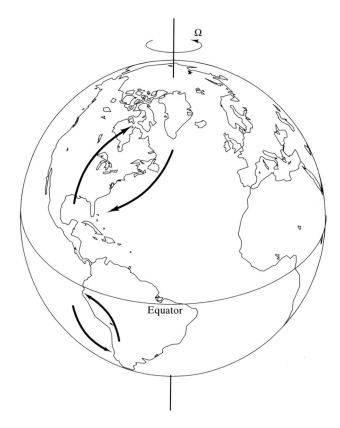

**Figure 9-7**  Deviation of winds in Northern and Southern hemispheres produced by the Coriolis force, (adapted from Williamson, 1973)

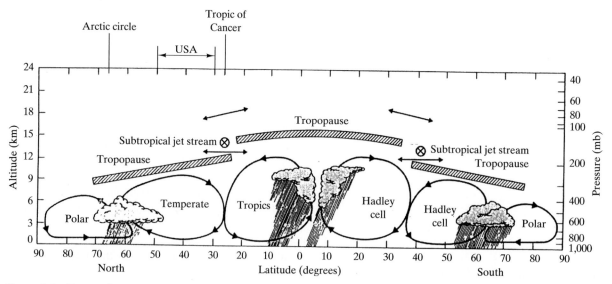

**Figure 9-8**  Tropospheric air circulation showing Hadley cells during the summer in the Northern and Southern Hemispheres that give rise to the Tropospheric subtropical jet stream (adapted from Williamson, 1973)

## EXAMPLE 9.2   CORIOLIS FORCE ON A MAN WALKING ON A ROTATING SURFACE

A high-speed amusement park ride consists of a horizontal circular surface (wheel) with a radius ($R$) of 10 m that rotates in a counterclockwise direction on a vertical axis at a speed ($\Omega$) of 20 rpm (1.045 rad/s). A 90-kg man tries to walk along the surface of the wheel. Estimate the force on the man under the following conditions.

- *Case I.* The man stands on the periphery of the wheel. What force does he have to exert to merely remain on the wheel?
- *Case II.* Beginning at the axis of rotation, the man walks outward in the radial direction at a constant speed $v_r = 2$ m/s (relative to the wheel).
  - **a.** From the vantage point of a stationary observer overhead, draw a graph of the man's trajectory until he reaches the edge of the wheel.
  - **b.** Relative to a stationary observer, compute the man's velocity after 1.0, 2.5, and 5.0 s.
  - **c.** What is the Coriolis force the man experiences at the times above?

### Solution

*Case I.* The centrifugal force on the man that must be overcome if the man is to remain on the periphery of the wheel is

$$\mathbf{F} = \hat{r}m\Omega^2 R = \hat{r}(90 \text{ kg})(1.045 \text{ rad/s})^2(10 \text{ m})$$

$$= 982.82 \text{ N} = (220.9 \text{ lb}_f)$$

Unless a force equal in magnitude but opposite in direction is exerted, the man will leave the wheel with a tangential velocity equal to

$$v_t = \Omega R = (1.045 \text{ rad/s})(10 \text{ m})$$

$$= 10.45 \text{ m/s} (23.4 \text{ mph})$$

It takes considerable strength merely to remain on the wheel, and if he files off the wheel he will strike something with damaging speed.

*Case II:* The person walking radially outward on the wheel is in an Eulerian reference system that rotates about the axis of the wheel. The stationary observer looking down on the man is in an inertial reference system and notices both the rotating wheel and the man attempting to move in a radial direction across the wheel. Figure E9-2 shows his trajectory.

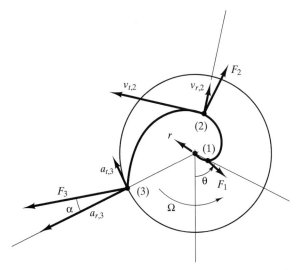

**Figure E9-2**   Coriolis acceleration and force on a 90-kg man walking ($v_r = 2$ m/s) across a rotating (20 RPM) surface ($R = 10$ m) at $t_1 = 1.0$ s, $t_2 = 2.5$ s and $t_3 = 5.0$ s

After 1 s ($t_1 = 1$ s) the man will be located at a radius ($r_1$) and an angle ($\theta_1$) displaced from his initial position.

$$r_1 = (v_r)(t_1) = (2 \text{ m/s})(1 \text{ s}) = 2 \text{ m}$$

$$\theta_1 = \Omega t_1 = (1.045 \text{ rad/s})(1 \text{ s}) = (1.045 \text{ rad})$$

$$(180°/\pi \text{ radians}) = 59.9°$$

$$v_{t,1} = \Omega r_1 = (1.045 \text{ rad/s})(2 \text{ m}) = 2.09 \text{ m/s}$$

$$v_{r,1} = 2 \text{ m/s}$$

The Coriolis acceleration can be found from Equation 9.7:

$$\mathbf{a} = -2(\boldsymbol{\Omega} \times \mathbf{v}) = -2[\hat{z}\Omega \times (\hat{r}v_r + \hat{\theta}v_t)]$$

$$= -\hat{\theta}2\Omega v_r + \hat{r}2\Omega v_t$$

$$= -\hat{\theta}2(1.045 \text{ rad/s})(2 \text{ m/s})$$

$$+ 2\hat{r}(1.045 \text{ rad/s})(2.09 \text{ m/s})$$

$$= -\hat{\theta}4.18 \text{ m/s}^2 + \hat{r}4.37 \text{ m/s}^2$$

The Coriolis force on the man is

$$\mathbf{F} = m\mathbf{a} = -\hat{\theta}ma_t + \hat{r}ma_r$$

$$= -\hat{\theta}(90 \text{ kg})(4.18 \text{ m/s}^2) + \hat{r}(90 \text{ kg})(4.37 \text{ m/s}^2)$$

$$= -\hat{\theta}376.2 \text{ N} + \hat{r}393.1 \text{ N}$$

The magnitude of the force is

$$F = \left(F_t^2 + F_r^2\right)^{1/2} = (296{,}054)^{1/2}$$
$$= 544.1 \text{ N } (122.3 \text{ lb}_f)$$

The table below summarizes the velocity, acceleration, and force for three periods of time.

| Time (s) | $r$(m) | $\theta$ | $v_r$(m/s) | $v_t$(m/s) | $a_r$(m/s²) | $a_t$(m/s²) | $F_r$(N) | $F_t$(N) | $F$(lb$_f$) |
|---|---|---|---|---|---|---|---|---|---|
| 1.0 | 2 | 59.9 | 2 | 2.09 | 4.37 | −4.18 | 393.1 | −376.2 | 122.3 |
| 2.5 | 5 | 149.2 | 2 | 5.22 | 10.92 | −4.18 | 982.8 | −376.2 | 236.6 |
| 5.0 | 10 | 299.5 | 2 | 10.45 | 21.84 | −4.18 | 1,965.6 | −376.2 | 449.9 |

Figure E9-2 shows the Coriolis force at the three periods of time. The Coriolis force has a radial and tangential component. The radial force is larger than the tangential force and becomes even larger as the radius increases. The net force acts at angle $\alpha = \tan^{-1}(v_r/v_t)$ to the radius and counter to the wheel's rotation. As the man travels outward, he will have to lean forward and to his left to offset the Coriolis force tending to topple him over. As a practical matter, he lacks the strength to reach the edge of the wheel.

Extrapolating this example to the atmosphere, it can be seen that air traveling south from the North pole will experience a force toward the west that produces a clockwise rotation to air in the northern hemisphere (Figure 9-7).

The *geostrophic wind* speed $(U_G)$ is the dominant wind speed at altitudes exceeding 500 m. It is dictated by a horizontal pressure gradient and Coriolis forces since viscous effects are negligible. As a simple illustration, we choose a point (latitude $\alpha$, longitude $\beta$) and assume that the air is steady, inviscid flow and that its density is constant. We then define a right-handed coordinate system on the earth in which $z$ is the vertical direction, $x$ is an eastward distance, and $y$ is a northward distance. The equations for the conservation of mass and momentum for such a frame of reference are (Holton, 1992)

*Mass:* $\qquad \nabla \cdot \mathbf{U} = 0 \qquad$ (9.15)

*Momentum:* $\quad \dfrac{D\mathbf{U}}{Dt} = \dfrac{\text{Grad } P}{\rho} + \mathbf{F}_c \quad$ (9.16)

$Ds/Dt$ is the *substantial derivative* of quantity $s$:

$$\frac{Ds}{Dt} = u\frac{\partial s}{\partial x} + v\frac{\partial s}{\partial y} + w\frac{\partial s}{\partial z} + \frac{\partial s}{\partial t} \quad (9.17)$$

and $u$, $v$, and $w$ are the air velocities in the $x$, $y$, and $z$ directions. For the atmosphere moving over the earth (radius $R_E$),

$$\frac{Du}{Dt} - \frac{uv}{R_E}\tan\alpha + \frac{uw}{R_E} = -\frac{1}{\rho}\frac{\partial P}{\partial x}$$
$$+ 2\Omega v \sin\alpha - 2\Omega w \cos\alpha + F_{fx} \quad (9.18)$$

$$\frac{Dv}{Dt} + \frac{u^2}{R_E}\tan\alpha + \frac{vw}{R_E} = -\frac{1}{\rho}\frac{\partial P}{\partial y}$$
$$- 2\Omega u \sin\alpha + F_{fy} \quad (9.19)$$

$$\frac{Dw}{Dt} - \frac{u^2 + v^2}{R_E} = -\frac{1}{\rho}\frac{\partial P}{\partial z}$$
$$+ \Omega u \cos\alpha + F_{fz} \quad (9.20)$$

where $F_f$ are frictional forces. The components of the Coriolis force in the $x$, $y$, and $z$ directions are

$$F_{cx} = -2\Omega(w\cos\alpha - v\sin\alpha) \quad \text{(a)}$$
$$F_{cy} = -2\Omega u \sin\alpha \quad \text{(b)} \quad (9.21)$$
$$F_{cz} = 2\Omega u \cos\alpha \quad \text{(c)}$$

In the geostrophic layer, the vertical velocity component $w$ is small with respect to the transverse $u$ and $v$ components and can be neglected. If the $x$-axis is oriented in the direction of the geostropic wind, $v = 0$. Equations 9.18 and 9.19 become

$$u\frac{\partial u}{\partial x} = -\frac{1}{\rho}\frac{\partial P}{\partial x} \quad \text{(a)}$$
$$\qquad\qquad\qquad\qquad\qquad (9.22)$$
$$0 = -2\Omega u \sin\alpha - \frac{1}{\rho}\frac{\partial P}{\partial y} \quad \text{(b)}$$

Since both $v$ and $w$ have been set equal to zero, the continuity equation requires that the velocity component $u$ (i.e., the *geostrophic wind speed* $U_G$) be constant. Equation 9.22 becomes

$$U_G = \frac{\partial P/\partial y}{2\Omega \sin \alpha} \qquad (9.23)$$

Thus the magnitude of the geostrophic wind depends on the transverse pressure gradient, which is dictated by meteorological conditions and latitude.

Another phenomenon affecting the behavior of plumes is the *Ekman spiral*. In the planetary boundary layer, airspeed increases with altitude and produces the classical boundary layer profile. Air moving in the boundary layer is influenced by three factors:

1. Pressure gradient
2. Viscous force
3. Coriolis force

Air moves from high to low pressure; thus the wind speed is parallel to the pressure gradient. The viscous force acts in the opposite direction to the velocity. Thus the pressure gradient and viscous force act along the same line. On the other hand, the Coriolis force is perpendicular to the wind direction and hence to the viscous force. As a result, the direction of the wind turns with height producing *wind directional shear* or a spiral known as the *Ekman spiral* (Figure 9-9). Typical (Hanna et al., 1982) observed angles between the surface wind and the geostropic or free wind are:

- 15 to 20° for neutral conditions
- 30 to 50° for stable conditions
- 5 to 10° for unstable conditions

Wind directional shear is very important for plume dispersion over long distances, where the top of the plume can move in directions differing as much as 40 or 50° from the bottom of the plume.

## 9.3 Lapse rate

The transport of pollutants over distances of the order of tens of kilometers and within the troposphere is influenced only slightly by this large-scale circulation. The major influence affecting pollutant transport is the stability of the troposphere, which is driven by the atmospheric temperature gradient.

To appreciate the significance of the temperature gradient, consider a stationary mass of air governed only by pressure forces and gravity. Admittedly, the atmosphere is not stationary, but viscous effects can be considered at a later time. A column of stationary air is shown in Figure 9-10. A free-body diagram of a differential volume $A\,dz$ yields

$$(P + dP)A + A\rho g\,dz = PA \qquad (9.24)$$

where the density ($\rho$) is given by the ideal gas law,

$$\rho = \frac{PM}{R_u T} = \frac{P}{RT} \qquad (9.25)$$

$M$ is the molecular weight of air, which in turn depends on the humidity; $R$ is the specific gas constant for air; $R = R_u/M$; and $T$ is the absolute temperature (K). To a first approximation, we neglect the moisture's contribution to the molecular weight and assume the air to be dry. Combining Equations 9.24 and 9.25 yields

$$\frac{dP}{dz} = -\rho g = -\frac{gP}{RT}$$
$$\frac{dP}{P} = -\frac{g}{RT}\,dz \qquad (9.26)$$

From Equation 9.26 it is seen that the pressure varies with altitude in ways related to how the temperature

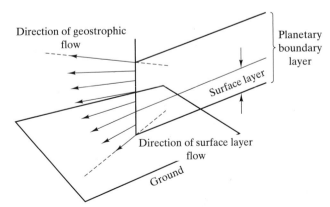

Figure 9-9 Ekman spiral (adapted from Seinfeld 1986)

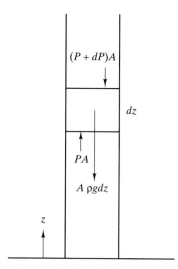

**Figure 9-10**    Quasistatic force balance in a column of air

varies with altitude. Consider three different types of temperature variations:

1. Isothermal atmosphere, $T(z) = T_0 =$ constant
2. Temperature and pressure related polytropically
3. Adiabatic (and reversible) variations

**Isothermal atmosphere.** If the temperature of air is constant up to an altitude $z$, Equation 9.26 can be integrated directly:

$$\int_{P_0}^{P} \frac{dP}{P} = \int_0^z \frac{-g}{RT} dz$$

$$\frac{P(z)}{P_0} = \exp\left(-\frac{gz}{RT}\right) \tag{9.27}$$

where $P_0$ is the ground-level ($z = 0$) pressure. Rarely is the atmosphere isothermal over an appreciable elevation or for an appreciable period of time. More commonly, the air temperature decreases since solar radiation heats the earth's surface and convection transports energy away from the earth.

**Polytropic atmosphere.** There are some atmospheric processes in which the pressure ($P$) and temperature ($T$) change with altitude in unique ways such that at any instant the air pressure ($P$), density ($\rho$), and absolute temperature ($T$) are related by

$$\frac{T(z)}{T_0} = \left(\frac{P(z)}{P_0}\right)^{(n-1)/n} = \left(\frac{\rho(z)}{\rho_0}\right)^{n-1} \tag{9.28}$$

and the subscript 0 refers to ground level ($z = 0$) or perhaps the typical 10-m altitude ($z = 10$ m) used by meteorologists. Substituting Equations 9.28 into Equation 9.26 and simplifying results in

$$\int_0^z dz = -\frac{R}{g} T_0 \int_{P_0}^{P} \frac{P_0^{(n-1)/n}}{g} P^{-1/n} dP$$

$$z = -\frac{RT_0}{g} \frac{n}{n-1} P_0^{(n-1)/n}\left(P^{(n-1)/n} - P_0^{(n-1)/n}\right) \tag{9.29}$$

If an expression for the temperature is needed, use Equation 9.28 to replace $P$ with

$$P = P_0\left(\frac{T}{T_0}\right)^{n/(n-1)} \tag{9.30}$$

and obtain an implicit expression for $z = f(T)$:

$$z = -\frac{R}{g}\left(\frac{n}{n-1}\right)(T - T_0) \tag{9.31}$$

The usefulness of Equation 9.31 is in differentiating $z$ with respect to $T$ and obtaining

$$\frac{dT}{dz} = -\frac{g}{R}\left(\frac{n-1}{n}\right) \tag{9.32}$$

The negative of the temperature gradient [i.e., $(-dT/dz)$] is called the *lapse rate* and represents the negative of the rate of change of temperature with altitude. The negative sign is introduced for convenience since for most conditions the temperature decreases with at altitude.

The standard (normal) lapse rate is the value of $-dT/dz$ averaged over the period of an entire year and takes into account yearly variations of temperature, pressure, and moisture. Meteorologists quantify the *standard (normal) lapse rate* as

$$-\left(\frac{dT}{dz}\right)_{\text{std atm}} = 6.6°C/km \ (3.5°F/1000 \text{ ft}) \tag{9.33}$$

Substituting this into Equation 9.32, one can generate a polytropic coefficient for the standard (normal) lapse rate:

$$n_{\text{std atm}} = 1.23 \tag{9.34}$$

**Dry adiabatic atmosphere.** Classical thermodynamics defines changes in temperature and pressure that occur in an adiabatic and reversible fashion as *isen-*

*tropic* changes. Such changes are those in which the entropy of the gas remains constant, and as such they represent an idealized limit in which the temperature and pressure change slowly while no energy is added or removed from the gas in the form of heat transfer.

There are occasions when the atmosphere changes adiabatically (i.e., when parcels of dry air rise or fall due to particular air movements). Under these conditions the real thermodynamic processes approach processes of constant entropy. During isentropic changes, the coefficient $n$ in Equation 9.32 becomes the isentropic constant $k$,

$$k = \frac{c_p}{c_v} \qquad (9.35)$$

where the thermodynamic coefficient $(c_p)$ is the specific heat at constant pressure and $c_v$ is the specific heat at constant volume. For dry air $k = 1.4$. Substituting $n = k = 1.4$ into Equation 9.32, one obtains the dry adiabatic temperature gradient,

$$\left(\frac{dT}{dz}\right)_{\text{dry adia}} = -\frac{g}{R}\left(\frac{k-1}{k}\right)$$

$$= -9.8°\text{C/km} \ (-5.4°\text{F}/1000 \text{ ft})$$

$$-\left(\frac{dT}{dz}\right)_{\text{dry adia}} \equiv \Gamma \qquad (9.36)$$

The symbol $\Gamma$ is used to define the *dry adiabatic lapse rate*. An alternative form of Equation 9.36 can be written using the fact that for an ideal gas with constant specific heats,

$$R = c_p - c_v$$

$$\frac{R}{c_p} = 1 - \frac{c_v}{c_p} = 1 - \frac{1}{k} = \frac{k-1}{k} \qquad (9.37)$$

Thus the dry adiabatic lapse rate can also be written as

$$\Gamma = -\left(\frac{dT}{dz}\right)_{\text{dry adia}} = \frac{g}{R}\frac{k-1}{k} = \frac{g}{c_p} \qquad (9.38)$$

In summary:

*Standard lapse rate:*

$$-\left(\frac{dT}{dz}\right)_{\text{std atm}} = 6.5°\text{C/km}$$

*Dry adiabatic lapse rate:* $\qquad (9.39)$

$$(\Gamma) = -\left(\frac{dT}{dz}\right)_{\text{dry adia}} = 9.8°\text{C/km}$$

Readers must remember that the lapse rate is the negative of the temperature gradient and that the symbol $\Gamma$ is used only to identify the dry adiabatic lapse rate. No particular symbol is used for other lapse rates. Readers must realize that for many atmospheric conditions it is impossible to predict a constant value of $n$ over a range of elevations. The previous discussion is useful as a point of reference to categorize atmospheric conditions and to compare the stability of one against another. In summary, then:

$$n = \begin{cases} 1 & \text{implies isothermal} \\ & T = T_0 \text{ and } \dfrac{dT}{dz} = 0 \quad \text{(a)} \\ 1.23 & \text{implies standard} \\ & \text{atmosphere and} \\ & \dfrac{dT}{dz} = -6.5°\text{C/km} \quad \text{(b)} \\ k = 1.4 & \text{implies dry adiabatic} \\ & \text{atmosphere and} \\ & \dfrac{dT}{dz} = -9.8°\text{C/km} \quad \text{(c)} \end{cases} \qquad (9.40)$$

---

**EXAMPLE 9.3 COMPUTING THE ATMOSPHERIC PRESSURE, $P(z)$, FROM MEASUREMENTS OF $T(z)$**

Atmospheric temperature measurements reveal that $dT/dz = 0.01$ °C/m, for $z < 500$ m. A positive value of $dT/dz$ is called an *inversion*. If the ground-level temperature and pressure are $T_0$ and $P_0$, write an equation specifying $P(z)$ for $z < 500$ m. The variation of temperature with altitude is $T(z) = T_0 + 0.01z$. From Equation 9.26, write

$$\frac{dP}{P} = -\frac{g\,dz}{RT} = -\frac{g}{R}\left(\frac{dz}{T_0 + 0.01z}\right)$$

$$\int_{P_0}^{P(z)} \frac{dP}{P} = -\frac{g}{R} \int_0^z \frac{dz}{T_0 + 0.01z}$$

$$\ln \frac{P}{P_0} = \frac{g}{0.01R} \ln\left(1 + \frac{0.01z}{T_0}\right)$$

$$= -\frac{9.8 \text{ m/s}^2}{(0.287 \text{ kJ kg}^{-1} \text{ K}^{-1})(0.01 \text{ K/m})}$$

$$\ln\left(1 + \frac{0.01z}{T_0}\right)(\text{kJ kPa}^{-1} \text{ m}^{-3})$$

$$(\text{kPa m}^2/10^3 \text{ N})(\text{N s}^2 \text{ kg}^{-1} \text{ m}^{-1})$$

$$= -3.414 \ln\left(1 + \frac{0.01z}{T_0}\right)$$

$$P(z) = P_0\left(1 + \frac{0.01z}{T_0}\right)^{-3.414}$$

**Moist saturated atmosphere.** Consider air containing water vapor at its dew point. If such moist air is caused to rise, its temperature will decrease and the water vapor will condense. To condense water vapor, energy equal to the enthalpy (heat) of vaporization $(h_{fg})$ must be added to the air and the lapse rate will be less than what it would be if the air was dry. To show why this is so, we write the first law of thermodynamics in the form

$$dq = dh - \frac{1}{\rho} dP \qquad (9.41)$$

where $dq$ is the heat transferred to a unit mass of air. Assuming that moist air can be treated as an ideal gas, the Equation 9.41 can be written as

$$\frac{dq}{dz} = c_p \frac{dT}{dz} - \frac{1}{\rho} \frac{dP}{dz} \qquad (9.42)$$

We introduce the *humidity ratio* $\omega$:

$$\omega = \frac{\text{mass of water vapor}}{\text{mass of dry air}} \qquad (9.43)$$

It can be shown that the relative humidity (RH) is related to the humidity ratio $(\omega)$:

$$\omega = \frac{0.622 \, (\text{RH}) P_v(T)}{P - P_v(T)} \qquad (9.44)$$

where $P$ is the total atmospheric pressure and $P_v(T)$ is the saturation pressure of water vapor at temperature $T$. To condense water vapor $(d\omega < 0)$, energy $(dq)$ must be transferred from the water vapor to the air. Thus

$$\frac{dq}{dz} = -\frac{d\omega}{dz} h_{fg} \qquad (9.45)$$

Combining Equations 9.42 and 9.45 gives

$$-h_{fg} \frac{d\omega}{dz} = c_p \frac{dT}{dz} - \frac{1}{\rho} \frac{dP}{dz} \qquad (9.46)$$

From Equation 9.26, $dP/dz$ can be replaced by $-\rho g$ and one obtains

$$-h_{fg} \frac{d\omega}{dz} = c_p \frac{dT}{dz} + g \qquad (9.47)$$

Rearrange and obtain

$$\left(-\frac{dT}{dz}\right)_{\text{moist air}} = \frac{g}{c_p} + \frac{h_{fg}}{c_p} \frac{d\omega}{dz} \qquad (9.48)$$

Replace $g/c_p$ using Equation 9.38 and the above becomes

$$\left(-\frac{dT}{dz}\right)_{\text{moist air}} = \Gamma + \frac{h_{fg}}{c_p} \frac{d\omega}{dz} \qquad (9.49)$$

The above reveals that the lapse rate $(-dT/dz)$ for a rising column of moist air in which condensation occurs $(d\omega/dz < 0)$ will be less than the dry adiabatic lapse rate. Since $h_{fg}/c_p$ is approximately 2440 K, a change in $d\omega/dz$ of $-0.0004$ m$^{-1}$ (0.4 km$^{-1}$) produces a moist air lapse rate of approximately 8.8°C/km.

### EXAMPLE 9.4   COMPUTING THE ATMOSPHERIC TEMPERATURE, $T(z)$, FROM RADIOSONDE DATA

A weather balloon (radiosonde) is sent aloft and transmits the following simultaneous pairs of temperature and pressure. The first data pair is obtained at the release point, $z = 30$ m. From the entire data set below, estimate the atmospheric temperature profile.

| P (kPa) | 102.3 | 101.2 | 100.0 | 96.8 | 96.6 | 90.9 | 87.8 | 84.0 | 72.5 | 70.0 |
|---------|-------|-------|-------|------|------|------|------|------|------|------|
| T (K)   | 284.2 | 282.0 | 285.2 | 288.2| 287.2| 286.2| 286.2| 285.8| 274.8| 273.0|

**Solution** On the basis of the data above, one does not know if the lapse rate is constant, if $n = k$, or even if a constant value of $n$ (Equation 9.32) is valid. Assume that the ideal gas relationship and the quasi-static model in Equation 9.26 are valid:

$$\frac{dP}{P} = -\frac{g}{RT}dz$$

Inspecting the $P$, $T$ data, it is apparent that the absolute temperature does not vary dramatically between signals. Thus the above can be integrated between two pairs of signals $i-1$ and $i$,

$$\ln\frac{P_i}{P_{i-1}} \simeq \frac{g}{RT_{avg}}(z_i - z_{i-1})$$

where $T_{avg} = \frac{1}{2}(T_i + T_{i-1})$. For $i=1$ the release point, $z_{1-1} = z_0 = 30$ m, the value of $z_1$ can be estimated:

$$z_1 = z_0 - \left(\ln\frac{P_1}{P_0}\right)\frac{RT_{avg}}{g}$$

Once $z_1$ is computed, the computation can be repeated between $i = 1$ and $i = 2$. Note that a new value of $T_{avg}$ has to be computed for each step. The calculation above can be accomplished with commercially available mathematical programs. Figure E9-4 is the result of the computation and shows the temperature versus altitude and the presence of a ground-based inversion, i.e. an increase in temperature with altitude. It also shows a substantial drop in temperature at 1500 m. This suggests the top of a well-mixed layer in which the temperature is fairly constant. The Mathcad program producing Figure E9-4 can be found in the textbook's Web site on the Prentice Hall Web page.

## 9.4 Atmospheric stability

*Stability* is an engineering term from control theory. A *stable* system is one which, if perturbed from a state of equilibrium, produces effects that nullify the perturbation and restore equilibrium. When perturbed, an *unstable* system produces changes that exacerbate the perturbation and do not restore the system to equilibrium. Figure 9-11 illustrates stable and unstable mechanical equilibria. Initially ($t = 0$), the marble is in equilibrium. It is then displaced to a new location at $t > 0$, whereupon gravity acts to move the marble. In case (a) the original equilibrium is restored, but in case (b) equilibrium is not restored.

Meteorologists distinguish three states of atmospheric stability at low altitudes (i.e., approximately $z < 100$ m): *unstable*, *neutral*, and *stable*. The stability of the atmosphere describes the vigor of vertical mixing. A *stable atmosphere* is one in which buoyancy returns a parcel of air to its original position after it has been displaced upward or downward in an isentropic fashion. An *unstable atmosphere* is one in which buoyancy increases the displacement of the parcel of air that has moved upward or downward in an isentropic

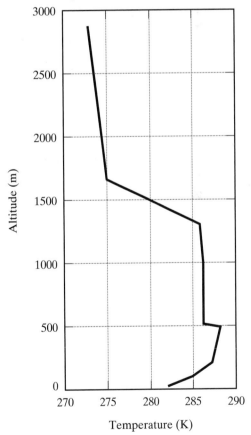

**Figure E9-4** Atmospheric temperature as a function of altitude computed from radiosonde data

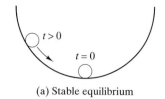

(a) Stable equilibrium

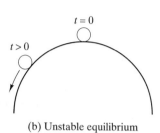

(b) Unstable equilibrium

**Figure 9-11**    Marble and bowl (a) stable equilibrium, (b) unstable equilibrium

fashion. A *neutral atmosphere* is one in which the lapse rate is equal to the dry adiabatic lapse rate.

Figure 9-12 shows a spherical parcel of air contained within an adiabatic, weightless, expandable balloon skin in equilibrium with the atmosphere at an altitude $z_e$. If the parcel is raised to altitude $z_2$, it expands instantly to allow the pressure of the air parcel inside to equilibrate with the pressure of the outside atmospheric pressure.

$$P_{\text{air parcel}} = P_{\text{atm}} \qquad (9.50)$$

As the pressure of the air parcel rises or falls, its temperature changes in an isentropic fashion since there is no heat transfer. Thus at elevation $z_2$ the temperature of the air parcel is

$$T_2(\text{airparcel}) = T_e \left[ \frac{P_2(\text{airparcel})}{P_e} \right]^{(k-1)/k} \qquad (9.51)$$

where the subscripts $e$ and 2 denote the initial and final location. Figure 9-12 depicts the variation in temperature of the parcel of air (dashed line) as it changes in an isentropic fashion and the actual atmospheric temperature variation (solid line). In the discussion that follows, the temperature gradients will not be discussed because all are negative and it is confusing to compare small or large negative terms. Rather, the lapse rate will be used. Readers should keep in mind that small lapse rates imply steep, nearly vertical lines on the $z$–$T$ graphs in Figure 9-12 while large lapse rates imply shallow, more tilted lines.

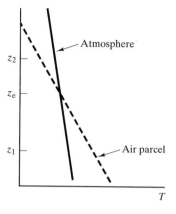

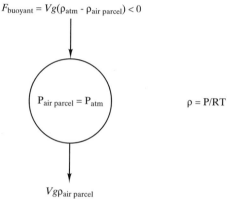

$$F_{\text{buoyant}} = Vg(\rho_{\text{atm}} - \rho_{\text{air parcel}}) < 0$$

$$P_{\text{air parcel}} = P_{\text{atm}} \qquad \rho = P/RT$$

$$Vg\rho_{\text{air parcel}}$$

Stable atmosphere:
$$T(z_2)_{\text{atm}} > T(z_2)_{\text{air parcel}}$$
$$\rho(z_2)_{\text{air parcel}} > \rho(z_2)_{\text{atm}}$$

**Figure 9-12**    Stable atmosphere, negative buoyant force restores an air parcel to its original location after it has been raised from $z_e$ to $z_2$

Figure 9-12 is identified as a *stable atmosphere* since the atmospheric lapse rate is less than it is in Figure 9-13, which is denoted as an *unstable atmosphere*. The phrases *stable* and *unstable* atmosphere become obvious when the reader determines the direction of the buoyant force acting on the parcel of air:

$$
\begin{aligned}
F_{\text{buoyant}} &= g(m_{\text{atm}} - m_{\text{air parcel}}) \\
&= Vg(\rho_{\text{atm}} - \rho_{\text{air parcel}})
\end{aligned}
\qquad (9.52)
$$

Consider a parcel of air at an altitude $z_e$ in which the temperature and pressure are identical to the surrounding atmosphere's temperature and pressure. Now elevate the parcel of air in a stable atmosphere from altitude $z_e$ to $z_2$. Equilibration of the pressure gives

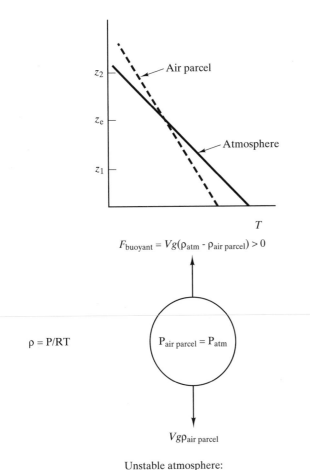

$$F_{\text{buoyant}} = Vg(\rho_{\text{atm}} - \rho_{\text{air parcel}}) > 0$$

$\rho = P/RT$

$P_{\text{air parcel}} = P_{\text{atm}}$

$Vg\rho_{\text{air parcel}}$

Unstable atmosphere:
$$T(z_2)_{\text{atm}} < T(z_2)_{\text{air parcel}}$$
$$\rho(z_2)_{\text{air parcel}} < \rho(z_2)_{\text{atm}}$$

**Figure 9-13** Unstable atmosphere, positive buoyant force does not restore an air parcel to its original location after it has been raised from $z_e$ to $z_2$

$$P_{\text{air parcel}}(z_2) = P_{\text{atm}}(z_2) \quad (9.53)$$

Figure 9-12 shows that the temperature of the air parcel is less than the surrounding air,

$$T_{\text{air parcel}}(z_2) < T_{\text{atm}}(z_2) \quad (9.54)$$

The mass of displaced air $(m_{\text{atm}})$ at the altitude $z_2$ is

$$m_{\text{atm}} = \frac{P_{\text{atm}}(z_2)V_{\text{atm}}(z_2)}{RT_{\text{atm}}(z_2)} \quad (9.55)$$

The mass of the air parcel $(m_{\text{air parcel}})$ is

$$m_{\text{air parcel}} = \frac{P_{\text{atm}}(z_2)V_{\text{atm}}(z_2)}{RT_{\text{air parcel}}(z_2)} \quad (9.56)$$

Since

$$T_{\text{atm}} > T_{\text{air parcel}}, \quad \text{then} \quad m_{\text{atm}} < m_{\text{air parcel}} \quad (9.57)$$

The weight of the displaced air is less than the weight of the air parcel and the parcel falls to $z_e$ where equilibrium is restored. If the air parcel is displaced downward to $z_1$ in a stable atmosphere, the reader should be able to prove that the weight of the displaced air is larger than the weight of the air parcel and a buoyant force causes the parcel rises to $z_e$ where equilibrium is restored.

Consider now the unstable atmosphere (Figure 9-13) and repeat the scenario above. The reader should be able to prove that if moved upward, the buoyant force acting on the air parcel now exceeds the weight of the air parcel, which causes it to continue rising of its own volition.

The lapse rate is used to denote stable and an unstable atmospheres. Furthermore, it is possible to differentiate very stable (or unstable) from mildly stable (or unstable) atmospheres. Consider lapse rates A through F in Figure 9-14. An atmosphere whose lapse rate corresponds to line D (dry adiabatic lapse rate) is called a *neutral atmosphere* and is of particular importance because it corresponds to the dry adiabatic lapse rate. I.e., for curve D (neutral; $n = k = 1.4$)

$$-\frac{dT}{dz} = \Gamma = 9.8°C/km \quad (9.58)$$

From the logic in the previous paragraphs it is possible to define all lapse rates lying to the left of line D as those for unstable atmospheres. Lapse rates A, B or C are also called *superadiabatic* lapse rates. All

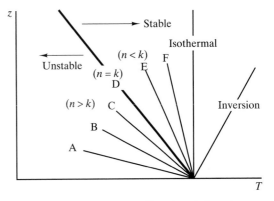

**Figure 9-14** Temperature profiles that define the atmospheric stability classes

lapse rates lying to the right of curve D are defined as stable. A vertical line corresponds to a unique isothermal atmosphere, and lapse rates to the right of the vertical correspond to atmospheres in which the temperature rises with altitude, $dT/dz > 0$. An atmosphere in which the temperature increases with altitude ($z$) is called an *inversion*. Thus

*Unstable atmosphere:* categories A, B, and C

*Neutral atmosphere:*   category D                    (9.59)

*Stable atmosphere:*    categories E, F, isothermal, and inversion

Since the temperature gradients of both stable and unstable atmospheres are negative, it is useful to introduce the concept of *potential temperature ($\Theta$) and the potential temperature gradient ($d\Theta/dz$)*. The potential temperature ($\Theta$) is the hypothetical temperature one would achieve if air at an actual (absolute) temperature and pressure ($T$, $P$) is compressed in an isentropic fashion to the ground-level pressure $P_0$.

$$\Theta = T\left(\frac{P_0}{P}\right)^{(k-1)/k} \qquad (9.60)$$

Taking the logarithm of both sides and differentiating with respect to altitude $z$, one obtains

$$\frac{1}{\Theta}\frac{d\Theta}{dz} = \frac{1}{T}\frac{dT}{dz} - \left(\frac{k-1}{k}\right)\frac{1}{P}\frac{dP}{dz} \quad (9.61)$$

Assuming that air is an ideal gas, $P = \rho R T$, the term $(k-1)/k$ can be shown to be equal to $R/c_p$. From Equation 9.26 and 9.37, the last term in Equation 9.61 can be expressed as

$$\left(\frac{k-1}{k}\right)\frac{1}{P}\frac{dP}{dz} = -\frac{g}{Tc_p} \qquad (9.62)$$

From Equation 9.36, the dry adiabatic lapse rate $\Gamma = -(dT/dz)_{\text{dry adiabatic}} = g/c_p = 9.8$ K/km and Equation 9.61 can be rearranged as

$$\begin{aligned}\frac{d\Theta}{dz} &= \frac{\Theta}{T}\left[\left(\frac{dT}{dz}\right)_{\text{actual}} + \Gamma\right] \\ &= \frac{\Theta}{T}\left[\left(\frac{dT}{dz}\right)_{\text{actual}} + 9.8 \text{ K/km}\right]\end{aligned} \quad (9.63)$$

Since the term $\Theta/T$ is always positive, it can be seen that $d\Theta/dz$ is positive for a stable atmosphere (categories E, F, isothermal, and inversion layers), negative for an unstable atmosphere (categories A, B, and C), and zero for a neutral atmosphere (category D).

*Stable atmosphere:*

$$\frac{d\Theta}{dz} = \left[\left(\frac{dT}{dz}\right)_{\text{actual}} + 9.8 \text{ K/km}\right] > 0$$

*Neutral atmosphere:*

$$\frac{d\Theta}{dz} = \left[\left(\frac{dT}{dz}\right)_{\text{actual}} + 9.8 \text{ K/km}\right] = 0 \quad (9.64)$$

*Unstable atmosphere:*

$$\frac{d\Theta}{dz} = \left[\left(\frac{dT}{dz}\right)_{\text{actual}} + 9.8 \text{ K/km}\right] < 0$$

Another useful parameter to differentiate stable from unstable atmospheres is the *Richardson number* (Ri):

$$\text{Ri} = \frac{g}{\Theta}\frac{d\Theta/dz}{(dU/dz)^2} \qquad (9.65)$$

The Richardson number is proportional to the rate of consumption of turbulent energy by buoyant forces divided by rate of production of turbulent energy by wind shear. Since the square of $dU/dz$ is always positive, the sign of the Richardson number is governed by the potential temperature gradient. Thus

*Unstable atmosphere:*

   (categories A, B, and C):   Ri $< 0$

*Neutral atmosphere:*

   (category D):          Ri $= 0$     (9.66)

*Stable atmosphere:*

   (categories E and F):      Ri $> 0$

Knowing how to identify the atmospheric stability category is important in modeling plumes. If sufficient meteorological data are available, users can compute the temperature gradient and choose the stability category with precision. In the absence of detailed data on atmospheric temperature and velocity variations with altitude, the stability category can be determined by a simple method based on inexpensive observations. The most widely used method was developed by Pasquill and modified later by Turner based on five classes of wind speed, three classes of daytime solar radiation, and two classes of nightime cloudiness. Table 9-1 was developed many years ago and continues to be in wide use today. The stability categories A to F are also known as *Pasquill stability categories.*

It should be emphasized that this method is valid in *dead weather* (i.e., when no significant meteorological changes are in progress). Thus Table 9-1 should be used with caution when frontal weather changes occur or at sun-up, when the nocturnal lapse rate gives way to the morning solar radiation.

## 9.5  *Wind-speed profile*

The wind speed in the planetary boundary layer increases with altitude. Figure 9-15 typifies what can be expected. It is seen that the profiles depend on solar radiation (i.e., day or night). The boundary layer profile can be approximated by the power-law formulation (Hanna et al., 1982)

$$U(z) = U_{10}\left(\frac{z}{10}\right)^p \quad \text{for } z < 200 \text{ m} \quad (9.67)$$

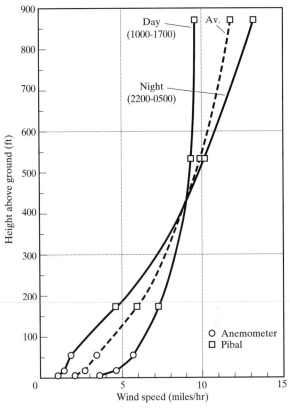

**Figure 9-15**  Planetary boundary layer velocity profiles. (Slade, 1968)

**Table 9-1**    Meteorological Conditions Defining Pasquill Stability Categories[a]

| ($U_{10}$) Surface wind speed at 10 m | Day, incoming solar radiation | | | Night, cloud cover, thickly overcast | |
|---|---|---|---|---|---|
| Class[b]: | Strong (1) | Moderate (2) | Slight (3) | $> \frac{1}{2}$ Low clouds (4) | $< \frac{3}{8}$ Clouds (5) |
| <2 (m/s) | A | A–B | B | Strongly stable | |
| 2–3 (m/s) | A–B | B | C | E | F |
| 3–5 (m/s) | B | B–C | D | D | E |
| 5–6 (m/s) | C | C–D | D | D | D |
| >6 (m/s) | C | D | D | D | D |

*Source:* Abstracted from Hanna et al. (1982).

[a] The neutral category, D, should be assumed for overcast conditions during the day or night. Category A is the most unstable and category F is the most stable, with category B moderately unstable and category E slightly stable.

[b] Class 1: clear skies, solar altitude greater than 60 degrees above the horizontal, typical of a sunny summer afternoon, very convective atmosphere; class 2: summer day with a few broken clouds; class 3: typical of a sunny fall afternoon, summer day with broken low clouds, or summer day with clear skies and solar altitude from only from only 15 to 35° above the horizontal; class 4: can be used for a winter day.

where $z$ is in meters, $U_{10}$ is the observed wind speed at 10 m, and the value of $p$ is selected from Table 9-2. The power-law expression is used by the EPA for several of its plume dispersion models, but it should not be used for altitudes above 200 m. For altitudes

---

**Table 9-2** Power-Law Exponents for Six Atmospheric Stability Categories

| Location | A | B | C | D | E | F |
|---|---|---|---|---|---|---|
| Urban ($p$) | 0.15 | 0.15 | 0.20 | 0.25 | 0.30 | 0.30 |
| Rural ($p$) | 0.07 | 0.07 | 0.10 | 0.15 | 0.35 | 0.55 |

*Source:* Abstracted from Turner (1994).

---

above 200 m, Hanna et al. (1982) recommend assuming a constant velocity,

$$U(z > 200\,\text{m}) = U_{200} \qquad (9.68)$$

There are occasions when engineers need to know the wind-speed profile close to the earth's surface. In these cases a logarithmic wind-speed profile law can be used:

$$U(z) = \frac{U^*}{k} \ln \frac{z}{z_0} \qquad (9.69)$$

where $k$ is the *von Kármán constant*, equal to 0.4, $U^*$ is the *friction velocity* (shear stress divided by the

density), and $z_0$ is a characteristic roughness height for the terrain over which the air flows.

$$U^* = \left( \frac{\tau}{\rho} \right)^{1/2} \qquad (9.70)$$

Table 9-3 defines values of $z_0$ and $U^*$ for several different land surfaces.

---

**Table 9-3** Typical Values of Parameters in the Logarithmic Wind-Speed Profile near Earth's Surface

| Type of surface | $z_0$ (cm) | $U^*$ (m/s) |
|---|---|---|
| Smooth mud flat; ice | 0.001 | 0.16 |
| Smooth snow | 0.005 | 0.17 |
| Smooth sea | 0.020 | 0.21 |
| Level desert | 0.030 | 0.22 |
| Snow surface; lawn to 1 cm high | 0.100 | 0.27 |
| Lawn grass, to 5 cm | 1–2 | 0.43 |
| Grass, to 60 cm | 4–9 | 0.60 |
| Fully grown root crops | 14 | 1.75 |

*Source:* Slade (1968).

---

When it is not possible to measure the wind speed, people will find it helpful to use the Beaufort scale of wind-speed equivalents (Table 9-4). The Beaufort scale relates the wind speed to common observable features of the wind. The Beaufort scale is very old but is recognized by the EPA as a useful way to specify that the wind speed is within a certain range of

---

**Table 9-4** Beaufort Scale of Wind-Speed Equivalents

| Description | Specifications | Wind speed at 10 m (mph) |
|---|---|---|
| Calm | Smoke rises vertically | Under 1 |
| | Direction of wind shown by smoke drift but not wind vanes | 1–3 |
| Light | Wind felt on face, leaves rustle, ordinary vane moved by wind | 4–7 |
| Gentle | Leaves and small twigs in constant motion, wind extends light flag | 8–12 |
| Moderate | Raises dust and loose paper, small branches are moved | 13–18 |
| Fresh | Small trees in leaf begin to sway, crested wavelets form on inland waters | 19–24 |
| | Large branches in motion, whistling heard in telegraph wires, umbrellas used with difficulty | 25–31 |
| Strong | Whole trees in motion, inconvenience felt in walking against wind | 32–38 |
| | Breaks twigs off trees, generally impedes progress | 39–46 |
| Gale | Slight structural damage occurs (chimney pots and slate removed) | 47–54 |
| | Trees uprooted, considerable structural damage occurs | 55–63 |
| Whole gale | Rarely experienced, accompanied by widespread damage | 64–75 |
| Hurricane | | Above 75 |

values [e.g., leaves that rustle (4 to 7 mph) as compared to leaves in constant motion (8 to 12 mph). Table 9-5 is a similar chart that relates indoor air speed and various detectable physiological responses.

## 9.6 *Mixing height*

For purposes of modeling the transport of pollutants for obtaining permits and prevention of significant deterioration (PSD) studies (Chapter 3), it is useful to define the *morning mixing height (depth)* $(z_{AM})$

and the *afternoon mixing height (depth)* $(z_{PM})$. The mixing height is the average thickness of the layer within which one can expect pollutants to mix for a particular geographic region over the course of a year. Mixing heights are not universal quantities and vary with details of the geographic region in question (i.e., proximity of bodies of water, terrain, annual rainfall, temperature, radiation intensity, etc.). Nonetheless mixing heights are frequently called for in EPA dispersion models and users will need to consult the annual meteorological data for particular

**Table 9-5**  Physiological Response to Moving Indoor Air[a]

| Response | Airspeed (ft/min) |
|---|---|
| Observable settling velocity of skin particles | 3 |
| Random indoor air movement, air at lower speed is considered "stale" | 20 |
| Exposed neck and ankles begin to sense air movement, upper limit of "comfort" at acceptable room temperatures | 50 |
| Skin on hands begins to sense air movement, moisture on skin increases sensitivity | 100 |
| Typical walking speed, eddy velocity of a person walking at a brisk speed | 250 |
| 8 h, upper comfort level of blown air for cooling, average outdoor air speed | 700 |
| 30 min upper comfort level of blown air for cooling, typical fan exit speed for free-standing fans | 1800 |
| 10 min upper comfort level of blown air for cooling | 3500 |

[a] In lieu of experimental measurements, the responses described above can be used to estimate the speed of moving indoor air.

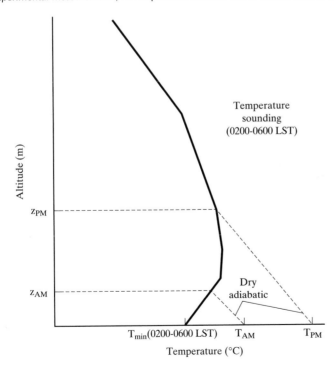

**Figure 9-16**  Computation of morning $(z_{AM})$ and afternoon $(z_{PM})$ mixing heights

locations of interest. Figure 9-16 shows how to calculate the morning $(z_{AM})$ and afternoon $(z_{PM})$ mixing heights from a graph of altitude $(z)$ versus temperature $(T)$, by the following steps.

1. For the locality and time of year in question, draw a representative atmospheric temperature profile obtained during the period 0200–0600 local standard time (LST; solid line in Figure 9-16).
2. Determine a representative morning ground-level temperature $(T_{AM})$ based on the lowest ground-level temperature during the period 0200–0600 local standard time.

$$T_{AM} = T_{min}(\text{during 0200–0600 local time}) + 5°C \tag{9.71}$$

3. Determine a representative afternoon temperature $(T_{PM})$ based on the highest ground-level temperature during the period 1200–1600 local standard time.

$$T_{PM} = T_{max}(\text{1200–1600 local time}) \tag{9.72}$$

4. From the points $T_{AM}$ and $T_{PM}$, draw lines corresponding to the dry adiabatic lapse rate.

The heights $z_{AM}$ and $z_{PM}$ are defined by the intersections of the lines created in step 4 and the actual atmospheric temperature profile defined in step 1.

Morning mixing heights $(z_{AM})$ typically vary between 300 and 900 m. Afternoon mixing heights $(z_{PM})$ typically vary between 600 and 4000 m in the summer and 600 and 1400 m during the winter. Due to the ameliorating effect of water, $T_{max}$ and $T_{min}$ tend to be closer in value for regions near bodies of water than for dry regions. Furthermore, regions near bodies of water have high relative humidity and have morning inversions that are weaker than dry regions, $(dT/dz)_{moist} < (dT/dz)_{dry}$. Thus:

- Morning mixing heights:

  $z_{AM}(\text{regions near water}) \approx 900$ m

  $z_{AM}(\text{regions on high dry plateaus}) \approx 300$ m

- Afternoon mixing heights:

  $z_{PM}(\text{regions near water}) \approx 600$ m

  $z_{PM}(\text{regions on high dry plateaus})$

  $\approx 1400$ m (winter), 4000 m (summer)

## 9.7 Actual atmospheric temperature variations

From day to day the atmospheric temperature gradient is affected by a variety of factors. Radiation from the sun to earth during the day and from the earth to space at night have a pronounced effect. Frontal movements of warm and cold air masses also affect the lapse rate. Of all the variations that are encountered, inversions warrant particular attention because they produce conditions that impede the dispersion of pollutants and can cause unusually large ground-level pollutant concentrations.

An *inversion* is the occurrence when the temperature increases with altitude. Inversions produce very stable air and are accompanied by very little vertical mixing because large buoyant forces restore air parcels to their original altitude. There are several types of inversions with which engineers should be familiar.

1. *Subsidence* is the name given to the condition when a huge mass of atmospheric air "sinks" (i.e., its altitude decreases rapidly). As the mass of air falls, its temperature rises due to the compression. Since the mass of air is huge, it does not have time to cool and the increased temperatures remain. Thus there will be a layer of warm air laying above cooler air.

2. *Frontal inversions* occur when, owing to its greater density, cool air slides under warm air and produces a layer of cool air below warm air. In a similar fashion, a moving warm front tends to override cooler air. In both cases an elevated inversion layer is produced. Elevated inversions may produce serious air pollution episodes. Pollutants produced below an elevated inversion layer cannot rise through it. On the other hand, pollutants issuing from tall stacks whose exits are above the inversion layer cannot travel downward and are thus lofted upward.

3. *Radiation inversions*, also known as *diurnal inversions*, are temperature variations driven by radiation heat transfer. A summary of low-altitude temperature profiles at various times during a 24-h period is shown in Figure 9-17. In the middle of the afternoon (1600) when the sun has heated the earth, the air just above the earth is warm but cools rapidly with altitude. The lapse rate is very large, the air is very unstable and there is a great deal of vertical mixing. As evening approaches (1800), the earth cools and cools the air

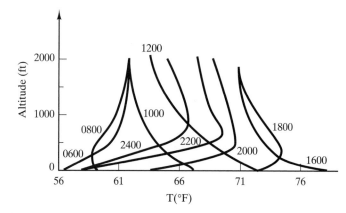

**Figure 9-17** Diurnal temperature variations produced by radiation heating and cooling. Such heat transfer produces a variety of temperature gradients in the surface boundary layer at different times throughout a 24hour period. (adapted from Slade, 1968)

above it. There will even be a time near sundown when a thin layer of air is isothermal and the air is very stable. As darkness approaches (2000) and passes into early morning (2400), the earth radiates to space and becomes cool (as home gardeners well know when their tomatoes freeze). By sun-up (0600), low earth temperatures produce a strong, thick, very stable ground-based inversion layer. As the sun heats the earth (0800), atmospheric stability decreases and ultimately becomes unstable in the heat of the day. The daily swings in temperature associated with diurnal temperature variations have a profound effect on low-altitude events such as open burning, domestic wood-burning stoves, industrial discharges, and unstable flying conditions for small aircraft.

## 9.8 Appearance of plumes

A process gas stream leaving an exhaust stack rises because of its momentum and buoyancy. The buoyant force is

$$
\begin{aligned}
F_{\text{buoyant}} &= Vg(\rho_{\text{atmosphere}} - \rho_{\text{air parcel}}) \\
&= (m_{\text{air parcel}})(a_{\text{buoyant}}) \\
&= Vg(\rho_a - \rho_s) = V\rho a_b
\end{aligned} \tag{9.73}
$$

which produces an bouyant acceleration $(a_b)$ given by

$$
a_b = \frac{g(\rho_a - \rho_s)}{\rho_s} \tag{9.74}
$$

where the subscript air parcel has been replaced by $s$ to relate it to the stack gas, and the subscript atmosphere has been shortened to $a$. Since the pressure of the plume gas and atmosphere are the same, the densities can be replaced by the reciprocals of the temperatures using the ideal gas law, $\rho = P/RT$, and the upward acceleration becomes

$$
a_b = \frac{g(T_s - T_a)}{T_a} \tag{9.75}
$$

Thus in judging the vigor with which plumes rise, readers should examine the temperature difference $(T_s - T_a)$ and the atmospheric stability. The plume temperature $(T_s)$ decreases as the plume rises and the slope $(dT_s/dz)$ is essentially the same as $(dT/dz)_{\text{dry adiabatic}}$ $(-9.8°C/km)$ if the plume is dry or less if the plume contains condensing water vapor.

Figure 9-18 shows pictorially several types of plumes. The names correspond to their appearances. The plumes are governed by the temperature difference $(T_s - T_a)$ and the atmospheric stability, as shown in Figure 9-19.

*Fanning* plumes (Figure 9-19a) occur in very stable air (categories E and F) or when there is a deep ground-based inversion. The plume rises rapidly at the stack outlet since the temperature difference $(T_s - T_a)$ there is large but $(T_s - T_a)$ decreases to zero very quickly with height. Once $T_s = T_a$, the plume is frozen at this altitude, because were it to depart from this altitude, the large temperature difference would produce a large force to restore the plume to its original altitude. Under these conditions vertical mixing is minimal and less than lateral mixing. Thus the plume acquires the appearance of a fan spreading laterally as it travels downwind. Fanning plumes are often observed at sun-up

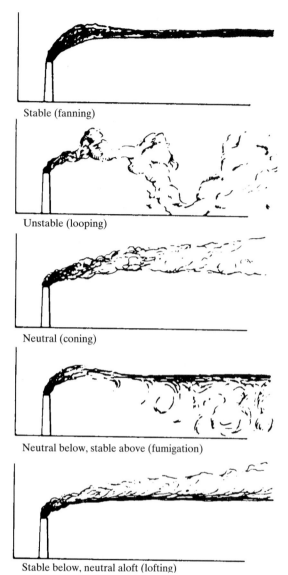

Stable (fanning)

Unstable (looping)

Neutral (coning)

Neutral below, stable above (fumigation)

Stable below, neutral aloft (lofting)

**Figure 9-18** Commonly encountered plumes (Slade, 1968)

when strong ground-based inversions are apt to occur.

*Looping* plumes (Figure 9-19b) occur in unstable air (category A, B, or C). Unstable air is characterized by vigorous vertical mixing. The temperature difference $(T_s - T_a)$ increases as the plume rises, which accentuates vertical mixing. As it rises the gases disperse and the plume does not persist very far downwind. Thus the plume moves about verti-

cally in a spastic fashion and the exhaust gases disperse rapidly. Looping plumes are often observed at the hottest part of the day when the lapse rate is superadiabatic.

*Coning* plumes (Figure 9-19c) occur in stable air (categories E and F) where there is little vertical mixing. The temperature difference $(T_s - T_a)$ is positive but it decreases with altitude and is equal to zero at some altitude. Once the plume rises and remains at this altitude, it is carried downwind by the prevailing wind. In a stable atmosphere vertical and lateral mixing are approximately the same. Thus as the plume travels downwind, it spreads vertically and laterally at nearly the same rate, which gives the plume the appearance of a cone. The plume retains its conical appearance for a considerable distance downwind of the stack. Coning plumes are often observed in the late morning.

Elevated inversions produce plumes of dramatic effect. The shape of the plume depends on the height of the discharge with respect to the base of the elevated inversion. Figure 9-19e depicts a variety of these effects by showing the temperature variations for plumes discharged at four different stack heights with respect to the base of the inversion layer. Plumes from a short stack $(h_{s,1})$ produce a coning plume at a low altitude, and its vertical movement is totally trapped by the inversion layer. If the plume drifted above the base of the inversion layer, it would encounter strong forces that would drive it downward. The elevated inversion layer traps pollutants below the inversion layer and can produce very high ground-level pollutant concentrations if the inversion layer is close to the ground and the wind speed is small. Plume location 1 and the corresponding elevated inversion produce conditions known as *fumigation* (Figure 9-19d), which was the cause of the Donora air pollution episode in 1948.

Plumes from stacks $h_{s,2}$, $h_{s,3}$, and $h_{s,4}$ reside above the base of the inversion layer and lead to desirable results. If the plumes were to depart from their equilibrium positions, strong forces would restore them to their equilibrium positions. For these plumes, the pollutants are trapped in or above the inversion layer, where they dissipate as the plume travels downwind. Ground-level pollutant concentrations will be low. If $T_s = T_a$ above the inversion

at $z_e$, $T_s = T_a$, $g_{b,e} = 0$ and plume ceases to rise

at $z_1 < z_e$, $T_s > T_a$, $g_{b,1} > 0$ and plume rises to $z_e$

at $z_2 > z_e$, $T_s < T_a$, $g_{b,2} < 0$ and plume falls to $z_e$

at $z_1$, $\Delta T_1 = (T_s - T_a) > 0$, $g_{b,1} > 0$
and the plume rises

at $z_2$, $\Delta T_2 = (T_s - T_a) > 0$, $g_{b,2} > 0$
and the plume rises

but $\Delta T_2 > \Delta T_1$ thus $g_{b,2} > g_{b,1}$ $dg_b/dz > 0$
buoyant force increases with altitude

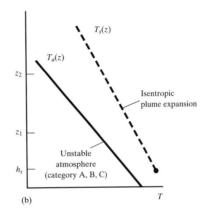

**Figure 9-19** (a) Fanning plume; horizontal mixing greatly exceeds vertical mixing and the plume travels downwind appearing (from above or below) like a fan. (b) Looping plume; considerable vigorous vertical mixing causing the plume to behave in an erratic fashion. (c) Coning plume; vertical and horizontal mixing are comparable and the plume travels downwind possessing a conical shape. (d) Fumigation; elevated inversion acts as a lid preventing upward dispersion and trapping pollutants below $z_e$, mild buoyant forces for $z < z_e$ produces modest vertical mixing and even allows downward mixing. (e) Virtue of building tall stacks to trap plumes within or above an elevated inversion, unfortunately economic conditions limit stack height and if elevated inversion is too high a plume may be trapped below the inversion.

(1) $h_{s,1}$: poor situation, plume trapped below the inversion layer, buoyancy restores plume to $z_1$, if the lapse rate is neutral below $z_1$, fumigation is possible,

(2) $h_{s,2}$: moderately good situation, fumigation might occur since plume is not likely to rise above $z_3$, but nearly neutral atmospheric conditions below $z_3$ may allow the plume to move downward,

(3) $h_{s,3}$: good situation, a fanning plume forms at $z_4$ and the inversion layer traps the plume near the top of the inversion,

(4) $h_{s,4}$: excellent condition, plume rises to an altitude $z_5$ above the top of the inversion layer, such a situation is called *lofting*.

at $z_e$, $\Delta T_e = (T_s - T_a) = 0$, $g_{b,e} = 0$
and plume spreads lateraly and vertically

at $z_2$, $\Delta T_2 = (T_s - T_a) < 0$, $g_{b,2} < 0$
and plume falls

at $z_1$, $\Delta T_1 = (T_s - T_a) > 0$, $g_{b,1} > 0$
and plume rises but $d\Delta T/dz$ = small
negative value such that $dg_b/dz$ is
small, horizontal and vertical
mixing are comparable

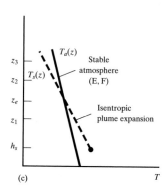

(c)

at $z_e$, $\Delta T_e = (T_s - T_a) = 0$, $g_{b,e} = 0$
and plume neither rises nor falls

at $z_3$, $\Delta T_3 = (T_s - T_a) < 0$, $g_{b,3} < 0$
and plume falls

at $z_2$, $\Delta T_2 = (T_s - T_a) > 0$, $g_{b,2} > 0$
and plume rises toward $z_e$

at $z_1$, $\Delta T_1 = (T_s - T_a)_1 \approx \Delta T_2 > 0$
thus $g_{b,1} \approx g_{b,2} \approx$ small and positive
thus plume rises slowly to $z_e$. Plume
can not pass beyond $z_e$ but may
diffuse downward owing
to small $g_b$ for $z < z_e$

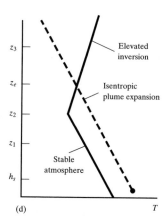

(d)

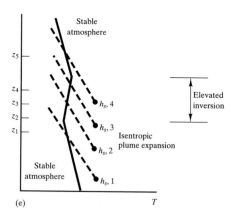

(e)

layer, the plume cannot travel downward. Such a plume is called a *lofting* plume.

The reader should be aware that the influence of buoyancy and the temperature difference $(T_s - T_a)$ that produces a buoyant acceleration $a_b$ is not the whole story. The analysis implicit in Figure 9-19 is based solely on the basis of the temperature difference. For a few moments after entering the atmosphere, the plume may behave in an adiabatic fashion. Within tens of seconds, however, heat transfer and mixing bring about an equality in temperature between stack gases and atmosphere. The appearance of a plume also is strongly influenced by the momentum with which it leaves the stack and the stability of the atmosphere it enters.

### EXAMPLE 9.5   PLUMES FROM WOOD-BURNING STOVES EARLY IN THE MORNING

At sun-up the air is motionless and there is a ground-based inversion, $dT/dz = 0.1$ C°/m, to an altitude of 120 m. Above 120 m the lapse rate is equal to the normal lapse rate. The ground-level air temperature is 10°C (283 K). Smoke (assume properties of dry air) from a poorly attended wood-burning stove leaves a chimney $(h_s = 5\text{ m})$ at a temperature $(T_s)$ of 20°C (293 K) and a negligible velocity. The smoke rises as dry adiabatic air (−9.8°C/km). Estimate the height to which you'd expect the smoke to rise.

*Solution*   The temperature of the smoke and atmosphere can be expressed as functions of altitude $(z)$:

$$T(z)_{\text{smoke}} = 293 - 0.0098(z - 5)$$

$$T(z)_{\text{atm}} = 283 + 0.1z$$

In the absence of significant air motion, the plume will rise to an altitude $(z_e)$ such that

$$T_{\text{smoke}}(z_e) = T_{\text{atm}}(z_e)$$

$$293 - 0.0098(z_e - 5) = 283 + 0.1z_e$$

$$293 - 283 + 0.049 = z_e(0.1 + 0.0098)$$

$$z_e = 91.5\text{ m}$$

The calculation above can be improved by including the condensation of water vapor that will be present in the smoke.

## 9.9  Gaussian plume model

The transport of contaminants downwind of their point of discharge is of great importance to engineers charged with the responsibility of ensuring that a process is in compliance with air pollution regulations. The pollutants may be gases or vapors or they may be particle matter of such small size that gravimetric settling can be neglected as a first approximation. The subject will be introduced in a simple direct manner to enable the ready grasp of the physical processes involved (Turner, 1979). With these fundamental principles readers will be able to make prudent estimates of ground-level concentrations using concepts that are recognized by professionals in the field. These fundamental concepts will enable engineers to acquire physical intuition so if they study more sophisticated presentations on the subject they will have a basis for examining the validity of predictions. The subject of *atmospheric diffusion or atmospheric dispersion* are serious subjects of study in the fields of meteorology, atmospheric physics, and turbulent fluid mechanics. Following the elementary study that follows, students are encouraged to consult any one of several outstanding texts on the subject (Seinfeld, 1986).

Consider a buoyant pollutant stream exiting a stack of geometric height $h_s$. Such a pollutant stream is called a *point source* and is depicted in Figure 9-20. The plume rises a distance, $\delta h$, called the *plume rise*, before buoyancy and its upward momentum cease. The stack gas is then transported downwind by the prevailing wind. While moving downwind, pollutants are dispersed vertically and horizontally. Dispersion in the vertical direction is governed largely by atmospheric stability. Dispersion in the horizontal plane is governed largely by molecular and eddy diffusion. The x-axis will always be oriented in the direction of the prevailing wind, the z-axis is vertically upward, and the y-axis is transverse to the wind. An *Eulerian* approach will be adopted in which properties of the plume will be described in terms of a coordinate system fixed with respect to the earth.

Define an elementary volume in the plume. Bulk motion and diffusion (or dispersion) transport pollutants into the upwind side of the volume and out the downwind side. Diffusion transports pollutants in the vertical and lateral directions as well. A source term

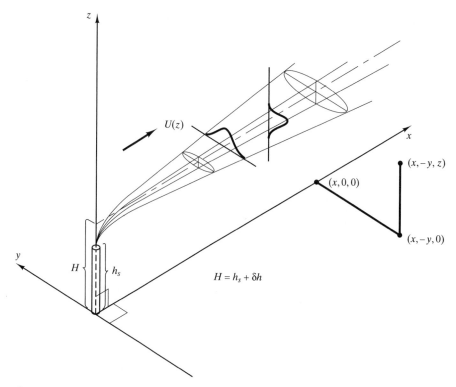

**Figure 9-20** Gaussian plume from an elevated source, effective stack height (H) is equal to the geometric stack height ($h_s$) plus the plume rise ($\delta h$) (adapted from Slade, 1968)

$(\dot{S}_s, \text{kg/s m}^3)$ will be included inside the control volume in the event that chemical reactions create (or consume) pollutants. A mass balance for the elementary volume yields

$$\frac{\partial c}{\partial t} = -\frac{\partial(cU)}{\partial x} + \dot{S}_s + \frac{\partial\left[\partial(cD_x)/\partial x\right]}{\partial x}$$

$$+ \frac{\partial\left[\partial(cD_y)/\partial y\right]}{\partial y} + \frac{\partial\left[\partial(cD_z)/\partial z\right]}{\partial z} \quad (9.76)$$

where $D_x$, $D_y$, and $D_z$ are the effective diffusion coefficients that incorporate both molecular diffusion and eddy diffusion resulting from atmospheric turbulence. Assume the following:

- Steady state, $\partial c/\partial t = 0$.
- Transport by bulk motion in the $x$-direction exceeds diffusion in the $x$-direction.
- Effective diffusion coefficients are constant.
- Wind speed in the $x$-direction does not vary with $x$.
- The plume species do not react (i.e., neglect the source term, $\dot{S}_s$).

Equation 9.76 becomes

$$U\frac{\partial c}{\partial x} = D_y\frac{\partial^2 c}{\partial y^2} + D_z\frac{\partial^2 c}{\partial z^2} \quad (9.77)$$

The above is solved subject to the following boundary conditions:

1. $c \to \infty$ as $x \to 0$. Concentration approaches infinity as the distance to the point source approaches zero (the infinite boundary condition is mathematical necessity, not a physical reality, when a finite source is reduced to a "point").
2. $c \to 0$ as $x, y, z \to \infty$. Concentration approaches zero as the distance from a point source approaches infinity.
3. $D_z\, \partial c/\partial z \to 0$ as $z \to 0$. There is no diffusion at ground level.
4. $\int_0^\infty \int_{-\infty}^\infty Uc(c, y, z)\, dy\, dz = \dot{m}_{i,s}$ at $x > 0$. Mass of pollutant transported downwind is constant and equal to what is emitted by the source.

In its general form Equation 9.77 is called *Gauss's equation* and its solution is called the *Gaussian func-*

*tion.* Since in this case it is used to describe the behavior of plumes, the resulting analytical description is called the *Gaussian plume model*. The solution to Equation 9.77 is

$$c_i(x, y, z) = \left( \frac{\dot{m}_{i,s}}{2\pi x (D_y D_z)^{1/2}} \right)$$
$$\exp\left( -\frac{U}{4x} \right)\left( \frac{y_2}{D_y} + \frac{z^2}{D_z} \right) \quad (9.78)$$

where $\dot{m}_{i,s}$ is called the *pollutant emission rate* and is equal to the mass flow rate (kg/s) at which molecular species $i$ is discharged from the stack. Note that $\dot{m}_{i,s}$ is the pollutant mass flow rate, which is not the same as the total mass flow rate of gas exiting the stack:

$$\dot{m}_{i,s} = c_i(\text{stack exit})Q_s \quad (9.79)$$

where $Q_s$ is the total actual volumetric flow rate of gas at the stack exit and $c_i$ is the actual pollutant concentration in the stack gas computed at the stack temperature and pressure $(T_s \text{ and } P_s)$. It is convenient to rearrange as follows:

$$c_i(x, y, z) = \frac{\dot{m}_{i,s}}{\pi U \sigma_y \sigma_z}$$
$$\exp\left\{ -\frac{1}{2}\left( \frac{y}{\sigma_y} \right)^2 - \frac{1}{2}\left( \frac{z}{\sigma_z} \right)^2 \right\} \quad (9.80)$$

where

$$\sigma_y^2 = \frac{2xD_y}{U} \qquad \sigma_z^2 = \frac{2xD_z}{U} \quad (9.81)$$

$\sigma_y$ and $\sigma_z$ are called *dispersion coefficients* in the transverse $(y)$ and vertical $(z)$ directions. It should be noted that the dispersion coefficients are not the same as the turbulent diffusion coefficients, owing to the inclusion of the downwind distance $x$ and the wind speed in Equation 9.81. Thus the dispersion coefficients increase with the distance $x$ as well as depending on the unstable nature of the atmosphere.

Values of dispersion coefficients obtained from experiments on air movement over flat rural terrain are shown in Figures 9-21 and 9-22 as functions of downwind distance $(x)$ and stability category (Turner, 1994). Because the accuracy of data taken from log-log graphs is poor, empirical expressions have been derived which lend themselves to numerical computation. These expressions may take the form

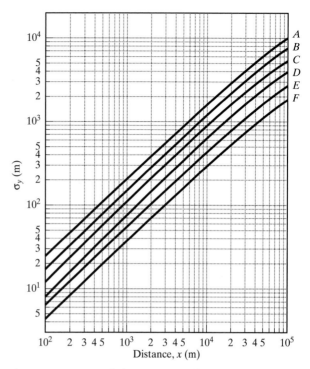

**Figure 9-21**  Lateral dispersion coefficient over flat rural terrain ($\sigma_y$) versus downwind distance (x) for different stability categories (Slade, 1968)

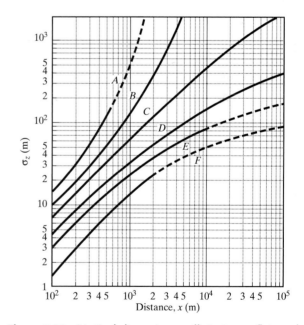

**Figure 9-22**  Vertical dispersion coefficient over flat rural terrain ($\sigma_z$) versus downwind distance (x) for different stability categories (Slade 1968)

**Table 9-6** Pasquill–Gifford Horizontal Dispersion Coefficient $(\sigma_y)$ for Rural Sites

$\sigma_y$ (m) = (1000/2.15)x tan($T$), where $x$ is the downwind distance in km and $T$, expressed in degrees, comes from the table below.

| Stability | Equation for $T$ (degrees) and $x$ (km) |
|---|---|
| A | $T = 24.167 - 2.5334 \ln(x)$ |
| B | $T = 18.333 - 1.8096 \ln(x)$ |
| C | $T = 12.5 - 1.0857 \ln(x)$ |
| D | $T = 8.3333 - 0.72382 \ln(x)$ |
| E | $T = 6.25 - 0.54287 \ln(x)$ |
| F | $T = 4.1667 - 0.36191 \ln(x)$ |

*Source:* Turner (1994).

$$\sigma_y = ax(1 + bx)^c \qquad (9.82)$$

$$\sigma_z = dx(1 + ex)^f$$

where a, b, c, d, e and f are constants and sometimes egual zero or some other function of $x$ and coefficients that account for the Pasquill stability categories. Shown in Tables 9-6 and 9-7 are values of $\sigma_y$ and $\sigma_z$, called the *Pasquill–Gifford coefficients*, suitable for rural sites. Shown in Table 9-8 are empirical expressions for $\sigma_y$ and $\sigma_z$ suitable for urban sites. It can be shown that for any stability category and downwind distance $(x)$,

$$\sigma_y(\text{urban}) > \sigma_y(\text{rural})$$

$$\sigma_z(\text{urban}) > \sigma_z(\text{rural}) \qquad (9.83)$$

**Table 9-7** Pasquill–Gifford Vertical Dispersion Coefficient $(\sigma_z)$ for Rural Sites
$^a$ $\sigma_z$ (m) $= \alpha x^\beta$, where $x$ in km.

| Stability | x (km) | α | β | $\sigma_z$ (m) at upper boundary[a] | Stability | x (km) | α | β | $\sigma_z$ (m) at upper boundary[a] |
|---|---|---|---|---|---|---|---|---|---|
| A | >3.11 | | | 5000 | E | >40.0 | 47.618 | 0.29592 | |
| | 0.5–3.11 | 453.85 | 2.1166 | | | 20.0–40.0 | 35.420 | 0.37615 | 141.9 |
| | 0.4–0.5 | 346.75 | 1.7283 | 104.7 | | 10.0–20.0 | 26.970 | 0.46713 | 109.3 |
| | 0.3–0.4 | 258.89 | 1.4094 | 71.2 | | 4.0–10.0 | 24.703 | 0.50527 | 79.1 |
| | 0.25–0.3 | 217.41 | 1.2644 | 47.4 | | 2.0–4.0 | 22.534 | 0.57154 | 49.8 |
| | 0.2–0.25 | 179.52 | 1.1262 | 37.7 | | 1.0–2.0 | 21.628 | 0.63077 | 33.5 |
| | 0.15–0.2 | 170.22 | 1.0932 | 29.3 | | 0.3–1.0 | 21.628 | 0.75660 | 21.6 |
| | 0.1–0.15 | 158.08 | 1.0542 | 21.4 | | 0.1–0.3 | 23.331 | 0.81956 | 8.7 |
| | <0.1 | 122.8 | 0.9447 | 14.0 | | <0.1 | 24.260 | 0.83660 | 3.5 |
| B | >35.0 | | | 5000 | F | >60.0 | 34.219 | 0.21716 | |
| | 0.4–35.0 | 109.30 | 1.0971 | | | 30.0–60.0 | 27.074 | 0.27436 | 83.3 |
| | 0.2–0.4 | 98.483 | 0.98332 | 40 | | 15.0–30.0 | 22.651 | 0.32681 | 68.8 |
| | <0.2 | 90.673 | 0.93198 | 20.2 | | 7.0–15.0 | 17.836 | 0.4150 | 54.9 |
| C | All x | 61.141 | 0.91465 | | | 3.0–7.0 | 16.187 | 0.4649 | 40.0 |
| D | >30.0 | 44.053 | 0.51179 | | | 2.0–3.0 | 14.823 | 0.54503 | 27.0 |
| | 10.0–30.0 | 36.650 | 0.56589 | 251.2 | | 1.0–2.0 | 13.953 | 0.63227 | 21.6 |
| | 3.0–10.0 | 33.504 | 0.60486 | 134.9 | | 0.7–1.0 | 13.953 | 0.68465 | 14.0 |
| | 1.0–3.0 | 32.093 | 0.64403 | 65.1 | | 0.2–0.7 | 14.457 | 0.78407 | 10.9 |
| | 0.3–1.0 | 32.093 | 0.81006 | 32.1 | | <0.2 | 15.209 | 0.81558 | 4.1 |
| | <0.3 | 34.459 | 0.86974 | 12.1 | | | | | |

*Source:* Turner (1994).

**Table 9-8**   Empirical Expressions for the Dispersion Coefficients for Urban Sites[a]

| Stability | $\sigma_y$ (m) | $\sigma_z$ (m) |
|---|---|---|
| A–B | $0.32x(1 + 0.0004x)^{-1/2}$ | $0.24x(1 + 0.0001x)^{1/2}$ |
| C | $0.22x(1 + 0.0004x)^{-1/2}$ | $0.20x$ |
| D | $0.16x(1 + 0.0004x)^{-1/2}$ | $0.14x(1 + 0.0003x)^{-1/2}$ |
| E–F | $0.11x(1 + 0.0004x)^{-1/2}$ | $0.08x(1 + 0.0015x)^{-1/2}$ |

*Source:* Griffiths (1994).
[a] Downwind distance $x$ measured in meters.

Plumes are generally emitted from stacks and Equation 9.80 must be modified to account for the fact that the source strength $\dot{m}_{i,s}$ is located at an elevated point called the *effective stack height* above the ground ($z = H$):

$$H \text{ (effective stack height)} = h_s + \delta h \quad (9.85)$$

where $h_s$ is the true or *geometric stack height* and $\delta h$ is called the *plume rise*. The plume rise is due to the momentum and buoyancy of the plume. The plume rise is the distance the plume rises before viscosity dissipates the plume's initial momentum and the plume temperature reaches equilibrium with the atmospheric temperature and buoyancy no longer exists. If the point source emits pollutants from a stack, an elevation H above the ground as show in Figure 9-20, the downwind concentration can be computed from:

$$c(x, y, z) = \frac{\dot{m}_{i,s}}{U\pi\sigma_y\sigma_z} \exp\left[-\frac{1}{2}\left(\frac{y}{\sigma_y}\right)^2 - \frac{1}{2}\left(\frac{z-H}{\sigma_z}\right)^2\right]$$

## 9.9.1   Ground-level concentrations from elevated point sources

Pollutants travel downwind and eventually reach the ground, where they may or may not be absorbed by soil, water, vegetation, and so on. Particles in plumes fall and collect on the ground. Pollutants such as $NO_x$, and $SO_2$ may be absorbed by the ground or water, whereas pollutants such as CO and VOCs are not absorbed and accumulate along the ground. The following material will be presented as three cases: (1) plumes containing particles, (2) plumes containing gaseous pollutants that are not absorbed (reflected) by the ground, and (3) plumes containing gaseous pollutants that are absorbed by the ground.

**Case 1: Plumes containing particles.** Equation 9.80 presumes that particles are absorbed by the ground. Large particles in plumes fall to earth with a velocity $v_t$ called the *settling velocity* as well as being carried downwind at the wind speed $U$. To account for the settling phenomenon, users may want to replace the constant effective stack height $H$ in Equation 9.85 by a value that decreases with the elapsed time and distance from the stack:

$$H(x) = H - (\delta t)(v_t) \quad (9.86)$$

Values of $v_t$ will be derived in Chapter 11; however, values of $v_t$ also can be obtained from Figure 11-19. The length of time ($\delta t$) a particle resides in a plume can be expressed as

$$\delta t = \frac{x}{U} \quad (9.87)$$

Thus a more accurate expression for the ground-level particle concentration from plumes can be obtained by replacing $H$ by $H(x)$ in Equations 9.86, where

$$H(x) = H - x\frac{v_t}{U} \quad (9.88)$$

For plumes containing particles,

$$c_p(x, y, z) = \frac{\dot{m}_p}{2U\pi\sigma_y\sigma_z}$$

$$\exp\left\{-\frac{1}{2}\left(\frac{y}{\sigma_y}\right)^2 - \frac{1}{2}\left[\frac{z - (H - v_t x/U)}{\sigma_z}\right]^2\right\} \quad (9.89)$$

where $v_t$ is the particle settling velocity. The ground-level concentration can be found from Equation 9.89 by setting $z = 0$. Since atmospheric diffusion causes the plume to spread in the lateral direction, it is also obvious that at any downwind distance $x$, the highest concentration always lies along the center-line of the plume, that is, to say where $y = 0$. Thus at any downwind distance $x$, the largest ground-level particle concentration from an elevated plume is given by the following.

$$c_p(x, 0, 0) = c_{p,GL}(x, 0, 0) = \frac{\dot{m}_p}{2U\pi\sigma_y\sigma_z}$$
$$\exp\left[-\frac{1}{2}\left(\frac{H - v_t x/U}{\sigma_z}\right)^2\right] \quad (9.90)$$

**Case 2: Plumes containing gaseous pollutants not absorbed (i.e., reflected) by the ground.** Many pollutants are not absorbed by the ground and accumulate along the ground. A mathematical way to account for the accumulation is to conceive of the superposition of the plume and its mirror image located a distance $z = -H$ below the ground, as shown in Figure 9-23. For plumes containing gaseous pollutants not absorbed (i.e., "reflected") by the ground,

$$c_i(x, y, z) = \frac{\dot{m}_{i,s}}{2U\pi\sigma_y\sigma_z}\left\{\exp\left[-\frac{1}{2}\left(\frac{y}{\sigma_y}\right)^2\right]\right\}$$
$$\left\{\exp\left[-\frac{1}{2}\left(\frac{z - H}{\sigma_z}\right)^2\right]\right.$$
$$\left. + \exp\left[-\frac{1}{2}\left(\frac{z + H}{\sigma_z}\right)^2\right]\right\} \quad (9.91)$$

At any downwind distance $x$, the largest ground-level concentration is found when $z = 0$ and $y = 0$. The ground-level pollutant concentrations for plumes containing gaseous pollutants not absorbed (i.e., "reflected") by the ground:

$$c_i(x, 0, 0) = c_{i,GL}(x, 0, 0)$$
$$= \frac{\dot{m}_{i,s}}{U\pi\sigma_y\sigma_z}\exp\left[-\frac{1}{2}\left(\frac{H}{\sigma_z}\right)^2\right] \quad (9.92)$$

**Case 3: Plumes containing gaseous pollutants absorbed by the ground.** For pollutants that are removed by the ground or water upon making contact, Equation 9.84 becomes

$$c_i(x, y, z) = \frac{\dot{m}_{i,s}}{2U\pi\sigma_y\sigma_z}$$
$$\exp\left\{-\frac{1}{2}\left[\left(\frac{y}{\sigma_y}\right)^2 + \left(\frac{z - H}{\sigma_z}\right)^2\right]\right\} \quad (9.93)$$

Once again, at any downwind distance $x$, the largest ground-level concentration is found when $z = 0$, and $y = 0$. Ground-level concentrations are given by

$$c_i(x, 0, 0) = c_{i,GL}(x, 0, 0) = \frac{\dot{m}_{i,s}}{2U\pi\sigma_y\sigma_z}$$
$$\exp\left[-\frac{1}{2}\left(\frac{H}{\sigma_z}\right)^2\right] \quad (9.94)$$

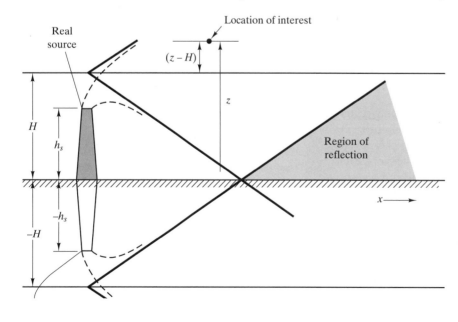

**Figure 9-23** Plume whose pollutants are not absorbed at ground-level. Introduce an imaginary source at $z = -H$ to simulate mathematically the accumulation of pollutants at ground-level (adapted from Wark and Warner, 1981)

## 9.9.2 Maximum ground-level pollutant concentrations from an elevated source

As the national ambient air quality standards (NAAQS; Chapter 3) concern the maximum allowable ground-level concentrations, engineers need to know the maximum ground-level concentration for a particular stack height ($H$) and emission rate ($\dot{m}_{i,s}$) and where this maximum value occurs. For a conservative estimate, assume the case in which pollutants are not absorbed (are reflected) by the ground or water. The location of the maximum ground-level concentration for gaseous pollutants is expressed as $x_{max}$; that is,

$$c_{i,GL}(x_{max}, 0, 0) \Rightarrow \text{the location at which } c_{i,GL}$$

$$\text{achieves its maximum value}\quad(9.95)$$

It is obvious that there is such a maximum value because at locations close to the stack, the plume is still overhead and the ground-level concentration is zero and at great distances from the stack the concentration approaches zero as the plume becomes diluted with ambient air. Since the concentration is a continuous function, it must reach a single maximum value somewhere between these two limits (Figure 9-24). One way to find the value and location of the maximum ground-level concentration is to substitute expressions for the dispersion coefficients as functions of $x$ (Equations 9.82 and 9.83), differentiate the concentration with respect to $x$, and set the derivative equal to zero. Since the empirical equations for the dispersion coefficients are not simple power laws, differentiation is a tedious process.

A shortcut can be achieved if it is assumed that the ratio of dispersion coefficients is a constant for any stability category:

$$\frac{\sigma_y}{\sigma_z} = \frac{ax^b}{f + cx^d} \approx \text{constant } C_1 \qquad (9.96)$$

Substituting the above into Equation 9.92 results in the following:

$$\frac{c_{i,GL}(x,0,0)U}{\dot{m}_{i,s}} = \frac{1}{\pi\sigma_y\sigma_z} \exp\left[-\frac{1}{2}\left(\frac{H}{\sigma_z}\right)^2\right]$$

$$= \frac{1}{\pi C_1 \sigma_z^2} \exp\left[-\frac{1}{2}\left(\frac{H}{\sigma_z}\right)^2\right] \quad (9.97)$$

Differentiate with respect to $\sigma_z$, set the derivative equal to zero, and obtain

$$\sigma_z = \frac{H}{\sqrt{2}} = \frac{H}{1.414} \qquad (9.98)$$

Substitute this value into Equation 9.92 for the ground-level concentration of unabsorbed gaseous pollutants:

$$\textit{Gaseous pollutants (reflection)}: \frac{c_{i,GL}U}{\dot{m}_{i,s}} = \frac{0.1171}{\sigma_y\sigma_z}$$
$$(9.99)$$

Equation 9.87 can be used for plumes containing particles, but the accuracy of the approximations leading to its derivation is poor and does not warrant a separate calculation for particle plumes. To determine the distance ($x_{max}$) at which this maximum ground-level

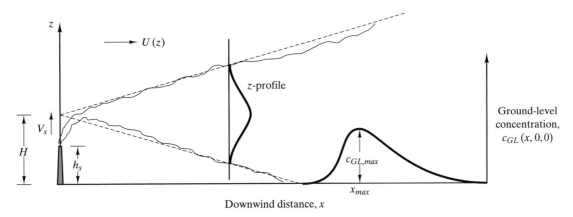

**Figure 9-24**   The ground-level pollutant concentration profile reaches a maximum value ($c_{GL, max}$) at a distance $x_{max}$ downwind of the stack

concentration occurs, use Equation 9.98 to determine $\sigma_z$ based on the effective stack height and then use Equation 9.82 or Figure 9-22 to determine the value of $x_{max}$.

Repeated application of Equation 9.92 for different downwind distances $(x)$, effective stack heights $(H)$, and Pasquill–Gifford stability categories pro-duces a series of graphs depicting ground-level pol-lutant concentrations $(c_{GL})$ versus $x$. If one examines only the location $(x_{max})$ and magnitude of the maxi-mum concentration $(c_{GL,\,max})$, it is possible to devise Figure 9-25. Users must be careful and use Figure 9-25 only for rural sites and gaseous pollutants that are not absorbed by the ground. Because the data are

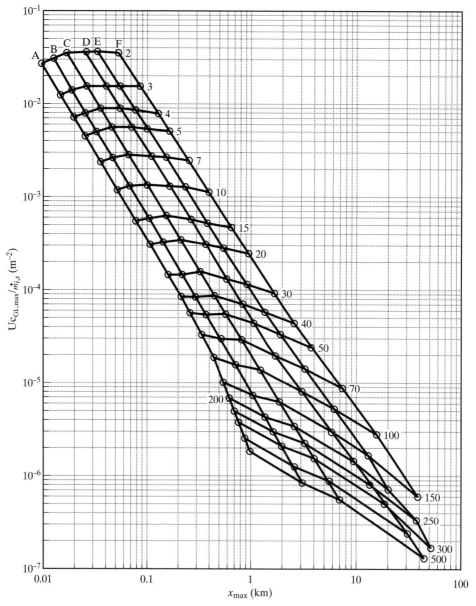

**Figure 9-25**  $(Uc_{GL,\,max}/\dot{m}_{i,\,s})$ versus $x_{max}$ for six stability classes and a range of effective stack heights (H = 2 to 500m) in rural sites assuming gaseous pollutants are not absorbed by the ground i.e. with reflection (Turner, 1994)

plotted in logarithmic fashion, the precision of data taken from the Figure 9-25 will be poor. Nonetheless, the figure can be used quickly and used as a check for more accurate calculations.

## EXAMPLE 9.6  GROUND LEVEL SO₂ CONCENTRATION DOWNWIND OF A STACK

Estimate the ground-level $SO_2$ concentration at points $(x)$ directly downwind of a 100-m stack $(h_s)$ on a coal-burning boiler that emits $SO_2$ at a rate predicted by the emission factors for an uncontrolled cyclone steam boiler. Assume an urban setting. Use Equation 9.92 with dispersion coefficients from Table 9-8 to compute the ground-level concentrations (ppb) for three atmospheric conditions. We call this the profile method.

- *Case 1*: cloudless, hot summer afternoon, strong radiation, lapse rate = 20 K/km, $U_{10}$ = 1.8 m/s
- *Case 2*: partially cloudy, windy afternoon, moderate radiation, lapse rate = 9.8 K/km, $U_{10}$ = 6.3 m/s
- *Case 3*: overcast evening, mild wind, lapse rate = 4 K/km, $U_{10}$ = 2.1 m/s

The coal consumption rate is 250 tons/day (5.79 lb$_m$/s) and the coal assay indicates 3% sulfur, HHV = 12,500 Btu/lb$_m$. Also, predict the location $(x, km)$ and magnitude (ppb) of the maximum ground-level $SO_2$ concentrations using Figure 9-25 alone and by using Equation 9.99 in conjunction with Figures 9.21 and 9.22. Note, however, that these two methods are appropriate only for rural sites. Compare these results with the maximum values predicted from the concentration profiles.

*Solution*  By a separate analysis to come (Example 9.7), it can be shown that the atmospheric stability and effective stack height $(H)$ are as follows. The wind speed at the effective stack height can be computed from Equation 9.67.

| Case | Stability | $H$ (m) | $U(H)$ (m/s) |
|------|-----------|---------|--------------|
| 1 | A | 518 | 2.37 |
| 2 | D | 178 | 9.70 |
| 3 | F | 179 | 10.26 |

The $SO_2$, emission factor for an uncontrolled cyclone furnace can be found from Table A-1.1 and is equal to 19.5 kg S₂/Mg coal where S is the sulfer concentration in coal written as a percent, (i.e. S=3). The emission rate of $SO_2$ noted earlier as $\dot{m}_{i,s}$ is

$$\dot{m}_{i,s} = (250 \text{ tons/day})(\text{day}/24\,\text{h})(\text{h}/3600\,\text{s})$$
$$(19.5\,\text{S kg } SO_2/\text{Mg coal})3(0.908\,\text{Mg/ton})$$
$$= 0.1537 \text{ kg/s}$$
$$= 153,700 \text{ mg/s}$$

Assuming that $SO_2$ is not absorbed by the ground, the ground-level concentration (in ppb) at points beneath the plume centerline can be found from Equation 9.92.

$$c_{GL}(x,0,0)(\text{ppm}) = \left[ \frac{382.8\dot{m}_{i,s}(\text{mg/s})_{SO_2}}{\pi \sigma_y \sigma_z U} \right]$$
$$\exp\left[ -\left(\frac{1}{2}\right)\left(\frac{H}{\sigma_z}\right)^2 \right]$$

where 382.8 converts the $SO_2$ concentration from mg/m³ into ppb.

- *Case 1: Stability category A. $H$ = 518 m, $U_{518}$ = 2.37 m/s. From Table 9-8,*

$$\sigma_z(\text{m}) = 0.32x(1 + 0.0004x)^{-1/2}$$
$$\sigma_z(\text{m}) = 0.24x(1 + 0.0001x)^{1/2}$$

- *Case 2: Stability category D. $H$ = 178 m, $U_{178}$ = 9.70 m/s. From Table 9-8;*

$$\sigma_y(\text{m}) = 0.16x(1 + 0.0004x)^{-1/2}$$
$$\sigma_z(\text{m}) = 0.14x(1 + 0.0003x)^{-1/2}$$

- *Case 3: Stability category F. $H$ = 179, $U_{179}$ = 10.26 m/s. From Table 9-8,*

$$\sigma_y(\text{m}) = 0.11x(1 + 0.0004x)^{-0.5}$$
$$\sigma_z(\text{m}) = 0.08x(1 + 0.0015x)^{-0.5}$$

For each stability category use the appropriate equation for $\sigma_y$ and $\sigma_z$, substitute values of $x$(m) into Equation 9.92 and compute $c_{GL}(x,0,0)$. Figure E9-6 shows the ground level $SO_2$ concentrations for these three stability classes. The Mathcad program producing Figure E9-6 can be found in the textbook Web site on the Prentice Hall Web page.

For a stable atmosphere, Figure E9-6c, the plume mixes slowly in the vertical direction; the maximum ground-level concentration is small and occurs several kilometers downwind of the stack. On the other hand,

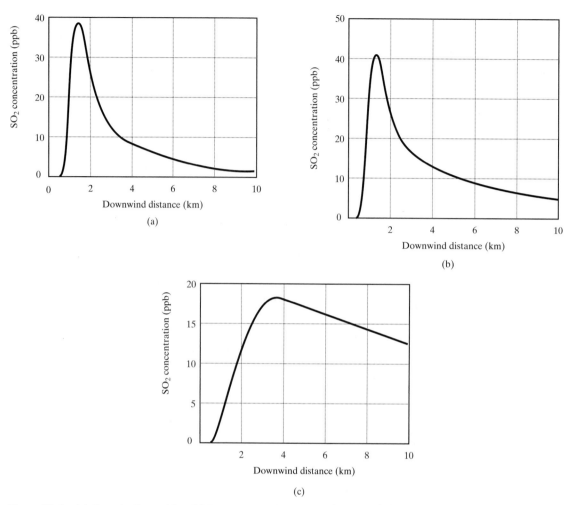

**Figure E9-6**    (a) Case 1, Ground-level $SO_2$ concentration versus downwind distance for stability category A. (b) Case 2, Ground-level $SO_2$ concentration versus downwind distance for stability category D. (c) Case 3, Ground-level $SO_2$ concentration versus downwind distance for stability category F

vertical mixing is larger for unstable and neutral atmospheres (Figures E9-6a and E9-6b, respectively) and the maximum ground-level concentrations are much higher and occur much closer to the stack. These results are included in the summary table near the end of the example.

*Maximum ground-level concentrations.* If readers are interested in estimating only the location $(x_{max})$ and value of the maximum ground-level concentrations $(c_{GL, max})$ for these three atmospheric conditions, they do not have to compute the entire ground-level concentration profile as done above. The location $(x_{max})$ and magnitude of the maximum ground-level $SO_2$ concentration

$(c_{GL, max})$ can be estimated directly by two methods. The verb *estimate* is used deliberately in this discussion because the accuracy of these methods is poor and different answers should be expected.

The direct use at Figure 9-25 is straight-forward. The results are shown in the summary table near the end of the example.

In the final method, from Equation 9.98 at $x_{max}$, $\sigma_z = H/1.414$ and from Equation 9.99, $c_{GL, max} = 0.1171 \dot{m}_{i, s}/U \sigma_y \sigma_z$.

- *Case 1: Stability category A.* $H = 518$ m. $\sigma_z = 518/1.414 = 366.3$ m. From Figure 9-22 $x_{max} \approx 900$ m, and from Figure 9-21, $\sigma_y = 200$ m.

$$c_{GL,\,max} = 0.1171\dot{m}_{i,\,s}/U\sigma_y\sigma_z = (0.1171)$$

$$(153{,}700\text{ mg/s})/(2.37\text{ m/s})$$

$$(200\text{ m})(366.3\text{ m})$$

$$= 0.10366\text{ mg/m}^3(39.7\text{ ppb})$$

- *Case 2: Category D.* $H = 178$ m. $\sigma_z = 178$ m $/1.414 = 125.8$ m. From Figure 9-22, $x_{max} \simeq 8$ km (crude estimate), and from Figure 9-21, $\sigma_y = 500$ m (crude estimate).

$$c_{GL,\,max} = 0.1171\dot{m}_{i,\,s}/U\sigma_y\sigma_z = (0.1171)$$

$$(153{,}700\text{ mg/s})/(9.70\text{ m/s})$$

$$(125.8\text{ m})(500\text{ m})$$

$$= 0.02949\text{ mg/m}^3(11.3\text{ ppb})$$

- *Case 3: Category F.* $H = 179$ m. $\sigma_z = 179$ m$/1.414 = 126.6$ m. From Figure 9-22, $x_{max} \simeq 100$ km (off scale), and from Figure 9-21, $\sigma_y = 2000$ m (off scale).

$$c_{GL,\,max} = 0.1171\dot{m}_{i,\,s}/U\sigma_y\sigma_z = (0.1171)$$

$$(153{,}700\text{ mg/s})/(10.26\text{ m/s})$$

$$(126.6\text{ m})(2000\text{ m})$$

$$= 0.00693\text{ mg/m}^3(2.6\text{ ppb})$$

In conclusion, three methods are available to readers if they are only interested in estimating $x_{max}$ and $c_{GL,\,max}$, as follows.

1. Plot $c_{GL}(x, 0, 0)$ versus $x$ using Equations 9.89 to 9.92 and determine the maximum value by inspection.
2. Use Figure 9-25.
3. Use Equations 9.98 and 9.99 in conjunction with Figures 9-21 and 9-22.

Predicted maximum ground-level concentrations based on the three methods were in good agreement for unstable air (Case 1) but not for Cases 2 and 3. The poor agreement among the methods for the case of stable air (Case 3) derives primarily from the unreasonable extrapolation of dispersion coefficient correlations required to obtain a numerical answer. Predicted ground-level concentrations based on dispersion coefficients for rural sites are lower for neutral air (Case 2) and stable air (Case 3) than those based on dispersion coefficients for urban sites. This is not surprising since planetary boundary layers over

| Case | Stability (method) | $x_{max}$ (km) | $c_{GL,\,max}U/\dot{m}_{i,\,s}$ (m$^{-2}$) | $c_{GL,\,max}$ (ppb) |
|------|--------------------|----------------|--------------------------------------------|----------------------|
| 1 | A | | | |
| | (profile) | 1.2 | | 39 |
| | (Figure 9-25) | 1 | $1.6\times10^{-6}$ | 39.7 |
| | (Equation 9.99) | 0.9 | | 39.7 |
| 2 | D | | | |
| | (profile) | 1 | | 40 |
| | (Figure 9-25) | 7 | $2\times10^{-6}$ | 12.1 |
| | (Equation 9.99) | 8 | | 11.3 |
| 3 | F | | | |
| | (profile) | 4 | | 18 |
| | (Figure 9.25) | 40 | $5\times10^{-7}$ | 2.9 |
| | (Equation 9.99) | >100 | | 2.6 |

urban terrain involve greater mixing than those over rural terrain and increase the downward transport of pollutants. Values of $\sigma_y$ and $\sigma_z$ from Table 9-8 have more precision than values from Figures 9-21 and 9-22 but Table 9-8 pertains to urban sites and Figures 9-21 and 9-22 pertain to rural sites.

To separate the effects of calculational methods and the urban/rural distinction, one could repeat the profile calculation using dispersion coefficients from correlations of rural data (Tables 9-6 and 9-7).

The entire analysis assumed that $SO_2$ was not absorbed by the ground. In truth, $SO_2$ may be absorbed by the ground and the ground-level concentrations may not be as large as predicted.

## 9.10  *Plume rise*

Exhaust gases exiting a stack are buoyant, possess momentum, and enter an atmosphere whose stability can be classified (Table 9-1). Figures 9-20 and 9-26 illustrate how the plume rises and travels downwind as it diffuses both laterally and vertically. The height to which the plume rises [i.e., *plume rise* $(\delta h)$] must be estimated to calculate the effective stack height $(H)$ used in the previous Gaussian dispersion equations:

$$H = h_s + \delta h \tag{9.100}$$

Expressions used to estimate the plume rise reflect its momentum and buoyancy and the stability of the atmosphere. Readers interested in a full discussion of these matters should review Turner (1994), Seinfeld (1986), and Hanna et al. (1982). Engineers using

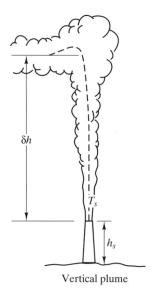

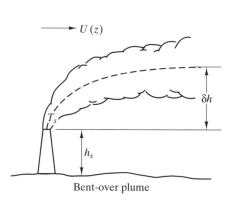

**Figure 9-26** Plume rise for different atmospheric categories (adapted from Slade, 1968)

Vertical plume

Bent-over plume

EPA dispersion models may be required to use one of several plume rise equations. For simplicity, the *Briggs equation* will be used in this book, since it contains terms that account for momentum, buoyancy, and atmospheric stability.

$$\delta h \ (\mathrm{m}) = \frac{114 \, C F^{1/3}}{U} \qquad (9.101)$$

where each of the terms above and the *required units* are defined as follows:

$$U(\mathrm{m/s}) = \text{wind speed at the geometric}$$
$$\text{stack height } (z = h_s) \qquad (9.102)$$

$$\textit{Buoyancy flux } F(\mathrm{m^4/s^3}) = \frac{g v_s D_s^2 (T_s - T_a)}{4 T_a}$$

where

$$v_s(\mathrm{m/s}) = \text{gas velocity at stack exit,}$$
$$v_s > 1.5U$$
$$D_s(\mathrm{m}) = \text{inside diameter of the stack exit}$$
$$g = 9.8 \ \mathrm{m/s^2}$$
$$T_s(\mathrm{K}) \text{ and } T_a(\mathrm{K}) = \text{gas temperature at stack exit and}$$
$$\text{ambient temperature at stack exit}$$

$$C = 1.58 - 41.4 \frac{\Delta \Theta}{\Delta z} \qquad (9.103)$$

where $\Delta \Theta / \Delta z$ (K/m) is the potential temperature difference. (Note that the units are K per meter!)

A plume's rise is also affected by the wake of the stack. In the stack's wake, pressures will be slightly subatmospheric to account for the eddies and vortices present in the wake. If the ratio of the stack gas exit velocity to the wind speed $(v_s/U)$ is less than 1.5, stack gases can be swept downward into the wake region. This phenomenon, called *downwash*, must be avoided if the ground-level concentration of stack gases is to be minimized. If downwash occurs, the dispersion coefficients in Tables 9-6 and 9-7 and the equations for the location of the maximum downwind concentration should be replaced by equations suggested by Bowman (1996).

### EXAMPLE 9.7 PLUME RISE FOR THREE ATMOSPHERIC STABILITY CLASSES

Consider the 100-m stack described in Example 9.6. Assume that the exhaust gas leaves the stack at a temperature $(T_s)$ of 450 K and with a velocity $(v_s)$ equal to 14.5 m/s. Assume that the atmospheric temperature $(T_a)$ at the stack exit is 289 K. The diameter of the stack exit $(D_s)$ is 1.8 m. Three atmospheric conditions will be considered:

- *Case 1*: cloudless, hot summer afternoon, strong radiation, lapse rate = 20 K/km, $U_{10} = 1.8$ m/s
- *Case 2*: partially cloudy, windy afternoon, moderate radiation, lapse rate = 9.8 K/km, $U_{10} = 6.3$ m/s

- *Case 3*: overcast evening, mild wind, lapse rate $= 4$ K/km, $U_{10} = 2.1$ m/s

**Solution**   The Pasquill atmospheric stability category can be found from Table 9-1. The wind speed at the stack height (100 m) is given by

$$U_{100} = U_{10}\left(\frac{100}{10}\right)^p$$

where $p$ is obtained from Table 9-2.

| Case | Stability | dT/dz (K/m) | p | $U_{100}$ (m/s) |
|------|-----------|-------------|------|-----------------|
| 1 | A | −0.020 | 0.07 | 2.11 |
| 2 | D | −0.0098 | 0.15 | 8.89 |
| 3 | F | −0.004 | 0.55 | 7.45 |

- *Case 1*: Stability category A

$$\frac{\Delta\theta}{\Delta z} = \left(\frac{dT}{dz}\right)_{actual} + \Gamma = (-20\text{ K/km}) + 9.8\text{ K/km}$$

$$= -10.2\text{ K/km} = -0.0102\text{ K/m}$$

$$C = 1.58 - 41.4\frac{\Delta\theta}{\Delta z}$$

$$= 1.58 - (41.4)(-0.012) = 2.002\text{ K/m}$$

$$F = \frac{gv_s D_s^2(T_s - T_a)}{4T_a}$$

$$= \frac{(9.8\text{ m/s}^2)(14.5)(1.8\text{ m})^2(450 - 289)}{4(289)}$$

$$= 64.12\text{ m}^4/\text{s}^3$$

$$\delta h = \frac{114CF^{1/3}}{U_{100}} = \frac{(114)(2.002)(64.12)^{1/3}}{2.11} = 417.7\text{ m}$$

The effective stack height is $H = 100 + 418 = 518$ m.

- *Case 2*: Stability category D

$$\frac{\Delta\theta}{\Delta z} = \left(\frac{dT}{dz}\right)_{actual} + \Gamma$$

$$= (-9.8\text{ K/km}) + 9.8\text{ K/km} = 0$$

$$C = 1.58 - 41.4\frac{\Delta\theta}{\Delta z} = 1.58\text{ K/m}$$

$$F = \frac{gv_s D_s^2(T_s - T_a)}{4T_a}$$

$$= \frac{(9.8\text{ m/s}^2)(14.5\text{ m/s})(1.8\text{ m})^2(450 - 289)}{4(289)}$$

$$= 64.12\text{ m}^4/\text{s}^3$$

$$\delta h = \frac{114\,CF^{1/3}}{U_{100}} = \frac{(114)(1.58)(64.12)^{1/3}}{8.99} = 78.2\text{ m}$$

The effective stack height is $H = 100 + 78.2 = 178.2$ m.

- *Case 3: Stability category F*

$$\frac{\Delta\theta}{\Delta z} = \left(\frac{dT}{dz}\right)_{actual} + \Gamma$$

$$= (-4.0\text{ K/km}) + 9.8\text{ K/km} = 5.8\text{ K/km}$$

$$= 0.0058\text{ K/m}$$

$$C = 1.58 - 41.4\frac{\Delta\theta}{\Delta z} = 1.58 - (41.4)(0.0058)$$

$$= 1.3399\text{ K/m}$$

$$F = \frac{gv_s D_s^2(T_s - T_a)}{4T_a}$$

$$= \frac{(9.8\text{ m/s}^2)(14.5\text{ m/s})(1.8\text{ m})^2(450 - 289)}{4(289)}$$

$$= 64.12\text{ m}^4/\text{s}^3$$

$$\delta h = \frac{114\,CF^{1/3}}{U_{100}} = \frac{(114)(1.3399)(64.12)^{1/3}}{7.45}$$

$$= 79.1\text{ m}$$

The effective stack height is $H = 100 + 79 = 179$ m.

| Case | Stability | H (m) | $U_{100}$ (m/s) |
|------|-----------|-------|-----------------|
| 1 | A | 518 | 2.11 |
| 2 | D | 178 | 8.89 |
| 3 | F | 179 | 7.45 |

Comparing the three values of plume rise, it can be seen that a neutral (D) and stable (F) atmospheres produce mild vertical mixing and a small value of $\delta h$. For a strong unstable atmosphere (A), there is vigorous vertical mixing and the plume rise is large.

## 9.11 Building exhaust stacks

One of the most common errors made in industrial ventilation is to place a makeup air inlet too close to an exhaust stack. The error arises generally because the locations of the inlet and exhaust are decided at different times by different people unaware of each other's actions. Second, errors are made because people are unaware of the size and consequences of *aerodynamic wakes* associated with buildings. Figures 9-27 and 9-28 illustrate the wakes and vortices one can expect from blocklike buildings on level terrain immersed in a deep terrestrial boundary layer. If the building lies in the wake of other buildings or if the building lies on terrain that is not level, the aerodynamic wakes are somewhat different but exist nonetheless. An aerodynamic wake is a region in which the local air velocities are not equal to free stream values. Wakes can be subdivided into (1) recirculation cavity or eddy and (2) turbulent shear region. A *recirculation eddy* is a region in which a

relatively fixed amount of air moves in a circular fashion and there is little air transported across the eddy boundaries. A *turbulent shear region* is one in which there is a net convective flow, but the turbulent shear stresses are larger than free stream values.

Relationships predicting the locations and velocity fields of wakes are not readily available, but alternatively, a series of empirical equations defining the boundaries of the wake regions are widely used (Hosker, 1982). In the discussion that follows it will be assumed that the building is a blocklike structure of height ($H$), cross-wind width ($W$), and length in the direction of the wind ($L$).

In selecting the location and dimensions of an air inlet or exhaust stack, two general principals should be followed:

1. Do not locate air inlets in recirculation regions or at other locations susceptible to contamination from exhaust gases.
2. The effective stack height above the building roof should be above the roof wake boundary.

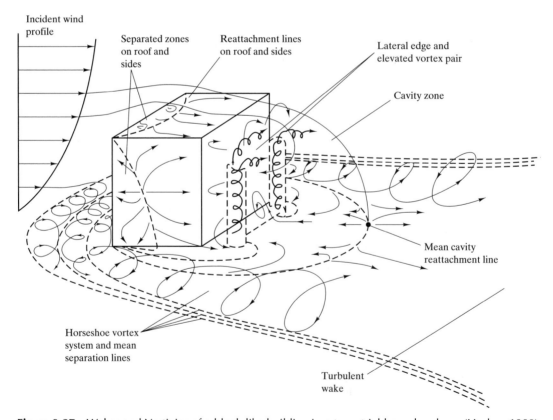

**Figure 9-27**  Wakes and Vorticies of a block-like building in a terrestrial boundary layer (Hosker, 1982)

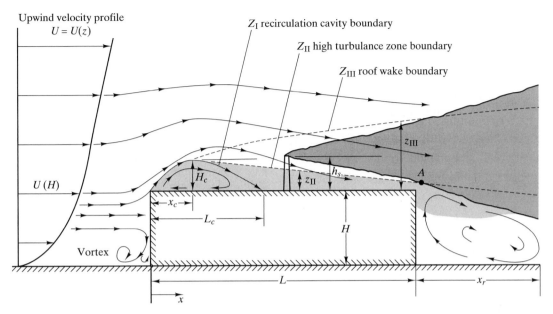

**Figure 9-28** Wakes and recirculation eddies of a two-dimensional flat building in a terrestrial boundary layer, (adapted from Hosker, 1982)

As *good engineering practice*, Section 123 of the Clean Air Act recommends that the geometric stack height (above ground level) should ensure that emissions from the stack do not result in excessive concentrations of any pollutant in the immediate vicinity of the stack as a result of atmospheric downwash, eddies, or wakes, which may be created by the stack, nearby structures, or nearby terrain. A safe engineering practice is a stack height ($h_s$, height above ground level)

$$h_s = H + 1.5z' \qquad (9.104)$$

where $z'$ is the smaller of $W$ or $H$. The stack height ($h_s$) is a geometric dimension. The effective stack height ($H$) is equal to the geometric stack height ($h_s$) plus the plume rise ($\delta h_s$), due to the buoyancy and the momentum of the exhaust gas (Wark and Warner, 1981; Hanna et al., 1982). If zoning laws or other considerations prevent Equation 9.104 from being followed, it will be necessary to estimate the location of the wakes. There are two regions to be concerned about: (1) the recirculation eddy and wake produced by the leading edge of the building roof, and (2) the large recirculation eddy directly downwind of the building.

Key dimensions of the *roof eddy* are shown in Figure 9-28 and can be expressed in terms of length parameters $L_L$, $L_S$, and $R'$, where

$$L_L \text{ larger of } H \text{ or } W \qquad \text{(a)}$$
$$L_S \text{ smaller of } H \text{ or } W \qquad \text{(b)} \qquad (9.105)$$
$$R' = L_S^{0.67} L_L^{0.33} \qquad \text{(c)}$$

The height ($H_c$), length ($L_c$), and center ($x_c$) of the roof eddy can be estimated from the following:

$$H_c = 0.22R' \qquad L_c = 0.9R' \qquad x_c = 0.5R' \quad (9.106)$$

The roof eddy is bounded above by a turbulent shear region and above that by a wake region. The height of the turbulent shear zone ($Z_{II}$) can be estimated from

$$\frac{Z_{II}}{R'} = 0.27 - 0.1\frac{x}{R'} \qquad (9.107)$$

where $x$ is measured in the downwind direction from the roof lip. The height of the roof wake ($Z_{III}$) can be estimated from

$$\frac{Z_{III}}{R'} = 0.28 \left( \frac{x}{R'} \right)^{0.33} \qquad (9.108)$$

The safest design is one in which the expanding plume remains above the roof wake boundary ($Z_{III}$). The manner in which the plume expands depends on the atmospheric stability, which is a function of wind speed and the solar radiation. If the plume cannot be kept above $(Z_{III})$, a plume that is kept in the turbulent shear region is the next best choice.

Directly downwind of the building a very large *building eddy* is formed. Contaminants trapped in this region are apt to enter building windows and affect people and vehicles on the ground. The size of the this eddy can be seen in Figure 9-28. The width and height of the eddy seldom exceed the building dimensions $W$ and $H$ by more than 50%, and the downwind length of the eddy $(x_r)$ can be estimated from the following:

$$\frac{x_r}{H} = \frac{A}{1 + B} \qquad (9.109)$$

For buildings in which $L/H < 1$,

$$A = -2.0 + 3.7\left(\frac{L}{H}\right)^{-0.33} \qquad (9.110)$$

$$B = -0.15 + 0.305\left(\frac{L}{H}\right)^{-0.33} \qquad (9.111)$$

For buildings in which $L/H > 1$, $A = 1.75$ and $B = 0.25$.

Predictions of ground-level concentrations from Gaussian plumes are valid for regions no closer than approximately 100 m of the stack. There is a paucity of information for regions closer to the stack where the plume is rising due to buoyancy and momentum. For neutrally buoyant plumes, Halitsky (1989) provides expressions to predict downwind concentrations for regions downwind of the short stack where the jet velocities have decayed and are essentially equal to the free stream values.

In the wake of a stack the pressure is lower than the free stream value, and under some conditions this may cause the exiting stack gases to fall rather than rise. Such a fall, called *downwash*, may be a serious consideration in the design of short stacks. Downwash can be prevented by keeping the ratio of the duct exit velocity $(v_s)$ to the wind speed $(U)$ greater than 1.5 (i.e., $v_s/U > 1.5$). If the ratio is less than 1.5, it is suggested (Hanna et al., 1982) that the downwash distance $(H_{dw})$ can be estimated from

$$\frac{H_{dw}}{D} = 2\left(\frac{v_s}{U} - 1.5\right) \qquad \text{if } \frac{v_s}{U} < 1.5 \quad (9.112)$$

where $D$ is the stack exit diameter. The negative value computed by Equation 9.112 implies that the downwash causes the plume centerline to fall rather than rise above the stack exit plane.

## 9.12 Instantaneous point source: puff diffusion

When there is an explosive, sudden, or very short-term release $(m_j)$ of a pollutant of molecular species $j$ from an elevated source $(H)$, the material is convected downwind as a "puff" by the prevailing wind $(U)$ and diffuses laterally and vertically. Figure 9-29 shows the trajectory of the puff with time and the instantaneous concentration at various downwind locations. The ground-level concentration at downwind point $P(x, y, 0)$ can be predicted by solving Equation 9.76. The ground-level concentration at any point $P(x, y, 0, t)$ for a pollutant that accumulates on the ground can be expressed by

$$c(x, y, 0, t) = \frac{m_j}{\sqrt{2}(\sigma_{xi}\sigma_{yi}\sigma_{zi})\pi^{1.5}}$$
$$\exp\left\{-\frac{1}{2}\left[\left(\frac{x - Ut}{\sigma_{xi}}\right)^2 + \left(\frac{y}{\sigma_{yi}}\right)^2 + \left(\frac{H}{\sigma_{zi}}\right)^2\right]\right\} \qquad (9.113)$$

Immediately after the release, the concentration at point $P$ is zero, but it begins to rise as the puff approaches point $P$, reaches a maximum, and then diminishes as the puff travels farther downwind. Of concern to health is not the time-varying concentration but the integral of the ground-level concentration over the time of exposure. The ground-level *dose* $D(x, y, 0)_j$ is defined as

$$D(x, y, 0)_j = \int_0^\infty c(x, y, 0)_j \, dt \qquad (9.114)$$

An expression for the ground-level dose is the following (Slade, 1968):

$$D(x, y, 0)_j = \frac{m_j}{\pi \sigma_{yi}\sigma_{zi}U}$$
$$\exp\left\{\frac{-1}{2}\left[\left(\frac{y}{\sigma_{yi}}\right)^2 + \left(\frac{H}{\sigma_{zi}}\right)^2\right]\right\} \qquad (9.115)$$

The coefficients $\sigma_{yi}$ and $\sigma_{zi}$ are instantaneous dispersion coefficients, which are not the same as the conventional time-averaged dispersion coefficients,

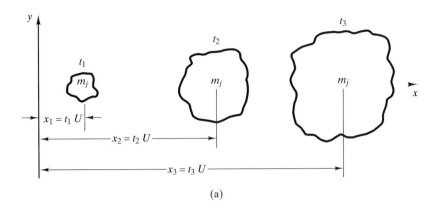

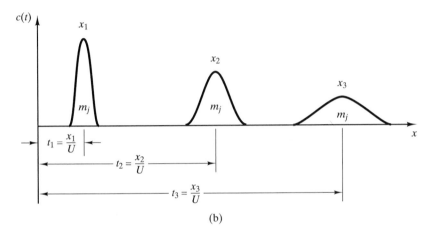

**Figure 9-29** (a) Location of an emission at three times, $t_1 < t_2 < t_3$, generated by an instantaneous point source ($m_j$). (b) Pollutant concentration versus time at three locations, $x_1 < x_2 < x_3$, generated by an instantaneous point source $m_j$.

evaluated by Figures 9-21 and 9-22 and/or data taken from Tables 9-6 to 9-8, owing to the fact that the instantaneous values need to be evaluated over the short lifetime of the puff. Shown in Table 9-9 are

**Table 9-9**   Suggested Values of Instantaneous Dispersion Coefficients

| Coefficient, m | Stability | Approximate function ($x$ in meters) |
|---|---|---|
| $\sigma_{yi}$ | Unstable | $0.14x^{0.92}$ |
| | Neutral | $0.06x^{0.92}$ |
| | Very stable | $0.02x^{0.89}$ |
| $\sigma_{zi}$ | Unstable | $0.53x^{0.73}$ |
| | Neutral | $0.15x^{0.70}$ |
| | Very stable | $0.05x^{0.61}$ |

*Source:* Slade (1968).

expressions that can be used to quantify these instantaneous dispersion coefficients.

### EXAMPLE 9.8   HCN CONCENTRATIONS AND DOSE DOWNWIND OF A BURSTING VESSEL

A ground-level storage tank, $D = 2$ m, contains 10 kg ($10^7$ mg) of HCN under high pressure. The nearest homes are 100 m from the tank. On a quiet evening ($U = 1.5$ m/s) an explosion occurs that ruptures the tank and allows HCN to enter the atmosphere. The plume of HCN drifts downwind, engulfing homes in its path. Compute and plot the concentrations and dose at locations downwind to define regions in which hazardous conditions exist.

***Solution***   From OSHA literature, the short-term exposure limit (STEL) of HCN is 4.7 ppm (5.17

mg/m$^3$). The STEL represents an exposure for 15 min (900 s) that constitutes hazardous conditions in the workplace for healthy adults. A corresponding dose is

$$\text{dose(STEL, 15 min)} = (5.17 \text{ mg/m}^3)(900 \text{ s})$$

$$= 4661 \text{ mg·s/m}^3$$

The reader should not put too fine a point on these values. While hydrogen cyanide (HCN) is a deadly gas, the body is capable of withstanding a certain dose without serious consequences. Thus it can be assumed that concentrations below 4.7 ppm for 15 min and concentrations and doses lower than 4661 mg·s/m$^3$ are not life threatening for healthy adults, whereas values larger than these are potentially life threatening. Using these data and Equations 9.113 and 9.115, it is possible to identify locations downwind of the bursting tank that are potentially dangerous.

If the air speed is low ($U < 2$ m/s), Table 9-1 indicates that the atmosphere can be considered very stable. From Table 9-9, the instantaneous dispersion coefficients are

$$\sigma_{yi} = \sigma_{xi} = 0.02x^{0.89} \qquad \sigma_{zi} = 0.05x^{0.61}$$

Since the wind speed is very small, the dispersion coefficient in the $x$-direction can be assumed to be the same as that in the $y$-direction. The equations can be solved with commercially available mathematical computational programs. Figure E9-8a shows how the concentration varies with time at three downwind locations along the centerline of the puff and Figure E9-8b shows the dose versus downwind distance along the centerline of the puff. The Mathcad program producing Figure E9-8 can be found in the textbook's Web site on the Prentice Hall Web page.

The graph of the concentration versus time reveals that the concentration rises suddenly as the HCN cloud passes particular locations, $x$. The peak concentration is inversely proportional to $x$, since the cloud diffuses out radially outward as it is convected downwind with the velocity $U$. If 4000 mg s/m$^3$ is the minimum dose defining potentially life-threatening

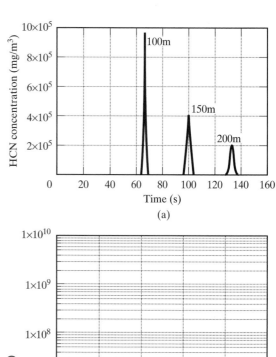

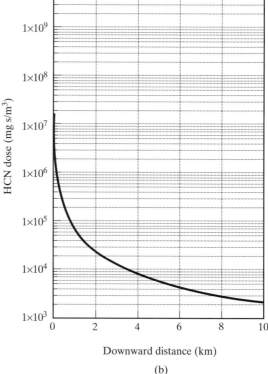

**Figure E9-8** (a) HCN concentration versus time at three locations (100, 150 and 200 m) directly downwind of an instantaneous release 10 kg of HCN. (b) HCN dose as a function of distance directly downwind of an instantaneous release of 10 kg HCN

conditions, the graph representing Equation 9.115 indicates that dangerous conditions exist at all downwind locations $x < 6500$ m (6.5 km).

# 9.13  Continuous elevated line source

When the source is distributed uniformly along a continuous elevated line in a cross wind (Figure 9-30), the downwind ground-level pollutant concentration can be found from Equation 9.76. The solution to the equation when there is no absorption by the ground can be expressed as (Slade, 1968)

$$c_i(x, 0) = \frac{2\dot{m}_i/L}{(2\pi)^{1/2}\sigma_z U_x} \exp\left[-\left(-\frac{1}{2}\right)\left(\frac{H}{\sigma_z}\right)^2\right]$$

(9.116)

where $\dot{m}_i/L$ is the pollutant emission rate per unit length, $x$ is in a direction perpendicular to the line source, and $U_x$ is the component of the cross wind perpendicular to the line. The equation should not be used if the angle (measured in the $x$–$y$ plane) between the wind's direction and the line source is less than 45 degrees.

# 9.14  Numerical dispersion models

The EPA has developed or adopted many air quality simulation models (dispersion models) that emphasize certain features such as pollutants that settle or react chemically, terrain, complex source locations, and averaging methods to account for hourly, monthly, and yearly averages. These models can be downloaded from the TTN electronic bulletin board discussed in Chapter 3. Table 9-10 is a list of several dispersion models used widely in PSD studies. A full discussion of these air quality simulation

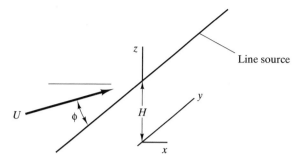

**Figure 9-30**  Atmospheric dispersion from a continuous elevated line source for wind inclined to the line source at an angle ϕ no less that 45° (angle ϕ measured in the x-y plane)

models is given by Turner et al. (1989). Turner (1994), has prepared for distribution a series of executable computer programs for a variety of dispersion calculations.

# 9.15  Closure

An integral part of air pollution control is estimating pollutant concentrations at points downwind of a source for different meteorological conditions. Preparing these estimates is important in the permitting process required by environmental regulatory agencies. The principles of meteorology and atmospheric dispersion are based on thermal science principles, which engineers should recognize from other courses. The models discussed in this chapter are elementary ones, but ones that provide a good understanding of phenomena that might later be treated with more sophisticated models.

**Table 9.10**   EPA Air Quality and Dispersion Models

| Model | Description |
| --- | --- |
| ISCLT (industrial source complex, long-term ) | Steady-state pollutant concentration from an industrial complex |
| ISCST (industrial source complex, short-term) | Short-term pollutant at concentrations from an industrial complex |
| UAM (urban airshed model) | Three-dimensional grid-based photochemical simulation for urban-scale domains |
| BLP (buoyant line and point source) | Gaussian plume model associated with aluminum reduction plants |
| CALINE3 (California line source) | Predicts CO concentrations near highways, given emissions, geometry, meteorology |
| CDM2 (climatological dispersion) | Long-term quasistable pollutant concentrations |

**Table 9.10**   *continued*

| Model | Description |
|---|---|
| CRSTER | Ground-level concentrations from up to 19 elevated stack emissions |
| EKMA | Ozone concentrations as a function of organics and $NO_x$ |
| MPTER (multiple-point-source model) | Gaussian plume model for multiple sources with terrain adjustments |
| RAM | Short-term Gaussian plume model for stable pollutants |
| CTDMPLUS | "Refined" air quality model for all stability categories for complex terrain applications |
| COMPLEX I | Multiple point source with terrain adjustments, program is a bridge between VALLEY and COMPLEX II |
| LONGZ-SHORTZ | Long- and short-term pollutant concentrations at a large number of receptors produced by emissions from multiple stack, building, and area sources |
| RTDM (rough-terrain diffusion model) | Ground-level concentrations in rough and flat terrain near one or more point sources |
| VALLEY | Steady-state Gaussian plume model for 24-h or annual concentrations from emissions for up to 50 point or area sources |
| PLUVUE | Transport, atmospheric diffusion, chemical conversion, optical effects, and surface deposition of point sources |
| SDM (shoreline dispersion model) | Ground-level concentrations from multiple, tall point sources near shorelines |
| MESOPUFF | Lagrangian transport model, diffusion and removal of pollutants from multiple point and area sources at transport distances exceeding 10 to 50 km |
| PTPLU-3 | Source dispersion model for 1-h ground-level concentrations |
| HIWAY-ROADWAY | Hourly concentrations of nonreactive pollutants within 200 m of highway |
| INPUFF | Integrated Gaussian and puff model for accidental short-term or continuous accidental stack emissions |
| MPTDS | Modification of MPTER to account for gravitational settling or deposition of pollutants |
| PAL (point, area, line source) | Transport of short-term stable pollutants |
| PBM (photochemical box model) | Single-cell, variable height for three urban pollutants, 1-day duration |
| PEM (pollutant episodic model) | Urban scale, ground-level concentrations for two gaseous or particle pollutants |

# *Nomenclature*

| Symbol | Description (Dimensions*) |
|---|---|
| $\rightleftharpoons$ | simultaneous forward and reverse reactions |
| $\rightarrow$ | direction of a chemical reaction |

*$F$, force; $L$, length; $M$, mass; $N$, mol; $Q$, energy; $t$, time; $T$, temperature.

| Symbol | Description (Dimensions*) |
|---|---|
| $=$ | empirical stoichiometric equation |
| $[=]$ | has the units of |
| [j] | concentration of species $j$ $\left(M/L^3, N/L^3\right)$ |
| $\mathbf{a}$ | acceleration $\left(L/t^2\right)$ |
| $A$ | area $\left(L^2\right)$ |
| $c$ | mass concentration $\left(M/L^3\right)$ |

| | | | |
|---|---|---|---|
| $C$ | constant used to predict plume rise (T/L) | $P_v$ | vapor pressure $(\text{F}/\text{L}^2)$ |
| $c_i$ | concentration of molecular species $i$ $(\text{M}/\text{L}^3)$ | $P_0$ | ground-level pressure $(\text{F}/\text{L}^2)$ |
| | | $q$ | heat added (Q) |
| $c_p$ | specific heat at constant pressure (Q/MT) | $Q$ | volumetric flow rate $(\text{L}^3/t)$ |
| | | $\hat{r}$ | unit vector in the radial direction |
| $c_v$ | specific heat at constant volume (Q/MT) | $R_E$ | radius of the earth (L) |
| $D_j$ | dose, product of concentration and time $(\text{Mt}/\text{L}^3)$ | $R_i$ | gas constant of molecular species $i$ $(R_i = R_u/M_i)$ |
| $D_x, D_y, D_z$ | atmospheric diffusion coefficients in $x$, $y$, and $z$ directions $(\text{L}^2/t)$ | $R_u$ | universal gas constant (Q/NT) |
| | | $s$ | distance in the windward direction (L) |
| $D_s$ | stack diameter (L) | $\dot{S}_s$ | source strength (M/t) |
| $F$ | coefficient used to predict plume rise $(\text{L}^4/t^3)$ | $T$ | temperature (T) |
| | | $T_0$ | ground-level temperature (T) |
| $\mathbf{F}$ | force | $U$ | velocity (L/t) |
| $F_c$ | Coriolis force (F) | $U_E, U_S, U_W$ | wind speed in the eastward, southern and westward directions (L/t) |
| $g$ | acceleration of gravity $(\text{L}/t^2)$ | | |
| $h$ | height (L) | $U_E$ | eastward velocity of the earth (L/t) |
| $H$ | effective stack height, $H = h_s + \delta h$, height of a building (L) | $U_G$ | geostrophic wind speed (L/t) |
| | | $U^*$ | friction velocity (defined by equation) (L/t) |
| $H_c$ | height off roof edge recirculation cavity | $V$ | volume $(L^3)$ |
| $H_{dw}$ | downwash distance (L) | $v_r, v_t$ | velocity in the radial and tangential directions (L/t) |
| $h_{fg}$ | enthalpy (heat) of vaporization (Q/M) | $v_s$ | velocity of gas at stack exit (L/t) |
| $h_s$ | stack height (L) | $v_t$ | particle settling velocity (L/t) |
| $k$ | isentropic constant $c_p/c_v$; von Kármán constant (0.4) | $w$ | width (L) |
| | | $W$ | width of a building perpendicular to the windward direction (L) |
| $L$ | length of a building in the windward direction (L) | | |
| | | $x, y, z$ | distances (L) |
| $M_i$ | molecular weight of species $i$ (M/N) | $x_c$ | distance to center of roof recirculation cavity |
| $L_c$ | length of roof edge recirculation cavity | $x_r$ | downwind length of building eddy |
| $\dot{m}_{i,s}$ | mass flow rate of species $i$ emitted by a stack (M/t) | $z_{\text{AM}}, z_{\text{PM}}$ | morning and afternoon mixing heights (L) |
| $m_j$ | mass of instantaneous release (M) | $z_0$ | roughness height (L) |
| $\dot{m}_s$ | source strength (M/t) | $Z_I$ | height of roof leading edge recirculation cavity |
| $\dot{m}_s/L$ | line source strength, emission rate per unit length (M/Lt) | $Z_{II}$ | height of turbulent shear zone |
| $n$ | thermodynamic constant of a polytropic process | $Z_{III}$ | height of roof wake |
| $p$ | exponent used to describe the atmospheric boundary layer velocity profile | *Greek* | |
| | | $\alpha$ | latitude, angle; constant |
| $P$ | pressure $(\text{F}/\text{L}^2)$ | $\beta$ | longitude; constant |

| | | | |
|---|---|---|---|
| $\Gamma$ | dry adiabatic lapse rate 9.8°C/km (T/L) | $(\cdot)_s$ | conditions at stack exit |
| $\delta h$ | plume rise (L) | $(\cdot)_S$ | southward |
| $\Theta$ | potential temperature (T) | $(\cdot)_{ss}$ | steady state |
| $\theta$ | angle | $(\cdot)_W$ | westward |
| $\hat{\theta}$ | unit vector in the tangential direction | $(\cdot)_x, (\cdot)_y, (\cdot)_z$ | components in the $x, y,$ and $z$ directions |
| $\rho$ | density $(M/L^3)$ | $(\cdot)_{10}$ | conditions measured at an altitude of 10 m[a] |
| $\rho_0$ | ground-level density $(M/L^3)$ | | |
| $\sigma_y, \sigma_z$ | atmospheric dispersion coefficients in the $y$ and $z$ directions (L) | $(\cdot)_\infty$ | far-field conditions |
| $\sigma_{y,i}, \sigma_{zi}$ | instantaneous dispersion coefficients (L) | $(\cdot)_p$ | particles |
| $\tau$ | shear stress $(F/L^2)$ | | |
| $\omega$ | humidity ratio | | |
| $\Omega$ | rotational speed $(t^{-1})$ | | |

*Subscript*

| | |
|---|---|
| $(\cdot)_a$ | ambient conditions |
| $(\cdot)_{atm}$ | atmospheric conditions |
| $(\cdot)_b$ | buoyant |
| $(\cdot)_E$ | eastward |
| $(\cdot)_{GL}$ | conditions measured at ground level |
| $(\cdot)_i$ | molecular species $i$ |
| $(\cdot)_{parcel}$ | pertaining to a parcel of air |

*Abbreviations*

| | |
|---|---|
| AM | morning, after sunrise |
| EPA | Environmental Protection Agency |
| GMT | Greenwich Mean Time |
| LST | local standard time |
| PM | evening, after sundown |
| PSD | prevention of significant deterioration |
| Ri | Richardson number |
| RH | relative humidity |

[a] The literature of air pollution meteorology commonly uses subscripts to denote heights at which certain parameters are evaluated whereas this text has used the more conventional parenthetical notation from mathematics. So that readers may pursue the underlying and advanced literature of air pollution meteorology more easily, we have made frequent use of their notation for Chapter 9 only. Hence $P_0 = P(0)$, $T_0 = T(0)$, $U_{10} = U(10)$, etc.

# References

Bowman, W. A., 1996. Maximum ground-level concentrations with downwash: the urban stability mode, *Journal of the Air and Waste Management Association*, Vol. 46, pp. 615–620

Davis, R. E., and Dolan, R., 1993. Nor'easters, *American Scientist*, Vol. 81, No. 5, September–October, pp. 428–439.

Griffiths, R. F., 1994. Errors in the use of the Briggs parameterization for atmospheric dispersion coefficients, *Atmospheric Environment*, Vol. 28, No. 17, pp. 2861–2865.

Halitsky, J., 1989. A jet plume model for short stacks, *Journal of the Air Pollution Control Association*, Vol. 39, No. 6, pp. 856–858.

Hanna, S. R., Briggs, G. A., and Hosker, R. P., 1982. *Handbook on Atmospheric Diffusion*, NTIS DE81009809 (DOE/TIC-22800), Springfield, VA.

Heinsohn, R. J., 1991. *Industrial Ventilation: Engineering Principles*, Wiley-Interscience, New York.

Holton, J. R., 1992. *An Introduction to Dynamic Meteorology*, 3rd ed., Academic Press, San Diego, CA.

Hosker, R. P. Jr., 1982. *Methods for Estimating Wake Flow and Effluent Dispersion near Simple Block-Like Buildings*, Report NUREG/CR-2521, ERL-ARL-108, National Oceanic and Atmospheric Administration, Washington, DC.

Jin, S., and Demerjian, K., 1993, A photochemical box model for urban air quality, *Atmospheric Environment*, Vol. 27B, No. 4, pp. 371–387.

Seinfeld, J. H., 1986. *Atmospheric Chemistry and Physics of Air Pollution*, J. Wiley, New York.

Slade, D. H., (Ed.), 1968. *Meteorology and Atomic Energy 1968*, U.S. Atomic Energy Commission, Air Resources

Laboratories, Research Laboratories, Environmental Sciences Services Administration, U.S. Department of Commerce, Washington, DC.

Turner, D. B., 1979. Atmospheric dispersion modeling, *Journal of the Air Pollution Control Association*, Vol. 29, No. 5, pp. 502–519.

Turner, D. B., 1994. *Workbook of Atmospheric Dispersion Estimates: An Introduction to Dispersion Modeling*, 2nd ed., Lewis Publishers, Boca Raton, FL.

Turner, D. B., Bender, L. W., Pierce, T. E., and Petersen, W. B., 1989. Air quality simulation models from EPA, *Environmental Software*, Vol. 4, No. 2, pp. 52–61.

Wark, K., and Warner, C. F., 1981. *Air Pollution*, Harper & Row, New York.

Williamson, S. J., 1973. *Fundamentals of Air Pollution*, Addison-Wesley, Reading, MA.

# Problems

**9.1.** The Coriolis acceleration causes a stationary body of water being drained from a small hole in the bottom of a container to acquire a certain rotation north of the equator and the opposite rotation south of the equator. Explain why this is the case and prove that the rotation is either clockwise or counterclockwise for locations north of the equator. To convince yourself of this fact, fill a container containing a small hole in its base with water. After the water has come to rest, open the bottom tap and note the direction of rotation acquired by the body of water above the hole.

**9.2.** Derive an expression that predicts the atmospheric pressure ($P$) at any altitude ($z$) if changes in pressure and temperature are related in a polytropic manner.

**9.3.** If the density of dry air is constant, show that the lapse rate is 34.14°C/km.

**9.4.** If the saturated (wet) adiabatic lapse rate is 8.8°C/km, show that the gradient of the humidity ratio ($\omega$, kg water vapor/kg dry air) is $-0.00041$ km$^{-1}$.

**9.5.** At sun-up the air is motionless and there is a ground-based inversion ($dT/dz = 0.03$°C/m) to an altitude of 500 m. Above 500 m the lapse rate is equal to the normal-lapse rate. The ambient temperature on the ground is 300 K. Smoke (assume air) from a poorly attended wood-burning stove leaves the chimney at 303 K with a negligible velocity and rises as dry adiabatic air. The chimney exit is 5 m above the ground. Estimate the height to which you'd expect the smoke to rise.

**9.6.** An unsavory industrial practice involves discarding unwanted volatile organic compounds (VOCs) by placing them in open vessels after dark, letting nature remove the unwanted VOCs by evaporation, and retrieving the empty vessels the next morning before they are noticed by the neighbors. To detect this illegal practice, state regulatory agencies require companies to account for VOCs entering and leaving a plant. Imagine that you live in a small community (1 km $\times$ 1 km) in a valley in which a small industry follows this practice. On one evening, 16,200 kg of MEK (methyl ethyl ketone, $M = 72$) evaporates over a period of 3 h. Using the box model, estimate how long it takes for the MEK concentration to reach 20 ppm. Assume that the wind speed is 0.5 m/s and that the community is bounded on two sides by hills 1000 m apart and that an evening inversion layer exists to a height of 50 m that traps the MEK vapor beneath it. Assume that the initial MEK concentration is zero.

**9.7.** Using the box model, derive an expression that predicts the wood smoke concentration ($PM_{10}$) as a function of time in a rural residential community lying in a valley in which the following occur.

- Wood smoke ($PM_{10}$) emission factor = 4 g/kg wood
- Wood burning rate = 0.1 kg/min
- 12,000 wood stoves
- Community area = 10 km $\times$ 10 km
- Morning mixing height = 100 m
- Wind speed = 2 m/s
- Upwind ambient smoke concentration = 20 $\mu$g/m$^3$

How long will it take before the concentration equals the annual primary ambient air quality standard?

**9.8.** The rendering process produces a high-intensity noxious odor that is difficult to control. Rendering consists of cooking animal parts and cooling the paste into fat and protein feed for animals and poultry. A rendering plant whose air pollution control practices leave much to be desired is located in a small rural community. The plant presently discharges noxious odors at 25°C from a short stack on its roof 5 m above the ground. The odors rise and cool at a rate equal to the dry adiabatic lapse rate. The owners wish to lengthen the stack and have asked you to model the process using the box model. On the basis of buoyancy alone, predict the height of the mixing layer and describe the type of plume you expect to see for the following stacks (heights measured from the ground):

$H_s$ = present stack (5 m)

= 15, 25, 35, 40 m

Odors are a serious problem between 6 A.M. and 9 A.M., with the following atmospheric temperature profiles:

| 6 A.M. | 9 A.M. |
|---|---|
| $U = 0.1$ m/s | $U = 1.0$ m/s |
| $T_{GL}(z = 0) = 24.5°C$ | $T_{GL}(z = 0) = 25.3°C$ |
| Linear, $0 \leq z \leq 50$ | Linear, $0 \leq z \leq 30$ m |
| $T(50\ m) = 24.8°C$ | $T(30\ m) = 24.7°C$ |
| Adiabatic lapse rate, $z > 50$ m | Linear, $30\ m \leq z \leq 50$ m |
| | $T(50\ m) = 24.8°C$ |
| | Adiabatic lapse rate, $z > 50$ |

**9.9.** A old and small coal-fired steam boiler provides steam for a county nursing home in a large rural community of 80,000 people. The plant is located on flat terrain and surrounded by one-story private homes. The power plant has the following characteristics:

- *Physical stack height*: $H_s = 25$ m (measured from the ground)
- *Stack exit diameter*: $D_s = 1.5$ m
- *Stack exit gas conditions*: $V_s = 20$ m/s, $T_s = 350$ K
- *Power plant dimensions*: $L = 20$ m, $H = 10$ m, $W = 15$ m

**(a)** Describe the stability of the atmosphere (using such parameters as the Pasquill letter code; words such as *stable, inversion, neutral*; the potential gradient; the Richardson number; etc.) that you expect under the following conditions.

(1) Sun-up in late fall, cloudless, clear cool morning with $U(10)$ essentially zero and $dT/dz = 0.005°C/m$

(2) High-noon, sunny, summer day with slight cloud cover and $U(10) = 3$ m/s

(3) Dark, overcast, windy late afternoon day in August when $U(10) = 7$ m/s

(4) Midmorning on a cloudless, cold January day when $U(10) = 1$ m/s

(5) Midnight on a cloudless evening with a full moon and $U(10) = 2$ m/s

(6) Noon on a clear sunny September day when $U(10) = 5$ m/s, 50% cloud cover

**(b)** If there is strong sunlight, $U(10) = 6$ m/s, $T$(ambient) = 300 K, a class C stability atmosphere, and $dT/dz = -11°C/km$, estimate the plume rise.

**(c)** The power plant is poorly run for a series of months and emits $SO_2$ to the atmosphere at a steady rate of 200 g/s.

The plume rise changes to 100 m and remains constant. Assume the same atmospheric conditions as part (b).

(1) Estimate the ground-level $SO_2$ concentration $(\mu g/m^3)$ at a point 1000 m directly downwind of the stack.

(2) Estimate the location $(x)$ where the ground-level $SO_2$ concentration is a maximum and compute the ground-level concentration at this location.

(3) Is the maximum ground-level concentration in compliance with the annual and 24-h average national ambient air quality standards for $SO_2$?

**9.10.** A new electric utility plant will use two circulating fluidized coal-burning steam boilers whose stacks are 200 f apart and lie in a plane perpendicular to the prevailing wind.

| Boiler 1 | Boiler 2 |
|---|---|
| Output = 685 MW | Output = 600 MW |
| Efficiency = 35% | Efficiency = 35% |
| $v_s = 23.7$ m/s | $v_s = 26.5$ m/s |
| $H_s = 336$ m | $H_s = 243$ m |
| $T_s = 413$ K | $T_s = 413$ K |
| $D_s = 7.3$ m | $D_s = 6.9$ m |

The particle emission rate for each stack is 0.5 kg/s, the particle density is 1700 kg/m³, and the mass mean particle diameter is 10 $\mu$m. For a rural setting, estimate the maximum ground-level particle concentration $(\mu g/m^3)$ and its unique location $(x, y)$ resulting from emissions from both of the stacks for stability classes A through F, assuming that wind speed and temperature at 10 m are 25°C and 4 m/s.

| Stability class | Lapse rate, $-dT/dz$ (°C/km) |
|---|---|
| A | 20 |
| B | 17 |
| C | 13 |
| D | 9.8 |
| E | 8 |
| F | 5 |

**9.11.** Describe a lapse rate that would result in no plume rise.

**9.12.** Butylamine vapor $(M = 73.2)$ and air are discharged through a stack of effective height $H$. Butylamine has a fishy, ammonia-like odor. The butylamine mass flow rate is 20 g/s. It is necessary that the ground-level concentration never exceed 1 mg/m³ anywhere downwind of the plant. If all atmospheric stability classes are equally proba- ble and the wind speed never falls below 2 m/s at the effec- tive stack height, what minimum effective stack height ($H$) is necessary?

**9.13.** A tunnel for automobile traffic passes under a river. The tunnel has a length $L$ (m) and cross-sectional area $A$ (m$^2$). A ventilation system withdraws air from the center of the tunnel at a volumetric flow rate $Q$ (m$^3$/h). Atmospheric air containing carbon monoxide (CO) at a concentration $c_0$ (mg/m$^3$) enters both ends of the tunnel at a volumetric flow rate $Q/2$. Automobiles traveling through the tunnel at a rate $N$ cars/h generate CO at a rate $G$ (g car$^{-1}$ km$^{-1}$). Write an expression for the steady-state CO concentration.

**(a)** Assume that the air within the tunnel can be mod- eled as well mixed throughout.

**(b)** Assume that the air within the tunnel can be mod- eled as plug flow in which the concentration varies only with length.

**9.14.** Data relayed from a weather balloon shows that the atmospheric pressure decreases linearly with altitude, $P(z) = P_0 - \alpha z$. Compute the lapse rate ($-dT/dz$).

**9.15.** To a first approximation, the ratio of the disper- sion coefficients ($\sigma_y/\sigma_z$) is a constant for each atmospheric stability class. If this is the case, show that the maximum ground-level pollutant concentration occurs at a downwind location ($x_{max}$), where

$$\sigma_z(x_{max}) = 0.707H_e$$

where $H_e$ is the effective stack height. In addition, show that the quantity $c_{max}U/Q$ is equal to $0.05857/\sigma_y\sigma_z$ when there is no ground-level reflection and to $0.1171/\sigma_y\sigma_z$ when there is ground-level reflection.

**9.16.** If there is a ground-level inversion, $dT/dz = \alpha$, derive an expression that shows how the pressure $P(z)$ varies with altitude and the ground-level temperature ($T_0$) and pressure $P_0$.

**9.17.** Paint fumes (assume toluene, $M = 174.2$) are col- lected by a paint booth inside a plant that manufactures wood furniture. The fumes are exhausted to the atmos- phere through a 15-m stack. Properties at the stack exit are:

- Toluene mass flow rate $= 10$ g/s
- Stack diameter $= 0.3$ m
- Stack temperature $= 25°$C
- Stack exit velocity of air and toluene $= 5$ m/s

**(a)** What is the plume rise on an overcast evening if the wind speed is 1 m/s and the ambient temperature is 20°C?

**(b)** How far downwind (m) of the stack will the ground- level concentration reach its maximum value?

**(c)** What is the maximum ground-level concentration ($\mu g/m^3$)?

**9.18.** An exhaust stack (effective stack height of 42 m) emits a plume containing sulfur dioxide. Is the maximum ground-level concentration for a very stable atmosphere greater or less than the maximum ground-level concentration for a very unstable atmosphere? Justify you answer quanti- tatively.

**9.19.** A weather balloon records a ground-based inver- sion up to 500 m within which $dT/dz = 0.002°$C/m. At ground level, $T_0 = 25°$C and $P_0 = 100$ kPa. For $z < 500$ m, write an expression for $P(z)$ as a function of $T_0, P_0, g$, and $R$ and the inversion lapse rate.

**9.20.** On an overcast evening when the wind speed $U_a(10$ m) is 2.5 m/s, $T_a(10$ m) $= 25°$C and the lapse rate is 8 C/km, a stack emits hydrogen sulfide (H$_2$S, $M = 34$) at a rate 0.1 g/s. Conditions at the stack exit are:

- $D_s = 1$ m
- $v_s = 10$ m/s
- $T_s = 30°$C
- $H_s = 10$ m

**(a)** Estimate the plume rise under these conditions.

**(b)** It is intended to replace the above stack to ensure that no one in the vicinity of the stack will detect the odor of H$_2$S on such an evening. If the H$_2$S odor detection level is 1 ppb, what effective stack height will be necessary? Hydrogen sulfide is not absorbed by the soil or vegetation.

# 10

# Capturing Gases and Vapors

---

In this chapter you will learn:

- To understand the physical principles on which control systems operate
- To apply these principles to design a separation process
- To select appropriate control systems for particular pollutants
- To express performance quantitatively
- To design for optimal performance

It must be emphasized that the most effective way to control pollution is to avoid producing pollutants in the first place. Every effort should be made to use materials and/or design industrial processes that do not generate pollutants. Design practices to minimize the generation of pollutants were covered in earlier chapters. This chapter is devoted to the selection and design of tailpipe devices to remove pollutant gases and vapors from a process gas stream. These processes are called pollution *control, capture,* or *removal*, whereas the term *separation* is more accu-

rate. All of these terms are used in this chapter. The physical separation models are presented in a rigorous but simplified form based on principles embodied in baccalaureate programs of chemical, mechanical, and environmental engineering. Some readers will be familiar with much of the material in this chapter, but few will have seen all of it.

In contrast with particle pollutants (covered in subsequent chapters) where separations are based on the large difference in density between particles and air, the capture of pollutant gases and vapors relies on the physical and chemical properties of the pollutant. In some cases the gas or vapor can be separated by purely physical means: for example, condensing a vapor. In other cases, the undesirable gas or vapor is adsorbed or oxidized before the process gas stream is discharged to the atmosphere. For example, volatile organic compounds (VOCs) can be oxidized to $CO_2$ and $H_2O$. Other undesirable gases or vapors can be reduced: for example, nitrogen oxides $(NO/NO_2)$ react with $NH_3$ to form molecular nitrogen $(N_2)$ and water. In all cases, the gas or vapor undergoes physical and/or chemical changes.

In this chapter the terms *gas* and *vapor* are used interchangeably. It is assumed that the polluted gas stream is an ideal mixture of ideal gases. Thus the perfect gas law and Dalton's law of partial pressures are taken for granted, with full recognition that more sophisticated models are available if needed. A process gas stream containing pollutant vapors will be referred to as a *gas stream* even if the principal constituent is air. If the term *air* is used, it will denote fresh air devoid of pollutants.

This chapter covers the removal of pollutant gases and vapors by the following processes:

- Condensation
- Adsorption,
- Absorption (wet scrubbers)
- Thermal oxidation processes
- Thermal reduction processes
- Flue gas desulfurization
- Biological processes
- UV and radical oxidation

Four examples will show how some of the foregoing methods can be used to design a control system to capture a gaseous pollutant from the industrial process gas stream described in Example 10.1. Condensation, adsorption, absorption, and thermal oxidation are considered in Examples 10.1, 10.2, 10.3 and 10.5, respectively. Section 10.6 contains a cost comparison of the three most practical methods.

# 10.1 Condensation

Condensation is not normally considered a separation process suitable for air pollution control. Rarely do pollutant vapors exist at partial pressures near their saturation values such that they can be condensed by cooling or increasing the pressure of the process gas stream. This is not to say that such occasions never occur, but when they do the process gas stream is likely to be an intermediate gas stream in a larger process rather than one discharged to the atmosphere and/or there are economic reasons to separate and reuse the vapor. The issue then is not air pollution control but process engineering. Examples of process gas streams where condensation is a feasible way to remove the vapor include:

- *Dry cleaners*: removing solvent (perchloroethylene) vapors in air used to dry freshly cleaned garments
- *Chemical manufacture*: removing solvent vapors from air used to dry a precipitate in a chemical process
- *Vent condensers*: using liquid nitrogen to reduce VOC concentration prior to its final removal by carbon adsorbers

Because condensers are rarely used and are almost always followed by an air pollution control device, a quantitative discussion of condensation will be omitted. If a condenser's performance cannot be predicted or if the manufacturer's performance data are not available, users can assume that the maximum partial pressure leaving the condenser is equal to the saturation pressure based on the condenser's cold temperature. This is the best performance a condenser can achieve. Figure 10-1 is a sketch of a typical condenser with a vapor–liquid separator.

## EXAMPLE 10.1   ENGINEERING DESIGN PROBLEM

Design a control device to reduce the emission of 1000 ppm diethylamine (DEA) in a 30,000-acfm process gas stream generated in the manufacture of textiles. The company has not been able to eliminate the emission by other methods and as a last resort wishes to install a pollution control device. The regional air pollution agency requires that the DEA emission not exceed either 50 tons/year or 500 $lb_m$/day, whichever is less. The process to be controlled normally runs 3600 h/yr (10 h/day, 360 days/yr). Specifically you are asked to recommend the important dimensions and operational parameters of the device and estimate the cost to overcome the pressure drop through the device assuming that electrical power costs \$0.07/kWh and that the fan efficiency is 75%.

*Solution*  Properties of the current process gas stream (input to control device):

$$Q = 30,000 \text{ acfm } (14.17 \text{ m}^3/\text{s}) \text{ at } 30°C,$$
$$101 \text{ kPa}$$
$$\dot{m} = 0.5694 \text{ kmol/s} = 16.45 \text{ kg/s}$$
$$36.24 \text{ lb}_m/\text{s}$$

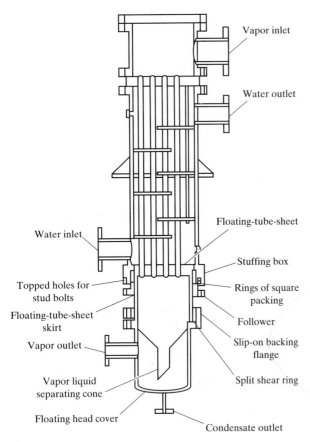

**Figure 10-1** Schematic diagram of a condenser (redrawn from Calvert and Englund, 1984)

$$c_0 = 1000 \text{ ppm} \left(2934.4 \text{ mg/m}^3\right)$$

$$\dot{m}_{\text{DEA, input}} = 41.58 \text{ g/s} \left(329.7 \text{ lb}_m/\text{h}\right)$$

$$3{,}297 \text{ lb}_m/\text{day} \left(593.46 \text{ tons/yr}\right)$$

The process DEA emissions are clearly in excess of the maximum allowable amounts. The company chooses to be conservative and decides to set as a design standard 50% lower than the two agency standards:

- *Daily emissions:* 250 lb$_m$/day = 3.15 g/s
- *Yearly emissions:* 25 tons/yr = 3.44 g/s

Thus the maximum allowable DEA emission rate after the control device will be chosen as 3.15 g/s. Such an emission rate amounts to an overall removal efficiency of

$$\eta = 1 - \frac{c_{\text{out}}}{c_{\text{in}}} = 1 - \frac{3.15}{41.58} = 0.924 \ (92.4\%)$$

Pertinent properties of DEA are the following:

- $(C_2H_5)_2NH$, $M = 73.1$
- Colorless liquid with a fishy, ammonia-like odor
- Irritates eyes, skin, and mucus membrane of the respiratory system
- Indoor workplace standard: 8-h threshold limit value 10 ppm
- Boiling point 132°F (55.5°C)
- Miscible in water
- Vapor pressure (21°C) 200 mm Hg
- Lower explosion limit 1.8%
- Freezing point −38.9°C

The vapor pressure of DEA at 21°C is 200 mm Hg, which corresponds to a mole fraction in air of 200/760 = 0.269 (269,000 ppm). Since the actual mole fraction is considerably below this value, there is little opportunity to condense diethylamine from the 30,000 acfm and satisfy the desired performance in an economical fashion. For $y_{\text{DEA}} = 0.001$, $P_{\text{DEA}} = (0.001)(760) = 0.76$ mm Hg. To reach this saturation pressure (and thus to condense DEA) the gas steam would have to be cooled below the freezing point of DEA, which is − 38.9°C (−38°F). One could do this with liquid $N_2$ and scrape off the frozen DEA from the cooling surface. But it is concluded that there are more economical ways to achieve compliance. For better solutions by other methods, see Examples 10.2, 10.3, and 10.5.

# 10.2 Adsorption

Adsorption is a process in which pollutant molecules separate themselves from the gas phase and attach themselves to the surface of solid adsorbent. Both physical and chemical characteristics play major roles. Adsorption is a preferred method of pollutant separation if:

- The pollutant is worth recovering (e.g., dry cleaning fluids).
- The concentration is very small (e.g., odors).
- The pollutant cannot be oxidized (e.g., radioactive gases from nuclear reactors).
- The pollutant is a poison (e.g., personal protection respirators for military and emergency personnel).

- Air in confined spaces is to be purified (e.g., submarines).

Adsorption is inadvisable if the process gas stream contains particles or other materials that will clog the adsorbent bed or coat the individual adsorbent particles. The adsorption of a flammable vapor from an airstream is inherently risky, owing to the chance that the adsorbed material and/or the adsorbent may catch fire. In addition, the adsorption process is exothermic. If the bed is large, substantial temperature rises may occur and steps must be taken to cool the bed.

Figure 10-2 illustrates the transport processes in adsorption. The pollutant is called an *adsorbate* or often simply *sorbate*, and the solid is called the *adsorbent*. Locations on the adsorbent surface where pollutant molecules adhere are called *active sites*. If a chemical bond occurs between the adsorbent and adsorbate, the process is called *chemisorption*. Adsorbents are very porous materials containing countless minuscule internal pores. The capacity of adsorbents is expressed by various indices, such as surface area or pore size distribution. *Surface area* is a relative term and refers to area occupied by adsorbing molecules in monolayer coverage on the surfaces within an adsorbent. The total surface area per unit mass of adsorbent is enormous, 0.1 to 1.0 km²/kg (100 to 1000 m²/g), or in more familiar terms, 20 to

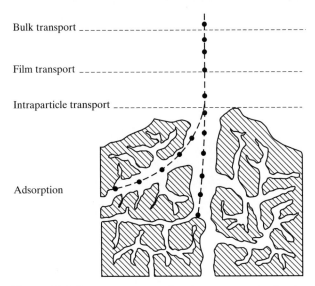

Bulk transport

Film transport

Intraparticle transport

Adsorption

**Figure 10-2** Transport process in microporous adsorbents

200 football fields per kilogram! Adsorbent particles can be made in a variety of sizes from as large as 1 cm to as small as 200 μm. Pore size distribution is related to the fraction of space within an adsorbent particle occupied by *micropores* [$D < 20$ angstroms (Å)], *mesopores* ($20 < D < 500$ Å) and *macropores* ($D > 500$ Å). Pore dimensions influence both the capacity and kinetics of adsorption. Adsorbents are grouped as follows:

1. *Activated alumina* is made from hydrated alumina, $Al_2O_3 \cdot nH_2O$, where $n$ is 1 to 3. Effective surface areas vary between 200 and 400 m²/g. Common forms are spheres 1 to 8 mm in diameter, granules, pellets 2 to 4 mm in diameter, and powders. Activated alumina is used to remove oxygenates and mercaptans from hydrocarbon feed streams and fluorides from water. Applications in the gas phase usually require preheating the adsorbent to 250°C. Alumina is widely used as a support material for catalysts and as a desiccant.

2. *Silica* ($SiO_2$) is available in a variety of forms: silica gel and porous borosilicate glass. The gel is a rigid but not crystalline ensemble of spherical microparticles of colloidal silica. The glass is open-celled and porous. The effective surface area of silica varies between 300 and 900 m²/g. Common forms of silica are beads 1 to 3 mm in diameter, granules 2 to 4 mm, and powder. Silica is used to separate hydrocarbons. Pretreatment for gas-phase adsorption may require preheating to 200°C. Other siliceous adsorbents include natural materials such as *fuller's earth* and *diatomaceous earth*.

3. *Zeolites* are generally aluminosilicates (i.e., stoichiometric compounds of silica and alumina). Compounds that have significant alumina content are usually hydrophilic, while compounds composed mainly of silica are hydrophobic. Zeolites are crystalline and possess micropores of uniform dimensions. These micropores are so uniform that they can discriminate between nearly identically sized molecules. For this reason, zeolites are also called *molecular sieves*. Virtually all commercial zeolite adsorbents are composites of very fine crystals held together by a binder. A common application of zeolites is to separate oxygen from air. Activation for gas-phase adsorption typically

requires heating to 300°C under full vacuum or an inert purge gas.

4. *Activated carbons* are derived from a variety of carbonaceous materials, including petroleum coke, hard wood, coal, peat, fruit pits, nutshells, and even recycled tires. Steam is commonly used to produce activated carbon by volatilizing of hydrocarbons and by pyrolysis. The pores of this carbonized material are either too small or constricted to be useful for adsorption. Superheated steam at 800 to 1000°C completes the process by enlarging the pores on all internal surfaces. The pore size distribution, surface area, and surface chemical composition affect strongly the capacity, selectivity, kinetics, cost, and regeneration characteristics of activated carbon. The range of surface areas in activated carbon vary from 300 to 1500 $m^2/g$, although activated carbon made from petroleum coke may exceed 3000 $m^2/g$. Common commercial forms of activated carbon include beads 1 to 3 mm in diameter, pellets 2 to 4 mm in diameter and powders. Activated carbons attract nonpolar molecules such as hydrocarbons, by van der Waals forces. Common applications for activated carbon are the removal of odors, taste, or hazardous organic compounds from drinking water, cleanup of gases containing VOCs, food decolorization, and pharmaceutical purification. Activated carbons can be impregnated with selective materials to adsorb a variety of unusual compounds (i.e., sulfuric acid removes ammonia and mercury, iron oxide removes hydrogen sulfide and mercaptans, zinc oxide removes hydrogen cyanide, and a combination of heavy metal salts removes phosgene, arsine, and nerve gas). Activation for gas-phase applications requires heating to 200°C. In Figure 10-3 are photographs of wood, coconut shell, and coal adsorbents.

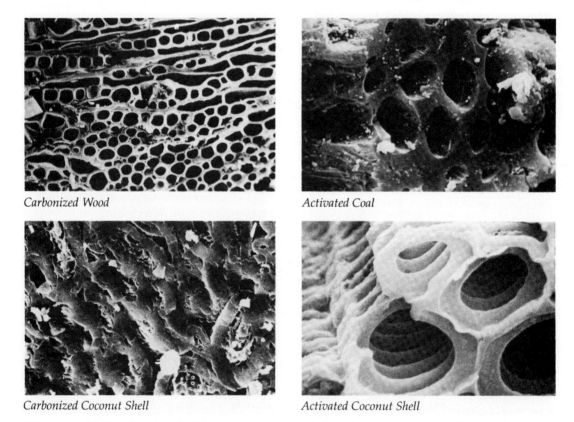

*Carbonized Wood*

*Activated Coal*

*Carbonized Coconut Shell*

*Activated Coconut Shell*

**Figure 10-3** Photographs showing the pore structure of carbonized wood and coconut shell and the enlarged pore structure of activated coal and activated coconut shell (with the permission of Barnebey and Sutcliffe Corp)

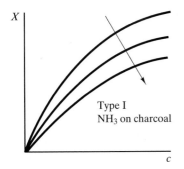

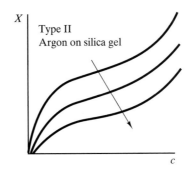

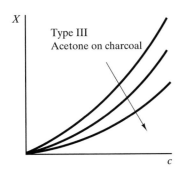

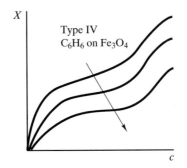

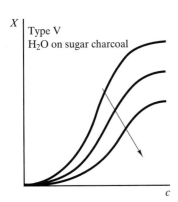

**Figure 10-4** Five types of equilibrium adsorption isotherms in which the arrows denote increasing temperature (redrawn from Calvert and Englund, 1984)

An important thermodynamic concept governing the extent of adsorption of a single component at equilibrium is the *adsorption isotherm* shown in Figure 10-4. It is important to understand the axes of adsorption isotherms:

$c$, the pollutant concentration in the gas phase, can be expressed in a variety of units, such as mg/m$^3$, ppm, mole fraction, or partial pressure.

$\chi$, the mass of adsorbate (pollutant) adsorbed per mass of adsorbent.

Figure 10-4 shows that adsorption isotherms may be concave upward or downward and they may have one or more inflection points. For many air pollution control applications, the pollutant gas-phase concentrations are thousands of parts per million or less. Thus, irrespective of the type of adsorbent, only the lower portion of the isotherm in the vicinity of the origin of the graphs is involved, and no inflection points exist there. A variety of theoretical equations are used to describe pure-component isotherms (i.e., Langmuir isotherm, Brunauer–Emmett–Teller isotherm, Redlich–Peterson isotherm, etc.) (Yu and Neretnieks, 1990). Because of its simplicity and reasonable accuracy, the empirical *Freundlich equation* is used in this book:

$$c = \alpha \chi^{\beta} \qquad (10.1)$$

Values of $\alpha$ and $\beta$ can be computed by numerical curve-fit programs based on adsorption isotherms obtained from adsorbent manufacturers or obtained experimentally. The Fruendlich equation is appropriate for type I adsorbents over a wide range of concentrations. The coefficient $\beta$ is unitless and is a small positive (or perhaps negative) number, depending on the type of adsorbent being used. The value of $\alpha$ has the same units as the pollutant concentration (i.e., partial pressure, mg/m$^3$, etc.).

If a fresh adsorbent is placed in an airstream containing a pollutant at concentration, $c_1$ (assumed to be constant), the absorbent will adsorb pollutant on its active sites until the value of $\chi_1$ corresponds to

$$c_1 = \alpha \chi_1^{\beta} \qquad (10.2)$$

If now the gas-phase concentration is reduced to a value $c_2$ (also assumed to be constant), pollutant will diffuse from the solid surface and enter the gas phase (i.e., it will desorb) until a new value of $\chi_2$ is achieved:

$$c_2 = \alpha \chi_2^{\beta} \qquad (10.3)$$

A similar argument can be given for adsorption and desorption processes when the temperature of the gas and adsorbent is decreased and increased. In general, low temperatures enhance equilibrium adsorption (i.e., higher values of $\chi$ for any value of $c$) and higher temperatures enhance desorption (or volatility). Of course, adsorption and desorption rates normally increase with temperature.

Adsorption air pollution control systems are designed as either *fixed beds* or *fluidized beds*. In fixed beds, polluted gas passes through a motionless bed of adsorbent particles until the bed is saturated (or nearly so), at which time the bed is replaced by new or regenerated adsorbent. In a fluidized bed, polluted air passes through adsorbent particles that behave as a quasi-fluid and are in a constant state of agitated motion inside a vessel. A portion of the particles are drawn off in a steady or periodic fashion, and fresh adsorbent particles are added, main-

taining a continuously active adsorbing medium. Figure 1-13 is an example of a commercially available fixed bed, and Figure 10-5 is an example of a fluidized adsorption system.

## 10.2.1 Adsorption of a single component in a fixed bed

The basic concept underlying the operation of a fixed-bed adsorption system is shown in Figure 10-6. Consider the steady flow of a gas containing a pollutant $(c_0)$ passing through a fresh porous adsorbent. For a long period of time $(t_1)$, polluted air enters the left-hand side of the bed $(x = 0)$, and pollutant-free air leaves the right-hand side of the bed $(x = L)$:

$$c(x = L) = c_L = 0 \qquad (10.4)$$

and

$$c(x = 0) = c_0 \qquad (10.5)$$

Inside the adsorbent bed at this instant $(t_1)$ there exists a zone between $x = x_1$ and $x = x_1 + \delta$ in which the pollutant concentration in the gas phase decreases from $c_0$ to zero [i.e., $c(x_1) = c_0, c(x_1 + \delta) = 0$]. The length $\delta$ is the *thickness of an adsorption wave, zone, or region*. A useful way to visualize an adsorption zone is to consider a photograph of the bed in which the wave is a region where the pollutant is being adsorbed. Upstream of the zone, $x < x_1$, the bed is saturated and the pollutant gas concentration, $c_0$, and the value of $\chi_0$ are related by Equation 10.1.

$$c_0 = \alpha \chi_0^{\beta} \qquad (10.6)$$

Downstream of the zone, $x > (x_1 + \delta)$, the gas stream contains no pollutant; hence $c = 0$ and $\chi = 0$.

At a subsequent time, $t_2 > t_1$, the zone has the same thickness $(\delta)$ but is displaced in the flow-wise direction, $x_2 > x_1$. Eventually the leading edge of the adsorption zone reaches the outlet of the adsorbent bed and the exit pollutant concentration $(c_L)$ rises steadily from zero and with the passage of time approaches $c_0$. At this time the bed is saturated (i.e., it can adsorb no additional pollutant). With respect to a stationary bed, the adsorption zone can be visualized as possessing a velocity $v_{ad}$ as it moves in the flow-wise direction.

Engineers are faced with two major design decisions: (1) they must identify adsorbents that will remove the pollutants, and (2) they need to select the bed length $(L)$ and cross-sectional area $(A)$ so that

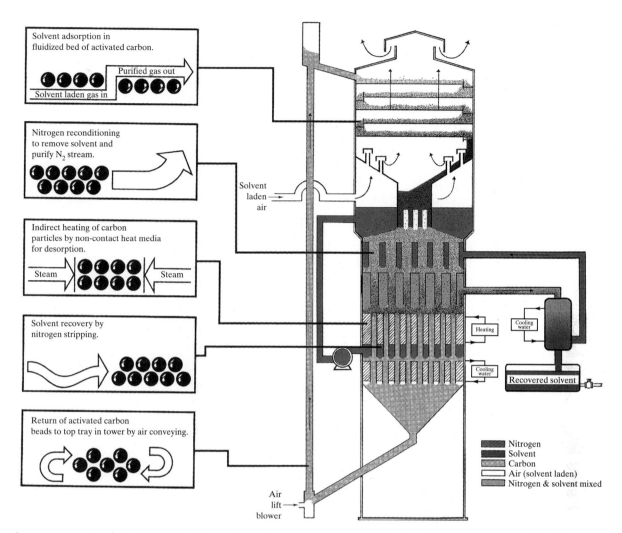

**Figure 10-5** A circulating bed adsorption system using PURASIV HR™ technology. A fluidized bed adsorbs solvents from air in the upper portion of the chamber and the activated carbon is regenerated with stream and nitrogen in the lower portion of the chamber. (Figure used with the permission of UOP, a partnership of Union Carbide Corp. and AlliedSignal, Inc.)

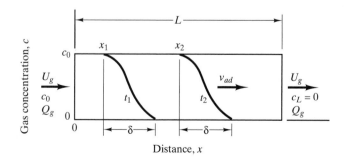

**Figure 10-6** Location of the adsorption zone at time periods $t_2 > t_1$ in a fixed bed adsorber

adsorption takes place within a time convenient to the user before the bed needs to be replaced or regenerated. Bed replacement time can be based on one of three methods.

## 10.2.2 Retention, saturation, and breakthrough times of a single component

**Retention time, $t_r$.** The mass of adsorbent $(m_{ad})$ is equal to the bed cross-sectional area $(A)$ times the bed length $(L)$ times the adsorbent bulk density $(\rho_{ad})$. Over a period of time $(t)$, the mass of pollutant that is adsorbed is equal to $Qtc_0$, where $Q$ is the actual volumetric flow rate. The *bulk density* is defined as the mass of the porous adsorbent divided by the volume (including pores) that it occupies. The retention time $(t_r)$ is the time required to saturate the adsorbent $(\chi_0)$. The value of $\chi_0$ is given by Equation 10.6. Thus

$$t_r = \frac{\chi_0 m_{ad}}{Qc_0} \qquad (10.7)$$

where

$$m_{ad} = AL\rho_{ad}$$

$$\chi_0 = \left(\frac{c_0}{\alpha}\right)^{1/\beta} \qquad (10.8)$$

and

$$t_r = \left(\frac{c_0}{\alpha}\right)^{1/\beta} \frac{m_{ad}}{Qc_0} \qquad (10.9)$$

The virtue of this method is that one can avoid computing the velocity $v_{ad}$ and thickness $(\delta)$ of the adsorption zone. The weakness in the method is that as the time $(t)$ approaches $t_r$ the adsorption zone protrudes through the bed exit and the exit pollutant concentration rises above zero. If the pollutant is not harmful or if the application allows it, the above method may be sufficient to determine when a bed needs to be replaced. In any event, the above method can be used as a crude way to estimate the bed dimensions. Once suitable bed dimensions have been selected, more accurate methods can be used to refine these estimates.

**Saturation time, $t_s$.** The time required to saturate a bed is equal to the time for an adsorption zone to travel the length of the bed,

$$t_s = \frac{L}{v_{ad}} \qquad (10.10)$$

This method requires users to compute the speed of the adsorption zone $(v_{ad})$. The method suffers from the same deficiency as the retention time, in that the pollutant concentration rises above zero as the zone begins to protrude from the bed exit.

**Breakthrough time, $t_b$.** A more accurate design practice is to design the bed so that its replacement occurs just prior to the arrival of the leading edge (right-hand edge, Figure 10-6) of the adsorption zone at the bed exit.

$$t_b = \frac{L - \delta}{v_{ad}} \qquad (10.11)$$

This method requires users to estimate the speed of the adsorption zone $(v_{ad})$ and the zone thickness $(\delta)$. If $L \gg \delta$, Equations 10.10 and 10.11 yield similar numerical results. Unfortunately, to minimize the pressure drop, engineers do not want a long bed, and if the zone thickness $(\delta)$ is just slightly smaller than the bed length $(L)$, $t_b$ will be considerably less than $t_s$. Engineers need the ability to predict the value of $\delta$ as a function of the mass transfer coefficient and gas *face* or *superficial* velocity (average gas velocity entering the bed), as it cannot be tabulated like the values of $\alpha$ and $\beta$.

## 10.2.3 Adsorption zone velocity and thickness

**Adsorption zone velocity, $v_{ad}$.** Figure 10-7 shows pollutant flowing into the zone. The mass flow of pollutant into the zone is equal to the rate at which it is adsorbed.

$$Qc_0 = \frac{\dot{m}_g}{\rho_g}c_0 = \rho_{ad}Av_{ad}\chi_0 \qquad (10.12)$$

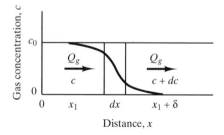

**Figure 10-7** Differential volume within an adsorption zone

where

$\dot{m}_g$ = mass flow rate of the process gas stream (pollutant plus air)

$\rho_g$ = average density of the process gas stream (pollutant plus air)

$\rho_{ad}$ = adsorbent bulk density (mass of solid/overall volume)

$c_0$ = actual pollutant concentration

Using Equation 10.8, replace $\chi_0$ and solve for the zone speed $(v_{ad})$:

$$v_{ad} = \frac{\dot{m}_g}{A\rho_g\rho_{ad}}\alpha^{1/\beta}c_0^{1-1/\beta}$$

$$= \frac{U_g}{\rho_{ad}}\alpha^{1/\beta}c_0^{1-1/\beta} \qquad (10.13)$$

where $U_g$ is the gas velocity at the inlet face of the adsorbent bed. Such a velocity is also called the *face velocity* or *superficial velocity* $(U_g)$. Face velocities are typically between 0.1 and 0.5 m/s.

$$U_g = \frac{Q_g}{A} = \frac{\dot{m}_g}{A\rho_g} \qquad (10.14)$$

Once the adsorption zone velocity has been computed, the saturation time can be computed by Equation 10.10.

**Adsorption zone thickness, δ.** Consider a differential volume $(A\,dx)$ within the adsorption zone as shown in Figure 10-7. The coordinates $(c, \chi)$ that exist in the differential volume are shown in Figure 10-8. A mass balance over the differential volume gives

$$cQ_g = (c + dc)Q_g + KA(c - c_e)\,dx \qquad (10.15)$$

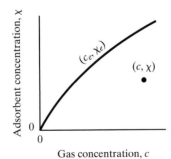

**Figure 10-8** Adsorption isotherm and coordinates (c,χ) associated with mass transfer from the gas phase to the adsorbent

where $K$ is the mass transfer coefficient and $c_e$ is the hypothetical equilibrium value of the gas-phase pollutant concentration corresponding to the actual adsorbent concentration $(\chi)$ that exists in the differential volume. The mass transfer coefficient $(K)$ is determined by independent calculations or by experiment. Values of $K$, depending on the molecular species to be adsorbed and the character of the adsorbent, are typically between 20 and 50 s$^{-1}$. Vahdat et al. (1994) suggest the following empirical equation:

$$K(\text{min}^{-1}) = \frac{14.4U_g^{0.5}}{D_{ad}^{1.5}} \qquad (10.16)$$

where $D_{ad}$ is the diameter (cm) of the adsorbent particles and $U_g$ is in the units of m/s. Replacing $c_e$ using Equation 10.1 yields

$$c_e = \alpha\chi^\beta \qquad (10.17)$$

Simplifying and integrating across the adsorption zone gives

$$\int_0^\delta dx = -\frac{\dot{m}_g}{KA\rho_g}\int_{c_0}^0 \frac{dc}{c - \alpha\chi^\beta} \qquad (10.18)$$

The expression above contains the local value of $\chi$ which varies throughout the adsorption zone. Consider the pollutant mass flow rate into the differential volume.

$$\frac{c\dot{m}_g}{\rho_g} = A\chi\rho_{ad}v_{ad} \qquad (10.19)$$

Replacing $v_{ad}$ by Equation 10.13 and solving for the local value of $\chi$ gives

$$\chi = c\alpha^{-1/\beta}c_0^{-(1-1/\beta)}$$

$$\alpha\chi^\beta = c^\beta c_0^{1-\beta} \qquad (10.20)$$

Substituting Equation 10.20 into Equation 10.18 gives

$$\int_0^\delta dx = -\frac{\dot{m}_g}{KA\rho_g}\int_{c_0}^0 \frac{dc}{c - c^\beta c_0^{1-\beta}} \qquad (10.21)$$

The integral above will be approximated because it becomes indeterminate (approaches infinity) at the lower limit. First, normalize the concentration by dividing by $c_0$, then note that there is little loss of accuracy if the upper and lower limits on the right-hand integral are replaced by 0.01 and 0.99.

$$\int_0^\delta dx = -\frac{\dot{m}_g}{KA\rho_g}\int_{0.99}^{0.01} \frac{d(c/c_0)}{c/c_0 - (c/c_0)^\beta} \qquad (10.22)$$

After integrating and rearranging, one arrives at an expression from which the adsorption zone thickness ($\delta$) can be calculated:

$$\delta \frac{KA\rho_g}{\dot{m}_g} = \delta \frac{KA}{Q_g} = \frac{\delta K}{U_g}$$

$$= 4.595 + \frac{1}{\beta - 1} \ln \frac{1 - 0.01^{\beta-1}}{1 - 0.99^{\beta-1}} \quad (10.23)$$

Equation 10.23 indicates that the thickness of the adsorption zone ($\delta$) is inversely proportional to the overall mass transfer coefficient ($K$). The higher the mass transfer coefficient, the thinner the adsorption zone. Since thin zones enable engineering designers to minimize the size of adsorbers, every opportunity should be explored to maximize the overall mass transfer coefficient. The breakthrough time can be evaluated by incorporating $\delta$ from Equation 10.23 and $v_{ad}$ from Equation 10.13 into Equation 10.11. Alternatively, one can obtain the breakthrough time from the widely used Wheeler equation (Vahdat et al., 1994):

$$t_b = \frac{W_e}{c_0 Q_g} \left( W_{ad} - \frac{\rho_{ad} Q_g}{K} \ln \frac{c_0}{c_e} \right) \quad (10.24)$$

where $W_e$ is the adsorption capacity (g of adsorbate/g of adsorbent), $W_{ad}$ is the mass of adsorbent $(AL\rho_{ad})$, and $c_e$ is the exit concentration.

The *adsorption capacity* is the maximum mass of the adsorbed species per mass of adsorbent. The adsorption capacity $(W_e)$ and mass transfer coefficient $(K)$ can be obtained from bench-top experiments in which $c_0$, $c_e$, and $Q_g$ are controlled and $t_b$ is measured for different values of $W_{ad}$. By plotting $t_b$ (ordinate) versus $W_{ad}$ (abscissa), the slope of the curve and abscissa intercept allow $W_e$ and $K$ to be determined. Mass transfer coefficients $(K_{ref})$ obtained for a reference molecular species $(M_{ref})$ at a particular temperature are related (Vahdat et al., 1994) to mass transfer coefficients for another molecular species, $M_j$, by

$$K_j = K_{ref} \left( \frac{M_{ref}}{M_j} \right)^{1/2} \quad (10.25)$$

## 10.2.4 Adsorbent replacement or regeneration

Whether an adsorbent is discarded or regenerated is largely dictated by economic considerations. Disposal in a landfill may be suitable if the adsorbent is inexpensive and if the adsorbate is not toxic, carcinogenic, or leachable. Regeneration occurs by the process of *desorption*, which can be achieved in a number of ways:

- Heating the bed
- Placing the bed under vacuum
- Stripping the bed with an inert gas
- Displacing the sorbate with a more adsorbable material
- Combinations of two or more of these process

It is useful to consider the *thermal regeneration* of granular activated charcoal as occurring in four steps.

1. *Drying* ($T < 200°C$). Remove water and highly volatile organics.
2. *Baking* ($200°C < T < 500°C$). Vaporize the less volatile adsorbates and decompose unstable adsorbates to form volatile fragments.
3. *Pyrolysis* ($500°C < T < 700°C$). Pyrolyze the nonvolatile adsorbates and adsorbate fragments to form carbonaceous char on the activated charcoal surface.
4. *Oxidation or reactivation* ($T > 700°C$). Oxidize the pyrolized residue using steam and/or $CO_2$ as the oxidizing agents.

Steam and $CO_2$ are often chosen as oxidizing agents because their reactions with carbon are endothermic, and reactivation can be managed by controlling the energy added to the regenerator. The quality of the regenerated charcoal is affected more by the oxidation step ($T > 700°C$) than the drying and baking steps. The oxidation step, known also as the *reactivation step*, is characterized by the following reactions:

*Regeneration with steam:*

$$H_2O + C_s \rightleftharpoons CO + H_2 \quad (10.26)$$

*Regeneration with carbon dioxide:*

$$CO_2 + C_s \rightleftharpoons 2CO \quad (10.27)$$

where $C_s$ is the carbonaceous char produced during pyrolysis of the skeletal activated charcoal. In addition, CO can react with steam in the *water-gas shift* reaction,

$$CO + H_2O \rightleftharpoons CO_2 + H_2 \qquad (10.28)$$

Under some circumstances, users may find it more economical to replace spent adsorbent or to contract with a vendor for regeneration of the adsorbent at the vendors facility. If the adsorption unit is sufficiently large, it is generally more economical to regenerate the adsorbent on site.

Regeneration is often facilitated by using parallel beds, one of which removes pollutant (a solvent in this case) while the other is being regenerated. In Figure 10-9 adsorbed solvent is desorbed by heated air and separated by condensation for reuse. Alternatively, a rotary configuration can be designed (Figure 10-10). If the solvent has little value and is combustible, it can be destroyed by thermal oxidation. In both cases, after regeneration, the bed needs to be cooled before it can be put back in operation. Repeated regeneration cycles subject the adsorbent to thermal shock that eventually causes it to fracture. This reduces the size of the adsorbent pellets, increases the bulk density of the bed, and affects the breakthrough time and pressure drop through the bed.

### 10.2.5 Regeneration time

During regeneration, a desorption zone progresses through the bed, usually in the opposite direction than the absorption zone. The speed of the desorption zone $(v_d)$ can be expressed by Equation 10.13, except

that the properties $\alpha_d$, $\beta_d$, gas density $(\rho_d)$ regeneration mass flow rate $(\dot{m}_d)$, and exit pollutant concentration $(c_d)$ are different than they were during the adsorption process (hence the use of the subscript $d$):

$$v_d = \frac{\dot{m}_d}{A\rho_d\rho_{ad}} \alpha_d^{\frac{1/\beta_d}{}} c_d^{1-1/\beta_d} \qquad (10.29)$$

During desorption, the coefficients $\alpha_d$ and $\beta_d$ exist at the elevated regeneration temperature. Unfortunately, the user has little idea of the exit concentration $c_d$, but it can be expressed in terms of the original inlet concentration. When desorption begins, the pollutant concentration is in equilibrium with the adsorbent as it was at the end of the adsorption portion of the cycle $(c_d \leqq c_o)$. Thus

$$c_d = \alpha_d \chi_0^{\beta_d} \qquad (10.30)$$

where

$$\chi_0 = \left(\frac{c_0}{\alpha}\right)^{1/\beta} \qquad (10.31)$$

Thus

$$c_d = \alpha_d \left(\frac{c_0}{\alpha}\right)^{\beta_d/\beta} \qquad (10.32)$$

Eliminate $c_d$ in Equation 10.29 and solve for the velocity of the desorption zone, $v_d$:

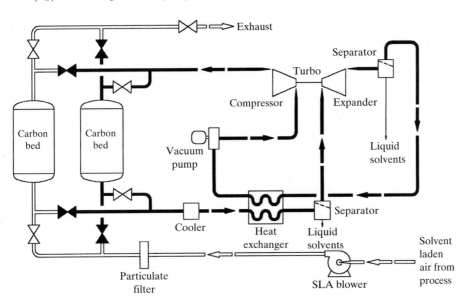

**Figure 10-9** Patented BRAYSORB ® dual-bed adsorption system using a reverse Brayton cycle produces useful mechanical work and simultaneously recovers adsorbate. For bed regeneration, air is recirculated through the loop indicated by the dark lines. Some make-up air is required. (Printed by permission of NUCON International Inc)

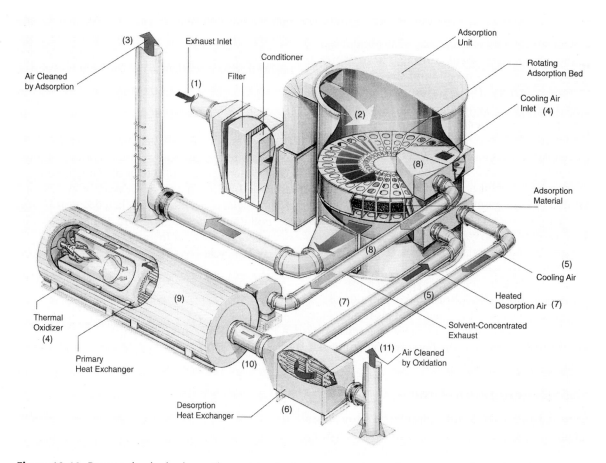

**Figure 10-10** Rotary wheel adsorber with provision for bed regeneration and recuperative thermal oxidation of the desorbed pollutant,
- The exhaust gas stream containing solvent enters at (1) and passes downward (2) through rotary adsorption bed where solvent is removed.
- The cleaned gas stream leaves the base of the rotary adsorption bed and is discharged to the atmosphere at (3).
- Clean cooling air (4) passes downward through the regenerated bed to cool the bed and prepare it for adsorption.
- Warm regeneration air leaves the regenerated bed (5), enters the desorption heat exchanger (6) where it is heated by hot exhaust leaving the thermal oxidizer (9).
- Heated desorption air (7) passes upward through the saturated adsorption bed to desorb solvent.
- Solvent-concentrated exhaust air leaves the bed (8) and enters the thermal oxidizer (9) where solvent is burned.
- Hot exhaust (10) from the thermal oxidizer enters the desorption heat exchanger (6) where it heats desorption air and finally exits at (11).
(Figure provided by Eisenmann Corp)

$$v_d = \frac{\dot{m}_d \alpha_d}{A \rho_d \rho_{ad}} \left( \frac{c_0}{\alpha} \right)^{(\beta_d - 1)/\beta} \quad (10.33)$$

$$t_d = \frac{L}{v_d} \quad (10.34)$$

The time to regenerate the adsorber $(t_d)$ can be found using either Equation 10.10 or 10.11. If the zone thicknesses $\delta$ and $\delta_d$ are much smaller than the bed length $(L)$, Equation 10.10 can be used as a first approximation. Thus the time to regenerate the adsorbent bed is

Since the adsorbent has to be cooled before it can be put back on line, the total time to regenerate can be expressed as

$$t_{\text{regeneration}} = t_d + t_{\text{cooling}} \quad (10.35)$$

## 10.2.6 Pressure drop through a fixed-bed adsorber

The pressure drop across the adsorber can be estimated using any number of empirical equations for flow through porous beds. If the Reynolds number, defined in terms of the average diameter $(D_p)$ of a granule in the bed,

$$\text{Re} = \frac{Q_g \rho D_p}{A\mu} \tag{10.36}$$

is less than 10, Crawford (1976) reports that the bed pressure drop can be estimated by the following:

$$\delta P = \frac{200\mu Q_g L f_f^2}{A D_p^2 \phi^2 (1 - f_f)^3} \tag{10.37}$$

The parameter $f_f$ is called the *packing density* or alternatively, *solids fraction*, and $\phi$ is an area ratio:

$f_f$ = bed bulk density $(\rho_{ad})$ divided by the density of the solid adsorbent, $f_f = \rho_{ad}/\rho_{absorbent}$

$\phi = \pi D_p^2/4 A_p$, where $A_p$ is the actual granule surface area

The units of $\delta P$ are dictated by the dimensions used of the variables in Equation 10.37.

### EXAMPLE 10.2 DESIGN OF A REGENERATIVE ADSORBER

Your supervisor wishes to install a rotary regenerative adsorber similar to Figure 10-10 to control the emission of diethylamine from the process described in Example 10.1. Recommend the dimensions and mass of activated carbon to meet the adsorption duty specified below and that can be regenerated by 3 kg/s of hot air at 400 K. The activated charcoal selected for the adsorber satisfies the Freundlich equation (Equation 10.1) and has the following properties:

- At 303 K:  $\alpha = 150$ kg/m$^3$,  $\beta = 2.4$
- At 400 K:  $\alpha_d = 200$ kg/m$^3$,  $\beta_d = 2.4$
- $f_f = 0.5$
- $D_p = 0.001$ m
- $\rho_{ad} = 300$ kg/m$^3$
- $K$ (mass transfer coefficient) $= 50$ s$^{-1}$
- $\phi = 0.6$

You have been instructed that working hours are 10 h/day, 3600 h/yr.

- No measurable VOC is to be emitted from the bed outlet.
- The breakthrough time should not exceed 1 h.
- The regeneration time should not exceed 4 h.
- The cooling time is equal to 2 h.

Since no VOC is allowed to pass from the bed outlet, VOC will only appear from leaks in the seals on the rotating bed. In the absence of leaks, the outlet VOC will be zero.

*Solution*  As a practical matter, the face velocity during adsorption should be between 0.1 and 0.5 m/s. At face velocities less that 0.1 m/s, the system becomes very large and at face velocities above 0.5 m/s the pressure drop across the bed is apt to be too large. The dimensions of the bed will be computed for both limiting face velocities and management will either select one of these face velocities or perform a study to optimize performance. No optimization will be undertaken in this example. From Example 10.1, the adsorber should remove DEA so that the discharge mass flow rate no larger than $\dot{m}_{DE} = 0.00315$ kg/s.

A volumetric flow rate of 30,000 acfm corresponds to an actual volumetric flow rate of 14.17 m$^3$/s. A diethylamine concentration of 1000 ppm corresponds to an actual mass concentration of 2934.4 mg/m$^3$ at 303 K, 101 kPa. The cross-sectional area $(A_{ad})$ of the adsorption chamber is equal to:

- $U_g = 0.1$ m/s: $A_{ad} = \dfrac{Q_g}{U_g} = \dfrac{14.17 \text{ m}^3/\text{s}}{0.1 \text{ m/s}} = 141.7$ m$^2$
- $U_g = 0.5$ m/s: $A_{ad} = 28.34$ m$^2$

The adsorption zone thickness ($\delta$) can be found from Equation 10.23:

$$\frac{\delta K}{U_g} = 4.595 + \frac{1}{\beta - 1} \ln \frac{1 - 0.01^{\beta - 1}}{1 - 0.99^{\beta - 1}}$$

- $U_g = 0.1$ m/s: $\delta = 0.0153$ m
- $U_g = 0.5$ m/s: $\delta = 0.076$ m

The velocity of the adsorption zone can be found from Equation 10.13.

$$v_{ad} = \frac{U_g}{\rho_{ad}} \alpha^{1/\beta} c_0^{(1 - 1/\beta)}$$

- $U_g = 0.1$ m/s: $v_{ad} = 0.0000896$ m/s $= 0.3226$ m/h
- $U_g = 0.5$ m/s: $v_{ad} = 0.0004479$ m/s $= 1.6124$ m/h

Since the breakthrough time cannot exceed 1 h, a value of 3600 s will be used for $t_b$. The thickness of the bed can be found from Equation 10.11:

$$t_b = \frac{L - \delta}{v_{ad}}$$

- $U_g = 0.1$ m/s: $L = 0.338$ m (1.11 ft)
- $U_g = 0.5$ m/s: $L = 1.689$ m (5.54 ft)

The volume of activated charcoal that is needed is equal to $AL\rho_{ad}$:

- $U_g = 0.1$ m/s: $m_{ad} = 14,360$ kg (15.81 tons)
- $U_g = 0.5$ m/s: $m_{ad} = 14,360$ kg (15.81 tons)

The pressure drop can be found from Equation 10.37 $(f_f = 0.6)$:

$$\delta P = \frac{200\, \mu Q_g L f_f^2}{A D_p^2 \phi^2 (1 - f_f)^3}$$

- $U_g = 0.1$ m/s: $\delta P = 0.693$ kPa
- $U_g = 0.5$ m/s: $\delta P = 17.32$ kPa

Electrical power to operate a 75% efficient fan to overcome the pressure drop is equal to

$$C_{\text{power}} = \frac{(\delta P)(Q)(0.07\$/\text{kWh})(3600\, \text{h/yr})}{0.75}$$

Thus

- $U_g = 0.1$ m/s: $C_{\text{power}} = \$3294/\text{yr}$
- $U_g = 0.5$ m/s: $C_{\text{power}} = \$82,451/\text{yr}$

During regeneration, the velocity of the desorption zone $(v_d)$ can be found from Equation 10.33:

$$v_d = \frac{\dot{m}_d \alpha_d}{A \rho_d \rho_{ad}} \left( \frac{c_0}{\alpha} \right)^{(\beta_d - 1/\beta)}$$

- $U_g = 0.1$ m/s: $v_d = 0.00002474$ m/s $= 0.0891$ m/h
- $U_g = 0.5$ m/s: $v_d = 0.000123$ m/s $= 0.4428$ m/h

and the time to desorb $(t_d)$ can be found from Equation 10.33.

$$t_d = \frac{L}{v_d}$$

- $U_g = 0.1$ m/s: $t_d = 3.81$ h
- $U_g = 0.5$ m/s: $t_d = 3.81$ h

In summary, adsorption wheels of two different diameters and lengths will satisfy the performance specifications. The speed of rotation should be such that the following events occur in one revolution:

$$\text{Time to adsorb} = 1\ \text{h}$$
$$\text{Time to desorb} = 3.81\ \text{h}$$
$$\text{Time to cool} = 2\ \text{h}$$
$$\text{Total cycle time} = 6.81\ \text{h}$$

Since the entire cycle (adsorption, desorption, and cooling) is to be completed in one revolution, the total surface area of the wheel $(A_{\text{wheel}})$ must satisfy

$$\frac{A_{\text{wheel}}}{6.81\ \text{h}} = \frac{A_{ad}}{1\ \text{h}} \quad \text{or} \quad A_{\text{wheel}} = 6.81 A_{ad}$$

Thus

- $U_g = 0.1$ m/s: $A_{\text{wheel}} = 965$ m$^2$, $D_{\text{wheel}} = 35$ m
- $U_g = 0.5$ m/s: $A_{\text{wheel}} = 193$ m$^2$, $D_{\text{wheel}} = 15.7$ m

The logic here assumes adsorbed-phase ideality and equilibrium control. Thus one may still expect surprises, especially if gas transport rates (manifested in the parameter $K$) dominate the uptake, as in synthetic zeolite (molecular sieve) adsorbents.

For the regenerative adsorber to operate properly, deliberate steps have to be taken to maintain seals so that the polluted gas does not short-circuit the bed and pass directly to the outlet duct. The larger wheel has a smaller pressure drop across the bed and may be easier to seal even though it has a larger seal. The smaller wheel will be more difficult to seal since the pressure drop is much larger. The thermal oxidizer that burns the desorbed diethylamine and the fan, and so on, are the same for both wheels. On the basis of capital costs, the smaller wheel has a slight advantage, but this is more than offset by its very large operating costs that are nearly 20 times larger.

## 10.2.7 Multicomponent adsorption

It is rare that a process gas stream contains only one pollutant. Removing multiple pollutants with a single adsorber may produce unusual results that engineers must anticipate. Since each component has a unique adsorption isotherm, one component will have a larger adsorption capacity $(W_e)$ than another. Consider a process gas stream containing two pollutant species 1 and 2 that enter the adsorber at concentrations $c_1(0)$ and $c_2(0)$. Figure 10-11 illustrates a

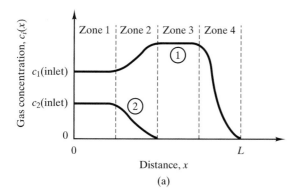

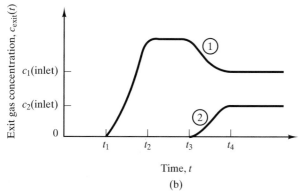

**Figure 10-11** (a) Gas phase pollutant concentration inside a fixed-bed adsorber during the adsorption of a binary mixture. (b) Concentration at the exit of a fixed-bed adsorber versus time for the adsorption of a binary mixture.

situation that engineers can expect. Figure 10-11a shows the instantaneous gas-phase pollutant concentrations of components 1 and 2 inside the bed at the instant $(t_1)$ when species 1 breaks through the exit plane of the bed. It should be noted that there is indicated a region within the bed where the gas-phase concentration of component 1 $(c_{1,\,max})$ exceeds the inlet concentration, $c_1(0)$! This phenomenon occurs when component 2 displaces component 1 from the adsorbent and reintroduces it to the process gas stream, causing its concentration to rise above its inlet value. As the two adsorption waves progress through the bed, component 2 continues to displace component 1 from the adsorbent. Over time, the profiles shown in Figure 10-11a move to the right, producing different break-through times for each species. Figure 10-11b illustrates how the gas-phase concentrations of components 1 and 2 at

the exit plane vary with time. Component 2 is said to be *more strongly adsorbed* than component 1 since the attraction between the adsorbent and component 2 is stronger than the attraction between component 1 and the adsorbent. To determine which component is more strongly adsorbed, engineers may use the concentrations of the incoming process gas stream $c_1(0)$ and $c_2(0)$ and the adsorption capacities of the pure components $(W_{e1})$ and $(W_{e2})$ at the respective concentrations. The component with the higher $W_e/c$ ratio will be adsorbed more strongly and the other will break-through first. Thus in Figure 10-11a and b, $W_{e2}/c_2(0) > W_{e1}/c_1(0)$.

The binary break-through behavior shown in Figure 10-11b is summarized below.

- *Component 1.* Break-through occurs at $t_1$.
  - $t_1 < t < t_2$. The exit concentration rises from zero to its inlet value and beyond, $0 < c_1(t) < c_{1,\,max} > c_1(0)$.
  - $t_2 < t < t_3$. The exit concentration achieves its maximum value, $c_1(t) = c_{1,\,max}$.
  - $t_3 < t < t_4$. The exit concentration decreases to its initial value, $c_1(0) < c_1(t) < c_{1,\,max}$.
- *Component 2.* Break-through occurs at $t_3$.
  - $t_3 < t < t_4$. The exit concentration increases to its original value, $0 < c_2(t) < c_2(0)$.

It is important for engineers to anticipate when the earliest breakthrough occurs, particularly if the leading component is more toxic than the other components. If the leading component is nonpolluting, engineers may want to design for the component that breaks through last. Analytical models of multicomponent adsorption are given by Mueller and DiToro (1993), Takeuchi and Shigeta (1991), Zwiebel et al. (1987), and Grant and Manes (1966). An excellent approximate mathematical model to predict break-through times is proposed by Vahdat et al. (1994).

## 10.3 Absorption

A *wet scrubber* is the generic name of an air pollution control device that uses the process of absorption to separate the pollutant from the process gas stream. Absorption is the physical (often augmented chemically) process in which the pollutant leaves the gas phase and becomes dissolved in the (scrubbing) liquid phase. Wet scrubbers are the mirror image to strippers discussed in Chapter 8, which use the process of desorption to transfer a pollutant from the liquid phase to the gas phase. Wet scrubbers will be grouped as follows:

- Spray chambers (counterflow, cross flow)
- Packed bed scrubbers (random or structured packing)
- Bubble plate and tray scrubbers
- Venturi scrubbers

Sketches of these systems are shown in Figures 1.9 and 1.10. Only a counterflow random packed bed scrubber will be modeled analytically in this chapter. There are many reference works on the subject (Perry et al, 1984), but this case will establish a common approach. Absorption is also affected by the pH of the scrubbing liquid. In some scrubbers it is desirable that the removed gas or vapor undergo a chemical reaction in the liquid phase and additives may be added to the scrubbing liquid.

Pollutants transferred to the scrubbing liquid need to be removed before the liquid is recycled. Thus scrubbers require equipment, storage vessels, additives, and so on, to treat the scrubbing liquid. In addition to air permits, users will need to obtain water permits for storage ponds and discharge permits for the transfer of treated liquid to a municipal waste water system.

A first step in the design is to select the scrubbing liquid. Water is the most commonly used material, but there are many processes in which other fluids are used. If water is chosen, the engineer should review the literature to determine the pollutant's solubility in water. Solubility is generally expressed as grams of pollutant soluble in 100 g of water. If the pollutant has a low solubility or is immiscible in water, another scrubbing fluid must be chosen if absorption is to be practical.

The physical principles governing absorption were presented in Chapter 8. On a $y_j$ (or $P_j$) versus $x_j$ diagram (see Figure 8-11), the characteristic feature about absorption is that the coordinates of the existing gas-and liquid-phase mole fractions lay above the equilibrium line, while in stripping the coordinates lay below the equilibrium line. For low concentrations, the absorption equilibrium can be expressed by Henry's law:

$$P_j = Hc_j \qquad\qquad\text{(a)}$$
$$P_j = Py_j = Hx_jc_L = H'x_j \qquad\text{(b)} \quad (10.38)$$
$$Hc_L = H' \qquad\qquad\text{(c)}$$
$$y_j = \frac{Hc_L}{P}x_j = \frac{H'}{P}x_j \qquad\text{(d)}$$

where

$H$ = Henry's law constant $(\text{atm m}^3\,\text{kmol}^{-1})$,
$H'$ = Henry's law constant (atm)
$P_j$ = partial pressure of molecular species $j$ in the gas phase
$P$ = total pressure of the gas phase,
$c_L$ = total molar concentration of the liquid phase
$c_j$ = molar concentration of $j$ in the liquid phase
$x_j$ = mole fraction of molecular species $j$ in the liquid phase
$y_j$ = mole fraction of molecular species $j$ in the gas phase

Values of $H'$ are given in Table A-8 for a variety of pollutants in water. There is no unanimity on how to express Henry's law nor on what units to ascribe to $H$. Readers can expect to see at least three forms of Henry's law used in the professional literature:

*Pressure and concentration:*

$$P_j = Hc_j \qquad\qquad\text{(a)}$$

*Pressure and mole fraction:*

$$P_j = (Hc_L)x_j = H'x_j$$
$$\text{where } H' = Hc_L \qquad\qquad\text{(b)} \quad (10.39)$$

*Mole fraction and mole fraction:*

$$y_j = \frac{P_j}{P} = \frac{Hc_L}{P}x_j = \frac{H'}{P}x_j = H''x_j$$

$$\text{where } H'' = \frac{H'}{P} = \frac{Hc_L}{P} \qquad\text{(c)}$$

If values of Henry's law constant can not be found, they can be estimated from solubility data provided that the concentrations are small. See Example 8.12 for details.

In air pollution work, pollutant concentrations are usually quite low and Henry's law is an acceptable representation of vapor–liquid equilibrium. For components present in large proportion, Raoult's law is usually more accurate. In general, for nonideal solutions of any concentration, methods exist for determining absorption equilibria (Perry et al, 1984).

## 10.3.1 Packed bed scrubber

Figure 10-12 shows the principal parts of a randomly packed fixed-bed scrubber. Not shown are ancillary components dealing with the treatment of the scrubbing liquid. The scrubbing liquid flows downward through the packing, where it encounters the process gas stream moving upward. Spray nozzles are not needed, but it is important that the liquid is distributed uniformly across the packing. Packing material is inert solid material that produces a large surface area per unit volume of the bed. Shown in Figure 10-13 are examples of random packing. The material may be plastic, ceramic, or metallic, since its function is to provide a large surface over which the liquid flows, enabling the gaseous pollutant to transfer to the liquid. Random packing should provide a bed of uniform porosity and should not fracture due to its own mass. The packing is supported by a structure that distributes the gas and liquid uniformly across the bed. Channeling, in which there is preferential passage of the gas in one location and the liquid in another location, is to be avoided.

The base of the scrubber is a sump that collects the scrubbing liquid prior to its treatment and recycling. The inlet gas header distributes the flow uniformly so that the gas velocity entering the bed is uniform. At the scrubber outlet, a *demister* or *mist eliminator* prevents particles (*carryover*) of the scrubbing liquid in the exit gas stream from leaving the scrubber. Demisters are separators that use the drop's inertia to cause them to contact the demister, coalesce, and fall downward into the bed.

## 10.3.2 Mass transfer within the packed bed

Engineers who design or select an absorption system are usually asked to recommend:

- The height ($Z$), diameter ($D$), and type of packing
- The molar flow rate of a scrubbing liquid ($L_s$)
- The pressure drop across the bed

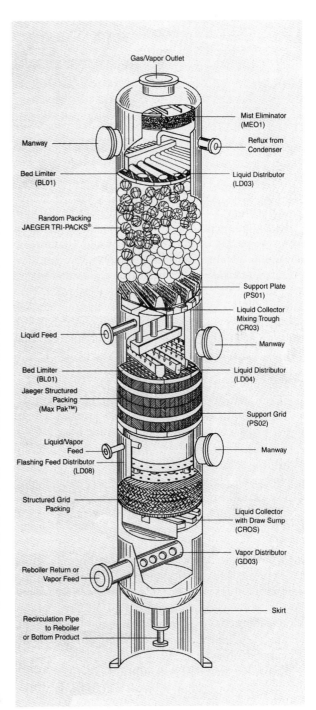

**Figure 10-12** Components of a counterflow wet scrubber: liquid collector, structured packing, liquid distributor, random packing and support plate, liquid distributor, mist eliminator. (With the permission of Jaeger Products, Inc.)

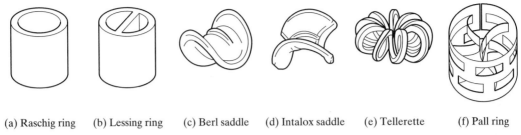

(a) Raschig ring    (b) Lessing ring    (c) Berl saddle    (d) Intalox saddle    (e) Tellerette    (f) Pall ring

**Figure 10-13** Types of random packing in a wet scrubber

so that a process can be brought into compliance with state regulations at a minimal cost. The total molar flow rate $(G_m)$ of the process gas stream (containing the pollutant) and the inlet pollutant concentration $(y_{j,\,inlet})$ are fixed by the industrial process the engineer is asked to control. The outlet pollutant concentration $(y_{j,\,outlet})$ can be no larger than a value specified by regulatory agencies or a more stringent value selected by the company. In the analysis that follows the following assumptions will be made:

- The only spatial variable is in the vertical (axial) direction, transverse gradients will be neglected.
- The liquid and gas velocities are uniform across the bed.
- No scrubbing fluid evaporates or is discharged from the scrubber exit.
- No carrier gas is absorbed in the scrubbing liquid.
- The exothermic nature of absorption will be neglected and the flow will be assumed to be isothermal.

In a particular application, one or more of these assumptions may be invalid. The last assumption (isothermality) rarely applies to commercial operations involving concentrated fluid streams and/or chemical reactions. It may be acceptable, however, for the very dilute streams found in many air pollution applications.

The inlet and outlet conditions of the packed bed are shown in Figure 10-14. The bottom of the bed is designated by the subscript 1 and the top of the bed by the subscript 2. Note that since this is a counterflow device, liquid enters at the top (2) of the bed while the gas enters at the bottom (1) of the bed. Readers are urged not to use the subscripts

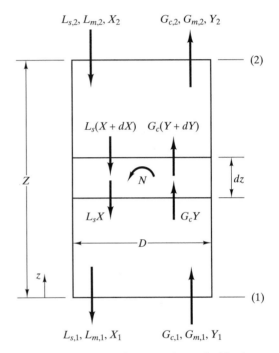

**Figure 10-14** Schematic diagram of a packed bed scrubber

(in) and (out) since they occur at different locations. Locations 1 and 2 and the notation developed forthwith are made clearer by reference to Figure 10-14.

It will be useful to describe the pollutant concentration by a new term called the *mole ratio* ($X_j$ and $Y_j$) which is related to the mole fraction ($x_j$ and $y_j$). Physically, $X_j$ and $Y_j$ represent the moles of molecular species $j$ per mole of scrubbing liquid and carrier gas, respectively. Hence

*Gas phase:*  $Y_j = \dfrac{y_j}{1 - y_j} = \dfrac{n_j/n_t}{1 - n_j/n_t}$ (a)

$$y_j = \dfrac{Y_j}{1 + Y_j}$$ (b)

(10.40)

*Liquid phase:* $X_j = \dfrac{x_j}{1 - x_j} = \dfrac{n_j/n_t}{1 - n_j/n_t}$ (c)

$$x_j = \dfrac{X_j}{1 + X_j}$$ (d)

where

$n_j$ = number of moles of molecular species $j$ in either the liquid or gas phase

$n_t$ = total number of moles in either the liquid or gas phase

$$= \sum n_j$$ (10.41)

$x_j = n_j/n_t$ = mole fraction in the liquid phase

$y_j = n_j/n_t$ = mole fraction in the gas phase

It is also useful to describe the gas flow rates by two new terms, *carrier gas molar flow rate* $(G_c)$ and the *total molar gas flow rate* $(G_m)$:

$$G_m = G_c + G_j$$ (10.42)

where $G_j$ is the molar flow rate of pollutant species $j$ in the gas phase. Thus $G_c$ is the molar flow rate of the uncontaminated carrier gas, such as air, entering or leaving the bed, while $G_m$ is the total molar flow rate of all species in the gas phase entering or leaving the bed. In a similar fashion, there are companion terms in the liquid phase,

$$L_m = L_s + L_j$$ (10.43)

where $L_m$ is the *total molar liquid flow rate* entering or leaving the bed, $L_j$ is the molar flow rate of pollutant species $j$ in the liquid phase, and $L_s$ is the *scrubbing liquid molar flow rate* of the pure scrubbing liquid, such as water.

If large amounts of the scrubbing liquid evaporate and/or if large numbers of liquid drops are carried out of the scrubber (called *carryover*, indicating faulty demister performance), $L_s$ will not be constant. If large amounts of the carrier gas are absorbed by the scrubbing liquid, $G_c$ will not be a constant. In this

analysis these events have been assumed not to occur; thus

$$L_{s,1} = L_{s,2} = L_s \quad \text{and} \quad G_{c,1} = G_{c,2} = G_c \quad (10.44)$$

The reader must keep in mind that since pollutant is transferred from the gas phase to the liquid phase, the total molar flow rate in the gas phase is larger at the inlet than at the outlet, and similarly, the total molar flow rate of liquid leaving the bed is larger than it is entering the bed. That is, both liquid and gas flows are largest at the bottom of the bed (point 1). Thus

$$G_{m,1} > G_{m,2} \quad \text{and} \quad L_{m,1} > L_{m,2} \quad (10.45)$$

The total number of moles of pollutant lost by the gas stream is equal to the total number of moles of pollutant gained by the scrubbing liquid:

$$y_1 G_{m,1} - y_2 G_{m,2} = x_1 L_{m,1} - x_2 L_{m,2} \quad (10.46)$$

Since the overall molar flow rates $(L_m$ and $G_m)$ vary between the bottom and the top of the packed bed, it is more useful to express the molar flow rates in terms of the molar flow rates of the pure scrubbing liquid $(L_s)$ and carrier gas $(G_c)$, which do not vary.

$$G_c(Y_1 - Y_2) = L_s(X_1 - X_2) \quad (10.47)$$

The *liquid-to-gas ratio* (also called *reflux ratio*) is

$$\frac{L_s}{G_c} = \frac{Y_1 - Y_2}{X_1 - X_2} \quad (10.48)$$

The molar flow rate of the carrier gas $(G_c)$ and inlet pollutant concentration $Y_1$ are known quantities fixed by the process to be controlled and $Y_2$ is dictated by the pollution regulations or company standards that need to be satisfied. The input concentration in the scrubbing liquid $(X_2)$ is an independent parameter fixed by the performance of the water treatment process. Consequently, there are two unknowns in Equation 10.48 (i.e., $X_1$ and $L_s$). Designers select $L_s$ and Equation 10.48 then determines the value of $X_1$.

Since the composition variables now being used are the mole ratios $X$ and $Y$, it is necessary to express Henry's law in these variables. Using Equations 10.38 and 10.40 gives

$$y_j = \frac{Hc_L}{P} x_j = H'' x_j$$

$$\frac{Y_j}{1 + Y_j} = H'' \frac{X_j}{1 + X_j} \qquad (10.49)$$

$$Y_j = H'' X_j \frac{1 + Y_j}{1 + X_j} = m X_j$$

The absorption equilibrium line on $X$–$Y$ axes may be redrawn as

$$Y_j = m X_j \qquad (10.50)$$

where

$$m = H'' \frac{1 + Y_j}{1 + X_j} \simeq H''$$

$$\text{if } 0 < X_j \ll 1 \quad \text{and} \quad 0 < Y_j \ll 1 \quad (10.51)$$

In this analysis, $m$ is the slope of the absorption equilibrium ($Y$–$X$) curve. The reader must be careful since even if $H$ is constant, $H'$ may not be if $c_L$ varies. Furthermore, depending on the range of the values of $X$ and $Y$, $m$ may not be a constant even if $H'$ and $H''$ are. In this analysis it will be assumed that the equilibrium curve is linear and that its slope ($m$) in $X$–$Y$ coordinates can be taken as constant.

Figure 10-15 is a $Y$–$X$ diagram on which the *equilibrium* line satisfies Equation 10.50 and values of $X$ and $Y$ satisfy Equation 10.48. Each line expressing the mass balance Equation 10.48 depends on a particular value of $L_s/G_c$ and is called an *operating line*.

There is no thermodynamic upper limit to the value $L_s/G_c$, although of course there are practical limits. Readers must realize that while the operating line is a straight line, it does not imply that the mole ratios ($X$, $Y$) vary linearly with height from the bottom to the top of the packed bed. Figure 10-16 indicates that a minimum value of $L_s/G_c$ is set by thermodynamics since the coordinates $(X_1, Y_1)$ can never lie below the equilibrium line, for to do so would indicate that pollutant can be no longer be transferred from the gas phase to the liquid phase. Recall that coordinates $(X, Y)$ lying below the equilibrium line indicate that pollutant gas is "stripped" or evaporated from the liquid, exactly opposite to what is intended in an absorption device. Thus the minimum value of $L_s/G_c$ corresponds to the value of $X_1$ that intersects the equilibrium line at the coordinates $(Y_1, X_1)$:

$$\left(\frac{L_s}{G_c}\right)_{\min} = \frac{Y_1 - Y_2}{(Y_1/m) - X_2} \qquad (10.52)$$

Figure 10-16 also shows the line corresponding to $(L_s/G_c)_{\min}$.

### 10.3.3 Mass transfer coefficients

Figure 10-17 depicts the contact between liquid and gas passing over an element of packing and the variations in mole ratios $X$ and $Y$ that exist inside each of the liquid and gas layers, respectively. The situation is identical to two-film mass transfer described in Chapter 8. Now consider a differential volume ($A \, dz$) in

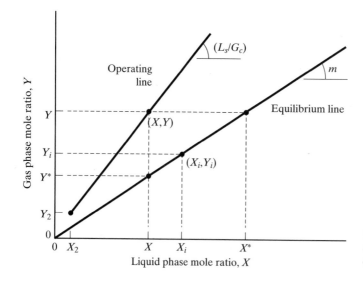

**Figure 10-15** Absorption equilibrium line and operating line

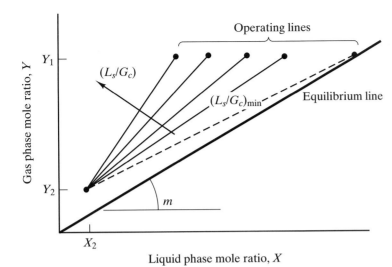

Gas phase mole ratio, $Y$

Operating lines

$(L_s/G_c)$

$(L_s/G_c)_{min}$

Equilibrium line

$Y_1$

$Y_2$

$X_2$

$m$

Liquid phase mole ratio, $X$

**Figure 10-16** Operating lines for different $(L_s/G_c)$ ratios, $(L_s/G_c) = (Y_1 - Y_2)/(X_1 - X_2)$

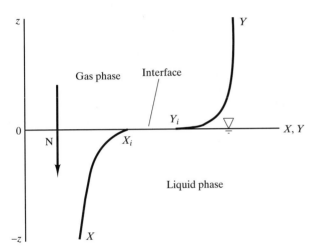

**Figure 10-17** Concentration profiles in liquid and gas phases associated with absorption

where

$k_X$ = liquid-phase mass transfer coefficient $\left(\text{kmol s}^{-1}\text{ m}^{-2}\right)$

$k_Y$ = gas-phase mass transfer coefficient $\left(\text{kmol s}^{-1}\text{ m}^{-2}\right)$

$X$ = actual liquid-phase mole ratio

$X_i$ = liquid-phase mole ratio at the gas–liquid interface

$Y$ = actual gas-phase mole ratio

$Y_i$ = gas-phase mole ratio at the gas–liquid interface

The variables $N$, $X$, and $Y$ are variables in the analysis. The variables at $(X_i, Y_i)$ are the values of the mole ratio that exist at the interface where vapor–liquid equilibrium occurs since the pollutant does not accumulate at the interface. Thus $X_i$ and $Y_i$ correspond to some unknown point on the equilibrium isotherm shown in Figure 10-15.

Because the values of $X_i$ and $Y_i$ are unknown and not easily measurable, it is useful to express the molar mass transfer rate in the alternative form on the far right-hand side of Equations 10.53a and b, which involve two new definitions:

$K_X$ = overall mass transfer coefficient for the liquid phase

the bed (Figure 10-14) and the molar transfer of pollutant $\left(N, \text{kmol s}^{-1}\text{ m}^{-2}\right)$ at some elevation $(z)$ inside the bed. The molar transfer rate $(N)$ inside the elemental volume can be expressed by the following:

$$N = k_X(X_i - X) = K_X(X^* - X) \quad \text{(a)}$$

$$N = k_Y(Y - Y_i) = K_Y(Y - Y^*) \quad \text{(b)} \quad (10.53)$$

$K_Y$ = overall mass transfer coefficient for the gas phase

$X^*$ = hypothetical (star state) liquid-phase mole ratio corresponding to the actual gas-phase mole ratio ($Y$)

$Y^*$ = hypothetical (star state) gas-phase mole ratio corresponding to the actual liquid-phase mole ratio ($X$).

The geometrical relationship between the star states and actual mole ratios is shown in Figure 10-18. The star states $X^*$ and $Y^*$ are hypothetical mole ratios that would exist at equilibrium with the actual mole ratios $Y$ and $X$. Based on the values of $X$ and $Y$, the star states may be computed by a graphical construction that can be expressed in mathematical terms of the Henry's law constant:

$$Y^* = mX \quad \text{and} \quad X^* = \frac{Y}{m} \quad (10.54)$$

The overall mass transfer coefficients $K_X$ and $K_Y$ will be developed by methods similar to those in Chapter 8. The analysis of absorption does not require the calculation of both $K_X$ and $K_Y$; thus only the overall gas-phase hypothetical mass transfer coefficient $\left(K_Y\right)$ will be computed. From Figure 10-18, the length $\left(Y - Y^*\right)$ can be expressed as

$$Y - Y^* = \left(Y - Y_i\right) + \left(Y_i - Y^*\right) \quad (10.55)$$

Using Equations 10.53 to 10-55, one can write

$$\frac{N}{K_Y} = \frac{N}{k_Y} + \left(mX_i - mX\right)$$

$$= \frac{N}{k_Y} + m\left(X_i - X\right)$$

$$= \frac{N}{k_Y} + \frac{mN}{k_X} \quad (10.56)$$

or

$$\frac{1}{K_Y} = \frac{1}{k_Y} + \frac{m}{k_X} \quad (10.57)$$

Since the slope of the equilibrium line ($m$) is known in terms of Henry's law constant, and since $k_X$ and $k_Y$ can be obtained from independent equations involving mass transfer properties, Equation 10.57 expresses the overall mass transfer coefficient $K_Y$ in terms of known parameters. Equations for the mass transfer coefficients $k_X$ and $k_Y$ are functions of the packing and the liquid and gas mass flow rates. Empirical equations for these mass transfer coefficients can be obtained from the manufacturers of packing material or from correlations in the research literature.

An instructive way to understand Equation 10.57 is to imagine that the mass transfer is proportional to

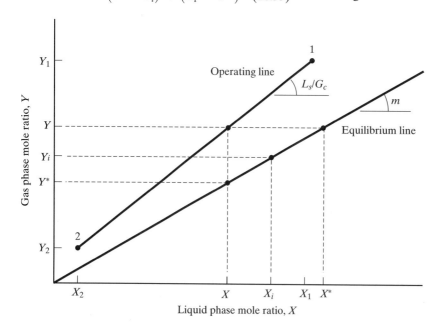

**Figure 10-18** Relationship between the star state and the actual state

a driving potential $(Y - Y^*)$ and inversely proportional to an *overall film resistance* $(R_m)$, that is, the sum of a gas-phase film resistance and a liquid-phase film resistance:

$$N = K_Y(Y - Y^*) = \frac{Y - Y^*}{1/K_Y}$$

$$= \frac{Y - Y^*}{1/k_Y + m/k_X} = \frac{Y - Y^*}{R_Y + R_X}$$

$$= \frac{Y - Y^*}{R_m} \qquad (10.58)$$

$$R_m = \frac{1}{K_Y} = R_Y + R_X \qquad (10.59)$$

$$R_Y = \frac{1}{k_Y} \qquad R_X = \frac{m}{k_X}$$

where the overall resistance to mass transfer $(R_m)$ is equal to the sum of the resistance to mass transfer in the gas phase $(R_Y)$ plus the resistance to mass transfer in the liquid phase $(R_X)$.

If we examine the consequences of choosing a scrubbing liquid that has a large or small value of the Henry's law constant (slope $m$ of the equilibrium line), we find the following:

**1.** *Small value of m.* [Pollutant highly soluble] A low value of $m$ produces a shallow slope of the equilibrium line in Figure 10-15, which implies that small values of $Y$ (which is precisely what is expected in air pollution) correspond to large values of $X$. Thus from Equation 10.48, the value of $L_s/G_c$ is very low, and only a small volumetric flow rate of scrubbing liquid will be required. The selection of such a scrubbing liquid is precisely what engineers desire (i.e., small amounts of scrubbing liquid can treat large volumes of gas). Under these conditions, $K_Y$ is essentially equal to $k_Y$, which means that the gas-phase film (resistance) controls the mass transfer. Consequently, the concentration gradient in the liquid phase is small compared to the concentration gradient in the gas phase.

**2.** *Large value of m.* [Pollutant moderately soluble] The arguments used above can now be reversed. A large value of $m$ implies a steep slope of the absorption line and a large value of $L_s/G_c$. A large volumetric flow rate of liquid will be required. This is precisely what engineers do not want. Such a liquid is an unattractive choice for a scrubber.

Returning to Equation 10.53, it is seen that the pollutant mass transfer rate,

$$N = K_Y(Y - Y^*) \qquad (10.60)$$

involves the variable $Y$ and the terms $K_Y$ and $Y^*$, which are now expressible in terms of independent equations.

### 10.3.4  Height of the packed bed

The simplified analysis that follows is valid for an equilibrium curve that can be represented by a straight line (i.e., $m$ is constant and an operating line that can be formulated from Equation 10.48),

$$Y = \frac{L_s}{G_c} X + Y_2 - \frac{L_s}{G_c} X_2 \qquad (10.61)$$

as seen in Figure 10-18. For equilibrium lines that are not straight and for a more detailed analysis, readers should consult Treybal (1980).

Consider a differential volume $(A\ dz)$ in the packed bed shown in Figure 10-14. A mass balance for the pollutant can be expressed by

$$YG_c = G_c(Y + dY) + Na_p A\ dz \qquad (10.62)$$

where $N$ is the molar transfer rate, $A$ is the cross-sectional area of the bed, and $a_p$ is a packing parameter that is equal to the total surface area of the packing per unit volume of the bed. Replace the molar transfer rate $(N)$ by Equation 10.58 and solve for $dz$:

$$\int_0^Z dz = -\int_{Y_1}^Y \frac{G_c}{a_p A K_Y} \frac{dY}{Y - Y^*} \qquad (10.63)$$

Assuming that the overall mass transfer coefficient $(K_Y)$ is constant throughout the packed bed,

$$Z = \frac{G_c}{a_p A K_Y} \int_{Y_2}^{Y_1} \frac{dY}{Y - Y^*} \qquad (10.64)$$

The term before the integral sign is defined as the *overall height of a transfer unit* (HTU),

$$\text{HTU} = \frac{G_c}{a_p A K_Y} \qquad (10.65)$$

The term has the units of length, which in many applications are on the order of 1 ft, which makes the HTU easy to visualize. As seen from Equation 10.65, the HTU incorporates the ratio of $G_C$ to $K_Y$, which makes the HTU less dependent on system flow rates than the more fundamental mass transfer coefficient.

Manufacturers of packing provide graphs of HTU as a function of gas and liquid flow rates for different chemical systems. Their clients can use these graphs in the design of absorption systems.

Figure 10-19 shows that the term $(Y - Y^*)$ in Equation 10.64 is the vertical distance between the operating and equilibrium lines at $X$. The integral is merely the reciprocal of this distance integrated in Equation 10.64 over the difference $(Y_1 - Y_2)$. For a straight equilibrium line of slope $m$, the integral can be evaluated in closed form.

$$Y - Y^* = Y - mX \qquad \text{(a)}$$

$$X = \frac{Y - Y_2 + X_2(L_s/G_c)}{L_s/G_c} \qquad \text{(b)}$$

$$ \text{(10.66)}$$

$$Y - Y^* = Y - m\frac{G_c}{L_s}\left(Y - Y_2 + X_2\frac{L_s}{G_c}\right) \qquad \text{(c)}$$

$$Y - Y^* = \alpha Y + \beta \qquad \text{(d)}$$

where

$$\alpha = 1 - m\frac{G_c}{L_s} \qquad \text{(a)}$$

$$\beta = m\frac{G_c}{L_s}\left(Y_2 - X_2\frac{L_s}{G_c}\right) \qquad \text{(10.67)}$$

$$= m\left(\frac{G_c}{L_s}Y_2 - X_2\right) \qquad \text{(b)}$$

The integral in Equation 10.64 can be expressed as

$$\int_{Y_2}^{Y_1} \frac{dY}{Y - Y^*} = \int_{Y_2}^{Y_1} \frac{dY}{\alpha Y + \beta}$$

$$= \frac{1}{\alpha}\ln\frac{\alpha Y_1 + \beta}{\alpha Y_2 + \beta} = \text{NTU} \quad \text{(10.68)}$$

The integral is unitless and is called the *number of transfer units* (NTUs). Combining Equations 10.64, 10.65 and 10.68 yields

$$Z = (\text{HTU})(\text{NTU}) \qquad \text{(10.69)}$$

Given knowledge of the height of a transfer unit (HTU), the height of an absorption column depends on the number of such transfer units (NTUs). The integration of Equation 10.68 to determine NTU can be performed conveniently on a graph such as Figure 10-20. The transfer units are stepped-off beginning at the top of the column (lowest mole ratios $Y_2$ and $X_2$) and drawing horizontal and vertical lines alternately between the operating and equilibrium lines, working up to the higher mole ratios, $Y_1$ and $X_1$. The number of horizontal lines (analogous to *equilibrium stages*) represents the number of transfer units.

An alternative method to evaluate the integral in Equation 10.64 is available if the equilibrium and operating lines are parallel, or nearly so. In this case $Y - Y^*$ is constant and its value can be taken directly from a graph like Figure 10-20:

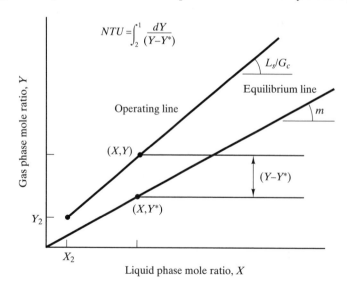

Figure 10-19  Diagram used to compute the number of transfer units

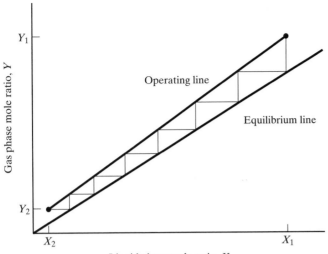

**Figure 10-20** Graphical method to compute NTU

$$\text{NTU} = \int_{Y_2}^{Y_1} \frac{dY}{Y - Y^*}$$

$$= \frac{1}{Y - Y^*} \int_{Y_2}^{Y_1} dY = \frac{Y_1 - Y_2}{Y - Y^*} \quad (10.70)$$

Equation 10.68 affords an opportunity to express the overall collection efficiency ($\eta$) as a function of the operating variables of the packed bed. Replace the value of $Y_2$ in Equation 10.68 by

$$Y_2 = (1 - \eta)Y_1 \quad (10.71)$$

After simplifying, Equation 10.68 can be written as

$$\eta = 1 - \frac{Y_2}{Y_1} = \left(1 - \frac{mX_2}{Y_1}\right) \frac{e^{(1-\theta)\text{NTU}} - 1}{e^{(1-\theta)\text{NTU}} - \theta} \quad (10.72)$$

where

$$\theta = \frac{H'}{P} \frac{G_c}{L_s} = m \frac{G_c}{L_s} \quad (10.73)$$

Using Equation 10.72, users can estimate an unknown exit pollutant concentration $(Y_2)$ if all the other operating parameters of the packed bed are known. Admittedly, $Y_2$ has to be known before a value of NTU can be determined accurately, but a crude trial-and-error process can be pursued in which an initial value of NTU can be estimated graphically assuming that $Y_2 \approx 0$. Substituting this value of NTU into Equation 10.72, a new value of $Y_2$ is obtained and the value of NTU is recomputed

graphically. The process can be repeated until satisfactory convergence is obtained. By this procedure the exit $Y_2$ can be estimated from a packed bed of known proportions ($Z$ and $A$) and operated at a known $L_s/G_c$ value.

### 10.3.5 Overall height of a transfer unit

Recalling Equation 10.57, multiplying each term by $G_c/a_p A$ and in addition multiplying the last term by $L_s/L_s$ gives

$$\frac{1}{K_Y} \frac{G_c}{a_p A} = \frac{1}{k_Y} \frac{G_c}{a_p A} + \frac{m}{k_X} \frac{G_c}{a_p A} \frac{L_s}{L_s}$$

$$\frac{G_c}{a_p A K_Y} = \frac{G_c}{a_p A k_Y} + \left(\frac{mG_c}{L_s}\right)\left(\frac{L_s}{a_p A k_X}\right) \quad (10.74)$$

The left-hand term corresponds to the overall height of a gas-phase transfer unit (HTU) in Equation 10.65:

$$\text{HTU} = \frac{G_c}{a_p A k_Y} + \left(\frac{mG_c}{L_s}\right)\left(\frac{L_s}{a_p A k_X}\right)$$

$$= \text{HTU}_c + \left(\frac{mG_c}{L_s}\right)\text{HTU}_s \quad (10.75)$$

where for the carrier gas (c) and the scrubbing liquid (s)

$$\text{HTU}_c = \frac{G_c}{a_p A k_Y} \quad (10.76)$$

$$\text{HTU}_s = \frac{L_s}{a_p A k_X} \quad (10.77)$$

Consequently, the overall height of a transfer unit (HTU) is the sum of the height of the gas-phase unit (HTU$_c$) and the height of the liquid-phase unit (HTU$_s$) multiplied by the slope of the equilibrium line and the ratio of carrier gas to liquid molar flow rates.

The height of the liquid (HTU$_s$) and gas (HTU$_c$) transfer units are given in terms of the individual mass transfer coefficients in the gas and liquid phases. The individual mass transfer coefficients $\left(k_x \text{ and } k_y\right)$ can be determined from dimensionless equations. Since the coefficients depend on the geometry of the packing, this can be a difficult task. Alternatively, vendors who sell packing provide such data. Typical of the data that can be expected is the following:

HTU$_s$ (height of a liquid transfer unit, ft)

$$= L_s/a_p \, Ak_X = \phi\left(\frac{L'}{\mu_L}\right)^\eta \mathrm{Sc}_L^{0.5} \quad (10.78)$$

where

$$\mathrm{Sc}_L = \text{liquid Schmidt number } \left(\frac{\mu}{D\rho}\right)_L$$

$L'\left(\mathrm{lb_m} \, \mathrm{h}^{-1} \, \mathrm{ft}^{-2}\right) = \text{superficial liquid mass flow rate}$

$$= \frac{L_s M_s}{A} \quad (10.79)$$

$$\mu_L = \text{liquid viscosity } \left(\mathrm{lb_m} \, \mathrm{ft}^{-1} \, \mathrm{h}^{-1}\right)$$

HTU$_c$ (height of a gas transfer unit, ft)

$$= G_c a_p \, Ak_Y = \frac{\alpha\left(\mathrm{Sc}_G\right)^{0.5}\left(G'\right)^\beta}{\left(L'\right)^\gamma} \quad (10.80)$$

where

$$\mathrm{Sc}_G = \text{carrier gas Schmidt number } \left(\frac{\mu}{D\rho}\right)_G$$

$G'\left(\mathrm{lb_m} \, \mathrm{h}^{-1} \, \mathrm{ft}^{-2}\right) = \text{superficial gas mass flow rate}$

$$= \frac{G_c M_c}{A} \quad (10.81)$$

Values of $\phi$, $\alpha$, $\beta$, $\eta$, and $\gamma$ and the range of L and G with which they are valid are shown in Table 10-1 for the packing by the selected vendor. The parameter $D$ in the Schmidt number is the molecular diffusion coefficient, and $M_s$ and $M_c$ are the molecular weights of the scrubbing liquid and carrier gas. Equations 10.78 through 10.81 are empirical equations devel-

oped by packing vendors for the different packings that they manufacture. The data in Table 10-1 require the expression of operating variables in specific, and perhaps unusual, units. Engineers have no alternative but to abide by these specifications. The equations above specify *superficial mass flow rates* $G'$ and $L'$ in the engineering units, $\mathrm{lb_m} \, \mathrm{h}^{-1} \, \mathrm{ft}^{-2}$.

If random packing is used for which the empirical data shown in Table 10-1 are not available, engineers may wish to estimate the liquid- and gas-phase mass transfer coefficients $\left(k_L \text{ and } k_G\right)$ by the commonly used Onda correlations, Equations 10.82. Dvorak et al. (1996) found that the Onda correlations are good predictors for mass transfer coefficients for large $\left(D_c > 5 \text{ cm}\right)$ random packing, although they tend to underpredict mass transfer at high gas flow rates and when the gas-film resistance is large.

$$k_G(\mathrm{m/s}) = 5.23\left(\frac{Q_G \rho_G}{a p \mu_G}\right)^{0.7}$$

$$\left(\frac{\mu_G}{\rho_G D_G}\right)^{0.33} \frac{a p D_G}{\left(a p D_c\right)^2} \quad (10.82)$$

$$k_L(\mathrm{m/s}) = 0.0051\left(\frac{Q_L \rho_L}{a p \mu_L}\right)^{0.67}$$

$$\left(\frac{\mu_L}{\rho_L D_L}\right)^{-0.5}\left(a D_c\right)^{0.4}\left(\frac{\rho_L}{\mu_L g}\right)^{-0.33}$$

where the terms above and their units are defined as follows:

$a_p = $ total packing surface area/overall bed volume $\left(\mathrm{m}^{-1}\right)$

$D_G, D_L = $ diffusion coefficients of the pollutant in gas and liquid phases $\left(\mathrm{m}^2/\mathrm{s}\right)$

$Q_G, Q_L = $ volumetric flow rates of gas and liquid $\left(\mathrm{m}^3/\mathrm{s}\right)$

$\mu_G, \mu_L = $ viscosities of gas and liquid N s m$^{-2}$

$\rho_G, \rho_L = $ densities of gas and liquid $\left(\mathrm{kg/m}^3\right)$

$A = $ packed bed cross-sectional area $\left(\mathrm{m}^2\right)$

$g = 9.8 \, \mathrm{m/s}^2$

$D_c = $ characteristic diameter of the packing (m)

A full discussion of methods to estimate the mass transfer coefficients and other parameters in the mass transfer realm is beyond the scope of this book and readers are encouraged to consult Hines and Maddox (1985).

**Table 10-1** Packing Constants

## a. Liquid-Phase HTU$_s$

| Packing | $\phi$ | $\eta$• | $L'\,(\mathrm{lb_m\ h^{-1}\ ft^{-2}})$ |
|---|---|---|---|
| Raschig rings | | | |
| $\frac{3}{8}$ in | 0.00182 | 0.46 | 400–15,000 |
| $\frac{1}{2}$ in. | 0.00357 | 0.35 | 400–15,000 |
| 1 in. | 0.0100 | 0.22 | 400–15,000 |
| $1\frac{1}{2}$ in. | 0.0111 | 0.22 | 400–15,000 |
| 2 in. | 0.0125 | 0.22 | 400–15,000 |
| Berl saddles | | | |
| $\frac{1}{2}$ in. | 0.00666 | 0.28 | 400–15,000 |
| 1 in. | 0.00588 | 0.28 | 400–15,000 |
| $1\frac{1}{2}$ in. | 0.00625 | 0.28 | 400–15,000 |
| Partion rings | | | |
| 3 in. | 0.0625 | 0.09 | 3000–14,000 |
| Spiral rings (stacked, staggered) | | | |
| 3-in. single spiral | 0.00909 | 0.28 | 400–15,000 |
| 3-in. triple spiral | 0.0116 | 0.28 | 3000–14,000 |

## b. Gas-Phase HTU$_c$

| Packing | $\alpha$ | $\beta$ | $\gamma$ | $G'\,(\mathrm{lb_m\ h^{-1}\ ft^{-2}})$ | $L'\,(\mathrm{lb_m\ h^{-1}\ ft^{-2}})$ |
|---|---|---|---|---|---|
| Raschig rings | | | | | |
| $\frac{3}{8}$ in. | 2.32 | 0.45 | 0.47 | 200–500 | 500–1500 |
| 1 in. | 7.00 | 0.39 | 0.58 | 200–800 | 400–500 |
| | 6.41 | 0.32 | 0.51 | 200–600 | 500–4500 |
| $1\frac{1}{2}$ in. | 17.30 | 0.38 | 0.66 | 200–700 | 500–1500 |
| | 2.58 | 0.38 | 0.40 | 200–700 | 1500–4500 |
| 2 in. | 3.82 | 0.41 | 0.45 | 200–800 | 500–4500 |
| Berl saddles | | | | | |
| $\frac{1}{2}$ in. | 32.40 | 0.30 | 0.74 | 200–700 | 500–1500 |
| | 0.81 | 0.30 | 0.24 | 200–700 | 1500–4500 |
| 1 in. | 1.97 | 0.36 | 0.40 | 200–800 | 400–4500 |
| $1\frac{1}{2}$ in. | 5.05 | 0.32 | 0.45 | 200–1000 | 400–4500 |
| Partion rings | | | | | |
| 3 in. | 6.50 | 0.58 | 1.06 | 150–900 | 3000–10,000 |
| Spiral rings (stacked, staggered) | | | | | |
| 3-in. single spiral | 2.38 | 0.35 | 0.29 | 130–700 | 3000–10,000 |
| 3-in. triple spiral | 15.60 | 0.38 | 0.60 | 200–1000 | 500–3000 |

*(continued)*

•these values of n are constants for Eq. 10.78 and should not be confused with the ?? symbol to denote overall collection effeciency,

**Table 10-1**    *(continued)*

**c. Packing Factor *F*, Porosity (ε), and Surface Area to Bed Volume $(a_p)$ for Random Packing**

| Packing | Normal size (in.) | | | | | | | | |
|---|---|---|---|---|---|---|---|---|---|
|  | $\frac{1}{4}$ | $\frac{3}{8}$ | $\frac{1}{2}$ | $\frac{5}{8}$ | $\frac{3}{4}$ | 1 | $1\frac{1}{4}$ | $1\frac{1}{2}$ | 2 |
| Raschig rings |  |  |  |  |  |  |  |  |  |
| Ceramic | 1600 | 1000 | 580 | 380 | 255 | 155 | 125 | 95 | 65 |
| ε | 0.73 | 0.68 | 0.63 | 0.68 | 0.73 | 0.73 | 0.71 | 0.74 | 0.74 |
| $a_p\,(\text{ft}^{-1})$ | 240 | 155 | 111 | 100 | 80 | 58 | 45 | 38 | 28 |
| Metal |  |  |  |  |  |  |  |  |  |
| $\frac{1}{32}$-in. wall | 700 | 390 | 300 | 170 | 155 | 115 |  |  |  |
| ε | 0.69 |  | 0.84 |  | 0.88 | 0.92 |  |  |  |
| $a_p\,(\text{ft}^{-1})$ | 236 |  | 128 |  | 83.5 | 62.7 |  |  |  |
| $\frac{1}{16}$-in. wall |  |  | 410 | 290 | 220 | 137 | 110 | 82 | 57 |
| ε |  |  | 0.73 |  | 0.78 | 0.85 | 0.87 | 0.90 | 0.92 |
| $a_p\,(\text{ft}^{-1})$ |  |  | 118 |  | 71.8 | 56.7 | 49.3 | 41.2 | 31.4 |
| Pall rings |  |  |  |  |  |  |  |  |  |
| Plastic |  |  |  | 97 |  | 52 |  | 40 | 25 |
| ε |  |  |  | 0.88 |  | 0.90 |  | 0.905 | 0.91 |
| $a_p\,(\text{ft}^{-1})$ |  |  |  | 110 |  | 63 |  | 39 | 31 |
| Metal |  |  |  | 70 |  | 48 |  | 28 | 20 |
| ε |  |  |  | 0.902 |  | 0.938 |  | 0.953 | 0.964 |
| $a_p\,(\text{ft}^{-1})$ |  |  |  | 131.2 |  | 66.3 |  | 48.1 | 36.6 |
| Intalox ceramic saddles | 725 | 330 | 200 |  | 145 | 98 |  | 52 | 40 |
| ε | 0.60 |  | 0.63 |  | 0.66 | 0.69 |  | 0.75 | 0.72 |
| $a_p\,(\text{ft}^{-1})$ | 274 |  | 142 |  | 82 | 76 |  | 44 | 32 |
| Berl saddles | 900 |  | 240 |  | 170 | 110 |  | 65 | 45 |
| ε | 0.60 |  | 0.63 |  | 0.69 | 0.69 |  | 0.75 | 0.72 |
| $a_p\,(\text{ft}^{-1})$ | 274 |  | 142 |  | 82 | 76 |  | 44 | 32 |

*Source:* Treybal (1980) and Wark and Warner (1981).

## 10.3.6  Pressure drop in packed beds

The power consumed by a fan needed to pass the gas through a packed bed can be expressed as

$$\text{power} = Q_g\,\delta P \qquad (10.83)$$

where $Q_g$ is the total volumetric gas flow rate and $\delta P$ is the overall pressure drop across the packed bed. The pressure drop depends on the type of packing, height $(Z)$ and diameter $(D)$ of the packed bed, and the total liquid $(L_s)$ and gas $(G_c)$ mass flow rates. Figure 10-21 typifies how the pressure drop per height $(\delta P/Z)$ varies with the volumetric gas flow rate for a particular type of packing. It is important for readers to understand that Figure 10-21 contains three flow domains.

**1.** At low values of $G_c$, $\delta P/Z$ increases with $G_c$ as if the packing were dry. It is only mildly dependent on $(L_s)$ since the liquid adheres to the packing surface and the void space in the packing is nearly constant.

**2.** At a particular value of $G_c$ called the *load point*, portions of liquid become airborne (called *holdup*), the packing void space becomes smaller, and the slope of the curve becomes steeper. As the value of $L_s$ increases, the value of $G_c$ at the load point decreases.

**3.** At values of $G_c$ above the load point, large amounts liquid (holdup) remain airborne and may

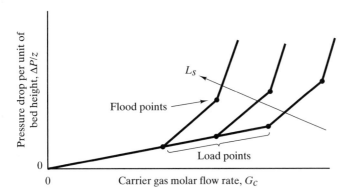

**Figure 10-21** Pressure drop in packed beds as function of the gas and liquid molar flow rates

even accumulate at the top of the packing. The packed bed is now said at be *flooded*. The value of $G_c$ at which flooding occurs is called the *flood point*. Typically, flooding produces a pressure drop of 2 to 3 in. of water/per foot of packing. The slope of the curve at values of $G_c$ above the flood point is quite steep.

As a practical matter, packed beds are operated above the load point but at 40 to 70% of the flood point. To achieve operation at 40 to 70% of flooding, the engineer will need to select the diameter of the packed bed carefully. Figure 10-22 illustrates the relationship between the pressure drop (inches of water per foot of packing) and the superficial liquid $\left(L', \text{lb}_m \text{ ft}^{-2} \text{s}^{-1}\right)$ and gas $\left(G', \text{lb}_m \text{ ft}^{-2} \text{s}^{-1}\right)$ flow rates. The term F refers to a *packing factor* which can be found in Table 10-1. Figure 10-22 compels users to use specific engineering units. The viscosity $\left(\mu_L\right)$ must be in centipoise.

$$1\text{cP} = 0.01 \text{ g cm}^{-1} \text{s}^{-1}$$

$$\mu_{\text{water}}(20°C) = 1 \text{ cP}$$

Figure 10-22 has been a staple in engineering textbooks for over twenty-five years. While more up-to-date data are available from vendors of packing, the newer formulations are similar to Figure 10-22. For purpose of initial computations, Figure 10-22 can be used with confidence.

## 10.3.7 Off-design performance

In previous sections we describe how to select the height, diameter, packing, and scrubbing mass flow rates to decrease the concentration of a pollutant to a certain value; in short, how to design a packed bed to achieve a particular collection efficiency. What remains to be discussed is how to predict *off-design per-*

*formance* for a packed bed already in existence. For example, if a packed bed of fixed dimensions and packing type is going to be used to remove a pollutant from a gas stream, how does an engineer predict the capture efficiency for a variety of gas and liquid mass flow rates? Assume that the following parameters are given:

- Packing, bed diameter, and height.
- The process gas stream total molar flow rate $G_m$.
- The inlet pollutant concentration $Y_1$.
- The concentration $X_2$ of the scrubbing liquid entering the scrubber.

The capture efficiency ($\eta$) is equal to

$$\eta = 1 - \frac{Y_2}{Y_1} \qquad (10.84)$$

To compute $Y_2$:

1. Select a scrubbing liquid molar flow rate $L_m$.
2. Compute HTU from Equation 10.75.
3. Calculate NTU from NTU = Z/HTU.
4. Substitute NTU into Equation 10.68, using Equations 10.67a and b, and solve by trial and error for $Y_2$.

While the dimensions and operating parameters of an individual packed bed scrubber depend on the specific application, the following are typical parameter values:

$L/G \approx$ 20 to 50 gpm/cfm

$L/A \approx$ 50 to 80% flood point

$Q_g/A \approx$ 3 ft/s (ceramic packing) to 6 ft/s (plastic packing)

HTU $\approx$ 1 to 1.7 ft

$\delta P/z \approx$ 0.25 in. H$_2$O/ft of packing for ceramic packing,

0.20 in. H$_2$O/ft packing for plastic packing

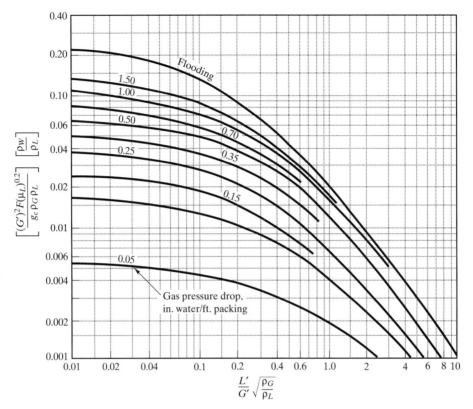

**Figure 10-22** Pressure drop in random packed towers as a function of the packing and liquid and gas mass flow rates. The variables G′ and L′ are expressed in $lb_m\ s^{-1}\ ft^{-2}$, $g_c = 32.2\ lb_m\ ft\ lb_f^{-1}\ s^{-2}$, $\mu_L$ is expressed in centipoise and $\rho_G$, $\rho_L$ and $\rho_W$ are earth densities of the gas, liquid and water expressed in $lb_m/ft^3$. F is a packing factor obtained from Table 10.1 (*permission has been granted by American Institute of Chemical Engineers*)

## EXAMPLE 10.3 DESIGN OF COUNTERFLOW PACKED BED ABSORBER

Propose the height $(Z)$, diameter $(D)$, liquid-to-gas flow rate ratio $(L_s/G_c)$, and pressure drop $(\delta P$, kPa) of a counterflow packed bed wet scrubber that uses water to remove diethylamine and satisfy the performance specifications described in Example 10.1. The company has decided to use clean water as the scrubbing liquid and to send the discharge water to another process that can tolerate dissolved diethylamine (DEA). The company wishes to use $1\frac{1}{2}$-in. ceramic Berl saddles as packing . The molar liquid-to-gas ratio $(L_s/G_c)$, is specified to be 30% larger than the minimum value. Estimate the overall pressure drop, if $(L_s/G_c)$, is to be 70% of the flood point. The densities of water and air are 62.4 $lb_m/ft^3$ and 0.0736 $lb_m/ft^3$, respectively.

**Solution** The first step is to determine the maximum emission rate that satisfies the state regulatory agency. From Example 10.1 the maximum tolerable diethylamine discharge rate $(\dot{m}_{DE})$ is 0.00315 kg/s. To satisfy this, the maximum allowable exit concentration is

$$c_2 = \frac{\dot{m}_{DEA}}{Q}$$

$$c_2(30°C, 101\ kPa) = \frac{0.00315\ kg/s}{14.17\ m^3/s}$$

$$= 0.000222\ kg/m^3 = 222\ mg/m^3$$

This corresponds to 76 ppm and an overall scrubber efficiency:

$$\eta = 1 - \frac{c_2}{c_1} = 1 - \frac{76}{1000} = 92.4\%$$

The gas concentration and flow rates at the base of the packed bed have the following values:

$$Y_1 = \frac{y_1}{1 - y_1} = \frac{0.001}{1 - 0.001} = 0.001$$

$$G_{c,1} = \left(14.17\,\text{m}^3/\text{s}\right)\left(1.1614\,\text{kg/m}^3\right)/\left(28.9\,\text{kg/kmol}\right)$$

$$= 0.5694\,\text{kmol/s}$$

$$\frac{\dot{m}_{\text{DEA}}}{M_{\text{DEA}}} = \frac{0.0458\,\text{kg/s}}{73.1\,\text{kg/kmol}} = 5.681 \times 10^{-5}\,\text{kmol/s}$$

$$G_{m,1} = 0.5694 + 4.301 \times 10^{-5} = 0.5694\,\text{kmol/s}$$

Thus at the inlet $y_1 = Y_1$ and $G_{c,1} = G_{m,1}$. Since the outlet contains even less diethylamine,

$$Y_2 = y_2 = 7.6 \times 10^{-5}$$

$$G_{c,2} = G_{m,2} = G_c = 0.5694\,\text{kmol/s}$$

The inlet DEA concentration in water is zero $\left(x_2 = X_2 = 0\right)$. The properties of the water at outlet are unknown and depend on the water mass flow rate that will be determined.

From the Table A-8, Henry's law constant $\left(H'\right)$ for diethylamine in water is $0.03658 \times 10^7\,\text{N/m}^2$. The equilibrium line in Figure E10-3 has the slope

$$Y = \left(H'/P\right)X = mX$$

$$m = \frac{H'}{P} = \frac{0.03658 \times 10^7\,\text{N/m}^2}{1.01 \times 10^5\,\text{N/m}^2} = 3.62$$

From Equation 10.52 the minimum liquid-to-gas ratio can be found:

$$\left(\frac{L_s}{G_c}\right)_{\min} = \frac{Y_1 - Y_2}{\left(Y_1/m\right) - X_2}$$

$$= \frac{0.001 - 0.000076}{0.001/3.62 - 0} = 3.334$$

The actual liquid-to-gas ratio has been selected to be 30% larger than the minimum; thus

$$\left(\frac{L_s}{G_c}\right)_{\text{actual}} = \left(1.3\right)\left(3.334\right)$$

$$= 4.3342\,\text{kmol water/kmol air}$$

From Equation 10.47,

$$X_1 = \frac{G_c}{L_s}\left(Y_1 - Y_2\right) + X_2$$

$$= \left(\frac{1}{4.3342}\right)\left(0.001 - 0.000076\right)$$

$$= 2.1318 \times 10^{-4}$$

The diameter of the packed bed is determined from Figure 10-22 by first finding the $G_c/A$ value at the flood point and then using 70% of this value, where $G'$ and $L'$ are defined as

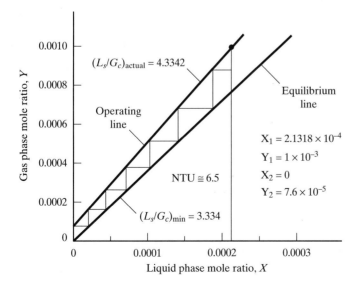

**Figure E10-3** Absorption (Y-X) diagram and graphical calculation of NTU

$$G'(\text{lb}_m\ \text{s}^{-1}\ \text{ft}^{-2}) = \frac{G_c M_c}{A}$$

$$= \frac{(0.5694\ \text{kmol/s})(28.9\ \text{km/kmol})(\text{lb}_m/0.454\ \text{km})}{A\ \text{ft}^2}$$

$$= 36.24\ \text{lb}_m\ \text{s}^{-1}/(A\ \text{ft}^2)$$

$$L'(\text{lb}_m\ \text{s}^{-1}\ \text{ft}^{-2}) = \frac{L_s M_s}{A}$$

$$= \frac{(2.4753\ \text{kmol/s})\ (18\ \text{km/kmol})\ (\text{lb}_m/0.454\ \text{km})}{A\ \text{ft}^2}$$

$$= 98.14\ \text{lb}_m\ \text{s}^{-1}/(A\ \text{ft}^2)$$

From the above, the coordinate on the (unitless) abscissa is

$$\frac{L'}{G'}\left(\frac{\rho_G}{\rho_L}\right)^{0.5} = \frac{98.14}{36.24}\left[\frac{1.1614}{1000}\right]^{0.5} = 0.0922$$

From Figure 10-22, the flood point curve has a coordinate equal to 0.15.

$$\frac{(G'_{\text{flood}})^2 F(\mu_L)^{0.2}}{g_c \rho_G \rho_L} = 0.15$$

Readers must be careful to use the units specifically called for in the caption of Figure 10-22. For ceramic $1\frac{1}{2}$-in. Berl saddles, Table 10-1 gives $F = 65$. Upon substitution and simplification, one obtains

$$\frac{(G'_{\text{flood}}\ \text{lb}_m\ \text{s}^{-1}\ \text{ft}^{-2})^2 (65)(1\ \text{cP})^{0.2}}{(32.2\ \text{lb}_m\ \text{ft}\ \text{lb}_f^{-1}\ \text{s}^{-2})(0.0736\ \text{lb}_m/\text{ft}^3)(62.4\ \text{lb}_m/\text{ft}^3)} = 0.15$$

$$(G'_{\text{flood}})^2 = \frac{0.15}{0.4395} = 0.3413$$

$$G'_{\text{flood}} = \frac{G}{A} = 0.5842\ \text{lb}_m\ \text{ft}^{-2}\ \text{s}^{-1}$$

The packing vendor recommends that the diameter be sized to operate at 70% flood point:

$$G'_{\text{design}} = (0.7)G'_{\text{flood}} = (0.7)(0.5842)$$

$$= 0.4089\ \text{lb}_m\ \text{ft}^{-2}\ \text{s}^{-1}$$

$$\dot{m}_c = (0.5694\ \text{kmol/s})$$

$$(28.9\ \text{kg/kmol})\ (\text{lb}_m/0.454\ \text{kg})$$

$$= 36.24\ \text{lb}_m/\text{s}$$

$$A_{\text{design}} = \frac{\dot{m}_c}{G'_{\text{design}}} = \frac{36.24\ \text{lb}_m/\text{s}}{0.4089\ \text{lb}_m\ \text{s}^{-1}\ \text{ft}^{-2}}$$

$$= 88.6\ \text{ft}^2$$

$$D_{\text{design}} = \left(\frac{4A}{\pi}\right)^{0.5} = 10.6\ \text{ft}\ (3.24\ \text{m})$$

Thus in subsequent computations,

$$L' = \frac{98.14\ \text{lb}_m/\text{s}}{88.6\ \text{ft}^2} = 1.11\ \text{lb}_m\ \text{s}^{-1}\ \text{ft}^{-2}$$

$$= 3988\ \text{lb}_m\ \text{h}^{-1}\ \text{ft}^{-2}$$

$$G' = 0.4089\ \text{lb}_m\ \text{s}^{-1}\ \text{ft}^{-2} = 1472\ \text{lb}_m\ \text{h}^{-1}\ \text{ft}^{-2}$$

To compute the height of the packed bed use Equation 10.69. The number of transfer units (NTUs) can be found from Equation 10.68.

$$\alpha = 1 - m\frac{G_c}{L_s} = 1 - \frac{3.62}{4.3342} = 0.1648$$

$$\beta = \frac{m}{L_s/G_c}\left(Y_2 - \frac{L_s}{G_c}X_2\right) = \frac{(3.62)(0.000076)}{4.3342}$$

$$= 6.348 \times 10^{-5}$$

$$\text{NTU} = \frac{1}{\alpha}\ln\frac{\alpha Y_1 + \beta}{\alpha Y_2 + \beta} = 6.626$$

As an alternative and more general method, the graphical construction (Figure E10-3), shows that NTU is equal to approximately 6.5.

The overall height of a transfer unit is given by Equation 10.75, and the height of liquid and gas units can be found from Equations 10.76 to 10.81. Packing costants $\phi$, $\eta$, $\alpha$, $\beta$, and $\gamma$ may be found in Table 10-1.

- *Height of a liquid unit.* The viscosity and Schmidt number for the liquid are taken to be 2.4 $\text{lb}_m^{-1}$ $\text{hr}^{-1}$ $\text{ft}^{-1}$ and 825.9.

$$\text{HTU}_s = \phi\left(\frac{L'}{\mu_L}\right)^{\eta}(\text{Sc}_L)^{0.5}$$

$$= (0.00625)\left(\frac{3988}{2.4}\right)^{0.28}(825.9)^{0.5}$$

$$= (0.00625)(7.98)(28.738)$$

$$= 1.43\ \text{ft}\ (0.437\ \text{m})$$

- *Height of a gas unit.* The Schmidt number for the gas ($\text{Sc}_c$) can be computed to be 1.806:

$$HTU_c = \frac{\alpha (Sc_c)^{0.5} (G')^\beta}{(L')^\gamma}$$

$$= \frac{(5.05)(1.806)^{0.5}(1472)^{0.32}}{(3988)^{0.45}}$$

$$= \frac{(5.05)(1.3438)(10.32)}{(41.72)}$$

$$= 1.68 \text{ ft } (0.512 \text{ m})$$

Note, the empirical data require $G'$ and $L'$ to be expressed in the units of $lb_m \ ft^{-2} \ h^{-1}$. The overall height of a transfer unit can be found from Equation 10.75:

$$HTU = HTU_c + \frac{mG_c}{L_s} HTU_s$$

$$= (0.512) + \left(\frac{3.62}{4.3342}\right)(0.437)$$

$$= 0.512 + 0.365 = 0.877 \text{ m } (2.88 \text{ ft})$$

The overall height $(Z)$ of the packed bed

$$Z = (NTU)(HTU) = (6.626)(0.877)$$

$$= 5.81 \text{ m } (19.1 \text{ ft})$$

To compute the overall pressure drop, use Figure 10-22 at the coordinates (note that $G'$ is expressed in $lb_m \ ft^{-2} \ s^{-1}$):

$$\frac{L'}{G'}\left(\frac{\rho_G}{\rho_L}\right)^{0.5} = 0.0922$$

$$\frac{(G')^2 F(\mu_L)^{0.2}}{g_c \rho_G \rho_L} = \frac{(0.4089)^2 (65)(1)^{0.2}}{(32.2)(0.0736)(62.4)}$$

$$= 0.0735$$

From Figure 10-22, $\delta P/z = 1.00$ in. $H_2O$ per foot of bed. The overall pressure drop,

$$\delta P = (1.00)(19.1 \text{ ft}) = 19.1 \text{ in. } H_2O$$

$$= (19.1 \text{ in. } H_2O)(\text{ft}/12 \text{ in.})$$

$$(1 \text{ atm}/33.9 \text{ ft } H_2O)(101 \text{ kPa/atm})$$

$$= 4.74 \text{ kPa}$$

The electrical power consumed during 1 year (3600 h) can be found from

$$C_{power} = (\delta P)(Q_g)(0.07 \text{ \$/kWh})(3600 \text{ h/yr})\left(\frac{1}{0.75}\right)$$

$$= \frac{(4.74 \text{ kPa})(14.17 \text{ m}^3/\text{s})(0.07)(3600)}{0.75}$$

$$= \$22,568/\text{yr}$$

# 10.4 Absorption and chemical reaction

There are many instances where pollutants are absorbed in a scrubbing liquid and subsequently undergo chemical reactions in the liquid phase. For example, soluble organics can be scrubbed from a gas stream and undergo oxidation with $H_2O_2$ or $KMnO_4$ dissolved in the scrubbing liquid. It is important for engineers to learn how to model such phenomena in order to design reactors. Modeling the absorption of gases in a reacting liquid involves a complex set of parallel and consecutive differential equations and is beyond the scope of this book. An excellent review of the literature can be found in De Leye and Froment (1986) and Zhukova et al. (1990). A commercially available computational program to solve these equations is available from Simulation Sciences Inc., Breva, California.

## 10.4.1 Absorption and chemical reaction in a well-mixed reactor

To simplify the process and gain physical understanding of absorption followed by chemical reaction (Sherwood et al., 1975), consider the well-mixed steady-flow reactor shown in Figure 10-23. The objective is to study how the steady-state liquid-phase concentration of pollutant depends on the rates of adsorption and reaction.

Scrubbing liquid containing no dissolved pollutant $(c_0 = 0)$ enters the reactor at a volumetric flow rate $Q_L \ (m^3/s)$. The carrier gas (air) containing pollutant enters the reactor at a molar flow rate $G_c \ (\text{kmol/s})$ through a *sparger* that produces bubbles of polluted air that are uniformly distributed within the reactor. The inlet pollutant mole fraction in the carrier gas is

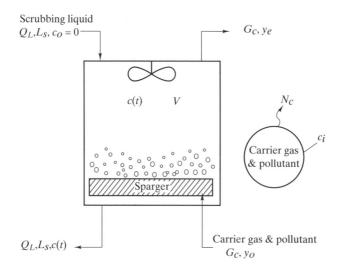

**Figure 10-23** Well mixed reactor in which the absorbed pollutant reacts chemically with the scrubbing liquid

$y_0$ (constant). The ratio of the total bubble surface area to the reactor volume is defined as $a_p$ $(m^2/m^3)$ and is assumed to be known.

The pollutant in the gas bubbles is absorbed by the scrubbing liquid. The mass transfer coefficient is $k_L$. The absorbed pollutant then reacts in the liquid phase. The reaction is assumed to be first order with respect to the pollutant concentration and the kinetic rate constant is designated as $k$. Not all of the pollutant is absorbed from the gas bubbles and the pollutant mole fraction in the carrier gas $(G_c)$ leaving the reactor is $y_e$. The scrubbing liquid exiting the reactor at a volumetric flow rate $Q_L$ contains pollutant at a steady state concentration $c_{ss}$.

The liquid-phase concentration of the pollutant at the interface of the bubble and scrubbing liquid is denoted as $c_i$. For the low concentrations encountered in air pollution, the maximum (equilibrium) value of $c_i$ is related to $y_e$ (the gas-phase pollutant mole fraction everywhere in the tank and therefore at the exit) by Henry's law,

$$c_i = y_e \frac{P c_L}{H'} \quad (10.85)$$

where $P$ is the total pressure, $H'$ is Henry's law constant, and $c_L$ is the total molar concentration of the scrubbing liquid. At steady state, $y_e$ and therefore $c_i$ will be constant. Writing a mass balance on the liquid phase alone gives the liquid-phase steady-state pollutant concentration. The balance comprises the following terms:

- Pollutant absorbed by the scrubbing liquid, $k_L a_p V(c_i - c)$
- Pollutant lost by chemical reactions in the scrubbing liquid, $kVc_{ss}$
- Pollutant transported from the reactor dissolved in the scrubbing liquid, $c_{ss}Q_L$

$$0 = k_L a_p V(c_i - c_{ss}) - Q_L c_{ss} - Vk c_{ss} \quad (10.86)$$

$$= k_L a_p c_i - c_{ss}\left(\frac{Q_L}{V} + k_L a_p + k\right) \quad (10.87)$$

Solving Equation 10.87 for the steady-state concentration $c_{ss}$ yields

$$c_{ss} = c_i \frac{k_L a_p}{(Q_L/V) + k + k_L a_p} \quad (10.88)$$

**1.** If $k_L a_p \gg (Q_L/V + k)$, $c_{ss} \simeq c_i$, which implies that the pollutant is absorbed in the liquid phase but does not react, and therefore its concentration achieves a concentration in equilibrium with the exiting gas. The result is that the chemical reagent has contributed little to the process.

**2.** If $k \gg (Q_L/V + k_L a_p)$, $c_{ss} \approx 0$ because the rapid reaction consumes all the pollutant that is absorbed into the liquid phase.

**3.** If $Q_L/V \gg (k + k_L a_p)$, then $c_{ss}$ depends on the balance between the liquid flow rate and the volume of the reactor. If $Q_L$ is large, $c_{ss} \approx 0$, because of the huge dilution effect. If $V$ is large, $c_{ss} \simeq c_i(k_L a_p V/Q_L)$. A long residence time provided by a large reactor volume would offset the effect of the low value of the mass transfer rate. This, however, is an impractical route to solving an air pollution problem.

**4.** For a practical absorption system, one would expect that the three terms in the denominator of Equation 10.88 should have comparable magnitudes.

**5.** All else being the same, a sparger that produces a large $a_p$ (many small bubbles compared to a few large bubbles) enhances the removal of pollutant from the carrier gas. (But alas, does one have to have an engineering degree to conclude this?)

## EXAMPLE 10.4   EFFECT OF CHEMICAL REACTIVITY ON GAS ABSORPTION IN A WELL-MIXED REACTOR

You have been asked by your supervisor to advise the company about the design of an air purification system. Polluted air is bubbled through ozonated water that oxidizes pollutants. What the company needs to know is how pollutant removal is influenced by:

- The ratio of scrubbing liquid flow rate to reactor volume, $Q_L/V$
- Sparger design (i.e., surface area of bubbles per volume of vessel, $a_p$)
- Mass transfer coefficient, $k_L$
- Reaction rate coefficient, $k$

To answer these questions, compute and plot $c_{ss}/c_i$ in the aqueous phase versus $k$ for the following values of $Q_L/V$, $a_p$, $k_L$, and $k$:

**a.** $125 < Q_L/V < 250 \text{ s}^{-1}$
**b.** $0.4 < k_L < 1.2 \text{ cm/s}$
**c.** $0.25 < a_p < 1.0 \text{ cm}^{-1}$
**d.** $10^{-4} < k < 10^4 \text{ s}^{-1}$

***Solution*** The solution calls for repetitive use of the equations. Several commercially available mathematical programs can be used for this purpose. The results of the computation are shown in Figure E10-4. The Mathcad program producing Figure E10-4 can be found in the textbook's Web site on the Prentice Hall Web page. The following conclusions can be drawn.

**1.** For $\log_{10} k < -2$, the concentration of pollutant in solution is governed solely by physical mass

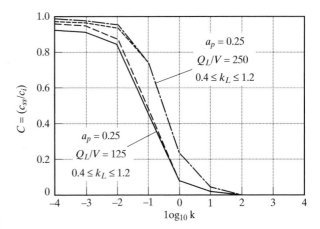

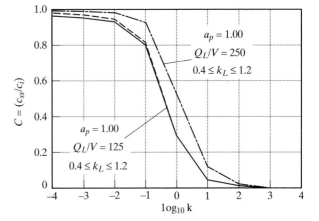

**Figure E10-4** Ratio of steady state concentration in the liquid phase to the maximum concentratrion calculated from Henry's law ($C = c_{ss}/c_i$) versus kinetic rate constant ($k$, $s^{-1}$) for $a_p = 0.25$ and $1.0$ (cm$^{-1}$) and two values of the mass transfer coefficient ($k_L$, cm/s) and ($Q_L/V$, s$^{-1}$)

transfer. The steady-state concentration $(c_{ss})$ is equal to the interface concentration $(c_i)$.

**2.** For $-2 < \log_{10} k < 2$, the concentration of pollutant in the liquid phase decreases as the rate of reaction $k$ becomes significant and the other parameters $(Q_L/V, k_L, \text{and } a_p)$ of the process become significant.

**3.** For $\log_{10} k > 2$, the concentration of pollutant in the liquid phase is very low, owing to vigorous chemical reaction occurring in the liquid phase.

**4.** For a constant absorption mass transfer coefficient $(k_L)$, the higher the value of $Q_L/V$ (i.e., short residence time), the higher the reaction rate. Similarly, for a constant residence time, the higher the mass transfer coefficient, $k_L$, the higher the reaction rate.

**5.** The value of $c_{ss}/c_i$ decreases with the kinetic rate constant $k$ but is less dependent on changes in $k_L$ and $Q_L/V$.

The most important information affecting in the design of such reactors is knowing the magnitude of the kinetic rate constant ($k$) describing the reactions in the liquid phase. If $\log_{10}k < -2$, the maximum removal of pollutant is governed by its concentration in the gas phase and its Henry's law constant. The operating parameters $Q_L/V$, $a_p$, and $k_L$ affect removal only to the extent that they increase physical absorption. If $\log_{10}k > 2$, the removal of pollutant from the gas phase is vastly improved by increasing any, or any combination of $Q_L/V$, $a_p$, and $k_L$, since vigorous chemical reaction consumes the pollutant as soon as it is absorbed in the aqueous phase. Spargers should be selected to produce large numbers of small bubbles at the same time as accommodating a large space velocity $Q_L/V$.

# 10.5 Thermal oxidation processes

The term *incineration* will not used to describe thermal oxidation because of its common association with municipal solid waste incineration. In this chapter it is assumed that the pollutant is a gas or vapor (although the theory can be used for the oxidation of limited amounts of liquid and solid matter). Municipal solid waste incineration is not discussed.

For there to be successful thermal oxidation, regulatory agencies require that the following conditions must be satisfied:

**1.** The pollutant must reside in the hot region for some agreed amount of *residence time*, typically 0.5 to 1.0 s.
**2.** The *average temperatures* in the hot region must be within certain limits, typically 1050 to 1250 K for direct flame oxidizers and 700 K for catalytic oxidizers.
**3.** The concentrations and mass emission rates of the exhaust products must be within certain limits.

Specific values of the residence time and average oxidation temperature depend on the pollutant being oxidized.

A principal drawback to thermal oxidizers is the generation of nitrogen oxides and possibly new pollutants. Whether new pollutants will be generated by oxidation is difficult to anticipate, although certain classes of waste streams are apt to produce certain type of pollutants. For example, it can be anticipated that oxidizing chlorinated compounds may produce an exhaust stream in which HCl or free chlorine are discharged to the atmosphere unless steps are taken to adsorb or absorb them. Thus engineers must take steps to ensure that thermal oxidizers are not new sources of pollution.

Thermal oxidation is a generic phrase and includes all thermal devices where air and an auxiliary fuel are used to produce a region of high temperature in which pollutants are oxidized. Thermal oxidation devices can be divided as follows (Figure 1.14):

- Afterburners
- Direct-flame oxidizers with energy recovery
  - Regenerative oxidizer
  - Recuperative oxidizer
- Catalytic oxidizers

*Afterburners* (Figure 7-3) are elementary combustors that have low initial and maintenance costs per cfm but high operating costs. Afterburners are attractive for small installations when the pollutant mass flow rate is small and the composition apt to vary with time. Oxidizers with heat recovery are attractive when the process gas stream is large and continuous and when transferring the energy of the exhaust to the incoming process gas stream is financially attractive. Thus per cfm treated, heat recovery oxidizers have a high initial cost but minimal operating costs.

Process gas streams associated with the laminating, surface finishing and coating, chemicals, and wood products industries are particularly difficult to control. Such streams contain solid particles and high-molecular-weight hydrocarbons that produce sticky condensates. Treating these process gas streams is particularly difficult and causes serious operating and maintenance problems. If the process or raw materials cannot be changed to prevent generating these pollutants, the only available remediation technology is *regenerative thermal direct flame oxidizers*, Figure 10-24.

The characteristic feature of regenerative oxidizers are beds of ceramic material through which flow

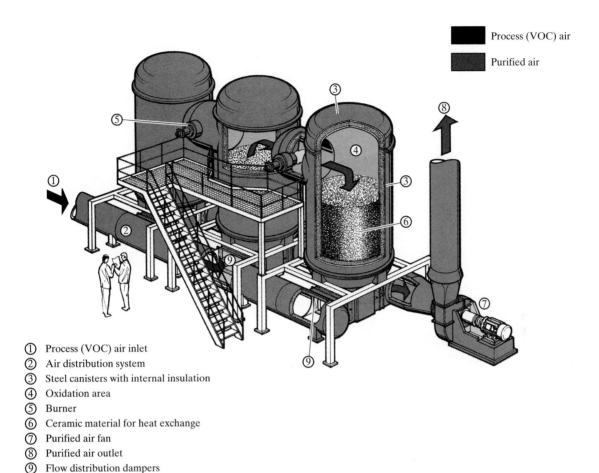

Process (VOC) air

Purified air

① Process (VOC) air inlet
② Air distribution system
③ Steel canisters with internal insulation
④ Oxidation area
⑤ Burner
⑥ Ceramic material for heat exchange
⑦ Purified air fan
⑧ Purified air outlet
⑨ Flow distribution dampers

**Figure 10-24**  Three-bed regenerative direct-flame thermal oxidizer for VOC control. (Figure used with the permission of Wahlco, Inc. Wahlco, Inc., VOC systems are manufactured under license from LTG Lufttechnische GmbH, Stuttgart, Germany.)

both hot exhaust products for one period of time and the cool inlet process gas stream for other periods of time. Thus the process gas stream flip-flops. The heat-resistant material in the regenerative oxidizer bed, stores energy from hot gases exiting the combustion chamber while another preheated regenerator releases heat to the cold gases entering the combustion chamber. This switching of the gas flow between regeneration beds is accomplished by large damper valves that alternately open and close. As a result of this switching, a pulse of untreated waste gas always escapes to the discharge side of the oxidizer and limits the overall oxidation efficiency to a maximum of 95%. Owing to the robust and large (1-in.) ceramic packing elements, regenerative oxidizers can

cope with gas streams containing sticky particulate matter or pollutants that might condense on the cooled bed and may coat the packing for a time. The packed beds are heated periodically to *bake-out* this particulate matter. If, however, the particulate matter is not combustible, the porosity of the packing may decrease over time.

*Recuperative oxidizers* (Figure 10-25), use high-efficiency air-to-air heat exchangers to recover energy. Heat is transferred from the hot exhaust to the cool incoming combustion air through a conducting surface that separates the gases (i.e., shell-tube or plate heat exchangers). Recuperative oxidizers tend to have lower capital costs than regenerative oxidizers for volumetric flow rates less than 40,000 cfm, yet

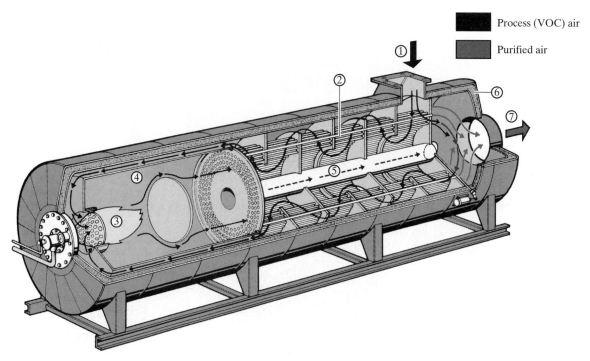

① Process (VOC) inlet
② Tubular heat exchanger for
   internal heat recovery
③ Conus burner
④ Combustion chamber
⑤ Internal bypass
⑥ Insulation
⑦ Purified air outlet

**Figure 10-25**  Recuperative direct-flame thermal oxidizer for VOC control. (Figure used with the permission of Wahlco, Inc. Wahlco, Inc., VOC systems are manufactured under license from LTG Lufttechnische GmbH, Stuttgart, Germany.)

have higher operating costs than regenerative oxidizers for all volumetric flow rates. Recuperative oxidizers are less forgiving than regenerative oxidizers since the narrow gas passages become irreparably fouled when handling dirty particulate-laden gas streams or gas streams containing pollutants that are apt to condense. The overall pollutant destruction efficiency of recuperative oxidizers is in general higher than in regenerative oxidizers. On the other hand, the efficiency of heat recovery in recuperative oxidizers is generally lower than in regenerative oxidizers. Recuperative oxidizers cannot be baked out as easily as regenerative oxidizers.

*Catalytic oxidizers* (Figure 10-26) contain catalytic surfaces that enable reactions to occur at lower temperatures, typically 300 to 800°F, than would be possible in the gas phase without the catalyst. The catalyst is not consumed in a reaction; it is merely the active surface agent that enables the chemical reaction to occur, whereas it would not occur, thermally. A catalyst is composed of a ceramic or metal substrate with a high surface area-to-volume ratio. On the surface of the substrate is a thin layer of catalytic material. Catalytic surfaces may be *deactivated* by exposure to certain materials. Deactivation may be irreversible or reversible.

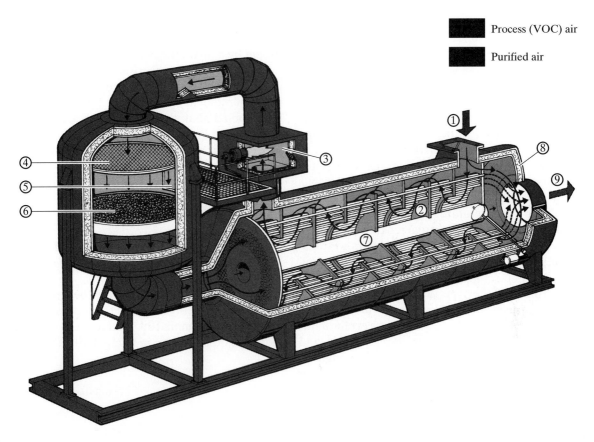

Process (VOC) air

Purified air

① Process (VOC) inlet
② Tube bundle heat exchangeer for internal heat recovery
③ Gas burner
④ Distribution grate
⑤ Assembly opening
⑥ Catalyst bed
⑦ Internal bypass
⑧ Insulation jacket
⑨ Purified air outlet

**Figure 10-26** Catalytic recuperative oxidizer for VOC control. (Figure used with the permission of Wahlco, Inc., Wahlco, Inc., VOC systems are manufactured under license from LTG Lufttechnische GmbH, Stuttgart, Germany.)

**1.** *Irreversible deactivation.* Poisons such as phosphorus, bismuth, lead, arsenic, antimony, mercury, iron oxides, tin, and silicon produce irreversible damage. Catalytic surfaces can also be damaged permanently if the temperatures are too high, causing the substrate or catalytic surface to sinter.

**2.** *Reversible deactivation.* Sulfur in a reducing environment, halogens, zinc, and solid organic materials produce damage that may be corrected by washing with detergent, acidic, or caustic fluids. It may be possible to remove organic materials deposited (blinding) on a catalytic surface by baking or burning.

Catalytic oxidizers (Figure 10-26), have low fuel costs because they operate at lower temperatures than direct flame devices but are useful mainly when the composition of the process gas steam is known and does not vary. Catalytic oxidizers are ideally suited for the oxidation of VOCs, but specially designed catalysts are required for the oxidation of halogenated hydrocarbons. Gases containing chlorine and sulfur can deactivate noble metal catalysts such as platinum or palladium. Lead, arsenic, and phosphorus are generally considered poisons for most oxidation catalysts. Because oxidation occurs at low temperature, catalytic oxidizers produce minimal amounts of nitrogen oxides. Catalytic oxidizers can incorporate heat exchangers to transfer energy from the exhaust stream to the incoming airstream.

Regardless of the type of oxidizer, designing a chamber in which pollutants reside for no less than a certain amount of time and achieve a certain minimum temperature is similar to the design of any kind of reaction chamber (internal combustion engines, jet engines, rocket engines, chemical reactors, steam boilers, etc.). Governing the design are the conservation equations of mass, momentum, and energy modified to include terms describing the formation and destruction of certain combustion species. Describing how to solve and manipulate these equations for this purpose is beyond the scope of this book. For the discussion that follows:

1. The reactor will be assumed to be well mixed such that the temperature is uniform.
2. The velocity field within the reactor will be neglected; consequently, the time (*mean residence time*) that the pollutant resides in the reaction chamber is equal to

$$t_r = \frac{V}{Q_g} \qquad (10.89)$$

The task to be undertaken in this section is to demonstrate how to estimate the fuel required to achieve a certain temperature in a steady-state, direct-flame recuperative thermal oxidizer. The analysis is also valid for the design of a catalytic oxidizer with heat recovery. This task will be under-

taken presuming that a portion of the energy in the exhaust products is transferred to the incoming gas stream by a high-efficiency heat exchanger.

To estimate the fuel costs of a direct-flame recuperative thermal oxidizer, engineers can employ basic thermodynamic concepts. For simplicity it will be assumed that the composition of the exhaust products can be specified and that all gases are ideal with constant specific heats. At a later time the computational accuracy can be improved by taking into account other reactions (e.g., dissociation) that occur at high temperatures.

You have been asked to estimate the cost of natural gas to oxidize diethylamine in the process gas stream described in Example 10.1. Consider the recuperative thermal oxidizer shown in Figure E10-5 that operates

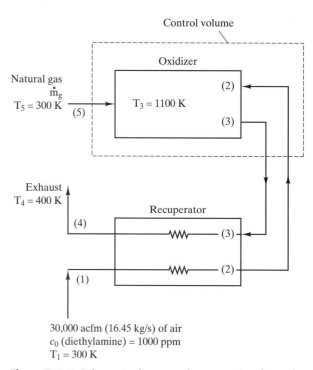

**Figure E10-5**  Schematic diagram of recuperative thermal oxidizer

described in Example 10.1. Consider the recuperative thermal oxidizer shown in Figure E10-5 that operates for 3600 h/yr. The state regulatory agency requires that the process gas stream be maintained at no less than 1100 K for at least 0.75 s and that the exhaust gas temperature must be no higher than 400 K. The oxidizer has a total exterior surface area of 2000 m² and will be located outdoors. The temperature ($T_s$) of the exterior surface of the oxidizer is 325 K. Heat transfer from the surface of the oxidizer to the atmosphere has an overall heat transfer coefficient ($h$) of 0.1 kJ s⁻¹ m⁻². The manufacturer of the thermal oxidizer guarantees that the efficiency of the heat exchanger, expressed as a thermal energy recovery (TER), is 87.5%. The overall pressure drop for the unit with this heat exchanger is 17 kPa. The fan efficiency is 75% and the cost of electricity is $0.0596/kWh.

$$\text{TER} = \frac{\dot{m}_a(h_2 - h_1)}{\dot{m}_a(h_3 - h_4)} = \frac{T_2 - T_1}{T_3 - T_4} = 0.875$$

$$T_2 = (0.875)(1100 - 400) + 300 = 912 \text{ K}$$

**Solution**  To estimate the natural gas mass flow rate, define the control volume to include only the oxidizer, and apply the first law of thermodynamics for an open system. Neglect changes in potential energy and kinetic energy:

$$\dot{Q} - \dot{W}_s = \dot{m}_3 h_3 - \dot{m}_2 h_2 - \dot{m}_g h_5$$

where $\dot{Q}$ is the rate of heat transfer into the control volume, $\dot{W}_s$ is output shaft power from the control volume extracted by a shaft, and $\dot{m}_g$ is the mass flow rate of natural gas. Since there is no shaft present in the system (Figure E10-5), the shaft power is zero. The natural gas mass flow rate is small compared to the mass flow rate of the process gas stream. Thus

$$\dot{m}_2 = \dot{m}_3 = \dot{m}_1 = \dot{m}_a$$

where $\dot{m}_a$ is the mass flow rate of the process gas stream. The equation above will be written in molar form:

$$\dot{Q} = \sum_i^{\text{products}} \dot{n}_i \bar{h}_i - \sum_j^{\text{reactants}} \dot{n}_j \bar{h}_j$$

Write the molar enthalpy of species $i$ at any $P$, $T$ as

$$\bar{h}_i(T, P) = \bar{h}_{f,i}^0(298 \text{ K}, 100 \text{ kPa}) + \delta \bar{h}_i$$

where $h_{f,i}^0$ is called the *enthalpy of formation* and is zero for all species $i$ in their elemental state at 298 K and 100 kPa. The enthalpy of formation for compounds is a thermodynamic property obtained from thermodynamic tables (e.g., Perry et al, 1984). Thus $h_{f,i}^0$ is zero for $O_2$, $N_2$, and so on, but is nonzero for $CO_2$, $CH_4$, and so on.

$$\delta \bar{h}_i = \bar{h}_i(T, P) - \bar{h}_{f,i}^0(298 \text{ K}, 100 \text{ kPa})$$

Thus

$$\dot{Q} = \sum_i^{\text{products}} \dot{n}_i \left( \bar{h}_{f,i}^0 + \bar{h}_i(T, P) - \bar{h}_{f,i}^0 \right)$$
$$- \sum_j^{\text{reactants}} \dot{n}_j \left( \bar{h}_{f,j}^0 + \bar{h}_j(T, P) - \bar{h}_{f,j}^0 \right)$$

For a first approximation, assume that gases are ideal and that the specific heat, $c_{p,i}$ is constant. Thus

$$\delta \bar{h}_i = \bar{c}_{p,i} \delta T$$

Thus

$$\dot{Q} = \sum_i^{\text{products}} \dot{n}_i \left[ \bar{h}_{f,i}^0 + \bar{c}_{p,i}(T - 298) \right]$$
$$- \sum_j^{\text{reactants}} \dot{n}_j \left[ \bar{h}_{f,j}^0 + \bar{c}_{p,j}(T - 298) \right]$$

Because the control volume was defined as the exterior shell of the thermal oxidizer, the only heat transfer is between the oxidizer's outer surface and the atmosphere. The large energy generation within the oxidizer is contained within it and hence not a part of the term $\dot{Q}$. The heat transfer ($\dot{Q}$) into the oxidizer from the atmosphere can be expressed as

$$\dot{Q} = A_s h(T_0 - T_s)$$

$$= (2000 \text{ m}^2)(0.1 \text{ kJ K}^{-1} \text{ m}^{-2} \text{ s}^{-1})(300 - 325)\text{K}$$

$$= -5000 \text{ kJ/s}$$

The negative sign indicates that heat is transferred from the oxidizer to the surroundings.

The stoichiometric equation describing the oxidation process needs to be postulated in terms of the known amounts of oxygen and nitrogen of the process gas stream and the unknown amount of natural gas that will be required. Since the amount of diethylamine is small, the energy released by its oxidization can be neglected. The mass flow rate of air is 16.45 kg/s and the molar flow rate of oxygen is 0.1196 kmol/s; hence

$$\dot{n}_a = 16.45 \text{ kg/s} = 0.5694 \text{ kmol/s}$$

$$\dot{n}(\text{oxygen}) = (0.5694 \text{ kmol/s})$$

$$(1 \text{ kmol O}_2/4.76 \text{ kmol air})$$

$$= 0.1196 \text{ kmol of O}_2/\text{s}$$

The stoichiometric equation can be written in its simplest form by neglecting the formation of CO and chemical dissociation. Once the problem has been analyzed, the reader can return to this equation and include additional products:

$$x\text{CH}_4 + 0.1196(\text{O}_2 + 3.76\text{N}_2) = x\text{CO}_2$$
$$+ 2x\text{H}_2\text{O(vapor)} + y\text{O}_2 + z\text{N}_2$$

where $x$ is the unknown molar flow rate of natural gas. The terms $y$ and $z$ can be evaluated by a balance of N and O atoms:

- Nitrogen (N): $(2)(3.76)(0.1196) = 2z$;
  $z = 0.4497$
- Oxygen (O): $(2)(0.1196) = 2x + 2x + 2y$;
  $y = 0.1196 - 2x$

The energy equation in the units of kJ/s can now be written as

$$\dot{Q} = -5000 = \sum_i^{\text{products}} \alpha_i(\bar{h}_f^0 + \delta\bar{h})_i$$
$$- \sum_j^{\text{reactants}} \beta(\bar{h}_f^0 + \delta\bar{h})_j$$

$$-5000 = \sum_i^{\text{products}} \alpha_i[\bar{h}_f^0 + \bar{c}_p(1100 - 298)]_i$$
$$- \sum_j^{\text{reactants}} \beta_j[\bar{h}_f^0 + \bar{c}_p(T_{\text{inlet}} - 298)]_j$$

where $\alpha_i$ and $\beta_j$ are the stoichiometric coefficients of the products and reactants, respectively:

$$\alpha_1 = \alpha(\text{CO}_2) = x$$

$$\alpha_2 = \alpha(\text{H}_2\text{O}) = 2x$$

$$\alpha_3 = \alpha(\text{O}_2) = y = 0.1196 - 2x$$

$$\alpha_4 = \alpha(\text{N}_2) = z = 0.4497$$

$$\beta_1 = \beta(\text{CH}_4) = x$$

$$\beta_2 = \beta(\text{O}_2) = 0.1196$$

$$\beta_3 = \beta(\text{N}_2) = z = 0.4497$$

Values of $\bar{c}_p$ can be determined from Perry et al (1984):

$$\bar{c}_p(\text{O}_2) = 29.49 \text{ kJ kmol}^{-1} \text{ K}^{-1}$$

$$\bar{c}_p(\text{N}_2) = 29.16 \text{ kJ kmol}^{-1} \text{ K}^{-1}$$

$$\bar{c}_p(\text{H}_2\text{O vapor}) = 33.71 \text{ kJ kmol}^{-1} \text{ K}^{-1}$$

$$\bar{c}_p(\text{CO}_2) = 37.04 \text{ kJ kmol}^{-1} \text{ K}^{-1}$$

$$\bar{c}_p(\text{CH}_4) = 36.06 \text{ kJ kmol}^{-1} \text{ K}^{-1}$$

To perform the summations above, it will be useful to express the equation above in tabular form.

**For reactants:**

| Species | $\beta_j$ | $\bar{h}_{f,j}^0$ (kJ/kmol) | $\beta_j[\bar{h}_{f,j}^0 + \bar{c}_{p,i}(T_{\text{inlet}} - 298)]$ |
|---|---|---|---|
| CH$_4$ | $x$ | $-74{,}873$ | $x(-74{,}873+0)$ |
| O$_2$ | 0.1196 | 0 | $0.1196[0+29.49(912-298)]$ |
| N$_2$ | 0.4497 | 0 | $0.4497[0+29.16(912-298)]$ |
| | | | $\Sigma = -74{,}873x+10{,}218$ |

**For products:**

| Species | $\alpha_i$ | $\bar{h}_{f,i}^0$ (kJ/kmol) | $\alpha_i[\bar{h}_{f,i}^0 + \bar{c}_{p,i}(1100-298)]$ |
|---|---|---|---|
| $CO_2$ | $x$ | $-393{,}522$ | $x[-393{,}522 + 37.04(1100-298)]$ |
| $H_2O$ | $2x$ | $-241{,}827$ | $2x[-241{,}827 + 33.71(1100-298)]$ |
| $N_2$ | $0.4497$ | $0$ | $0.4497[0 + 29.16(1100-298)]$ |
| $O_2$ | $0.1196 - 2x$ | $0$ | $(0.1196;1\text{-}2x)[0 + 29.49(1100-298)]$ |

$$\sum = -840{,}701x + 13{,}344$$

Substituting the values and solving for $x$ yields

$$-5000 = (13{,}344 - 840{,}701x) - (10{,}218 - 74{,}837x)$$

$$= 3126 - 765{,}828x$$

$$x = 0.0106 \text{ kmol/s}$$

Thus 0.0106 kmol/s of $CH_4$ are needed to burn with 0.1196 kmol/s of $O_2$. The fuel–air ratio (F/A) expressed in moles is

$$F/A = \left(\frac{0.0106 \text{ mol } CH_4}{0.1196 \text{ mol } O_2}\right)\left(\frac{1 \text{ mol } O_2}{4.76 \text{ mol air}}\right)$$

$$= \frac{(0.0106)}{(4.76)(0.1196)} = 0.0186$$

The reader may be surprised at this value because it is well below the stoichiometric F/A and even below the lean combustible limit (0.0525) for $CH_4$. The computed result is not an aberration when it is recalled that the air enters the reactor at an elevated temperature (912 K), which means that combustion can occur at F/A considerably leaner than what is required if air enters at STP. In addition, the stoichiometric equation used for this analysis ignores dissociation reactions that occur. The fuel cost for thermal oxidation can now be computed.

$$\dot{m}(CH_4) = (0.0106 \text{ kmol/s})(16 \text{ kg/kmol})$$

$$= 0.1696 \text{ kg/s}$$

Assuming that natural gas has a higher heating value of 55,496 kJ/kg and costs $5/10^6$ Btu and

that the process operates for 3600 h/yr; the cost of using natural gas is \$115,560/yr. Without energy recovery, the yearly fuel cost would be even larger. The estimated fuel cost neglects the fuel to consumed to operate a pilot flame to keep the system hot when the plant is not in operation. It also neglects the energy consumed to bring the oxidizer up to temperature before the plant operations begin each morning and to sustain a pilot flame that heats the combustion chamber during the evening.

The volume of the combustion chamber can be estimated by Equation 10.89. Assuming a cylindrical combustion chamber in which the diameter and height are equal ($D = H$), a residence time of 0.75 s requires that

$$V = tQ = (0.75 \text{ s})(\text{min}/60 \text{ s})(30{,}000 \text{ ft}^3/\text{min})$$

$$= (D)\frac{\pi D^2}{4}$$

$$D = 7.2 \text{ ft}$$

The cost of electricity to operate the system is

$$C_{\text{power}} = \frac{(\delta P)(Q)(\$0.0596/\text{kWh})t}{\eta}$$

$$= \frac{(17 \text{ kPa})(14.17 \text{ m}^3/\text{s})(3600 \text{ h/yr})(\$0.0596/\text{kWh})(1)}{0.75}$$

$$= \$68{,}220/\text{yr}$$

## 10.6  Summary of operational costs of three methods to remove diethylamine

The following is a summary of the fuel and fan costs ($Q_g \, \delta P$) if adsorption, absorption, or thermal oxidation were used to remove the dietheylamine from the process gas stream described in Example 10.1.

The table considers only two of many operational costs of an air pollution control system. For example, raw material and/or regeneration costs have not been addressed for adsorption and absorption sys-

| Control system | Physical process | Fan cost ($/yr) | Fuel cost ($/yr) | Total |
|---|---|---|---|---|
| Activated charcoal | Adsorption | | | |
| $U_a = 0.1$ m/s | | 3,294 | | |
| $U_a = 0.5$ m/s | | 82,451 | | |
| Packed bed wet scrubber | Absorption | 22,568 | | |
| Thermal oxidizer | Combustion | 68,220 | 115,560 | $183,780 |

tems. In Chapter 13 a method will be presented to estimate the total initial cost (TIC) and the total annual cost (TAC) to operate these three systems following procedures required for an EPA RACT analysis.

## 10.7  Thermal reduction of $NO_x$

Nitrogen oxides are generated by high-temperature combustion processes in which nitrogen contained in fuel and ambient diatomic nitrogen are converted to NO and $NO_2$ (hereafter called $NO_x$). A large portion of the nitrogen oxides can be eliminated by designing and operating combustors in ways that do not produce the oxides in the first place. Nonetheless, a certain amount will remain that may have to be removed by a tailpipe device. Nitrogen dioxide ($NO_2$) is very soluble in water, whereas NO, $N_2O$, $SO_2$, and $CO_2$ have limited solubility, and CO is virtually insoluble in water. Because the exhaust from high-temperature combustion processes contains NO frozen at the peak temperature of the process (see Chapter 7 for details), the task of removing the oxides of nitrogen is a difficult one.

Thermal reduction processes are classified as either *selective catalytic reduction* (SCR) or *selective noncatalytic reduction*. Both processes use ammonia ($NH_3$) or ammonia-forming compounds such as urea ($NH_2CONH_2$) to reduce $NO_x$ to $N_2$.

**Ammonia**

- Equilibrium:

$$2NO_2 + 4NH_3 + O_2 \rightleftharpoons 3N_2 + 6H_2O \qquad \text{(a)}$$

$1070 \text{ K} < T < 1200 \text{ K}$

$$4NO + 4NH_3 + O_2 \longrightarrow 4N_2 + 6H_2O \qquad \text{(b)}$$

$$\text{(10.90)}$$

$T > 1370 \text{ K}$

$$4NH_3 + 5O_2 \longrightarrow 4NO + 6H_2O \qquad \text{(c)}$$

**Urea**

- Hydrolysis reactions

$$NH_2CONH_2 + H_2O \rightleftharpoons NH_4COONH_2$$

$$\rightleftharpoons 2NH_3 + CO_2 \qquad \text{(10.91)}$$

- Pyrolysis reactions

$$NH_2CONH_2 \longrightarrow NH_3 + HNCO \qquad \text{(a)}$$

$$8HNCO + 4NO + 3O_2 \longrightarrow 6N_2O$$

$$+ 8CO + 4H_2O \qquad \text{(b)}$$

$$\text{(10.92)}$$

$$4HNCO + 2NO + 2O_2 \longrightarrow 3N_2$$

$$+ 4CO_2 + 2H_2O \qquad \text{(c)}$$

Reactions occurring at high temperature ($T > 1370$ K) are counterproductive and result in removal efficiencies between 30 and 50%, while reactions occurring at lower temperatures with catalysts achieve removal efficiencies in excess of 80%.

### 10.7.1  Selective catalytic reduction

The choice of catalyst depends on other materials present in the gas stream. For example, gas streams containing SO$_3$ form ammonium bisulfate (NH$_4$HSO$_4$), which is a sticky material that coats all internal surfaces. If safety considerations allow, ammonia gas stored under pressure can be injected into the exhaust stream. In other instances, aqueous ammonia (NH$_4$OH) is stored under pressure and vaporized to generate ammonia gas, which is injected into the exhaust stream. Three types of catalysts are commonly used. Figure 10-27 shows an application in which both direct ammonia injection and catalysts are used to remove NO$_x$ from the exhaust of an electric utility boiler.

**1.** *Noble metal* catalysts containing platinum supported on Al$_2$O$_3$ operate at low temperatures (490 to 555 K) but are adversely affected by SO$_2$. Noble metal catalysts are limited to natural gas and light-fuel oil applications. Spent noble metal catalysts can be recovered. Thus noble metal does not present a solid waste disposal problem. Care must be taken to clean the catalysts routinely because solid (NH$_4$)$_2$SO$_4$ and NH$_4$HSO$_4$ corrode the catalyst and inhibit heat transfer (Figure 10-28).

**2.** *Base metal* catalysts, typically containing vanadium deposited on a high-surface-area anatase titania (TiO$_2$) or vanadia (V$_2$O$_5$) support, operate in the temperature range 590 to 670 K. Vanadia (V$_2$O$_5$) catalysts are attractive because they are highly active and insensitive to sulfur compounds in the exhaust gas. Some catalysts containing tungsten or molybdenum promoters assist in the oxidation of SO$_2$. Vanadia catalysts are mildly toxic and present a modest disposal problem. Particulate matter can blind a catalyst and certain base metal oxides (i.e., K$_2$O, CaO,

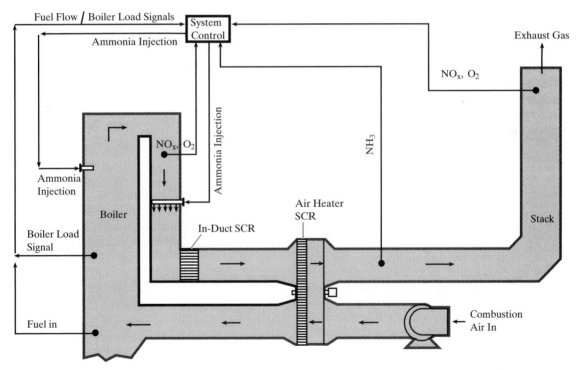

**Figure 10-27**  Schematic diagram of staged thermal reduction of NO$_x$ for an electric utility boiler showing nonselective catalytic ammonia injection (NSCR) and in-duct and air-heater applications of selective catalytic reduction (SCR) (with the permission of *Wahlco*)

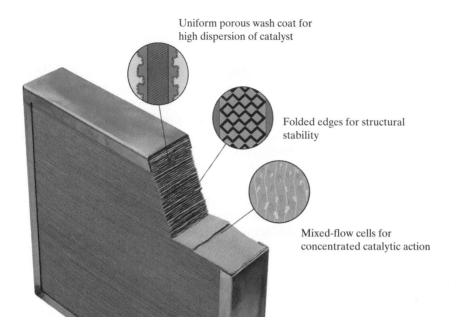

Uniform porous wash coat for
high dispersion of catalyst

Folded edges for structural
stability

Mixed-flow cells for
concentrated catalytic action

**Figure 10-28** Metal core
catalytic converters using
noble or base metal
catalysts. (Figure courtesy
Engelhard Corp.)

MgO, $P_2O_5$, etc.) and acid gases can poison the catalyst.

**3.** *Zeolite* catalysts contain zeolite crystals which may be subjected to ion exchange with base metal cations such as iron and copper to promote chemical activity. Zeolite catalysts operate at higher temperatures (645 to 755 K) and require higher $NH_3$ concentrations, which may result in the inadvertent appearance of unreacted ammonia in the exhaust stream. Zeolite materials are relatively benign and their disposal may present a modest problem, depending on the metal cation in the zeolite.

Selective catalytic reduction (SCR) catalysts operate in relatively narrow temperature bands. At the lower end of the temperature band, larger amounts of catalysts are needed to achieve a particular removal efficiency. At the upper end of the temperature band, the oxidation of ammonia competes with the SCR reactions, resulting in lower $NO_x$ removal efficiencies. When temperatures exceed the narrow band, the catalyst itself suffers permanent damage because of thermal sintering.

Aside from the temperature band, another factor affecting $NO_x$ removal is the geometric structure of the catalyst. The structure is homogeneous if the monolithic catalyst is fabricated from a catalyst paste extruded as a honeycomb. Gas streams containing particles tend to clog these catalysts. Nonhomogeneous structures consist of a thin layer of catalyst deposited on a support structure such as metal plates or ceramics. Such nonhomogeneous catalysts are used with particle-laden gas streams or gas streams containing poisons. The ammonia injection rate should follow closely a 1:1 molar relationship with $NO_x$. Excess ammonia generally passes through the system and appears as ammonia in the outlet gas stream.

Removing nitrogen oxides by selective catalytic reduction (SCR) is attractive for clean exhaust streams containing nothing but nitrogen oxides. Rarely is this the case, however, since most exhaust streams contain particles that will clog a catalyst and traces of rare molecular species that poison catalysts.

## 10.7.2  Selective noncatalytic reduction

$NO_x$ may be reacted thermally, without catalysts, in which case ammonia gas or compounds containing ammonia are introduced directly into the exhaust stream at higher temperatures. The stoichiometric reactions involving NO and $NH_3$ are very sensitive to temperature. At temperatures below 1070 K, the ammonia may not react and unwanted ammonia will appear in the exhaust gas. The process is called selective since the temperature must be controlled to ensure that the ammonia reacts with NO rather than oxygen. Catalysis is appropriate for selected "clean" process gas streams such as reciprocating engines, natural gas-fired combustion turbines, and so on. Catalytic reduction of nitrogen oxides is not appropriate for treatment of the exhaust from municipal incinerators, pulverized coal-fired boilers, circulating fluidized bed combustors, and so on, because of a wide variety of materials that cause blinding and/or deactivation of catalysts.

# 10.8  Flue gas desulfurization

In Chapter 7, fluidized-bed combustors are described. In them, pulverized limestone or powdered lime is mixed with coal to react directly with sulfur, forming $CaSO_4$ and $CaSO_3$ particles in the combustion zone. The particles remain in the bed and are removed with the ash or are carried over with the fly ash and are removed by particle collectors downstream. If the $SO_2$ cannot be removed in the combustion zone, it must be removed by either of two *flue gas desulfurization* methods, *wet scrubbing* and *dry scrubbing*. Lime and limestone are the principal reagents used to remove $SO_2$, but MgO and $Na_2CO_3$ can also be used.

## 10.8.1  Dry scrubbing

In dry scrubbing a lime or limestone slurry is sprayed into the hot gas stream and $CaSO_3$ and $CaSO_4$ are formed as airborne particles which are captured by a fabric filter. When $Ca(OH)_2$ is sprayed into the hot process gas stream, the water evaporates and dry particles are captured by a fabric filter. The $SO_2$ reacts with the calcium in the dust cake to form $CaSO_4$ and $CaHSO_4$, which is removed along with the fly ash as dry material. Dolomite and other natural materials can also be used. The baghouse needs to be oversized because the total particle concentration in the gas stream is large, but this cost is more than offset by reducing the huge cost in materials handling, disposal, land, and personnel required by wet scrubbing. There are two variations of the basic process, both called dry scrubbing.

**1.** *Spray dryer (semidry scrubbing).* The reagent, typically $Ca(OH)_2$, is mixed with water into a slurry. The slurry is injected into a spray dryer vessel as atomized droplets that mix with the flue gas. The evaporating slurry drops react with $SO_2$ at the drop boundaries. The reaction continues in the dry state on the dust cake on the filter (Figure 10-29) after the aerosol leaves the spray dryer. Ultimately, $SO_2$ is converted to a dry salt which is removed along with fly ash in the baghouse.

**2.** *Dry injection (dry–dry system or direct injection system).* With direct injection systems, the reagents are either trona (naturally occurring $Na_2CO_3$) or nahcolite (naturally occurring $NaHCO_3$). The reagent is injected into the flue gas in a dry state upstream of the baghouse. The reagent aerosol reacts with $SO_2$ as a suspended solid (Garding and Svedberg, 1988). The dry reagent is collected by the baghouse filters, where it becomes an element of the dust cake. $SO_2$ passing through the dust cake reacts with the reagent to form a salt. In summary, the dust cake becomes a porous bed of reagent that coverts $SO_2$ to a salt. The salt and particulate matter are removed during the normal baghouse cleaning cycle.

$$2NaHCO_3(s) \longrightarrow Na_2CO_3(s) \\ + CO_2(g) + H_2O(g) \qquad\qquad \text{(a)}$$

$$Na_2CO_3(s) + SO_2(g) \longrightarrow Na_2SO_3(s) \\ + CO_2(g) \qquad\qquad \text{(b)} \quad \text{(10.93)}$$

$$Na_2SO_3(s) + \tfrac{1}{2}O_2(g) \longrightarrow Na_2SO_4(s) \quad \text{(c)}$$

The only difference between dry and semidry scrubbing is the way the reagent is introduced to the flue gas. The adsorption and reaction processes of the reagent are the same while in the dust cake. Dry scrubbing eliminates the absorption/reaction processes that occur in a huge wet scrubber. Thus a wet scrubber and its ancillary water treatment steps are avoided. Figure 10-29 illustrates semidry scrubbing using a spray dryer.

## 10.8.2  Wet scrubbing

Figure 10-30 illustrates the essential features of wet scrubbing, using limestone as the principal removal agent. Limestone and lime slurries are used to remove $SO_2$ from an exhaust gas stream in wet scrubbing. Limestone is a naturally formed mineral con-

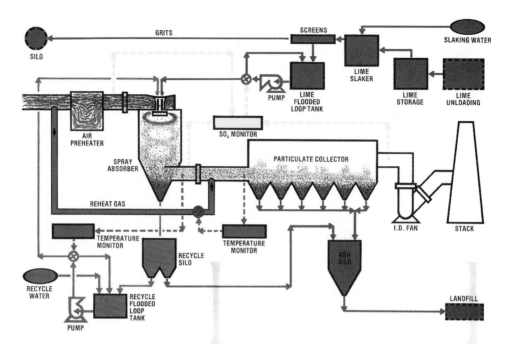

**Figure 10-29** Schematic diagram of the GE Dry Flue Gas Desulfurization System using lime to remove $SO_2$. The lime reagent enters the system as a slurry. Hot flue gas evaporates water from the slurry and $SO_2$ is adsorbed on the remaining particles of dry reagent. Flyash and the dry reagent are collected by a fabric filter. Additional $SO_2$ is adsorbed as the flue gas passes through the dust cake containing the dry reagent particles. (Figure used by permission of General Electric Environmental Services, Inc.)

sisting of a mixture of $CaCO_3$ and a variety of inert siliceous compounds. Lime (CaO) is obtained from limestone by heating, called *calcining*:

$$CaCO_3 + heat \longrightarrow CaO + CO_2 \quad (10.94)$$

A limestone slurry is basically a suspension of limestone particles in water, as limestone is only slightly soluble in water (0.00153 g per 100 g of $H_2O$ at 273 K). When lime is added to water, calcium hydroxide, called *slaked lime*, is formed:

$$CaO + H_2O \longrightarrow Ca(OH)_2 \quad (10.95)$$

Calcium hydroxide is considerably more soluble than limestone (0.185 g per 100 g of water at 273 K). and dissociates to calcium and hydroxyl ions. The degree of dissociation of calcium hydroxide decreases with temperature (0.078 g per 100 g water at 373 K):

$$Ca(OH)_2 \rightleftharpoons Ca^{2+} + 2OH^- \quad (10.96)$$

The following mechanisms apply to flue gas desulfurization using limestone and lime slurries.

**Wet scrubbing mechanism using limestone slurry ($CaCO_3$)**

$$SO_2 + H_2O \longrightarrow SO_2 \cdot H_2O \quad \text{(a)}$$

$$SO_2 \cdot H_2O \longrightarrow H^+ + HSO_3^- \quad \text{(b)}$$

$$H^+ + CaCO_3 \longrightarrow Ca^{2+} + HCO_3^- \quad \text{(c)}$$

$$(10.97)$$

$$Ca^{2+} + HSO_3^- + 2H_2O \longrightarrow CaSO_3 \cdot 2H_2O + H^+ \quad \text{(d)}$$

$$H^+ + HCO_3^- \longrightarrow CO_2 \cdot H_2O \quad \text{(e)}$$

$$CO_2 \cdot H_2O \longrightarrow H_2O + CO_2 \quad \text{(f)}$$

The overall reaction for the mechanism above is

$$CaCO_3 + SO_2 + 2H_2O \rightleftharpoons CaSO_3 \cdot 2H_2O + CO_2$$

$$(10.98)$$

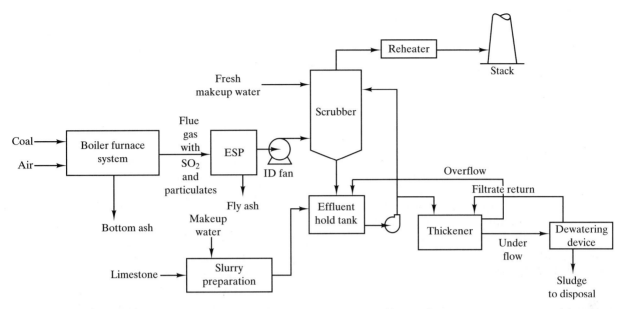

**Figure 10-30** Schematic diagram of a wet $SO_2$ scrubber. Flyash is removed by an electrostatic precipitator or fabric filter and the remaining gas passes through a wet scrubber (packed bed, venturi, etc.) that uses an alkaline reagent slurry. After the absorbed $SO_2$ is converted into a sulfate or bisulfate, the scrubber slurry is thickened, filtered and discharged to a disposal pond as wet sludge.

Excess $O_2$ in the flue gas becomes dissolved in the slurry, which oxidizes the sulfite ion to a sulfate ion. The conversion increases as the pH is decreased:

$$SO_3^{2-} + \tfrac{1}{2}O_2 \longrightarrow SO_4^{2-} \qquad (10.99)$$

When this reaction occurs, some of the $CaSO_3$ is converted to $CaSO_4$:

$$CaSO_3 \cdot 2H_2O + \tfrac{1}{2}O_2 \longrightarrow$$
$$CaSO_4 \cdot 2H_2O \downarrow \quad (gypsum) \quad (10.100)$$

**Wet scrubbing mechanism using lime (CaO)**

$$SO_2 + H_2O \longrightarrow SO_2 \cdot H_2O \qquad (a)$$

$$SO_2 \cdot H_2O \longrightarrow H^+ + HSO_3^- \qquad (b)$$

$$CaO + H_2O \longrightarrow Ca(OH)_2 \qquad (c)$$
$$(10.101)$$

$$Ca(OH)_2 \longrightarrow Ca^{2+} + 2OH^- \qquad (d)$$

$$Ca^{2+} + HSO_3^- + 2H_2O \longrightarrow CaSO_3 \cdot 2H_2O + H^+ \qquad (e)$$

$$H^+ + OH^- \longrightarrow H_2O \qquad (f)$$

The overall reaction can be written as

$$CaO + SO_2 + 2H_2O \rightleftharpoons CaSO_3 \cdot 2H_2O \quad (10.102)$$

In the presence of dissolved oxygen and at low pH, the sulfite is converted to sulfate:

$$CaSO_3 \cdot 2H_2O + \tfrac{1}{2}O_2 \longrightarrow$$
$$CaSO_4 \cdot 2H_2O \downarrow \quad (gypsum) \quad (10.103)$$

After passage through the scrubber the scrubbing liquid is sent to a retention tank, where the $CaSO_3$, $CaSO_4$, and unreacted $CaCO_3$ are precipitated as a waste sludge that is discarded. The principal problem encountered with lime and limestone scrubbing is scaling and plugging of the scrubbing equipment. The low pH results in corrosion unless special materials are used.

## 10.8.3 Desulfurization in fluidized combustors

Pulverized limestone or dolomite mixed with coal in a fluidized-bed combustor (FBC; Figure 1.15) can convert $SO_2$ to solid sulfate particles (Badin and Frazier, 1985). The chemical sorbency of $SO_2$ by limestone occurs in two steps:

$$CaCO_3 + heat\ (825–900\ C) \longrightarrow CaO + CO_2$$

$$CaO + SO_2 + \tfrac{1}{2}O_2 \longrightarrow CaSO_4 \qquad (10.104)$$

Calcium sulfite ($CaSO_3$) and calcium sulfide (CaS) sometimes form in reducing regions of the FBC but then oxidize to form sulfate in the oxidizing region of the combustor. The sorbency of $SO_2$ by dolomite also occurs in two steps:

$$CaCO_3 \cdot MgCO_3 + \text{heat (730–760 C)} \longrightarrow$$
$$[CaCO_3 + MgO] + CO_2$$

$$[CaCO_3 + MgO] + SO_2 + \tfrac{1}{2}O_2 \longrightarrow$$
$$[CaSO_4 + MgO] + CO_2 \quad (10.105)$$

At 850°C, the reactions for limestone tend to be thermally balanced, the first reaction being endothermic while the second is exothermic. Because of metal oxide impurities in naturally formed limestone and dolomite, a large range of sulfate conversion can be expected. Sorbency can be improved by the addition of chlorides of potassium and sodium.

# 10.9  Bioscrubbers, biofilters, and trickle-bed reactors

Bioreactors comprise a generic category including three types of pollution control devices which employ microorganisms to metabolize pollutants. It is possible to divide bioreactors into categories based on where the microorganisms reside. Two terms in common use and the reactor types that correspond to them are:

- *Suspended-growth*: bioscrubbers
- *Fixed-film bioreactors*: trickle-bed reactors and biofilters

Hydrocarbons and odorous emissions can be removed by biological microorganisms in pollution control systems called biofilters and bioscrubbers. The terms are sometimes used interchangeably, but in this book *biofilters* pertain to gas-phase systems in which the microorganisms exist in a moist thin film surrounding a solid substrate and *bioscrubbers* pertain to liquid-phase absorbers in which the microorganisms are suspended in the scrubbing liquid after absorption has occurred. The attraction of biological processes is that they occur at ambient temperature and pressure, consume very little energy, and produce no nitrogen oxides such as accompany thermal oxidation.

Biological control systems use immobilized microorganisms and natural processes of biological degradation. *Biodegradation* reactions oxidize hydrocarbons to produce alcohols, which then react to form aldehydes, followed by organic acids and eventually $CO_2$ and $H_2O$. Even if biodegradation is not complete, the partially oxidized products may still be discharged to the atmosphere, depending on their concentrations, molecular configurations, and state regulations. Biofilters are most effective when the hydrocarbon concentration is below 1 mg/m³. At concentrations greater than 10 mg/m³, biofilters should be followed by charcoal adsorbers. Bioscrubbers are used generally when biological degradation products such as acids, $H_2S$, and ammonia would harm a biofilter bed.

Complex hydrocarbon chains containing many carbon bonds require a longer time to biodegrade than simple alcohols, such as ethanol which contains only two carbons. Longer biodegradation times require a larger system with a longer residence time. Industrial processes where biological systems have been used to control odors with efficiencies above 90% include wastewater treatment, flavor and fragrance, food processing, and fermentation. Compounds that have been controlled include ammonia, hydrogen sulfide, ethanol, isobutyl alcohol, acetone, methyl ethyl ketone, ethyl acetate, benzene, toluene, styrene, phenol, chlorophenols, methylene chloride, petroleum distillates, and mineral spirits; or more generally, alcohols, aldehydes, ketones, esters, aromatics, amines, organic acids, organic sulfides and mercaptans.

Figure 10-31 is a schematic diagram of a bioscrubber. A *bioscrubber* is essentially a packed-bed scrubber in which the microorganisms are suspended in the scrubbing liquid. The liquid is called a *suspended-growth slurry*. The suspended material is called a *floc*. The inert packing serves the traditional purpose of increasing the contact surface area per volume of the bed and enhancing the transfer of the pollutant from the gas to the liquid phase. Bioscrubbers are equipped with a well-mixed slurry tank that is aerated to provide oxygen. The tank is stirred to maintain the biomass (activated sludge) in suspension.

Figure 10-32 is a schematic diagram of a *biofilter* and Figure 10-33 is a sketch of an actual biofilter. It consists of one or more beds of biologically active material such as peat, compost, municipal waste, wood chips, bark, leaves, or even soil, all of which provide nutrients to the microorganisms. The particles of biologically active material increase the wet-

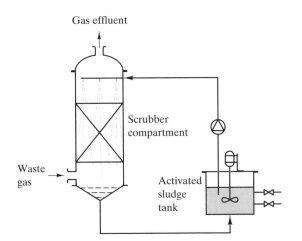

**Figure 10-31** Bioscrubber containing a random packed bed and an activated sludge tank. The scrubber removes the pollutant and biodegradation occurs in the activated sludge tank. [figures taken from Ottengraf 1986]

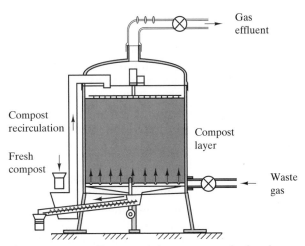

**Figure 10-32** Biofilter containing a compost bed and mechanical compost recirculation system (patented by Kneer 1976) [figure taken from Ottengraf 1986]

**Figure 10-33** To reduce compacting compost, to improve the distribution of humidified air, and to maintain a desired bed temperature, it is advantageous to place the biofilter in an enclosure containing two beds in series. (Reprinted with the permission of Ambient Engineering, Inc.)

ted surface area per volume of bed and thus perform a function similar to inert packing. The biological material will henceforth be called *compost*. Beds are typically 1 m high in order not to produce too large a pressure drop. Waste gas enters the base of the filter. Sufficient time exists for the pollutants to diffuse into the biologically active layer (biofilm) attached to the compost particles. Aerobic reactions occur in the biofilm, where microorganisms metabolize pollutants. If metabolism is complete and no undesirable products are generated, the end products are water vapor, $CO_2$, and microbial biomass. Sulfur-bearing and chlorinated organic compounds generate inorganic acids. It is essential that the inlet waste gas be saturated with water vapor (relative humidity should never be less than 90%); otherwise, water will be evaporated from the biomass medium, killing the microorganisms. Biofilters are also equipped with water sprays to add water to the bed.

In a *trickle-bed reactor* (Figure 10-34) liquid flows slowly (trickles) through the bed to contact a fixed biofilm attached to column packing (Diks and Ottengraff, 1991). The packing elements are larger (5 to 10 cm) than those used in bioscrubbers. Following the microbial action in the trickle bed, the solid product is removed in a settling tank.

Biological methods have found wide use in Europe (Hartmans and Tramper, 1991; Derkix et al., 1989; Kirchner et al., 1985; Livingston and Chase, 1989) and Japan (Cho et al., 1991a, b; Hirai et al., 1990) as economically effective ways to treat odors

in waste gases from wastewater treatment and food processing plants, as well as chemical and pharmaceutical manufacture. A summary of U.S. activity can be found in Leson and Winer (1991), Utgikar et al. (1991), Overcamp et al. (1993), and Shareefdeen et al. (1993). Deshusses et al. (1995a, b) propose a novel analytical model to describe the dynamic behavior of biofilters when the concentration and/or volumetric flow rate vary with time. The model is also capable of describing the performance of biofilters in handling a waste gas steam containing multiple molecular species.

The bacteria used are aerobic bacteria, mostly common *mesophiles*, which feed on both organic and inorganic compounds. Such bacteria function between 65 and 105°F and require heating and thermal insulation in cold climates. Activated sludges from municipal wastewater plants are widely used in bioreactors. Used also are natural microorganisms in peat, humus, mushroom compost, and soil. To metabolize chlorinated and sulfur-bearing compounds, special microorganisms have been developed. A problem facing all biological methods is selecting microorganisms that tolerate and function metabolically in a gas stream containing numerous materials. For the microorganisms to function properly, chemicals may have to be added to control pH.

## 10.9.1 Bioscrubbers

Bioscrubber operation is composed of two processes: absorption in the packed bed and regeneration when microorganisms metabolize pollutants in the activated sludge tank. The absorption process is identical to that described in Section 10.3.1. The activated well-mixed sludge tank performs three functions:

1. mechanically suspend the sludge and maintain well-mixed conditions
2. aerate the sludge to provide oxygen
3. foster contact between dissolved oxygen and sludge

Consider Figure 10-35, depicting a bed of thickness $Z$. The total rate at which pollutants are transferred from the carrier gas to the scrubbing liquid ($N_j$, kmol/s) is equal to the rate at which pollutants are absorbed by the scrubbing liquid:

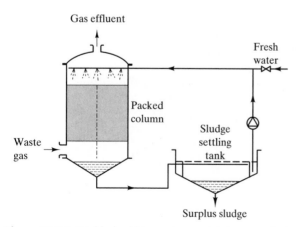

**Figure 10-34** Trickle-bed bioreactor containing a packed bed and a settling tank to remove sludge

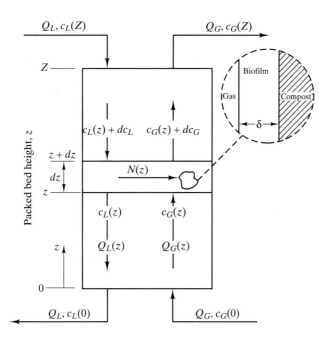

**Figure 10-35** Schematic diagram of the packed bed in a biofilter or trickle-bed reactor

$$N_j = Q_G[c_{G,j}(0) - c_{G,j}(Z)]$$

$$= Q_L[c_{L,j}(0) - c_{L,j}(Z)] \quad (10.106)$$

where $Q_G$ and $Q_L$ are the gas and liquid volumetric flow rates, $c_{G,j}$ and $c_{L,j}$ are the molar concentrations of pollutant ($j$) in air and water, and $Z$ is the height of the packed bed. The rate $(r_j)(\text{kmol m}^{-3}\,\text{s}^{-1})$ at which pollutant (species $j$) is consumed by the microorganism   is given by the *Monod, or Michaelis and Menten*, expression:

$$r_j = \left(\frac{\mu_m}{Y_j}\right)\frac{Xc_{L,j}}{c_{L,j} + K_{sj}} \quad (10.107)$$

where $\mu_m(\text{h}^{-1})$ is the maximum growth rate, $Y_j$ the cell yield coefficient $\left(\text{kg}_j\,\text{kmol}_j^{-1}\,\text{h}^{-1}\right)$, $K_{sj}\left(\text{kmol/m}^3\right)$ is the *Monod or Michaelis–Menten* constant of component $j$, $X$ is the active microorganism concentration $\left(\text{kg/m}^3\right)$, and $c_{L,j}$ is the molar concentration of pollutant $j$ in water $\left(\text{kmol/m}^3\right)$. The rate $\left(N_j, \text{kmol/s}\right)$ at which pollutant ($j$) is absorbed by the liquid in the packed bed  is expressed as

$$N_j = AK_La_pZ\Delta c_{\text{ln}} \quad (10.108)$$

where $a_p$ is the ratio of packing surface area to volume of bed and $\Delta c_{\text{ln}}$ is the log-mean concentration difference:

$$\Delta c_{\text{ln}} = \frac{\Delta c(0) - \Delta c(Z)}{\ln[\Delta c(0)/\Delta c(Z)]} \quad (10.109)$$

Shown in Figure 10-36 are the equilibrium and operating lines from which the terms $\Delta c(Z)$ and $\Delta c(0)$

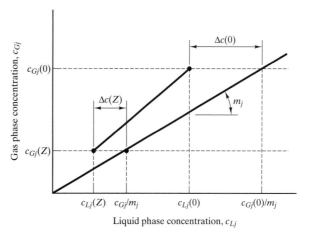

**Figure 10-36** Equilibrium and operating lines for a bioscrubber

(called the *driving forces* at elevation $z = Z$ and $z = 0$) are defined as

$$\Delta c(Z) = \frac{c_{G,j}(Z)}{m_j} - c_{L,j}(Z) \quad \text{(a)} \quad (10.110)$$

$$\Delta c(0) = \frac{c_{G,j}(0)}{m_j} - c_{L,j}(0) \quad \text{(b)}$$

and $m_j$ is Henry's law constant:

$$P_j = Hc_{L,j} \quad \text{(a)}$$

$$c_{G,j}R_uT = Hc_{L,j} \quad \text{(b)}$$

$$c_{G,j} = \left(\frac{H}{R_uT}\right)c_{L,j} = \frac{H'}{c_L}c_{L,j} \quad \text{(c)} \quad (10.111)$$

$$c_{G,j} = m_j c_{L,j} \quad \text{(d)}$$

where $c_L$ is the molar concentration of water and $m_j$ is a form of the Henry's law constant, defined as

$$m_j = \frac{H'}{R_uTc_L} \quad (10.112)$$

Beginning with Equation 10.110,

$$\Delta c(Z) - \Delta c(0) = \left[\frac{c_{G,j}(Z)}{m} - c_{L,j}(Z)\right]$$

$$- \left[\frac{c_{G,j}(0)}{m} - c_{L,j}(0)\right]$$

$$(10.113)$$

$$= \frac{1}{m_j}[c_{G,j}(Z) - c_{G,j}(0)]$$

$$+ [c_{L,j}(Z) - c_{L,j}(0)]$$

Introduce Equation 10.106:

$$\Delta c(Z) - \Delta c(0) = \frac{N_j}{Q_L} - \frac{N_j}{m_j Q_G}$$

$$= N_j\left(\frac{1}{Q_L} - \frac{1}{m_j Q_G}\right)$$

$$= \frac{N_j}{Q_L}\left(1 - \frac{Q_L}{m_j Q_G}\right) \quad (10.114)$$

$$= -\frac{N_j}{Q_L(E - 1)}$$

where

$$E = \frac{Q_L}{m_j Q_G} \quad (10.115)$$

The overall mass transfer coefficient $K_L$ is related to the individual mass transfer coefficients in the liquid and gas phases by

$$\frac{1}{K_L} = \frac{1}{k_L} + \frac{m_i}{k_G} \quad (10.116)$$

The units on $K_L$, $k_L$, and $k_G$ are m/s. The total number of moles of pollutant oxidized in the activated sludge tank are

$$R_j = \int_0^{V_R} r_j\,dV = \frac{V_RX(\mu_m/Y_j)c_{L,j}(Z)}{c_{L,j}(Z) + K_{sj}} \quad (10.117)$$

where $V_R$ is the volume of the activated sludge tank. For a bioscrubber at steady state, the total mass transfer rate $N_j$ in the packed bed is equal to the steady-state conversion in the sludge tank, $R_j$:

$$R_j = N_j = AK_La_pZ\Delta c_{ln}$$

$$= V_RX\frac{\mu_m}{Y_j}\frac{c_{L,j}(Z)}{c_{L,j}(Z) + K_{sj}} \quad (10.118)$$

The bioscrubber removal efficiency

$$\eta = 1 - \frac{c_{G,j}(Z)}{c_{G,j}(0)} = 1 - \frac{N_j}{Q_Gc_{G,j}(0)} \quad (10.119)$$

The analytical model presumes that the rate-determining step is the microbial elimination reaction. If the cell concentration is very high, the rate may shift toward a diffusion-controlled regime.

### 10.9.2 Trickle-bed reactors

A trickle-bed reactor (Figure 10-34) consists of a packed bed of compost particles having a relatively low surface area per volume of bed $(a_p)$. A large void space is needed because biological growth tends to fill the void volume. Recirculating liquid containing essential nutrients is supplied continuously at the top of the column and distributed uniformly across the bed. Recirculated liquid flows downward wetting the fixed biofilm attached to the compost particles. The waste gas enters the base of the column and travels upward countercurrent to the liquid. Water-soluble pollutants are absorbed in the liquid film through which they diffuse into the biofilm, where they are metabolized by microorganisms. Surface films slough off periodically and are removed from the recirculating liquid in the sludge settling tank before the liquid is returned to the top of the column. Due to evaporation, the cleaned gas is nearly saturated and fresh water has to be supplied to the recirculating liquid stream. In contrast to bioscrubbers, the process of absorption and metabolism in trickle-bed reactors occurs in the biofilm surrounding the packing rather than in the aerated sludge tank. The analytical model describing this simultaneous absorption and metabolism is the same as it is for biofilters, described in the next section.

### 10.9.3   Biofilters

In fixed-film biofilters, it is assumed that the compost particle is surrounded by a layer of water containing microorganisms. The *fixed film* will also be called a *biofilm*. In the discussion that follows, two kinetic expressions for the rate of metabolism will be studied (Ottengraf and Van Den Oever, 1983). Figure 10-37 is a schematic diagram of the biofilm showing the liquid-phase pollutant concentration for these cases:

- **Case 1:** first-order kinetics $\left(r_j = k_1 c_{L,j}\right)$
- **Case 2:** zeroth-order kinetics $\left(r_j = k_0\right)$
  **2a.** Reaction limited
  **2b.** Diffusion limited

Figure 10-35 shows an elemental volume in the beds of biofilters or trickle-bed reactors.

When air passes upward (increasing values of $z$) through the compost particles, oxygen, water-soluble

gases, and vapors in the waste gas are absorbed in the biofilm and produce the concentration gradients shown in Figure 10-37. The products of metabolism $\left(\text{notably } CO_2\right)$ diffuse out through the biofilm into the waste gas (Ottengraf and Konings, 1991). Nutrients needed by the microorganisms may be added to the waste gas stream or circulating liquid, but generally they are present in the packing material (i.e., humus, peat, compost, and other biological material). As a practical matter, the packing material should be circulated and a fixed amount continually replaced by fresh material. The size of the compost particles in fixed-film reactors is on the order of several millimeters. The thickness of the biofilm $(\delta)$ is assumed to be considerably smaller than the diameter of the compost particle. This allows modeling of the biofilm with variations in only one dimension $(x)$ perpendicular to the biofilm surface. It is also assumed that:

- Pollutant concentration in the gas stream varies only in the vertical direction [i.e., $c_{G,j}(z)$].
- Pollutant concentration in the biofilm varies with respect to $x$ and $z$ [i.e., $c_{L,j}(x, z)$].

Consider just the biofilm at a height $z$ in the bed surrounding a compost particle (Figure 10-37). Ottengraf (1986) has shown that mass transfer through the gas film is rapid and the overall mass transfer process is controlled by the rate at which material diffuses into the biofilm (i.e., single-film

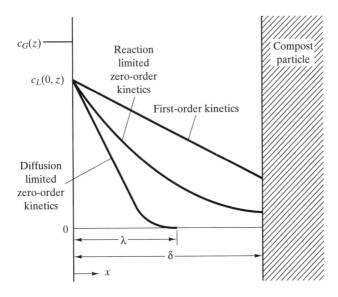

**Figure 10-37**
Concentration profiles in biofilm for different reaction orders.

mass transfer). The film can be considered a liquid sheet of thickness δ whose transverse dimension is $x$. Outside the biofilm the pollutant concentration in the gas is $c_{G,j}(z)$ since the concentration is assumed to be uniform at any height $z$ in the bed (i.e., $\partial c_{G,j}/\partial x = 0$). Inside the biofilm the pollutant concentration is $c_{L,j}(x, z)$. The differential equation describing the variation of pollutant concentration in the biofilm is

$$D_j \frac{\partial^2 c_{L,j}}{\partial x^2} - r_j = 0 \qquad (10.120)$$

where $D_j$ is the diffusion coefficient of species $j$ in the liquid biofilm and $r_j$ is the rate at which pollutants are consumed . Assuming that equilibrium exists at the gas–liquid interface, Henry's law provides

*Boundary condition 1:*

$$(x = 0, z = z): \quad c_{L,j}(0, z) = \frac{c_{G,j}(0, z)}{m_j} \qquad (10.121)$$

where $m_j$ is the slope of the equilibrium line.

**Case 1: First-order kinetics.** If the pollutant concentration in the liquid phase is small such that if $K_{sj} \gg c_{L,j}(x, z)$ throughout the biofilm, the reaction rate equation 10.107 reduces to

$$r_j = \left(\frac{\mu_m}{Y_j}\right) \frac{X}{K_{sj}} c_{L,j}(x, z) = k_1 c_{L,j}(x, z) \ (10.122)$$

where $k_1(\text{h}^{-1})$ is called a *first-order* reaction rate constant:

$$k_1 = \left(\frac{\mu_m}{Y_j}\right) \frac{X}{K_{sj}} \qquad (10.123)$$

The second boundary condition exists at the biofilm–particle interface where there is no diffusion.

*Boundary condition 2:*

$$(x = \delta, z = z): \quad \left(\frac{\partial c_{L,j}}{\partial x}\right)_{\delta, z} = 0 \ (10.124)$$

When the differential equation (10.120) is solved subject to these boundary conditions, the following expression for the concentration within the biofilm is obtained:

$$\frac{c_{L,j}(x, z)}{(c_{G,j}(z)/m_j)} = \frac{\cosh 1[\Phi_1(1 - \sigma)]}{\cosh \Phi_1} \qquad (10.125)$$

where $\Phi_1$ is called the *Thiele number* for the first-order kinetics,

$$\Phi_1 = \delta \left(\frac{k_1}{D_j}\right)^{1/2} \qquad \text{(a)} \quad (10.126)$$

$$\sigma = \frac{x}{\delta} \qquad \text{(b)}$$

At any height $z$, the mass transfer into the biofilms of all particles in an elemental volume $A \ dz$ is (Figure 10-35)

$$N(z) = \left[-D_j \left(\frac{\partial c_{L,j}}{\partial x}\right)_{0,z}\right] a_p A \ dz \qquad \text{(a)}$$
$$\qquad (10.127)$$
$$= \left[\frac{D_j}{\delta} \frac{c_{G,j}(z)}{m_j} \Phi_1 \tanh \Phi_1\right] a_p A \ dz \quad \text{(b)}$$

Pollutant removed from the gas phase is equal to the mass transferred to the biofilm. A mass balance for an element of the gas column at $z = z$ results in

$$Q_G c_{G,j}(z) = Q_G[c_{G,j}(z) + dc_{G,j}(z)] + N(z) \quad \text{(a)}$$
$$\qquad (10.128)$$

$$dc_{G,j}(z) = -\frac{N(z)}{Q_G}$$
$$= -\frac{D_j}{\delta} \frac{c_{G,j}(z)}{m_j} \frac{a_p}{U_G} (\Phi_1) \tanh \Phi_1 \ dz \quad \text{(b)}$$

where $U_G$ is the average gas velocity, $U_G = Q_G/A$. Integrating Equation (10.128) over the height of the bed $(Z)$, one obtains

$$\int \frac{dc_{G,j}}{c_{G,j}} = \int \left(\frac{D_j}{\delta} \frac{a_p}{U_G} \frac{\Phi_1}{m_j}\right) \tanh \Phi_1 \ dz \quad (10.129)$$

$$\frac{c_{G,j}(Z)}{c_{G,j}(0)} = \exp\left(-\frac{Z}{m_j} \frac{a_p D_j}{\delta U_G} \Phi_1 \tanh \Phi_1\right)$$

The removal efficiency of a biofilter satisfying first-order kinetics,

$$\eta = 1 - \frac{c_{G,j}(Z)}{c_{G,j}(0)} = 1 - \exp\left(-\frac{Z}{m_j} \frac{K_1}{U_G}\right)$$

$$= 1 - \exp\left(-\frac{V_{bed} K_1}{m_j Q_G}\right) \qquad (10.130)$$

The term $K_1(\text{h}^{-1})$, called a *reaction unit*, is

$$K_1 = \frac{a_p}{\delta} D_j \Phi_1 \tanh \Phi_1 \qquad (10.131)$$

$$\frac{c_{L,j}(x,z)}{c_{G,j}(z)/m_j} = 1$$

The parameter $\Phi_0$ i
$\sigma = x/\delta$. Since $c_L($
be solved for this co

$$\lambda(z)$$

At $z = z$, the molar
in an element of $A$

$$N(z) = $$

$$= $$

Pollutant removed
mass transferred to
an element of the g

$$Q_G c_{G,j}(z) = Q_G [c$$

$$dc_{G,j} = -\frac{N}{C}$$

$$= -k_0\lambda$$

Integrating Equatic
bed $(Z)$ yields

$$\frac{c_{G,j}(Z)}{c_{G,j}(0)} = \left[1 - \right.$$

$$= \left[1 - \right.$$

where $K_0$ is given b
The biofilter rer
ited zeroth-order k

$$\eta = 1 - \frac{c_{G,j}(Z)}{c_{G,j}(0)}$$

$$= 1 - \left[1 - \frac{Z}{U_G}\right.$$

Removal by the be

$$\frac{K_0 Z}{U_G c_{G,j}(0)} \geqq 2$$

## EXAMPLE 10.6   IMPROVING BIOFILTER EFFICIENCY

A biofilter composed of local materials (bark, leaves, and humus) has been designed to remove foul-smelling odors from a wastewater treatment plant. The bed was designed to be 1 m thick and have a cross-sectional area of 100 m². Waste gases $(Q_G)$ enter the bed through a system of distribution ducts beneath the bed to ensure uniform gas velocity across the bed face. Bench-top experiments have confirmed that the reaction kinetics are first-order because the value of $K_{sj}$ is considerably larger than the concentration of the odorous compounds in the liquid phase. The present system has the following operating properties:

$$Z = 0.6 \text{ m} \qquad U_G = 100 \text{ m/h}$$
$$\eta = 35\% \qquad V_{bed} = 60 \text{ m}^3$$
$$A = 100 \text{ m}^2 \qquad Q_G = 10^4 \text{ m}^3/\text{h}$$

Without changing the compost material show how the efficiency varies with respect to the face velocity $(U_G)$ and bed height $(Z)$ and recommend changes in $Z$ and $U_G$ that can increase the efficiency to 90%.

**Solution**   For first-order kinetics, the removal efficiency is equal to

$$\eta = 1 - \exp\left(-\frac{Z}{m_j}\frac{K_1}{U_G}\right) = 1 - \exp\left(-\frac{V_{bed}K_1}{m_j Q_G}\right)$$

The first-order rate constant and Henry's law constant will not change with changes in bed depth $(Z)$ or superficial velocity $(U_G)$. Thus from the measured efficiency of 35%, $K_1/m_j$ is found to be equal to 71.8 h$^{-1}$. The efficiency can be computed for various values of $Z/U_G$. Figure E10-6 shows how the efficiency varies for different values of $Z$ and $U_G$. An efficiency of 90% can be obtained for the following beds of different depths, cross-sectional areas, and face velocities:

$$Z = 0.4 \text{ m} \qquad U_G = 12.5 \text{ m/h} \qquad A_{bed} = 800 \text{ m}^2$$
$$Z = 1.0 \text{ m} \qquad U_G = 31.5 \text{ m/h} \qquad A_{bed} = 318 \text{ m}^2$$

While both beds have the same volume, the thicker bed can achieve the desired efficiency but at a higher operating cost, owing to the larger pressure drop. The cross-sectional area of the thinner bed is large and requires additional land, which may or may not be available. The design of the thinner bed may also be costly, owing to the difficulty of distributing the

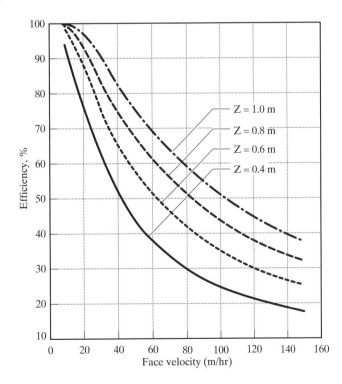

**Figure E10-6** Efficiency of a bioflter to consume odors from a wastewater treatment plant versus bed face velocity, $U_G = Q/A_{bed}$, and for several bed depths

air and gas flov
tional area.

---

**Case 2: Zeroth**
ations when the
phase is large
Under these cc
10.107) reduces

$$r_j =$$

The constant
*zeroth-order* rat
with different i
display zeroth-

**2a.** *Reaction-li*
lutants diff
than the re
**2b.** *Diffusion-l*
lutants diff
react so rap
packing ma
the biofilm

**Case 2a: Rea**
Under these cc
tion is

and the bounc
10.121) remain:
(10.120) shows
the biofilm:

$$\frac{c_{L,j}(x,z)}{c_{G,j}(z)/m_j}$$

The parameter
order kinetics:

At $z = z$, the m
an element of t

$$N(z) = -\Big[$$

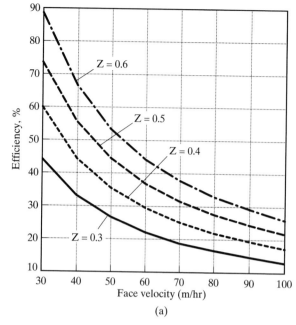

(a)

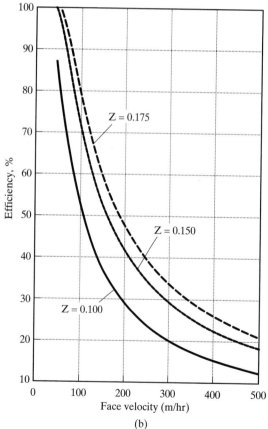

(b)

**Figure E10-7** (a) Efficiency of a bioflter to consume MEK vapor versus bed face velocity, $U_G = Q/A_{bed}$, for various bed depths, when the reaction is zeroth-order but limited by reaction kinetics
(b) Efficiency of a bioflter to consume MEK vapor versus bed face velocity, $U_G = Q/A_{bed}$, for various bed depths, when the reaction is zeroth-order but limited by diffusion

$Z = 1.0$ m: $\quad U_G = 48$ m/h $\quad A_{bed} = 20.4$ m$^2$

Thus the original bed ($Z = 0.6$ m) could be retained, but the gas velocity would have to decrease to 29.4 m/h.

*Case 2: Diffusion-limited*

$$\eta = 1 - \left[1 - \frac{Z}{U_G}\sqrt{\frac{K_0 D_i a_p}{2\delta m_j c_{G,j}(0)}}\right]^2$$

From the performance data at 35%, $K_0 D_j a_p / 2\delta m_j$ can be computed, and from this the efficiency at other $U_G$, $Z$, and $c_{G,j}(0)$ can be predicted. From Figure E10.7b, the following are three of many design conditions that should satisfy the desired performance:

$Z = 0.1$ m: $\quad U_G = 4.8$ m/h $\quad A_{bed} = 208$ m$^2$

$Z = 0.15$ m: $\quad U_G = 7.2$ m/h $\quad A_{bed} = 139$ m$^2$

$Z = 0.175$ m: $\quad U_G = 8.5$ m/h $\quad A_{bed} = 118$ m$^2$

If the kinetics are limited by diffusion, it is seen that a large bed is needed, accompanied by a low gas velocity which allows considerable time for the MEK to diffuse into the biofilm. The dramatically different values for $U_G$ and $A_{bed}$ for the same $Z = 0.1$ m, even though both products $U_G A_{bed} = Q = 1000$ m$^3$/h, suggest the criticality of a clear understanding of the kinetics that govern the process. Bench-top experiments must be conducted to reconcile the details of the kinetics.

Experimental studies indicate that the transient response of biological systems to variations in the inlet concentration and flow rate is slow. Consequently, biological systems are best suited in dedicated facilities where the composition and concentration of the inlet gas stream does not vary with time and the process operates 7 days a week for extended periods of time. Biological systems are ideally suited to remove odors from large volumetric flow rates from steady-state processes such as fermentation, wastewater, and food processing.

# 10.10 Ultraviolet–ozone oxidation

Several advanced technologies are currently drawing attention in the United States, Europe, and Asia (Peral and Ollis, 1992.). The new technologies are commonly called *advanced oxidation technologies* because oxidation of organic compounds is accomplished under unusual circumstances. Two examples are discussed: *ultraviolet–ozone oxidation* and *supercritical water oxidation* (SCWO).

It is ironic that reactions initiated by ultraviolet light and ozone which produce urban air pollution can also be used in a controlled fashion at STP to oxidize organic compounds in process gas streams. Since the UV/O$_3$ technique is isothermal, there are no fuel costs and no production of nitrogen oxides. Ozone gas is added to the process gas stream and the entire mixture is then subjected to UV radiation, (180 - 400 nm). Ozone can be generated by cold plasma discharge or photolysis of dry O$_2$ subjected to UV below 200 nm. The sources of ultraviolet radiation vary, but many presently employ inexpensive low-wattage mercury vapor lamps. The technology is based on concepts used for decades to oxidize organic compounds in water (Figure 10-38). The goal of the UV/O$_3$ technique is to produce OH· and HO$_2$· radicals that initiate reactions to *mineralize* (convert to CO$_2$ and H$_2$O) hydrocarbons and organic compounds. Ultraviolet–ozone oxidation techniques contain some, or all, of the following reactions that occur simultaneously:

- Photolysis of organic compounds
- Oxidation of organic compounds using ozone
- O$_3$, H$_2$O$_2$, photolysis, and the formation of OH· and HO$_2$·
- Photocatalysis using titania

## 10.10.1 Photolysis of organic compounds

Most aliphatic and aromatic hydrocarbons are not photolytic, while aldehydes and ketones are photolytic with varying degrees of reactivity. The objective of photolysis is to decompose organic compounds into radicals that ultimately form CO$_2$ and H$_2$O.

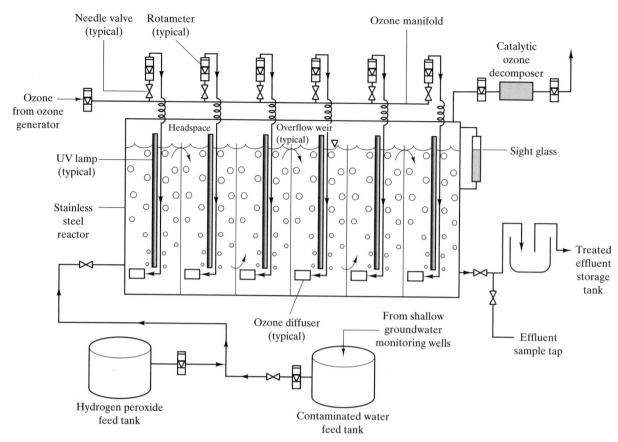

**Figure 10-38** UV/O$_3$ sparging system for water purification, taken from ULTROX diagram in EPA Report, EPA/540/A5-89/012

**1.** Aldehydes

*Formaldehyde*:
$$HCHO + h\nu \ (\lambda < 370 \text{ nm})$$
$$\rightarrow HCO\cdot + H\cdot \quad (a)$$

$$HCHO + h\nu \ (\lambda > 320 \text{ nm})$$
$$\rightarrow CO + H_2 \quad (b) \quad (10.150)$$

*Acetaldehyde*:
$$CH_3CHO + h\nu \ (\lambda < 300 \text{ nm})$$
$$\rightarrow CH_3\cdot + HCO\cdot \quad (a)$$

$$CH_3CHO + h\nu \ (\lambda < 260 \text{ nm})$$
$$\rightarrow CH_4 + CO \quad (b) \quad (10.151)$$

**2.** Ketones

*Acetone*:
$$CH_3COCH_3 + h\nu \ (260 \text{ nm} < \lambda < 300 \text{ nm})$$
$$\rightarrow CH_3\cdot + CH_3CO\cdot \quad (10.152)$$

*Methyl ethyl ketone*:
$$CH_3COC_2H_5 + h\nu \ (260 \text{ nm} < \lambda < 300 \text{ nm})$$
$$\rightarrow CH_3CO\cdot + C_2H_5\cdot \quad (a) \quad (10.153)$$

$$CH_3COC_2H_5 + h\nu \ (260 \text{ nm} < \lambda < 300 \text{ nm})$$
$$\rightarrow CH_3\cdot + C_2H_5CO\cdot \quad (b)$$

**3.** Alcohols

*Ethanol*:
$$C_2H_5OH + h\nu \rightarrow C_2H_4 + H_2O \quad (a)$$
$$C_2H_5OH + h\nu \rightarrow CH_3CHO + H_2 \quad (b) \quad (10.154)$$
$$C_2H_5OH + h\nu \rightarrow HCHO + CH_4 \quad (c)$$
$$C_2H_5OH + h\nu \rightarrow H\cdot + C_2H_5O\cdot \quad (d)$$
$$\hookrightarrow CH_3\cdot + HCHO$$

## 10.10.2 Oxidation of organic compounds using ozone

Reactions of ozone with alkenes produce aldehydes, ketones, and acids. For example, the reaction with ethene has a rate constant equal to

$$k(cm^3 \, molecule^{-1} \, s^{-1}) = 120 \times 10^{15}$$
$$\exp\left(-5232 \, cal \, gmol^{-1} \, (RT)^{-1}\right) \quad (10.155)$$

*Ethene:* $O_3 + C_2H_4 \rightarrow HCHO + H_2COO^*$

(intermediate species)

$$H_2COO^* + M \rightarrow H_2COO + M$$

$$H_2COO^* \rightarrow CO + H_2O \qquad (a)$$

$$H_2COO^* \rightarrow CO_2 + H_2 \qquad (b) \quad (10.156)$$

$$H_2COO^* \rightarrow CO_2 + H\cdot + H\cdot \qquad (c)$$

$$H_2COO^* \rightarrow HCOOH \qquad (d)$$

## 10.10.3 $O_3$, $H_2O_2$ photolysis, and OH· and HO$_2$· formation

Ozone gas reacts with UV radiation over a wide range of wavelengths ($\lambda$). The reader is urged to review atmospheric photochemical reactions in Chapter 6. Ozone generated externally is added to the process gas stream and when irradiated, produces $O(^1D)\cdot$ radicals, which in the presence of water vapor produce OH· radicals. Within the reactor, ozone is formed by

$$O_2 + O(^3P)\cdot + M \rightarrow O_3 + M \quad (10.157)$$

and removed by

$$O_3 + h\nu(\lambda < 320 \, nm) \rightarrow O_2 + O(^1D)\cdot \quad (a)$$
$$O(^1D)\cdot + H_2O(vapor) \rightarrow OH\cdot + OH\cdot \quad (b)$$
$$\qquad\qquad\qquad\qquad\qquad\qquad\qquad (10.158)$$
$$O(^1D)\cdot + H_2O(liquid) \rightarrow H_2O_2 \quad (c)$$
$$O_3 + h\nu(\lambda < 300 \, nm) \rightarrow O_2 + O(^3P)\cdot \quad (d)$$

The net effect is to generate OH· and O· that sustain reactions to form $H_2O_2$, which is photolytic and leads to the production of two OH· radicals for every $O_3$ that is consumed.

Radicals OH·, HO$_2$·, and $H_2O_2$ are interconnected through a series of reactions involving organics, $O_2$, and ultraviolet radiation. The major source

of OH· is the photolysis of $O_3$ in the presence of water vapor:

$$O_3 + h\nu(\lambda < 320 \, nm) \rightarrow O_2 + O(^1D)\cdot \quad (a)$$
$$\qquad\qquad\qquad\qquad\qquad\qquad\qquad (10.159)$$
$$O(^1D)\cdot + H_2O \rightarrow OH\cdot + OH\cdot \quad (b)$$

The photolysis of HONO and $H_2O_2$ also produces OH·:

$$HONO + h\nu(\lambda < 400 \, nm) \rightarrow OH\cdot + NO \quad (a)$$
$$\qquad\qquad\qquad\qquad\qquad\qquad\qquad (10.160)$$
$$H_2O_2 + h\nu(\lambda < 360 \, nm) \rightarrow OH\cdot + OH\cdot \quad (b)$$

Possible sources of HONO are reactions with NO and $NO_2$ that may be present in the process gas stream. Depending on the process under control, $H_2O_2$ may be formed from HO$_2$· radicals:

$$HO_2\cdot + HO_2\cdot \rightarrow H_2O_2 \quad (10.161)$$

Conversion of HO$_2$· to OH· is possible by

$$HO_2\cdot + NO \rightarrow OH\cdot + NO_2 \quad (10.162)$$

Formaldehyde is a principal by-product of organic oxidation. It is also the major source of HO$_2$· via reactions of H· and HCO· with $O_2$:

$$HCHO + h\nu \rightarrow H\cdot + HCO\cdot \quad (a)$$
$$H\cdot + O_2 \rightarrow HO_2\cdot \quad (b) \quad (10.163)$$
$$HCO\cdot + O_2 \rightarrow HO_2\cdot + CO \quad (c)$$

Any process that produces HCO· or H· is a source of HO$_2$·. As seen above, HO$_2$· is produced by the photolysis of aldehydes. Alkoxy radicals also produce HO$_2$·:

$$RCH_2O + O_2 \rightarrow RCHO + HO_2\cdot \quad (a)$$
$$RCHO + h\nu \rightarrow R\cdot + HCO\cdot \quad (b) \quad (10.164)$$
$$HCO\cdot + O_2 \rightarrow HO_2\cdot + CO \quad (c)$$

## 10.10.4 Titania photocatalysis

The anatase crystalline structure of titania $(TiO_2)$ is an unusual compound because electrons are displaced when the compound is subjected to UV radiation ($\lambda < 400 \, nm$). Titania is not alone in this regard; to varying degrees $SnO_2$ and ZnO react similarly. This displacement of electrons makes it possible to generate OH· and participate in the reactions above. The oxidation of organic compounds proceeds in the following general way:

alkanes $\xrightarrow[\text{TiO}_2]{h\nu}$ alcohols $\xrightarrow[\text{TiO}_2]{h\nu}$ alkenes $\xrightarrow[\text{TiO}_2]{h\nu}$

aldehydes and/or ketones $\xrightarrow[\text{TiO}_2]{h\nu}$ $CO_2 + H_2O$   (10.165)

Because of this capability, titania is called a *photocatalyst*. Excitation of $TiO_2$ with UV promotes the migration of an electron into the conduction band $(e_{CB}^-)$ and a positive "hole" in the valance band $(h_{VB}^+)$. Figure 10-39 illustrates the band structure of a photocatalyst. The photogenerated electron reduces $O_2$ molecules to a form of *superoxide ion* $(O_2^-)$, while the photogenerated "hole" reacts with surface-bound compounds or ions such as $OH^-$. In a water-free environment, the photogenerated "hole" reacts with the superoxide ion to form $(O\cdot)$. In humid air, $(OH\cdot)$ is produced.

If the fluid in contact with the photocatalytic surface is water, surface "holes" react with $H_2O$ or $(OH^-)$ to form $(OH\cdot)$:

$$TiO_2 + h\nu \to h_{VB}^+ + e_{CB}^- \qquad (a)$$
$$h_{VB}^+ + H_2O\,(\text{liquid}) \to OH\cdot + H^+ \qquad (b) \quad (10.166)$$
$$h_{VB}^+ + OH^-\,(\text{surface}) \to OH\cdot \qquad (c)$$

and excess electrons in the conduction band react with molecular oxygen to form superoxide ions:

$$e_{CB}^- + O_2 \to (O_2^-) \qquad (10.167)$$

which may later react with water to form additional $OH\cdot$:

$$2(O_2^-)\cdot + 2H_2O \to OH\cdot + OH\cdot + OH^- + O_2 \qquad (10.168)$$

or produce $OH\cdot$ by

$$(O_2^-)\cdot + H^+ \to HO_2\cdot \qquad (a)$$
$$HO_2\cdot + HO_2\cdot \to H_2O_2 + O_2 \qquad (b) \quad (10.169)$$
$$(O_2^-)\cdot + HO_2\cdot \to HO_2^- + O_2 \qquad (c)$$
$$HO_2^- + H^+ \to H_2O_2 \qquad (d)$$

Hydroxyl radicals are formed by cleaving $H_2O_2$ by any of the following reactions:

$$H_2O_2 + e_{CB}^- \to OH\cdot + OH^- \qquad (a)$$
$$H_2O_2 + (O_2^-) \to OH\cdot + OH^- + O^2 \qquad (b) \quad (10.170)$$
$$H_2O_2 + h\nu(\lambda \approx 260\ \text{nm}) \to OH\cdot + OH\cdot \qquad (c)$$

Thus when treating a liquid wastewater stream, there is much to be gained by adding $H_2O_2$ to the input wastewater stream. In the absence of $TiO_2$, Peyton and Glaze (1988) propose the following mechanism for $UV/O_3$ processes in the aqueous phase:

$$O_3 + H_2O + h\nu \to H_2O_2 + O_2 \qquad (a)$$
$$H_2O_2 + h\nu \to OH\cdot + OH\cdot \qquad (b)$$
$$H_2O_2 \rightleftharpoons HO_2^- + H^+ \qquad (c)$$
$$O_3 + HO_2^- \to O_3^- + HO_2\cdot \qquad (d)$$
$$HO_2\cdot \rightleftharpoons O_2^- + H^+ \qquad (e)$$
$$O_3 + O_2^- \to O_3^- + O_2 \qquad (f)$$
$$O_3^- + H^+ \rightleftharpoons HO_3 \qquad (g) \quad (10.171)$$
$$HO_3 \to OH\cdot + O_2 \qquad (h)$$
$$OH\cdot + H_2O_2 \to H_2O + HO_2\cdot \qquad (i)$$
$$OH\cdot + O_3 \to O_2 + HO_2\cdot \qquad (j)$$
$$OH\cdot + OH\cdot \to H_2O_2 \qquad (k)$$
$$HO_2\cdot + HO_2\cdot \to H_2O_2 + O_2 \qquad (l)$$
$$H_2O + HO_2\cdot + O_2^- \to H_2O_2 + O_2 + OH^- \qquad (m)$$

## 10.10.5 Photolytic kinetics

Photochemical reactions occurring in the gas phase are modeled as reactions involving molecular species or free radicals. The rate constants for photochemical reactions are expressed as

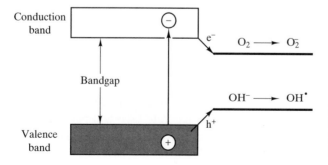

**Figure 10-39** Charge transfer in a photocatalytic material (TiO2) subjected to UV radiation

$$k = \int_{\lambda_1}^{\lambda_2} \sigma_A(\lambda, T)\varphi_A(\lambda, T)I(\lambda)\, d\lambda \quad (10.172)$$

where $\sigma_A(\lambda, T)$ is the *absorption cross section* of molecule $A$ at wavelength $\lambda$ and temperature $T$, $\varphi_A(\lambda, T)$ is the *quantum yield* (a probability) that molecule $A$ reacts when subjected to radiation of wavelength $(\lambda)$ and temperature $(T)$, and $I(\lambda)\, d\lambda$ is the *actinic flux* (photons $cm^{-2}\, s^{-1}$) in the band of wave lengths $d\lambda$. Absorption cross sections and quantum yields are parameters obtained from the literature of photochemistry. The parameters are unique to each molecular species and are generally discontinuous functions of wavelength. The actinic flux is a function of how bulbs are configured in the reactor and is inversely proportional to the radial distance from the bulb surface. Since the actinic flux decreases rapidly in regions not near the bulb surface, it is apparent that bulbs should be packed closely.

The kinetic expressions describing how OH•(or any other radical or molecular species) is generated (or consumed) at a photocatalytic surface depend on the local value of the actinic flux. Ostensibly, one wants an equation expressing how fast free radicals (or any other species) are generated or consumed. When titania $(TiO_2)$ particles are suspended in the fluid, photolysis can be modeled as a homogeneous source or sink within the fluid. When $TiO_2$ is deposited on surfaces in the reactor, photolysis is modeled as a surface source or sink boundary condition for the overall set of ordinary differential equations describing reactions within the reactor. Species produced on a surface diffuse into the gas stream and react with appropriate species in the gas phase. The diffusion and bulk motion of the surface-generated species are described by the conventional Navier–Stokes equations.

The rates of reaction on a photocatalytic surface are not expressible by the law of mass action and Arrhenius rate constant used to describe rates of reaction in the gas phase. Reactions that consume species $i$ on a catalytic surface are described by a *Langmuir–Hinshelwood* rate expression, for example

$$-r_i = k\frac{Kc_i}{1 + Kc_i} \quad (10.173)$$

where $k$ is the kinetic rate constant and $K$ is the Langmuir equilibrium constant. Since the photolytic reactions occur in the boundary layers surrounding the UV bulbs and $TiO_2$ surfaces (Mallery and Heinsohn, 1996), the design of UV reactors must maximize the ratio of the surface area of the boundary layers per volume of the reactor. Alternatively, to maximize $TiO_2$ catalysis, photolytic reactors need to maximize the ratio of surface area receiving UV radiation to the volume of reactor. The task is similar to the objective of arranging tubes in heat exchangers and condensers. The professional literature (Webb, 1993) on the subject should be consulted.

Ultraviolet radiation is produced by low-pressure mercury vapor reactions occurring inside quartz tubes. Quartz is chosen since it does not absorb the UV radiation. Thus a second consideration affecting the design of UV photoreactors is to eliminate deposits that might form on the outer surface of the UV bulbs from chemical reactions or particles that contact the tubes, absorb the radiation, and prevent photolysis.

## EXAMPLE 10.8  GAS PHASE PHOTOLYTIC REACTOR

An airstream at STP containing 100 ppm of ethanol vapor $(c_0)$ enters a photolytic reactor 6 cm high $(H)$, 1 m wide $(W)$, and of variable length $(L)$ (Figure E10-8a). The average velocity $(U_0)$ is 1 m/s. Inside the duct, 100-W low-pressure Hg UV bulbs 1 m long will be placed with their axes perpendicular to the flow. The bulbs are 1 cm in diameter. Laying between the bulbs in a staggered array are 1-cm cylinders coated with $TiO_2$. The bulbs and $TiO_2$ cylinders are spaced 2 cm between centers. The upper and lower surfaces of the duct are also coated with $TiO_2$. You have been asked to prepare an analytical model of a photolytic reactor to remove ethanol and predict the efficiency of the reactor as a function of its length $(L)$. From the kinetic data reported by Nimlos et al. (1996), the absorption coefficient $(K)$ is 0.49 $ppm^{-1}$. From the experimental data reported by these authors, a surface photolytic kinetic rate constant $(k/n_t)$ can be computed and is equal to 0.06 ppm m $s^{-1}$.

*Solution*  For modeling purposes, consider a short length of duct, $L = 4.46$ cm, containing nine bulbs and four $TiO_2$-coated cylinders (Figure E10-8a). The total surface area per unit volume of reactor $(a_s)$ is

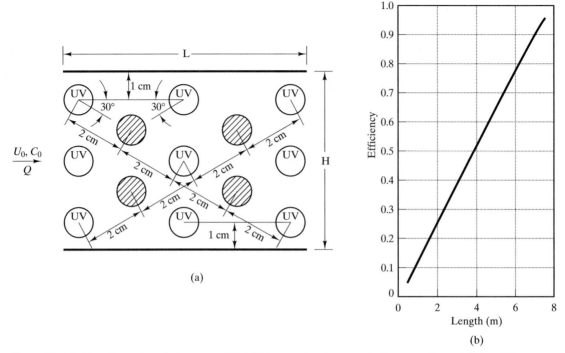

**Figure E10-8** Gas phase photolytic reactor (b) Efficiency of a photocatalytic reactor for oxidizing airborne ethanol versus reactor length, for an array of 100-W, low-pressure Hg UV bulbs and $TiO_2$ coated surfaces

$$a_s = \frac{(4)\pi(0.01)(1)\ \text{m}^2 + (2)(1)(0.0446)\ \text{m}^2}{(0.0446)(1)(0.06)\ \text{m}^3}$$

$$= \frac{0.1256 + 0.0892}{0.002676} = 80.3\ \text{m}^{-1}$$

The energy required per volume of reactor is

$$P_v = \frac{(9\ \text{bulbs})(100\ \text{W/bulb})}{0.002676} = 336{,}322\ \text{W/m}^3$$

$$= 336\ \text{kW/m}^3$$

To model the removal of ethanol, consider an elemental volume and perform a mass balance, where $y$ is the ethanol mole fraction and $N$ is the total molar flow rate.

$$Ny = N(y + dy) + \frac{kKy}{(1 + Ky)a_s A\ dL}$$

$$\frac{dy(1 + Ky)}{Ky} = -\frac{a_s Ak}{N}\ dL$$

Integrating over a length of duct ($L$) when the concentration changes from $y_0$ to $y(L)$ gives

$$\int_{y_0}^{y(L)} \frac{1 + Ky}{Ky}\ dy = -\int_0^L \frac{s_s Ak}{N}\ dL$$

$$-\frac{a_s kAL}{N} = \frac{1}{K}\ln\frac{y(L)}{y_0} + [y(L) - y_0]$$

The overall removal efficiency is

$$\eta = 1 - \frac{y(L)}{y_0}$$

Substituting in the above yields

$$\frac{1}{K}\ln(1 - \eta) - \eta y_0 = -\frac{a_s kAL}{N} = -\frac{a_s kV}{N}$$

$$= -\frac{a_s kAL}{U A n_t} = -a_s \frac{k}{n_t}\frac{L}{U}$$

where $n_t$ is the total molar concentration. Figure E10-8b is a graph of $\eta$ versus $L$ which shows that for the data above, the efficiency is linearly proportional to $L$.

To achieve 95% removal would require a reactor approximately 8 m long ($V = 0.4\ \text{m}^3$) with 81 bulbs.

The electrical power consumed is 1.27 kW/cfm. If operated for 3600 h/yr, the electrical costs (0.0596 $/kWh) would be $272/cfm-yr. These costs are considerably higher than the costs associated with an activated charcoal adsorber, wet scrubber, or thermal oxidizer (see Chapter 13 for details about these three air pollution control systems).

## 10.10.6 Applications

Using ozone to disinfect drinking water been practiced in Europe for decades and is an attractive alternative to the chlorinating techniques practiced in the North America (Glaze, 1987). Ozone can be combined with UV, hydrogen peroxide, and so on, to produce additional benefits. Ultraviolet–ozone techniques can remove chlorinated organic compounds from ground water (Kutsuna et al., 1993, 1994). The EPA SITE program was a comprehensive study of new techniques to remove pollutants from process gas streams and groundwater (Lewis et al., 1992; Lewis and Parker, 1994). Ultraviolet–ozone techniques demonstrated the ability to oxidize chlorinated organics in the aqueous phase or to oxidize organics stripped from groundwater. Obee and Brown (1995) proposed using UV and $TiO_2$ to oxidize indoor air pollutants.

Ultraviolet radiation is produced by cylindrical bulbs containing a selected gas that radiates at particular UV, wavelengths. Commercially available low-wattage mercury vapor bulbs, approximately 5 ft long and $\frac{1}{2}$ in. in diameter are typically used. In many design configurations, the gas flow is parallel to the bulb's axis, but in some configurations the flow is perpendicular to the bulb axis. Since the actinic flux $\left(\text{photons cm}^{-2}\,\text{s}^{-1}\right)$ from a cylindrical bulb is inversely proportional to the distance from the bulb, bulbs should be packed close together for maximum effect. If titania is used, it is important to maximize the surface area per unit volume of reactor. In aqueous systems, titania particles can be suspended as a slurry (Mills and Hoffmann, 1993; Pacheco and Yellowhorse, 1992). In the gas phase the situation is far more complicated, particularly if the volumetric flow rate is large. Bench-top experiments with fluidized beds (Dibble and Raupp, 1992), porous titania frits (Nimlos et al., 1993), or gauze impregnated with titania surrounding each cylindrical bulb have been

tried. A novel approach using diffuse radiation from optical fibers is described by Matsunaga and Okochi (1995).

Figure 10-40 is a schematic diagram of a $UV/O_3$ system that illustrates the essential features of the process (Anon., 1991). The system contains three distinct components which treat the process gas stream.

**1.** *Gas-phase photo-oxidation.* In the first component, photolytic compounds react with UV to form secondary products, which in turn may, or may not, be photolytic. Simultaneously, free radicals O•, OH•, and $HO_2$• are generated when UV photolyzes $O_3$ or $H_2O_2$ in the presence of $O_2$ and $H_2O$ (vapor).

**2.** *Scrubbing with ozonated water.* The absorption of the VOCs in the packed beds is similar to that discussed earlier. Ozone $\left(O_3\right)$ and $H_2O_2$ in the scrubbing liquid (and perhaps even UV light radiation) causes chemical reactions to occur in the aqueous phase.

**3.** *Granular activated charcoal.* Pollutants are adsorbed on the charcoal according to concepts described earlier. The desorption process, however, is vastly different. Regeneration with ozonated air initiates chemical oxidation reactions on the charcoal surface. Thus an oxidation reaction occurs on the surface rather than the purely physical desorption process discussed earlier.

Other commercial systems do not have three components necessarily, since they may incorporate several functions in one component.

# 10.11 Supercritical water oxidation

Supercritical water oxidation (SCWO) refers to the oxidation of organic compounds dissolved in water raised above its critical temperature. Water, or any other liquid or gas is supercritical if its temperature and pressure exceed the critical values listed in Table A-9. In the case of water, the critical temperature and pressure are 374°C and 22.13 MPa. Supercritical water enhances transport properties that sustain chemical reactions that ordinarily occur rapidly only at higher temperatures. Under supercritical conditions the organic compounds and oxygen are all present in the single, dense, aqueous phase minimizing

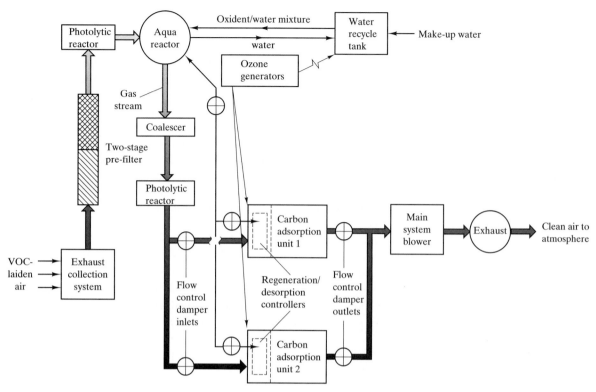

**Figure 10-40** Schematic diagram of UV-radical oxidation system. Air containing VOC's passes through a photolytic reactor which adds ozone and subjects the mixture to UV radiation. The gas then passes through a wet scrubber using water containing ozone. The gas then passes through activated carbon that is regenerated with ozonated air.

mass transfer resistance and providing relatively fast reaction rates. Under supercritical conditions, nonpolar hydrocarbons become soluble and salts become insoluble. Supercritical oxidation is attractive economically for aqueous waste streams with organic concentrations from 1 to 20%. At lower organic concentrations, an auxiliary fuel is required. With the addition of oxygen, hydrogen peroxide, or other suitable oxidants $(KMnO_4, Cu–H_2O_2,$ etc.), hazardous organics can be oxidized in the aqueous phase. In many respects oxidation is similar to combustion in the gas phase, but it occurs at much lower temperatures and the oxidation products are totally contained within the aqueous medium since many of the oxidation products are soluble in supercritical water. Heavy metals are converted to their oxides. Sulfur and phosphorus are converted to sulfates and phosphates. The unique properties of supercritical water makes SCWO attractive for oxidizing hazardous organic wastes, which would be more costly to oxidize by direct flame oxidizers. Several types of waste streams are amenable for treatment by SCWO:

- PCB-contaminated oil, waste solvents
- Groundwater contaminated with organics
- Stored wastes containing sludges or other oils
- Soil contaminated by spills or burial of organics
- Pharmaceutical wastes, mixed chemical wastes
- DOE hazardous and mixed wastes
- Spent carbon adsorbents, DOD wastes (propellants, explosives, etc.), and medical wastes

Commercial processes typically take aqueous wastes at STP and compress them to 3400 psig, whereupon compressed air is metered into the system. Exothermic reactions increase the water temperature to between 600 and 620°C. Caustic soda can be injected onto the system to neutralize acids formed by oxidation. This forms salts, which sink to the bottom of the reactor. Gaseous products of oxi-

$k_1$

$k_0$

$K_0$

$K_1$

$L$

$L'$

$L_j$

$L_m$

$L_s$

$m$

$\dot{m}$

$\dot{m}_a$

$\dot{m}_{ad}$

$\dot{m}_d$

$\dot{m}_{DE}$

$\dot{m}_g$

$\dot{m}_g$

$m_j$

$M_s, M_c$

$N$

$\dot{n}_i$

$n_j$

$n_t$

$N_t$

NTU

$P$

$P_G, P_L$

$P_j$

dation, primarily CO, $CO_2$, and nitrogen leave the reactor. No exhaust stacks are needed and virtually no $NO_x$ is generated, although small amounts of $N_2O$ may be formed. The reactor effluent is quenched by direct contact with recycle water, which dissolves the small amount of salt remaining in solution. A handful of SCWO processes are commercially available and others are being prepared for pilot-scale demonstrations. Supercritical water oxidation is a developing, not a mature technology, under active study in the United States, Europe, and Japan. Shown in Figure 10-41 is a diagram of a SCWO process treating water containing long-chain organics and amines formerly treated by direct flame thermal oxidizers.

Whereas SCWO is effective in the destruction of original organic compounds, incomplete conversion may produce low-molecular-weight partial oxidation products such as formic acid, acetic acid, benzyl alcohol, and benzoic acid. In all cases, conversion of

organic compounds to $CO_2$ is enhanced by the addition of heterogeneous catalysts such as CuO, $MnO_2/CeO$, $Al_2O_3/V_2O_5$, or $Cr_2O_3/Al_2O_3$ (Ding et. al., 1995). In the absence of catalysts, supercritical oxidation proceeds along parallel reaction paths. One path is direct oxidation to form stable products such as $CO_2$ and $N_2$. The second path leads to the formation of low-molecular-weight intermediates, which are then slowly oxidized to the end products (Savage et al., 1995). In experiments with phenol, Ding et al. (1995) report that catalysts enhanced the yield ratio (ratio of $CO_2$ to the carbon in phenol) from 0.3–0.5 to 0.5–0.98.

## *10.12 Closure*

Separating pollutant gases and vapors from a process gas stream lies at the heart of environmental engineering. Application of the principles

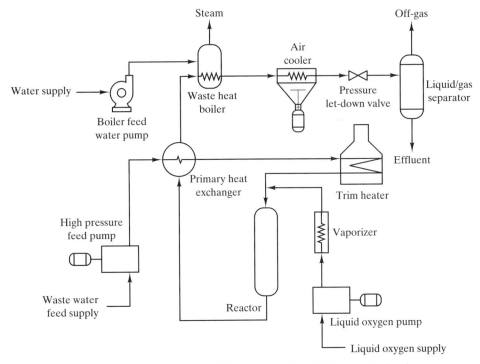

**Figure 10-41** Schematic diagram of a SCWO system of the ECO Waste Technologies process. Long-chain organics and amines are compressed and heated to supercritical state whereupon these impurities are oxidized. The liquid effluent contains less than 5 ppm organic carbon and less than 1 ppm ammonia and is suitable for discharge to a publicly owned treatment system. (redrawn from Chemical Engineering Progress, April 1995)

of the the
chemical
dents in a
discipline
stand hov
operate
removal e
tions engi

*Nomei*

*Symbol*

$\rightleftharpoons$

$\rightarrow$

$=$

$[=]$

$[j]$

$A$

$a, a_p$

$A_s$

$c$

$c, c_j$

$c_e$

$c_L$

$\bar{c}_{p,i}$

$C_{\text{power}}$

$c_0$

$c(x)$

$D$

$D$

$D_c$

$D_G, D_L$

* F, forc
t, time; T, t

---

**Table 11-2**   Dynamic Shape Factors ($X$) Averaged over All Orientations

| Shape | $X$ |
|---|---|
| Sphere | 1.00 |
| Cube | 1.08 |
| Cylinder ($L/D = 4$) | |
|   Axis horizontal | 1.32 |
|   Axis vertical | 1.07 |
| Ellipsoid, across polar axis, | 1.20 |
|   major/minor diameters = 4 | |
| Parallelepiped, square base; | |
|   height to base: | |
|     0.25 | 1.15 |
|     0.50 | 1.07 |
|     2.00 | 1.16 |
|     3.00 | 1.22 |
|     4.00 | 1.31 |
| Clusters of spheres | |
|   Chain of two | 1.12 |
|   Chain of three | 1.27 |
|   Three compact | 1.15 |
|   Chain of four | 1.32 |
|   Four compact | 1.17 |

*Source:* Abstracted from Fuchs (1964) and Strauss (1966).

where $D_{p,p}$ and $D_{s,p}$ are called the *equivalent projected area diameter* and *equivalent surface area diameter*, respectively. $D_{p,p}$ is the diameter of a sphere with the same projected area (i.e., cross-sectional area of the particle normal to the direction of flow) and $D_{s,p}$ is the diameter of a sphere with the same surface area as the actual particle. The *Sauter mean diameter* is the ratio of the total volume of all particles in an aerosol divided by the total surface area of all the particles. For this reason the Sauter diameter can be thought of as a mean volume-surface diameter.

Spherical particles whose density is equal to the density of water are called *unit density spheres*. For the most part, the density of a particle is the density of the compound of which it is composed. In the event the particle contains voids, is a loose feathery agglomerate, or is a composite material, the density is more difficult to define. Details of how to cope with these circumstances can be found in Hinds (1982).

## 11.2.3  Size distribution

Another physical property to distinguish one group of particles from another is size distribution. To select the appropriate particle control device to satisfy EPA performance standards, it is necessary that the size distribution at the source be known. An aerosol in which all particles have the same diameter is called a *monodisperse aerosol*. An aerosol in which particles have different diameters is called a *polydisperse aerosol*. If two aerosols have the same mass concentration (mg/m$^3$) but different size distributions, the dynamic behavior of one aerosol will be quite different from the other. The particle mass concentration is defined as the mass of particles per volume of carrier gas. For simplicity, the particle mass concentration will be expressed by the symbol $c$ without any subscript.

### EXAMPLE 11.1   CLOUDS OF WATER DROPLETS

To illustrate some important features about size distributions, consider four different monodisperse aerosols consisting of H$_2$O drops. The mass concentration of each aerosol is 0.5233 mg/m$^3$.

| Aerosol | $D_p(\mu\text{m})$ | $n(\text{particles/m}^3)$ | $c(\text{mg/m}^3)$ |
|---|---|---|---|
| A | 1,000 | 1 | 0.5233 |
| B | 100 | $10^3$ | 0.5233 |
| C | 10 | $10^5$ | 0.5233 |
| D | 1 | $10^7$ | 0.5233 |

Thus aerosol A, consisting of one rain drop (1000 μm) suspended in a cubic meter of air, produces the same mass concentration as aerosol D, consisting of $10^7$ fog droplets in a cubic meter of air! Obviously, mass concentration alone does not satisfactorily differentiate one aerosol from another. Other factors, such as the average mass, average diameter, or median diameter, can be used to discriminate one polydisperse aerosol from another.

To illustrate the difference between the *arithmetic mean* $(D_{p,am})$ of an ensemble and the *median* diameter $(D_{p,\text{median}})$ of the ensemble, consider an aerosol containing one particle 1000 μm in diameter and $10^7$

particles 1 μm in diameter. The total number of particles $(n_t)$ is

$$n_t = 10,000,001$$

The total mass concentration $(c)$ is

$$c = 0.5233 + 0.5233 = 1.046 \text{ mg/m}^3$$

The arithmetic mass mean diameter, $D_{p,am}(\text{mass})$, is the diameter 10,000,001 particles would have to have to produce a mass concentration of 1.046 mg/m³:

$$c = \frac{n_t \pi \rho_p [D_{p,am}(\text{mass})]^3}{6}$$

$$1.046 \text{ mg/m}^3 = (10,000,001)(10^{-9} \text{ mg/m}^3)$$

$$\pi [D_{p,am}(\text{mass})]^3$$

$$D_{p,am}(\text{mass}) = 7.258 \text{ μm}$$

There are no particles this size, but the single 1000-μm particle has enormous influence on the mass of the aerosol. With respect to the 10,000,001 particles in the aerosol, the median particle with respect to number, $D_{p,median}(\text{number})$, is 1 μm, since 50% of the particles is equal to or larger than this size:

$$D_{p,median}(\text{number}) = D_{p,50}(\text{number}) = 1\text{μm}$$

---

Monodisperse aerosols, composed of particles of a single size, are rarely encountered in engineering practice. If the physical properties of the particles in two polydisperse aerosols are the same, one way to distinguish one aerosol from another is to compare the size distributions of the two aerosols. What follows is a synopsis of a commonly encountered statistical description, called the *log-normal distribution*. For a systematic study of particle size distributions, readers should consult texts devoted to the subject (Cadle, 1965; Hinds, 1982; Willeke and Baron, 1993; Waters et al., 1991).

Suppose that an aerosol is analyzed on the basis of size and found to have the distribution shown in Table 11-3. The *arithmetic mean* diameter based on number, $D_{p,am}(\text{number})$, *variance*, $\sigma^2$, and *standard deviation*, $\sigma$, can be computed from the following:

$$D_{p,am} = D_p (\text{number arithmetic mean})$$

$$= \sum_i \frac{n_i D_{p,i}}{n_t} \tag{11.17}$$

$$D_{p,am} = \frac{1}{n_t} \int_0^\infty n(D_p) \, dD_p \tag{11.18}$$

$$\sigma^2 = \sum_i \frac{n_i}{n_t} (D_{p,am} - D_{p,i})^2 \tag{11.19}$$

The variable $n(D_p)$ is the *number distribution function*, and the quantity $n(D_p)dD_p$ represents the number of particles between sizes $D_p$ and $(D_p + dD_p)$. The variable $n_i$, on the other hand, is the number of particles in the size range $i$ where the midrange particle has a diameter, $D_{p,i}$. The symbol $n_t$ is the total number of particles:

$$n_t = \sum_i n_i = \int_0^\infty n(D_p) \, dD_p \tag{11.20}$$

The reader must be very careful to distinguish between the discrete number concentration within a certain range $n_i$, which has the units number of particles, and the continuous size distribution function $n(D_p)$, which has the units number of particles/μm. The variables $n_i$ and $n(D_p)$ may also be defined in terms of a number concentration, number of particles/m³ and number of particles/m³·μm, respectively.

Figure 11-5 is a graph of the discrete number fraction $n_i/n_t$ versus particle diameter $D_p$ and shows that the distribution is not a Gaussian distribution. Many particle size distributions encountered in air pollution can be described as *log-normal*; that is, if the discrete number fraction $n_i/n_t$ is plotted against $(\ln D_p)$ as shown in Figure 11-6, the resulting distribution has the behavior associated with a normal distribution. It will be useful to generate an analytical function to describe a log-normal distribution since it can be incorporated in equations and computer programs to calculate parameters needed in air pollution control. Log-normal distributions can be characterized by two properties, geometric mean particle diameter based on number $(D_{p,gm})$ and the geometric standard deviation $(\sigma_g)$.

**Table 11-3**   Particle Size Distribution

| Class boundary (μm) | Class width (μm) | Class midpoint (μm) | Particles per class | Percent particles in class | Percent of particles less than upper interval size |
|---|---|---|---|---|---|
| *Frequency distribution of 1000 particles classified according to classes of equal linear size* | | | | | |
| 0–0.5 | 0.5 | 0.25 | 16 | 1.6 | 1.6 |
| 0.5–1.0 | 0.5 | 0.75 | 159 | 15.9 | 17.5 |
| 1.0–1.5 | 0.5 | 1.25 | 235 | 23.5 | 41.0 |
| 1.5–2.0 | 0.5 | 1.75 | 200 | 20.0 | 61.0 |
| 2.0–2.5 | 0.5 | 2.25 | 133 | 13.3 | 74.3 |
| 2.5–3 0 | 0.5 | 2.75 | 97 | 9.7 | 84.0 |
| 3.0–3.5 | 0.5 | 3.25 | 55 | 5.5 | 89.5 |
| 3.5–4.0 | 0.5 | 3.75 | 36 | 3.6 | 93.1 |
| 4.0–4.5 | 0.5 | 4.25 | 24 | 2.4 | 95.5 |
| 4.5–5.0 | 0.5 | 4.75 | 15 | 1.5 | 97.0 |
| 5.0–5.5 | 0.5 | 5.25 | 10 | 1.0 | 98.0 |
| 5.5–6.0 | 0.5 | 5.75 | 6 | 0.6 | 98.6 |
| 6.0–6.5 | 0.5 | 6.25 | 5 | 0.5 | 99.1 |
| 6.5–7.0 | 0.5 | 6.75 | 3 | 0.3 | 99.4 |
| 7.0–7.5 | 0.5 | 7.25 | 2 | 0.2 | 99.6 |
| 7.5–8.0 | 0.5 | 7.75 | 1 | 0.1 | 99.7 |
| 8.0–8.5 | 0.5 | 8.25 | 1 | 0.1 | 99.8 |
| 8.5–9 0 | 0.5 | 8.75 | 1 | 0.1 | 99.9 |
| 9.0–9.5 | 0.5 | 9.25 | 1 | 0.1 | 100.0 |
| *Frequency distribution of 1000 particles classified according to size classes of equal logarithmic diameter size width* | | | | | |
| 0.3–0.4 | 0.1 | 0.35 | 6 | 0.6 | 0.6 |
| 0.4–0.53 | 0.13 | 0.46 | 15 | 1.5 | 2.1 |
| 0.53–0.71 | 0.18 | 0.62 | 41 | 4.1 | 6.2 |
| 0.71–0.94 | 0.24 | 0.82 | 88 | 8.8 | 15.0 |
| 0.94–1.28 | 0.31 | 1.1 | 150 | 15.0 | 30.0 |
| 1.28–1.68 | 0.42 | 1.5 | 180 | 18.0 | 48.0 |
| 1.68–2.23 | 0.57 | 2.0 | 205 | 20.5 | 68.5 |
| 2.23–3.00 | 0.75 | 2.6 | 156 | 15.6 | 84.1 |
| 3.0–4.0 | 1.0 | 3.5 | 91 | 9.1 | 93.2 |
| 4.0–5.3 | 1.32 | 4.6 | 44 | 4.4 | 97.6 |
| 5.3–7.1 | 1.63 | 6.2 | 18 | 1.8 | 99.4 |
| 7.1–9.4 | 2.32 | 8.2 | 5 | 0.5 | 99.9 |
| 9.4–12.8 | 3.13 | 11.0 | 1 | 0.1 | 100.0 |

If an aerosol has a log-normal size distribution, the normalized *number distribution function* can be expressed as

$$\frac{n(D_p)}{n_t} = \frac{1}{D_p \ln \sigma_g \sqrt{2\pi}} \exp \left\{ -\left(\frac{1}{2}\right) \left[ \frac{\ln\left(D_p/D_{p,gm}\right)}{\ln \sigma_g} \right]^2 \right\} \quad (11.21)$$

The term $\sigma_g$, called the *geometric standard deviation*, is related to the variance

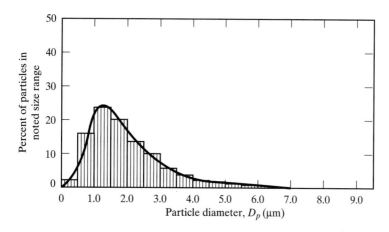

**Figure 11-5** Number distribution versus particle diameter for data in Table 11.3

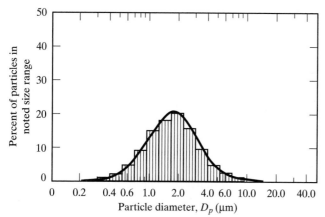

**Figure 11-6** Number distribution versus logarithm of particle diameter in Table 11.3

$$\sigma_g = \exp\left\{\sqrt{\ln\left[1 + \left(\frac{\sigma}{D_{p,am}}\right)^2\right]}\right\} \quad (11.22)$$

The *geometric mean diameter* based on number $\left(D_{p,gm}\right)$ is related to the arithmetic mean diameter based on number:

$$D_{p,gm} = D_{p,am} \exp\left[-\frac{(\ln\sigma_g)^2}{2}\right] \quad (11.23)$$

The *cumulative number distribution function* $N(D_p)$ is the total number of particles smaller than $D_p$. The cumulative distribution function will be written as $N$ in the material that follows, to simplify equations. The normalized cumulative distribution function for a log-normal distribution can be expressed in terms of the error function (erf):

$$\frac{N(D_p)}{n_t} = \int_0^{D_p} \frac{n(D_p)}{n_t}\, dD_p \quad (11.24)$$

$$= \frac{1}{2}\left\{1 + \mathrm{erf}\left[\frac{\ln(D_p/D_{p,gm})}{1.414\,\ln\sigma_g}\right]\right\}$$

The fraction of particles between any two particle sizes, $D_{p,1}$ and $D_{p,2}$, can be found by subtracting the values of the respective normalized cumulative number distribution functions:

$$\frac{N(D_{p,2})}{n_t} - \frac{N(D_{p,1})}{n_t} \quad (11.25)$$

There exists a unique particle diameter $(D_{p,50})$, called the *median particle diameter*, such that when the value is substituted in Equation 11.24, the value of the normalized cumulative number distribution function is one-half (i.e., $N/n_t = 0.5$), and

$$\mathrm{erf}\left[\frac{\ln(D_{p,50}/D_{p,gm})}{1.414\,\ln\sigma_g}\right] = 0 \quad (11.26)$$

After simplification the above reduces to

$$D_{p,gm} = D_{p,50} \quad (11.27)$$

Thus for a log-normal size distribution, the median particle diameter based on number is equal to the geometric mean particle diameter based on number, and the terms *median* and *geometric* mean can be used interchangeably. There also exists another unique particle diameter $(D_{p,\,84.1})$, having the value

$$\ln \frac{D_{p,\,84.1}}{D_{p,\,gm}} = \ln \sigma_g \qquad (11.28)$$

$$\sigma_g = \frac{D_{p,\,84.1}}{D_{p,\,gm}} = \frac{D_{p,\,84.1}}{D_{p,\,50}}$$

When $D_{p,\,84.1}$ is substituted into Equation 11.24, we obtain

$$\frac{N(D_{p,\,84})}{n_t} = \frac{1}{2}\left[1 + \text{erf}\,(0.707)\right] = 0.841 \quad (11.29)$$

The term $D_{p,\,84.1}$ corresponds to a particle whose normalized cumulative number distribution function is 84.1% (i.e., 84.1% of the particles are smaller than $D_{p,\,84.1}$. A similar analysis can be performed to show that

$$\sigma_g = \frac{D_{p,\,50}}{D_{p,\,15.9}} \qquad (11.30)$$

where $D_{p,\,15.9}$ corresponds to a particle whose normalized cumulative number distribution function is 15.9%. On the basis of number, 34% of the total number of particles are between $D_{p,\,gm}$ and $D_{p,\,84.1}$. Similarly, 34% of the particles are between $D_{p,\,gm}$ and $D_{p,\,15.9}$.

Readers can determine whether a size distribution is log-normal by trying to describe the data by the equations above, or by plotting the data on special graph paper, called *log probability paper*, designed for this purpose. The graph paper contains a conventional logarithmetic axis for the particle diameter and a specially scaled axis for the cumulative distribution. Particle size distributions which are log-normal are straight lines on such graph paper and the values of $D_{p,\,50}$ and $\sigma_g$ can be computed using Equations 11.27 and 11.28. Plotted in Figure 11-7 are the data in Table 11-3. From the graph it is reasonable to conclude that the distribution is log-normal.

$$D_{p,\,50}(\text{number}) = D_{p,\,g}(\text{number}) = 1.4\ \mu\text{m}$$

$$D_{p,\,84}(\text{number}) = 2.7\ \mu\text{m}$$

$$\sigma_g = \frac{D_{p,\,84}}{D_{p,\,50}} = \frac{2.7}{1.4} = 1.93$$

From Table 11-3 one can compute the arithmetic mean diameter based on number, $D_{p,\,am}(\text{number})$, to be 2.05 μm and the standard deviation ($\sigma$) to be 1.16. From Equation 11.22 the geometric standard deviation, $\sigma_g$, is computed to be 1.69, which is reasonably close to the 1.93 calculated (left) from Figure 11-7, considering the experimental accuracy of the data.

The discussion to this point has concerned only the *number* of particles of particular sizes. Many instruments used to analyze aerosols are based on light scattered by particles. Light scattering depends on the square of the diameter. Other instruments discriminate among particles on the basis of their masses, which depend on the cube of the particle diameters. Consequently, there are particle area distributions and particle mass distributions. Equations 11.17 to 11.30 have counterparts based on mass in which the normalized number distribution function $n/n_t$ is replaced by a normalized *mass distribution function*, $m(D_p)/m_t$, where $m_t$ is the total mass of the aerosol and in which the normalized cumulative number distribution, $N(D_p/n_t)$, is replaced by the normalized *cumulative mass distribution function*, $G(D_p/m_t)$. Again the mass distribution function, $m(D_p)/m_t$, versus particle diameter can be plotted similarly to Figure 11-7. If the size distribution is log-normal, Hatch and Choate devised a conversion equation in 1929 that relates the number geometric mean diameter and the mass geometric mean diameter. Equation 11.31 is used frequently and is called the *Hatch–Choate equation* (Hinds, 1982):

$$\ln \frac{D_{p,\,gm}(\text{mass})}{D_{p,\,gm}(\text{number})} = 3(\ln \sigma_g)^2 \qquad (11.31)$$

$$\ln D_{p,\,gm}(\text{mass}) = \ln D_{p,\,gm}(\text{number}) + 3(\ln \sigma_g)^2$$

The proof of this statement is derived from the following. Define the mass distribution function $m(D_p)$ as

$$m(D_p) =$$

$$\frac{\pi}{6} \frac{n_t D_p^3}{D_p \ln \sigma_g \sqrt{2\pi} \exp\left\{-[\ln(D_p/D_{p,\,g})]^2/2(\ln \sigma_g)^2\right\}} \qquad (11.32)$$

Using the identity $D_p^3 = \exp(3 \ln D_p)$, expanding the exponential portion of Equation 11.32, and completing the square within the exponent, the above can be written as

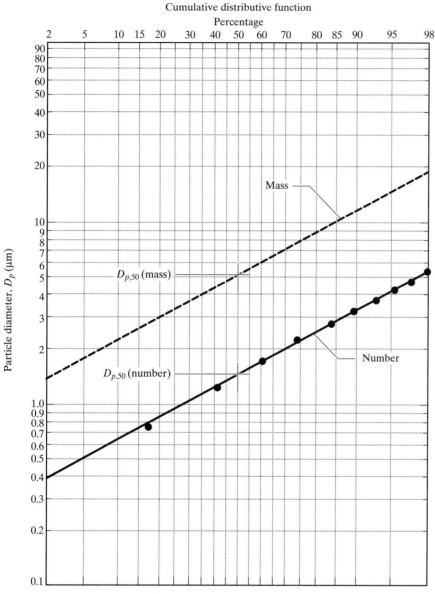

**Figure 11-7** Size distribution for data in Table 11.3

$$m(D_p) = \frac{n_t}{D_p \ln \sigma_g \sqrt{2\pi}} \exp\left[3\ln D_{p,gm} + \frac{9}{2}(\ln \sigma_g)^2\right]$$

$$\exp\left[-\frac{\ln D_p - \left[\ln D_{p,gm} + 3(\ln \sigma_g)^2\right]^2}{2(\ln \sigma_g)^2}\right]$$

(11.33)

While Equation 11.33 is complicated, it should be noted that $\ln \sigma_g$ appears in ways that indicate that Equation 11.33 is yet another log-normal expression

that has the same geometric standard deviation as Equation 11.21. Consequently, if $n(D_p)$ is log-normal, $m(D_p)$ is also log-normal with the *same geometric standard deviation* $\sigma_g$. The logarithm of the median particle size with respect to mass is related to the median particle size with respect to number by Equation 11.31. Since $D_p$ has the units of length and one cannot obtain the logarithm of a dimensional quantity, the reader should think of $\ln D_p$ as $\ln (D_p/1 \ \mu m)$ where 1 $\mu m$ is a "reference" particle size even though it is not stated explicitly. Thus in using

Equation 11.31, introduce $D_p$ in micrometers ($\mu$m). From Equation 11.31 one can compute the geometric mean diameter based on mass, $D_{p,gm}(\text{mass})$, to be 5.12 $\mu$m. With this value, the mass distribution shown in Figure 11-7 can be drawn. From the mass distribution one can estimate the percent of the mass between particle diameters $D_{p,1}$ and $D_{p,2}$. For example, the percent of the mass that is inhalable $(D_p < 10 \ \mu\text{m})$ is 85% and the percent of the mass that is respirable $(D_p < 2 \ \mu\text{m})$ is 7%. Thus the percent of the mass between 2 and 10 $\mu$m is 78%.

The relationship between the mass of particles per unit volume of air $(c)$ and the number of particles per unit volume, *number concentration ($n$)*, can be found from

$$c(\text{mass/volume}) =$$
$$\frac{n(\text{number/volume})\pi\rho_p[D_{p,am}(\text{mass})]^3}{6} \quad (11.34)$$

The arithmetic mean diameter based on mass can be found from Table 11-3 by compiling a column based on $D_{p,i}^3$:

$$D_{p,am}(\text{mass}) = \left[\sum \frac{n_i D_{p,i}^3}{n_t}\right]^{1/3} \quad (11.35)$$

It is obvious that the arithmetic average diameter based on mass is not the same as the arithmetic average based on number:

$$D_{p,am}(\text{number}) = \sum \frac{n_i D_{p,i}}{n_t} \quad (11.36)$$

since a few large particles have considerably greater mass than the same number of smaller particles. The median particle size signifies only that particles smaller than this size constitute 50% of the total number of particles, while those larger than the median size constitute the other 50% of the particles. Figure 11-5 shows clearly that the median particle size is not the same as the arithmetic average particle size.

Computing the cumulative distribution may mask whether the aerosol has one mode or several modes Whitby et al. 1978. For example, if an aerosol consists of two slightly overlapping log-normal distributions, the cumulative distribution of the entire aerosol will not be log-normal, and users may not be able to discern the separate modes. They might even apply a "best fit" for a single mode.

To reveal the existence of different modes and their modal diameters, it is common practice to plot the data as

$$\frac{(N/n_t)_{i+1} - (N/n_t)_i}{\ln D_{p,i+1} - \ln D_{p,i}} \quad (11.37)$$

versus the mean diameter in the interval, where $N_i$ is the cumulative number distribution. Why these variables reveal different modes in a distribution can be seen by differentiating the normalized cumulative distribution function with respect to $\ln D_p$:

$$\frac{N(D_p)}{n_t} = \int_0^{\ln D_p} \frac{n}{n_t} d(\ln D_p) \quad (11.38)$$

$$\frac{n}{n_t} = \frac{d(N/n_t)}{d(\ln D_p)}$$

or in finite form,

$$\frac{(N/n_t)_{i+1} - (N/n_t)_i}{\ln D_{p,i+1} - \ln D_{p,i}} \quad (11.39)$$

Thus peaks in the value of the function of $d(N/n_t)/d(\ln D_p)$ correspond to the existence of modes.

Of special concern to human health are fine particles, $D_p \leqq 2.5 \ \mu$m. It is instructive to identify the atmospheric particles that fall in this category. Figure 11-8a depicts the size distribution of particles in the atmosphere. *Primary particles* generated by biogenic and anthropogenic processes enter the atmosphere directly as particles (soot, fly ash, sea salt, pollen fungi, soil, etc.). *Secondary particles* are generated by gas-to-particle conversions occurring in the atmosphere. Secondary organic aerosols occur when a parent organic molecule reacts with OH$\cdot$ and/or $O_3$ to yield products that partition themselves between gas and aerosol phases. The concentration of OH$\cdot$ determines the rate of oxidation of $SO_2$ and $NO_2$ to sulfuric and nitric acids and thence to aerosol sulfate and nitrate. Figure 11-8b summarizes these gas-to-particle conversions. The vertical dashed lines indicate conversions that form droplets that leave the atmosphere by gravimetric settling. The horizontal dashed lines indicate conversions that form aerosols that remain in suspension for a considerable time. Examine the following portions of Figure 11-8b.

- *Top left.* OH$\cdot$, HO$_2\cdot$ and $H_2O_2$ sustain themselves through gas-phase photolytic reactions (Equations 6.15 to 6.21).

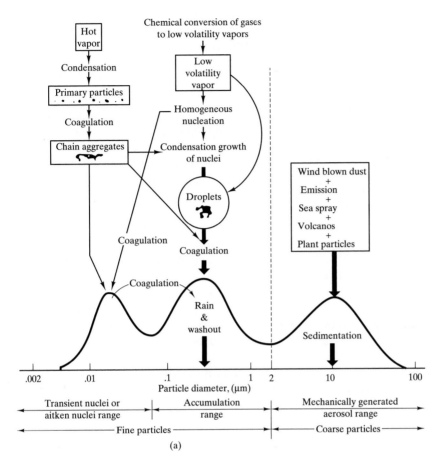

(a)

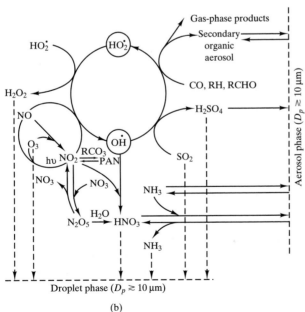

(b)

**Figure 11-8** (a) Formation and removal mechanisms of atmospheric particles (b) Hydroxyl radical reactions that convert $SO_2$ and $NO_2$ to sulfuric and nitric acid droplets and gas-phase precursors to sulfate and nitrate secondary particles. (Meng et al., 1997)

- *Top right.* CO, hydrocarbons with seven or more carbon atoms (RH) and aldehydes (RCOH) react with OH· and HO₂· to generate semi-volatile organic aerosols and gas-phase products (Figs 6-6 and 6-10).
- *Middle right.* SO₂ in combination with OH· and HO₂· produces particulate H₂SO₄ through nucleation or condensation on existing aerosols such as clouds and fog (Equation 6.10).
- *Lower left.* NO, O₃, and NO₂ sustain themselves via atmospheric photochemical reactions. In the presence of the products of olefin reactions, NO₂ produces peroxyacetyl nitrate (PAN) (Figure 6-6):

$$NO_2 + CH_3C(O)OO\cdot$$

$$\rightarrow CH_3CO(O)OONO_2(PAN) \quad (11.40)$$

In the presence of OH·, NO₂ forms gas-phase HNO₃,

$$OH\cdot + NO_2 \rightarrow HNO_3(g) \quad (11.41)$$

that deposits by adsorption or reaction on surfaces or by settling of a hydrated species, HNO₃·$n$H₂O. Such processes are often referred to as *dry deposition.*

- *Lower right.* Ubiquitous NH₃ reacts with HNO₃(g) to produce particulate NH₄NO₃, which can leave the atmosphere via rain as *wet deposition* or by gravimetric settling of dry particles (*dry deposition*):

$$NH_3(g) + HNO_3(g) = NH_4NO_3(particulate) \quad (11.42)$$

Gas-phase photochemistry determines the relative amounts of HNO₃ and PAN. Airborne NH₃ has a direct influence on airborne nitrate. At low NH₃ concentration, more nitrate remains as HNO₃(g), whereas at higher NH₃ concentrations, HNO₃(g) is converted to particulate NH₄NO₃.

Figure 11-9 is the size distribution for an urban aerosol and confirms the existence of accumulation and primary mode. When Lee (1972) plotted the data on log-probability paper only, he concluded that atmospheric aerosols had only a single mode. Unfortunately, the log-probability graph masked the existence of these two modes. Readers should always test data using Equation 11.39 to determine whether the data have a single or multiple modes.

The reader must also develop an appreciation of particle distributions based on mass and number. A

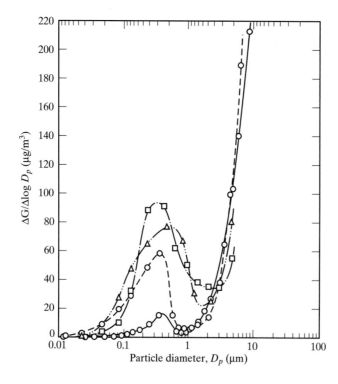

**Figure 11-9** Atmospheric particle size distribution in four communities (redrawn from Whitby et al, 1974)

few large particles can have an enormous influence on the mass distribution, since mass varies with the cube of the diameter. To dramatize this fact consider the following Example 11.2.

## EXAMPLE 11.2 PARTICLE SIZE DISTRIBUTION

Air in a foundry is dusty because of handling sand used to make molds, shaking castings out of the sand molds, and so on. A sample of the workplace air was drawn through a filter at a rate of 0.15 L/min for a period of 100 s. The sampled air contained 240 particles, which were counted and sized optically on the basis of diameter. Table 11-4 shows the results.

a. Can the size distribution be described as log-normal?
b. What is the mass concentration ($c_0$, mg/m$^3$) of particles in the workplace sample assuming that the particle density ($\rho_p$) is 1800 kg/m$^3$?

### Solution

a. Figure E11-2a shows that the distribution appears to be log-normal, although a more accurate test can be made when the data in Table 11-4 are plotted on log probability paper. Figure E11-2b shows that the assumption of log normality appears to be accurate. A further test can be performed when the predicted median particle diameter $D_{p, 50}$ and geometric standard deviation $\sigma_g$ are compared with the plotted data.
b. The arithmetic mean diameter based on number, variance and standard deviation can be computed from the data in Table 11-4.

$$D_{p, am} = \sum_i \frac{n_i D_{p, i}}{n_t} = \frac{7782}{240} = 32.4 \ \mu m$$

$$\sigma^2 = \sum_i \frac{n_i (D_{p, am} - D_{p, i})^2}{n_t}$$

$$= \frac{42,519.2}{240} = 177.2 \ \mu m^2$$

$$\sigma = 13.3 \ \mu m$$

From these values, the geometric standard deviation and geometric mean diameter based on number are predicted to be

$$\sigma_g = \exp \left\{ \left[ \ln \left( 1 + \frac{177.2}{(32.4)^2} \right) \right]^{1/2} \right\} = 1.484$$

$$D_{p, gm}(\text{number}) = 32.4 \exp \left[ -\frac{(\ln 1.484)^2}{2} \right]$$

$$= 30.0 \ \mu m$$

The predicted median particle diameter $D_{p, 50}$ of 30.0 μm is seen to agree well with that obtained from the plotted data. The slope of the predicted log-normal distribution can be found in terms of the predicted values of $D_{p, 84}$ and $D_{p, 16}$:

$$D_{p, 84} = (1.484)(30) = 44.5 \ \mu m$$

$$D_{p, 16} = \frac{30}{1.484} = 20.2 \ \mu m$$

**Table 11-4** Foundry Dust Size Distribution for Example 11.2

| Class | Range | $D_{p, i}$ | $n_i$ | $n_i/n_t$ | $N/n_t$ | $n_i D_{p, i}$ | $n_i (D_{p, am} - D_{p, i})^2$ | $n_i D_{p, i}^3/n_t$ | $g(D_{p, i})$ |
|---|---|---|---|---|---|---|---|---|---|
| 1 | 5–6 | 5.5 | 0 | 0 | 0 | 0 | 0 | 0 | 0 |
| 2 | 6–9 | 7.5 | 0 | 0 | 0 | 0 | 0 | 0 | 0 |
| 3 | 9–13 | 11.0 | 2 | 0.008 | 0.008 | 22 | 915.9 | 11 | 0.0002 |
| 4 | 13–18 | 15.5 | 29 | 0.121 | 0.129 | 450 | 8,282.7 | 450 | 0.0083 |
| 5 | 18–26 | 22.0 | 54 | 0.225 | 0.354 | 1,188 | 5,840.6 | 2,396 | 0.0441 |
| 6 | 26–37 | 32.0 | 84 | 0.350 | 0.704 | 2,688 | 13.4 | 11,469 | 0.2110 |
| 7 | 37–52 | 42.5 | 54 | 0.225 | 0.929 | 2,295 | 5,508.5 | 17,272 | 0.3178 |
| 8 | 52–73 | 62.5 | 14 | 0.058 | 0.987 | 875 | 12,684.1 | 14,241 | 0.2620 |
| 9 | 73–103 | 88.0 | 3 | 0.012 | 0.999 | 264 | 9,274.1 | 8,518 | 0.1567 |
| | | | 240 | | | 7,782 | 42,519.3 | 54,357 | |

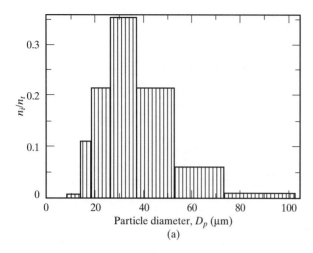

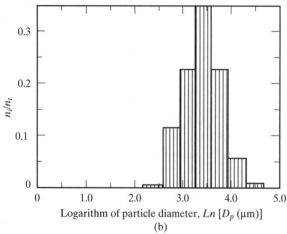

**Figure E11-2** (a) Number distribution versus particle diameter for Example 11.2. (b) Number distribution versus the logarithm of particle diameter, for Example 11.2

When these values are plotted on Figure E11-2c, the agreement is also found to be good. The geometric mean diameter based on mass is found to be

$$\ln D_{p,gm}(\text{mass}) = \ln 30 + 3(\ln 1.484)^2$$

$$D_{p,gm}(\text{mass}) = 47.8 \; \mu m$$

The total mass of particles in the sample is

$$m_t = \frac{240\pi\rho_p}{6} \sum \frac{n_i D_{p,i}^3}{n_t}$$

$$= \frac{240\pi}{6}(1800 \; \text{kg/m}^3)(54{,}357) \; \mu m^3 (m/10^6 \; \mu m)^3$$

$$= 1.2288 \times 10^{-8} \; \text{kg}$$

The volume of the air sample is

$$V = Qt = (0.15 \; \text{L/min})(100/60 \; \text{min})(1 \; \text{m}^3/1000 \; \text{L})$$

$$= 0.000250 \; \text{m}^3$$

The particle concentration in the air sample is

$$c_0 = \frac{m_t}{V}$$

$$= \left(\frac{1.2288 \times 10^{-8} \; \text{kg}}{2.5 \times 10^{-4} \; \text{m}^3}\right)(10^6 \; \text{mg/kg})$$

$$= 49.2 \; \text{mg/m}^3$$

## 11.2.4 Physical explanation of log-normal distribution

Ott (1990) gives a straightforward explanation of why many natural processes produce log-normal distributions. His explanation is called *successive random dilutions*. These processes consist of a series of random events that follow one another over a period of time (e.g., grinding operations where the size of particles is reduced by successive grinding operations, or where particles in a monodisperse aerosol grow by the process of agglomeration). If the successive dilutions involve fixed amounts, the final state can be predicted with certainty. Such predictions are called *deterministic*. If the successive dilutions involve randomness, the final state can only be predicted using probability theory. Random processes are called *stochastic*.

To illustrate the concept of successive random processes, consider the concentration of a contaminant (particles, vapor, or gas) in an enclosure of volume $V_0$ (Figure 11-10). At time zero, the contaminant concentration is $c_0$. Fresh air (containing no contaminant) enters the enclosure at an arbitrary volumetric flow rate $Q_i$ for an elapsed time $t_i$, and simultaneously, an exhaust system removes contaminated air from the room at the same volumetric flow rate for the same elapsed time. At the end of each step a circulating fan distributes the contaminant uniformly so that at any instant the concentration is spatially uniform. The enclosure contains no contaminant sources or sinks and there are no chemical reactions. Consider the following sequence of events.

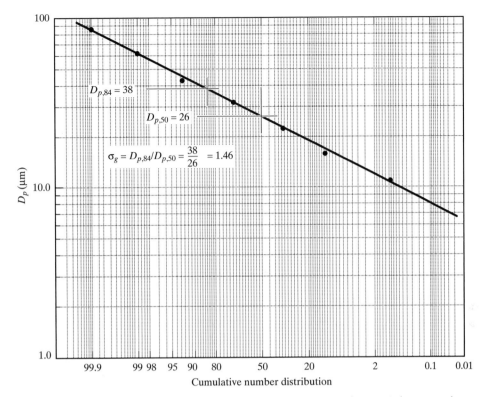

**Figure E11-2c** Testing whether the size distribution in Example 11.2 is log-normal

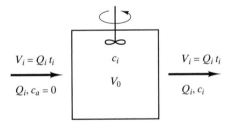

**Figure 11-10** Successive random dilutions, i = dilution number, c(0) = $c_0$ > 0

1. Add fresh air at a volumetric flow rate $Q_1$ to the enclosure over an elapsed time $t_1$ and simultaneously remove contaminated air at concentration $c_0$ from the enclosure at a volumetric flow rate $Q_1$ over the elapsed time $t_1$.
2. Wait for well-mixed conditions to occur and measure the concentration $c_1$.
3. Repeat the process, add fresh air at a volumetric flow rate $Q_2$ to the enclosure over an elapsed time $t_2$, and simultaneously remove contaminated air at concentration $c_1$ from the enclosure at a volumetric flow rate $Q_2$ over the elapsed time $t_2$.
4. Wait for well-mixed conditions to occur and measure the concentration $c_2$.
5. Repeat the first four steps $m$ times.

Initially ($i = 0$), the concentration is $c_0$ and the mass of contaminant in the enclosure is $m_0$:

$$c_0 = \frac{m_0}{V_0} \qquad (11.43)$$

Following the first dilution ($i = 1$), the concentration in the enclosure is

$$c_1 = \frac{m_1}{V_0} = \frac{m_0 - Q_1 t_1 c_0}{V_0}$$

$$= \frac{V_0 c_0 - V_1 c_0}{V_0} = c_0 \left(1 - \frac{V_1}{V_0}\right) \qquad (11.44)$$

where $V_1$ is the volume of air added and removed from the enclosure $(V_1 = Q_1 t_1)$. Define $D_1$ as the *dilution factor:*

$$D_1 = 1 - \frac{V_1}{V_0} \qquad (11.45)$$

Thus

$$c_1 = c_0 D_1 \qquad (11.46)$$

Following the second dilution ($i = 2$), the concentration in the enclosure is $c_2$, where

$$c_2 = \frac{m_2}{V_0} = \frac{m_1 - Q_2 t_2 c_1}{V_0}$$

$$= \frac{V_0 c_1 - V_2 c_1}{V_0} = c_1 \left(1 - \frac{V_2}{V_0}\right) \quad (11.47)$$

where $V_2$ is the volume of contaminated air added and removed from the enclosure and $D_2$ is the dilution factor:

$$D_2 = 1 - \frac{V_2}{V_0} \qquad (11.48)$$

The process is repeated $m$ times and the dilution ratios are selected randomly. The concentration after $m$ dilutions is

$$c_m = c_0 D_1 D_2 \cdots D_m \qquad (11.49)$$

Taking the logarithm of the above gives

$$\ln c_m = \ln c_0 + \left(\ln D_1 + \ln D_2 + \cdots + \ln D_m\right)$$
$$(11.50)$$

Using the additive form of the central limit theorem, Ott showed that since each dilution factor $D_i$ is an independent random variable, the sum is also an independent random variable. Random variables have frequency distributions that are called *Gaussian* or *normal*. Thus since the logarithm of the concentration is normally distributed, it can be said that $c_m$ is log-normally distributed.

Graphic proof of why successive dilutions produce log-normal distributions can be obtained by making 1000 computations (henceforth called a *trial*) each involving five dilutions ($m = 5$). In each trial select the dilution factors ($D_i$) by a random number generator, keeping in mind that $1.0 \geqq D_i > 0$. Now tabulate the results of the 1000 trials and draw a graph of the number of trials in which the final concentration has a particular value versus the frequency that this value is achieved (i.e., a histogram or frequency plot). The graph will be log-normal and resemble Figures

E11-2a and b. Compute the cumulative distribution and plot the cumulative distribution versus the final concentration on log-probability paper. One will find that the results can be approximated in a reasonable manner by a straight line. The analysis of 1000 trials of five dilutions each is approximately log-normal since the number of dilutions and the number of trials are not infinite. If the number of dilutions and trials are increased and the computation above repeated, the log-normal fit improves. In conclusion, the example shows that if a process consists of a succession of random events, the final result has a frequency distribution that can be characterized as log-normal.

## 11.3  Overall collection efficiency

Particles are removed from a gas stream by a collection device in which particles either settle or contact and adhere to the surfaces of other bodies. The removal process is traditionally characterized by a *grade*, or *fractional efficiency*, in which the percent of particles of a particular size that are removed $\eta(D_p)$ is expressed as function of the particle size $D_p$ and other operating parameters.

$$\eta(D_p) = 1 - \frac{c(D_p)_f}{c(D_p)_i} \qquad (11.51)$$

where the subscripts $i$ and $f$ mean *initial* and *final or exit* states, respectively. Fractional efficiency curves are obtained from experiment or predicted from first principles. Deriving these curves is a large subject in itself and will be developed later. What is important now is how such data and knowledge of the aerosol size distribution can be used to predict the overall collection efficiency. Typical grade efficiency curves are shown in Figure 11-11. In general, high-efficiency collectors have higher $\eta(D_p)$ at any given $D_p$ and set of inlet conditions.

If an aerosol enters a device or process with a size distribution that can be expressed statistically, the aerosol will leave with another size distribution in which there is a higher percentage of small particles. The overall mass concentration is lower than when it entered, but the size distribution is skewed toward the smaller particles. The *overall removal efficiency* by the process is equal to

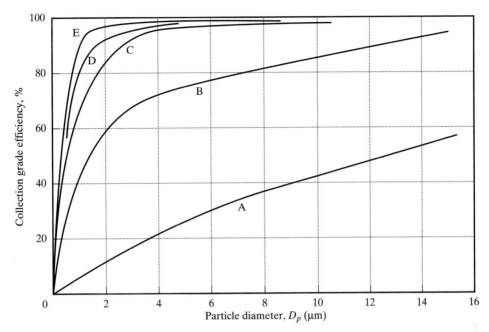

**Figure 11-11**  Typical collection grade efficiencies of different dust collectors; A - low-efficiency high-flow rate cyclones, B - high-efficiency, low flow rate cyclones, C - spray towers, D - electrostatic precipitators, E - venturi scrubbers

$$\eta(\text{overall}) = 1 - \frac{c_f}{c_i} \qquad (11.52)$$

where $c_f$ and $c_i$ are the overall mass concentrations at the exit and inlet of the device. The size distribution can be described by the number fraction or mass fraction of particles of a given size. For this analysis it is preferable to use the *mass fraction*. The variable $g(D_{p,j})_i$ is called the inlet mass fraction and is defined as the concentration of particles of a particular size $j$ at the inlet $c(D_{p,j})_i$ divided by the total mass concentration at the inlet $c_i$:

$$g(D_{p,j})_i = \frac{c(D_{p,j})_i}{c_i} \qquad (11.53)$$

The *overall collection efficiency* $\eta$ (overall) can be expressed as

$$\eta = 1 - \sum_j \frac{c(D_{p,j})_f}{c_i} \qquad (11.54)$$

where $c(D_{p,j})_f$ refers to the concentration of particles within size range $j$ that appear at the exit of the device. The fractional efficiency defined in Equation 11.51 can be rearranged as follows:

$$c(D_{p,j})_f = c(D_{p,j})_i \left[1 - \eta(D_{p,j})\right] \quad (11.55)$$

Combining Equations 11.54 and 11.55 and simplifying yields

$$\eta = 1 - \sum_j \left[1 - \eta(D_{p,j})\right] \frac{c(D_{p,j})_i}{c_i}$$

$$\eta = 1 - \sum_j \frac{c(D_{p,j})_i}{c_i} + \sum_j \frac{\eta(D_{p,j})c(D_{p,j})_i}{c_i}$$

$$\eta = 1 - 1 + \sum_j \eta(D_{p,j})g(D_{p,j})_i$$

$$\eta = \sum_j \eta(D_{p,j})g(D_{p,j})_i \qquad (11.56)$$

Thus the overall collection efficiency is equal to the mass fraction for a certain range of particles times the fractional efficiency for that size range summed over all the particle size ranges.

To compare collectors whose efficiencies are very close to 100%, it is useful to use the term *penetration* (*P*), defined as $1 - \eta$. Thus to compare two collectors whose efficiencies are 99.5% and 99.8%, emphasis can be gained by saying the penetration of one collector is 0.5%, while the penetration of the other is 0.2%.

## EXAMPLE 11.3 OVERALL REMOVAL EFFICIENCY OF A POLYDISPERSE AEROSOL

It has been proposed to remove particles from the air described in Example 11.2 with a particle collection device whose fractional (grade) efficiency is given below.

| $D_{p,j}(\mu m)$ | $\eta(D_{p,j})$ | $D_{p,j}(\mu m)$ | $\eta(D_{p,j})$ |
|---|---|---|---|
| 5.5 | 0.42 | 32 | 0.80 |
| 7.5 | 0.50 | 42.5 | 0.83 |
| 11 | 0.60 | 62.5 | 0.93 |
| 15.5 | 0.68 | 88 | 0.98 |
| 22 | 0.72 | | |

Will this device be able to bring the workplace air into compliance with the OSHA standard that specifies that the maximum allowable concentration for nonrespirable nuisance dust is 15 mg/m³?

**Solution** The overall dust concentration was found to be 49.2 mg/m³. The minimum overall removal efficiency is

$$\eta_{min} = 1 - \frac{c_{max, allowable}}{c_0} = 1 - \frac{15}{49.2}$$

$$= 0.6948\,(69.5\%)$$

Using the mass fraction given in Table 11-4, compute the overall removal efficiency using Equation 11.56:

$$\eta = \sum \eta(D_{p,i})g(D_{p,i}) = 0.00012 + 0.00564$$

$$+ 0.03175 + 0.1688 + 0.2637$$

$$+ 0.2437 + 0.1536$$

$$= 0.8687\,(86.9\%)$$

Consequently, the collector is capable of satisfying the OSHA standard, although it remains to be shown that all the air in workplace can be cleaned with this efficiency, that pockets of unusually high dust concentration do not exist, and so on.

## 11.3.1 Collectors in series

Consider a collection system consisting of several ($n$) collectors arranged in series as shown in Figure 11-12a. The overall collection efficiency is

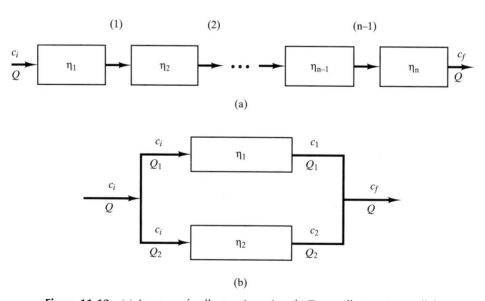

(a)

(b)

**Figure 11-12** (a) A group of collectors in series. (b) Two collectors in parallel

$$\eta(\text{overall}) = 1 - \frac{c_f}{c_i}$$

$$= 1 - \frac{c_1}{c_i}\frac{c_2}{c_1}\frac{c_3}{c_2}\cdots\frac{c_{n-1}}{c_{n-2}}\frac{c_f}{c_{n-1}} \qquad (11.57)$$

The efficiency of each collector can be expressed as

$$\eta_1 = 1 - \frac{c_1}{c_i} \qquad \frac{c_1}{c_i} = 1 - \eta_1 \qquad (11.58)$$

$$\eta_2 = 1 - \frac{c_2}{c_1} \qquad \frac{c_2}{c_1} = 1 - \eta_2 \qquad (11.59)$$

$$\eta_n = 1 - \frac{c_f}{c_{n-1}} \qquad \frac{c_f}{c_{n-1}} = 1 - \eta_n \qquad (11.60)$$

Combining yields

$$\eta(\text{overall}) = 1 - (1 - \eta_1)(1 - \eta_2)\cdots(1 - \eta_n) \qquad (11.61)$$

Thus the overall collection efficiency is not the arithmetic average of the separate collection efficiencies, but unity minus the product of the penetrations ($P = 1 - \eta$) of each collector in series:

$$\eta(\text{overall}) = 1 - (P_1 P_2 \cdots P_n) \qquad (11.62)$$

## 11.3.2 Collectors in parallel

Consider two collectors arranged in parallel as shown in Figure 11-12b. The efficiencies and the volumetric flow rates through each collector are not, in general, the same, although the concentration entering each collector is the same. The contaminant mass flow rate leaving each collector is

$$\begin{aligned} c_1 Q_1 &= (1 - \eta_1)Q_1 c_i \\ c_2 Q_2 &= (1 - \eta_2)Q_2 c_i \end{aligned} \qquad (11.63)$$

The overall collection efficiency of the parallel configuration is

$$\eta(\text{overall}) = 1 - \frac{c_1 Q_1 + c_2 Q_2}{Q c_i}$$

$$= 1 - \frac{(1 - \eta_1)Q_1 c_i + (1 - \eta_2)Q_2 c_i}{Q c_i}$$

$$= 1 - \left[(1 - \eta_1)\frac{Q_1}{Q} + (1 - \eta_2)\frac{Q_2}{Q}\right]$$

$$\qquad (11.64)$$

Thus the overall collection efficiency is not the arithmetic average of each collector's efficiency but depends on the manner in which the volumetric flow rates are split as well as the magnitudes of $\eta_1$ and $\eta_2$.

### EXAMPLE 11.4 MOST ECONOMICAL RETROFIT OF A PARTICLE COLLECTION SYSTEM

A coal-fired electric utility boiler produces an exhaust gas stream of 30,000 acfm at 400 K. Fly ash particles are removed with a cyclone particle collector that has an overall capture efficiency of 75%. New state environmental regulations require that 95% of the particles be removed. Your supervisor asks you to consider three options and recommend the one with the lowest capital cost.

1. Replace the existing 75% collector with a new 95% collector.
2. Add one or more new collectors in series with the existing collector.
3. Devise a parallel configuration in which a fraction of the flow passes through the existing collector and the remainder passes through a new collector.

*Solution* The capital cost for new collectors is as follows:

| $\eta$(%) | Capital cost ($/scfm) |
|---|---|
| 75 | 30 |
| 80 | 40 |
| 85 | 46 |
| 90 | 52 |
| 95 | 58 |
| 99 | 70 |
| 99.5 | 80 |

The standard volumetric flow rate (scfm) is [30,000(298/400)] or 22,350 scfm.

*Option 1.* The cost to replace the existing collector with a new 95% collector is

$$\text{cost}_{\text{replacement}} = (\$58/\text{scfm})(22,350\text{ scfm}) = \$1,296,300$$

*Option 2.* The overall collection efficiency for the series configuration is given by Equation 11.61. The new collector must have an efficiency ($\eta_2$) such that

$$\eta(\text{overall}) = 0.95 = 1 - (1 - 0.75)(1 - \eta_2)$$
$$= 1 - (0.25)(1 - \eta_2) = 0.75 + 0.25\eta_2$$
$$\eta_2 = \frac{0.20}{0.25} = 0.80(80\%)$$

The cost of an 80% collector placed in series with the existing collector is

$$\text{cost}_{\text{series}} = (\$40/\text{scfm})(22{,}350\ \text{scfm}) = \$894{,}000$$

*Option 3.* The overall efficiency for a parallel configuration is given by Equation 11.64, in which $\eta_1 = 0.75$ and $\eta_2$ is equal to any one of the efficiencies above. To achieve an overall efficiency of 95%, there will be a unique flow spilt $(f_1 = Q_1/Q)$ and $(f_2 = Q_2/Q)$ for each configuration, where $1 = f_1 + f_2$.

$$\eta(\text{overall}) = 0.95 = 1 - [f_1(1 - \eta_1)$$
$$+ f_2(1 - \eta_2)]$$
$$= 1 - [(1 - f_2)(0.25) + f_2(1 - \eta_2)]$$
$$= 1 - 0.25 + 0.25f_2 - f_2 + f_2\eta_2$$
$$f_2 = \frac{0.20}{\eta_2 - 0.75}$$

The following is a summary of the cost for each of the collectors above arranged in a parallel configuration.

| $\eta_2(\%)$ | $f_2$ | $Q_2$ (scfm) | Cost$_{\text{parallel}}$ |
|---|---|---|---|
| 75 | ∞ (physically impossible) | | |
| 80 | 4 (physically meaningless, cannot achieve an overall efficiency of 95%) | | |
| 85 | 2 (physically meaningless, cannot achieve an overall efficiency of 95%) | | |
| 90 | 1.33 (physically meaningless, cannot achieve an overall efficiency of 95%) | | |
| 95 | 1.0 (same as option 1) | | $1,296,300 |
| 99 | 0.833 | 18,617.55 | $1,303,228 |
| 99.5 | 0.8163 | 18,244.30 | $1,459,544 |

It is seen that a series configuration (option 2) has the lowest capital cost. Other considerations must also be considered. Since this is a retrofit of an existing facility, there must be space for the second collector in series with the first. If there is no room for a second collector to follow the first, the parallel configuration (option 3) may have to be considered. Even option 3 has its drawback. Since such a large percentage of the flow will have to be rerouted to it, considerable space must be available.

## 11.4 Equations of particle motion

To design particle collection devices and to predict their performance, engineers must be able to predict the trajectories of particles passing through the devices. The objective of this section is to establish the equations used to predict the trajectories of particles. If the number of particles per unit volume of gas is below a certain value, it will be assumed that particles move through the carrier gas independently of each other and that the particles do not influence the gas velocity field. The assumption is valid if:

**1.** Particles do not collide with one another.
**2.** Particles do not pass through each other's wake.

A useful rule of thumb that quantifies these two assumptions for a monodisperse aerosol is that the average distance between particles is at least 10 times the particle diameter. Assuming that eight particles are located at the corners of a cube $L \times L \times L$, one finds that $L/D_p > 10$ when

$$\frac{4\pi\rho_p}{3c} = \frac{8}{n_t D_p^3} < 1000 \qquad (11.65)$$

where $c$ is the mass of particles per volume of gas and $n_t$ is the number of particles per volume of gas. Table 11-5 illustrates these upper limits and indicates that the number of particles per unit volume of gas has to be exceedingly large for the particles to influence the gas flow field. For water droplets

**Table 11-5**   Particle Concentrations beyond Which Particles Influence the Flow Field

| $D_p(\mu m)$ | $n$ (particles/m$^3$) |
|---|---|
| 1 | $8 \times 10^{15}$ |
| 10 | $8 \times 10^{12}$ |
| 100 | $8 \times 10^{9}$ |

$(\rho_p = 1000 \text{ kg/m}^3)$, applying Equation 11.65 leads to a particle concentration ($c$) of 4.2 kg/m³ of air, corresponding to the upper limit of 1000. For most problems in air pollution, particle concentrations of water and other particulate pollutants are hundreds of times smaller than 4.2 kg/m³. Consequently, the motion of the carrier gas is independent of the motion of the particles.

To describe the motion of an aerosol, one must first compute the velocity field of the carrier gas and then compute the motion of the particles. The motion of the carrier gas can be expressed analytically if one is so fortunate as to have a system of simple geometry, or it may be established experimentally and the data stored numerically as a data file. If an analytical expression for the velocity field exists, one can compute the particle trajectories explicitly. For most industrial applications, only experimentally measured velocity data are available and the computer will be needed to compute the particle trajectories.

The density of a particle is approximately 1000 times greater than the density of air. Thus the force of buoyancy on a particle is negligible with respect to its weight. The motion of a *single spherical particle* is given by

$$\frac{\pi D_p^3}{6}\rho_p\frac{d\mathbf{v}}{dt} = -c_D\frac{\rho}{2C}\frac{\pi D_p^2}{4}(\mathbf{v} - \mathbf{U})|\mathbf{v} - \mathbf{U}|$$

$$- \frac{\pi D_p^3}{6}\rho_p\mathbf{g} \tag{11.66}$$

where $\mathbf{g}$ is the acceleration of gravity. To illustrate the use of this equation, examples of increasing complexity will be undertaken.

Consider the general motion of a spherical particle traveling through a moving two-dimensional gas stream in which the air velocity ($U$) varies. Let the particle have an initial velocity $v(0)$. Figure 11-2 depicts such motion. The *equation describing a particle's acceleration* reduces to

$$\frac{d\mathbf{v}}{dt} = \hat{j}g - \frac{3c_D}{4C}\frac{\rho}{\rho_p D_p}(\mathbf{v} - \mathbf{U})|\mathbf{v} - \mathbf{U}| \tag{11.67}$$

$$\mathbf{v} - \mathbf{U} = \hat{i}(v_x - U_x) + \hat{j}(v_y - U_y)$$

$$= \hat{i}\,v_{rx} + \hat{j}\,v_{ry} \tag{11.68}$$

$$|\mathbf{v} - \mathbf{U}| = \sqrt{v_{rx}^2 + v_{ry}^2} \tag{11.69}$$

The above reduces to a pair of coupled differential equations:

$$\frac{dv_x}{dt} = -\frac{3c_D}{4C}\frac{\rho}{\rho_p D_p}v_{rx}\left(v_{rx}^2 + v_{ry}^2\right)^{1/2} \tag{11.70}$$

$$\frac{dv_y}{dt} = -g - \frac{3c_D}{4C}\frac{\rho}{\rho_p D_p}v_{ry}\left(v_{rx}^2 + v_{ry}^2\right)^{1/2} \tag{11.71}$$

$$c_D = 0.4 + \frac{24}{\text{Re}} + \frac{6}{\left(1 + \text{Re}^{1/2}\right)} \tag{11.72}$$

$$\text{Re} = \frac{\rho D_p\left(v_{rx}^2 + v_{ry}^2\right)^{1/2}}{\mu} \tag{11.73}$$

If the particle's motion is entirely within the Stokes regime, replacing $c_D$ by 24/Re *uncouples equations* 11.70 and 11.71,

$$\frac{dv_x}{dt} = -\frac{v_x - U_x}{\tau C} \tag{11.74}$$

$$\frac{dv_y}{dt} = -g - \frac{\left(v_y - U_y\right)}{\tau C} \tag{11.75}$$

where $\tau$ has the units of time and is called the *relaxation time:*

$$\tau = \frac{\rho_p D_p^2}{18\mu} \tag{11.76}$$

Thus if $U_x$ and $U_y$ are constant or known functions of $x$ and $y$, Equations 11.74 and 11.75 can be solved to predict $v_x$ and $v_y$ as functions of time and known initial conditions, $v_x(0)$ and $v_y(0)$.

To illustrate this kind of Stokes flow, consider the motion shown in Figure 11-13, in which particles enter an airstream that is moving to the right with a

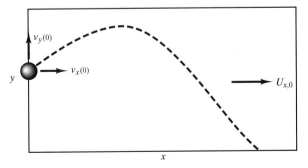

**Figure 11-13**  Particle trajectory in a horizontal streaming flow

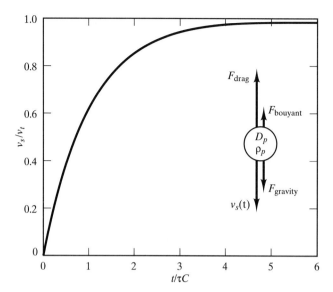

**Figure 11-14**   Speed of freely falling particle versus time, Stokes flow

$$c_D = \frac{24}{\text{Re}} = 24\frac{\rho D_p v_s}{\mu} \qquad (11.86)$$

Only the vertical component of Equation 11.75 is to be solved. After substitution and simplification,

$$\frac{dv_s}{dt} = g - \frac{v_s}{\tau C} \qquad (11.87)$$

where

$$\tau = \frac{\rho_p D_p^2}{18\mu} \qquad (11.88)$$

The term $\tau$ is called the *relaxation time* since it possesses the units of time and because it is customary to use this name when it appears in first-order differential equations such as Equation 11.87. If the particle starts from rest, the downward velocity is

$$v_s(t) = Cg\tau\left[1 - \exp\left(-\frac{t}{\tau C}\right)\right] \qquad (11.89)$$

When $t \gg \tau$, a steady-state condition occurs that is equivalent to setting the left-hand side of Equation 11.87 to zero. Under these conditions the downward velocity is constant and is called the *settling, fall*, or *terminal velocity* $(v_t)$. Thus,

$$v_t = \tau g C \qquad \text{for Re} < 1 \qquad (11.90)$$

The variation of $v_s$ with time is shown in Figure 11-15. It is seen that it increases rapidly with time and

achieves 63.2% of $v_t$ after an elapsed time $t = \tau$ and 99.9% of its final value in $t = 7\tau$ for $C = 1$. Since the particle starts from rest, the maximum velocity is the terminal velocity. If the Reynolds number is always to be less than unity, the results above pertain to particles and fluid such that

$$\text{Re} = \frac{\rho D_p v_t}{\mu} < 1$$

$$= \frac{\rho \rho_p D_p^3 g C}{18\mu^2} < 1 \qquad (11.91)$$

The variation of the settling velocity $(v_t)$ for spherical particles of different size and density in air at STP is shown in Figure 11-15. In the Stokes regime, the slope of the lines on the lower left of the log-log graph is 2, indicating that the settling velocity varies with the square of the particle diameter as implied by Equations 11.88 and 11.90. Table 11-6 shows values of the settling velocity $(v_t)$ for several particles of specific gravity 1.0 (i.e., *unit density spheres* in air at STP). It should be noted that the values are small for particles smaller than 100 μm which have Reynolds numbers (based on their settling velocities) less than 1.0 and therefore within the Stokes regime. In air pollution problems, characteristic times are on the order of seconds or minutes; consequently, the transient portion of vertical fall is of negligible importance. From a practical point of

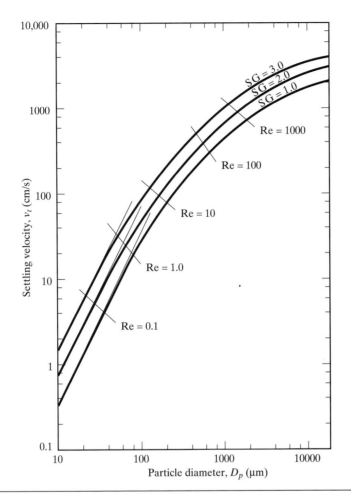

**Figure 11-15** Gravimetric settling velocity of spheres in air at STP

**Table 11-6** Gravimetric Settling of Unit Density Spheres in Air at STP

| $D_p$ (μm) | $\tau$ (s) | $C$ | $v_t$ (m/s) | $Re_t$ |
|---|---|---|---|---|
| 0.01 | $3.069 \times 10^{-10}$ | 22.2 | $6.5 \times 10^{-8}$ | $4.3 \times 10^{-11}$ |
| 0.10 | $3.069 \times 10^{-8}$ | 2.9 | $8.5 \times 10^{-7}$ | $5.7 \times 10^{-9}$ |
| 1.00 | $3.069 \times 10^{-6}$ | 1.2 | $3.6 \times 10^{-5}$ | $2.4 \times 10^{-6}$ |
| 10.00 | $3.069 \times 10^{-4}$ | 1.0 | $3.1 \times 10^{-3}$ | $2.1 \times 10^{-3}$ |
| 100.00 | $3.069 \times 10^{-2}$ | 1.0 | 0.3 | 2 |
| 1000.00 | 3.069 | 1.0 | 4.0 | 267 |

view the vertical velocity for these particles (relative to air) can always be set equal to the settling velocity.

Suppose that the particle is very large and the Reynolds number is always above 1000 (i.e., *Newtonian flow regime*). The slip factor, $C$, is unity and will be omitted. The drag coefficient $(c_D)$ is constant and will be set equal to 0.4. The steady-state vertical velocity (i.e., settling velocity) can be found by setting the left-hand side of Equation 11.71 equal to zero and manipulating the remaining variables. Then for Re > 1000,

$$v_t = \left[ \left( \frac{4}{3} \right) \frac{\rho_p D_p g}{0.4\rho} \right]^{1/2} \qquad (11.92)$$

A general expression for the settling velocity for particles of any diameter and Reynolds number can also be found by setting the left-hand side of Equation 11.71 equal to zero and solving the simultaneous equations:

$$v_t = \left( \frac{4 C \rho_p D_p g}{3 \rho c_D} \right)^{1/2} \tag{11.93}$$

$$c_D = 0.4 + \frac{24}{Re} + \frac{6}{1 + Re^{1/2}} \tag{11.94}$$

$$Re = \frac{\rho D_p v_t}{\mu} \tag{11.95}$$

## EXAMPLE 11.6 SETTLING VELOCITY

Write a numerical program to compute the settling velocity of a particle of any diameter and density falling in quiescent air at an arbitrary temperature and pressure. One program will use the modified Euler method and the other will make use of Mathcad functions. Compute the settling velocity for particles of density of 3000 kg/m$^3$ falling through air at 1 atm and 300 K. The particle diameter varies from 40 μm to 1200 μm. The density of air can be found from the ideal gas law, where $R(air) = 0.287$ kJ kg$^{-1}$ K$^{-1}$. The

viscosity of air is given as a function of temperature by the power-law expression in Table A-10.

**Solution** The settling velocity can be computed by solving Equations 11.93 to 11.95 using trial-and-error techniques in several commercially available mathematical programs. Figure E.11-6 shows the settling velocities for a wide range of particle diameters. The Mathcad program producing Figure E11-6 can be found in the textbook's Web site on the Prentice Hall Web page.

Comparing Figures 11-15 and E11-6, one finds that there is good agreement. To compute the Stokes flow settling velocity of particles of arbitrary density falling in fluids at arbitrary temperatures (hence viscosity), users can find $v_t$ from Figure 11-15, for STP and several densities. Such a value can then be corrected to the temperature (hence viscosity) and particle density of interest via Equation 11.88, which applies rigorously only to Stokes flow. This is an alternative to merely solving Equations 11.88 and 11.90 for the desired settling velocity. As an approximation for non-Stokes flow $(D_p > 2$ μm$)$ the den-

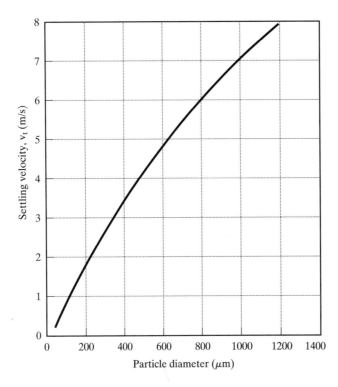

**Figure E11-6** Settling velocity versus particle diameter for particles with a density of 3,000 kg/m$^3$ falling freely in air at STP.

sity and viscosity dependence of Equation 11.88 can be used to correct the settling velocity obtained at STP from Figure 11-15.

### EXAMPLE 11.7  FALLING BULLETS

Great concern is exercised in the United States to instruct people to use firearms properly. On the other hand, TV, movies, newspapers, and magazines often show people firing pistols and rifles into the air, with reckless abandon. Rarely, if ever, does anyone seem to care about harm that can be done when the bullets fall to earth. Estimate the settling velocity of a 9-mm bullet, $\rho_p = 8000 \text{ kg/m}^3$.

**Solution** For a start, assume that $c_D$ is constant (0.4) and after the settling velocity $(v_t)$ is computed, return to check that the Reynolds number is sufficiently large. Using Equation 11.92, we obtain

$$v_t = \left(\frac{\rho_p D_p g}{0.3\rho}\right)^{0.5}$$

$$= \left[\frac{(8000 \text{ kg/m}^3)(0.009 \text{ m})(9.8 \text{ m}^2/\text{s})}{(0.3)(1.2 \text{ kg/m}^3)}\right]^{0.5}$$

$$= 44.138 \text{ m/s } (99.01 \text{ mph})$$

Ouch, beware of falling bullets! The final Reynolds number is

$$\text{Re} = \frac{\rho D_p v_t}{\mu} = 26{,}483$$

From Figure 11-3 it can be seen that assuming that $c_D = 0.4$ is valid.

---

Consider two particles, A and B, that have the same settling velocity:

- *Particle A:* perfect sphere of density 1000 kg/m³
- *Particle B:* nonspherical particle of unknown density

The *aerodynamic diameter* $(D_a)$ is the diameter assigned to a particle of unknown density and shape that possesses the same settling velocity as a sphere of water. If the unknown particle has a settling velocity $v_t$(observed) within the Stokes regime,

$$D_a = \left[\frac{18v_t \text{ (observed) } \mu}{\rho_w g C(D_a)}\right]^{1/2} \quad (11.96)$$

where $C(D_a)$ is the Cunningham factor, based on the aerodynamic diameter, and $\rho_w$ is the density of water. If the unknown particle happens to be a sphere of diameter $(D_p)$, then

$$D_a = D_p\left[\frac{C(D_p)}{C(D_a)}\frac{\rho_p}{\rho_w}\right]^{1/2} \quad (11.97)$$

where $C(D_p)$ and $C(D_a)$ are the Cunningham factors based on the actual and aerodynamic diameters and $\rho_p$ is the actual particle density. For particles larger than 1 μm, the Cunningham slip factors approach unity and

$$D_a = D_p\left(\frac{\rho_p}{\rho_w}\right)^{1/2} \quad (11.98)$$

### 11.5.2  Quiescent water

For particles in a liquid medium, settling is often called *sedimentation*. Sedimentation is described by the same model used for settling in a gaseous medium except that buoyancy cannot be neglected. As a result, the equations above can be used, but the particle density should be replaced by the difference between the particle density and the density of water.

## 11.6  Horizontal motion in quiescent air

To illustrate the dominating effect of viscosity in reducing the relative motion between a particle and the carrier gas, consider the horizontal velocity $v_x(t)$ of a sphere in quiescent air where the initial Reynolds number is less than 1.0. Since subsequent Reynolds numbers will be smaller than the initial value, the particle's motion will be entirely within the Stokes regime. The differential equation for the horizontal velocity component is

$$\frac{dv_x}{dt} = -\frac{v_x}{\tau C} \quad (11.99)$$

If a particle enters quiescent air with an initial velocity $v_x(0)$, the horizontal velocity component at a subsequent time is

$$v_x(t) = v_x(0) \exp\left(-\frac{t}{\tau C}\right) \quad (11.100)$$

The horizontal displacement can be found from

$$\int_0^{x(t)} dx = \int_0^t v_x\, dt$$

$$= \tau C v_x(0)\left[1 - \exp\left(-\frac{t}{\tau C}\right)\right] \quad (11.101)$$

The maximum horizontal displacement is called the *penetration or pulvation distance or stopping distance* (*l*) and is found by allowing $t \gg \tau$. Thus

$$l = v_x(0)\tau C \quad (11.102)$$

Table 11-7 is a compilation of penetration distances for several different-sized water drops possessing initial Reynolds numbers less than 1.0. It can be seen that penetration distances are small, indicating that viscosity damps relative motion very quickly.

**Table 11-7** Penetration Distance of Unit Density Spheres in Quiescent Air at STP[a]

| $D_p$ (μm) | Penetration distance (m) |
|------------|--------------------------|
| 0.01 | $68.18 \times 10^{-10}$ |
| 0.10 | $8.79 \times 10^{-8}$ |
| 1.00 | $3.57 \times 10^{-6}$ |
| 10.00 | $3.12 \times 10^{-4}$ |

[a] Initial velocity $U_0 = 1$ m/s.

## 11.7 Gravimetric settling in chambers

If one is interested in estimating the particle concentration in a chamber as a function of time, a crude model of the *sedimentation* process can be used. The model assumes that the particles have only two velocity components:

**1.** *Vertical component:* equal to the settling velocity
**2.** *Horizontal component:* equal to the carrier gas velocity

The assumptions ignore transient behavior and take the particles to be in equilibrium with the carrier gas everywhere inside the chamber. It is often difficult to predict or measure small gas velocities in a chamber. Thus it is useful to consider upper and lower limits to the particle concentration, knowing that reality lies somewhere in between. Define the upper limit as one

for which mixing is a maximum (well-mixed turbulent model) and the lower limit for the case of no mixing (laminar model).

- *Laminar model (no mixing).* All particles of the same size fall at a uniform speed equal to the settling velocity, and there is no mixing mechanism to redistribute particles as they fall.
- *Turbulent model (well-mixed).* All particles of the same size fall at a uniform speed equal to the settling velocity, but there is an idealized mixing mechanism that redistributes the remaining particles completely so that even though the concentration decreases with time, it is always uniform within the volume.

Figure 11-16 illustrates the laminar and well-mixed models. In the laminar model all particles of the same size fall uniformly such that if $c(D_p)_0$ is the initial concentration of particles of diameter $D_p$ in the chamber, then at any subsequent time there will be no particles of that size above a certain height $y(D_p)$ and a concentration $c(D_p)_0$ below this height. The same argument applies to particles of other sizes except the values of $y(D_p)$ will be different because they settle at different velocities. The rate of change of mass of particles of this size within the chamber is equal to the rate of deposition.

### 11.7.1 Laminar settling model

Consider the chambers in Figure 11-16 to have volume $V$ and horizontal cross-sectional area $A$.

$$\frac{d[Ayc(D_p)_0]}{dt} = Ac(D_p)_0 \frac{dy}{dt}$$

$$= -v_t Ac(D_p)_0$$

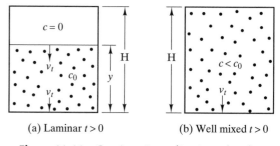

(a) Laminar $t > 0$      (b) Well mixed $t > 0$

**Figure 11-16** Gravimetric settling in a chamber

$$\int_0^y dy = -\int_0^t v_t \, dt \qquad (11.103)$$

$$y = H - tv_t$$

Let the average concentration of particles be $\bar{c}(D_p)$:

$$\bar{c}(D_p) = \frac{Ayc(D_p)_0}{V} = \frac{Ac(D_p)_0(H - tv_t)}{V}$$

$$= c(D_p)_0\left(1 - \frac{tv_t}{H}\right) \qquad (11.104)$$

The average concentration of these particles will decrease linearly with time until a time $t_c$ elapses:

$$t_c = \frac{H}{v_t} \qquad (11.105)$$

and all of particles of the size whose settling velocity is $v_t$ have settled to the floor.

### 11.7.2 Well-mixed settling model

On the other hand, suppose the well-mixed model, Figure 11-16b, is valid. As particles of a particular size fall to the floor with a velocity $v_t$, a "magic mixer" redistributes the remaining particles throughout the chamber. The rate of change of mass of particles of this size within the chamber is equal to the rate of deposition:

$$\frac{d[c(D_p)V]}{dt} = V\frac{dc(D_p)}{dt} = -v_t c(D_p)A$$

$$\frac{c(D_p)}{c(D_p)_0} = \frac{\bar{c}(D_p)}{c(D_p)_0} = \exp\left(-\frac{tv_t}{H}\right) \qquad (11.106)$$

Since the well-mixed model presumes that the concentration is the same throughout the enclosure at any instant of time, the term $c(D_p)$ in Equation 11.106 also expresses the average concentration, $\bar{c}(D_p)$.

The well-mixed model predicts that the average concentration decreases exponentially and the laminar model predicts that it decreases linearly. In the well-mixed model $\bar{c}(D_p)/c(D_p)_0 = 0.368$ at $t = t_c$, while in the laminar model it is zero. Only after $t = 6.9t_c$ does the average concentration decrease to 0.001 of its initial value in the well-mixed model.

The laminar model leads one to believe that the dust will be removed more quickly than is realistic

because it ignores unavoidable thermal currents, drafts, diffusion, and so on, that redistribute particles. The laminar model also predicts an infinite concentration gradient at an interface (see Figure 11-16a), which cannot exist in nature. The well-mixed model overestimates the time to clean the air because it exaggerates the mixing mechanisms. For small, inhalable particles, it is, however, the more realistic model to use. Certainly, it is a more conservative model.

## 11.8 Gravimetric settling in ducts

Gravimetric settling in ducts, also called *sedimentation*, can be analyzed using the same concepts of a laminar model and a well-mixed model. Figure 11-17 illustrates the laminar and well-mixed models for flow in a horizontal duct of rectangular cross section ($A = WH$).

### 11.8.1 Laminar settling model

Assume that the average duct velocity ($U_0$) is constant (i.e., *plug flow* in the gas phase only). In the laminar model all particles of the same size fall at their settling velocity $v_t$, and each possesses the horizontal velocity of the carrier gas (i.e., $v_x = U_0$). Thus at a downstream distance $x$, all particles of the same diameter have fallen the same distance and the uppermost particles have fallen a distance ($H - y$),

$$H - y = tv_t = \frac{x}{U_0}v_t \qquad (11.107)$$

The average concentration of particles of a certain size $\bar{c}(D_p)$ distance $x$ from the inlet can be written as

$$\bar{c}(D_p) = \frac{yc(D_p)_0}{H} \qquad (11.108)$$

Combining yields

$$\frac{\bar{c}(D_p)}{c(D_p)_0} = 1 - \frac{x}{H}\frac{v_t}{U_0} \qquad (11.109)$$

The collection efficiency $\eta$ is,

$$\eta = 1 - \frac{\bar{c}(D_p)}{c(D_p)_0} = \frac{x}{H}\frac{v_t}{U_0} \qquad (11.110)$$

Equation 11.110 also applies to gravimetric settling in fully established flow between parallel plates. See

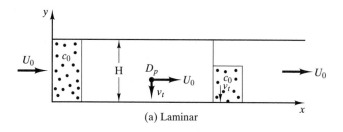

(a) Laminar

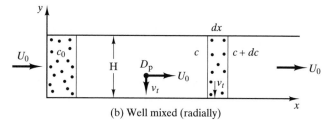

(b) Well mixed (radially)

**Figure 11-17**   Gravimetric settling in a horizontal duct of rectangular cross section

Flagan and Seinfeld (1988) for the derivation. At a *critical distance* (downstream) $L_c = HU_0/v_t$, the collection efficiency will be 100% and the duct will contain no particles of the size defined by the settling velocity. The authors also show that the laminar collection efficiency given by Equation 11.110 is valid for all laminar flow irrespective of the velocity profile as long as the profile does not change in the flow-wise direction [i.e., as long as the flow is *fully established*, $\partial(\cdot)/\partial x = 0$].

## 11.8.2  Well-mixed settling model

Following the assumptions of the well-mixed model used in Section 11.7, it can be assumed that the concentration $c(D_p)$ of particles of a particular diameter $(D_p)$ is uniform in an element $(A\,dx)$. The difference in the mass concentration of particles entering and leaving the elemental volume is equal to the rate of deposition within the volume. Thus

$$c(D_p)AU_0 = [c(D_p) + dc(D_p)]AU_0$$
$$+ c(D_p)v_t W\,dx$$

$$\int_{c(D_p)_0}^{c(D_p)} \frac{dc(D_p)}{c(D_p)} = -\int_0^x \frac{v_t}{U_0 H}\,dx \qquad (11.111)$$

$$\frac{c(D_p)}{c(D_p)_0} = \exp\left(-\frac{v_t}{U_0}\frac{x}{H}\right)$$

Using the definition of collection efficiency above, and realizing that for the well-mixed model $\bar{c}(D_p) = c(D_p)$, we have

$$\eta = 1 - \exp\left(-\frac{v_t}{U_0}\frac{x}{H}\right)$$

$$= 1 - \exp\left(-\frac{v_t}{U_0}\frac{x}{H}\frac{W}{W}\right)$$

$$= 1 - \exp\left(-\frac{v_t A_s}{Q}\right) \qquad (11.112)$$

where $A_s$ is the area of lower collecting surface $(A_s = xW)$ and $Q$ is the volumetric flow rate $(Q = U_0 HW)$. The collection efficiency can also be written in terms of the critical length, $L_c = HU_0/v_t$:

$$\eta = 1 - \exp\left(-\frac{x}{L_c}\right) \qquad (11.113)$$

A comparison of the collection efficiencies for laminar and well-mixed models shows differences similar to those concluded for settling in chambers.

1. The laminar model overestimates deposition because it ignores turbulence that mixes and redistributes particles. The well-mixed model exaggerates mixing but nevertheless provides a more accurate and conservative design estimate of deposition.

2. At a critical downstream distance $L_c$, the well-mixed model predicts a collection efficiency of 63.2% while the laminar model predicts 100%. At a downstream distance of $6.9L_c$, the well-mixed model predicts a collection efficiency of 99.9%.

It is common practice in industrial ventilation to design for duct velocities of 3500 to 4500 fpm (17.8 - 22.9 m/s) to minimize gravimetric settling. The expressions above can be used to examine the settling that one can expect at these velocities. Table 11-8 is a compilation of the duct lengths (expressed as $x/H$) at which 1% and 10% of unit density spheres settle.

It can be seen that 3500 to 4500 fpm is not an exaggerated recommendation. For particles less than 100 μm, settling is minimal in duct lengths up to $30H$; however, deposition is serious for larger particles. Thus it is a wise practice to provide a particle collector (cyclone or gravity dropout chamber) whenever a ventilation system is connected to a source that generates particles, particularly if a portion contains large particles. If this practice is not followed, a ventilation duct may unknowingly become a particle collector, which after a period of time reduces the duct cross-sectional area and alters the pressure drop and volumetric flow rate for which the system of ducts was designed. The volumetric flow rate in the duct will decrease and, in extreme cases, the flow may cease.

If the flow in the rectangular duct is fully established, it can be shown that the expressions for the collection efficiency do not change for either model. It can also be shown that if the duct cross-sectional area is circular, the expression for the collection efficiency for the well-mixed model is the same, but the duct diameter ($D$) should replace the duct height ($H$). Anand and McFarland (1989) analyzed particle deposition for turbulent flow in transport lines of circular cross-sectional area at arbitrary angles of inclination. They found that there is an optimum inside diameter at which deposition is a minimum that is independent of the duct length but depends on particle diameter, flow rate, and angle of inclination.

## 11.8.3 Applications

Gravimetric settling is not considered an efficient air pollution control method. Indeed, particles carried in a duct will settle, but this is not a desired situation since the settled dust has to be removed periodically because it adds unwanted weight (vertical load) to ducts not designed to suffer such loading. Nonetheless, duct systems are often equipped with *dropout boxes* (Figure 1-4a) at key junctions to allow large particles to be collected prior to entry of the process gas stream into conventional particle removal systems. In addition, dropout boxes with access doors provide the opportunity for workers to enter duct systems to perform routine maintenance and to remove settled dust. The pressure drop and fan costs associated with dropout boxes are negligible.

One application for gravimetric settling is the field of industrial hygiene, where one may only be interested in knowing the concentration of respirable dust to which workers are exposed. Devices using gravimetric settling are called *elutriators* and allow nonrespirable dust to settle while allowing the respirable dust to pass through and be captured or measured.

### EXAMPLE 11.8 GRAVIMETRIC SETTLING IN A DUCT WITH A CONVERGING CROSS SECTION

Consider the flow ($Q$, m³/s) of a dust-laden gas stream (inlet dust concentration $c_0$, g/m³) passing through a duct of constant width $W$ with a horizontal

**Table 11-8** Gravimetric Settling of Unit Density Spheres in Ducts[a]

| $D_p$ (μm) | Efficiency = 1% | | Efficiency = 10% | |
| | $v_t$ (m/s) | $(x>H)_{lam} = (x>H)_{turb}$ | $(x>H)_{lam}$ | $(x>H)_{turb}$ |
| --- | --- | --- | --- | --- |
| 10 | $3.1 \times 10^{-3}$ | 74.6 | 745.8 | 785.8 |
| 50 | $7.3 \times 10^{-2}$ | 3.12 | 31.2 | 32.9 |
| 100 | $2.5 \times 10^{-1}$ | 0.92 | 9.2 | 9.7 |
| 200 | 0.7 | 0.33 | 3.3 | 3.4 |
| 600 | 2.5 | 0.09 | 0.9 | 1.0 |
| 1000 | 4.0 | 0.06 | 0.6 | 0.6 |

[a] Duct velocity = 22.86 m/s (4500 fpm), air at STP.

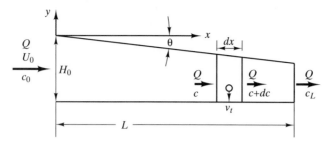

(a) Well mixed settling model, $\eta = 1 - \exp[-(v_t/U_0)(L/H_0)]$

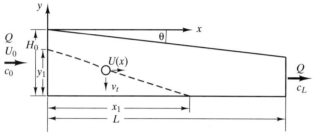

(b) Laminar settling model, $\eta = (v_t/U_0)(L/H_0)\,[1 - (L/2H_0)\tan\theta]$

**Figure E11-8** Gravimetric settling in a converging duct of width W.

floor and a roof that is inclined downward at an angle $\theta$ to the horizontal (Figure E11-8):

$$H(x) = H_0 - x\tan\theta$$

Predict the particle removal efficiency using the well-mixed settling model and the laminar settling model. Compare and discuss the results.

**Solution**

*Well-mixed settling model.* Write a mass balance for a differential element downstream of the inlet.

$$Qc = Q(c + dc) + v_t cW\,dx$$

$$\frac{dc}{c} = -\frac{v_t W}{Q}\,dx$$

Integrate over a length $L$ and obtain

$$\frac{c_L}{c_0} = \exp\left(-\frac{v_t LW}{Q}\right)$$

Since the volumetric flow rate is constant even though $U$ and $H$ vary with $x$, the overall efficiency

$$\eta = 1 - \frac{c_L}{c_0} = 1 - \exp\left(-\frac{v_t}{U_0}\frac{L}{H_0}\right)$$

It should be noted that this is the same expression that would have been obtained if the duct was of constant height or if the duct diverged rather than converged. It is clear that the well-mixed model fails to account for the geometry of the duct, which is contrary to an engineer's intuition.

*Laminar settling model.* Assume that the gas velocity has no vertical component $(U_y = 0)$ and that $U_x$ does not vary in the vertical direction but varies in the $x$-direction, owing to the reduction in duct height. Thus

$$U(x) = \frac{Q}{WH(x)} = \frac{Q}{W(H_0 - x\tan\theta)}$$

Take note of a particle at the inlet that exists at a height $y_1$, and track its trajectory until it settles to the duct floor at a downstream distance $x_1$. From the figure it can be seen that the rate at which particles settle to the floor over the distance $x_1 \geqq x \geqq 0$ is

$$y_1 c_0 W U_0$$

Over the distance $x_1$, the collection efficiency is

$$\eta = \frac{y_1 c_0 W U_0}{H_0 c_0 W U_0} = \frac{y_1}{H_0}$$

The particle falls through the distance $y_1 = v_t t_1$; thus

$$\eta = \frac{v_t t_1}{H_0}$$

where $t_1$ is equal to

$$
\begin{aligned}
t_1 &= \int_0^{t_1} dt = \int_0^{x_1} \frac{dx}{U(x)} \\
&= \int_0^{x_1} \frac{W(H_0 - x \tan \theta) \, dx}{Q} \\
&= \frac{W x_1 [H_0 - (x_1/2) \tan \theta]}{Q}
\end{aligned}
$$

For a duct length $L$, the overall efficiency can be expressed as

$$
\begin{aligned}
\eta &= W \frac{v_t L}{H_0} \frac{H_0 - (L/2) \tan \theta}{Q} \\
&= W \frac{v_t L}{H_0} \frac{H_0 - (L/2) \tan \theta}{U_0 H_0 W} \\
&= \frac{v_t}{U_0} \frac{L}{H_0} \left( 1 - \frac{L}{2 H_0} \tan \theta \right)
\end{aligned}
$$

*Discussion.* The laminar settling model contains terms that account for the slope of the duct roof and thus satisfies an engineer's intuition. In the limit, as the angle $\theta$ approaches zero, the expression above also approaches Equation 11.110. In conclusion, the laminar settling model seems best suited to describe gravimetric settling in a converging duct despite the simplistic assumptions about the gas velocity.

## 11.9 Clouds

Consider an ensemble of particles suspended in air, henceforth called a *cloud*. Clouds can be produced by explosions, smoke generators, insecticide mists, or discharges from chimneys, automobiles, and so on. A common example is exhaled tobacco smoke. It will be assumed that the cloud has discernible boundaries and that it rises or falls relative to the surrounding air with a detectable velocity $(v_c)$. It will be useful if practitioners can predict the cloud settling velocity. In the analysis that follows it will be assumed that the cloud contains monodisperse particles.

It is observed that clouds with a small particle number concentration $(n_p)$ settle with a velocity similar to the settling velocity of the individual particles $(v_t)$. It is also observed that when either the particle number concentration is large or the cloud is large (or both), the cloud settles with a much larger velocity $(v_c)$.

From the concept of ideal (potential) flow it can be shown that the velocity field produced by one particle does not influence the flow field produced by another particle if the distance between particles exceeds 10 particle diameters. Of course, particle motion is dictated by viscous effects and one must broaden the analysis to account for the influence of the wakes produced by particles. If the wake of one particle influences the motion of another particle, one can anticipate unique dynamics of the bulk aerosol. The equations that govern the flow of an aerosol with a large particle number concentration is a topic in multiphase flow. The subject is complex because the equations describing the motion of the aerosol and the carrier gas are coupled.

The behavior of concentrated aerosols is called *colligative behavior* (Phalen et al., 1994). Colligative behavior can occur for concentrated aerosols within a confined space or in an unbounded environment. In an unbounded environment, the concentrated aerosol is called a *cloud*. In lieu of such a detailed analysis, Fuchs (1964) and Hinds (1982) describe a simple way to predict the settling velocity of a cloud. Consider the clouds shown in Figure 11-18 that satisfy the following:

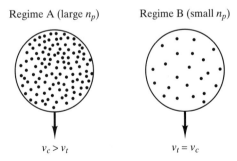

Regime A (large $n_p$)      Regime B (small $n_p$)

$v_c > v_t$        $v_t = v_c$

**Figure 11-18** Clouds with large and small particle concentrations, assume monodisperse particles (a) large particle concentration $n_p$, cloud settling velocity larger than particle settling velocity, $v_c > v_t$, (b) small particle concentration $n_p$, cloud settling velocity equals individual particle settling velocity, $v_c = v_t$.

- The cloud contains monodisperse particles of diameter $(D_p)$ and density $(\rho_p)$, where $\rho_p = 1000$ kg/m³.
- The particle number concentration $(n_p)$ is uniform within the cloud.
- The cloud is spherical $(D_c)$.
- The temperatures of the particles and the air inside and outside the cloud are the same.
- The surrounding air is motionless.
- When the particle number concentration $(n_p)$ is sufficiently large, it causes the cloud to settle as an entity with a velocity $(v_c)$ greater than the individual particle settling velocity $(v_t)$.

If the air and particles within the cloud settle at a velocity $v_c > v_t$, then with respect to the cloud, the ambient air passes around the cloud as if its outer boundary was an impervious spherical envelope. Under these conditions, aerodynamic drag on the cloud is the drag produced by air passing around the spherical envelope. On the other hand, if the cloud settles with a velocity $v_t$, the ambient air passes through the envelope and the drag on the cloud is the sum of the drag on each individual particle. The settling velocity $v_c$ of a solid sphere of diameter $D_c$ whose mass is equal to the mass of the particles within is as follows:

$$n_p \frac{\pi D_c^3}{6} \frac{\pi D_p^3}{6} g\rho_p = c_D \frac{\pi D_c^2}{4} \frac{\rho v_c^2}{2}$$

$$v_c = \left( \frac{2\pi g}{9 c_D} \frac{\rho_p}{\rho} n_p D_c D_p^3 \right)^{1/2} \tag{11.114}$$

There may be occasions when processes occur within the cloud that change $D_p$, or $n_p$, or $D_c$, and users will want to anticipate whether the cloud's motion might change from one regime to the other. For these situations it is useful to define

$$\beta = \frac{v_c}{v_t} \tag{11.115}$$

Upon substitution, and assuming that the individual particles are governed by Stokes flow,

$$\beta = \frac{\left[ (2\pi g / 9 c_D)(\rho/\rho_p) n_p D_c D_p^3 \right]^{1/2}}{\tau_p g C}$$

$$= \frac{6\mu}{C} \left( \frac{2\pi n_p}{g c_D} \frac{D_c}{D_p} \frac{1}{\rho \rho_p} \right)^{1/2} \tag{11.116}$$

where $C$ is the slip factor and $\tau_p$ is the relaxation time (Equation 11.88). Thus, when $\beta > 1$, the cloud settles with a velocity $v_c$, and if $\beta \leqq 1$, the cloud settles with a velocity $v_t$.

When the cloud is large, the Reynolds number of the spherical cloud exceeds 1000 and Figure 11-3 shows that one can assume that $c_D = 0.4$. It will be useful to use the product of the number concentration and cloud diameter $(n_p D_c)$ as a variable. Assuming that the cloud particles are water $(\rho_p = 1000$ kg/m³), Table 11-9 indicates the maximum value of $n_p D_c$ at which the cloud at STP settles with a velocity equal to $v_t$.

**Table 11-9**    Clouds Whose Settling Velocities are Equal to the Particle Settling Velocities

| $D_p$ (µm) | $v_t$ (m/s) | Maximum $n_p D_c$ (particles/m²) |
|---|---|---|
| 0.1 | $8.5 \times 10^{-7}$ | $1.91 \times 10^{14}$ |
| 1.0 | $3.6 \times 10^{-4}$ | $1.91 \times 10^{12}$ |
| 10.0 | $3.6 \times 10^{-3}$ | $1.91 \times 10^{10}$ |

To acquire a frame of reference that relates particle number concentration $(n_p)$ and mass concentration $(c)$, it can be shown that typical atmospheric clouds have a particle mass concentration of 1 g/m³ $(10^{-3}$ kg/m³). If a cloud contains a monodisperse aerosol of unit-density spheres of diameter, $D_p = 1$ µm, the relationship between number and mass concentration is as follows:

| $D_p$ (µm) | $c$ (kg/m³) | $n_p$ (particles/m³) |
|---|---|---|
| 1.0 | 0.1 | $2 \times 10^{14}$ |
| 1.0 | $5.233 \times 10^{-2}$ | $10^{14}$ |
| 1.0 | $5.233 \times 10^{-4}$ | $10^{12}$ |
| 1.0 | $5.233 \times 10^{-6}$ | $10^{10}$ |

Figure 11-19 shows cloud settling velocities $(v_c)$ as a function of particle size $(D_p)$ for different values of $n_p D_c$. As a result of these calculations it can be seen that small, highly dense aerosol clouds may settle with a velocity considerably larger than one would expect based on the size of the individual particles. On the other hand, large clouds with modest particle number concentrations also settle faster than one expects based on the size of particles.

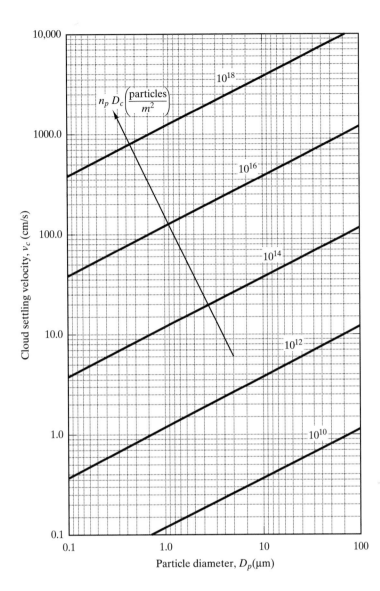

**Figure 11-19** Cloud settling velocity as a function of particle size and concentration

The analysis is a useful approximation, but users must be aware of several physical processes affecting cloud behavior that the model ignores. From everyday observations (watch smokers blow smoke rings) users should recall that clouds behave as an entity early in their lifetime but with the passage of time, clouds tend to be governed by the motion of individual particles, as particles coagulate and/or diffuse into the atmosphere, and the cloud breaks up.

**1.** The boundaries of the cloud expand and the particle number concentration decreases near the boundaries.

**2.** The assumption that the air is motionless inside the cloud's envelope ignores that fact that there is circulation of air even within small clouds.

**3.** As a cloud of particles falls, it loses its spherical shape and the drag coefficient changes. The cloud may even divide into two or more fragments.

**4.** The atmosphere is never motionless and there will be air currents whose velocity exceeds $v_t$ and may even exceed $v_c$.

**5.** Particles coagulate, thus increasing $D_p$ and decreasing $n_p$ and $D_c$. Coagulation inhibits settling. It also decreases $\beta$ and may bring about a change of flow regimes. Alternatively, $D_p$ may decrease due to evaporation, thus reducing the settling velocity. It also increases β and may bring about a change of flow regimes.

## 11.10 Stokes number

The *Stokes number* is an important dimensionless parameter in particle dynamics, just as the *Reynolds number* is important in fluid dynamics. Rather than use the *Buckingham Pi theorem* to show the uniqueness of the Stokes number, it is easier to transform the equation for particle motion into a dimensionless equation and allow the Stokes number to emerge through the derivation. This process is called *inspectional analysis*, as contrasted with *dimensionless analysis*. To begin inspectional analysis, it is necessary to define a characteristic velocity $U_0$ and a characteristic length $L$. The characteristic time is equal to $L/U_0$. The characteristic velocity and length are merely two terms relevant to the problem that the engineer can easily identify and assign numerical values to. Typical values might be the velocity of an aerosol entering a device and the width or diameter of the inlet. The choice of $U_0$ and $L$ is generally dictated by the phenomenon being modeled.

Consider a spherical particle traveling with a velocity **v** through a carrier gas that has a velocity **U**

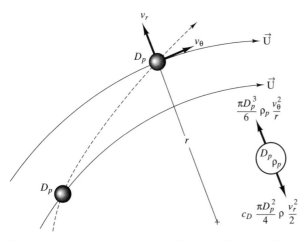

**Figure 11-20** Quasi-static equilibrium of a particle in curvilinear flow

(see Figure 11-20). The equation of motion for the particle is

$$\frac{\pi D_p^3}{6} \rho_p \frac{d\mathbf{v}}{dt} = -\frac{\rho c_D}{2C} \frac{\pi D_p^2}{4}(\mathbf{v} - \mathbf{U})|\mathbf{v} - \mathbf{U}|$$

$$- \mathbf{g}\rho_p \frac{\pi D_p^3}{6} \qquad (11.117)$$

Gravity will be neglected since inertial and viscous effects are much larger. The Reynolds number is defined in terms of the relative velocity (Equation 11.4).

$$\mathrm{Re} = \frac{\rho D_p |\mathbf{v} - \mathbf{U}|}{\mu}$$

Substituting the Reynolds number into Equation 11.67 and simplifying yields

$$\frac{d\mathbf{v}}{dt} = -\left(\frac{3}{4}\right)\frac{c_D \mu \mathrm{Re}}{C D_p^2 \rho_p}(\mathbf{v} - \mathbf{U}) \qquad (11.118)$$

To reduce Equation 11.118 to a dimensionless equation, we multiply and/or divide the velocity, length, and time by the known characteristic velocity $U_0$, length $L$ and time $L/U_0$.

$$\frac{d(\mathbf{v}/U_0)U_0}{d[t/(L/U_0)](L/U_0)} = -\frac{3}{4}\frac{c_D \mu \mathrm{Re}}{C D_p^2 \rho_p}\left(\frac{\mathbf{v}}{U_0} - \frac{\mathbf{U}}{U_0}\right)U_0$$

$$\frac{d\mathbf{v}^*}{dt^*} = -\frac{18\mu}{D_p^2 \rho_p}\frac{L}{U_0}\frac{c_D \mathrm{Re}}{24C}(\mathbf{v}^* - \mathbf{U}^*)$$

$$(11.119)$$

where the dimensionless particle and carrier gas velocities are $\mathbf{v}^* = \mathbf{v}/U_0$ and $\mathbf{U}^* = \mathbf{U}/U_0$ and the dimensionless time is $t^* = t/(L/U_0)$. A dimensionless parameter called the *Stokes number* (ψ) may be defined as

$$\psi = \frac{D_p^2 \rho_p}{18\mu}\frac{U_0}{L} = \frac{\tau C U_0}{L} \qquad (11.120)$$

The Stokes number is the ratio of the penetration distance $(\tau U_0 C)$ (see Equation 11.102) to the characteristic distance $(L)$. Equation 11.119 can now be rewritten as

$$\frac{d\mathbf{v}^*}{dt^*} = -\frac{c_D \mathrm{Re}}{24\psi}(\mathbf{v}^* - \mathbf{U}^*) \qquad (11.121)$$

Since the differential equation is dimensionless, solutions to widely different problems are the same provided that the initial values of the dimensionless variables and the

$$\text{scaling parameter} = c_D \frac{\text{Re}}{24\psi}$$

are the same. For example, provided that the scaling parameters of the particles are the same, the dynamics of dust entering a personnel dust sampler will be similar to the dynamics of rain entering a jet engine during takeoff. The dynamics of dust generated by trucks traveling on unpaved roads will be similar to the movement of welding fume from a workbench. In general, if two different physical phenomena (A and B) are such that the following is true:

$$\left( c_D \frac{\text{Re}}{24\psi} \right)_A = \left( c_D \frac{\text{Re}}{24\psi} \right)_B \qquad (11.122)$$

the motion of the particles written in dimensionless terms will be equivalent.

## 11.11 Inertial deposition in curved ducts

It is important to understand the deposition of particles in a curved duct (bend) since it occurs in many situations of practical interest. In the field of health, inhaled particles pass through a number of bifurcations (divisions) as the airflow divides again and again in passing through the bronchial tree. This is important since the highest incidence of carcinogenic tumors occur in the upper generations of the bronchial system. In terms of particle sampling, industrial dust control and pneumatic transfer of powders, the impaction of particles in bends through inertial deposition has important practical consequences. In the analysis that follows, gravitational settling will be neglected. The reader should consult Balashazy et al. (1990) for an analysis, including gravitational settling.

As an aerosol flows through a curved duct (Figure 11-21), one can expect a particle's inertia to cause it to move radially outward. The velocity of the carrier gas can be obtained from the solution of the Navier–Stokes equations. For purposes of illustration, assume that the duct cross-section is rectangular (constant width $W$) and that the tangential velocity is given by

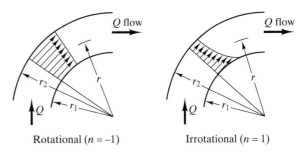

**Figure 11-21** Velocity profiles for rotational and irrotational flow in a curved duct

$$U_\theta r^n = \text{constant} = C_n \qquad (11.123)$$
$$U_r = 0$$

where the value of $n$ lies between (or equal to) $+1$ and $-1$. When $n = 1$ the flow is irrotational (**Curl U** $= 0$), and when $n = -1$, the flow is called *solid body rotation* (constant vorticity). For lack of a better phrase, flow fields that can be described by Equation 11.123, where $-1 < n < 1$, will be called *n-degree rotational* since irrotational implies the specific condition (**Curl U** $= 0$), while rotational can be used more broadly.

*Irrotational:* $\qquad n = 1 \qquad \text{constant} = rU_\theta$
$$(11.124)$$

*Solid body rotation:* $\quad n = -1 \quad \text{constant} = \dfrac{U_\theta}{r}$
$$(11.125)$$

*n-Degree rotational:* $\quad -1 < n < 1 \text{ constant} = U_\theta r^n$
$$(11.126)$$

Figure 11-21 shows the velocity profiles for rotational and irrotational flow. If the tangential velocity is given by Equation 11.123, application of the continuity equation requires that $\partial U_r/\partial r = 0$. If, in addition, the flow is irrotational, then $U_r$ is zero.

The volumetric flow rate through a curved duct of constant rectangular cross section and constant radius is

$$Q = \int_{r_1}^{r_2} W U_\theta \, dr \qquad (11.127)$$

where $W$ is the width of the $r$–$\theta$ plane. Upon substituting Equations 11.124 to 11.126 into 11.127, the constants can be evaluated and the tangential velocities expressed as:

- *Irrotational flow, $n = 1$*

$$C_n = \frac{Q}{W \ln\left(r_2/r_1\right)} \tag{11.128}$$

$$U_\theta = \frac{Q/r}{W \ln\left(r_2/r_1\right)} \tag{11.129}$$

- *Solid body rotation, $n = -1$*

$$C_n = \frac{2Q}{W(r_2^2 - r_1^2)} \tag{11.130}$$

$$U_\theta = \frac{2Qr}{W(r_2^2 - r_1^2)} \tag{11.131}$$

- *$n$-Degree rotational flow, $-1 < n < 1$*

$$C_n = \frac{(1 - n)Q}{W\left(r_2^{1-n} - r_1^{1-n}\right)} \tag{11.132}$$

$$U_\theta = \frac{(1 - n)Q}{W\left[(r_2)^{1-n} - (r_1)^{1-n}\right]r^n} \tag{11.133}$$

Rather than solving the full set of equations describing the motion of particles in a moving flow field, a useful approximation is available by assuming:

1. Quasistatic equilibrium
2. Stokes flow
3. Negligible gravimetric settling

Consistent with the assumption of quasistatic equilibrium is the assumption that the tangential velocity of the particle and the gas are equal. In the tangential direction there will be no relative motion between the particle and the carrier gas (Figure 11-20). In the radial direction it will be assumed that the *centrifugal force* of the particle is equal to the viscous drag. Since the gas velocity has no radial component $(U_r = 0)$, the particle's relative velocity is its absolute velocity in the radial direction $(v_r)$. Thus the particle velocity has the following tangential and radial components:

*Tangential direction $(\theta)$:*

$$v_\theta = U_\theta = \frac{C_n}{r^n} \tag{11.134}$$

*Radial direction $(r)$:*

$$\rho_p \frac{\pi D_p^3}{6} \frac{U_\theta^2}{r} = c_D \frac{\pi D_p^2}{4} \frac{\rho v_r^2}{2} \tag{11.135}$$

$$v_r = U_\theta \sqrt{\frac{\rho_p}{\rho} \frac{4}{3c_D} \frac{D_p}{r}}$$

It will be assumed that the particle's Reynolds number, based on the relative velocity, is less than unity such that Stokes flow can be assumed.

$$c_D = \frac{24}{\text{Re}} \tag{11.136}$$

$$\text{Re} = \frac{\rho D_p v_r}{\mu} \tag{11.137}$$

At any point in the flow field, the particle radial and tangential velocity components are:

*Tangential $(\theta)$:*  $v_\theta = U_\theta = \dfrac{C_n}{r_m} \tag{11.138}$

*Radial $(r)$:*  $v_r = \dfrac{\tau\left(U_\theta\right)^2}{r} \tag{11.139}$

Deposition of particles on the outer wall of the bend will be modeled in a fashion similar to deposition in horizontal ducts: for the laminar model and for the well-mixed model (turbulent model).

## 11.11.1  Laminar model

Figure 11-22 depicts the particle concentration downwind from the inlet of the bend. It will be useful to find the angle $\theta$ (in *radians*) at which a particle enter-

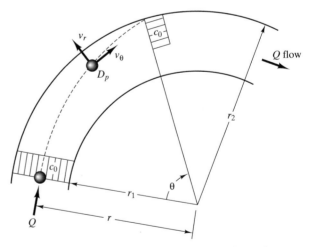

**Figure 11-22**   Particle motion in a curved duct of rectangular cross section, laminar flow model

ing the bend at a radius $r$, strikes the outer wall. Such an expression can be found as follows:

$$\frac{v_\theta}{v_r} = \frac{r\,d\theta/dt}{dr/dt} = r\frac{d\theta}{dr}$$

$$\frac{r}{\tau U_\theta} = \frac{r^{1+n}}{\tau C_n} \tag{11.140}$$

where $C_n$ is given by Equations 11.128, 11.130, and 11.132 and must be evaluated in terms of other flow parameters. For irrotational and $n$-degree rotational flow ($n \neq -1$)

$$\int_0^\theta d\theta = \int_r^{r_2} \frac{r^n\,dr}{C_n\tau}$$

$$r = \left[r_2^{n+1} - (n+1)\theta\tau C_n\right]^{1/(1+n)} \tag{11.141}$$

For rotational flow ($n = -1$)

$$\int_0^\theta d\theta = \int_r^{r_2} \frac{dr}{r C_n\tau}$$

$$r = r_2 \exp(-\tau\theta C_n) \tag{11.142}$$

Define collection efficiency ($\eta$) as

$$\eta = \frac{r_2 - r}{r_2 - r_1} = \frac{1 - (r/r_2)}{1 - (r_1/r_2)} \tag{11.143}$$

For irrotational flow ($n = 1$),

$$\eta_{n=1} = \frac{1 - \sqrt{1 - 2\theta Q\tau/Wr_2^2\ln(r_2/r_1)}}{1 - r_1/r_2} \tag{11.144}$$

For solid body rotation ($n = -1$),

$$\eta_{n=-1} = \frac{1 - \exp\left[-2Q\theta\tau/W(r_2^2 - r_1^2)\right]}{1 - r_1/r_2} \tag{11.145}$$

Equations 11.144 and 11.145 can be rearranged in terms of an average Stokes number ($\psi_a$) defined in terms of an average velocity, $U_{\text{ave}} = Q/W(r_2 - r_1)$

$$\psi_a = \frac{\tau U_{\text{ave}}}{r_2} = \frac{\tau}{r_2}\frac{Q}{W(r_2 - r_1)} \tag{11.146}$$

For irrotational flow ($n = 1$), the efficiency becomes

$$\eta_{n=1} = \frac{1 - \sqrt{1 - 2\psi_a\theta\left[(r_2 - r_1)/r_2\ln(r_2/r_1)\right]}}{1 - r_1/r_2} \tag{11.147}$$

and for solid body rotation ($n = -1$),

$$\eta_{n=-1} = \frac{1 - \exp\left\{-2\psi_a\theta\left[r_2/(r_1 + r_2)\right]\right\}}{1 - r_1/r_2} \tag{11.148}$$

A similar development can be performed for $n$-rotational flows [i.e., that are neither irrotational ($n = 1$) nor solid body rotation ($n = -1$)]. The efficiency can be shown to be

$$\eta_n =$$
$$\frac{1 - \left[1 - (1 - n^2)\theta r_2(r_2 - r_1)\psi_a/r_2^{1+n}(r_2^{1-n} - r_1^{1-n})\right]^{1/(1+n)}}{1 - r_1/r_2} \tag{11.149}$$

A useful parameter to consider is the critical angle ($\theta_c$, in *radians*) at which all particles entering the bend strike on the outer wall. For the laminar model and for irrotational flow, the critical angle is

$$\theta_c(\text{radians}) = (r_2 + r_1)\frac{\ln(r_2/r_1)}{2r_2\psi_a} \tag{11.150}$$

## 11.11.2 Well-mixed model

The well-mixed model assumes that within the flow field there exists a mixing mechanism that redistributes particles remaining in the duct after those along the outer radius strike and stick to the outer wall. For this reason, the particle concentration varies with angle $\theta$ (radians) but not the radius. To analyze the flow, begin by constructing a mass balance for an element of the flow as shown in Figure 11-23:

$$cQ = Q(c + dc) + cv_r(r_2)Wr_2\,d\theta \tag{11.151}$$

$$\int_{c_0}^{c_\theta} \frac{dc}{c} = -\int_0^\theta \frac{v_r(r_2)Wr_2}{Q}\,d\theta$$

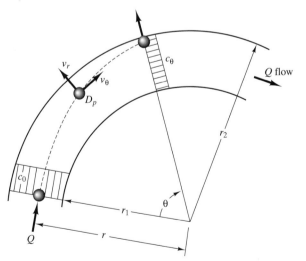

**Figure 11-23** Particle motion in a curved duct of rectangular cross section, well mixed flow model

where $v_r(r_2)$ is the particle radial velocity (Equation 11.139) evaluated at the outer radius $(r_2)$. Define the collection efficiency as

$$\eta = 1 - \frac{c_\theta}{c_0} \qquad (11.152)$$

The right-hand side of Equation 11.152 can be found by integrating Equation 11.151, replacing $v_r(r_2)$ by Equation 11.139, simplifying, and introducing the Stokes number based on the average velocity (Equation 11.146). As a consequence of these substitutions and simplifications, the efficiency can be written as

$$\eta = 1 - \exp(-K\theta\psi_a) \qquad (11.153)$$

where

$$K(\text{irrotational}) = \frac{r_2 - r_1}{r_2[\ln(r_2/r_1)]^2} \qquad (11.154)$$

$$K(\text{solid body rotation}) = \frac{4r_2^3}{(r_2 + r_1)(r_2^2 - r_1^2)} \qquad (11.155)$$

$$K(n\text{-degree rotational}) = \frac{(1-n)^2(r_2 - r_1)r_2^{1-2n}}{(r_2^{1-n} - r_1^{1-n})^2} \qquad (11.156)$$

A summary of Equations 11.146 to 11.156 is given in Table 11-10.

The assumption of quasi-static equilibrium is valid for a large class of flows when one recalls the analysis of Section 11.4. The time required for small particles to achieve equilibrium is very small and only several times longer than the relaxation time. Thus it is reasonable to assume that viscosity acts rapidly on particles flowing in a curved duct. The assumption of Stokes flow is also not restrictive since it is only the particle's radial velocity component that produces the relative velocity. The radial velocity is small since it depends on the relaxation time $(\tau)$. The restrictive assumption in the above analysis is that the gas velocity is given by

$$U_\theta r^n = \text{constant} \qquad (11.123)$$

The velocities entering the bend depend on conditions upstream of the inlet. If air enters the bend from the atmosphere, the inlet velocity profile will be uniform. If a long rectangular duct precedes the bend, the velocity profile entering the bend will be fully established, or well on its way to becoming so. In either case, the velocity profile in the bend would not be given by Equation 11.123. The velocity profile is also dictated by the pressure distribution in the bend.

---

**Table 11-10** Impaction of Particles on Outer Wall of Curved Duct Having a Rectangular Cross Section [a]

| | | Impaction efficiency | |
|---|---|---|---|
| **Flow** | $U_\theta r^n = \text{const.}$ | **Laminar model** | **Well-mixed model** |
| Irrotational $(n=1)$ | $\dfrac{Q/r}{W \ln(r_2/r_1)}$ | $\dfrac{1 - \left[1 - \dfrac{2\theta(r_2 - r_1)\psi_a}{r_2 \ln(r_2/r_1)}\right]^{1/2}}{1 - r_1/r_2}$ | $1 - \exp\left[-\dfrac{\theta(r_2 - r_1)\psi_a}{r_2[\ln(r_2/r_1)]^2}\right]$ |
| Solid-body rotation $(n=-1)$ | $\dfrac{2Qr}{W(r_2^2 - r_1^2)}$ | $\dfrac{1 - \exp\left[-\dfrac{2\theta r_2 \psi_a}{r_2 - r_1}\right]}{1 - r_1/r_2}$ | $1 - \exp\left[-\dfrac{4\theta r_2^3 \psi_a}{(r_2 + r_1)(r_2^2 - r_1^2)}\right]$ |
| $n$-Rotational $(-1 \le n < 1)$ | $\dfrac{(1-n)Q}{W(r_2^{1-n} - r_1^{1-n})r^n}$ | $\dfrac{1 - \left[1 - \dfrac{(1-n^2)\theta r_2(r_2 - r_1)\psi_a}{r_2^{1+n}(r_2^{1-n} - r_1^{1-n})}\right]^{1/(1+n)}}{1 - r_1/r_2}$ | $1 - \exp\left[-\dfrac{(1-n)^2\theta(r_2 - r_1)r_2^{1-2n}\psi_a}{(r_2^{1-n} - r_1^{1-n})^2}\right]$ |

[a] $\psi_a = \dfrac{\tau}{r_2}\dfrac{Q}{W(r_2 - r_1)}$.

The radial and tangential pressure gradients for irrotational and rotational flow are quite different. A thorough analysis of particle motion depends on solving the equations of motion of the air in the bend.

## EXAMPLE 11.9 PARTICLE CLASSIFIER

A company processes agricultural materials, grains, corn, rice, and so on. One of the processes is a milling operation. Significant fugitive dust is produced. Enclosures and exhaust air ($Q$, cfm) are needed to capture the dust. A classifier is needed to separate no less than 50% of the particles larger than 100 μm

$(D_p > 100 \mu m)$ which then are returned for reprocessing. The smaller particles are removed by filters (baghouse). Your supervisor suggests constructing a simple device consisting of a 180° elbow of rectangular cross section containing louvers on the outside surface (Figure E11-9a). The volumetric flow rate of air in the elbow is $Q$. Centrifugal force sends large particles in the radial direction where they pass through the louvers and are drawn off by a slip stream and removed by other means. Your supervisor asks you to compute the grade efficiency curves (similar to Figure 11-11) that will enable operators to select the proper volumetric flow rate $Q$ to achieve a

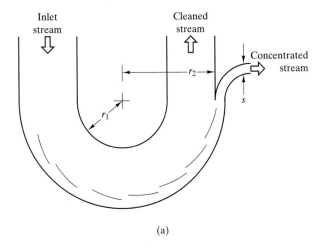

(a)

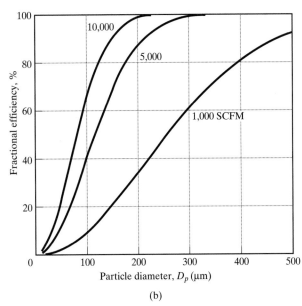

(b)

**Figure E11-9** (a) Centrifugal particle classifier. (b) Fractional efficiency of a centrifugal particle classifier for three volumetric flow rates. Classifier dimensions: $r_1 = 0.3$ m, $r_2 = 0.7$ m, W = 0.4 m s = 0.07 cm

certain removal efficiency ($\eta$). Assume that the gas flow is irrotational and well mixed. Plot the results for a classifier whose dimensions are

$$r_1 = 0.3 \text{ m} \qquad r_2 = 0.7 \text{ m}$$

$$W = 0.4 \text{ m} \qquad s = 0.07 \text{ cm}$$

and which separates unit density particles ($\rho_p = 1000$ kg/m$^3$) traveling in an air stream at 300 K.

**Solution** From Equations 11.146 to 11.154,

$$\eta = 1 - \exp(-K\theta\psi_a)$$

$$K = \frac{r_2 - r_1}{r_2[\ln(r_2/r_1)]^2} = \frac{0.7 - 0.3}{0.7^2[\ln(0.7/0.3)]^2} = 0.7959$$

$$\theta = \pi$$

$$\psi_a = \frac{\tau}{r_2} \frac{Q}{W(r_2 - r_1)} = 8.9286\tau Q$$

$$\tau = \frac{\rho_p D_p^2}{18\mu}$$

$$\eta = 1 - \exp(-7.1062\tau Q)$$

Figure E11-9b shows the fractional (grade) efficiency at three volumetric flow rates. The Mathcad program producing Figure E11-9b can be found in the textbook Web site on the Prentice Hall Web page. Clearly, the lowest volumetric flow rate classifies particles poorly.

The use of the well-mixed model is a reasonable selection since Reynolds numbers in the elbow are surely large enough to establish turbulent flow. The assumption of irrotational flow is an assumption that needs to be justified. The next level of sophistication is to use commercially available *computational fluid dynamics* (CFD) computer programs to predict the trajectories of particles in the three dimensional velocity field of the 180° elbow.

## 11.12 Closure

If the number of particles per unit volume of gas is low (valid for most problems in air pollution) the motion of each particle is not affected by the motion or wake of another particle. This fortunate state of affairs accommodates the wonderful simplification of modeling the motion of each particle separately. The motion of a particle is governed by Newton's second law, in which the drag force can be expressed as a function of a conventional drag coefficient (modified by the Cunningham slip factor for very small particles) and the square of the relative velocity. Computers facilitate calculations that are repeated for each particle (or small range of particles) in an aerosol. Further simplification occurs for small particles in the Stokes regime, in which the drag coefficient can be expressed in simple terms. The range of particle diameters in air pollution can often be described as log-normal, and the log-normal size distribution function can be used in conjunction with the equations of particle motion to describe the motion of an entire aerosol. Gravitational and inertial forces exerted on particles can be included in the equations of motion, which allows engineers to model the performance of mechanical particle collectors.

## Nomenclature

| Symbol | Description (Dimensions*) |
|---|---|
| [=] | has the units of |
| [j] | concentration of species $j$ ($M/L^3$, $N/L^3$) |

\* F, force; L, length; M, mass; N, mols; Q, energy; t, time; T, temperature.

| | |
|---|---|
| = | stoichiometric equation |
| A | area, constant defined by equation ($L^2$) |
| b | constant defined by equation |
| B | constant defined by equation |
| c | mass concentration ($M/L^3$) |

| | | | |
|---|---|---|---|
| $C$ | Slip factor (Cunningham correction factor) | $K$ | constant defined by equation |
| $c_D$ | particle drag coefficient | $L_c$ | penetration or stopping distance (L); penetration (pulvation) distance (L) |
| $c_i, c_0$ | inlet or free-stream mass concentration $(M/L^3)$ | $L$ | characteristic length (L) |
| $c_L, c_H, c_f$ | final or exit mass concentrations $(M/L^3)$ | $L_c$ | critical length (L) |
| $C_n$ | constant used to describe curvilinear flow | $\dot{m}$ | mass flow rate (M/t) |
| $C_1, C_2$ | constants defined by equation | $M, M_i$ | molecular weight, molecular weight of species $i$ (M/N) |
| $c(D_p), c(D_p)_0, c(D_p)_a$ | local, initial and average mass concentration of particles of diameter $D_p$ $(M/L^3)$ | $m_t$ | total mass of particles (M) |
| | | $m(D_p)$ | mass distribution function $(L^{-1})$ |
| $\bar{c}(D_p)$ | average dust concentration in an enclosure $(M/L^3)$ | $n(D_p)$ | number distribution function $(L^{-1})$ |
| | | $n$ | number of molecules per volume of gas $(L^{-3})$; exponent for curvilinear flow |
| $c(D_{p,j})$ | concentration of particles in the $j$th interval $(M/L^3)$ | $n_p$ | number of drops per volume of air, number particles per volume in a cloud $(L^{-3})$ |
| $D$ | constant defined by equation, dilution ratio | | |
| $D$ | particle diffusion coefficient $(L^3/t)$ | $n_i$ | number of particles of diameter $D_{p,i}$ |
| $D_a$ | aerodynamic diameter defined by equation (L) | $N_i$ | cumulative distribution for particles less than $D_{p,i}$ |
| $D_c$ | diameter of the collecting particle, cloud diameter (L) | $n_t$ | total number of particles; (N) total number of mols |
| $D_{e,p}$ | equivalent volume diameter, defined by equation (L) | $N(D_p)$ | cumulative number distribution function |
| | | $P$ | pressure $(F/L^2)$; penetration $(P = 1 - \eta)$ |
| $d_n$ | collision diameter (L) | $Q$ | volumetric flow rate $(L^3/t)$ |
| $D_p$ | particle diameter (L) | $Q_a, Q_s$ | volumetric flow rates of air and scrubbing fluid $(L^3/t)$ |
| $D_{p,am}$ | arithmetic mean diameter (L) | | |
| $D_{p,gm}$ | geometric mean diameter (L) | $r$ | radius (L)$v$ |
| $D_{p,j}$ | mean particle diameter in the $j$th interval (L) | $R$ | $D_p/D_f$ |
| | | $R_u, R_j$ | universal gas constant, gas constant for a specific molecular species $j$ (Q/NT) |
| $D_{p,p}$ | equivalent projected area diameter, defined by equation (L) | | |
| | | $t$ | time (t) |
| $D_{p,16}, D_{p,50}, D_{p,84}$ | particle diameters at which the cumulative distributions are 16%, 50%, and 84% (L) | $s$ | width of louvers (L) |
| | | $t^*$ | dimensionless time |
| | | $T$ | temperature (T) |
| $D_{s,p}$ | equivalent surface area diameter, defined by equation (L) | $t_c$ | critical time (t) |
| $F$ | force on a particle (F) | $\mathbf{U}, U_i$ | gas velocity (L/t) |
| $F_{drag}$ | drag force (F) | $U^*, v^*$ | dimensionless gas and particle velocities |
| $g$ | acceleration of gravity $(L/t^2)$ | $U_a$ | air velocity (L/t) |
| $G$ | cumulative mass distribution function | $U_{ave}$ | average velocity (L/t) |
| $g(D_{p,j})$ | mass fraction of particles in size range $j$ | $U_r, U_\theta$ | radial and tangential gas velocity (L/t) |
| $H$ | height of a duct, enclosure (L) | $U_0$ | characteristic velocity (L/t) |
| $J$ | constant defined by equation | $\mathbf{v}, v_i$ | particle velocity (L/t) |
| $k$ | Boltzmann constant (Q/T) | $|\mathbf{v} - \mathbf{U}|$ | relative velocity (L/t) |

| | | | |
|---|---|---|---|
| $\|\vec{v} - \vec{U}\|$ | relative velocity (L/t) | $\sigma, \sigma_g$ | standard deviation, geometric standard deviation |
| $V$ | volume (L$^3$) | | |
| $v_c$ | velocity of the collecting particle (L/t) | $\tau$ | relaxation time (t) |
| $v_c$ | cloud setting velocity (L/t) | $\chi$ | particle dynamic shape factor |
| $V_p$ | particle volume (L$^3$) | $\psi$ | Stokes number |
| $v_r, v_\theta$ | radial and tangential particle velocity (L/t) | | |

*Subscripts*

| | | | |
|---|---|
| $v_s(t)$ | instantaneous settling velocity (L/t) |
| $v_t$ | terminal (gravimetric, steady state) settling velocity (L/t) |
| $v_{t,c}$ | settling velocity of collecting particle (L/t) |
| $W$ | width of a duct (L) |
| $x, y, z$ | coordinates (L) |

| | |
|---|---|
| $(\cdot)_a$ | property of air, average property |
| $(\cdot)_{ave}$ | average |
| $(\cdot)_b$ | bleed air |
| $(\cdot)_c$ | critical property |
| $(\cdot)_d$ | drop property |
| $(\cdot)_e$ | exit condition |
| $(\cdot)_f$ | final condition, filter property, streaming flow property |
| $(\cdot)_g$ | geometric |
| $(\cdot)_i$ | initial condition, inlet condition |
| $(\cdot)_L$ | condition at distance $L$ |
| $(\cdot)_p$ | particle property |
| $(\cdot)_{p,j}$ | particle in size range $j$ |
| $(\cdot)_r$ | radial component |
| $(\cdot)_w$ | water, values relevant to inlet, width $2w$ |
| $(\cdot)_x, (\cdot)_y, (\cdot)_z, (\cdot)_i$ | vector components |
| $(\cdot)_0$ | initial or incoming properties, STP |
| $(\cdot)_\phi$ | component in angular direction |

*Greek*

| | |
|---|---|
| $\alpha$ | turning angle |
| $\beta$ | constant defined by equation |
| $\delta t$ | incremental time step (t) |
| $\eta$ | efficiency (mass removed/mass entering) |
| $\eta(D_p)$ | fractional or grade efficiency |
| $\theta$ | angle defined by figure |
| $\theta_c$ | critical angle in curvilinear flow |
| $\lambda$ | carrier gas mean free path |
| $\mu$ | viscosity (Ft/L$^2$) |
| $\upsilon$ | kinematic viscosity ($\mu/\rho$) (L$^2$/t) |
| $\pi$ | 3.14159 |
| $\rho, \rho_p, \rho_w$ | density of carrier gas, particle and water (M/L$^3$) |

*Abbreviations*

| | |
|---|---|
| Kn | Knudsen number |
| Re | Reynolds number |
| STP | standard temperature and pressure |

# References

Anand, N. K., and McFarland, A. R., 1989. Particle deposition in aerosol sampling lines caused by turbulent diffusion and gravimetric settling, *American Industrial Hygiene Association Journal*, Vol. 50, No. 6, pp. 307–312.

Balashazy, I., Martonen, T. B., and Hofmann W., 1990. Simultaneous sedimentation and impaction of aerosols in two-dimensional channel bends, *Aerosol Science and Technology*, Vol. 13, pp. 20–34.

Cadle, R. D., 1965. *Particle Size*, Reinhold, New York.

Cheng, Y. S., Yeh, H. C., and Allen, M. D., 1988. Dynamic shape factor of a plate-like particle, *Aerosol Science and Technology*, Vol. 8, pp. 109–123.

Crawford, M., 1976. *Air Pollution Control Theory*, McGrawHill, New York.

Davies, C. N. (Ed.), 1966. *Aerosol Science*, Academic Press, San Diego, CA.

Flagan, R. C., and Seinfeld, J. H., 1988. *Fundamentals of Air Pollution Engineering*, Prentice Hall, Upper Saddle River, NJ.

Fuchs, N. A., 1964. *The Mechanics of Aerosols*, Pergamon Press, Terrytown, NY.

Hidy, G. M., and Brock, J. R., 1970. *The Dynamics of Aerocolloidal Systems*, Pergamon Press, Terrytown, NY.

Hinds, W. C., 1982. *Aerosol Technology*, Wiley-Interscience, New York.

Jenning, S. G., 1988. The mean free path, *Journal of Aerosol Science*, Vol. 19, No. 2, pp. 159–166.

Lapple, C.E., 1961. "The Size of Common Aerosols" *Stanford Research Institute Journal*, Vol. 5, pp. 322–325.

Lee, R. E., Jr., 1972. The size of suspended particle matter in air, *Science*, Vol. 178, pp. 567–575.

Lee, C. T., and Leith, D., 1989. Drag force on agglomerated spheres in creeping flow, *Journal of Aerosol Science*, Vol. 20, No. 5, pp. 503–513.

Leith, D., 1987. Drag on nonspherical objects, *Aerosol Science and Technology*, Vol. 6, pp. 153–161.

Meng, Z., Dabdub, D., and Seinfeld, J. H., 1997. Chemical coupling between atmospheric ozone and particulate matter, *Science*, Vol. 277, July 4, pp. 116–119.

Ott, W. R., 1990. A physical explanation of the lognormality of pollutant concentrations, *Journal of the Air and Waste Management Association*, Vol. 40, pp. 1378–1383.

Phalen, R. F., Oldam, M. J., Mannix, R. C., and Schum, G. M., 1994. Cigarette smoke deposition in the tracheobronchial tree: evidence for colligative effects, *Aerosol Science and Technology*, Vol. 20, pp. 215–226.

Strauss, W., 1996. *Industrial Gas Cleaning*, Pergamon Press, Tarrytown, NY.

Waters, M. A., Selvin, S., and Pappaport, S. M., 1991. A measure of goodness-of-fit for the lognormal model applied to occupational exposures, *American Industrial Hygiene Association Journal*, Vol. 52, No. 11, pp. 493–502.

Whitby, K. T., Charlson, R. E., Wilson, W. E., and Stevens, R. K., 1974. The size of particle matter in air, *Science*, Vol. 183, pp. 1098–1100.

Whitby, K. T., and Sverdrup, G. M., "Advances in Environmental Science and Technology", Vol. 10, p. 477, 1980.

Willeke, K., and Baron, P. A., (Eds.), 1993. *Aerosol Measurement*, Van Nostrand Reinhold, New York.

# Problems

**11.1.** A puff of smoke from a wood-burning stove is initially 5 cm in diameter, $D_c(0)$, and contains a unit density aerosol $(D_p)$ 0.1 $\mu$m in diameter. Initially the particle number concentration, $n_c(0)$, is $10^{15}$ particles/m³. It is observed that the diameter of the smoke puff increases steadily at the rate of 1 cm/min. Assuming that no smoke particles escape, plot the cloud velocity versus time.

**11.2.** High-speed stamping machines used to manufacture small electrical devices require a fluid to be sprayed (as a mist) on the dies for lubrication and cooling. Unfortunately, an aerosol is formed around the machine that irritates the surface of the nasal cavity of the operator. From time to time the aerosol is analyzed and found to be lognormal. The results of a recent analysis are shown below. Compute the median particle size based on number and mass. The total particle concentration is 10,000 particles/cm³ and the density of the fluid is 1.2 g/cm³. Estimate the aerosol mass concentration (mg/m³).

| Range of $D_p$ ($\mu$m) | Number of particles/range |
|---|---|
| 0–2 | 0 |
| 2–8 | 0.0359 |
| 8–14 | 0.1999 |
| 14–20 | 0.2642 |
| 20–30 | 0.2940 |
| 30–45 | 0.1556 |
| 45–60 | 0.0365 |
| 60–80 | 0.0139 |

**11.3.** An aerosol produced during a bagging operation is log-normal with a mass median diameter of 10 $\mu$m and a geometric standard deviation of 2.5. What is the median particle size based on number? If the specific gravity of the dust is 2.5 and the number concentration is 100,000 particles/cm³, what is the mass concentration (mg/m³)?

**11.4.** A thin layer of finely ground cornstarch is placed on the surface of newly printed pages leaving a four-roll, color offset printing press. The purpose is to keep the pages from sticking together. Unfortunately, some of the cornstarch becomes airborne and settles on the rolls, ink reservoir, fountain (wetting) solution, and so on. Ultimately, it is transferred to the rolls and produces what printers humorously call "hickeys" on the printed page. A sample of the air above the press is passed through a filter and analyzed under the microscope. The number of particles (by number) within size intervals is as follows:

| Particle interval $D_p$ (μm) | Number of particles/interval |
|---|---|
| 2–3 | 10 |
| 3–4 | 0 |
| 4–5 | 0 |
| 5–6 | 30 |
| 6–7 | 100 |
| 7–8 | 210 |
| 8–9 | 450 |
| 9–10 | 910 |
| 10–11 | 850 |
| 11–12 | 800 |
| 12–13 | 440 |
| 13–14 | 230 |
| 14–15 | 90 |
| 15–16 | 30 |
| 16–17 | 20 |

Is the size distribution log-normal? What is the median size with respect to number and with respect to mass?

**11.5.** Paint particles from the exhaust of a spray booth in an autobody repair shop are sampled. The mass of particles per unit volume of air is found to be 2 g/m³ at STP. The particle size distribution is found to be:

| Particle size range (μm) | Number of particles |
|---|---|
| 2–3 | 1 |
| 3–4 | 1 |
| 4–5 | 1 |
| 5–6 | 3 |
| 6–7 | 10 |
| 7–8 | 21 |
| 8–9 | 45 |
| 9–10 | 91 |
| 10–11 | 85 |
| 11–12 | 80 |
| 12–13 | 44 |
| 13–14 | 23 |
| 14–15 | 9 |
| 15–16 | 3 |
| 16–17 | 2 |

**(a)** Draw a histogram.

**(b)** Compute the arithmetic mean particle diameter and standard deviation.

**(c)** Compute the geometric standard deviation and geometric mean particle diameter.

**(d)** Plot the actual cumulative size distribution based on diameter.

**(e)** Using the data in part (c), plot the cumulative size distribution assuming a log-normal size distribution.

**(f)** Compute the mass median diameter.

**(g)** If the particle density is 1800 kg/m³, how many particles are there per cubic meter of air ?

**11.6.** Particles (SG = 1.5) are generated by brick curing in a baking oven. The exhaust from the oven is discharged to the atmosphere. A sample of the exhaust gas is taken and the particle concentration is found to be 20,000 particles/cm³. A light-scattering instrument records the following size distribution:

| Class boundaries (μm) | Class midpoint (μm) | Particles per class |
|---|---|---|
| 0.30–0.40 | 0.35 | 6 |
| 0.40–0.53 | 0.46 | 15 |
| 0.53–0.71 | 0.62 | 41 |
| 0.71–0.94 | 0.82 | 88 |
| 0.94–1.28 | 1.10 | 150 |
| 1.28–1.68 | 1.50 | 180 |
| 1.68–2.33 | 2.00 | 205 |
| 2.33–3.00 | 2.66 | 156 |
| 3.00–4.00 | 3.50 | 91 |
| 4.00–5.30 | 4.60 | 44 |
| 5.30–7.10 | 6.20 | 18 |
| 7.10–9.40 | 8.20 | 5 |
| 9.40–12.80 | 11.00 | 1 |

**(a)** Test to see if the size distribution is log-normal by plotting the data on log-probability paper.

**(b)** Find:

**1.** The arithmetic mean diameter based on number

**2.** The geometric mean diameter based on number

**3.** The median particle diameter based on number

**4.** The median particle diameter based on mass

**(c)** What is the mass concentration of "fine" particles $\left( \text{i.e., } D_p < 2.5 \mu\text{m} \right)$.

**(d)** Find the particle concentration (mg/m³) in the process gas stream. Is the stream in compliance with the new source performance standards (0.05 g/dry ft³ at STP)?

**(e)** A particle removal system with the following fractional efficiency (based on mass) is suggested. What is the overall particle removal efficiency, and what is the exit mass concentration?

| $D_p$ (µm) | $\eta(D_p)$ |
|---|---|
| 0 | 0 |
| 0.5 | 0.10 |
| 1.0 | 0.19 |
| 2.0 | 0.32 |
| 3.0 | 0.42 |
| 4.0 | 0.52 |
| 5.0 | 0.60 |
| 6.0 | 0.68 |
| 7.0 | 0.76 |
| 8.0 | 0.82 |
| 9.0 | 0.87 |
| 10.0 | 0.92 |
| 12.0 | 0.97 |
| 14.0 | 0.99 |

**11.7.** The size distribution shown in Table 11-4 was obtained by drawing workplace air (5°C, 101 kPa) through a filter for 100 s at a volumetric flow rate of 0.15 L/min and then counting and sizing the particles on special filter paper. A total of 240 particles were collected and counted. If the particle density is 1800 kg/m³, compute the particle mass concentration (mg/m³) in the sampled air. The company management wishes to install a cyclone air cleaner to bring the workplace in compliance with the inert dust OSHA PEL of 15 mg/m³. The grade efficiency of the collection device is shown below. What is the particle concentration (mg/m³) leaving the device? Will it be in compliance with the OSHA PEL?

| $D_p$ (µm) | $\eta(D_p)$ (%) |
|---|---|
| 0 | 0 |
| 10 | 60 |
| 20 | 72 |
| 30 | 78 |
| 40 | 82 |
| 50 | 86 |
| 60 | 90 |
| 70 | 93 |
| 80 | 98 |

**11.8.** An aerosol has the size distribution (based on number) shown in Table 11-3. If the mass concentration of the entire aerosol is 100 mg/m³, what is the mass concentration between 5 µm $< D_p <$ 20 µm?

**11.9.** The aerosol described in Example 11.2 flows through a horizontal duct 23.3 m long and a 1 m $\times$ 1 m square cross-sectional area. The particle concentration entering the duct is 100 mg/m³. After 2000 h, what is the mass of particles that have settled on the floor of the duct?

**11.10.** A baghouse (Figure 1-7) contains 72 identical bags. The total volumetric flow rate is 7200 scfm. The volumetric flow rate into each bag is the same. The particle collection efficiency of each bag is 99%. Over time the cell plate corrodes and allows 10% of the inlet flow (720 scfm and particles) to bypass the bags and mix with the air leaving the bags.

(a) What is the new overall efficiency of the baghouse?

(b) If $k$ bags break, allowing all the dusty air to pass to the clean side of the baghouse, show that the rate of change of the overall collection efficiency ($\eta$) with respect to the number of broken bags ($k$) is given by

$$\frac{d\eta}{dk} = -\frac{\eta_b}{N_b}$$

**11.11.** A duct carries 25°C, 101 kPa air at 1 m³/s and water drops of 100 µm diameter. The duct makes a 90° turn upward (Figure 11-23) and some of the drops impact the outer elbow wall and are removed from the air stream. The dimensions of the duct are $R_1 = 0.5$ m, $R_2 = 1.0$ m, and width $W = 1.0$ m. The viscosity of air $\mu(25°C)$ is $1.8 \times 10^{-5}$ kg/m·s. Assuming well-mixed, irrotational flow, what percentage of the water drops will be removed?

**11.12.** A widely used experimental technique to study the velocity, temperature, and concentration in a moving gas stream uses a laser to excite small particles injected into the stream. By choosing unique particles and lasers, velocities can be studied (laser-Doppler velocimetry), or temperatures and concentrations can be determined by analyzing the emission spectra. It has been suggested that large particles can be used, but others believe that gravimetric settling will result in bogus information. You have been asked to study particle motion for the following conditions:

- Duct diameter $(D) = 0.10$ m
- Volumetric flow rate $(Q) = 0.07$ m³/min
- Initial particle velocity $\left[ v_x(0) = 10\, U_{avg} \right]$
- $T = 25°C$, $P = 1$ atm
- Particle density $(\rho_p) = 1200$ kg/m³

(a) If the air velocity is uniform and constant and $U_r = 0$, $U_\theta = 0$, plot a graph of $D_p$ versus $L/D$ for $D_p = 10$, 100, and 1000 µm, where $L$ is the horizontal distance a particle travels before it encounters the duct wall.

(b) Repeat part (a) if the flow is fully established; that is, the axial air velocity is

$$\frac{U_x(r)}{U_{ave}} = 2\left[1 - \left(\frac{r}{R}\right)^2\right] \qquad U_{ave} = \frac{Q}{A}$$

**(c)** A polydisperse aerosol is injected along the axis at a mass flow rate 10 mg/min, $v_x(0) = 10U_{ave}$. The mass median diameter is 15 μm and the geometric standard deviation is 1.2. If the air flow velocity is fully established as in part (b), plot the mass fraction of the deposited particles as a function of $L/D$.

**11.13.** A remote sensor monitors an aerosol in a foundry using electric arc furnaces and determines that:

- The particle size distribution is log-normal.
- The geometric mean diameter based on number is 20 μm.
- The geometric standard deviation is 1.8.
- The overall concentration is 1500 particles/cm³.

On the basis of mass $(\rho_p = 8000 \text{ kg/m}^3)$, what percent of the particles are greater than 40 μm in diameter?

**11.14.** The concentration of particles in a quarry is 25 mg/m³. The density of the particles is 1200 kg/m³. The size distribution is log-normal. The geometric standard deviation is 2.0 and the mass median diameter is 10 μm. What is the concentration of particles (mg/m³) between 5 and 25 μm?

**11.15.** Contaminant collectors 1 and 2 are arranged in series and remove smoke and fume from a stream of air $(Q_t)$ taken from the workplace. The efficiencies of both collectors are the same $(\eta_1 = \eta_2 0.60)$.
**(a)** What is the overall efficiency $(\eta_0)$ of the pair of units?
**(b)** The company wishes to increase the overall efficiency by 95% and plans to buy an additional collector $(A)$ with an efficiency $(\eta_a)$. Which configuration below requires the smallest amount of air to be diverted through the new unit $(Q_A)$, and what is the minimum $Q_a/Q_t$.

**1.** Units A and 1 in parallel followed by unit 2.
**2.** Unit 1 followed by units 2 and A arranged in parallel.
**3.** Unit A in parallel with units 1 and 2 arranged in series.

**11.16.** Two pollution control devices were installed in series many years ago. The collection efficiency of each device is 75% and is independent of the volumetric flow rate. Company management wishes to increase the overall collection efficiency to 95% by purchasing a new device with a collection of efficiency of 95%. The new device is to be installed in parallel with the two devices in series. What fraction of the flow should be diverted through the new device? Would it make any difference if the new device were placed in parallel with the first device?

**11.17.** Compute the terminal velocity of the following "particles" in air at STP.
**(a)** Baseball $(D_p = 9 \text{ cm}, \rho_p = 950 \text{ kg/m}^3)$
**(b)** Ping-Pong ball $(D_p = 3 \text{ cm}, \text{overall density} = 2 \text{ kg/m}^3)$
**(c)** Agricultural dust $(D_p = 5 \text{ μm}, \rho_p = 2 \text{kg/m}^3)$

**11.18. (a)** How long will it take the following particles to achieve 34% of their steady-state terminal settling velocity if they have an initial velocity (downward) of 100 m/s in quiescent air at −10°C and 0.9 atm.

**1.** $D_p = 1$ μm, $\rho_p = 980$ kg/m³
**2.** $D_p = 10$ μm, $\rho_p = 980$ kg/m³
**3.** $D_p = 100$ μm, $\rho_p = 980$ kg/m³
**4.** $D_p = 25$ μm, $\rho_p = 3000$ kg/m³

**(b)** If the particles in part (a) have an initial velocity in the horizontal direction of 100 m/s in quiescent air at $-0°C$ and 0.9 atm, what are the displacement and elapsed time before their velocities have decreased to 0.001 m/s?

**11.19.** Compute the penetration distance of a unit density sphere, $D_p = 500$ μm, injected into still air at STP with a velocity of 100 m/s.

**11.20.** A 1-mm sphere (density 8000 kg/m³) is injected into still air (1 atm, 25°C) with an initial velocity of 50 m/s inclined at 60° to the horizontal. How far will it travel in the horizontal direction? What maximum height will it attain?

**11.21.** A spherical particle $(D_p = 10 \text{ μm}, 1800 \text{ kg/m}^3)$ is injected into still air at STP in a horizontal direction with a velocity of 5 m/s. After 10 s, find the $x$ and $y$ displacement from the injection point.

**11.22.** A 6-mm water drop is given an initial downward velocity of 30 m/s into an airstream which is traveling uniformly to the right at 5 m/s. What is the steady-state vertical component of the particle velocity? How long does take for the drop to acquire a downward velocity of 15 m/s?

**11.23.** An aerosol consists of glass spheres $(\rho_p = 2724 \text{ kg/m}^3)$ in an upward-moving stream of hot air. The air temperature and pressure are 1000 K and 0.8 atm. For particles of the following diameters: 1.0, 10.0, and 100.0 μm.
**(a)** What is the gas viscosity (kg m$^{-1}$ s$^{-1}$)?
**(b)** What is the Cunningham correction factor $(C)$?
**(c)** What are the aerodynamic diameters of the particles?
**(d)** What gas velocity will levitate the particles?

**11.24.** A water drop 20 μm in diameter is traveling in an irrotational flow field where the irrotational constant $C(C = rU_\theta$, Eq. 11.123) is equal to 10 m²/s. If the tangential velocity of the particle is equal to the tangential air velocity, what is the radial velocity of the particle at a radius of 1.0 m?

**11.25.** Consider the gravimetric settling of particles on the floor of a long horizontal duct of square cross section through which an aerosol flows at a uniform velocity $U_0$. Assume that settling is governed by the well-mixed model. Define $R$ as

$$R = \frac{\text{local rate of settling at a distance } x \text{ from the inlet}}{\text{total inlet mass flow rate/duct cross-sectional area}}$$

Show that a graph of $R$ versus $v_t/U_0$ for constant values of $x/H$ have maximum values and that the maximum values of $R$ occur when $v_t/U_0 = H/X$.

**11.26.** An aerosol of concentration $c_0$ enters a horizontal duct of width $W$ and height $H$. The volumetric flow rate is $Q$. What is the total rate of deposition ($D$, kg/s) over the entire length of duct ($L$)?

**11.27.** An aerosol of concentration $c_0$ enters a horizontal duct of width $W$ and height $H$. The volumetric flow rate is $Q$. Show that the rate of deposition ($D$, kg/s) over the entire length of duct varies with respect to the length ($L$) as follows:

$$\frac{dD}{dL} = v_t c_0 W \exp\left(\frac{LWv_t}{Q}\right)$$

**11.28.** Air from an enclosure containing a foundry shake-out process is to be ducted to a baghouse to remove particles of molding sand. The dust concentration $c_0$ entering the 0.1 m $\times$ 0.1 m duct is 10 g/m$^3$. The horizontal duct is 50 m long.

**(a)** Foundry dust settles to the bottom of the duct and collects in a layer in which the bulk density ($\rho_b$) is 5000 kg/m$^3$. Assuming that the volumetric flow rate ($Q$) is constant and that settling proceeds according to the well-mixed model, derive an expression for the thickness ($h$) of the deposited dust layer as functions of time ($t$), distance downstream of the inlet ($x$), volumetric flow rate ($Q$), dust settling velocity ($v_t$) and duct width ($W$).

**(b)** Assuming that the fan maintains a constant volumetric flow rate ($Q = 0.1$ m$^3$/s), how long will it take to block one-half of the cross-sectional area ? Where will this occur ?

**(c)** Assuming that the fan maintains a constant volumetric flow rate ($Q = 0.1$ m$^3$/s), draw a graph of the friction head loss ($h_f$) versus time where

$$h_f = f\left(\frac{L}{W}\right)\frac{U^2}{2g} = \frac{f(L/W)Q^2}{2gA^2}$$

The term $f$ is the friction factor, which for a first approximation can be given by its laminar value, 64/Re, where Re is the Reynolds number:

$$\mathrm{Re} = \frac{\rho WU}{\mu} = \frac{4\rho Q}{\pi W \mu}$$

**(d)** Repeat part (c) but assume that the friction factor has to be taken from the Moody chart, which can be described by the empirical equation

$$\frac{1}{f^{1/2}} = -2.0\log_{10}\left(\frac{\epsilon/W}{3.7} + \frac{2.51}{\mathrm{Re}\,f^{1/2}}\right)$$

where $\varepsilon$ is the duct roughness factor (0.2 mm).

**(e)** Repeat part (d) but take into account that as dust fills the duct, the volumetric flow rate decreases in accordance with the fan operating curve:

$$h_f(\mathrm{m}) = h_p(\mathrm{m})$$

$$\int_0^L \frac{f}{W}\frac{U^2}{2g}\,dx = 75 - 52Q^2$$

where $Q$ has the units m$^3$/s. Draw a graph of volumetric flow rate versus time.

**11.29.** A settling chamber is to be constructed in the form of $N$ horizontal circular trays (see Figure 1-4c). The inner radius is $R_1$ and the outer radius is $R_2$. Air enters the center of the apparatus, divides equally between the $N$ trays, and flows radially outward. The distance separating the trays is $h$; the overall height of the unit is $H$.

**(a)** Assuming the well-mixed model, show that the fractional efficiency can be expressed as

$$\eta(D_p) = 1 - \exp\left[-\frac{g\rho_p D_p^2}{18Q\mu}\frac{H\pi}{h}(R_2^2 - R_1^2)\right]$$

**(b)** Assuming that the flow is reversed and travels radially inward, compute the fractional gravimetric collection efficiency.

**11.30.** A horizontal duct of length ($L$), width ($W$), and height ($H$) is used as a gravimetric settling device.

**(a)** Using the well-mixed model, draw curves of the fractional efficiency $\eta(D_p)$ versus particle diameter $D_p$ of unit density spheres for three volumetric flow rates.

$$\eta(D_p) = 1 - \frac{c(D_p)_a}{c(D_p)_0}$$

$$H = 0.5\,\mathrm{m} \qquad W = 1.0\,\mathrm{m} \qquad L = 10\,\mathrm{m}$$

$$Q = 1.0, 1.25, \text{ and } 1.5\,\mathrm{m}^3/\mathrm{s}$$

**(b)** An aerosol of unit density spheres enters the duct at an overall concentration of 100 mg/m$^3$ and volumetric flow rate of 1.25 m$^3$/s. The entering aerosol has a log-normal size distribution in which the mass median diameter is 30 $\mu$m and the geometric standard deviation is 1.3. What is the overall mass concentration (mg/m$^3$) leaving the duct and what is the mass median particle size (mm)?

**11.31.** Air carrying dust particles flows between two horizontal parallel plates separated by a distance $H$. The velocity profile is given by

$$\frac{U}{U_0} = 1 - 4\left(\frac{2z - H}{2H}\right)^2$$

where $z$ is measured from the bottom plate and $U_0$ is the velocity along the midplane. At $x = 0$ the particle concentration at all points between the plates is equal to $c_0$ and the particle velocity in the flow-wise direction is equal to the local gas velocity. Assume that gravimetric settling can be described as laminar and that at any downstream location the particle velocity in the flow-wise direction is equal to the local air velocity. Write a general expression for the collection efficiency and show that all the particles will be removed within a critical downstream distance $x_c$ given by

$$x_c = \frac{HU_0}{v_t}$$

**11.32.** The flow through a 90° elbow is irrotational and well mixed. Show that at the exit the concentration of particles of a particular diameter $c(D_p)$ varies with particle size as follows:

$$\frac{dc}{dD_p} = -c_0\beta D_p \exp\left(\frac{-\beta D_{p2}}{2}\right)$$

where

$$\beta = \frac{\pi Q \rho_p}{18\mu W\left(R_2 \ln R_2/R_1\right)^2}$$

**11.33.** [*Design Problem*] A duct (0.3 m high by 0.6 m wide) is placed on the roof of a building for makeup air to a newly renovated optics laboratory. Unfortunately, the installers ignored an unpaved access road directly beneath the side of the building where the duct inlet was located. Vehicles using the access road produce fugitive dust that enters the makeup air duct. The dust drawn into the duct has a concentration of 10 mg/m³, a density 1500 kg/m³ and a log-normal size distribution (mass median diameter of 20 μm and geometric standard deviation of 2.0). The duct is 50 m long and has a 90° elbow at the inlet ($r_1 = 1$ m). What is the concentration and size distribution of particles enter-

ing the laboratory if the average velocity in the duct is 10 m/s?

**11.34.** [*Design Problem*] Compute and plot the trajectories of coal dust particles traveling over an infinitely long mound of coal having a semicircular cross sectional area of radius ($a$). Assume that the velocity field around the mound can be described by the stream function

$$\psi = U_0 r\left[1 - \left(\frac{a}{r}\right)^2\right]\sin\theta$$

where the radius $r$ is measured from the center of the mound, the positive $x$-axis corresponds to a value of $\theta$ equal to zero. Air with a uniform velocity $(U_0)$ approaches the mound in the negative $x$-direction. Begin the analysis with particles at points along a vertical line $x = 1.5$, $y = h$, where the particle velocity $v_x = -1.5U_0$. Some coal dust particles will impact the mound and others will pass over it to settle to the ground downwind of the mound. Analyze the particle trajectories and determine where each particle impacts either the mound or the ground. Carry out the computations for particles originating at values of $h$ from 0 to $1.5a$.

**(a)** Plot the trajectory of a particle whose diameter is 50 μm.

**(b)** Plot a graph of $h/a$ versus the normalized displacement $x/a$ at which the particle impacts a solid surface. Repeat the calculation for several particle sizes. Assume the following physical conditions:

- $\rho_p = 1000$ kg/m³
- $U_0 = 1$ m/s
- $D_p = 20, 30, 50$, and 100 μm
- $a = 10$ m

**(c)** Repeat part (b) but omit the mound. Discuss the significance of the mound in enhancing the collection of particles.

**(d)** A polydisperse particle stream having a log-normal size distribution (mass median diameter 30 μm, geometric standard deviation $= 1.2$) enters the airstream at $x = 1.5a$, $y = a$ with an initial velocity $\vec{v}_p = -\hat{i}\, 1.5$ m/s. Plot the mass fraction versus $y$ at the point $x = -1.5a$.

# 12
# Capturing Particles

I n this chapter you will learn:

- To describe the physical phenomena underlying particle removal systems
- To model the performance of particle removal systems

## 12.1 Cyclone collectors

Equations 11.151 to 11.153 can be used to estimate the particle collection efficiency of a large class of particle collection devices called *cyclones*. Shown in Figures 12-1 and 12-2 are schematic diagrams of *straight-through* and *reverse-flow* cyclone collectors. Cyclone collectors can be made of inexpensive materials to capture high-temperature aerosols and liquid particles. Cyclones have no moving parts and may have a modest pressure drop. As a rule, cyclones do not remove very small particles, although they can be designed to do so. Traditionally, cyclones are used to remove coarse particles and find use inside coal-fired boilers,

inlets to military gas turbines for helicopters, and inlet air cleaners for diesel engines of earth moving equipment. Shown in Figures 1-5 and 1-6 are reverse-flow and straight-through cyclones arranged in parallel within a collector so that large volumetric flow rates can be cleaned. Cyclones are also used in powder technology, in which case one is interested in increasing the concentration of particles in an airstream. Cyclones used for this purpose are called *concentrators*.

It is necessary to bleed air from straight-through cyclones to remove the particles that migrate radially outward. If used as a concentrator, bleed air transports particles downstream to the next step in a process. If used as an air cleaning device, bleed air reduces the amount of clean air available for use. If a reverse-flow cyclone is mounted with its conical hopper beneath the outlet, it is not necessary to bleed air to remove particles. Particles migrating radially outward strike the cyclone walls and slide downward into the hopper, where they can be removed later as a batch process or removed continually by hopper valves. In either event, only a minimal amount of bleed air is required.

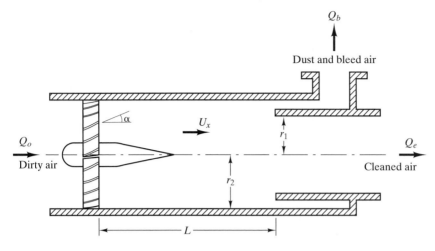

**Figure 12-1** Straight-through cyclone, bleed air is required to remove particles

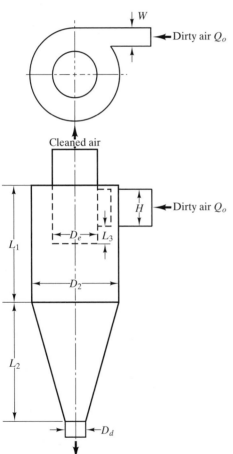

**Figure 12-2** Standard reverse-flow cyclone with tangential entry ($L_1 = 2D_2$, $L_2 = 2D_2$, $H = D_2/2$, $W = D_2/4$, $D_e = D_2/2$, $L_3 = D_2/8$, $D_d$ = arbitrary but usually $D_2/4$)

Consider the straight-through cyclone of length $L$ shown in Figure 12-1. The helical turning vanes (angle $\alpha$) impart rotation to the air such that

$$\tan \alpha = \frac{U_\theta}{U_x} \qquad (12.1)$$

Flow through the cyclone can be conceived as a wrapped-around bend of total angular displacement, $\theta_t$ (radians), where

$$\theta_t = \frac{U_\theta(r_2)}{r_2} \frac{L}{U_x} = \frac{L}{r_2} \tan \alpha \qquad (12.2)$$

and $r_2$ is the outer radius of the straight-through cyclone. The *number of turns* $(N_e)$ is equal to $\theta_t/2\pi$. The width of the bend $(W)$ can be scaled as follows:

$$W = \frac{L}{N_e} = \frac{U_x 2\pi r_2}{U_\theta(r_2)} = \frac{2\pi r_2}{\tan \alpha} \qquad (12.3)$$

Flow through cyclones is turbulent but neither irrotational nor of constant vorticity. A realistic estimate is difficult to make, but Strauss (1966) assumed the flow to be $n$-degree rotational, with $n$ equal to 0.5 for the reverse-flow cyclone. Assuming that $n = 0.5$ can also be used for straight-through cyclones and that an overall collection efficiency for straight-through cyclones can be given by Equations 11.154 and 11.156, we obtain

$$\eta = 1 - \exp - \left[ \frac{L(\tan \alpha)^2 \tau Q}{8\pi r_2^3 (r_2 + r_1 - 2\sqrt{r_1 r_2})} \right] \quad (12.4)$$

where $r_1$ and $r_2$ are the inner and outer radii of the outlet annulus.

A reverse-flow cyclone in Figure 12-2 can also be conceived as a wrapped-around bend of angle θ where θ is obtained from Equation 12.3. The width (*W*) of the bend can be approximated as the width of the inlet duct. The overall collection efficiency can be expressed by Equation 11.153 with *K* given by Equations 11.154 to 11.156 with $r_2$ replaced by the radius of the exit duct $(D_e/2)$.

Reverse cyclones have been used for decades in series with other collectors of higher efficiency. Placed upstream of high-efficiency filters or electrostatic precipitators, cyclones reduce the particle mass concentration by removing large particles and allow the high-efficiency collector to remove smaller particles. With the advent of *computational fluid dynamics* (CFD) modeling programs, which accurately predict particle trajectories inside cyclones, it is believed that the grade efficiency of cyclones can be improved. Figure 12-3 illustrates the capabilities of CFD to predict the velocity field of air in a particular reverse flow cyclone. Since CFD methods improve steadily with time, there is reason to be optimistic that the performance of cyclones to capture smaller particles will be improved.

Shown in Figure 12-2 is what is called a *standard reverse-flow cyclone*. The configuration is called standard, because the dimensions of the collector are scaled to the outer diameter $D_2$ (i.e., large and small collectors are geometrically similar). Over the years, many researchers have proposed different scale values, yet they are all remarkably alike. The choice of the scale factors and the grade efficiency for these configurations have received a great deal of attention, see for example, (Crawford, 1976; Fuchs, 1964; Strauss, 1966; Licht, 1980; Danielson, 1973; Boysan et al., 1982; Moore and McFarland, 1996). The grade efficiency of these configurations is expressed in forms similar to Equations 11.151 to 11.153. Calvert and Englund (1984) summarize the scale values and performance equations of several of the standard designs.

For simplicity, the *Lapple standard cyclone* will be used in this text as a representative example of the configurations presented by Calvert and Englund (1984). The scale factors of the Lapple cyclone are:

$$L_1 = L_2 = 2D_2 \qquad W = L_3 = D_d = \frac{D_2}{4}$$

$$\text{(12.5)}$$

$$H = D_e = \frac{D_2}{2} \qquad N_e = \frac{1}{H}\left(L_1 + \frac{L_2}{2}\right)$$

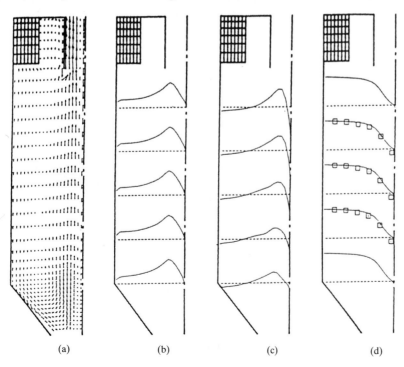

(a)      (b)      (c)      (d)

**Figure 12-3** Predicted velocity and pressure distribution in a reverse-flow cyclone; (a) maximum component of the velocity, (b) tangential velocity, (c) axial velocity, (d) pressure distribution with experimental data. (Boysan, et al., 1982)

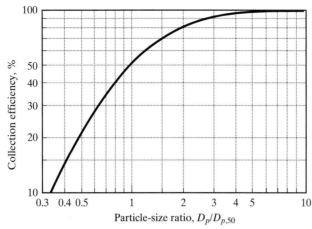

**Figure 12-4** Dimensions and collection efficiency of a standard reverse-flow cyclone (adapted from Perry and Chilton, 1973)

An attraction for Lapple standard cyclones is that their performance can be described by a single *grade (fractional) efficiency* curve shown in Figure 12-4. The dimensionless abscissa is the actual particle diameter $(D_p)$ divided by the *cut diameter*, which is defined as the diameter of a particle that can be removed with a collection efficiency of 50%. The value of the cut diameter is given by

$$D_{p,50} = \sqrt{\frac{9\mu H W^2}{2\pi N_e Q \rho_p}} \qquad (12.6)$$

An acceptable pressure drop is generally less than 20 cm of water.

## 12.1.1 Applications

The pressure drop across a cyclone is proportional to the square of the volumetric flow rate. Calvert and Englund (1984) suggest an equation for the pressure drop of the Lapple standard cyclone, which reduces to

$$\Delta P(\text{cm H}_2\text{O}) = 40.96\,\rho\left(\frac{Q}{WH}\right)^2 \qquad (12.7)$$

where $\rho$ is the actual gas density $(\text{g/cm}^3)$, $Q$ is the actual volumetric flow rate $(\text{m}^3/\text{s})$, and $W, H$ are the inlet dimensions (m).

The removal of particles in a cyclone requires maximizing a particle's radial velocity (i.e., $v_r = U_\theta^2/r$). To achieve high removal efficiency, engineers must either maximize the gas volumetric flow rate $(Q)$ through the cyclone or reduce the diameter $(D_2)$ of the cyclone. Since the energy loss associated with the pressure drop varies with the cube of the gas volumetric flow rate $(Q)$, maximizing the flow rate can be quite costly. Reducing the cyclone diameter yet coping with a large volumetric flow rate requires using many small cyclones arranged in parallel (Figure 1-5). The latter is the most economical course of action.

A second important use of inertial separation is the removal of liquid droplets from a process gas steam. *Mist eliminators* or *demisters* are the names given to the class of collectors that achieve this function. Mist eliminators contain a labyrinth of passages producing curvilinear flow that causes liquid drops to acquire a radial velocity. The drops impact a solid surface, coalesce and form a liquid film that is removed by other means. Mist eliminators are used in all wet scrubbers (Figure 1-9) because the scrubbing fluid contains undesirable material that should not be emitted to the atmosphere. There are a large variety of mist eliminators (Figures 12-5 and 12-6) but all have the common feature of causing the liquid droplets to impact a solid surface, thus removing them from the gas stream.

**EXAMPLE 12.1   FRACTIONAL EFFICIENCY OF A REVERSE-FLOW CYCLONE**

Compute and plot the fractional efficiency of a standard reverse-flow cyclone that is 2 cm in diameter $(D_2)$ to collect unit density $(\rho_p = 1000\,\text{kg/m}^3)$ spherical particles as a function of the volumetric flow rate at STP, $1 \times 10^{-7}$ to $5 \times 10^{-7}$ $\text{m}^3/\text{s}$ at STP.

$D_2 = 0.02\,\text{m} \quad H = 0.01\,\text{m}$

$W = 0.005\,\text{m} \quad \mu(\text{STP}) = 1.81 \times 10^{-5}\,\text{kg m}^{-1}\,\text{s}^{-1}$

**Figure 12-5**   FLEXIMESH ® mist eliminators (figure used with the permission of Koch Engineering Co. Inc.)

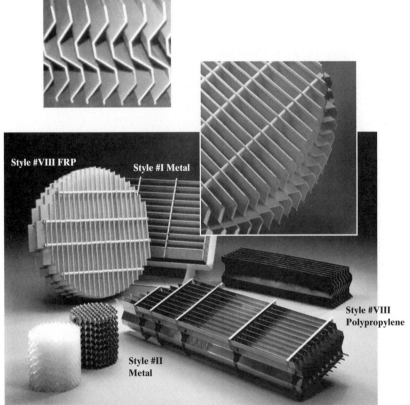

**Figure 12-6** FLEXICHEVRON ® mist eliminators (figure used with the permission of Koch Engineering Co. Inc.)

**Solution** The fractional efficiency will be based on Equation 12.5 and 12.6 and Figure 12-4,

$$N_e = \frac{1}{H}\left[L_1 + \frac{L_2}{2}\right] = \frac{2}{D_2}[2D_2 + D_2] = 6$$

$$D_{p,50} = \left[\left(\frac{9}{2\pi N_e}\right)\frac{\mu H W^2}{Q\rho_p}\right]^{1/2}$$

$$= \left[\frac{9}{12\pi}\frac{(1.8\times10^{-5})(\text{kg m}^{-1}\text{ s}^{-1})(0.01)(0.005)^2(\text{m}^3)}{Q(\text{m}^3/\text{s})1000\ (\text{kg/m}^3)}\right]^{1/2}$$

The fractional efficiency is shown graphically in Figure 12-4, but to use the curve analytically it will be necessary to express it by an empirical equation. The following form is selected:

$$\eta = ax^2 + bx + \frac{c}{1+x}$$

where $x = D_p/D_{p,50}$. The coefficients $a$, $b$, and $c$ can be found by empirical curve-fitting programs in commercially available mathematical computational programs ($a = -0.067$, $b = 0.531$, $c = -0.025$). Once the coefficients are determined, the equation can be used to generate the fractional efficiency curve for the cyclone described above. Figure E12-1 is the curve generated by this process. The figure shows that the fractional efficiency is strongly dependent on the volumetric flow rate. The Mathcad program producing Figure E12-1 can be found in the textbook's Web site on the Prentice Hall Web page.

---

## 12.1.2 Cascade impactors

To size and select particle collectors, it is necessary that engineers describe the size distribution statistically using the techniques described in Section 11.2.3. Ultimately, however, one needs to classify particles on the basis of mass, since compliance with air pollution regulations is based on the mass emitted by a process. The mass distribution can be calculated from the size distribution based on number using Equation 11.31.

A simple experimental method to measure the size distribution based on mass is the *cascade impactor* shown schematically in Figure 12-7. The device consists of consecutive stages in which the aerosol impacts a series of horizontal surfaces that hold filter paper. A sample of the aerosol is drawn into the device by a vacuum pump at a prescribed volumetric flow rate $Q$ for a measured time $t$. Stage (i) removes all particles in the range $(D_{p,\text{high}}$ and $D_{p,\text{low}})_i$ and stage (i + 1) removes smaller particles over a range $(D_{p,\text{high}}$ and $D_{p,\text{low}})_{i+1}$. The selective removal continues until at the last stage the remaining particles are removed by a filter. Figure 12-8 shows the removal efficiency of each stage

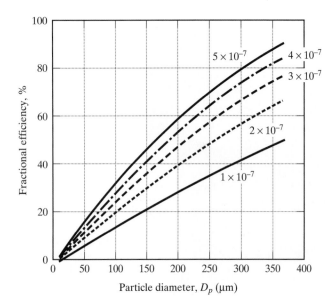

**Figure E12-1** Fractional efficiency of a standard reverse-flow cyclone versus particle diameter for various volumetric flow rates

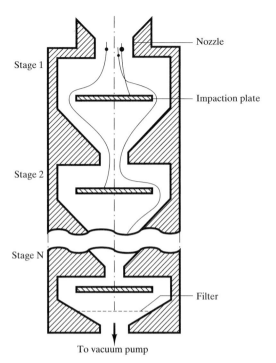

**Figure 12-7** Schematic diagram of a cascade impactor showing the trajectories of particles having two different diameters (redrawn from Willeke and Baron, 1993)

of a cascade impactor (Marple et al., 1993) designed to separate particles over the range, 0.1 $\mu$m < $D_p$ < 10 $\mu$m. After the impactor is taken apart, the mass of particles collected on each stage is determined and a table similar to Table 11-3 is con-

structed except that the discriminating parameter is the mass of particles on each stage. The mass fraction and cumulative mass fraction for each stage is computed. The data can also be plotted on log-probability graph paper to test for log-normality. The mass concentration of the entire aerosol is equal to total mass collected on all the stages divided by the total volume of the sampled gas, $Qt$. The data yield the mass distribution, which must be combined with the collector fractional (grade) efficiency to compute the overall collection efficiency via the equations of Section 11.3.

The cascade impactor is a rugged instrument and can be used to measure the size distribution of ambient aerosols or aerosols in process gas streams for various temperatures, pressures, and velocity profiles. The accuracy of the measurement requires obtaining an aerosol sample that is same as that which exists in the process gas stream. Obtaining a representative sample is ensured if:

**1.** The velocity of the aerosol entering the sampling probe is equal to the velocity of the process gas stream.
**2.** The gas velocity is perpendicular to the plane of the sampling probe.

Criterion 1 is called *isokinentic sampling* and criterion 2 is called *isoaxial sampling*. Both are illustrated in Figure 12-9.

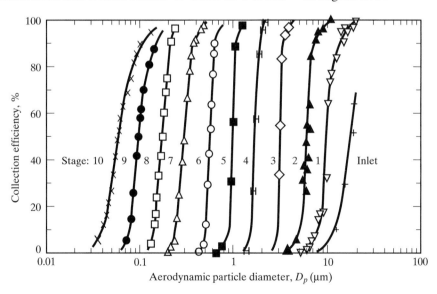

**Figure 12-8** Particle collection efficiency for each stage of a micro-orifice uniform deposit cascade impactor.(Marple, et al., 1991)

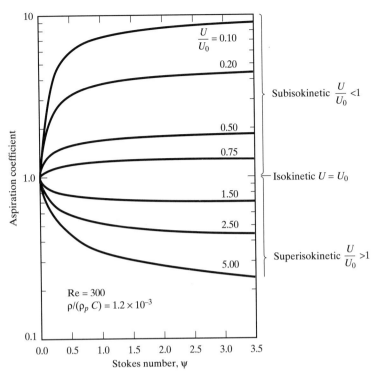

**Figure 12-11** Aspiration coefficienet $(c/c_0)$ versus Stokes number and several different velocity ratios $(U/U_0)$ (Heinsohn, 1991)

where

$$R_e = \frac{\rho D_i U_0}{\mu}$$

$$\psi = \frac{U_0 \tau}{D_i}$$

$D_i = $ diameter of sampling probe

$\tau = $ relaxation time

Equation 12.12

*Interception* accounts for the fact that even though the particle's center of gravity may not collide with a collecting surface, collision still occurs if the distance between the path of the center of gravity and surface is less than the particle's radius. *Diffusion* accounts for the fact that in addition to inertial effects, particles exhibit Brownian movement and very small particles will migrate toward a collecting surface and be removed from the gas stream. In a subsequent section a spherical particle's diffusion coefficient will be shown to be inversely proportional to its diameter such that above 1 μm, diffusion is negligible. For convenience, all three processes (impaction, interception, and diffusion) occur simultaneously and hereafter will be subsumed under the phrase *impaction*.

Impaction occurs between small particles and large bodies as long as there is relative motion between the two, for example, raindrops falling through quiescent dusty air, a high-speed aerosol passing through slower-moving water droplets in a venturi scrubber, and so on. For generality, consider the impaction between a small particle $(D_p)$ possessing the velocity of the carrier gas and a large collecting body $(D_c)$ moving through the gas stream at a different velocity (Figure 12-12). Assume both bodies $(D_p$ and $D_c)$ are spheres. Throughout this section the term *small particle* refers to the small contaminant particle $(D_p)$ that is to be removed by the larger *collecting particle* of scrubbing liquid denoted by $D_c$.

It will be assumed that if a small particle impacts a large particle, it will stick and be removed from the gas stream. The consequent growth in size and mass of the collecting particle will be neglected. The effectiveness of the removal process is expressed as the *single-drop collection* or *particle collection efficiency* $(\eta_d)$, defined as the rate (by mass) with which particles are removed by impacting with the collecting drop (sphere $D_c$) divided by the mass rate of flow of particles in a stream tube of cross-sectional area equal to that of the collecting sphere:

$$\eta_d = \frac{\text{mass removal rate}}{c_0 v_r (\pi D_c^2 / 4)} \qquad (12.8)$$

where $v_r$ is the speed of the gas stream relative to the collector and $c_0$ is the concentration of contaminant particles in the gas stream approaching the collector (Figure 12-13). If the collecting particle is at rest, the relative velocity, $v_r$, is equal to the gas velocity, $U_0$. Figure 12-14 shows the stream tube (radius $r_1$) defined by particles that collide with the collector

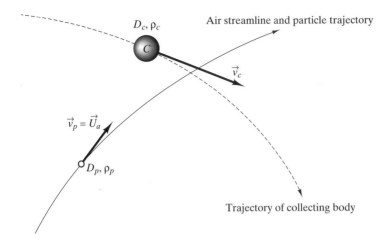

**Figure 12-12**   Impaction between a small particle and a collector particle (Heinsohn, 1991)

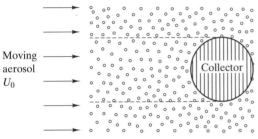

**Figure 12-13**   Stream tube used to define the single-particle, single-fiber collection efficiency (Heinsohn, 1991)

because their center of gravity *impacts* the collector or because the outer edge of the particle collides with (*intercepts*) the collector. The single-particle, sphere-on-sphere removal efficiency is the percent of the particles in the stream tube defined by the collector (radius $R_c$) that are removed by the processes shown in Figure 12-14:

$$\eta_d = \frac{c_0 \pi r_I^2 U_0}{c_0 \pi R_c^2 U_0} = \left(\frac{r_I}{R_c}\right)^2 \qquad (12.9)$$

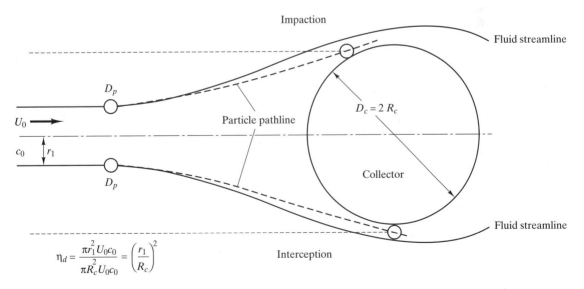

**Figure 12-14**   Sphere-on-sphere, single particle collection efficiency including both impaction and interception

Similar expressions can be defined for impaction of spheres on cylinders or any pair of impacting bodies.

Calvert and Englund (1984) recommend that for potential flow and viscous flow in which the Stokes numbers exceed 0.2, the single particle collection efficiency of *spheres impacting on spheres* can be approximated by

$$\eta_d = \left(\frac{\psi}{\psi + 0.7}\right)^2 \qquad (12.10)$$

where

$$\psi = \frac{\tau_p |\mathbf{v}_r|}{D_c} \qquad (12.11)$$

$$\tau_p = \frac{\rho_p D_p^2}{18\mu} \qquad (12.12)$$

The velocity of particle relative to the collector,

$$\mathbf{v}_r = \mathbf{v}_p - \mathbf{v}_c \qquad (12.13)$$

where $\mathbf{v}_p$ is the velocity of the particle and $\mathbf{v}_c$ is the velocity of the collector. A graph of the single-particle collection efficiency for spheres impacting on spheres shows that the $\eta_d$ is zero at $\psi = 0.083$ and approaches unity asymptotically for large values of $\psi$.

## EXAMPLE 12.2 SINGLE-DROP COLLECTION EFFICIENCY OF RAINDROPS FALLING THROUGH DUSTY AIR

Consider raindrops falling through quiescent dusty air at STP. Compute and plot the single-particle collection efficiency $(\eta_d)$ as a function of the diameter of the drop $(D_c)$ and a range of dust particles $(D_p)$.

*Solution* Equation 12.10 predicts the single-particle collection efficiency as a function of the Stokes number. To eliminate confusion, the water drop will be called the collector $(D_c, v_c)$ and the dust will be called the particle $(D_p)$. Since the dust particle is motionless, the relative velocity is equal to the settling velocity of the rain drop $(v_c)$, which in turn varies with its diameter $(D_c)$ (i.e., Equations 11.93 to 11.95). The relaxation time $(\tau_p)$ is a function of the dust particle diameter $(D_p)$ (Equation 12.12). To compute the efficiency, calculate the drop settling velocity $(v_c)$, for particular $D_c$ and then the efficiency for a particular value of $D_p$. Repeat the process for various combinations of $D_c$ and $D_p$. Figure E12-2 shows the fractional efficiency of a raindrop falling through dusty air. The Mathcad program producing Figure E12-2 can be found in the textbook's Web site on the Prentice Hall Web page.

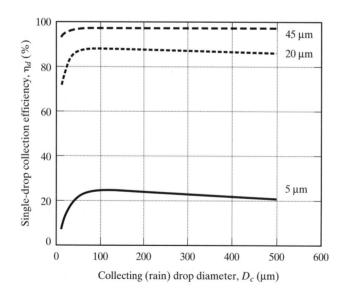

**Figure E12-2** Ability of rain to remove airborne dust. Single-drop collection efficiency of various size rain drops (μm) falling freely through still air containing dust particles, $D_p = 5$, 20 and 45 μm, $\rho_c = \rho_p = 1{,}000$ kg/m³

Readers may be surprised to see the efficiency reach a maximum and then fall for large drop diameters. The explanation turns on how the settling velocity varies with drop diameter. For small drop diameters, the Reynolds numbers are small, Stokes flow exists, and the settling velocity increases with the square of the drop diameter. Thus the Stokes number increases linearly with $D_c$ and the efficiency rises with $D_c$. For large values of $D_c$, the Reynolds numbers are large and the drag coefficient approaches a constant (0.4). Thus the settling velocity increases with the square root of the drop diameter $D_c$, and the Stokes numbers are inversely proportional to the square root of the drop diameter $D_c$. Thus as $D_c$ increases, the Stokes numbers fall and the efficiency falls as well.

## 12.2.1  Spray chambers

To illustrate impaction as a method for control of small particles, consider the impaction between a falling stream of (collecting) water drops $\left(D_c\right)$ and an ascending stream of small particles $\left(D_p\right)$. In nature this occurs when falling rain drops impact dust suspended in the air and the process is called *scavenging* or *washout*. Two industrial processes in which scavenging is used are (1) a spray chamber to remove small particles of cutting fluid generated by a high-speed punch and (2) a water spray

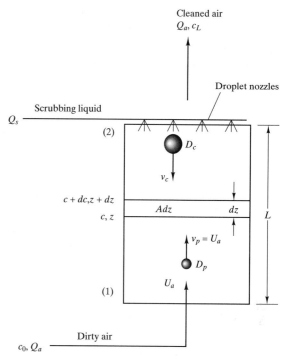

**Figure 12-16**   Schematic diagram of a spray chamber illustrating an infinitesimal well-mixed volume (Heinsohn, 1991)

to control dust generated by a conveyor that discharges material to a stockpile. The overall effectiveness of a *counterflow spray chamber* (Figures 12-15 and 12-16) can be modeled using an

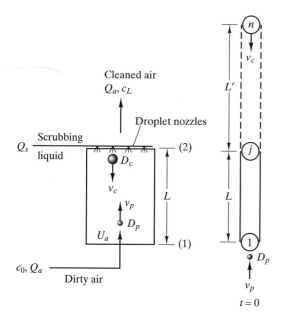

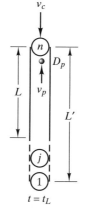

**Figure 12-15**   Schematic diagram of a spray chamber illustrating the number of encounters between the particles and the falling drops (Heinsohn, 1991)

approach suggested by Crawford (1976) and Calvert (1972). Assume the following:

- The height of the control region is $L$. The number of liquid drops per unit volume, the diameter of the drops $(D_c)$, and the absolute velocity $(\mathbf{v}_c)$ of the falling drops are constant.
- The diameter of the small particles $(D_p)$ is constant.
- The concentration of collecting droplets $(n_c)$ is constant, the particle concentration varies only with $z$ [i.e., $n(z)$], and the number of encounters between particles and collecting drops is uniform.
- The volumetric flow rate of air $(Q_a)$ and spray liquid $(Q_s)$ are constant.
- Gravimetric settling of the small particles is negligible; thus it is assumed that the velocity of the small particles $(\mathbf{v}_p)$ is equal to the upward constant air velocity $(\mathbf{U}_a)$.

$$\mathbf{v}_p = \mathbf{U}_a \qquad (12.14)$$

- The velocity of impaction is the velocity of the particle relative to the collector:

$$\mathbf{v}(\text{impaction}) = \mathbf{v}_p - \mathbf{v}_c = \mathbf{U}_a - \mathbf{v}_c \qquad (12.15)$$

Because the air and particles flow upward and the drops fall downward, the impaction velocity is the sum of the airspeed and the drop speed, which in turn is equal to the settling velocity $(v_{t,c})$ of the collector particle in quiescent air:

$$v_{t,c} = U_a + v_c \qquad (12.16)$$

The number of collecting drops per unit volume of air $(n_c)$ is easy to understand but is a troublesome parameter to determine. The mass flow rate of the scrubbing liquid is a parameter controlled by the engineer and can be expressed by the continuity equation

$$\dot{m}_c = \bar{\rho}_c \mathbf{A} \cdot \mathbf{v}_c \qquad (12.17)$$

where $\bar{\rho}_c$ is a bulk density, i.e. the mass of drops per volume of duct, $\mathbf{v}_c$ is the velocity of the collecting drop, $\mathbf{A}$ is a cross-sectional area:

$$\mathbf{A} \cdot \mathbf{v}_c = A_i v_{c,i} = A_x v_{c,x} + A_y v_{c,y} + A_z v_{c,z} \qquad (12.18)$$

$v_{c,i}$ is the component of the collecting drop velocity in the $i$th direction and $A_i$ is the component of the duct area whose unit normal in *the outward direction*

is defined by the direction of flow. The bulk density can be constructed from the following:

$$\bar{\rho}_c = \frac{\text{mass of drops}}{\text{volume of duct}}$$

$$= (\text{mass of drop/volume of a drop})(\text{volume of a drop/drop})(\text{number of drops/volume of duct})$$

$$= \rho_c \left( \frac{\pi D_c^3}{6} \right) n_c \qquad (12.19)$$

The volumetric flow rate of the scrubbing liquid $(Q_s)$ is equal to the mass flow rate of the collecting liquid divided by the density of the collecting liquid $(\rho_c)$:

$$\frac{\dot{m}_c}{\rho_c} = Q_s = \frac{\pi D_c^3}{6} n_c A_i v_{c,i} \qquad (12.20)$$

For the spray chamber the collector velocity is $v_{c,i}$, the collecting drop velocity is $v_c$, and the area $A_{c,i}$ is the cross-sectional area of the chamber $(A)$:

$$Q_s = n_c \frac{\pi D_c^3}{6} v_c A$$

$$n_c = \frac{6 Q_s}{v_c A \pi D_c^3} \qquad (12.21)$$

Multiplying and dividing by the volumetric flow rate of air $(Q_a)$ where, $Q_a = U_a A$

$$n_c = \frac{Q_s}{Q_a} \frac{6}{\pi D_c^3} \frac{U_a A}{v_c A}$$

$$n_c = \frac{Q_s}{Q_a} \frac{U_a}{v_c} \frac{6}{\pi D_c^3}$$

$$n_c = \frac{Q_s}{Q_a} \frac{6}{\pi D_c^3} \frac{U_a}{v_{t,c} - U_a} \qquad (12.22)$$

Since the only spatial variable is the height, drops being collected can be imagined to travel one behind the other in a column of height $(L + L')$. Impaction occurs as small particles rise through a height $L$ and encounter $n$ drops. The single-drop impaction efficiency $(\eta_d)$ is constant since the velocities are constant and can be calculated from Equation 12.10. The parameter $n$ is the number of times that small particles encounter collector drops as the small particles travel upward through a height $L$. The overall removal of small particles can be imagined as the flow of an aerosol through a stream tube containing $n$ collectors in series, each with a single-particle collection efficiency $(\eta_i)$. Thus the overall collection efficiency for the column can be constructed as follows:

$$\eta = 1 - \frac{c_L}{c_0} = 1 - \frac{c_L}{c_n}\frac{c_n}{c_{n-1}}\cdots\frac{c_3}{c_2}\frac{c_2}{c_1}\frac{c_1}{c_0} \qquad (12.23)$$

where $n$ is the number of collectors (number of encounters). Each of the ratios, $c_2/c_1$, and so on, can be expressed in terms of the single-particle collection efficiency $\eta_d$. Thus

$$\eta = 1 - \frac{c_L}{c_0} = 1 - (1 - \eta_d)^n \qquad (12.24)$$

where $c_L$ and $c_0$ are the small-particle concentrations at the exit and entrance of the chamber.

The total number of encounters can be deduced using Figure 12-15. It must be kept in mind that as the small particles travel upward, $n$ drops travel downward, such that at the instant a particle has risen through the height $L$, drop number $n$ is being encountered. The total number of encounters is

$$n = n_c \frac{\pi D_c^2}{4}(L + L')$$

$$\qquad (12.25)$$

$$n = \frac{Q_s}{Q_a}\frac{6}{\pi D_c^3}\frac{U_a}{v_c}\frac{U_a}{v_c}\frac{\pi D_c^2}{4}(v_c t_L + U_a t_L)$$

where $t_L$ is the time it takes for the small particles to rise a height $L$ which is also the time it takes for drop number n to fall through a height $L'$.

$$t_L = \frac{L}{U_a} = \frac{L'}{v_c} \qquad (12.26)$$

Thus

$$n = \left(\frac{3}{2}\right)\frac{Q_s}{Q_a}\frac{U_a}{v_c}\frac{t_L}{D_c}(v_c + U_a)$$

$$\qquad (12.27)$$

$$n = \left(\frac{3}{2}\right)\frac{Q_s}{Q_a}\frac{v_{t,c}}{v_c}\frac{L}{D_c}$$

The overall collection efficiency of spray chambers is inherently low. Since the collecting drops fall by gravity alone, the relative velocity between falling collector drops and rising small particles is low which in turn produces a low single particle collection efficiency.

An alternative model of a vertical (counterflow) spray chamber can be formulated. A schematic diagram of a vertical spray chamber of height $L$ is shown in Figure 12-16. It will be assumed that the particle concentration varies only in the vertical direction ($z$-direction) and that there are no variations in the transverse direction. Consider a differential volume $A\,dz$. The conservation of mass for particles can be written as

$$cAU_a = (c + dc)AU_a$$

$$\qquad (12.28)$$

$$+ \eta_d c(v_c + U_a)n_c \frac{\pi D_c^2}{4}A\,dz$$

where $n_c$ is the number of collection drops per volume of the carrier gas. The quantity $n_c$ is given by Equation 12.21. Combine Equations 12.21 and 12.28 and simplify:

$$\frac{dc}{c} = -\eta_d \frac{v_c + U_a}{v_c}\left(\frac{3}{2}\right)\frac{Q_s}{Q_a}\frac{dz}{D_c} \qquad (12.29)$$

where $Q_s$ and $Q_a$ are the volumetric flow rates of the scrubbing fluid and carrier gas. The ratio $Q_s/Q_a$ is called by a number of names, *liquid-to-gas ratio* and/or the *reflux ratio*. The settling velocity of the collecting drop is sufficiently large to overcome the upcoming gas velocity. Thus the absolute velocity of the collecting drop $(v_c)$ plus the carrier gas velocity $(U_a)$ is equal to the settling velocity $(v_{t,c})$ of the collecting drop:

$$v_{t,c} = v_c + U_a \qquad (12.30)$$

Integrating Equation 12.29 over the height of the spray tower $(L)$, one obtains the following expression for the overall collection efficiency of a vertical, counterflow spray chamber:

$$\eta = 1 - \exp\left[-\eta_d \frac{v_{t,c}}{v_c}\left(\frac{3}{2}\right)\frac{Q_s}{Q_a}\frac{L}{D_c}\right] \qquad (12.31)$$

Equations 12.31 and 12.24 predict the overall collection efficiency of a vertical spray chamber. Differences between the two expressions can be resolved if both the exponential term in Equation 12.31 and the power term in Equation 12.24 are expanded in a Taylor series. Since the number of collecting drops $(n)$ over the height of the tower $(L)$ is large, it can be shown that both expressions yield similar numerical results. If the model introduced at the beginning of this section is repeated but the encounters between particles and collecting droplets are described by a Poisson distribution, an alternative form of Equation 12.24 will emerge in an exponential form that is compatible with Equation 12.31.

## EXAMPLE 12.3 FRACTIONAL EFFICIENCY OF A COUNTERFLOW SPRAY CHAMBER

Compute and plot the fractional efficiency of a counterflow spray chamber that removes unit density dust particles $(D_p)$ from an airstream traveling upward with a velocity $(U)$ of 1 ft/s (0.305 m/s). The scrubbing liquid is water and the diameter $(D_c)$ of the collecting particles is 1000 μm. The spray tower is 5 m high $(L)$ and 1 m in diameter.

**Solution** For collecting drops 1000 μm in diameter, Figure 11-15 indicates that the settling velocity $(v_{t,c})$ is 4.0 m/s. The settling velocity is the speed with which the dust particle and the collecting particle impact one another. Since the gas velocity is 0.305 m/s, the velocity of the collecting drops $(v_c)$ is 3.695 m/s. The single-drop collection efficiency $(\eta_d)$ is given by Equation 12.10:

$$\eta_d = \left( \frac{\psi}{\psi + 0.7} \right)^2$$

where

$$\psi = \frac{\tau_p v_{t,c}}{D_c}$$

The overall collection efficiency of the spray chamber is given by Equation 12.31:

$$\eta = 1 - \exp\left[ -\eta_d \left( \frac{3}{2} \right) \frac{Q_s}{Q_a} \frac{L}{D_c} \frac{v_{t,c}}{v_c} \right]$$

$$= 1 - \exp\left[ -\eta_d \frac{Q_s}{Q_a} \left( \frac{5}{0.001} \right) \left( \frac{4}{3.562} \right) \right]$$

$$= 1 - \exp\left( -5614.8 \eta_d \frac{Q_s}{Q_a} \right)$$

The fractional efficiency can be computed by repetitive calculations using mathematical computational programs. The efficiencies are modest. For a constant tower height $(L)$, the efficiency is strongly dependent on the reflux ratio, $Q_s/Q_a$, e.g., for $D_p = 50$ μm: Figure E12-3 shows the fractional efficiency for five drop diameters. The Mathcad program producing

| $\eta(D_p = 50 \text{ μm})$ | $Q_s/Q_a$ | gpm water/1000 cfm air |
|---|---|---|
| 0.42 | $1 \times 10^{-4}$ | 0.748 |
| 0.56 | $1.5 \times 10^{-4}$ | 1.122 |
| 0.66 | $2 \times 10^{-4}$ | 1.496 |
| 0.74 | $2.5 \times 10^{-4}$ | 1.870 |
| 0.80 | $3.0 \times 10^{-4}$ | 2.244 |

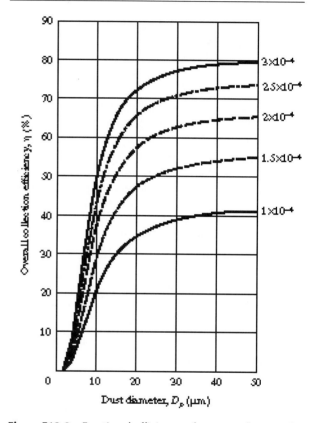

**Figure E12-3** Fractional efficiency of a counterflow gravity spray chamber. Dust removal efficiency versus reflex ratio for collecting water drops, $D_c = 1000$ μm. Dust density, $\rho_p = 1,000 \text{ kg/m}^3$. Air at STP traveling upward with a velocity, $U_0 = 1$ ft/s

Figure E12-3 can be found in the textbook's Web site on the Prentice Hall Web page.

## 12.2.2 Transverse packed bed scrubber

In industrial plants there is often an insufficient vertical distance to accommodate a tall vertical spray chamber. Under these conditions, engineers often install a *cross-flow (transverse) spray cham-*

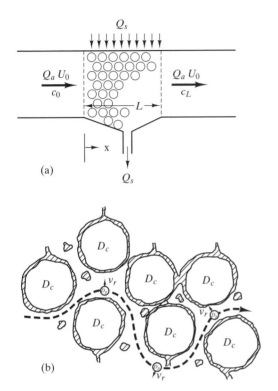

**Figure 12-17** (a) Transverse packed bed scrubber to remove particles. (b) Curvilinear air flow through a packed bed that generates particle velocities ($v_r$) that cause particles to impact wetted packing material

*ber* Figures 1-11 and 12-17a). In a cross-flow spray chamber, the gas and particles flow in the horizontal direction and the scrubbing liquid is sprayed downward (i.e., transverse to the carrier gas flow). By modeling the device in a manner similar to the above, and assuming that well-mixed conditions exist vertically in a unit volume $A\,dx$, where $x$ is in the direction of the flow and $A$ is the chamber cross-sectional area perpendicular to that direction. The overall collection efficiency of the device can be expressed by an equation similar to Equation 12.31.

The aerosol flows horizontally with a velocity $(U_0)$ and encounters a packed bed of inert solid material with bed porosity $(\varepsilon)$. A scrubbing liquid $(Q_s)$ flows over and around the packing $(D_c)$ and removes particles that impact the packing. See Figure 12-17b. The scrubbing liquid is treated in a secondary process to remove the captured particles. The purpose of the

packing is to provide curved surfaces that create centrifugal forces enabling particles to acquire radial velocities $(v_r)$ and strike the surface of the wetted packing. A second function of the packing is to maximize the total surface area per volume of the duct $(a_p)$. Common packing material is shown in Figure 10-13 and properties are shown in Table 10-1.

To model the performance of a transverse packed bed scrubber it is assumed that the aerosol enters the bed with uniform velocity $(U_0)$ and particle concentration $(c_0)$. We assume also that as the aerosol passes through the labyrinth of open passages in the bed, particles acquire radial velocities given by Equation 11.139:

$$v_r = \frac{\tau U_\theta^{\,2}}{r} \tag{12.32}$$

where $r$ is equal to the radius of the packing element $(D_c/2)$, $\tau$ is the relaxation time (Equation 11.76), and $U_\theta$ is the gas velocity inside the bed. Since the packing blocks the gas flow, the gas velocity inside the bed is larger than the approach velocity $(U_0)$:

$$U_\theta = \frac{U_0}{\varepsilon} \tag{12.33}$$

Consider an element of the packing $(A\,dx)$ and write an equation for the conservation of mass of particles,

$$Q_a c = Q_a(c + dc) + v_r a_p c A\,dx$$

$$\frac{dc}{c} = -\frac{v_r a_p A\,dx}{Q_a}$$

$$= -\frac{\tau}{D_c}\left(\frac{U_0}{\varepsilon}\right)^2 a_p\left(\frac{A\,dx}{U_0 A}\right)$$

$$= -\frac{\tau a_p U_0}{D_c \varepsilon^2}dx \tag{12.34}$$

Integrating over the length of the packed bed $(L)$, the removal efficiency can be written as

$$\eta = 1 - \frac{c_L}{c_0} = 1 - \exp\left(-\frac{\tau Q_a}{A}\frac{a_p}{\varepsilon^2}\frac{L}{D_c}\right)$$

$$= 1 - \exp\left(-\psi\frac{a_p}{\varepsilon^2}L\right) \tag{12.35}$$

where the Stokes number is defined as

$$\psi = \frac{\tau U_0}{D_c} = \frac{\tau Q_a}{A D_c} \qquad (12.36)$$

The volumetric flow rate of the scrubbing liquid $(Q_s)$ does not enter the calculation unless it is of such magnitude that globs of liquid slough off the packing and reduce the porosity $(\varepsilon)$ of the bed. Readers should consult Section 10.3.6 for details about computing the pressure drop across the bed and how the pressure drop varies with the volumetric flow rate of gas $(Q_a)$ and liquid $(Q_s)$.

### EXAMPLE 12.4   PARTICLE REMOVAL EFFICIENCY OF A TRANSVERSE PACKED BED WET SCRUBBER

Compute and plot the fractional efficiency of a 1-ft-thick $(L = 12 \text{ in.})$ packed bed composed of 1-in. ceramic Raschig rings $(D_c = 0.025 \text{ m})$. The particles are unit density spheres $(\rho_p = 1000 \text{ kg/m}^3)$ and the gas velocity $(U_0)$ is equal to 0.5 ft/s $(0.152 \text{ m/s})$, 0.75 ft/s $(0.229 \text{ m/s})$, and 1.0 ft/s $(0.305 \text{ m/s})$.

***Solution***   Table 10-1 indicates that $a_p = 58 \text{ ft}^{-1}$ and $\varepsilon = 0.73$:

$$\tau_p = \frac{\rho_p D_p^2}{18 \mu}$$

$$\psi = \frac{\tau_p U_0}{D_c}$$

$$\eta = 1 - \exp\left[ -\psi \left( \frac{a_p}{\varepsilon^2} \right) L \right]$$

$$= 1 - \exp\left[ -\psi \left( \frac{58 \text{ ft}^{-1}}{0.73^2} \right) (1 \text{ ft}) \right]$$

$$= 1 - \exp(-108.9\psi)$$

The fractional efficiency can be computed by repetitive calculations that can be performed by mathematical programs. Figure E12-4 shows the fractional efficiency for three values of the air velocity $U_0$. The Mathcad program producing Figure E12-4 can be found in the textbook's Web site on the Prentice Hall Web page.

---

The transverse packed bed scrubber is ideally suited for removing droplets of liquids that are miscible in water or solid particles that can easily be separated from water. The scrubber is poorly suited for removing sticky solid particles or liquids that are apt to form a sticky gum in the presence of water. The packing can tolerate a certain amount of surface coating, but the scrubber should not be used if the coating buildup reduces the bed porosity.

Packing increases the surface area per scrubber volume and improves the particle removal efficiency

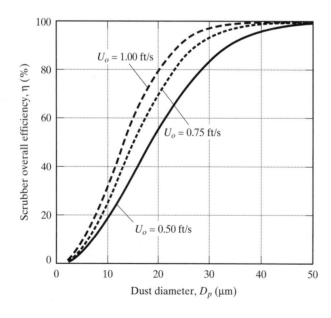

**Figure E12-4**   Dust removal efficiency of a transverse-flow, packed bed scrubber that uses water.

in a significant manner. Examine below the space and water requirements of the spray tower and the packed bed to remove 30-$\mu$m particles with an efficiency of 80%.

- *Spraychamber* $\left[\eta(30\ \mu\text{m}) = 0.80,\ \text{required reflux ratio},\ R = Q_s/Q_a = 3 \times 10^{-4}\right]$

$$Q_s = RQ_a = \left(3 \times 10^{-4}\right)\left(0.239\ \text{m}^3/\text{s}\right)\left(3.28\ \text{ft/m}\right)^3$$
$$\left(60\ \text{s/min}\right)\left(\text{gal}/0.13368\ \text{ft}^3\right) = 1.136\ \text{gpm}$$

$$\text{tower volume/cfm} = \frac{AL}{AU_0} = \frac{L}{U_0}$$

$$= \left(5\ \text{m}/0.305\ \text{m/s}\right)\left(60\ \text{s/min}\right) = 983.6\ \text{ft}^3/\text{cfm}$$

- *Transverse scrubber* $\left[\eta(30\ \mu\text{m}) = 0.80,\ \text{required gas velocity} = 0.152\ \text{m/s}\right]$

$Q_s$ = variable, only enough to bathe the packing and remove the deposited particles; $Q_s$ could easily be less than 1.136 gpm

packing volume/cfm =

$$\frac{AL}{AU_0} = \frac{L}{U_0} = \frac{\left(1\ \text{ft}\right)\left(\text{m}/3.28\ \text{ft}\right)}{\left(0.152\ \text{m/s}\right)\left(60\ \text{s/min}\right)} = 120\ \text{ft}^3/\text{cfm}$$

The packed bed tower is eight times smaller per cfm than the spray tower and offers the opportunity to use considerably less water.

## 12.2.3 Venturi scrubber

A second example in which impaction is used to remove particles from a gas stream is the venturi scrubber. Figures 1-9 and 12-18 are sketches of venturi scrubbers and the downstream cyclone separator used to remove particles of the collecting liquid after they leave the scrubber. Some venturi scrubbers have a constant throat cross-sectional area. In some cases it is desirable to vary the throat cross section. Figure 12-19 shows three ways to control the throat cross-sectional area. Figure 12-20 depicts the throat of a venturi scrubber that produces a large relative velocity and high overall collection efficiency. The operation of a venturi scrubber is based on the following:

1. Air containing the undesired small particles is brought to a high velocity by passing it through a

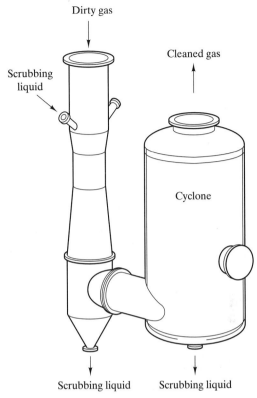

**Figure 12-18** Venturi scrubber with cyclone collector to remove drops of a scrubbing liquid

constricted area (throat) nozzle. Velocities of 50 to 100 m/s are common.

2. Drops of the collecting liquid are injected into the aerosol at the inlet of the throat section. The collecting liquid may be produced by spray nozzles or generated as the high-velocity gas shears small drops from the scrubbing liquid that flows down the sides of the venturi. The drops have a low velocity as they enter the gas stream but accelerate as they travel downstream and approach (if not achieve) the gas velocity.

3. The high-speed small particles impact the slower-moving collecting drops and are removed from the gas steam. The single-drop collection efficiency decreases as the collecting drops accelerate down the throat and the relative velocity between the particles and collectors decreases.

4. The collecting drops containing the impacted small particles are removed from the gas stream by a cyclone or some other conventional particle removal system.

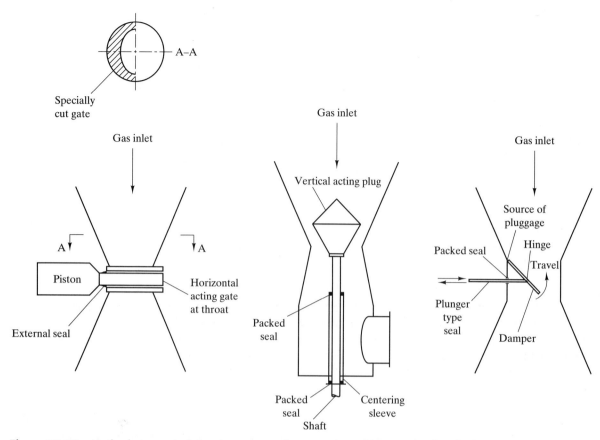

**Figure 12-19** Methods to control the throat area of a venturi scrubber. (Printed with the permission of D. R. Technology)

The *overall collection efficiency* of a venturi scrubber can be modeled using equations developed earlier if the following assumptions are made:

- The diameter of the collecting drops $(D_c)$ and the diameter of the dust particles $(D_p)$ are constant.
- Steady-state, laterally well-mixed conditions exist at any value of $y$.
- Only spatial variations in the direction of flow $(y)$ will be considered, variations transverse to this direction are assumed to be zero.
- The velocity of the small particles $(\mathbf{v}_p)$ equals the air velocity $(\mathbf{U}_a)$ and is constant throughout the throat.
- The evaporation of the scrubbing liquid is neglected and the pressure and temperature of the gas in the throat section are constant.

Writing a mass balance for an elemental volume:

$$cAU_a = (c + dc)AU_a + v_r c \frac{\pi D_c^2}{4} \eta_d n_c A \, dy$$

$$\frac{dc}{c} = -\eta_d \frac{v_r}{v_c} \frac{3}{2} \frac{Q_s}{Q_a} \frac{dy}{D_c} \tag{12.37}$$

where $\eta_d$ is the single-drop collection efficiency for an *accelerating drop* given by Equation 11.66 and $v_r$ is the relative velocity given by Equation 11.67. The collector drop velocity $v_c(y)$ increases from $v_c(0)$ at the throat inlet and may ultimately reach the gas velocity at some point downstream of the inlet. At any point downstream of the inlet, the collector velocity $v_c(y)$ can be found by solving the equation of motion (Equations 11.67 to 11.73), which because the distance $y$ is now measured in the downward direction (Figure 11-18) becomes

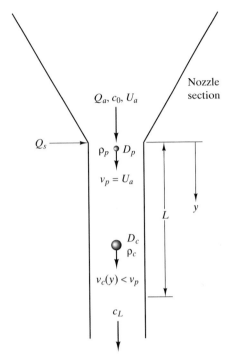

**Figure 12-20** Schematic diagram of a venturi scrubber throat section (Heinsohn, 1991)

$$\frac{dv_c}{dt} = g - \frac{3c_D}{4} \frac{\rho}{\rho_c D_c}(v_c - U_a)|v_c - U_a|$$

$$= g - \frac{3c_D}{4} \frac{\rho}{\rho_c D_c} v_r(v_c - U_a) \qquad (12.38)$$

where

$$v_r = |v_c - U_a|$$

$$c_D = 0.4 + \frac{24}{Re} + \frac{6}{1 + Re^{1/2}}$$

$$Re = \frac{\rho D_c v_r}{\mu}$$

and

$$\frac{dv_c}{dt} = Av_c + D \qquad (12.39)$$

where

$$A = -\frac{3c_D}{4} \frac{\rho v_r}{\rho_c D_c}$$

$$D = -AU_a + g$$

It must be emphasized that only collector drops accelerate. Thus only the equations of motion for the collecting drop need to be written and solved. The small particles that one wishes to remove travel at a constant velocity equal to the gas velocity. The equation for the collector velocity is a nonlinear, second-order differential equation. Numerical methods will be required to solve this equation and the equation to predict the overall collection efficiency. An elementary algorithm for this purpose consists of the following.

1. Using the flow condition at the throat inlet, compute the single-particle collection efficiency. Next, select a small increment of distance $\Delta y$ in the direction of flow and compute a new concentration using Equation 12.37.
2. At this new location $y = \delta y$, compute the elapsed time $(\Delta t = \Delta y/U_a)$ and using the computer program developed in Example 11.5, compute the new velocity of the collector drop. Be careful to use a small value of $\delta y$ such that the time increment is sufficiently small. With this new collector velocity, compute the new relative velocity, Stokes number, and single-drop collection efficiency. Because the collector drop accelerates rapidly, very small time steps will be needed near the inlet of the throat section, but progressively larger increments can be used as the collector particle approaches the throat exit.
3. Using the values computed in step 2, return to step 1.
4. Repeat the process until the overall displacement of the collector particle is equal to the length of the throat.

**EXAMPLE 12.5   PERFORMANCE OF A VENTURI SCRUBBER**

Consider the venturi scrubber depicted schematically in Figure 12.20 in which L = 1 m, that operates under the following conditions:
Gas velocity, density and viscosity: $U_a = 60$ m/s, $\rho_a = 1.161$ kg/m³, $\mu = 1.511 \times 10^{-5}$ kg m⁻¹ s⁻¹
Inlet particle density, concentration and diameter: $c_0 = 0.1$ mg/m³, $\rho_p = 1,500$ kg/m³, 1 μm < $D_p$ < 25 μm
Collecting drop density and diameter:
$\rho_c = 1000$ kg/m³,  $D_c = 200$ μm, 400 μm and 600 μm

convince themselves that the liquid is clean, has a low viscosity, and that the material collected will not clog the filter.

## 12.4 Electrostatic recipitators

Electrostatic precipitators (ESPs) are commonly used to remove particles from process gas streams having large steady volumetric flow rates (Oglesby and Nichols 1970; White 1963). Batch processes and processes in which the gas properties vary considerably with time do not lend themselves to ESPs because the collection efficiency of an ESP is strongly dependent on small changes in these properties. A common ESP design is the *single-stage plate-wire* configuration shown in Figure 1.8. Figure 12.36 is a top view of a single-stage plate-wire ESP. Single-stage ESPs use a single set of electrodes to produce the corona and the collecting electric field to remove particles. Figure 12.37 is a diagram of a cylinder-wire ESP, and Figure 12.38 is a diagram of an ESP in which the collecting electrodes are cleaned with water. A *two-stage* ESP has one set of electrodes to produce the corona and another set of electrodes to collect the particles. Two-stage ESPs produce less $O_3$ than a single-stage ESP, which is advantageous if the device is to be used in an indoor environment. Two-stage devices also offer users the opportunity to optimize particle charging and particle collecting independently. The key features of a single-stage (see Figure 1.7) ESP are:

- Flow straighteners and baffle-plate are located upstream of the charging and collecting sections to distribute the flow uniformly.
- Vertically aligned corona wires are spaced several inches apart mounted midway between vertically aligned collecting plates that are nearly 1 ft apart,

and dead weights that keep the corona wires under tension and vertical alignment.
- The corona wires are maintained at a negative voltage of several thousand volts to generate a corona that provides electrons for transfer to the airborne dust particles.
- The collecting plates are grounded and provide the oppositely charged electrodes that attract the negatively charged dust particles. Each collecting plate is several feet wide and 10 to 20 ft high.
- Electrical connections at the top of the ESP maintain each corona wire at a desired voltage.
- Mechanical "rappers" on the top of the ESP strike the collection plate and dislodge the collected dust; the gas velocity between the plates is low such that dislodged clumps of dust fall vertically to dust hoppers below and are not resuspended in the gas stream.
- Modular design divides the ESP into several parallel paths, each path consisting of several modules in series; modules in parallel operate at similar voltages and rapping frequency while modules in series might operate with different voltages and rapping frequencies.

To illustrate the essential features that govern the collection of charged particles, consider the flow of electrically charged particles passing between the parallel plates shown in Figure 12.39. The plates are of height $H$ (separation distance) and of length $L$ in the flow-wise direction. An external power supply establishes a uniform electrical field $(E_\infty)$ between the plates. The upper plate is the cathode and the lower plate is the anode. Gravimetric settling will be neglected. Dust particles of diameter $D_p$ possessing a negative charge $q_p$ enter the region with a velocity equal to the gas velocity $U_a$.

$$v_p = U_a \qquad (12.70)$$

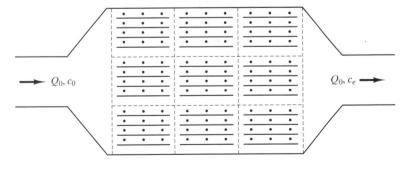

**Figure 12-36** Single-stage, plate-wire electrostatic precipitator, thee modules in parallel, three modules in series

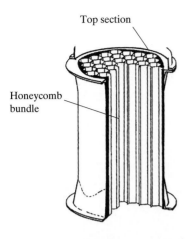

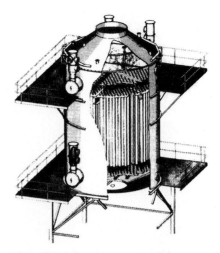

**Honeycomb bundle**                    **Corrosion-resistant wet-type precipitator**

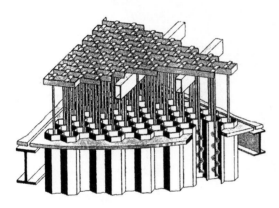

**Honeycomb bundle and discharge system**

**Figure 12-37**   Cylinder-wire (tubular) electrostatic precipitator. Tubular precipitators consist of nonconducting parallel vertical tubes, each with a central corona electrode. The tubes are grounded and serve as support for a conducting liquid film of collected particles. Tubular designs are well-suited to collect small corrosive liquid particles from a process gas stream. (Figure courtesy of the Lurgi Company, Germany)

Charged particles experience a transverse electrical force $F_{el}$ in the downward direction:

$$F_{el} = q_p E_\infty \qquad (12.71)$$

that produces a downward velocity called the *drift velocity* (w). For simplicity, assume a state of quasi-equilibrium in which the downward electrical force equals the drag force. Thus,

$$F_{el} = F_{drag}$$
$$q_p E_\infty = \frac{c_D \pi D_p^2}{4C} \frac{\rho w^2}{2} \qquad (12.72)$$

Assuming Stokes flow and recognizing that the relative velocity is the drift velocity, one finds that the value of the drift velocity w is,

$$c_D = \frac{24}{Re} = \frac{24}{\rho D_p w / \mu}$$
$$w = \frac{q_p E_\infty C}{3\pi\mu} \frac{1}{D_p} \qquad (12.73)$$

To predict the overall collection efficiency $(\eta_{overall})$, assume that the flow is well mixed laterally and write a mass balance on an element between the plates of length dx,

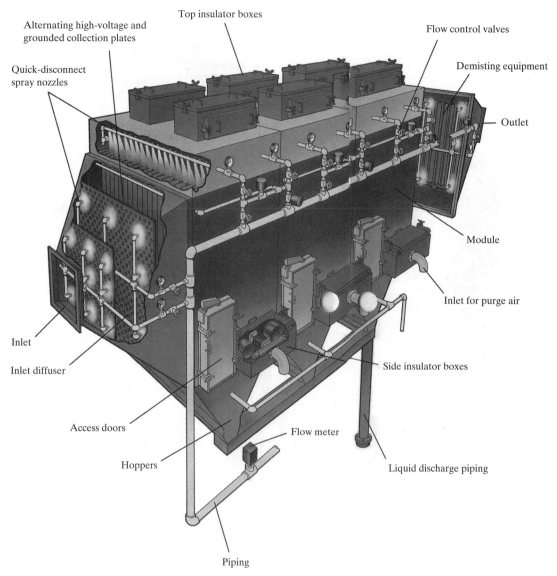

**Figure 12-38** Wet plate electrostatic precipitator. Alternating high-voltage and grounded collection plates produce the corona and transverse electric fields to remove sticky particles and condensed vapors. Wet collectors are well-suited to collect troublesome sticky organic condensibles, resinous materials and binders generated by forming and curing fiberglass and resins, turpenes, and fatty acids generated by burning and curing wood. Water nozzles continuously spray the high-voltage and grounded collecting plates. (Illustration courtesy of United McGill, Corp.)

$$Qc = Q(c + dc) + wcW \, dx$$

$$\int_{c_0}^{c_L} \frac{dc}{c} = -\int_0^L \frac{wW}{Q} \, dx \qquad (12.74)$$

where $W$ is the depth of the configuration perpendicular to the paper.

Integrating over a length of duct $x = L$ yields

$$\frac{c_L}{c_0} = \exp\left(-\frac{wA_s}{Q}\right) \qquad (12.75)$$

where $A_s$ is the total area of the collecting electrode $(A_s = WL)$. The overall collection efficiency $\eta_{\text{overall}}$ is

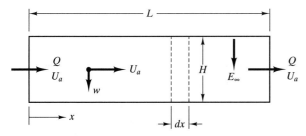

**Figure 12-39**  Collecting section of a parallel-plate electrostatic precipitator

$$\eta_{overall} = 1 - \frac{c_L}{c_0} = 1 - \exp\left(-\frac{wA_s}{Q}\right) \qquad (12.76)$$

This expression is called the *Deutsch equation*. The reader must be careful and realize that the term $Q$ is the *actual volumetric flow rate*, evaluated at the local temperature and pressure inside the ESP:

$$Q = U_a HW \qquad (12.77)$$

where $U_a$ is the actual gas velocity. When the Deutsch equation is placed alongside the equation for laminar settling in a horizontal duct (Equation 11.109), it is seen that the expressions are similar except that the drift velocity $(w)$ has replaced the gravimetric settling velocity $(v_t)$.

Even though Equation 12.76 was derived for the simplified geometry of Figure 12.39, it is nevertheless used for ESPs of complicated geometry because of its simplicity and clear statement about the actual volumetric flow rate $(Q)$ and area of the collecting plates

$(A_s)$. To account for actual ESPs' complex design, the drift velocity is replaced by an empirical parameter (possessing the units of velocity) derived from experiment called the *precipitation parameter or migration velocity*. The symbol $(w)$ is nevertheless retained. Table 12-2 lists typical values of the precipitation parameter for a variety of particles removed by ESPs.

Equation 12.77 enables engineers to understand how the collection efficiency depends on the collecting plate area $(A_s)$, temperature, the mass flow rate of the process gas stream, and the electrical properties of the collected dust. The relationship between the precipitation parameter and other operating parameters is not obvious. To appreciate this relationship, consider the physical phenomena that dictate the value of the drift velocity (Equation 12.73): (1) how particles acquire charge, and (2) how space charge and the electrical resistance of the collected dust affect the electric field intensity.

## 12.4.1  Particle charging

The reader should review the subject of electricity and magnetism, particularly the field of electrostatics.

$$\nabla \cdot \mathbf{E} = \frac{q_v}{\varepsilon_0} \qquad \mathbf{E} = -\nabla V \qquad (12.78)$$

where $\varepsilon_0$ is the permitivity of free space. Equation 12.73 shows that one of the parameters affecting the drift velocity is the charge on a dust particle, $q_p$. To

**Table 12-2**   Typical Migration Velocities (cm/s) for Plate-Wire Electrostatic Precipitators

| Process (ESP temperature) | Design Efficiency (%): | | | |
| | 95 | 99 | 99.5 | 99.9 |
|---|---|---|---|---|
| Bituminous coal fly ash (300°F) | 12.6 | 10.1 | 9.3 | 8.2 |
| Subbituminous coal fly ash in a tangentially fired boiler (300°F) | 17.0 | 11.8 | 10.3 | 8.8 |
| Cement kiln (600°F) | 1.5 | 1.5 | 1.8 | 1.8 |
| Glass plant (500°F) | 1.6 | 1.6 | 1.5 | 1.5 |
| Incinerator fly ash (250°F) | 15.3 | 11.4 | 10.6 | 9.4 |
| Iron/steel sinter plant dust with mechanical precollector (300°F) | 6.8 | 6.2 | 6.6 | 6.3 |
| Lime dust | 1.5 | 1.5 | 1.5 | 1.5 |

*Source*: Abstracted from Vatavuk (1990).

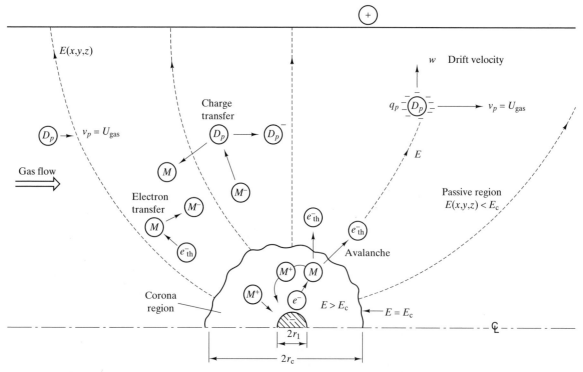

**Figure 12-40** Particle charging mechanism in a plate-wire electrostatic precipitator. High-energy electrons (e) are generated in the *active corona region*. The electrons travel outward, encounter neutral molecules (M) and transfer their charge in the *passive region*. Charged ions (M⁻) encounter dust particles $D_p$, transfer their charge to dust particles in the passive region. Charged particles $(D_p^-)$ travel to the anode where they adhere to the collecting plate and transfer their charge to the external circuit

understand the process by which uncharged dust particles acquire negative charges, consider a single negatively charged wire located between two grounded plates (Figure 12.40). Two distinct regions can be defined.

**1.** *Corona region.* A corona is a thin annular region close to the negatively charged wire (cathode), $r < r_c$. in which the electric field intensity is very large, $E > E_c$. The radius $r_c$ is called the corona radius and $E_c$ is the electric field intensity capable of ionizing gas molecules. Crawford (1976) states that the electric field $E_c$ at the outer radius of the corona $r_c$ can be expressed as

$$E_c(\text{V/m}) = 3 \times 10^6 f \frac{T_0}{T} \frac{P}{P_0}$$
$$+ 0.03 \left( \frac{T_0}{T} \frac{P}{P_0} \frac{1}{r_c} \right)^{1/2} \qquad (12.79)$$

$$r_c = r_w + 0.02(r_w)^{1/2} \qquad (12.80)$$

where $r_w$ is the radius of the corona wire (m), $r_c$ is in meters, and $E_c$ is in the units of V/m and $T_0$ and $P_0$ refer to temperature and pressure at STP. The parameter $f$ is equal to 1 for a smooth corona wire and greater than 1 if the corona wire is rough. Recall that the voltage gradient is equal to the electric field strength; thus small-diameter wires charged to a large voltage produce large gradients close to their surface. Within the corona, $r < r_c$, an electron acquires very large velocity, and when it strikes a neutral gas molecule, it dislodges an electron from the molecule. The positive gas ion moves toward the cathode, where it becomes neutral. The original and newly formed electrons are free to ionize another gas molecule or travel outward toward the anode. Within the corona, electrons are generated (i.e., an *avalanche* of electrons) that travel outward toward the anode. At a certain radius $(r_c)$ the electric field strength decreases to a value $(E_c)$ such that electrons are no longer sufficiently energetic to ionize neutral molecules but now merely attach themselves to gas molecules.

**2. *Passive region.*** In the passive region, $r > r_c$, $E < E_c$, electrons attach themselves to gas molecules through a process called *electron transfer*. Per unit volume of gas, the number of gas molecules is much larger than the number of dust particles; thus electrons are more likely to attach themselves to gas molecules and produce negative ions than to attach themselves to dust particles. Negatively charged gas ions travel outward toward the anode. Along the way they encounter dust particles and transfer their negative charges to dust particles through a process called *charge transfer*. It must be recalled (see Figure 11.1) that a dust particle is many orders of magnitude larger in diameter than a gas molecule. Through countless encounters with gas ions, dust particles acquire a negative charge and begin their outward migration toward the anode at a velocity equal to the drift velocity ($w$). The charge a particle acquires has an upper limit called the *saturation charge* ($q_{ps}$) because as it becomes charged it produces its own electric field that ultimately repels gas ions traveling toward it.

Particles acquire charge by two distinct physical processes, *field charging* (also called *bombardment charging*) and *diffusion charging*.

**Field Charging.** Imagine a dust particle subjected to a stream of "bombarding" ions. The rate at which charge is transferred to the dust particle is given by

$$\frac{dq_p}{dt} = \frac{Nq_e}{4\varepsilon_0} \kappa q_{ps} \left(1 - \frac{q_p}{q_{ps}}\right)^2 \quad (12.81)$$

where $q_{ps}$ is called the *saturation charge* (in coulombs):

$$q_{ps} = \frac{3\kappa}{(\kappa + 2)} \pi \varepsilon_0 D_p^2 E_\infty \quad (12.82)$$

and

$N$ = number of gas ions per unit volume

$q_e$ = coulombs per ion (charge on one electron = $-1.6029 \times 10^{-19}$ C)

$\varepsilon_0$ = permitivity of free space ($8.85 \times 10^{-12}$ C²/N·m²)

$\kappa$ = dielectric constant of the dust particle

$E_\infty$ = electric field driving the stream of ions

$D_p$ = diameter of a dust particle

The time it takes to acquire charge is given by

$$t = \frac{(4\varepsilon_0/\kappa Nq_e)(q_p/q_{ps})}{1 - q_p/q_{ps}} \quad (12.83)$$

For electric fields encountered in ESPs, the parameter $4\varepsilon_0/Nq_eK$ is on the order of 0.01 to 0.1 s. The typical time ($t$) a dust particles resides inside an ESP is on the order of tens of seconds. Thus for design purposes, one can assume that if field charging is the dominant charging mechanism, dust particles always possess their saturation charge ($q_{ps}$).

**Diffusion charging.** As a continuum, gas ions possess both a velocity dictated by their charge ($Nq_e$) and the local electric field intensity ($E_\infty$). Individual gas molecules also possess velocities given by a Gaussian distribution dictated by the gas temperature, and there are a small number of ions possessing sufficiently large velocities as to be unaffected by the repelling electric field on a charged dust particle. Charging that occurs by the process of molecular diffusion is called *diffusion charging*. The rate at which a particle acquires charge by diffusion is given by

$$\frac{dq_p}{dt} = Nq_e \sqrt{\frac{kT\pi}{2m}} D_p^2 \exp\left(-\frac{q_e q_p}{2\pi\varepsilon_0 kT D_p}\right) \quad (12.84)$$

$$q_p = \frac{2\pi\varepsilon_0 kT D_p}{q_e} \ln\left(1 + \frac{ND_p q_e^2 t}{2\varepsilon_0 \sqrt{2m\pi kT}}\right) \quad (12.85)$$

where $m$ is the mass of the ion and $k$ is the Boltzmann constant.

It is seen that charge acquired by the process of diffusion charging has no upper limit, whereas charge acquired by the process of field charging has an upper limit called the saturation charge. Thus at any time, the charge on a dust particle ($q_p$) can be thought of as the sum of Equations 12.82 and 12.85. Since field charging depends on the square of the particle diameter, whereas diffusion charging depends on the diameter to the first power, diffusion charging will be important for small particles ($D_p < 2$ μm), while field charging will be more important for larger particles ($D_p > 2$ μm). It should be noted that both particle charging mechanisms depend on the ion density ($Nq_e$).

| | | | |
|---|---|---|---|
| $D_c$ | diameter of the collecting particle, cloud diameter $(L)$ | $k$ | Boltzmann constant $(Q/T)$ |
| $D_{e,p}$ | equivalent volume diameter, defined by equation $(L)$ | $K_1$ | residual drag $(M/L^2)$ |
| | | $K_2$ | dust cake specific resistance $(t^{-1})$ |
| $D_f$ | fiber diameter $(L)$ | $l$ | penetration or stopping distance $(L)$ |
| $D_f$ | characteristic diameter of a single fiber $(L)$ | $L$ | characteristic length, length of ESP collecting electrode $(L)$ |
| $D_i$ | diameter of inlet nozzle of sampling probe | $L, L'$ | height of spray chamber, hypothetical height $(L)$ |
| $D_n$ | collision diameter $(L)$ | $L_c$ | critical length $(L)$ |
| $D_p$ | particle diameter $(L)$ | $L_f$ | length of filter fiber per unit volume of filter $(L^{-2})$ |
| $D_{p,am}$ | arithmetic mean diameter $(L)$ | | |
| $D_{p,gm}$ | geometric mean diameter $(L)$ | $L(t)$ | dust cake thickness $(L)$ |
| $D_{p,j}$ | mean particle diameter in the $j$th interval $(L)$ | $L_1, L_2, L_3$ | characteristic lengths of Lapple cyclone $(L)$ |
| $D_{p,min}$ | diameter of the most penetrating particle in a filter $(L)$ | $\dot{m}$ | mass flow rate $(M/t)$ |
| | | $M, M_i$ | molecular weight, molecular weight of species $I$ $(M/N)$ |
| $D_{p,p}$ | equivalent projected area diameter, defined by equation $(L)$ | $m_t$ | total mass of particles $(M)$ |
| $D_{p,16}, D_{p,50}, D_{p,84}$ | particle diameter for cumulative distributions 16%, 50%, and 84% $(L)$ | $m(D_p)$ | mass distribution function $(L^{-1})$ |
| $D_{s,p}$ | equivalent surface area diameter, defined by equation $(L)$ | $n$ | number of encounters between particle and falling drop; number of molecules per volume of gas $(L^{-3})$; exponent for curvilinear flow |
| $D_2, D_c, D_d$ | characteristic diameters of Lapple cyclone $(L)$ | | |
| $E_\infty$ | electric field intensity acting on a charged particle $(V/L)$ | $N$ | number of ions/volume $(L^{-3})$ |
| | | $n_c$ | number of drops per volume of air, number particles per volume in a cloud $(L^{-3})$ |
| $E_c$ | Electric field intensity at the outer edge of a corona $(V/L)$ | | |
| $F$ | drag force on a particle $(F)$ | $N_e$ | number of turns |
| $f_d$ | solids fraction of dust cake | $n_i$ | number of particles of diameter $D_{p,i}$ |
| $F_{drag}$ | drag force $(F)$ | $N_i$ | cumulative distribution for particles less than $D_{p,i}$ |
| $F_{el}$ | force on a charged particle due to electric field $(F)$ | $n_t$ | total number of particles |
| | | $n_t$ | total number of moles $(N)$ |
| $f_f$ | solids fraction of filter | $n(D_p)$ | number distribution function $(L^{-1})$ |
| $g$ | acceleration of gravity $(L/t^2)$ | $N(D_p)$ | cumulative number distribution function |
| $G$ | cumulative mass distribution function | | |
| $g(D_p)_j$ | mass fraction of particles in size range $j$ | $P$ | pressure $(F/L^2)$ penetration $(P = 1 - \eta)$ |
| $H$ | height of a duct, enclosure, ESP collecting electrode, filter thickness height Lapple cyclone inlet $(L)$ | $\delta P_c$ | pressure drop across dust cake $(F/L^2)$ |
| | | $\delta P_d$ | overall pressure drop across dust-caked filter $(F/L^2)$ |
| $I$ | electrical current in an ESP $(C/t)$ | $\delta P_f$ | pressure drop across the filter $(F/L^2)$ |
| $J$ | constant defined by equation | $Q$ | volumetric flow rate $(L^3/t)$ |

| | | | |
|---|---|---|---|
| $Q_a, Q_s$ | volumetric flow rates of air and scrubbing fluid $(L^3/t)$ | $w$ | drift velocity, precipitation parameter $(L/t)$; width of a duct $(L)$ |
| $q_p$ | electrical charge per particle $(C)$ | $w(t)$ | dust cake mass per area of filter $(M/L^2)$ |
| $q_{ps}$ | saturation charge on a particle $(C)$ | $x, y, z$ | coordinates $(L)$ |
| $q_v$ | space charge density $(C/L^3)$ | | |
| $r$ | radius $(L)$ | *Greek* | |
| $R$ | $D_p/D_f$, gas constant | | |
| $r_c$ | corona radius $(L)$ | $\epsilon$ | porosity |
| $R_c$ | radius of collecting drop $(L)$ | $\alpha$ | turning angle |
| $R_{dc}$ | dust cake electrical resistance $(Vt/C)$ | $\Delta t$ | incremental time step $(t)$ |
| $R_u, R_j$ | universal gas constant, gas constant for a specific molecular species $j$ $(Q/NT)$ | $\varepsilon_0$ | permitivity of free space $(C^2/ML^2)$ |
| | | $\eta$ | efficiency (mass removed/mass entering) |
| $r_w$ | radius of the corona wire in an ESP $(L)$ | $\eta_d$ | single drop (particle) collection efficiency |
| $s$ | distance separating collecting plates in an ESP $(L)$ | $\eta_f$ | single fiber collection efficiency |
| $S$ | filter drag $(F/L^2)$ | $\eta_I, \eta_R, \eta_D, \eta_{DR}$ | single-fiber collection efficiencies of impaction, interception, diffusion, combination of diffusion and interception |
| $t$ | time $(t)$ | | |
| $T$ | temperature $(T)$ | | |
| $t^*$ | dimensionless time | $\eta(D_p)$ | fractional or grade efficiency |
| $t_c$ | critical time $(t)$ | $\theta_c$ | critical angle in curvilinear flow |
| $U, U_i$ | gas velocity $(L/t)$ | $\theta_t$ | total angular displacement |
| $U_a$ | air velocity $(L/t)$ | $\kappa$ | dielectric constant |
| $U_{ave}$ | average velocity $(L/t)$ | $\lambda$ | carrier gas mean free-path $(L)$ |
| $U^*, v^*$ | dimensionless gas and particle velocities $(L/t)$ | $\mu$ | viscosity $(Ft/L^2)$ |
| | | $\nu$ | kinematic viscosity $\mu/\rho$ $(L^2/t)$ |
| $U_0$ | characteristic velocity $(L/t)$ | $\pi$ | 3.14159 |
| $U_\theta,$ | tangential velocity component $(L/t)$ | $\rho, \rho_p, \rho_w$ | density of carrier gas, particle and water $(M/L^3)$ |
| $v, v_i$ | particle velocity $(L/t)$ | | |
| $V$ | voltage $(V)$ | $\rho_{dc}$ | dust cake resistivity |
| $V$ | volume $(L^3)$ | $\sigma, \sigma_g$ | standard deviation, geometric standard deviation |
| $v_c$ | velocity of the collecting particle $(L/t)$ | | |
| $V_c$ | voltage at edge of corona | $\tau$ | relaxation time $(t)$ |
| $V_{dc}$ | voltage drop across the dust cake $(V)$ | $\chi$ | particle dynamic shape factor |
| $V_p$ | particle volume $(L^3)$ | $\psi$ | Stokes number |
| $v_r$ | relative velocity, $(L/t)$ | | |
| $v_r, v_{r,i}$ | relative velocity, $(v, U)$ $(L/t)$ | *Subscripts* | |
| $v_s(t)$ | instantaneous settling velocity $(L/t)$ | $(\cdot)_a$ | property of air, average property |
| $v_t$ | terminal (gravimetric, steady state) settling velocity $(L/t)$ | $(\cdot)_{ave}$ | average |
| | | $(\cdot)_b$ | bleed air |
| $v_{t,c}$ | settling velocity of collecting particle $(L/t)$ | $(\cdot)_c$ | collecting particle, critical property, corona |
| | | $(\cdot)_d$ | drop property |
| $W$ | width of a bend $(L)$, width of Lapple cyclone inlet $(L)$ | $(\cdot)_{DA}$ | dry air |
| | | $(\cdot)_e$ | exit condition |

$( \cdot )_f$    final condition, filter property, streaming flow property

$( \cdot )_i$    initial condition, inlet condition

$( \cdot )_L$    condition at distance $L$

$( \cdot )_{\text{overall}}$    overall

$( \cdot )_p$    particle property

$( \cdot )_{p,j}$    particle in size range $j$

$( \cdot )_r$    radial component, relative component

$( \cdot )_s$    scrubbing liquid property

$( \cdot )_w$    water, values relevant to inlet, width $2w$

$( \cdot )_x, ( \cdot )_y, ( \cdot )_z, ( \cdot )_i$    vector components

$( \cdot )_0$    initial or incoming properties, STP

$( \cdot )_\theta$    component in angular direction

*Abbreviations*

Kn    Knudsen number

Ku    Kuwabara number

Pe    Peclet number

Re    Reynolds number

$\text{Re}_f$    Reynolds number defined in terms of the difference between cyclone inlet and outlet diameters

STP    standard temperature and pressure

# References

Billings, C. E., and Wilder, J., 1970. *Handbook of Fabric Filter Technology*, Vols. I and II, PB200648, PB-200649, U.S. Government Printing Office, Washington, DC.

Boysan, F., Ayers, W. H., and Swithenbank, J. A., 1982. A fundamental mathematical modeling approach to cyclone design, *Transactions of the Institution of Chemical Engineering*, Vol. 60, pp. 222–230.

Brockmann, J., 1993. Sampling and transport of aerosols, Chapter in 6, *Aerosol Measurement*, Willeke, K., and Baron, P., Van Nostrand Reinhold, New York, pp. 77–108.

Calvert, S., 1972. *Scrubber Handbook*, PB-213016, U.S. Government Printing Office, Washington, DC.

Calvert, S., and Englund, H. M., (Eds.), 1984. *Handbook of Air Pollution Technology*, Wiley-Interscience, New York.

Carr, R. C., and Smith, W. B., 1984. Fabric filter technology for utility coal-fired power plants, *Journal of the Air Pollution Control Association*, Vol. 31, No. 4, pp. 399–412.

Crawford, M., 1976. *Air Pollution Control Theory*, McGraw-Hill, New York.

Danielson, J. A., 1973. *Air Pollution Engineering Manual*, EPA Publication AP-40, U.S. Government Printing Office, Washington, DC.

Davies, C. N. (Ed.), 1966. *Aerosol Science*, Academic Press, San Diego, CA.

Elliott, G. K., and Withers, C. J., 1991. Ceramic filters for high-temperature furnace gas applications, in *Smelter Process Gas Handling and Treatment*, Smith, T. J. A., and Newman, C. J., (Eds.), The Minerals, Metals & Materials Society, London, England, pp. 175–184.

Fuchs, N. A., 1964. *The Mechanics of Aerosols*, Pergamon Press, Elmsford, NY.

Gregg, W., and Griffin, J., 1991. Baghouse control of particulate pollution, *Pollution Engineering*, April, Vol. 23, pp. 82–83.

Heinsohn, R. J., 1991. *Industrial Ventilation: Engineering Principles*, Wiley-Interscience, New York.

Hidy, G. M., and Brock, J. R., 1970. *The Dynamics of Aerocolloidal Systems*, Pergamon Press, Tarrytown, NY.

Hinds, W. C., 1982. *Aerosol Technology*, Wiley-Interscience, New York.

Lee, K. W., and Liu, B. Y. H., 1980. On the minimum efficiency and the most penetrating particle size for fibrous filters, *Journal of the Air Pollution Control Association*, Vol. 30, No. 4, pp. 377–381.

Licht, W., 1980. *Air Pollution Control Engineering*, Marcel Dekker, New York.

Marple, V. A., Rubow, K. L., and Behm, S. M., 1991. A multi-orifice uniform deposit impactor (MOUDI), *Aerosol Science and Technology*, Vol. 14, pp. 434–436.

Marple, V. A., Rubow, K. L., and Olson, B. A., 1993. Inertial, gravitational, centrifugal and thermal collection tecniques, Chapter 11 in *Aerosol Measurement*, Willeke, K., and Baron, P. A., (Eds.), Van Nostrand Reinhold, New York, pp. 206–228.

Meritt, R. L., and Vann Bush, P., 1997. Status and future of baghouses in the utility industry, *Journal of the Air & Waste Management Association*, Vol. 47, pp. 704–709.

Moore, M. E., and McFarland, A. R., 1993. Performance modeling of single-inlet aerosol sampling cyclones, *Environmental Science and Technology*, Vol. 27, No. 9, pp. 1842–1848.

Moore, M. E., and McFarland, A. R., 1996. Design methodology for multiple inlet cyclones, *Environmental Science and Technology*, Vol. 30, pp. 271–276.

Oglesby, S., and Nichols, G. B., 1970. *A Manual of Electrostatic Precipitator Technology, Parts I and II*, PB 196380 and 196381, National Technical Information Service, Springfield, VA.

Perry, R. H., and Chilton, C. H., 1973. *Chemical Engineers Handbook*, 5th ed., McGraw-Hill, New York.

Pontius, D. H., and Smith, W. B., 1988. Method for computing flow distribution and pressure drop in multi-chamber baghouses, *Journal of the Air Pollution Control Association*, Vol. 38, No. 1, pp. 39–45.

Strauss, W., 1966. *Industrial Gas Cleaning*, Pergamon Press, Tarrytown, NY.

Vatavuk, W. M., 1990. *OAQPS Control Cost Manual*, 4th ed., EPA 450/3-90-006, Office of Air Quality Planning and Standards, Research Triangle Park, NC.

White, H. T., 1963. *Industrial Electrostatic Precipitation*, Addison-Wesley, Reading, MA.

Willeke, K., and Baron, P. A., (Eds.), 1993. *Aerosol Measurement*, Van Nostrand Reinhold, New York.

# Problems

**12.1.** Diesel engines used in construction vehicles are equipped with an air cleaner consisting of a number of standard reverse-flow cyclones arranged in parallel followed by a cartridge paper filter. The total volumetric flow rate of air is 10 scfm. It is necessary to remove all particles $(\rho_p = 1500 \text{ kg/m}^3)$ larger than 10 μm with an efficiency no less than 80%. Each reverse flow cyclone has an overall diameter $(D_2)$ of 2 cm. How many cyclones are needed?

**12.2.** Consider raindrops $(D_c)$ falling through air containing suspended dust particles $(D_p)$. Show that the change in the single-drop collection efficiency $(\eta_d)$ varies with $D_p$ in the following manner:

$$\frac{d\eta_d}{dD_p} = 2.8 \frac{\eta_d}{D_p(0.7 + \psi)}$$

where $\psi$ (Stokes number) and $\eta_d$ are defined by Eqs. 12.10 to 12.12.

**12.3.** An aerosol $(D_p = 3 \text{ μm}, \rho_p = 1700 \text{ kg/m}^3)$ enters the standard cyclone described in Example 12.1 at a velocity of 15 m/s and temperature 350 K. If the inlet concentration is 100 mg/m$^3$, what exit concentration can be expected?

**12.4.** Hot exhaust (assume air) from a fluidized-bed combustor contains large particles:

- *$D_p$, $\rho_p$*: 50 μm, 2500 kg/m$^3$
- *Gas temperature and pressure*: 1000°C, 1000 kPa

What is the particle settling velocity in such a gas? It has been suggested that 80% of these particles can be removed by a bank of reverse-flow cyclones. Assuming that collection occurs at the temperature and pressure above, how many cyclones of the dimensions below should be installed in parallel? The total gas mass flow rate is 1 kg/s.

- *Outer diameter*: 10 cm

- *Exit diameter*: 5 cm
- *Height and width of inlet*: 5 cm and 2.5 cm

**12.5.** Predict the fractional efficiency curve for an air cleaning unit consisting of a straight-through cyclone and a reverse-flow cyclone that removes the particles from the bleed flow $(Q_b)$ of the straight-through cyclone and returns the air to the air exiting the straight-through cyclone. Assume that a fan draws a total volumetric flow rate $(Q_0)$ into the unit and produces a bleed flow of 10% $(Q_b/Q_0 = 0.2)$. Assume also that the flow through both devices is $n$-degree rotational with $n = 0.5$. Assume that the particles are unit-density spheres and that air is at STP and the following dimensions of the cyclones apply.

| Straight-through cyclone | Reverse-flow cyclone |
|---|---|
| $D_2 = 5$ cm | $D_2 = 10$ cm |
| $D_1 = 4.5$ cm | $D_1 = 5$ cm |
| $L = 50$ cm | $L_1 = L_2 = 20$ cm |
| Vane angle $= \pi/4$ | $W = 2.5$ cm |
| | $H = 5$ cm |

Draw the fractional efficiency curve for $Q_0 = 0.05, 0.075,$ and 0.10 m$^3$/s.

**12.6.** Derive an expression for the gravimetric settling efficiency of a horizontal duct of rectangular cross section in which the base is horizontal but the roof slopes downward at an angle $\alpha$ in the flow-wise direction. The duct width $(b)$ is constant. The flow enters the duct at a Reynolds number of 100,000 and the particles are large such that the drag coefficient is constant. Show the collection efficiency as a function of some (but not necessarily all) of the following:

- Volumetric flow rate $(Q)$

- Flow-wise distance $(x)$
- Angle $\alpha$
- Local average velocity $[U(x)]$
- Duct width $(b)$ and initial height $(h_0)$
- Dust diameter and density
- Air properties

**12.7.** A spray chamber uses water to capture small particles produced in grinding marble. Draw the fractional efficiency curves for $Q_a/Q_s$ of 1000 and 5000 if the spray chamber has the following properties:

$$D_c = 1 \text{ mm} \qquad L = 5 \text{ m}$$
$$U_a = 3 \text{ m/s} \qquad c_0 = 1000 \text{ mg/m}^3$$
$$\rho_p = 4000 \text{ kg/m}^3 \qquad \rho_c = 1000 \text{ kg/m}^3$$
$$D_p = 2, 5\ 10, 20, 40 \ \mu\text{m}$$

**12.8.** Show that the single-drop collection efficiency $(\eta_d)$ varies with the relative velocity as follows:

$$\frac{d\eta_d}{dv_r} = \frac{\psi}{v_r} \frac{1.4\psi}{(\psi + 0.7)^3}$$

**12.9.** Spherical water drops $(D_c = 1 \text{ mm})$ are sprayed on dust generated when a conveyor discharges coal $(\rho = 1500 \text{ kg/m}^3)$ to a stockpile. Assume that the air and coal dust velocities are negligible. The coal dust concentration is 100 mg/m³. Estimate the single-drop collection efficiency assuming that the drops fall at their settling velocity. If there are 1000 drops/m³ of air, what is the initial rate of removal of coal dust $(\text{mg m}^{-3} \text{sec}^{-1})$?

**12.10.** Consider a transverse flow scrubber through which an aerosol passes in a horizontal direction. Assume that there is only one spatial variable (in the direction of flow) and that the following are known:

- *Duct dimensions*: height $H$, length $L$, depth $W$
- *Scrubbing liquid volumetric flow rate*: $Q_s$ that produces a uniform stream of drops of diameter $(D_c)$
- *Uniform air velocity*: $U_0 = Q_a/HW$
- *Aerosol*: $D_p, \rho_p$, inlet and outlet concentration $c(0)$ and $c(L)$
- *Drop velocity*: $v_d = iU_0 - jv_t$, where $v_t$ is the constant settling velocity, where $v_t \ll U_0$
- *Drop diameter*: $D_c$

Assume that the single-drop collection efficiency $(\eta_d)$ between drop and aerosol particle is known and constant. Show that the overall collection efficiency $(E_0)$ is equal to

$$\eta = 1 - \exp\left[ -\eta_d \left( \frac{3}{2} \right) \frac{Q_s}{Q_a} \frac{H}{D_c} \frac{v_t L}{v_t L + U_a H} \right]$$

**12.11.** You are employed by a firm that makes bituminous concrete (roadway asphalt). The process produces a sticky aerosol of concentration $(c_0)$ 2.0 g/m³ at STP and it has been decided to design a venturi scrubber to bring the process into compliance with state air pollution regulations. You have been asked to conduct a preliminary analysis of the overall collection process and to produce a series of graphs showing how the overall collection efficiency varies with the operating parameters:

- Drop diameter $(D_c)$.
- Throat length $(L_t)$.
- Ratio of the volumetric flow rates of gas to liquid $(G/S)$.

Process parameters:

- Particle density $(\rho_p) = 1500 \text{ kg/m}^3$.
- Throat diameter-0.6 m.
- Throat gas velocity $(U_a) = 15 \text{ m/s}$.
- Gas temperature and pressure-25°C, 1 atm.
- Scrubbing liquid is water and enters the throat with a negligible velocity; once particles impact water drops, they are removed, and liquid particles do not impact one another.

**(a)** Develop an analytical model that predicts the overall collection efficiency by impaction. State all assumptions carefully and fully. Develop a finite difference technique to integrate the differential equations so that the overall collection efficiency can be computed for a venturi throat of arbitrary length. Compute the velocity of a drop $(D_c = 0.25, 1.0, 2.0, \text{and } 4.0 \text{ mm})$ versus throat length.

**(b)** Using Eqs. 12.37 to 12.39 and the following:

$$\eta_d = \left( \frac{\psi}{\psi + 0.7} \right)^2$$

compute the single-drop collection efficiency as function of throat length for $D_c = 1$ and 2 mm and $D_p = 2 \ \mu\text{m}$.

**(c)** Plot the overall collection efficiency as a function of the length of the throat $(L_t)$ assuming that

$$D_c = 1 \text{ mm} \qquad D_p = 2 \ \mu\text{m} \qquad G/S = 9000$$

for $L_t = 0.1$ to 10.0m.

**(d)** Plot the overall collection efficiency as a function of length assuming that

$$D_c = 1 \text{ mm} \qquad D_p = 2 \ \mu\text{m}$$

for $G/S = 3000, 9000, \text{ and } 15,000$.

**(e)** Plot the overall collection efficiency as a function of the drop diameter $(D_c)$ assuming that

$$D_p = 2 \ \mu\text{m} \qquad L_t = 3.0 \text{ m} \qquad G/S = 9000$$
$$D_c = 0.25, 1.0, 2.0, 3.0, 4.0 \text{ mm}$$

**12.12.** Fog particles $(D_p = 10 \ \mu m)$ in air pass through a window screen $(D_c = 1000 \ \mu m)$ with a velocity 1.0 m/s. Some of the particles are collected by the screen. Which of the following fog–air aerosols will be collected with the same efficiency?

| Particle | $D_p$ (μm) | $D_c$ (μm) | $U$ (m/s) |
|----------|-----------|-----------|-----------|
| (a) | 20 | 1000 | 0.5 |
| (b) | 20 | 2000 | 1.0 |
| (c) | 10 | 2000 | 4.0 |
| (d) | 10 | 2000 | 2.0 |

**12.13.** Show that the overall collection efficiency $(\eta)$ of a filter bed of thickness $(H)$ consisting of spherical particles of diameter $D_c$ is the following:

$$\eta(\text{overall}) = 1 - \exp\left[-\eta_d \frac{f_s}{1 - f_s}\left(\frac{3}{2}\right)\frac{H}{D_c}\right]$$

The bed has the following properties:

- $U_0 =$ approach velocity $= Q/A$, where $Q$ is the volumetric flow rate through a bed of frontal area $A$
- $\eta_d =$ single-particle collection efficiency, collection efficiency of each bed particle $D_c$ to collect particles of diameter $D_p$
- $f_s =$ solids fraction, fraction of bed composed of solid material, the solids fraction is equal to (1 - porosity)

**(a)** Derive an expression for the overall collection efficiency in terms of the parameters above.

**(b)** Derive an expression for the rate of change of the overall collection efficiency $(\eta)$ with respect to the face velocity $(U_0)$, $d\eta/dU_0$, if the single-particle collection efficiency is governed by

$$\eta_d = \left(\frac{\psi}{\psi + 0.7}\right)^2$$

where $\psi = \dfrac{U_0 \tau}{D_c}$ and $\tau = \dfrac{\rho_p D_p^2}{18 \ \mu}$.

**12.14.** You wish to install a baghouse to capture particles removed from the air in a foundry. The baghouse manufacturer claims that the filter resistance $(K_1)$ is 10,000 N s m$^{-3}$ and estimates that the dust cake resistance $(K_2)$ will be 80,000 s$^{-1}$. The dust concentration entering the unit is 5 g/m$^3$. The recommended air-to-cloth ratio is 1.0 m/min. If the baghouse is to operate at a maximum pressure drop of 500 N/m$^2$, how often should the bags be cleaned?

**12.15.** A horizontal duct of square cross section ($W$ by $W$) and length $(L)$ carries an aerosol at a volumetric flow rate $(Q)$. The inlet dust concentration is $c_0$. The duct is

mounted only at the inlet end and has no other supports. Dust settles to the floor of the duct. After a while a sufficient amount of dust accumulates to produce a critical bending moment $M_c$ that fractures the duct mounting. Write an expression to predict the time $(t_c)$ when the mounting fails.

**12.16.** A jewelry manufacturer asks you to design an inertial collector followed by a filter to capture small particles of silver generated by cutting tools. The capture efficiency of the collector should be 80%. The inertial collector consists of $N$ individual standard reverse-flow cyclones (Figure 12-2) arranged in parallel (Figure 10-5). How many cyclone collectors will be needed? The following characterize the particle and the gas stream.

| Particles | Gas stream (air) | Individual cyclones |
|-----------|------------------|---------------------|
| $D_p = 25$ μm | $Q$ (actual) $= 0.001$ m$^3$/s | $D_2 = 4$ cm |
| $\rho_p = 10{,}500$ kg/m$^3$ | $P = 100$ kPa, $T = 300$ K | $N_e = 4$ |

**12.17.** Consider a cylindrical filter for an off-road vehicle. Air containing particles at an inlet concentration $c_0$ is drawn inward in a radial direction through the filter of height $(H)$, outer diameter $(D_2)$, and inner diameter $(D_1)$. The volumetric flow rate of air is $Q$. The filter is composed of fibers of a diameter $D_f$ and has a solid fraction $\alpha$. The single-fiber collection efficiency is $\eta_f$. Show that the fractional efficiency of the filter can be expressed as

$$\eta(D_p) = 1 - \exp\left[\left(-\frac{2\eta_f}{\pi}\right)\left(\frac{\alpha}{1 - \alpha}\right)\left(\frac{D_2 - D_1}{D_f}\right)\right]$$

**12.18.** An electrostatic precipitator with a collecting area of 5000 ft$^2$ was designed to remove 95% of the fly ash particles from a 250°F exhaust gas stream from a municipal incinerator. Your supervisor suggests that the same design can be used to remove 95% of the particles from a 590 K, 101 kPa, 150-kg/min gas stream from a cement kiln.

**(a)** What capture efficiency do you believe the original design will achieve when used for the cement kiln?

**(b)** If your supervisor's expectation is to be achieved, what should be the area of the collecting surface?

**12.19.** An electrostatic precipitator is used to remove particles from the exhaust of a pulverized coal furnace in an electric power generating station. The collection efficiency is 95%. Management decides to increase the collection efficiency to 98% and to increase the exhaust gas temperature from 350 to 380°K to increase plume buoyancy. The exhaust gas mass flow rate remains the same. Estimate the change in area of the collecting electrodes needed to accomplish these tasks.

**12.20.** The dust concentration in the air of a foundry increases during the day. The workplace is of height $H$ and floor area $A$. Circulating fans within the room ensure that the dust concentration varies only with time and not location:

$$c(x, y, z, t) = c(t)$$

At the end of the workday the concentration is $c_0$. Assuming that settling is governed by the well-mixed model, write an expression that predicts the concentration at any instant of time after the workday ends:

$$c(t) f(c_0, D_p, \rho_p, t, \mu, H, A)$$

and the total mass of dust that settles on the floor $m_p(t)$ over an elapsed time

$$m(t)_p = f(c_0, D_p, \rho_p, \mu, H, A)$$

**12.21.** A coal-fired steam boiler is equipped with a particle collector that has an overall efficiency of 75%. The volumetric flow rate of exhaust gas is 30,000 scfm. In the near future 95% fly ash removal will be required. Your supervisor asks that you consider three options and recommend the option with the lowest cost.

- *Option 1.* Replace the entire 75% collector with a new collector.
- *Option 2.* Add one or more new collectors in series with the 75% collector.

- *Option 3.* Devise a parallel configuration in which a fraction (that you specify) of the flow passes through a the 75% collector and the remainder passes through the new collector.

| Overall efficiency (%) | Capital cost ($/scfm) |
|---|---|
| 75 | 30 |
| 80 | 40 |
| 85 | 46 |
| 90 | 52 |
| 95 | 58 |
| 99 | 70 |
| 99.5 | 80 |

**12.22.** An electrostatic precipitator with a collecting area of 5000 ft$^2$ was designed to remove 95% of the fly ash particles from a 250°F exhaust gas stream from a municipal incinerator. Your supervisor believes that the same design can be used to remove 95% of the particles from a 590°F, 101 kPa, 150 kg/min gas stream from a cement kiln.

**(a)** What capture efficiency do you believe the original design will achieve when used for the cement kiln?

**(b)** If your supervisor's expectation is to be achieved, what should be the area of the collecting surface?

# 13

# Cost of Air Pollution Control Systems

In this chapter you will learn:

- How to estimate the initial cost of an air pollution control system and its components,
- How to estimate the annual cost to operate the air pollution control system

In many instances the selection of an air pollution control system is based on the lowest cost per ton of pollutant removed. The objective of the chapter is to instruct engineers on how to estimate the initial and annual costs of several air pollution control systems. The first things that must be determined are the dimensions, materials, and operating parameters of the system. Next, engineers should compute the data listed in Tables 13-1 and 13-2. If engineers have explicit information from vendors, they can omit much of the material in the subsequent sections and complete Tables 13-1 and 13-2 directly. Unfortunately, some or all of this information is often lacking. The sections in this chapter present ways to estimate typical costs of the control systems and their

components in the absence of explicit data provided by vendors.

Two pollution control systems to remove particles will be studied. The costs associated with a reverse-flow fabric filter and an ESP to control particles from the lime kiln analyzed in Examples 12.7 and 12.8 will be estimated. The performance data for the lime kiln are:

- *Reverse-flow fabric filter and ESP*: $Q = 50,000$ scfm, $T_{inlet} = 325\,°F$, $c_{p,inlet} = 10g/m^3$, minimum removal efficiency $= 99\%$, operating time $= 8640$ h/yr

Three control systems will be studied for removal of gaseous pollutants. The costs associated with a carbon adsorber, a packed bed scrubber, and a recuperative thermal oxidizer to control diethylamine vapor in the process gas stream analyzed in Examples 10.2, 10.3, and 10.5 will be estimated. The performance data for diethylamine removal are:

- *Carbon adsorber, packed bed scrubber and recuperative thermal oxidizer*: $Q = 30,000$ scfm, $T_{inlet} = 30\,°C$,

---

**Table 13-1**    Total Initial Capital Cost (TIC) of an Air Pollution Control System

---

The total initial capital cost is equal to the cost of equipment plus the cost to install, erect, and start up an air pollution system.

1. *Initial direct cost* (IDC) = initial equipment costs (IEC) + initial facilities cost (IFC)

    1.1  Initial equipment cost (IEC)

| | |
|---|---|
| [Control system + auxiliary equipment] (C) | C |
| Instrumentation (10% C) | 0.10C |
| Sales tax (3% C) | 0.03C |
| Freight (5% C) | 0.05C |
| Initial equipment cost (IEC) = | 1.18C |

    1.2  Initial facilities cost (IFC))

| | |
|---|---|
| Site preparation | Variable |
| Building | Variable |
| Foundations and supports ($\alpha$% IEC) | $\alpha$ (IEC) |
| Air handling (fans and ductwork) and erection ($\beta$% IEC) | $\beta$ (IEC) |
| Electrical ($\gamma$% IEC) | $\gamma$ (IEC) |
| Piping ($\theta$% IEC) | $\theta$ (IEC) |
| Insulated ductwork ($\delta$% IEC) | $\delta$ (IEC) |
| Painting (1% IEC) | 0.01 (IEC) |
| Initial facilities cost (IFC) = | *sum of the above* |

Total initial direct cost (IDC) = 1.18C + *sum of the above*

2. *Initial installation cost* (IIC)

| | |
|---|---|
| Engineering ($\varepsilon$ % IEC) | $\varepsilon$ (IEC) |
| Contractor fees (10% IEC) | 0.1 (IEC) |
| Construction and field expenses ($\eta$% IEC) | 0.01 $\eta$ (IEC) |
| Startup (2% IEC) | 0.02 (IEC) |
| Performance tests (1% IEC) | 0.01 (IEC) |
| Contingencies (3% IEC) | 0.03 (IEC) |
| Total initial installation cost (IIC) = | *sum of the above* |

Total initial cost:    TIC = IDC + IIC

| System | $\alpha$ | $\beta$ | $\gamma$ | $\delta$ | $\theta$ | $\varepsilon$ | $\eta$ |
|---|---|---|---|---|---|---|---|
| Thermal oxidizer | 8% | 14% | 4% | 1% | 2% | 10% | 10% |
| Adsorption | 8% | 14% | 4% | 1% | 2% | 10% | 5% |
| Fabric filter | 4% | 50% | 8% | 7% | 2% | 10% | 20% |
| ESP | 4% | 50% | 8% | 2% | 2% | 20% | 20% |
| Scrubber | 12% | 40% | 1% | 1% | 30% | 10% | 10% |

---

$c_{\text{inlet}} = 1000$ ppm $(2934.4 \text{ mg/m}^3)$, minimum efficiency = 92.4%, operating time = 3600 h/yr, 360 days/yr

For all the collectors studied the costs will be compared on the basis of cost per scfm and cost per ton of pollutant.

## 13.1  Total initial cost and total annual cost

*Cost* is a slippery word; there is the initial cost to buy an air pollution control system that will last for many years, there is the cost of a component that lasts for only a few years (or part of a year), and then there is

---

**Table 13-2**   Total Annual Cost ($TAC_{1995}$) of an Air Pollution Control System

---

1. *Annual direct cost* (ADC)
    Operating labor (application specific)
        Operator (h/yr, typically $12/h)
        Supervisor (15% operator)
    Operating materials (application specific)
    Maintenance
        Labor (h/yr, typically $13.20/h)
        Material (application specific)

Operations and maintenance (O&M) = sum of the above

    Replacement
        Labor (h/yr, typically $13.20/h)
        Materials (application specific, include capital
            recovery costs, transportation)
    Utilities
        Natural gas ($3.95/ 1000 ft$^3$, ft$^3$/yr)
        Electricity ($0.07/kW h, h/yr)
        Steam
        Cooling water
        Compressed air
    Wastes (application specific)
        Labor
        Disposal fees

Total annual direct cost (ADC) = sum of the above

2. *Annual indirect cost* (AIC)
    Overhead (60% O&M)
    Administrative charges (2% TIC)
    Insurance (1% TIC)
    Property tax (1% TIC)
    Capital recovery cost (CRC):
        product of the capital recovery factor (CRF)
        and total initial cost (TIC)
            CRC = (CRF)(TIC)

Total annual indirect cost (AIC) = sum of the above

3. *Annual recovery cost* (ARC)
    Revenue generated by the sale of recovered material
*Total annual cost*:   TAC = ADC + AIC − ARC

---

a continuing cost that reflects the money to finance and operate the system that is used by accountants in the preparation of annual budgets. It is advisable not to use the single word *cost*. For clarity, modify "costs" with an additional word. In this chapter it will be the practice to use the following language: *initial cost* and *annual cost*. The cost analysis presented in this chapter is recommended by the EPA and used by state regulatory agencies to compare alternative air pollution control technologies. The calculation of initial and annual cost, the specific items included within them, and the procedures to estimate these costs are estimated are drawn from Vatavuk (1990). The fundamental concepts of engineering economics can be found in Grant and Ireson (1970).

To simplify procedures, the EPA Office of Air Quality Planing and Standards (OAQPS) has prepared spreadsheets called (CO$T-AIR) that are available on both the Control Technology Center (CTC) and Clean Air Act Amendments (CAAA) *telnet* electronic bulletin boards in TTNBBS.RTPNC.EPA.GOV. These spread sheets enable engineers to compute and manipulate data for various control technologies described in the examples in this chapter.

The *initial cost* is the purchase price of an item (i.e., the final price of a item arrived by negotiation). *Annual costs* are incorporated in yearly budgets as expenditures that must be offset by revenue in order for the company to remain profitable. The *Total Initial Cost* will be abbreviated as TIC (some call this the total capital investment). Table 13-1 lists the items included in TIC. All the items are portions of the first-cost, up-front cost, initial expenditure, and so on, that institutions need to make to consummate the purchase and to put a pollution control system into operation. The *Total Annual Cost* will be abbreviated as TAC. Table 13-2 lists the items included in TAC. All the items are yearly expenditures to operate the pollution control system plus a capital recovery cost (CRC). The *Capital Recovery Cost* (CRC) represents a way of distributing TIC over many years. However, the CRC is not merely the TIC divided by the life span of the system. It also reflects the fact that institutions may define as an expense the interest revenue it would have obtained if money had not been spent to meet the total initial cost (TIC).

Tables 13-1 and 13-2 organize costs in ways used by industry and the EPA. Different institutions may use different titles and abbreviations for these cost categories. The titles and abbreviations used in this book are chosen for clarity and to minimize confusion.

### 13.1.1  Capital recovery cost

*Capital recovery cost* (CRC) is the annualized cost of an original purchase that is to be replaced in *n* years. Capital recovery costs include interest not received because money has been spent on items that are purchased every few years rather than every year. The capital recovery cost can be computed as the product of the *capital recovery factor* (CRF) and the total initial cost (TIC):

$$CRC = (CRF)(TIC) \qquad (13.1)$$

where the *capital recovery factor* (CRF) is

$$CRF = \frac{i(i+1)^n}{(1+i)^n - 1} \qquad (13.2)$$

and *i* is the investment rate of return, or *nominal annual interest rate*, and *n* is the useful lifetime (years).

### 13.1.2  Inflation

If users wish to reflect the effect of inflation, the *nominal interest rate* can be computed from the following:

$$i = (1 + i_r)(1 + r) - 1 \qquad (13.3)$$

where, *i* is the nominal interest rate, $i_r$ is the real or posted interest rate, and *r* is the annual rate of inflation.

Users may have only manufacturer-listed equipment costs that are several years old. For purposes of calculation, the current total initial cost $[TIC(t)]$ is related to its value *n* years earlier $(TIC_0)$ by the following,

$$TIC(t) = TIC_0(1 + r)^n \qquad (13.4)$$

where *r* is the annual inflation rate and *n* is the elapsed time in years.

In Sections 13.3 to 13.8 we discuss each generic class of air pollution control system and elaborate on unique items to be included in TIC and TAC.

### 13.1.3  Cost indices

A more accurate way to account for costs that change with time is to use compiled indexed cost data published in the professional literature. Cost or price indices are needed to record changes in costs owing to inflation (which may vary from year to year, or even quarter to quarter) and advances in technology in which the costs of some control systems change more rapidly than others. Indeed, technological advancements may even cause costs of some equipment to decrease over time. The Consumer Price Index (CPI) and the Producer Price Index (PPI) compiled monthly by the U.S. Bureau of Labor and Statistics are used widely in the United States. Unfortunately, the mixture of goods and services do not correlate well with air pollution activities. Relevant to air pollution are the *Vatavuk Air Pollution Control Index* (VAPCCI) and the *Chemical Engineering Plant Cost Index* (CE) (Vatavuk, 1995), which can be found in the monthly publication *Chemical Engineering* (see Table 13-3). The CE index is compiled annually, while the VAPCCI is a more finely tuned index for specific generic classes of air pollution control equipment and is compiled quarterly. The VAPCCI is the preferred index since it accurately reflects specific changes in the cost of generic classes of air pollution systems. The CE index is based on broader classes of equipment used in the chemical industry. To estimate the cost of a piece of equipment at a particular time, $C(t)$, using costs published at an earlier time $(C_0)$, use the following equation:

$$C(t) = C_0 \left[ \frac{\text{Index}(t)}{\text{Index}_0} \right]_j \qquad (13.5)$$

where the subscript $j$ refers to the index for a particular control device $j$ obtained from the professional literature. Indices obtained from different publications covering different time periods can be used with one another, but users must be careful doing so.

**Table 13-3**   Economic Indicators

**a. Chemical Engineering (CE) Plant Cost Annual Index**

| Date | Annual CE index | Date | Annual CE index |
|------|-----------------|------|-----------------|
| 1986 | 318.4 | 1991 | 361.3 |
| 1987 | 323.8 | 1992 | 358.2 |
| 1988 | 342.5 | 1993 | 359.2 |
| 1989 | 355.4 | 1994 | 368.1 |
| 1990 | 357.6 | 1995 | 381.9 |

**b. Vatavuk Air Pollution Control Cost Index (VAPCCI)**

| Year | Quarter | Catalytic incinerators | Packed-column absorbers | Regenerative thermal oxidizers | Thermal incinerators | Venturi scrubbers |
|------|---------|------------------------|-------------------------|--------------------------------|----------------------|-------------------|
| 1989 | 1 | 100.0 | 100.0 | 100.0 | 100.0 | 100.0 |
|      | 2 | 100.3 | 100.5 | 100.2 | 100.7 | 100.5 |
|      | 3 | 100.4 | 101.0 | 100.6 | 101.0 | 102.0 |
|      | 4 | 100.9 | 101.5 | 101.2 | 101.8 | 102.5 |
| 1990 | 1 | 102.7 | 103.4 | 101.6 | 106.9 | 102.9 |
|      | 2 | 103.0 | 103.8 | 103.9 | 108.1 | 103.3 |
|      | 3 | 103.8 | 104.2 | 104.3 | 108.2 | 103.7 |
|      | 4 | 104.2 | 104.6 | 104.6 | 109.3 | 104.1 |
| 1991 | 1 | 106.8 | 106.0 | 105.9 | 112.6 | 106.0 |
|      | 2 | 107.0 | 106.4 | 105.9 | 112.7 | 106.4 |
|      | 3 | 107.1 | 106.8 | 107.2 | 112.8 | 106.8 |
|      | 4 | 107.1 | 107.2 | 107.9 | 113.3 | 107.2 |
| 1992 | 1 | 100.2 | 107.6 | 108.0 | 115.3 | 107.6 |
|      | 2 | 100.2 | 108.0 | 108.1 | 115.6 | 108.0 |
|      | 3 | 100.2 | 108.4 | 109.3 | 115.7 | 109.4 |
|      | 4 | 100.2 | 108.9 | 109.6 | 115.8 | 109.8 |
| 1993 | 1 | 104.2 | 109.0 | 114.7 | 117.9 | 110.0 |
|      | 2 | 104.2 | 109.1 | 114.8 | 118.3 | 110.1 |
|      | 3 | 104.2 | 109.3 | 115.0 | 118.3 | 110.3 |
|      | 4 | 104.2 | 109.4 | 109.7 | 118.5 | 110.4 |

*Source: (a) Chemical Engineering* (monthly); (b) Vatavuk (1995).

**EXAMPLE 13.1 ESTIMATED COST OF A PACKED COLUMN ABSORBER IN 1995 BASED ON THE COST IN 1990**

Estimate the cost in the fourth quarter of 1995 of a wet scrubber that was listed as costing $100,000 in 1990.

**Solution** Table 13-3 dose not have VAPCCI data for the period 1990 to the 4$^{th}$ quarter of 1995, however it contains CE data for this period.

$$C(1995) = C(1990) \left[ \frac{CE(1995)}{CE(1990)} \right]$$

$$= 100,000 \left[ \frac{381.9}{357.6} \right] = 106,795 \ (6.7\% \text{ more})$$

It would be more accurate to use the VAPCCI index for the entire period, but in the absence of such data, engineers should not hesitate to mix the VAPCCI and CE indices, realizing that this is only an estimate. *Chemical Engineering* April 1995 reports VAPCCI index is 106.2 for the 4$^{th}$ quarter 1995 and 100.0 for the 1$^{st}$ quarter of 1994. The CE indices can be found from Table 13-3.

$$C(4^{th} \ qtr \ 1995) = C(1990) \left[ \frac{CE(1994)}{CE(1990)} \right]$$

$$\left[ \frac{VAPCCI(4^{th} \ qtr \ 1995)}{VAPCCI(1^{st} \ qtr \ 1994)} \right]$$

$$= \$100,000 \left[ \frac{368.1}{357.6} \right] \left[ \frac{106.2}{100} \right]$$

$$= \$109,318 \ (9.3\% \text{ more})$$

## 13.2  Utility costs

In the absence of specific information relating to the cost of basic equipment, fuel, raw materials, disposal, and so on, engineers may estimate costs for the following categories:

- *Basic equipment cost*
  - Fans and blowers
  - Electric motors
  - Pumps
- *Equipment electrical power cost*
  - Fans and blowers
  - Refrigeration
  - Pumps
- *Basic energy cost*
  - Natural gas
  - Steam
  - Cooling water
  - Makeup water
  - Compressed air
- *Disposal cost*
  - Wastewater
  - Hopper dust and sludge

In economic analyses, engineers need to know the cost of these items. In lieu of specific information from vendors, the following expressions can be used. The costs reported below were obtained from Vatatuk (1990) and updated to 1995 using the date the original data were compiled and the CE index (Equation 13.5).

### 13.2.1  Basic equipment cost

**Fans and blowers.** The cost of a fan or blower is typically included in the purchase price of an air pollution control system (see Table 13-1). If fans have to be purchased separately and a specific price is not known, one can estimate the 1995 cost of a *centrifugal fan*, with backward inclined bladed fan of diameter $D$ (inches) from the following:

$$C(\$)_{\text{fan cen, 1995}} = 69.4 D_{\text{fan}}^{1.38} \quad (13.6)$$

**Electric motor.** If the cost of a fan (or pump) does not include an *electric motor*, one can estimate the 1995 cost of a electric motor from the following:

$$C(\$)_{\text{el motor, 1995}} = 124.7 \text{hp}^{0.821} \quad (13.7)$$

**Pump.** In lieu of knowing specific prices for a pump (and electric motor), one can use the following relationship and volumetric flow rate:

$$C(\$)_{\text{pump, 1995}} = (\$19.2/\text{gpm})(L) \quad (13.8)$$

where $L$ is the liquid flow rate, gpm.

## 13.2.2  Equipment electrical power cost

**Electric power for fans and blowers.**  To estimate the *electric power for fans and blowers*, users must know the volumetric flow rate $(Q)$, pressure drop $(\delta P)$, and operating time $(t)$.

$$\text{Power}_{\text{fan}}(kW) = 1.17 \times 10^{-4} \frac{Q\,\delta P}{\eta} \quad (13.9)$$

where, $Q$ is in scfm, $\delta P$ is the flange-to-flange pressure drop in inches of water, and $\eta$ is the overall efficiency, $0.30 < \eta < 0.70$. The annual cost of electricity to operate fans and blowers can be computed from

$$C(\$)_{\text{fan}} = (p_{\text{el}})(t_{\text{op}})(\text{Power}_{\text{fan}}) \quad (13.10)$$

where $t_{\text{op}}$ is the fan operating time, h/yr, and $p_{\text{el, 1995}}$ is the electric power cost. Readers should use values appropriate to a geographical area. A typical value is \$0.07/kWh.

**Electric power for refrigeration.**  To determine the annual cost of electricity for *refrigeration*, $C(\$)_{\text{rf}}$, the user must know the temperature and refrigeration capacity that is needed:

$$C(\$)_{\text{rf, 1995}} = \frac{(R)(E)(t_{\text{op}})(p_{\text{el, 1995}})}{\eta} \quad (13.11)$$

where

$R$ = refrigeration in tons, 1 ton = 12,000 Btu/h cooling
$t_{\text{op}}$ = system operating time, h/year
$p_{\text{el, 1995}}$ = electric power cost, a typical value is \$0.07/kWh
$\eta$ = overall efficiency of refrigeration unit, a typical value is 0.85
$E$ = electric power cost per ton of refrigeration

Typical values for $E$ depend on the cooling temperature as follows.

| Cooling temperature (°F) | Electric power, $E$ (kW/ton) |
| --- | --- |
| 40 | 1.3 |
| 20 | 2.2 |
| −20 | 4.7 |
| −50 | 5.0 |
| −100 | 11.7 |

**Electric power for pumps.**  The annual cost *of electricity to pump water*, $C(\$)_{\text{pump}}$, can be estimated from the following:

$$\text{Power}_{\text{pump}}(kW) = \frac{(0.746)(0.01512)(L)}{\eta}$$

$$C(\$)_{\text{pump, 1995}} = (p_{\text{el, 1995}})(t_{\text{op}})\,\text{Power}_{\text{pump}} \quad (13.12)$$

where

$t_{\text{op}}$ = pump operating time, h/yr
$L$ = liquid flow rate, gpm
$\delta P$ = pump pressure drop, ft of water
$\eta$ = pump-motor efficiency, typically $0.40 < \eta < 0.70$
$p_{\text{el, 1995}}$ = electric power rate, typically \$0.07/kWh

## 13.2.3  Basic energy cost

**Natural gas.**  The annual *cost of natural gas*, $C(\$)_{\text{gas}}$, is equal to the yearly consumption times the cost of the natural gas:

$$C(\$)_{\text{gas, 1995}} = (Q_{\text{gas}})(t_{\text{op}})(3.95 \times 10^{-3}) \quad (13.13)$$

where $Q_{\text{gas}}$ is the volumetric flow rate of gas scfm and $t_{\text{op}}$ is the operating time, min/yr. Readers should use costs appropriate to a geographical region. A typical cost of natural gas in 1995 is $\$3.95 \times 10^{-3}/\text{ft}^3$.

**Steam.**  To regenerate adsorbers with steam, one typically uses 3 to 4 $\text{lb}_{\text{m}}$ of steam/$\text{lb}_{\text{m}}$ of VOC (i.e., $\dot{m}_s = 3.5\dot{m}_p$). Thus the annual cost of process *steam*, $C(\$)_{\text{steam}}$, can be estimated from the following:

$$C(\$)_{\text{steam, 1995}} = (3.5\ \text{lb}_{\text{m}}\ \text{steam/lb}_{\text{m}}\ \text{pollutant})$$
$$= (\dot{m}_p)(p_{\text{s, 1995}})(t_{\text{op}}) \quad (13.14)$$

where $p_{\text{s, 1995}}$ is the cost to produce steam (a typical value is 0.007 $\$/\text{lb}_{\text{m}}$ steam), $\dot{m}_p$ is the pollutant mass flow rate ($\text{lb}_{\text{m}}$/h), and $t_{\text{op}}$ is the steam operating time (hours of steam use per year).

**Cooling water.**  The annual cost to produce *cooling water*, $C(\$)_{\text{cw}}$, to regenerate adsorbers, operate a condenser, and so on, is related directly to the amount of steam used. The cost can be estimated from the following:

$$C(\$)_{\text{cw, 1995}} = (3.43\ \text{gal water/lb}_{\text{m}}\ \text{steam})$$
$$(\dot{m}_s)(p_{\text{cw, 1995}})(t_{\text{op}}) \quad (13.15)$$

where

3.43 = typical ratio of gallons of cooling water per $lb_m$ of steam

$\dot{m}_s$ = steam mass flow rate ($lb_m$/h), typically equal to 3.5 $\dot{m}_p$

$\dot{m}_p$ = pollutant mass flow rate ($lb_m$/h)

$p_{cw, 1995}$ = cost to produce cooling water in the units of \$/gal of water, typically $1.8 \times 10^{-4} < p_{cw} < 3.6 \times 10^{-4}$

$t_{op}$ = cooling water operating time, hours of water use per year

**Make up water (public water).** In 1995 the cost of public water $\left(p_{pw, 1995}\right)$ for a scrubber, cooling water, and so on, was approximately \$0.24/1000 gal.

**Compressed air.** In the absence of knowing the annual cost to produce *compressed air*, $C(\$)_{ca}$, the annual cost can be estimated from the following:

$$C(\$)_{ca, 1995} = (Q_{ca})(t_{op})(\$1.9 \times 10^{-4}) \quad (13.16)$$

where $Q_{ca}$ is the volumetric flow rate of compressed air (scfm), $t_{op}$ is the compressed air operating time (min/yr), and \$1.9 $\times 10^{-4}$ is the typical cost in 1995 to produce a standard (1 atm, 300 K) cubic foot of compressed air.

### 13.2.4 Disposal cost

**Wastewater disposal.** The cost to treat and dispose of *wastewater* $\left(p_{ww, 1995}\right)$ is between \$1.2 and \$2.4 per 1000 gal depending on the extent of treatment. If solid material is removed by filtration, an additional cost is incurred to dispose of the sludge.

**Disposal of hopper dust or sludge.** If solid material is removed from wastewater, it must be dried and disposed of. The typical cost $\left(p_{sw, 1995}\right)$ to dispose of dried sludge or hopper dust is \$24 to \$36/ton plus transportation costs of \$0.60/ton·mile unless the material is hazardous, in which case the costs are higher.

In Sections 13.3 to 13.8 we describe how to conduct an engineering economic analysis for five air pollution control systems:

**1.** Fabric filters (reverse-flow baghouse)
**2.** Electrostatic precipitators (plate-wire ESP)
**3.** Adsorption (activated charcoal adsorber)
**4.** Absorption (countercurrent packed bed wet scrubber)

**5.** Direct-flame oxidizer (thermal oxidation with recuperative heat exchanger)

The first economic analysis will compare the costs of using a baghouse and an ESP to control the particles produced by the lime kiln described in Examples 12.7 and 12.8. A second economic analysis will compare the costs of using a charcoal adsorber, wet scrubber, and thermal oxidizer to control the emissions of diethylamine described in Examples 10.2, 10.3, and 10.5.

## 13.3 Fabric filters

The engineer must select the type of cleaning method best suited for the composition and temperature of the process gas stream and the particles that are to be removed (i.e., shaker, reverse-flow, or pulse-jet baghouse). Following this, Table 12.1 should be consulted to select a gas-to-cloth ratio appropriate for the particles to be removed. From this, the total fabric surface area can be computed. The overall pressure drop and cleaning cycle also need to be chosen. The costs described in Table 13-4 were obtained from Vatavuk (1990) and updated to 1995 using the CE Index (Equation 13.5) and the date the original data were compiled.

### 13.3.1 Equipment cost

The total equipment cost $C(\$)$ is the sum of the equipment cost of the housing, $C(\$)_{housing}$, plus the cost of the bags, $C(\$)_{bag}$.

$$C(\$)_{fabric\ filter} = C(\$)_{housing} + C(\$)_{bags} \quad (13.17)$$

The bag cost is based on the fabric surface area $A_c$:

$$C(\$)_{bag} = \left(A_c\ ft^2\right)\left(cost/ft^2\right) \quad (13.18)$$

For the analysis that follows only the cost of nomex bags of typical fabric weight will be reported. The costs of other fabric materials and other cleaning methods can be found in Vatavuk (1990).

**Table 13-4** Fabric Filter Costs

| Method of cleaning | Bag diameter | Typical bag cost ($/ft$^2$, 1995) |
|---|---|---|
| Pulse jet, removal from the top | 4.12 in. $< D_{bag} < 5\frac{1}{8}$ in. | 2.25 |
| | 6 in. $< D_{bag} < 8$ in. | 1.87 |
| Shaker, strap at top | $D_{bag} = 5$ in. | 1.53 |
| Reverse air with rings | $D_{bag} = 8$ in. | 2.06 |
| | $D_{bag} = 11\frac{1}{2}$ in. | 2.03 |

The cost of the baghouse housing depends on the method of cleaning, plus additional costs if stainless steel is used and if the baghouse is to be insulated.

- *Intermittent shaker (in 1995 dollars):*

$$C(\$)_{shaker} = C(\$)_{housing} + C(\$)_{bag} \quad (13.19)$$

where,

$$C(\$)_{housing} = C(\$)_{metal} + C(\$)_{stainless} + C(\$)_{insulation}$$
$$C(\$)_{metal} = 4{,}239 + 8.08 A_c(\text{ft}^2)$$
$$C(\$)_{stainless} = 13{,}853 + 3.96 A_c(\text{ft}^2)$$
$$C(\$)_{insulation} = 2{,}249 + 0.54 A_c(\text{ft}^2)$$

- *Reverse flow (in 1995 dollars):*

$$C(\$)_{reverse\ flow} = C(\$)_{housing} + C(\$)_{bag} \quad (13.20)$$

where,

$$C(\$)_{housing} = C(\$)_{metal} + C(\$)_{stainless} + C(\$)_{insulation}$$
$$C(\$)_{metal} = 35{,}086 + 8.41 A_c(\text{ft}^2)$$
$$C(\$)_{stainless} = 16{,}349 + 6.32 A_c(\text{ft}^2)$$
$$C(\$)_{insulation} = 1{,}355 + 0.96 A_c(\text{ft}^2)$$

- *Pulse jet (in 1995 dollars):*

$$C(\$)_{pulsejet} = C(\$)_{housing} + C(\$)_{bag} \quad (13.21)$$

where,

$$C(\$)_{housing} = C(\$)_{metal} + C(\$)_{stainless} + C(\$)_{insulation}$$
$$C(\$)_{metal} = 11{,}620 + 6.66 A_c(\text{ft}^2)$$
$$C(\$)_{stainless} = 12{,}580 + 5.41 A_c(\text{ft}^2)$$
$$C(\$)_{insulation} = 1{,}713 + 1.12 A_c(\text{ft}^2)$$

## 13.3.2 Total annual cost

- *Annual direct cost* (ADC)
  - *Labor*: operation = 6 h/day, maintenance = 2 h/day
  - *Materials*: operation = 0, maintenance = 100% maintenance labor
  - *Material replacement*: bags typically replaced every 2 years; include capital recovery costs

$$\text{CRC} = \text{CRF}\left[C(\$)_{bags} + C(\$)_{labor}\right]$$

where $C(\$)_{bags}$ and $C(\$)_{labor}$ are the cost of bags and labor to replace bags.

$$C(\$)_{labor} = 10\text{–}20 \text{ min/bag reverse-flow,}$$
$$= 5\text{–}10 \text{ min/bag for the pulse jet}$$

- *Utilities* (see Section 13.2)
  - *Electric power for fans*: typical $\delta P = 6$ to 7 in. $H_2O$ for reverse-flow baghouses
  - *Compressed air (dried and filtered) for pulse jet*: typically 2 scfm/1000 ft$^3$ acfm of flue gas
- *Waste disposal*: $24/ton on site $< 2$ miles, off-site $0.60$ ton$^{-1}$ mile$^{-1}$ plus tipping fees if applicable

- *Annual indirect cost* (AIC). Capital recovery cost should reflect that the cost of bags are itemized separately:

$$\text{CRC} = \text{CRF}\left[\text{TIC} - C(\$)_{bags} - C(\$)_{labor}\right] \quad (13.22)$$

The typical lifetime of the entire baghouse is 20 years.

**EXAMPLE 13.2   REMOVAL OF PARTICLES FROM THE EXHAUST OF A LIME KILN USING A FABRIC FILTER**

Consider Examples 12.7 and 12.8, in which a fabric filter and an ESP were sized to remove particles from the exhaust of a lime kiln. The exhaust gas leaves the kiln at 325°F (162.8°C) at a volumetric flow rate of 50,000 scfm (23.61 m³/s), and contains particles with a mass concentration of 10 g/m³ at STP. It is desired to remove 99% of the particles. Because the particles collected can be used for other purposes, a vendor will buy the delivered particles for $6/ton. Since lime dust is alkaline, the baghouse will use stainless steel. Management has decided to install a reverse-flow baghouse containing five modules and each module will be cleaned every 60 min (i.e., 3600 s/cycle). The bags are 15 ft long and 10 in. in diameter (39.09 ft²/bag). The maximum pressure drop in the ducts between the baghouse and exhaust fan will be 5 in. of $H_2O$. The installation requires the preparation of the site ($50,000) and the construction of a building ($200,000).

Performance and design data are as follows:

- Operating time $(t_{op}) = 8640$ h/yr
- Inlet concentration $= 10$ g/m³
- Outlet concentration $= 0.1$ g/m³
- Removal efficiency $= 99\%$
- Mass of lime particles collected $= 7776$ tons/yr
- Electrical power $= \$0.07/kWh$
- $Q_{exhaust} = 50,000$ scfm
- $T_{exhaust} = 325°F$ (162.8°C)
- Bag cleaning cycle $= 1$ h (60 min)
- Collected lime dust worth $6/ton delivered
- Pressure drop in ducts $(\delta P_{duct}) = 5$ in. $H_2O$
- Fan efficiency $= 65\%$
- The preparation of the site and construction of new building $= \$50,000$ and $\$200,000$ respectively
- Bag lifetime $= 2$ yr, baghouse lifetime $= 20$ years, inflation $= 3\%$, CRF interest $= 10\%$

From Table 10-12, a typical gas-to-cloth ratio for a lime kiln is 0.7 m/min (0.01167 m/s) and the dust cake constants can be assumed to be $K_1 = 350$ N min m⁻³ and $K_2 = 9 \times 10^4$ s⁻¹.

*Equipment cost* $(C_{1995})$. The actual volumetric flow rate (162.8°C, 1 atm) is

$$Q(\text{acfm}) = Q(\text{scfm}) \frac{162.8 + 273}{298}$$

$$= 73,120 \text{ acfm } (34.53 \text{ m}^3/\text{s})$$

The total bag surface area is

$$A_{\text{bags}} = (34.53 \text{ m}^3/\text{s}) \left( \frac{0.7 \text{ m/min}}{60 \text{ m/s}} \right)$$

$$= 2959 \text{ m}^2 (31,842 \text{ ft}^2)$$

and the total number of bags is

$$N_{\text{bags}} = 31,842 \text{ ft}^2/39.09 \text{ ft}^2/\text{bag} = 814 \text{ bags}$$

The purchase price of the baghouse $(C_{1995})$ is

$$C_{\text{bags}} = (\$2.03/\text{ft}^2)(31,842 \text{ ft}^2) = \$64,639$$

$$C_{\text{housing}} = C_{\text{metal}} + C_{\text{stainless}} + C_{\text{insulation}}$$

$$= [35,086 + (8.41)(31,842)]$$

$$+ [16,349 + (6.32)(31,842)]$$

$$+ [1355 + (0.96)(31,842)]$$

$$= 302,877 + 217,590 + 31,923 = \$552,390$$

$$C(\$)_{\text{reverse-flow}} = C_{\text{bags}} + C_{\text{housing}} = 64,639 + 552,390$$

$$= \$617,029$$

*Utility costs.* The overall pressure drop is

$$\delta P_{\text{overall}} = K_1 U_0 + K_2 c_0 U_0^2 t + \delta P_{\text{duct}}$$

$$= (350 \text{ N min m}^{-3})(\text{Pa m}^2 \text{N}^{-1})(1 \text{ atm}/10^5 \text{Pa})$$

$$(0.7 \text{ m/min})(33.91 \text{ ft } H_2O/\text{atm})(12 \text{ in./ft}) +$$

$$(9 \times 10^4/\text{s})(10 \text{ g/m}^3)(0.01167 \text{ m/s})^2$$

$$(3600 \text{ s})(\text{kg}/1000 \text{ g})(\text{N s}^2 \text{ kg}^{-1} \text{ m}^{-1})$$

$$(\text{atm}/10^5 \text{ N m}^{-2})(33.91)(12)(\text{in. } H_2O/\text{atm})$$

$$+ 5.0 \text{ in. } H_2O$$

$$= 0.99 + 1.79 + 5.0 = 7.78 \text{ in. } H_2O$$

The electrical power cost for the fan is

$$P(\text{kW}) = \frac{(1.17 \times 10^{-4})(50,000 \text{ scfm})(7.78 \text{ in.} H_2O)}{0.65}$$

$$= 70.02 \text{ kW}$$

$$C(\$)_{\text{fan power}} = (\$0.07/\text{kWh})(70.02 \text{ kW})(8640 \text{ h/yr})$$

$$= \$42,348$$

*Waste disposal.* The particle collection rate is

$$\dot{m}_p = \eta Q c_{\text{p, into baghouse}} = (0.99)(10 \text{ g/m}^3)(23.61 \text{ m}^3/\text{s})$$

$$(60 \text{ s/min})(\text{lb}_m/454 \text{ g})(\text{ton}/2000 \text{ lb}_m)$$

$$= 0.015 \text{ ton/min} = 7776 \text{ tons/yr}$$

The cost of hauling is

$$C(\$)_{hauling} = (\$24/ton)(0.015 \ ton/min)$$
$$(60 \ min/h)(8640 \ h/yr) = \$186,624$$

The annual recovery cost (ARC) of the lime dust is

$$ARC = (\$6/ton)(7776 \ tons/yr) = \$46,656$$

---

#### Total Initial Cost $(TIC_{1995})$ for a Fabric Filter

1. *Initial direct cost* $(IDC_{1995})$

    1.1 Initial equipment cost $(IEC_{1995})$

    | | |
    |---|---:|
    | Baghouse purchase price (C) | 617,029 |
    | Instrumentation (10% C) | 61,703 |
    | Sales tax (3% C) | 18,510 |
    | Freight (5% C) | 30,851 |

    Initial equipment cost $(IEC_{1995}) = 728,093$

    1.2 Initial facilities cost $(IFC_{1995})$

    | | |
    |---|---:|
    | Site preparation | 50,000 |
    | Building | 200,000 |
    | Foundations and supports (4% IEC) | 29,123 |
    | Air handling and erection (50% IEC) | 364,046 |
    | Electrical (8% IEC) | 58,247 |
    | Piping (2% IEC) | 14,562 |
    | Insulation ductwork (7% IEC) | 50,966 |
    | Painting (1% IEC) | 7,280 |

    Initial facilities cost $(IFC_{1995}) = 774,224$

*Total initial direct cost* $(IDC_{1995}) = (IEC + IFC)_{1995} =$
$728,093 + 774,224 = \$1,502,317$

2. *Initial installation cost* $(IIC_{1995})$

    | | |
    |---|---:|
    | Engineering (10% IEC) | 72,809 |
    | Contractor fees (10% IEC) | 72,809 |
    | Construction and field expenses (20% IEC) | 145,619 |
    | Start-up (2% IEC) | 14,562 |
    | Performance (1% IEC) | 7,280 |
    | Contingencies (3% IEC) | 21,842 |

    *Total initial installation cost* $(IIC_{1995}) = \$334,921$

*Total initial cost:* $TIC_{1995} = (IDC + IIC)_{1995} =$
$1,502,317 + 334,921 = \$1,837,238$

---

#### Total Annual Cost $(TAC_{1995})$ for a Fabric Filter

1. *Annual direct cost* (ADC)
   (24 h/day, 360 days/yr, 8640 h/yr)
   Operations

    | | |
    |---|---:|
    | Operator (6 h/day, 360 days/yr, $12.00/h) | 25,920 |
    | Supervisor (15% operator) | 3,880 |

Maintenance

| | |
|---|---:|
| Labor (3 h/day, 360 days/yr, $13.20/h) | 14,256 |
| Material (100% labor) | 14,256 |

Operations and maintenance (O&M) = 58,312

Bag replacement (bags are replaced every 2 years)

| | |
|---|---:|
| Labor (3 h/day, 360 days/yr, $13.20/h) | 14,256 |
| Materials $(C_{bags, 1995}) = (64,639)$ | 64,639 |
| Bag replacement cost $= (C_{bags} + labor)$ $(CRF, 2 \ yr, 10\%)$ | |
| $= (64,639 + 14,256)(0.5762)$ | 45,459 |
| Utilities, electric power for fans | 42,348 |
| Waste disposal (assume 99 % removal) | 186,624 |

*Total annual direct* $(ADC_{1995}) =$
$58,312 + 45,459 + 42,348 + 186,624 = \$332,743$

2. *Annual indirect cost* $(AIC_{1995})$

    | | |
    |---|---:|
    | Overhead (60% O&M) | 34,987 |
    | Administrative charges (2% TIC) | 36,744 |
    | Insurance (1% TIC) | 18,372 |
    | Property tax (1% TIC) | 18,372 |
    | Capital recovery cost (CRC) (assume 20 years, 10%, CRF = 0.1174) | |
    | $CRC = (CRF)[TIC - (C_{bags} + C_{labor})_{1995}]$ | |
    | $= (0.1174)[1,837,238 - (64,639 + 14,256)]$ | 206,429 |

    *Total annual indirect cost* $(AIC_{1995}) = \$314,904$

3. *Annual recovery cost* $(ARC_{1995})$      46,656

*Total annual cost:* $TAC_{1995} = (ADC + AIC - ARC)_{1995}$

$$= 332,743 + 314,904 - 46,656 = \$600,991$$

---

*Summary: fabric filter annual costs*

| | | |
|---|---|---:|
| Annual cost/scfm = 600,991/50,000 | | $12.02/scfm |
| Annual cost/ ton of particles removed: | | |
| 600,991/7776 tons/yr | | $77.29/ton |

---

# 13.4 Electrostatic precipitator

The engineer must select the type of electrostatic precipitator (ESP) suitable for the particles to be removed, the temperature, and the composition of the process gas stream. Table 12.2 should be used to select the value of the migration velocity. In some instances the manufacturer will perform field tests using a slip stream of the actual process gas stream to determine the migration velocity by experimental means. From this value of the migration velocity, the area of the collecting plates can be estimated using the Deutsch equation (Equation 12-76). Next, the engineer must select the number of modules in parallel and number modules in series and the rapping frequency. The costs shown below were obtained from Vatavuk (1990) and updated to 1995 using the CE index (Eq 13.5) and the date the original data were compiled.

## 13.4.1 Equipment cost

Shown below is an empirical equation to estimate the flange-to-flange cost (1995 dollars) of a dry, rigid-electrode, plate-wire precipitator that includes the essential auxiliary equipment and the following options:

- Inlet and outlet nozzles and diffuser plates
- Hopper heater and level gauges
- Weather enclosure and access stairs
- Structural supports
- Insulation

$$A_p < 50,000 \text{ ft}^2 \quad C(\$)_{ESP} = 1050.6 A_p^{0.6276}$$
$$A_p > 50,000 \text{ ft}^2 \quad C(\$)_{ESP} = 98.93 A_p^{0.8431} \quad (13.23)$$

See Vatavuk (1990) for similar expressions for other types of ESPs.

## 13.4.2 Total annual cost

- *Annual direct cost* $(ADC_{1955})$
  - *Operations*: operator = 3 h/day, supervisor 15% operator, coordinator = 33.3% operator
  - *Maintenance*:
    labor = $4864 minium *cost*; $0.0973/ft² for $A_p > 50,000$ ft², material = 1% EEC
  - *Utilities* (see Section 13.2)
    - *Electricity to produce corona and operate rappers*
    $(\$2.287 \times 10^{-3}/\text{ft}^2) A_p(\text{ft}^2) t(\text{h/yr})(\$0.07/\text{kWh})$
    - *Electricity to operate fans*: typically 0.1 in. $H_2O < \delta P_{ESP} < 0.4$ in. $H_2O$
  - *Waste disposal*: $24/ton on site < 2 miles, $0.60/ton-mile off-site plus tipping fees, if applicable
- *Annual indirect cost* $(AIC_{1955})$. The typical lifetime of an ESP is 20 years.

## EXAMPLE 13.3   REMOVAL OF PARTICLES FROM THE EXHAUST OF A LIME KILN USING AN ESP

Estimate the cost of removing 99% of the particles from the lime kiln described in Example 13.2 and compare the costs to those for a fabric filter. Assume that the "flange-to-flange" pressure drop across the ESP is 0.3 in. $H_2O$. From Table 12.2 a migration velocity $(w)$ of 0.05 ft/s (1.524 cm/s) will be used.

Performance and design data are as follows:

- Operating time $(t_{op})$ = 8640 h/yr
- Electrical power, $0.07/kWh
- $Q_{exhaust}$ = 50,000 scfm
- $T_{exhaust}$ = 325°F (162.8°C)
- Inlet particle concentration = 10 g/m³
- Outlet particle concentration = 0.1 g/m³
- Overall collection efficiency $(\eta)$ = 99%
- Lime dust captured = 7776 tons/yr
- Pressure drop in duct $(\delta P\text{duct})$ = 5 in. $H_2O$
- Fan efficiency = 65%
- Preparation of the site and construction of new buildings: $50,000 and $200,000, respectively
- ESP lifetime = 20 yr, inflation = 4%, CRF interest = 10%

*Equipment cost* $(C_{1995})$. Using the Deutsch equation, the surface area $(A_p)$ is

$$\eta = 1 - \exp\left(-\frac{A_p w}{Q}\right)$$

$$\ln(1 - \eta) = \ln(1 - 0.99) = \ln(0.01) = -4.605$$

$$= -\frac{A_p(0.05 \text{ ft/s})(60 \text{ s/min})}{73,120 \text{ acfm}}$$

$$A_p = 112,239 \text{ ft}^2$$

The "flange-to-flange" purchase price $(C_{ESP\ 1995})$

$$C(\$)_{ESP\ 1995} = (98.93)(A_p \text{ ft}^2)^{0.8431}$$
$$= (98.93)(112,239 \text{ ft}^2)^{0.8431}$$
$$= \$1,791,035$$

*Utility costs.* The overall pressure drop is

$$\delta P_{overall} = \delta P_{ESP} + \delta P_{duct} = 0.30 \text{ in. } H_2O$$
$$+ 5.0 \text{ in. } H_2O = 5.3 \text{ in. } H_2O$$

The electrical power for the fan

$$\text{Power}_{\text{fan}}(\text{kW}) =$$

$$\frac{(1.17 \times 10^{-4})(50{,}000 \text{ scfm})(5.3 \text{ in. H}_2\text{O})}{0.65} = 47.7 \text{kW}$$

The total electrical power consumed to operate the fans and to produce the corona and rappers is:

$$C(\$)_{\text{electrical, 1995}} = \left[C(\$)_{\text{fan}} + C(\$)_{\text{ESP}}\right]_{1995}$$

$$= (\$0.07/\text{kWh})(8640 \text{ h/yr})(47.7 \text{ kW})$$

$$+ (1.94 \times 10^{-3})(112{,}239 \text{ ft}^2)$$

$$(8640 \text{ h/yr})(\$0.07/\text{kWh})$$

$$= 28{,}848 + 131{,}691 = \$160{,}539$$

*Waste disposal.* The particle collection rate is

$$\dot{m}_p = \eta Q c_{p,\text{ inlet to ESP}} = (0.99)(10 \text{ g/m}^3)(23.61 \text{ m}^3/\text{s})$$

$$(60 \text{ s/min})(\text{lb}_m/454 \text{ g})(\text{ton}/2000 \text{ lb}_m)$$

$$= 0.015 \text{ ton/min} = 7776 \text{ tons/yr}$$

The cost of hauling is

$$C(v)_{\text{hauling}} = (\$24/\text{ton})(0.015 \text{ ton/min})$$

$$(60)(8640 \text{ h/yr}) = \$186{,}624$$

The annual recovery cost (ARC) of the lime dust is

$$\text{ARC}_{1995} = (\$6/\text{ton})(7776 \text{ tons/yr}) = \$46{,}656$$

---

## Total Initial Cost $\left(\text{TIC}_{1995}\right)$ for an ESP

1. *Initial direct cost* $\left(\text{IDC}_{1995}\right)$
   1.1 Initial equipment cost $\left(\text{IEC}_{1995}\right)$

   | | |
   |---|---:|
   | Electrostatic precipitator purchase price (C) | 1,791,035 |
   | Instrumentation (10% C) | 179,104 |
   | Sales tax (3% C) | 53,731 |
   | Freight (5% C) | 89,552 |

   Initial equipment cost $\left(\text{IEC}_{1995}\right) = 2{,}113{,}422$

   1.2 Initial facilities cost $\left(\text{IFC}_{1995}\right)$

   | | |
   |---|---:|
   | Site preparation | 50,000 |
   | Building | 200,000 |
   | Foundations and supports (4% IEC) | 84,537 |
   | Air handling and erection (50% IEC) | 1,056,711 |
   | Electrical (8% IEC) | 169,073 |
   | Piping (1% IEC) | 21,134 |
   | Insulation for ducts (2% IEC) | 42,268 |
   | Painting (2% IEC) | 42,268 |

   Initial facilities cost $\left(\text{IFC}_{1995}\right) = 1{,}665{,}991$

*Total initial direct cost* $\left(\text{IDC}_{1995}\right)$
$$= \left(\text{IEC} + \text{IFC}\right)_{1995} = 2{,}113{,}422 + 1{,}665{,}991$$
$$= \$3{,}779{,}413$$

2. *Initial installation cost* $\left(\text{IIC}_{1995}\right)$

   | | |
   |---|---:|
   | Engineering (20% IEC) | 422,684 |
   | Construction and field expenses (20% IEC) | 422,684 |
   | Contractor fees (10% IEC) | 211,342 |
   | Startup (1% IEC) | 21,134 |
   | Performance test (1% IEC) | 21,134 |
   | Model studies (2% IEC) | 42,268 |
   | Contingencies (3% IEC) | 63,402 |

   *Total initial installation cost* $\left(\text{IIC}_{1995}\right) = 1{,}204{,}648$

*Total initial cost:* $\text{TIC}_{1995}$
$$= \left(\text{IDC} + \text{IIC}\right)_{1995} = 3{,}779{,}413 + 1{,}204{,}648$$
$$= \$4{,}984{,}061$$

---

## Total Annual Cost $\left(\text{TAC}_{1995}\right)$ for an ESP

1. *Annual direct cost* (ADC) (8640 h/yr, 360 days/yr)
   Operations

   | | |
   |---|---:|
   | Operator (3 h/day, 360 days/yr, $12.00/h) | 12,960 |
   | Supervisor (15% operator) | 1,944 |
   | Coordinator (33.3% operator) | 4,315 |

   Maintenance

   | | |
   |---|---:|
   | Labor ($0.0973/ft$^2$)(112,239 ft$^2$) | 10,920 |
   | Material (1% IEC) | 21,134 |

   Operations and maintenance (O&M) = 51,273

   Utilities

   | | |
   |---|---:|
   | Electricity for fan | 28,848 |
   | Electricity for corona and rappers | 131,691 |
   | Waste disposal (same as fabric filter) | 186,624 |

   *Total annual direct cost* $\left(\text{ADC}_{1995}\right) = \$398{,}436$

2. *Annual indirect cost* $\left(\text{AIC}_{1995}\right)$

   | | |
   |---|---:|
   | Overhead (60% O&M) | 30,764 |
   | Administrative charges (2% TIC) | 99,681 |
   | Property tax (1% TIC) | 49,840 |
   | Insurance (1% TIC) | 49,840 |
   | Capital recovery cost (CRC) (assume 20 yr, 10%, CRF = 0.1175) | |
   | CRC = (CRF)(TIC) = (0.1175)(4,984,061) | 585,627 |

   *Total annual indirect cost* $\left(\text{AIC}_{1995}\right) = \$815{,}752$

3. *Annual recovery cost* $(ARC_{1995})$

$$ARC_{1995} = (\$6/ton)(7776 \text{ tons/yr} \qquad 46,656$$

*Total annual cost:* $\quad TAC_{1995} = (ADC + AIC - ARC)_{1995}$

$$= (398,436 + 815,752 - 46,656) = \$1,167,532$$

Summary: ESP annual costs

Annual cost/scfm $= 1,167,532/50,000 \qquad \$23.35/\text{scfm}$

Annual cost/ ton of particles removed

$$= 1,167,532/7776 \qquad \$150.14/\text{ton}$$

---

# 13.5  Comparison of costs to remove particles from the exhaust of a lime kiln

Control system performance is as follows:

- $Q = 50,000$ scfm ($23.61$ m$^3$/s)
- $c_0 =$ inlet particle concentration 10 g/m$^3$
- $T_0 =$ inlet gas temperature 325°F (162.8°C)
- $\eta =$ removal efficiency 99%
- $t_{op} =$ operation time 8640 h/yr

| Cost | Fabric filter | ESP |
|---|---|---|
| TIC$_{1995}$ ($) | 1,837,238 | 3,779,413 |
| TAC$_{1995}$ ($) | 600,991 | 1,167,532 |
| Annual cost ($/scfm) | 12.02 | 23.35 |
| Annual cost, ($/ton of particles removed) | 77.29 | 150.14 |

We observe that the TIC and TAC of the electrostatic precipitator are approximatly double those for the fabric filter system. Accordingly, one would choose elec- trostatic precipitation only to remove particles that can not be removed adequately by a fabric filtration.

# 13.6  Carbon adsorber

Carbon adsorption is an attractive control method if the pollutant is very toxic and it is necessary to guarantee that no pollutant escapes removal. Adsorption is also attractive if the gas or vapor is present in very low concentrations or if the recovered material has financial value. Adsorption should be avoided if the process gas steam contains particles that can clog the adsorber. The costs shown below were obtained from Vatavuk (1990) and updated to 1995 using the CE index (Eq 13.5) and the date the original data were compiled.

## 13.6.1  Equipment cost

The cost of a carbon adsorber depends on decisions about the bed geometry and regeneration cycle. Once the mass of carbon $(M_c)$ has been selected, the basic initial equipment cost (IEC) can be estimated from the following.

$$C(\$)_{ca, 1995} = 6.49[C_c + (N_A + N_D + N_C)C_v]Q^{-0.133} \qquad (13.24)$$

where

$Q =$ flue gas volumetric flow rate (acfm)
$\qquad 4,000 < Q < 500,000$ acfm

$C_c(\$) =$ carbon cost $= p_c M_c$

$\quad M_c =$ mass of carbon (lb$_m$)

$\quad p_c =$ cost of carbon, use pollutant specific values, typically \$2.00/lb$_m$

$N_A =$ number of adsorbing beds

$N_D =$ number of desorbing beds

$N_C =$ number of beds being cooled

$C_v(\$) =$ cost of vessel $= 302\, S_v^{0.776}$

$\quad S_v =$ vessel surface area (ft$^2$)
$\qquad (97 < S_v < 2100$ ft$^2)$

## 13.6.2  Annual direct cost

- *Labor*: operator $= 0.5$ h/shift, maintenance $= 0.5$ h/shift
- *Materials*: operation $= 0$, maintenance $= 0$
- *Material replacement*: carbon replacement (pollutant specific, typically 2 to 5 years)
- *Capital recovery costs* (CRC), *and tax and transportation* (i.e., tax and transportation factor $= 1.08$)
  - CFC $=$ CRF $(1.08C_c + C_{\text{replacement}})$ where $C_c$ and $C_{\text{replacement}}$ are the cost of the initial and replacement carbon
  - Carbon replacement: labor hours $= 3.3 \times 10^{-3}$ h/lb$_m$ carbon, or labor cost $= \$0.06/\text{lb}_m$
- *Utilities* (see Section 13.2)
  - Steam, cooling water
  - Electric power fan costs, calculate the pressure drop across the bed $(\delta P)$ directly

- Waste disposal: carbon is typically regenerated at a cost to be negotiated; disposal costs are typically $39 to $72, plus transportation for a 150-lb$_m$ canister of carbon disposed in a landfill.

## 13.6.3  Annual indirect cost

The capital recovery cost (CRC) should reflect the fact that the capital costs of the carbon costs are itemized separately. The typical lifetime of a carbon adsorber unit is 10 years.

$$CRC = CRF\left[TIC - \left(1.08C_c + C_{replacement}\right)\right] \quad (13.25)$$

**EXAMPLE 13.4   REMOVAL OF DIETHYLAMINE WITH A REGENERATIVE CARBON ADSORBER**

Estimate the total initial cost (TIC) and total annual cost (TAC) of the rotary regenerative adsorber described in Example 10.2. It will be assumed that a face velocity of 0.1 m/s will be used in order to minimize the pressure drop across the bed.
Performance and design data are as follows:

- $Q = 30,000$ acfm at 30°C, 101 kPa
- Operating time $(t_{op})$, 3,600 h/yr, 360 days/yr, 10 h/day
- Electrical power, $0.07/kWh
- Inlet VOC concentration 1000 ppm
- Inlet VOC mass flow rate $(\dot{m}_p) = 41.58$ g/s (593.46 tons/yr)
- 100% removal efficiency, mass of VOC removed = 593.46 tons/yr (3297 lb$_m$/day)
- $U_a$ (face velocity) = 0.1 m/s
- Bed surface area $(A_s) = 965$ m$^2$, diameter = 35 m
- Bed depth $(Z) = 0.338$ m,
- Mass of adsorbing carbon = 15.8 tons
- Overall pressure drop across the bed $(\delta P) = 0.693$ kPa (2.79 in. H$_2$O)
- Fan efficiency $(\eta) = 75\%$
- Preparation of the site and construction of a new building will cost $50,000 and $200,000 respectively.
- The carbon has a lifetime of 5 years and the adsorber assembly has a lifetime of 10 years. Waste carbon will be placed in 150 lb$_m$ canisters that cost $ 50/canister to transport.

*Equipment cost* $(C_{1995})$. For computational purposes it will be assumed that regeneration is achieved using steam, which is later cooled with cooling water

to separate the pollutant from the condensed steam. The total number of adsorbing ($N_A$), desorbing ($N_D$) and cooling ($N_C$) units is

$$N_{total} = N_A + N_D + N_C = 1 + 3.81 + 2 = 6.81$$

Based on the time for adsorption (1 h), desorption (3.24 h) and cooling (2 hrs) (total = 6.24), it will be assumed that the total mass of carbon $(M_c)$ and the total cost of this carbon $(C_c)$ is as follows:

$$M_c = (6.81)(15.8 \text{ tons/unit})$$
$$= 107.6 \text{ tons } (215,200 \text{ lb}_m)$$
$$C_{c, 1995} = (\$2/\text{lb}_m)(107.6)(2000) = \$430,400$$

The total surface area of the vessel $(S_v)$ is equal to 6.81 times the surface area of each adsorbing portion which will be treated as a right circular cylinder of diameter $(D)$ and height $(L)$.

$$S_v = 6.81(2A_s + \pi DL)$$
$$= 6.81[2(965) + \pi(35)(0.338)]$$
$$= (6.81)(1967.1) = 13,396 \text{ m}^2 \ (144,120 \text{ ft}^2)$$

The cost of the adsorbing portion of the vessel $(C_v)$ is

$$C_v = 302\left(\frac{S_v}{6.81}\right)^{0.776} = (302)\left(\frac{144,120}{6.81}\right)^{0.776}$$
$$= (302)(21,163)^{0.776} = \$686,522$$

The equipment cost $(C)$ can be estimated from

$$C_{1995} = 6.49\left[C_c + C_v(N_A + N_D + N_C)\right]Q^{-0.133}$$
$$= 6.49\left[430,000 + (6.81)(686,522)\right]$$
$$(30,000^{-0.133}) = 6.49(5,105,215)(0.2538)$$
$$= \$8,409,116$$

It must be noted that if separate carbon canisters were used, it would be easier to identify the number of beds: $N_A$, $N_D$, and $N_C$.

*Utility costs.* The power for the fans is

$$Power_{fans} = \frac{(1.17 \times 10^{-4})Q\delta P}{\eta}$$
$$= \frac{(1.17 \times 10^{-4})(30,000)(2.79)}{0.75}$$
$$= 13.06 \text{ kW}$$

The cost of the electric power for the fans is

$$C(\$)_{fans,1995} = (\$0.07/kWh)(10\ h/day)$$
$$(360\ days/yr)(13.06\ kW)$$
$$= \$3291.12$$

The mass of steam for regeneration is

$$\dot{m}_s = (3.5\ lb_m\ steam/lb_m\ VOC)$$
$$(3297\ lb_m\ VOC/day)(360\ days/yr)$$
$$= 4{,}154{,}220\ lb_m/yr$$

The cost of the steam is

$$C(\$)_{steam,1995} = (4{,}154{,}220\ lb_m/yr)$$
$$(\$0.007/lb_m\ steam) = 29{,}079$$

The cost of the cooling water to condense steam and separate diethylamine for reuse is

$$C(\$)_{cooling\ water,\ 1995} = (3.43\ gal\ water/lb_m\ steam)$$
$$(\dot{m}_s\ lb_m/yr)(p_{cw}\ \$/gal)$$
$$= (3.43)(4{,}154{,}220)$$
$$(2.5 \times 10^{-4}) = \$3562$$

---

## Total Initial Cost $(TIC_{1995})$ for a Carbon Adsorber

1. *Initial direct cost* $(IDC_{1995})$
   1.1 Initial equipment cost $(IEC_{1995})$

   | | |
   |---|---|
   | Purchase cost rotary regenerative adsorber (C) | $8,409,116 |
   | Instrumentation (10% C) | 840,912 |
   | Sales tax (3% C) | 252,273 |
   | Freight (5% C) | 420,456 |
   | Initial equipment cost $(IEC_{1995})$ = | $9,922,757 |

   1.2 Initial facilities cost $(IFC_{1995})$

   | | |
   |---|---|
   | Site preparation | 50,000 |
   | Building | 200,000 |
   | Foundations and supports (8% IEC) | 793,820 |
   | Air handling and erection (14% IEC) | 1,389,186 |
   | Electrical (4% IEC) | 396,910 |
   | Piping (2% IEC) | 198,455 |
   | Insulated ductwork (1% IEC) | 99,228 |
   | Painting (1% IEC) | 99,228 |
   | Initial facilities cost $(IFC_{1995})$ = | 3,226,827 |

   *Total initial direct cost*
   $$(IDC_{1995}) = (IEC + IFC)_{1995} = \$13{,}149{,}583$$

2. *Initial Installation Cost* $(IIC_{1995})$

   | | |
   |---|---|
   | Engineering (10% IEC) | 992,275 |
   | Contractor fees (10% IEC) | 992,275 |
   | Construction and field expenses (5% IEC) | 496,138 |
   | Start-up (2% IEC) | 198,455 |
   | Contingencies (3% IEC) | 297,682 |
   | Total initial installation cost $(IIC_{1995})$ = | $2,976,825 |

---

*Total initial cost:* 
$$TIC_{1995} = (IDC + IIC)_{1995}$$
$$= 9{,}922{,}757 + 2{,}976{,}825$$
$$= \$12{,}899{,}582$$

---

## Total Annual Cost $(TAC_{1995})$ for a Carbon Adsorber

1. *Annual direct cost* (ADC)
   (10 h/day, 1 shift/day, 360 days/yr)

   | | |
   |---|---|
   | Operations | |
   | Operator (0.5 h/shift, $12.00/h) | 2,160 |
   | Supervisor (15% operator) | 324 |
   | Operating materials | 0 |
   | Maintenance | |
   | Labor (0.5 h/shift, $13.20/hr) | 2,376 |
   | Material (100% labor) | 2,376 |
   | Operations and maintenance (O&M) = | 7,236 |

   Replacement: (assume carbon replaced every 5 years, capital recovery factor (CRF, 5 yr, 10%) = 0.2638)

   | | | |
   |---|---|---|
   | Labor: | CRF [($0.05/lb_m) (107.6) (2000) lb_m carbon] = (0.2638)(0.05)(107.6)(2,000) | 2823 |
   | Materials: | CRF $(1.08\ C_c + C_r)$ = 0.2638 [(1.08)(430,400) + 430,400] | 236,162 |

   | | |
   |---|---|
   | Utilities | |
   | Fan power | 3,291 |
   | Steam generation | 29,079 |
   | Cooling water | 3562 |

   Waste disposal (assume $50/150 lb_m, CRF based on 5 years, 10% = 0.2638)

   | | |
   |---|---|
   | Waste removal = CFC[(50/150)(107.6) (2000) lb_m] = (0.2638) [71,733] = | 18,923 |
   | Total annual direct cost $(ADC_{1995})$ = | $297,785 |

2. *Annual indirect cost* $(AIC_{1995})$

   | | |
   |---|---|
   | Overhead (60% O&M) | 4,342 |
   | Administrative charges ((2% TIC) | 257,992 |
   | Insurance ((1% TIC) | 128,996 |
   | Property tax (1% TIC) | 128,996 |

   Capital recovery cost (CRC) CRF based on 10 years, 10% = 0.1628
   $$CRC = CRF\left[TIC - (1.08 C_c + C_{replacement})\right]$$

$$= (0.1682)[12,899,582$$
$$- ((1.08)(430,400) + 430,400)]$$

$$= (0.1682)(12899582 - 895,232) = 2,019,134$$

*Total annual indirect cost* $(\text{AIC}_{1995}) = \$2,539,460$

3. *Annual recovery cost* $(\text{ADC}_{1995})$          0

*Total annual cost*:
$$\text{TAC}_{1995} = (\text{ADC} + \text{AIC} - \text{ARC})_{1995}$$
$$= 297,785 + 2,539,460 - 0 = \$2,837,245$$

---

Summary: regenerative adsorber annual costs
Annual cost/scfm $= \$2,837,245/30,000 = \$94.57/\text{scfm}$
Annual cost/ ton of VOC $= \$2,837,245/593.46 =$
$$\$4780.85/\text{ton VOC}$$

# 13.7  Scrubbers

Scrubbers are an attractive control technique if the pollutant is soluble in water and the treatment of the scrubbing water does not result in prohibitive costs. Prior to estimating the costs of scrubbers, users must compute the flue gas volumetric flow rate, liquid flow rate, height and diameter of the packed bed, and the flange-to-flange pressure drop. The costs shown below were obtained from Vatavuk (1990) and updated to 1995 using the CE index (Equation 13.5) appropriate for the original data.

## 13.7.1  Equipment cost

The purchase price of the basic scrubber tower is

$$C(\$)_{s,\,1995} = C(\$)_t + C(\$)_p + C(\$)_{\text{aux}} \quad (13.26)$$

$$C(\$)_{t,\,1995} = \text{tower cost} = (C_f)137.9S_t$$

and $S_t$ is the total surface area of tower $(\text{ft}^2)$ and $C_f$ is a cost factor that depends on the material used to construct the tower:

| Tower material | $C_f$ |
|---|---|
| 304 stainless steel | 1.10–1.75 |
| Polypropylene | 0.80–1.10 |
| Polyvinyl chloride | 0.50–0.80 |

The surface area of the tower $(S_t)$, including space for the packing and headers, can be estimated from

$$S_t = \pi D_t \left( H_t + \frac{D_t}{2} \right) \quad (13.27)$$

where $D_t$ is the diameter of the tower. The height of the tower $H_t$ can be estimated from the following:

$$H_t(\text{ft}) = (1.4Z + 1.02D_t + 2.81) \quad (13.28)$$

where $Z$ is the height of the packing (ft).

---

$$C(\$)_{p,\,1995} = \text{cost of packing materials, } (\$/100 \text{ ft}^3) \text{ in lots of } > 100 \text{ ft}^3$$

---

- 1 inch, 304 stainless Pall, Raschig, Ballast rings      78-119
- 1 inch, ceramic Raschig rings, Berl saddles      31-43
- 1 inch, polypropylene Pall, Ballast rings, flexisaddles      14-41
- 2 inch ceramic Raschig, rings, Berl saddles      12-36
- 2 inch polypropylene Pall, Ballast rings, flexisaddles      6-23

$C(\$)_{\text{aux}} =$ cost of auxiliary equipment such as pumps, demisters, controls for pH, chemical addition, solvent flow, etc.

The cost of a circulating pump for the scrubbing liquid is

$$C(\$)_{\text{pump},\,1995} = (\$19.2/\text{gpm})(L) \quad (13.29)$$

where $L$ is the volumetric flow rate of the water in gpm

## 13.7.2  Total annual cost

*Annual direct cost* $(\text{ADC}_{1995})$

- *Operating labor*: operator 0.5 h per shift typically $12.00/h, supervisor 15% of operator,
- *Operating materials*
  - Chemical for solvent treatment (pH, chlorine, ozone, etc.)
  - Makeup solvent (typically 0.1 to 10%) to account for solvent discharged with output gas (carryover)
- *Maintenance*:
  - *Labor* 0.5 h per shift, typically $13.20/h
  - *Materials*: 100% maintenance labor
- *Utilities*: (see Section 13.2)
  - Electric power for pump
  - Electric power for fan
  - Makeup scrubbing liquid
  - Disposing of a portion of the scrubbing liquid ($1.2/1000 gal)
  - Removal of solids from the scrubbing liquid
  - Disposal of the removed solids (typically $20 to $30/ton, depends on the toxicity of the material) plus transportation of $0.60/ton-mile

## 13.7.3  Annual indirect cost

The capital recovery cost can be computed assuming that a typical lifetime of a scrubber unit is 15 years.

### EXAMPLE 13.5  REMOVAL OF DIETHYLAMINE WITH A COUNTERFLOW PACKED-BED ABSORBER

Consider Example 10.3 in which a packed-bed scrubber was designed to remove diethylamine using water as the scrubbing fluid. The objective of this analysis is to estimate the total initial cost (TIC) and total annual cost (TAC) of the design.
Performance and design data are as follows:

- Operating time $(t_{op})$ 3,600 h/yr, 360 days/yr, 10 h/day
- Electrical power $0.059/kWh
- Packing $1\frac{1}{2}$ in. ceramic Berl saddles
- Height of the packed bed $(Z)$ = 5.82 m (19.1 ft)
- Diameter of the packed bed $(D_t)$ = 3.24 m (10.6 ft)
- Volume of the packed bed $(Z\pi D_t^2/4)$ = 1685 ft$^3$
- Tower height
  $(H_t)$ = $1.4Z + 1.02D_t + 2.81$ = 40.36 ft

- Tower surface area
  $(S_t)$ = $\pi D_t(H_t + D_t/2)$ = 1520 ft$^2$
- Volumetric flow rate of polluted air $(Q)$ = 30,000 acfm
- Diethylamine mass flow rate entering scrubber = 41.58 g/s (3297 lb$_m$/day, 593.46 tons/yr)
- $\dot{m}_s$ scrubbing liquid volumetric flow rate $(L)$ = 98.14 lb$_m$/s (706 gpm)
- Tower material = 304 stainless steel
- Auxiliary costs (demister, controls for pH, water flow rate; displays, etc.) = $15,000
- Chemical for water treatment = $2000/yr
- Fresh water makeup = 5%
- Disposal of wastewater = 1%
- Sludge disposal = 0
- $\delta P$ (pressure drop across bed) = 4.74 kPa (19.1 in. H$_2$O)
- $\delta P$ (pressure drop across pump) = 25 psia (57.5 ft H$_2$O)
- Efficiency $(\eta)$ of fan and water pump = 70%
- Lifetime of scrubber = 15 years.
- Assumed costs of preparing the site and a building to house the scrubber are $200,000 and $50,000 respectively
- Outlet VOC mass flow rate is the maximum allowable mass flow rate = 44.9 tons/yr
- VOC removed by scrubber = 548.5 tons/yr $(\eta = 92.4\%)$

*Equipment cost* $(C_{1995})$. The purchase price $(C)$ of the scrubber is

$$C_{1995} = [C(\$)_p + C(\$)_t + C(\$)_{aux}]_{1995}$$

For ceramic $1\frac{1}{2}$ in. Berl saddles,

$$C(\$)_{p,1995} = (\$36/100 \text{ ft}^3)(1685 \text{ ft}^3) = \$607$$

For a 304 stainless steel tower $(C_f = 1.75)$,

$$C(\$)_{t,1995} = 138(C_f)S_t = (138)(1.75)(1520 \text{ ft}^2)$$
$$= \$367,080$$

Since the cost of the fan and motor are incorporated in the initial facilities costs for air handling and erection $(C)$, the only items in $C_{aux}$ are $15,000 for the demister and the controls for pH, water flow rate, plus the cost of the circulating water pump and its electric motor. The cost for the pump and motor is

$$C(\$)_{\text{water pump, 1995}} = (\$19/\text{gpm})(706\ \text{gpm})$$
$$= \$13,414$$

Thus

$$(C)_{1995} = (607 + 367,080 + 15,000 + 13,414)$$
$$= \$396,101$$

*Utility costs.* The cost of (5%) makeup water is

$$C(\$)_{\text{make-up water, 1995}} = (0.05)(706\ \text{gpm})(60)$$
$$\frac{(3600)\text{hr}}{\text{yr}}(\$0.24/1000\ \text{gal}) = \$1830$$

The cost of disposing of 1% of the scrubbing liquid is

$$C(\$)_{\text{waste disposal, 1995}} = (0.01)(706\ \text{gpm})(60)$$
$$\frac{(3600)\text{hr}}{\text{yr}}(\$1.5/1000\ \text{gal}) = \$2287$$

The power and electricity for the fan are

$$\text{Power}_{\text{fan}}(\text{kW}) =$$
$$\frac{(1.17 \times 10^{-4})(30,000\ \text{acfm})(19.1\ \text{in. H}_2\text{O})}{0.7}$$
$$= 95.8\ \text{kW}$$

$$C(\$)_{\text{electricity for fan, 1995}} =$$
$$(\$0.07/\text{kWh})(10)\text{h/day}(360)\text{day/yr}$$
$$(95.8\ \text{kW}) = \$24,142$$

The power and electricity for the circulating pump are

$$\text{Power}_{\text{pump}}(\text{kW}) = \frac{(0.746)(0.0152)(706)}{0.7} = 11.4\ \text{kW}$$

$$C(\$)_{\text{electricity for pump, 1995}} =$$
$$(11.4\ \text{kW})(10\ \text{h/day})(360\ \text{days/yr})$$
$$(\$0.07/\text{kWh}) = \$2,873$$

---

### Total Initial Cost $(\text{TIC}_{1995})$ for a Wet Scrubber

1. *Initial direct cost* $(\text{IDC}_{1995})$

   1.1 Initial equipment cost $(\text{IEC}_{1995})$

   | | |
   |---|---|
   | Scrubber purchase cost (C) | $396,101 |
   | Instrumentation (10% C) | 39,610 |

   | | |
   |---|---|
   | Sales tax (3% C) | 11,883 |
   | Freight (5% C) | 19,805 |
   | Initial equipment cost $(\text{IEC}_{1995})$ = | 467,399 |

   1.2 Initial facilities cost $(\text{IFC}_{1995})$

   | | |
   |---|---|
   | Site preparation | 50,000 |
   | Building | 200,000 |
   | Foundations and supports (12% IEC) | 56,088 |
   | Air handling and erection (40% IEC) | 186,960 |
   | Electrical (1% IEC) | 4,674 |
   | Piping (30% IEC) | 140,220 |
   | Insulated ductwork (1% IEC) | 4,674 |
   | Painting (1% IEC) | 4,674 |
   | Initial facilities cost $(\text{IFC}_{1995})$ = | 647,290 |

   *Total initial direct cost* $(\text{IDC}_{1995}) = (\text{IEC} + \text{IFC})_{1995}$
   $$= 467,399 + 647,290 = 1,114,689$$

2. *Initial installation cost* $(\text{IIC}_{1995})$

   | | |
   |---|---|
   | Engineering (10% IEC) | 46,740 |
   | Contractor fees (10% IEC) | 46,740 |
   | Construction and field expenses (10% IEC) | 46,740 |
   | Start-up (2% IEC) | 9,348 |
   | Performance tests (1% IEC) | 4,674 |
   | Contingencies (3% IEC) | 14,022 |
   | *Total initial installation Cost* $(\text{IIC}_{1995})$ = | $168,264 |

   *Total initial cost:* $\text{TIC}_{1995} = (\text{IDC} + \text{IIC})_{1995} =$
   $$(1,114,689 + 168,264) = \$1,282,953$$

---

### Total Annual Cost $(\text{TAC}_{1995})$ for a Wet Scrubber

1. *Annual direct cost* (ADC)
   (10 h/day, 1 shift/day, 360 days/yr)

   | | |
   |---|---|
   | Operations | |
   | Operator (0.5 h/shift, $12.00/h) | 2,160 |
   | Supervisor (15%, operator) | 324 |
   | Operating materials | |
   | Chemicals for water treatment | 2,000 |
   | Fresh water makeup (5%) | 1,555 |
   | Waste water disposal (1%) | 1,555 |
   | Maintenance | |
   | Labor (0.5 h/shift, $13.20/h) | 2,376 |
   | Materials (100% labor) | 2,376 |
   | Operations and maintenance (O&M) = | 12,346 |
   | Utilities | |
   | Electricity for fan | 24,142 |
   | Electricity for pump | 2,873 |
   | *Total annual indirect costs* $(\text{ADC}_{1995})$ = | $27,015 |

2. *Annual Indirect Cost* $(AIC_{1995})$

| | |
|---|---:|
| Overhead (60% O&M) | 7,408 |
| Administrative charges (2% TIC) | 25,659 |
| Insurance (1% TIC) | 12,830 |
| Property taxes (1% TIC) | 12,830 |

Capital recovery cost (CRC) (assume 15 years, 10%, CRF = 0.1315)

$$CRC = (CRF)(TIC) = (0.1315)(1,282,953) = 168,708$$

Total annual indirect cost $(AIC_{1995})$ = $227,435

3. *Annual recovery cost* (ARC)                   0

*Total annual cost*:

$$TAC_{1995} = (ADC + AIC + ARC)_{1995}$$
$$= (27,015 + 227,435) = \$254,450$$

---

*Summary: wet scrubber annual costs*

Annual cost/scfm = 254,450/30,000 = $8.48/scfm

Annual cost/ton VOC = 254,450/548.5 = $563.90/ton

# 13.8  *Thermal oxidizers*

Thermal oxidizers are an attractive air pollution control system if one needs assurance that combustible pollutants will be totally destroyed. If the destruction temperatures are high, nitrogen oxides will be produced and that may violate local air pollution regulations. In addition, one must be certain that the products of combustion are not hazardous materials. A process gas stream free from particulate matter may be oxidized using a catalytic or recuperative oxidizer. If the process gas stream contains particles, oxidation can be achieved with a regenerative heat exchanger which uses a packed bed heat exchanger that can tolerate particles. Before costs can be estimated it will be necessary to determine the dimensions of the oxidizer and estimate the fuel flow rate. The costs shown below were obtained from Vatavuk (1990) and updated to 1995 using the CE index (Equation 13.5) appropriate for the data.

## 13.8.1  Equipment cost

There are three types of thermal oxidizers; recuperative, regenerative and catalytic oxidizers. The purchased cost is a function of the percent heat recovery (HR) and can be estimated from the following.

- *Recuperative oxidizer* $(500 < Q < 50,000$ scfm$)$

  HR = 0%      $C(\$)_{ox,rcup\ 1995} = 11,478Q^{0.2355}$

  HR = 35%     $C(\$)_{ox,rcup\ 1995} = 14,661Q^{0.2609}$

  HR = 50%     $C(\$)_{ox,rcup\ 1995} = 19,017Q^{0.2502}$

  HR = 70%     $C(\$)_{ox,rcup\ 1995} = 23,796Q^{0.2500}$

  (13.30)

- *Regenerative oxidizer* (HR = 95%): $(10,000 < Q < 100,000$ scfm$)$

$$C(\$)_{ox,rgen\ 1995} = 245,746 + 12.9Q \quad (13.31)$$

- *Fixed bed catalytic oxidizer* $(2000 < Q < 50,000$ scfm$)$

  HR = 0%      $C(\$)_{fx,cat\ 1995} = 1232Q^{0.5471}$

  HR = 35%     $C(\$)_{fx,cat\ 1995} = 3638Q^{0.4189}$  (13.32)

  HR = 50%     $C(\$)_{fx,cat\ 1995} = 1354Q^{0.5575}$

  HR = 70%     $C(\$)_{fx,cat\ 1995} = 1609Q^{0.5527}$

- *Fluidized bed catalytic oxidizer* $(2000 < Q < 25,000$ scfm$)$

  HR = 0%      $C(\$)_{fb,cat\ 1995} = 94,552 + 14.7Q$

  HR = 35%     $C(\$)_{fb,cat\ 1995} = 98,566 + 16.3Q$

  HR = 50%     $C(\$)_{fb,cat\ 1995} = 96,559 + 17.6Q$

  HR = 70%     $C(\$)_{fb,cat\ 1995} = 93,548 + 21.4Q$

  (13.33)

## 13.8.2  Total annual cost

*Annual direct cost* $(ADC_{1995})$

- *Labor*: operation = 0.5 h/shift, maintenance = 0.5 h/shift
- *Materials*: operation = 0, maintenance = 0
- *Material replacement*:
  - Catalyst = 100% every 2 year, cost is pollutant specific, include capital recovery costs at 10%, and transportation and sales tax typically 8% of catalyst cost

- Recuperative/regenerative = 0
- *Utilities* (see Section 13.2)
- Fuel
- Electric power for fans, typical pressure drop
- Waste disposal = 0

| Equipment | δP (in. of H₂O) |
|---|---|
| Direct flame, HR = 0% | 4 |
| Catalytic | |
| Fixed bed, HR = 0% | 6 |
| Fluidized bed HR, = 0% | 6–10 |
| Heat exchangers | |
| HR = 35% | 4 |
| HR = 50% | 8 |
| HR = 70% | 15 |

## 13.8.3 Annual indirect cost

The typical lifetime of catalysts is 2 years, the typical lifetime of an overall thermal oxidizer is 10 years.

### EXAMPLE 13.6 REMOVAL OF DIETHYLAMINE WITH A DIRECT FLAME THERMAL OXIDIZER WITH HEAT RECOVERY

Estimate the total initial cost (TIC) and total annual cost (TAC) of the direct flame thermal oxidizer with 87.5% heat recovery discussed in Example 10.5 and used to oxidize diethylamine in a 30,000 scfm airstream. As a result of the analysis, the following operating parameters are known:

- Operating hours $(t_{op})$ 3600 h/yr, 1 shift/day, 10 h/day, 360 days/yr
- Electrical power, $0.07/kWh
- Inlet diethylamine (VOC) concentration = 1000 ppm
- Inlet VOC mass flow rate = 41.58 g/s
- Destruction efficiency, assumed to be 100%, VOC removal rate = 593.46 tons/yr
- Natural gas, $3.95/1000 ft³
- $Q$ = 30,000 scfm at 30°C, 101 kPa
- Mass flow rate of gas $(CH_4)$ during normal operating hours = 0.1696 kg/s
- $CH_4$ density (STP) = 0.648 kg/m³ (0.0404 lb$_m$/ft³)
- $CH_4$ volumetric flow rate during normal operating hours = 555 scfm
- Internal volume of oxidizer = 246.8 ft³

- Fan efficiency = 70%
- Life time of thermal oxidizer = 15 years
- Overall pressure drop $(\delta P)$ = 15 in. H₂O
- Site preparation and construction of a new building, $50,000 and $200,000, respectively

*Equipment cost* $(C_{1995})$. The purchase price of a recuperative thermal oxidizer can be found from the following:

$$C(\$)_{\text{thermal oxidizer 1995}} = 23,796Q^{0.25}$$

$$= 23,796(30,000)^{0.25} = \$313,173$$

While the thermal oxidizer controls emissions for 10 h/day, it is necessary to maintain a minimum temperature inside the oxidizer even when there are no emissions. Thus for 14 h a day, a pilot flame using 100 cfm of air burns natural gas at the theoretical fuel–air ratio (0.0072 lb$_m$ natural gas/lb$_m$m air). Consequently, two fuel costs and two electric power fan costs must be included in the total annual cost (TAC)

- *Normal operation, 10 hours a day.* The power and cost of the electrical power to operate the fan is

$$\text{Power}_{\text{fan}} = \frac{(1.17 \times 10^{-4})(30,000)(15)}{(0.7)} = 75.21 \text{ kW}$$

$$C(\$)_{\text{fan, 1995}} = (75.21 \text{ kW})(10 \text{ h/day})(360 \text{ day/yr})$$
$$(\$0.07/\text{kWh}) = \$18,953$$

The cost of natural gas during normal operating hours can be found from the following:

$$C(\$)_{\text{gas, 1995}} = (\$3.95/1000 \text{ ft}^3 \text{ at STP})(555 \text{ scfm})$$
$$(60 \text{ min/h})(10 \text{ h/day})(360 \text{ day/yr}) = \$473,526$$

- *Pilot flame operation, 14 hours a day.* The power and cost of the electrical power to operate the fan is

$$\text{Power}_{\text{fan, pilot flame}} =$$
$$\frac{(1.17 \times 10^{-4})(100 \text{ scfm})(15 \text{ in. H}_2\text{O})}{(0.7)}$$
$$= 0.251 \text{ kW}$$

$$C(\$)_{\text{fan, pilot flame, 1995}} = (0.251 \text{ kW})(14)(360 \text{ h/yr})$$
$$(\$0.07/\text{kWh}) = \$88$$

The volumetric flow rate of gas for the pilot flame can be found from the following:

$$\frac{(100 \text{ scfm air})(0.0072 \text{ lb}_m \text{ gas/lb}_m \text{ air})(0.075 \text{ lb}_m \text{ air/ft}^3)}{(0.0404 \text{ lb}_m \text{ gas/ft}^3)}$$

$$= 1.34 \text{ scfm}$$

The cost of gas to operate the pilot flame is

$$C(\$)_{\text{gas, pilot flame, 1995}} = (\$3.95 \times 10^{-3}/\text{ft}^3 \text{ gas})$$

$$(1.34 \text{ scfm})(14 \text{ h/day})(60 \text{ min/h})(360 \text{ days/yr})$$

$$= \$1,600$$

---

### Total Initial Cost $(\text{TIC}_{1995})$ for a Thermal Oxidizer

1. *Initial direct cost* $(\text{IDC}_{1995})$

   1.1  Initial equipment cost $(\text{IEC}_{1995})$

| | |
|---|---:|
| Oxidizer purchase price (C) | $313,173 |
| Instrumentation (10% C) | 31,317 |
| Sales tax (3% C) | 9,395 |
| Freight (5% C) | 15,658 |
| Initial equipment cost $(\text{IEC}_{1995}) = $ | 369,544 |

   1.2  Initial facilities cost $(\text{IFC}_{1995})$

| | |
|---|---:|
| Site preparation | 50,000 |
| Building | 200,000 |
| Foundations and supports (8% IEC) | 29,563 |
| Air handling and erection (14% IEC) | 51,736 |
| Electrical (4% IEC) | 14,781 |
| Piping (2% IEC) | 7,391 |
| Insulated ductwork (1% IEC) | 3,695 |
| Painting (1% IEC) | 3,695 |
| Initial facilities cost $(\text{IFC}_{1995}) = $ | 360,861 |

*Total initial direct cost* $(\text{IDC}_{1995}) = (\text{IEC} + \text{IFC})_{1995} =$
$$\$369,544 + 360,861 = \$730,405$$

2. *Initial installation Cost* $(\text{IIC}_{1995})$

| | |
|---|---:|
| Engineering (10% IEC) | 36,954 |
| Contractor fees (10% IEC) | 36,954 |
| Construction and field expenses (10% IEC) | 36,954 |
| Start-up (2% IEC) | 7,391 |
| Performance tests (1% IEC) | 3,695 |
| Contingencies (3% IEC) | 11,086 |
| *Total initial installation cost* $(\text{IIC}_{1995}) = $ | $133,034 |

*Total initial cost:*   $\text{TIC}_{1995} = (\text{IDC} + \text{IIC})_{1995} =$
$$730,405 + 133,034 = \$863,439$$

---

### Total Annual Cost $(\text{TAC}_{1995})$ for a Thermal Oxidizer

1. *Annual direct cost* $(\text{ADC}_{1995})$ (10 h/day, 1 shift/day, 360 days/yr)

   Operating labor

| | |
|---|---:|
| Operator (0.5 h/shift, $12.00/h) | 2,160 |
| Supervisor (15% labor) | 324 |
| Operating materials | 0 |
| Maintenance | 0 |
| Operations and maintenance (O&M) = | 2,484 |
| Replacement | 0 |

Utilities
| | |
|---|---:|
| Gas (normal operation + pilot flame) | |
| = 473,526 + 1600 | 475,126 |
| Electricity (normal operation + pilot flame) = | |
| 18,953 + 88 | 19,041 |
| Wastes | 0 |
| Total annual direct cost $(\text{ADC}_{1995}) = $ | $494,167 |

2. *Annual indirect cost* $(\text{AIC}_{1995})$

| | |
|---|---:|
| Overhead (60% O&M) | 1,490 |
| Administrative charges (2% TIC) | 17,268 |
| Insurance (1% TIC) | 8,634 |
| Property tax (1% TIC) | 8,634 |

   Capital recovery cost (CRC) (assume 10 years, 10%, CRF = 0.1628)

$$\text{CRC} = (\text{CRF})(\text{TIC}) = (0.1628)(863,435) = 140,567$$

*Total annual indirect cost* $(\text{AIC}_{1995}) = \$176,593$

3. *Annual recovery cost* $(\text{ARC}_{1995})$       0

*Total annual cost:*

$$\text{TAC}_{1995} = (\text{ADC} + \text{AIC} - \text{ARC})_{1995} =$$

$$(494,167 + 176,593 - 0) = \$670,760$$

---

*Summary: thermal oxidizer annual costs*
Annual cost/scfm = 670,760/30,000 = $22.36/scfm
Annual cost/ton VOC = 670,760/593.46
$$= \$1130.25/\text{tons VOC}$$

## 13.9  Comparison of costs to remove diethylamine

Control system performance is as follows:

$$Q = 30,000 \text{ SCFM}$$

$c_0$ = inlet DEA concentration

$\eta$ = removal efficiency, 92.4% minimum

$t_{op}$ = operation time 3600 h/yr

| Cost | Carbon adsorber | Wet scrubber | Thermal oxidizer |
|---|---|---|---|
| $TIC_{1995}$ ($) | $12.9 \times 10^6$ | $1.3 \times 10^6$ | $0.9 \times 10^6$ |
| $TAC_{1995}$ ($) | $2.8 \times 10^6$ | $0.3 \times 10^6$ | $0.7 \times 10^6$ |
| Annual cost ($/SCFM) | 94.57 | 8.48 | 22.36 |
| Annual cost ($/ton DEA removed) | 4780.85 | 463.90 | 1130.25 |

The cost data above are misleading because different presumptions are involved in the design of the different DEA removal devices. The carbon adsorber appears to be the most expensive by far to install and operate. Its design was based upon 100% removal efficiency, however. Also, the cost of regeneration of the adsorption beds has been included in the calculation. This cost is of similar magnitude to the cost of a different mode of regeneration that would allow the ready retrieval of the DEA solvent for reuse.

By comparison, the costs of the wet scrubber are an order of magnitude lower than those for adsorption. The required removal efficiency in this case, however, was 92.4% and not 100%. Furthermore, the DEA collected with the scrubbing water was merely discharged as a waste. Thus the current analysis ignores the considerable cost of cleaning up the wastewater before discharge or reuse, as well as the cost of makeup DEA for the upstream process. In general, the cost of recovering a strong chemical reagent (such as DEA) exceeds the costs of separation processes in which such solvents are used. If the same demands were to be made of the wet scrubber as for the adsorption system, the costs would increase substantially, but probably not to the level of the adsorption system. Like electrostatic precipitation, adsorption is a high-cost operation especially suited to specific difficult separations.

From the table above, thermal oxidation which (like adsorption) achieves 100% removal efficiency now appears to be an attractive economic alternative. One should keep in mind, however, that the generation of $NO_x$, which accompanies such high-temperature operations, may not be tolerable, except for the destruction of particular materials such as pathogens. It is not a likely choice for a valuable agent such as DEA, which should be reused rather than destroyed. Clearly, a refined and more comprehensive design and economic analysis is called for.

## 13.10 Closure

As important as designing the most efficient air pollution control system to meet certain air pollution regulations is the task of estimating the cost of the system. Indeed, complying with RACT requires one to select an air pollution control strategy based on the lowest cost per ton of pollutant removed. For this purpose the EPA requires engineers to use the procedures described in this chapter. The EPA might be persuaded to use other economic models the engineer believes to be more appropriate, but for minimal competency, engineers must be able to use the procedures described in this chapter. The essence of the costing procedures is estimating the total initial cost (TIC) and the total annual cost (TAC). Knowledge of these procedures will be helpful if the need to learn more sophisticated costing procedures arises.

## NOMENCLATURE

| Symbol | Description (Dimensions*) |
|---|---|
| [=] | has the units of |

*F, force; kWh, kilowatthour; L, length; M, mass; N, mols; Q, energy; t, time; T, temperature.

| | |
|---|---|
| $[j]$ | concentration of species j $(M/L^3, N/L^3)$ |
| $A_c$ | surface area of a filter bag $(L^2)$ |
| $A_p$ | surface area of collecting plates in an ESP $(L^2)$ |
| $A_s$ | surface area of adsorbing carbon $(L^2)$ |

$C_f$    material cost factor for scrubber tower

$C(\$)_{j,\,date}$    yearly cost of a device $j$ on a particular date $(\$)$

$c_p, c_0$    pollutant mass concentration $(M/L^3)$

$C_v$    cost of a scrubber vessel $(\$)$

$D_{bag}$    diameter of a filter bag $(L)$

$D_{fan}$    diameter of a fan or blower $(L)$

$D_t$    diameter of a scrubber tower $(L)$

$E$    electric power cost per ton of refrigeration (kW/ton refrigeration)

$H_t$    height of scrubber tower $(L)$

$i$    nominal annual interest rate $(\%)$

$i_r$    real or posted interest rate $(\%)$

$IEC(\$)_j$    initial equipment cost of device $j$ $(\$)$

$K_1, K_2$    constants related to filtration pressure drop (units depend on equation)

$L$    liquid volumetric flow rate $(L^3/t)$; thickness of an adsorber bed $(L)$

$M_c$    mass of carbon adsorber $(M)$

$\dot{m}_p$    pollutant mass flow rate $(M/t)$

$\dot{m}_s$    mass flow rate of steam $(M/t)$

$n$    time, useful lifetime $(t)$

$N_A, N_D, N_c, N_{bags}$    number of adsorbing, desorbing, and cooling vessels and bags

$p_c$    cost of carbon $(\$/M)$

$p_{cw}$    cost to produce cooling water $(\$/L^3)$

$p_{el}$    cost of electrical power $(\$/kwh)$

$(Power_j)$    electrical power associated with device $j$ (kwh)

$Q$    volumetric flow rate of the process gas stream $(L^3/t)$

$Q_{ca}$    volumetric flow rate of compressed air $(L^3/t)$

$r$    annual rate of inflation $(\%)$

$R$    refrigeration in tons

$S_t$    surface area of a scrubbing tower $(L^2)$

$S_v$    vessel surface area $(L^2)$

$T$    temperature $(T)$

$t_{op}$    operating time $(t/t)$

$U_a$    face velocity $(L/t)$

$U_0$    gas velocity approaching a filter $(L/t)$

$w$    precipitation parameter, migration velocity $(L/t)$

$Z$    height of a packed bed, thickness of an adsorption bed $(L)$

*Greek*

$\delta P$    pressure drop $(\text{in. or ft } H_2O, kPa)$

$\eta$    efficiency $(\%)$

*Subscripts*

$(\cdot)_{aux}$    auxiliary

$(\cdot)_{bag}$    bag house

$(\cdot)_{ca}$    carbon adsorber

$(\cdot)_{cw}$    cooling water

$(\cdot)_{duct}$    duct

$(\cdot)_{el}$    electrical

$(\cdot)_{ESP}$    electrostatic precipitator

$(\cdot)_{fabric\ filter}$    fabric filter

$(\cdot)_{fan,\ cen}$    centrifugal fan

$(\cdot)_{fb}$    fluidized bed

$(\cdot)_{fb,\ cat}$    fluidized bed catalyst

$(\cdot)_{fx,\ cat}$    fixed bed catalyst

$(\cdot)_{fr}$    fixed bed

$(\cdot)_{gas}$    gas

$(\cdot)_{housing}$    housing

$(\cdot)_{inlet}$    inlet

$(\cdot)_{insulation}$    insulation

$(\cdot)_{op}$    operating

$(\cdot)_{ox,\ cup}$    recuperative oxidizer

$(\cdot)_{ox,\ re}$    regenerative oxidizer

$(\cdot)_p$    pollutant

$(\cdot)_{pulse\ jet}$    pulse jet

$(\cdot)_{pump}$    pump

$(\cdot)_{rf}$    refrigeration

$(\cdot)_s$    steam

$(\cdot)_{stainless}$    stainless steel

$(\cdot)_t$    packing tower

$(\cdot)_0$    initial state

*Abbreviations*

acfm    actual cubic feet per minute $(L^3/t)$

ADC    annual direct cost $(\$/t)$

AIC    annual indirect cost $(\$/t)$

ARC    annual recovery cost $(\$/t)$

| | | | |
|---|---|---|---|
| CE | Chemical Engineering Plant Cost Index | IIC | initial installation cost ($) |
| CRC | capital recovery cost | O&M | annual operations and maintenance cost ($/t) |
| CRF | capital recovery factor | ppm | parts per million |
| ESP | electrostatic precipitator | scfm | standard (1 atm, 298 K) cubic feet per minute $(L^3/t)$ |
| gpm | gallons per minute | | |
| hp | horsepower | TAC | total annual cost $($/t)$ |
| HR | heat recovery (%) | TIC | total initial cost ($) |
| IDC | initial direct cost ($) | VAPCCI | Vatavuk Air Pollution Control Cost Index |
| IEC($)$_j$ | initial equipment cost of device $j$ ($) | | |
| IFC | initial facilities cost ($) | VOC | volatile organic compound |

# REFERENCES

Grant, E. L., and Ireson, W. G., 1970. *Principles of Engineering Economy*, 5th ed., Ronald Press, New York.

Vatavuk, W. M., 1990. *OAQPS Control Cost Manual*, 4th ed., EPA 450/3-90-006, Office of Air Quality Planning and Standards, Research Triangle Park, NC.

Vatavuk, W. M., 1995. Air pollution control escalates equipment costs, *Chemical Engineering*, December, pp. 88–95.

# PROBLEMS

**13.1.** Estimate the total initial cost (TIC) and total annual cost (TAC) of the activated charcoal adsorber designed in Problem 10.10c. Use a spreadsheet to save time and improve accuracy.

**13.2.** Estimate the total initial cost (TIC) and total annual cost (TAC) of the packed bed absorber designed in Problem 10.10d. Use a spreadsheet to save time and improve accuracy.

**13.3.** Estimate the total initial cost (TIC) and the total annual cost (TAC) of the direct-flame oxidizer designed in Problem 10.10e. Use a spreadsheet to save time and improve accuracy.

**13.4.** Estimate the total initial cost (TIC) and total annual cost (TAC) of the electrostatic precipitator designed in Problem 12.18. Use a spreadsheet to save time and improve accuracy.

**13.5.** Estimate the total initial cost (TIC) and total annual cost (TAC) of the baghouse designed in Problem 12.14. Use a spreadsheet to save time and improve accuracy.

# Appendix

---

**Table A-1.1**  AP-42 Emission Factors for Uncontrolled [a] Bituminous and Subbituminous Coal Furnaces (kg pollutant/Mg coal)

| Combustor | Particles | $SO_2$[b] | $NO_x$ | CO | VOC[c] | $CH_4$ |
|---|---|---|---|---|---|---|
| Pulverized | | | | | | |
|   Dry bottom | 5A[d] | 19.5S[e] | 10.5 | 0.3 | 0.04 | 0.015 |
|   Wet bottom | 3.5A | 19.5S | 17 | 0.3 | 0.04 | 0.015 |
| Cyclone furnace | 1A | 19.5S | 18.5 | 0.3 | 0.04 | 0.015 |
| Spreader stoker | 30[f] | 19.5S | 7 | 2.5 | 0.04 | 0.015 |
| Overfed stoker | 8 | 19.5S | 3.25 | 3 | 0.04 | 0.015 |
| Underfed stoker | 7.5 | 15.5S | 4.75 | 5.5 | 0.65 | 0.4 |
| Hand-fired | 7.5 | 15.5S | 1.5 | 45 | 5 | 4 |

*Source:* Abstracted from U.S. EPA, AP-42.

[a] Emissions as they leave boiler prior to any control devices or settling in breaching.

[b] Includes $SO_2$, $SO_3$, and gaseous sulfates.

[c] Volatile organic compounds excluding methane.

[d] A is the ash content expressed as a percent (%) by mass (i.e., ash $= 8\% \Rightarrow A = 8$).

[e] S is the sulfur content expressed as a percent (%) by mass (i.e., sulfur $= 2.1\% \Rightarrow S = 2.1$).

[f] Emissions entering breeching before control device.

**Table A-1.2**  AP-42 Emission Factors for Uncontrolled [a] Anthracite Furnaces (kg pollutant/Mg coal)

| Combustor | Particles | SO$_2$ [b] | NO$_x$ | CO | VOC | CH$_4$ |
|---|---|---|---|---|---|---|
| Pulverized, coal fired | c | 19.5S | 9 | c | c | c |
| Traveling grate stoker | 4.6 | 19.5S | 5 | 0.3 | c | c |
| Hand-fired stoker | 5 | 19.5S | 1.5 | c | c | c |

*Source:* Abstracted from U.S. EPA, AP-42.

[a] Emissions as they leave boiler prior to any control devices or settling in breaching.

[b] Includes SO$_2$, SO$_3$, and gaseous sulfates.

[c] Emissions same as bituminous combustion.

**Table A-1.3**  AP-42 Emission Factors for Uncontrolled [a] Emissions from Fuel Oil Furnaces (kg pollutant/1000 L fuel)

| Boiler [b] | Particles | SO$_2$ | SO$_3$ | CO | NO$_x$ [c] | VOC [d] | CH$_4$ |
|---|---|---|---|---|---|---|---|
| Utility (>100 MBtu/h) residual oil | e | 19S [f] | 0.34S | 0.6 | 8.0 | 0.09 | 0.03 |
| Industrial (10 to 100 MBtu/hr) | | | | | | | |
|     Residual oil | e | 19S | 0.24S | 0.6 | 6.6 | 0.034 | 0.12 |
|     Distillate oil | 0.24 | 17S | 0.24S | 0.6 | 2.4 | 0.024 | 0.006 |
| Commercial (0.5 to 10 MBtu/h) | | | | | | | |
|     Residual oil | e | 19S | 0.24S | 0.6 | 6.6 | 0.14 | 0.057 |
|     Distillate oil | 0.24 | 17S | 0.24S | 0.6 | 2.4 | 0.04 | 0.026 |
| Residential (<0.5 MBtu/h) distillate oil | 0.3 | 17S | 0.24S | 0.6 | 2.2 | 0.085 | 0.214 |

*Source:* Abstracted from U.S. EPA, AP-42.

[a] Emission prior to any control device.

[b] Boiler defined in terms of firing rate in MBtu/h.

[c] NO$_x$ related to fuel-nitrogen, more accurate expression, kg NO$_x$/1000 L $= 2.75 + 50(N)^2$, where $N$ is the fuel-nitrogen expressed as a percent (%) by mass.

[d] Volatile organic compounds, excluding methane.

[e] Residual oil, No. 6, particles (kg/1000 L) $= 1.25S + 0.38$.

[f] S is the sulfur content of the fuel expressed as a percent (%) by mass (i.e., 2.1% sulfur $\Rightarrow$ S $= 2.1$).

**Table A-1.4**  AP-42 Emission Factors for Uncontrolled [a] Emissions from Natural Gas Furnaces (kg pollutant/10$^6$ m$^3$ gas)

| Boiler [b] | Particles | SO$_2$ | NO$_x$ | CO | VOC [c] | CH$_4$ |
|---|---|---|---|---|---|---|
| Utility (>100 MBtu/h) | 16–80 | 9.6 | 8800 | 640 | 23 | 4.8 |
| Industrial (10 to 100 MBtu/h) | 16–80 | 9.6 | 2240 | 560 | 44 | 48 |
| Domestic and commercial (<10 MBtu/h) | 16–80 | 9.6 | 1600 | 320 | 84 | 43 |

*Source:* Abstracted from U.S. EPA, AP-42.

[a] Emission prior to any control device.

[b] Boiler defined in terms of firing rate in MBtu/h.

[c] Volatile organic compounds, excluding methane.

**Table A-1.5**    AP-42 Emission Factors for Uncontrolled Refuse Incinerators (kg pollutant/Mg refuse)[a]

| Incinerator | Particles | $SO_2$ | CO | $NO_2$[b] | Organics[c] |
|---|---|---|---|---|---|
| Municipal multiple chamber | 15 | 1.25 | 17.5 | 1.5 | 0.75 |
| Industrial | | | | | |
|    Multiple chamber | 3.5 | 1.25 | 5 | 1.5 | 1.5 |
|    Single chamber | 7.5 | 1.25 | 10 | 1 | 7.5 |
| Domestic single chamber with primary burner | 3.5 | 0.25 | neg[d] | 1 | 1 |
| Pathological | 4 | neg | neg 1.5 | neg | |

*Source:* Abstracted from U.S. EPA, AP-42.

[a] Emissions exiting incinerator prior to controls.

[b] Oxides of nitrogen reported as $NO_2$.

[c] Organic compounds reported as $CH_4$.

[d] neg, Negligible.

**Table A-2.1**    Emissions From Aircraft (kg per landing and takeoff cycle)[a]

| Aircraft (no engines) | Particles | $SO_x$[b] | CO | $NO_x$[c] | HC[d] |
|---|---|---|---|---|---|
| Short-, medium-, long-range, jumbo jets | | | | | |
|    Boeing 707 (4) | 2.05 | 1.94 | 119.12 | 11.64 | 99 |
|    Boeing 727 (3) | 0.53 | 1.48 | 25.38 | 13.44 | 6.09 |
|    Boeing 737 (2) | 0.35 | 0.99 | 16.92 | 8.96 | 4.06 |
|    Boeing 747 (4) | 2.36 | 3.25 | 117.76 | 37.76 | 43.96 |
|    DC 10-30 (3) | 0.10 | 2.26 | 53.01 | 22.17 | 21.36 |
| Commuter and freighter, turboprop | | | | | |
|    Beech 99 (2) | 0.08 | 3.25 | 0.37 | 2.3 | |
|    DeHavilland Twin Otter (2) | 0.08 | 3.25 | 0.37 | 2.3 | |
|    Lockheed L188 Electra (4) | 0.83 | 22.12 | 19.65 | 8.91 | |
|    Lockheed L100 Hercules (4) | 0.83 | 22.12 | 19.65 | 8.91 | |
| Business jets | | | | | |
|    Cessna Citation (2) | 0.18 | 8.85 | 0.91 | 3.05 | |
|    Learjet 24D (2) | 0.38 | 40.26 | 0.72 | 3.82 | |
| General aviation, piston | | | | | |
|    Cessna 150 | 0 | 3.77 | 0.01 | 0.10 | |
|    Cessna Skymaster | 0 | 15.01 | 0.06 | 0.52 | |
|    Piper Warrior | 0 | 6.52 | 0.01 | 0.12 | |
| Military | | | | | |
|    B-52H (8) | 42.67 | 4.64 | 228.65 | 24.06 | 229.41 |
|    C-130 (4) | 1.98 | 0.73 | 14.68 | 4.35 | 9.20 |
|    C-5A (4) | 1.87 | 1.74 | 37.25 | 36.11 | 12.74 |
|    C-130 (4) | 1.98 | 0.73 | 14.68 | 4.35 | 9.20 |
|    F-14 Tomcat (2) | 11.0 | 0.56 | 18.09 | 3.46 | 7.87 |
|    F-15A (1) | 0.20 | 1.06 | 24.68 | 13.58 | 1.22 |
|    F-16 (1) | 0.10 | 0.53 | 12.34 | 6.97 | 0.61 |
|    UH-1H Huey (1) | 0.09 | 0.70 | 0.54 | 1.15 | |
|    HH-3 Green Giant (2) | 0.18 | 0.20 | 6.14 | 1.37 | 3.08 |

*Source:* Abstracted from Bare (1988).

[a] Total emission between time aircraft descends through 3500 ft in its approach to land and when it ascends through 3500 ft during takeoff.

[b] All sulfur oxides and sulfuric acid reported as $SO_2$.

[c] All nitrogen oxides reported as $NO_2$.

[d] Total hydrocarbons (volatile organics, including unburned hydrocarbons and organic pyrolysis products).

**Table A-2.2**  AP-42 Emission Factors for Gasoline
and Diesel Industrial Engines (20–250 hp)[a]

| Pollutant (rate) | Gasoline | Diesel |
|---|---|---|
| CO (g hp$^{-1}$ h$^{-1}$) | 199 | 3.03 |
| Exhaust HC (g hp$^{-1}$ h$^{-1}$) | 6.68 | 1.12 |
| Evaporative HC (g/h) | 62.0 | |
| Crankcase HC (g/h) | 38.3 | |
| NO$_x$ (g hp$^{-1}$ h$^{-1}$) | 5.16 | 14.0 |
| Aldehydes (g hp$^{-1}$ h$^{-1}$) | 0.22 | 0.21 |
| SO$_x$ (g hp$^{-1}$ h$^{-1}$) | 0.268 | 0.931 |
| Particles (g hp$^{-1}$ h$^{-1}$) | 0.327 | 1.00 |

*Source:* Abstracted from U.S. EPA, AP-42.

[a] Small engines for such uses as forklift trucks, mobile refrigeration units, electrical generators, pumps, and portable drilling equipment.

**Table A-2.3**  Emission Factors for Automobiles and Trucks

| | Automobile emissions | | Truck emissions | |
|---|---|---|---|---|
| Pollutant | Uncontrolled vehicles prior to 1968 (g/km) | Reduction in new vehicles (%) | Spark ignition (g/km) | Diesel (g/km) |
| NO and NO$_2$ | 2.5 | 75 | 7 | 12 |
| CO | 65 | 95 | 150 | 17 |
| Unburned HC | 10 | 90 | 17[a] | 3 |
| Particles | 0.5[b] | 40 | neg[c] | 0.5 |

*Source:* Heywood (1988).

[a] 95% exhaust emissions and 5% evaporative emissions.

[b] Diesel engines only, particulates from spark ignition engines are negligible.

[c] neg, Negligible.

**Table A-3.1**  AP-42 Hydrocarbon Emissions for Loading and Transportation of Liquid Petroleum Products (kg/1000 L fuel)

| Activity | Gasoline | Crude | JP-4 | Kerosene | No. 2 | No 6 |
|---|---|---|---|---|---|---|
| Loading of tank cars and trucks | | | | | | |
| Submerged balanced | 0.6 | 0.4 | 0.18 | 0.002 | 0.001 | 0.00001 |
| Splash | 1.4 | 0.8 | 0.5 | 0.005 | 0.004 | 0.00004 |
| Splash-balanced | 1.0 | 0.6 | 0.3 | NA | NA | NA |
| Transit | | | | | | |
| Loaded | 0.001–0.009 | | | | | |
| Empty | 0.013–0.44 | | | | | |
| Marine vessels | | | | | | |
| Tanker loading | 0.08 | 0.06 | 0.0006 | 0.0006 | $5.0 \times 10^{-6}$ | |
| Barge loading | 0.20 | 0.14 | 0.0016 | 0.0014 | $1.1 \times 10^{-5}$ | |
| Tanker ballast | 0.01 | 0.07 | | | | |

*Source:* Abstracted from U.S. EPA, AP-42.

[a] NA, not applicable.

**Table A-3.2**  AP-42 Emission Factors for Fugitive Nonmethane Hydrocarbons from Piping Systems

| Fitting | Emission ($lb_m$/h) |
|---|---|
| Valves | |
| Gas–vapor stream | 0.047 |
| Light liquid/two-phase stream | 0.023 |
| Heavy liquid stream | 0.0007 |
| Pump seals | |
| Light liquid stream | 0.26 |
| Heavy liquid stream | 0.045 |
| Flanges (all) | 0.00058 |
| Gas–vapor | 0.0005 |
| Light liquid/two-phase stream | 0.0005 |
| Heavy liquid stream | 0.0007 |
| Compressor seals | |
| Hydrocarbon service | 0.98 |
| Hydrogen service | 0.10 |
| Drains (all) | 0.070 |
| Light liquid/two-phase stream | 0.085 |
| Heavy liquid stream | 0.029 |
| Relief valves (all) | 0.19 |
| Gas–vapor stream | 0.36 |
| Light liquid/two-phase stream | 0.013 |
| Heavy liquid stream | 0.019 |

*Source:* Abstracted from U.S. EPA, AP-42

**Table A-4.1** AP-42 Emission Factors for Uncontrolled Metallurgical Processes (kg/Mg)

### a. Aluminum Production

| Process | Particulate | Fluoride |
|---|---|---|
| Bauxite grinding | 3.0 | neg [a] |
| Aluminum hydroxide calcining | 100.0 | neg |
| Anode baking furnace | 1.5 | 0.45 |
| Prebake cell | 47.0 | 12.0 |
| Vertical Soderberg stud cell | 39.0 | 16.5 |
| Horizontal Soderberg stud cell | 49.0 | 11.0 |

### b. Coke Manufacture

| Process | Particles | $SO_2$ | CO | VOC [b] | $NO_x$ | $NH_3$ |
|---|---|---|---|---|---|---|
| Coal crushing with cyclone | 0.055 | | | | | |
| Coal preheating | 1.75 | | | | | |
| Wet coal charging larry car | 0.24 | 0.01 | 0.3 | 1.25 | 0.015 | 0.01 |
| Door leak | 0.27 | | 0.3 | 0.75 | 0.005 | 0.03 |
| Coke pushing | 0.58 | | 0.035 | 0.1 | | 0.05 |
| Quenching dirty water | 2.62 | | | | | |
| Combustion stack | 0.234 | 2.0 | | | | |
| Coke handling with cyclone | 0.003 | | | | | |

### c. Iron and Steel Mills

| Process | Particles |
|---|---|
| Sintering windbox, leaving grate | 5.56 (kg/Mg finished sinter) |
| Sinter discharge | 3.4 (kg/Mg finished sinter) |
| Casthouse roof monitor | 0.3 (kg/mg hot metal) |
| Taphole and trough | 0.15 (kg/Mg hot metal) |
| Hot metal desulfurization | 0.55 (kg/Mg hot metal) |
| Basic oxygen furnace (BOF) | |
|   Top blown | 14.25 (kg/Mg steel) |
|   Charging, at source | 0.3 (kg/Mg steel) |
|   Tapping, at source | 0.46 (kg/Mg steel) |
|   Hot metal transfer, at source | 0.095 (kg/Mg steel) |
|   BOF roof monitor, all sources | 0.25 (kg/Mg steel) |
| Electric arc furnace | |
|   Melting and refining carbon steel | 19.0 (kg/Mg steel) |
|   Charging, tapping and slagging at roof monitor | 0.7 (kg/Mg steel) |
|   Melting, refining, charging tapping and slagging | |
|     Alloy steel | 5.65 (kg/Mg steel) |
|     Carbon steel | 25.0 (kg/Mg steel) |
| Open hearth | |
|   Melting and refining | 10.55 (kg/Mg steel) |
|   Teeming leaded steel | 0.405 (kg/Mg steel) |
| Teeming unleaded steel | 0.035 (kg/Mg steel) |
| Machine scarfing | 0.05 (kg/Mg steel) |

**Table A-4.1** *(continued)*

**d. Gray Iron**

| Process | Particles | CO | SO$_2$ | NO$_x$ | VOC | Lead |
|---|---|---|---|---|---|---|
| Furnace | | | | | | |
| Cupola | 6.9 | 73 | 0.6S [c] | | | 0.05–0.6 |
| Electric arc | 6.3 | 0.5–19 | neg | | 0.02–0.3 | 0.03–0.15 |
| Electric induction | 0.5 | neg | neg | neg | neg | 0.005–0.05 |
| Reverberatory | 1.1 | neg | neg | neg | neg | 0.006–0.07 |
| Scrap and charge handling | 0.3 | | | | | |
| Magnesium treatment | 0.9 | | | | | |
| Inoculation | 1.5–2.5 | | | | | |
| Pouring, cooling | 2.1 | | | | | |
| Shakeout | 1.6 | | | | | |
| Cleaning, finishing | 8.5 | | | | | |
| Sand handling | 1.8 | | | | | |
| Core making, baking | 0.6 | | | | | |

*Source:* Abstracted from U.S. EPA, AP-42.

[a] neg, Negligible.

[b] Expressed as methane.

[c] S is the sulfur content expressed as a percent by mass.

**Table A-5.1** AP-42 Emission Factors for Uncontrolled Emissions from Mineral Processes

**a. Asphalt**

| Production | (g/Mg asphalt) |
|---|---|
| Sulfur oxides (as SO$_2$) | 146S [a] |
| Nitrogen oxides (as NO$_2$) | 18 |
| VOC | 14 |
| CO | 19 |
| Polycyclic organics | 0.013 |
| Aldehydes, total | 10 |
| Formaldehyde | 0.075 |
| 2-Methylpropanal | 0.65 |
| 1-Butanal | 1.2 |
| 3-Methylbutanal | 8.0 |

**b. Brick Manufacturing (kg/Mg)**

| Process | Particles | SO$_2$ | CO | VOC | CH$_4$ | NO$_x$ | Fluorides |
|---|---|---|---|---|---|---|---|
| Raw material | | | | | | | |
| Drying | 35 | | | | | | |
| Grinding | 38 | | | | | | |
| Storage | 17 | | | | | | |
| Brick dryer | 0.006A [b] | 0.55S | | | | 0.33 | |
| Tunnel kiln (coal) | 0.34A | 3.65S | 0.71 | 0.005 | 0.006 | 0.73 | 0.5 |

**Table A-5.1**   *(continued)*

### c. Concrete Batching (kg/Mg of material)

| Process | Particles |
|---|---|
| Sand and aggregate transfer to elevated bin | 0.014 |
| Cement unloading to elevated vessel to silo (bucket elevator) | 0.12 |
| Weigh hopper loading | 0.01 |
| Truck loading | 0.01 |
| Mixer loading | 0.02 |
| Vehicle traffic (unpaved road) | 4.5 kg vehicle$^{-1}$ km$^{-1}$ |
| Wind erosion from sand and aggregate storage stock pile | 3.9 kg ha$^{-1}$ day$^{-1}$ |

### d. Glass Manufacture (kg/Mg)

| Process | Particles | $SO_2$ | $NO_x$ | VOC | CO | Lead |
|---|---|---|---|---|---|---|
| Melting furnace | 0.7 | 1.7 | 3.1 | 0.1 | 0.1 | |
| Flat glass | 1.0 | 1.5 | 4.0 | <0.1 | <0.1 | |
| Pressed glass | 8.7 | 2.8 | 4.3 | 0.2 | 0.1 | |
| Lead glass manufacturing | | | | | | 2.5 |

### e. Crushed Stone Plants (g/Mg stone)

| Process | TSP particles [c] |
|---|---|
| Wet quarry drilling | 0.4 |
| Batch drop, truck unloading | 0.17 |
| Truck loading | |
|   Conveyor | 0.17 |
|   Front end loader | 29.0 |
| Conveyor, tunnel belt | 1.7 |

### f. Lime Manufacture (kg/Mg)

| Process | Particles | $SO_2$ | $NO_x$ | VOC | CO | Lead |
|---|---|---|---|---|---|---|
| Rotary kiln | 180 | [d] | 1.4 | neg | 1 | neg |
| Product coolers | 20 | neg[e] | neg | neg | neg | neg |
| Closed truck loading | | | | | | |
|   Limestone | 0.38 | neg | neg | neg | neg | neg |
|   Lime | 0.15 | neg | neg | neg | neg | neg |

**Table A-5.1**   *(continued)*

**g. Surface Coal Mining**

| Process | TSP particles (units) [f] | |
|---|---|---|
| Truck loading | $0.580/M^{1.2}$ | (kg/Mg) |
| Bulldozing | | |
|    Coal | $35.6s^{1.2}/M^{1.3}$ | (kg/h) |
|    Overburden | $2.6s^{1.2}/M^{1.3}$ | (kg/h) |
| Dragline, overburden | $0.0046\ d^{1.1}/M^{0.3}$ | (kg/m$^3$) |
| Scraper (travel mode) | $9.6 \times 10^{-6}\ s^{1.3}\ W^{2.4}$ | (kg/VkT) |
| Grading | $0.0034U^{2.5}$ | (kg/VkT) |
| Vehicle traffic (light, medium duty) | $1.63/M^{4.0}$ | (kg/VkT) |
| Haul truck | $0.0019\ (wh)^{3.4}L^{0.2}$ | (kg/VkT) |
| Active coal storage pile (wind erosion and maintenance) | $1.8u$ | (kg hectare$^{-1}$ h$^{-1}$) |

*Source:* Abstracted from U.S. EPA, AP-42.

[a] S is the sulfur content in coal used as a percent by mass.

[b] A is the ash content in coal used as a percent by mass.

[c] TSP, total suspended particulate.

[d] neg, Negligible.

[e] $M$, material moisture content (%); $s$, material silt content (%); $d$, drop height (m); $W$, vehicle weight (Mg); VKT, vehicle kilometers traveled; $U$, vehicle speed (km/h); $wh$, number of wheels; $L$, road surface silt loading (g/m$^2$); $u$, wind speed (m/s).

**Table A-6.1**  AP-42 Emission Factors for Fugitive Emissions from Vehicles Traveling on Roadways

| Roadway | TSP particles (units) [a] |
|---|---|
| Passenger car, paved road | 5.0 (g/V-mile) |
| Trucks, paved roads | |
|    10-Wheel, gasoline | 12.70 (g/V-mile) |
|    12-Wheel, diesel | 15.40 (g/V-mile) |
|    18-Wheel, diesel | 22.50 (g/V-mile) |
|    Industrial paved roads truck traffic | $0.022(I)(4/n)(s/10)(L/280)(W/2.7)^{0.7}$ (kg/VkT) |
|    Unpaved roads | $1.7\,k(s/12)(U/48)(W/2.7)^{0.7}[(365-p)/365]$ (kg/VkT) |

$n$ = number of traffic lanes;

$s$ = surface material silt content (%), mean values below

      (copper smelters, 19%

      iron and steel, 12.5%;

      asphalt, 3.3%;

      concrete batching, 5.5%;

      sand and gravel, 7.1%);

$L$ = surface dust loading (kg/km), mean values below

      (copper smelters, 15.9 kg/km;

      iron and steel, 0.495 kg/km;

      asphalt, 14.9 kg/km;

      concrete batching, 1.7 kg/km;

      sand and gravel, 3.8 kg/km);

$U$ = vehicle speed (km/h);

$k$, = multiplier, emission rate composed of particles less than $D_p$,

      $k = 0.8$ for $D_p < 30\ \mu m$

      $k = 0.5$ for $D_p < 15\ \mu m$

      $k = 0.36$ for $D_p < 10\ \mu m$

      $k = 0.20$ for $D_p < 5\ \mu m$)

$W$ = vehicle weight Mg

$p$ = mean number of days with greater than or equal to 0.01 in. of precipitation, typical values

      $p = 150$, New England and mid-Atlantic states

      $p = 120$, south

      $p = 100$, midwest

      $p = 180$, northwest

      $p = 80$ southwest

VkT, V-mile = vehicle kilometer traveled, vehicle mile traveled

$I$ = industrial augmentation factor

      $I = 7.0$, industrial roadway carrying traffic which enters from unpaved roads

      $I = 3.5$, industrial roadway with unpaved shoulders on which 20% vehicles temporarily travel on shoulders

      $I = 1.0$, traffic does not travel on unpaved roads

* abstracted from U.S. EPA, AP-42; Orlemann, 1983 and U.S. EPA/625/5-87/022, 1987

**Table A-7** Emission Factors for Indoor Processes, Activities, and Furnishings

**a. Cigarettes**

| Pollution factor | Mainstream (inhaled) | Sidestream (smoldering) |
|---|---|---|
| Duration (s) | 20 | 550 |
| Tobacco burned (mg) | 347 | 441 |
| Particles/cigarette | $1.05 \times 10^{12}$ | $3.5 \times 10^{12}$ |

| | Emissions (mg/cigarette) | |
|---|---|---|
| Contaminant | Mainstream | Sidestream |
| Particles | | |
|   Unfiltered cigarettes | | |
|     Tar | 20.8 | 44.1 |
|     Nicotine | 0.92 | 1.69 |
|   Filtered cigarettes | | |
|     Tar | 10.2 | 34.5 |
|     Nicotine | 0.46 | 1.27 |
| CO | 18.3 | 86.3 |
| $NH_3$ | 0.16 | 7.4 |
| $NO_x$ | 0.014 | 0.051 |
| HCN | 0.24 | 0.16 |
| Acrolein | 0.084 | 0.825 |

**b. Gas Ranges and Kerosene Space Heaters at Two Firing Rates ($\mu$g/kcal)**

| Compound | Gas range (2500 kcal/h) | Space heater (2800 kcal/h) |
|---|---|---|
| CO | 890 | 632 |
| $CO_2$ | 209,000 | 200,000 |
| NO | 31 | 76 |
| $NO_2$ | 85 | 46 |
| $SO_2$ | 0.8 | |
| HCHO | 7.1 | |

**Table A-7**  *(continued)*

**c. Formaldehyde HCHO**

| Material | Emission rate (mg m$^{-2}$ day$^{-1}$) |
|---|---|
| UF-foam insulation | 1–50 |
| Plywood (UF-bonded) | 1–34 |
| Hardwood paneling (UF) | 1–34 |
| Particleboard (std, UF) | 2–34 |
| Fiberglass ceiling panel | 2.8 |
| 100% cotton drapery fabric | 0.2–0.7 |
| Paper cups and plates | 0.33–0.7 |
| Fiberglass insulation | 0.45 |
| Latex-backed fabric | 0.19 |
| Foam-backed carpet | 0.12 |
| Nylon upholstery fabric | 0.018 |

**d. Common Household Aerosols in 1970**

| Material | Emission rate (g/month) |
|---|---|
| Deodorant spray | 112–140 |
| Hairspray | 84–112 |
| Shaving foam | 84–112 |
| Air fresheners | 28–56 |
| Disinfectant sprays | 112 |
| Furniture polish | 56 |
| Dust sprays | 28–56 |
| Oven cleaners | 84 |

**e. Small Stoves**

| Source | Emission factor (g/kg fuel) |
|---|---|
| Wood | |
|   Particles | 1.6–6.4 |
|   Carbon monoxide | 100 |
| Bituminous coal | |
|   Particles | 10.4 |
|   Carbon monoxide | 116 |
| Anthracite coal | |
|   Particles | 0.5 |
|   Carbon monoxide | 21 |

*Source:* (a) Abstracted from Commission on Indoor Air Pollution (1981), Lofroth et al. (1989); (b) abstracted from Meyer (1982); (c), (d) abstracted from Meyer (1983); (e) abstracted from Butcher and Ellenbecker (1982).

**Table A-8**    Henry's Law Constant ($H'$) and Diffusion Coefficients of Pollutants in Air and Water[a,b]
Unit convention: $D = \alpha(10^{-5} \text{ m}^2/\text{s}) \Rightarrow D = \alpha \times 10^{-5} \text{ m}^2/\text{s}$

| Substance | SG | M | Henry's law constant, $H'$ ($10^7$ N/m$^2$) | Diffusion coefficient, $D$ Air ($10^{-5}$ m$^2$/s) | Diffusion coefficient, $D$ Water ($10^{-9}$ m$^2$/s) |
|---|---|---|---|---|---|
| Acetaldehyde | 0.79 | 44 | 0.56[Y] | | |
| Acetic acid | 1.05 | 60 | 0.00067[Y] | 1.06 | 1.19 |
| Acetone | 0.79 | 56 | 0.024[Y] | 0.83 | 1.16 |
| Acetonitrile | 0.78 | 41 | 0.011[Y] | | 1.26 |
| Acetylene | | 26 | 14.2[Y] | 1.7 | 2.0 |
| Acrylic acid | | 72 | 0.00023[Y] | | |
| Aldrin | 1.60 | 364.9 | 0.0008 | | |
| Allyl alcohol | 0.85 | 58 | 0.0031[Y] | | |
| Ammonia | | 17 | 0.03 | 2.2 | 2.0 |
| Aniline | 1.02 | 93 | 0.75 | 0.92 | |
| Benzene | 0.88 | 78 | 3.05[S] | 0.77 | 1.02 |
| Benzoic acid | | 122 | | | 1.00 |
| Benzyl alcohol | | 94 | | | 0.82 |
| Biphenyl | | 154 | 0.48[Y] | | |
| Bromine | | 160 | 0.747 | 1.0 | 1.3 |
| Butadiene | | 50 | 286.2[Y] | | |
| n-Butane | | 58 | 514.6[Y] | 0.96[V] | 0.89 |
| n-Butanol | 0.81 | 74 | | 0.89 | 0.77 |
| Carbon dioxide | | 44 | 12.3[Y] | 1.5 | 2.0 |
| Carbon disulfide | 1.26 | 76 | 10.8[Y] | 0.89[P] | |
| Carbon monoxide | | 28 | 640.6[Y] | 2.0 | 2.0 |
| Carbon tetrachloride | 1.59 | 154 | 16.5[Y] | 0.62 | 0.82 |
| Carbonyl sulfide | | 60 | 28.4[Y] | 1.3 | 1.5 |
| Chlorine | | 71 | 6.82 | 1.2 | 1.5 |
| Chlorine dioxide | 1.60 | 67.5 | 0.541 | | |
| Chlorobenzene | | 113 | 2.0[S] | 0.62 | 0.86[M] |
| Chloroform | 1.48 | 119 | 2.66[L] | 0.87 | 0.92 |
| Cyanogen | | 52 | 2.98[Y] | | |
| Cycloheptane | | 98 | 52.5[Y] | | |
| Cyclohexane | | 84 | 18.0[S] | 0.86[P] | |
| Cyclohexene | | 82 | 25.3[Y] | | |
| Cyclopentane | | 70 | 105.3[Y] | | |
| Cyclopentene | | 68 | 35.7[Y] | | |
| Cyclopropane | | 42 | 42.8[Y] | | |
| Dibromochloropropane | | | 0.021 | 0.69 | 0.72 |
| 1,1-Dichloroethylene | 1.27 | 97 | 106.7 | | |
| Dichloromethane | | 85 | 1.39[Y] | | |
| Dieldrin | 1.75 | 380.9 | 0.011 | | |
| Diethylamine | 0.71 | 73 | 0.03658 | 0.88 | 0.97 |
| Ethane | | 30 | 281[S] | 1.5 | 1.4 |
| Ethyl acetate | 0.90 | 88 | | 0.72 | 1.00 |
| Ethyl alcohol | | 46 | 0.00456[Y] | 1.02 | 0.84 |

(continued)

## Table A-8 *(continued)*

| Substance | SG | M | Henry's law constant, $H'$ ($10^7$ N/m²) | Diffusion coefficient, $D$ — Air ($10^{-5}$ m²/s) | Water ($10^{-9}$ m²/s) |
|---|---|---|---|---|---|
| Ethyl benzene | 0.87 | 106 | 4.44[S] | 0.66[P] | 0.81[R] |
| Ethyl bromide | 0.46 | 109 | 40.48 | | |
| Ethyl formate | 0.92 | 74 | | 0.84[P] | |
| Ethylene | | 28 | >4769[L] | 1.6 | 1.5 |
| Ethylene dibromide | 2.17 | 188 | 6432[L] | 0.81 | 0.89 |
| Ethylene dichloride | 1.24 | 99 | 0.61[L] | | |
| Ethylene glycol | 1.49 | | | | 1.16 |
| Formaldehyde | 1.08 | 30 | | | |
| Formic acid | 1.22 | 46 | 0.00062[Y] | 1.31[P] | 0.69[P] |
| Furfural | 1.16 | 96 | | | 1.04[R] |
| Glycerol | | | | | 0.82 |
| Glycine | | | | | 1.06 |
| Heptane | | 100 | 1514[Y] | 0.71[R] | |
| Hexachloroethane | 2.09 | 236.7 | 1.404 | | |
| Hexane | 0.66 | 0.68 | 86 | 944[S] | 0.8[R] |
| Hexylbenzene | | 162 | 12.0[Y] | | |
| Hydrogen cyanide | 0.69 | 27 | 0.064 | 1.5 | 1.8 |
| Hydrogen sulfide | | 34 | 5.52 | 1.7 | 1.6 |
| Isobutyl acetate | 0.80 | 116 | | 0.61 | |
| Isopropyl alcohol | | 60 | | 1.07[V] | 0.87[R] |
| Mercury | 13.6 | 201 | 6.1[L] | | |
| Methane | | 16 | 357[Y] | 2.2 | 1.8 |
| Methyl acetate | 0.93 | 74 | | 0.84[P] | |
| Methyl alcohol | 0.79 | 32 | 0.0039[Y] | 1.33 | 0.84 |
| Methyl bromide | 1.73 | 95 | 7.20[L] | | |
| Methyl chloride | | 51 | 13.3[L] | 1.3 | 1.5 |
| Methyl chloroform | 1.34 | 133 | 0.346 | 0.78 | 0.81 |
| Methyl formate | 0.98 | 60 | | 0.87[P] | |
| Methylene chloride | 1.33 | 85 | 1.67 | | |
| Napthalene | 1.15 | 128 | 0.634[L] | 0.51[P] | |
| Nitric oxide | | 30 | 291.0 | 2.0 | 2.4 |
| Nitrobenzene | 1.20 | 123 | 0.0122[L] | 0.86[V] | |
| Nitrogen | | 28 | 853.5 | | |
| Nitrous oxide | | 44 | 22.7 | 1.5 | 1.8 |
| Octane | 0.70 | 114 | 1667[S] | | |
| Oxalic acid | 1.65 | 90 | 1.53 | | |
| Oxygen | | 32 | 434.4 | | |
| Ozone | | 48 | 46.4 | | 2.0 |
| Pentane | 0.63 | 72 | 709.5[Y] | | |
| Perchloroethylene | | 166 | 4.60[L] | 0.74 | 0.76 |
| Phenol | 1.06 | 94.1 | 0.0004 | | |
| Phosgene | 1.43 | 99 | | 0.80 | |
| Phosphine | | 34 | 398[S] | 1.6 | |
| Propane | | 44 | 384.1[Y] | 0.88 | 0.97 |

*(continued)*

|  | | | | Diffusion coefficient, $D$ | |
|---|---|---|---|---|---|
| Substance | SG | M | Henry's law constant, $H'$ ($10^7$ N/m$^2$) | Air ($10^{-5}$ m$^2$/s) | Water ($10^{-9}$ m$^2$/s) |
| Ethyl benzene | 0.87 | 106 | 4.44[S] | 0.66[P] | 0.81[R] |
| Propylene | | 42 | 57.3 | | 1.1 |
| *n*-Propyl acetate | 0.84 | 102 | | 0.67[P] | |
| Propyl alcohol | 0.81 | 60 | | 0.85[P] | 1.1[P] |
| Pyridine | 0.98 | 79 | | | 0.58 |
| Styrene | | 104 | 1.47[Y] | | |
| Sulfur dioxide | | 64 | 0.485 | 1.3 | 1.7 |
| Tetrachloroethane | | 166 | 15.07[Y] | | |
| Tetrachloroethylene | 1.62 | 165.8 | 10.42 | | |
| Toluene | 0.87 | 92 | 3.72[S] | 0.71 | 0.844[M] |
| 1,1,2-Trichloroethane | 1.44 | 133.4 | 0.4161 | | |
| Trichloroethylene | 1.46 | 131 | 6.54[Y] | 0.78 | 0.81 |
| Triethylamine | 0.73 | 101 | 0.007[L] | | |
| Urethane | | | | 1.06 | |
| Vinyl chloride | | 62 | 1331[L] | | |
| Water | 1.00 | 18 | | 2.64 | |
| Xylene | 0.86 | 106 | 2.78[S] | | |

*Source:* Abstracted from Crawford (1976) except where noted by the following subscripts: [L] Lyman et al. (1990); [M] Mackay and Yuen (1983); [P] Perry and Chilton (1973); [R] Reid et al. (1977); [S] converted from data in Mackay and Shiu (1981); [V] Vargaftik (1975); [Y] Yaws et al. (1991).

[a] See also http://chemfinder.camsoft.com/.

[b] Note Crawford does not attach temperatures to the given data. Most data are for temperatures from 0 to 30 °C. If greater accuracy is warranted, more sophisticated predictive techniques from the chemical engineering literature should be employed.

**Table A-9** Critical Temperature, Pressure, and PEL Values
for Common Pollutants

| Name | Formula | $T_c$ (°C) | $P_c$ (atm) | TWA-PEL[a] (ppm) |
|---|---|---|---|---|
| Acetaldehyde | $C_4H_4O$ | 187.8 | 54.7 | 100 |
| Acetic acid | $C_2H_4O_2$ | 321.6 | 57.1 | 10 |
| Acetone | $C_3H_6O$ | 235.5 | 47 | 750 |
| Acetonitrile | $C_2H_3N$ | 274.7 | 47.7 | 42 |
| Aniline | $C_6H_7N$ | 425.6 | 52.3 | 2[b] |
| Benzene | $C_6H_6$ | 288.9 | 48.6 | 10[c] |
| Benzyl chloride | $C_6H_5Cl$ | 359.2 | 44.6 | 1 |
| Boron triflouride | $BF_3$ | −12.3 | 49.2 | 1 (ceiling) |
| Carbon disulfide | $CS_2$ | 279 | 78 | 4[c] |
| Carbon tetrachloride | $CCl_4$ | 283.4 | 45.6 | 2[c] |
| Diethylamine | $C_4H_{11}N$ | 223.3 | 36.6 | 10 |
| Dimethylamine | $C_2H_7N$ | 164.6 | 52.4 | 10 |
| Ethylene oxide | $C_2H_5O$ | 195.8 | 71 | 1 |
| Hydrogen chloride | HCl | 51.4 | 82.1 | 5 (ceiling) |
| Hydrogen cyanide | HCN | 183.5 | 48.9 | 4.7 (STEL) |
| Hydrogen sulfide | $H_2S$ | 100.4 | 88.9 | 10 |
| Methyl alcohol | $CH_4O$ | 240 | 78.5 | 200 |
| Methyl mercaptan | $CH_4S$ | 196.8 | 71.4 | 0.5 |
| Methylamine | $CH_5N$ | 156.9 | 40.2 | 10 |
| Methylene chloride | $CH_2Cl_2$ | 237 | 60 | d |
| Naphthalene | $C_{10}H_8$ | 474.8 | 40.6 | 10 |
| Nitric oxide | NO | −93 | 64 | 25 |
| Ozone | $O_3$ | −5.2 | 67 | 0.1 |
| Phenol | $C_6H_6O$ | 421.1 | 60.5 | 5[b] |
| Propylene oxide | $C_3H_6O$ | 209 | 48.6 | 20 |
| Styrene | $C_8H_8$ | 374.4 | 39.4 | 50 |
| Toluene | $C_7H_8$ | 320.8 | 41.6 | 100 |
| Triethylamine | $C_6H_{15}N$ | 258.9 | 30 | 10 |

*Source:* Abstracted from ACGIH (1988); CFR (29) Part 1910.1000 (1989), and Weast (1975).

[a] PEL as of 1989. PEL are reviewed on a regular basis and readers should always use currently approved values.

[b] Additional entry through skin.

[c] See 29 CFR Part 1910.1000 for ceiling and maximum transitory values.

[d] In process of rulemaking.

**Table A-10** Thermophysical Properties of Air and Water
Unit convention: $\mu = \alpha(10^{-5} \text{ kg/ms}) \Rightarrow \mu = \alpha \times 10^{-5}$ kg/ms

**a. Air**

| $T$ (°C) | $c_p$ (kJ kg$^{-1}$ K$^{-1}$) | $\rho$ (kg/m³) | $\mu$ ($10^{-5}$ kg m$^{-1}$ s$^{-1}$) | $\nu$ ($10^{-5}$ m²/s) | $k$ ($10^{-2}$ W m$^{-1}$ °C$^{-1}$) | Pr |
|---|---|---|---|---|---|---|
| 20 | 1.0061 | 1.2042 | 1.817 | 1.509 | 2.564 | 0.713 |
| 30 | 1.0064 | 1.1644 | 1.865 | 1.601 | 2.638 | 0.712 |
| 40 | 1.0068 | 1.1273 | 1.911 | 1.696 | 2.710 | 0.710 |
| 50 | 1.0074 | 1.0924 | 1.957 | 1.792 | 2.781 | 0.709 |
| 60 | 1.0080 | 1.0596 | 2.003 | 1.890 | 2.852 | 0.708 |
| 70 | 1.0087 | 1.0287 | 2.047 | 1.990 | 2.922 | 0.707 |
| 80 | 1.0095 | 0.9996 | 2.092 | 2.092 | 2.991 | 0.706 |
| 90 | 1.0103 | 0.9721 | 2.135 | 2.196 | 3.059 | 0.705 |
| 100 | 1.0113 | 0.9460 | 2.178 | 2.302 | 3.127 | 0.704 |
| 110 | 1.0123 | 0.9213 | 2.220 | 2.410 | 3.194 | 0.704 |
| 120 | 1.0134 | 0.8979 | 2.262 | 2.519 | 3.261 | 0.703 |
| 130 | 1.0146 | 0.8756 | 2.303 | 2.631 | 3.328 | 0.702 |
| 140 | 1.0159 | 0.8544 | 2.344 | 2.744 | 3.394 | 0.702 |
| 150 | 1.0172 | 0.8342 | 2.384 | 2.858 | 3.459 | 0.701 |
| 160 | 1.0186 | 0.8150 | 2.424 | 2.975 | 3.525 | 0.701 |
| 170 | 1.0201 | 0.7966 | 2.463 | 3.093 | 3.589 | 0.700 |
| 180 | 1.0217 | 0.7790 | 2.503 | 3.213 | 3.654 | 0.700 |
| 190 | 1.0233 | 0.7622 | 2.541 | 3.334 | 3.718 | 0.699 |
| 200 | 1.0250 | 0.7461 | 2.579 | 3.457 | 3.781 | 0.699 |

**b. Saturated Water**

| $T$ (°C) | $c_p$ (kJ kg$^{-1}$ K$^{-1}$) | $\rho$ (kg/m³) | $\mu$ ($10^{-3}$ kg m$^{-1}$ s$^{-1}$) | $\nu$ ($10^{-6}$ m²/s) | $k$ (W m$^{-1}$ °C$^{-1}$) | Pr |
|---|---|---|---|---|---|---|
| 20 | 4.182 | 998.3 | 1.003 | 1.004 | 0.5996 | 6.99 |
| 25 | 4.180 | 997.1 | 0.8908 | 0.8933 | 0.6076 | 6.13 |
| 30 | 4.180 | 995.7 | 0.7978 | 0.8012 | 0.6150 | 5.42 |
| 35 | 4.179 | 994.1 | 0.7196 | 0.7238 | 0.6221 | 4.83 |
| 40 | 4.179 | 992.3 | 0.6531 | 0.6582 | 0.6286 | 4.34 |
| 45 | 4.182 | 990.2 | 0.5962 | 0.6021 | 0.6347 | 3.93 |
| 50 | 4.182 | 998.0 | 0.5471 | 0.5537 | 0.6405 | 3.57 |
| 55 | 4.184 | 985.7 | 0.5043 | 0.5116 | 0.6458 | 3.27 |
| 60 | 4.186 | 983.1 | 0.4668 | 0.4748 | 0.6507 | 3.00 |
| 65 | 4.187 | 980.5 | 0.4338 | 0.4424 | 0.6553 | 2.77 |
| 70 | 4.191 | 977.7 | 0.4044 | 0.4137 | 0.6594 | 2.57 |
| 75 | 4.191 | 974.7 | 0.3783 | 0.3881 | 0.6633 | 2.39 |
| 80 | 4.195 | 971.6 | 0.3550 | 0.3653 | 0.6668 | 2.23 |
| 85 | 4.201 | 968.4 | 0.3339 | 0.3448 | 0.6699 | 2.09 |
| 90 | 4.203 | 965.1 | 0.3150 | 0.3264 | 0.6727 | 1.97 |

**Table A-11** Odor Threshold and OSHA PEL Values for Common Industrial Materials

| Chemical (M) | CAS | Odor[a] (ppm) | Description | PEL(ppm) |
|---|---|---|---|---|
| Acetaldehyde (44.1) | 75–07–0 | 2.3 | Green, sweet, fruity | 100 |
| Acetic acid (60.1) | 64–19–7 | 102 | Sour, vinegar-like | 10 |
| Acetic anhydride (102) | 108–24–1 | | | 5[b] |
| Acetone (58.1) | 67–64–1 | 680 | Minty chemical, sweet | 750 |
| Acetonitrile (41.1) | 75–05–8 | 42 | Ether-like | 40 |
| Acrolein (56.1) | 107–02–8 | 16 | Burnt, sweet | 0.1 |
| Acrylontrile (53.1) | 107–13–1 | 36 | Onion-garlic pungent | c |
| Aldrin (364.9) | 309–00–2 | 0.03 | Mild chemical odor | 0.02 |
| Allyl alcohol (58.1) | 107–18–6 | 2 | Mustard, pungent | 2 |
| Allyl chloride (76.5) | 107–05–1 | 24 | Green, garlic, oniony | 1 |
| Allyl glycidyl ether (114.2) | 106–92–3 | 9.4 | Sweet | 5 |
| Ammonia (17.0) | 7664–41–7 | 57 | Pungent, irritating | 50 |
| Amyl acetate ( ) (130.2) | 628–63–7 | 7 | Fruity, banana, pear | 100 |
| Aniline (93) | 62–53–3 | | | 2 |
| Arsine (78.0) | 7784–42–1 | 0.6 | Garlic-like | 0.05 |
| Benzene (78.1) | 71–43–2 | 85 | Sweet, solventy | c |
| Benzyl chloride (126.6) | 100–44–7 | 0.3 | Solventy | 1 |
| Biphenyl (154) | 92–52–4 | | | 0.2 |
| Boron trifluoride (68) | 7637–07–2 | | | 1[b] |
| Bromine (158.8) | 7726–95–6 | 3.7 | Bleachy, penetrating | 0.1 |
| Butyl acetate (N) (116.2) | 123–86–4 | 20 | Fruity | 150 |
| Butyl alcohol (N) (74.1) | 71–36–3 | 50 | Sweet | 50 |
| Butyl alcohol (sec) (74.1) | 78–92–2 | 43 | Strong, pleasant | 100 |
| Butyl alcohol (tert) (74.1) | 75–65–0 | 72- | Camphor-like | 100 |
| Butyl amine (N) (73.2) | 109–73–9 | 2 | Fishy, ammonia-like | 5[b] |
| Butyl mercaptan (90.2) | 109–79–5 | $9 \times 10^{-4}$ | Repulsive | 0.5 |
| Carbon disulfide (76.1) | 75–15–0 | 74 | Sweet, disagreeable | 4 |
| Carbon monoxide (28.0) | 630–08–0 | | No odor | 35 |
| Carbon tetrachloride (153.8) | 56–23–5 | 239 | Sweet, pungent | 2 |
| Chloracetaldehyde (78.5) | 107–20–0 | 0.9 | Sharp, irritating | 1.0[b] |
| Chlorine (70.9) | 7782–50–5 | 5 | Bleachy, pungent | 0.5 |
| Chlorine dioxide (67.5) | 10049–04–4 | 0.1 | Sharp, pungent | 0.1 |
| Chlorobenzene (112.6) | 108–90–7 | 61 | Sweet, almond-like | 75 |
| Chlorobromomethane (129.4) | 74–97–5 | 398 | Sweet | 200 |
| Chloroform (119.4) | 67–66–3 | 205 | Pleasant odor | 2 |
| Chloropicrin (164.4) | 76–06–2 | 1 | Sharp, penetrating | 0.1 |
| Cresol (108.2) | 1319–77–3 | 5 | Sweet, creosote tar | 5 |
| Crotonaldehyde (70.1) | 123–73–9 | 1 | Pungent, suffocating | 2 |
| Cumene (120.2) | 98–82–8 | 1.3 | Sharp, aromatic | 50 |
| Cyclohexane (84.2) | 110–82–7 | 0.4 | Sweet, aromatic | 300 |
| Cyclohexanol (100.2) | 108–93–0 | 98 | Camphor-like | 50 |
| Cyclohexanone (98.2) | 108–94–1 | 100 | Sweet, pepperminty | 25 |
| Decaborane (122) | 17702–41–9 | | | 0.05 |
| Diacetone alcohol (116.2) | 123–42–2 | 101 | Sweet | 50 |
| Diazomethane (42) | 334–88–3 | | | 0.2 |
| Diborane (27.7) | 19287–45–7 | 3.5 | Repulsively sweet | 0.1 |
| Dichlorobenzene (p) (147.0) | 106–46–7 | 30 | Mothballs | 75 |

(*continued*)

**Table A-11** *(continued)*

| Chemical (M) | CAS | Odor[a] (ppm) | Description | PEL (ppm) |
|---|---|---|---|---|
| Dichloroethane (99.0) | 75–34–3 | 200 | Chloroform-like | 100 |
| Dichloroethyl ether (143) | 111–44–4 | | | 5 |
| Dichloroethylene (97.0) | 540–59–0 | 500 | Acrid, ethereal | 200 |
| Diethylamine (73.1) | 109–89–7 | 38 | Fishy, ammonia-like | 10 |
| Diisobutyl ketone (142.3) | 108–83–8 | 0.3 | Sweet, ester | 25 |
| Diisopropyl amine (101.2) | 108–18–9 | 0.8 | Fishy, amine | 5 |
| Dimethylamine (45.1) | 124–40–3 | 30 | Fishy, ammonia-like | 10 |
| Dioxane (1,4–) (88.1) | 123–91–1 | 171 | Ether-like | 25 |
| Epichlorohydrin (92.5) | 106–89–8 | 21 | Chloroform-like | 2 |
| Ethanolamine (61.1) | 141–43–5 | 4.3 | Ammonia | 3 |
| Ethyl acetate (88.1) | 141–78–6 | 185 | Fruity, pleasant | 400 |
| Ethyl acrylate (100.1) | 140–88–5 | 8 | Earthy, acrid, plastic | 5 |
| Ethyl alcohol (46) | 64–17–5 | | | 1000 |
| Ethyl amine (45.1) | 75–04–7 | 215 | Sharp, ammonia-like | 10 |
| Ethyl benzene (106.2) | 100–41–4 | 200 | Aromatic | 100 |
| Ethyl bromide (109.0) | 74–96–4 | 200 | Ether-like | 200 |
| Ethyl chloride (65) | 75–00–3 | | | 1000 |
| Ethyl ether (74.1) | 60–29–7 | 1 | Sweet, ether-like | 400 |
| Ethyl formate (74.1) | 109–94–4 | 327 | Fruity, irritating | 100 |
| Ethyl mercaptan (62.1) | 75–08–1 | 0.4 | Garlic | 0.5 |
| Ethyl silicate (208.3) | 78–10–4 | 85 | Alcohol-like, sharp | 10 |
| Ethylene dibromide (187.9) | 106–93–4 | 10 | Mild, sweet | c |
| Ethylene dichloride (99.0) | 107–06–2 | 109 | Sweet | 1 |
| Ethylene oxide (44.1) | 75–08–1 | 778 | Sweet, olefinic | c |
| Flourine (38.0) | 7782–41–4 | 4 | Pungent, irritating | 0.1 |
| Formaldehyde (30.0) | 50–00–0 | 60 | Hay-like, pungent | 3[b] |
| Formic acid (46.0) | 64–18–6 | 20 | Pungent, penetrating | 5 |
| Furfural (96.1) | 98–01–1 | 5 | Almonds | 2 |
| Heptane (100.2) | 142–82–5 | 313 | Gasoline-like | 400 |
| Hydrazine (32.1) | 302–01–2 | 3 | Fishy, ammonia-like | 0.1 |
| Hydrochloric acid (36.5) | 7647–01–0 | 33 | Irritating, pungent | 5[b] |
| Hydrofluoric acid (20.0) | 7664–39–3 | 0.2 | Strong, irritating | 3 |
| Hydrogen bromide (80.9) | 10035–10–6 | 2 | Sharp irritating | 3 |
| Hydrogen cyanide (27.0) | 74–90–8 | 4.5 | Bitter, almond-like | 4.7[d] |
| Hydrogen peroxide (34.0) | 7722–84–1 | 108 | Slightly sharp | 1 |
| Hydrogen selenide (81.0) | 7783–07–5 | 3.6 | Decayed horseradish | 0.05 |
| Hydrogen sulfide (34.1) | 7783–06–4 | 0.01 | Rotten eggs | 10 |
| Iodine (254) | 7553–56–2 | | | 0.1[b] |
| Isobutyl acetate (116) | 110–19–0 | | | 150 |
| Isobutyl alcohol (74) | 78–83–1 | | | 50 |
| Isopropyl acetate (102.2) | 108–21–4 | 365 | Fruity | 250 |
| Isopropyl alcohol (60.1) | 67–63–0 | 200 | Pleasant | 400 |
| Isopropylamine (59.1) | 75–31–0 | 199 | Pungent, ammonia-like | 5 |
| Isopropyl ether (102.2) | 108–21–3 | 302 | Sweet, sharp, ether | 500 |
| Maleic anhydride (98.1) | 108–31–6 | 0.5 | Acrid | 0.25 |
| Mercury vapor (201) | 7439–97–6 | | | 0.05 mg/m$^3$ |
| Methyl acetate (74.1) | 72–20–9 | 302 | Fragrant, fruity | 200 |

*(continued)*

**Table A-11** *(continued)*

| Chemical (M) | CAS | Odor[a] (ppm) | Description | PEL (ppm) |
|---|---|---|---|---|
| Methyl acrylate (86.1) | 96–33–3 | 20 | Sharp, sweet, fruity | 10 |
| Methyl alcohol (32.1) | 67–56–1 | 20,485 | Sweet | 200 |
| Methylamine (31.1) | 74–89–5 | 9.5 | Fishy, pungent | 10 |
| Methyl bromide (95.0) | 74–83–9 | 1031 | Sweetish | 5 |
| Methyl cellosolve (76.1) | 109–86–4 | 93 | Mild, nonresidual | c |
| Methyl chloride (50.5) | 74–87–3 | 10 | Sweet, etheral | 50 |
| Methyl chloroform (133.4) (1,1,1-Trichloroethane) | 71–55–6 | 698 | Chloroform-like | 350 |
| Methylcyclohexane (98.2) | 108–87–2 | 500 | Faint, benzene-like | 400 |
| Methyl cyclohexanol (114.2) | 25639–42–3 | 505 | Weak, like coconut oil | 50 |
| Methyl ethyl ketone (72.1) 2-Butanone | 78–93–3 | 50 | Sweet, acetone-like | 200 |
| Methyl formate (60.1) | 107–31–3 | 2802 | Pleasant | 100 |
| Methyl iodide (142) | 74–88–4 | | | 2 |
| Methyl isocyanate (102) | 624–83–9 | | | 0.02 |
| Methyl mercaptan (48.1) | 74–93–1 | 0.04 | Sulfidy | 0.5 |
| Methyl methacrylate (100.1) | 80–62–6 | 0.3 | Arid, fruity, sulfidy | 100 |
| Methyl stryene ($\alpha$) (118.2) | 98–83–9 | 48 | Sweet, sharp | 50 |
| Methylene chloride (84.9) | 75–09–2 | 623 | Sweet, chloroform-like | c |
| Morpholine (87.1) | 110–91–8 | 0.13 | Fishy, amine | 30 |
| Naphthalene (128.2) | 91–20–3 | 24 | Mothball, tar-like | 10 |
| Nickel carbonyl (170.7) | 13463–39–3 | 3 | Musty | 0.001 |
| Nitric acid (63) | 7697–37–2 | | | 2 |
| Nitric oxide (30) | 10102–43–9 | | | 25 |
| Nitrobenzene (123.1) | 98–95–3 | 1.9 | Shoe polish, pungent | 1 |
| Nitroethane (75.1) | 79–24–3 | 202 | Mild, fruity | 100 |
| Nitrogen dioxide (46.0) | 10102–44–0 | 5 | Sweetish, acrid | 1[d] |
| Nitromethane (61.0) | 75–52–5 | 100 | Mild, fruity | 100 |
| Nitropropane (1–) (89.1) | 108–03–2 | 297 | Mild, fruity | 25 |
| Nitrotoluene (137) | 88–72–2 | | | 2 |
| Octane (114.2) | 111–65–9 | 259 | Gasoline-like | 300 |
| Oxygen difluoride (54.0) | 7783–41–7 | 0.4 | Foul | 0.05 |
| Ozone (48.0) | 10028–15–6 | 0.5 | Pleasant, clover-like | 0.1 |
| Pentaborane (63.1) | 19624–22–7 | 1 | Strong, pungent | 0.005 |
| Pentane (N) (72.2) | 109–66–0 | 1018 | Gasoline-like | 600 |
| Perchloroethylene (165.8) (tetrachloroethylene) | 127–18–4 | 70 | Chlorinated solvent | 25 |
| Perchlorylflouride (102.5) | 7616–94–6 | 11 | Sweet | 3 |
| Phenol (94.1) | 108–95–2 | 5.8 | Medicinal, sweet | 5 |
| Phenyl ether (170.2) | 101–84–8 | 0.1 | Disagreeable | 1 |
| Phosgene (98.9) | 75–44–5 | 1 | Musty hay, green corn | 0.1 |
| Phosphine (34.0) | 7803–51–2 | 2.6 | Oniony, mustard, fishy | 0.3 |
| Phosphoric acid | 7664–38–2 | | | 1 mg/m$^3$ |
| Propyl acetate (N) (102.2) | 109–50–4 | 26 | Sweet, ester | 10 |
| Propyl alcohol (60.1) | 71–23–8 | 61 | Sweet, alcohol | 200 |
| Propyl nitrate (N) (105.1) | 627–13–4 | 49 | Ether-like | 25 |
| Propylene dichloride (113.0) | 78–87–5 | 131 | Sweet | 75 |

*(continued)*

**Table A-11** *(continued)*

| Chemical (M) | CAS | Odor[a] (ppm) | Description | PEL (ppm) |
|---|---|---|---|---|
| Propylene oxide (58.1) | 75–56–9 | 211 | Sweet, alcoholic | 20 |
| Pyridine (79.1) | 110–86–1 | 4.6 | Burnt, sickening | 5 |
| Quinone (108.1) | 106–51–4 | 0.1 | Acrid | 0.1 |
| Styrene (104.2) | 100–42–5 | 202 | Solvexty, rubber-like | 50 |
| Sulfur dioxide (64.1) | 7446–09–5 | 4.7 | Irritating, pungent | 2 |
| Sulfur monochloride (135.0) | 10025–67–9 | 2.1 | Nauseating | 1[b] |
| Sulfuric acid (98.1) | 7664–93–9 | 0.2 | Pungent, irritating | 2 |
| Tetrachloroethane (167.9) | 79–34–5 | 5 | Sickly sweet | 1 |
| Tetrahydrofuran (72.1) | 109–99–9 | 60 | Ether-like | 200 |
| Toluene (92.1) | 108–88–3 | 40 | Rubbery, mothballs | 100 |
| Toluene 2,4 diisocyanate (174.2) | 584–84–9 | 2.4 | Sweet, fruity, acrid | 0.02[b] |
| Trichloroethylene (TCE) trichloroethene (131.4) | 79–01–6 | 403 | Solventy | 25 |
| Trichloropropane (147.4) | 96–18–4 | 50 | Strong, acrid | 10 |
| Triethylamine (101.2) | 121–44–8 | 0.3 | Fishy, amine | 10 |
| Vinyl toluene (118.2) | 25013–15–4 | 50 | Strong, disagreeable | 100 |
| Xylene (106.2) | 1330–20–7 | 40 | Sweet | 100 |
| Xylidene (121.2) | 1300–73–8 | 0.005 | Weak, amine-like | 2 |

*Source:* Odor thresholds abstracted from Ruth (1986); PEL values abstracted from CFR 1910.1000.

[a] Lowest concentration consistently detected by a panel of experts.

[b] Ceiling value.

[c] In process of rule-making.

[d] STEL value.

**Table A-12**  Vapor Pressure[a] and OSHA PEL Values[b] for Industrial Volatile Liquids

| Substance | TWA-PEL (ppm) | M | SG | 1 mm | 5 mm | 10 mm | 20 mm | 40 mm | 60 mm | 100 mm | 200 mm | 400 mm | 760 mm |
|---|---|---|---|---|---|---|---|---|---|---|---|---|---|
| Acetaldehyde | 100 | 44 | 0.79 | −81.5 | −65.1 | −56.8 | −47.8 | −37.8 | −31.4 | −22.6 | −10.0 | +4.9 | 20.2 |
| Acetic acid | 10 | 60 | 1.05 | −17.2 | +6.3 | 17.5 | 29.9 | 43.0 | 51.7 | 63.0 | 80.0 | 99.0 | 118.1 |
| Acetic anhydride | 5[c] | 102 | 1.08 | 1.7 | 24.8 | 36.0 | 48.3 | 62.1 | 70.8 | 82.2 | 100.0 | 119.8 | 139.6 |
| Acetone | 750 | 58 | 0.79 | −59.4 | −40.5 | −31.1 | −20.8 | −9.4 | −2.0 | +7.7 | 22.7 | 39.5 | 56.5 |
| Acrolein | 0.1 | 56 | 0.84 | −64.5 | −46.0 | −36.7 | −26.3 | −15.0 | −7.5 | +2.5 | 17.5 | 34.5 | 52.5 |
| Allyl alcohol[d] | 2 | 58 | 0.85 | −20.0 | +0.2 | 10.5 | 21.7 | 33.4 | 40.3 | 50.0 | 64.5 | 80.2 | 96.6 |
| Aniline[d] | 2 | 93 | 1.02 | 34.8 | 57.9 | 69.4 | 82.0 | 96.7 | 106.0 | 119.9 | 140.1 | 161.9 | 184.4 |
| Benzene[e] | 10 | 78 | 0.88 | −36.7 | −19.6 | −11.5 | −2.6 | +7.6 | 15.4 | 26.1 | 42.2 | 60.6 | 80.1 |
| Biphenyl[e] | 0.2 | 154 | | 70.6 | 101.8 | 117.0 | 134.2 | 152.5 | 165.2 | 180.7 | 204.2 | 229.2 | 254.9 |
| Carbon disulfide[e] | 4 | 76 | 1.26 | −73.8 | −54.3 | −44.7 | −34.3 | −22.5 | −15.3 | −5.1 | +10.4 | 28.0 | 46.5 |
| Carbon tetrachloride[e] | 2 | 154 | 1.59 | −50.0 | −30.0 | −19.6 | −8.2 | +4.3 | 12.3 | 23.0 | 38.3 | 57.8 | 76.7 |
| Chlorobenzene | 75 | 113 | 1.11 | −13.0 | +10.6 | 22.2 | 35.3 | 49.7 | 58.3 | 70.7 | 89.4 | 110.0 | 132.2 |
| Chloroform | 2 | 119 | 1.48 | −58.0 | −39.1 | −29.7 | −19.0 | −7.1 | +0.5 | 10.4 | 25.9 | 47.7 | 61.3 |
| 1-Chloroprene | 10 | 89 | 0.96 | −81.3 | −63.4 | −54.1 | −44.0 | −32.7 | −25.1 | −15.1 | +1.3 | 18.0 | 37.0 |
| Cyclohexane | 300 | 84 | 0.78 | −45.3 | −25.4 | −15.9 | −5.0 | +6.7 | 14.7 | 25.5 | 42.0 | 60.8 | 80.7 |
| Dimethylamine | 5 | 45 | 0.67 | −87.7 | −72.2 | −64.6 | −56.0 | −46.7 | −40.7 | −32.6 | −20.4 | −7.1 | +7.4 |
| Dimethylaniline | 10 | 121 | 0.96 | 29.5 | 56.3 | 70.0 | 84.8 | 101.6 | 111.9 | 125.8 | 146.5 | 169.2 | 193.1 |
| Ethyl acetate | 400 | 88 | 0.90 | −43.4 | −23.5 | −13.5 | −3.0 | +9.1 | 16.6 | 27.0 | 42.0 | 59.3 | 77.1 |
| Ethyl acrylate[d] | 5 | 100 | 0.92 | −29.5 | −8.7 | +2.0 | 13.0 | 26.0 | 33.5 | 44.5 | 61.5 | 80.0 | 99.5 |
| Ethyl alcohol | 1000 | 46 | | −31.3 | −12.0 | −2.3 | +8.0 | 19.0 | 26.0 | 34.9 | 48.4 | 63.5 | 78.4 |
| Ethylamine | 10 | 45 | 0.69 | −82.3 | −66.4 | −58.3 | −48.6 | −39.8 | −33.4 | −25.1 | −12.3 | +2.0 | 16.6 |
| Ethyl benzene | 100 | 106 | 0.97 | −9.8 | +13.9 | 25.9 | 38.6 | 52.8 | 61.8 | 74.1 | 92.7 | 113.8 | 136.2 |
| Ethyl chloride | 1000 | 65 | 0.92 | −89.8 | −73.9 | −65.8 | −56.8 | −47.0 | −40.6 | −32.0 | −18.6 | −3.9 | +12.3 |
| Ethyl formate | 100 | 74 | 0.92 | −60.5 | −42.2 | −33.0 | −22.7 | −11.5 | −4.3 | +5.4 | 20.0 | 37.1 | 54.3 |
| Ethyl mercaptan | 0.5 | 62 | 0.84 | −76.7 | −59.1 | −50.2 | −40.7 | −29.8 | −22.4 | −13.0 | +1.5 | 17.7 | 35.0 |
| Ethylene dibromide[d] | f | 188 | 2.17 | −27.0 | +4.7 | 18.6 | 32.7 | 48.0 | 57.9 | 70.4 | 89.8 | 110.1 | 131.5 |
| Ethylene dichloride | 1 | 99 | 1.24 | −44.5 | −24.0 | −13.6 | −2.6 | +10.0 | 18.1 | 29.4 | 45.7 | 64.0 | 82.4 |
| Formaldehyde[e] | 3 | 30 | 1.08 | | | −88.0 | −79.6 | −70.6 | −65.0 | −57.3 | −46.0 | −33.0 | −19.5 |
| Formic acid | 5 | 46 | 1.22 | −20.0 | −5.0 | +2.1 | 10.3 | 24.0 | 32.4 | 43.8 | 61.4 | 80.3 | 100.6 |
| Furfural[d] | 2 | 96 | 1.16 | 18.5 | 42.6 | 54.8 | 67.8 | 82.1 | 91.5 | 103.4 | 121.8 | 141.8 | 161.8 |
| Heptane | 400 | 100 | 0.68 | −34.0 | −12.7 | −2.1 | +9.5 | 22.3 | 30.6 | 41.8 | 58.7 | 78.0 | 98.4 |
| Hexachloroethane[d] | 1 | 237 | 2.09 | 32.7 | 49.8 | 73.5 | 87.6 | 102.3 | 112.0 | 124.2 | 143.1 | 163.8 | 185.6 |
| Hexane | 50 | 86 | 0.68 | −53.9 | −34.5 | −25.0 | −14.1 | −2.3 | +5.4 | 15.8 | 31.6 | 49.6 | 68.7 |
| Isobutyl acetate | 150 | 116 | 0.87 | −21.2 | +1.4 | 12.8 | 25.5 | 39.2 | 48.0 | 59.7 | 77.6 | 97.5 | 118.0 |

(continued)

673

## Table A-12  (continued)

| Substance | TWA-PEL (ppm) | M | SG | Temperature (°C) | | | | | | | | | |
|---|---|---|---|---|---|---|---|---|---|---|---|---|---|
| | | | | 1 mm | 5 mm | 10 mm | 20 mm | 40 mm | 60 mm | 100 mm | 200 mm | 400 mm | 760 mm |
| Isobutyl alcohol | 50 | 74 | 0.80 | -9.0 | +11.6 | 21.7 | 32.4 | 44.1 | 51.7 | 61.5 | 75.9 | 91.4 | 108.0 |
| Isopropyl acetate | 250 | 102 | 0.87 | -38.3 | -17.4 | -7.2 | +4.2 | 17.0 | 25.1 | 35.7 | 51.7 | 69.8 | 89.0 |
| Isopropyl alcohol | 400 | 60 | 0.79 | -26.1 | -7.0 | +2.4 | 12.7 | 23.8 | 30.5 | 39.5 | 53.0 | 67.8 | 82.5 |
| Methacrylic acid | 20 | 86 | 0.93 | 25.5 | 48.5 | 60.0 | 72.7 | 86.4 | 95.3 | 106.6 | 123.9 | 142.5 | 161.0 |
| Methyl acetate | 200 | 74 | 0.96 | -57.2 | -38.6 | -29.3 | -19.1 | -7.9 | -0.5 | +9.4 | 24.0 | 40.0 | 57.8 |
| Methyl acrylate[d] | 10 | 86 | 0.96 | -43.7 | -23.6 | -13.5 | -2.7 | +9.2 | 17.3 | 28.0 | 43.9 | 61.0 | 80.2 |
| Methyl alcohol | 200 | 32 | 0.79 | -44.0 | -25.3 | -16.2 | -6.0 | +5.0 | 12.1 | 21.2 | 34.8 | 49.9 | 64.7 |
| Methylamine | 10 | 31 | 0.70 | -95.8 | -81.3 | -73.8 | -65.9 | -56.9 | -51.3 | -43.7 | -32.4 | -19.7 | -6.3 |
| Methyl chloride | 50 | 51 | | | -99.5 | -92.4 | -85.9 | -76.0 | -70.4 | -63.0 | -51.2 | -38.0 | -24.0 |
| Methyl formate | 100 | 60 | 0.98 | -74.2 | -57.0 | -48.6 | -39.2 | -28.7 | -21.9 | -12.9 | +0.8 | 16.0 | 32.0 |
| Methyl iodide[d] | 2 | 142 | 2.28 | | -55.0 | -45.8 | -35.6 | -24.2 | -16.9 | -7.0 | +8.0 | 25.3 | 42.4 |
| Methylene chloride | [f] | 85 | 1.33 | -70.0 | -52.1 | -43.3 | -33.4 | -22.3 | -15.7 | -6.3 | +8.0 | 24.1 | 40.7 |
| Naphthalene | 10 | 128 | 1.15 | 52.6 | 74.2 | 85.8 | 101.7 | 119.3 | 130.2 | 145.5 | 167.7 | 193.2 | 217.9 |
| Nitrobenzene[d] | 1 | 123 | 1.20 | 44.4 | 71.6 | 84.9 | 99.3 | 115.4 | 125.8 | 139.9 | 161.2 | 185.8 | 210.6 |
| Nitroethane | 100 | 75 | 1.05 | -21.0 | +1.5 | 12.5 | 24.8 | 38.0 | 46.5 | 57.8 | 74.8 | 94.0 | 114.0 |
| Nitromethane | 100 | 61 | 1.14 | -29.0 | -7.9 | +2.8 | 14.1 | 27.5 | 35.5 | 46.6 | 63.5 | 82.0 | 101.2 |
| 2-Nitrotoluene[d] | 2 | 137 | | 50.0 | 79.1 | 93.8 | 109.6 | 126.3 | 137.6 | 151.5 | 173.7 | 197.7 | 222.3 |
| Octane | 300 | 114 | 0.70 | -14.0 | +8.3 | 19.2 | 31.5 | 45.1 | 53.8 | 65.7 | 83.6 | 104.0 | 125.6 |
| Perchloroethylene | 25 | 166 | | | | | | | | | | | |
| Phenol[d] | 5 | 94 | 1.06 | 40.1 | 62.5 | 73.8 | 86.0 | 100.1 | 108.4 | 121.4 | 139.0 | 160.0 | 181.9 |
| n-Propyl acetate | 200 | 102 | 0.84 | -26.7 | -5.4 | +5.0 | 16.0 | 28.8 | 37.0 | 47.8 | 64.0 | 82.0 | 101.8 |
| n-Propyl alcohol[d] | 200 | 60 | 0.81 | -15.0 | +5.0 | 14.7 | 25.3 | 36.4 | 43.5 | 52.8 | 66.8 | 82.0 | 97.8 |
| Propylene oxide | 20 | 58 | 0.83 | -75.0 | -57.8 | -49.0 | -39.3 | -28.4 | -21.3 | -12.0 | +2.1 | 17.8 | 34.5 |
| Pryidine | 5 | 79 | 0.98 | -18.9 | 2.5 | 13.2 | 24.8 | 38.0 | 46.8 | 57.8 | 75.0 | 95.6 | 115.4 |
| Styrene, monomer | 50 | 104 | 0.91 | -7.0 | +18.0 | 30.8 | 44.6 | 59.8 | 69.5 | 82.0 | 101.3 | 122.5 | 145.2 |
| Toluene | 100 | 92 | 0.87 | -26.7 | -4.4 | +6.4 | 18.4 | 31.8 | 40.3 | 51.9 | 69.5 | 89.5 | 110.6 |
| Trichloroethylene | 50 | 131 | 1.46 | -43.8 | -22.8 | -12.4 | -1.0 | +11.9 | 20.0 | 31.4 | 48.0 | 67.0 | 86.7 |
| 1, 2, 3-Trichloropropane | 10 | 147 | 1.39 | 9.0 | 33.7 | 46.0 | 59.3 | 74.0 | 83.6 | 96.1 | 115.6 | 137.0 | 158.0 |
| Xylene | 100 | 106 | 0.86 | -3.8 | +20.2 | 32.1 | 45.1 | 59.5 | 68.8 | 81.3 | 100.2 | 121.7 | 144.4 |

[a] Vapor pressure measured in mm Hg (Perry et al. 1984).
[b] PEL values are reviewed on regular basis and readers should always use currently approved values.
[c] Ceiling value.
[d] Additional entry through skin.
[e] See CFR Part 1910.1000 for ceiling and maximum transitory values.
[f] In process of rule-making.

**Table A-13**  Fresh Air Requirements for Ventilation

| | Estimated maximum persons/1000 ft² | Cfm/person | Cfm/ft² |
|---|---|---|---|
| Commercial facilities | | | |
| Commercial laundry | 10 | 25 | |
| Commercial dry cleaner | 30 | 30 | |
| Bar, cocktail lounge | 100 | 30 | |
| Cafeteria, fast food | 100 | 20 | |
| Auto repair room | | | 1.5 |
| Office space | 7 | 20 | |
| Office conference room | 50 | 20 | |
| Office corridors | | | 0.05 |
| Office smoking lounge | 70 | 60 | |
| Elevators | | | 1.0 |
| Warehouses | 5 | | 0.5 |
| Shipping and receiving | 10 | 1.5 | |
| Beauty shop | 25 | 25 | |
| Supermarket | 8 | 15 | |
| Clothier, furniture | | | 0.3 |
| Spectator areas | 150 | 15 | |
| Gymnasium playing floor | 30 | 20 | |
| Auditorium | 150 | 15 | |
| Laboratory | 30 | | |
| Meat processing | 10 | 15 | |
| Transportation vehicles | 150 | 15 | |
| Photography darkrooms | 10 | | 0.5 |
| Duplicating, printing | | | 0.5 |
| Hospital operating room | 20 | 30 | |
| Hospital patient room | 10 | 25 | |
| Education laboratory | 30 | 20 | |
| Education classroom | 50 | 15 | |
| Prison cell | 20 | 20 | |
| Private home | | | |
| Living area | 0.35 air changes/h but not less than 1.5 cfm/person | | |
| Kitchen | 100 cfm intermittent or 25 cfm continuous or operable windows | | |
| Separate garage | 100 cfm per car | | |

*Source:* Abstracted from ASHRAE (1989).

---

**Table A-14**   Modified Euler Method

---

Suppose that one wants to find the value of the continuous function $U(t)$ at arbitrary values of $t$ where the function satisfies the nonlinear differential equation,

$$\frac{dU}{dt} = UA(t) + B(t) \tag{1}$$

and the quantities $A$ and $B$ are known analytical functions of $t$. Assume that the initial value of $U(0)$ is known, that is,

$$U(t_0) = U(0) = \text{known} \tag{2}$$

Express the function $U(t)$ as a Taylor series:

$$U(t + \delta t) = U(t) + \frac{\partial U}{\partial t}\delta t + \frac{\partial^2 U}{\partial t^2}\frac{\delta t^2}{2} + \cdots \tag{3}$$

where $\delta t$ is a small increment of time and the function $U(t)$ and its derivatives are evaluated at time $t$. Eliminate the second- third- and all higher-order derivatives and replace the first derivative by Eq. 1:

$$U(t + \delta t) = U(t) + \delta t[U(t)A(t) + B(t)] \tag{4}$$

Since dropping the higher-order derivatives introduces error, replace $U(t)$ inside the brackets on the right-hand side by the average value of the function between $t$ and $t + \delta t$:

$$U(t + \delta t) = U(t) + \delta t\left[A(t)\frac{U(t) + U(t + \delta t)}{2} + B(t)\right] \tag{5}$$

Rearrange and obtain

$$U(t + \delta t) = \frac{U(t)[1 + (\delta t/2)A(t)] + \delta t\,B(t)}{1 - (\delta t/2)A(t)} \tag{6}$$

To obtain the value of $U$ at the end of a time period $t_n$ [i.e., $U(t_n)$], begin by evaluating Eq. 6 at $t_1$:

$$t_1 = t_0 + \delta t$$

$$U(t_1) = \frac{U(t_0)[1 + (\delta t/2)A(t_0)] + B(t_0)\,\delta t}{1 - (\delta t/2)A(t_0)} \tag{7}$$

Now evaluate Eq. 6 at $t_2$ but use $U(t_1)$, $A(t_1)$, and $B(t_1)$ on the right-hand side of Eq. 6:

$$t_2 = t_1 + \delta t$$

$$U(t_2) = \frac{U(t_1)[1 + (\delta t/2)A(t_1)] + \delta t\,B(t_1)}{1 - (\delta t/2)A(t_1)} \tag{8}$$

Repeat the process until you have computed $U(t_n)$:

$$t_n = t_{n-1} + \delta t \tag{9}$$

The magnitude of the time step $(\delta t)$ is selected by the user. The value should be small such that

$$\frac{\delta t}{2}A(t) < 1 \tag{10}$$

and the denominator in Eq. 6 never changes sign.

---

**Table A-15**  SI Prefixes, Factors, and Symbols

| Prefix | Factor | Symbol |
|--------|--------|--------|
| pico   | $10^{-12}$ | p |
| nano   | $10^{-9}$  | n |
| micro  | $10^{-6}$  | $\mu$ |
| milli  | $10^{-3}$  | m |
| centi  | $10^{-2}$  | c |
| deci   | $10^{-1}$  | d |
| deca   | $10^{1}$   | da |
| hecto  | $10^{2}$   | h |
| kilo   | $10^{3}$   | k |
| mega   | $10^{6}$   | M |
| giga   | $10^{9}$   | G |
| tera   | $10^{12}$  | T |

**Table A-16**  Common Physical Constants and Conversions

| | |
|---|---|
| Avogadro's number | $6.02 \times 10^{23}$ molecules/gmol |
| Energy | $1\ J = 1\ kg\ m^2\ s^{-2}$<br>$1\ Btu = 778.16\ ft \cdot lb_f = 1.055 \times 10^{10}\ ergs = 252\ cal = 1.055\ kJ$<br>$1\ cal = 4.186\ J$<br>$1\ erg = 1\ g\ cm^2\ s^{-2}$<br>$1\ kJ = kPa\ m^3$ |
| Force | $1\ N = 1\ kg\ m\ s^{-2} = 10^5\ dyn$<br>$1\ dyn = 1\ g\ cm\ s^{-2}$<br>$1\ lb_f = 4.448 \times 10^5\ dyn = 4.448\ N$ |
| Length | $1\ m = 100\ cm = 1000\ mm = 3.280\ ft = 39.37\ cm$<br>$1\ \mu m = 10^{-6}\ m = 10^4\ Å$<br>$1\ mile = 5280\ ft = 1609.344\ m$ |
| Mass | $1\ kg = 1000\ g = 2.2046\ lb_m = 6.8521 \times 10^{-2}\ slug$<br>$1\ slug = 1\ lb_f\ s^2\ ft^{-1} = 32.174\ lb_m$<br>$1$ (U.S.) $ton = 2000\ lb_m$<br>$1\ metric\ ton = 1000\ kg = 1.102$ (U.S.) ton<br>$1\ ounce$ (troy) $= 3.110347 \times 10^{-2}\ kg$<br>$1\ lb_m = 7000\ grains$ |
| Power | $1\ Watt = 1\ J/s = 1\ kg\ m^2\ s^{-3}$<br>$1\ hp = 550\ ft\ lb_f\ s^{-1} = 746\ W = 2545\ Btu/h$<br>$1\ MW = 10^3\ kW = 10^6\ W$<br>$1\ Btu/s = 1.055\ kW$ |
| Pressure | $1\ atm = 14.696\ lb_f/in^2 = 760\ torr = 101.325\ kPa = 33.91\ ft\ H_2O$<br>$1\ kPa = 1000\ Pa = 1,000\ n/m^2 = kJ/m^3$<br>$1\ mm\ Hg = 1\ torr = 0.01934\ lb_f/in^2$<br>$1\ bar = 10^6\ dyn/cm^2 = 10^5\ N/m^2 = 14.504\ lb_f/in^2$ |

*(continued)*

| | **Table A-16** *(continued)* |
|---|---|
| Temperature | $T(C) = [T(F) - 32]/1.8$ <br> $T(K) = T(C) + 273.15$ <br> $T(R) = 1.8\,T(K) = T(F) + 459.67$ |
| Time | $1\ h = 60\ min = 3600\ s$ <br> $1\ ms = 10^{-3}\ s$ <br> $1\ \mu s = 10^{-6}\ s = 1000\ ns$ |
| Universal <br> gas constant | $8.314\ J\ gmol^{-1}\ K^{-1} = 8.314\ kJ\ kmol^{-1}\ K^{-1}$ <br> $0.082\ L\ atm\ gmol^{-1}\ K^{-1}$ <br> $1.987\ Btu\ lbmol^{-1}\ R^{-1} = 1.987\ cal\ gmol^{-1}\ K^{-1}$ <br> $1545.33\ ft\ lb_f\ lbmol^{-1}\ R^{-1}$ |
| Volume | $1\ gal\ (liquid) = 0.13368\ ft^3 = 3.785\ L$ <br> $1\ L = 10^{-3}\ m^3 = 1000.028\ cm^3$ <br> $1\ barrel\ (petroleum) = 42\ gal$ <br> $1\ fluid\ ounce = 2.957352 \times 10^{-5}\ m^3$ |

Molecular weight of air = 28.97

Gravitational constant $g_c = 32.2\ lb_m\ ft\ lb_f^{-1}\ s^{-2}$

# References

ACGIH, 1988. *Industrial Ventilation: A Manual of Recommended Practice*, 20th ed., American Conference of Governmental and Industrial Hygienists, Committee on Industrial Ventilation, Lansing, MI.

ASHRAE, 1989. *Ventilation for Acceptable Indoor Air Quality*, ASHRAE Standard 62-1989, American Society of Heating, Refrigerating, and Air-Conditioning Engineers, Atlanta, GA.

Bare, J. C., 1988. Indoor air pollution source database, *Journal of the Air Pollution Control Association*, Vol. 38, No. 5, pp. 670–671.

Butcher, S. S., and Ellenbecker, M. J., 1982. Particulate emissions for small wood stoves and coal stoves, *Journal of the Air Pollution Control Association*, Vol. 32, No. 4, pp. 380–384.

Committee on Indoor Air Pollution. "Indoor Air Pollution", National Academy Press, Washington, DC, 1981.

Crawford, M., 1976. *Air Pollution Control Theory*, McGraw-Hill, New York.

Heywood, J. B., 1988. "Internal Combustion Engine Fundamentals" McGraw-Hill, NY.

Lofroth, G., Burton, R. M., Forehand, L., Hammond, S. K., Seila, R. L., Zwerdlinger, R. B., Lawtas, J., "Characterization of Environmental Tobacco Smoke", Env. Science Tech., Vol. 23, No. 5, pp. 610–614, 1989.

Lyman, W. J., Reehl W. F., and Rosenblatt D. H., "Handbook of Chemical Property Estimation Methods", American Chemical Society, Wash. DC. 1990.

Mackay, D., and Shiu, W. Y., 1981. A critical review of Henry's law constants for chemicals of environmental interest, *Journal of Physical and Chemical Reference Data*, Vol. 10, No. 4, pp. 1175–1191.

Mackay, D., and Yeun, A. T. K., 1983. Mass coefficient correlations for volatilization of organic solutes from water, *Environmental Science and Technology*, Vol. 17, No. 4, pp. 211–217.

Meyer, B., 1983. *Indoor Air Quality*, Addison-Wesley, Reading, MA.

Orlemann, J. A., Kalman, T. J., Cummings, J. A., Lin, E. Y., Jutze, G. A., Zoller, J. M., Wunderle, J. A., Zieieniewski, J. L., Gibbs, L. L., Loudin, D. J., Pfetzing, E. A., and Ungers, L. J., 1983. *Fugitive Dust Control Technology*, Noyes Data Corporation, Park Ridge, NJ.

Perry, R. H., Green, D. W., and Maloney, J. O., 1981. *Perry's Chemical Engineers' Handbook*, 6th ed., McGraw-Hill, New York.

Ruth, J. H., 1986. Odor thresholds and irritation levels of several chemical substances: a review, *Journal of the American Industrial Hygiene Association*, Vol. 47, March, pp. A-142 to A-151.

U.S. EPA, AP-42. *Compilation of Pollutant Emission Factors*, Vol. I, *Stationary Point and Area Sources*, PB87-150959 Supplement A, October 1986, PB82-101213 Third Edition Supplement 12, 1981; PB81-244097 Third Edition Supplement 9; PB81-178014 Third Edition Supplement 11, October 1980; PB80-199045 Third Edition Supplement 10, February 1980; National Technical Information Service, Springfield, VA.

U.S. EPA, 1977, *User's Guide: Emission Control Tecnologies and Emission Factors for Unpaved Road Fugitive Emissions*, EPA /625/5-87/022, September, U.S. Government Printing Office, Washington, DC.

Vargaftik, N. B., 1975. *Tables on the Thermophysical Properties of Liquids and Gases*, 2nd ed., Hemisphere, Washington, DC.

Weast, R. C., 1975. *Handbook of Chemistry and Physics*, 58th ed., CRC Press, Boca Raton, FL.

Yaws, C., Yang, H. C., and Pan, X., 1991. Henry's law constants for 362 organic compounds in water, *Chemical Engineering*, Vol. 98, No. 11, pp. 179–185.

# INDEX

**Appendix, (A*x-y*)**    **Example, (E*x.y*)**    **Figure, (F*x-y*)**    **Table, (T*x-y*)**

## A

Absorbers, 453 (*see also* Scrubbers)
 bed pressure drop, 466, 468
  (**F10-22**)
 costs, 643, 644 (**E13.5**)
 design, 468 (**E10.3**)
 dry scrubbing, 486 (**F10-29**)
 efficiency, 467
 equilibrium line, 458
 height of packed bed, 461
 limestone and lime scrubbing, 487
  (**F10-30**)
 load point, flood point, holdup,
  467 (**F10-21**)
 mass transfer packed bed, 456
  (**F10-14**)
 mass transfer coefficient, 458
 off-design performance, 467
 operating line, 458, 459 (**F10-16**)
 random packing, 456 (**F10-13**)
 with chemical reaction, 471, 472
Absorption, 18
 cross section, 503
 isotherm, 356 (**F8-11**)
 spectra of atmospheric gases, 77
  (**F2-7**)
 uptake efficiency, 151
Accelerating pace of social change,
 61, 64 (**F2-1**)
Acetaldehyde, 246, 500
Acetylene, 236, 312
Acetyl radical (HCO●), 246
ACFM, 28
ACGIH, 12
Acid rain, 101, 187
 chemical and physical processes,
  187, 228
 effect on soil and feeder roots,
  193 (**F5-3**)
Acidification of bodies of water, 192
Acinus, 134
Acoustic power, 208

ACQR, 93
Actinic flux, 503
Activated
 alumina, 441
 charcoal, 18, 441, 442 (**F10-3**)
Activation energy, 44
Active sites, 441
Activity coefficients, 35, 352
Actual cubic feet per minute,
 conversion ACFM to SCFM, 28
Acute effects, 11
Adduct, 174
Adiabatic lapse rate, 399
Administrative controls, 5
Adsorber,
 cost, 640, 641 (**E13.4**)
 design, 451 (**E10.2**)
 dual bed, 448 (**F10-9**)
 regeneration, 25 (**F1-13**)
 rotary wheel, 450 (**F10-10**)
Adsorption, 18, 440
 adsorbate, adsorbent, 441
 adsorbent surface area, 441
 bed thickness, 446
 capacity, 448
 efficiency, 151, 149 (**F 4-26**)
 face, superficial velocity, 447
 isotherm, 443 (**F10-4**)
 macropores, micropores,
  mesopores, 441
 mass transfer, 441
 multicomponent mixtures, 453,
  (**F10-11**)
 physical explanation, 441
 zone thickness, 447
 zone velocity, 446
 zone wave, 445 (**F10-6**)
Advanced oxidation
 technology, 18, 500, 506
Aerodynamic
 building wake, 426
 diameter, 236, 521, 543
 mass resistance in leaves, 263

characteristic Weibel
 model, 135 (**F4-7**)
Aerosols, 527 (see also particles)
 accumulation mode, 527
  (**F11-8**)
 anthropogenic and biogenic
  sources, 240 (**T6-8**)
 classification of size, 239
  (**F6-9**), 516 (**F 11-1**)
 fine & coarse, 162, 238
 light extinction due to, 198
 mean free path, 518
 nuclei mode, 527 (**F11-8**)
 primary & secondary, 162, 526,
  527 (**F11-8**)
 urban air, 246 (**T6-9**)
Afterburner, thermal oxidizer, 25
 (**F1-14**)
Agglomeration, 238 (see also
 coagulation)
Aggregate risk, 178
AIHA, 51
Air basin, 114
Air fuel ratio,
 relation to combustion
  products, 290, 341
Air pollution (EPA definition), 4
Air Pollution Control Act of 1955, 90
 Extension, 90
 summary of legislation, 91, 92,
  (**T3-1**)
Air pollution episodes, 117
 criteria, 118 (**T3-10**)
 response, 119 (**T3-11**)
Air quality (EPA) models, 431
 (**T9-10**)
Air Quality:
Act (1967), 93
 control regions (AQCR), 93
 criteria, 93
 EPA model (SCRAM), 112, 431
 regions (AQR), Class I, II &
  III, 94

review, 106 (**E3.2**)
standards (AQSs), 7
Air standard cycle, 285
Air toxics, 101
Air-to-cloth ratio, 601 (**T12-1**)
Airshed or air basin model, 246
Airway resistance, 139
Airway irritation, 164
Albedo, 74
Alcohols, 313
Aldehydes (RCHO), 232, 236, 246, 247, 500
Alert, air pollution episode, 118 (**T3-10**), 119 (**T3-11**)
Aliphatic ketones, 312
Alkyl peroxy radical [$RO_2\bullet$], 247
Altering process, 5
Alkanes (paraffins), 313
Alkenes (se also olefins) 231, 247
Alkynes (acetylenes), 313
Alkyl radicals ($R\bullet$), 247
Allergens, 164, 169
    house pets, 166
    household pests, 167
    plants, 167
Allergies, 166
Alveoli, 133
    alveolar ducts, 134
    alveolar sacs, 134
    alveolar membrane, 133, 145 (**F4-18**)
    alevolar capillary barrier, 150 (**F4-25**)
    alveolar ventilation rate, 138, 141
    $CO_2$ concentration, 147 (**F4-22**)
    clearance, 161
    $CO_2$ & ($O_2$) concentration, 147
    diagram, 133 (**F4-5**)
    gas exchange, 147 (**F4-20**), 148 (**F4-23**)
    macrophage, 133 (**F4-5**)
    surface area, 135 (**F4-5**)
Ambient monitoring (AMTIC), 112
American Conference of Governmental and Industrial Hygienists (ACGIH), 12
Ames, 15
Ammonia,
    atmospheric behavior, 229
    atmospheric concentrations, 220 (**T6-1**), 222 (**T6-2**)
    injection for NO control, 482, 483 (**F10-27**)
    selective catalytic $NO_x$ reduction, 482, 483 (**F10-27**)
Ammonium nitrate, 229
Ammonium sulfate and bisulfate, 158
Anatomic dead space, 133, 137, 141 (**F4-14**)
Anemic hypoxia, 160
Annual cost (see Total Initial Cost and Total Annual Cost)
Anthropocentric, 60

Anthropogenic processes, 217
Antigen, 166
AP-42 Emission Factors, 336, 652-663 (**A-1** to **A-7**)
Appearance of plumes, 203-205, 409
AQS, 7
Area source, 101
Arithmetic (linear) progression, 62, 64
Arithmetic mean
    particle diameter, 520, 521
    transient time, 140
Arrhenius equation, 44
Art, effects of pollutants, 165
Arteries, arterioles, 135
Aromatic fuels, 312
Asbestos removal, 14
Asbestosis, 165, 166
Asphyxiant, 169
Aspiration efficiency, 575, 576 (**F12-11**)
Asthma, 164, 167
Asynchronous lung function, 143
Asymmetric
    lungs, pendelluft, 142 (**F4-15**), 143 (**F4-16**)
    velocity profiles, 144 (**F4-17**)
Atmospheric dispersion (see Dispersion)
Atmospheric detectors, 4
Atmospheric,
    composition, 220 (**T6-1**), 222 (**T6-2**)
    $CO_2$, $CH_4$ versus time, 63 (**F2-2**)
    diffusion, 413
    dispersion, 386, 413
    generalized reaction mechanism, 247 (**F6-10**), 248 (**T6-10**)
    interfacial transport, 254
    layers, 389, 390 (F9-2)
    photochemical reactions, 232 (**F6-6**), 499
    reactivity, 313 (**T7-5**)
    stability, 401-405
    temperature versus altitude, 218 (**F6-1**)
    temperature versus time, 80 (**F2-9**)
    water vapor concentration, 219 (**E6.2**)
Atmospheric stability
    definition, 401
    neutral, 402
    stable, 402
    unstable, 402
    superadiabatic, 403
Attainment and nonattainment regions, 97
Audit, environmental, 11, 118, 112
Automotive high emitters, 105
Automotive emissions, 97, 105 (**F3-1**), 319 (**T7-6**)
Avalanche, electron ,612

Average molecular weight and density, 31, 664, 673
Avogadro's number, 31, 677 (**A-16**)

**B**

Back-corona, 615 (**F12-43, -44**)
Bacteria, 167
BAT, 96
BACT, 97
Bagassosis, 167
Baghouse, 20
    cost, 634
    pressure drop, 600, 605
    reverse-flow, shaker, pulse-jet, 592-594
    types, 21 (**F1-7**)
Bake-out regenerative thermal oxidizer, 475
Base metal catalyst, 483
Beaufort scale, 406 (**T9-4**)
Benefits and costs, 7 (**F1-2**), 17
Benzene, 174 (**E4.5**)
Best Available Control Technology (BACT), 97, 112
Bioavailability, 169
Biodegradation, 488
Biodiversity, 60
Biofilm, 493 (**F10-37**)
Biogenic sulfur compounds, 228
Biological Exposure Index (BEI), 169, 170 (**E4.4**)
Biological
    mercury cycle, 194 (**F 5-4**)
    monitoring, 169
    oxidation, 18
Biomass conversion, 75
Bioscrubbers, biofiltration, 18, 26, 493
    equilibrium & operating lime, 491 (**F10-36**)
    overall mass transfer coefficient, 492
    trickle bed reactor, 490 (**F10-34**)
    1st order kinetics, 495 (**E10.6**)
    0th order kinetics, 497 (**E10.7**)
    reaction limited
    diffusion limited
Biosphere, 73
Biot number, 363
Biotransformation, 171
Blackbody radiation model, 73 (**F2-6**)
Black gypsum crust, stone monuments, 197
BLIS, 112
Black lung disease, 168
Blood flow during exercise, 142 (**T4-2**)
Blow-by, 315 (**F7-16**)
Bodies of water, 192
Body burden, 169

Bohr model, 141 (**F4-14**)
  extended model, 149 (**F4-24**)
Bombardment charging, 613
Bore of a cylinder, 316
Boundary layer,
  planetary, 387, 389
  surface velocity profile, 405 (**F9-15**)
Box model, 386 (**F9-1**), 387 (**E 9.1**)
Branching airways, 132 (**F4-2**)
  aerodynamic characteristics, 135
  particle removal, 161, 162
  Weibel model, 134 (**F4-6**)
Branching reactions, 291
Brayton cycle, 285, 287 (**F7-6**)
Breakthrough time of adsorber, 446
Briggs equation, 424 (**E 9.7**)
Broadband noise, 207
Bronchi:
  primary and secondary, 132, 133
  right and left, 134
  terminal, 132
Bronchoalveolar lavage (BAL), 164
Bronchial wall, layers, 150
  (**F4-25**), 165 (a)
Bronchiole, 133
Bronchitis, 165 (**F4-36**)
Bronchi: primary, secondary and
  terminal, 132
Bubble-cap scrubbers, 22, 24 (**F1-12**)
Buckingham Pi Theorm, 552
Building
  exhaust stacks, 427
  wakes, 426 (**F9-27**)
Bulk density, 446
Bureau of Labor Statistics (BLS), 8
Byssinosis, 167

**C**

CAAA, 90-94, 100-105
Calculations
  Atmospheric dispersion
    Gaussian plume model, 420
      (**E9.6**)
    plume rise, 424 (**E9.7**)
    puff dispersion, 429 (**E9.8**)
    radiosonde (P,T) data, 400
      (**E9.4**)
  Cancer risk, 179 (**E4.6**)
  Chemical equilibrium,
    $CO/CO_2/O_2$, 296 (**E7.3**)
    $SO_2/SO_3/O_2$, 41 (**E1.9**)
    $NO/NO_2/O_2$, 301 (**E7.5**)
    water-gas shift, 37 (**E1.7**)
  Chemical kinetics
    $1^{st}$, $2^{nd}$ order, 45
    Leighton mechanism, 251
      (**E6.7**)
    $CO/CO_2$, 299 (**E7.4**)
    $NO/NO_2$, 306 (**E7.6**)

pseudo steady state
  assumption, 47
$SO_2/SO_3$, 311
simultaneous, 45
Control of gases, removal
  efficiency
    absorption and chemical
      reaction, 473 (**E10.4**)
    biofiltration, 495, 497
      (**E10.6 & 10.7**)
    packed bed scrubbing, 468
      (**E10.3**)
    regenerative adsorption, 451
      (**E10.2**)
    thermal oxidation with energy
      recovery, 478 (**E10.5**)
Control of particles, removal
  efficiency
    collectors in series and parallel,
      478 (**E11.4**)
    cyclone collectors, 570 (**E12.1**)
    electrostatic precipitator, 617
      (**E12.8**)
    filtration, 606 (**E12.7**)
    inertial deposition in curved
      ducts, 557 (**E11.9**)
    settling in ducts, 547 (**E11.8**)
    spray chambers, 582 (**E12.3**)
    transverse packed bed
      scrubber, 584 (**E12.4**)
    venturi scrubber, 587 (**E12.5**)
Cost of control systems
  carbon adsorber, 641 (**E13.4**)
  costs that varying over time,
    632 (**E13.1**)
  counter-flow scrubber, 644
    (**E13.5**)
  electrostatic precipitator, 638
    (**E13.3**)
  fabric filtration, 635 (**E13.2**)
  thermal oxidizer, 647 (**E13.6**)
  utilities, 632 (**E13.2**)
Earth black body radiation
  model, 74 (**E2.5**)
Emission factor, 336 (**E8.4**)
Exponential growth
  Malthusian dilemma, 65 (**E2.2**)
Gas uptake in lung, 152 (**E4.1**)
Leaks
  flashing, 347 (**E8.19**)
  gaseous leaks, 372 (**E8.18**)
  liquid leaks, 370 (**E8.17**)
Mass and volumetric flow rate, 29
  (**E1.4**)
  cooling hot gas stream, 602
    (**E12.6**)
Mass balance
  $CO_2$ and $SO_2$, 332 (**E8.2**)
  sulfur, 27 (**E1.3**)
Mass transfer rates
  nitrogen dioxide ($NO_2$) into
    water, 259 (**E6.9**)
  nitrogen dioxide ($NO_2$) uptake
    by potato, 262 (**E6.10**)

sulfur dioxide ($SO_2$) uptake by
  leaves, 265 (**E6.11**)
Noise, several sources, 209 (**E5.3**)
Pollutant concentration &
  emission rate, 31 (**E1.5**)
Pollutant generation rate
  drop evaporation, 365 (**E8.16**)
  empirical equations, 338 (**E8.5**)
  evaporation in confined space,
    361 (**E8.15**)
  evaporation in stagnant air, 344
    (**E8.8**)
  flux chamber, 335 (**E8.3**)
  single film evaporation, 348
    (**E8.10**)
  single film evaporation of
    several components, 352
    (**E8.11**)
  two-film evaporation, 359
    (**E8.14**)
Risks
  benzene and leukemia, 174
    (**E4.5**)
  voluntary and involuntary, 16
    (**E1.2**)
Small particles
  particle trajectory of  in moving
    air, 538 (**E11.5**)
  settling velocity, 542 (**E11.6**),
    543 (**E11.7**)
  size distribution, 529 (**E11.2**)
Thermophysical properties
  diffusion coefficient, 341 (**E8.7**)
  equilibrium coefficient, 40
    (**E1.8**)
  Raoult's law and Henry's law,
    357 (**E8.12**) & **E8.13**)
  vapor pressure, 339 (**E8.6**)
Well mixed model
  box model, 387 (**E9.1**)
  flux chambers, 335 (**E8.3**)
Calcining, 486
Capillaries, 135
Capital Recovery Cost
  (CRC), 630
Capital Recovery Factor
  (CRF), 630
Cancer, 166
  initiation, promotion and
    progression, 173
Carbon cycle (global), 226 (**F6-3**)
Carbon adsorber costs, 640
Carbon dioxide ($CO_2$),
  air and oceans, 226 (**E6.3**)
  atmospheric concentration, 220
    (**T6-1**)
  atmosphere versus ancient time,
    63, 237
  atmosphere versus time, 223
    (**F6-2**)
  $CO/CO_2/O_2$ equilibrium, 296
    (**E7.3**)
  confined space, 160 (**E4.3**)

formation in
  combustion, 291-193
rate of absorption in ocean, 266
  (**E6.12**)
Carbon disulfide ($CS_2$), 228
Carbon monoxide (CO), 172
  atmospheric chemistry, 227
  atmospheric concentration, 220
    (**T6-1**)
  combustion of $CH_4$, 241 (**E6.5**)
  formation in
    combustion, 291-293
  global budget, 226 (**F6-3**)
  toxicity, 173 (**F4-40**)
  rate of oxidation of CO
    to $CO_2$, 240 (**E7.4**)
Carbon monoxide-carbon dioxide
  ($CO/CO_2$)
  equilibrium, 296 (**E7.3**)
  kinetics in flames, 299 (**E7.4**)
Carbon, oceanic transport, 81 (**F2-10**)
Carbonic acid, 187
Carbon tetrachloride, 235
Carboxyhemoglobin, 173
Carbonyls, 312
Carbonyl sulfide (COS), 228
Carcinogen potency factor, 169, 177
  (**F12-7 & F12-8**)
Cardiotoxin, 169
Carnia, 133
Carnial ridges, 163
Carryover, 455
CAS, 669 (**A-11**)
Cascade impactor, 551
Case-by-case, top-down permit
  review process, 113 (**T3-5**)
Catalytic
  base metal, noble metal 483
  converters, 317
  converters for internal combustion
    engines, 318 (**F7-18**)
  dual catalyst reactions, 318 (**F7-19**)
  metal core, 484 (**F10-28**)
  oxidation, 18, 25 (**F1-14**), 474
  recuperative oxidizer, 477 (**F10-26**)
  types of Selective Catalytic
    Reduction (SCR), 483
  zeolite, 484
Ceiling (concentrations), 121
Cellular damage in lung, 164
Centrifugal force, 553
Ceramic filters, 603
CFB, 323
CFC, 234
CFR, 111
Chain-branching reactions, 292
Change, absolute & relative, 64
Change the product, control
  strategy, 5
Charge transfer, 612
Chemical element balances, 27
  (**E1.3**), 332 (**E8.2**)
Chemicals of commerce, 10

Chemical Abstract Service
  (CAS), 12, 120
Chemical Engineering Plant Cost
  Index (CE), 630
  631 (**T13-3**), 632 (**E13.1**)
Chemical equilibrium, 35, 46
Chemical Information Service
  (CIS), 12, 120
Chemical kinetics, 43
  first-order, second-order, 45
  mechanisms, 44
  rate constants, 44
Chemical oxidation, 18
Chemical Safety and Hazard
  Investigation Board, 101
Chemisorption, 441
Chloroflurohydrocarbons
  (CFC), 234
  atmospheric concentrations, 220
    (**T6-1**)
  effect on stratospheric ozone,
    252, 253 (**E6.8**)
Chloroform, 235
Chloroplasts cells of plants, 261
Chlorosis, 188
Choked flow, 371
Chronic effects, 11
CI, 285
Cilia, 132
  components, 132 (**F4-3**)
  on bronchial wall, 132 (**F4-4**)
Circulation;
  blood, 159
  global atmospheric, 389-394
  oceanic, 81
Circulatory hypoxia, 160
CIS, 12, 120
Class notes, **http://www.engr.psu.edu
  /cde/me470**
Classes of organic compounds, 313
Clausius-Clapeyron equation, 339
Clean Air Act:
  of 1963, 90
  Amendments (CAAA) of
    1966,1970, 1977, 90-94
  Amendments (CAAA)
    of 1990, 100-104
Clearance, 161
Closed system, 47 (**F1-20**)
Clouds, 520 (**E11.1**), 549, 550, 551
Coal combustion
  emissions of $SO_2$ and $NO_x$, see
    Emission factors
Coagulation, 527 (**F11-8**)
Coarse particles, 162, 238
Code of Federal Regulations
  (CFR), 111
Coefficient of Haze (COH), 201
Collagen, 136, 165
Collection efficiency, (see also
  efficiency)
  collectors in series and parallel,
    534, 535 (**E11.4**)

flow in horizontal ducts, 546
  (**F11-17**)
flow in curved ducts, 553
impaction sphere-on-sphere, 577,
  578 (**E12.2**)
impaction sphere-on cylinder,
  595, 597 (**F12-27**)
polydisperse aerosol, 534 (**E11.3**)
overall, 532
Colligative behavior, 549
Combustion controls, 313
Combustion, staged, 320
Combustors, types, 281
Command and control, 96
Compartmental models of lung, 149
Compliance of lungs, 137 (**F4-10**)
Compliant surface coating, 116, 117
  (**T3-9**)
Compost, 490
Compressed air cost, 634
Compressibility factor, 30, 339
Compression ignition engine (CI), 285
Computer, air quality models
  (SCRAM), 112, 431
Concentration, 29-31
  average or total molar, 32
  gas and vapor at STP and actual
    T and P, 30
  partial molar concentration, 32
  particles, 29, 31 (**E 1.5**)
Concentrators, 567
Condensation, 440 (**F10-1**)
Conditioning of inhaled air, 155, 156
Conducting airway, 132
Confined space, 361
Coning plume, 410 (**F9-18**), 412
  (**F9-19c**)
Continuity equation, 387, 342
Continuum flow regime, 518
Conservation, 5
Constricted lungs, 139 (**F4-13**)
Contact scrubbers, 590 (**F12-21**)
Continuous-stirred tank reactor
  (CSTR), 34
Control
  system (definition), 4
  technology documents, 93
  volume and surface, 148
Control of emissions (see Emission
  control)
Control strategy for combustors,
  air pollution controls (APC), 313
  combustion controls, 313
Convective diffusion, 355
Convection heat transfer,
  relationships, 349 (**T8-1**)
Cooling water cost, 623
Cooling gas streams, 602 (**E12.6**)
Coriolis force & acceleration, 392-
  395, 395 (**E9.2**)
Corona, 612
  electron avalanche, 612
Cost-benefit ratio, 10
Cost indices, 630

Cost of utilities, 632
Costs, see
    Total Annual Cost (TAC), 629
    Total Initial Cost (TIC), 628
Coughing, 133
Counterflow scrubber efficiency,
    spray chamber particle
        removal, 579, 581
    packed bed gas absorption, 455,
        467, 468 (**E10.3**)
Crankcase emissions, 315 (**F7-16**)
CRC, 630
CRF, 630
Criteria (EPA)
    documents, 90
    pollutants, 93
Critical temperature and
    pressure, 647 (**A-9**)
Cross-flow spray chamber, 582
Cumulative
    mass distribution function, 524
    number distribution, 523
Cunningham correction factor, 519
    (**T11-1**)
Curvilinear flow, 552-555
Curved ducts, 553, 556 (**T11-10**)
Cyanosis, 160
Cycles of atmospheric gases (see
    formation and fate cycles)
Cycloparaffins (cyclanes), 313
Cyclones,
    straight-through, 18, 20 (**F1-6**),
        568 (**F12-1**)
    reverse-flow, 19 (**F1-5**), 568
        (**F12-2**), 569 (**F12-3**)
    fractional efficiency, 569 (**F12-3**)

**D**

Dalton's law of partial pressure, 29
Darcy's law, 600
Daughters of Radon, 234
Dead weather, 405
    (dB) Decibels, 209
Demister, 23, 455, 570, 571
Denitrification processes, 224
Density,
    average or total, 31
    liquid pollutants, 664 (**A-8**), 673
        (**A-12**)
    molar, 330
    moist air, 30
    partial, 31
Deposition velocity, 256, 257 (**T6-11**)
Design defects or errors, 17
Desorption, 346, 355
Detection (odor) threshold, 206
Deutsch equation, 590
    precipitation parameter, 611
        (**T12-2**)
Deviation, standard, 521
Diameter (see particle size)

arithmetic mean, 520
    geometric number mean, 523
    geometric mass mean, 524
    median, 520
Diatomaceous earth adsorbent, 441
Diesel cycle, 285, 287 (**F7-6**)
Diffusion
    flames, 289, 290 (**F7-10**)
    charging, 613
Diffusion coefficient, 339, 664
    (**A-8**), 341 (E8.7), 340 (**F8-3**)
Diffusion-limited biofilms, 496
Diffusion through stagnant air, 343
Dimensionless analysis, 552
Dimethylmercury, 194
Dimethyl sulfide ($CH_3SCH_3$), 228
Direct flame oxidizer, 18, 474
    regenerative, recuperative,
        catalytic, 475 (**F10-24**), 476
        (**F10-25**), 477 (**F10-26**)
Directory factor (sound), 208
Dispersion, (see also Pasquill-
    Gifford coefficients)
    coefficients, 415, (**F9-21** &
        **F9-22**), 416 (**T9-6**) & (**T9-7**), 417
        (**T9-8**)
    EPA air quality models, 431
        (**T9-10**)
    Gaussian plume model, 413, 414
        (**F9-20**), 421 (**E 9.6**)
    puff, instantaneous point source,
        429 (**E 9.8**)
Disposal costs, wastewater, sludge,
    hopper dust, 634
Distributed parameter model, 153
    (**F4-27**)
Diurnal,
    hydrocarbon emissions, 315
    temperature variations, 409 (**F9-17**)
DOE, 49
Doldrums, 390 (**F9-3**), 393
Donora, 90
Dose, total and dose rate, 168, 172
Dose-response characteristics, 168,
    172 (**F4-39**)
Doubling time,
    exponential growth, 64
    linear growth, 62
Double exponential rate, 69
Downwash, 428
Downwelling, 81
Drag coefficient, 517 (**F11-3**)
Drift velocity, 611 (**T12-2**)
Drop evaporation, 362, 363 (**F8-13**),
    365 (**E8.16**)
Dropout boxes, 18, 547
Dry adiabatic
    atmosphere, 399
    lapse rate, 399
Dry
    injection, 485
    deposition, 528
    scrubbing, 485, 486 (**F10-29**)

Dual-bed catalytic converter, 318
    (**F7-18**)
    reactions, 318 (**F7-19**)
Dust, fugitive, 115 (**T3-8**)
Dust cake, 21, 595, 596
    back corona, 615
    filter loading, 600
    resistivity, 616
    thickness on filter, 595, 596, 600
Dust mites, 167
Dynamic viscosity, 518, 668 (**A-10**)
Dynamic shape factor, 520 (**T11-2**)
Dyspnea, 160, 164

**E**

Earth
    effective radiation temperature, 74
    human carrying capacity, 66
    mean surface temperature, 74
    radiation model, 73 (**F2-6**)
Ecoaudits, 122
Ecology, 59
Economic incentives, 60, 96
Ecosystem, 59
Eddy cell model, 258
Edema, 136, 164
Effective stack height, 417, 424
Efficiency of collection and
    removal, (overall) 532
    absorbers, 467, 468 (**E10.3**), 473
        (**E10.4**)
    adsorbers, 440, 451 (**E10.2**)
    biofilters & bioscrubbers, 495
        (**E10.6**), 497 (**E10.7**)
    cyclones,
        reverse-flow, 570 (**E12.1**), 568
            (**F12-2**), 569 (**F12-3**), 570 (**F12-4**)
        straight-through, 568 (**F12-1**)
    counter-flow scrubbers
    gases and vapors, 467
    particles, 579 (**F12-15**) &
        (**F12-16**), 582 (**E12.3**)
    electrostatic precipitators, 611,
        617 (**E12.8**)
    filters, 606 (**E12.7**)
    gravitation settling in ducts, 547
        (**E11.8**)
    inertial deposition in curved
        ducts, 556 (**T11-10**), 557 (**E11.9**)
    polydisperse aerosol, 534 (**E11.3**)
    series and parallel, 534 (**F11-12**),
        535 (**E11.4**)
    single drop, sphere-on-sphere,
        577, 578 (**E12.2**)
    single fiber, sphere-on-cylinder,
        577, 595, 597
    spray chambers (particles), 579,
        582 (**E12.3**)
    thermal oxidation, 478 (**E10.5**)
    overall and fractional efficiency,
        533, 534 (**E11.3**)

transverse-flow scrubbers, 583, 584 (**E12.4**)
uptake of gases in lung, 151
venturi scrubber, 585, 587 (**E12.5**)
EGR, 319
Ehrlich Index, 171 (**F4-38**)
Ekman spiral, 397 (**F9-9**)
Electron transfer, 612 (**F12-40**), 613
Electronic
  air cleaners, 617
  bulletin boards, 111
Electric field, 614
  space charge, 615
Electric motor cost, 632
Electrostatic precipitators
  (ESP), 18, 22, 608
  corona, 612
  cost, 637, 638 (**E13.3**)
  cylinder-wire, 609 (**F12-37**), 614
   (**F12-41**)
  drift velocity, 609
  efficiency (Deutsch equation), 611 (**F12.39**)
  electric field, 614
  electron avalanche, 612
  migration velocity, 611 (**T12-2**)
  particle charging, 612 (**F12-40**)
  plate wire, 608 (**F12-36**), 617
   (**E12.8**)
  precipitation parameter, 611
   (**T12-2**)
  single-stage, plate-wire, 614
  space charge, 614
  wet-plate, 610 (**F12-38**)
Elevated line source, 431
El Nino, 80
Elimination, 5
Elutriators, 547
Emergency, air pollution episode, 118 (**T3-10**) 119 (**T3-11**)
Emissions
  anthropogenic, 222 (**T6-2**)
  area and point source, 330
  automotive, 314 (**F7-15**), 319
   (**T7-6**)
  credit or allowance, 101
  elevated point source, 413-425
  elevated line source, 431
  emission rate, 31 (**E1.5**)
  fugitive, 115 (**T3-7**), 330
  instantaneous point source, 429
  natural, 222 (**T6-2**)
  odors, 114
  offsets, 96
  particles, 240 (**T6-8**)
  pollution allowance, 108
  reduction credit (ERC), 108
  trading plan, 101
  test methods (EMTIC), 112
  visible, 115 (**T3-7**)
Emission Factors (CHIEF), 112, 336
Emission Factors (**AP-42**), 336
  (**E8.4**), 652-663 (**A-1 to A-7**)
Emission standards,

particulate, 115 (**T3-7**)
sulfur dioxide ($SO_2$), 95 (**T3-3**), 114 (**T3-6**)
Emission sources,
  fugitive, 115 (**T3-8**)
  point, area, line, 330
Emission spectra of gases, 77 (**F2-7**)
Emphysema, 164, 165 (**F4-35**), 152 (**E4.1**)
Empirical Kinetic Modeling
  Approach (EKMA), 246
  diagram, 247 (**F6-10**)
  mechanism, 248 (**T6-10**)
  ozone isopleths, 250 (**F6-12**)
  pollutant concentration versus
   time, 249
Enclosing a process, 5
Energy
  consumption rate, 67, 69
  exchange, sun, earth & space, 78
  exchange with earth, sun and
   space, 78 (**F2-8**)
  feed back, 79
  per capita consumption versus
   time, 63 (**F2-3**)
  per capita consumption & growth
   rate, 69
  relation to GNP, 70 (**F2-4**)
  total energy consumption growth
   rate, 69
  world consumption rate versus
   time, 71 (**F2-5**)
Engineering Economics (see Total
  Annual Cost & Total Initial Cost)
Engineering controls, 5
Entropy of formation, 39
Environmental:
  audit, 11
  disease, causes, 9
  epidemiology and toxicology, 13
  handbooks, 12
  Hearing Board, 112
  management system, 112
  Protection Agency (EPA), 93
  Quality Board, 112
EPA, 93
  air quality models, 431 (**T9.10**)
  dispersion (air quality) models, 431 (**T9-10**)
  opacity assessment, EPA
   Method 5, 203
  reference methods, 331
  emission rate, EPA Method 5, 575 (**F12-10**)
Epidermic cells of plants, 261
Epiglotis, 133
Episode, levels
  definitions, 118 (**T3-10**)
  response, 118 (**T3-11**)
Epoxides, 171
Equilibrium,
  chemical, 35
  $CO/CO_2/O_2$, 296 (**E7.3**)
  $SO_2/SO_3/O_2$, 41 (**E1.9**)

$NO/NO_2/O_2$, 301 (**E7.5**)
stable and unstable, 402 (**F9-11**)
water gas shift, 37 (**E1.7**)
Equilibrium
  of combustion species, 298 (**T7-3**)
  line (adsorption), 458
  isotherm (absorption), 356
  multicomponent, 356, (**F8-11**)
Equilibrium constant, 35-37
  effect of temperature, pressure, diluent, 41
  calculation from thermal data, 40 (**E1.8**)
  chart of different chemicals, 38 (**F1-19**)
Equivalence ratio, 288, 314
Equitorial doldrums, 390, 393 (**F9-3**)
Enthalpy of formation, 39, 479
ERC, 108
Erosion of stone monuments, 196 (**F5-6**), 197
Errors, design and production, 17
ESP, 18,22, 608
Ethyl tertiary butyl ether
  (ETBE), 316
Eulerian reference frame, 46, 391
Euler method (modified), 538, 676 (**A-14**)
Evaporative
  cooling, 362, 603
  hydrocarbon emissions, 315
Evaporation,
  confined space, 361 (**E8.15**)
  droplets, 363 (**F8-13**), 356 (**E8.16**)
  multiple species in moving air, 352 (**E8.11**)
  single species in moving air, 347, 348 (**E8.10**)
  stagnant air, 343 (**F8-5**), 344 (**E8.8**)
  two-film, 355 (**F8-8**), 356 (**F8-9**), 359 (**E 8.14**)
Exceedances per year, 106
Exercise, 142 (**T4-2**)
Excess lifetime cancer risk, 178
Exhaust gas recirculation (EGR), 319
Exponential
  energy consumption growth rate, 69
  geometric change, 64
  growth, 60
  growth rate, 64, (**E2.1**)
Express warrantee, 17
Extended Bohr model, 149 (**F4-24**)
Extractable organic material
  (EOM), 176
Extrathoracic (upper) airways, 131
Eye irritation, 169

**F**

Fabric filters (see filters, filtration)

Face velocity, 446, 600
Fall (settling) velocity, 539, 541
  (**F11-15**)
Fan and blower cost, 632
  power cost, 633
Fanning plume, 409, 410 (**F9-18**),
  411 (**F9-19a**)
Farmers lung, 167
Fatality rates 14 (**T1-2**)
Fatigue, olfaction, 206
Federal
  enforcement, 104
  implementation plan (FIP), 101
  Register, 110
Feedback, positive and negative
  greenhouse effect, 79
$FEV_1$, 139
FGD, 485
Fibers solid fraction, 575, (see also
  porosity)
Fibrosis, 165
Fick's law, 341
Field charging, 613
Filling station losses, 315
Film resistance
  overall liquid-phase film, 359
  overall gas-phase film, 461
Filtration, filters, 21 (**F1-7**), 591
  analytical model, 598 (**F12-30**)
  air-to-cloth ratio, 600, 601 (**T12-1**)
  cost, 634, 635 (**E13.2**)
  dust cake and filter, 595, 596
  dust cake thickness, 600
  dust cake loading, 600
  efficiency, 599
  felt, woven fabric, 595 (**F12-25**)
  filter drag, 601
  HEPA, 603
  high-temperature, 602
  impaction, interception, diffusion,
    597 (**F12-27**)
  most penetrating particle, 599
  penetration, 598 (**F12-28**) & (**F12-
    29**), 599
  pressure drop, 602 (**F12-31**)
  residual drag, 600
  reverse flow, 21 (**E1.7**)
  shaker, reverse-flow, pulse-jet, 21,
    592-594
  specific resistance, 600
  superficial velocity, 600
  types, 21 (**F1-7**)
  wet, 606
Filtration,
  design problem, 606 (**E12.7**)
  high-temperature, 602, 604
    (**F12-32**)
  particles, 18
Fine particles, 162, 238
FIP, 101
First generation respiratory
  passage, 133
Fish, effects of pollution, 191
Fixed bed adsorption, 444

Fixed film bioreactors, 488
Flames,
  diffusion, 290 (**F7-10**)
  flame holder, 290
  flat flame, 291 (**F7-10**) & (**F7-11**)
  flat flame, 291 (**F7-11**)
  premixed, 291 (**F7-11**)
  zones, 291
Flash boiling, 368, 369, 372, 373, 374
  (**E.8.19**)
Floc, 488
Flood point, flooding, 467 (**F10-21**)
Flow, plug, 34 (**F1-18**)
Flow regimes
  packed bed , 467 (**F10-21**)
  particle motion, 517
  mean free path, 518
Flue gas
  desulfurization, 485
  recirculation, 319
Fluidized-bed
  bubbling, 26 (**F1-15**)
  circulating fluidized-bed reactor
    (CFB), 323
  classes, 322 (**F7-22**)
  incineration, 26
  waste combustion, 31
Flux chambers, 334 (**F8-2**), 335
  (**E8.3**)
Forced,
  expiratory volume at 1 second
    ($FEV_1$), 139
  vital capacity (VC), 139
Forecast, air pollution episode, 118
  (**T3-10**), 119 (**T3-11**)
Force,
  buoyant, 409
  centrifugal (radial), 552
  Coriolis, 391
  drag, 515
Forest canopy, 265
Form drag, 517
Formaldehyde, 236, 243, 245, 663
  photolysis, 500
Formation and fate cycles
  (global), 59, 217
  carbon compounds, 226 (**F6-3**)
  chloroflourocarbons (CFC), 253
  mercury, 193, 194 (**F5-4**)
  methane oxidation, 241 (**E6.5**)
  nitrogen compounds, 231 (**F6-5**)
  ozone 243 (see EKMA model)
  sulfur compounds, 229 (**F6-4**)
Formyl radical ($HCO\bullet$), 243
Four-stroke cycle, 288 (**F7-8**)
Four-layer barrier, 150, 155
Fractional conversion, 34
Fractional efficiency, definition
Free-energy change, 39
Free-molecular flow regime, 518
Free-radical chemistry in flames, 289
Freely falling particles, 539, 540
Freons, 235
Fresh air requirements, 675 (**A-13**)

Freundlich equation, 444
Friction velocity, 255, 359, 406
Fuel
  blending, 319
  $NO_x$, 301, 308
Fugacity, coefficient, 36
Fugitive emissions, 115 (**T3-8**)
Fuller's earth adsorbent, 441
Fully alveoleted alveolar duct, 134
Fume, 29
Fumigation, 410, 412 (**F9-19d**)
Function,
  cumulative distribution, 523
  Gaussian distribution, 414
  number distribution, 521, 522
Functional residual capacity, 138


**G**

Gametoxin, 169
Gas-to-cloth ratio, 601 (**T12-1**)
Gas-phase transfer
  coefficient, 360, 460
Gas transport in lung, 146 (**F4-19**)
  composition in and out of
    lung, 146, 147
Gas turbine
  can burner, 321
  cycle, 285
  staged combustion, 321 (**F7-21**)
  stripping from water, 346 (**F8-6**)
Gas and vapor control systems,
  classes, 18
Gaussian plume model, 414 (**F9-20**)
  down wind pollutant
    concentration, 417-420
  elevated point source, 417-419
  maximum ground-level
    concentration, 419, 420
  particles, 417
  with and without reflection, 418
    (**F9-23**), 421 (**E9.6**)
General circulation, atmospheric, 389
Geometric
  (exponential) change, 64
  mean diameter, 523
  stack height, 417
  standard deviation, 522
Geostrophic
  layer, 389, 390 (**F9-2**)
  wind, 396
Geothermal energy, 73, 75
Gibbs free energy
  change of reaction, 35
  of formation, 39
Global:
  atmospheric circulation, 389-394
  combustion kinetics, 293
    (**T7-1**), 298
  oceanic circulation, 81

temperature change, 80 (**E2.3**), (**F2-9**)
Global sources and sinks, natural processes, 222 (**T6-2**)
Global formation and fate cycles, 59
  carbon, 226 (**F6-3**)
  mercury, 194 (**F5-4**)
  nitrogen, 231 (**F6.5**)
  particles, 240 (**T6-8**)
  photochemical oxidants, 232 (**F6-6**)
  ozone, 243, 246
  sulfur, 229 (**F6-4**)
Good engineering practice, 18
Gravimetric settling, 18
  classification, 19 (**F1-4**), 544
  chambers, 544 (**F11-16**)
  clouds, 549
  flow in ducts, 546 (**F11-17**), 547 (**T11-8**)
  laminar model, 544
  turbulent (well mixed) model, 544, 547 (**E11.8**)
Greenhouse
  effect, 72, 74-76
  feedback,
    water, ice-snow, clouds, ocean-atmosphere, 79
  gases, 77 (**F2-7**)
  heating, 74
Ground-level concentrations, elevated point sources, 419
  maximum, 420 (**F9-25**)
Growth laws,
  arithmetic (linear), 62
  geometric (exponential), 64

**H**

Haber's law, 171
Hadley cell, 390, 394 (**F9-8**)
Halogens,
  atmospheric concentration, 220 (**T6-1**)
  atmospheric reactions, 188
Halogenated hydrocarbons, 234
  hydrofluorocarbons (HFC),
  hydrochlorofluorocarbons (HCFC)
HAPs, 102, (**T3-4**)
Hardin, 66
Hatch-Choate equation, 524
Hay fever, 164
Hazard, 6
  acute, 11
  chronic, 11
Hazardous Air Pollutants (HAP), 12, 101, 102 (**T3-4**)
Haze, (see smog, 89, visibility, 198)
HC, abbreviation for hydrocarbon
HCFC, 234

Health hazards, percent known, 11 (**F1-3**)
Hearing impairment, 210
Hearing loss, 212 (**F5-16**)
Heat transfer
  in respiratory system, 145
  of evaporation, 339, 364
  relationships, 349 (**T8-1**)
Hedonic tone (odors), 207
Height of a transfer unit (HTU), 461
  overall, liquid, gas, 463
Hematopoietic toxin, 169
Hemoglobin, 172
  carboxy-, 172
Henry's law, 356 (**F8-11**), 357 (**E8.12**) & (**E8.13**), 454
  coefficients, 664 (**A-8**)
HEPA, 603
Hepatoxin, 169
HERP Index (Human Exposure Dose-Rodent Potency Dose, 15 (**T1-3**)
HFC, 234
High Efficiency Particulate (HEPA) filter, 603
High-frequency ventilation, 142
High-temperature filtration, 602
Histamine, 167
Histotoxic hypoxia, 160
Holdup, 466
Hopper dust disposal, 634
Horse latitudes, 390 (**F9-3**), 393
Hot-soak hydrocarbon emissions, 315
Household pests (allergens), 167
Housekeeping, 5
House pets (allergens), 166
HTU, 461
Human carrying capacity, 66
Human Exposure Dose/Rodent Potency Dose (HERP), 15 (**T1-3**)
Humidity, relative
  definition, 220 (**E6.2**), 400
  effect on lapse rate, 400
Hydrocarbon (HC)
  atmospheric concentrations, 220 (**T6-1**), 222 (**T6-2**)
  atmospheric reactions, 242
  emissions of auto engines, 290 (**F7-90**), 314 (**F7-15**)
  emission factors, 112, 336, 631
  reactivities, 252 (**F6-13**)
  urban air, 238 (**T6-7**)
Hydrogen chloride (HCl)
  stratospheric reactions, 253 (**F6-14**)
Hydrogen peroxide ($H_2O_2$), 220 (**T6-1**)
  atmospheric concentration, 220 (**T6-1**)
  photolysis, 501
Hydrogen sulfide ($H_2S$), 228
Hydroperoxyl radical [$HO_2\bullet$], 242
  reactions, 247 (**F6-10**), 248 (**T6-10**)
Hydrolase, 163

Hydrophobic and hydrophilic pollutants, 188
Hydroxyl radical [$OH\bullet$], 242
  in atmosphere, 227 (**E6.4**)
  in combustion, 289
Hygroscopic particles, 198
Hypercapnia, 160
Hypoxia:
  anemic hypoxia, 160
  circulatory hypoxia, 160
  histotoxic hypoxia, 160
  hypoxic hypoxia, 159
  hypoxemia, 165
Hysteresis, 137 (**F4-10**)

**I**

Ideal
  gas law, 29
  solution, 31, 350
Ignition temperature, 291
Impact noise, 207
Impact wet scrubbers, 590 (**F12-21**)
Impaction of particles,
  between particles, 577
  cascade impactors, 573 (**F12-7**) & (**F12-8**)
  curved ducts, 556 (**T11-10**)
  nasopharyngeal region, 161
  sphere-on-sphere, 576-578, 578 (**E12.2**)
  sphere-on-cylinder, 597 (**F12-27**)
Incinerator, two-stage, 285 (**F7-4**)
  thermal oxidation, 474
Inflation, 630
Indictor plants,189
Indoor workplace pollutant standards, 121
Industrial revolution, 62, 68
Inertial, particle
  deposition in curved ducts, 553, 556 (**T11-10**)
  nasopharyngeal deposition, 161
  separation, 18
  sphere-on-sphere collection efficiency, 576, 577
  sphere-on-cylinder collection efficiency, 597 (**F12-27**)
Inertial reference frame, 391
Inhalable particles, 162, 238
Injury statistics, 8 (**T1-1**)
Inspectional analysis, 552
Instantaneous point source, 428
International Standards Organization (ISO-14000), 121
Internet web pages, 49
  EPA, 111
  CFR Title 40, 111
  Prentice Hall,
    **http://www.prenhall.com**

textbook,
  **http://www.engr.psu.edu/cde/m e470**
class notes,
  Mathcad tutorial,
  solutions to textbook Examples that use Mathcad,
errata
thermophysical properties of chemicals
  **http://www.uic.edu/~mansoori/T hermodynamic.Data.and.Proper ty_html**
Interception, 577 (**F12-14**)
Interfacial, 223
  equilibrium, 258, 259
  flux calculations, 224, 254, 259 (**E6.9**)
Internal combustion engine,
  air standard cycle, 285
  emissions from, 319 (**T7-6**)
  compression ratio, 286 (**E7.1**)
  fuel-air ratio, 290, 314
  emissions reduction,
  catalytic converters, 318 (**F7-18**)
  combustion controls, 313
  crankcase emission control, 315 (**F7-16**)
  cylinder dimensions, 316
  evaporative hydrocarbon emissions, 315
  exhaust gas recirculation (EGR), 315
  reformulated gasoline, 316
International Standards Organization (ISO), 121
Interstitium, interstitial fluid, 136 (**F4-8**)
Inversion, 399, 403 (**F9-14**), 408
  frontal & radiation, 408
Involuntary risk, 6
Ion transfer in soil, 193 (**F5-3**)
Irrotational flow, 553
ISO, 121
Isoaxial sampling, 574
Isokinetic sampling, 573, 574, 576
Isooctane combustion, equilibrium products, 290 (**F7-9**)
Isoprene & terpene, 236
Isothermal
  saturation boundary, 157
  atmosphere, 398

**J**

Jet engine, 287 (**F7-7**)
Just noticeable change in visibility, (JNC), 198

**K**

Ketones, 313, 500
Kiln, 284 ( **F7-3**)
Kinematic viscosity, 518
Kinetic mechanisms
  acid rain, 5-1, 188, 228
  ammonia, 242
  carbon monoxide, 241 (**E6.5**)
  carbonic acid ($H_2CO_3$), 187
  CFC's, 242
  flames, 191-193
  hydrochloric acid (HCl), 188
  Leighton mechanism, 251 (**E6.7**)
  mercury, 194
  methane ($CH_4$), 240, 243, 245
  methanol, 245
  nitric acid ($HNO_3$), 188, 241, 259
  oxides of carbon, 299 (**E7.4**)
  oxides of nitrogen ($NO_x$), 245, 259
  oxides of sulfur, 241
  oxygen & ozone, 243
  stratosphere, 188, 244
  troposphere, 245
  urban air photochemical mechanism (EKMA), 232 (F6-6), 247 (F6-10), 248 (**T6-10**)
  Zeldovitch, 303-306
Knock-out drum, 607 (**F12-35**)
Knudsen number, 518
Kuwabara factor, 596

**L**

LAER, 97, 112
Lagrangian point of view, 46
La Nina, 80
Laminar,
  flow in curved ducts, 554
  flow in horizontal ducts, 545
  lungs, 143
Langmuir-Hinshelwood kinetics, 503
Lapple standard cyclone, 568 (**F12-2**)
  fractional efficiency, 570 (**F12-4**)
  pressure drop, 570
Lapse rate, 397
  computation, 399 (**E9.3**), 400 (E9.4)
  dry adiabatic, 399
  dry air, 397
  isothermal, 398
  lapse rate moist saturated air, 400
  moist air, 400
  normal, 398
  stable, neutral & unstable, 403
  polytropic, 398
Larynx, 131
Law of mass action, 44
Layers of the atmosphere, 390

Laughing gas ($N_2O$), 221
Le (Lewis Number), 348
LeChatelier's principle, 309
Leaf area index, 265
Leaves of plants
  cross sectional diagram, 262 (**F6-19**)
  epidermic, palisade, mesophyll, chloroplast cells, 262
  overall mass transfer & resistance, 264
  water loss , 263 (**F6-20**)
Leaks
  flashing, 368 (**F8-14**), 372 (**E8.19**), 373
  gases, 371 (**F8-17**) & (**F8-18**), 372 (**E8.18**)
  liquids 369 (**F8-16**), 370 (**E8.17**), 375 (**E8.20**), 376
Legislation and regulations, 91 (**T3-1**)
Leighton model, 251 (**E6.7**)
Length of fiber per volume, 599
Levels of offenses to regulations, Summary, Misdemeanors, Consent, 118
Lewis Number (Le), 348
Liability, negligence and strict liability, 17
Life-cycle assessments, 122
Light scattering and opacity, 199
Limits to Growth, 66
Line source, 431 (**F9-30**)
Linear (arithmetic) progression, 62
Liquid-to-gas ratio, 457, 581
Liquid-phase mass transfer coefficient, 258
Lipids, 136
Load point, 467 (**F10-21**)
Lofting plume, 410 (**F9-18**), 411 (**F9-19**)
Logarithmic wind-speed profile, 406 (**T9-3**)
Log-mean partial pressure ratio, 343
Log-normal particle size distribution, 521, 530
Log-probability graph paper, 525 (**F11-7**)
Local toxins,
  irritant, corrosive, pulmonary, ocular, 168
Looping plume, 410, 411 (**F9-18**) & (**F9-19b**)
Los Angles, visibility, 199 (**F5-7**)
Loss coefficient, 370
Low-$NO_x$ burners, 319, 320 (**F7-20**), 321 (**F7-21**)
Low oxygen, effects, 160 (**T4-4**)
Lowest Achievable Emission Rate (LAER), 97, 112
Luminous flame
  chemical kinetics, 291
  zone, 291 (**F7-11**)

Lumped-parameter earth radiation model, 73 (**F2-6**)
Lumped heat capacity assumption, 364
Lung disease, 128 (**T4-1**)
  irritation, cell damage, allergies, fibrosis, oncogenesis, 163
Lung function,
  clearance, 133
  compliance, 137
  during exercise, 142 (**T4-2**)
  normal, obstructed, constructed, 139 (**F4-13**)
Lung volumetric flow rates, 128 (**F4-12**)
  modes of air motion, 146 (**F4-19**)
Lymphatic system, 136, 137 (**F4-9**)
Lysosome, 163

# M

MACT, 101
Macrophage, 133 (**F4-5**), 134, 163
Macropore, absorbent, 441
Macroscopic scale, 389
Maintenance, 5
Major source, 101
Make-up water cost, 634
Malthus, 65, (**E2.2**)
Managerial structure, 5
Marine boundary layer, 245
Market financial incentives, 96, 101
Mast cells, 167
Mass
  balance, 27 (**E1.3**), 332 (**E8.2**)
  concentration, 29
  distribution function, 523
  fraction, 32 (**E1-6**)
Mass and volumetric flow rates, 28, 29 (**E1.4**)
Mass transfer,
  coefficients, adsorption, 460
  coefficient, stagnant air, 343
  exchange $CO_2$ to ocean, 266 (**E6.12**)
  in respiratory system, 145, 156
  through stagnant air, 343
  to soil, 257 (**F6-16**), 266
  to vegetation, 261
  to bodies of water, 258
Material Safety Data Sheets (MSDS), 12, 120
Mathcad tutorial and solutions to text book Examples, 49
  **http://www. engr.psu.edu/cde/me470**
Maximum Achievable Control Technology (MACT), 101
Maximum lifetime individual risk, 178
Mean,
  arithmetic, 520

earth surface temperature, 74
free path (air STP), 518
geometric, 523
Median particle diameter, 520
Mercury
  biological cycle in environment, 193, 194 (**F5-4**)
  coal combustion, 195
  poisoning, 193
Mesoscale, mesoscopic scale 389
  mesopore, adsobent, 441
  mesophiles, 490
Mesophyll cells of plants, 261
Mesophyllic resistance, 264
Mesosphere & mesopause, 218
Mesothelioma, 166
Methane
  atmospheric chemistry, 240
  atmospheric concentration, 220 (**T6-1**)
  concentration versus ancient time, 63, 237
  global cycle, 226 (**F6-3**)
  oxidation kinetics, 243
Metal surfaces, effect of pollutants, 197
Method (EPA) 9, 201
Method (EPA) 5, 575 (**F12-10**)
Methyl radical, $[CH_3 \bullet]$, 242
Michaelis-Menten kinetics, 491
Methyl mercapatan ($CH_3SH$), 228
Methyl tertiary butyl ether (MTBE), 316
Microscopic scale, microscale, 389
Micropore, adsorbent, 441
Mie scattering, 198
Migration velocity, 611 (**T12-2**)
Minamata Bay (mercury poisoning), 193
Mineralize, 499
Minor head loss, 370
Minute respiratory rate, minute volume, 138, 141
Misrepresentation, 17
Mist, 238, 515
Mist eliminator, 455, 570
  cheverons, 571 (**F12-6**)
  knockout drum, 607
  mesh, 571 (**F12-5**)
Mixing height, 407 (**F9-16**)
Mixture
  ideal liquids, 31
  perfect gases, 30
Mobile combustors, 281
Modeling, distributed and compartmental, 148
Modified Euler method, 676 (**A-14**)
Moist air, density, 30
Molar concentration, total, average, partial, 32
Mole
  fraction, 30, 32
  ratio, 450
Molecular weight

average, 31, 32 (**E1.6**)
  pollutants, 664 (**A-8**), 673 (**A-12**)
Molecular sieves, 441
Moments, first and second, 140
Monod kinetics, 491
Monodisperse aerosol, 520
Monomethylmercury (ion), 193
Monuments, effects of pollutants, 196
Motor Vehicle
  Air Pollution Control Act 90
  Exhaust Study Act (1960), 90
MSDS, 12, 120
Mucociliary
  clearance, 161
  escalation, 132
Mucus, mucosal layer, mucous membrane, 132
Multi-clone dust collectors, 19 (**F1-5**)
Multicomponent
  adsorption, 452
  equilibrium, 356 (**F8-11**)
Multihit model, 174 (**E1.5**)
Multistage model, 174
Municipal waste incinerator, 284 (**F7-4**)
Mutagen, 169
Mutagenic Emission Factors, 177 (**T4-6**)

# N

NAAQS, 93
Narrow band noise, 207
NASA Chemical Equilibrium Code, 40
Nasal turbinates, 131
Nasopharyngeal region, 131
  particle deposition, 161
National Ambient Air Quality Standards (NAAQS), 93 (**T3-2**), 121
National Emission Standards for Hazardous Air Pollutants (NESHAP), 94
National Institute of Occupational Safety and Health (NIOSH), 12
National Safety Council (statistics), 7
Natural gas cost, 633
Necrosis, 188
Negligence, 18
Nephrotoxin, 169
Neurotoxin, 169
Neutral atmosphere, 402
New source
  definition, 94
  modified, 94, 104
New Source Review (NSR), 113 (**T3-5**)
NESHAP, 94

New Stationary Source
  Performance Standards (NSPS), 94
  (**T3-3**)
Newtonian flow regime, 517, 541
NIH, 49
NIOSH, 12
Nitric oxide/nitrogen dioxide
  ($NO/NO_2$)
  equilibrium, 301 (**E7.5**)
  equilibrium constant, 301 (**T7-4**)
Nitric oxide/nitrogen dioxide (*cont.*)
  $NO/NO_2/O_2$ equilibrium, 301
    (**T7.4**)
  NO generation rate, 306 (**E7.6**)
  reduction, 482
Nitric acid ($HNO_3$)
  atmospheric chemistry, 188, 245
Nitrification in soil and air, 229
Nitrogen ($N_2$),
  atmospheric concentration, 220
    (**T6-1**)
  fixation, 228
  global cycle, 231 (**F6-5**)
Nitrogen dioxide ($NO_2$)
  air-water absorption, 259 (**E 6.9**)
  atmospheric concentrations, 220
    (**T6-1**)
  atmospheric reactions, 232
    (**F6-6**), 247 (**F6-10**), 248
    (**T6-10**)
  equilibrium with nitric oxide, 301
    (**E7.5**)
  in engine exhaust, 319 (**T7-6**)
  photolysis, 247 (**F6-10**), 248 (**T6-10**)
  plant uptake, 262 (**E6.10**)
Nitrogen oxides ($NO_x$)
  atmospheric reactions, 247
    (**F6-10**), 248 (**T6-10**)
  control of in stationary
    combustion
  exhaust gas recirculation (FGR),
    315, 319
  fuel blending, 319
  low $NO_x$ burners, 319
  staged combustion, 319
  emission factors, 336
  global budget, 222 (**T6-2**)
  global cycle, 231 (**F6-5**)
  mechanism of formation in
    combustion, 306 (**E7.6**)
  fuel, 301, 308
  prompt, 308
  thermal, 290
Nitrous oxide ($N_2O$)
  role in stratospheric
    chemistry, 229
Noise, 207
  common sources, 211 (**F5-15**)
  OSHA standard, 210 (**T5-2**)
  several sources, 209 (**E5.3**)
Nominal interest rate, 630
Nonattainment regions, 100
  top five areas, 105 (**F3-2**)
Non-ideal solutions, 352

Nonmethane hydrocarbon
  (NMH), 114
Nonselective reduction
  (NSCR), 482, 483
Normal (quiet) breathing, 137
  boiling temperature, 339
  lapse rate, 399
  minute respiratory rate, 138
  tidal volume, 138
Noxious, 6
NSCR, 482, 483
NSPS, 95
Nucleation, nuclei, 238
Number
  distribution function, 523
  transfer units (NTU), 462
    (**F10-19**), 463 (**F10-20**)
  turns ($N_e$), 568
Nu (Nusselt) Number, 347


**O**

Obstructed lungs, 139 (**F4-13**)
Ocean-atmosphere interaction, 79
  $NO_2$ absorption, 259 (**E6.9**)
  $CO_2$ exchange, 266 (**E6.12**)
Oceanic carbon, 81
Occupational Safety and Health
  Administration (OSHA), 12
Odor
  index, 207
  dilution factor, 207
  standards, 115 (**T3-8**)
  quantification, quality,
    acceptability, intensity, 207
  threshold values, 669 (**A-11**)
Office of Technology
  Assessment, 8
Offsets, 96
Oil steam boiler, 286 (**F7-5**)
Olefine (see alkenes), 231, 247,
  312, 313
Olfaction, 206
One-hit model, 174
Opacity of plumes, 198
  assessing, 203
  opacity and particle
    concentration, 200 (**E5.1**)
  steam plumes, 202
Open system, 47 (**F1-20**)
Operations and Maintenance
  (O&M), 629 (**T13-2**)
Operating line, 458 (**F10-15**), 459
  (**F10-16**)
Optical path length, 198
Order-of-magnitude
  analysis, 289, 303
Organic compounds,
  classes, 313 (**F7-14**)
  nitrogen compounds, 308 (**F7-12**)
  sulfur compounds, 308

OSHA, 121
Otto cycle, 285, 287 (**F7-6**), 288
  (**F7-8**)
Over firing, 320
Overall
  collection efficiency, 532
  combustion mechanism, 293 (**T7-1**)
  mass transfer to a liquid, 359
  absorption in soil, 460
  transfer to soil & water, 257
  mass transfer resistance to
    leaves, 264
  several collectors, 534
Oxidation, advanced (UV), 18, 499
Oxides of nitrogen (see $NO_2$
  and NO)
Oxygen deficient atmosphere, 160
  (**T4-4**)
Oxygenated (oxygenates)
  hydrocarbons, 312
Ozone,
  adsorption in respiratory system,
    154, 155
  air quality standard, 93 (**T3-2**)
  background, troposphere, 222
    (**T6-2**), 220 (**T6-1**)
  dose to mucus layer, 154 (**F4-28**)
  dose to tissue, 155 (**F4-29**)
  exposure, 164
  health consequences
  isopleths, ridge, 250 (**F6-12**)
  oxygen radical, $O(^1D)\bullet$ &
    $O(^3P)\bullet$, 242
  photolysis, 244
  reactions in stratosphere, 253
    (**F6-14**)
  toxicity, 173 (**F4-41**)
Ozone, in atmosphere (also see
  EKMA)
  depletion potentials, 253 (**T6-6**)
  deterioration, 244 (**E6.6**)
  generalized kinetic mechanism,
    247 (**F6-10**), 248 (**T6-10**)
  pollutant concentration versus
    time, 249 (**F6-11**)
  ozone isopleth, 250 (**F6-12**)
  ozone hole, 244 (**E6.6**)


**P**

Pacific warm pool, 81
Packed-bed scrubbers, 22, 23, 455
Packing constants, 465 (**T10-1**)
Packing, 456 (**F10-13**)
Painted surfaces and pollution, 196
  (**F5-5**)
Palisade cells of plants, 261
PAN, 230, 247-249
Paper, log-probability, 525
  (**F11-7**), 531
Paracelsus, xii, 168

Paraffins (alkanes), 312, 313
Parallel, collector configuration, 534
Partial pressure, 30, 330
Partially alveolated respiratory bronchiole, 134
Particle charging (ESP), 612 (**F12-40**)
  bombardment charging, 613
  charge transfer, 613
  corona, 612
  diffusion charging, 613
  electron avalanche, 612
  electron transfer, 613
  field charging, 613
  passive region, 612
  saturation charge, 613
  space charge, 615 (**F12-42**)
Particle collection,
  gravitational, 544
  wet scrubbing, 579, 583, 585
  filtration, 591
  electrostatic precipitation, 608
  impaction, interception and diffusion, 577, 597
Particle control systems, classes, 18
Particle sizes, 239 (**F6-9**)
  aerodynamic diameter, 162
  arithmetic mean diameter, 520
  coarse, 162
  cumulative number distributive function, 523
  geometric standard deviation, 522
  geometric mean, 523
  Hatch-Choat equation, 524
  Lapple chart, 516 (**F11-1**)
  inhalable & respirable, 162
  mass mean diameter, 524
  median diameter, 520
  number distribution function, 521, 522
  number mean diameter, 524
  $PM_{10}$, 98
  primary, 162
  Sauter diameter, 520
  secondary, 162
  standard deviation, 521
  surface area equivalent diameter, 520
  volume equivalent diameter, 520
Particle size distribution, 520 (**E11.1**), 523 (**F11-5**) & (**F11-6**), 525 (**F11-7**), 529 (**E11.2**)
  arithmetic mean, 520
  atmospheric, 528 (**F11-9**)
  based on mass and number, 524
  classifier, 557 (**E11.9**)
  geometric mean ($D_{p,gm}$), 523
  log-normal, 521
  median ($D_{p,50}$), 520, 523
  standard deviation and variance, 521
Particle emission rates, 6-53 (**T 6-8**)
Particle, health effects

deposition in lung, 161 (**F4-33**), 162 (**F4-34**)
  inhalable, respirable, coarse, fine, secondary, 162, 238
  primary and secondary, 238, 526 (**F11-8a**)
  size of common particles, 239 (**F6-9**)
  size of common particles, 516 (**F11-1**)
  trajectory, 38 (**E11.5**)
Particle motion, 536
  curvilinear flow, 552
  drag force, 515
  equations of motion, 537
  particle-particle impaction, 577
  settling velocity, 541 (**F11-15**)
Particle sampling,
  EPA Method 5, 574-576
  cascade impactors, 572, 573 (**F12-7**)
Particulate emission standards, 115 (**T3-7**)
Parts per million (ppm), 30
Partitioning, 356
Pasquill-Gifford coefficients, 416 (**T9-6**) & (**T9-7**), 417 (**T9-8**)
  stability categories, 405 (**T9-1**), 416 (**T9-6**) & (**T 9-7**), 498 (**T9-8**), 429 (**T9-9**)
Passive region, 613
Path mean free, 518
Pathways for toxin movement, 168 (**F4-37**)
Peclect number, 596
PEL, 7, 121
Pendelluft, 142 (**F4-15**), 143 (**F4-16**)
Penetration
  distance, 544 (**T11-7**)
  efficiency, 533
  of filter, 600
Per capita energy consumption rate, 63, 69
Perchloroethylene, 235
Perfusion, 135
  during exercise, 142 (**T4-2**)
Permeability, 136, 145
Permissible Exposure Limits (PEL), 7, 121, 667 (**A-9**), 669 (**A-11**), 673 (**A-12**)
Permits, 104
  case-by-case, top-down review, 113 (**T3-5**)
  New Source Review (**NSR**), 113 (**T3-5**)
  steam boiler, 106 (**E3.2**)
Peroxyacetyl radical, [$RC(O)O_2\bullet$], 247
Peroxyacetyl nitrate (PAN), 230
  atmospheric reactions, 247 (**F6-10**), 248 (**T6-10**)
Persons at risk, fatality rates, 14 (**T1-2**)
Perspectives about risk, 13

competing and entwined interests, 17
pH scale of common materials, 188
pH isopleths in US, 189 (**F5-2**)
Phagocytosis, 134, 163
Pharmacokinetics, 169
Pharynx, 131
Phenols, 312
Photochemical oxidants, atmospheric concentrations, 222 (**T6-2**)
Photochemical reactions, 230, 232 (**F6-6**), 240
Photochemical smog (see Smog) 223
Photocatalysis, 501, 503 (**E10.8**)
  of atmospheric species, 506 (**F10-40**)
Photolytic kinetics, 502
Physiological response, moving air, 407 (**T9-5**)
Physical constants, 677 (**A-16**)
Pink noise, 207
Pinocytosis, 163
Planetary boundary layer, 389, 390 (**F-2**), 405 (**F9-15**)
Planck's law, 76
Plant allergens, 167
Plants,
  effect of pollutants, 190 (**T 5-1**)
  leaf pollutant uptake, 261
  $NO_2$ uptake, 262 (**E6.10**)
  $SO_2$ uptake by canopy of leaves, 265 (**E6.11**)
Pleura, 137
Plug flow, 545
Plug flow reactor, 34 (**F1-18**)
Plume,
  combined particle and steam, 203
  classification, 409-411, 413 (**E9.5**)
  differentiating steam from particle plumes, 203
  rise, 413, 417, 423, 424 (**E9.7**) & (**F9-26**)
$PM_{10}$ standard, 98
$PM_{2.5}$ standard, 106
PMN, 12, 120
Pneumonia, 165
Point mutation, 174
Polar easterlies, westerlies, fronts, 390 (**F9-3**), 393
Pollens, 167
Pollution, Air
  allowance, 108
  abatement strategy, 5
  control systems, classes, 18
  definition, 4
  effect on soil, 192
  effect on bodies of water, 192
  effect on stone surfaces, 196
  effect on painted surfaces, 197
  hydrocarbons in urban air, 238 (**T6-7**), 246 (**T6-9**)
  lung diseases, 128 (**T4-1**)

Pollution, Air (*cont.*)
  measurement methods, 331
  solubility and uptake in
    water, 224, 254, 258
  sources and sinks, 222 (**T6-2**)
Polycyclic aromatic compounds, 238
  (**T6-7**)
Polydisperse aerosol, 520
Polytropic atmosphere, 398
Pond lilies, 64
Population growth rates, 68, 69
Population versus time, 63 (**F2-2**)
Porosity (see solids fraction), 597
Post-flame
  chemical kinetics, 243, 292
  zone, 291 (**F7-11**)
Potential
  to emit, 97, 94
  emissions, 94, 113
Potential temperature &
  gradient, 404
Power law exponents, 405 (**T9-2**)
Pr (Prandtl Number), 347
Precipitation parameter, 611 (**T12-2**)
Pre-exponential factor, 44
Preflame
  chemical kinetics, 291, 243
  zone, 291
Prefixes (SI), 677 (**A-15**)
Premanufacture Notice
  (PMN), 12, 120
Premixed flames, 290, 291 (**F7-11**)
Preservation, 5
Pressure drop,
  adsorber bed, 451
  baghouse modules, 605
  baghouse, 592-594
  filter, 602 (**F12-31**)
  packed bed, 466, 467 (**F10-21**),
    408 (**F10-22**)
Prevention of Significant
  Deterioration (PSD), 96, 108
  increment consumption, 109
  significant level, 109
Prevention of air pollution, 3, 5
Primary particles, 238, 526, 527
  (**F11-8**)
Probability graph paper, 525
  (**F11-7**)
Process dump, 368
Process rate, 44
Products and reactants, 35
Production errors, 17
Prompt NO, 301, 308
Propaganda, 13
Properties, thermophysical,
  **http://www.uic.edu/~mansoori/Th
  ermodynamic.Data.and.Propert
  y_html**
Proteoglycan, 136
Proximate cause, 17
Pseudo steady state assumption, 46
  applied to NO/NO₂ reactions,
  PSD, 96

Puff diffusion, 428 (**F9-29**), 429
  (**T9-9**)
Pulmonary:
  fibrosis, 165
  inflammation, 163
  region, 133, 134
Pulsatile flow, 144 (**F4-17**)
Pulse-jet filtration, 594 (**F12-23**)
Pulvation distance, 544
Pulverized coal steam boilers, 282
  (**F7-1**)
  staged combustion, 320 (**F7-20**)
Pump cost, 632
  power cost, 633
Pure diffusion, 342
Pyrolize, 291

**Q**

Quad, 70
Quantum yield, 503
Quench layer, reciprocating
  engine, 316
Quiet breathing, 133, 137

**R**

RACT, 98, 112
Radiation (effective) temperature
  of earth, 74
Radicals (free), 242
Radiosonde, 400 (**E9.4**)
Radon, radioactive decay, 231, 233
  (**F6-7**)
Rain, drop, 238, 515
Rain-out, Rain-dust impaction, 558
  (**E12.2**)
Random packed scrubber, 23, 455
Raoult's law, 350
Rate, 33-34
  of change, 33
  of disappearance, 33
  of formation, 33
  process rate, 33
  of reaction, 33
Rayleigh scattering, 198
Reactants and products, 35
Reactions,
  first-order, second-order,
    simultaneous, 45
  photochemical, 230, 232
    (**F6-6**), 240
Reaction-limited biofilm, 493
  (**F10-37**), 497 (**E10.7**)
Reactivity of hydrocarbons, 252
  (**F6-13**), 313 (**T7-5**)
Reasonable Available Control
  Technology (RACT), 98, 112
Receptors, 4

Reciprocating engine, 285
Reciprocal space velocity, 35
Recirculating eddy, 426 (**F9-27**), 427
  (**F9-28**)
Recognition (odor) threshold, 206
Recuperative thermal oxidizer, 476
  (**F10-25**)
Reduction of NOₓ, 482
Reformulated gasoline (RFG), 316
Reflux ratio, 457, 581
Refrigeration, power cost, 633
Regeneration of adsorber, 448
  (**F10-9**), 450 (**F10-10**)
Regenerative thermal oxidizer, 475
  (**F10-24**)
Regime, of flow, continuum, free
  molecular, transition, 518
Registry of Toxic Effects of
  Chemical Substances (RTECS), 11
Regulations and legislation, 89
  historical record, 91 (**T3-1**)
Reid Vapor Pressure (RVP), 316
Relative humidity, mol fraction, 220
  (**E6.2**)
Relaxation time, 537
Renewable fuel usage, 317
Reproductive toxins, 169
Residence time, thermal
  oxidizer, 474
Residual
  filter drag, 600
  risk, 104
  volume, 137
Respirable particles, 162, 238
Respiratory
  airspace, 133
  bronchiole, 134
  components, 131, (**F4-1**)
  generations, 134
  heat and mass transfer, 145
  water loss, 159 (**E4.2**)
RTECS, 119
Retention time of adsorber, 446
Reverse-flow
  cyclone, standard, 568 (**F12-2**)
  filtration, 20, 593 (**F12-22b**)
Revertants, 176
Re (Reynolds Number), 347
Reynolds,
  analogy, 348
  numbers in bronchial system, 135
Richardson number, 404
Ringelmann scale, 89
Risk, 6
  analysis, 176
  benzene and lukemia, 174 (**E4.5**)
  fatality rates, 7, 8 (**T1-1**), 14 (**T1-2**)
  involuntary and voluntary, 7
    (**F1-2**), 16 (**E1.2**)
  management and assessment, 9
  perspective, 13
  risk versus benefit, 7 (**F1-2**)
  rule-making, 110 (**F3-3**)
  unit risk factors, 178 (**T4-7**)

Roll cell, 258 (**F6-18**)
Roof eddy, 427
Rotary kiln incinerator, 284 (**F7-3**)
Rotary wheel adsorber, 450 (**F10-10**)
Rotational flow, 533
Roughness height, 256, 406
Rule making process, 110 (**F3-3**)
Runge-Kutta, 538
Running (hydrocarbon) losses, 315

## S

Salmonella, 176
Sampling particles,
    cascade impactor, 573 (**F12-7**)
    EPA Method 5, 574-576
Saturation
    charge, 613
    pressure, 330
    time of adsorber, 446
Sauter particle diameter, 520
Sc (Schmidt Number), 347
Scattering of light, particles, 199
    coefficient, 198
    effect of light intensity, 202 (**F5-9**)
    forward, back, 90-degree, 201
    Mie scattering, 198
    Rayleigh scattering, 198
    scattering angle, 201 (**F5-8**)
Scavenging, 579
    precipitation, dry deposition,
        aerosol, 223
SCFM, 28
SCR, 483
Scrubbers, 18, 23 (**F1-9**) & (**F1-10**),
    24 (**F1-11**) & (**F1-12**)
    contact, 590
    counter-flow, 23 (**F1-10**)
    packed bed, 23 (**F1-10**), 456
        (**F10-13**)
    transverse, 24 (**F1-11**), 583
    venturi, 23 (**F1-9**), 585
Scrubbing ($SO_2$), 487 (**F10-30**), 486
    (**F10-29**)
Secondary
    particles, 162, 238, 526, 527
        (**F10-27**)
    organic aerosols, 317
Sedimentation, 543
Selective Catalytic Reduction
    (SCR), 482, 483 (**F10-27**)
    catalytic reduction, types, 483
        (**F10-27**)
Series, collector configuration, 534
    (**F11-12**)
Settling
    chambers, 18, 19 (**F1-4**)
    velocity, 417, 540 (**F11-14**), 541
        (**T11-6**), 542 (**E11.6**), 543
        (**E11.7**)
Shaker fabric filter, 20, 592 (**F12-22a**)

Shape factor, 520 (**T11-2**)
Sherwood number (Sh), 347
Short-Time Exposure Limit
    (STEL), 121
SI, 308
SIP, 93
Silent Spring, 90
Silica adsorbent, 441
Single-drop collection
    efficiency, 577-578
Single-fiber collection
    efficiency, 577, 595, 597
Single-film theory, 350
Site-specific toxins, 169
Size distribution, particles, 516
SI prefixes, factors, symbols, 677
    (**A-15**)
Skin flames, 167
Skylight, 201
Slaked lime, 486
Slip
    flow regime, 518
    slip factor, 519
    velocity, 321, 322 (**F7-22**)
Sludge disposal cost, 634
Slurry, limestone and lime, 487
Smog, 89, see also haze, visibility,
    198, 199 (**F5-7**), 223, 238, 515
Smoke reading, 202
Sneezing, 133
Solar energy, 75
Soil,
    effects of acid rain, 192
    mass transfer pollutants from
        air, 266
Solid body
    rotation, 553
    fiber fraction, 597
Solvent recovery system, 25
    (**F1-13**)
Solvent mixture, 32 (**E1.6**)
Sound level,
    common sources, 211 (**F5-15**)
    free-field loudness, 210 (**F5-13**)
    frequency response, 210 (**F5-14**)
    pressure, intensity, power, 208
Spark ignition (SI )engine,
    catalytic converters, 318
        (**F7-18**) & (**F7-19**)
    equilibrium exhaust products,
        290 (**F7-9**)
    exhaust composition vs
        equivalence ratio, 314 (**F7-15**)
    exhaust composition, 319 (**T7-6**)
    four-stroke cycle, 288 (**F7-8**)
    importance of peak cycle
        temperature, 308
    NO frozen equilibrium, 308
    reactivity of a exhausts, 313 (**T7-5**)
    sulfur- and nitrogen-bearing
        fuel, 308
    wall quenching, 316 (**F7-17**)
Specific
    gravity 664 (**A-8**), 673 (**A-12**)

heat, 39
    resistance, filter, 600
Spectral emissive power, 76
Sphere on sphere impaction, 577
    (**F12-13**) & (**F12-14**)
Spray chambers (particle removal)
    counter-flow, 579 (**F12-16**)
    transverse flow, 583 (**F12-17**)
$SO_2/SO_3$, sulfur dioxide/Sulfur
    trioxide
    combustion kinetics, 311
    equilibrium, 308 (**E7.7**)
    equilibrium computation, 41
        (**E1.9**)
    flue gas desulfurization, 485
Soak, (hot) auto emissions, 315
Solar intensity, constant ($S_0$), 74
Solvent recovery adsorption, 25
    (**F1-13**)
Sorbate, 441
Sources
    area, 101
    emission rates, (see EPA
        Emission Factors)
    major, 101
    modified, 94, 104
    new, 94
    uncontrolled, 94
Sound,
    intensity, pressure, power
        level, 208
    OSHA standards, 210 (**T5.2**)
    point source, 208
    pressure level, 209 (**E5.2**)
    several sources, 209 (**E5.3**)
    speed, 207
Southern oscillation, 80
Space charge, 641
Spectral absorptivity (of gases), 77
    (**F2-7**)
Sphere-on-sphere impaction, 577
    (**F12-13**) & (F12-14)
Spills, 337
Spirometry, 137
    lung volumes, 138 (**F4-11**)
    terminology, 137, 138
Spray dryer (scrubbing), 485
Spraying transfer efficiency, 117
    (**T3-9**)
Spray chamber (wet scrubber), 579
Stability categories, atmosphere,
    401-405
Stack height
    building wakes and eddies, 428
    down wash, 428
    effective, 417
    geometric, 417
    good engineering practice, 427
    plume rise, 423
Staged combustion in gas turbines,
    320, 321 (**F7-21**)
Standard
    cubic feet per minute (SCFM), 28
    lapse rate, 398

Standard (*cont.*)
mortality ratio (SMR), 174
(**E4.5**)
reverse-flow cyclone,
568 (**F12-2**)
standard deviation, size
distribution, 521
temperature and pressure
(STP), 28
Standards, air quality, 93 (**T3-2**)
Star state, 358 (**F8-12**), 460
(F10-18)
State Implementation Plans
(SIP), 93
Stationary combustor controls, 281
fuel blending, 316, 319
exhaust gas recirculation, 317
low $NO_x$ burners, 319
staggered combustion., 319
Statistics,
environmental disease, 9
occupational injury, 8 (**T1-1**)
Steady-state approximation, 46
see also pseudo steady state, 46
STEL,
Steam, cost, 633
Steam engine, 78
Steam plumes, 202
attached & detached, 203
back lighting, 205 (**F5-12**)
combined with particles, 204
(**F5-11b**)
identification, 203 (**F5-12**)
Stoichiometric
equation, coefficients, 27, 35
proportions, 289
Stokes
flow regime, 517
number, 552
settling velocity, 538
Stoker boilers, 283 (**F7-2**)
Stomatal resistance, 263
Stone surfaces, effects of pollutants,
196 (**F5-6**)
Stopping distance, 544 (**T11-7**)
STP, 28
Stratified water model, 258
Stratosphere & stratopause, 218
Stratospheric ozone layer, 252
ClO kinetics, 253 (**E6.8**)
kinetic mechanisms, 244
Nobel prize, 244 (**E6.6**)
reactions, 253 (**F6-14**)
Strict liability, 17
Stripping, 346, 355 (**F 8-6**)
Stroke (reciprocating engines)
optimization, 316
square design, 316
stroke-bore ratio, 316
Subsidence, 390, 408
Substantial derivative, 46, 396
Substitution, control strategy, 5
Subpolar lows, 390 (**F9-3**)

Subtropical jet stream, 389, 390
(**F9-8**)
Successive random dilutions, 530
Sulfur dioxide ($SO_2$)
coal-fired power plant, 98 (**E3.1**)
emission factors, 652-663
(**A-1** to **A-7**)
emission standards, 114 (**T3-6**)
$SO_2/SO_3/O_2$ equilibrium, 308
(**E7.7**)
uptake by vegetation, 265
(**F6-21**)
Sulfur trioxide ($SO_3$)
kinetics with $SO_2$ and $O_2$, 311
Sulfur cycle,
global cycle, 229 (**F6-4**)
principal fluxes, 230 (**T6-5**)
Sulfur oxides from combustion
removal, flue gas
desulfurization, 485
removal, scrubbing, 486, 487
Sulfates,
global fluxes, 229 (**F6-4**)
deposition velocity, 256, 257
(**T6-11**)
Surface,
renewal theory 258
soiling and deterioration, 195
Surface coating, 5
compliant and noncompliant
coatings, 116
control strategy, 5 (**E1.1**), 116
(**E3.3**)
Surfactant, 134
Superadiabatic lapse rate, 403
Supercritical water oxidation
(SCWO), 505, 507
Superficial velocity (see also face
velocity)
adsorber, 446
filter, 600
mass flow rates, 464
Suspended-growth biofilters, 488
Switching point, 131
System International (SI) prefixes,
677 (**A15**)
System, open and closed, 47
Swamp gas ($CH_4$), 221


**T**


TAC, 629
Taylor-type dispersion, 144
TCE, 235
Technical Conference on Air
Pollution (1950), 90
Technology Transfer Network
Bulletin Boards (TTNBBS), 111
Telnet, 111
Temperature
atmospheric profile, 218 (**F6-1**)

depression, 364
potential, 404
Teratogen, 169
Termination, air pollution episode,
118 (**T3-10**), 119 (**T3-11**)
Termites, 236
Terminal settling velocity, 540, 541
(**F11-15**)
Terpenes, 235
Thermal
cyclic efficiency, 68, 284
destruction, gas and vapor, 18
energy recovery (TER), 479
fixation of atmospheric
nitrogen, 301
NO, 292
reduction of $NO_x$, 482
regeneration of activated
charcoal, 448
Thermal oxidizer, 23, 25 (**F1-14**), 474
afterburners, 25, 474
average temperature, 474
categories, 25 (**F1-14**), 463
costs, 646, 647 (**E13.6**)
design, 478 (**E10.5**)
catalytic, 25, 477 (**F10-26**)
recuperative, 25, 476 (**F10-25**)
regenerative, 25, 475 (**F10-24**)
residence time, 474, 478
types, 25 (**F1-14**)
with heat recovery, 475-478
Thermal reduction of $NO_x$, 482
selective catalytic reduction
(SCR),
selective non-catalytic reduction
(NSCR)
Thermocline, 81, 258 (**F6.17**)
Thermodynamic
activities, 35
equilibrium, 35
Thermophysical properties ($c_p$, Pr,
$\rho$, $\mu$, $\nu$, k)
air and water, 668 (**A-10**)
chemical, Web site
**http://www.uic.edu/~mansoori/Th**
**ermodynamic.Data.and.Propert**
**y_html**
Thermosphere, 218 (**F6-1**), 219
Thiele Number, 494
Threshold Limit Values (TLV), 121,
667 (**A-9**), 669 (**A-11**), 673 (**A-12**)
TIC, 628
Tidal volume, 137
Time
regeneration of adsorber, 450
relaxation, 537
$SO_2/SO_3$ reactions, 311-312
Time-weighted averages, 121
Time Weighted Average
Threshold Limit Values (TWA-
TLV), 7, 12, 121
Tissue gel, 136
Titania photocatalysis, 501
TLV, 667, 669, 673

Total Annual Cost (TAC), 629
(**T13-2**)
  Annual Direct Cost (ADC)
  Operations & Maintenance
  (O&M)
  Annual Indirect Cost (AIC)
  Annual recovery Cost (ARC)
Total Initial Cost (TIC), 628
(**T13-1**)
  Initial Direct Cost (IDC)
  Initial Facilities Cost (IFC)
  Initial Installation Cost (IIC)
  Initial Equipment Cost (IEC)
Tort, 17
Top-down process, 98
Total
  dose, 172
  head loss, 370
  lung capacity, 138
  suspended particulate (TSP), 6
Toxic Substances Control Act
  (TSCA), 11, 119
Toxicity, chronic, acute, systemic,
  local, 168
Toxicants producing lung disease,
  128 (**T4-1**)
Toxicology, 13, 159
Toxins, site-specific and
  reproductive, 169
Trachea, tracheobronchial
  region, 132
  particle removal, 161
Trade winds, 393 (**F9-3**)
Tragedy of the Commons, 66
Trajectory of a particle, 517, 537
Transfer efficiency, spraying, 117
(**T3-9**)
Transfer unit
  height, 461
  number, 462
Transition flow regime, 517
Transmissometers, 202
Transplacental carcinogen, 169
Transport reactor, 321
Transverse
  flow scrubbers, 22, 24 (**F1-11**)
  packed bed scrubbers, 582, 583,
  584 (**E12.4**)
Trichloroethylene (TCE), 235
Trickle-bed reactors, 490 (**F10-34**)
Troposphere & tropopause, 219
Troposphere chemistry, 245
Trumpet lung model, 153 (**F4-27**)
TSCA, 11, 119
TTNBBS, 111
Turbulent shear region, 426, 427
(**F9-28**)
Turbulent flow
  curved ducts, 555
  horizontal ducts, 546
TWA, 7, 12, 121
Tumor dose, 50% ($TD_{50}$) 15
Two-film mass transfer, 354, 355
(**F8-8**)

Two-stage municipal
  incinerator, 285
Two-step overall reaction, 293
Tuberculosis, 164

### U

Ultraviolet-ozone oxidation, 18,
  499, 503 (**E10.8**)
Uncontrolled emissions, 94
Unit,
  conversions, 677 (**A-16**)
  density spheres, 520, 540
  risk factor, 178 (**T4-7**)
Universal gas constant, 28
Unreasonably dangerous, 18
Unstable,
  atmosphere, 403
  equilibrium, 401
Upwelling, 81
Upper (extrathoracic) airways, 131
Uptake,
  absorption efficiency of lung, 149
  effect of emphysema, 152 (**E4.1**)
  $NO_2$ by potato, 262 (**E6.10**)
  $SO_2$ by vegetation, 230 (**T6-5**),
  264 (**F6-21**)
  in water and alfalfa, 224 (**T6-4**)
Urban air
  hydrocarbon concentrations, 238
  (**T6-7**)
  particle size distribution, 528
  (**F11-8**) & (**F11-9**)
  pollutant concentrations, 246
  (**T6-9**)
  pollution model, 246
US death statistics, 7
Utility costs, 632
Uvula, 133

### V

van der Waals forces, 442
Vapor-pressure, 330, 339 (**E8.6**),
  673 (**A-12**)
Variance
  air pollution regulations, 118
  log-normal distribution, 521
  particle size distribution, 521
Vatavuk Air Pollution Control
  Index (VAPCCI), 630,
  631 (**T13-3**)
Vegetation, effects of pollutants
  interfacial mass transfer, 224, 261
  ion transfer in soil, 193 (**F5-3**)
  mass transfer resistances:
  stomatl, aerodynamic,
  mesophyllic, 264
  $NO_2$ uptake, 262 (**E6.10**)
  $SO_2$ uptake, 265 (**F6-21**)
  visual effects, 190 (**T5-1**)
  water vapor transfer, 263 (**F6-20**)

Vehicle emissions versus time, 105
(**F3-1**)
Velocity
  deposition, 256, 257 (**T6-11**)
  drift, 611
  friction, 255
  migration, 611 (**T12-2**)
  molar average, 342
  terminal, settling, 541 (**F11-15**)
Veins, venules, 135
Velocity profile,
  planetary boundary layer, 389,
  390 (**T9-2**), 405 (**F9-15**)
  inspiration and expiration, 144
  (**F4-17**)
Vent condenser, 439
Ventilation (rates)
  fresh air requirements,
  675 (**A-13**)
  lung, 141 (**F4-14**)
  rate during exercise, 142 (**T4-2**)
Ventilation perfusion ratio, 141
  during exercise, 142 (**T4-2**)
Venturi scrubber, 25 (**F1-9**), 585
(**F12-18**)
  efficiency, 587 (**E12.5**)
  flow control, 586 (**F12-19**)
  modeling, 587 (**F12-20**)
Viscosity, 518 (**F11-4**)
  air and water, 668 (**A-10**)
Viscous drag, 515
Visibility, 198
  assessing opacity, 203
  range, 198
  reduction, 199 (**F5-7**)
Visible emissions, 115 (**T3-8**)
  EPA Method 9, 203
Vesticle, 163
Vital capacity, 137
VOC,
Void fraction, 597
Volatile organic compound
  (VOC), 114, 236
Volume
  equivalent particle diameter, 519
  volumetric and mass flow
  rates, 28
  total lung, 138 (**F4-12**)
Voluntary risk, 6
VonKarman constant, 256, 406

### W

Wall quenching in internal
  combustion engine, 316 (**F7-17**)
Warning, air pollution episode, 118
  (**T3-10**), 119 (**T3-11**)
Washout, 579
Waste disposal, 5
Wastewater disposal cost, 634
Water, feed back, 79
Water loss,

leaf canopy, 263 (**F6-20**)
respiration, 159 (**E4.2**)
Water-gas shift reaction, 40,
equilibrium, 37 (**E1.7**)
Water vapor, concentration in
air, 219 (**E6.2**)
Water and pollutants, 192
mass transfer coefficients, 257
(**F6-16**)
roll cells, 258 (**F6-18**)
solubility and uptake, 224, 259
(**E6.9**)
Weathering of soil minerals, 192
Web page,
Code of Federal Regulations
(CFR), 111
pollution control agencies, 111
textbook,
**http://www.engr.psu.edu/cde/m
e470**
class notes,
Mathcad tutorial,
solution to text Examples that
use Mathcad,

errata
Prentice Hall, **http://prenhall.com**
thermophysical properties,
**http://www.uic.edu/~mansoori/Th
ermodynamic.Data.and.Propert
y_html**
Weber-Fechner law, 207
Weibel symmetric model,
134 (**F4-6**)
Well-mixed flow,
curved ducts, 555
horizontal ducts, 546
Well-mixed reactor, 34 (**F1-17**)
Wet collection
deposition velocity, 257
filtration, 606
impact scrubbers, 590
counter flow (particles), 579
(**F12-15**) & (**F12-16**), 582 **E12.3**)
wet deposition, 528
scrubbers (gases), 18, 453, 485,
487 (**F10-30**)
Wheeler equation, 444
White noise, 207

Wilke-Chang Equation, 340
Wind speed profile, 405 (**F9-15**)
logarithmic, 406 (**T9-3**)
power-law exponent, 405 (**T9-2**)
Wind;
directional shear, 397
energy, 75
Work;
accident statistics, 8 (**T1-1**)
rules, 5
Womersley number, 145
Workplace air standards, 121
World population vs time,
63 (**F2-2**)

## X,Y & Z

Zeldovitch mechanism, 303
Zeolites, 441, 484
Zero-emission vehicles
(ZEV), 105
Zero-order biofilter kinetics, 493